T0276189

Undergraduate Lecture Notes in Physics

Series editors

N. Ashby, University of Colorado, Boulder, USA

W. Brantley, Department of Physics, Furman University, Greenville, USA

M. Deady, Physics Program, Bard College, Annandale-on-Hudson, USA

M. Fowler, Dept of Physics, Univ of Virginia, Charlottesville, USA

M. Hjorth-Jensen, Dept. of Physics, University of Oslo, Oslo, Norway

M. Inglis, Earth &Space Sci, Smithtown Sci Bld, SUNY Suffolk County Community College, Long Island, USA

H. Klose, Humboldt University, Oldenburg, Germany

H. Sherif, Department of Physics, University of Alberta, Edmonton, Canada

Undergraduate Lecture Notes in Physics (ULNP) publishes authoritative texts covering topics throughout pure and applied physics. Each title in the series is suitable as a basis for undergraduate instruction, typically containing practice problems, worked examples, chapter summaries, and suggestions for further reading.

ULNP titles must provide at least one of the following:

- An exceptionally clear and concise treatment of a standard undergraduate subject.
- A solid undergraduate-level introduction to a graduate, advanced, or non-standard subject.
- A novel perspective or an unusual approach to teaching a subject.

ULNP especially encourages new, original, and idiosyncratic approaches to physics teaching at the undergraduate level.

The purpose of ULNP is to provide intriguing, absorbing books that will continue to be the reader's preferred reference throughout their academic career.

More information about this series at
http://www.springer.com/series/8917

Wolfgang Demtröder

Mechanics and Thermodynamics

Wolfgang Demtröder
Kaiserslautern, Germany
demtroed@rhrk.uni-kl.de

ISSN 2192-4791 ISSN 2192-4805 (electronic)
Undergraduate Lecture Notes in Physics
ISBN 978-3-319-27875-9 ISBN 978-3-319-27877-3 (eBook)
DOI 10.1007/978-3-319-27877-3
Library of Congress Control Number: 2016944491

© Springer International Publishing Switzerland 2017
This work is subject to copyright. All rights are reserved by the Publisher, whether the whole or part of the material is concerned, specifically the rights of translation, reprinting, reuse of illustrations, recitation, broadcasting, reproduction on microfilms or in any other physical way, and transmission or information storage and retrieval, electronic adaptation, computer software, or by similar or dissimilar methodology now known or hereafter developed.
The use of general descriptive names, registered names, trademarks, service marks, etc. in this publication does not imply, even in the absence of a specific statement, that such names are exempt from the relevant protective laws and regulations and therefore free for general use.
The publisher, the authors and the editors are safe to assume that the advice and information in this book are believed to be true and accurate at the date of publication. Neither the publisher nor the authors or the editors give a warranty, express or implied, with respect to the material contained herein or for any errors or omissions that may have been made.

Printed on acid-free paper

This Springer imprint is published by Springer Nature
The registered company is Springer International Publishing AG
The registered company address is: Gewerbestrasse 11, 6330 Cham, Switzerland

Preface

The present textbook represents the first part of a four-volume series on experimental Physics. It covers the field of Mechanics and Thermodynamics. One of its goal is to illustrate, that the explanation of our world and of all natural processes by Physics is always the description of models of our world, which are formulated by theory and proved by experiments. The continuous improvement of these models leads to a more detailled understanding of our world and of the processes that proceed in it.

The representation of this textbook starts with an introductory chapter giving a brief survey of the history and development of Physics and its present relevance for other sciences and for technology. Since experimental Physics is based on measuring techniques and quantitative results, a section discusses basic units, techniques for their measurements and the accuracy and possible errors of measurements.

In all further chapters the description of the real world by successively refined models is outlined. It begins with the model of a point mass, its motion under the action of forces and its limitations. Since the description of moving masses requires a coordinate system, the transformation of results obtained in one system to another system moving against the first one is described. This leads to the theory of special relativity, which is discussed in Chap. 3. The next chapter upgrades the model of point masses to spatially extended rigid bodies, where the spatial extension of a body cannot be ignored but influences the results. Then the deformation of bodies under the influence of forces is discussed and phenomena caused by this deformation are explained. The existence of different phases (solid, liquid and gaseous) and their relation with external influences such as temperature and pressure, are discussed.

The properties of gases and liquids at rest and the effects caused by streaming gases and liquids are outlined in Chap. 7 and 8.

Many insights in natural phenomena, in particular in the area of atomic and molecular physics could only be explored after sufficiently good vacua could be realized. Therefore Chap. 9 discusses briefly the most important facts of vacuum physics, such as the realization and measurement of evacuated volumina.

Thermodynamics governs important aspects of our life. Therefore an extended chapter about definitions and measuring techniques for temperatures, heat energy and phase transitions should emphazise the importance of thermodynamics. The three principle laws ot thermodynamics and their relevanve for energy transformation and dissipation are discussed.

Chapter 11 deals with oscillations and waves, a subject which is closely related to acoustics and optics.

While all foregoing chapters discuss classical physics which had been developed centuries ago, Chap. 12 covers a modern subject, namely nonlinear phenomena and chaos theory. It should give a feeling for the fact, that most phenomena in classical physics can be described only approximately by linear equations. A closer inspection shows that the accurate description demands nonlinear equations with surprising solutions.

A description of phenomena in physics requires some minimum mathematical knowledge. Therefore a brief survey about vector algebra and vector analysis, about complex numbers and different coordinate systems is provided in the last chapter.

A real understanding of the subjects covered in this textbook can be checked by solving problems, which are given at the end of each chapter. A sketch of the solutions can be found at the end of the book.

For further studies and a deeper insight into special subjects some selected literature is given at the end of each chapter.

The author hopes that this book can transfer some of his enthusiasm for the fascinating field of physics. He is grateful for any comments and suggestions, also for hints to possible errors. Every e-mail will be answered as soon as possible.

Several people have contributed to the realization of this book. Many thanks go the Dr. Schneider and Ute Heuser, Springer Verlag Heidelberg, who supported and encouraged the authors over the whole period needed for translating this book from a German version. Nadja Kroke and her team (le-tex publishing services GmbH) did a careful job for the layout of the book and induced the author to improve ambiguous sentences or unclear hints to equations or figures. I thank them all for their efforts.

Last but not least I thank my wife Harriet, who showed much patience when her husband disappeared into his office for the work on this book.

Kaiserslautern, December 2016 Wolfgang Demtröder

Contents

Introduction and Survey

1

Chapter 1

© Springer International Publishing Switzerland 2017

W. Demtröder, *Mechanics and Thermodynamics*, Undergraduate Lecture Notes in Physics, DOI 10.1007/978-3-319-27877-3_1

The name "Physics" comes from the Greek ("$\varphi\upsilon\sigma\iota\kappa\eta$" = nature, creation, origin) which comprises, according to the definition of *Aristotle* (384–322 BC) the theory of the material world in contrast to *metaphysics*, which deals with the world of ideas, and which is treated in the book by Aristotle after (*Greek: meta*) the discussion of physics.

Definition

The modern definition of physics is: Physics is a basic science, which deals with the fundamental building blocks of our world and the mutual interactions between them.

The goal of research in physics is the basic understanding of even complex bodies and their composition of smaller elementary particles with interactions that can be categorized into only four fundamental forces. Complex events observed in our world should be put down to simple laws which allow not only to explain these events quantitatively but also to predict future events if their initial conditions are known.

In other words: Physicists try to find laws and correlations for our world and the complex natural events and to explain all observations by a few fundamental principles.

Note, however, that complex systems that are composed of many components, often show characteristics, which cannot be reduced to the properties of these components. The amalgamation of small particles to larger units brings about new and unforeseen characteristics, which are based on cooperative processes. *The whole is more than the sum of its parts (Heisenberg 1973, Aristotle; metaphysics VII).* Examples are living biological cells, which are composed of lifeless molecules or molecules with certain chemical properties consisting of atoms that do not show these properties of the molecule.

The treatment of such complex systems requires new scientific methods, which have to be developed.

This should remind enthusiastic physicists, that physics alone might not explain everything although it has been very successful to expands the borderline of its realm farther and farther in the course of time.

1.1 The Importance of Experiments

The more astronomically oriented observations of ancient Babylonians brought about a better knowledge of the yearly periods of the star sky. The epicycle model of *Ptolemy* gave a nearly quantitative description of the movements of the planets. However, modern Physics in the present meaning started only much later with *Galileo Galilei* (1564–1642, Fig. 1.1), who performed as the first physicist well planned experiments under defined conditions, which could give quantitative answers to open questions. These experiments can be performed at any time under conditions chosen by the experimentalist independent of external influences. This distinguishes them from the observations of natural phenomena, such as thunderstorms, lightening or volcanism, which cannot be influenced. This freedom of choosing the conditions is the great advantage of experiments, because all perturbing external influences can be partly or even completely eliminated (e. g. air friction in experiments on free falling bodies). This facilitates the analysis of the experimental results considerably.

Experiments are aimed questions to nature, which yield under defined conditions definite answers.

The goal of all experiments is to find reasons and causes for all phenomena observed in nature, to see connections between the manifold of observations and to categorize them under a common law. Even more ambitious is the quantitative prediction of future experimental results, if the initial conditions of the experiments are known.

A physical law connects measurable quantities and concepts. Its clear form is a mathematical equation.

Such mathematical descriptions give a clearer insight into the relations between different physical laws. It can reduce the manifold of experimental findings, which might seem at first glance uncorrelated but turn out to be special cases of the same general law that is valid in all fields of physics.

Examples

1. Based on many careful measurements of planetary orbits by *Tycho de Brahe* (1546–1601), *Johannes Kepler* (1571–1630) could postulate his three famous laws for the quantitative description of distances and movements of the planets. He did not find the cause for these movements, which was discovered only later by Isaac Newton (1642–1727) as the gravitational force between the sun and the planets. However, Newton's gravitation law did not only describe the planetary orbits but all movements of bodies in gravitational fields. The problem to unite the gravitational force with the other forces (electromagnetic, weak and strong force) has not yet been solved, but is the subject of intense current research.
2. The laws of energy and momentum conservation were only found after the analysis of many experiments in different fields. Now they explain and unify many experimental findings. Such a unified summary of different physical laws and principles to a consistent general description is called a *physical theory*. ◀

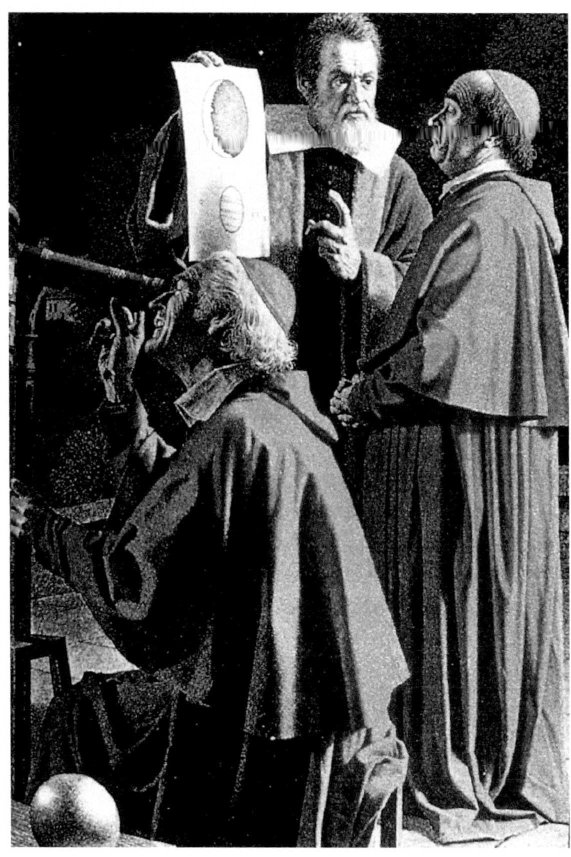

Figure 1.1 *Left:* Galileo Galilei. *Right:* Looking of Cardinales through Galilio's Telescope

Its range of validity and predictive capability is checked by experiments.

Since the formulation of a theory requires a mathematical description, a profound knowledge of basic mathematics is indispensible for every physicist.

1.2 The Concept of Models in Physics

The close relation between theory and experiments is illustrated by the following consideration:

If a free falling body in a vacuum container at the surface of the earth is observed one finds that the fall time over a definite distance is independent of the size or form of the body and also independent of its weight. In contrast to this result is the fall of a body in any fluid, instead of vacuum where the form of the body does play a role because here perturbing influences, such as friction often cannot be neglected. Neglecting these perturbations one can replace the body by the **model of a point mass**. With other words: In these experiments the falling body behaves like a point mass, because its size does not matter. The theory can now give a complete description of the movement of point masses under the influence of gravitational forces and it can predict the results of corresponding future experiments (see Chap. 2).

Now the experimental conditions are changed: For a body falling in water the velocity and fall time do depend on size and weight of the body, because of friction and buoyancy. In this case the model of a point mass is no longer valid and has to be broadened to the **model of spatially extended rigid bodies** (see Chap. 5). This model can predict and quantitatively explain the movements of extended rigid bodies under the influence of external forces.

If we now further extend our experimental condition and let a massive body fall onto a deformable elastic steel plate, our rigid body model is no longer valid but we must include in our model the deformation of the body, This results in the **model of extended deformable bodies**, which describes the interaction and the forces between different parts of the body and explains elasticity and deformation quantitatively (see Chap. 6).

The theory of phenomena in our environment is always the description of a model, which describes the observations. If new phenomena are discovered which are not correctly represented by the model, it has to be broadened and refined or even completely revised.

The details of the model depend on the formulation of the question asked to nature and on the kind of experiments which should be explained. Generally a single experiment tests only certain statements of the model. If such an experiment confirms these statements, we say, that nature behaves in this experiment like the model predicts, i. e. nature gives the same answer to selected experiments as the model.

Since theory can in principle calculate all properties of an accepted model it often gives valuable hints, which experiments could best test the validity of the model.

Such a cooperation and mutual inspiration of theoretical and experimental physics contribute in an outstanding way to the progress in physical knowledge.

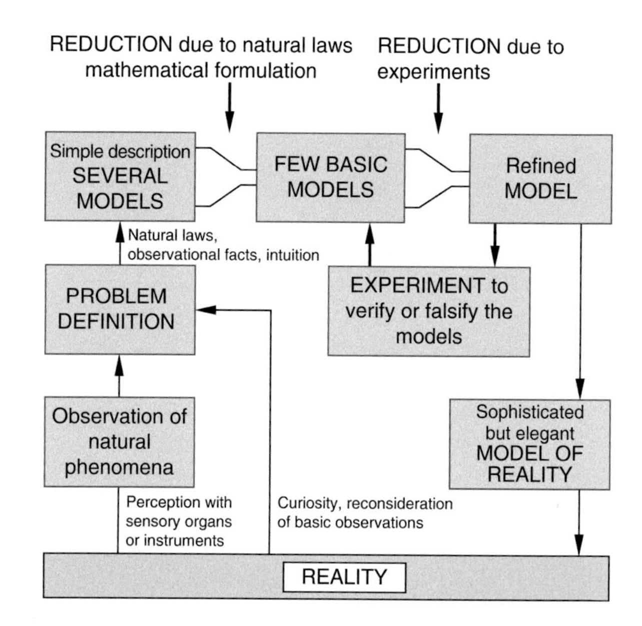

Figure 1.2 Schematic representation of the way, how scientists gain information on nature

An impressive example is the development of *quantum chromodynamics*. This modern theory describes the substructure of particles, which had been regarded as elementary, such as protons, neutrons and mesons, but are really composed of still smaller particles, the **quarks**. Theoretical predictions about the possible masses of unstable particles, composed of these quarks, which appear as resonances in the collision cross sections, allowed the experimentalists to restrict their search which is like the search for a needle in the haystack, to the predicted energy range, which facilitated their efforts considerably.

The model concept for the description of observations in nature is in particular obvious in the world of microphysics (atomic, molecular and nuclear physics), because here the particles cannot be seen with the naked eye and therefore a vivid picture cannot be given. Attempts to transfer vivid models useful in macrophysics to microphysics have often led to misunderstandings and wrong ideas. One example is the particle-wave dualism for the description of microparticles (see Vol. 3).

Figure 1.2 comprises the discussion above. One example shall illustrate the development and refinement of models in physics. The explanation of lightning by Greek philosophers was the god Zeus who flung flashes to the earth while he was in a furious mood. Modern models explain lightning by the separation of positive and negative electrical charges by charged water drops floating in turbulent air, leading to large electric voltages between different clouds or between clouds and earth with resulting strong discharges. This modern model is based on many detailed observations with high speed photographic instruments and on experimental simulations of lightning in high voltage laboratories where discharges can be observed under controlled conditions.

The goal of sciences is the understanding of natural phenomena observed under different conditions and to categorize their explanations under a common law. It is assumed, that the observed reality exists independent of the observer. However, the experiments performed in order to reproduce the observations demand nevertheless characteristic features of the observing subject, such as imagination for the planning of decisive experiments, an open mind for new ideas, etc. Many ideas turn out to be wrong. They can be already excluded by comparison with former experiments. Such ideas which do not contradict already existing knowledge can contribute to a working hypothesis. Even such a hypothesis might be only partly correct and has to be modified by the results of further experiments. If all these results confirm the working hypothesis it can become a **proved theory**, which allows us to summarize many observations to a general law (see Fig. 1.3).

This procedure where a theory is built up from many experimental results is called the **inductive method**.

In theoretical physics often a reverse procedure is chosen. The starting point are fundamental basic equations such as Newton's law of gravitation or the Maxwell equations or symmetry laws. From these general laws the outcome of possible experiments is predicted (**deductive method**).

Both procedures have their justification with advantages and drawbacks. They supplement each other.

An important aspect which one should keep in mind is summarized in the following fundamental statement:

Physics describes objective and as accurate as possible the reality of the material world. For human beings this is, however, only a small section of the world we experience, as a specific example illustrates: From the standpoint of physics a painting can be described, by giving for each point (x, y) the reflectivity $R(\lambda, x, y)$, which depends on the wavelength λ, the spectrum of

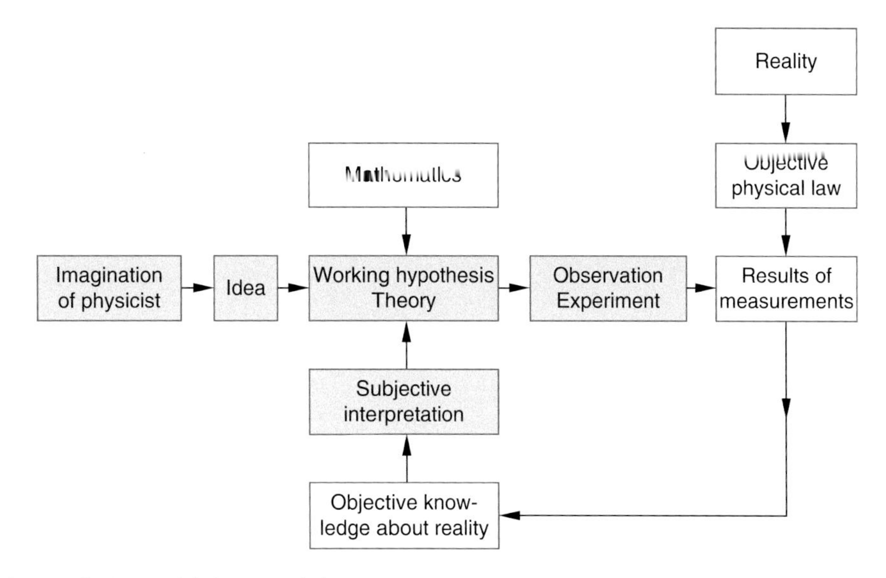

Figure 1.3 Schematic diagram of gaining insight into natural phenomena

the illuminating radiation source and the angles of incidence and observation direction. A computer which is fed with these characteristic input data can reproduce the painting very accurately.

Nevertheless this physical description lacks an essential part of the painting, which is in the mind of the observer. When looking at the painting a human being might remember other similar paintings which he compares with the present painting, even if these other paintings are not present but only in the mind they still change the subjective impression of the observer. The subject of the painting may induce cheerful or sad feelings in the mind of the observer, it may call back remembrances of former events or impressions which are related to this painting. All these different influences will determine the judgement about the painting, which therefore might be different for different observers.

All these aspects are not the realm of physics, because they are subjective, although they are essential for the quality of the painting as judged by human beings and they represent an important part of the "reality" as perceived by us.

These remarks should warn physicists, not to forget that our fascinating science is only competent for the description of the material basis of our world. Although the other nonmaterial realms are based on the material world their description and understanding reaches far beyond physics. The question, how living cells are built from inanimate molecules and how the human mind is related to the structure of the brain are still pending but exciting problems, which might be solved in the future. This is related to the question whether the human brain is more than a highly developed computer, which is the subject of hot discussions between the supporter of artificial intelligence and biologists.

For more detailed discussions of these questions, the reader is referred to the literature [1.1a–1.6].

1.3 Short Historical Review

The historical development of physics can be roughly divided into three periods:

- The natural philosophy in ancient times
- The development of classical physics
- The modern physics.

1.3.1 The Natural Philosophy in Ancient Times

The investigation of natural phenomena and the efforts to explain them by rational arguments started already 4000 years ago. The astronomical observation of the Babylonian and the Egyptian scientists were important for the prediction of annual occurrences, such as the Nile flood or the correct time for sowing. The Greek philosophers produced many ideas for the explanation of the observed natural phenomena. All these ideas were treated within the framework of general philosophy. For example, the textbook on Physics ($\varphi\upsilon\sigma\iota\kappa\eta\ \alpha\kappa\varrho o\alpha\sigma\iota\varsigma =$ lectures on physics) by Aristotle contains mainly philosophical considerations about space and time, movements of bodies and their causes.

Probably the most important achievement of Greek philosophy was the overcoming of the widespread mythology, where the life of mankind was governed by a hierarchy of gods, whose mood was not predictable and everybody had to win the liking of gods by sacrificing precious gifts to them. Most Greek philosophers abandoned the belief, that the world was a playing ground for gods, demons and ghosts who generated thunderstorm, floods, sunshine or disastrous droughts just according to their mood (see Homer's Odyssey).

The Greek philosophers believed that all natural phenomena obeyed eternal unchanging laws which were not always obvious

Figure 1.4 Aristotle. With kind permission of the "Deutsches Museum"

because of the complex nature but which were independent of men or gods. This means that it is, at least in principle, possible to find such laws merely by human reason.

Example

A solar eclipse is no longer described by a monster that engulfs the sun, but by the temporarily blocking of the sunlight by the moon. This changes the solar eclipse from an accidental event to a predictable occurrence. ◄

Famous representatives of Greek philosophy were *Thales from Milet* (624–546 BC), who discovered magnetism and frictional electricity, but could not correctly explain his findings. *Empedocles* (495–435 BC) assumed that fire, water, air and soil formed the four basic elements, which can mix, divide and build compositions from which all other material is composed. The mathematical aspect of natural phenomena was introduced by *Pythagoras* (572–492 BC) and his scholars who assumed that numbers and mathematical relations between these numbers reflect the reality. They made acoustic experiments with striking chords of different lengths and measured the resulting tones. However, they erroneously generalized their results to other fields such as the movement of the planets.

Anaxagoras (499–428 BC) was the first to postulated that the world consists of many infinitely small different particles. The force which keeps them together is the Nus (= world spirit). *Leucippus* (489–428 BC) and his student *Democritus* (455–370 BC) followed these ideas and refined this hypothesis. *Democritus* assumed that the world consists of atoms ($\alpha\tau o\mu o\varsigma$ = indivisble), very small indivisible identical particles, which move forever in an infinite empty space. The different forms of matter differ only by the number and arrangement of atoms of which they are composed. This hypothesis comes close to our present understanding of the atomic composition of the different elements in the periodic table (see Sect. 1.4).

The doctrine of the "atomists" was declined by *Plato* (427–347 BC) and *Aristotle* (Fig. 1.4) since it contradicted their view of a continuous world. Since these two philosophers had such a great reputation the atomistic theory was forgotten for nearly 2000 years.

Aristotle (384–322 BC) (Fig. 1.4) regarded nature as the forever moving and developing universe, where at the beginning a "divine mover" was assumed who started the whole world. The planets move apparently without obvious mover and therefore Aristotle assumed that they do not consist of the four earthly elements fire, air, water and soil but of a fifth "divine element" which he called "Ether". This ether should be massless and elastic and should penetrate the whole world, including rigid bodies.

Archimedes (287–212 BC) studied in Alexandria, the centre of science at that time. Later he moved to Syracuse on Sicily. He was the greatest mathematician, physicist and technical expert of his time. He succeeded to calculate the area and the perimeter of a circle, the surfaces of spheres, cones and cylinders and he solved third order equations. As a physicist he determined the centre of mass for bodies of different shape, he found the lever principle, calculated the buoyancy of bodies in water (Archimedes' principle), he built a planetarium and measured star positions and proved the curvature of the sea surface. He was famous for his technical achievements. He invented and constructed about 40 different machines, such as the worm gear drive, catapults, hydraulic levers for lifting ships and many machines used for warfare.

In spite of great success in many fields the Greek philosophers could not reach natural science in the present sense, because they did not accept the experiment as the touchstone for every theory. They believed that an initial observation was sufficient and that all subsequent conclusions and knowledge could be achieved by pure thinking without further confirming or disproving experiments.

This rather speculative procedure has influenced, due to the great impact of Aristotle's generally accepted teaching, many generations of philosophers for more than 1500 years. Even when *Galilei Galileo* observed through his telescope the four moons of Jupiter, most philosophers and high members of the church did not believe him, because his observation contradicted the theory of Aristotle, who taught that the planets were fixed on crystal spheres moving with the planet around the earth. If moons circled around Jupiter they had to penetrate these crystal spheres and would smash them. Therefore, the moons should be impossible. Even when Galilei offered to the sceptics to look through the telescope (Fig. 1.1b) many of them refused and said: "Why should we look and be deceived by optical illusions when we are sure about Aristotle's statements".

Although some inconsistencies in Aristotle's teaching had been found before, Galilee was the first to disprove by his observations and experiments the whole theory of the shining example of Greek philosophy, in particular when he also advertised the new astronomy of *Copernicus*, which brought him many enemies and even a trial before the catholic court.

1.3.2 The Development of Classical Physics

One may call Galileo the first physicist in the present meaning. He tried as the first scientist to prove or disprove physical theories by specific well-planned experiments. Famous examples are his experiments on the movement of a body with constant acceleration under the influence of gravity. He also considered how large the accuracy of his experimental results must be in order to decide between two different versions for the description of such movements. He therefore did not choose the free fall (it is often erroneously reported, that he observed bodies falling from the Leaning tower in Pisa). This could never reach the required accuracy with the clocks available at that time. He chose instead the sliding of a body on an inclined plane with an angle α against the horizontal. Here only the fraction $g \cdot \sin\alpha$ acts on the body and thus the acceleration is much smaller.

His astronomical observations (phases of Venus, Moons of Jupiter) with a self-made telescope (after he had learned about its invention by the optician *Hans Lipershey* (1570–1619) in Holland) helped the Copernican model of the planets circling around the sun instead of the earth, finally to become generally accepted (in spite of severe discrepancies with the dogmatic of the church and heavy oppression by the church council).

The introduction of mathematical equations to physical problems, which comprises several different observations into a common law, was impressively demonstrated by *Isaac Newton* (Fig. 1.5). In his centennial book "*Philosophiae Naturalis Principia Mathematica*" he summarizes all observations and the knowledge of his time about mechanics (including celestial mechanics = astronomy) by reducing them to a few basic principles (principle of inertia, actio = reactio, the force on a body equals the time derivative of his momentum and the gravitational law).

Supported by progress of mathematics in the 17th century (analytical geometry, infinitesimal calculus, differential equations) the mathematical description of physical observations becomes more and more common. Physics emancipates from Philosophy and develops its own framework using mathematical language for the clear formulation of physical laws. For example classical mechanics experiences its complete and elegant mathematical form by *J. L. de-Lagrange* (1736–1813) and *W. R. Hamilton* (1805–1865) who reduced all laws for the movement of bodies under arbitrary forces to a few basic equations.

Figure 1.5 Sir Isaac Newton. With kind permission of the "Deutsches Museum München"

Contrary to mechanics which had developed already in the 18th century to a closed complete theory the knowledge about the structure of matter was very sketchy and confused. Simultaneously different hypotheses were emphasized: One taken form the ancient Greek philosophy, where fire, water, air and soil were assumed as the basic elements, or from the alchemists who favoured mercury, sulphur and salt as basic building blocks of matter.

Robert Boyle (1627–1591) realized after detailed experiments that simple basic elements must exist, from which all materials can be composed, which however, cannot be further divided. These elements should be separated by chemical analysis from their composition. Boyle was able to prove that the former assumption of elements was wrong. He could, however, not yet find the real elements.

A major breakthrough in the understanding of matter was achieved by the first critically evaluated quantitative experiments investigating the mass changes involved in combustion processes, published in 1772 by *A. L. de Lavoisier* (1743–1794). These experiments laid the foundations of our present ideas about the structure of matter. Lavoisier and *John Dalton* (1766–1844) recognised metals as elements and postulated like Boyle that all substances were composed of atoms. The atoms were now, however, not just simple non-divisible particles, but had specific characteristics which determined the properties of the composed substance. *Karl Wilhelm Scheele* (1724–1786) found that air consisted of nitrogen and oxygen.

Antoine-Laurent Lavoisier furthermore found that the mass of a substance increased when it was burnt, if all products of the combustion process were collected. He recognized that this mass increase was caused by oxygen which combined with the substance during the burning process. He formulated the law of mass conservation for all chemical processes. Two elements can combine in different mass ratios to form different chemical products where the relative mass ratios always are small integer numbers.

The British Chemist *John Dalton* was able to explain this law based on the atom hypothesis.

Examples

1. For the molecules carbon monoxide and carbon dioxide the mass ratio of oxygen combining with the same amount of carbon is 1 : 2 because in CO one oxygen atom and in CO_2 two oxygen atoms combine with one carbon atom.
2. For the gases N_2O (Di-Nitrogen oxide), NO (nitrogen mono oxide), N_2O_3 (nitrogen trioxide), and NO_2) nitrogen dioxide) oxygen combines with the same mass of nitrogen each time in the ratio 1 : 2 : 3 : 4. ◄

Dalton also recognized that the relative atomic weights constitute a characteristic property of chemical elements. The further development of these ideas lead to the periodic system of elements by *Julius Lothar Meyer* (1830–1895) and *Dimitri Mendelejew* (1834–1907), who arranged all known elements in a table in such a way that the elements in the same column showed similar chemical properties, such as the alkali atoms in the first column or the noble gases in the last column.

Why these elements had similar chemical properties was recognized only much later after the development of quantum theory.

The idea of atoms was supported by *Amedeo Avogadro* (1776–1856), who proposed in 1811 that equal volumes of different gases at equal temperature and pressure contain an equal number of elementary particles.

A convincing experimental indication of the existence of atoms was provided by the *Brownian motion*, where the random movements of small particles in gases or liquids could be directly viewed under a microscope. This was later quantitatively explained by Einstein, who showed that this movement was induced by collisions of the particles with atoms or molecules.

Although the atomic hypothesis scored indisputable successes and was accepted as a working hypothesis by most chemists and physicists, the existence of atoms as real entities was a matter of discussion among many serious scientists until the end of the 19th century. The reason was the fact that one cannot see atoms but had only indirect clues, derived from the macroscopic behaviour of matter in chemical reactions. Nowadays the improvement of experimental techniques allows one to see images of single atoms and the theoretical basis of atomic theory leaves no doubt about the real existence of atoms and molecules.

The theory of heat began to become a quantitative science after thermometers for the measurement of temperatures had been developed (air-thermoscope by Galilei, alcohol thermometer 1641 in Florence, mercury thermometer 1640 in Rome). The Swedish physicist *Anders Celsius* (1701–1744) introduced the division into 100 equal intervals between melting point (0 °C) and boiling point (100 °C) of water at normal pressure. *Lord Kelvin* (1824–1907) postulated the absolute temperature, based on gas thermometers and the general gas law. On this scale the zero point $T = 0\,\mathrm{K} = -273.15\,°\mathrm{C}$ is the lowest temperature which can be closely approached but never reached (see Chap. 10).

Denis Papin (1647–1712) investigated the process of boiling and condensation of water vapour (Papin's steam pressure pot). He built the first steam engine, which *James Watt* (1736–1819) later improved to reliable technical performance. The terms ***amount of heat*** and ***heat capacity*** were introduced by the English physicist and chemist *Joseph Black* (1728–1799). He discovered that during the melting process heat was absorbed which was released again during solidification.

The more precise formulation of the theory of heat was essentially marked by establishing general laws. *Robert Mayer* (1814–1878) postulated the first law of the theory of heat, which states that for all processes the total amount of energy is conserved. *Nicolas Carnot* (1796–1832) started 1831 after some initial errors a fresh successful attempt to describe the conversion of heat into mechanical energy (Carnot's cycle process). This was later more precisely formulated by *Rudolf Clausius* (1822–1888) in the **second law of heat theory**.

A real understanding of heat was achieved, when the kinetic gas theory was formulated. Here the connection between heat properties and mechanical energy was for the first time clearly formulated. Since the dynamical properties of molecules moving around in a gas were related to the temperature of a gas, the heat theory was now called **thermodynamics**, which was formulated by several scientists (Clausius, Avogadro, Boltzmann) (see Fig. 1.6). They proved under the assumption that gases consist of many essentially free atoms or molecules, which move randomly around and collide with each other, that the heat energy of a gas is equivalent to the kinetic energy of these particles. The Austrian physicist *Joseph Loschmidt* (1821–1895) found that under normal pressure the gas contains the enormous number of about $3 \cdot 10^{19}$ atoms per cm^3.

Optics is one of the oldest branches of physics which was already studied more than 2000 years ago where the focussing of light by concave mirrors was used to ignite a fire. However, only in the 17th century optical instruments and their imaging properties were studied systematically. A milestone was the fabrication of lenses and the invention of telescopes. *Willibrord Snellius* (1580–1626) formulated his law of refraction (see Vol. 2, Chap. 9), Newton found the separation of different colours when white sun light passed through a prism. The explanation of the properties of light was the subject of hot discussions. While Newton believed that light consisted of small particles (in our present model these are the *photons*) the experiments on interference and diffraction of light by *Grimaldi*

Figure 1.6 Ludwig Boltzmann. With kind permission from Dr. W. Stiller Leipzig

Figure 1.7 James Clerk Maxwell. With kind permission from the American Institute of Physics, Emilio Segre Visual hives, College Park MD

(1618–1663), *Christiaan Huygens* (1629–1695), *Thomas Young* (1773–1829) and *Augustin Fresnel* (1788–1827) decided the dispute in favour of the wave theory of light. *Melloni* showed 1834 that the laws for visible light could be extended into the infrared region and *Max Felix Laue* (1879–1960) and *William Bragg* (1862–1942) demonstrated the wave character of X-rays, which had been discovered by *Conrad Roentgen* (1845–1923), by their famous experiments on X-ray diffraction in crystals.

The velocity of light was first estimated by *Ole Rømer* (1644–1710) by astronomical observations of the appearance time of Jupiter moons and later more precisely determined by Huygens. With measurements on earth *Jean Foucault* (1819–1868) and *Armand Fizeau* (1819–1896) could obtain a rather accurate value for the velocity of light.

William Gilbert (1544–1603) was called "the father of electricity". He investigated the magnetic field of permanent magnets and measured the magnetic field of the earth with the help of magnetic needles. He made extensive experiments on friction electricity and divided the different materials into electrical and non-electrical substances. He built the first electroscop and measured the forces between charged particles. *Stephen Gray* (1670–1736) discovered the electrical conductivity of different materials and made detailed experiments on electric induction. He made electricity very popular by spectacular demonstrations.

Charles Augustin Coulomb (1736–1806) built the first electrometer, constructed the Coulomb torsion balance and formulated the famous Coulomb law for the forces between charged particles. *Benjamin Franklin* (1706–1790) recognized that lightening is not a fire but an electrical discharge and constructed the first lightning conductor. *Luigi Galvani* (1737–1798) discovered the stimulation of nerves by electrical currents (frog's leg experiments); and the contact voltage between different conductors, which lead to the construction of batteries (Galvanic element). *Allessandro Volta* (1745–1827) continued the experiments of Galvani and he categorized the different metals in an electrochemical series.

Hans *Christean Oersted* (1777–1851) discovered the magnetic field of an electric current. *Andre Marie Ampere* (1775–1836) coined the terms "***electrical current***" and ***electrical voltage***. By many detailed experiments, he established modern electrodynamics.

Michael Faraday (1791–1867) performed basic experiments on the relations between electric currents and magnetic fields (Faraday's induction law). He prepared the foundations for the development of alternating currents and their applications.

James Clerk Maxwell (1831–1879) (Fig. 1.7) summarized all known results of former experiments by a few basic equations (Maxwell's equations) and gave them a general mathematical formulation, which represents the basis for electrodynamics and optics. Their solutions are electro-magnetic waves, which found a brilliant confirmation by the experiments of *Heinrich Hertz* (1857–1894), who showed that these waves were transversal and propagate in space with the velocity of light.

1.3.3 Modern Physics

At the end of the 19th century, all problems in physics seemed to be solved and many physicists believed, that a closed theory describing all known facts could be realized in the near future.

This optimistic opinion changed, however, in a dramatic way, induced by the following experimental findings.

- The Michelson experiment (see Sect. 3.4) showed without doubt, that the velocity of light is constant, independent of the direction or the velocity of the observer. This result was in sharp contrast to former concepts and induced Albert Einstein (Fig. 1.8) to formulate his theory of special relativity (see Sect. 3.6).
- Experimentally found deviations from the theoretically expected spectral intensity distribution of the thermal radiation of hot bodies, as calculated by *Stephan Boltzmann* and *Wilhelm Wien*, could not be explained by classical physics. This discrepancy led *Max Planck* (1858–1947) (Fig. 1.9) to the conclusion of quantized energy of radiation fields. This bold assumption, which could perfectly reproduce the experimental results, represented the beginning of quantum theory that was later on imbedded in a concise mathematical framework by *Erwin Schrödinger* (1887–1961) and *Werner Heisenberg* (1901–1976) (see Vol. 3). The concept of energy quanta was further supported experimentally by the photoelectric effect, which was quantitatively explained by Einstein, who received the Nobel Prize for his theory of the photo-effect (not for his theory of relativity!).
- New experimental techniques allowed investigating the structure of atoms and molecules. The light emitted from atoms or molecules could be sent through a spectrograph and showed discrete lines, indicating that it has been emitted from discrete energy levels. Through the development of spectral analysis by *Gustav Robert Kirchhoff* (1824–1887) and *Robert Bunsen* (1811–1899) it was found that atoms of a specific element emitted spectral lines with wavelengths characteristic for this element. The results could not be explained by classical physics but needed quantum theory for their interpretation. Today the physics of atomic electron shells and their energy levels can be completely described by a closed theory called **quantum-electrodynamics**.

Figure 1.9 Max Planck. With kind permission of the "Deutsches Museum München"

Figure 1.8 Albert Einstein. With kind permission of the "Deutsches Museum München"

This illustrates that always in the history of natural sciences new experimental results forced physicists to revise former concepts and to formulate new theories which, however, should include proved earlier results. In most cases the old theories were not completely abandoned but their validity range was restricted and more precisely characterized. For example the classical physics is perfectly correct for the description of the motion of macroscopic bodies or for many applications in daily life, while for the description of the micro-world of atoms and molecules it may completely fail and quantum theory is necessary.

The properties of atomic nuclei could be only investigated after appropriate detectors had been developed. Nuclear physics is therefore a rather new field where most of the results were obtained in the 20th century. The substructure of atomic nuclei and the physics of elementary particles could start after particle accelerators could be operated and many results in this field have been achieved only recently.

This short historical review should illustrate that many concepts which today are taken for granted, are not as old and have been accepted only after erroneous ideas and a long way of successive corrections, guided by new experiments. It is worthwhile for every physicist to look into some original papers and follow the gradual improvements of concepts and representation of results.

More extensive literature about the historical development of physics and about bibliographies of physicists can be found in the references [1.6–1.14c].

1.4 The Present Conception of Our World

As the result of all experimental and theoretical investigations our present model of the material world has been established (Fig. 1.10). In this introduction, we will give only a short summary. The subject will be discussed more thoroughly in Vol. 3 and 4 of this textbook series.

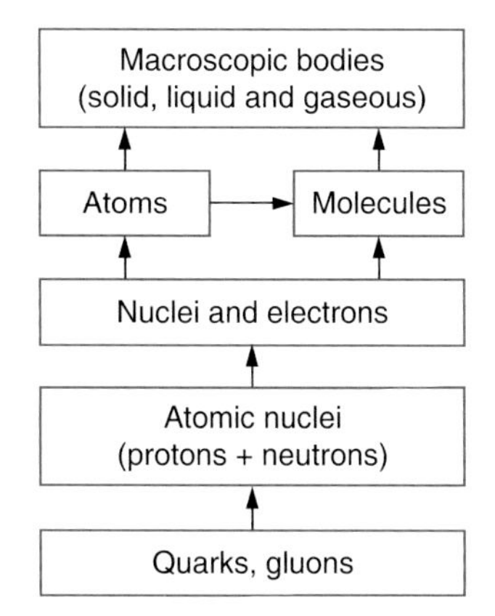

Figure 1.10 Build up of our material world (H. J. Jodl [1.14b])

Elementary Particles

The entire material world known up to now is composed of only a few different particles. The three most important are the electron (e^-), the proton (p^+) and the neutron (n). All other elementary particles (muons, π-Mesons, Kaons, Λ-particles etc.) exist after their production only a very short time (10^{-6}–10^{-15} s). They convert either spontaneously or by collisions into other particles which finally decay into p^+, e^-, neutrinos or photons $h \cdot \nu$. Although neutrinos are stable particles they show such a small interaction with matter that they are difficult to detect and they therefore play no role in daily life.

Recent experiments and theoretical consideration have shown, that the particles p^+, n, mesons and hyperons, which had been regarded as elementary, show a substructure (see Vol. 4), According to our present understanding they consists of smaller particles, called "quarks", which occur in 6 different species.

All building blocks of matter can be divided into two groups:

1. the quarks, which build up the heavy particles (*baryons*), such as proton, neutron, mesons and hyperons
2. the light particles (*leptons*) electron, myon and neutrino.

Each of these two groups consists of three families of elementary particles, which are listed in Tab. 1.1. For each of these particles there exists an anti-particle with equal mass but opposite charge. For instance the anti-particle of the electron e^- is the positron e^+, the proton p^+ has as anti-particle the anti-proton p^- and the anti-neutron has the same mass and the charge zero as the neutron.

According to present theories the interaction between the particles can be described by the exchange of "interaction particles", which are called the quanta of the interaction field. For example the quanta of the electromagnetic field, which determine the interaction between charged particles are the ***photons*** $h \cdot \nu$.

The quanta of the strong interaction between nucleons are called ***gluons***. The ***gravitons*** are the quanta of the gravitational field. Our present knowledge is that there exist only four different kinds of interaction, which are summarized in Tab. 1.2.

An essential goal of present research is to reduce the four types of interaction to one common force (grand unification). The reduction of the manifold of different particles to two groups of elementary particles was in a certain sense successful, because the classification into two groups with three families in each group gives a rather simple arrangement. However, the number of 24 different particles together with their antiparticles is still large and adding the 15 interaction quanta the total number of elementary particles is 39. Whether the "grand unification" will allow a further reduction or a simpler ordering scheme is still an open question.

This field of research is very interesting because it ventures to the limit where matter and energy might become indistinguishable. It is also closely related to processes occurring at the very beginning of our universe where elementary particles and their interaction played a major role in the extremely hot fireball during the first seconds of the big bang.

Table 1.1 The three families of Leptons and Quarks

Leptons				Quarks			
Name	Symbol	Mass MeV/c^2	Charge	Name	Symbol	Mass MeV/c^2	Charge
Electron	e^-	0.51	−1	Up	u	≈ 300	2/3
Electron neutrino	ν_e	$< 10^{-5}$	0	Down	d	≈ 306	−1/3
Myon	μ^-	−105.66	−1	Charm	c	≈ 1200	2/3
Myon neutrino	ν_μ	$< 10^{-4}$	0	Strange	s	≈ 450	−1/3
Tau-lepton	τ	1840	−1	Top	t	$1.7 \cdot 10^5$	2/3
Tau-neutrino	ν_τ	$< 10^{-4}$	0	Bottom	b	≈ 4300	−1/3

Table 1.2 The four types of interaction between particles (known up to now) and their field quanta. There are 8 gluons, 2 charged (W^+ and W^-) W bosons, 1 neutral boson (Z^0) and probably only 1 graviton with spin $I = 2$

Interaction	Field quantum	Rest mass MeV/c^2
Strong interaction	Gluons	0
El. magn. interaction	Photons	0
Weak interaction	W bosons	81,000
	Z bosons	91,010
Gravitational interaction	Gravitons	0

Atomic Nuclei

Protons and Neutrons can combine to larger systems, the atomic nuclei. The smallest nucleus is the proton as the nucleus of the hydrogen atom. The largest naturally existing nucleus is that of the uranium atom with 92 protons and 146 neutrons. Its diameter is about 10^{-14} m. Besides the nuclei found in nature there are many artificially produced nuclei, which are however, generally not stable but decay into other stable nuclei. Nearly every atom has many isotopes with nuclei differing in the number of neutrons. Meanwhile there is a wealth of information about the strong attractive forces, which keep the protons and neutrons together in spite of the repulse electrostatic force between the positively charged protons.

Atoms

Atomic nuclei together with electrons can form stable atoms, where for neutral atoms the number of electrons equals the number of protons. The smallest atom is the hydrogen atom, which consists of one proton and one electron. The diameter of atoms ranges from $5 \cdot 10^{-11}$ m to $5 \cdot 10^{-10}$ m and is about 10,000 times larger than that of the nuclei, although the mass of the nuclei is about 2000 times larger than that of the electrons. The electrons form a cloud of negative charge around the nucleus. The electro-magnetic interaction between electrons and protons has been investigated in detail and there is a closed theory, called **quantum electrodynamics**, which describes all observed phenomena of atomic physics very well.

The chemical properties of the different atoms are completely determined by the structure of the atomic electron shell. This is illustrated by the periodic system of the elements (Mendelejew 1869, Meyer 1870), where the elements are arranged in rows and columns and ordered according to the number of electrons of the atoms (see Vol. 3). With each new row a new electron shell starts. In each column the number of electrons in the outer shell (valence electrons) is equal and the chemical properties of the elements in the same column are similar. A real understanding of the periodic table could only be reached 60 years later after the quantum theory of atomic structure had been developed.

Molecules

Two or more atoms can combine to form a molecule, where the atoms are held together by electro-magnetic forces. The magnitude of the binding energy depends mainly on the electron density between the nuclei. Biological molecules such as proteins or DNA-molecules may consist of several thousand atoms and have diameters up to 0.1 μm, which is about 1000 times larger than the hydrogen atom. Molecules form the basis of all chemical and biological substances. The properties of these substances depend on the kind and structure of the molecules, such as the geometrical arrangement of the atoms forming the molecule.

Macroscopic Structures, Liquids and Solid Substances

Under appropriate conditions many equal or different atoms can form large macroscopic bodies which can contain a huge number of atoms. Depending on temperature they can exist in the solid or liquid phase. The interaction between the atoms is in principle known (el. magn. forces) but difficult to calculate because of the enormous number of participating atoms ($10^{22}/\text{cm}^3$). Most theoretical treatments therefore use statistical methods. Up to now many characteristics of macroscopic bodies can be calculated and understood from their atomic structure but a general exact theory of liquids and solids, which can explain also finer details, is still not available. Therefore approximations are used where each approximate model can describe special features quite well but others less satisfactorily. Examples are the band structure model, which can explain the electrical conductivity but not as well the elastic properties.

Structure and Dynamics of Our Universe

In our universe all of the constituents discussed so far are present.

- *Free elementary particles* (p^+, n, e^-, photons $h\nu$, also short lived mesons in the cosmic radiation, in the atmosphere of stars and in hot interstellar clouds, in the hot fireball during some minutes after the big bang, of our universe).

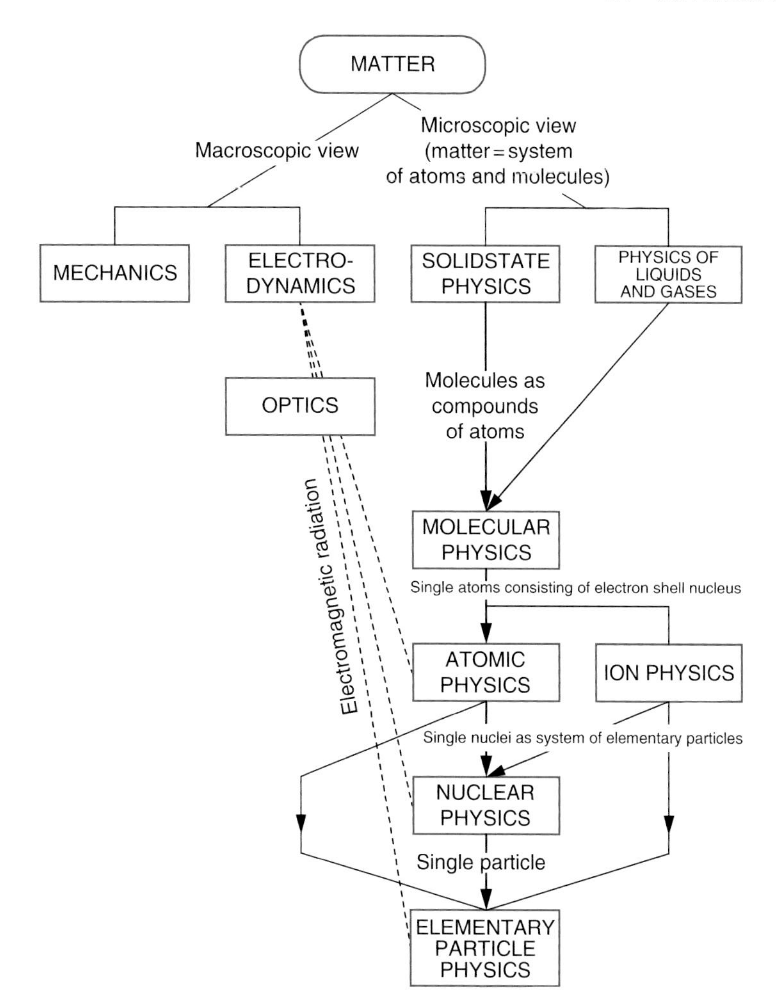

Figure 1.11 Family tree of physics (with kind permission of Dr. H. J. Jodl) [1.14b]

- *Atomic nuclei* in the inner part of stars, in neutron stars and in hot gas clouds.
- *Atoms* in atmospheres of planets and stars and in the interstellar medium.
- *Molecules* in molecular clouds, in comet tails, in interstellar space, in the atmospheres of cold stars and of planets.
- *Solid and liquid macroscopic bodies* (in planets and moons, in meteorites).

For the understanding of the origin and the development of our universe the interactions between these particles have to be known. Although in the early stage of the universe and later on in the interior of stars all four kinds of interaction played a role, gravitation is by far the most important force between celestial bodies such as stars, planets and moons.

Systematic Hierarchy of Physics

The systematic building up principle from small to larger entities discussed so far would suggest to start studies of physics with elementary particles and then proceed gradually to larger systems. However, since the theoretical treatment of elementary particles and nuclear physics is rather difficult, it is advisable from the didactical point of view to go the opposite way, We therefore start with classical physics of macroscopic bodies and proceed then to smaller structures like atoms, molecules, nuclei and elementary particles (see Fig. 1.11). The Physics courses therefore start with classical mechanics and thermodynamics (Vol. 1), continue with electrodynamics and optics (Vol. 2) and then with a basic knowledge of quantum mechanics treat the physics of atoms, molecules, solid and liquid states (Vol. 3) to arrive finally at nuclear physics, elementary particle physics and astrophysics (Vol. 4).

There exist a large number of good books on the subjects treated in this section [1.14b–1.19], which discuss in more detail the questions raised here. In order to gain a deeper understanding of how all this knowledge has been achieved, a more thorough study of basic physics, its fundamental laws and the experimental techniques, which test the developed theories, is necessary. The present textbook will help students with such studies.

1.5 Relations Between Physics and Other Sciences

Since physics deals with the basic elements of our material world it represents in principle the foundations of every natural science. However, until a few decades ago the scientific methods in chemistry, biology and medicine were more empirically oriented. Because of the complex nature of the objects studied in these sciences it was not possible to start the investigations "ab initio" in order to understand the atomic structure of large complex molecules and biological cells to say nothing of the human body and its complex reactions as the research object in medicine. Therefore, in former years a more phenomenological method was preferred.

With refined experimental techniques developed in recent years (electron microscopy, (Fig. 1.12), tunnel microscopy, x-ray structural analysis, neutron diffraction, nuclear magnetic resonance tomography and laser spectroscopy) in many cases it became possible to uncover the atomic structure even of complex molecules such as the DNA (Fig. 1.13). Here physics was helpful in a twofold way: First of all physicists developed, often in cooperation with engineers, the experimental equipment and secondly it provided the theoretical understanding for the atomic basis of the research objects. Therefore the differences in the research methods become less and less important and the cooperation between researchers of different fields is rapidly increasing, indicated by the growing number of interdisciplinary research projects. For example the essential question of the relation between molecular structure and chemical binding is attacked in common efforts by experimental chemists, theoretical quantum chemists and physicists. Overstated one may say that chemistry is applied quantum theory and therefore a branch of physics.

Due to the complex diagnostic techniques in medicine the cooperation between physicists and medical doctors has enormously increased as will be outlined in the next section.

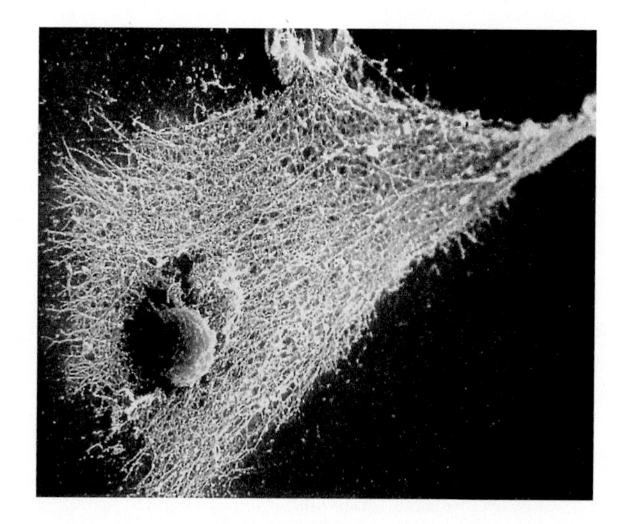

Figure 1.12 Scavenger Cells visualized with an electron microscope

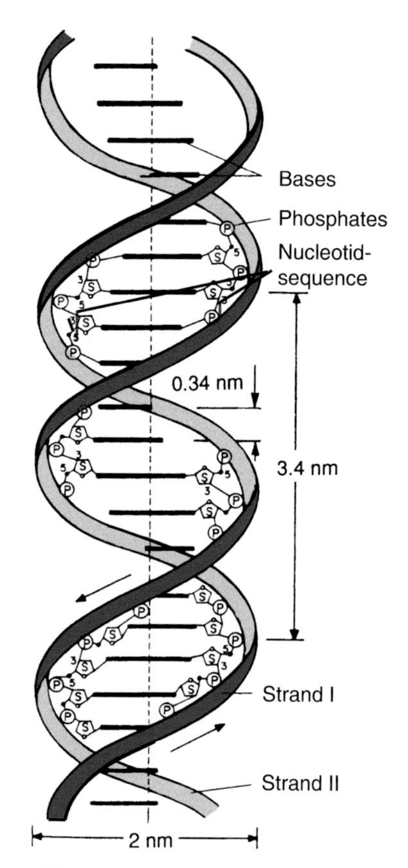

Figure 1.13 Double Helix of DNA (deoxyribonucleic acid)

1.5.1 Biophysics and Medical Physics

Meanwhile biophysics has developed to an independent branch of physics. Some of the many research projects are the physical processes in living cells, e. g. the energy balance during cell processes, the ion transport through cell membranes, the penetration of bacteria and viruses into cells, the different steps of photosynthesis or the visual process. The very sensitive detection techniques for the detection of single molecules, developed in physics laboratories, allow the tracing of single laser excited molecules on their way from outside a cell through membrane channels into the cell interior. In particular the realization of ultra short laser pulses down to below a femtosecond (10^{-15} s) opens for the first time the possibility to view ultrafast processes such as molecular isomerisation.

In recent years, medical physics has been established at many universities and research institutes. The development of new diagnostic techniques and therapy methods are based on experimental techniques invented and optimized in physics laboratories and on new insights about the interaction between radiation and tissue. Examples of such new methods are ultrasonic diagnostics with improved spatial resolution, nuclear magnetic resonance tomography, thermography or laser-induced cell fluorescence. One specific example is the localization of brain tumours by optical coherence tomography and methods for their operation with laser techniques, which are investigated in cooperation between laser physicists and neurosurgeons. [1.20a–1.23b]

1.5.2 Astrophysics

For ages the closest relation with physics had the astronomy, which tried to determine the positions of stars, the movement of planets and the prediction of eclipses. Modern astronomy goes far beyond this type of problems and looks for information about the composition of stars, conditions for their birth and the different stages of their development. It turns out that nearly all branches of physics are necessary in order to solve these problems. Therefore, this part of modern astronomy is called *astrophysics*. The cooperation with physicists who measure in the laboratory processes relevant for the understanding of star atmospheres and the energy production in the interior of stars has greatly improved our knowledge in astrophysics (see Vol. 4). One of the results is for example, that in the universe the same elements are present as can be found on earth and that the same physical laws are valid as known from experiments on earth. The correct interpretation of many astrophysical observations could only be given, because laboratory experiments had been performed which could give unambiguous decisions between several possible explanations of astrophysical phenomena.

The following facts have contributed essentially to the impressive progress in astronomy.

- The development of new large telescopes in the optical, near infrared and radio region, of satellites and space probes (Fig. 1.14) and sensitive detectors.
- New and deeper knowledge in the fields of atomic, nuclear and elementary particle physics, in plasma physics and magneto-hydrodynamics.
- Faster computers for the calculation of more complex models for the present composition, the birth, evolution and final stages of stars [1.24a–1.24c].

Figure 1.14 Last inspection of the Giotto-space probe before its journey to the comet Halley (with kind permission of the European Space Agency ESA)

1.5.3 Geophysics and Meteorology

Although geophysics and meteorology have developed into autonomous disciplines, they are completely based on fundamental physical laws. In particular, in meteorology it is evident how important fundamental physical processes are, such as the interaction of light with atoms and molecules, collisions between electrons, ions, atoms and molecules or light scattering by aerosols and dust particles. Without the detailed understanding of these and other processes the complex preconditions for the local and global climate could not be calculated within a climate model. However, it turns out, that in spite of the knowledge of these basic processes it is often not possible to give a reliable long term weather forecast, because already tiny changes of the present status of the atmosphere could result in huge changes of its future development. The system shows a chaotic behaviour. This astonishing feature has lead to a new branch of physical and mathematical sciences, called chaos research (see Chap. 12). [1.25–1.30b]

1.5.4 Physics and Technology

The application of physical research has pushed the development of our industrial society in a way, which can hardly be overestimated. Examples are the inventions of the steam engine, the electromotor, research on semiconductors, which form the basis of computers, information technology, such as the telephone and extremely fast optical communication over glass fibres, Lasers and their various applications, precision measuring techniques down into the nanometre range. This connection between applied physics and technology has received new impetus through the urgent problems of energy crisis, lack of raw materials, global warming, which have to be solved within a limited time. Urgent problems are, for example

- the development of new energy sources, such as nuclear fusion, which demands a profound knowledge of plasma physics under extreme conditions,

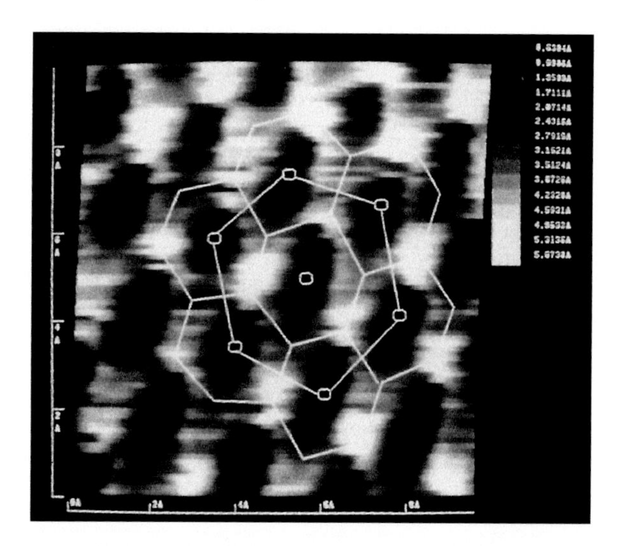

Figure 1.15 Hexagonal structure of a graphite surface, visualized by a tunnel microscope (M. Müller, H. Öchsner, TU Kaiserslautern)

- the optimization of wind converters,
- the development of solar cells with sufficiently high efficiencies,
- increasing the conversion efficiency from heat into electrical energy,
- improving the transport efficiency of energy.

Further examples are the development of reliable electrically driven cars with new designs of batteries, hydrogen technology, magnetically levitated railways (trans-rapid), development of "-clean air cars" etc.

Of particular interest for many branches of industry is the research on new materials such as *met-glasses* (amorphous metals with particular properties such as high tensile strength), compound materials or amorphous semiconductors, which have found meanwhile numerous applications. Surface science (Fig. 1.15) has given the basic understanding for corrosion processes, catalytic effects and the properties of thin films in optics and for the creation of very hard surfaces of tools, which decrease the wear and tear of such tools considerably.

One should keep in mind that for densely populated countries such as Germany, which do not have sufficient raw material at their disposal, technological innovations and inventions of new products as well as progress in environmental protection are essential for a better and safe life in the future. Here physicists encounter great challenges and new ideas and a critical but pragmatic way of thinking are demanded, characteristics, which are trained during the physics education. [1.30a–1.30b]

1.5.5 Physics and Philosophy

Since its beginning in the Greek period, physics always had a close relation to philosophy (see Sect. 1.3). Already for the Greek philosophers recognition in natural sciences gave new directions to the philosophical way of thinking. The essential goal of modern physics is the understanding and the detailed description of our world and the reduction of many observations to a few general laws. The essential point is, that the human consciousness and the attitude against the human surroundings are changed by this new knowledge. The fascinating question, how cognitive faculty is received by communication with other thinking persons and whether the structured mind which allows to process this information to form a unique world view, had been already formed prenatal had been extensively discussed by the great philosopher *Immanuel Kant* (1724–1804) in his famous book *"Kritik der reinen Vernunft"*.

Nowadays biophysicists and neurologists try to understand by well aimed experiments the connection between specified parts of the brain and the storage of information which we receive from outside. All these progress in natural sciences has influenced philosophical theories. Although the approach to this subject is often different for philosophers and scientists, an intense discussion between the representatives of the two camps could remove many misunderstandings and could lead to a more extensive view of our world. If such discussions should be fruitful, both sides have to learn more about the way of thinking and arguing of the other side. The study of physics and its way of arguing can shape the way we are looking onto our world and represents an essential part of our culture.

An important aspect of such cooperation is the critical evaluation of ethical questions related to scientific research, which have found more and more concern in our society. Since the developments in physics and their applications, essentially change our daily life, physicists have to think about the consequences of their scientific results. The research itself is unbiased and value-free. Ethical problems arise when the results of basic research are applied in such a way, that society might be damaged by such applications. For instance, the discovery of nuclear fission by Otto Hahn could be used for peaceful applications as well as to build an atomic bomb; lasers can be used for health treatment in medicine or as laser weapons.

People who demand social relevance for every research projects forget that this is a question of possible applications, which can often not be predicted from basic research. There are many examples where basic research was done without any ideas of possible benefit for the public, such as the beginning of solid state physics, low temperature physics, semiconductor research. [1.31–1.35]

1.6 The Basic Units in Physics, Their Standards and Measuring Techniques

Since any objective description of nature demands quantitative relations between measurements of different objects, which can be expressed by numbers, one has to define units for the results of measurements. This means that every numerical result of a measurement must be expressed in multiples of such units. One needs a scale that can be compared with the measured quantity.

To measure always means to compare two quantities!

There are several possibilities for choosing units. For the length unit for instance one may use units which are given by nature such as foot or the distance between two atoms in a crystal; for the time unit the time interval between two successive heart beats, or the time between two culminations of the sun. A better choice of physical units is to use arbitrary but suitable units, which are conveniently adapted to daily life. Such units have to be defined by standards with which they can be always compared (calibration).

Every standard has to meet the following demands:

- It must be possible to compare with sufficient accuracy the quantity in question with the standard.
- The standard must be reproducible with the demanded accuracy.
- The production and the safekeeping of the standard and the comparison with measurable elements must be possible with justifiable expenditure.

According to these demands ulna, foot or heartbeat period are not good standards, because they are dependent on the person who measures them. They may change with time and are not general constants.

The **quality** of a measurement is judged according to the following aspects:

- How reliable is the measurement?
  Here the experimental apparatus plays an important role, the interpretation of the experimental results by the observer; his ability and experience (see for instance temperature estimations guided by our senses (Chap. 10 or "optical illusions" Vol. 2)).
- How accurate is the measurement, i.e. how large is the maximum possible error of the result?
- Are measurements performed under different experimental conditions reproducible?

Of course, each physical quantity cannot be measured more accurately than the accuracy of the normal's measurement. Therefore such a normal should be chosen which is so accurately defined that it does not represent a limitation for the accuracy of the measurement. For many measurements, a stopwatch or a micrometre-screw might not be accurately enough and should not be used as normal.

The question is now how many basic units are necessary to describe all physical quantities. Since all physical processes go off in space and time one certainly needs basic units for length and time. We will see that all physical quantities can be derived from three basic units for length, time and mass. One would therefore need in principle only these three basic units. It turns out, however that it is useful to add four more basic units for the temperature, the mole fraction of material, for the strength of an electric current and the luminous intensity of radiation sources, because many derived units can be simpler expressed when these four additional units are included [1.37–1.39].

In the following we will discuss the different basic units and also give a short outline of the historical development of this units and their increasing accuracy. This shall illustrate how new measurement techniques have improved the quality of a measurement and asked for new and better standards that could meet the demands for higher accuracy and reproducibility.

1.6.1 Length Units

As length unit the metre (m) was chosen in 1875 which was originally meant as the 1/10,000,000 fraction of the equator quadrant (¼ of the earth circumference). The prototype as the primary standard was kept in Paris. In order to maintain this normal as reproducible as possible, it was realized by the distance between two markers on a platinum-iridium rod with a low thermal expansion coefficient. The rod was kept in a box at 0 °C. More precise later measurements of the earth circumference showed that the metre deviated from the original definition by about 0.02%. The comparison of length standards with this prototype was only possible with a relative uncertainty of 10^{-6}. This means that it is only possible to detect a deviation of larger than 1/1000 mm. This does not meet modern requirements of accuracy.

Therefore in 1960 a new length standard was defined by the wavelength λ of the orange fluorescence line of a discharge lamp filled with the krypton isotope 86 (Fig. 1.16), where the conditions in the krypton lamp (pressure, discharge current and temperature) were fixed. The metre was defined as $1{,}650{,}763.73 \cdot \lambda$. The wavelength λ can be measured with an uncertainty of 10^{-8}, which is 100 times more accurate than the comparison with the original metre standard in Paris.

With increasing accuracy of measurements this standard was again abandoned and a new standard was chosen, which was based on a completely new definition. Since time can be measured much more accurate than length, the length standard was

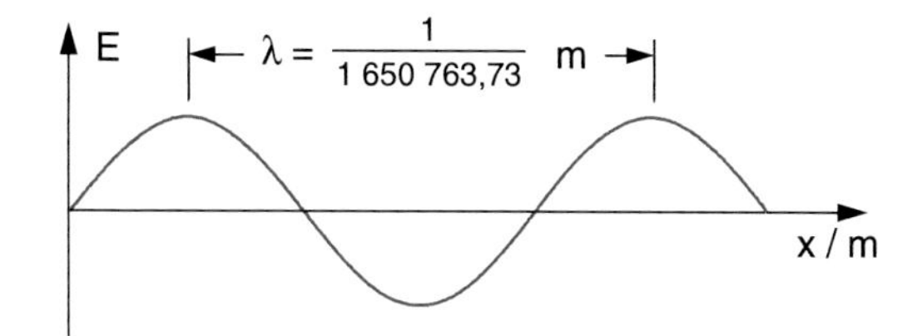

Figure 1.16 The old definition of the length unit, based on the wavelength of a Krypton line (valid from 1960–1983)

Table 1.3 Range of actual lengths in our world

Object	Dimension/m
Radius of the electron	$\leq 10^{-18}$
Radius of the proton	10^{-15}
Distance between atoms in solids	10^{-10}
Thickness of the skin of a soap bubble	10^{-7}
Mean distance between air molecules at 10^5 Pa	10^{-6}
Radius of the earth	$6 \cdot 10^6$
Distance earth–moon	$4 \cdot 10^8$
Distance earth–sun	$1.5 \cdot 10^{11}$
Diameter of the solar system	10^{14}
Distance to the nearest star	$4 \cdot 10^{16}$
Diameter of our galaxy	$3 \cdot 10^{20}$
Extension of the universe	$3 \cdot 10^{25}$

related to time measurements via the velocity c of light. The weighted average of the most precise measurement of the speed of light in vacuum is now defined as

$$c = 299{,}792{,}458\,\text{m/s}\,.$$

This means that the speed of light is no longer a result of new measurements but ***is defined*** as a fixed value.

Definition

The length unit 1 m is now fixed by the following definition:

One metre is the length of the path that is travelled by light in vacuum during the time interval $1/299{,}792{,}485$ s.

From the relation $c = \nu \cdot \lambda$ between speed of light c, frequency ν and wavelength λ of an electro-magnetic wave the wavelength λ of any spectral line can now be determined from the frequency ν (which can be measured with a much higher accuracy than wavelengths) and the defined speed of light (see Sect. 1.6.2 and 1.6.4).

The order of magnitude of length-scales in physics covers the enormous range from 10^{-18} m for the size of elementary particles to 10^{+25} m for the radius of the present universe (Tab. 1.3). It is therefore appropriate to give metre scales in powers of ten. For specific powers a shorthand notation is used, e. g. 10^{-6} m = 1 micrometer (µm); 10^3 m = 1 kilometer (km). These shorthand notations are listed in Tab. 1.4.

In astronomy, the distances are very large. Therefore, appropriate units are used. The astronomical unit AU is the mean distance between earth and sun. The new and more exact definition, adopted 1976 by the International Astronomical Union is the following:

Table 1.4 Labels for different orders of magnitude of length units

1 attometer	= 1 am	$= 10^{-18}$ m
1 femtometer	= 1 fm	$= 10^{-15}$ m
1 picometer	= 1 pm	$= 10^{-12}$ m
1 nanometer	= 1 nm	$= 10^{-9}$ m
1 micrometer	= 1 µm	$= 10^{-6}$ m
1 millimeter	= 1 mm	$= 10^{-3}$ m
1 centimeter	= 1 cm	$= 10^{-2}$ m
1 dezimeter	= 1 dm	$= 10^{-1}$ m
1 kilometer	= 1 km	$= 10^{3}$ m
Often used units in		
– atomic and nuclear physics		
1 fermi = 1 femtometer		$= 10^{-15}$ m
1 X-unit	= 1 XU	$= 1.00202 \cdot 10^{-13}$ m
1 Ångström	= 1 Å	$= 10^{-10}$ m
– astronomy:		
1 astronomical unit		= 1 AU
≈ mean distance earth–sun		$\approx 1.496 \cdot 10^{11}$ m
1 light year	= 1 ly	$= 9.5 \cdot 10^{15}$ m
1 parsec	= 1 pc	$= 3 \cdot 10^{16}$ m = 3.2 ly

Definition

1 AU is the distance to the centre of the sun, which a hypothetical body with negligible mass would have, if it moves on a circle around the sun in 365.256 8983 days.

One light-year (1 ly) is the distance, which light travels in 1 year. An object has a distance of one parsec (1 pc) if the astronomical unit seen from this object appears under an angle of one second of arc (1″) (Fig. 1.17). The distance d of a star, where this angle is α is $d = 1\,\text{AU}/\tan\alpha$. With $\tan 1'' = 4.85 \cdot 10^{-6}$ we obtain

$$1\,\text{pc} = 2.06 \cdot 10^5\,\text{AU} = 3.2\,\text{ly}\,.$$

Note: In some countries other non-metric length units are in use: 1 inch = 2.54 cm = 0.0245 m and 1 yard (1 yd) = 0.9144 m, 1 mile (1 mi) = 1609.344 m.

However, **in this textbook only SI units are used.**

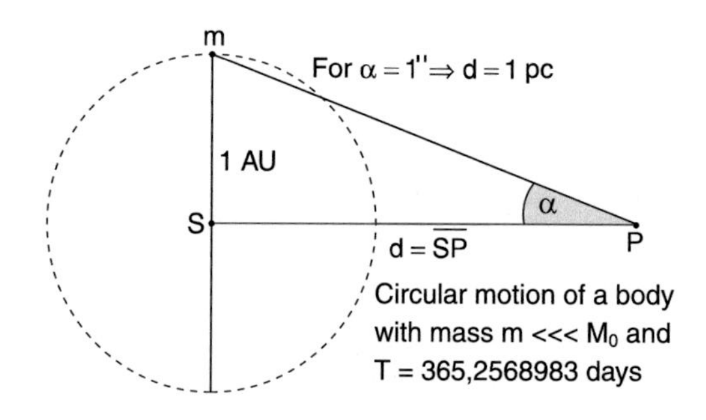

Figure 1.17 Definition of the astronomical units 1 AU and 1 pc

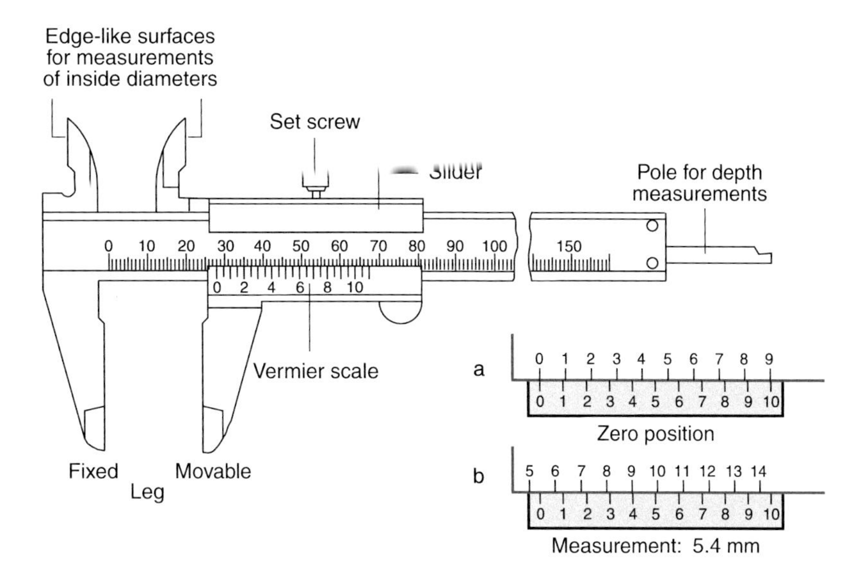

Figure 1.18 Caliper gauge with vernier scale

1.6.2 Measuring Techniques for Lengths

For measuring of lengths in daily life secondary standards are used which are not as accurate as the primary standards but are more readily usable. The accuracy of such standards is adapted to the application for which they are constructed. One simple example is the sliding vernier (Fig. 1.18). Its accuracy is based on the nonius principle. The upper scale is divided into millimetres, the lower scale has 10 scale divisions for 9 mm, which means that every division is 9/10 mm. For the situation in Fig. 1.18b the division mark 9 mm on the upper scale coincides with the division mark 4 on the lower scale. The distance D between the two fold limbs is then

$$D = (9 - 4 \cdot 9/10)\,\mathrm{mm} = 5.4\,\mathrm{mm}\,.$$

The uncertainty of the measurement is about 0.1 mm.

Higher accuracies can be reached with a micrometer screw (Fig. 1.19) where a full turn of the micrometer drum corresponds to a translation of 1 mm. If the scale on the drum is divided into 100 divisions each division mark corresponds to 0.01 mm. The shackle is thermally isolated in order to minimize thermal expansion, With differential micrometer screws, which have two coaxial drums turning into opposite directions, where one drum produces a translation of 1 mm per turn, the other of -0.9 mm in the backward direction, one full turn corresponds now to 0.1 mm. This allows an accuracy of $0.001\,\mathrm{mm} = 1\,\mu\mathrm{m}$. This is about the accuracy limit of mechanical devices.

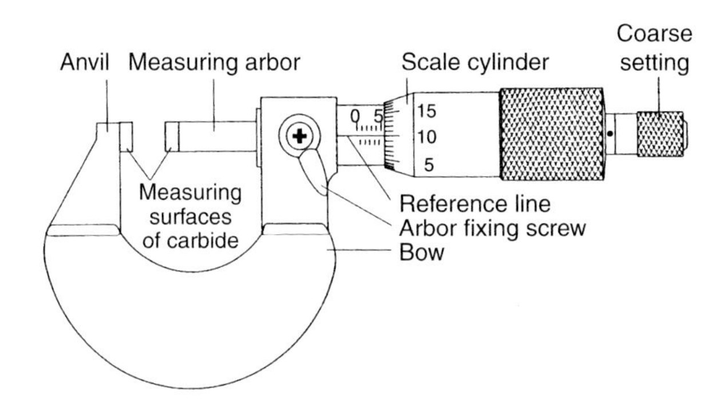

Figure 1.19 Micrometer caliper

More accurate length measurements are based on optical techniques. For distances below 1 m interferometric methods are preferable (see Vol. 2) where lasers (see Vol. 3) are used as light sources. Here distances are compared to the wavelength of the light source. Modern interferometers reach accuracies of $\lambda/100$. With a wavelength of $\lambda = 500\,\mathrm{nm}$ an accuracy of $5\,\mathrm{nm} = 5 \cdot 10^{-9}\,\mathrm{m}$ can be achieved.

Larger distances can be measured via the travel time of a light pulse. For instance the distance of the retro-reflector which the astronauts have positioned on the moon, can be measured within a few cm using laser pulses with 10^{-12} s pulse width (LIDAR technique see Fig. 1.20). Measuring this distance from different locations on earth at different times even allows to detect continental drifts of the earth crust plates [1.41–1.42].

For the exact location of planes, ships or land vehicles the global positioning system GPS has been developed. Its principle is illustrated by Fig. 1.21.

The navigator, who wants to determine his position, measures simultaneously the phases of radio signals emitted from at least four different satellites. The radio signals on frequencies at 1575 MHz and 1227 MHz are modulated. This allows to determine unambiguously the distances d_i from the receiver to the satellites S_i from the measured phase differences ϕ_i. From these four distances d_i the position (x, y, z) of the receiver can be determined with an uncertainty of only a few cm if relativistic effects (see Sect. 3.6) are taken into account! In order to achieve this accuracy, the frequencies of the radio signals must be kept stable within 10^{-10}. This can be realized with atomic clocks which

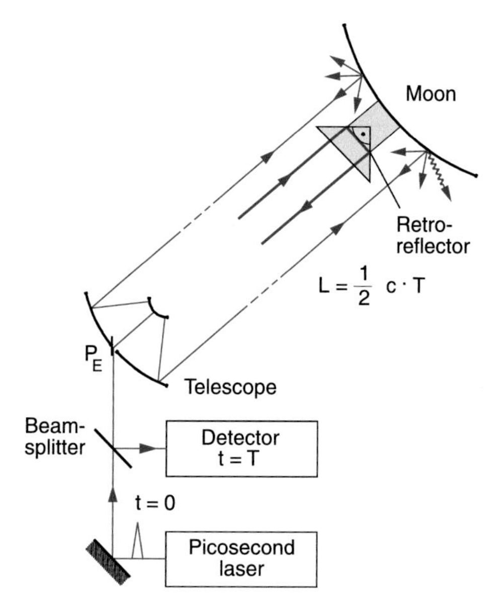

Figure 1.20 Measurement of the distance Earth–Moon with the LIDAR-technique

reach a relative stability $\Delta\nu/\nu = 10^{-14}$. The exact position of the satellites is fixed by radio signals from several stations and receivers at selected precisely known locations on earth. The European Space Agency has launched several satellites for the realization of a new GPS System called Galileo with predicted higher accuracy.

Also a more precise value of the astronomical unit 1 AU can be obtained by measuring the travel time of short light pulses. A radar pulse is sent from the earth to Venus where it is reflected. The time delay between sending and receiving time is measured for a time of closest approach of Venus to Earth. which gives

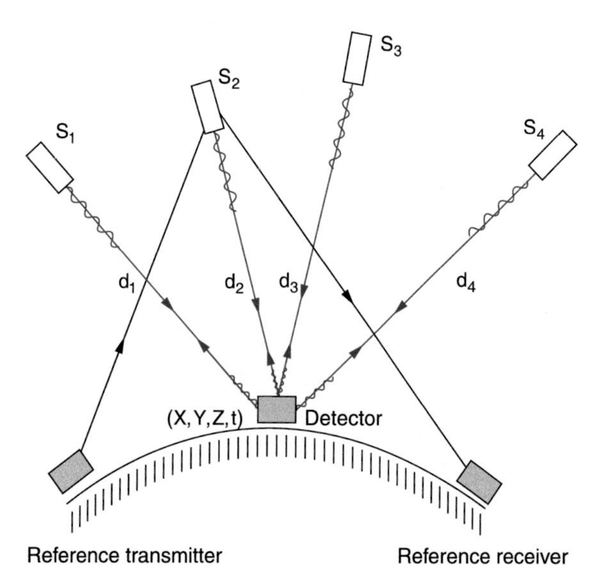

Figure 1.21 Principle of the Global Positioning System GPS

a precise value of the distance between Earth and Venus. From the angle between the radii Earth–Sun and Earth–Venus at the time of the measurement the distance Earth–Sun can be obtained by trigonometric relation in the triangle Earth–Venus–Sun and using Kepler's 3rd law (see Sect. 2.9).

As the result of many different measurements, which became more and more accurate, the Astronomical Union has recommended in 2012 to take the average of these measurements as the **definition** of the Astronomical Unit:

$$1\,\mathrm{AU}^{\mathrm{def}} = 149{,}597{,}870{,}700\,\mathrm{m}\ .$$

1.6.3 Time-Units

The unit of time is the second (1 s). Its initial definition was

$$1\,\mathrm{s} = 1/(60 \cdot 60 \cdot 24)\,\mathrm{d} = (1/86{,}400) \text{ of a solar day}\ ,$$

where a ***solar day*** is defined as the time between two lower culminations of the sun i.e. between two successive midnights.

When the earth rotates around its axis with the angular velocity ω one sun day is $d = (2\pi + \alpha)/\omega$, where the additional angle α is due to the revolution of the earth around the sun. On the other hand a **sidereal day** (= time between two culminations of a star) is $d = 2\pi/\omega$ and therefore shorter by $1/365$ d (Fig. 1.22a). 365.25 solar days correspond to 366.25 sidereal days.

Later it was found that the period of a solar day showed periodic and erratic changes, which can amount up to 30 s per day. (Fig. 1.22b) These changes are caused by the following effects:

- A yearly period due to the non-uniform movement of the earth on an ellipse around the sun (Fig. 1.23 and Sect. 2.9). The velocity v_2 around the perihelion (minimum distance between earth and sun) is larger than v_1 around the aphelion (maximum distance). Since the revolution of the earth around the sun and the rotation of the earth around its axis have the same rotation sense, a solar day is longer around the perihelion than around the aphelion.
- A half-year period due to the inclination of the earth axis against the ecliptic (the plane of the earth's movement around the sun), which causes a variation of the sun culmination at a point P on earth (Fig. 1.24).

In order to eliminate the effect of such changes on the definition of the second, a fictive "mean sun" is defined which (seen by an observer on earth) moves with uniform velocity (= yearly average) along the earth equator. The time between two successive culmination points of this fictive sun defines the mean solar day $\langle d\rangle$. This gives the definition of the **mean solar second**

$$1\,\mathrm{s} = (1/86{,}400)\langle d\rangle\ .$$

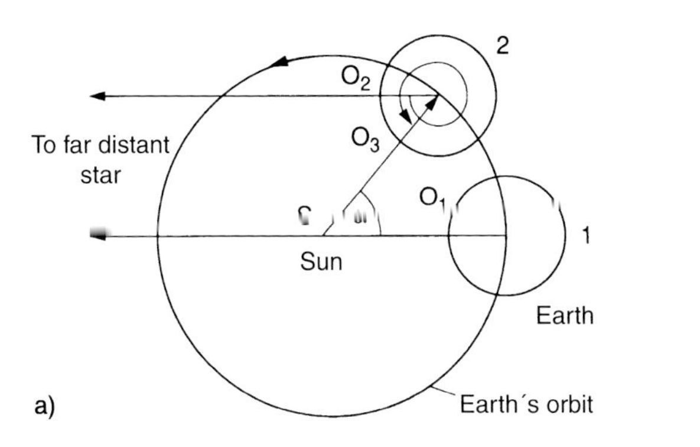

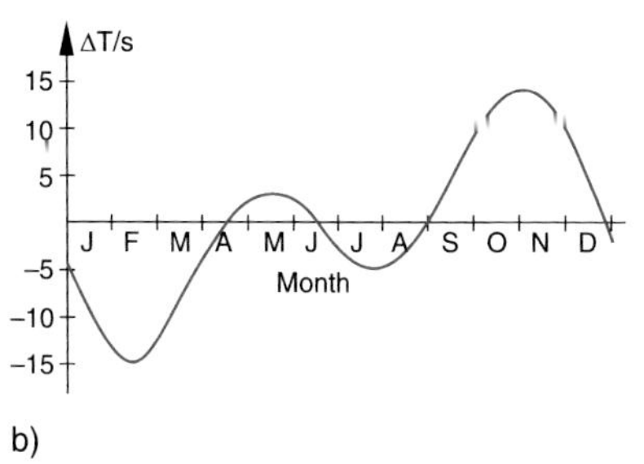

Figure 1.22 **a** Difference between solar day and sidereal day, **b** Difference between the true and the mean solar time

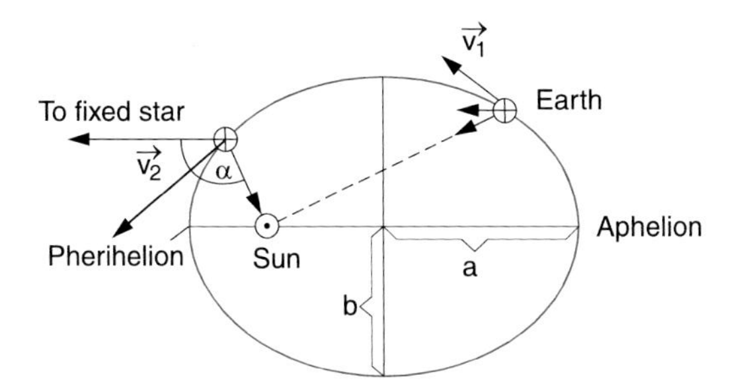

Figure 1.23 Changing velocity of the earth during one revolution on its elliptical path around the sun

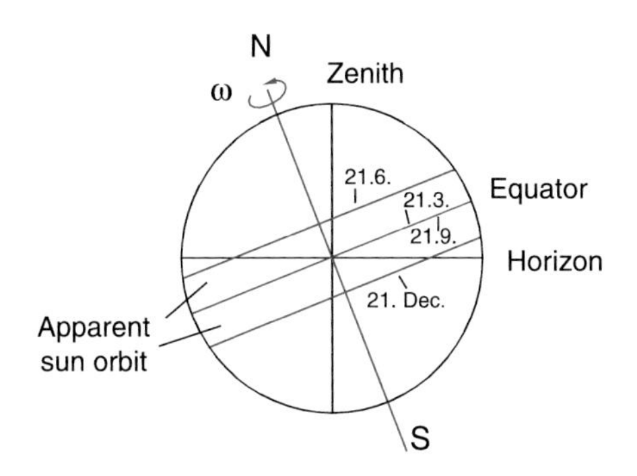

Figure 1.24 Variation of the point of culmination of the sun with a half-year period, due to the inclination of the earth axis

With the development of modern precise quartz clocks it was found that even this mean solar day showed periodic and irregular variations due to changes of the earth's moment of inertia caused by melting of glaciers at the poles, falling of leaves in autumn, volcano eruptions, earth quakes, and turbulent movements of material in the liquid part of the earth's interior. The deviations from the mean sun day amount up to 10 milliseconds per day and cause a relative deviation of $10^{-2}/85{,}400 \approx 10^{-7}$ per day. Therefore the astronomers no longer use the earth rotation as a clock but rather the time span of the **tropical year**. This is the revolution period of the earth around the sun between two successive spring equinoxes, which are the intersection point of the ecliptic and the equator plane vertical to the earth's axis (Fig. 1.25). This tropical year equals the annual period of the mean sun on its way along the earth's equator.

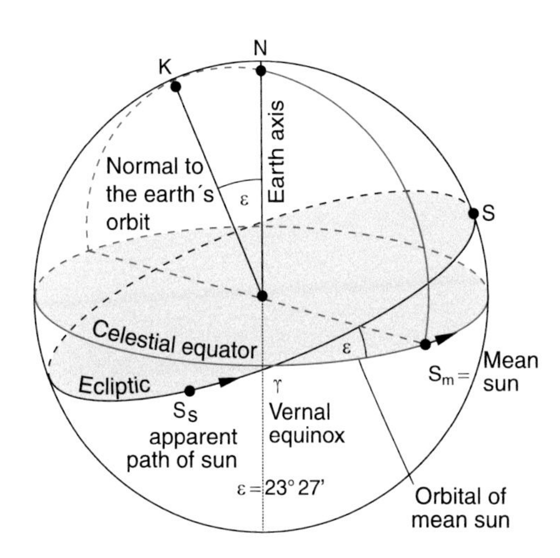

Figure 1.25 Definition of the tropical year

Since even the tropical year suffers in the course of time small variations, the astronomers introduced 1960 the **ephemeris time**, based on tables which give the calculated positions of sun, moon and planets at a given time [1.24d].

The astronomical definition of the second is now 1 s = period of the tropical year 1900 divided by 31,556,925.9747.

For daily use, quartz clocks are more convenient and therefore more useful secondary time standards. Their essential part is a quartz rod of definitive length, which is excited by an external electric high frequency field to length oscillations (see Vol. 2). If the exciting frequency is tuned to the resonance frequency of the quartz rod, the oscillation amplitude reaches a maximum. By appropriate feedback the system becomes a stable self sustaining oscillator which does not need an external frequency source. The relative frequency deviation of good quartz clocks are $\Delta\nu/\nu \leq 10^{-9}$. The second is then counted by the number of oscillation periods per time. Of course, the quartz clocks need a calibration with primary time standards.

The subdivisions of the second and longer time periods are listed in Tab. 1.5.

A better time standard which is still valid up to now is the **caesium atomic clock**. Its principle is illustrated in Fig. 1.26.

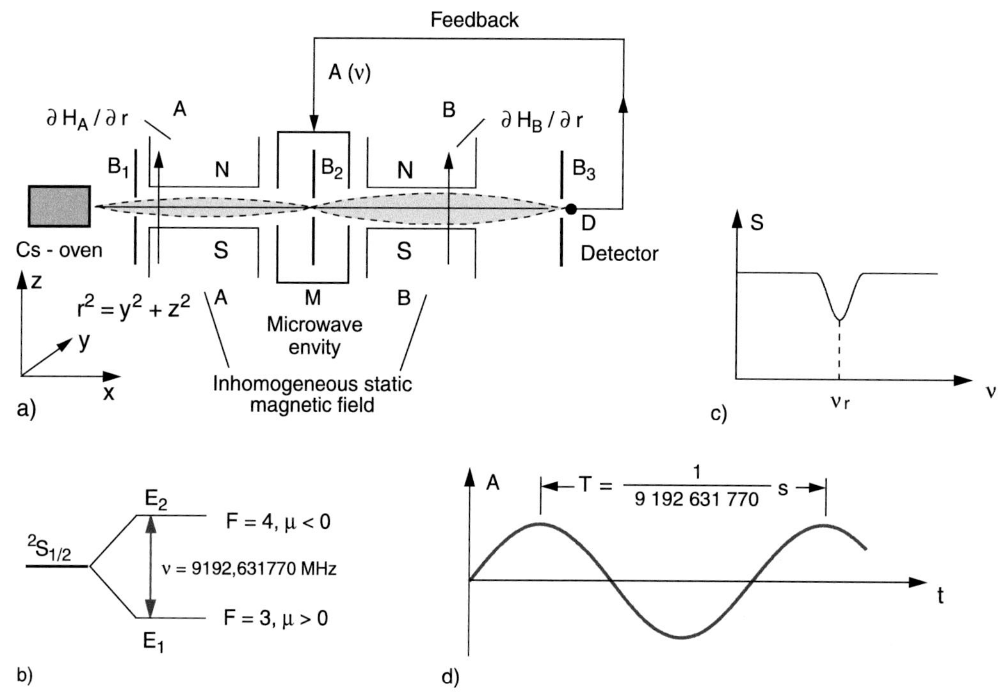

Figure 1.26 Caesium atomic clock. **a** Experimental arrangement; **b** level scheme of the hyperfine-transition; **c** detector signal as a function of the microwave frequency; **d** Definition of the second as a multiple of the oscillation period T

Table 1.5 Labelling of subdivisions of the second or of longer time intervalls

Subdivisions of second		
1 millisecond	$= 1\,\text{ms}$	$= 10^{-3}\,\text{s}$
1 mikrosecond	$= 1\,\mu\text{s}$	$= 10^{-6}\,\text{s}$
1 nanosecond	$= 1\,\text{ns}$	$= 10^{-9}\,\text{s}$
1 picosecond	$= 1\,\text{ps}$	$= 10^{-12}\,\text{s}$
1 femtosecond	$= 1\,\text{fs}$	$= 10^{-15}\,\text{s}$
1 attosecond	$= 1\,\text{as}$	$= 10^{-18}\,\text{s}$
Larger time units		
1 hour	$= 1\,\text{h}$	$= 3.6 \cdot 10^3\,\text{s}$
1 day	$= 1\,\text{d}$	$= 8.64 \cdot 10^4\,\text{s}$
1 year	$= 1\,\text{a}$	$= 3.15 \cdot 10^7\,\text{s}$

Cs-atoms evaporate through a hole in an oven into a vacuum tank. Several apertures collimate the evaporating atoms and form a collimated atomic beam which passes through a microwave resonator M placed between two six pole magnets A and B. They act on atoms with a magnetic moment like an optical lens and focus the atomic beam onto the detector D where the focusing characteristics depend on the hyperfine structure level of the atoms. If the resonator is excited on the frequency $\nu = (E_2 - E_1)/h$ which corresponds to the transition between the two hyperfine levels $F = 3 \to F = 4$ in the $S_{1/2}$ electronic ground state of Cs (Fig. 1.26b) (see Vol. 3), the atoms can absorb the microwave radiation and are transferred from the $F = 3$ level into the $F = 4$ level. In this level they have a different magnetic moment and are therefore defocused in the magnetic field B. They cannot reach the detector D and the measured signal decreases (Fig. 1.26c). When the microwave frequency ν is tuned over the resonance at $\nu = 9{,}192{,}631{,}770\,\text{s}^{-1}$ a dip in the signal $S(\nu)$ appears which is transferred by a feedback circuit to the microwave generator and keeps its frequency exactly on resonance. The frequency stability of the microwave generator is now determined by the atomic transition frequency and serves as a very stable clock, called **atomic clock**. The achieved frequency stability of modern versions of the Cs-clock is $\Delta\nu/\nu = 10^{-15}$.

The new definition of the second, which is still valid today, is: 1 s is the time interval of 9,192,631,770.0 oscillation periods of the Cs clock.

Table 1.6 gives a survey about the time scales of some natural phenomena, which extend from 10^{-23} to 10^{+18} s.

The new definition of the second shows that the time measurement is put down to frequency measurements. The frequency of any oscillating system is the number of oscillation periods per second. Its metric unit is $[1\,\text{s}^{-1}]$ or $[1\text{ hertz} = 1\,\text{Hz}]$. Larger units are

- 1 kilohertz $= 1\,\text{kHz} = 10^3\,\text{s}^{-1}$,
- 1 Megahertz $= 1\,\text{MHz} = 10^6\,\text{s}^{-1}$,
- 1 Gigahertz $= 1\,\text{GHz} = 10^9\,\text{s}^{-1}$,
- 1 Terahertz $= 1\,\text{THz} = 10^{12}\,\text{s}^{-1}$.

Smaller units are

- 1 Millihertz $= 1\,\text{mHz} = 10^{-3}\,\text{s}^{-1}$,
- 1 Microhertz $= 1\,\mu\text{Hz} = 10^{-6}\,\text{s}^{-1}$.

Table 1.6 Time scales occuring in natural phenomena

Natural phenomenon	Period/s
Transit time of light over the diameter of an atomic nucleus	10^{-23}
Revolution period of electron in the hydrogen atom	10^{-15}
Transit time of electrons in old tv-tubes	10^{-7}
Oscillation period of tuning fork	$2.5 \cdot 10^{-3}$
Time for light propagation sun–earth	$5 \cdot 10^{2}$
1 day	$8.64 \cdot 10^{4}$
1 year	$3.15 \cdot 10^{7}$
Time since the first appearance of homo sapiens	$2 \cdot 10^{13}$
Rotational period of our galaxy	10^{16}
Age of our earth	$1.6 \cdot 10^{17}$
Age of universe	$5 \cdot 10^{17}$

1.6.4 How to measure Times

For the measurement of times periodic processes are used with periods as stable as possible. The number of periods between two events gives the time interval between these events if the time of the period is known. Devices that measure times are called **clocks**.

Quartz Clocks: Modern precision clocks are quartz clocks with a frequency instability $\Delta\nu/\nu \leq 10^{-9}$. This means that they deviate per day from the exact time by less than 10^{-4} s.

Atomic Clocks: For higher accuracy demands atomic clocks are used, which are available as portable clocks (Rubidium clocks with $\Delta\nu/\nu \leq 10^{-11}$) or as a larger apparatus fixed in the lab e. g. the Cs clock with $\Delta\nu/\nu \leq 10^{-15}$.

As world-standard Cs-clocks are used at several locations (National Institute of Standards and Technology NIST in Boulder, Colorado, Physikalisch-Technische Bundesanstalt PTB in Braunschweig, Germany and the National Physics Laboratory in Teddington, England) which are connected and synchronized by radio signals. Two of such clocks differ in 1000 years by less than 1 millisecond [1.44a–1.44b].

Frequency stabilized Lasers: A helium-Neon laser with a frequency of 10^{14} Hz can be locked to a vibrational transition of the CH_4 molecule and reaches a stability of 0.1 Hz, which means a relative stability $\Delta\nu/\nu \leq 10^{-15}$ comparable to the best atomic clocks [1.45]. With the recently developed optical frequency comb (see Vol. 3) stabilities $\Delta\nu/\nu \leq 10^{-16}$ could be achieved [1.46]. It is therefore expected, that the Cs-standard will soon be replaced by stabilized lasers as frequency and time standards.

The time resolution of the human eye is about 1/20 s. For the time resolution of faster periodic events stroboscopes can be used. These are pulsed light sources with a tuneable repetition frequency. If the periodic events are illuminated by the light source, a steady picture is seen, as soon as the repetition frequency equals the event frequency. If the two frequencies differ the appearance of the event is changing in time the faster the more the two frequencies differ.

Periodic and non-periodic fast events can be observed with high speed cameras, which reach a time resolution down to 10^{-8} s; with special streak cameras even 10^{-12} s can be achieved. Faster events, such as the rearrangement of the atomic electron shell after excitation with fast light pulses or the dissociation of molecules which occur within femtoseconds ($1\,\text{fs} = 10^{-15}$ s) can be time- resolved with special correlation techniques using ultrafast laser pulses with durations down to 10^{-16} s.

1.6.5 Mass Units and Their Measurement

As the third basic unit the mass unit is chosen. The mass of a body has always a fixed value, even if its form and size is altered or when the aggregation state (solid, liquid or gaseous) changes as long as no material is lost during the changes. The mass is the cause of the gravitational force and for the inertia of a body, which means that all bodies on earth have a weight and if they are moving, magnitude and direction of their velocity is not changing as long as no external force acts on the body (see Sect. 2.6).

As mass unit the kilogram is defined as the mass of a platinum-iridium cylinder, which is kept as the primary mass standard in Paris. (Fig. 1.27)

Initially the kilogram should have been the mass of a cubic decimetre of water at 4 °C (at 4 °C water has its maximum density). Later more precise measurements showed, however, that the mass of 1 dm^3 water was smaller by $2.5 \cdot 10^{-5}\,\text{kg} = 0.025\,\text{g}$ than the primary standard.

In Tab. 1.7 the subunits of the kilogram, which are used today, are listed. For illustration in Tab. 1.8 some examples of masses which exist in nature are presented.

Figure 1.27 Standard kilogram of platin-iridium, kept under vacuum in Paris (https://en.wikipedia.org/wiki/Kilogram#International_prototype_kilogram)

Chapter 1

Table 1.7 Subdivisions and multiples of the kilogram

Unit	Denotion	Mass/kg
1 gram	= 1 g	10^{-3}
1 milligram	= 1 mg	10^{-6}
1 microgram	= 1 µg	10^{-9}
1 nanogram	= 1 ng	10^{-12}
1 pikogram	= 1 pg	10^{-15}
1 ton		10^{3}
1 megaton		10^{9}
1 atomic mass unit	= 1 AMU	$1.6605402 \cdot 10^{-27}$

Table 1.8 The masses of particles and bodies found in nature

Body	Mass/kg
Electron	$9.1 \cdot 10^{-31}$
Proton	$1.7 \cdot 10^{-27}$
Uranium nucleus	$4 \cdot 10^{-25}$
Protein molecule	10^{-22}
Bacterium	10^{-11}
Fly	10^{-3}
Man	10^{2}
Earth	$6 \cdot 10^{24}$
Sun	$2 \cdot 10^{30}$
Galaxy	$\sim 10^{42}$

Masses can be measured either by their inertia or they weight, since both properties are proportional to their mass and unambiguously defined (see Sect. 2.6). The inertia of a mass is measured by the oscillation period of a spring pendulum. Here the mass measurement is reduced to a time measurement.

The weight of a mass is determined by comparison with a mass normal on a spring balance or a beam balance and therefore reduced to a length measurement. Today balances are available with a lower detection limit of at least 10^{-10} kg (magnetic balance, electromagnetic balance, quartz fibre microbalance).

Note: In some countries non-metrical units are used: 1 pound = 0.453 kg.

1.6.6 Molar Quantity Unit

As already mentioned in the beginning of this section in addition to the three basic units for length, time and mass four further units (molar quantity, temperature, electric current and luminous intensity of a radiation source) are introduced because of pragmatic reasons. Strictly speaking they are not real basic units because they can be expressed by the three basic units.

Definition

The unit of molar quantity is the mol, which is defined as follows:

1 mol is the amount of a substance that consist of as many particles as the number N of atoms in 0.012 kg of the carbon nuclide ^{12}C.

These particles can be atoms, molecules, ions or electrons. The number N of particles per mol with the numerical value $N = 6.02 \cdot 10^{23}/\text{mol}$, is called **Avogadro's number** (*Amedeo Avogadro* 1776–1856).

Example

1 mol helium has a mass of 0.004 kg, 1 mol copper corresponds to 0.064 kg, one mol hydrogen gas H_2 has the mass $2 \cdot 0.001\,\text{kg} = 0.002\,\text{kg}$. ◄

1.6.7 Temperature Unit

The unit of the temperature is 1 Kelvin (1 K). This unit can be defined by the thermo-dynamic temperature scale and can be reduced to the kinetic energy of the molecules (see Sect. 10.1.4). Because of principal considerations and also measuring techniques, which are explained in Chap. 10, the following definition was chosen:

1 Kelvin is the fraction (1/273.16) of the thermodynamic temperature of the triple point of water.

The triple point is that temperature where all three phases of water (ice, liquid water and water vapour) can simultaneously exist (Fig. 1.28).

There are plans for a new definition of 1 K which is independent on the choice of a special material (here water). It reads:

1 Kelvin is the temperature change which corresponds to a change $\Delta(kT) = 1.3806505 \cdot 10^{-23}$ Joule of the thermal energy kT, where $k = 13{,}806{,}505 \cdot 10^{-23}\,\text{J/K}$ is the Boltzmann constant.

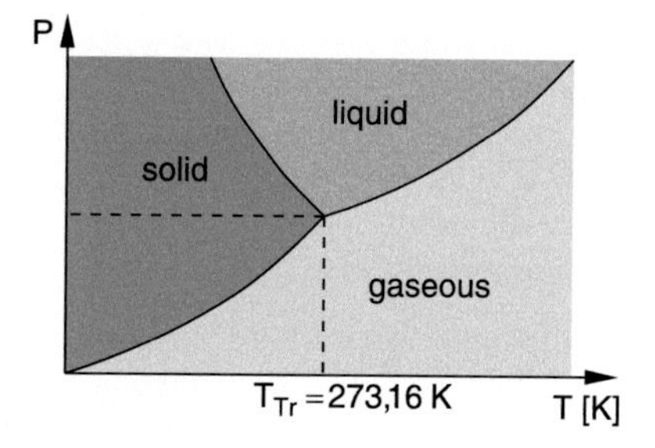

Figure 1.28 Phase diagram and triple point of water

New very accurate measurements of the Boltzmann constant allow a much better definition of the temperature T with an uncertainty of $\Delta T/T \leq 8 \cdot 10^{-6}$.

1.6.8 Unit of the Electric Current

The unit of the electric current is 1 Ampere (1 A) (named after Andre-Marie Ampère 1775–1836). It is defined as follows:

1 Ampere corresponds to a constant electric current through two straight parallel infinitely long wires with a distance of 1 m which experience a mutual force of $2 \cdot 10^{-7}$ Newton per m wire length (Fig. 1.29).

The definition of the electric current unit is therefore based on the measurement of the mechanical quantities length and force (see Vol. 2)

1.6.9 Unit of Luminous Intensity

The luminous intensity of a radiation source is the radiation power emitted into the solid angle 1 Sterad $= 1/(4\pi)$. It could be defined in Watt/Sterad, which gives the radiation power independent of the observing human eye. However, in order to characterize the visual impression of the light intensity of a light source, the spectral characteristics of the radiation must be taken into account, because the sensitivity of the human eye depends on the wavelength. Therefore the definition of the light intensity is adapted to the spectral sensitivity maximum of the eye at a wavelength $\lambda = 555$ nm. The luminosity unit is called 1 **candela** (1 cd).

1 cd is the radiation power of (1/6839)W/Sterad emitted by a source at the frequency 540 THz ($\lambda = 555$ nm) into a selected direction.

Note: 1. The luminous intensity of a source can differ for different directions.
2. The definition of the candela is related to the radiation power in Watt/Sterad, which shows that the candela is not a basic unit.

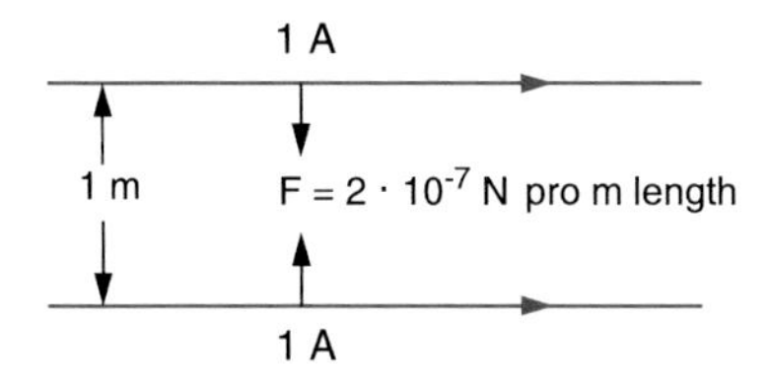

Figure 1.29 Illustration how the unit of the electric current is defined

1.6.10 Unit of Angle

Plane angles are generally measured in degrees of arc. The full angle of a circle is 360°. The subdivisions are minutes of arc ($1° = 60'$) and seconds of arc ($1' = 60'' \rightarrow 1° = 3600''$). Often it is convenient to use dimensionless units by reducing angle measurements to length measurements of the arc length L of a circle, which corresponds to the angle α (Fig. 1.30).

The circular measure (radian) of the angle α is defined as the ratio L/R of circular arc L and radius R of the circle. The unit of this dimensionless quantity is 1 radian (rad) which is realized for $L = R$. Since the total circumference of the circle is $2\pi R$ the angle $\alpha = 360°$ in the unit degrees corresponds to $\alpha = 2\pi$ in the units radian = rad.

The conversion from radians to degrees is

$$1 \text{ rad} = \frac{360°}{2\pi} = 57.296° = 57°17'45'' .$$

While the plane angle $\alpha = L/R$ cuts the arc with length L out of a circle with radius R the solid angle $\Omega = A/R^2$ is the angel of a cone that cuts the area $A = \Omega R^2$ out of a full sphere with area $4\pi R^2$ and radius R (Fig. 1.31). The dimensionless unit of the solid angle is 1 steradian (1 sr) for which $A = R^2$.

Definition

1 sr is the solid angle of a cone which cuts an area $A = 1\text{ m}^2$ out of the unit sphere with $R = 1$ m.

Since the total surface of a sphere is $4\pi R^2$ the total solid angle around the centre of the sphere with $A = 4\pi R^2$ is $\Omega = 4\pi$.

The three planes xy, xz, yz through the positive coordinate axis $+x$, $+y$, $+z$ cut a sphere around the origin $(0,0,0)$ into 8 octands, The solid angle of one octand is

$$\Omega = \frac{1}{8} \cdot 4\pi = \frac{1}{2}\pi \ sr .$$

Note: The numerical values of the units for the basic physical quantities discussed so far have been often adapted by *the International Comission for Weights and Measures* (CIPM for

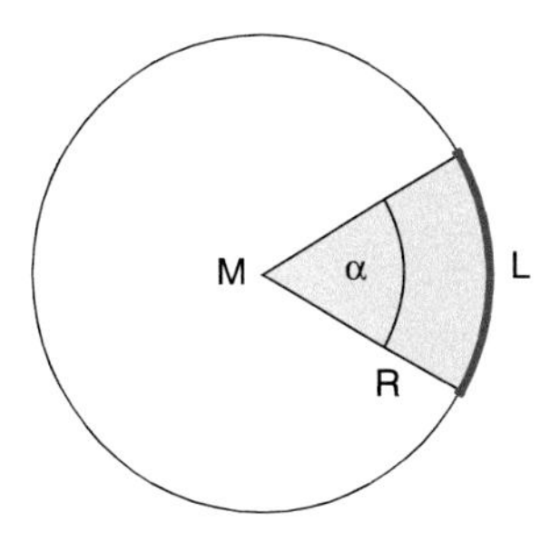

Figure 1.30 To the definition of the radian $\alpha = L/R$

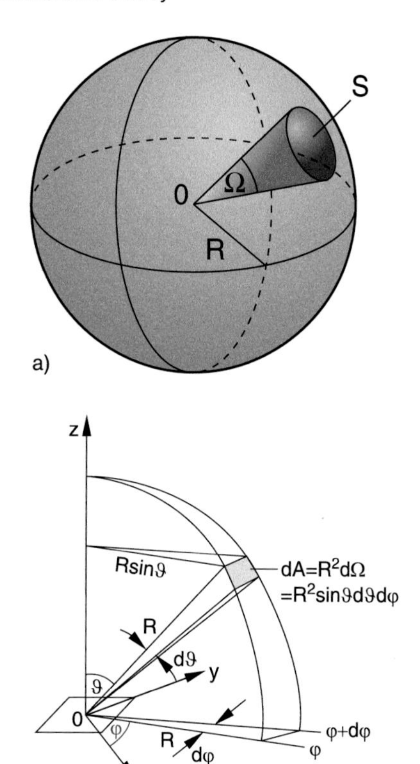

Figure 1.31 **a** To the definition of the solid angle Ω; **b** Illuration of the solid angel element $\mathrm{d}\Omega = \mathrm{d}A/r^2$

the French *comite international des poids et measures*) in order to take into account the results of new and more accurate measurements. At present, considerations are made to reduce all quantities to combinations of fundamental constants in order to give them more accurate and time independent values. This has been realized up to now only for the length unit which is defined through the fixed speed of light and the frequency of the Cs-clock. This might be soon generalized to all physical units in order to get a system of time-independent values for the units which do not need to be corrected in future times.

One example is the mass unit. There are many efforts in several laboratories to create a better and more accurately defined mass normal. One realistic proposal is a large silicon single crystal in form of a polished sphere, where the atomic distances in the crystal have been precisely measured with X-ray interferometry. This allows the determination of the total number of atoms in the crystal and the mass of the crystal can be related to the mass of a silicon atom and is therefore reduced to atomic mass units and the Avogadro constant [1.48a]. Although it has been shown, that such a mass normal would be more accurate ($\Delta\,\mathrm{m/m} \le 10^{-8}$) and would represent a durable mass standard, it has not yet been internationally acknowledged.

Similar considerations are discussed for the temperature unit 1 K which might be reduced to the Boltzmann constant k (see above).

1.7 Systems of Units

As has been discussed in Sect. 1.6 the three basic quantities and their units in physics are

- length with the unit 1 Meter = 1 m
- time with the unit 1 second = 1 s
- mass with the unit 1 kilogram = 1 kg

with four additional quantities

- molar quantity with the unit 1 mole = 1 mol
- temperature with the unit 1 Kelvin = 1 K
- electric current with the unit 1 Ampere = 1 A
- radiation luminosity with the unit 1 candela = 1 cd

where these four quantities can be reduced in principle to the three basic quantities and are therefore no real basic quantities.

All other quantities in physics can be expressed by these 3 basic quantities with the additional 4 quantities for convenient use. This will be shown for each derived quantity in this textbook when the corresponding quantity is introduced.

Each physical quantity is defined by its unit and its numerical value. For instance the speed of light is $c = 2.9979 \cdot 10^8$ m/s or the earth acceleration $g = 9.81\,\mathrm{m/s^2}$ etc.

In a physical equation all summands must have the same units.

These units or the products of units are called the dimension of a quantity. The check, whether all summands in a equation have the same dimension is called dimensional analysis. It is a very helpful tool to avoid errors in conversion of different systems of units.

Each physical quantity can be expressed in different units, for example, times in seconds, minutes or hours. The numerical value differs for the different units. For instance the velocity $v = 10\,\mathrm{m/s}$ equals $v = 36\,\mathrm{km/h}$. In order to avoid such numerical conversions one can use a definite fixed system of units.

If the three basic units are chosen as

- 1 m for the length unit,
- 1 s for the time unit,
- 1 kg for the mass unit.

The system is called the *mks-system*. If the unit Ampere for the electric current is added, the system is called the mksA.-system, often named the **SI-System** after the French nomenclature *System International d'Unites*. It has the very useful advantage that for the conversion from mechanical into electrical and magnetic units all numerical conversion factors have the value 1. All basic units and also the units derived from them are called **SI units**.

In theoretical physics often the *cgs system* is used, where the basic units are 1 cm (instead of 1 m), 1 Gramm (instead of 1 kg) and only the time unit is 1 s as in the SI-system. According to international agreements from 1972 only the SI-system should be used. **In this textbook exclusively SI units are used.**

For a more detailed representation of the subject the reader is referred to the literature [1.37–1.39,1.50].

1.8 Accuracy and Precision; Measurement Uncertainties and Errors

Every measurement has in different ways uncertainties which can be minimized by a reliable measuring equipment and careful observation of the measurement. The most important part in the measuring process is an experienced and critical experimenter, who can judge about the reliability of his results. The final results of an experiment must be given with error limits which show the accuracy of the results. There are two different kinds of possible errors: *Systematic* and *statistical* errors.

1.8.1 Systematic Errors

Most systematic errors are caused by the measuring equipment, as for instance a wrong calibration of an instrument, ignoring of external conditions which can influence the results of the measurement (temperature change for length measurements, lengthening of the string of a threat pendulum by the pendulum weight or air pressure changes for measurements of optical path length). Recognizing such systematic errors and their elimination for precision measurements is often difficult and demands the experience and care of the experimental physicist. Often the influence of systematic errors on the experimental results is underestimated. This is illustrated by Fig. 1.32, which shows the results of measurements of the electron mass during the time from 1950 up to today with the error bars given by the authors. Due to improved experimental techniques the error bars become smaller and smaller in the course of time. The dashed line gives the value that is now accepted. One can clearly see, that all the error bars given by the authors are too small because the systematic error is much larger.

The electron mass can be only determined by a combination of different quantities. For example, from the deflection of electrons in magnetic fields one can only get the ratio e/m of electron charge e and electron mass m. According to the CODATA publication of NIST the value accepted today is $m_e = 9.10938291(40) \cdot 10^{-31}$ kg, where the number in brackets gives the uncertainty of the last two digits.

1.8.2 Statistical Errors, Distribution of Experimental Values, Mean Values

Even if systematic errors have been completely eliminated, different measurements of the same quantity (for instance the falling time of a steel ball from the same heights) do not give the same results. The reasons are inaccurate reading of meters, fluctuations of the measured quantity, noise of the detection system etc. The measured results show a distribution around a mean value. The width of this distribution is a measure of the quality of the results. It is illustrative to plot this distribution of measured values x_i in a histogram (Fig. 1.33), where the area of the rectangles represents the number $n_i \Delta x = \Delta n_i$ of measurements which have given a value within the interval from $x_i - \Delta x/2$ to $x_i - \Delta x/2$.

The mean value $\overline{x}$ of n measurements is chosen in such a way that the sum of the squares of the deviations $(\overline{x} - x_i)$ from the

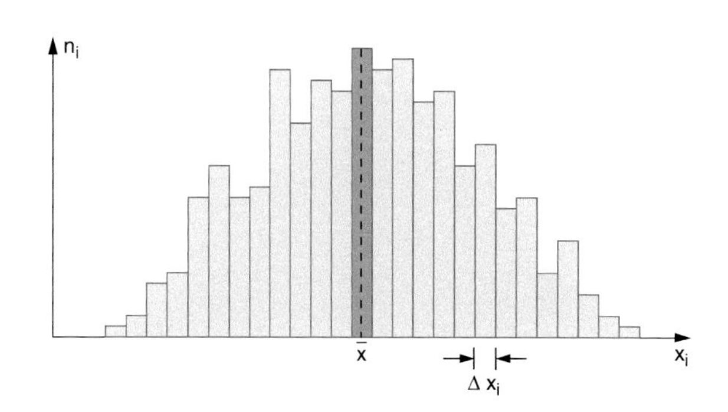

Figure 1.33 Typical histogram of the statistical distribution of measured values x_i around the mean value $\overline{x}$

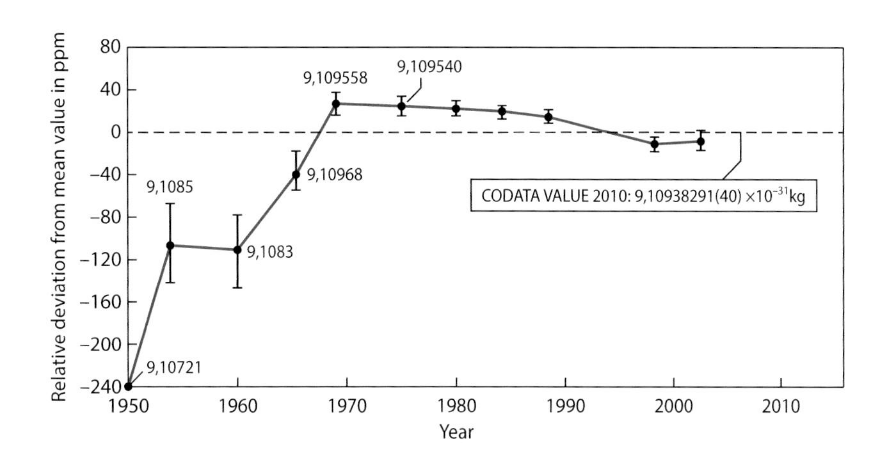

Figure 1.32 Historical values of measurements of the electron mass in units of 10^{-31} kg, demonstrating the underestimation of measuring uncertainties. The relative deviations Δ m/m from the best value accepted today are plotted in units of 10^{-6} (ppm = parts per million)

mean value become a minimum, i.e.

$$S = \sum_{i=1}^{n} (\overline{x} - x_i)^2 = \text{Minimum} . \tag{1.1}$$

For the derivative follows:

$$\frac{dS}{d\overline{x}} = 2 \cdot \sum_{i=1}^{n} (\overline{x} - x_i) = 0 .$$

This gives for the mean value

$$\overline{x} = \frac{1}{n} \sum_{i=1}^{n} x_i , \tag{1.2}$$

the arithmetic mean value of all measured results. Because $\sum(\overline{x} - x_i) = 0$ the arithmetic mean is at the centre of the symmetric distribution, which means that the sum of the positive deviations equals the sum of the negative ones. Contrary to this symmetric distribution of values with statistical errors the systematic errors cause deviations in one direction.

The question is now how much the mean value deviates from the true, but generally unknown value of the measured quantity. We will now prove, that after elimination of all systematic errors the arithmetic mean converges against the true value x_w with increasing number of measurements. This means:

$$x_w = \lim_{n\to\infty} \frac{1}{n} \sum_{i=1}^{n} x_i . \tag{1.3}$$

Since it is impossible to perform infinitely many measurements the true value generally remains unknown!

We define the absolute error of the measured value x_i as the difference

$$e_i = x_w - x_i \tag{1.4}$$

and the absolute error of the mean value as the difference

$$\varepsilon = x_w - \overline{x} . \tag{1.5}$$

The mean values of these errors are

$$\langle e \rangle = (1/n) \sum e_i ; \quad \langle e^2 \rangle = (1/n) \sum e_i^2 .$$

From (1.2) it follows

$$\varepsilon = x_w - \overline{x} = \frac{1}{n} \sum_{i=1}^{n} (x_w - x_i) = \frac{1}{n} \sum_{i=1}^{n} e_i . \tag{1.6}$$

The absolute error ε of the arithmetic mean $\overline{x}$ equals the arithmetic mean $\langle e_1 \rangle = \frac{1}{n} \sum e_i$ of the absolute errors of the individual results x_i.

From (1.6) we obtain by squaring

$$\varepsilon^2 = \frac{1}{n^2} \Big(\sum_i e_i \Big)^2 = \frac{1}{n^2} \sum_i e_i^2 + \frac{1}{n^2} \sum_i \sum_{j \neq i} e_i e_j \approx \frac{1}{n^2} \sum_i e_i^2 . \tag{1.7}$$

The double sum converges for $n \to \infty$ towards zero because for any fixed number j it follows from (1.3)

$$\lim_{n\to\infty} \frac{1}{n} \sum_{i=1}^{n} e_i = x_w - x_w = 0 .$$

Since for statistical errors the deviations e_i and e_j are uncorrelated.

The quantity

$$\sigma = \sqrt{\langle e^2 \rangle} = \sqrt{\frac{\sum (x_w - x_i)^2}{n}} \tag{1.8a}$$

is named **standard deviation** or *root mean square deviation*. It equals the square root of the squared arithmetic mean $\langle e^2 \rangle$

$$\langle e^2 \rangle = \frac{1}{n} \sum e_i^2 = \frac{1}{n} \sum_{i=1}^{n} (x_w - x_i)^2 \tag{1.8b}$$

The smaller quantity

$$\sigma_m = \sqrt{\varepsilon^2} = \sqrt{\frac{1}{n^2} \sum e_i^2} = \frac{1}{n} \sqrt{\sum_i (x_w - x_i)^2} \tag{1.8c}$$

is the **mean error of the arithmetic mean $\overline{x}$**.

From (1.8a)–(1.8c) we can conclude

$$\sigma_m = \frac{\sigma}{\sqrt{n}} . \tag{1.9}$$

The mean error of the arithmetic mean equals the mean error of the individual measurements divided by the square root of the total number n of measurements.

In the next section it will be shown that σ approaches a constant value for $n \to \infty$. Equation 1.9 then implies, that $\lim \sigma_m = 0$, which means that the arithmetic mean $\bar{x}$ approaches the true value x_w for a sufficiently large number n of measurements.

1.8.3 Variance and its Measure

Since for a finite number n of measurements the true value of the measured quantity is generally unknown, also the absolute errors and the mean errors σ and σ_m cannot be directly determined. We will now show how σ and σ_m are related to quantities that can be directly measured.

We introduce instead of the unknown deviations $e_i = x_w - x_i$ of the measured values from the true value x_w the deviations $v_i = \bar{x} - x_i$ from the mean value, which contrary to e_i are known values.

According to (1.4) and (1.5) we can express the v_i by the quantities e_i and ε.

$$\begin{aligned} v_i &= \bar{x} - x_i \\ &= x_w - x_i - (x_w - \bar{x}) \\ &= e_i - \varepsilon \, . \end{aligned} \tag{1.10}$$

The mean square deviation of the measured values x_i from the arithmetic mean $\bar{x}$ can then be written as

$$\begin{aligned} s^2 &= \frac{1}{n}\sum_i v_i^2 = \frac{1}{n}\sum_i (e_i - \varepsilon)^2 \\ &= \frac{1}{n}\left[\sum_i e_i^2 - \left(\frac{2\varepsilon}{n}\sum_i e_i\right) + \varepsilon^2\right] \\ &= \frac{1}{n}\sum_i \left(e_i^2 - \varepsilon^2\right) , \end{aligned} \tag{1.11}$$

because according to (1.6) $\varepsilon = (1/n)\sum e_i$. The comparison with (1.8a,b,c) yields the relation

$$s^2 = \frac{1}{n}\sum_i \left(e_i^2 - \varepsilon^2\right) = \sigma^2 - \sigma_m^2 \, . \tag{1.12}$$

From the equations (1.8b), (1.9) and (1.12) it follows

$$\begin{aligned} s^2 &= \left(\frac{1}{n} - \frac{1}{n^2}\right)\sum_i (x_w - x_i)^2 \\ &= \frac{n-1}{n^2}\sum_i (x_w - x_i)^2 \\ &= (n-1)\sigma_m^2 = \frac{n-1}{n}\sigma^2 \, . \end{aligned}$$

For the standard deviation of the individual results x_i we obtain the mean deviation of the arithmetic mean value

$$\sigma^2 = \frac{n}{n-1}s^2 \to \sigma = \sqrt{\frac{\sum (\bar{x} - x_i)^2}{n-1}} , \tag{1.13}$$

which can be obtained from measurements and is therefore a known quantity.

For the mean deviation of the arithmetic mean (also called standard deviation of the arithmetic means) we get

$$\sigma_m^2 = \frac{1}{n-1}s^2 \to \sigma_m = \sqrt{\frac{\sum (\bar{x} - x_i)^2}{n(n-1)}} . \tag{1.14}$$

Example

For 10 measurements of the period of a pendulum the following values have been obtained:
$T_1 = 1.04$ s; $T_2 = 1.01$ s; $T_3 = 1.03$ s; $T_4 = 0.99$ s; $T_5 = 0.98$ s; $T_6 = 1.00$ s; $T_7 = 1.01$ s; $T_8 = 0.97$ s; $T_9 = 0.99$ s; $T_{10} = 0.98$ s.

The arithmetic mean is $\overline{T} = 1.00$ s. The deviations $x_i = T_i - \overline{T}$ of the values T_i from the mean $\overline{T}$ are
$x_1 = 0.04$ s; $x_2 = 0.01$ s; $x_3 = 0.03$ s; $x_4 = -0.01$ s; $x_5 = -0.02$ s; $x_6 = 0.00$ s; $x_7 = 0.01$ s; $x_8 = -0.03$ s; $x_9 = -0.01$ s; $x_{10} = -0.02$ s. This gives

$$\Sigma(T_i - \langle T \rangle)^2 = \Sigma x_i^2 = 46 \cdot 10^{-4}\,\mathrm{s}^2 \, .$$

The standard deviation is then

$$\sigma = \sqrt{(46 \cdot 10^{-4}/9)} = 2.26 \cdot 10^{-2}\,\mathrm{s}$$

and the standard deviation of the arithmetic mean is

$$\sigma_m = \sqrt{(46 \cdot 10^{-4}/90)} = 0.715 \cdot 10^{-2}\,\mathrm{s} \, .$$ ◄

1.8.4 Error Distribution Law

In the histogram of Fig. 1.33 the resolution of the different measured values depends on the width Δx_i of the rectangles. All values within the interval Δx_i are not distinguished and regarded to be equal. If Δn_i is the number of measured values within the interval Δx_i and k the total number of intervals Δx_i we can write Eq. 1.2 also as

$$\bar{x} = \frac{1}{n}\sum_{i=1}^{k} \Delta n_i \cdot x_i \quad \text{with} \quad \sum_{i=1}^{k} \Delta n_i = n \, . \tag{1.15}$$

The histogram in Fig. 1.33 can be obtained in a normalized form when we plot the fraction n_i/n ($n_i = \Delta n_i/\Delta x_i$ and $n = \sum \Delta n_i$),

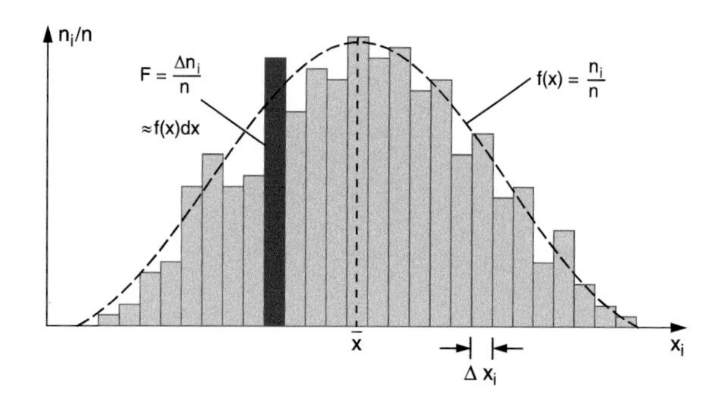

Figure 1.34 Normalized statistical distribution and distribution function of measured data

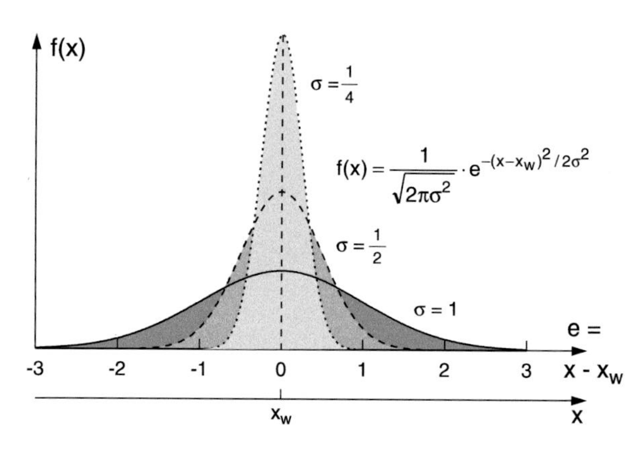

Figure 1.35 Error distribution function (Gaussian distribution) around the true value x_w for different standard deviations σ

which represents the number of measured values within the unit interval $\Delta x_i = 1$ (Fig. 1.34). The heights of the rectangles give these fractions. The quantity $\Delta n_i/n$ can be regarded as the probability that the measured values fall within the interval Δx_i. With increasing number n of measurements we can decrease the width of the intervals Δx_i which means that the total number k of all intervals increases. For $\Delta x_i \to 0$ the number $k \to \infty$ and $\Delta n_i \to 0$ but the fraction $\Delta n_i/\Delta x_i$ approaches a finite value. The sum $n = \sum n_i \Delta x_i$ which represents the total number of measured values, stays of course constant. The discontinuous distribution of the histogram in Fig. 1.34 converges against a continuous function $f(x)$, which is shown in Fig. 1.34 as black dashed curve. The function $f(x)$ is defined as

$$f(x) = (1/n)\lim(\Delta n_i/\Delta x_i) = (1/n)\cdot \mathrm{d}n/\mathrm{d}x\,; \qquad (1.16a)$$

$f(x)$ is the continuous **distribution function**. The product $f(x)\cdot \mathrm{d}x$ gives the probability to find a measured value in the interval from $x - \mathrm{d}x/2$ to $x + \mathrm{d}x/2$. From (1.16a) and $\sum n_i \Delta x_i$ follows the normalization

$$\int f(x)\mathrm{d}x = \lim\left[(1/n)\sum n_i \Delta x_i\right] = 1\,. \qquad (1.16b)$$

This means that the probability to find a measured value somewhere within the total x-range must be of course 100% = 1, because it has to be somewhere in this range.

The integral $\int f(x)\mathrm{d}x$ represents the area under the black curve which is normalised to 1 because the ordinate in Fig. 1.34 is given as the normalized quantity n_i/n.

The standard deviation σ is a measure for the width of the distribution $f(x)$. Its square σ^2 gives, as for the discontinuous distribution (1.8b), the mean square deviation of the arithmetic mean from the true value x_w, which determines the centre of the symmetric curve $f(x)$

$$\sigma^2 = \langle e^2\rangle = \int_{-\infty}^{+\infty} (x_w - x)^2 f(x)\,\mathrm{d}x\,. \qquad (1.17)$$

The quantity σ^2 is named the **variance**.

If only statistical errors contribute, the normalized distribution of the measured values can be described by the normalized Gauss-function

$$f(x) = \frac{1}{\sqrt{2\pi\sigma^2}}\, e^{-(x-x_w)^2/2\sigma^2}\,, \qquad (1.18)$$

which has its maximum at $x = x_w$. The inflection points of the curve $f(x)$ are at $x = x_w \pm \sigma$. The full width between the inflection points where $f(x) = f(x_w)/e$ is therefore 2σ. The distribution $f(x)$ is symmetrical around its centre at x_w (Fig. 1.35). For infinitely many measurements the arithmetic mean x becomes x_w.

When the standard deviation has been determined from n measurements, the probability $P(\sigma)$ that further measured values fall within the interval $x = x_w \pm \sigma$ and are therefore within the standard deviation from the true value. It is given by the integral

$$P\left(|x_w - x_i| \le \sigma\right) = \int_{x_w-\sigma}^{x_w+\sigma} f(x)\mathrm{d}x\,. \qquad (1.19)$$

When inserting (1.18) the integral can be solved and yields the numerical values

$$\begin{aligned} P\left(e_i \le \sigma\right) &= 0.683 \quad (68\%\ \text{confidence range})\\ P\left(e_i \le 2\sigma\right) &= 0.954 \quad (95\%\ \text{confidence range})\\ P\left(e_i \le 3\sigma\right) &= 0.997 \quad (99.7\%\ \text{confidence range})\,. \end{aligned}$$

The results of a measurement are correctly given with the 68% confidence range as

$$x_w = \bar{x} \pm \sigma\,. \qquad (1.20)$$

This means that the true value falls with a probability of 68% within the uncertainty range from $\bar{x} - \sigma$ to $\bar{x} + \sigma$ around the arithmetic mean, if all systematic errors has been eliminated. The relative **accuracy** of a measured value x_w is generally given as $\sigma/\bar{x}$.

Cautious researchers extend the uncertainty range to $\pm 3\sigma$ and can than state that their published result lies with the probability of 99.7%, which means nearly with certainty within the given limits around the arithmetic mean. The result is then given as

$$x_w = \bar{x} \pm 3\sigma = \bar{x} \pm 3\cdot\sqrt{\frac{\sum (x_i - \bar{x})^2}{n-1}}\,. \qquad (1.21)$$

Since the arithmetic mean is more accurate than the individual measurements often the uncertainty range is given as the standard deviation σ_m of the arithmetic mean which is smaller than σ. The result is then given as

$$x_w = \bar{x} \pm \sigma_m = \bar{x} \pm \sqrt{\frac{\sum (x_i - \bar{x})^2}{n(n-1)}} . \quad (1.22)$$

Example

For our example of the measurements of the periods of a pendulum the result would be given with the 69% confidence range as

$$T_w = \langle T \rangle \pm \sigma = (1.000 \pm 0.025)\,\mathrm{s}$$

and for the 99.7% confidence range as

$$T_w = \langle T \rangle \pm 3\sigma = (1.000 \pm 0.075)\,\mathrm{s} .$$

For the standard deviation σ_m of the arithmetic mean one gets

$$T_w = \langle T \rangle \pm \sigma_m = (1.0000 \pm 0.0079)\,\mathrm{s} .$$

The relative uncertainty of the true value is then with a probability of 68%

$$\Delta T_w / T_w = 7.9 \cdot 10^{-3} = 0.79\% .$$ ◀

Remark. For statistical processes where the measured quantity is an integer number $x_i = n_i$ that statistically fluctuates (for instance the number of electrons emitted per sec by a hot cathode, or the number of decaying radioactive nuclei per sec) one obtains instead of the Gaussian function (1.18) a Poisson distribution

$$f(x) = \frac{\bar{x}^x}{x!} e^{-\bar{x}} \qquad x = \text{integer number} . \quad (1.23)$$

1.8.5 Error Propagation

If a quantity $y = f(x)$ depends in some way on the measured quantity x, the uncertainty $\mathrm{d}y$ is related to $\mathrm{d}x$ by (Fig. 1.36)

$$\mathrm{d}y = \frac{\mathrm{d}f(x)}{\mathrm{d}x} \mathrm{d}x . \quad (1.24)$$

When the quantity x has been measured n-times its standard deviation is

$$\sigma_x = \sqrt{\frac{\sum (\bar{x} - x_i)^2}{n-1}} ,$$

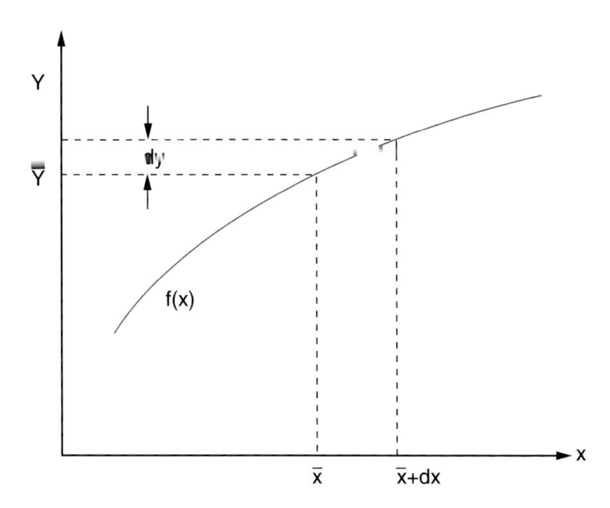

Figure 1.36 Error propagation for a function $y = f(x)$

which results in the standard deviation of the y_i values

$$\sigma_y = \sqrt{\frac{\sum (\bar{y} - y_i)^2}{n-1}} = \sqrt{\frac{\sum \left(f(\bar{x}) - f(x_i) \right)^2}{n-1}}$$
$$= \left(\frac{\mathrm{d}f(x)}{\mathrm{d}x} \right)_{\bar{x}} \cdot \sigma_x . \quad (1.25)$$

Often the value of a quantity, which is not directly accessible to measurements, and its uncertainty should be known. Examples are the density of a body which can be calculated as the ratio of mass and volume of the body, or the acceleration of a moving mass which is determined from measurements of distances and times.

The question is now: What is the accuracy of a quantity $f(x, y)$, if the uncertainties of the measurements of x and y are known.

Assume one has made n measurements of the quantity x from which the uncertainty range of the arithmetic mean is determined as

$$\bar{x} \pm \sigma_x = \bar{x} \pm \sqrt{\frac{\sum v_i^2}{n-1}} \quad \text{with} \quad v_i = x_i - \bar{x}$$

and m measurements of the quantity y with the mean

$$\bar{y} \pm \sigma_y = \bar{y} \pm \sqrt{\frac{\sum u_k^2}{m-1}} \quad \text{with} \quad u_k = y_k - \bar{y} ,$$

one obtains the quantity

$$\begin{aligned} f_{ik} &= f(x_i, y_k) = f(\bar{x} + v_i, \bar{y} + u_k) \\ &= f(\bar{x}, \bar{y}) + v_i \left(\frac{\partial f(x,y)}{\partial x} \right)_0 \\ &\quad + u_k \left(\frac{\partial f(x,y)}{\partial y} \right)_0 + \ldots \end{aligned} \quad (1.26)$$

by a Taylor expansion, where $(\partial f / \partial x)_0$ is the partial derivative for the values x, y. Often the deviations v_i and u_k are so small that the higher powers in the expansion can be neglected. The mean value of all f_{ik} is then

$$\begin{aligned}\overline{f} &= \frac{1}{n\cdot m}\sum_i\sum_k f_{ik} = \frac{1}{n\cdot m}\sum_{i=1}^{n}\sum_{k=1}^{m}\Big[f(\overline{x},\overline{y})\\ &\quad + v_i\frac{\partial f}{\partial x}(\overline{x},\overline{y}) + u_k\frac{\partial f}{\partial y}(\overline{x},\overline{y})\Big]\\ &= \frac{1}{n\cdot m}\Big[n\cdot m\cdot f(\overline{x},\overline{y}) + m\sum_i v_i\frac{\partial f}{\partial x}\\ &\quad + n\sum_k u_k\frac{\partial f}{\partial y}\Big] = f(\overline{x},\overline{y})\ ,\end{aligned} \tag{1.27}$$

because $\partial f/\partial x|_{x,y}$ is constant and $\sum v_i = \sum u_i = 0$.

The arithmetic mean $\overline{f}$ of all values f_{ik} equals the value $f(\overline{x},\overline{y})$ of the function $f(x,y)$ for the arithmetic means $\overline{x},\overline{y}$ of the measured values $x_i y_k$.

In books about error calculus [1.53a–1.55] it is shown, that the standard deviation of the derived quantity f is related to the standard deviations σ_x and σ_y of the measured values x_i, y_k by

$$\sigma_f = \sqrt{\sigma_x^2\left(\frac{\partial f}{\partial x}\right)^2 + \sigma_y^2\left(\frac{\partial f}{\partial y}\right)^2}\ . \tag{1.28}$$

The mean uncertainties σ_x and σ_y propagate to the uncertainty σ_f of the derived mean $f(x,y)$. The 68% confidence range of the true value $f_w(x,y) = f(x_w,y_w)$ is then

$$f_w(x,y) = f(\overline{x},\overline{y}) \pm \sqrt{\sigma_x^2\left(\frac{\partial f}{\partial x}\right)^2 + \sigma_y^2\left(\frac{\partial f}{\partial y}\right)^2}\ . \tag{1.29}$$

With the inequality $\sqrt{a^2+b^2} \le |a| + |b|$ the uncertainty (1.29) can be also written as

$$\Delta f = f_w - f(\overline{x},\overline{y}) \le \left|\sigma_x\frac{\partial f}{\partial x}\right| + \left|\sigma_y\frac{\partial f}{\partial y}\right|\ . \tag{1.30}$$

Examples

1. The length L is divided into two sections x and y with $L = x + y$ which are separately measured (Fig. 1.37a). The final result of L is then, according to (1.27) and (1.28) with $\partial f/\partial x = \partial f/\partial y = 1$,

$$\overline{L} = \overline{x} + \overline{y} \pm \sqrt{\sigma_x^2 + \sigma_y^2}\ .$$

This means: the mean error of a sum (or a difference) equals the square root of the sum of squared errors of the measured values.

$x_w = \overline{x} \pm \sigma_x$, $y_w = \overline{y} \pm \sigma_x$

x y

L

$\overline{L} = \overline{x} + \overline{y} \pm \sqrt{\sigma_x^2 + \sigma_y^2}$

a)

$\overline{A} = \overline{x}\cdot\overline{y} \pm \sqrt{(\overline{y}\cdot\sigma_x)^2 + (\overline{x}\cdot\sigma_y)^2}$

$A = x \cdot y$ y

x

b)

Figure 1.37 **a** Mean error of a length measurement, that consists of two individual measurements x and y; **b** Error propagation for the measurement of an area $x \cdot y$

2. The area $A = x \cdot y$ of a rectangle shall be determined for the measured side lengths x and y. The true values of x and y are

$$\begin{aligned}&x_w = \overline{x} \pm \sigma_x\ ,\quad y_w = \overline{y} \pm \sigma_y\ ,\\ &\frac{\partial A}{\partial x}(\overline{x},\overline{y}) = \overline{y}\ ,\quad \frac{\partial A}{\partial y}(\overline{x},\overline{y}) = \overline{x}\ ,\\ &\overline{A} = \overline{x}\cdot\overline{y} \pm \sigma_{xy}\\ &\quad = \overline{x}\cdot\overline{y} \pm \sqrt{(\overline{y}\cdot\sigma_x)^2 + (\overline{x}\cdot\sigma_y)^2}\ .\end{aligned}$$

The relative error of the product $A = x \cdot y$

$$\frac{\sigma_{xy}}{A} = \sqrt{\left(\frac{\sigma_x}{\overline{x}}\right)^2 + \left(\frac{\sigma_y}{\overline{y}}\right)^2}$$

equals the Pythagorean sum of the relative errors of the two factors x and y.

3.

$$\begin{aligned}&y = \ln x\ ;\quad x = \overline{x} \pm \sigma_x \Rightarrow \frac{\partial y}{\partial x} = 1/x\\ &\overline{y} = \ln\overline{x} \pm \sigma_x/\overline{x}\end{aligned}$$

The mean absolute error of the logarithm of a measured value x equals the relative error of x. ◄

1.8.6 Equalization Calculus

Up to now we have discussed the case, where the same quantity has been measured several times and how the arithmetic means of the different measured values and its uncertainty can be obtained. Often the problem arises that a quantity $y(x)$, which depends on another quantity x shall be determined for different values of x and the question is how accurate the function $y(x)$ can be determined if the measured values of x have a given uncertainty.

Example

1. A falling mass passes during the time t the distance $d = \frac{1}{2}g \cdot t^2$ and its velocity $v = g \cdot t$ is measured at different times t_i.

2. The change of the length $\Delta L = L_0 \cdot \alpha \cdot \Delta T$, a long rod with length L and thermal expansion coefficient α experiences for a temperature change ΔT, is measured at different temperatures T. ◀

In our first example distances and velocities are measured at different times. The goal of these measurements is the accurate determination of the earth acceleration g. In the second example length changes and temperatures are measured in order to obtain the thermal expansion coefficient α as a function of temperature T.

The relation between $y(x)$ and x can be linear (e. g. $v = g \cdot t$), but may be also a nonlinear function (e. g. a quadratic or an exponential function). Here wee will restrict the discussion to the simplest case of linear functions, in order to illustrate the application of equalization calculus to practical problems.

This will become clear with the following example.

Example

We consider the linear function

$$y = ax + b$$

and will answer the question, how accurate the constants a and b can be determined when y is calculated for different measured values of x.

Solution

It is often the case that the values x can be measured more accurately than y. For instance for the free fall of a mass the times can be measured with electronic clocks much more accurately than distances or velocities. In such cases the errors of x can be neglected compared to the uncertainties of y. This reduces the problem to the situation depicted in Fig. 1.38. The measured values $y(x)$ are given by points and the standard deviation by the length of the error bars.

The question is now, how it is possible to fit a straight line to the experimental points in such a way that the uncertainties of the constants a and b become a minimum.

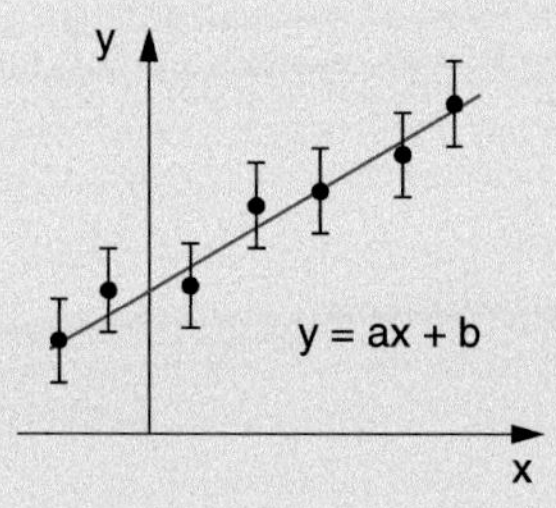

Figure 1.38 Equalization calculus for the function $y = ax + b$, when the values x_i can be measured much more accurate than the values y_i

This is the case if the sum of the squared deviations reaches a minimum.

$$S = \sum (y_i - ax_i - b)^2 \qquad (1.31)$$

Differentiating (1.31) gives the two equations. (Note that a and b are here the variables!)

$$\frac{\partial S}{\partial a} = -2\sum_{i=1}^{n} x_i(y_i - ax_i - b) = 0 \qquad (1.32a)$$

$$\frac{\partial S}{\partial b} = -2\sum_{i=1}^{n} (y_i - ax_i - b) = 0\,. \qquad (1.32b)$$

Rearranging yields

$$a \cdot \sum_i x_i^2 + b \cdot \sum_i x_i = \sum_i x_i y_i \qquad (1.33a)$$

$$a \cdot \sum_i x_i + b \cdot n = \sum_i y_i\,. \qquad (1.33b)$$

The last equation is matched exactly for the point $(\overline{x}, \overline{y})$ with the mean coordinates

$$\overline{x} = (1/n)\sum x_i; \quad \overline{y} = (1/n)\sum y_i\,.$$

Inserting these values into (1.33b) yields after division by the number n the relation

$$a \cdot \overline{x} + b = \overline{y}\,.$$

This proves that the point $(\overline{x}, \overline{y})$ fulfils the equation and is located in Fig. 1.38 exactly on the red straight line.

From (1.33b) one obtains for the slope b of the straight line

$$b = y - ax = (1/n)\sum y_i - (a/n)\sum x_i\,.$$

Inserting this into (1.33a) gives with the abbreviation

$$d = n \cdot \sum x_i^2 - \left(\sum x_i\right)^2,$$

the constants a and b as

$$a = \frac{n\left(\sum x_i y_i\right) - \left(\sum x_i\right)\left(\sum y_i\right)}{d}, \qquad (1.34a)$$

$$b = \frac{\left(\sum x_i^2\right)\left(\sum y_i\right) - \left(\sum x_i\right)\left(\sum x_i y_i\right)}{d}. \qquad (1.34b)$$

The true constants a and b give the true values $y_w(x_i) = ax_i + b$ within the 68% confidence limits $y_i \pm \sigma_y$ around the mean value $\overline{y}$. From (1.18) and (1.19) one obtains the probability $P(y_i)$ to find the measured value y_i

$$P(y_i) \propto \frac{1}{\sigma_y} e^{-(y_i - ax_i - b)^2/2\sigma_y^2}\,. \qquad (1.35)$$

The uncertainties of the constants a and b can be obtained according to the error propagation rules. The results are

$$\sigma_a^2 = \frac{n \cdot \sigma_y^2}{d}, \quad \sigma_b^2 = \frac{\sigma_y^2 \sum x_i^2}{d}. \qquad (1.36)$$

The full width between the two points $P(y_w)/e$ is $\sigma_y \cdot \sqrt{2}$.

For more information on error analysis and regression fits see [1.53a–1.56]. ◀

Summary

- Physics deals with the basic building blocks of our world, their mutual interactions and the synthesis of material from these basic particles.
- The gain of knowledge is pushed by specific experiments. Their results serve for the development of a general theory of nature and to confirm or contradict existing theories.
- Experimental physics started in the 16th century (e. g. Galilei, Kepler) and led to a more and more refined and extensive theory, which is, however, even today not yet complete and consistent.
- All physical quantities can be reduced to three basic quantities of length, time and mass with the basic units 1 m, 1 s, and 1 kg. For practical reasons four more basic quantities are introduced for molar mass (1 mol), temperature (1 K), electric current (1 A) and the luminous power (1 cd).
- The system of units which uses these basic 3 + 4 units is called SI-system with the units 1 m, 1 s, 1 kg, 1 mol, 1 K, 1 A and 1 cd.
- Every measurement means the comparison of the measured quantity with a normal (standard).
- The *length standard* is the distance which light travels in vacuum within a time interval of (1/299,792,458) s. The *time standard* is the transition frequency between two hyperfine levels in the Cs atom measured with the caesium atomic clock. The present *mass standard* is the mass of the platinum-iridium kilogram, kept in Paris.
- Each measurement has uncertainties. One distinguishes between systematic errors and statistical errors. The mean value of n independent measurements with measured values x_i is chosen as the arithmetic mean

$$\overline{x} = \frac{1}{n}\sum_{i=1}^{n} x_i ,$$

which meets the minimum condition

$$\sum_{i=1}^{n} (\overline{x} - x_i)^2 = \text{minimum}.$$

If all systematic errors could be eliminated the distribution of the measured values x show the statistical Gaussian distribution

$$f(x) \propto e^{-(x-x_w)^2/2\sigma^2} ,$$

about the most probable value, which equals the true value x_w. The half-with of the distribution between the points $f(x_w)/e = f(x_w \pm \sigma)$ is $\sigma \cdot \sqrt{2}$ Within the range $x = x_w \pm \sigma$ fall 68% of all measured values. The standard deviation σ of individual measurements is

$$\sigma = \sqrt{\frac{\sum(\overline{x} - x_i)^2}{n-1}} ,$$

the standard deviation of the arithmetic means is

$$\sigma_m = \sqrt{\frac{\sum(\overline{x} - x_i)^2}{n(n-1)}} .$$

The true value x_w lies with the probability of 68% within the interval $x_w \pm \sigma$, with a probability of 99.7% in the interval $x_w \pm 3\sigma$. The Gaussian probability distribution for the measured values x_{oi} has a full width at half maximum of

$$\Delta x_{1/2} = 2\sigma\sqrt{2 \cdot \ln 2} = 2.35\sigma .$$

Problems

1.1 The speed limit on a motorway is 120 km/h. An international commission decides to make a new definition of the hour, such that the period of the earth rotation about its axis is only 16 h. What should be the new speed limit, if the same safety considerations are valid?

1.2 Assume that exact measurements had found that the diameter of the earth decreases slowly. How sure can we be, that this is not just an increase of the length of the meter standard?

1.3 Discuss the following statement: "The main demand for a length standard is that its length fluctuations are smaller than length changes of the distances to be measured".

1.4 Assume that the duration of the mean solar day increases by 10 ms in 100 years due to the deceleration of the earth rotation. a) After which time would the day length be 30 hours? b) How often would it be necessary to add a leap second in order to maintain synchronization with the atomic clock time?

1.5 The distance to the next star (α-Centauri) is $d = 4.3 \cdot 10^{16}$ m. How long is the travelling time of a light pulse from this star to earth? Under which angle appears the distance earth-sun from α-Centauri? If the accuracy of angular measurements is $0.1''$ what is the uncertainty of the distance measurement?

1.6 A length L is seen from a point P which is 1 km (perpendicular to L) away from the centre of L, under an angle of $\alpha = 1°$. How accurate can the length be determined by angle measurements from P if the uncertainty of α is $1'$?

1.7 Why does the deviation of the earth orbit from a circle cause a variation of the solar day during the year? Give some arguments why the length of the mean solar day can change for different years?

1.8 How many hydrogen atoms are included in 1 kg of hydrogen gas?

1.9 How many water molecules H_2O are included in 1 litre water?

1.10 The radius of a uranium nucleus ($A = 238$) is $8.68 \cdot 10^{-15}$ m. What is its mean mass density?

1.11 The fall time of a steel ball over a distance of 1 m is measured 40 times, with an uncertainty of 0.1 s for each measurement. What is the accuracy of the arithmetic mean?

1.12 For which values of x has the error distribution function $\exp[-x^2/2]$ fall to 0.5 and to 0.1 of its maximum value?

1.13 Assume the quantity $x = 1000$ has been measured with a relative uncertainty of 10^{-3} and $y = 30$ with $3 \cdot 10^{-3}$. What is the error of the quantity $A = (x - y^2)$?

1.14 What is the maximum relative error of a good quartz clock with a relative error of 10^{-9} after 1 year? Compare this with an atomic clock ($\Delta\nu/\nu = 10^{-14}$).

1.15 Determine the coefficients a and b of the straight line $y = ax + b$ which gives the minimum squared deviations for the points $(x, y) = (0, 2); (1, 3); (2, 3); (4, 5)$ and $(5, 5)$. How large is the standard deviation of a and b?

References

1.1a. R.D. Jarrard, *Scientific Methods.* An online book: https://webct.utah.edu/webct/RelativeResourceManager/288712009021/Public%20Files/sm/sm0.htm
1.1b. K. Popper, *The Logic of of Scientific Discovery.* (Routledge 2002)
1.2. R. Feynman, *The Character of Physical Laws* (Modern Library, 1994)
1.3. W. Heisenberg, *Physics and Philosophy* (Harper Perennial Modern Thoughts, 2015)
1.4. A. Franklin, Phys. Persp. **1**(1), 35–53 (1999)
1.5. L. Susskind, G. Hrabovsky, *The theoretical Minimum. What you need to know to start doing Physics* (Basic Books, 2013)
1.6. J.Z. Buchwald, R. Fox (eds.), *The Oxford Handbook of the History of Physics* (2014)
1.7. A. Einstein, *The evolution of Physics From early Concepts to Relativity and Quanta* (Touchstone, 1967)
1.8. D.C. Lindberg, E. Whitney, *The Beginning of western Sciences* (University of Chicago Press, 1992)
1.9. https://en.wikipedia.org/wiki/History_of_physics
1.10. H.Th. Milhorn, H.T. Milhorn, *The History of Physics* (Virtualbookworm.com publishing)
1.11a. W.H. Cropper, *Great Physicists: The Life and Times of Leading Physicists from Galileo to Hawking* (Oxford University Press, 2004) https://en.wikipedia.org/wiki/Special:BookSources/0-19-517324-4

1.11b. John L. Heilbron, *The Oxford Guide to the History of Physics and Astronomy* (Oxford University Press, 2005) https://en.wikipedia.org/wiki/Special:BookSources/0-19-517198-5
1.11c. J.Z. Buchwald, I.B. Cohen (eds.), *Isaac Newton's natural philosophy* (Cambridge, Mass. and London, MIT Press, 2001)
1.12. J.T. Cushing, *Philosophical Concepts in Physics: The historical Relation between Philosophy and Physics* (Cambridge Univ. Press, 2008)
1.13a. E. Segrè, *From Falling Bodies to Radio Waves: Classical Physicists and Their Discoveries* (W.H. Freeman, New York, 1984) https://en.wikipedia.org/wiki/Special:BookSources/0-7167-1482-5, https://www.worldcat.org/oclc/9943504
1.13b. E. Segrè, *From X-Rays to Quarks: Modern Physicists and Their Discoveries* (W.H. Freeman, San Francisco, 1980) https://en.wikipedia.org/wiki/Special:BookSources/0-7167-1147-8, https://www.worldcat.org/oclc/237246197+56100286+5946636
1.13c. G. Gamov, *The Great Physicists from Galileo to Einstein* (Dover Publ., revised edition 2014) ISBN: 978-0486257679
1.14a. P. Fara, *Science, A Four Thousand Year History* (Oxford Univ. Press, 2010) ISBN: 978-0199580279
1.14b. H.J. Jodl, in: E. Lüscher, H.J. Jodl, (eds.), *Physik, Gestern, Heute, Morgen* (München, Heinz Moos-Verlag 1971)
1.14c. N.N., *From Big to Small: The Hierarchy Problem in Physics (String Theory)* (what-when-how – In Depth Tutorials and Information) http://what-when-how.com/string-theory/from-big-to-small-the-hierarchy-problem-in-physics-string-theory/
1.15. St. Weinberg, *The Discovery of Subatomic Particles* (Scientifique American Library Freeman Oxford, 1984)
1.16a. G. Gamov: *One, Two, Three ... Infinity. Facts and Speculations of Science* (Dover Publications, 1989) ISBN: 978-0486256641
1.16b. G. Gamov: *Mr. Tomkins in Paperback* (Cambridge Univ. Press, reprint 2012) ISBN: 978-1107604681
1.17. N. Manton, *Symmetries, Fields and Particles.* (University of Cambridge), http://www.damtp.cam.ac.uk/user/examples/3P2.pdf
1.18. F. Close, *Particle Physics: A Very Short Introduction.* (Oxford Univ. Press, 2004)
1.19. R. Oerter, *The Theory of Almost Everything: The Standard Model, the Unsung Triumph of Modern Physics.* (Plume, Reprint edition, 2006)
1.20a. T. Plathotnik, E.A. Donley, U.P. Wild, *Single Molecule Spectroscopüy.* Ann. Rev. Phys. Chem. **48**, 181 (1997)
1.20b. N.G. Walter, *Single Molecule Detection, Analysis, and Manipulation,* in *Encyclopedia of Analytical Chemistry,* ed. by R.A. Meyers (John Wiley & Sons Ltd, 2008)
1.21. E. Schrödinger, *What is life?* (Macmillan, 1944)
1.22a. D. Goldfarb, *Biophyiscs Demystified.* (Mc Grawhill, 2010) ISBN: 978-0071633642
1.22b. R. Glaser, *Biophysics: An Introduction.* (Springer, Berlin, Heidelberg, 2012) ISBN: 978-3642252112
1.23a. K. Nouri (ed.), *Lasers in Dermatology and Medicine.* (Springer, Berlin, Heidelberg, 2012) ISBN: 978-0-85729-280-3
1.23b. K. Nouri (ed.), *Laser Applications in Medicine International Journal for Laser Treatment and Research.* (Elsevier, Amsterdam)
1.24a. St. Hawking, *The Universe in a Nutshell.* (Bantam Press, 2001)
1.24b. L.M. Krauss, *A Universe from Nothing.* (Atria Books, 2012)
1.24c. E. Chaisson, St. McMillan, *Astronomy Today.* (Addison Wesley, 2010)
1.24d. T. Dickinson, A. Schaller, T. Ferris, *Night watch. A practical Guide to viewing the Universe.* (Firefly Books)
1.24e. B.W. Carroll, D.A. Ostlie, *An Introduction to Modern Astrophysics.* (Pearson Education, 2013)
1.25. A.E. Musset, M. Aftab Khan, S. Button, *Looking into the Earth: An Introduction to Geological Geophysics,* 1st edn. (Cambridge University Press, 2000) ISBN: 978-0521785747
1.26a. C.D. Ahrens, *Meteorology Today: Introduction to Wheather, Climate and the Environment.* (Brooks Cole, 2012) ISBN: 978-0840054999
1.26b. St.A. Ackerman, J.A. Knox, *Meteorology: Understanding the Atmosphere,* 4th edn. (Jones & Bartlett Learning, 2013)
1.27. P.J. Crutzen, Pure Appl. Chem. **70**(7), 1319–1326 (1998)
1.28. P. Saundry, *Environmental physics.* (The Encyclopedia of Earth, 2011), http://www.eoearth.org/view/article/152632
1.29a. F.K. Lutgens, E.J. Tarbuck, D.G. Tasa, *Essentials of Geology,* 11th edn. (Prentice Hall) ISBN: 978-0321714725
1.29b. W.S. Broecker, *How to Build a Habitable Planet.* (Eldigio Press, 1988)
1.30a. R.A. Müller, *Physics and Technology für Future Presidents.* (Princeton Univ. Press, 2010)
1.30b. M.L. Forlan, *Modern Physics and Technology.* (World Scientific, Singapore, 2015)
1.31. L. Sklar, *Philosophy of Physics.* (Oxford University Press, 1990) ISBN: 978-0198751380
1.32. Th. Brody, *The Philosophy behind Physics.* (Springer, Berlin, Heidelberg, 1994)
1.33. St. Cole, Am. J. Sociol. **89**(1) 111–139 (1983)
1.34a. B.R. Cohen, Endeavour **25**(1) 8–12 (2001)
1.34b. R.M. Young, *Science and Humanities in the understanding of human nature.* (The Human Nature Review), http://human-nature.com/rmyoung/papers/pap131h.html
1.35. W. Heisenberg, *Physics and Beyond: Encounters and Conversations.* (Harper & Row, 1971)
1.36a. P. Becker, *The new kilogram is approaching.* (PTB-news, 3/2011), http://www.ptb.de/en/aktuelles/archiv/presseinfos/pi2011/pitext/pi110127.html
1.36b. S. Hadington, *The kilogram is dead, long live the kilogram!* (The Royal Society of Chemistry, 2011), http://www.rsc.org/chemistryworld/News/2011/October/31101101.asp
1.37. N.N., *SI based Units.* (Wikipedia), https://en.wikipedia.org/wiki/SI_base_unit

1.38. N.N., https://www.bing.com/search?q=Phyiscal%20basis%20for%20SI%20units&pc=cosp&ptag=ADC890F4567&form=CONMHP&conlogo=CT3210127
1.39. B.W. Petley, *The fundamental physical constants and the frontiers of measurements.* (Adam Hilger, 1985)
1.40. N.N., *History of the Meter.* (Wikipedia), https://en.wikipedia.org/wiki/History_of_the_metre
1.41. J. Levine, Ann. Rev. Earth Planet Sci. **5** 357 (1977)
1.42. J. Müller et al., *Lunar Laser Ranging. Recent Results.* (AG Symposia Series, Vol 139, Springer, 2013)
1.43a. K.M. Borkowski, Journal of the Royal Astronomical Society of Canada **85** 121 (1991)
1.43b. http://en.wikipedia.org/wiki/Tropical_year
1.44a. N.N., *Atomic clock.* (Wikipedia), https://en.wikipedia.org/wiki/Atomic_clock
1.44b. N.N., *Clock network.* (Wikipedia), https://en.wikipedia.org/wiki/Clock_network
1.45. A. deMarchi (ed.), *Frequency Standards and Metrology.* (Springer, Berlin, Heidelberg, 1989)
1.46. S.A. Diddams, T.W. Hänsch et al., Phys. Rev. Lett. **84** 5102 (2000)
1.47. Ch. Gaiser, Metrologia **48** 382 (2011)
1.48a. N.N., *Kilogram* (Wikipedia), https://en.wikipedia.org/wiki/Kilogram
1.48b. N.N., *Mole (unit)* (Wikipedia), https://en.wikipedia.org/wiki/Mole_(unit)
1.49. V.V. Krutikov, *Measurement Techniques.* (Springer, Berlin, Heidelberg, 2013)
1.50. Journal of Measurement Techniques (Springer Berlin, Heidelberg)
1.51. K.T.V. Grattan (ed.), *Measurement: Journal of International Measuring Confederation (IMEKO)* (Elsevier, Amsterdam), ISSN: 0263-2241
1.52. L. Marton (ed.), *Methods of experimental Physics.* (29 Vol.) (Academic Press, New York, 1959–1996) (continued as Experimental Methods in the Physical Sciences up to Vol. 49, published by Elsevier Amsterdam in 2014), ISBN 13: 978-0-12-417011-7
1.53a. J.R. Taylor, *An Introduction to Error Analysis:The Study of Uncertainties in Physical measurements,* 2nd ed. (University Science Books, 1996)
1.53b. D. Roberts, *Errors, discrepancies, and the nature of physics.* (The Physics Teacher, March 1983) a very useful introduction
1.54. J.W. Foremn, *Data Smart: Using Data Science to transform Information into Insight.* (Wiley, 2013)
1.55. R.C. Sprinthall, *Basic statistical Analysis,* 9th ed. (Pearson, 2011)
1.56. J. Schmuller, *Statistical Analysis with Excel for Dummies,* 3rd ed. (For Dummies, 2013)

Mechanics of a Point Mass

2

© Springer International Publishing Switzerland 2017
W. Demtröder, *Mechanics and Thermodynamics*, Undergraduate Lecture Notes in Physics, DOI 10.1007/978-3-319-27877-3_2

As has been discussed in the previous chapter, the theoretical description of the physical reality often proceeds by successively refined models which approach the reality more and more with progressive refinement. In this chapter the motion of bodies under the influence of external forces will be depicted by the model of point masses, which neglects the spatial form and extension of bodies, which might influence the motion of these bodies.

2.1 The Model of the Point Mass; Trajectories

For many situations in Physics the spatial extension of bodies is of no importance and can be neglected because only their masses play the essential role. Examples are the motion of the planets around the sun where their size is very small compared with the distance to the sun. They can be described as point masses.

The position $P(t)$ of a point mass in the three-dimensional space can be described by its coordinates, which are defined if a suitable coordinate-system is chosen. These coordinates are $\{x, y, z\}$ in a Cartesian system, $\{r, \vartheta, \varphi\}$ in a spherical coordinate system and $\{\varrho, \vartheta, z\}$ in cylindrical coordinates (see Sect. 13.2).

The motion of a point mass is described as the change of its coordinates with time, for example in Cartesian coordinates

$$\left.\begin{array}{l} x = x(t) \\ y = y(t) \\ z = z(t) \end{array}\right\} \equiv \boldsymbol{r} = \boldsymbol{r}(t) \,,$$

where the position vector $\boldsymbol{r} = \{x, y, z\}$ combines the three coordinates x, y and z (Sect. 13.1).

Note: Vectors are always marked as bold letters.

The function $\boldsymbol{r}(t)$ represents a trajectory in a three-dimensional space, which is passed by the point mass in course of time (Fig. 2.1). The representation $\boldsymbol{r} = \boldsymbol{r}(t)$ is called parameter representation because the coordinates of the point $P(t)$ depend on the parameter t.

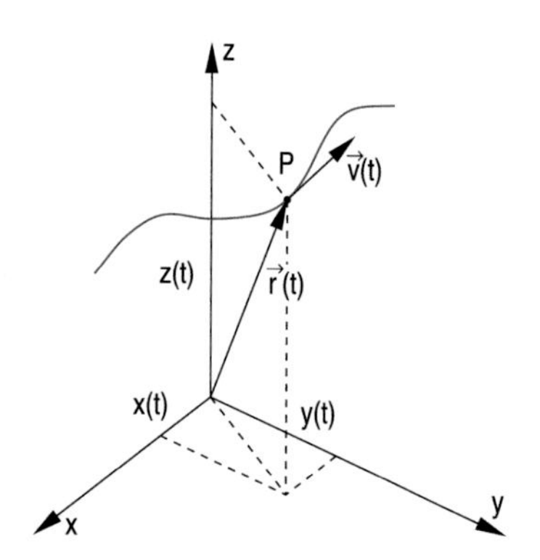

Figure 2.1 Illustration of a trajectory

The motion performed by $P(t)$ on its trajectory is called **translation**. Contrary to the point mass bodies with extended size can also perform **rotations** (Chap. 5) and **vibrations** (Chap. 6).

Note: The model of a point mass moving on a well-defined trajectory fails in micro-physics for the motion of atoms or elementary particles described correctly by quantum mechanics (Vol. 3), where position and velocity cannot be precisely given simultaneously. Instead of a precisely defined trajectory where the point mass can be find at a specific time with certainty at a well-defined position, only probabilities $\mathfrak{P}(x, y, z, t)\mathrm{d}x\mathrm{d}y\mathrm{d}z$ can be given for finding the point mass in a volume $\mathrm{d}V = \mathrm{d}x\mathrm{d}y\mathrm{d}z$ around the position (x, y, z). Strictly speaking a geometrical exact trajectory does not exist in the framework of quantum mechanics.

Examples

1. **Motion on a straight line**

$$x = a \cdot t \,, \quad y = b \cdot t \,, \quad z = 0 \,.$$

Elimination of t gives the usual representation $y = (b/a)x$ of a straight line in the (x, y)-plane.
The point mass moves in the x, y-diagram on the straight line with the slope (b/a) (Fig. 2.2).

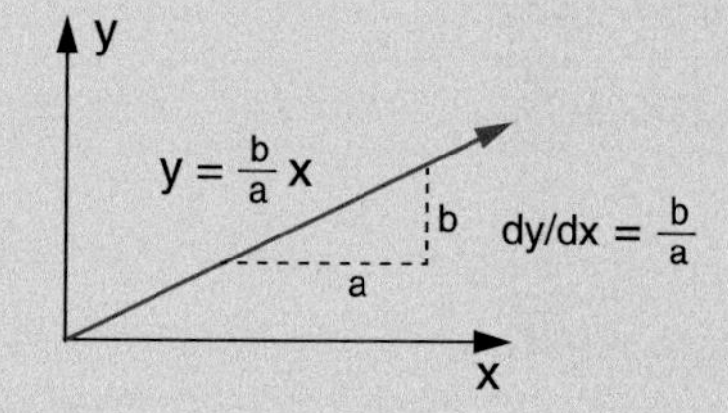

Figure 2.2 Motion on a straight line in the x-y plan

Motions where one of the coordinates are time-independent constants are named **planar motions**, because they are restricted to a plane (in our example the x, y-plane)

2. **Planar circular motion**
We can describe this motion by the coordinates R and φ (Fig. 2.3), where R is the radius of the circle and $\varphi(t)$ the angle between the x-axis and the momentary radius vector $\boldsymbol{R}(t)$. From Fig. 2.3 the relations

$$x = R \cdot \cos \omega t \,, \quad y = R \cdot \sin \omega t \,,$$
$$R = \text{const} \,, \qquad \omega = \mathrm{d}\varphi/\mathrm{d}t \,.$$

can be derived. Squaring of x and y yields

$$x^2 + y^2 = R^2(\cos^2 \omega t + \sin^2 \omega t) = R^2 \,,$$

which is the equation of a circle with radius R. The point mass m with the coordinates $\{x, y, 0\}$ moves with the angular velocity $\omega = \mathrm{d}\varphi/\mathrm{d}t$ and the velocity $v = R \cdot \omega$ on a circle in the x, y-plane.

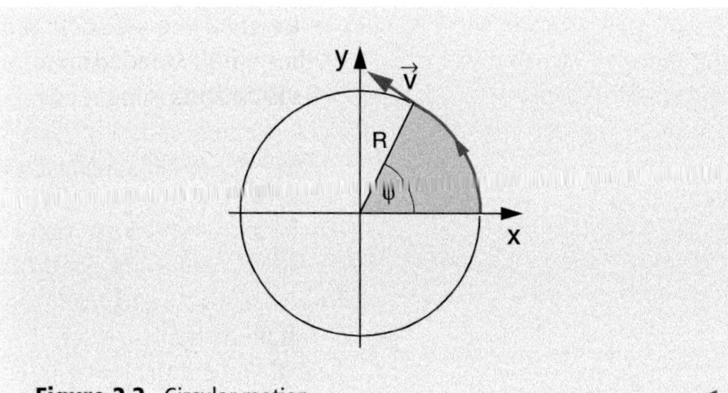

Figure 2.3 Circular motion ◄

Note: The point mass moves relative to a chosen coordinate system (in our case a plane system with the origin at $x = y = 0$). The description of this motion depends on the choice of the reference frame (coordinate system) (see Chap. 3).

Example

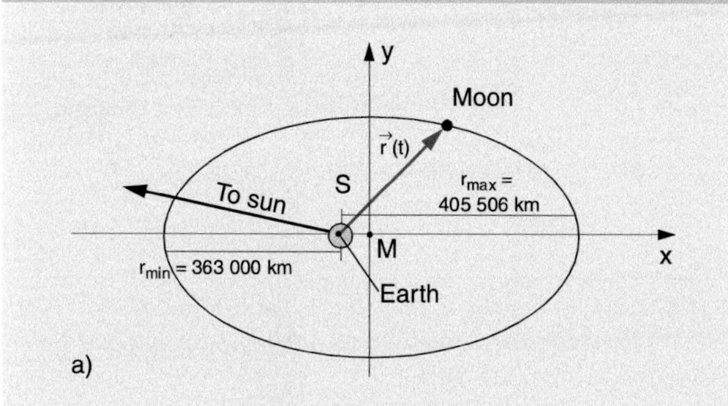

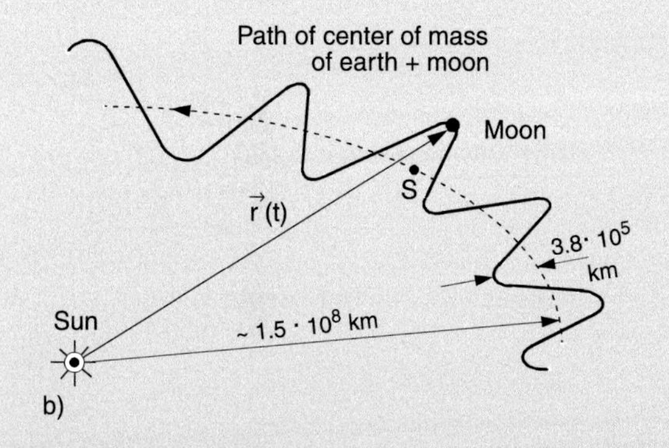

Figure 2.4 Part of the moon trajectory described in two different coordinate systems. **a** Origin in the mass centre of the moon-earth system, located in the focal point of the ellipse; **b** origin in the centre of the sun. The deviations from the elliptical path of the mass centre earth-moon are here exaggerated in order to illustrate these deviations. In reality the orbit of the moon around the sun is always concave, i. e. the curvature radius always points towards the sun. The orbital plane of the moon is inclined against that of the earth

The orbital motion of the moon around the earth is approximately an ellipse if $\boldsymbol{r}(t)$ is measured in a coordinate system with the origin in the centre of mass of the earth-moon system.(Fig. 2.4a). If one chooses, however, the centre of the sun as origin, the trajectory is much more complex (Fig. 2.4b), because now two motions are superimposed: the orbit around the centre of mass and the motion of the centre of mass around the sun. ◄

2.2 Velocity and Acceleration

For a uniformly moving point mass the position vector

$$\boldsymbol{r} = \boldsymbol{v} \cdot t \quad \text{with} \quad \boldsymbol{v} = \{v_x, v_y, v_z\} = \textbf{const} , \qquad (2.1)$$

increases linearly with time. This means that in equal time intervals Δt equal distances $\boldsymbol{\Delta r}$ are covered.

The ratio $\boldsymbol{v} = \boldsymbol{\Delta r}/\Delta t$ is the velocity of the point mass. The unit of the velocity is $[v] = 1\,\text{m/s}$.

A motion where the magnitude and the direction of the velocity vector $\boldsymbol{v}$ is constant, i. e. does not change with time, is called **uniform rectilinear motion** (Fig. 2.5). In Cartesian coordinates with the unit vectors $\hat{e}_x, \hat{e}_y, \hat{e}_z$, the velocity vector $\boldsymbol{v}$ can be written as

$$\boldsymbol{v} = v_x\hat{\boldsymbol{e}}_x + v_y\hat{\boldsymbol{e}}_y + v_z\hat{\boldsymbol{e}}_z \quad \text{or} \quad \boldsymbol{v} = \left\{v_x, v_y, v_z\right\} .$$

Equation 2.1 reads for the components of v as

$$x = v_x t\,; \quad y = v_y t\,; \quad z = v_z t\,. \qquad (2.1a)$$

Example

Uniform motion along the x-axis:

$$v_x = v_0 = \text{const}; \quad v_y = v_z = 0 \to \boldsymbol{v} = \{v_0, 0, 0\} .$$

The trajectory is the x-axis and the motion is $x = v_0 t$. ◄

In general the velocity will not be constant but can change with time its magnitude as well as its direction. Let us regard a point mass m, which is at time t in the position P_1 (Fig. 2.6). Slightly

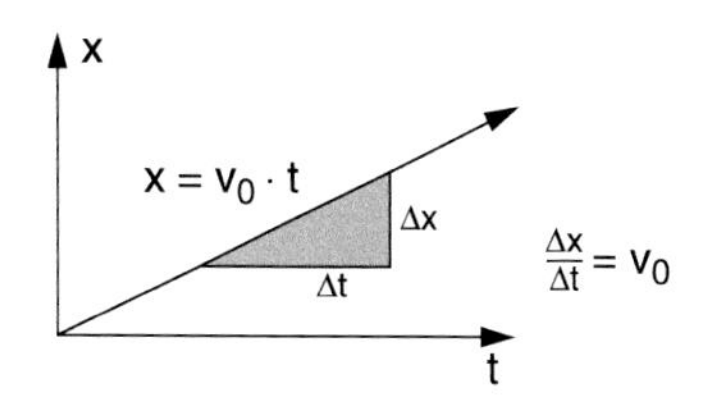

Figure 2.5 Uniform motion on a straight line

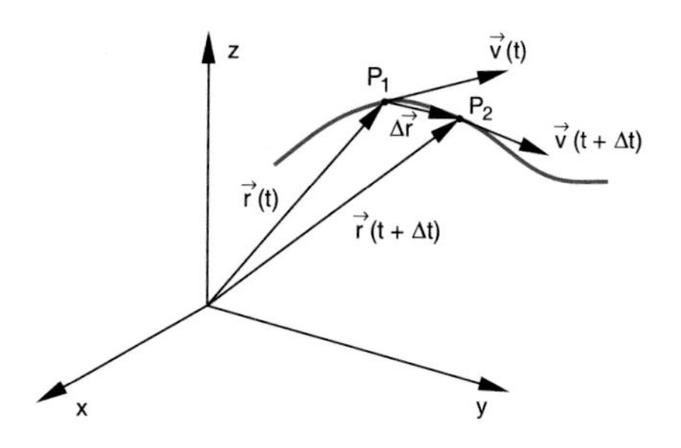

Figure 2.6 Non-uniform motion on an arbitrary trajectory in space

later at the time $t + \Delta t$ it has proceeded to the point P_2. The ratio

$$\frac{\overrightarrow{P_1P_2}}{(t+\Delta t)-t} = \frac{\boldsymbol{r}(t+\Delta t)-\boldsymbol{r}(t)}{\Delta t} = \frac{\Delta \boldsymbol{r}}{\Delta t} = \overline{\boldsymbol{v}}$$

is the average velocity $\overline{\boldsymbol{v}}$ over the distance $\overline{P_1P_2}$.

For $\Delta t \to 0$ the two points P_1 and P_2 merge together and we define as the momentary velocity $\boldsymbol{v}(t)$ the limiting value

$$\boldsymbol{v}(t) = \lim_{\Delta t\to 0} \frac{\boldsymbol{r}(t+\Delta t)-\boldsymbol{r}(t)}{\Delta t} = \frac{\mathrm{d}\boldsymbol{r}}{\mathrm{d}t} = \dot{\boldsymbol{r}}\,,$$

which equals the time derivative of the function $\boldsymbol{r}(t)$. In order to distinguish this time derivative $\mathrm{d}r/\mathrm{d}t = \dot{r}(t)$ from the spatial derivative $y'(x) = \mathrm{d}y/\mathrm{d}x$ the time derivative is marked by a point instead of an apostrophe.

Since the derivative $\mathrm{d}f/\mathrm{d}x$ of a function $f(x)$ gives the slope of the curve $f(x)$ at the point $P(x, y)$ the velocity v has at any point the direction of the tangent (Fig. 2.6). Its magnitude is in Cartesian coordinates:

$$v = |\boldsymbol{v}| = \sqrt{v_x^2 + v_y^2 + v_z^2} = \sqrt{\dot{x}^2 + \dot{y}^2 + \dot{z}^2}\,. \tag{2.2}$$

Examples

1. Linear accelerated motion

$$z = a \cdot t^2 \to v_z = \dot{z} = 2a \cdot t\,.$$

For $a = \text{const}$ the velocity increases linearly with time. For $a = -g/2$ this describes the free fall with the initial velocity $v_z(t = 0) = 0$ (see Sect. 2.3.1). Here only the magnitude, not the direction of the velocity changes with time.

2. Uniform circular motion

$$\left.\begin{aligned} x &= R\cdot\cos\omega t &\Rightarrow\quad \dot{x} &= -R\cdot\omega\cdot\sin\omega t\\ y &= R\cdot\sin\omega t &\Rightarrow\quad \dot{y} &= R\cdot\omega\cdot\cos\omega t\\ z &= 0 &\Rightarrow\quad \dot{z} &= 0 \end{aligned}\right\}$$

$$\to |\boldsymbol{v}| = \sqrt{\dot{x}^2 + \dot{y}^2 + \dot{z}^2} = R\cdot\omega\,.$$

For $\omega = \text{const}$ the magnitude of $\boldsymbol{v}$ does not change, only its direction. ◀

We will now discuss the time dependence of the velocity $\boldsymbol{v}$ in more detail: Let us regard a point mass with the velocity $\boldsymbol{v}(t)$ at the point P_1 of the curve $v(t)$. At a slightly later time $t + \Delta t$ the point mass has arrived at P_2 and has there generally a different velocity $\boldsymbol{v}(t + \Delta t)$ (Fig. 2.7). We define the mean acceleration $\overline{\boldsymbol{a}}$ as

$$\overline{\boldsymbol{a}} = \frac{\boldsymbol{v}(t+\Delta t)-\boldsymbol{v}(t)}{\Delta t}\,.$$

Analogous to the definition of the momentary velocity the momentary acceleration is the limit

$$\begin{aligned} \boldsymbol{a}(t) &= \lim_{\Delta t\to 0}\frac{\boldsymbol{v}(t+\Delta t)-\boldsymbol{v}(t)}{\Delta t} = \frac{\mathrm{d}\boldsymbol{v}}{\mathrm{d}t} = \dot{\boldsymbol{v}}(t) = \ddot{\boldsymbol{r}}(t)\\ \boldsymbol{a}(t) &= \dot{\boldsymbol{v}}(t) = \ddot{\boldsymbol{r}}(t) \end{aligned} \tag{2.3}$$

The acceleration $\boldsymbol{a}(t)$ is the first time derivative $\mathrm{d}\boldsymbol{v}/\mathrm{d}t$ of the velocity $\boldsymbol{v}(t)$ and the second derivative $\mathrm{d}^2\boldsymbol{r}/\mathrm{d}t^2$ of the position vector $\boldsymbol{r}(t)$. $\boldsymbol{a}(t) = \{a_x, a_y, a_z\}$ is a vector and has the dimensional unit $[a] = [1\,\mathrm{m/s^2}]$.

2.3 Uniformly Accelerated Motion

A motion with $\boldsymbol{a} = \mathbf{const}$ where the magnitude and the direction of a do not change with time is called uniformly accelerated motion. It is described by the equation

$$\ddot{\boldsymbol{r}}(t) = \boldsymbol{a} = \mathbf{const}\,. \tag{2.4}$$

Equation 2.4 is named *differential equation* because it is an equation between the derivative of a function and other quantities (here the constant vector $\boldsymbol{a}$).

The vector equation (2.4) can be written as the corresponding three equations for the components

$$\begin{aligned} \ddot{x}(t) &= a_x\\ \ddot{y}(t) &= a_y\\ \ddot{z}(t) &= a_z\,. \end{aligned}$$

The equation of motion (2.4) is readily solvable. The velocity if obtained by integrating (2.4) which yields:

$$\boldsymbol{v}(t) = \dot{\boldsymbol{r}}(t) = \int \boldsymbol{a}\,\mathrm{d}t = \boldsymbol{a}\cdot t + \boldsymbol{b}\,. \tag{2.5}$$

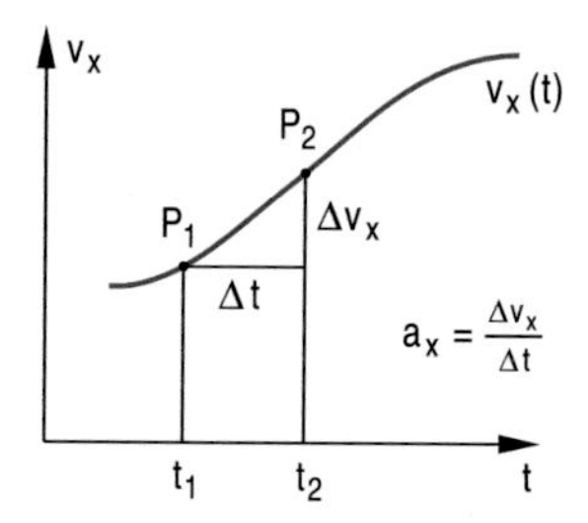

Figure 2.7 Definition of acceleration

The integration constant $\boldsymbol{b}$ ($\boldsymbol{b}$ is a vector with constant components) can be defined by choosing the initial conditions for the motion. For $t = 0$ is $\dot{\boldsymbol{r}}(0) = \boldsymbol{v}(0) = \boldsymbol{b}$. I. e. the constant $\boldsymbol{b}$ gives the initial velocity $\boldsymbol{v}(0) = \boldsymbol{v}_0$.

Further integration of (2.5) gives the trajectory $\boldsymbol{r}(t)$

$$\boldsymbol{r}(t) = \tfrac{1}{2}\boldsymbol{a}t^2 + \boldsymbol{v}_0 t + \boldsymbol{c} \quad \text{with} \quad \boldsymbol{c} = \boldsymbol{r}(0) = \boldsymbol{r}_0 \,. \tag{2.6}$$

This vector-equation can be written for the 3 components

$$\begin{aligned} x(t) &= \tfrac{1}{2}a_x \cdot t^2 + v_{0x}t + x_0 \,, \\ y(t) &= \tfrac{1}{2}a_y \cdot t^2 + v_{0y}t + y_0 \,, \\ z(t) &= \tfrac{1}{2}a_z \cdot t^2 + v_{0z}t + z_0 \,. \end{aligned} \tag{2.6a}$$

One should realize the following statement:

All functions $f(x) + c$ with arbitrary constants c have the same derivative $y' = f'(x)$ because the derivative of a constant is zero. This implies:

All functions $f(x) + c$, which represent an infinite parametric curve family, are solutions of the differential equation $y' = f'(x)$. Therefore infinitely many position vectors $r(t)$ are found for the same velocity $v(t)$. Only the initial conditions select one specific position vector.

We will illustrate this by several examples in the next sections.

2.3.1 The Free Fall

We choose the vertical direction as the z-axis. A body experiences in the gravitational field of the earth the acceleration

$$\begin{aligned} a_x &= a_y = 0 \,, \\ a_z &= -g = -9.81\,\text{m/s}^2 \,, \end{aligned}$$

where the numerical value is obtained from experiments.

When a body at rest falls at time $t = 0$ from the height h, the initial conditions are $x(0) = y(0) = 0$: $z(0) = h$; $v_x(0) = v_y(0) = v_z(0) = 0$.

With these initial conditions the system of equations (2.6a) reduces to

$$z(t) = -\tfrac{1}{2}gt^2 + h \,. \tag{2.7}$$

The derivative gives $v_z(t) = -g \cdot t$. The motion $z(t)$ plotted in the z-t-plane represents a parabola (Fig. 2.8). For $t = \sqrt{2h/g}$ the body has reached the ground at $z = 0$. The falling time for the distance h is

$$t_\text{fall} = \sqrt{2h/g} \,, \tag{2.8}$$

and the final velocity at $z = 0$ is $v_\text{max} = \sqrt{2hg}$.

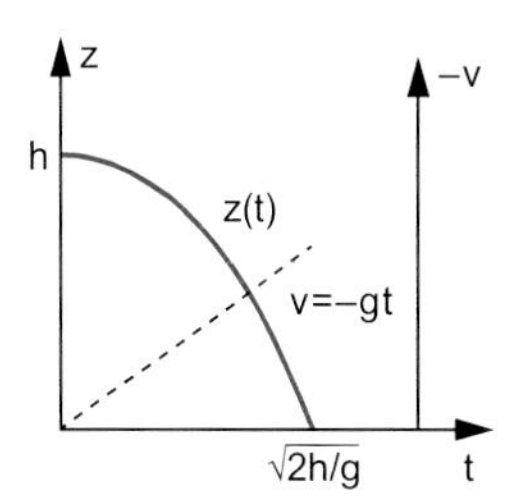

Figure 2.8 Path-time function $z(t)$ (*red curve*) and velocity-time function (*dotted line*)

2.3.2 Projectile Motion

As starting point we choose $x(0) = y(0) = 0$; $z(0) = h$; and the z-axis is again the vertical direction, while the x-axis marks the horizontal direction, so that the trajectory for the projectile is in the x-z-plane (Fig. 2.9). The initial velocity should be $\boldsymbol{v}_0 = \{v_{0x}, 0, v_{0z}\}$. The acceleration is $\boldsymbol{a} = \{9, 0, -g\}$. Equation 2.6 becomes then

$$\begin{aligned} x(t) &= v_{0x}t \,, \\ y(t) &= 0 \,, \\ z(t) &= -\tfrac{1}{2}gt^2 + v_{0z}t + h \,. \end{aligned}$$

The motion is therefore a superposition of a uniform straight motion into the x-direction and a uniformly accelerated motion into the z-direction. For $v_{0z} = 0$ we obtain the special case of the horizontal throw and for $v_{0x} = 0$ the vertical throw.

Elimination of $t = x/v_{0x}$ yields the projectile parabola

$$z(x) = -\frac{1}{2}\frac{g}{v_{0x}^2}x^2 + \frac{v_{0z}}{v_{0x}}x + h \,. \tag{2.9}$$

The value $x = x_s$ where the maximum occurs is found for $\mathrm{d}z/\mathrm{d}x = 0$.

$$x_\text{S} = \frac{v_{0x} \cdot v_{0z}}{g} = \frac{v_0^2 \cdot \sin\varphi \cdot \cos\varphi}{g} \,. \tag{2.10}$$

For a given value of the initial velocity v_0 the maximum of x_s is achieved for $\varphi = 45°$. In order to calculate the projectile range

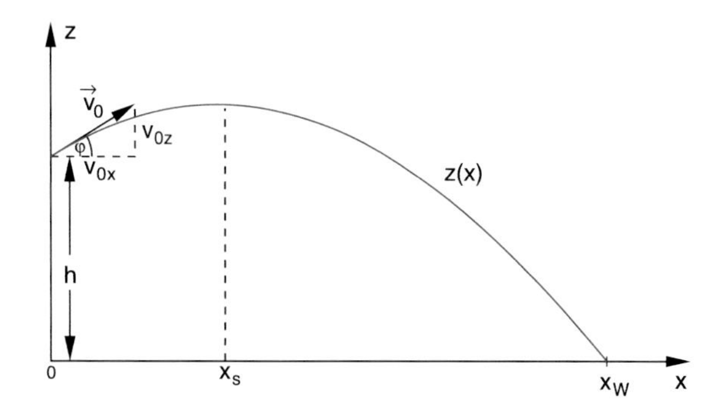

Figure 2.9 Projectile motion

x_w we solve (2.9) for $z(x_w) = 0$. This gives

$$x_W = \frac{v_{0x} \cdot v_{0z}}{g} \pm \left[\left(\frac{v_{0x} \cdot v_{0z}}{g} \right)^2 + \frac{2v_{0x}^2}{g} \cdot h \right]^{1/2} . \quad (2.11)$$

Since $x_w > 0$ only the positive sign is possible. With the relation $v_{z_0} \cdot v_{x_0} = \frac{1}{2} v_0^2 \cdot \sin 2\varphi$ we can transform (2.11) into

$$x_W = \frac{v_0}{2g} \sin 2\varphi \left[v_0 + \left(v_0^2 + \frac{2gh}{\sin^2 \varphi} \right)^{1/2} \right] . \quad (2.12)$$

The optimum angle φ_{opt} for achieving the largest throwing range for a given initial velocity v_0 is achieved when $dx_w/d\varphi = 0$. This gives

$$\varphi_{opt} = \arcsin \left(\frac{1}{\sqrt{2 + 2gh/v_0^2}} \right) . \quad (2.13)$$

For the special case $h = 0$ (2.13) simplifies because of $\arcsin(\sqrt{2}/2) = \pi/4$ to $\varphi_{opt} = 45°$ (see the detailed derivation of (2.13) in the solution of Problem 2.5c).

2.4 Motions with Non-Constant Acceleration

While the differential equation for motions with constant acceleration is elementary integrable this might not be true for arbitrary time dependent accelerations. We will at first treat the simple example of the uniform circular motion, where the magnitude of the acceleration is constant but not the direction.

2.4.1 Uniform Circular Motion

For the uniform circular motion equal distances are gone for equal time intervals. This means that the magnitude of the velocity $\boldsymbol{v}$ is constant and the component a_φ of the acceleration $\boldsymbol{a} = \{a_r, a_\varphi\}$ in the direction of v must be therefore zero.

The path length Δs on the circle arc for the angle $\Delta\varphi$ is $\Delta s = R \cdot \Delta\varphi$ (Fig. 2.10a). The magnitude of the velocity is then

$$v = \frac{ds}{dt} = R \cdot \frac{d\varphi}{dt} = R \cdot \omega .$$

The quantity $\omega = d\varphi/dt$ is the **angular velocity** with the dimension $[\omega] = [\text{rad/s}]$.

The acceleration is now

$$\boldsymbol{a} = \frac{d\boldsymbol{v}}{dt} = \frac{d}{dt}(v\hat{\boldsymbol{e}}_t) = \frac{dv}{dt}\hat{\boldsymbol{e}}_t + v\frac{d\hat{\boldsymbol{e}}_t}{dt}$$
$$= v\frac{d\hat{\boldsymbol{e}}_t}{dt} \quad \text{because} \quad v = \text{const} .$$

Because $\hat{\boldsymbol{e}}_t^2 = 1 \rightarrow 2\hat{\boldsymbol{e}}_t \cdot d\hat{\boldsymbol{e}}_t/dt = 0$.

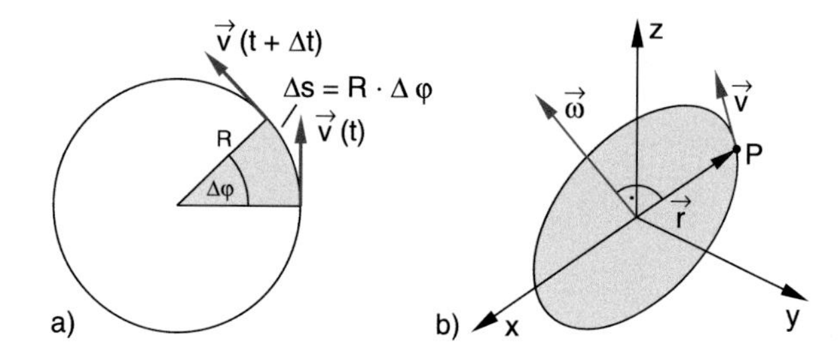

Figure 2.10 a uniform circular motion, b Illustration of the angular velocity

The scalar product of two vectors becomes zero, if either at least one of the vectors is zero or if the two vectors are orthogonal. Since $\hat{e}_t \neq \boldsymbol{0}$ and $d\hat{e}_t/dt \neq 0$ it follows

$$\frac{d\hat{\boldsymbol{e}}_t}{dt} \perp \hat{\boldsymbol{e}}_t .$$

This means that the acceleration $\boldsymbol{a}$ is orthogonal to the velocity $\boldsymbol{v}$ which is collinear with $\hat{\boldsymbol{e}}_t$. The vector $d\hat{e}_t/dt$ gives the angular velocity of the tangent to the circle. Since the radius vector R is orthogonal to the vector $\boldsymbol{v}$ both vectors turn with the angular velocity $\omega = d\varphi/dt$. This means that the magnitude is $|d\hat{e}_t/dt| = \omega$. This gives for the acceleration

$$\boldsymbol{a} = v \cdot \frac{d\hat{\boldsymbol{e}}_t}{dt} = R \cdot \omega^2 \hat{\boldsymbol{e}}_a = -R\omega^2 \hat{\boldsymbol{r}} , \quad (2.14)$$

where the unit vector $\boldsymbol{e}_a = -\boldsymbol{R}/R$ always points into the direction towards the centre of the circle, and $\hat{r} = \boldsymbol{r}/|\boldsymbol{r}|$ points into the opposite direction.

Proof

$$\boldsymbol{r} = \begin{Bmatrix} R \cdot \cos\omega t \\ R \cdot \sin\omega t \end{Bmatrix}$$

$$\boldsymbol{v} = \begin{Bmatrix} -R \cdot \omega \cdot \sin\omega t \\ R \cdot \omega \cdot \cos\omega t \end{Bmatrix}$$

$$\boldsymbol{a} = \begin{Bmatrix} -R\omega^2 \cos\omega t \\ -R\omega^2 \sin\omega t \end{Bmatrix} = -\omega^2 \cdot \boldsymbol{r} = -R\omega^2 \cdot \hat{\boldsymbol{r}} .$$ ◀

The vector of the acceleration for the uniform circular motion

$$\boldsymbol{a} = -R\omega^2 \hat{\boldsymbol{r}} \quad \text{with} \quad |\boldsymbol{a}| = R \cdot \omega^2$$

is called *centripetal-acceleration* because it points towards the centre of the circle (Fig. 2.11).

If also the orientation of the plane in the three-dimensional space should be defined, it is useful to define a vector $\boldsymbol{\omega}$ of the angular velocity which is vertical to the plane of motion (Fig. 2.10b) and has the magnitude $\omega = |\boldsymbol{\omega}| = d\varphi/dt = v/R$.

Figure 2.11 Rollercoaster, where the superposition of centripetal acceleration and gravity changes along the path and inluences the feelings of the passenger (with kind permission of Foto dpa)

2.4.2 Motions on Trajectories with Arbitrary Curvature

In the general case the velocity $\boldsymbol{v}$ will change its magnitude as well as its direction with time. However, the momentary velocity $\boldsymbol{v}(t)$ at time t is always the tangent to the trajectory in the point $P(t)$, while the acceleration $\boldsymbol{a}(t)$ can have any arbitrary direction (Fig. 2.12). The acceleration can be always composed of two components $a_t = \mathrm{d}v/\mathrm{d}t \cdot \hat{e}_t$ along the tangent to the curve (*tangential acceleration*) and a_n in the direction of the normal to the tangent, i. e. perpendicular to a_t (*normal acceleration*).

For $\boldsymbol{v} = v \cdot \hat{\boldsymbol{e}}_t$ where $\hat{\boldsymbol{e}}_t$ is the unit vector tangential to the trajectory, the acceleration $\boldsymbol{a}$ is

$$\boldsymbol{a} = \frac{\mathrm{d}\boldsymbol{v}}{\mathrm{d}t} = \frac{\mathrm{d}v}{\mathrm{d}t} \cdot \hat{\boldsymbol{e}}_t + v\frac{\mathrm{d}\hat{\boldsymbol{e}}_t}{\mathrm{d}t} = \boldsymbol{a}_t + \boldsymbol{a}_n \,. \tag{2.15}$$

The change of the magnitude of the velocity is described by a_t while the change of the direction of v is described by a_n.

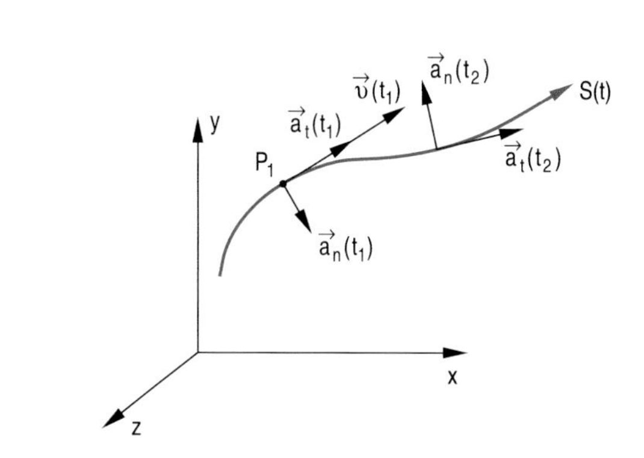

Figure 2.12 Tangential and normal acceleration

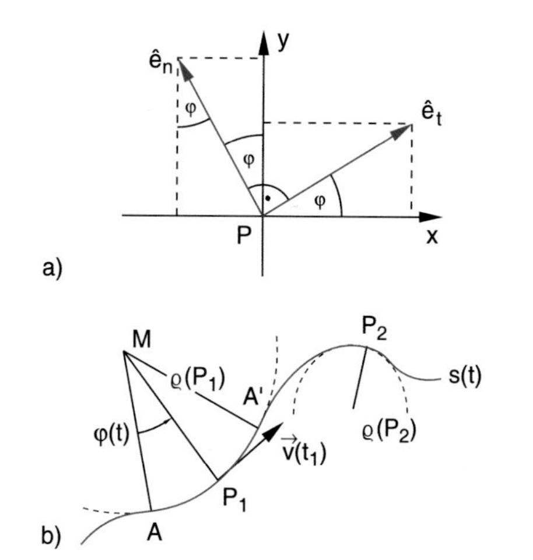

Figure 2.13 **a** Derivation of the normal acceleration. **b** Local radius of curvature of a trajectory with arbitrary curvature ϱ

For $a_n = 0$ the trajectory is a straight line, where the body moves with changing velocity if $a_t \neq 0$. For $a_t = 0$ the point mass moves with a constant velocity $|\boldsymbol{v}|$ on a curve which is determined by $\boldsymbol{a}_n(t)$. For the free fall of Sect. 2.3.1 is $a_n = 0$ and $a_t =$ const., while for the uniform circular motion $a_t = 0$ and $a_n =$ const.

For the motion on trajectories with arbitrary curvature the acceleration can be obtained as follows: We choose the x-y-plane as the plane of the two vectors $\boldsymbol{v}(t)$ and $\boldsymbol{a}(t)$, which implies that all vectors have zero z-components.

According to Fig. 2.13a the two mutually vertical unit vectors $\hat{e}_t$ and $\hat{e}_n$ can be composed as

$$\begin{aligned}
\hat{e}_t &= \cos\varphi\hat{e}_x + \sin\varphi\hat{e}_y \\
\hat{e}_n &= \cos(\varphi + \frac{\pi}{2})\hat{e}_x + \sin(\varphi + \frac{\pi}{2})\hat{e}_y \\
&= -\sin\varphi\hat{e}_x + \cos\varphi\hat{e}_y
\end{aligned}$$

There we get

$$\begin{aligned}
\frac{d\hat{e}_t}{dt} &= -\sin\varphi\frac{d\varphi}{dt}\hat{e}_x + \cos\varphi\frac{d\varphi}{dt}\hat{e}_y \\
&= \frac{d\varphi}{dt}\hat{e}_n
\end{aligned}$$

The normal acceleration is therefore

$$\boldsymbol{a}_n = v\frac{d\varphi}{dt}\hat{e}_n .$$

We regard in Fig. 2.13b an infinitesimal section between the points A and A' of an arbitrary curve and approximate this section by a circular arc AA' with the center of curvature M. Shortening the section $\overline{AA'}$ more and more, i.e. the points A and A' converge towards the point P_1 the curve section $\overline{AA'}$ approaches more and more the circular arc with radius $\overline{MP_1}$. The radius $\varrho = \overline{MP_1}$ is the radius of curvature of the curve in the point P_1.

For the small section of the curve we get

$$ds = \varrho d\varphi \tag{2.16a}$$

$$\frac{d\varphi}{dt} = \frac{d\varphi}{ds}\frac{ds}{dt} = \frac{d\varphi}{ds}v = \frac{1}{\varrho}v . \tag{2.16b}$$

The acceleration vector becomes

$$\boldsymbol{a} = \frac{dv}{dt}\hat{e}_t + \frac{v^2}{\varrho}\hat{e}_n \tag{2.16c}$$

Examples

1. Assume a motion on a straight line experiences the acceleration $a(x) = b \cdot x^4$.
   Calculate the velocity $v(x)$ for the initial condition $v(0) = v_0$.

Solution

$$a = \frac{dv}{dt} = \frac{dv}{dx}\cdot\frac{dx}{dt} = \frac{dv}{dx}\cdot v , \quad \int_{x_0}^{x} a dx = \int_{v_0}^{v} v dv .$$

Inserting a and integration yields

$$\tfrac{1}{5}b\left(x^5 - x_0^5\right) = \tfrac{1}{2}\left(v^2(x) - v_0^2\right) .$$

Resolving this equation for $v(x)$ gives

$$v(x) = \sqrt{\tfrac{2}{5}b\left(x^5 - x_0^5\right) + v_0^2} .$$

2. The open parachute of a parachutist experiences, due to air friction, a negative acceleration besides the acceleration by gravity.

$$a = -b \cdot v^2 \quad \text{with} \quad b = 0.3\,\text{m}^{-1} .$$

   a) What is his constant final velocity v_e?
   b) What is the time-dependent velocity $v(t)$, if the parachutist opens his parachute only after $t_0 = 10$ s free fall for which friction can be neglected?

Solution

a) A constant final velocity is reached, when the total acceleration becomes zero. This is the case when

$$g - b \cdot v_e^2 = 0 \to v_e = \sqrt{g/b} = 5.7\,\text{m/s} .$$

b) The equation of motion after the parachute is opened is with the z-axis in the vertical direction

$$\ddot{z} = g - b \cdot \dot{z}^2 .$$

With $v = \mathrm{d}z/\mathrm{d}t$ and $\mathrm{d}v/\mathrm{d}t = \mathrm{d}^2z/\mathrm{d}t^2$ we obtain

$$\mathrm{d}v/\mathrm{d}t = b - b \cdot v^2 ,$$

which leads to the equation

$$\int_{v_0}^{v} \frac{\mathrm{d}v}{g - bv^2} = \frac{1}{g} \int_{v_0}^{v} \frac{\mathrm{d}v}{1 - v^2/v_e^2} = \int_{t_0}^{t} \mathrm{d}t' = t - t_0 .$$

We substitute $v/v_e = x$, for $x > 1$ i. e. for $v > v_e$ we get

$$\int \frac{\mathrm{d}x}{1 - x^2} = \frac{1}{2} \ln \frac{x+1}{x-1}$$
$$\rightarrow t - t_0 = \frac{1}{2} \frac{v_e}{g} \ln \frac{v + v_e}{v - v_e} + C .$$

For $t = t_0 \rightarrow v = v_0 = g \cdot t_0 = 98.1\,\mathrm{m/s}$. This gives for the integration constant C the value

$$C = -\frac{1}{2} \frac{v_e}{g} \ln \frac{v_0 + v_e}{v_0 - v_e}$$
$$\rightarrow t - t_0 = \frac{1}{2} \frac{v_e}{g} \ln \left[\frac{v + v_e}{v - v_e} \frac{v_0 - v_e}{v_0 + v_e} \right] .$$

Eliminating v from this equation for v yields

$$v(t) = v_e \frac{d \cdot e^{c(t-t_0)} + 1}{d \cdot e^{c(t-t_0)} - 1} \quad \text{with}$$
$$d = \frac{v_0 + v_e}{v_0 - v_e} \quad \text{and} \quad c = 2g/v_e .$$

The velocity decreases from the initial value $v(t_0) = v_0$ at t_0 exponentially to the final value v_e for $t = \infty$. However, already after $t - t_0 = 2v_e/g = 1.16\,\mathrm{s}$ the velocity has reached 96.7% of its final value. ◀

2.5 Forces

We will now discuss the question, **why** a body performs that motion that we observe, why for instance the earth moves around the sun on an elliptical trajectory, or why a stone in a free fall moves on a vertical straight line to the ground.

Newton recognized that the cause for changes of a body's velocity must be interactions of the body with its surroundings. These can be long range interactions such as the gravitational interaction between the sun and the earth, or short range interaction which work for example in collisions between colliding billiard balls, or even ultrashort range strong interactions between neutrons in an atomic nucleus. All such interactions are described by the concept of **forces**. When a body changes its state of motion we say that a force acts upon the body.

If, for instance, two bodies collide we say: Each of the two bodies has exerted during the collision a force onto the other body, which causes a change of the state of motion for both bodies.

A body without any interaction with its surroundings (or for which the vector sum of all forces is zero), is called a *free body*. A free body does not change its state of motion. Strictly speaking there are in reality no free bodies without any interaction (because we would not see them). However, in many cases the interaction is so small, that we can neglect it. Examples are atoms in a tank where a very good vacuum has been established, or a sliding carriage on a nearly frictionless horizontal air track. Such free bodies move uniformly on a straight trajectory. For such cases the model of a free body is justified.

2.5.1 Forces as Vectors; Addition of Forces

Since velocity changes which are caused by forces are vectors, also forces must be described by vectors, i. e. they are defined by their magnitude and their direction.

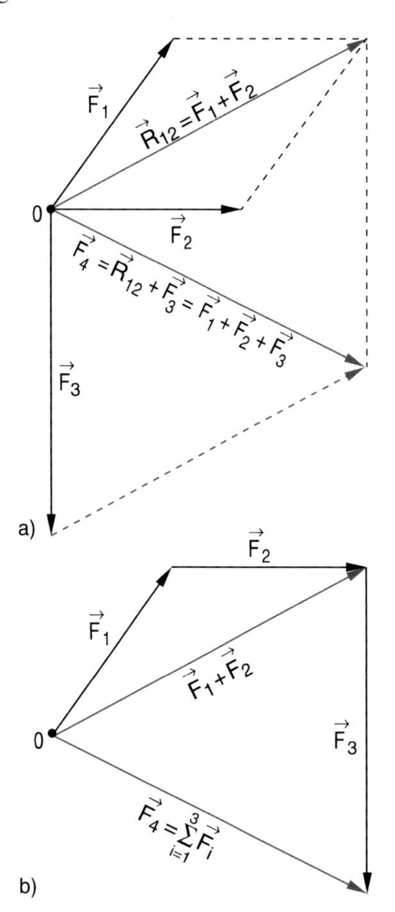

Figure 2.14 Vector sum of forces. **a** all forces act on the same point, **b** equivalent representation of the vector sum

Note: When forces act on extended bodies, also the point of origin is important (see Sect. 5.4).

A force, as any vector, can be reduced to the sum of its components. This reduction depends on the chosen coordinate system. For example in Cartesian coordinates the vector and its components are $\boldsymbol{F} = \{F_x, F_y, F_z\}$. If we choose the coordinate system in such a way that the z-direction points into the direction of $\boldsymbol{F}$, the component representation becomes $\boldsymbol{F} = \{0, 0, F_z = F\}$ with $F = |\boldsymbol{F}|$. Often the solution of a problem can be essentially simplified by choosing the optimum coordinate system (see Sect. 2.3.2). If several forces act on a body the total force is the vector sum of the individual forces (superposition principle)

$$\boldsymbol{F} = \sum_i \boldsymbol{F}_i \,.$$

This vector equation is equivalent to the three equations for the components

$$F_x = \sum_i F_{ix} \quad F_y = \sum_i F_{iy} \quad F_z = \sum_i F_{iz} \,.$$

The addition of several vectors is illustrated in Fig. 2.14a and b. Both ways to add vectors are equivalent, because the origin of the vectors can be shifted. If $\sum \boldsymbol{F}_i = \boldsymbol{0}$ the total force is zero and the body remains in its constant state of motion (either at rest or in a uniform motion on a straight line.

Examples

1. A body with mass m rests on a friction-free sloped plane (Fig. 2.15). The gravitational force can be regarded as the vecor sum of the two forces $\boldsymbol{F}_\perp$ perpendicular to the sloped plain and $\boldsymbol{F}_\parallel$ parallel to this plane. $\boldsymbol{F}_\perp$ exerts a force onto the surface of the plane and causes an opposite force $\boldsymbol{N}$ of equal magnitude by the elastic response of the surface. Only the force $\boldsymbol{F}_\parallel$ can cause an acceleration of the body. It can be compensated by an opposite force $\boldsymbol{Z}$ in order to reach a zero total force and keep the body at rest on the sloped plane. This situation can be described by the equation

$$m \cdot \boldsymbol{g} = \boldsymbol{F}_\parallel + \boldsymbol{F}_\perp = -(\boldsymbol{Z} + \boldsymbol{N}) \,.$$

Attractive force $\boldsymbol{Z}$ and elastic force $\boldsymbol{N}$ compensate the gravitational force and the body remains at rest.

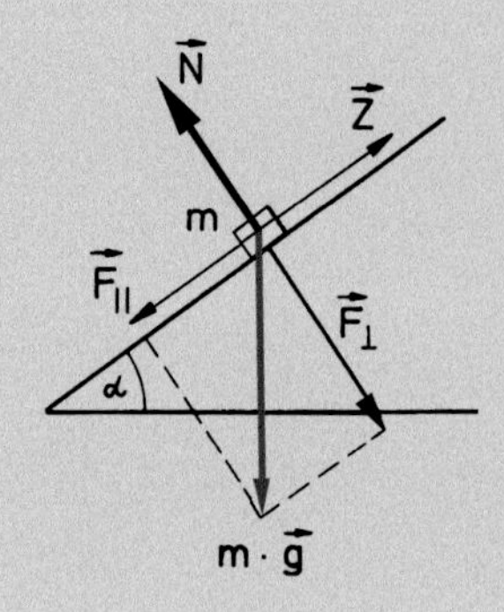

Figure 2.15 Equilibrium of forces for a body on an inclined plane

2. A circular pendulum is a mass m hold by a string which is fixed at a point P. The mass can move on a circle in the x-y-plane while the string movement forms the surface of a cone (Fig. 2.16). The total force $\boldsymbol{F} = m \cdot \boldsymbol{g} + \boldsymbol{F}_{\mathrm{el}}$ as the sum of gravitational force and elastic force of the stressed string acting on the mass m always points towards the centre of the circle in the x-y-plane and acts as centripetal force which causes the circular motion of m.

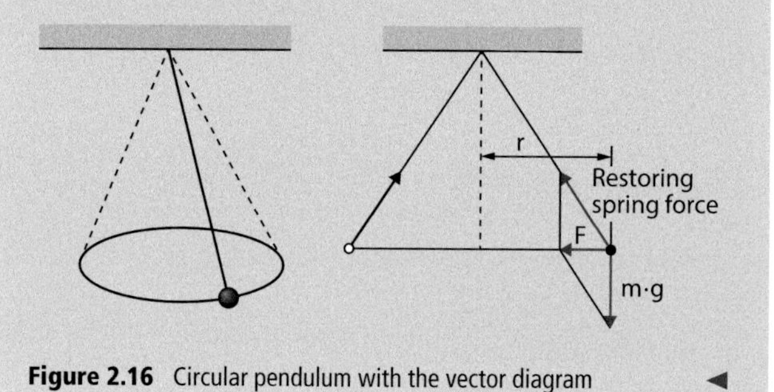

Figure 2.16 Circular pendulum with the vector diagram ◀

2.5.2 Force-Fields

Often the force acting on a body depends on the location. If it is possible to unambiguously assign to each point (x, y, z) a force with defined magnitude and direction the spatial force function $\boldsymbol{F}(x, y, z)$ is called a **force-field**. Its components depend on the chosen coordinate system:

$$\begin{aligned} &\boldsymbol{F}(\boldsymbol{r}) = \boldsymbol{F}(x, y, z) && \text{in Cartesian coordinates, or} \\ &\boldsymbol{F}(r, \vartheta, \varphi) && \text{in spherical coordinates, or} \\ &\boldsymbol{F}(r, \varphi, z) && \text{in cylinder coordinates.} \end{aligned}$$

In a graphical representation the direction of the force is illustrated by "force-lines" where the force at any point (x, y, z) is the tangent to the force-line (Fig. 2.17).

If the force has for any point in space only a radial component with a magnitude which depends on the distance r to the centre $r = 0$ the force field is centro-symmetric and is called a *central force field*. It can be written as

$$\boldsymbol{F} = f(r) \cdot \hat{\boldsymbol{r}} \,,$$

where $\hat{\boldsymbol{r}} = \boldsymbol{r}/|\boldsymbol{r}|$ is the unit vector in radial direction. The sign of the scalar function $f(r)$ is: $f(r) < 0$ if the forces point to the centre and $f(r) > 0$ is it points from the centre away.

Surfaces where the force field has the same magnitude are called *equipotential surfaces*. (see Sect. 2.7.5)

Central force fields are spherical symmetric.

Examples

1. **Central force fields**

a) ***Gravitational force field of the earth*** (Fig. 2.17a)
F depends on the distance for the earth's centre. For the idealized case that the earth can be described by a homogeneous sphere with spherical symmetric mass distribution (see Fig. 2.9) the gravitational force is for $r > R$ (R = radius of the earth)

$$\boldsymbol{F} = -G\frac{m \cdot M}{r^2}\hat{\boldsymbol{r}}$$

(M = mass of earth, m = mass of body, G = gravitation constant, unit vector $\hat{\boldsymbol{r}} = \boldsymbol{r}/|r|$)

b) ***Force field of a positive electric charge*** Q (Fig. 2.17b).
In the electric force field of an electric charge Q the force on a small test charge q is

$$\boldsymbol{F} = \frac{1}{4\pi\varepsilon_0}\frac{q \cdot Q}{r^2}\hat{\boldsymbol{r}}\,;$$

(ε_0 = dielectric constant see Vol. 2). The spherical symmetric force field has the same form as the gravitational force field.

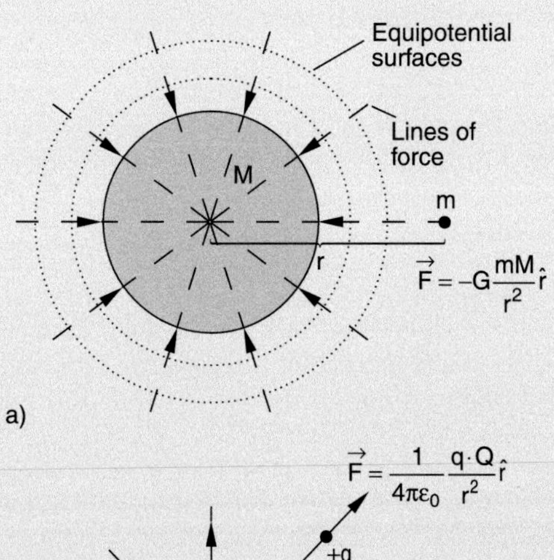

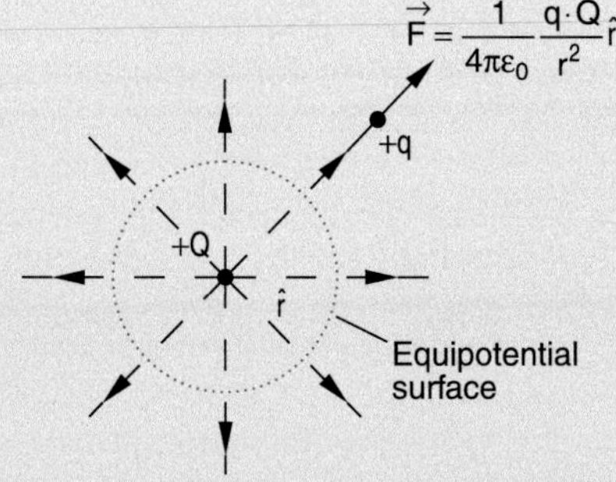

Figure 2.17 Spherical symmetric force fields **a** gravitational force field of a mass M (attractive force) and **b** electric force field of a positive charge Q and repulsive force on a positive test charge

2. **Non central force fields**

a) ***Dipole force field***
The force field in the surrounding of two charges $+Q$ and $-Q$ with equal magnitude but opposite sign is no longer spherical symmetric. The force on a test charge not only depends on the distance from the centre of the two charges but also on the angle ϑ of the position vector against the connecting line of the two charges (Fig. 2.18). The calculation of the force field gives (see Sect. 1.5 of Vol. 2)

$$\boldsymbol{F} = \boldsymbol{F}_1 + \boldsymbol{F}_2 = \frac{q \cdot Q}{4\pi\varepsilon_0}\left[\frac{1}{r_1^2}\hat{\boldsymbol{r}}_1 - \frac{1}{r_2^2}\hat{\boldsymbol{r}}_2\right].$$

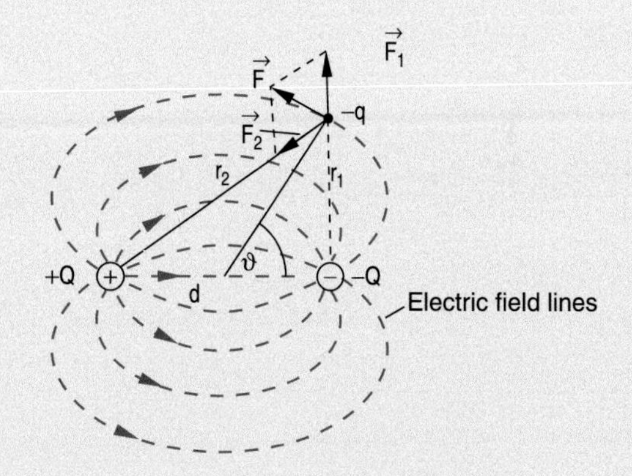

Figure 2.18 Force field of an electric dipole and the force on a negative test charge

b) ***Force field of a planetary system***
At each position $\boldsymbol{r}$ the gravitational forces on a test mass exerted by the sun, the planets and the moons superimpose. The force field $\boldsymbol{F}(\boldsymbol{r}) = \sum \boldsymbol{F}_i$ is very complex. It even can be zero at certain points in space, for example at a point N between earth and moon (neutral point) where the opposite gravitational forces from earth and moon just compensate (Fig. 2.19).

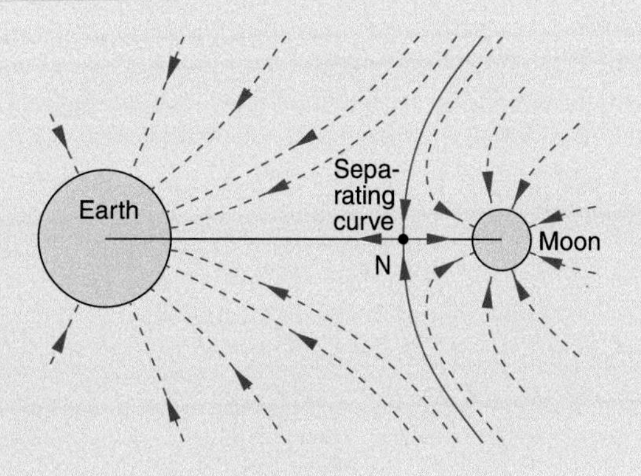

Figure 2.19 Gravitational field between earth and moon

c) ***Homogeneous force field of a parallel plate capacitor***
For a voltage V between the plates with a distance d the force on an electric charge $+q$ is $\boldsymbol{F} = +q \cdot (V/d) \cdot \hat{e}_z$ vertical to the plates and pointing from the positively charged plate to the negative one (Fig. 2.20). The force $\boldsymbol{F}$ has at any point inside the capacitor the same magnitude and direction. Such a force field is called *homogeneous*.
Within a small volume also the gravitational field of the earth can be treated as a homogeneous force field as long as the vertical extension Δz of this volume is very small compared to the radius R of the earth. The force on a mass is then $\boldsymbol{F} = m \cdot \boldsymbol{g}$, where $|\boldsymbol{g}| = 9.81\,\mathrm{m/s^2}$ is the earth gravitational acceleration which remains constant in a small volume.

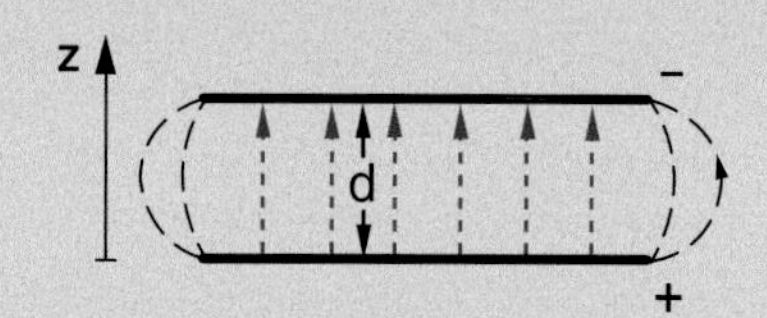

Figure 2.20 Homogeneous force field for electric charges inside a parallel plate capacitor ◄

2.5.3 Measurements of Forces; Discussion of the Force Concept

Forces can be measured due to their effect on the deformation of elastic bodies (see Chap. 6). One example is the spring balance (Fig. 2.21). Here the elongation of a spring under the influence of a force is measured. Its displacement $x - x_0$ from the equilibrium position x_0 is proportional to the acting force

$$F_x = -D(x - x_0)\ . \tag{2.17}$$

If the spring constant $D = F/\Delta x$ is known, the determination of the force $\boldsymbol{F}$ is reduced to a length measurement $\Delta x = x - x_0$. The spring constant D can be obtained from measurements of the oscillation period of the spring balance. After a mass m has been

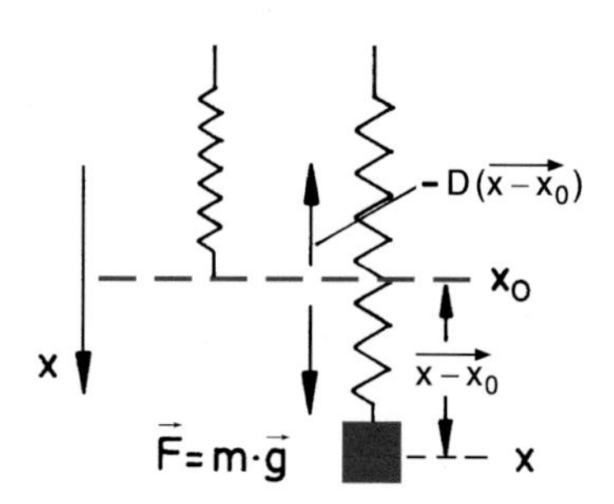

Figure 2.21 Spring balance for the measurement of forces

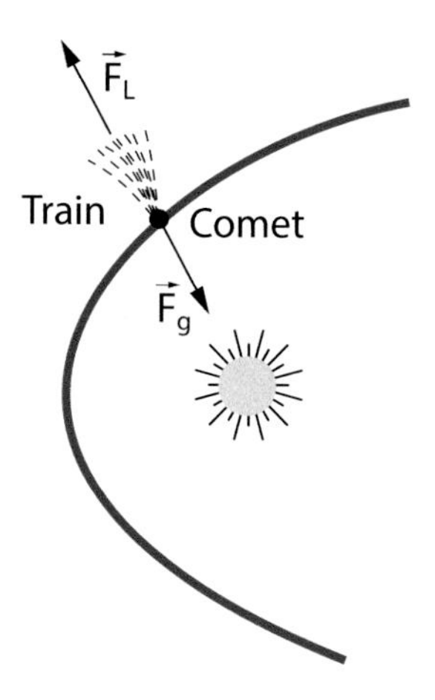

Figure 2.22 Interaction between sun and comet as an example for the far distance effect of forces

displaced from its equilibrium position x_0 and then released, it performs oscillations around x_0 (see Sect. 2.9.7).

Often forces can act on bodies without physical contact between them. Examples are the gravitational force between sun and earth or between sun and a comet (Fig. 2.22). In the latter case the comet is attracted by the sun due to the gravitational force and vice versa. Its tail is repelled because of the radiation pressure and the sun wind which is exerted by particles (protons and electrons) emitted from the sun.

Even if there is no direct contact between two bodies we say that a force acts on each body which causes the change of its motional state i. e. its velocity with time. Also for the investigation of atomic collision processes the information on the forces between the colliding atoms is obtained from the observed change of the velocities of the two collision partners (see Sect. 4.3). Here the change of the momentum $\mathrm{d}\boldsymbol{p}/\mathrm{d}t$ is used to determine the force. This explanation goes beyond the ordinary meaning of forces as directly perceptible phenomena as for instance the physical strength.

In all cases the force is a synonym for the interaction between bodies. The range of distances between the interacting bodies can reach from 10^{-17} m to infinity.

The question, what the real cause for this interaction is and whether it is transferred between the interacting bodies infinitely fast or with a finite speed can be up to now only partly answered and is the subject of intense research but is not yet fully understood. Theoretical predictions claim a finite transfer time which equals the speed of light. The description of the interaction between very fast moving bodies has therefore to take into account this finite transfer time (retardation, see Sect. 3.5). For velocities which are small compared to the speed of light this effect can be neglected (realm of non-relativistic physics).

We will now discuss more quantitatively the relations between forces and the change of motional states of bodies.

2.6 The Basic Equations of Mechanics

The mathematical description of the motion of bodies under the influence of forces can be reduced to a few basic equations. These equations are based on assumptions (axioms) which are suggested by experiments. They were first postulated by Isaac Newton in his famous multi-volume opus "*Philosophiae naturalis principia mathematica*" which was published in the years 1687–1726 [2.1].

2.6.1 The Newtonian Axioms

For the introduction of the force model and its relation with the state of motion of bodies Newton started from three basic assumptions which were taken from daily experience. They are called the *three Newtonian axioms* (sometimes also Newton's three laws).

First Newtonian Axiom

Each body remains in the state of rest or of uniform motion on a straight line as long as no force is acting on it.

As the measure for the state of motion of a body with mass m we define the **momentum**

$$\boldsymbol{p} = m \cdot \boldsymbol{v} .$$

The momentum $\boldsymbol{p}$ is a vector parallel to the velocity $\boldsymbol{v}$ and has the dimension $[p] = [\mathrm{kg \cdot m \cdot s^{-1}}]$. A particle on which no force is acting is called a **free particle**.

With this definition Newton's first law can be formulated as

The momentum of a free particle is constant in time.

This means: always when a particle changes its state of motion a force is acting on it, i. e. it interacts with other particles or it is moving in a force field (Fig. 2.23).

Second Newtonian Axiom

Since we attribute a force to any change of momentum we define the force $\boldsymbol{F}$ as

$$\boldsymbol{F} = \frac{\mathrm{d}\boldsymbol{p}}{\mathrm{d}t} . \tag{2.18}$$

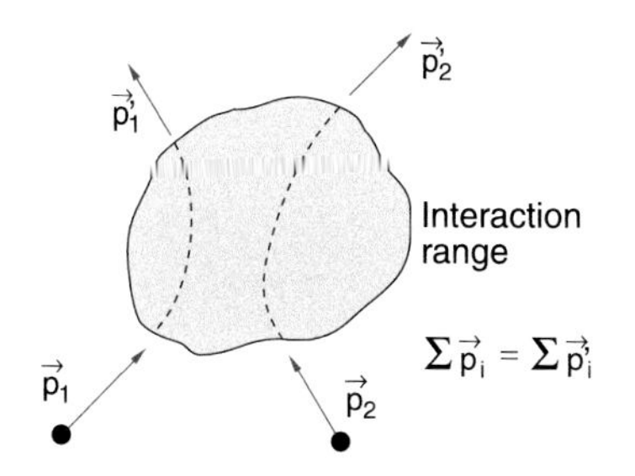

Figure 2.23 Forces as cause for a change of momentum

With $\boldsymbol{p} = m \cdot \boldsymbol{v}$ we can write this in the form

$$\boldsymbol{F} = m \cdot \frac{\mathrm{d}\boldsymbol{v}}{\mathrm{d}t} + \frac{\mathrm{d}m}{\mathrm{d}t} \cdot \boldsymbol{v} . \tag{2.18a}$$

The second term describes a possible change of the mass m with the velocity of the particle. There are many situations where this second term becomes important, for instance when a rocket is accelerated by the expulsion of fuel (see Sect. 2.6.3) or when a particle is accelerated to very high velocities, comparable to the velocity c of light, where the relativistic mass $m(v)$ increase with velocity, cannot be neglected (see Sect. 4.4.1).

Example

A freight train moves with the velocity v in the horizontal x-direction (Fig. 2.24). It is loaded continuously with sand from a stationary reservoir above the train. The mass increase per time $\mathrm{d}m/\mathrm{d}t$ is assumed to be constant. When friction can be neglected the total force onto the train is zero. The equation of motion is then

$$0 = m \cdot \mathrm{d}v/\mathrm{d}t + A \cdot v \tag{2.18b}$$

with $m = m_0 + A \cdot t$. Integration yields

$$\ln \frac{v}{v_0} = \ln \frac{m_0}{m_0 + A \cdot t}$$

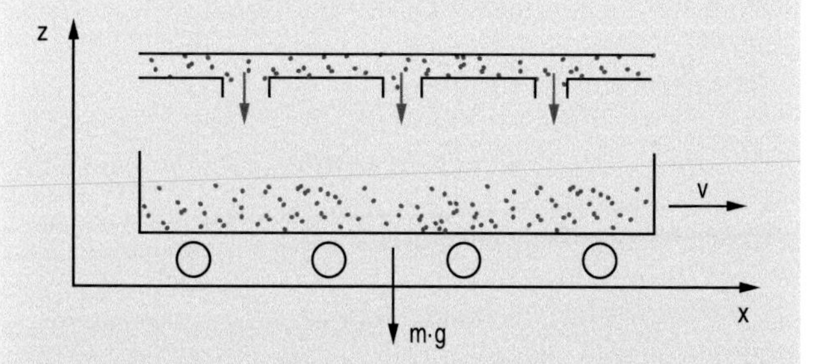

Figure 2.24 Example to Eq. 2.18a

with the solution

$$v(t) = v_0 \frac{1}{1 + (A/m_0) \cdot t} . \tag{2.18c}$$

With $m_0 = 1000$ tons and $\mathrm{d}m/\mathrm{d}t = A = 1$ ton/s the train velocity $v(t) = v_0(1 + 1 + 10^{-3}\,t)^{-1}$ the velocity slows down to $v_0/2$ in 1000 s. ◄

Chapter 2

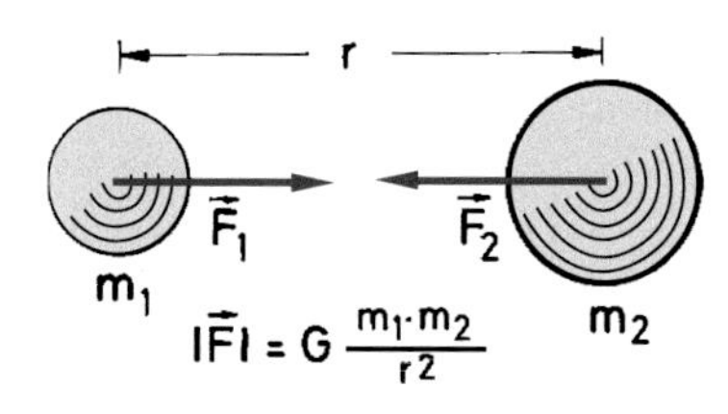

Figure 2.25 actio = reactio for the example of gravitational forces $F_1 = -F_2$ between two masses

If the mass m is constant ($dm/dt = 0$) Eq. 2.18b takes the simple form

$$F = m \cdot a \quad \text{with} \quad a = \frac{dv}{dt} . \tag{2.18d}$$

The unit of the force is $[F] = 1\,\text{kg} \cdot \text{m} \cdot \text{s}^{-2} = 1$ Newton $= 1$ N.

Third Newtonian Axiom

When two bodies interact with each other but not with a third partner the force acting on the first body has equal magnitude but opposite direction as the force on the second body (Fig. 2.25). Newton's formulation in Latin was

$$\textit{actio} = \textit{reactio}$$
$$F_1 = -F_2 .$$

We will apply Newton's axioms to a system of two masses m_1 and m_2 which interact with each other, i. e. they collide, but are otherwise completely isolated from their surroundings. Such a system is called a *closed system*.

Since there are no external forces on a closed system we can conclude in analogy to a free particle that the total momentum of the system remains constant:

$$p_1 + p_2 = \textbf{const} . \tag{2.19a}$$

Differentiating this equation yields

$$\frac{dp_1}{dt} + \frac{dp_2}{dt} = 0 \Rightarrow F_1 = -F_2 . \tag{2.19b}$$

This axiom can be proved experimentally with two equal spring balances (Fig. 2.26a), which are connected to each other at one end. If one pulls at the two other ends into opposite directions they show that on each spring balance the same force is acting.

Another experimental verification is shown in Fig. 2.26b where a spring is compressed by two equal masses on an air track which are hold together by a string. If the string is burnt by a candle, the two masses are pushed by the expanding spring to opposite sides and slide on the air track with equal velocities, which means that they have equal but opposite momenta. The velocities can be accurately measured by photoelectric barriers.

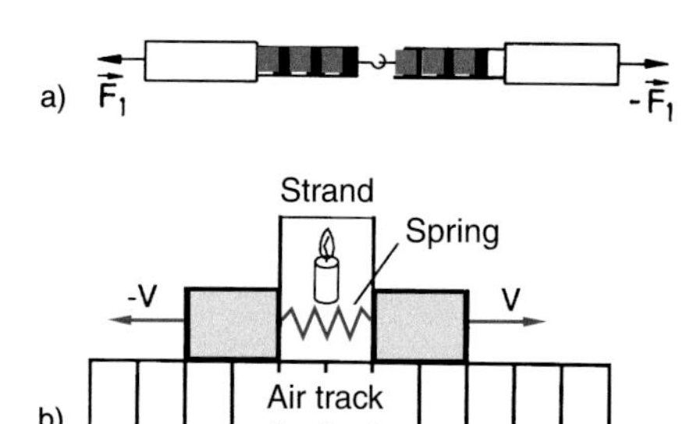

Figure 2.26 Experiment to prove the 3. Newtonian law **a** with two equal spring balances, **b** with two equal masses on an air track

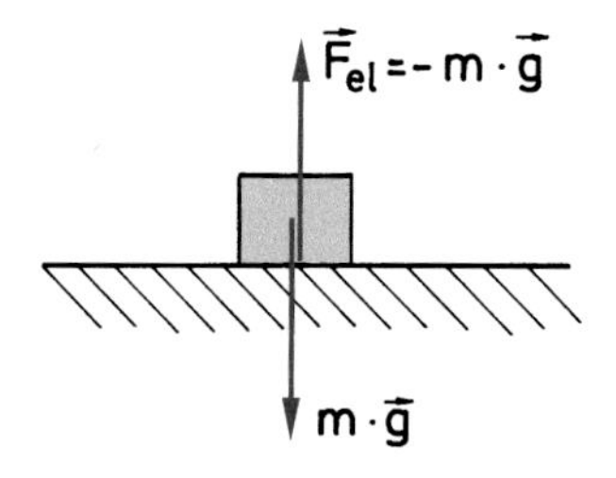

Figure 2.27 The gravitational force $F = m \cdot g$ of a mass m on a solid surface is compensated by the antiparallel deformation force of the solid surface

Newton's third law can be also proved for resting bodies. A mass m resting on a solid surface acts with the gravitational force $F_1 = m \cdot g$ on the surface which is deformed and responds with an equal but opposite elastic force $F_{el} = -F_1 = -mg$ (Fig. 2.27).

2.6.2 Inertial and Gravitational Mass

The property of bodies to remain in their state of motion when left alone (i. e. when no force is acting on them) is called their **inertia**. Since the accelerating force is proportional to the mass of the body its mass can be regarded as the cause of the inertia and is therefore called the **inertial mass** m_{inertial}. Newton's second law means this inertial mass. There are many demonstration experiments which illustrate this inertia. Assume, for example, a glass of water standing on a sheet of paper. If the paper is pulled suddenly away, the glass remains a rest without moving, because of its inertia.

There is another property of masses which is the gravitational force ($F_{\text{grav}} = m \cdot g$ on the earth surface). This force is also called the weight of the mass. Experiments measure the weight of a mass of 1 kg as

$$F_{\text{grav}} = 1\,\text{kg} \cdot 9.81\,\text{m/s}^2 = 9.81\,\text{N} .$$

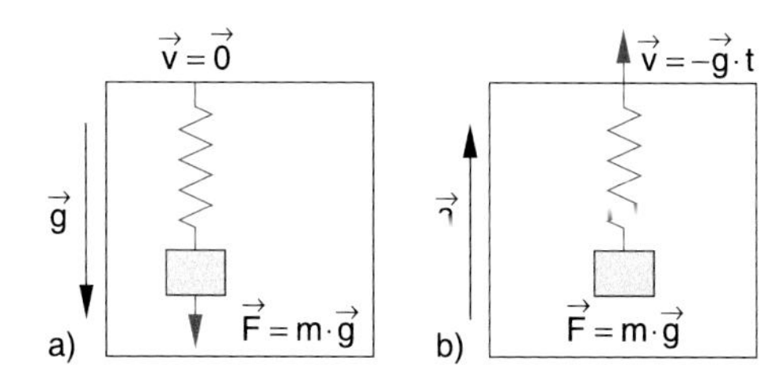

Figure 2.28 Einstein's Gedanken-experiment for the equivalence of gravitational and inertial mass **a** in the homogeneous gravitational field of the earth; **b** in a gravitation-free space inside an accelerated lift

Note: The gravitational force is always present when the mass m is attracted by another mass M and it is proportional to the product $m \cdot M$ (see Sect. 2.9.2).

The question is now: Are these two properties related to the same mass i. e. is $m_{\text{inertial}} = m_{\text{grav}}$?

Many detailed and accurate measurements for many different masses have proved that within the relative uncertainty of 10^{-10} there is no measurable difference between m_{inertial} and m_{grav}.

Starting from this experimental result Einstein has postulated the general *equivalence principle* **that inertial and gravitational masses are always equal**.

By the following "Gedanken-experiment" he has shown, that it doesn't make sense to distinguish between inertial and gravitational masses:

An observer in a closed lift measures a mass m hanging on a spring balance (Fig. 2.28). He cannot distinguish, whether the elevator is resting in a gravitational field with the gravitational force $\boldsymbol{F}_{\text{grav}} = m \cdot \boldsymbol{g}$ on the mass m (Fig. 2.28a) or whether the elevator moves upwards with the velocity $\boldsymbol{v} = -\boldsymbol{g}t$ and the acceleration $-\boldsymbol{g}$ in a force-free surrounding (Fig. 2.28b). Both situations lead to the same elongation of the spring balance. Any further experiment performed inside the closed elevator leads to the same results for the two situations (a) and (b).

For instance when the observer in the elevator throws a ball in the horizontal direction the trajectory of the ball is for both situations a parabola (see Fig. 2.9).

We will therefore no longer distinguish between inertial and gravitational mass and call it simply the mass m of a body which has the two characteristic features of inertia under acceleration and weight in gravitational fields.

Note: The question what the mass of a body really means is up to date not answered, although great efforts are undertaken to solve this problem.

2.6.3 The Equation of Motion of a Particle in Arbitrary Force Fields

Integration of Newton's equation of motion $\boldsymbol{F} = m\cdot \mathrm{d}\boldsymbol{v}/\mathrm{d}t$ yields the equations

$$\boldsymbol{v}(t) = \frac{1}{m}\int \boldsymbol{F}\mathrm{d}t + \boldsymbol{C}_1 , \tag{2.20a}$$

$$\begin{aligned}\boldsymbol{r}(t) &= \int \boldsymbol{v}(t)\,\mathrm{d}t + \boldsymbol{C}_2\\ &= \frac{1}{m}\int\left[\int \boldsymbol{F}\mathrm{d}t\right]\mathrm{d}t + \int \boldsymbol{C}_1\,\mathrm{d}t + \boldsymbol{C}_2\end{aligned} \tag{2.20b}$$

For the velocity $\boldsymbol{v}(t)$ and the position vector $\boldsymbol{r}(t)$ with the integration constants $\boldsymbol{C}_1$ and $\boldsymbol{C}_2$ which are fixed by the initial conditions (e.g. $\boldsymbol{v}(t=0) = \boldsymbol{v}_0$ and $\boldsymbol{r}(t=0) = \boldsymbol{r}_0$).

Whether these equations are analytical solvable depends on the form of the force $\boldsymbol{F}$ which can be a function of position $\boldsymbol{r}$, velocity $\boldsymbol{v}$ or time $\boldsymbol{t}$. We will illustrate this by some examples.

Constant Forces

For the most simple case of constant forces $\boldsymbol{F} = \mathbf{const}$, which do not depend on time nor on the position or velocity of the particle the integration of (2.20) immediately gives

$$\begin{aligned}\boldsymbol{F} &= m\cdot\boldsymbol{a} = \mathbf{const}\,,\\ \boldsymbol{v}(t) &= \boldsymbol{a}t + \boldsymbol{C}_1 \quad \text{with} \quad \boldsymbol{C}_1 = \boldsymbol{v}_0 = \boldsymbol{v}(t=0)\,,\\ \boldsymbol{r}(t) &= \tfrac{1}{2}\boldsymbol{a}t^2 + \boldsymbol{v}_0 t + \boldsymbol{r}_0 \quad \text{with} \quad \boldsymbol{r}_0 = \boldsymbol{r}(t=0)\,.\end{aligned} \tag{2.21}$$

The trajectory of the particle can be directly determined, if the initial conditions are known. It is advisable to choose the coordinate system in such a way that the force coincides with one of the coordinate axes.

Example

The motion of a particle under the influence of the constant force $\boldsymbol{F} = \{0, 0, -mg\}$ pointing into the $-z$-direction gives the three equations for the 3 components of the force

$$\begin{aligned}\ddot{x} &= 0 \;\Rightarrow \dot{x} = A_x \qquad\;\;\; \Rightarrow x = A_x t + B_x\\ \ddot{y} &= 0 \;\Rightarrow \dot{y} = A_y \qquad\;\;\; \Rightarrow y = A_y t + B_y\\ \ddot{z} &= -g \Rightarrow \dot{z} = -gt + A_z \Rightarrow z = -\tfrac{1}{2}gt^2 + A_z t + B_z\,.\end{aligned} \tag{2.22}$$

These equations describe every possible motion of particles under the influence of the earth gravitation in a volume which is small compared with the dimensions of the earth where the gravitational force can be regarded as constant. From the many possible solutions of (2.22) the initial conditions with fixed values of A and B select special solutions (see examples in Sect. 2.3). ◀

Chapter 2

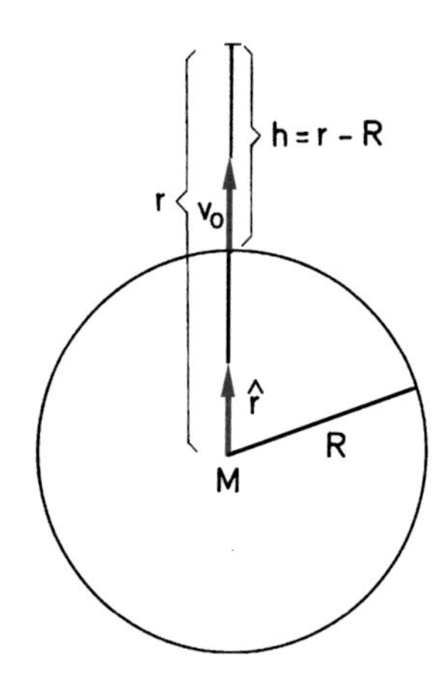

Figure 2.29 Launch of a body from the earth surface

Forces $F(r)$ that depend on the Position

As an example of position dependent forces we choose the gravitational force

$$F(r) = -G\frac{m \cdot M}{r^2}\hat{r} .$$

The minus-sign indicates that the attractive force points into the direction of $-r$.

The acceleration a has in this central force field only a radial component $a_r = -G \cdot M/r^2$. For vertical motions the velocity becomes in spherical coordinates $v = \{v_r, 0, 0\}$ and its magnitude is $|v| = v = v_r$. Our problem therefore becomes one-dimensional. From the relation

$$a = \frac{dv}{dt} = \frac{dv}{dr} \cdot \frac{dr}{dt} = \frac{dv}{dr} \cdot v ,$$

it follows: $v \cdot dv = -(G \cdot M/r^2)dr$.

Integration yields

$$\frac{1}{2}v^2 = \frac{GM}{r} + C_1 . \tag{2.23}$$

Let us discuss the case that a projectile is fired from the earth surface ($r = R$) upwards in vertical direction with the initial velocity v_0 (Fig. 2.29). The integration constant C_1 then becomes

$$C_1 = \frac{1}{2}v_0^2 - \frac{GM}{R} = \frac{1}{2}v_0^2 - g \cdot R ,$$

because $a(R) = -g = -G(M/R^2)$. This gives

$$\frac{1}{2}v^2 = \frac{gR^2}{r} + \frac{1}{2}v_0^2 - g \cdot R . \tag{2.24}$$

At the maximum vertical height $r = r_{max}$ the velocity becomes $v(r_{max}) = 0$ and we obtain from (2.24)

$$r_{max} = \frac{R}{1 - (v_0^2/2Rg)} . \tag{2.25}$$

For the initial velocity $v_0 \to \sqrt{2Rg}$ the maximum vertical height r_{max} becomes infinity and the projectile can leave the earth. This velocity is called *the escape velocity*. Inserting the numerical values for R and g gives

$$v_0 \geq v_2 = \sqrt{2Rg} = 11.2\,\text{km/s} . \tag{2.26a}$$
(escape velocity)

The velocity v_2 is often named the 2nd cosmic velocity while the first cosmic velocity v_1 is the velocity of a projectile which is fired in horizontal direction and orbits around the earth on a circle closely above the earth surface. From the relation

$$\frac{v_1^2}{R} = \frac{GM}{R^2} \to v_1 = \sqrt{\frac{GM}{R}} = \sqrt{g \cdot R} \tag{2.26b}$$

the numerical value of v_1 becomes (when neglecting the earth rotation) $v_1 = v_2/\sqrt{2} \approx 7.9\,\text{km/s}$.

Note: The general case of arbitrary motion in a central force field is treated in Sect. 2.9.

Time-dependent Forces

There are many situations where the force on a particle changes with time. One simple example is a mass hanging on a spring, which is induced to vertical oscillations, or a comet moving on a parabolic trajectory through the solar system. We will illustrate the solution of the equation of motion for time dependent forces by two numerical examples.

Examples

1. Assume the time dependent force $F = b \cdot t + c$ with $b = 120\,\text{N/s}$ and $c = 40\,\text{N}$, which points into the x-direction, is acting on the mass $m = 10\,\text{kg}$. For $t = 0$ the mass should be at $x = 5\,\text{m}$ with a velocity $v(0) = 6\,\text{m/s}$. Calculate the position $x(t)$.

Solution

The straight motion proceeds along the x-axis. The acceleration is $a = F/m$ and the velocity

$$v(t) = \frac{1}{m}\int_0^t F(\tau)d\tau = \frac{b}{2m}t^2 + \frac{c}{m}t + v_{0x} ;$$

$$x(t) = \int v_x d\tau = \frac{b}{6m}t^3 + \frac{c}{2m}t^2 + v_{0x}t + x_0$$
$$= (2t^3 + 2t^2 + 6t + 5)m \quad \text{with } t \text{ in s} .$$

2. What is the final velocity of a mass m initially at rest ($v(0) = 0$) which experiences a force $F(t) = A \cdot \exp[-a^2t^2]$?

Chapter 2

Solution

$$m \cdot v(t = \infty) = \int F \,\mathrm{d}t = A \int \mathrm{e}^{-a^2 t^2} \,\mathrm{d}t = \frac{A\sqrt{\pi}}{2a}\,;$$

$$v_\infty = \frac{A}{2} \frac{\sqrt{\pi}}{a \cdot m}\,.$$

◀

Acceleration of a Rocket

In the example for position dependent forces we have assumed that the projectile starts with the initial velocity $v_0 > 0$. In reality it starts with $v_0 = 0$. However, the velocity $v > 0$ is reached within a short distance that is very small compared with the earth radius R. We will now study the acceleration during the start phase of the rocket in more detail. Within this small distance $d \ll R$, which the rocket passes during its acceleration, we can fairly assume the earth acceleration g to be constant.

During the burning phase the rocket is continuously accelerated by the recoil momentum of the propellant hot gases (Fig. 2.30).

With v' we denote the velocity of the propellant gases relative to the surface of the earth which represents our reference coordinate system, and with v the rocket velocity in this system. The escaping gas mass per second is $\Delta m/\Delta t$. The momentum of the rocket at time t is $p(t) = m \cdot v$. At time $t + \Delta t$ the mass of the rocked has been reduced by $-\Delta m$ (which equals the mass of the expanding gas during this time interval) and its velocity has increased by Δv while the gases have transported the momentum $\Delta m \cdot v'$. The total momentum of the system rocket + gas is then with $\Delta m > 0$

$$\boldsymbol{p}(t + \Delta t) = (m - \Delta m)(\boldsymbol{v} + \Delta \boldsymbol{v}) + \Delta m \cdot \boldsymbol{v}'\,. \tag{2.27a}$$

During the time interval Δt the momentum of the system has changed by

$$\begin{aligned} \Delta \boldsymbol{p} &= \boldsymbol{p}(t + \Delta t) - \boldsymbol{p}(t) \\ &= m \cdot \Delta \boldsymbol{v} + \Delta m(\boldsymbol{v}' - \boldsymbol{v}) - \Delta m \cdot \Delta v\,. \end{aligned} \tag{2.27b}$$

Figure 2.30 Acceleration of a rocket

For the limit $\Delta t \to 0$; $\Delta m/\Delta t \to \mathrm{d}m/\mathrm{d}t$ is $\lim_{\Delta t \to 0}(\Delta m \cdot \Delta v/\Delta t) = 0$.

Since the time derivative $\mathrm{d}p/\mathrm{d}t$ of the momentum equals the force $F_g = m \cdot \boldsymbol{g}$ of gravity acting on the rocket we obtain

$$\frac{\mathrm{d}\boldsymbol{p}}{\mathrm{d}t} = m\frac{\mathrm{d}\boldsymbol{v}}{\mathrm{d}t} + \frac{\mathrm{d}m}{\mathrm{d}t}(\boldsymbol{v}' - \boldsymbol{v}) = m \cdot \boldsymbol{g}\,. \tag{2.27c}$$

The velocity $\boldsymbol{v}'$ of the propellant gases *relative to the earth* depends on the velocity $\boldsymbol{v}$ of the rocket. For $|\boldsymbol{v}| < |\boldsymbol{v}'|$ the direction of $\boldsymbol{v}'$ is downwards, for $|\boldsymbol{v}| > |\boldsymbol{v}'|$ it is upwards. It is therefore better to introduce the velocity $\boldsymbol{v}_\mathrm{e} = \boldsymbol{v}' - \boldsymbol{v}$ of the propellant gases *relative to the rocket*, which is independent of $\boldsymbol{v}$ and constant in time. This converts Eq. 2.27c into

$$m \cdot \frac{\mathrm{d}\boldsymbol{v}}{\mathrm{d}t} + \frac{\mathrm{d}m}{\mathrm{d}t}\boldsymbol{v}_\mathrm{e} = m \cdot \boldsymbol{g}\,. \tag{2.27d}$$

With $\boldsymbol{v} = \{0, 0, v_z\}$, $\boldsymbol{v}_\mathrm{e} = \{0, 0, v_\mathrm{e}\}$, $\boldsymbol{g} = \{0, 0, -g\}$ this equation becomes after division by m and multiplication by $\mathrm{d}t$

$$\mathrm{d}v = -v_\mathrm{e}\frac{\mathrm{d}m}{\mathrm{d}t} - g \cdot \mathrm{d}t\,, \tag{2.27e}$$

Integration from $t = 0$ up to $t = T$ (propellant time of the rocket) yields

$$v(T) = v_0 + v_\mathrm{e} \ln \frac{m_0}{m} - gT\,, \tag{2.28}$$

where $v_0 = v(t = 0)$.

Numerical Example

Launching of a *Saturn rocket* with $m_0 = 3 \cdot 10^6\,\mathrm{kg}$; $v_\mathrm{e} = 4000\,\mathrm{m/s}$, $T = 100\,\mathrm{s}$, $v_0 = 0$. Final mass at $t = T$ is $m(T) = 10^6\,\mathrm{kg}$, which means that the mass of the fuel is $2 \cdot 10^6\,\mathrm{kg}$. Equation 2.28 yields

$$\begin{aligned} v(T = 100\,\mathrm{s}) &= 0 + 4000\,\mathrm{m/s} \cdot \ln 3 - 9.81\,\mathrm{m/s^2}\,100\,\mathrm{s} \\ &= 3413.5\,\mathrm{m/s}\,. \end{aligned}$$

◀

The heights $z(t)$ of the rocket during its burning time for constant loss of mass $q = \mathrm{d}m/\mathrm{d}t = \mathrm{const}$ is readily obtained. With $m(t) = m_0 - q \cdot t$, Eq. 2.28 becomes

$$v(t) = v_0 - v_\mathrm{e} \ln\left(1 - \frac{q}{m_0}t\right) - gt\,;$$

$$z(t) = v_0 t - v_\mathrm{e} \int \ln\left(1 - \frac{q}{m_0}t\right)\mathrm{d}t - \frac{1}{2}gt^2 + C_0\,,$$

and integration yields

$$z(t) = v_0 \cdot t - v_\mathrm{e} \int \ln\left(1 - \frac{q}{m_0}t\right)\mathrm{d}t - \frac{1}{2}gt^2 + C_0\,.$$

The integration constant is $C_0 = 0$ (because $z(0) = 0$).

Since $\int \ln x\mathrm{d}x = x\ln x - x$ the integration gives

$$z(t) = (v_0 + v_e)\,t + v_e\left[\frac{m_0}{q} - t\right]\ln\left(1 - \frac{q}{m_0}t\right) - \frac{1}{2}gt^2\,. \tag{2.29}$$

Numerical Example

For our example above we obtain with $q = 2 \cdot 10^4\,\mathrm{kg/s}$, $v_0 = 0$; $v_e = 4000\,\mathrm{m/s}$ $T = 100\,\mathrm{s}$

$$\begin{aligned} z(T) &= (4 \cdot 10^5 + 2 \cdot 10^3 \cdot \ln 0.33 - 4.9 \cdot 10^4\,\mathrm{m} \\ &= (400 - 219.7 - 49)\,\mathrm{km} = 131\,\mathrm{km}\,, \\ v(T) &= \left[-4 \cdot 10^3 \cdot \ln(0.33) - 981\right]\,\mathrm{m/s} = 3413\,\mathrm{m/s}\,, \end{aligned}$$

◀

This example illustrates that with $z(T) \ll R$ the earth acceleration does not change much and can be regarded as constant. It further demonstrates that with a single stage the escape velocity $v = 11200\,\mathrm{m/s}$ of the rocket cannot be achieved with reasonable fuel masses. It is therefore necessary to use multistage rockets.

Numerical Example

After the end of the burning time T_1 of the first stage the velocity of the rocket in our example is $v(T_1) = 3400\,\mathrm{m/s}$. The second stage starts with a mass $m(T_1) = 9 \cdot 10^5\,\mathrm{kg}$ (the fuel tank with $m = 10^5\,\mathrm{kg}$ has been pushed off) including $m = 7 \cdot 10^5\,\mathrm{kg}$ for the fuel. The burning time is again 100 s and the final mass $m(T_2) = 2.10^5\,\mathrm{kg}$. According to (2.28) the final velocity v is

$$\begin{aligned} v(T_1 + T_2) &= (3400 + 4000\ln(9/2) - 9.81 \cdot 100) \\ &= 8435\,\mathrm{m/s}\,. \end{aligned}$$

The third stage starts with a velocity $v = 8435\,\mathrm{m/s}$ and a mass $m = 1.8 \cdot 10^5\,\mathrm{kg}$ (the fuel tank of the 2nd stage with $m = 2 \cdot 10^4\,\mathrm{kg}$ has been pushed off). With $T_3 = 100\,\mathrm{s}$ we obtain the final velocity

$$\begin{aligned} v(T_{\mathrm{final}}) &= (8400 + 4000\ln 7.2 - 9.8 \cdot 100)\,\mathrm{m/s} \\ &= 15{,}000\,\mathrm{m/s} > v_{\mathrm{escape}}\,. \end{aligned}$$

◀

Note: For the second and third stage one should, strictly speaking, take into account the decrease of the earth acceleration g with increasing z. Instead of the constant g one should use the function $g(z) = G \cdot M/r^2$ with $r = z + R$ and M = mass of the earth. With the approximation $(1 + z/R)^{-2} \approx 1 - 2z/R$ one obtains instead of (2.27e) the equation

$$\mathrm{d}v = -v_e\frac{\mathrm{d}m}{m} - g(1 - 2z/R)\mathrm{d}t\,. \tag{2.30}$$

This equation illustrates that even for $z = 100\,\mathrm{km}$ the correction term $2z/R$ for g amounts only to 3%. This means for the calculation of the velocity v only a correction of 1%, because the term $g \cdot T$ in Eq. 2.28 represents only about $1/3v$.

The integration of (2.28) is now more tedious but an approximation is still possible, if the function (2.29) is inserted for $z(t)$.

2.7 Energy Conservation Law of Mechanics

In this section we will discuss the important terms "*work*", "*power*", "*kinetic and potential energy*" before we can formulate the **energy conservation law of mechanics**.

2.7.1 Work and Power

If a point mass m proceeds along the path element $\boldsymbol{\Delta r}$ in a force field $\boldsymbol{F}(\boldsymbol{r})$ (Fig. 2.31), the scalar product

$$\Delta W = \boldsymbol{F}(\boldsymbol{r}) \cdot \boldsymbol{\Delta r} \tag{2.31a}$$

is called **the *mechanical work***, due to the action of the force F on the point mass m.

The work is a scalar quantity!

Written in components of the vectors $\boldsymbol{F}$ and $\boldsymbol{r}$ Eq. 2.31a reads

$$\Delta W = F_x\Delta x + F_y\Delta y + F_z\Delta z\,. \tag{2.31b}$$

The unit of work is [work] = [force · length] = $1\,\mathrm{N} \cdot m$ = 1 Joule = 1 J.

Remark. In the *cgs-system* the unit is [W] = 1 dyn · cm = 1 erg = 10^{-7} J.

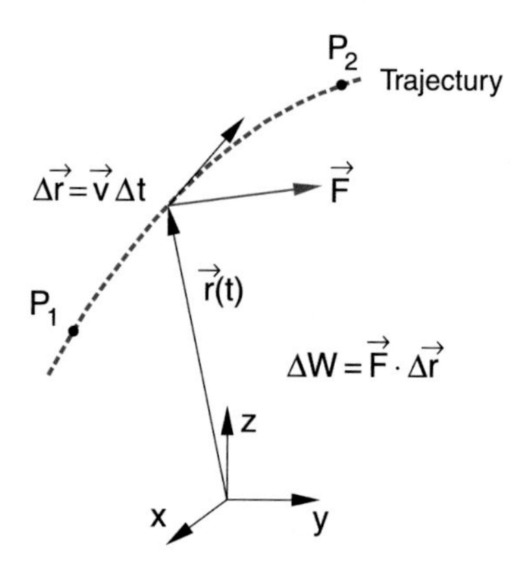

Figure 2.31 Definition of work

If the point mass moves under the action of the force F from point P_1 to point P_2 the total work on this path is the sum $W = \sum \Delta W_i$ of the different contributions $\Delta W_i = \boldsymbol{F}(\boldsymbol{r}_i) \cdot \Delta \boldsymbol{r}_i$ which converges in the limit $\Delta r_i \to 0$ to the integral

$$W = \int_{P_1}^{P_2} \boldsymbol{F} \cdot \mathbf{dr} \,. \tag{2.32a}$$

The integral is called *line-integral* or *curvilinear integral*. Because of the relation $\boldsymbol{F} \cdot \mathbf{dr} = F_x \mathrm{d}x + F_y \mathrm{d}y + F_z \mathrm{d}z$ it can be reduced to a sum of simple Rieman integrals:

$$\int \boldsymbol{F} \cdot \mathbf{dr} = \int_{x_1}^{x_2} F_x \mathrm{d}x + \int_{y_1}^{y_2} F_y \mathrm{d}y + \int_{z_1}^{z_2} F_z \mathrm{d}z \,, \tag{2.32b}$$

which can be readily calculated if the force is known (see the following examples). In Equation 2.32 is $W > 0$ for $\boldsymbol{F} \cdot \mathbf{dr} > 0$ i.e. if the force $\boldsymbol{F}$ has a component in the direction of the movement. In this case the mass m is accelerated. According to this definition the work is positive if the energy of the mass m is increased. Work which is performed by the mass on other systems decreases its energy and is therefore defined as negative (see Sect. 2.7.3).

If $\boldsymbol{F}$ is perpendicular to $\boldsymbol{r}$ (and therefore also to the velocity $\boldsymbol{v}$) the work is $W = 0$, because then the scalar product $\boldsymbol{F} \cdot \mathbf{dr} = \mathbf{0}$.

The work per time unit

$$P = \frac{\mathrm{d}W}{\mathrm{d}t} \tag{2.33a}$$

is called the **power** $\boldsymbol{P}$. Its unit is $[P] = 1\,\mathrm{J/s} = 1\,\mathrm{Watt} = 1\,\mathrm{W}$.

$$\begin{aligned} P &= \frac{\mathrm{d}}{\mathrm{d}t} \int_{t_0}^{t} \boldsymbol{F}(\boldsymbol{r}(t'), t') \cdot \dot{\boldsymbol{r}}(t') \mathrm{d}t' \\ &= \boldsymbol{F}(\boldsymbol{r}(t), t) \cdot \boldsymbol{v}(t) = \boldsymbol{F} \cdot \boldsymbol{v} \,. \end{aligned} \tag{2.33b}$$

Remark. In daily life the electrical work is defined in kWh. With $1\,\mathrm{J} = 1\,\mathrm{Ws}$ the relation is $1\,\mathrm{kWh} = 3.6 \cdot 10^6\,\mathrm{Ws}$.

Examples

1. Uniform circular motion under the action of a radial constant force. Here $\boldsymbol{v}$ always points in the direction of the tangent to the circle, but the force is always radial, i. e. $\boldsymbol{F} \perp \boldsymbol{v}$. The scalar product $\boldsymbol{F} \cdot \boldsymbol{v} = 0$ and therefore the work is zero.
2. A mass is moved with constant velocity without friction on a horizontal plane. (motion on a straight line). The gravitational force is always perpendicular to the motion, $\to \boldsymbol{F} \cdot \mathbf{dr} = 0$. The work is zero.
3. The work performed by a mountaineer against the gravitational force (man + pack = 100 kg), who climbs up the Matterhorn ($\Delta z = 1800\,\mathrm{m}$) is $W = \int F_g \mathrm{d}z = -m \cdot g \cdot \Delta z = 10^2 \cdot 9.81 \cdot 1.8 \cdot 10^3\,\mathrm{kg \cdot m^2/s^2} = -17.6 \cdot 10^5\,\mathrm{J} \approx 0.5\,\mathrm{kWh}$.
   The work is negative, because the force is antiparallel to the direction of the movement. The mountaineer produces energy by burning his food and converts it into potential energy thus decreasing its internal energy. The prize for the electrical equivalent of 0.5 kWh is about 10 Cents!
4. In order to expand a coil spring one has to apply a force $\boldsymbol{F} = -\boldsymbol{F}_\mathbf{r}$ opposite to the restoring spring force $\boldsymbol{F}_\mathbf{r} = -D(\boldsymbol{x} - \boldsymbol{x}_0)$ which is proportional to the elongation $(x - x_0)$ of the spring from its equilibrium position x_0. The work which has to be applied is

$$\begin{aligned} W &= \int F_x \mathrm{d}x = D \int (x - x_0) \mathrm{d}x \\ &= \tfrac{1}{2} D (x - x_0)^2 \,. \end{aligned}$$

   This is equal to the area A in Fig. 2.32a between the x-axis and the straight line $F = D(x - x_0)$.

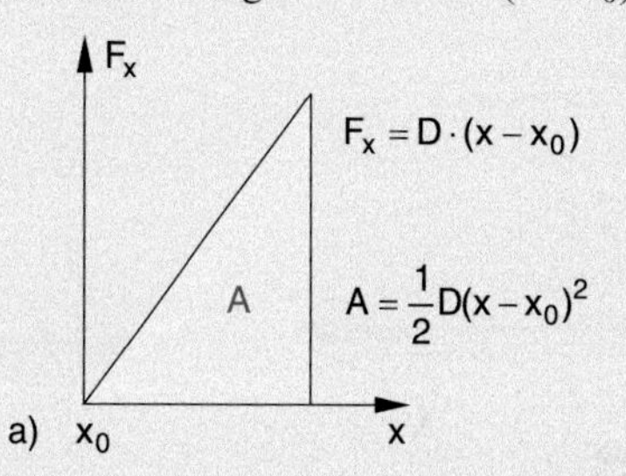

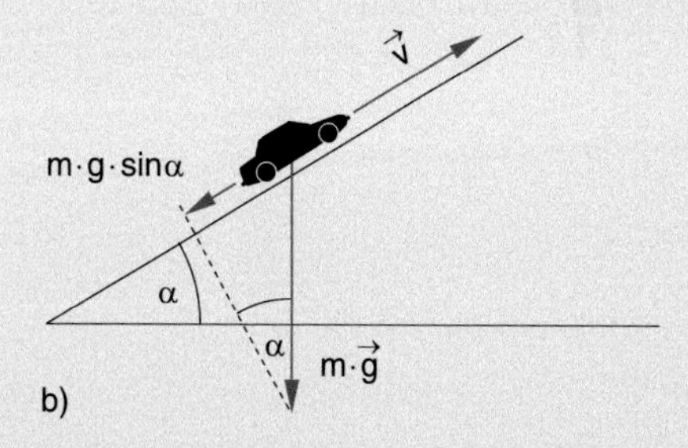

Figure 2.32 **a** Work for expanding a spring, **b** work of a car climbing up a slope

5. A car ($m = 1000\,\mathrm{kg}$) moves with constant velocity of 48 km/h on a straight line with a slope of 5° against the horizontal (Fig. 2.32b). What is the work the engine has to produce within 5 min, if friction effects can be neglected?
   The force in the direction of the motion is

$$F = -F_g \cdot \sin\alpha = m \cdot g \cdot \sin\alpha \,.$$

   The distance which the car moves within 5 min is

$$s = 48\,\mathrm{km} \cdot 5/60 = 4\,\mathrm{km} = 4000\,\mathrm{m} \,.$$

The work is then with $1\,\text{kWh} = 10^3 \cdot 3.6 \cdot 10^3\,\text{Ws} = 3.6 \cdot 10^6\,\text{J}$

$$W = 4 \cdot 10^3 \cdot 9.81 \cdot \sin 5° \cdot 10^3\,\text{N} \cdot m$$
$$= 3.4 \cdot 10^6\,\text{J} \approx 1\,\text{kWh}\,.$$

The power is

$$P = \frac{\mathrm{d}W}{\mathrm{d}t} = \frac{3.4 \cdot 10^6\,\text{J}}{300\,\text{s}} \approx 1.13 \cdot 10^4\,\text{W}$$
$$= 11.3\,\text{kW}\,.$$

◀

2.7.2 Path-Independent Work; Conservative Force-Fields

We regard a force field $\boldsymbol{F}(\boldsymbol{r})$ that depends only on the position $\boldsymbol{r}$ but not on time. When a mass m is moved from point P_1 to point P_2 on the path (a) (Fig. 2.33) the work necessary for this motion is

$$W_a = \int \boldsymbol{F} \cdot \mathrm{d}\boldsymbol{r}_\mathrm{a}\,.$$

On the path (b) it is

$$W_b = \int \boldsymbol{F} \cdot \mathrm{d}\boldsymbol{r}_\mathrm{b}\,.$$

If for arbitrary paths (a) and (b) always $W_\mathrm{a} = W_\mathrm{b}$ we name the integral *path-independent* and the force field $F(\boldsymbol{r})$ *conservative*.

With other words:

In conservative force fields the work necessary to move a mass m from a point $P(r_1)$ to a point $P(r_2)$ is independent of the path between the two points.

If we move the mass from P_1 to P_2 and back to P_1 the total work is then zero.

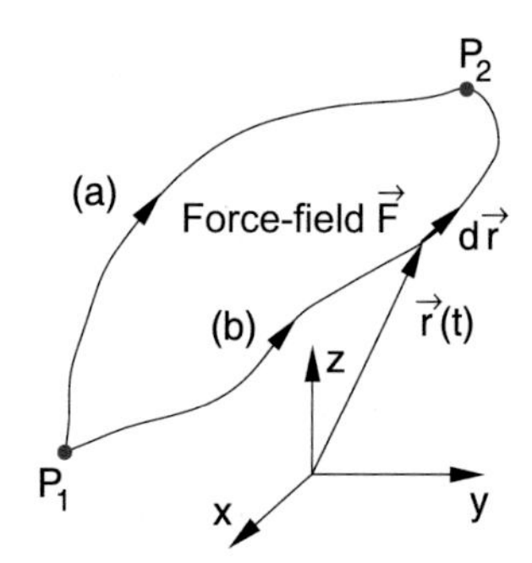

Figure 2.33 Path-independent work in a conservative force field

In conservative force fields the work for moving a mass m on a closed loop is zero.

$$\begin{aligned} W_\mathrm{a} - W_\mathrm{b} &= \int_{P_1}^{P_2} \boldsymbol{F} \cdot \mathrm{d}\boldsymbol{r}_\mathrm{a} - \int_{P_1}^{P_2} \boldsymbol{F} \cdot \mathrm{d}\boldsymbol{r}_\mathrm{b} \\ &= \int_{P_1}^{P_2} \boldsymbol{F} \cdot \mathrm{d}\boldsymbol{r}_\mathrm{a} + \int_{P_2}^{P_1} \boldsymbol{F} \cdot \mathrm{d}\boldsymbol{r}_\mathrm{b} \\ &= \oint \boldsymbol{F} \cdot \mathrm{d}\boldsymbol{r} = 0\,. \end{aligned} \qquad (2.34)$$

The work depends only on initial and final position of the motion, not on the chosen path between them.

In Vector-Analysis it is proved that the equivalent condition for a conservative force field $F(r)$ is $\mathbf{curl}\,\boldsymbol{F} = \mathbf{0}$ (theorem of Stokes). For the definition of $\mathbf{curl}\,\boldsymbol{F}$ see Sect. 13.1. It is

$$\begin{aligned} \mathbf{curl}\,\boldsymbol{F} &= \mathbf{rot}\,\boldsymbol{F} = \nabla \times \boldsymbol{F} \\ &= \left\{ \frac{\partial F_z}{\partial y} - \frac{\partial F_y}{\partial z}, \frac{\partial F_x}{\partial z} - \frac{\partial F_z}{\partial x}, \frac{\partial F_y}{\partial x} - \frac{\partial F_x}{\partial y} \right\}\,. \end{aligned}$$

Conservative force fields are a special case of force fields $\boldsymbol{F}(\boldsymbol{r})$ that depend only on the position $\boldsymbol{r}$, not on time or velocity.

Note: Not every force field $\boldsymbol{F}(\boldsymbol{r})$ is conservative! (see Example below)

Examples

Conservative Force Fields

1. A homogeneous force field $\boldsymbol{F}(\boldsymbol{r}) = \{0, 0, F_z\}$ with $F_z = \text{const}$ (Fig. 2.34a) is conservative because

$$\begin{aligned} \boldsymbol{F} \cdot \mathrm{d}\boldsymbol{r} &= F_z \mathrm{d}z \to W = \int \boldsymbol{F} \cdot \mathrm{d}\boldsymbol{r} \\ &= \int_{z_1}^{z_2} F_z \mathrm{d}z = -\int_{z_2}^{z_1} F_z \mathrm{d}z \to \oint \boldsymbol{F} \cdot \mathrm{d}\boldsymbol{r} = 0\,. \end{aligned}$$

2. Every time-independent *central force field*, written in spherical coordinates (see Sect. 13.1) as $\boldsymbol{F} = \{F_r, F_\vartheta = 0, F_\varphi = 0\}$, which depends only on the distance r from the centre $r = 0$ and not on the angles ϑ and φ is conservative.
It can be written as $\boldsymbol{F}(\boldsymbol{r}) = f(r) \cdot \hat{\boldsymbol{r}}$, where $f(r)$ is a scalar function of r (Fig. 2.34b).

$$\int \boldsymbol{F} \cdot \mathrm{d}\boldsymbol{r} = \int_{r_1}^{r_2} F_r \mathrm{d}r = -\int_{r_2}^{r_1} F_r \mathrm{d}r \Rightarrow \oint \boldsymbol{F} \cdot \mathrm{d}\boldsymbol{r} = 0\,.$$

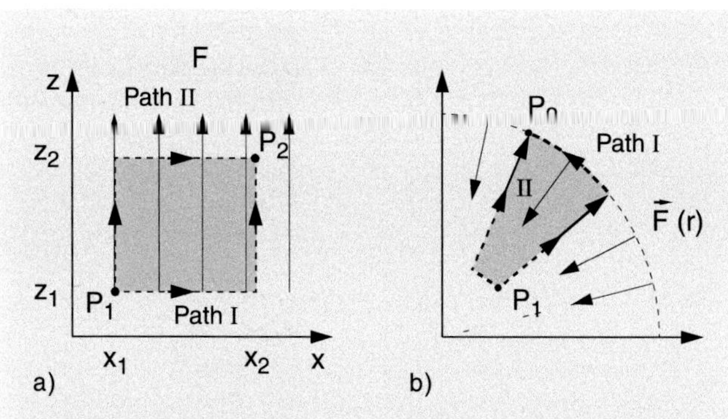

Figure 2.34 Examples for conservative force fields. **a** Homogeneous field, **b** central field

Non-conservative Force Fields

1. Position-dependent non-central force field

$$F(r) = ye_x + x^2e_y .$$

The work one has to expend for moving a body from point $P_1 = \{0, 0, 0\}$ to point $P_2 = \{2, 4, 0\}$ is

$$W = \int_0^P F \cdot dr = \int_{x=0}^{2} F_x dx + \int_{y=0}^{4} F_y dy$$

$$= \int_{x=0}^{2} y\,dx + \int_{y=0}^{4} x^2 dy .$$

We choose two different paths (Fig. 2.35):
(a) along the straight line $y = 2x$
(b) along the parabola $y = x^2$.
On the path (a) is $y = 2x \Rightarrow x^2 = (y/2)^2$

$$\int F \cdot dr_a = \int_0^2 2x dx + \int_0^4 \left(\frac{y}{2}\right)^2 dy$$

$$= x^2\Big|_0^2 + \frac{y^3}{12}\Big|_0^4 = 4 + \frac{16}{3} = 28/3 ,$$

On the path (b) is $y = x^2$.

$$\int F \cdot dr_b = \int_0^2 x^2 dx + \int_0^4 y\,dy$$

$$= \frac{1}{3}x^3\Big|_0^2 + \frac{1}{2}y^2\Big|_0^4 = \frac{8}{3} + 8 = \frac{32}{3} .$$

$\Rightarrow \oint F \cdot dr \neq 0$. The force field is not conservative!

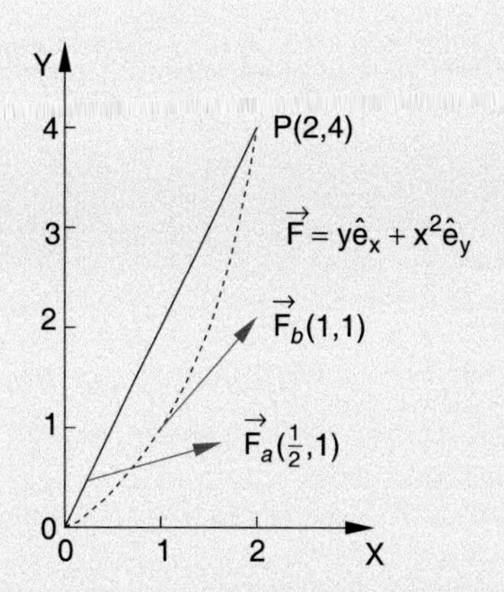

Figure 2.35 Movement in a non-conservative force field $F(r) = y \cdot e_x + x^2 \cdot e_y$

2. For time-dependent force fields the integral cannot be path-independent, because the force field varies during the travel of the body and therefore the work expended for the different paths is generally different.
3. If the force depends on the velocity of the body (for instance the friction for a body moving through a medium or on a surface, or the Lorentz-force $F = q \cdot (v \times B)$ on a charge q moving with the velocity v in a magnetic field B) such fields are generally not conservative because the velocity differs generally on the different paths. For friction forces F_f the force is for small velocities v proportional to $v (F_f \sim v)$, when the body moves slowly through a liquid. For large velocities is $F_f \sim v^3$ for example when a body moves through turbulent air. For all friction forces heat is produced and therefore the mechanical energy cannot be preserved. In all these cases $\oint F \cdot dr \neq 0$ (see also Sect. 6.5) ◄

Time-dependent or velocity-dependent forces are generally not conservative.

2.7.3 Potential Energy

When a body is moved in a conservative force field from a starting point $P_1(r_1)$ to another point $P_2(r_2)$ the work expended or gained during this movement does not depend on the path between the two points. If P_0 is a fixed point P_0 and $P(r)$ has an arbitrary position r the work solely depends on the initial point P_0 and the final point $P(r)$. It is therefore a function of $P(r)$ with respect to the fixed point P_0. This function is called the **potential energy $E_p(P)$** of the body.

The work

$$\Delta W = \int_{P_1}^{P_2} F\,dr \overset{\text{Def}}{=} -\left(E_p(P_2) - E_p(P_1)\right) , \qquad (2.35a)$$

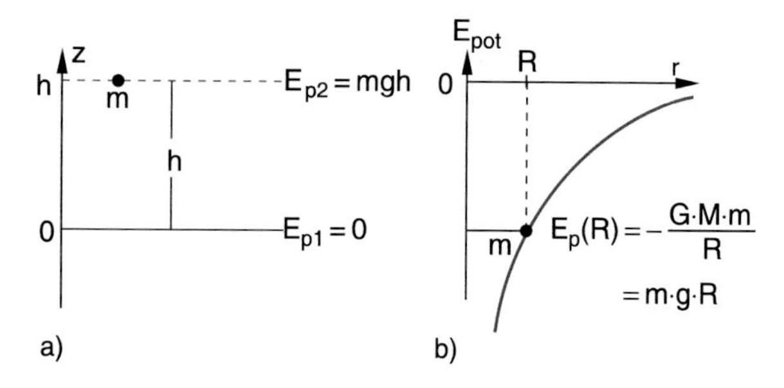

Figure 2.36 Different possibilities to choose the zero of the potential energy: **a** $E_p(z = 0) = 0$; **b** $E_p(r = \infty) = 0$

Chapter 2

which the force $\boldsymbol{F}(\boldsymbol{r})$ accomplishes on the body when it is moved between two points P_1 and P_2 is equal to the difference of the potential energies in these two points. For $\boldsymbol{F} \cdot \mathrm{d}\boldsymbol{r} > 0$ the force is directed into the direction of the motion. The potential energy difference $\Delta E_p = E_p(P_1) - E_p(P_2)$ is then negative. This means, that the mass m can deliver the work ΔW but looses potential energy.

One example is the free fall in the gravitational field of the earth, when a mass m falls from the height h with potential energy $m \cdot g \cdot h$ to the ground with $h = 0$. When we lift the mass m from $h = 0$ to $h > 0$ against the gravitational force, the scalar product $\boldsymbol{F} \cdot \mathrm{d}\boldsymbol{r}$ is negative and the potential energy increases (Fig. 2.36a). The work spend on the body to lift it against the force results in an increase of the potential energy. A body with a positive potential energy can convert this potential energy again into work. An example is water falling down through pipes and drives a turbine which drives maschines and produces electricity.

Note:

1. The sign of work and potential energy difference in (2.35a) has been chosen in such a way, that for $\boldsymbol{F} \cdot \mathrm{d}\boldsymbol{r} < 0 \rightarrow \Delta W < 0$ but $\Delta E_p > 0$, i. e. one has to spend work in order to move the body against the force which increases its potential energy. Work which the body can deliver to its surrounding for $\boldsymbol{F} \cdot \mathrm{d}\boldsymbol{r} > 0$ decreases its potential energy.
2. The defined zero $E_p = 0$ for the potential energy is not fixed by the definition (2.35a). If we choose the fixed reference point P_0 as the zero point of the potential energy and define $E_p(P_0) = 0$, then the absolute value of the potential energy in point P is given by

$$W = \int_{P_0}^{P} \boldsymbol{F} \cdot \mathrm{d}\boldsymbol{r} = -E_p(P) \,. \tag{2.35b}$$

For our example of the free fall we can choose $h = 0$ as the reference point with $E_p(0) = 0$. In many cases where a body can be moved to very large distances from the earth (for instance space crafts) it is more convenient to choose $r = \infty$ as the reference point for $E_p(\infty) = 0$. We then have the definition

$$\int_{P}^{\infty} \boldsymbol{F} \cdot \mathrm{d}\boldsymbol{r} = E_p(P) - E_p(\infty) = E_p(P) \,, \tag{2.35c}$$

the potential energy $E_p(P)$ is then negative for $F \cdot dr < 0$. It is equal to the work one has to spend in order to bring the body from the point P to infinity. For instance the potential energy of a mass m in the gravitational field of the earth $F_g = -GMm/r^2$ at a distance $r = R$ from the centre of the earth is then

$$E_p(R) = -GMm/R \,, \tag{2.35d}$$

where G is Newton's constant of gravity and M is the mass of the earth (Fig. 2.36b).
3. The work which one has to spend on the body (for $F \cdot dr < 0$ or which can be gained from the body (for $F \cdot dr > 0$) when it is moved from point P_1 to point P_2 is of course independent of the choice of the zero point because it depends only on the difference $\Delta E_p = E(P_1) - E_p(P_2)$ of the potential energies.

Examples

1. A body with mass m is lifted in the constant gravitational force field $F = \{0, 0, -mg\}$ from $z = 0$ to $z = h$, where $h \ll R$ (earth radius). The necessary work to achieve this lift is

$$W = \int \boldsymbol{F} \cdot \mathrm{d}\boldsymbol{r} = -\int_0^h m \cdot g \,\mathrm{d}z$$
$$= -m \cdot g \cdot h = E_p(0) - E_p(h) \,.$$

If we choose $E_p(z = 0) = 0$ the potential energy for $z = h$ is $E_p(h) = +mgh$ (Fig. 2.37a). The work applied to the mass m appears as potential energy.

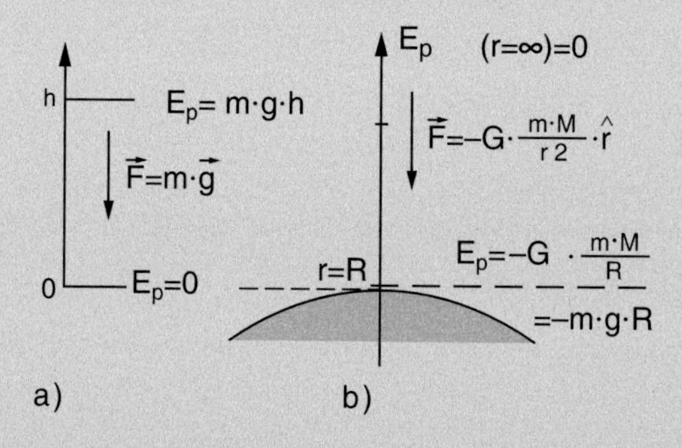

Figure 2.37 **a** Approximately homogeneous gravitational force field as small section of the spherical field of the earth in **b**. The selection of the definition $E_p = 0$ is $E_p(z = 0) = 0$ in case **a** and $E_p(r = \infty) = 0$ for case **b**

2. In an attractive force field, such as the gravitational field of the earth $\boldsymbol{F} = -(GMm/r^2)\boldsymbol{e}_r$ a mass m is moved from $r = R$ (earth surface) to $r = \infty$. In this case is $\boldsymbol{F} \cdot \mathrm{d}\boldsymbol{r} < 0$. The necessary work is negative:

$$W = -\int_r^\infty \frac{GMm}{r^2} \hat{\boldsymbol{r}} \mathrm{d}\boldsymbol{r} = -\int_r^\infty \frac{GMm}{r^2} \mathrm{d}r$$
$$= -\frac{GMm}{r} = E_p(r) \,. \tag{2.35e}$$

$E_p(r)$ is negative because $E(r = \infty) = 0$. To raise the mass m work has to be applied, which is converted to the increase of potential energy (Fig. 2.37b).
For repulsive potentials (e.g. the Coulomb potential of two positive electrical charges q_1 and q_2)

$$\boldsymbol{F} = (q_1 \cdot q_2/r^2)\boldsymbol{e}_r$$

the potential energy is positive and one wins work when the charge separation increases, while the potential energy decreases.
When a body with mass m should be moved from the earth surface $r = R$ to $r = \infty$ one needs the work $W = -GMm/R$. With $g = GM/R^2$ this can be written as $W = -mgR$.
Numerical example: With $g = 9.81\,\text{m/s}^2$, $R = 6371\,\text{km}$, the work to launch a mass of 100 kg is $W = 6.25 \cdot 10^9\,\text{J} = 1736\,\text{kWh}$. ◄

2.7.4 Energy Conservation Law in Mechanics

Multiplying the Newton equation

$$\boldsymbol{F} = m \cdot \frac{\mathrm{d}\boldsymbol{v}}{\mathrm{d}t}$$

scalar with the velocity $\boldsymbol{v}$ and integrating over time yields

$$\int \boldsymbol{F} \cdot \boldsymbol{v}\mathrm{d}t = m \int_{t_0}^{t_1} \frac{\mathrm{d}\boldsymbol{v}}{\mathrm{d}t} \cdot \boldsymbol{v}\,\mathrm{d}t\,. \tag{2.36}$$

The integral on the left hand side gives with $v = \mathrm{d}\boldsymbol{r}/\mathrm{d}t$

$$\int \boldsymbol{F} \cdot \boldsymbol{v}\mathrm{d}t = \int_{P_0}^{P_1} \boldsymbol{F} \cdot \mathrm{d}\boldsymbol{r} = E_p(P_0) - E_p(P_1)\,,$$

where the last equality is valid for conservative force fields.

The right hand side of (2.36) gives

$$m \cdot \int \frac{\mathrm{d}\boldsymbol{v}}{\mathrm{d}t} \cdot \boldsymbol{v}\,\mathrm{d}t = m \int_{v_0}^{v_1} \boldsymbol{v} \cdot \mathrm{d}\boldsymbol{v} = \frac{m}{2}v_1^2 - \frac{m}{2}v_0^2\,.$$

The expression

$$E_{kin} = mv^2/2 \tag{2.37}$$

is called the **kinetic energy** of a body with mass m and velocity $v = |\boldsymbol{v}|$.

The integral $\int \boldsymbol{F} \cdot \mathrm{d}\boldsymbol{r}$ represents the work W which is supplied to the body. The statement of Eq. 2.36 can therefore be formulated as:

$$\Delta E_{kin} = \Delta W\,. \tag{2.38a}$$

The increase of kinetic energy of a body is equal to the work supplied to this body.

In conservative force fields $\int \boldsymbol{F} \cdot \mathrm{d}\boldsymbol{r}$ is equal to the change of potential energy. Then Eq. 2.36 states:

$$E_p(P_0) + E_{kin}(P_0) = E_p(P) + E_{kin}(P) = E\,. \tag{2.38b}$$

When a body is moved in a conservative force field from a point P_0 to a point P the total mechanical energy E (sum of potential and kinetic energy) is conserved, i. e. it has for all positions in the force field the same amount.

Examples

1. For the free fall starting from $z = h$ with the velocity $v(h) = 0$ we choose $E_p(h = 0) = 0$. For arbitrary z the following equations hold:

$$E_p(z) = -\int_0^z -mg\mathrm{d}z = mgz\,.$$

With $v = g \cdot t$ and $s = h - z = \frac{1}{2}gt^2 \rightarrow \frac{1}{2}v^2 = \frac{1}{2}g^2t^2 = g(h-z)$ (see Sect. 2.3).
This gives

$$E_{kin}(z) = \tfrac{1}{2}mv^2 = m \cdot g \cdot (h - z)\,.$$

The sum $E_p(z) + E_{kin}(z) = mgh$ is independent of z and for all z equal to the total energy $E = mgh$.

2. A body with mass m oscillates in the x-direction, driven by the force $F = -D \cdot x$. For each point of its path the total energy is $E = E_p(x) + E_{kin}(x) = \text{const.}$ For $x = 0$ the potential energy is zero. In the upper turning points for $x = \pm x_m$ the velocity is zero and therefore $E_{kin} = 0$. (Fig. 2.38).

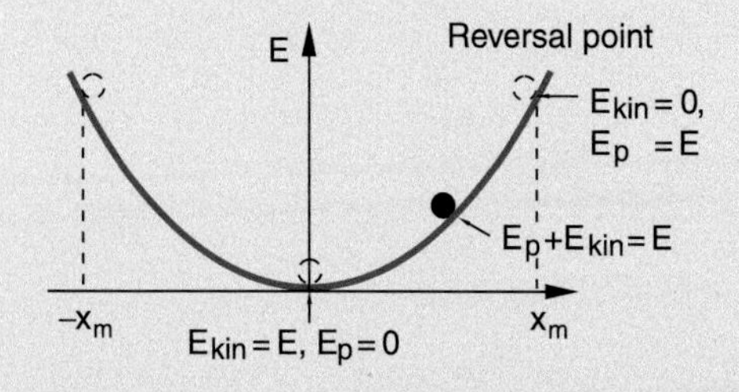

Figure 2.38 Example of energy conservation for a harmonic motion

The oscillation can be described by

$$x = x_m \sin \omega t \rightarrow v = \mathrm{d}x/\mathrm{d}t = x_m\omega \cos \omega t\,.$$

The potential energy is $E_p = \int Dx\mathrm{d}x = \frac{1}{2}Dx^2 = \frac{1}{2}Dx_m^2 \sin^2 \omega t$. The kinetic energy is $E_{kin} = \frac{1}{2}mv^2 =$

$\frac{1}{2}m \cdot x_m^2 \omega^2 \cos^2 \omega t$. From the Newton equation $F = ma = m \cdot \mathrm{d}^2x/\mathrm{d}t^2$ we obtain by comparison with $F = -Dx$ the relation $D = m \cdot \omega^2$. Inserting this into the expression for the potential energy we get

$$E = E_\mathrm{p} + E_\mathrm{kin} = \tfrac{1}{2} m x_m^2 \omega^2 (\sin^2 \omega t + \cos^2 \omega t) = \tfrac{1}{2} m \cdot x_m^2 \omega^2 ,$$

which is independent of x. ◀

2.7.5 Relation Between Force Field and Potential

If a body in a conservative force field is moved from the point P by an infinitesimal small distance Δr to a neighbouring point P' (Fig. 2.39) the potential energy changes by the amount

$$\Delta E_\mathrm{p} = \frac{\partial E_\mathrm{p}}{\partial x}\Delta x + \frac{\partial E_\mathrm{p}}{\partial y}\Delta y + \frac{\partial E_\mathrm{p}}{\partial z}\Delta z , \qquad (2.39)$$

where the partial derivative $\partial E/\partial x$ means that for the differentiation of the function $E(x, y, z)$ the two other variables are kept fixed (see Sect. 13.1.6).

The movement of the body from P to P' requires the work

$$\Delta W = \boldsymbol{F} \cdot \Delta \boldsymbol{r} = -\Delta E_p , \qquad (2.40)$$

where $\boldsymbol{F}$ is an average of $\boldsymbol{F}(P)$ and $\boldsymbol{F}(P')$. The comparison between (2.39) and (2.40) yields

$$F\Delta r = F_x\Delta x + F_y\Delta y + F_z\Delta z = -\frac{\partial E_\mathrm{p}}{\partial x}\Delta x - \frac{\partial E_\mathrm{p}}{\partial y}\Delta y - \frac{\partial E_\mathrm{p}}{\partial z}\Delta z .$$

Since this equation holds for arbitrary paths, i. e. arbitrary values of Δx, Δy, Δz it follows that

$$F_x = -\frac{\partial E_\mathrm{p}}{\partial x}; \quad F_y = -\frac{\partial E_\mathrm{p}}{\partial y}; \quad F_z = -\frac{\partial E_\mathrm{p}}{\partial z} . \qquad (2.41)$$

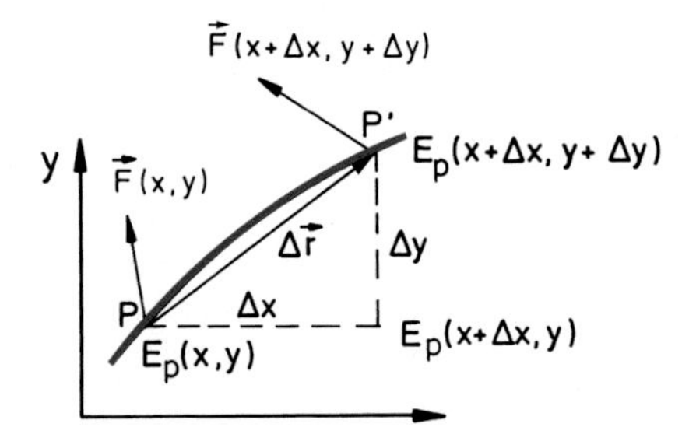

Figure 2.39 Relation between force and potential

Defining the gradient of the function $E_\mathrm{p}(x, y, z)$ as

$$\mathbf{grad} E_\mathrm{p} \overset{\mathrm{Def}}{=} \left\{ \frac{\partial E_\mathrm{p}}{\partial x}, \frac{\partial E_\mathrm{p}}{\partial y}, \frac{\partial E_\mathrm{p}}{\partial z} \right\} , \qquad (2.42)$$

the relations (2.41) for the components of F can be combined into the vector equation

$$\boldsymbol{F} = -\mathbf{grad} E_\mathrm{p} = -\boldsymbol{\nabla} E_\mathrm{p} , \qquad (2.41\mathrm{a})$$

where the symbol ∇ = *nabla* (∇ has the form of an old Egyptian string instrument called nabla) is an abbreviation to make the equation more simple to write.

The potential energy E_p of a body with mass m in the gravitational field of a mass M depends on both masses. However, for $m \ll M$ (for instance a mass m in the gravitational field of the earth with $M \gg m$) the small contribution of m to the gravitational field can be neglected. In such cases it is possible to define a function $V(P)$ for each point P, called the gravitational potential

$$V(P) \overset{\mathrm{Def}}{=} \lim_{m\to 0} \left(\frac{1}{m} E_\mathrm{p}(P) \right) ; \qquad (2.42\mathrm{a})$$

which is the potential energy pro unit mass m in the limit of $m \to 0$ in the gravitational field of M. $V(P)$ is a scalar function which depends only on the position of P and on the mass M that generates the gravitational field.

The gravitational potential of the earth is for instance

$$V(r) = -G \cdot M_\mathrm{E}/r ,$$

where r is the distance from the centre of the earth.

The gravitational field strength is defined as

$$\mathcal{G} = -\mathbf{grad} V . \qquad (2.43)$$

The force on a mass m is then

$$\boldsymbol{F}_\mathrm{G} = -m \cdot \mathcal{G} . \qquad (2.44)$$

For the gravitational field of a spherical symmetric mass M one obtains

$$\mathcal{G} = G\frac{M}{r^2}\hat{\boldsymbol{r}} , \qquad (2.43\mathrm{a})$$

and for the force on a body with mass m in this field Newton's gravitational law

$$\boldsymbol{F}_\mathrm{G} = -G\frac{m \cdot M}{r^2}\hat{\boldsymbol{r}} . \qquad (2.44\mathrm{a})$$

These definitions are completely equivalent to their pendants in electrostatics: The electrical potential of an electric charge Q and the Coulomb law (see Vol. 2, Sect. 1.3).

2.8 Angular Momentum and Torque

Assume a point mass moving with the momentum $\boldsymbol{p} = m \cdot \boldsymbol{v}$ on an arbitrary path $\boldsymbol{r} = \boldsymbol{r}(t)$ (Fig. 2.40). We define its *angular momentum* L with respect to the coordinate origin $\boldsymbol{r} = \boldsymbol{0}$ as the vector product

$$\boldsymbol{L} = (\boldsymbol{r} \times \boldsymbol{p}) = m \cdot (\boldsymbol{r} \times \boldsymbol{v}) . \tag{2.45}$$

Note, that $\boldsymbol{L}$ is perpendicular to $\boldsymbol{r}$ and $\boldsymbol{v}$!

In Cartesian coordinates $\boldsymbol{L}$ has the components (see Sect. 13.4)

$$\begin{aligned} L_x &= yp_z - zp_y ; \quad L_y = zp_x - xp_z ; \\ L_z &= xp_y - yp_x . \end{aligned} \tag{2.46}$$

If the body moves in a plane but on an arbitrarily curved path we can compose the velocity in any point of the path of a radial component $\boldsymbol{v}_\mathrm{r} \parallel \boldsymbol{r}$ and a tangential component $\boldsymbol{v}_\varphi \perp \boldsymbol{r}$ using polar coordinates r and φ (Fig. 2.40). This gives the relations:

$$\begin{aligned} \boldsymbol{L} &= m \cdot [\boldsymbol{r} \times (\boldsymbol{v}_\mathrm{r} + \boldsymbol{v}_\varphi)] \\ &= m \cdot (\boldsymbol{r} \times \boldsymbol{v}_\varphi) \quad \text{because} \quad \boldsymbol{r} \times \boldsymbol{v}_\mathrm{r} = 0 . \end{aligned}$$

The value of $\boldsymbol{L}$ is

$$|\boldsymbol{L}| = m \cdot r^2 \cdot \frac{\mathrm{d}\varphi}{\mathrm{d}t} \quad \text{because} \quad |\boldsymbol{r} \times v_\varphi| = r^2 \cdot \frac{\mathrm{d}\varphi}{\mathrm{d}t} = r^2 \cdot \omega .$$

These equations describe the following facts:

For planar motions the angular momentum L always points into the direction of the plane-normal perpendicular to the plane (Fig. 2.40). The vector product $(r \times v)$ forms a right-handed screw.

When the angular momentum is constant, the motion proceeds in a plane perpendicular to the angular momentum vector.

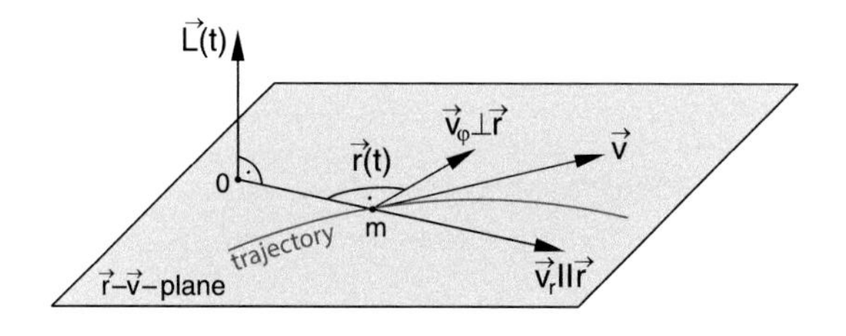

Figure 2.40 Angular momentum $\boldsymbol{L}$ referred to an arbitrarily chosen origin $\boldsymbol{0}$ for a plain motion of a point mass m

Example

For the uniform circular motion the constant angular momentum points into the direction of the axis through the circle centre perpendicular to the circular plane, i. e. into the direction of the angular velocity vector $\boldsymbol{\omega}$ (Fig. 2.11).

$$|\boldsymbol{L}| = L = m \cdot r \cdot v \cdot \sin(\boldsymbol{r}, \boldsymbol{v}) = m \cdot r \cdot v = m \cdot r^2 \cdot \omega .$$

$$\sin(\boldsymbol{r}, \boldsymbol{v}) = 1 \quad \text{because} \quad \boldsymbol{r} \perp \boldsymbol{v} . \tag{2.47}$$

For the uniform circular motion is $r = \text{constant}$ and $v = \text{constant} \rightarrow \boldsymbol{L} = \text{constant}$.

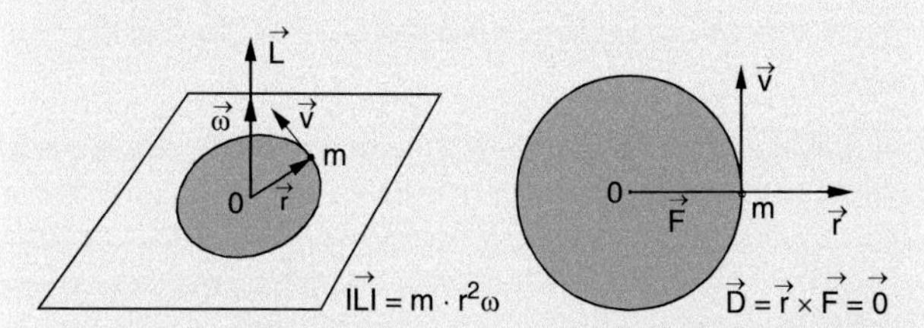

Figure 2.41 Constant angular momentum of the uniform circular motion ◄

Differentiating (2.45) with respect to time we obtain

$$\begin{aligned} \frac{\mathrm{d}\boldsymbol{L}}{\mathrm{d}t} &= \left[\frac{\mathrm{d}\boldsymbol{r}}{\mathrm{d}t} \times \boldsymbol{p}\right] + \left[\boldsymbol{r} \times \frac{\mathrm{d}\boldsymbol{p}}{\mathrm{d}t}\right] \\ &= (\boldsymbol{v} \times \boldsymbol{p}) + (\boldsymbol{r} \times \dot{\boldsymbol{p}}) = (\boldsymbol{r} \times \dot{\boldsymbol{p}}), \quad \text{because} \quad \boldsymbol{v} \parallel \boldsymbol{p} , \\ \frac{\mathrm{d}\boldsymbol{L}}{\mathrm{d}t} &= (\boldsymbol{r} \times \boldsymbol{F}), \quad \text{because} \quad \boldsymbol{F} = \frac{\mathrm{d}\boldsymbol{p}}{\mathrm{d}t} . \end{aligned} \tag{2.48}$$

The vector product

$$\boldsymbol{D} = (\boldsymbol{r} \times \boldsymbol{F}) \tag{2.49}$$

is the **torque of the force** around the origin $\boldsymbol{r} = \boldsymbol{0}$ acting on the mass m at the position $\boldsymbol{r}$. Equation 2.48 can then be written as

$$\frac{\mathrm{d}\boldsymbol{L}}{\mathrm{d}t} = \boldsymbol{D} . \tag{2.49a}$$

The change of the angular momentum $\boldsymbol{L}$ with time is equal to the torque $\boldsymbol{D}$.

In other words: If the torque on a mass is zero, its angular momentum remains constant.

Note the equivalence between linear momentum $\boldsymbol{p}$ and angular momentum $\boldsymbol{L}$:

$$\frac{\mathrm{d}\boldsymbol{p}}{\mathrm{d}t} = \boldsymbol{F}, \quad \frac{\mathrm{d}\boldsymbol{L}}{\mathrm{d}t} = \boldsymbol{D} , \tag{2.50}$$

$\boldsymbol{p} =$ **constant** for $\boldsymbol{F} = \boldsymbol{0}$ and $\boldsymbol{L} =$ **constant** for $\boldsymbol{D} = \boldsymbol{0}$.

In central force fields $\boldsymbol{F}(r) = f(r) \cdot \hat{r}$ the torque $\boldsymbol{D} = r \times \boldsymbol{F} = \boldsymbol{0}$ because $\boldsymbol{F} \parallel \boldsymbol{r}$. Therefore the angular momentum is constant

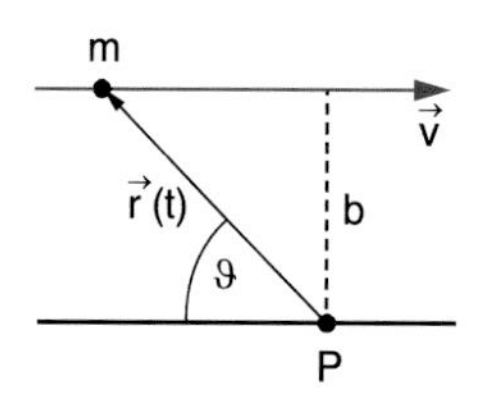

Figure 2.42 Illustration of angular momentum of a body moving on a straight line with respect to a point P which does not lie on the straight line

for all motions in a central force field. **This implies that all trajectories are in a plane, perpendicular to the angular momentum vector.**

Note: Angular momentum and torque are always defined **with respect to a selected point** (for instance the origin of the coordinate system). Even a body moving on a straight line can have an angular momentum with respect to a point, which is not on the straight line.

In Fig. 2.42 the amount L of the angular momentum $\boldsymbol{L}$ of the mass m moving with the constant velocity v on a straight line is with respect to the point P

$$L = m \cdot r \cdot v \cdot \sin\vartheta = m \cdot b \cdot v$$

where b (*called the impact parameter*) is the perpendicular distance of P from the straight line.

Figure 2.43 Tycho de Brahe (1546–1601) (with kind permission of "Deutsches Museum")

2.9 Gravitation and the Planetary Motions

In the previous section we have learned that in central force fields the angular momentum $\boldsymbol{L}$ is constant in time. The motion of a body therefore proceeds in a plane perpendicular to $\boldsymbol{L}$. The orientation of the plane is determined by the initial conditions (for instance by the initial velocity v_0) and is then fixed for all times. The most prominent example are the motions of the planets in the central gravitational field of the sun which we will now discuss.

2.9.1 Kepler's Laws

Based on accurate measurements of planetary motions (in particular the motion of Mars) by *Tycho de Brahe* (Fig. 2.43) *Johannes Kepler* (Fig. 2.44) could show, that the heliocentric model of *Copernicus* allowed a much simpler explanation of the observations than the old geocentric model of Ptolemy where the earth was the centre and the planets moved around the earth in complex trajectories (epicycles).

Figure 2.44 Johannes Kepler (1571–1630) (with kind permission of "Deutsches Museum")

Kepler assumed at first circular trajectories because such motions seemed to him as perfect in harmony with God's creation. However, this assumption led to small inconsistencies between

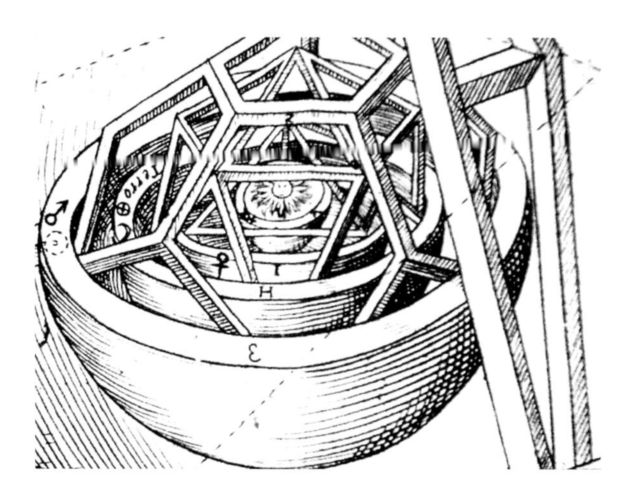

Figure 2.45 Initial model of Kepler illustrating the location of the planets at the corners of regular geometric figures (with kind permission of Prof. Dr. Ron Bienek)

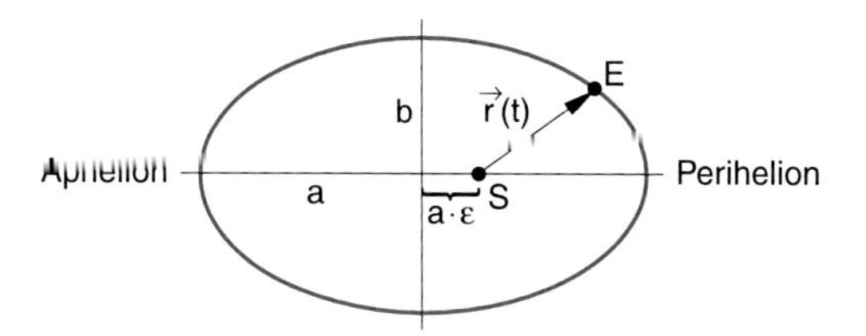

Figure 2.46 Kepler's first law

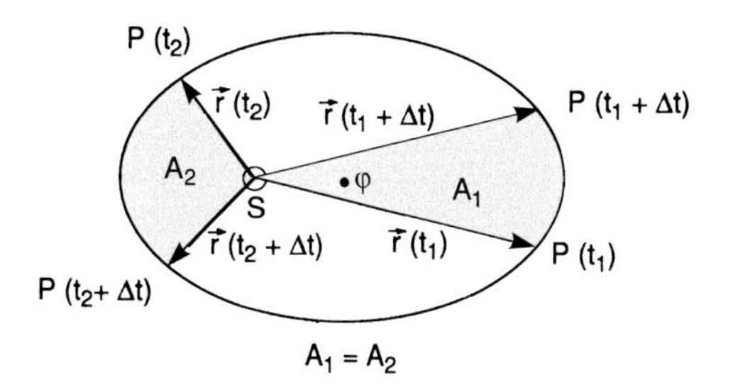

Figure 2.47 Kepler's second law. S: sun, φ: center of ellipse

calculated and observed motions of the planets which exceeded the error limits of the observations. After a long search with several unsuccessful models (for instance a model where the planets were located at the corners of symmetric figures which rotate around a centre (Fig. 2.45). Kepler finally arrived at his famous three laws which were published in his books: *Astronomia Nova* (1609) and *Harmonices Mundi Libri V* (1619).

Kepler's first law

The planets move on elliptical trajectories with the sun in one of the focal points (Fig. 2.46).

Kepler's second law

The radius vector from the sun to the planet sweeps out in equal time intervals equal areas (Fig. 2.47).

Kepler's third law

The squares of the full revolution times T_i of the different planets have the same ratio as the cubes of the large half axis a_i of the elliptical paths.

$$T_1^2/T_2^2 = a_1^3/a_2^3 \quad \text{or} \quad T_i^2/a_i^3 = \text{constant}\,,$$

where the constant is the same for all planets.

The 2. Kepler's law tells us that the areas A_i in Fig. 2.47 is for equal time intervals Δt always the same, i. e. the area $A_1 = SP(t_1)P(t_1 + \Delta t) = A_2 = SP(t_2)P(t_2 + \Delta t)$. For sufficiently small time intervals $\mathrm{d}t$ we can approximate the arc length $\mathrm{d}s = P_1P_2 = v\mathrm{d}t$ in Fig. 2.48b by the straight line P_1P_2. The area of the triangle SP_1P_2 is then

$$\mathrm{d}A = \tfrac{1}{2} \cdot |r \times v| = \tfrac{1}{2}|r| \cdot |v| \cdot \sin\alpha = \tfrac{1}{2} \cdot \frac{|L|}{m}\,. \tag{2.51}$$

Kepler's second law therefore states that the angular momentum of the planet is constant. Kepler's first law postulates that the motion of the planets proceeds in a plane. Since the angular momentum is perpendicular to this plane it follows that also the direction of $\boldsymbol{L}$ is constant.

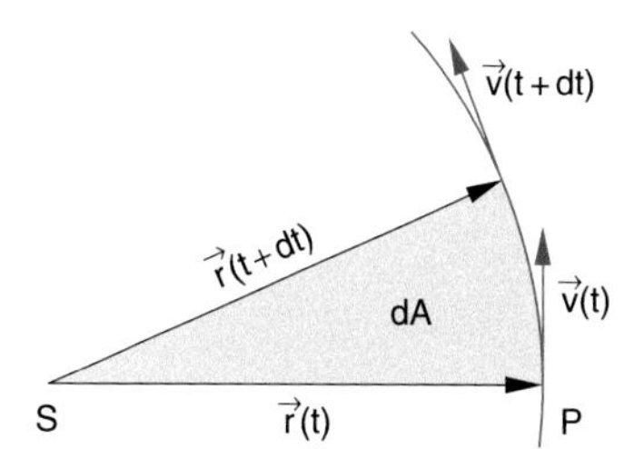

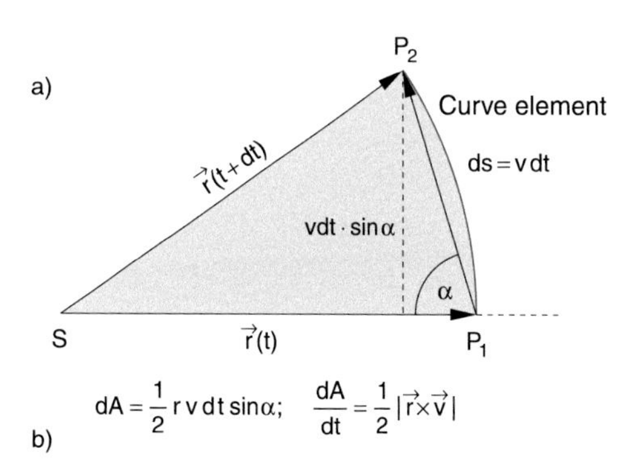

Figure 2.48 Kepler's second law as conservation of angular momentum. **a** schematic representation of the equal area law. **b** calculation of the area covered by the radius vector in the time interval $\mathrm{d}t$

2.9.2 Newton's Law of Gravity

Newton came to the conclusion that the free fall of a body as well as the motion of the planets have a common cause: the gravitational attraction between two masses. In order to find a quantitative formulation of the gravitational force he started his considerations with Kepler's laws. Since the angular momentum of the planetary motion is constant the force field has to be a central force field

$$\boldsymbol{F}(\boldsymbol{r}) = f(r) \cdot \hat{r} .$$

The gravitational force which acts on a body with mass m at the surface of the earth with mass M (which is equal to its weight) is proportional to m. According to the principle *actio = reactio* and also because of symmetry principles the equal but opposite force acting on M should be also proportional to the mass M of the earth (Fig. 2.25). It is therefore reasonable to postulate that the gravitational force is proportional to the product $m \cdot M$ of the two masses. We therefore can write for the force between two masses m_1 and m_2

$$\boldsymbol{F}_g = G \cdot m_1 \cdot m_2 \cdot f(r) \cdot \hat{\boldsymbol{r}} . \tag{2.52a}$$

The proportionality factor G is the *Newtonian gravitational constant*.

The function $f(r)$ can be determined from Kepler's third law. Since (2.52a) must be also valid for circular orbits we obtain for the motion of a planet with mass m around the sun with mass $M_\odot$ the equation

$$G \cdot m \cdot M_\odot \cdot f(r) = m \cdot \omega^2 \cdot r , \tag{2.52b}$$

because the gravitational force is the centripetal force which causes the circular motion of the planet with the angular velocity $\omega = v/r$. The revolution period of the planet is $T = 2\pi/\omega$. For the orbits of two different planets Kepler's third law postulates:

$$T^2/r^3 = \text{const} .$$

With $\omega = 2\pi/T$ this gives $\omega^2 \cdot r^3 = \text{const}$ or $\omega^2 \sim r^{-3}$.

Inserting this into (2.52b) yields $f(r) \sim r^{-2}$.

We then obtain Newton's law of gravity

$$F_g(r) = -G \cdot \frac{m \cdot M_\odot}{r^2} \hat{\boldsymbol{r}} . \tag{2.52c}$$

The minus sign indicates that the force is attractive.

The gravitational force

$$F(r) = -G \cdot \frac{m_1 \cdot m_2}{r^2} \hat{\boldsymbol{r}}$$

acts not only between sun and planets but also between arbitrary masses m_1 and m_2 separated by the distance r. However, the force between masses realized in the laboratory is very small and it demands special very sensitive detection techniques in order to measure it. The gravitational constant G can be determined from such experiments in the lab. Among all physical constant it is that with the largest uncertainty. Therefore many efforts are undertaken to determine G with new laser techniques which should improve the accuracy [2.5a–2.5b]. The present accepted numerical value is

$$G = 6.67384(80) \cdot 10^{-11}\,\text{N} \cdot \text{m}^2/\text{kg}^2$$

with a relative uncertainty of $1.2 \cdot 10^{-4}$.

Note: The gravitational force is always attractive, never repulsive! This differs from the static electric forces between two charges Q_1 and Q_2

$$F(r) \sim Q_1 \cdot Q_2/r^{2+} ,$$

which can be attractive or repulsive, depending on the sign of the charges Q_i.

2.9.3 Planetary Orbits

Since the gravitational force field is conservative the sum of potential and kinetic energy of a planet is constant. Because it is a central field also the angular momentum $\boldsymbol{L} = \boldsymbol{r} \times \boldsymbol{p}$ is constant. This can be used to determine the orbit of a planet which proceeds in a plane with constant orientation perpendicular to $\boldsymbol{L}$. We use polar coordinates r and φ with the centre of the sun as coordinate origin (Fig. 2.49).

The kinetic energy is

$$\begin{aligned} E_{\text{kin}} &= \frac{m}{2} v^2 = \frac{m}{2}\left(v_r^2 + v_\varphi^2\right) \\ &= \frac{m}{2}\left(\dot{r}^2 + r^2\dot{\varphi}^2\right) . \end{aligned} \tag{2.53}$$

The amount $L = |\boldsymbol{L}|$ of the angular momentum $\boldsymbol{L}$ is

$$L = mr^2\dot{\varphi} = \text{const} . \tag{2.54}$$

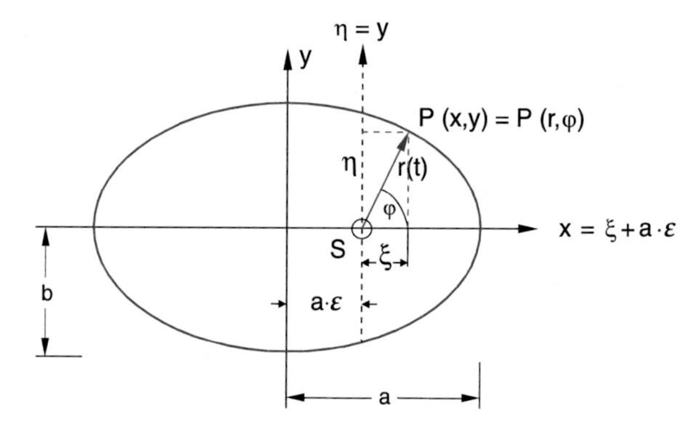

Figure 2.49 Elliptical orbit in Cartesian and in polar coordinates

Conservation of energy demands

$$E_p + \frac{m}{2}\dot{r}^2 + \frac{L^2}{2mr^2} = E = \text{const} , \qquad (2.55)$$

where E and L^2 are temporally constant. Resolving (2.55) for $\mathrm{d}r/\mathrm{d}t$ gives

$$\frac{\mathrm{d}r}{\mathrm{d}t} = \sqrt{\frac{2}{m}\left(E - E_p - \frac{L^2}{2mr^2}\right)} . \qquad (2.56)$$

For the angular variable $\varphi(t)$ one gets from (2.54)

$$\frac{\mathrm{d}\varphi}{\mathrm{d}t} = \frac{L}{mr^2} . \qquad (2.57)$$

Division of (2.57) by (2.56) yields

$$\frac{\mathrm{d}\varphi}{\mathrm{d}r} = \frac{L}{mr^2}\left[\frac{2}{m}\left(E - E_p - \frac{L^2}{2mr^2}\right)\right]^{-1/2} ,$$

integration gives

$$\begin{aligned}\int \mathrm{d}\varphi &= \varphi - \varphi_0 \\ &= \frac{L}{m}\int \frac{\mathrm{d}r}{r^2\sqrt{\frac{2}{m}\left(E - E_p - L^2/(2mr^2)\right)}} . \end{aligned} \qquad (2.58)$$

This allows to get the polar representation of the orbit in the following way:

With $E_p = -G \cdot M \cdot m/r$ the integral in (2.58) belongs to the type of elliptical integrals with the solution for the initial condition $\varphi(0) = \varphi_0 = 0$ (see integral compilation [2.6a–2.6b]):

$$\varphi = \arccos\left(\frac{L^2/r - Gm^2M}{\sqrt{(Gm^2M)^2 + 2mE \cdot L^2}}\right) . \qquad (2.59)$$

With the abbreviations

$$a = -\frac{GmM}{2E} \quad \text{and} \quad \varepsilon = \sqrt{1 + \frac{2EL^2}{G^2m^3M^2}} , \qquad (2.59a)$$

Eq. 2.59 can be written as

$$\varphi = \arccos\left(\frac{a(1-\varepsilon^2) - r}{\varepsilon \cdot r}\right) . \qquad (2.59b)$$

Solving for r gives

$$r = \frac{a\left(1-\varepsilon^2\right)}{1 + \varepsilon \cdot \cos\varphi} . \qquad (2.60)$$

This is the equation of a conic section (ellipse, hyperbola or parabola) in polar coordinates with the origin in the focal point S [2:6]. The minimum distance $r_{min} = a(1 - \varepsilon)$ is obtained for $\cos\varphi = +1$, the maximum distance $r_{max} = a(1 + \varepsilon)$ for $\cos\varphi = -1$. For the shortest distance (perihelion) and the largest distance (Aphelion) from the sun the derivative $\mathrm{d}r/\mathrm{d}t = 0$. Inserting this into (2.56) gives

$$E - \frac{GmM}{r} - \frac{L^2}{2m \cdot r^2} = 0 .$$

The solutions of this equation are

$$r_{min,max} = -\frac{GmM}{2E} \pm \left[\frac{G^2m^2M^2}{4E^2} + \frac{L^2}{2mE}\right]^{1/2} . \qquad (2.61)$$

We distinguish between three cases:

a) $\boldsymbol{E < 0}$.
For $E < 0$ is the constant $a = -GmM/(2E) > 0$ and $\varepsilon < 1$. The orbit is an ellipse with the major axis a and the excentricity ε. This can be readily seen from (2.60), when the transformation $\xi = r \cdot \cos\varphi$ and $\eta = r \cdot \sin\varphi$ to Cartesian coordinates with the origin in the focal point S is applied. This gives

$$a\left(1-\varepsilon^2\right) - \varepsilon\xi = \sqrt{\xi^2 + \eta^2} . \qquad (2.61a)$$

When we shift the origin $\{0, 0\}$ from S into the centre of the ellipse with the transformation $x = \xi + a\varepsilon$ and $y = \eta$ we obtain from (2.61a) the well-known equation for an ellipse in Cartesian coordinates

$$\frac{x^2}{a^2} + \frac{y^2}{b^2} = 1 \quad \text{with} \quad b^2 = a^2\left(1-\varepsilon^2\right) . \qquad (2.61b)$$

For the special case $\varepsilon = 0 \Rightarrow a = b$ the orbit becomes a circle with $r = \text{const}$. From (2.54) it follows because of $L = \text{const}$ that $\mathrm{d}\varphi/\mathrm{d}t = const$ the planet proceeds with uniform velocity around the central mass M.

For a negative total energy $E < 0$ the planet proceeds on an elliptical orbit (Kepler's first law).

b) $\boldsymbol{E = 0}$.
For $E = 0$ one immediately obtains from (2.59)

$$r = \frac{L^2}{Gm^2M(1 + \cos\varphi)} . \qquad (2.62)$$

This is the equation of a parabola [2.6a, 2.6b] with the minimum distance $r_{min} = L^2/(2Gm^2M)$ from the focal point for $\varphi = 0$.

c) $\boldsymbol{E > 0}$.
Since in (2.61) the distance r has to be positive ($r > 0$) for $E > 0$ only the positive sign before the square root is possible. Therefore only one r_{min} exists and the orbit extends until infinity ($r = \infty$). For $E > 0 \Rightarrow \varepsilon > 0$ (see (2.59a)). The orbit is a hyperbola.

In Tab. 2.1 the relevant numerical data for all planets of our solar system are compiled, where the earth moon is included for comparison.

Table 2.1 Numerical values for the orbits of all planets in our solar system. The earth moon is included for comparison

Name	Symbol	Large semi axis a of orbit			Revolution period T	Mean velocity	Numerical excentricity	Inclination of orbit	Distance from earth	
		In AU	In 10^6 km	In light travel time t		In km s^{-1}			Minimum in AU	Maximum in AU
Mercury	☿	0.39	57.9	3.2 min	88 d	47.9	0.206	7.0°	0.53	1.47
Venus	♀	0.72	108.2	6.0 min	225 d	35.0	0.007	3.4°	0.27	1.73
Earth	♁	1.00	149.6	8.3 min	1.00 a	29.8	0.017	–	–	–
Mars	♂	1.52	227.9	12.7 min	1.9 a	24.1	0.093	1.8°	0.38	2.67
Jupiter	♃	5.20	778.3	43.2 min	11.9 a	13.1	0.048	1.3°	3.93	6.46
Saturn	♄	9.54	1427	1.3 h	29.46 a	9.6	0.056	2.5°	7.97	11.08
Uranus	⛢	19.18	2870	2.7 h	84 a	6.8	0.047	0.8°	17.31	21.12
Neptun	♆	30.06	4496	4.2 h	165 a	5.4	0.009	1.8°	28.80	31.33
Earth moon	☾	0.00257	0.384	1.3 s	27.32 d	1.02	0.055	5.1°	356410 km	406740 km

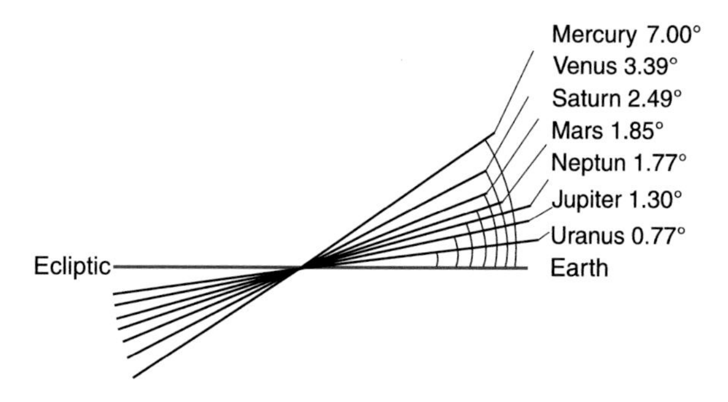

Figure 2.50 Inclination angles of the orbital planes for the different planets against the earth ecliptic

Remark.

1. Pluto is since 2006 no longer a planet but is now listed according to a decision of the International Astronomical Union in the group of *dwarf planets*. To this group also belong Ceres, Eris and about 200 additional dwarf planets in the Kuiper belt far beyond the orbit of Neptune.
2. The orientation of the orbital plane of a planet depends on the initial conditions when the solar system was created from a rotating gas cloud [2.7]. Since these initial conditions were different for the different planets the orbital planes are slightly inclined against each other (Fig. 2.50). Furthermore the gravitational interaction between the planets is small compared to the interaction with the sun, but not completely negligible. This disturbs the central force field and leads over longer time periods to a change of the orientation of the orbital planes.
3. For more accurate calculations (which are necessary for astronomical predictions) one has to take into account that the sun is not exactly located in a focal point of the ellipse. Because $M_\odot$ is not infinite, the sun and the planets move around the common centre of mass, which is, however, not far away from the focal point because $M_\odot \gg m$ [2.8]. For more accurate calculations one has to replace the mass m of a planet by the reduced mass $\mu = m \cdot M_\odot/(m + M_\odot)$ (see Sect. 4.1) where $M_\odot$ is 700 times larger than the mass of all planets ($M_\odot \approx 700 \cdot \sum m_i$). The constant a in Eq. 2.60 has to be replaced by

$$a = -\frac{G\mu M}{2E} = -\frac{GmM^2}{2E \cdot (m+M)} .$$

4. For the accurate calculation of the planetary orbits one has to take into account the interactions between the planets. Because of the small deviations from a central force field the angular momentum is no longer constant but shows slight changes with time.
5. Most of the comets have been formed within our solar system. They therefore have a negative total energy $E < 0$ and move on elongated elliptical orbits with $a \gg b$.

2.9.4 The Effective Potential

The radial motion of a body in a central force field, i. e. the solution of Eq. 2.56, can be illustrated by the introduction of the *effective potential.*

We decompose the kinetic energy in (2.53) into a radial part $(m/2)\dot{r}^2$ which represents the kinetic energy of the radial motion, and an angular part $\frac{1}{2}m \cdot r^2(\mathrm{d}\varphi/\mathrm{d}t)^2$ which stands for the kinetic energy of the tangential motion at a fixed distance r. The second part can be expressed by the angular momentum L

$$E_{\text{kin}}^{\text{tan}} = \frac{1}{2}mr^2\dot{\varphi}^2 = \frac{L^2}{2mr^2} \qquad (2.63)$$

(see (2.55)). Since for a given constant L this part depends only on r but not on the angle φ or on the radial velocity $\dot{r}$, it is added to the potential energy E_p, which also depends only on r. The sum

$$E_\mathrm{p}^{\text{eff}} = E_\mathrm{p}(r) + \frac{L^2}{2mr^2} \qquad (2.64)$$

is the effective potential energy. Often the effective potential

$$V_\mathrm{p}^{\text{eff}} = E_\mathrm{p}^{\text{eff}}/m$$

is introduced which is the potential energy per mass unit. The part $L^2/(2m \cdot r^2)$ is called the centrifugal potential energy

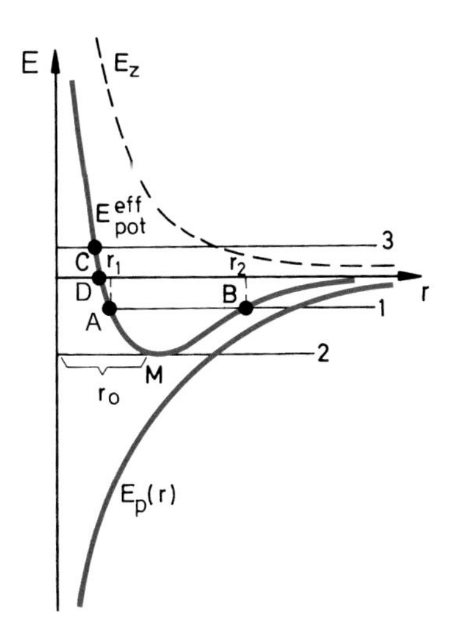

Figure 2.51 Effective potential energy $E_p^{eff}(r)$ as the sum of potential energy and centrifugal energy

and $L^2/(2m^2r^2)$ the *centrifugal potential*, while the radial part $E_p(r)/m$ is the *radial potential*.

The kinetic energy of the radial motion is then

$$E_{kin}^{rad} = \frac{1}{2}m\dot{r}^2 = E - E_p^{eff} , \qquad (2.65)$$

where E is the constant total energy.

In the gravitational force field is

$$E_p^{eff} = -G \cdot \frac{mM}{r} + \frac{L^2}{2mr^2} . \qquad (2.66)$$

Both parts are depicted in Fig. 2.51. The centrifugal-term E_z decreases with increasing r as $1/r^2$ and is for large r negligible while for small values of r it can overcompensate the negative radial part to make the total energy positive.

The minimum of E_p^{eff} is obtained from $dE_p^{eff}/dr = 0$. This gives

$$r_0 = \frac{L^2}{Gm^2M} . \qquad (2.67)$$

The kinetic energy of the radial motion $E_{kin}(r) = E - E_p^{eff}(r)$ at the distance r from the centre is indicated in Fig. 2.51 as the vertical distance between the horizontal line E = constant and the effective potential energy. The body can only reach those intervals $\Delta r = r_{min} - r_{max}$ of r where $E - E_p^{eff} > 0$.

These intervals depend on the total energy E, as is illustrated in Fig. 2.51.

- $\boldsymbol{E < 0}$ but $\boldsymbol{E_{kin}^{rad} > 0}$. (horizontal line 1)
  The body moves between the points $A(r_{min})$ and $B(r_{max})$. They correspond to the radii $r = a(1 \pm \varepsilon)$ for the motion of planets on an ellipse around the sun.
- $\boldsymbol{E < 0}$ but $\boldsymbol{E_{kin}^{rad} = 0}$ (horizontal line 2)
  The orbital path has a constant radius r_0, which means it is a circle. In the diagram of Fig. 2.51 the body always remains at the point M in the minimum of E_p^{eff}.
- $\boldsymbol{E > 0}$ and $E_{kin}^{rad} < |E_p^{eff}(r = \infty)|$ (horizontal line 3)
  The body has the minimum value of r in the point C, where $E_{kin}^{rad} = 0$. It can reach $r = \infty$. Its orbit is a hyperbola.
- $\boldsymbol{E = 0}$
  From (2.65) it follows that $E_{kin}^{rad} = -E_p^{eff}$. The body reaches the minimum distance r_{min} in the point D on the curve $E(r)$. Here is $E_{kin}^{rad} = 0$ and $E_p^{eff} = 0$. It can reach $r = \infty$, where $E_{kin}^{rad} = 0$. The orbit is a parabola.

2.9.5 Gravitational Field of Extended Bodies

In the preceding sections we have discussed the gravitational field generated by point masses. We have neglected the spatial extension of the masses and have assumed that the total mass is concentrated in the centre of each body. This approximation is justified for astronomical situations because the distance between celestial objects is very large compared to their diameter.

Example

The radius of the sun is $R_\odot = 7 \cdot 10^8$ m, the mean distance sun–earth is $r = 1.5 \cdot 10^{11}$ m, i. e. larger by the factor 210! ◄

We will now calculate the influence of the spatial mass distribution on the gravitational field. We start with the field of a hollow sphere in a point P outside the sphere (Fig. 2.52). The hollow sphere should have the radius a and the wall-thickness $da \ll a$.

A disc with the thickness dx cuts a circular ring with the breadth $ds = dx/\sin\vartheta$ and the diameter $2y$. The mass of this ring (thickness da and breadth ds) is for a homogeneous mass density ϱ

$$\begin{aligned} dM &= 2\pi y\varrho \cdot ds \cdot da \\ &= 2\pi a \cdot \varrho \cdot dx \cdot da \quad \text{because} \quad y = a \cdot \sin\vartheta . \end{aligned}$$

All mass elements dM of this ring have the same distance to the point P. Therefore the potential energy of a small probe mass m in the gravitational field generated by dM is

$$dE_p = -G \cdot m \cdot dM/r = -G \cdot m \cdot 2\pi a \cdot \varrho \cdot da \cdot dx/r .$$

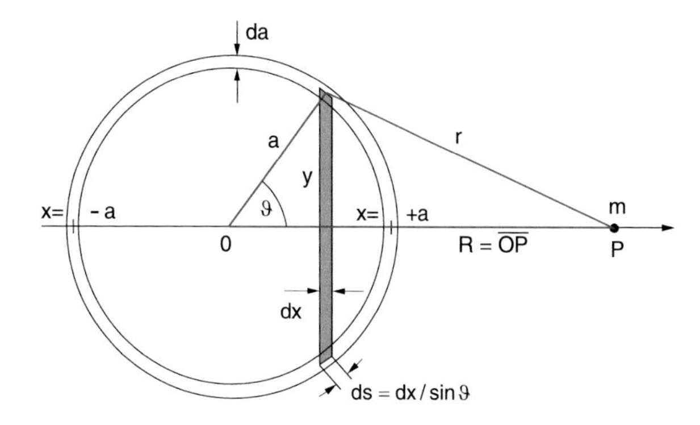

Figure 2.52 Potential and gravitational field-strength of a hollow sphere

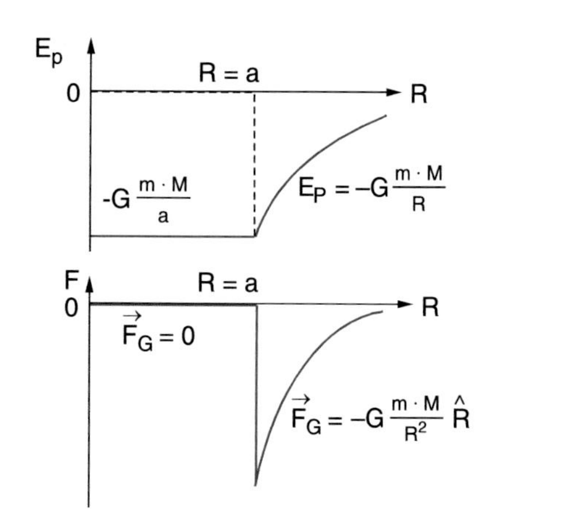

Figure 2.53 Potential energy of a sample mass m and gravitational field strength in the gravitational field of a hollow sphere with mass M

The gravitational field of the total mass M is obtained by integrating over x from $x=-a$ to $x=+a$.

$$E_p = -2\pi\varrho G m a \cdot da \int\limits_{x=-a}^{+a} \frac{dx}{r}. \tag{2.68}$$

From Fig. 2.52 the relations

$$\begin{aligned} r^2 &= y^2 + (R-x)^2 = y^2 + x^2 + R^2 - 2Rx \\ &= a^2 + R^2 - 2Rx; \quad r\,dr = -R\,dx \end{aligned}$$

can be verified. This yields

$$\begin{aligned} E_p &= \frac{2\pi\varrho a\,da \cdot m}{R} G \int\limits_{r=R+a}^{R-a} dr \\ &= -G \cdot \frac{m \cdot M}{R}, \end{aligned} \tag{2.69}$$

because $M = 4\pi a^2 \cdot \varrho da$ is the mass of the hollow sphere.

The gravitational force on the mass m is

$$\begin{aligned} \boldsymbol{F}_G &= -\mathbf{grad} E_p \\ &= -\frac{dE_p}{dR}\hat{\boldsymbol{R}} = -G \cdot \frac{m \cdot M}{R^2} \cdot \hat{\boldsymbol{R}}. \end{aligned} \tag{2.70}$$

The gravitational field of a hollow sphere with mass M is outside the sphere exactly the same as if the mass M is concentrated in the centre of the sphere (Fig. 2.53).

For $R < a$ the calculation proceeds in the same way. Only the upper limit of the integration changes. For $x = +a$ the limit becomes $r = a - R$ as can be seen from Fig. 2.52. With

$$\int\limits_{r=a+R}^{r=a-R} dr = -2R$$

the potential energy becomes

$$E_p = -G\frac{m \cdot M}{a} = \text{const} \quad \text{for} \quad R \le a. \tag{2.71}$$

The gravitational force in the inner volume of the hollow sphere is then

$$\boldsymbol{F} = -\mathbf{grad} E_p = \mathbf{0} \quad \text{for} \quad R < a. \tag{2.72}$$

In the inner volume of the hollow sphere there is no gravitational field. The force on a test mass m is zero. The contributions from the different parts of the hollow sphere cancel each other. In Fig. 2.53 the potential energy $E_p(R)$ and the force $F(R)$ are shown inside and outside of the hollow sphere.

A homogeneous full sphere can be composed of many concentric hollow spheres. Its mass is

$$M = \int\limits_{a=0}^{R_0} \varrho \cdot 4\pi a^2 da.$$

For a test mass outside the sphere ($R > R_0$) we obtain from (2.69)

$$\begin{aligned} E_p &= -G\frac{4\pi}{R}\varrho m \int\limits_0^{R_0} a^2\,da = -G\frac{4\pi}{3R}R_0^3 \varrho m \\ &= -G\frac{m \cdot M}{R}. \end{aligned} \tag{2.71a}$$

For a point inside the sphere ($R < R_0$) we perform the integration in two steps over the ranges $0 \le a \le R$ and $R \le a \le R_0$. From the Eqs. 2.71 and 2.71a the potential energy can be derived as

$$\begin{aligned} E_p &= -4\pi\varrho G m \left[\int\limits_{a=0}^{R} \frac{a^2\,da}{R} + \int\limits_{a=R}^{R_0} a\,da \right] \\ &= -4\pi\varrho G m \left[\frac{R^2}{3} + \frac{1}{2}R_0^2 - \frac{1}{2}R^2 \right]; \end{aligned} \tag{2.73}$$

since $M = (4/3) \cdot \varrho\pi R_0^3$ this becomes

$$E_p = \frac{GMm}{2R_0^3}\left(R^2 - 3R_0^2\right). \tag{2.74}$$

The physical meaning of the two steps for the integration is the following: For a test mass in the point P(R) only the mass elements of the sphere with $r \le R$ contribute to the total gravitational force while the contributions of all mass elements with $r \ge R$ exactly cancel each other. The second term in (2.73) gives a constant part to the potential energy and therefore no contribution to the force. From (2.71) and (2.74) one obtains the force (Fig. 2.54 lower part)

$$\begin{aligned} \boldsymbol{F} &= -G\frac{Mm}{R^2}\hat{\boldsymbol{r}} \quad \text{for} \quad R \ge R_0 \\ \boldsymbol{F} &= -\frac{GMm}{R_0^3}R\hat{\boldsymbol{r}} \quad \text{for} \quad R \le R_0. \end{aligned} \tag{2.75}$$

Remark. The earth is not a sphere with homogeneous density

1. Because it is an oblate spheroid due to the rotation of the earth which deforms the plastic earth crust [2.10].

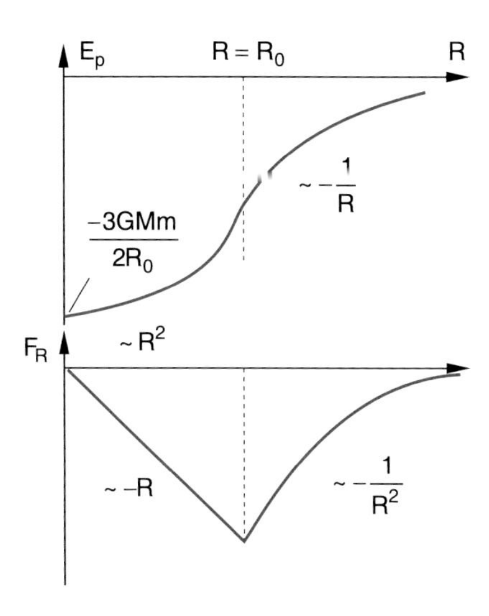

Figure 2.54 Potential energy E_p and gravitational force F of a sample mass m in the gravitational field of a full sphere with mass M

2. Because the density increases towards the centre. Therefore the mass $M(R)$ inside a sphere with radius $R < R_0$ increases with R only as R^n (with $n < 3$, Fig. 2.55). The earth acceleration g measured in a deep well therefore decreases with $r^q (q < 1)$ [2.11].

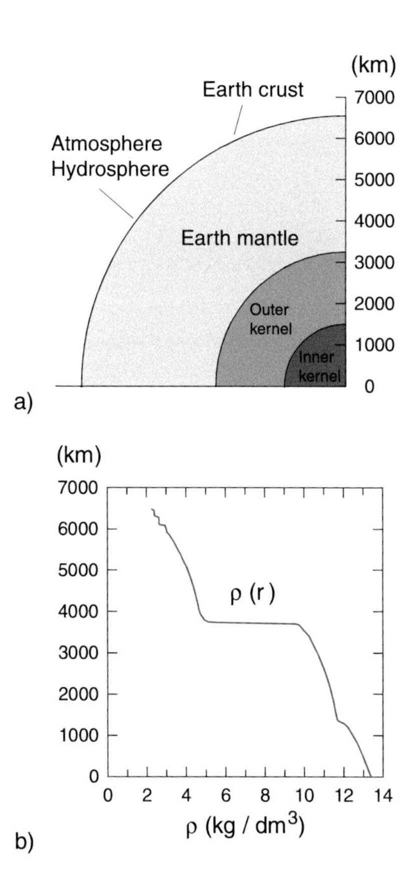

Figure 2.55 **a** Radial cut through the earth showing the different layers. **b** radial density function $\varrho(r)$

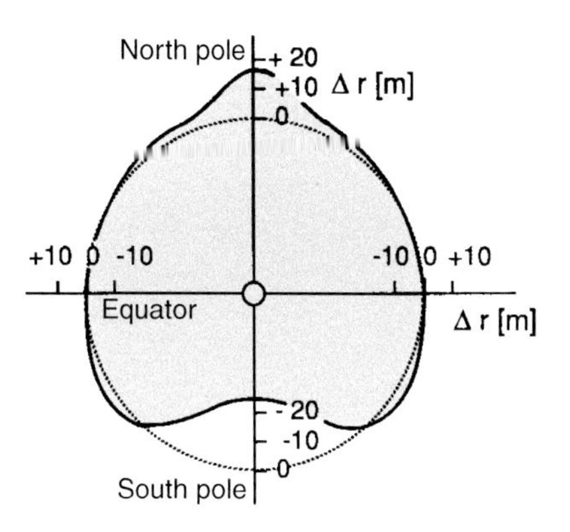

Figure 2.56 The shape of the earth as geoid. The deviation of the geoid from a spheroid with $(a - b)/a = 1/298.25$ (*dotted curve*) is shown 80 000 times exaggerated. Even the geoid gives only the approximate shape of the real earth

3. The mass distribution is not exactly spherical symmetric. The gravitational field of the earth is therefore not exactly a central force field. This implies that the angular momentum of a satellite, orbiting around the earth is not really constant. Measurements of the change of the orbital plane with time (the position $r(t)$ of a satellite can be determined with RADAR techniques with an uncertainty of a few cm!) allows the determination of the mass distribution $\varrho(\vartheta, \varphi)$ in the earth [2.9a, 2.9b].
4. The equipotential surfaces of the earth form a *geoid* (Fig. 2.56). One of these surfaces, which coincides with the average surface of the oceans is defined as the normal zero surface. All heights on earth are given with respect to this surface.

2.9.6 Measurements of the Gravitational Constant G

Measurements of the planetary motions allow only the determination of the product $G \cdot M_\odot$ of gravitational constant G and mass of the sun. The absolute value of G has to be measured by laboratory experiments. Such experiments were at first performed 1797 by *Henry Cavendish* and later on repeated by several scientists with increased accuracy [2.12a–2.14], where *Lorand Eötvös* (1848–1919) was especially of high repute because of his very careful and extensive precision experiments [2.2].

Most of these experiments use a **torsion balance** (Fig. 2.57). A light rod (1) with length $2L$ and two small lead balls with equal masses m hangs on a thin wire. Two large masses $M_1 = M_2 = M$ are placed on a rotatable rod (2), which can be turned to the two positions (a) or (b). Due to the gravitational force between m and M the light rod (1) is clockwise turned for the position (a) and counter-clockwise for the position (b) by an angle φ where the retro-driving torque

$$D_r = \frac{\pi}{2} G^* \frac{d^4}{16l} \cdot \varphi \tag{2.76}$$

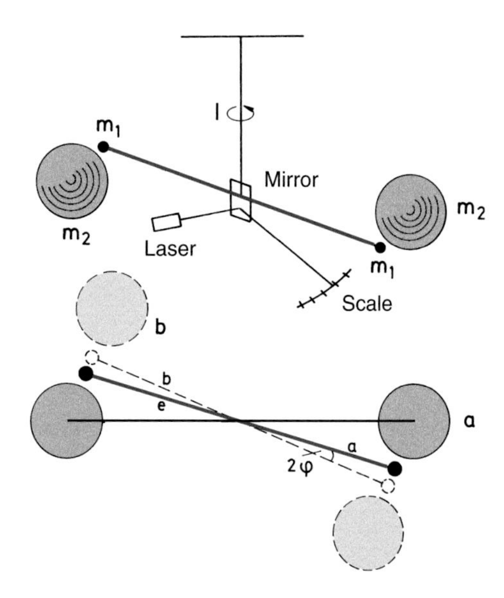

Figure 2.57 Eötvös' torsion balance for measuring Newton's gravitational constant G

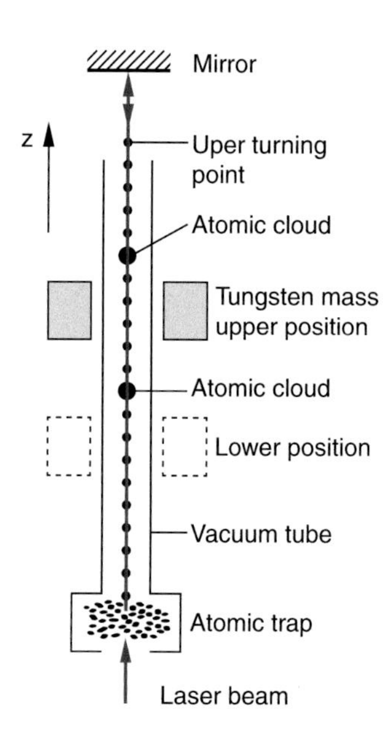

Figure 2.58 Atom interferometer for the measurement of the Newtonian gravitational constant G [2.13b]

of the twisted wire just compensates the torque with the amount $2L \cdot F_g$ generated by the gravitational forces

$$F_g = G \cdot \frac{m \cdot M}{r^2} = G \cdot \frac{16\pi^2}{9r^2} \varrho^2 R_1^3 R_2^3 . \tag{2.77}$$

Here G^* is the torsion module of the wire, d its diameter and l its length, ϱ the mass density of the spheres, R_1 and R_2 their radii and r the distance between their centres. In the equilibrium position, where the two torques cancel, we have the condition $D_r = 2L \cdot F_g$. This gives for the gravitational constant

$$G = \frac{9G^*}{64\pi} \frac{r^2 (d/2)^4}{l \cdot L \cdot \varrho^2 R_1^3 R_2^3} \cdot \Delta\varphi . \tag{2.78}$$

In order to maximize the force F_g, the density ϱ should be as high as possible, because the distance r between the masses m and M cannot be smaller than $r_{\min} = R_1 + R_2$. The measurement of φ is performed by placing a mirror at the turning point of the rod with the masses m, which reflects a laser beam by an angle 2φ. On a far distant scale the deflection of the laser spot is a measure for the angle φ.

The most accurate measurement proceeds as follows: The masses M are turned into the position (a). The system now performs oscillations around the new equilibrium position φ_1 which can be determined as the mean of the turning points of the oscillations. Now the masses M are turned into the new position (b). Again oscillations start around the new equilibrium position φ_2, which is determined in the same way. The difference $\Delta\varphi = \varphi_1 - \varphi_2$ than gives according to (2.78) the gravitational constant G.

Equation 2.78 tells us, that the diameter d of the wire should be as small as possible. New materials, such as graphite composites, have a large tear strength. They can carry the masses m even for small values of d. This increases the sensitivity.

In recent years new methods for measuring G have been developed. Most of them are based on optical techniques. We will just discuss one of them: A collimated beam of very cold atoms (laser-cooled to $T < 1\,\mu$K) is sent upwards through an evacuated tube (Fig. 2.58). At the heights $z = h$ where $\frac{1}{2}mv^2 = mgh$ they reach their turning point where they fall down again. A large tungsten mass surrounds the tube and can be shifted upwards or downwards. Above the mass the atoms experience during their upwards motion an acceleration $-(g + \Delta g)$ due to the gravitational attraction by the earth (g) and the mass (Δg). Below the mass their acceleration is $-(g - \Delta g)$. These accelerations are measured via atom interferometry [2.13b].

Figure 2.59 gives the results of many experiments in the course of time, using different measuring techniques. This illustrates, that the error bars are still large but the differences between the results of many experiments are even larger, indicating the underestimation of systematic errors. The value accepted today

$$G = 6.67384(80)\,\mathrm{m^3\,kg^{-1}\,s^{-2}}$$

is the weighted average of the different measurements where the number in the brackets give the standard deviation σ (see Sect. 1.8.2). The relative error is $1.2 \cdot 10^{-4}$ which illustrates that among all universal constants G is the one with the largest uncertainty.

2.9.7 Testing Newton's Law of Gravity

In order to test the validity of the $1/r^2$ dependence of the gravitational force (2.52) several precision experiments have been performed [2.13d]. An interesting proposal by Stacey [2.17] is based on the following principle: In the vertical tunnel within

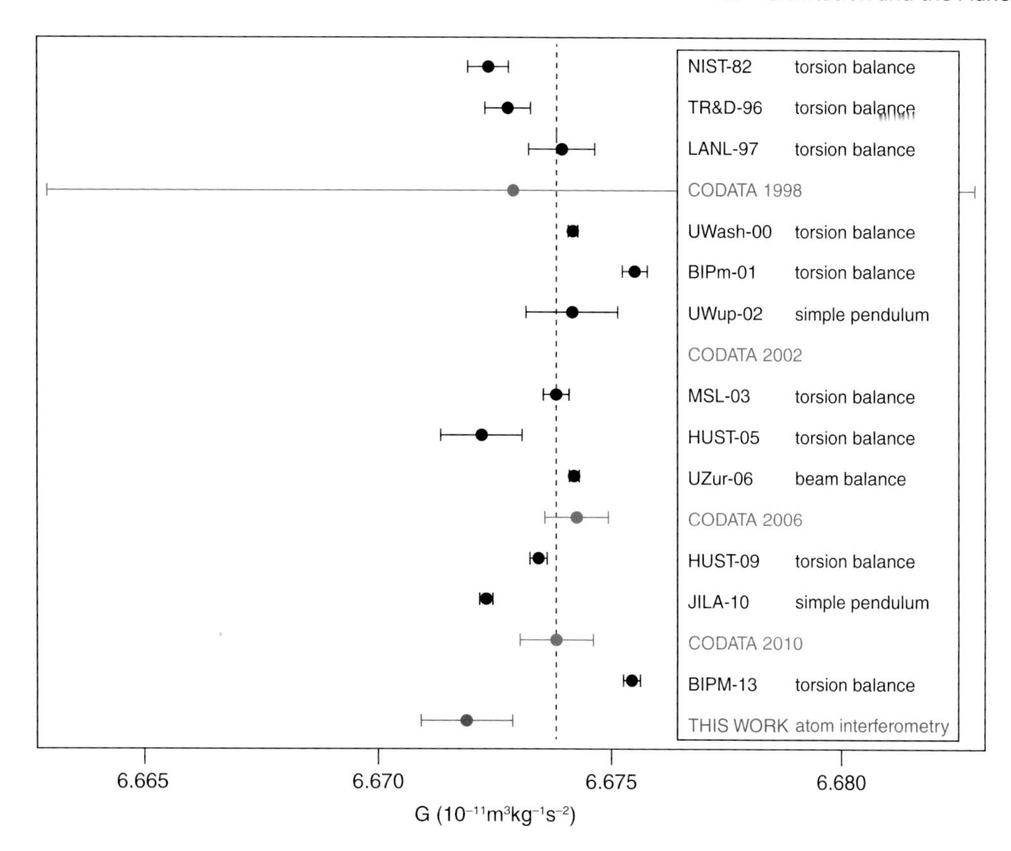

Figure 2.59 Results of different measurements of the Newtonian gravitational constant G [2.13b]

a large water reservoir a sensitive gravitation-balance is placed, where two masses m are hold at different heights, one above the water level and one below (Fig. 2.60). When the water level is lowered by Δh, the change of the gravitational force differs for the two masses. For the lower mass it increases by

$$\delta F_G = G \cdot m \cdot 2\pi\varrho \cdot \Delta h \qquad (2.78a)$$

because the water above the mass decreases, while the water below the mass stays constant. For the upper mass the force decreases because the distance between the mass and the water surface increases (see problem 2.34).

There is still an open question concerning the exact validity of the r^{-2} dependence in Newton's gravitation law over astronomical distances. Astronomical observations of the rotation of galaxies showed, that the visible mass distribution in the galaxy could not explain the differential rotation $\omega(R)$ as the function of the distance R from the galaxy centre, if Newton's law is assumed to be valid. There are two different explanations of this discrepancy: Either the $1/r^2$ dependence of F_G is not correct over large distances, or there exists invisible matter (*dark matter*) which interacts with the visible matter only by gravitation and therefore changes the gravitational force of the visible matter.

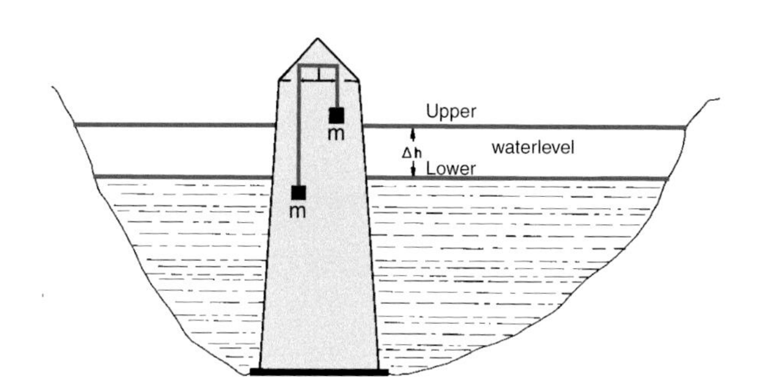

Figure 2.60 Possible method for measuring the $1/r^2$ dependence of the gravitational force

Such very difficult precision experiments have a great importance for testing fundamental physical laws. There are many efforts to develop theories which reduce the four fundamental forces (see Tab. 1.2) to a common origin and to understand more deeply the difference between energy and matter. One example of such precision experiments are tests of possible differences between gravitational and inertial mass as has been performed by *Eötvös* 1922 and *Dicke* 1960 and many other scientists.

Here the inertial mass is measured for different materials by the oscillation period of a gravitational torsion balance [2.18a]. The results obtained up to now show that the ratio m_{in}/m_g of inertial mass to gravitational mass does not differ from 1 within the error limits. For two different materials A and B a possible difference

$$\eta(A, B) = [m_{\text{in}}/m_g]_A - [m_{\text{in}}/m_g]_B < 10^{-12}$$

must be very small and lies below the detection limit of 10^{-12} with the presently achievable accuracy.

Table 2.2 Mass and mean density of sun, planets and the earth-moon

Planet	Symbol	Mass/earth mass	Mean density $\bar{\varrho}$ in 10^3 kg/m^3
Sun	☉	$3.33 \cdot 10^5$	1.41
Mercury	☿	0.0558	5.42
Venus	♀	0.8150	5.25
Earth	♁	1.0	5.52
Mars	♂	0.1074	3.94
Jupiter	♃	317.826	1.314
Saturn	♄	95.147	0.69
Uranus	♅	14.54	1.19
Neptun	♆	17.23	1.66
Moon	☾	0.0123	3.34

From the revolution period $T = 2\pi/\omega$ of a satellite around the earth (e.g. the moon or an artificial satellite) the mass M of the earth can be determined. For a circular motion the gravitational force is equal to the centripetal force

$$m \cdot \omega^2 \cdot r = G \cdot mM/r^2 .$$

With the known gravitational constant G and the measured distance r of the satellite from the earth centre the mass of the earth is obtained from

$$M = \omega^2 \cdot r^3/G .$$

The experimental value is

$$M = 5.974 \cdot 10^{24}\,\text{kg} .$$

From measurements of the gravity acceleration g on the earth surface the equation

$$m \cdot g = G \cdot m \cdot M/R^2$$

yields the earth radius R. From M and R the mean density $\varrho = 3M/(4\pi R^3)$ can be derived.

A comparison of the densities of the different planets (Tab. 2.2) illustrates that the inner planets (Mercure, Venus, Earth and Mars) formed of rocks have comparable densities around $\varrho = 5\,\text{g/cm}^3$, while the outer gas planets and the sun have much lower densities. These differences give hints to the formation process of our solar system [2.7] (see Volume 4).

2.9.8 Experimental Determination of the Earth Acceleration *g*

The most accurate determination of g can be performed by measuring the oscillation period of a pendulum. This pendulum consist of a sphere with the mass m suspended by a string with length L (measured between suspension point A and the centre C of the sphere). If the mas of the string is negligibly small compared to m and the radius R of the sphere small compared

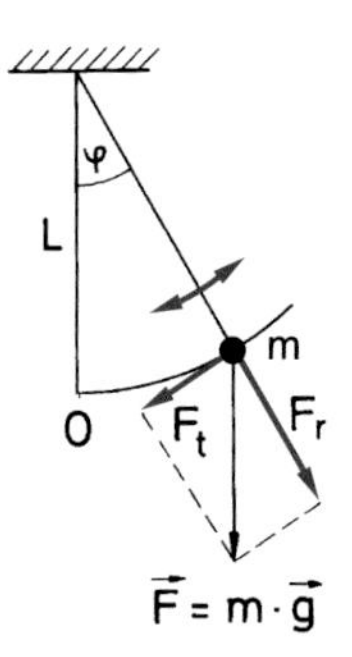

Figure 2.61 Measuring the free fall acceleration g with a pendulum

with L this device is called a *mathematical pendulum*. The motion of the pendulum under the influence of gravity can be best calculated when the force $\boldsymbol{F}_g = m \cdot \boldsymbol{g}$ is decomposed into the two components (Fig. 2.61):

- a radial component F_r in the direction of the string, which generates in the string an equal but opposite restoring force. Since the total force component in this direction is zero, it does not contribute to the acceleration.
- a tangential component $F_t = -m \cdot g \cdot \sin\varphi$ which causes a tangential acceleration $a_t = -g \cdot \sin\varphi$.

The pendulum represents an example of a position dependent force which is **not** a central force. The angular momentum is therefore not preserved. However, if the initial velocity for $\varphi \neq 0$ lies in the plane of the components F_r and F_t the motion remains in this plane. It can be therefore described by planar polar coordinates z and φ. The equation of motion reads

$$m \cdot g \cdot \sin\varphi = -m \cdot L \cdot \varphi . \quad (2.79a)$$

Expanding $\sin\varphi$ into a Taylor-series

$$\sin\varphi = \varphi - \frac{\varphi^3}{3!} + \frac{\varphi^7}{7!} - \cdots .$$

The higher order terms can be neglected for small elongations φ. For example is for $\varphi = 10° = 0.17\,\text{rad}$ the term $\varphi^3/3! = 8.2 \cdot 10^{-4}$ which means that the second term is already smaller by the factor 208 than the first term. The error in the approximation $\sin\varphi \approx \varphi$ is for $\varphi = 10°$ only $< 0.5\%$.

The equation of motion (2.79a) is then in the approximation $\sin\varphi \approx \varphi$

$$\ddot{\varphi} = -(g/L)\varphi . \quad (2.79b)$$

With the initial condition $\varphi(0) = 0$ the solution is

$$\varphi(t) = A \cdot \sin(\sqrt{g/L} \cdot t) . \quad (2.80)$$

The pendulum performs a periodic oscillation with the oscillation period

$$T = 2\pi \cdot \sqrt{L/g} . \quad (2.81)$$

Measuring the time for 100 periods with an uncertainty of 0.1 s allows the determination of T with an error of 10^{-3} s. The largest uncertainty comes from the measurement of the length L. The errors for L and T in the determination of

$$g = \frac{4\pi^2 \cdot L}{T^2}$$

give a total error of g according to

$$\left|\frac{\Delta g}{g}\right| \leq 2\left|\frac{\Delta T}{T}\right| + \frac{\Delta L}{L} .$$

Example

$\Delta T/T = 5 \cdot 10^{-5}$, $\Delta L/L = 10^{-3}$ for $L = 1$ m. $\Rightarrow$ $\Delta g/g = 1.1 \cdot 10^{-3}$. ◄

For a more accurate solution of (2.79a) we use the energy conservation law (see Sect. 2.7), which saves one integration. From Fig. 2.62 we see that

$$E_p = m \cdot g \cdot L \cdot (1 - \cos\varphi)$$
$$E_{kin} = \tfrac{1}{2} m \cdot v^2 = \tfrac{1}{2} mL^2 \cdot \dot{\varphi}^2 .$$

The constant total energy is

$$\begin{aligned} E = E_p + E_{kin} &= \frac{m}{2} L^2 \dot{\varphi}^2 + mgL(1 - \cos\varphi) \\ &= mgL(1 - \cos\varphi_0) . \end{aligned}$$

Where φ_0 is the angle at the turning point where $E_{kin} = 0$. Solving for φ gives

$$\frac{d\varphi}{dt} = \sqrt{\frac{2g(\cos\varphi - \cos\varphi_0)}{L}} .$$

Integration yields

$$\sqrt{\frac{L}{2g}} \int_{\varphi=0}^{\varphi_0} \frac{d\varphi}{\sqrt{\cos\varphi - \cos\varphi_0}} = \int_{t=0}^{T/4} dt = T/4 . \qquad (2.82)$$

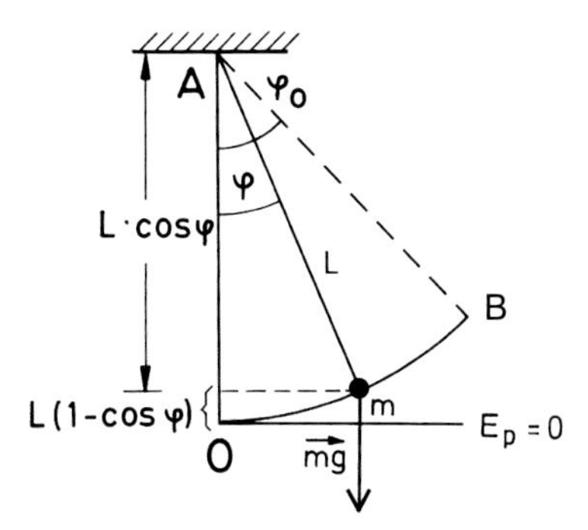

Figure 2.62 Illustration of the integration of the pendulum equation based on the energy conservation

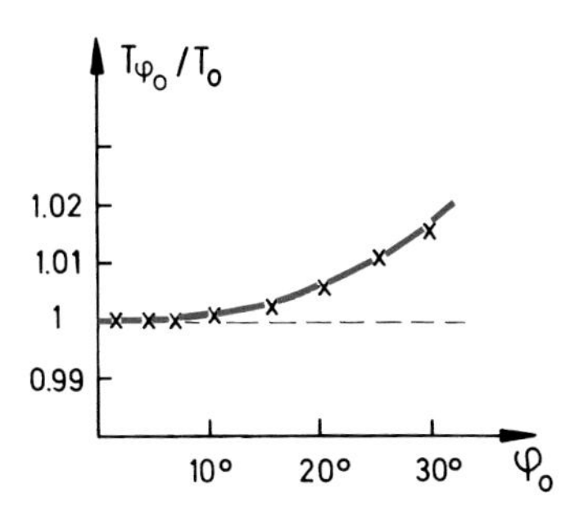

Figure 2.63 Dependence of the oscillation period on the deflection of the pendulum

With the substitution $\sin\xi = \sin(\varphi/2)/\sin(\varphi_0/2)$ the integral can be reduced to an elliptical integral

$$T = 4\sqrt{L/g} \int_0^{\pi/2} \frac{d\xi}{\sqrt{1 - k^2 \sin^2\xi}} \qquad (2.83)$$
$$\text{with} \quad k = \sin(\varphi_0/2) ,$$

which can be solved by a Taylor expansion of the integrand [2.18b]. The result is

$$T(\varphi_0) = 2\pi \sqrt{\frac{L}{g}} \left(1 + \frac{1}{16}\varphi_0^2 + \cdots\right) . \qquad (2.84)$$

For the accurate determination of T the oscillation period is measured as a function of the elongation φ_0 and the measured values are extrapolated towards $\varphi_0 = 0$ (Fig. 2.63).

If the shape of the earth is approximated by a spheroid the dependence of g on the latitude $\beta = 90° - \vartheta$ can be approximated by the formula

$$\begin{aligned} g(\beta) \approx g_e \big(1 &+ 0.0053024 \sin^2\beta \\ &- 5.8 \cdot 10^6 \sin^2 2\beta\big) \end{aligned} \qquad (2.85)$$

where $g_e = g(\beta = 0) = 9.780327\,\text{m/s}^2$ is the earth acceleration at the equator. This formula takes into account, that g is diminished by the centrifugal acceleration of the rotating earth which depends on β (see Sect. 3.2). Because of the inhomogeneous mass distribution of the earth additional local changes of g appear which are not considered in (2.85).

Instead of the pendulum nowadays modern gravimeters are used for the determination of g. These are sensitive spring balances which had been calibrated with a precision pendulum. The restoring force $\boldsymbol{F} = -D(x - x_0)$ is determined by measuring the displacement from the equilibrium position by a calibrated mass m and gets the local variation of the earth acceleration g according to [2.19]

$$m \cdot g = -D(x - x_0) .$$

Recently two identical satellites were launched which orbit around the earth on identical paths with an angle distance $\Delta\varphi$.

This distance can be measured very accurately (within a few millimetres) by the time laser pulses need to travel from one satellite to the other and back. Local variations of the gravity cause a different local acceleration which changes the distance $d = R \cdot \Delta\varphi$ between the satellites. This allows the determination of even tiny changes of the gravity force [2.20a, 2.20b, 2.20c].

Summary

- A body with mass m can be described by the model of a point mass as long as its spatial extensions are small compared to its distance to other bodies.
- The motion of a body is described by a trajectory $\boldsymbol{r}(t)$, which the body traverses in the course of time. Its momentary velocity is $\boldsymbol{v}(t) = \dot{\boldsymbol{r}} = \mathrm{d}\boldsymbol{r}/\mathrm{d}t$ and its acceleration is $a(t) = \mathrm{d}v/\mathrm{d}t = \mathrm{d}^2r/\mathrm{d}t^2$.
- Motions with $\boldsymbol{a}(t) = \boldsymbol{0}$ are called uniform straight-line motions. Magnitude and direction of the velocity are constant.
- For the uniform circular motion the magnitude $|\boldsymbol{a}(t)|$ is constant, but the direction of $\boldsymbol{a}(t)$ changes uniformly with the angular velocity $\boldsymbol{\omega}$.
- A force acting on a freely movable body causes an acceleration and therefore a change of its state of motion.
- A body is in an equilibrium state if the vector sum of all acting forces is zero. In this case it does not change its state of motion.
- The state of motion of a body with mass m and velocity $\boldsymbol{v}$ is defined by the momentum $\boldsymbol{p} = m \cdot \boldsymbol{v}$.
- The force $\boldsymbol{F}$ acting on a body is defined as $F = \mathrm{d}\boldsymbol{p}/\mathrm{d}t$ (2. Newton's law).
- For two bodies with masses m_1 and m_2 which interact with each other but not with other bodies the 3. Newtonian law is valid: $F_1 = -F_2$ (F_1 is the force acting on m_1, F_2 acting on m_2).
- The work executed by the force $\boldsymbol{F}(\boldsymbol{r})$ on a body moving along the trajectory $\boldsymbol{r}(t)$ is the scalar quantity $W = \int \boldsymbol{F}(\boldsymbol{r})\mathrm{d}\boldsymbol{r}$.
- Force fields where the work depends only on the initial point P_1 and the final point P_2 but **not** on the choice of the path between P_1 and P_2 are called **conservative**. For such fields is $\mathbf{rot}\,\boldsymbol{F} = \boldsymbol{0}$. All central force fields are conservative.
- To each point P in a conservative force field a potential energy $E_\mathrm{p}(P)$ can be attributed. The work $\int F(r)\mathrm{d}r = E(P_1) - E(P_2)$ executed on a body to move it from P_1 to P_2 is equal to the difference of the potential energies in P_1 and P_2. The choice of the point of zero energy is arbitrarily. Often one chooses $E(r = 0) = 0$ or $E(r = \infty) = 0$.
- The potential energy $E(P)$ and the force $F(r)$ in a conservative force field are related by $\boldsymbol{F}(\boldsymbol{r}) = -\mathbf{grad}E_\mathrm{p}$.
- The kinetic energy of a mass m moving with the velocity v is $E_\mathrm{kin} = \frac{1}{2}mv^2$.
- In a conservative force field the total energy $E = E_p + E_\mathrm{kin}$ is constant (law of energy conservation).
- The angular momentum of a mass m with momentum $\boldsymbol{p}$, referred to the origin of the coordinate system is $\boldsymbol{L} = \boldsymbol{r} \times \boldsymbol{p} = m \cdot (\boldsymbol{r} \times \boldsymbol{v})$. The torque acting on a body in a force field $\boldsymbol{F}(\boldsymbol{r})$ is $\boldsymbol{D} = \boldsymbol{r} \times \boldsymbol{F}$. It is $\boldsymbol{D} = \mathrm{d}\boldsymbol{L}/\mathrm{d}t$.
- All planets of our solar system move in the central force field $F(r) = -G \cdot (m \cdot M/r^2)\hat{\boldsymbol{r}}$ of the sun. Therefore their angular momentum is constant. Their motion is planar. Their trajectories are ellipses with the sun in one focal point.
- The gravitational field of extended bodies depends on the mass distribution. For spherical symmetric mass distributions with radius R the force field outside the body ($r > R$) is exactly that of a point mass, inside the body ($r < R$) the force $F(r)$ increases for homogeneous distributions linearly with r from $\boldsymbol{F} = 0$ at the centre $r = 0$ to the maximum value at $r = R$.
- The free fall acceleration $\boldsymbol{g}$ of a body with mass m equals the gravitational field strength $\boldsymbol{G} = \boldsymbol{F}/m$ at the surface $r = R$ of the earth with mass M. With Newton's law of gravity $\boldsymbol{g}$ can be expressed as $\boldsymbol{g} = G \cdot (M/R^2)\hat{\boldsymbol{r}}$ (G = gravitational constant). It can be determined from the measured oscillation period $T = 2\pi\sqrt{L/g}$ of a pendulum with length L, or with gravitational balances.

Problems

2.1 A car drives on a road behind a foregoing truck (length of 25 m) with a constant safety distance of 40 m and a constant velocity of 80 km/h. As soon as the driver can foresee a free distance of 300 m he starts to overtake. Therefore he accelerates with $a = 1.3\,\mathrm{m/s^2}$ until he reaches a velocity of $v = 100\,\mathrm{km/h}$. Can he safely overtake? How long are time and path length of the overtaking procedure if he considers the same safety distance after the overtaking? Draw for illustration a diagram for $s(t)$ and $v(t)$.

2.2 A car drives half of a distance x with the velocity $v_1 = 80\,\mathrm{km/h}$ and the second half with $v_2 = 40\,\mathrm{km/h}$. Estimate and calculate the mean velocity $\langle v \rangle$ as the function of v_1 and v_2. Make the same consideration if $x_1 = 1/3x$ and $x_2 = 2/3x$.

2.3 A body moves with constant acceleration along the x-axis. It passes the origin $x = 0$ with $v = 6\,\mathrm{cm/s}$. 2 s later it arrives at $x = 10\,\mathrm{cm}$. Calculate magnitude and direction of the acceleration.

2.4 An electron is emitted from the cathode with a velocity v_0 and experiences in an electric field over a distance of 4 cm a constant acceleration $a = 3 \cdot 10^{14}\,\mathrm{m/s^2}$, reaching a velocity of $7 \cdot 10^6\,\mathrm{m/s}$. How large was v_0?

2.5 A body is thrown from a height $h = 15\,\mathrm{m}$ with an initial velocity $v_0 = 5\,\mathrm{m/s}$

a) upwards,
b) downwards.

Calculate for both cases the time until it reaches the ground.

c) Derive Eq. 2.13.

2.6 Give examples where both the magnitude and the direction of the acceleration are constant but the body moves nevertheless not on a straight line. Which conditions must be fulfilled for a straight line?

2.7 A car crashes with a velocity of 100 km/h against a thick tree. From which heights must it fall down in order to experience the same velocity when reaching the ground? Compare this with two equal cars with velocities of 100 km/h crashing head on against each other.

2.8

a) A body moves with constant angular velocity $\omega = 3\,\mathrm{rad/s}$ on a vertical circle in the x-z-plane with radius $R = 1\,\mathrm{m}$ in the gravity field $F = \{0, 0, -g\}$ of the earth. How large are its velocities at the lowest and the highest point on the circle? How large is the difference between the two values? Could you relate this to the potential energy?
b) A body starts with $v_0 = 0$ from the point $A(z = h)$ in Fig. 2.64 on the frictionless looping path. How large are velocities and accelerations in the points B and C of the circular path with radius R? What is the maximum ratio R/h to prevent that the body falls down in B? How large is then the velocity $v(B)$?

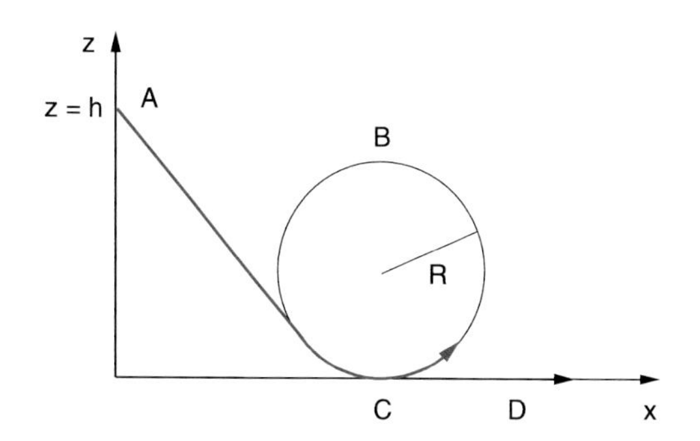

Figure 2.64 Looping path (Probl. 2.8 b)

2.9 How large is the escape velocity

a) of the moon ($d = 384\,000\,\mathrm{km}$) in the gravitational field of the earth?
b) of a body on the surface of the moon in the gravitational field of the moon?

2.10 What is the minimum fuel mass of a one stage rocket with a payload of 500 kg for a horizontal launch at the equator to bring the rocket to the first escape velocity of $v_1 = 7.9\,\mathrm{km/s}$ when the velocity of the propellant gas relative to the rocket is $v_e = 4.5\,\mathrm{km/h}$

a) in the east direction
b) in the west direction?

2.11 Check the energy conservation law for the examples given in the text. Show, that (2.26) follows directly from the condition $E_{\mathrm{kin}} \geq E_{\mathrm{p}}$, i. e. $\frac{1}{2}mv^2 \geq m \cdot g \cdot R$.

2.12 A rocket to the moon is launched from a point at the equator. How much energy is saved compared to a vertical launch, when it is shot in the eastern direction under 30° against the horizontal?

2.13 A wooden cylinder (radius $r = 0.1\,\mathrm{m}$, heights $h = 0.6\,\mathrm{m}$) is vertically immersed in water with 2/3 of its length which is its equilibrium position. Which work has to be performed when it is pulled out of the water? How is the situation if the cylinder lies horizontally in the water? How deep does it immerse?

2.14 A body with mass $m = 0.8\,\mathrm{kg}$ is vertically thrown upwards. In the heights $h = 10\,\mathrm{m}$ its kinetic energy is 200 J. What is the maximum heights it can reach?

2.15 A spiral spring of steel with length $L_0 = 0.8$ m is expanded by the force $F = 20$ N to a length $L = 0.85$ m. Which work is needed to expand the spring to twice its initial length, if the force is always proportional to the expansion $\Delta L = L - L_0$?

2.16 What is the minimum initial velocity of a body at a vertical launch from the earth when it should reach the moon?

2.17 What is the distance of a geo-stationary satellite from the centre of the earth? Which energy is needed to launch it? How accurate has its distance to the earth centre be stabilized in order to maintain its position relative to a point on earth within 0.1 km/d?

2.18 What is the change of potential, kinetic and total energy of a satellite when its radius r on a stable circular orbit around the earth centre is changed? What is the ratio E_{kin}/E_p? Does it depend on r? Express the total energy E by m, g, r and the mass M_E of the earth. Are these quantities sufficient or are more needed?

2.19 Prove, that the force $\boldsymbol{F} = m \cdot g \cdot \sin\varphi \cdot \boldsymbol{e}_t$ for the mathematical pendulum is conservative and that for arbitrary values of φ conservation of energy $E_{kin} + E_p = $ const holds.

2.20 Assume one is able to measure the length $L = 10$ m of a pendulum within 0.1 mm and the period T within 10 ms. How many oscillation periods have to be measured in order to equalize the contribution of ΔL and ΔT to the accuracy of g? How large is then the uncertainty of g?

2.21 How much accuracy is gained for the determination of G with the gravity balance if the large masses M are increased by a factor of 10? How accurate has the measurement of the angle φ to be in order to determine G with an accuracy of 10^{-4}? Give some physical reasons for the limits of the accuracy of φ.

2.22 The comet Halley has a period of 76 years. His smallest distance to the sun is 0.59 AU. How large is its maximum distance to the sun and what is the eccentricity of its elliptical orbit? Hint: Look for a relation between T and $r_{min} = a(1-\varepsilon)$ and $r_{max} = a(1+\varepsilon)$.

2.23 Assume that the gravity acceleration at the equator of a rotating planet is $11.6\,\mathrm{m/s^2}$, the centripetal acceleration $a = 0.3\,\mathrm{m/s^2}$ and the escape velocity for a vertical launch 23.6 m/s. At the heights $h = 5000$ km above the surface is $g = 8.0\,\mathrm{m/s^2}$. What are the radius R and the mass M of the planet. How fast is it rotating? Which planet meets these requirements?

2.24 The gravitational force exerted by the sun onto the moon is about twice as large as that exerted by the earth. Why is the moon still circling around the earth and has not escaped?

2.25 Which oscillation period would a pendulum have on the moon, if its period on the earth is 1 s?

2.26 A vertical straight tunnel is cut through the earth between opposite points A to B on the earth surface.

a) Show that without friction a body released in A performs a harmonic oscillation between A and B.
b) What is the oscillation period?
c) Compare this value with the period of a satellite, which circles around the earth closely above the surface.
d) A straight tunnel is cut between London and New York. What is the travel time of a train without friction and extra driving force (besides gravity) which starts in London with the velocity $v_0 = 0$? How much does the time change, if $v_0 = 40$ m/s?

2.27 Calculate the distance earth-moon from the period of revolution of the moon $T = 27$ d (mass of the earth is $M = 6 \cdot 10^{24}$ kg).

2.28 Saturn has a mass $M = 5.7 \cdot 10^{26}$ kg and a mean density of $0.71\,\mathrm{g/cm^3}$. How large is the gravitational acceleration on its surface?

2.29 How large is the relative change of the gravity acceleration g between a point on the earth surface and a point with $h = 160$ km above the surface?

2.30 How large is the change Δg of the earth acceleration due to the attraction by

a) the moon and
b) the sun?

Compare the two changes and discuss them. How large is the relative change $\Delta g/g$?

2.31 Two spheres made of lead with masses $m_1 = m_2 = 20$ kg are suspended by two thin wires with length $L = 100$ m where the suspension points are 0.2 m apart. What is the distance between the centres of the spheres, when the gravitational field of the earth is assumed to be spherical symmetric?

a) without
b) with the gravitational force between the two masses.

2.32 Based on the energy conservation law determine the velocity of the earth in its closest distance from the sun (Perihelion) and for the largest distance (aphelion). How large is the difference Δv to the mean velocity? Discuss the relation between the eccentricity of the elliptical orbit and Δv.

2.33 A satellite orbiting around the earth has the velocity $v_A = 5$ km/s in the aphelion and $v_P = 7$ km/s in its perihelion. How large are minor and major half axes of its elliptical orbit?

2.34 Prove the equation (2.78a).

References

2.1. F. Cajori (ed.), *Sir Isaac Newton's Mathematical Principles of Natural Philosophy and His System of the World. (Principia).* (University of California Press, Berkely, 1962)

2.2. R.V. Eötvös, D. Pekart, F. Fekete, Ann. Phys. **68** 11 (1922)

2.3. A.P. French, *Special Relativity Theory.* (MIT Interoductor Series Norton & Company, 1968)

2.4. A. Beer, P. Beer (ed.) *Kepler. Four Hundred Years.* (Pergamon, Oxford, 1975)

2.5a. J. Stuhler, M. Fattori, T. Petelski, G.M. Tino, J. Opt. B: Quantum Semiclass. Opt. **5** 75 (2003)

2.5b. A. Bertoldi et al., Europ Phys. Journal D **40**, 271 (2006)

2.6a. B.R. Martin, C. Shaw *Mathematics for Physicists. Manchester Physics Series.* (Wiley, London, 2015)

2.6b. A. Jeffrey, D. Zwillinger, *Table of Integrals, Series and Products.* 8th ed. (Elsevier Oxford, 2014)

2.7. M.A. Seeds, D. Backman, *Foundations of Astronomy,* 13th ed. (Cangage Learning, 2015)

2.8. E. Chaisson, St. McMillan, *A Beginners Guide to the Universe.* 7th ed (Benjamin Cummings Publ. Comp., 2012)

2.9a. W.M. Kaula, Satellite Measurements of the Earth's Gravity Field. in *Methods of Experimental Physics.* ed. by R. Celotta, J. Levine, Ch.G. Sammis, Th.L. Henyey, Vol. 24, part B (Academic Press, San Diego, 1987), p.163

2.9b. Ch. Hirt et al., Geophys. Research Lett. **40**(16), 4279 (2013)

2.9c. http://en.wikipedia.org/wiki/Gravity_of_Earth

2.10. R.H. Rapp, F. Sanso (ed.) *Determination of the Geoid.* (Springer, Berlin, Heidelberg, 1991)

2.11. C.M.R. Fowler *The Solid Earth: An Introduction to Global Geophysics,* 2nd ed. (Cambridge Univ. Press, Cambridge, 2004)

2.12a. https://en.wikipedia.org/wiki/Gravitational_constant

2.12b. B. Fixler, G.T. Foster, J.M. McGuirk, M.A. Kasevich, Science **315**(5808), 74 (2007)

2.13a. H.V. Parks, Phys. Rev. Lett. **105**, 110801 (2010)

2.13b. G. Rosi et al., Nature **510**, 518 (2014)

2.13c. C. Moskowitz, *Puzzling Measurement of "Big G" Gravitational Constant.* Scientific American, Sept. 18, 2013

2.13d. T. Quinn, Nature **408**, 919 (2000)

2.14. P.J. Mohr, B.N. Taylor, Rev. Mod. Physics **80**, 633 (2008)

2.15. NIST: Reference on constants, units and uncertainties. http://physics.nist.gov/cuu/

2.16a. C.C. Speake, T.M. Niebauer et al., Phys. Rev. Lett. **65**, 1967 (1990)

2.16b. C.W. Misner, K.S. Thorne, J.A. Wheeler, *Gravitation.* (Freeman, San Franscisco, 1973)

2.17. F. Stacey, G. Tuck, Phys. World **1**(12), 29 (1988)

2.18a. C.B. Braginski, V.I. Panov, Sov. Phys. JETP **34**, 464 (1971)

2.18b. https://en.wikipedia.org/wiki/Elliptic_Intergral

2.19. R. Celotta, J. Levine, Ch.G. Sammis, Th.L. Henyey (ed.), *Methods od experimental Physics,* Vol. 24 (Academic Press, San Diego, 1987)

2.20a. https://en.wikipedia.org/wiki/Geographic_information_system

2.20b. http://www.stevenswater.com/telemetry_com/leo_info.aspx

2.20c. Ch. Hwang, C.K. Shum, J. Li (ed.), *Satellite Altimetry for Geodesy, Geophysics and Oceanography.* (Springer, Berlin, Heidelberg, 2004)

Moving Coordinate Systems and Special Relativity

3

© Springer International Publishing Switzerland 2017
W. Demtröder, *Mechanics and Thermodynamics*, Undergraduate Lecture Notes in Physics, DOI 10.1007/978-3-319-27877-3_3

For the description of the location and the velocity of a body in a three-dimensional space one needs a coordinate system where the position vectors $\boldsymbol{r}(t)$ and its time derivative $\mathrm{d}\boldsymbol{r}/\mathrm{d}t = \boldsymbol{v}(t)$ are defined. Of course are all physical processes independent of the choice of the coordinate system. However, their mathematical formulation can be much simpler in a suitable coordinate system than in other systems. It is therefore essential to choose that system which allows the optimum description of a process and to find the transformation equations to change from one to another coordinate system.

For example is the coordinate system connected with the earth which moves around the sun, the best choice for the description of measurements on earth. For astronomical observations the results of such measurements must be transformed into a galactic coordinate system which has its origin in the galactic centre and moves with the rotating galaxy, in order to eliminate the complex motion of the earth relative to the galactic centre. For coordinate systems at rest these transformations impose no problems. The situation is different for systems which move against each other.

In this chapter we will discuss question which arise for transformations between moving coordinate systems when physical processes are described in different systems. It turns out that many concepts derived from daily life experience which were taken for granted, had to be revised. The mathematical framework for such revisions is the special relativity theory developed by *Albert Einstein*, which will be briefly treated in this chapter.

3.1 Relative Motion

An observer, sitting in the origin O of a coordinate system looks at two objects A and B with the coordinates $\boldsymbol{r}_\mathrm{A}$ and $\boldsymbol{r}_\mathrm{B}$ and the relative distance

$$\boldsymbol{r}_\mathrm{AB} = \boldsymbol{r}_\mathrm{A} - \boldsymbol{r}_\mathrm{B} \, , \tag{3.1}$$

which move with the velocities

$$\boldsymbol{v}_\mathrm{A} = \frac{\mathrm{d}\boldsymbol{r}_\mathrm{A}}{\mathrm{d}t} \quad \text{and} \quad \boldsymbol{v}_\mathrm{B} = \frac{\mathrm{d}\boldsymbol{r}_\mathrm{B}}{\mathrm{d}t}$$

relative to the coordinate system O (Fig. 3.1). The velocity of A relative to B is then

$$\boldsymbol{v}_\mathrm{AB} = \frac{\mathrm{d}\boldsymbol{r}_\mathrm{AB}}{\mathrm{d}t} = \boldsymbol{v}_\mathrm{A} - \boldsymbol{v}_\mathrm{B} \, , \tag{3.2a}$$

while the velocity of B relative to A

$$\boldsymbol{v}_\mathrm{BA} = \boldsymbol{v}_\mathrm{B} - \boldsymbol{v}_\mathrm{A} = -\boldsymbol{v}_\mathrm{AB} \, . \tag{3.2b}$$

This illustrates that position vector and velocity do depend on the reference system.

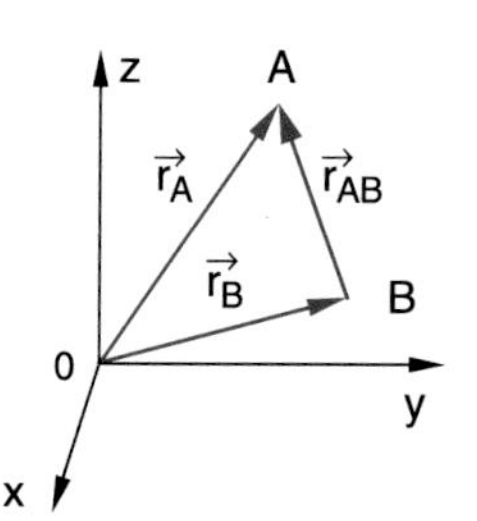

Figure 3.1 Definition of the relative distance

3.2 Inertial Systems and Galilei-Transformations

Two observers B and B′ sit in the origins O and O' of two coordinate systems $S(x, y, z)$ and $S'(x', y', z')$ which move against each other with the constant velocity $\boldsymbol{u}$ (Fig. 3.2). Both observers measure the motion of an object A. which has the position vector $\boldsymbol{r}(x, y, z)$ in the system S and $\boldsymbol{r}'(x', y', z')$ in the system S'.

As can be erived from Fig. 3.2 it is

$$\boldsymbol{r}' = \boldsymbol{r} - \boldsymbol{u} \cdot t \, , \tag{3.3}$$

which can be written for the components as

$$\left\{ \begin{aligned} x'(t) &= x(t) - u_x \cdot t \\ y'(t) &= y(t) - u_y \cdot t \\ z'(t) &= z(t) - u_z \cdot t \\ t' &= t \end{aligned} \right\} \, , \tag{3.3a}$$

where $t = t'$ means that both observers use synchronized equal clocks for their time measurements. This is **not** obvious and is generally not true if the velocity $\boldsymbol{u}$ approaches the velocity of light (see Sect. 3.4). For the velocity of A the two observers find

$$\boldsymbol{v} = \frac{\mathrm{d}\boldsymbol{r}}{\mathrm{d}t} \quad \text{and} \quad \boldsymbol{v}' = \frac{\mathrm{d}\boldsymbol{r}'}{\mathrm{d}t} \, . \tag{3.4}$$

From (3.3) follows

$$\boldsymbol{v}' = \boldsymbol{v} - \boldsymbol{u} \, . \tag{3.5}$$

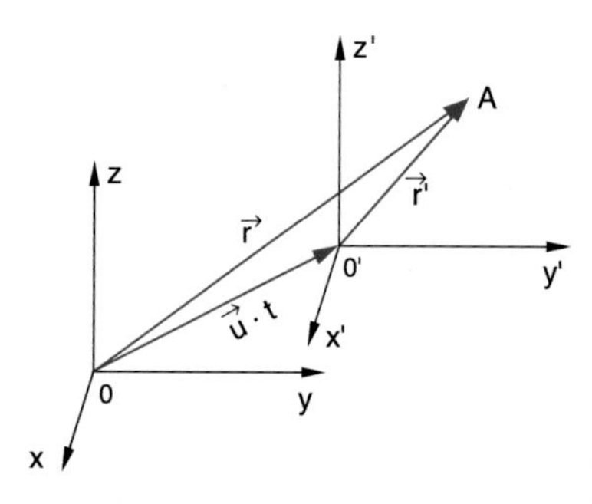

Figure 3.2 The coordinates of a point A, described in two different systems O and O' which move against each other with the constant velocity $\boldsymbol{u}$

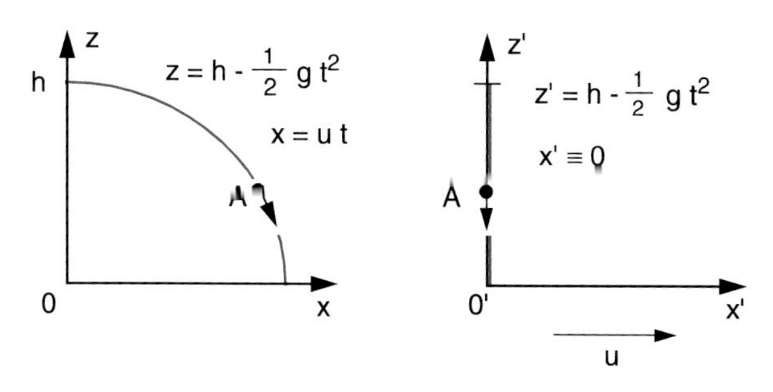

Figure 3.3 Description of the free fall in two different inertial systems

The acceleration a of A can be derived from (3.5) as

$$\boldsymbol{a}' = \frac{\mathrm{d}\boldsymbol{v}'}{\mathrm{d}t} = \frac{\mathrm{d}\boldsymbol{v}}{\mathrm{d}t} = \boldsymbol{a} \,. \tag{3.6}$$

Both observers in the systems which move with constant velocity $\boldsymbol{u}$ against each other, measure the same value for the acceleration $\boldsymbol{a}$. Because the force on a body with mass m is $\boldsymbol{F} = m \cdot \boldsymbol{a}$ both observers come to the same conclusion about the force acting on A and find the same relations for dynamical processes in the two systems.

Such systems which move with a constant relative velocity $\boldsymbol{u}$ against each other are named **inertial systems**.

Between the quantities $\boldsymbol{r}$, $\boldsymbol{v}$ and t for the motion of an object A measured in two different inertial systems the Galilei transformations pertain

$$\begin{aligned} \boldsymbol{r} &= \boldsymbol{r}' + \boldsymbol{u}t \,, \\ \boldsymbol{v} &= \boldsymbol{v}' + \boldsymbol{u} \Rightarrow \boldsymbol{a} = \boldsymbol{a}' \quad \text{and} \quad \boldsymbol{F} = \boldsymbol{F}' \,, \\ t &= t' \,, \end{aligned} \tag{3.7}$$

where $u = |\boldsymbol{u}| \ll c$ is the constant velocity of S against S'.

Because of $\boldsymbol{F} = \boldsymbol{F}'$ both observers measure the same forces and derive identical physical laws. This can be illustrated by the example of the free fall observed in the two systems S and S' moving with the velocity $u = u_x$ in the x-direction against each other (Fig. 3.3):

A body A which is released at the heights $z = h$ falls down in the system S' along the z'-axis ($x' = y' = 0$), which moves with the velocity u against the z-axis in the system S. For the observer O' in S' the motion of A appears as vertical free fall. For the observer O in S the body A starts at $z = z' = h$ with the velocity $v(h) = u$ in the x-direction, which bends down into the $-z$-direction because of the gravitation. The trajectory of A is for O a parabola (horizontal throw see Sect. 2.3.2). However, both observes measure the same fall acceleration $\boldsymbol{g} = \{0, 0, -g\}$ and the same fall times. They derive the same law for the free fall.

All inertial systems are equivalent for the description of physical laws.

In other words: An observer siting in a train who does not look out through the window cannot decide by arbitrary many experiments whether he sits in a train at rest or in a train moving against another reference system with constant velocity.

3.3 Accelerated Systems; Inertial Forces

If the two observers sit in two systems which move against each other with a velocity $\boldsymbol{u}(t)$ changing with time resulting in an acceleration $\boldsymbol{a} = \mathrm{d}\boldsymbol{v}/\mathrm{d}t$ they measure for the motion of a body A relative to their system different accelerations and therefore conclude that different forces act on A.

The observer in an accelerated system can, however, ascertain that his system moves accelerated against another system. If he takes into account this acceleration he comes to the same conclusions about physical laws for the observed motion of the body A as an observer in an inertial system.

We will discuss this for two different accelerated motions:

a) rectilinear motion of S against S' with constant acceleration
b) rotation of S against S' around the common origin $0 = 0'$.

Remark. In the following sections we will always assume that the observers O and O' sit in the origins 0 and 0′ of the systems S and S'.

The discussion of the description of physical processes in accelerated coordinate systems leads to the introduction of special forces (inertial forces), which are often confusing students. Therefore these forces will be discussed as vivid as possible in order to illustrate that these forces are no real forces but are only necessary, when the observer in the accelerated system does not take into account the acceleration of his system.

3.3.1 Rectilinear Accelerated Systems

If the origin 0′ of the system S' moves along the x-axis of S with the time dependent velocity $u(t) = u_0 + a \cdot t (\boldsymbol{a} = a_x \boldsymbol{e}_x$ with $a_x = \mathrm{d}u/\mathrm{d}t = \mathrm{d}^2x/\mathrm{d}t^2)$ against S, only the magnitude of the velocity changes not its direction (Fig. 3.4). An example is an observer in a train accelerating on a straight track.

For a body A with the coordinates (x', y', z') in the system S' the observer in S measures the coordinates $x = u_0 t + \frac{1}{2}at^2 + x'$, $y = y'$, $z = z'$, if for $t = 0$ the two origins of S and S' coincide and the relative velocity $u(t)$ between S and S' at time $t = 0$ is u_0. The velocity of A is then $v' = \{v'_x, v'_y, v'_z\}$ for O' and $v = \{v_x = u_0 + a \cdot t + v'_x, v_y = v'_y, v_z = v'_z\}$ for O.

The description of different situations by O (sitting in a system S at rest) and O' (sitting in the accelerated System S') shall be illustrated by three examples. **Note** that S' is no inertial system!

Chapter 3

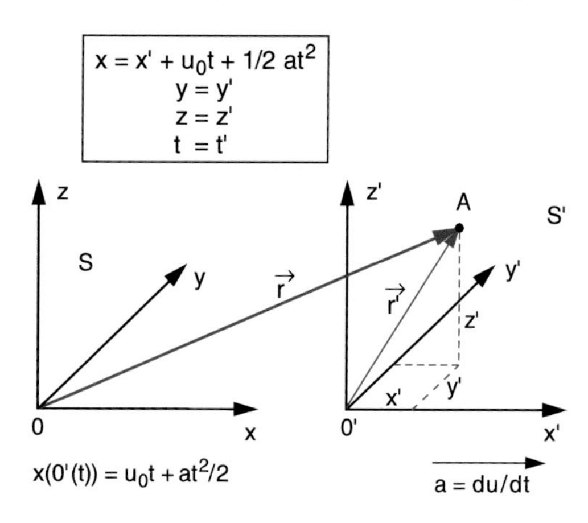

Figure 3.4 Coordinates of point A in a system S with origin O and a system S' with origin O', that moves against O with the acceleration a in x-direction

Examples

1. The observer O' is sitting on a carriage at a fixed table with plain tabletop. On the tabletop rests a ball A without friction (Fig. 3.5a). If the system S' is accelerated to the left (i. e. in $-x$-direction), both observers O (in the system S at rest) and O' see that the ball moves accelerated towards O'. Both O and O' make the same observation but interpret this in a different way:
   O' says: The ball moves accelerated towards me. Therefore a force $\boldsymbol{F} = m \cdot \boldsymbol{a}$ must act on the ball.
   O says: The system S' moves with the acceleration $-\boldsymbol{a}$ to the left, while the ball does not participate in the acceleration and stays at rest. This means: Not the ball is accelerated towards O', but O' is accelerated towards the resting ball. Therefore no force is acting on A.

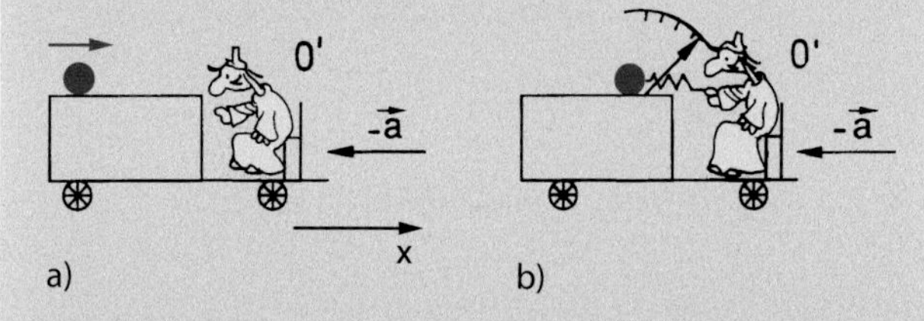

Figure 3.5 **a** A freely movable ball; **b** a ball fixed to a spring balance, both on a table that is accelerated into the $-x$-direction with constant acceleration a

Note: If O' knows that his system S' is accelerated, he also knows that the ball must stay at rest, because it is frictionless and therefore not linked with the table, which means that it will not participate in the motion of the table. In order to explain his observation of the acceleration $\boldsymbol{a}$ of the ball he introduces a force $\boldsymbol{F} = m \cdot \boldsymbol{a}$ which he calls *fictitious force* (often named *pseudo-force*), because he knows that this is not a real force but merely the description of a virtual acceleration $\boldsymbol{a}$ of the ball when its motion is described in a reference system which itself is accelerated with the acceleration $-\boldsymbol{a}$. Often the notation "*inertial force*" is used in order to point to the *inertial mass* of the ball which prevents it to follow the acceleration of the table.

2. The observer O' connects the ball with an elastic spring scale and holds the other end with his hand (Fig. 3.5b). If the system S' is now accelerated with the acceleration $-\boldsymbol{a}$ to the left O' observes that the spring is compressed. The spring balance measures the force $\boldsymbol{F}_1 = -m \cdot \boldsymbol{a}$. He must apply an equal but opposite force $\boldsymbol{F}_2 = +m \cdot \boldsymbol{a}$ in order to keep the ball at rest.
   O' says: The total force $\boldsymbol{F} = \boldsymbol{F}_1 + \boldsymbol{F}_2$ acting on the ball is zero in accordance with my observation that the ball rests.
   The observer O in the rest system S says: Since the ball is now connected with the table in S' it participates in the acceleration $-\boldsymbol{a}$ of S'. The observer O' has to apply the force $\boldsymbol{F} = -m \cdot \boldsymbol{a}$ in order to transfer the same acceleration $-\boldsymbol{a}$ to the ball as the system S' and to keep the ball at rest relative to the system S'.
3. A mass m in an elevator is suspended by a spring balance (Fig. 3.6). If the elevator moves with the acceleration $\boldsymbol{a} = \{0, 0, -a\}$ downwards (Fig. 3.6a) the spring balance measures the force $\boldsymbol{F} = m(\boldsymbol{g} - \boldsymbol{a})$, if the elevator moves upwards with the acceleration $+a$ the balance measures $\boldsymbol{F} = m(\boldsymbol{g} + \boldsymbol{a})$ where $\boldsymbol{g} = \{0, 0, -g\}$ is the earth acceleration. The observer O', sitting in the elevator, says: The body is at rest. Therefore the total force acting on it must be zero. The total force $F = F_1 + F_2 + F_3$ (Fig. 3.6c) is the sum of

$$\begin{aligned} \boldsymbol{F}_1 &= m \cdot \boldsymbol{g} && = \text{the weight of the mass } m \\ \boldsymbol{F}_2 &= -m(\boldsymbol{g} - \boldsymbol{a}) && = \begin{array}{l}\text{opposite force of} \\ \text{the spring balance}\end{array} \\ \boldsymbol{F}_3 &= -m \cdot \boldsymbol{a} && = \text{inertial force} \end{aligned} .$$

   O' must introduce the inertial force F_3 in order to explain his observation.
   The observer O outside the elevator at rest says: The body with mass m is connected with the elevator. It therefore participates in the acceleration of the elevator. This demands the force $\boldsymbol{F} = m \cdot \boldsymbol{a}$. The total force acting on the body is the sum of its weight $\boldsymbol{F}_1 = m \cdot \boldsymbol{g}$ and the restoring force $\boldsymbol{F}_2 = -m \cdot (\boldsymbol{g} - \boldsymbol{a})$ of the spring balance. Which gives, as expected the total force $\boldsymbol{F} = m \cdot \boldsymbol{g} - m \cdot (\boldsymbol{g} - \boldsymbol{a}) = m \cdot \boldsymbol{a}$.
   If the suspension cable of the elevator is ruptured and the elevator goes down in a free fall its acceleration is $a = g$. For O' the total force remains $\sum F_i = 0$ while for O the total force becomes $F = m \cdot g$.
   These examples illustrate, that the inertial forces are introduced only for measurements in accelerated coordinate systems if the acceleration of the system is not taken into account. They are therefore also called *fictitious forces* or *pseudo-forces*. A transformation to an inertial system lets all pseudo-forces vanish. This

means an observer O in an inertial system does not need any pseudo-force for the explanation of the observed physical processes.

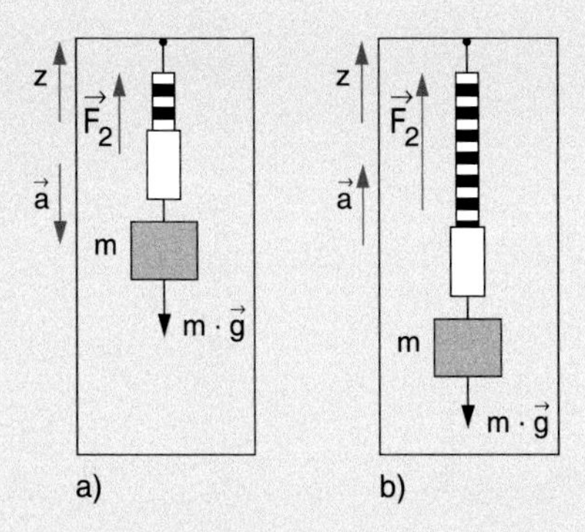

O'		O	
$\vec{F}_1 = m \cdot \vec{g}$	$\vec{F}_2 = -m(\vec{g} - \vec{a})$ $\vec{F}_3 = -m \cdot \vec{a}$	$\vec{F}_1 = m \cdot \vec{g}$	$\vec{F}_2 = -m(\vec{g} - \vec{a})$
$\Sigma \vec{F}_i = 0$		$\Sigma \vec{F}_i = m \cdot \vec{a}$	

c)

Figure 3.6 Elevator experiment. Description of the forces acting on a mass m, that hangs on a spring balance in an elevator accelerated downwards in **a** and upwards in **b**. In **c** the forces are listed as observed by O' in the elevator (*left hand side*) and by O at rest outside the elevator (*right hand side*) ◀

3.3.2 Rotating Systems

We regard two coordinate systems $S(x, y, z)$ and $S'(x', y', z')$ with the unit vectors $\hat{\boldsymbol{e}}_x$, $\hat{\boldsymbol{e}}_y$, $\hat{\boldsymbol{e}}_z$ and $\hat{\boldsymbol{e}}_{x'}$, $\hat{\boldsymbol{e}}_{y'}$, $\hat{\boldsymbol{e}}_{z'}$ of the coordinate axes and a common origin $0 = 0'$. S' rotates against S with the constant angular velocity $\boldsymbol{\omega} = \{\omega_x, \omega_y, \omega_z\}$ around $0 = 0'$ (Fig. 3.7). S' is therefore no inertial system. We assume that for all times $0 = 0'$.

A point A should have at time t in the system S the position vector

$$\boldsymbol{r}(t) = x(t) \cdot \hat{\boldsymbol{e}}_x + y(t) \cdot \hat{\boldsymbol{e}}_y + z(t) \cdot \hat{\boldsymbol{e}}_z \tag{3.8}$$

and the velocity

$$\boldsymbol{v}(t) = \frac{\mathrm{d}x}{\mathrm{d}t}\hat{\boldsymbol{e}}_x + \frac{\mathrm{d}y}{\mathrm{d}t}\hat{\boldsymbol{e}}_y + \frac{\mathrm{d}z}{\mathrm{d}t}\hat{\boldsymbol{e}}_z \,. \tag{3.9}$$

The same point A has in the system S' the position vector

$$\boldsymbol{r}'(t) = \boldsymbol{r}(t) = x'\hat{\boldsymbol{e}}_{x'} + y'\hat{\boldsymbol{e}}_{y'} + z'\hat{\boldsymbol{e}}_{z'} \,. \tag{3.10}$$

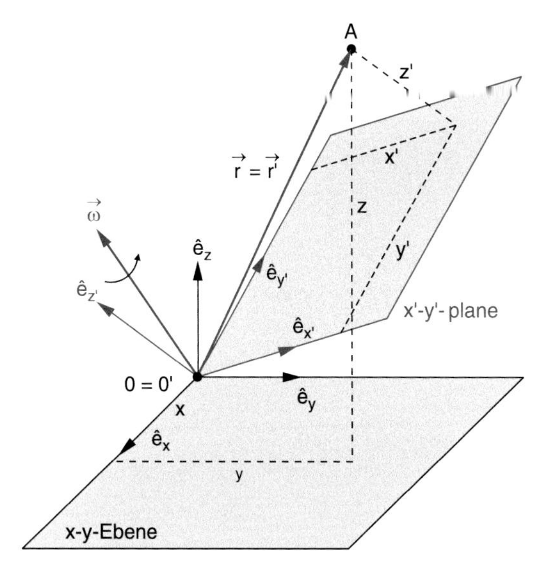

Figure 3.7 A system S', that rotates around the axis $\boldsymbol{\omega}$ against S. Both systems have the same origin $O = O'$

Note: $\boldsymbol{r} = \boldsymbol{r}'$ means that we regard the same vector in both systems with the same magnitude but different components.

If the observer O' does not take into account that his system rotates, he will define as the velocity of A in his system

$$\boldsymbol{v}'(t) = \frac{\mathrm{d}\boldsymbol{r}'}{\mathrm{d}t} = \frac{\mathrm{d}x'}{\mathrm{d}t}\hat{\boldsymbol{e}}_{x'} + \frac{\mathrm{d}y'}{\mathrm{d}t}\hat{\boldsymbol{e}}_{y'} + \frac{\mathrm{d}z'}{\mathrm{d}t}\hat{\boldsymbol{e}}_{z'} \,. \tag{3.11}$$

However, if the observer O in the inertial system S describes the velocity of A in the coordinates of S'. he knows that the axis of S' are rotating and therefore not constant in time. He therefore must write:

$$\begin{aligned} \boldsymbol{v}(x', y', z') &= \left(\frac{\mathrm{d}x'}{\mathrm{d}t}\hat{\boldsymbol{e}}_{x'} + \frac{\mathrm{d}y'}{\mathrm{d}t}\hat{\boldsymbol{e}}_{y'} + \frac{\mathrm{d}z'}{\mathrm{d}t}\hat{\boldsymbol{e}}_{z'}\right) \\ &\quad + \left(x'\frac{\mathrm{d}\hat{\boldsymbol{e}}_{x'}}{\mathrm{d}t} + y'\frac{\mathrm{d}\hat{\boldsymbol{e}}_{y'}}{\mathrm{d}t} + z'\frac{\mathrm{d}\hat{\boldsymbol{e}}_{z'}}{\mathrm{d}t}\right) \\ &= \boldsymbol{v}' + \boldsymbol{u} \,. \end{aligned} \tag{3.12}$$

The endpoints of the unit vectors $\hat{\boldsymbol{e}}_{x'}$, $\hat{\boldsymbol{e}}_{y'}$, $\hat{\boldsymbol{e}}_{z'}$ perform a circular motion with the angular velocity ω around $0 = 0'$. Their velocity is then

$$\frac{\mathrm{d}\hat{\boldsymbol{e}}_{x'}}{\mathrm{d}t} = \boldsymbol{\omega} \times \hat{\boldsymbol{e}}_{x'}; \quad \frac{\mathrm{d}\hat{\boldsymbol{e}}_{y'}}{\mathrm{d}t} = \boldsymbol{\omega} \times \hat{\boldsymbol{e}}_{y'}; \quad \frac{\mathrm{d}\hat{\boldsymbol{e}}_{z'}}{\mathrm{d}t} = \boldsymbol{\omega} \times \hat{\boldsymbol{e}}'_z \,. \tag{3.13}$$

Inserting this into (3.12) the second term in (3.12) becomes

$$\begin{aligned} \boldsymbol{u} &= (\boldsymbol{\omega} \times \hat{\boldsymbol{e}}_{x'})x' + (\boldsymbol{\omega} \times \hat{\boldsymbol{e}}_{y'})y' + (\boldsymbol{\omega} \times \hat{\boldsymbol{e}}_{z'})z' \\ &= \boldsymbol{\omega} \times (\hat{\boldsymbol{e}}_{x'}x' + \hat{\boldsymbol{e}}_{y'}y' + \hat{\boldsymbol{e}}_{z'}z') \\ &= \boldsymbol{\omega} \times \boldsymbol{r}' = \boldsymbol{\omega} \times \boldsymbol{r}, \quad \text{because } \boldsymbol{r} \equiv \boldsymbol{r}' \,. \end{aligned}$$

We therefore get the transformation between the velocity v of the point A measured by O in the system S and the velocity v' measured by O' in the system S'

$$\boldsymbol{v} = \boldsymbol{v}' + (\boldsymbol{\omega} \times \boldsymbol{r}) \,. \tag{3.14}$$

Note: v' is the velocity measured by O', if he does not take into account, that his system S' rotates with the angular velocity $\boldsymbol{\omega}$, while $\boldsymbol{v}$ in (3.9) is the velocity in the resting system S and $\boldsymbol{v}$ in (3.14) the velocity of A measured by O but expressed in the coordinates of the rotating system S'.

The acceleration a can be obtained by differentiating (3.14). The result is

$$a = \frac{\mathrm{d}\boldsymbol{v}}{\mathrm{d}t} = \frac{\mathrm{d}\boldsymbol{v}'}{\mathrm{d}t} + \left(\boldsymbol{\omega} \times \frac{\mathrm{d}\boldsymbol{r}}{\mathrm{d}t}\right) , \qquad (3.15)$$

because we have assumed that $\omega = \text{const}$. The observer O' gets the result for a, expressed in the coordinates of his system S':

$$\begin{aligned} \frac{\mathrm{d}\boldsymbol{v}'}{\mathrm{d}t} &= \left(\hat{\boldsymbol{e}}_{x'}\frac{\mathrm{d}v'_x}{\mathrm{d}t} + \hat{\boldsymbol{e}}_{y'}\frac{\mathrm{d}v'_y}{\mathrm{d}t} + \hat{\boldsymbol{e}}_{z'}\frac{\mathrm{d}v'_z}{\mathrm{d}t}\right) \\ &+ \left(\frac{\mathrm{d}\hat{\boldsymbol{e}}_{x'}}{\mathrm{d}t}v'_x + \frac{\mathrm{d}\hat{\boldsymbol{e}}_{y'}}{\mathrm{d}t}v'_y + \frac{\mathrm{d}\hat{\boldsymbol{e}}_{z'}}{\mathrm{d}t}v'_z\right) \qquad (3.16) \\ &= \boldsymbol{a}' + (\boldsymbol{\omega} \times \boldsymbol{v}') , \end{aligned}$$

where a' is again the acceleration of A measured by O' in the system S'. We therefore obtain with (3.15)

$$a = \frac{\mathrm{d}\boldsymbol{v}}{\mathrm{d}t} = \boldsymbol{a}' + (\boldsymbol{\omega} \times \boldsymbol{v}') + (\boldsymbol{\omega} \times \boldsymbol{v}) .$$

Inserting for v the expression (3.14) we finally obtain from (3.15)

$$\boldsymbol{a} = \boldsymbol{a}' + 2(\boldsymbol{\omega} \times \boldsymbol{v}') + \boldsymbol{\omega} \times (\boldsymbol{\omega} \times \boldsymbol{r}) , \qquad (3.17)$$

and for $\boldsymbol{a}'$

$$\begin{aligned} \boldsymbol{a}' &= \boldsymbol{a} + 2(\boldsymbol{v}' \times \boldsymbol{\omega}) + \boldsymbol{\omega} \times (\boldsymbol{r} \times \boldsymbol{\omega}) \\ &= \boldsymbol{a} + \boldsymbol{a}_\mathrm{C} + \boldsymbol{a}_\mathrm{cf} . \end{aligned} \qquad (3.18)$$

While the observer in his resting system S measures the acceleration $\boldsymbol{a} = \mathrm{d}\boldsymbol{v}/\mathrm{d}t$, the observer O' in his rotating system S' has to add additional terms for the acceleration in order to describe **the same motion of A**. These are

the Coriolis-acceleration

$$\boldsymbol{a}_\mathrm{C} = 2(\boldsymbol{v}' \times \boldsymbol{\omega}) , \qquad (3.19a)$$

the centrifugal acceleration

$$\boldsymbol{a}_\mathrm{cf} = \boldsymbol{\omega} \times (\boldsymbol{r} \times \boldsymbol{\omega}) . \qquad (3.20a)$$

Special Cases: If the point A moves parallel to the rotation axis we have $\boldsymbol{v} \parallel \boldsymbol{\omega}$ and therefore the Coriolis acceleration becomes $\boldsymbol{a}_\mathrm{C} = \boldsymbol{0}$. The Coriolis acceleration appears only, if $\boldsymbol{v}'$ has a component perpendicular to $\boldsymbol{\omega}$. When we choose the z-axis as the direction of $\boldsymbol{\omega}$ (Fig. 3.8), both the Coriolis acceleration a_C and the centrifugal acceleration a_cf lie in the x-y-plane. The centrifugal acceleration points outwards in the radial direction. The direction of the Coriolis acceleration depends on the direction of the velocity v' in the coordinate system (x', y', z'). Since the v'_z-component does not contribute to a_C only the projection $v_\perp = \{v'_x, v'_y\}$ is responsible for the determination of the vector

$$\boldsymbol{a}_\mathrm{C} = \boldsymbol{\omega} \cdot \{v'_y, -v'_x, 0\} .$$

The vector $\boldsymbol{a}_\mathrm{C}$ is perpendicular to $\boldsymbol{v}_\perp$ as can be immediately seen when forming the scalar product $\boldsymbol{a}_\mathrm{C} \cdot \boldsymbol{v}'_\perp$.

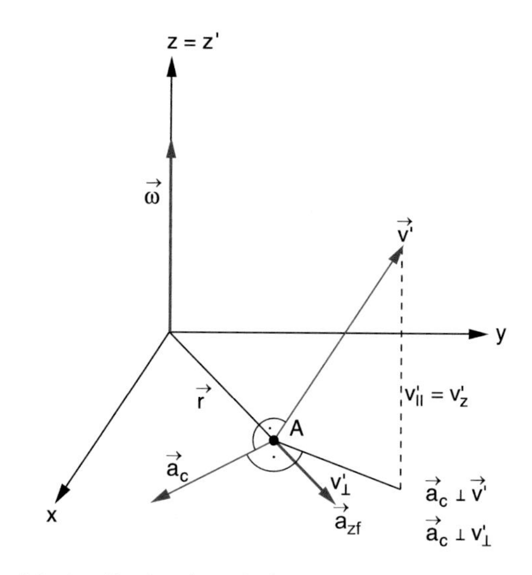

Figure 3.8 Centrifugal- and Coriolis-force acting on a mass m in $A(x, y, z = 0)$ described in a system S', that rotates with constant angular velocity ω around the z-axis

3.3.3 Centrifugal- and Coriolis-Forces

According to Newton's laws accelerations are caused by forces. Therefore the observer O', who measures in his rotating system S' additional accelerations has to introduce additional forces based on the equation $\boldsymbol{F} = m \cdot \boldsymbol{a}$ These are the Coriolis force

$$\boldsymbol{F}_\mathrm{C} = 2\,\mathrm{m} \cdot (\boldsymbol{v}' \times \boldsymbol{\omega}) , \qquad (3.19b)$$

and the centrifugal force

$$\boldsymbol{F}_\mathrm{cf} = m \cdot \boldsymbol{\omega} \times (\boldsymbol{r} \times \boldsymbol{\omega}) . \qquad (3.20b)$$

Both forces are *inertial forces* or *virtual forces* because they are not real forces due to interactions between bodies. They have only to be introduced if the rotation of the coordinate axes of the rotating system S' are not taken into account. If the same motion of the body A are described in an inertial system S or in the rotating system S' where the rotation of the coordinate axes in considered, these forces do not appear.

We will illustrate these important facts by some examples.

Examples

1. A mass m is attached to one end of a string with length L while the other end is connected to the end of a bar with length d which rotates with the angular velocity ω around a vertical axis fixed to the centre of a rotating table (Fig. 3.9). In the equilibrium position the string forms an angle α against the vertical direction, where α depends on ω, d and L. The observer O in the resting frame S and the observer O' sitting on the rotating table describe their observations as follows:

O says: Since m moves with constant angular velocity ω on a circle with radius $r = d + L \cdot \sin\alpha$ a centripetal force $\boldsymbol{F}_{cp} = -m \cdot \omega^2 \cdot \boldsymbol{r}$ acts on m which is the vector sum of its weight $m \cdot g$ and the restoring force F_r of the string (Fig. 3.9a).

O' says: Since m is resting in my system S' the total force on m must be zero, i. e. $\sum F_i = 0$. The vector sum $m \cdot g + F_r$ has to be compensated by the centrifugal force $\boldsymbol{F}_{cf} = +m\omega^2 \boldsymbol{r}$ (Fig. 3.9b). He has to introduce the virtual force F_{cf} if he does not take into account the rotation of his system.

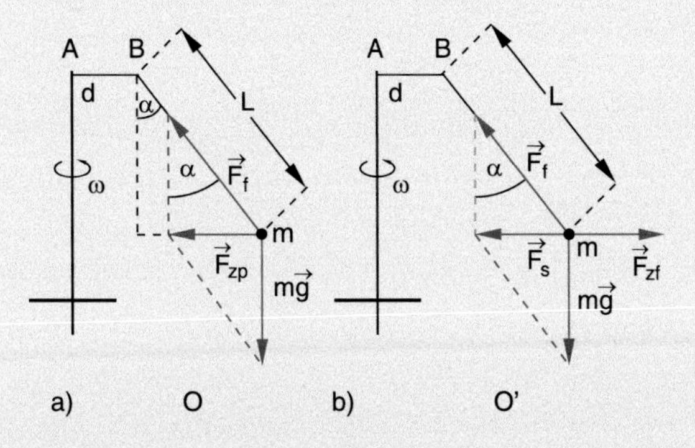

Figure 3.9 Forces on a rotating string pendulum, described by the observer O at rest and O' rotating with the pendulum

2. In a satellite, circling around the earth with constant angular velocity ω experiments are performed concerning the "weightlessness" (Fig. 3.10). For example an astronaut can freely float in his satellite without touching the walls.
The observer O' in the satellite (i. e. the astronaut) says: I know that the gravity force

$$\boldsymbol{F}_g = -(GmM/r^2)\hat{\boldsymbol{e}}_r$$

acts on me, where r is the distance to the centre of the earth. It is compensated by the opposite centrifugal force

$$\boldsymbol{F}_{cf} = +m\omega^2 \cdot r \cdot \hat{\boldsymbol{e}}_r \,.$$

The total force acting on me is zero and therefore I can freely float.

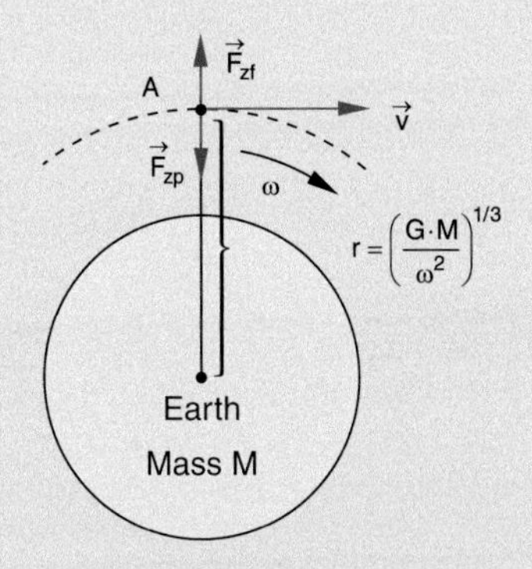

Figure 3.10 Force-free conditions in a satellite orbiting around the earth

Note: The state of the astronaut should be better called "force-free" instead of "weightlessness".
The observe O in a resting system S (for example the galactic coordinate system) says: The gravity force F_g acts as centripetal force on both the satellite and the astronaut. Both are therefore forced to move on a circle around the earth. The acceleration $\boldsymbol{a} = -(GM/r^2)\hat{\boldsymbol{e}}_r$ is the same for the astronaut and the satellite and the difference of the accelerations is zero. Therefore the astronaut can freely float in his satellite.
Note: Both observes can describe consistently the situation of the astronaut, however the observer S' has to introduce the inertial force F_{cf} if he does not take into account the accelerated motion of his space ship.

3. A sled moves with constant velocity v on a linear track and writes with a pen on a rotating disc (Fig. 3.11). The marked line on the rotating disc is curved where the curvature depends on the velocity v of the sled, the perpendicular distance d of the track from the centre of the disc and the angular velocity ω of the rotating disc. The two observers O and O' describe the observed curve as follows:
O says: The sled moves with constant velocity on a straight line, as can be seen from the marked line outside the disc. Therefore no force is acting on the sled and its acceleration is zero. The curved path marked on the disc is due to the fact that the disc is rotating.
O' says: I observe a curved path. Therefore a force has to act on the sled. By experiments with different values of v, ω and d he finds:
For $d = 0$ is $|\boldsymbol{a}'| \propto v' \cdot \omega$; $\boldsymbol{a} \perp \boldsymbol{v}'$ and $\boldsymbol{a} \perp \omega$.
For $d \neq 0$ is $a = c_1\omega + c_2\omega^2$ with $c_1 \propto v$ and $c_2 \propto r$, where r is the distance of the sled from the centre of the disc. The quantitative analysis of his measurements gives the result:

$$\boldsymbol{a}' = 2(\boldsymbol{v}' \times \boldsymbol{\omega}) + \boldsymbol{\omega} \times (\boldsymbol{r} \times \boldsymbol{\omega}) \,,$$

which is consistent with (3.18) and shows that the acceleration of the sled measured by O' is the sum of centrifugal and Coriolis accelerations.

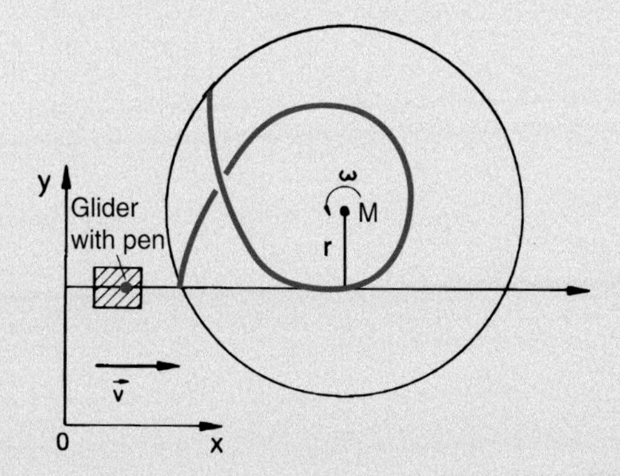

Figure 3.11 Experimental illustration of the inertial forces. A glider, moving on a straight line above a rotating disc writes with a pen its path on the rotating disc which appears as a curved trajectory

This example illustrates clearly that the two accelerations and the corresponding forces are only virtual, because the sled moves in fact with constant velocity on a straight line and therefore experiences no real forces.

4. A hollow sphere filled with sand hangs on a string which is connected to a fixed suspension point and swings in the fixed x-z-plane of a resting system S, driven by the gravity force $\boldsymbol{F}_g = m \cdot \boldsymbol{g}$ with $\boldsymbol{g} = \{0, 0, -g\}$.
   Below the swinging pendulum is a rotating table in the x-y-plane which rotates with the angular velocity ω around a vertical axis through the minimum position of the pendulum.
   If the sand flows through a small hole in the hollow sphere it draws for $\omega = 0$ a straight line on the table while for $\omega \neq 0$ a rosette-like figure is drawn (Fig. 3.12) with a curvature which depends on the ratio of oscillation period T_1 of the pendulum to rotation period T_2 of the rotating table.
   The two observers give the following explanations:
   O says: The x-z-oscillation plane remains constant because the driving force $F_g = m \cdot g \cdot \sin\alpha$ (see Sect. 2.9.7) lies always in the x-z-plane and therefore the motion must stay in this plane. The projection of the trajectory onto the x-y-plane should be a straight line. The curved trajectory drawn on the rotating table is caused by the rotation and not by an additional force.
   O' says sitting on the rotating table: I see a curved path which must be caused by forces, which depend on ω, v' and r. Its form can be explained by the centrifugal and the Coriolis forces. My careful measurements prove that the paths is due to the action of the total acceleration $a' = a_{\mathrm{cf}} + a_{\mathrm{C}}$ in accordance with Eq. 3.18.

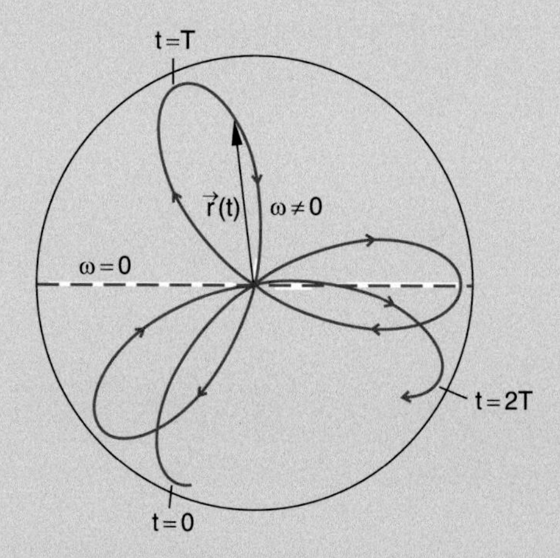

Figure 3.12 Apparent trajectory written on a rotating disc by a sand pendulum than oscillates in a constant plane

5. Foucault pendulum. Since our earth is a rotating system, the path drawn by a linearly swinging pendulum onto the ground must show curved lines as discussed in example 4). However, because of the slow earth rotation ($\omega = 7.3 \cdot 10^{-5}\,\mathrm{s}^{-1}$) the curvature is very small. Using a pendulum with a large length L and a corresponding large oscillation period T the rotation of the earth under the linearly swinging pendulum could be first demonstrated 1850 by Leon Foucault (1819–1868) who used a copper ball ($m = 28\,\mathrm{kg}$) suspended by a 67 m long string ($T = 16.4\,\mathrm{s}$). The turn of the oscillation plane against the rotating ground occurs with the angular velocity $\omega_s = \omega \cdot \sin\varphi$ where φ is the geographic latitude of the pendulum (Fig. 3.13). in Kaiserslautern with $\varphi = 49°$ the pendulum plane turns in 1 h by 11°32′, which can be readily measured. Using shadow projection of the pendulum string defining the oscillation plane this turn can be quantitatively measured within a physics lecture.

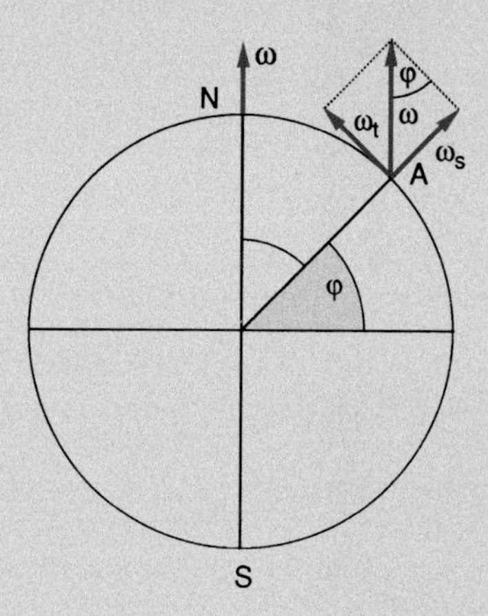

Figure 3.13 Explanation of the turning plane of oscillation of the Foucault-pendulums on the surface the rotating earth

6. An impressive demonstration of the Coriolis force is provided by the motion of cloud formations around a low pressure region as for instance realized by tornados or typhoons (Fig. 3.14). For an observer on the rotating earth looking from above onto the ground the wind does not blow radially into the low pressure region but rotates on the northern hemisphere anticlockwise around it, on the southern hemisphere clockwise. Around a high pressure region the rotation is clockwise on the northern hemisphere and anticlockwise on the southern.

   Note: If a small balloon which floats in the air is used as indicator of the wind flow an observer on earth would see the balloon moving on one of the lines in Fig. 3.14. An observer O at a fixed position outside the earth, would however see, that the balloon moves on a straight line radially into the centre of the deep pressure region or out from the centre of a high pressure region. These centres are fixed at a point on earth and rotate with the earth.

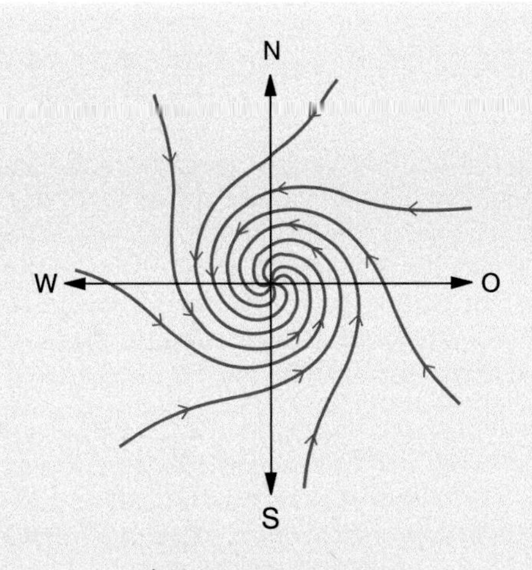

a) Northern hemisphere

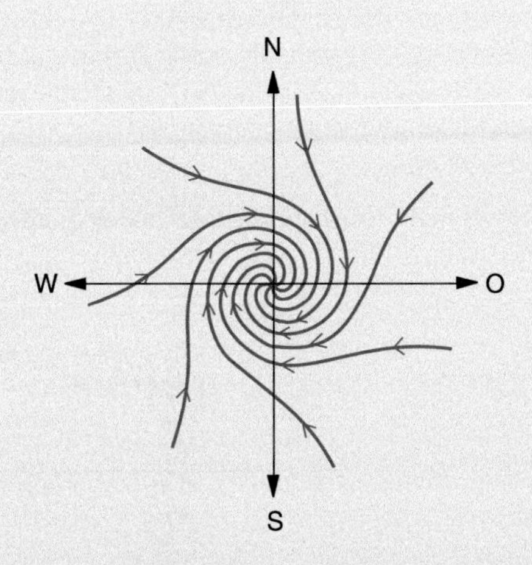

b) Southern hemisphere

Figure 3.14 Stream-lines of the air around a deep-pressure region. **a** on the northern hemisphere; **b** on the southern hemisphere. On the northern hemisphere the Coriolis force acts (seen in the wind direction from above) in the right direction against the radial force of the pressure gradient, on the southern hemisphere in the left direction. **c** Satellite photo of the "death-hurricane" north of Hawaii (with kind permission of NASA photo HP 133)

3.3.4 Summary

Inertial forces (virtual forces) have to be introduced, if the motion of bodies are described in accelerated coordinate systems. These forces are not caused by real interactions between bodies but only reflect the acceleration of the coordinate system. They do not appear if the same motion is described in an inertial system.

In rotating systems with a fixed centre the inertial forces are centrifugal and Coriolis forces. In systems with arbitrarily changing velocities further inertial forces have to be introduced.

3.4 The Constancy of the Velocity of Light

We consider a body A which has the velocity $\boldsymbol{v}$ measured in the system S but the velocity $\boldsymbol{v}'$ in a system S', which moves itself with the velocity $\boldsymbol{u}$ against the resting system S. According to the Galilei transformations the different velocities are related through the vector sum (Fig. 3.15)

$$\boldsymbol{v} = \boldsymbol{v}' + \boldsymbol{u} \,. \tag{3.21a}$$

Therefore one might suggest, that also the velocity of light, emitted from a light source which is fixed in a system S' moving with the velocity $\boldsymbol{u}$ against the system S, should be measured in the system S as the vector sum

$$\boldsymbol{c} = \boldsymbol{c}' + \boldsymbol{u} \,, \tag{3.21b}$$

where c' is the velocity measured by O' in his system S'. This means that the observer O should measure the velocity $\boldsymbol{c}_1 = c' + u$ if c' and u have the same direction, and $c_2 = c' - u$ if they have opposite directions.

Very careful measurements performed 1881 by Albert Abraham Michelson and Edward Morley [3.2a, 3.3] and later on by many other researchers [3.4a, 3.4b] produced evidence that the velocity of light is independent of the relative velocity u between source and observer. For example measurements of the velocity of light from a star at different times of the year always

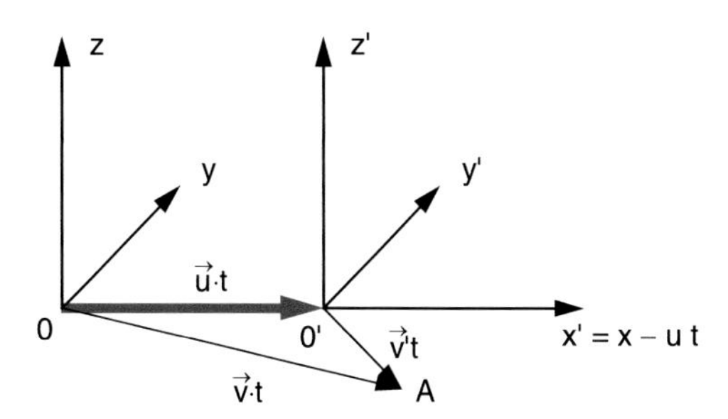

Figure 3.15 Galilei transformations of velocities in two inertial systems

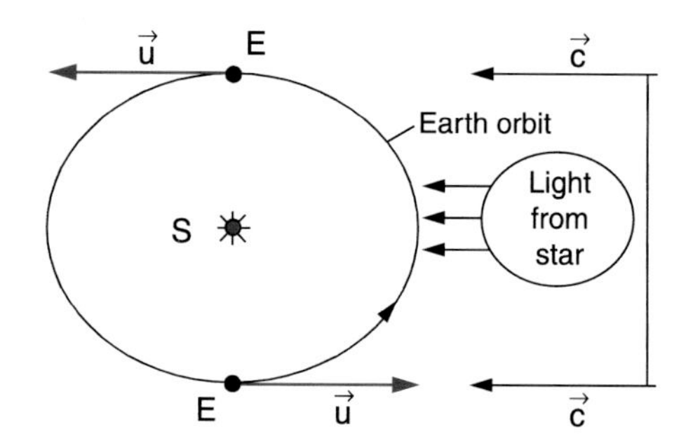

Figure 3.16 Experimental possibility to prove the constant velocity of light, by measuring the velocity of light from a distant star at two different days with a time interval of half a year, when the earth on its way around the sun moves towards the star and away from it

brought the same result although the earth moved with a velocity of 30 km/s at one time of the year against the star and half a year later away from the star (Fig. 3.16). This result was very surprising and brought about many discussions but induced the formulation of the theory of special relativity by Albert Einstein.

According to these unambiguous experimental results we must conclude:

The velocity of light is the same in all inertial systems, independent of their velocity against the light source.

The Galilei-transformations (3.7) which appear very plausible apparently fail for very large velocities. It turns out, that in particular the assumption $t = t'$ in Eq. 3.3a needs a critical revision. It must be precisely defined what "simultaneity" means for two events at different locations. The question is: How does one measure the times of two events at different locations?

To illustrate this point we regard in Fig. 3.17 two systems S and S' where light pulses are emitted from the points A and B in the system S and from A′ and B′ in the system S'. If the two systems do not move against each other (Fig. 3.17a) the situation is clear: The observers O and O' measure the arrival time of the two light pulses in O resp. O' and can decide, whether the pulses had been sent from A and B resp. from A′ and B′ simultaneously or at different times. For the first case they arrive in O or O' simultaneously. Both observers come to the same result.

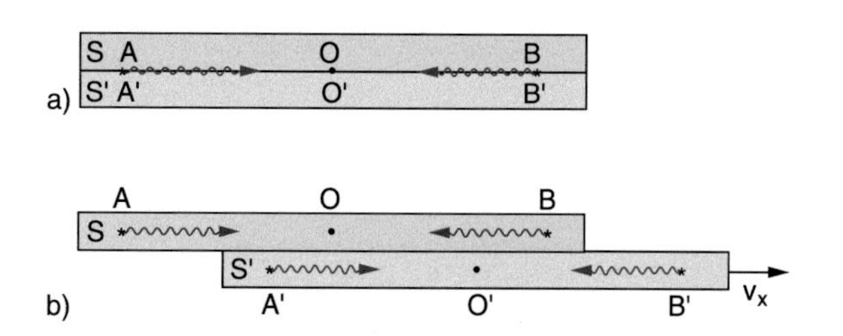

Figure 3.17 Illustration of the problem of measurements of simultaneous events in A and B resp. in A′ and B′ in two different systems: **a** which are at rest, **b** which move against each other

The situation is more difficult, if S' moves with the velocity v_x against S (Fig. 3.17b). We assume, that at time $t = 0$ the origins of both systems coincide, i. e. $O = O'$ and therefore also A and A′ as well as B and B′ coincide. If now at $t = 0$ two pulses are emitted form A and B in S and from A′ and B′ in S' the observer O in the rest frame measures their arrival times in O. During the light travel time Δt for the pulses from A or B the system S' has moved over the distance $\Delta x = v_x \cdot \Delta t$ to the right side in Fig. 3.17b. The pulses from B′ therefore arrive earlier in O' than those from A′. Therefore O' concludes that the pulses from B′ had been sent earlier than those from A′.

Now we will take the standpoint of O', who assumes that his system S' is at rest and that S moves with the velocity $-v_x$ to the left in Fig. 3.17b. He now defines the Simultaneity of the events in A′ and B′ if he receives the light pulses at O' simultaneously. Now the pulses from A arrive for O earlier that those from B. This illustrates that the definition of simultaneity depends on the system in which the pulses are measured. The reason for this ambiguity is the finite velocity of light. If this velocity would be infinite, the problem of simultaneity would not exist because then the travel time for the signals from the two points A and B would be always zero.

The question is now: what are the true equations for the transformation between different inertial systems?

3.5 Lorentz-Transformations

We regard two inertial systems $S(x, y, z)$ and $S'(x', y', z')$ with parallel axes and with $O(t = 0) = O'(t' = 0)$ which move with the constant velocity $v = \{v_x, 0, 0\}$ against each other in the x-direction (Fig. 3.18). Assume, that a short light pulse is emitted at $t = 0$ from $O = O'$. The observer O measures that the pulse has reached the point A after a time t. He describes his observation by the equation

$$\boldsymbol{r} = \boldsymbol{c} \cdot t \quad \text{or:} \quad x^2 + y^2 + z^2 = c^2 \cdot t^2 . \tag{3.22a}$$

The observer O' in S' measures that the pulse has arrived in A after the time t'. he therefore postulates:

$$\boldsymbol{r}' = \boldsymbol{c} \cdot t' \quad \text{or:} \quad x'^2 + y'^2 + z'^2 = c^2 \cdot t'^2 . \tag{3.22b}$$

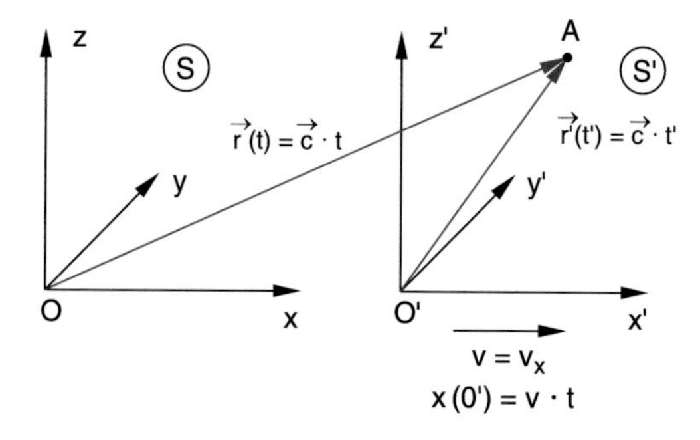

Figure 3.18 Schematic diagram for deriving the Lorentz-transformations

Both observers know about the result of the Michelson experiment. They therefore assume the same velocity of light c. The coordinate x of the origin O' measured in the system S is

$$x(O') = v \cdot t \quad \text{for} \quad x' = 0 \,.$$

Since the coordinate x' refers to the system S' the transformation to the coordinates x(A) of the point A, expressed in the system S must depend on the argument $(x - v \cdot t)$. We make the ansatz

$$x' = k(x - v \cdot t) \,, \tag{3.23}$$

where the constant k has to be determined. At time $t = 0$ the two observers were at the same place $x = x' = 0$ and have simultaneously started their clocks, i. e. $t(x = 0) = t'(x' = 0) = 0$. However, the time measurements for $t > 0$ are not necessarily the same for the two observers, because they are no longer at the same place but move against each other with the velocity v. The simplest transformation between t and t' is a linear transformation

$$t' = a(t - bx) \,, \tag{3.24}$$

where the constants a and b have again to be determined. Inserting (3.23) and (3.24) into (3.22b) yields with $y = y'$ and $z = z'$

$$\begin{aligned} &k^2\left(x^2 - 2vxt + v^2t^2\right) + y^2 + z^2 \\ &\quad = c^2a^2\left(t^2 - 2bxt + b^2x^2\right) \,. \end{aligned}$$

Rearrangement gives

$$\begin{aligned} &\left(k^2 - b^2a^2c^2\right)x^2 - 2\left(k^2v - ba^2c^2\right)xt + y^2 + z^2 \\ &\quad = \left(a^2 - k^2v^2/c^2\right)c^2t^2 \,. \end{aligned}$$

This has to be identical with (3.22a) for all times t and all locations x. Therefore the coefficients of x and t have to be identical. This gives the equations

$$\left.\begin{aligned} k^2 - b^2a^2c^2 &= 1 \\ k^2v - ba^2c^2 &= 0 \\ a^2 - k^2v^2/c^2 &= 1 \end{aligned}\right\} \Rightarrow \quad \begin{aligned} a = k &= \frac{1}{\sqrt{1-v^2/c^2}} \\ b &= v/c^2 \,. \end{aligned} \tag{3.25}$$

Inserting the expressions for a, b and k into (3.23) and (3.24) gives the special **Lorentz-Transformations**

$$\begin{aligned} x' &= \frac{x - vt}{\sqrt{1 - v^2/c^2}} \,, \quad y' = y \,, \quad z' = z \\ t' &= \frac{t - vx/c^2}{\sqrt{1 - v^2/c^2}} \,, \end{aligned} \tag{3.26}$$

between the coordinates (x, y, z) and (x', y', z') of two inertial systems which move against each other with the constant velocity $\boldsymbol{v} = \{v, 0, 0\}$. These equations were first formulated 1890 by Hendrik Lorentz [3.5]. They show, that for $v \ll c$ the Lorentz transformations converge towards the Galilei transformations (because for $v^2/c^2 \ll 1 \rightarrow \sqrt{1 - v^2/c^2} \approx 1$), which are therefore a special approximation for small velocities v:

Example

For $v = 10$ km/s (36 000 km/h) is $v/c \approx 3 \times 10^{-5}$ and $(1 - v^2/c^2)^{-1/2} \approx 1 + \frac{1}{2}v^2/c^2 = 1 + 10^{-10}$. The difference between Galilei and Lorentz transformations is then only $5 \cdot 10^{-10}$ and therefore smaller than the experimental uncertainty. ◀

With the abbreviation $\gamma = (1 - v^2/c^2)^{-1/2}$ the Lorentz transformations can be written in the clear form

$$\begin{aligned} x' &= \gamma(x - vt) & x &= \gamma(x' + vt') \\ y' &= y & y &= y' \\ z' &= z & z &= z' \\ t' &= \gamma(t - vx/c^2) & t &= \gamma(t' + vx'/c^2) \end{aligned} \,. \tag{3.26a}$$

Note: The Lorentz transformations have, compared to the Galilei transformations, only one additional assumption: The constancy of the velocity of light and its independence of the special inertial system, which was used in (3.22a,b), where both observers anticipate the same value of the velocity of light.

We will now discuss, how the velocity u of a body A, measured in the system S transforms according to (3.26) into the velocity u' of A, measured by O' in S'.

For O pertains:

$$u_x = \frac{\mathrm{d}x}{\mathrm{d}t}; \quad u_y = \frac{\mathrm{d}y}{\mathrm{d}t}; \quad u_z = \frac{\mathrm{d}z}{\mathrm{d}t} \,, \tag{3.27}$$

while for O' applies

$$\boldsymbol{u}' = \{u'_x, u'_y, u'_z\} = \left\{\frac{\mathrm{d}x'}{\mathrm{d}t'}, \frac{\mathrm{d}y'}{\mathrm{d}t'}, \frac{\mathrm{d}z'}{\mathrm{d}t'}\right\} \,.$$

Using (3.26) and considering that $x = x(t)$ depends on t, we get:

$$\begin{aligned} u'_x &= \frac{\mathrm{d}x'}{\mathrm{d}t'} = \frac{\mathrm{d}x'}{\mathrm{d}t} \cdot \frac{\mathrm{d}t}{\mathrm{d}t'} = \frac{\mathrm{d}x'}{\mathrm{d}t} \Big/ \frac{\mathrm{d}t'}{\mathrm{d}t} \\ &= \frac{\gamma\left(\frac{\mathrm{d}x}{\mathrm{d}t} - v\right)}{\gamma\left(1 - \frac{v}{c^2}\frac{\mathrm{d}x}{\mathrm{d}t}\right)} = \frac{u_x - v}{1 - \frac{v \cdot u_x}{c^2}} \,. \end{aligned}$$

Solving for u_x gives the back-transformation

$$u_x = \frac{u'_x + v}{1 + u'_x v/c^2} \,. \tag{3.28a}$$

In the same way one obtains

$$u'_y = \frac{u_y}{\gamma\left(1 - vu_x/c^2\right)}; \quad u_y = \frac{u'_y}{\gamma\left(1 + vu'_x/c^2\right)} \,, \tag{3.28b}$$

$$u'_z = \frac{u_z}{\gamma\left(1 - vu_x/c^2\right)}; \quad u_z = \frac{u'_z}{\gamma\left(1 + vu'_x/c^2\right)} \,. \tag{3.28c}$$

These equations demonstrate that the velocity components u_y and u_z perpendicular to the velocity $v = v_x$ of S' against S

transforms differently from the component u_x parallel to v_x. For $v_x \cdot u \ll c^2$ one obtains again the Galilei transformations.

If the body A moves parallel to the velocity v i. e. parallel to the x-axis and therefore also to the x'-axis, we have $u_y = u_z = 0 \Rightarrow u = u_x$, the Lorentz transformations simplify to

$$u' = \frac{u - v}{1 - vu/c^2} \,. \tag{3.28d}$$

For $u = c$ we get

$$u' = \frac{c - v}{1 - v/c} \equiv c \,, \tag{3.28e}$$

which means that O and O' measure the same value for the light velocity in accordance with the results of the Michelson experiment.

3.6 Theory of Special Relativity

Starting with the results of the Michelson experiment and the Lorentz transformations Einstein developed 1905 his theory of special relativity [3.5–3.8], which is based on the following postulates:

- All inertial systems are equivalent for all physical laws
- The velocity of light in vacuum has the same value in all inertial systems independent of the motion of the observer or the light source.

For the comparison of measurements of the same event by two observers in two different systems S and S', which move against each other the time definition plays an important role. The Lorentz-transformations (3.26) show that also the time has to be transformed when changing from S to S'. We will therefore at first discuss the relativity of simultaneity. The presentation in Sect. 3.6 follows in parts the recommendable book by French [3.8].

3.6.1 The Problem of Simultaneity

We will now treat the problem of simultaneity in different inertial systems in some more detail. We regard three points A, B and C which rest in the system S and have equal distances, i. e. AB = BC = Δx. In an x-t-coordinate-system with rectangular axis. For $t = 0$ the three points are located on the x-axis (Fig. 3.19a). In the x-t-diagram the points A, B and C proceed in the course of time on vertical straight lines since they are fixed in the system S and have therefore constant positions x. At time $t = 0$ a light pulse is emitted from point B. The light pulse, however proceeds on an inclined straight line with an inclination angle α_1 with $\tan\alpha_1 = t_1/\Delta x$. This line intersects the vertical position lines of A and C at the points A_1 and C_1. The connecting line through A_1 and C_1 is the horizontal line $t = t_1$. Since the light travels with the same speed c in all directions the pulses reach the points A and C at the same time $t_1 = \Delta x/c$.

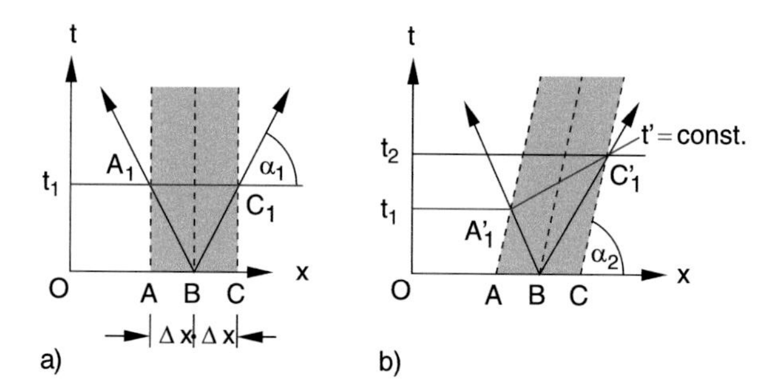

Figure 3.19 Illustration of the different results in a space-time diagram, when measuring simultaneous events in two different systems that move against each other

Now we regard the same situation in the system S' which moves with the velocity $v = v_x$ against S (Fig. 3.19b). The points A, B and C should rest in the system S', they therefore move with the velocity v_x against the system S and pass in the x-t-diagram of S inclined straight lines with the inclination angle α_2 and the slope $\tan\alpha_2 = \mathrm{d}t/\mathrm{d}x = 1/v$. The light pulses travel with the same velocity c as in S. At time $t = 0$ both systems S and S' should coincide. The light pulse, emitted from B at $t = 0$ now reaches the two points A and C for the observer O **not simultaneously** but in A at $t = t_1$ and in C at $t = t_2$, which correspond with the intersection points A'_1 and C'_1 in Fig. 3.19b. The reason is that A propagates towards the light pulse but C from it away.

Since for the observer O' in S' the points A, B and C are resting in S' the events A'_1 and C'_1 (we define as event the arrival of the light pulse in the point A'_1 or C'_1) has to occur simultaneously, equivalent to the situation for S in Fig. 3.19a because all inertial systems at rest are completely equivalent. In the x'-t'-diagram the line through the points A'_1 and C'_1 has to be a line $t' = \text{const}$ i. e. it must be parallel to the x'-axis (Fig. 3.20). One has to choose for the moving system S' other x'- and t'-axes which are inclined against the x- and t-axes of the system S. The x'-axis ($t' = 0$) and the t'-axis ($x' = 0$) are generally not perpendicular to each other.

One obtains the t'-axis in the following way: If O' moves with the velocity $v = v_x$ against O he propagates in the system S' along the axis $x' = 0$ which is the t'-axis (because he is in his

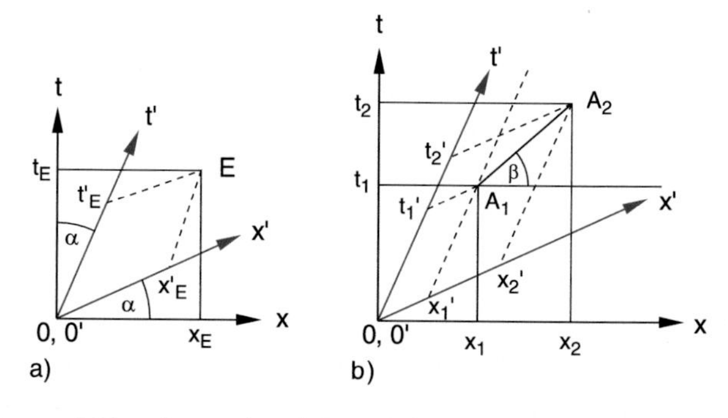

Figure 3.20 **a** Space axis and time axis in a moving inertial system S' are inclined by the angle α against the axes in a system S at rest. **b** Definition of the velocity u of a point A in the two systems S and S'

Chapter 3

system S' always resting at the origin $x' = 0$). In the system S this axis is $x = v \cdot t$ which is inclined against the t-axis $x = 0$ by the angle α with $\tan\alpha = v/c$. The slope of the t'-axis against the x-axis in the system S is $dt/dx = 1/v$.

Any event E is completely defined by its coordinates (x, t) in S or (x', t') in S'.

Note, however, that for the same event E the spatial and time coordinates (x_E, t_E) for O in S are different from (x'_E, t'_E) for O' in S' (Fig. 3.20)

For each observer the simultaneity of two events at different spatial points depends on the coordinate system in which the events are described.

We regard a point mass A which moves with the velocity u_x against O and with u'_x against O'. Its velocity is determined by O and O' by measuring the coordinates $x_1(t_1)$ and $x_2(t_2)$ in S resp. $x'_1(t'_1)$ and $x'_2(t'_2)$ in S' (Fig. 3.20b).

$$O \text{ obtains: } \quad u_x = \frac{x_2 - x_1}{t_2 - t_1} \, ,$$

$$O' \text{ obtains: } \quad u'_x = \frac{x'_2 - x'_1}{t'_2 - t'_1} \, .$$

The velocity u_x is represented in S by the reciprocal slope $\Delta x/\Delta t = u_x$ of the straight line A_1A_2, In S' however by $u'_x = \Delta x'/\Delta t'$. One can see already from, Fig. 3.20b that $u_x \neq u'_x$, which is quantitatively described by Eq. 3.28.

3.6.2 Minkowski-Diagram

The relativity of observations and their dependence of the reference system can be illustrated by space-time-diagrams as shown in Fig. 3.20. Each physical event which occurs at the location $\boldsymbol{r} = \{x, y, z\}$ at time t can be represented by a point in the four-dimensional space-time $\{x, y, z, t\}$. For simplicity we will restrict the following to one spatial dimension x and the relative motion of S' against S should occur only in the x-direction. Then the four-dimensional representation reduces to a two-dimensional one. Furthermore the time axis t is changed to $c \cdot t$ in order to have the same physical dimension [m] for both axes. Such a depiction is called ***Minkowski-diagram*** (Fig. 3.21).

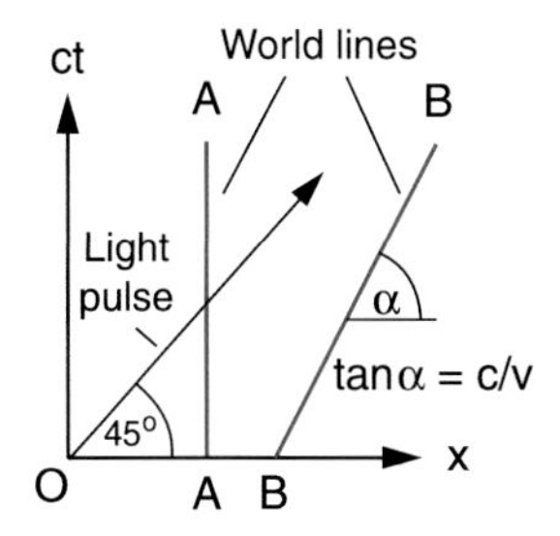

Figure 3.21 Minkowski diagram showing the world lines of a point A resting in the system, of a point B moving in the system with the velocity u and a light pulse emitted at time $t = o$ from the origin

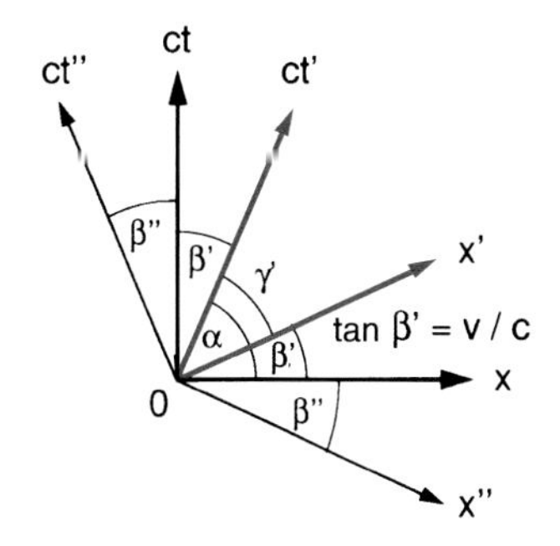

Figure 3.22 Minkowski diagram of the axes (x, t) of a system S at rest, of the axes (x', t') of a system S' that moves with the velocity v against S and of the axes (x'', t'') of a system S'', moving with $-v$ against S

A body A at rest propagates in an orthogonal (x, ct) diagram on a vertical line while a body B with the constant velocity v relative to O propagates on a sloped straight line with the slope $c \cdot \Delta t/\Delta x = c/v$. A light pulse which is emitted from $x = 0$ at $t = 0$ and propagates with the velocity c into the x-direction traverses on a straight line with the inclination of 45° against the x-axis because the slope is $\tan\alpha = c/c = 1$. It is represented by the diagonal in an orthogonal (x, ct)-diagram. Such lines for moving bodies or for light pulses are called *world lines* or *space-time-lines*, which can be also curved. Two events A and B occur in the system S simultaneously, if their points in the Minkowski-diagram lie on the line $t = t_1$ parallel to the x-axis (Fig. 3.19). The ct'-axis in S' is the world line of O'.

We had already discussed in the preceding section that the axis of two inertial systems S and S', which move against each other with the constant velocity v_x are inclined against each other. If the x- and the ct-axes in system S are orthogonal the ct'-axis has the slope $\tan\alpha = c/v_x$ against the x-axis. Also the x'-axis is inclined against the x-axis. According to the Lorentz-transformations the relation $t' = 0 \Rightarrow t = v \cdot x/c^2$ must be satisfied (Fig. 3.22). Its slope against the x-axis is therefore $dt'/dx = \tan\beta' = v/c$. The angle between the x' and the ct' axes is $\gamma = \alpha\text{-}\beta' = \arctan(c/v) - \arctan(v/c)$.

For illustration also a third system S'' is shown in Fig. 3.22, which moves with the velocity $v = -v_x$ against S. The slope of the x''-axis against the x-axis is now $\tan\beta'' = -v_x/c$. the angle between ct''-axis and ct-axis is also β''. The ct''-axis forms an angle $\delta = 2(\beta' + \beta'') + \gamma > 90°$ against the x''-axis.

3.6.3 Lenght Scales

Not only the inclination of the axis but also their scaling is different in the systems S, S' and S''. Since the velocity of light is the same in all inertial systems (which implies $c = dx/dt =$

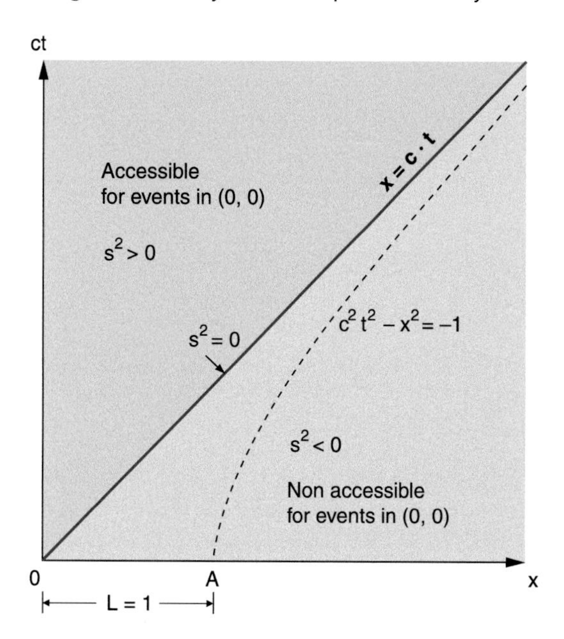

Figure 3.23 Illustration of the invariant s^2

$c' = \mathrm{d}x'/\mathrm{d}t' = \text{const}$) the quantity

$$s^2 = (ct)^2 - x^2 = (ct')^2 - x'^2 \tag{3.29}$$

must be equal in all inertial systems. This can be also seen from the Lorentz-equations (3.26). The quantity s^2 is therefore invariant under transformation between different inertial systems. For $s^2 = 0$ the world-line $x = \pm ct$ of a light pulse is obtained. For the motion of a body with velocity $v < c$ starting at $t = 0$ and $x = 0$ it follows $x^2(t) < (ct)^2 \Rightarrow s^2 > 0$.

In the (x, ct)-diagram no points with $x^2 > (ct)^2 (s^2 < 0)$ can be reached by signals emitted by O at $t = 0$. The area in Fig. 3.23 with $s^2 < 0$ is *non-accessible*, while all points with $s^2 > 0$ can be reached by such signals.

Such an invariant quantity like s^2 can be used to fix the scale length in Minkowski-diagrams. If we allow also imaginary values of s, the square s^2 can be also negative. For $s^2 = -1$ we obtain from (3.29) for all inertial systems (i. e. for S as well as for S') the hyperbola

$$x^2 - (ct)^2 = x'^2 - (ct')^2 = 1 ,$$

which is drawn in Fig. 3.23. It intersects in the system S the x-axis ($t = 0$) in the point A at $x = 1$. This defines the scale length $L = 1$ for the system S.

Also in the system S' is $x' = 1$ for $t' = 0$, which gives the scale length $L' = 1$ for the observer O'. However, for the observer O in S the length L' appears as $L \neq 1$ as can be seen from Fig. 3.24 where $L = O$A but $L' = 0$B. Each observer measures for the length in his own system another value than for the length in a system moving against his system. This seems very strange but is a consequence of the problem of simultaneity, because in order to measure the lengths L and L', O has to measure the endpoints 0 and A or 0 and B simultaneously, i. e. at the same time t, while O' measures them at the same time t'.

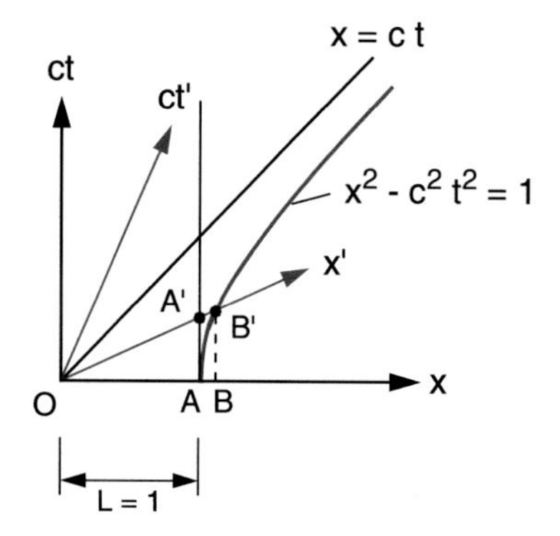

Figure 3.24 A yardstick with length OA that rests in the system S, appears shortened in the system S' moving against S

This shows that length standards in different inertial systems can be in fact different. If O in his system S measures distances in another system S', moving against S, he uses a larger scale, which means that the length of distances appears shorter.

3.6.4 Lorentz-Contraction of Lengths

One of the surprising results of the Lorentz-transformations is the contraction of the length of bodies in Systems S' moving against the observer in a rest frame S. In the foregoing section we have already indicated that this contraction is caused by the change of the length scale L' and that it can be ascribed to the problem of simultaneity.

Assume a rod with the endpoints P'_1 and P'_2 rests in the moving system S'. The coordinates x'_1 and x'_2 therefore move in the course of time on straight lines parallel to the t'-axis (Fig. 3.25). The observer O' measures at time t'_1 the length

$$L' = P'_1P'_2 = x'_2(t'_1) - x'_1(t'_1) .$$

For the observer O in S the rod resting in S' moves with the system S' with the velocity v in the x-direction. In order to determine the length of the rod, O has to measure the endpoints x_1

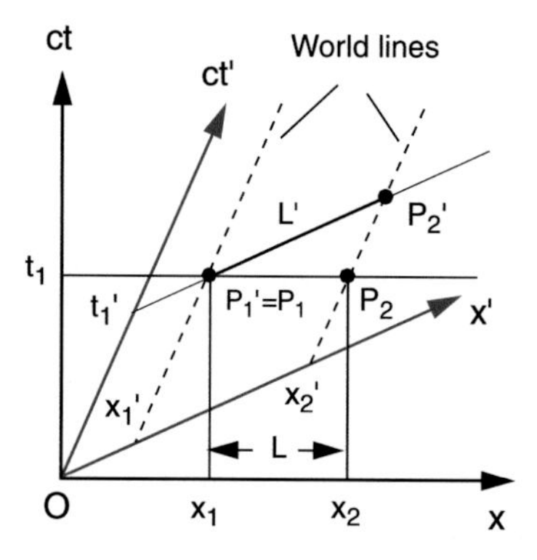

Figure 3.25 Graphical illustration of the Lorentz-contraction of a yard stick with length L resting in the moving system S', when O in S measures the length L', expressed in the Minkowski diagram of S'

and x_2 simultaneously, i.e. for $t = t_1$. These endpoints are for $t = t_1$ at the intersection points $P_1(t_1) = P'_1$ and $P_2(t_1) \neq P'_2$ of the world-lines $x'_1(t)$ and $x'_2(t)$ with the horizontal line $t = t_1$ in Fig. 3.25. For O is therefore the length of the rod

$$L = P_1 P_2 = x_2 - x_1 ,$$

where x_1 and x_2 are the vertical projections of P_1 and P_2 onto the x-axis $t = 0$ (Fig. 3.25).

Since $\Delta x'$ differs from Δx the two lengths L and L' are different. Because the scale lengths s and s' are different one cannot directly geometrically compare the length of the rod measured in S and S' from Fig. 3.25, but has to use the Lorentz-transformations.

$$\begin{aligned} x'_1 &= \gamma(x_1 - vt_1); \qquad x'_2 = \gamma(x_2 - vt_2) \\ \Rightarrow x'_2 - x'_1 &= \gamma(x_2 - x_1) \quad \text{for} \quad t_1 = t_2 \\ \Rightarrow L' &= \gamma \cdot L \Rightarrow L < L'; \quad \text{because} \quad \gamma > 1 . \end{aligned} \tag{3.30}$$

The lengths of a moving rod seems for an observer to be shorter than that of the same rod at rest.

- The contraction does not depend on the sign of the velocity $v = \pm v_x$.
- The contraction is really relative as can be seen from the following example: Two rods should have the same length $L_1 = L_2$ if both are resting in the same system S. Now L_2 is brought into a moving system S' where it rests relative to the origin O' of S'. For the observer O the length L_2 seems to be shorter then L_1 but for O' L_1 seems to be shorter than L_2. This implies that the Lorentz contraction is symmetric. This is no contradiction, because the different length measurements are due to the different observations of simultaneity as has been discussed before.
  Each observer can only make statements of events and times with respect to his own system S. If he transfers measurements of events in moving systems S' to his own system S, he has to take into account the relative velocity of S' against S and must use the Lorentz transformations. Then O and O' come to the same results.

Note that both observers O and O' come to consistent results for measurements in their own system and in the other system which moves against their own system, if they use consequently the Lorentz-transformations.

The answer to the often discussed question whether there is a "real contraction" depends on the definition of "real". The only information we can get about the length of the rod is based on measurements of the distance between its endpoints. For rods moving against the observer the locations of the two endpoints have to be measured simultaneously, which gives the results discussed above.

The relativity of the contraction can be visualized in the Minkowski diagram of Fig. 3.26. We regard again two iden-

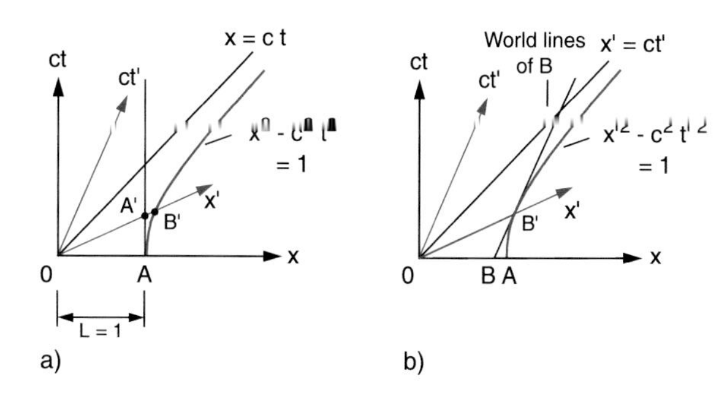

Figure 3.26 Relativity of the Lorentz contraction: **a** The yardstick $OA = 1$ rests in S, **b** the yardstick $OB = 1$ rests in the moving system S'

tical yardsticks with the scale $L = 1$, which rest in the system S resp. S'. The yardstick in S has for the observer O the endpoints O and A with the distance $OA = 1$. The world line for O is the ct-axis $x = 0$ and for A the parallel vertical line $x = 1$. In Fig. 3.26a also the world line $x = c \cdot t$ of a light pulse and the hyperbola $x^2 - c^2t^2 = 1$ are drawn. The intersection of the hyperbola with the x-axis $t = 0$ defines the scale $L = 1$, in the system S.

How is the situation in the system S'? The world line of A intersects the x'-axis $ct' = 0$ in the point A$'$. Therefore the distance OA$'$ is for O the length $L' = 1$ of the yardstick. However, for O' in his system S' the length of the yardstick is $x' = 1$ given by the distance OB$'$ where B$'$ is the intersection point of the parabola $x'^2 - (ct')^2 = 1$. For O is the length of the moving yardstick therefore smaller than for O', who regards the stick resting in his system.

Note that the parabola is the same in both systems S and S' (see Sect. 3.6.3).

For O' is the scale of O which he measures as OA$'$ shorter than his own scale OB$'$ this means that it appears for O' shorter.

Now we take a scale OB$'$ which rests in the system S' and has there the length $x' = 1$ because B$'$ is the intersection point of the parabola $x'^2 - (ct)^2 = 1$ with the x'-axis $ct' = 0$ (Fig. 3.26b). The world line of O' is the ct'-axis $x' = 0$ and that of the point B the line through B$'$ parallel to the ct' axis. This line intersects the x-axis in the point B. The observer O measures the length of the scale $x' = 1$ as the distance OB which is shorter than the distance OA with $x = 1$. Now the scale $x' = 1$ of the observer O' is shorter for the observer O.

This illustrates that the length contraction is due to the different prolongation of the scale which is caused by the different simultaneity for measurements of the endpoints by O and O'.

Note that both observers O and O' come to contradiction-free statements concerning measurements in their own system and in the other system if they use the Lorentz-transformations.

Chapter 3

3.6.5 Time Dilatation

We regard a clock, which rests in the origin O of system S. We assume that this clock sends two light pulses at times t and $t+\Delta t$ with a time delay Δt of the second pulse. An observers O at the location x_0 in the system S receives the light pulses at times t_1 and t_2, at the event points A and B of his world line $x = x_0 =$ const (Fig. 3.27). For O the time interval between the two pulses is

$$\Delta t = \mathrm{AB} = t_2 - t_1 \ .$$

An observers O' sitting at $x' = x_0'$ in the system S' which moves with the velocity v against S receives the light pulses at the intersection points A′ and B′ of his world line $x' = x_0'$ with the two axes $x' = ct'$ and $x' = c(t' + \Delta t')$ which are observed at times t_1' and t_2' measured with his clock in S'.

The observer O in S knows, that these times t_1' and t_2' are transformed into his measured times t_1 and t_2 by the Lorentz transformations

$$t_1' = \gamma \frac{t_1 - v \cdot x_0}{c^2} t_2' = \gamma \frac{t_2 - v \cdot x_0}{c^2} \ .$$

According to these equations he determines the time difference in the moving system S' as

$$\Delta t' = t_2' - t_1' = \gamma \cdot \Delta t \ . \qquad (3.31)$$

Since $\gamma = (1 - v^2/c^2)^{-1/2} > 1$ the observer O at rest measures for the moving system S' a longer time interval $\Delta t'$ between the two pulses than the moving observer O'. Because the clock resting in S moves for the observer S' he measures, that this clock runs slower than his own clock. This can be expressed by: **Moving clocks run slower**. Equivalent to the length contraction also the time dilatation is caused by the different observations of simultaneity in the systems S and S'. This effect increases with increasing velocity v and reaches essential values only for velocities v close to the velocity c of light. However this time dilatation can be measured with very precise clocks already for smaller velocities. For example, if two clocks are synchronized in Paris and transported by a fast plane (such as the concorde with $v = 2400\,\mathrm{km/h} = 667\,\mathrm{m/s} \Rightarrow \gamma = 1 + 8.9 \cdot 10^{-12}$) to New York the difference between Δt and $\Delta t'$ during a flight time of 3 hours is $8.9 \cdot 10^{-9}\,\mathrm{s} = 8.9\,\mathrm{ns}$.

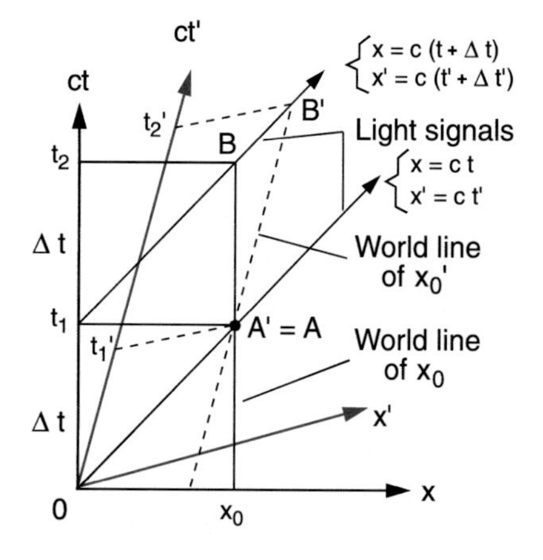

Figure 3.27 Minkowski diagram for illustration of the time dilatation. Two signals with the time difference $\Delta t = t_2 - t_1$ in the resting system S reach the moving observer O' in S' with the time difference $\Delta t' = \gamma \cdot \Delta t$

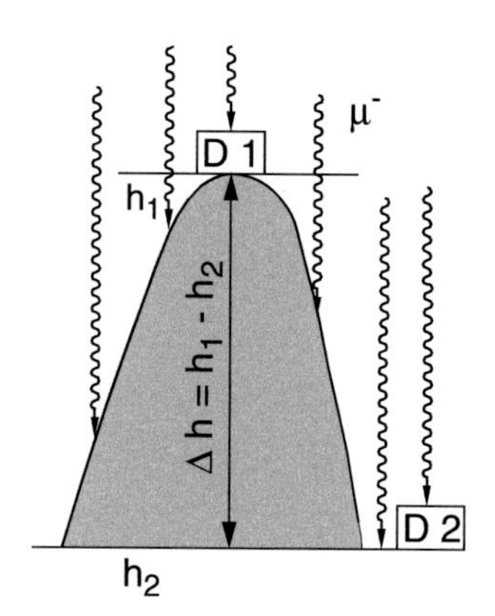

Figure 3.28 Measurement of the lifetime of relativistic muons with two detectors at different heights h_1 and h_2 above sea level

A much more precise measurement of the time dilatation can be obtained with faster moving clocks. Examples of such fast moving clocks are fast elementary particle such as electrons, protons or muons which move with velocities $v \approx c$.

The cosmic radiation (electrons and protons with very high energy) produce in the upper earth atmosphere at collisions with the atomic nuclei of the atmospheric molecules muons μ^- with velocities $v \approx c$ which reach nearly the velocity of light. Part of these muons reach the earth surface, while part of them decay during their flight through the atmosphere according to the scheme

$$\mu^- \xrightarrow{\tau} e^- + \nu_\mu + \overline{\nu_e} \qquad (3.32)$$

into an electron and two neutrinos (see Vol. 4). The lifetime of decelerated resting muons can be precisely measured as $\tau = 5 \cdot 10^{-6}\,\mathrm{s}$.

In order to measure the lifetime τ' of fast flying muons the rate of muons, incident onto a detector is measured at different altitudes above sea level, for instance at the altitude $h = h_1$ on the top of a mountain and at $h = h_2$ at the bottom of the mountain (Fig. 3.28). For a mean decay time τ' of the muons moving with the velocity v the relative fraction $\mathrm{d}N/N$ decays during the time interval $\mathrm{d}t/\tau'$

$$\mathrm{d}N = -a \cdot (N/\tau')\mathrm{d}t \ .$$

Integration yields

$$N(h_2) = N(h_1) \cdot e^{-\Delta t/\tau'} \quad \text{with} \quad \Delta t = \frac{h_1 - h_2}{v} \ ,$$

where the factor $a < 1$ takes into account the scattering of muons by the atmospheric molecules. This factor can be calculated from known scattering data. The often repeated measurements clearly gave essentially higher lifetimes $\tau' = 45 \cdot 10^{-6}\,\mathrm{s}$

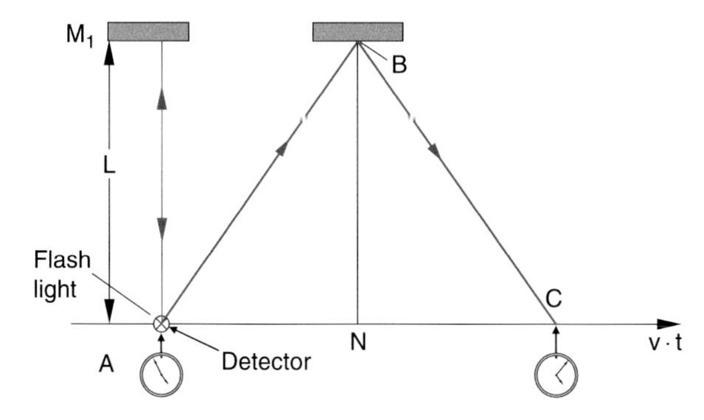

Figure 3.29 Einstein's "light-clock" for illustration of the time dilatation

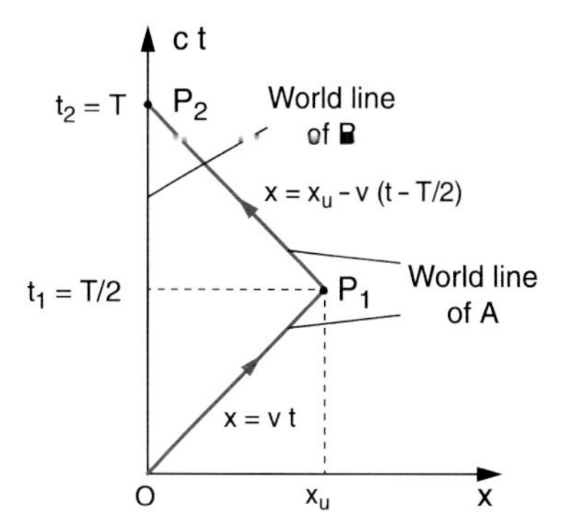

Figure 3.30 Minkowski diagram for the twin paradox

of the moving muons than $\tau = 5 \cdot 10^{-6}$ s for muons at rest. From $\tau' = \gamma \cdot \tau$ follows $\gamma = 9 \rightarrow v = 0.994c$. The muons have a velocity v which is smaller than c by only 6‰.

Meanwhile many short-lived elementary particles can be produced and accelerated to high velocities. Comparing the lifetimes of these particles at rest and while moving, unambiguously confirms the time dilatation postulated by the Lorentz transformations.

The relativistic time dilatation can be illustrated by an "gedanken-experiment" proposed by Einstein using a light pulse clock (Fig. 3.29). The system consists of a box with length L. On one side a flash lamp is mounted at the point A and on the opposite end a mirror M_1. The flash lamp emits a short light pulse and starts a clock. The light pulse reflected by M_1 is received by a detector which stops the clock. The time interval $\Delta t_0 = 2L/c$ is used as time scale in the system S in which the light clock rests.

Now we let the system S move with the velocity v relative to a system S' in a direction perpendicular to the length L. For the observer O' in S' the light pulse now travels from A to B and is reflected to C. With $AN = NC = v \cdot \Delta t/2$ it follows from Fig. 3.29

$$\begin{aligned} \mathrm{AB} + \mathrm{BC} &= 2 \cdot \left[L^2 + \left(v \frac{\Delta t'}{2} \right)^2 \right]^{1/2} = c \cdot \Delta t' \\ \Rightarrow \Delta t' &= \frac{2L}{(c^2 - v^2)^{1/2}} \,. \end{aligned} \tag{3.33}$$

The observer O measures $\Delta t = 2L/c$. The comparison between Δt and $\Delta t'$ gives:

$$\Delta t' = \frac{\Delta t}{(1 - v^2/c^2)^{1/2}} = \gamma \cdot \Delta t \,, \tag{3.34}$$

which turns out to be identical with (3.31).

3.6.6 The Twin-Paradox

No other problem of special relativity has aroused so many controversial discussions as the twin paradox (often called the clock-paradox), discussed by Einstein in his first paper 1905 about relativity. It deals with the following situation:

Two clocks which are synchronized show equal time intervals when sitting in the same system S at rest. One of the clocks is taken by O' on a fast moving spacecraft and returns after the travel time T (measured by O in S) back to the other clock which always had stayed in the system S. A comparison of the two clocks shows that the moved clock is delayed, that means that it shows a smaller value T' than T [3.8–3.11].

This "gedanken-experiment" has meanwhile be realized and the time dilatation has been fully verified (see previous section). For manned space missions this means that an astronaut A after his return to earth after a longer journey through space is younger than his twin brother B who has stayed at home. The "gained" time span is, however, for velocities of spacecrafts which can be realized up to now, very small and therefore insignificant. Nevertheless an understanding of the twin paradox is of principal significance because it illustrates the meaning of relativity, which is often used in a popular but wrong way.

We have discussed in the previous section that the time dilatation is relative, i. e. for each of the two observers O and O' the time scale of the other seems to be prolonged. Why is it then possible to decide unambiguously that A and not B is younger after his return?

The essential point is that A is not strictly in an inertial system, even if he moves with constant velocity, because at his returning point he changes the system from one that moves with the velocity $+v$ into one that moves with $-v$ against B. This shows that the measurements of A and B are not equivalent.

In order to simplify the discussion we will categorize the journey of A into three sections, which are illustrated in the Minkowski diagram of the resting system S of B in Fig. 3.30.

- A starts his journey from $x = 0$ at time $t = t' = 0$, reaches in a negligibly small time interval his final speed v until he arrives at his point of return $P_1(x_r, T/2)$ after the time $t_1 = T/2$.
- At time $t_1 = T/2$ he decelerates to $v = 0$ and accelerates again to $-v$. This should all happen within a time interval which is negligibly small compared with the travel time T.
- Astronaut A flies with $v_2 = -v$ back to B and reaches B in $x = 0$ at $t_2 = T$.

While the world line of B in Fig. 3.30 is the vertical line $x = 0$, A follows the line $x = v \cdot t \rightarrow ct = (c/v)x$ until the point of return P_1 from where he travels on the line $x = x_r - v(t - T/2) \Rightarrow ct = (c/v)(x_r - x) + cT/2$, until the point $P_2(0, T)$ where he meets with B.

From (3.29) we obtain for the invariant

$$\mathrm{d}s^2 = c^2\mathrm{d}t^2 - \mathrm{d}x^2 = c^2\mathrm{d}t'^2 - \mathrm{d}x'^2 .$$

This yields the different travel times for A and B: For B is always $\mathrm{d}x = 0$. We therefore get for the total distance s in the Minkowski diagram:

$$s = \int \mathrm{d}s = c \cdot \int \mathrm{d}t = c \cdot T .$$

For the moving astronaut A the resting observer B measures on the way OP_1: $\mathrm{d}x = v \cdot \mathrm{d}t \rightarrow \mathrm{d}s^2 = c^2\mathrm{d}t^2 - v^2\mathrm{d}t^2$, which gives for the total path

$$\int \mathrm{d}s = \sqrt{c_2 - v^2} \int \mathrm{d}t = \frac{c \cdot T}{2\gamma} = \frac{cT'}{2} ,$$

and on the way P_1P_2 back: $\mathrm{d}x = -v\mathrm{d}t$:

$$\int \mathrm{d}s = \sqrt{c^2 - v^2} \int \mathrm{d}t = \frac{c \cdot T}{2\gamma} = \frac{cT'}{2} .$$

The total travel time measured by B in S for the system S' of his twin A is then $T' = T/\gamma < T$. This result can be also explained by the Lorentz contraction: For A is the path L shortened by the factor γ. Therefore the travel time T for A is shorter by the factor γ since A as well as B measure the same velocity v of A relative to B.

The asymmetry of the problem can be well illustrated by regarding light pulses sent at constant intervals by A to B and by B to A. Both the observer B and the astronaut A send these light signals at the frequency f_0 measured with their clocks. The sum of the sent pulses at a frequency $f_0 = 1/s$ gives the total travel time in seconds (Fig. 3.31).

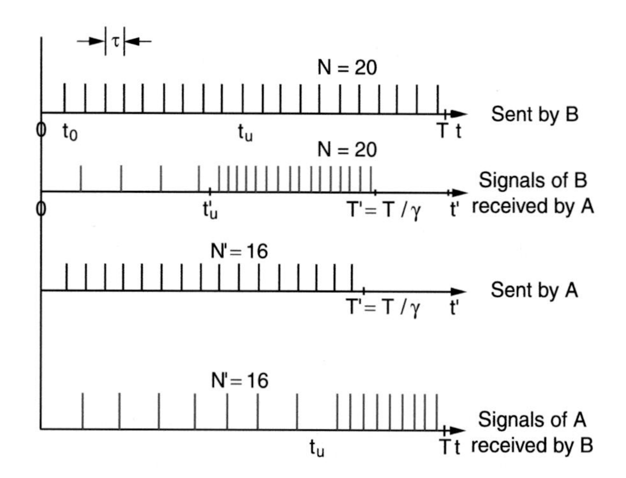

Figure 3.31 Illustration of the twin paradox, using the signals sent and received by A and B

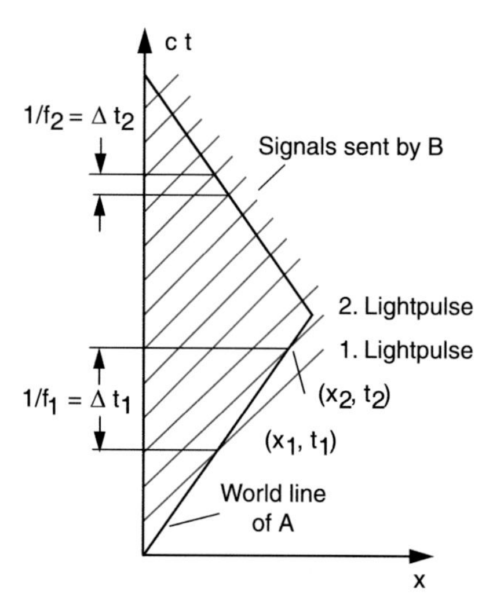

Figure 3.32 Doppler-effect of the signal frequency illustrated in the Minkowski diagram

While A moves away from B both observers receive the pulses at a lower frequncy f_1 because each successive pulse has to travel a longer way than the preceding pulse. The asymmetry occurs at the turning point P_1. While A on his way back now receives the pulses with a higher frequency f_2 directly after he turns around, B receives the pulses from A with the frequency f_2 only at times $\Delta t \leq c \cdot x_0$ after the return time. He receives the same total number N' of pulses as has been sent by A but he receives for a longer time signals with lower frequency than A. Therefore he measures a longer travel time for A than A himself.

This is illustrated in Fig. 3.31 for $v = 0.6c$. B sends during the travel time of A altogether 20 pulses, which are all received by A. While B sends his signals at constant time intervals τ_0, A receives them on the outbound trip with larger time intervals τ_1 i. e. lower frequency and on the way back with shorter time intervals τ_2, i. e. higher frequency. Measured in the system S' of A the travel time T' is shorter by the factor γ. The astronaut A sends during this time only $N' = N/\gamma = 16$ pulses which are all received by B. Since B receives the signal sent by A at the return time only delayed, he receives signals with the larger interval τ_1 (lower frequency) for a longer time and only after the time $t'_r + L/2c$ the signals with the shorter intervals τ_2.

This is further illustrated by the Minkowski diagram of Fig. 3.32 which explains the relativistic Doppler-effect. The astronaut A is at time $t = 0$ at the point $(x = 0, t = 0)$ in the Minkowski diagram. He moves, measured by B on the line

$$x = v \cdot t \rightarrow c \cdot t = \frac{c}{v} x .$$

The observer B sitting always at $x = 0$ sends light pulses at a repetition frequency f_0. A pulse sent by B at time t_0 travels in the (x, ct)-diagram on lines with a 45° slope which intersects the world line of A in the point (x_1, t_1), where it is received by A. The next pulse is sent by B at time $t = t_0 + \tau = t_0 + 1/f_0$ and reaches A at (x_2, t_2). According to Fig. 3.31 the following

Table 3.1 Measurement of multiple physical quantities of resident and traveler (according to [3.7])

Physical quantity	Measurement of B (resident)	Measurement of A (traveler)
Total travel time	$T = \frac{2L}{v}$	$T' = \frac{2L}{\gamma v}$
Total number of sent signals	$f \cdot T = \frac{2fL}{v}$	$f \cdot T' = \frac{2fL}{\gamma v}$
Reversal time of A	$t_\mathrm{u} = \frac{L}{v} + \frac{L}{c} = \frac{L}{v}(1+\beta)$	$t'_\mathrm{u} = \frac{L}{\gamma v}$
Number of received signals with frequency $f' \left(f' = f \cdot \left(\frac{1-\beta}{1+\beta}\right)^{1/2}\right)$	$f' t_\mathrm{u} = f \cdot \left(\frac{1-\beta}{1+\beta}\right)^{1/2} \cdot \frac{L}{v}(1+\beta)$ $= \frac{fL}{v}(1-\beta^2)^{1/2}$	$f' t'_\mathrm{u} = f \cdot \left(\frac{1-\beta}{1+\beta}\right)^{1/2} \cdot \frac{L}{v}(1-\beta^2)^{1/2}$ $= \frac{fL}{v}(1-\beta)$
Travel time after reversal	$t_2 = \frac{L}{v} - \frac{L}{c} = \frac{L}{v}(1-\beta)$	$t'_2 = \frac{L}{\gamma v} = \frac{L}{v}\frac{1}{(1-\beta^2)^{1/2}}$
Number of received signals with frequency $f'' = f \cdot \left(\frac{1+\beta}{1-\beta}\right)^{1/2}$	$f'' t_2 = f \cdot \left(\frac{1+\beta}{1-\beta}\right)^{1/2} \cdot \frac{L}{v}(1-\beta)$ $= \frac{fL}{v}(1-\beta^2)^{1/2}$	$f'' t'_2 = f \cdot \left(\frac{1+\beta}{1-\beta}\right)^{1/2} \cdot (1-\beta^2)^{1/2}$ $= \frac{fL}{v}(1+\beta)$
Total number of received signals $N = f' t_\mathrm{u} + f'' t_2$, $N' = f' t'_\mathrm{u} + f'' t'_2$	$N = f' t_\mathrm{u} + f'' t_2 = \frac{2fL}{v}(1-\beta^2)^{1/2}$ $= \frac{2fL}{\gamma \cdot v}$	$N' = f' t'_\mathrm{u} + f'' t'_2 = \frac{2fL}{v}$
Conclusion regarding the time measured by the other	$T' = \frac{2L}{\gamma v}$	$T = \frac{2L}{v}$
	$\beta = v/c,\ \gamma = (1-\beta^2)^{-1/2}$	

relations apply:

$$x_1 = c \cdot (t_1 - t_0) = x_0 + v \cdot t_1$$
$$x_2 = c \cdot (t_2 - t_0 - \tau) = x_0 + v \cdot t_2\,.$$

Subtraction of the first from the second equation yields

$$t_2 - t_1 = \frac{c \cdot \tau}{c - v}; \qquad x_2 - x_1 = \frac{v \cdot c \cdot \tau}{c - v}\,.$$

Figure 3.32 illustrates that for A the time intervals τ' are longer on the outward flight than on the return flight. Astronaut A measures in his system S' according to the Lorentz transformations

$$\tau' = t'_2 - t'_1 = \gamma \cdot \left[(t_2 - t_1) - \frac{v}{c^2}(x_2 - x_1)\right]$$
$$= \gamma \cdot (1 + \beta) \cdot \tau\,, \quad \text{with } \beta = v/c\,.$$

With $\gamma = (1 - \beta^2)^{-1/2}$ this becomes

$$\tau' = \tau \left(\frac{1+\beta}{1-\beta}\right)^{1/2} \Rightarrow f' = \frac{1}{\tau'} = f_0 \left(\frac{1-\beta}{1+\beta}\right)^{1/2}.$$

Astronaut A measures therefore on the outward flight the smaller repetition frequency f_1 which is smaller than f_0 by the factor $[(1-\beta)/(1+\beta)]^{1/2}$ and on the return flight with the velocity $-v$ he measures the higher repetition rate $f_2 = [(1+\beta)/(1-\beta)]^{1/2} f_0$.

In Tab. 3.1 the different measurements of A and B are summarized. The table shows again, that the total number of pulses sent by B is equal to the number received by A but different from the number sent by A. The last line in Tab. 3.1 makes clear, that B can conclude the travel time measured by A from the number of pulses received from A and vice versa can A conclude the time measured by B. **Both observers are therefore in complete agreement in spite of the different times measured in their systems. This shows that there are no contradictions in the description of the twin paradox.** Observer B knows, that the travel time T' measured by A is shorter than the time measured by himself because A is sitting in a moving system, and A knows that B measures in his resting system a longer time.

3.6.7 Space-time Events and Causality

Since the speed of light is the upper limit for all velocities with which signals can be transmitted from one space-time point (x_1, t_1) to another point (x_2, t_2) all space-time events can be classified into those which can be connected by signals and those which cannot. In the first case an event in (x_2, t_2) can be caused by an event in (x_1, t_1).

In the Minkowski diagram of Fig. 3.33 the two diagonal lines $x = \pm c \cdot t$ are the worldliness of light signals passing through

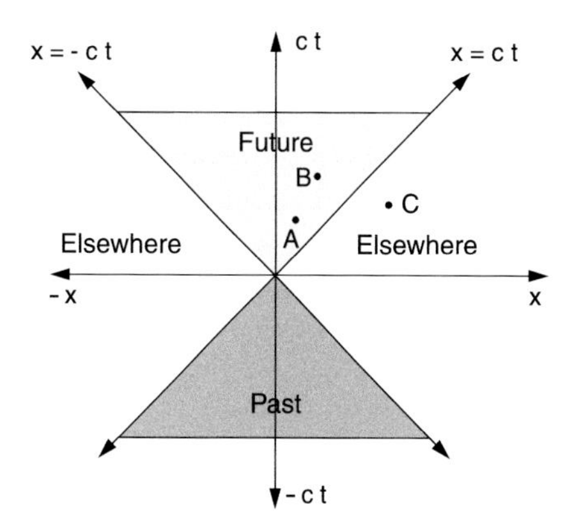

Figure 3.33 Two-dimensional Minkowski diagram with the shaded areas for past and future and the white areas for non-accessible space-time point (r, t)

Chapter 3

the point ($x = 0, t = 0$). These world lines divided the space-time into different regions: All regions with ($x, t > 0$) with $|x|| \leq ct$ represent *the future* seen from ($x = 0, t = 0$). They can be reached by signals sent from $(0, 0)$, while the region with ($x, t < 0$) form the *past*.

This can be also expressed in the following way: All events in space-time point (x, t) can be causally connected with each other, i. e. an event in (x_2, t_2) can be caused by an event in (x_1, t_1) if both points lie in the red shaded regions in Fig. 3.33. this means signals can be transferred between these points and interactions between bodies in these points are possible. For instance the event A can influence the event B in Fig. 3.33 but not the event C.

An observer in the space-time point (x, t) with $|x| \leq |ct|$ can never receive a signal from points in the white regions with $|x| > |ct|$. We call these regions therefore "elsewhere".

In a three-dimensional space-time diagram (x, y, ct) the surfaces $x^2 + y^2 = c^2t^2$ form a cone called the light-cone. Past and future are inside the cone. "elsewhere" is outside. In a four-dimensional space-time diagram (x, y, z, ct) this light cone becomes a hyper-surface.

Very well written introductions to the special relativity and its consequences without excessive Mathematics, which are also understandable to undergraduate students can be found in [3.7–3.11].

Summary

- For the description of motions one needs a coordinate system. Coordinate systems in which the Newtonian Laws can be formulated in the form, discussed in Sect. 2.6 arte called inertial systems. Each coordinate system which moves with constant velocity v against another inertial system is also an inertial system.
- The transformation of coordinates (x, y, z), of time t and of velocity v and therefore also of the equation of motion from one to another inertial system is described by the Lorentz transformations. They are based on the constancy of the speed of light c, confirmed by experiments, which is independent of the chosen inertial system and has the same value in all inertial systems. For small velocities $v \ll c$ the Lorentz transformations approach the classical Galilei transformations.
- The description of motions in accelerated systems demand additional accelerations, which are caused by "inertial or virtual" forces. In a rotating system with constant angular velocity these are the Coriolis force $\boldsymbol{F}_\mathrm{C} = 2\,\mathrm{m}(\boldsymbol{v}' \times \boldsymbol{\omega})$ which depends on the velocity v' of a body relative to the rotating system, and the centrifugal force $\boldsymbol{F}_\mathrm{cf} = m \cdot \boldsymbol{\omega} \times (\boldsymbol{r} \times \boldsymbol{\omega})$ which is independent of v'.
- The theory of special relativity is based on the Lorentz transformations and discusses the physical effects following from these equations when the motion of a body is described in two different inertial systems which move against each other with constant velocity v. An essential point is the correct definition of simultaneity of two events. Many statements of special relativity can be illustrated by space-time diagrams (x, ct) (Minkowski diagrams), as for instance the length-contraction or the time-dilatation. Such diagrams show that these effects are relative and symmetric, which means that each observers measures the lengths in a system moving against his system contracted and the time prolonged. The description of the two observers O and O' are different but consistent. There is no contradiction.
- For the twin-paradox an asymmetry occurs, because the astronaut A changes its inertial system at the point of return. It is therefore possible to attribute the time dilation unambiguously to one of the observers.
- The statements of special relativity have been fully confirmed by numerous experiments.

Problems

3.1 An elevator with a cabin heights of 2.50 m is accelerated with constant acceleration $a = -1\,\mathrm{m/s^2}$ starting with $v = 0$ at $t = 0$. After 3 s a ball is released from the ceiling.

a) At which time reaches it the bottom of the cabin?
b) Which distance in the resting system of the elevator well has the ball passed?
c) Which velocity has the ball at the time of the bounce with the bottom in the system of the cabin and in the system of the elevator well?

3.2 From a point A on the earth equator a bullet is shot in horizontal direction with the velocity $v = 200\,\mathrm{m/s}$.

a) in the north direction
b) in the north-east direction 45° against the equator
c) In the north-west direction 135° against the equator

What are the trajectories in the three cases described in the system of the rotating earth?

3.3 A ball hanging on a 10 m long string is deflected from its vertical position and rotates around the vertical axis with $\omega = 2\pi \cdot 0.2\,\mathrm{s}^{-1}$. What is the angle of the string against the vertical and what is the velocity v of the ball?

3.4 In the edge region of a typhoon over Japan (geographical latitude $\varphi = 40°$) the horizontally circulating air has a velocity of 120 km/h. What is the radius of curvature r of the path of the air in this region?

3.5 A fast train ($m = 3 \cdot 10^6$ kg) drives from Cologne to Basel with a velocity of $v = 200\,\mathrm{km/h}$ exactly in north-south direction passing 48° latitude. How large is the Coriolis force acting on the rail? Into which direction is it acting?

3.6 A body with mass $m = 5$ kg is connected to a string with $L = 1$ m and rotates
a) in a horizontal plane around a vertical axis
b) in a vertical plane around a horizontal axis
At which angular velocity breaks the string in the cases a) and b) when the maximum tension force of the string is 1000 N?

3.7 A plane disc rotates with a constant angular velocity $\omega = 2\pi \cdot 10\,\mathrm{s}^{-1}$ around an axis through the centre of the disc perpendicular to the disc plane. At time $t = 0$ a ball is launched with the velocity $v = \{v_r, v_\varphi\}$ with $v_r = 10\,\mathrm{m/s}$, $v_\varphi = 5\,\mathrm{m/s}$ (measured in the resting system) starting from the point A ($r = 0.1$ m, $\varphi = 0°$). At which point (r, φ) does the ball reach the edge of the disc?

3.8 A bullet with mass $m = 1$ kg is shot with the velocity $v = 7\,\mathrm{km/s}$ from a point A on the earth surface with the geographical latitude $\varphi = 45°$ into the east direction. How large are centrifugal and Coriolis force directly after the launch? At which latitude is its impact?

3.9 Two inertial systems S and S' move against each other with the velocity $v = v_x = c/3$. A body A moves in the system S with the velocity $\boldsymbol{u} = \{u_x = 0.5c, u_y = 0.1c, u_z = 0\}$. What is the velocity vector $\boldsymbol{u}'$ in the system S' when using
a) the Galilei transformations and
b) the Lorentz transformations?
How large is the error of a) compared to b)?

3.10 A meter scale moves with the velocity $v = 2.8 \cdot 10^8\,\mathrm{m/s}$ passing an observer B at rest. Which length is B measuring?

3.11 A space ship flies with constant velocity v to the planet Neptune and reaches Neptune at its closest approach to earth. How large must be the velocity v if the travel time, measured by the astronaut is 1 day? How long is then the travel time measured by an observer on earth?

3.12 Light pulses are sent simultaneously from the two endpoints A and B of a rod at rest. Where should an observer O sit in order to receive the pulses simultaneously? Is the answer different when A, B and O moves with the constant velocity v? At which point in the system S an observer O' moving with a velocity v_x against S receives the pulses simultaneously if he knows that the pulses has been sent in the system S simultaneously from A and B?

3.13 At January 1st 2010 the astronaut A starts with the constant velocity $v = 0.8c$ to our next star α-Centauri, with a distance of 4 light years from earth. After arriving at the star, A immediately returns and flies back with $v = 0.8c$ and reaches the earth according to the measurement of B on earth at the 1st of January 2020. A and B had agreed to send a signal on each New Year's Day. Show that B sends 10 signals, but A only 6. How many signals does A receive on his outbound trip and how many on his return trip?

3.14 Astronaut A starts at $t = 0$ his trip to the star Sirius (distance 8.61 light years) with the velocity $v_1 = 0.8c$. One year later B starts with the velocity $v_2 = 0.9c$ to the same star. At which time does B overtake A, measured
a) in the system of A,
b) of B and
c) of an observer C who stayed at home?
At which distance from C measured in the system of C does this occur?

References

3.1. https://en.wikipedia.org/wiki/Fictitious_force
3.2a. A.A. Michelson, E.W. Morley, Am. J. Sci. **34**, 333 (1887)
3.2b. R. Shankland, Am. J. Phys. **32**, 16 (1964)
3.3. A.A. Michelson, *Studies in Optics*. (Chicago Press, 1927)
3.4a. A. Brillet, J.L. Hall, in Laser Spectroscopy IV, Proc. 4th Int. Conf. Rottach Egern, Germany, June 11–15 1979. (Springer Series Opt. Sci. Vol 21, Springer, Berlin, Heidelberg, 1979)
3.4b. W. Rowley et al., Opt. and Quant. Electr. **8**, 1 (1976)
3.5. A. Einstein, H.A. Lorentz, H. Minkowski, H. Weyl, *The principles of relativity*. (Denver, New York, 1958)
3.6. R. Resnik, *Introduction to Special Relativity*. (Wiley, 1968)
3.7. E.F. Taylor, J.A. Wheeler, *Spacetime Physics: Introduction to Special Relativity*, 2nd ed. (W.H. Freeman & Company, 1992)
3.8. A.P. French, *Special Relativity Theory*. (W.W. Norton, 1968)
3.9. D.H. Frisch, H.J. Smith, Am. J. Phys. **31**, 342 (1963)
3.10. N.M. Woodhouse, *Special Relativity*. (Springer, Berlin, Heidelberg, 2007)
3.11. C. Christodoulides, *The Special Theory of Relativity: Theory, Verification and Applications*. (Springer, Berlin, Heidelberg, 2016)

4 Systems of Point Masses; Collisions

© Springer International Publishing Switzerland 2017
W. Demtröder, *Mechanics and Thermodynamics*, Undergraduate Lecture Notes in Physics, DOI 10.1007/978-3-319-27877-3_4

In the preceding chapters we have discussed the motion of a single particle and its trajectory under the influence of external forces. In this chapter we will deal with systems of many particles, where besides possible external forces also interactions between the particles play an important role.

4.1 Fundamentals

At first we introduce several expressions and definitions of fundamental terms and notations for systems of many particles.

4.1.1 Centre of Mass

We consider N point masses with position vectors r_i and define as the centre of mass the point with the position vector

$$\boldsymbol{R}_S = \frac{\sum_i m_i \boldsymbol{r}_i}{\sum_i m_i} = \frac{1}{M}\sum_i m_i \boldsymbol{r}_i \,, \tag{4.1}$$

where $M = \sum m_i$ is the total mass of all N particles (Fig. 4.1).

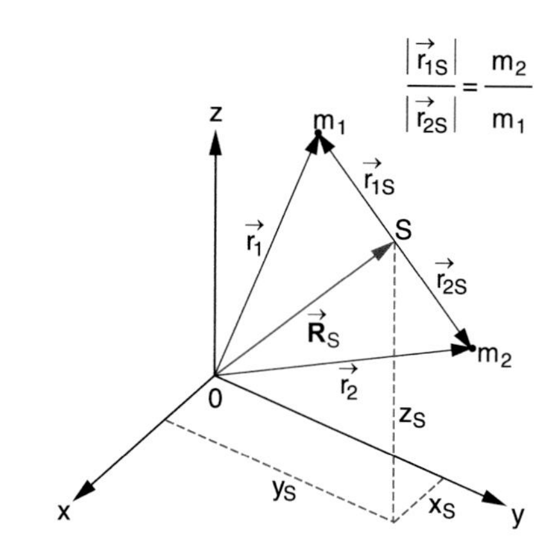

Figure 4.1 Definition of center of mass

When the masses m_i move with the velocities $\boldsymbol{v}_i = \mathrm{d}\boldsymbol{r}_i/\mathrm{d}t$ we define the velocity v_S of the centre of mass as

$$\boldsymbol{v}_S = \frac{\mathrm{d}\boldsymbol{R}_S}{\mathrm{d}t} = \frac{1}{M}\sum_i m_i \boldsymbol{v}_i \,. \tag{4.2a}$$

With the momenta $\boldsymbol{p}_i = m_i \cdot \boldsymbol{v}_i$ (4.2a) can be also expressed by the total momentum $\boldsymbol{P} = \sum \boldsymbol{p}_i$ as

$$\boldsymbol{P} = M\boldsymbol{v}_S \,. \tag{4.2b}$$

If no external forces are acting on the particles, we need to regard only internal forces, i. e. interactions between the particles. Such a system without external forces is called a *closed system.*

From the Newtonian law $\boldsymbol{F}_{ik} = -\boldsymbol{F}_{ki}$ it follows: $\sum_i \sum_{k \neq i} \boldsymbol{F}_{ik} = \boldsymbol{0}$. **In a closed system the vector sum of all forces is zero.**

With $F_i = \sum_{k \neq i} F_{ik}$ and $F_i = \mathrm{d}\boldsymbol{p}_i/\mathrm{d}t$ the total momentum of the system

$$\boldsymbol{P} = \sum \boldsymbol{p}_i = \mathbf{const} \,. \tag{4.3}$$

Since P is the momentum of the centre of mass we can state:

The centre of mass of a closed system moves with constant momentum. This implies that its velocity does not change.

If an external total force $\boldsymbol{F} \neq \boldsymbol{0}$ acts onto the system we can write

$$\boldsymbol{F} = \frac{\mathrm{d}}{\mathrm{d}t}\sum \boldsymbol{p}_i = \frac{\mathrm{d}\boldsymbol{P}}{\mathrm{d}t} \,, \tag{4.4}$$

With the acceleration of the centre of mass $\boldsymbol{a}_S = d\boldsymbol{v}_S/dt$ we obtain

$$\boldsymbol{F} = M\boldsymbol{a}_S \,. \tag{4.5}$$

The centre of mass of an arbitrary system of particles moves in the same way as a body with the total mass $M = \sum m_i$ would move under the action of the external force $\boldsymbol{F}$.

Often it is useful to choose a coordinate system with the centre of mass as origin, which moves with the velocity $\boldsymbol{v}_S$ of the centre of mass against the fixed laboratory system. Such a system is called the **centre of mass system (CM-system).**

The position vectors $\boldsymbol{r}_i$ in the lab-system are related to the position vectors $\boldsymbol{r}_{iS}$ in the CM-system (Fig. 4.1) by

$$\boldsymbol{r}_i = \boldsymbol{r}_{iS} + \boldsymbol{R}_S \,. \tag{4.6a}$$

Inserting into (4.1) gives

$$\begin{aligned}\sum_i m_i \boldsymbol{r}_{iS} &= \sum_i m_i (\boldsymbol{r}_i - \boldsymbol{R}_S) \\ &= \sum_i m_i \boldsymbol{r}_i - \boldsymbol{R}_S \sum_i m_i = \boldsymbol{0} \,,\end{aligned}$$

$$\sum m_i \boldsymbol{r}_{iS} = \boldsymbol{0} \tag{4.6b}$$

This implies that in the CM-system the position vector $\boldsymbol{R}_S$ of the centre-of-mass is $\boldsymbol{R}_S = (1/M)\sum m_i \boldsymbol{r}_{iS} = \boldsymbol{0}$.

The relation between the velocity $\boldsymbol{v}_i$ in the lab-system and $\boldsymbol{v}_{iS}$ in the CM-system is

$$\boldsymbol{v}_i = \boldsymbol{v}_{iS} + \boldsymbol{V}_S \,, \tag{4.6c}$$

which can be verified by differentiation of (4.6a). For the momenta we therefore get

$$\sum_i m_i \boldsymbol{v}_{iS} = \sum_i \boldsymbol{p}_{iS} = \boldsymbol{0}\,. \tag{4.6d}$$

The sum of all momenta in the CM-system is always zero.

For a closed system of two masses m_1 and m_2 the total kinetic energy in the lab-system is

$$\begin{aligned} E_{\text{kin}} &= \tfrac{1}{2}m_1 v_1^2 + \tfrac{1}{2}m_2 v_2^2 \\ &= \tfrac{1}{2}\left(m_1 v_{1S}^2 + m_2 v_{2S}^2\right) + \tfrac{1}{2}\left(m_1 + m_2\right) V_S^2 \\ &\quad + \left(m_1 \boldsymbol{v}_{1S} + m_2 \boldsymbol{v}_{2S}\right)\cdot \boldsymbol{V}_S \,. \end{aligned} \tag{4.7a}$$

The last term is zero because $\boldsymbol{p}_{1S} + \boldsymbol{p}_{2S} = \boldsymbol{0}$ and we obtain:

$$E_{\text{kin}} = E_{\text{kin}}^{(S)} + \tfrac{1}{2}MV_S^2 \,. \tag{4.7b}$$

In the Lab-system the kinetic energy of a closed system can be written as the sum of $E_{\text{kin}}^{(S)}$ in the CM-system plus the kinetic energy of the total mass M concentrated in the center of mass S (translational energy of the system).

The total motion of the closed system can be divided into a uniform motion of S with the constant velocity V_S and a relative motion of the two particles against S.

4.1.2 Reduced Mass

We consider two particles with masses m_1 and m_2 which interact with each other due to the forces $\boldsymbol{F}_{12} = -\boldsymbol{F}_{21}$. Without other external forces the equations of motion read:

$$\frac{\mathrm{d}\boldsymbol{v}_1}{\mathrm{d}t} = \frac{\boldsymbol{F}_{12}}{m_1}\,; \quad \frac{\mathrm{d}\boldsymbol{v}_2}{\mathrm{d}t} = \frac{\boldsymbol{F}_{21}}{m_2}\,. \tag{4.8a}$$

Subtraction yields

$$\frac{\mathrm{d}}{\mathrm{d}t}\left(\boldsymbol{v}_1 - \boldsymbol{v}_2\right) = \left(\frac{1}{m_1} + \frac{1}{m_2}\right)\boldsymbol{F}_{12}\,, \tag{4.8b}$$

where $\boldsymbol{v}_{12} = \boldsymbol{v}_1 - \boldsymbol{v}_2$ is the relative velocity of the two particles.

Introducing the **reduced mass**

$$\mu = \frac{m_1 m_2}{m_1 + m_2} \tag{4.9}$$

and rewrite Eq. 4.8b we get

$$\boldsymbol{F}_{12} = \mu \frac{\mathrm{d}\boldsymbol{v}_{12}}{\mathrm{d}t}\,. \tag{4.10}$$

This means: For the relative motion of the two particles the equation of motion is completely analogous to Newton's equation (2.18a) for a single particle with the mass μ. This shows the usefulness of defining the reduced mass.

The kinetic energy E_{kin}^{S} of the two particles in the CM-system

$$\begin{aligned} E_{\text{kin}}^{(S)} &\overset{\text{Def}}{=} \sum_i \frac{m_i}{2} v_{iS}^2 \\ &= \frac{1}{2}\sum m_i v_i^2 - \frac{1}{2}MV_S^2 \,. \end{aligned} \tag{4.11a}$$

is the difference of E_{kin} in the lab-system and the kinetic energy of the CM.

Inserting $\boldsymbol{v}_S = (1/M)\sum m_i \boldsymbol{v}_i$ gives with (4.9)

$$E_{\text{kin}}^{(S)} = \frac{1}{2}\mu v_{12}^2 \,. \tag{4.11b}$$

The kinetic energy of a closed system of two particles in the CM-system equals the kinetic energy of a single parrtivle with the reduced mass μ which moves with the relative velocity $\boldsymbol{v}_{12}$.

This important relations can be summarized as:

The relative motion of two particles under the influence of their mutual interaction $\boldsymbol{F}_{12} = -\boldsymbol{F}_{21}$ can be reduced to the motion of a single particle with the reduced mass μ driven by the force $\boldsymbol{F}_{12}$.

This is illustrated in Fig. 4.2 where two masses $m_1 = m$ and $m_2 = 1.5m$ move around their centre of mass S which moves itself with the velocity V_S. An example of such a system is a double-star system, where two stars with different masses circulate around their common CM (see Vol. 4).

4.1.3 Angular Momentum of a System of Particles

We consider two point masses m_1 and m_2 with their mutual interaction forces

$$\boldsymbol{F}_{12} = -\boldsymbol{F}_{21}$$

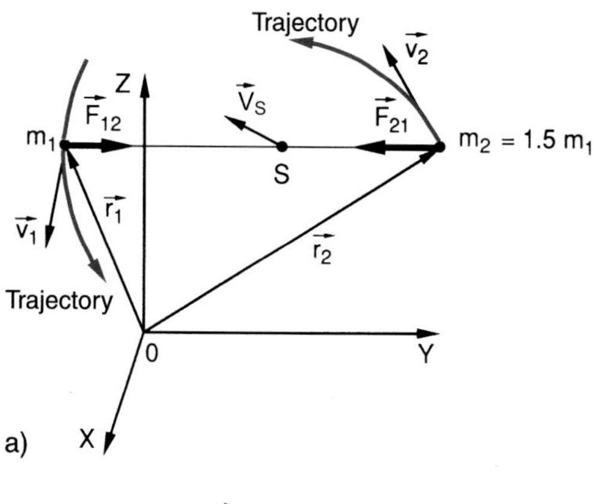

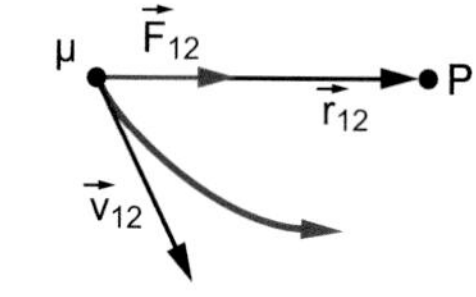

Figure 4.2 **a** Velocity V_S of the CM of a system of two masses with velocities v_i; **b** Reduction of the relative motion of two masses m_i to the motion of a single particle with the reduced mass μ under the action of the force F_{12}

and the external forces $\boldsymbol{F}_1$ acting on m_1 and $\boldsymbol{F}_2$ acting on m_2. The torques on the two masses with respect to the origin 0 of the coordinate system are

$$\boldsymbol{D}_1 = \boldsymbol{r}_1 \times (\boldsymbol{F}_1 + \boldsymbol{F}_{12}) \ ,$$
$$\boldsymbol{D}_2 = \boldsymbol{r}_2 \times (\boldsymbol{F}_2 + \boldsymbol{F}_{21}) \ ,$$

and the total torque of the system is then (Fig. 4.3)

$$\boldsymbol{D} = (\boldsymbol{r}_1 \times \boldsymbol{F}_1) + (\boldsymbol{r}_2 \times \boldsymbol{F}_2) + (\boldsymbol{r}_1 - \boldsymbol{r}_2) \times \boldsymbol{F}_{12} \ .$$

Since the direction of the internal forces $\boldsymbol{F}_{12} = -\boldsymbol{F}_{21}$ lies in the direction of the connecting line $\boldsymbol{r}_{12} = (\boldsymbol{r}_1 - \boldsymbol{r}_2)$ the last term vanishes and the total torque

$$\boldsymbol{D} = (\boldsymbol{r}_1 \times \boldsymbol{F}_1) + (\boldsymbol{r}_2 \times \boldsymbol{F}_2) \tag{4.12}$$

becomes the vector sum of the torques on the individual particles. Without external forces the total torque on the system is zero!

The total angular momentum L of the system with respect to the origin 0 is

$$\boldsymbol{L} = (\boldsymbol{r}_1 \times \boldsymbol{p}_1) + (\boldsymbol{r}_2 \times \boldsymbol{p}_2) \ , \tag{4.13}$$

and we obtain, analogous to the Eq. 2.48 for a single particle:

$$\frac{\mathrm{d}\boldsymbol{L}}{\mathrm{d}t} = (\boldsymbol{r}_1 \times \boldsymbol{F}_1) + (\boldsymbol{r}_2 \times \boldsymbol{F}_2) = \boldsymbol{D} \ . \tag{4.14}$$

The derivation of these equations and the situations discussed for a system of two particles can be readily generalized to a system of many particles. This gives the important statement:

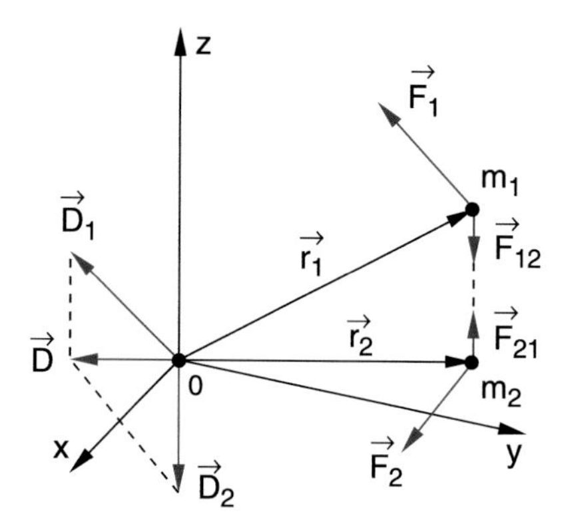

Figure 4.3 Torques acting on a system of two masses under the influence of external forces

The time derivative of the total angular momentum of a system of particles referred to an arbitrary point is equal to the total torque exerted onto the system and referred to the same point.

For the special case that no external forces are present the torque is zero and therefore the angular momentum L is constant.

The total angular momentum of a closed system of particles is constant.

Using CM-coordinates we can divide the angular momentum (4.13) according to (4.6a)

$$\begin{aligned}\boldsymbol{L} &= m_1 (\boldsymbol{r}_{1S} + \boldsymbol{R}_S) \times (\boldsymbol{v}_{1S} + \boldsymbol{V}_S) \\ &\quad + m_2 (\boldsymbol{r}_{2S} + \boldsymbol{R}_S) \times (\boldsymbol{v}_{2S} + \boldsymbol{V}_S) \ .\end{aligned}$$

For many particles this reads

$$\begin{aligned}\boldsymbol{L} &= \sum_i m_i (\boldsymbol{r}_{iS} + \boldsymbol{R}_S) \times (\boldsymbol{v}_{iS} + \boldsymbol{V}_S) \\ &= M (\boldsymbol{r}_S \times \boldsymbol{V}_S) + \sum_i m_i (\boldsymbol{r}_{iS} \times \boldsymbol{v}_{iS}) \\ &\quad + \sum_i m_i (\boldsymbol{R}_S \times \boldsymbol{v}_{iS}) + \sum_i m_i (\boldsymbol{r}_{iS} \times \boldsymbol{V}_S) \ .\end{aligned}$$

The terms $\sum_i m_i (\boldsymbol{R}_S \times \boldsymbol{v}_S)$ and $\sum_i m_i (\boldsymbol{r}_{iS} \times \boldsymbol{v}_S)$ are zero according to (4.6d) and (4.6b) and it follows:

$$\boldsymbol{L} = M (\boldsymbol{r}_S \times \boldsymbol{V}_S) + \sum_i m_i (\boldsymbol{r}_{iS} \times \boldsymbol{v}_{iS}) \ . \tag{4.14a}$$

The first term

$$\boldsymbol{L}_{0S} = M (\boldsymbol{R}_S \times \boldsymbol{V}_S) \tag{4.15a}$$

is the angular momentum of the total mass contracted in the CM referred to the origin of the coordinate system. The second term gives the total angular momentum referred to the CM.

For a system of two particles we can transform $\boldsymbol{L}_S$ because of $\sum_i m_i \boldsymbol{v}_{iS} = 0$ into

$$\begin{aligned} \boldsymbol{L}_S &= \sum \boldsymbol{L}_{iS} = (\boldsymbol{r}_{1S} \times \boldsymbol{p}_{1S}) + (\boldsymbol{r}_{2S} \times \boldsymbol{p}_{2S}) \\ &= (\boldsymbol{r}_{1S} - \boldsymbol{r}_{2S}) \times \boldsymbol{p}_{1S} = \boldsymbol{r}_{12} \times \mu \boldsymbol{v}_{12} , \end{aligned} \tag{4.15b}$$

(with $\boldsymbol{p}_{iS} = \mu \boldsymbol{v}_{12}$). This follows from (4.6d) and (4.10). We can therefore state:

The angular momentum $\boldsymbol{L}_S$ of a system of two particles in the CM is equal to the angular momentum of a single particle with the reduced mass μ and the position vector $\boldsymbol{r}_{12} = \boldsymbol{r}_1 - \boldsymbol{r}_2$.

Examples

1. The relative motion of the earth-moon system around their common center of mass S (Fig. 4.4) can be reduced to the motion of a single body with reduced mass $\mu = m_E \cdot m_{Mo}/(m_E + m_{Mo}) \approx 0.99 m_{Mo}$ in the central gravitational force field between earth and moon around the centre M of the earth. The centre of mass is located inside the earth 4552 km away from the centre M because the mass of the moon $m_{Mo} \approx 0.01 m_E$ is small compared with the earth mass. In the CM-system earth and moon describe nearly circular elliptical orbits around the common CM with radii

$$r_E = (m_{Mo}/(m_E + m_{Mo}))\, r_{EMo} \approx 0.01 r_{EMo}$$

and

$$r_{Mo} = (m_E/(m_E + m_{Mo}))\, r_{EMo} \approx 0.99 r_{EMo} ,$$

where r_{EMo} is the distance between earth and moon. In a coordinate system which is referred to the centre of our galaxy the lunar orbit is a complicated curve, shown in Fig. 4.4b where the deviations from the path of the CM are exaggerated in order to elucidate the situation. This complicated motion can be composed of

a) the motion of the moon around the CM of the earth-moon system
b) the motion of the CM around the centre of mass of the solar system, which is located inside the sun, because $M_\odot > 10^3 \cdot \sum m_{\text{Planets}}$.
c) the motion of the CM of the solar system around the centre of our galaxy.
d) The exact calculation of the lunar orbit has to take into account the simultaneous gravitational attraction of the moon by the earth and the sun, which changes with time because of the changing relative position of the three bodies. Because of this "perturbation" the lunar orbit is not exactly an ellipse around the CM. Although there is no analytical solution for the exact orbit, very good numerical approximations have been developed [4.1b].

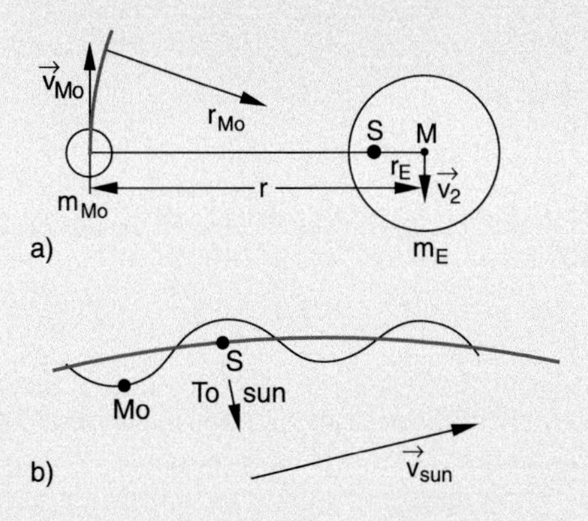

Figure 4.4 **a** Motion of the moon in the CM-system earth–moon. **b** Motion of the moon and the CM in the galactic coordinate system where the sun also moves

2. The hydrogen atom is a two-body system of an electron with mass m_e and proton with mass m_p. Because $m_p = 1836 m_e$ the reduced mass is $\mu = 0.99946 m_e \approx m_e$. In a classical picture proton and electron circulate around the CM. With the mean distance r_{pe} between proton and electron the CM lies $(1/1836) r_{pe}$ from the centre of the proton. The motion of the two particles can be separated into the translation of the CM with the velocity V_S and the motion of a particle with mass μ with the relative velocity v_{pe} around the CM. The total kinetic energy of the H-atom is then:

$$E_{\text{kin}} = \tfrac{1}{2}\left(m_p + m_e\right) V_S^2 + \tfrac{1}{2}\mu v_{pe}^2 .$$

For velocities of the H-atom which correspond to thermal energies at room temperature the first term (≈ 0.03 eV) is very small compared to the second term of the "internal" energy (≈ 10 eV). ◀

4.2 Collisions Between Two Particles

This section is of great importance for the understanding of many phenomena in Atomic and Nuclear Physics, because an essential part of our knowledge about the structure and dynamics of atoms and nuclei arises from investigations of collision processes.

When two particles approach each other they are deflected due to the interaction forces between them. The deflection occurs in the whole spatial range where the forces are noticeable (Fig. 4.5). Due to this interaction both particles change their momentum and often also their energy. However, conservation laws demand that momentum and energy of the total system are always preserved.

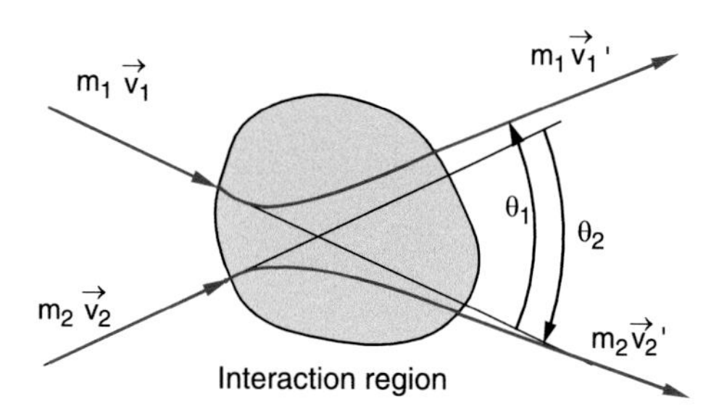

Figure 4.5 Schematic illustration of a collision with the asymptotic scattering angles θ_1 an θ_2

The exact form of the trajectory of the particles inside the interaction zone can be determined only if the exact interaction potential is known. However, it is possible to make definite statements about magnitude and direction of the particle momenta after the collision in a great distance from the interaction zone. These statements are based solely on the conservation of momentum and energy. We will illustrate this in more detail in the following section.

4.2.1 Basic Equations

Although the total energy of the two colliding partners is preserved during collisions, part of the translational energy is often converted into other forms of energy, as for instance potential energy or thermal energy. From (4.3) it follows however, that the total momentum of the collision partners is always retained.

The basic equations for collision processes between two particles with velocities v which are small compared to the velocity c of light (non-relativistic collisions) can be written as:

conservation of momentum (Fig. 4.6)

$$\boldsymbol{p}_1' + \boldsymbol{p}_2' = \boldsymbol{p}_1 + \boldsymbol{p}_2 \tag{4.16}$$

conservation of energy

$$\frac{p_1'^2}{2m_1'} + \frac{p_2'^2}{2m_2'} = \frac{p_1^2}{2m_1} + \frac{p_2^2}{2m_2} + U \tag{4.17}$$

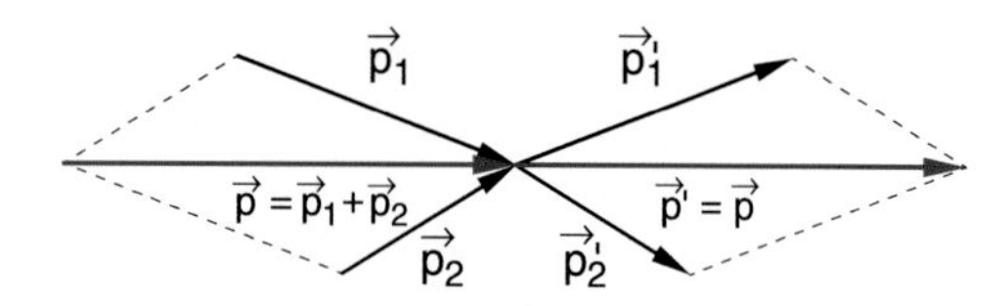

Figure 4.6 Conservation of total momentum at the collision of two particles

where p_i' is the momentum of particle i after the collision and U is that part of the initial energy that had been converted into internal energy of one or both of the collision partners and is therefore missing in the kinetic energy after the collision ($U < 0$). If internal energy of the colliding partners has been transferred into kinetic energy we get $U > 0$.

The Eq. 4.16 and 4.17 describe the collision process completely in that sense, that relations between magnitude and direction of the individual momenta of the particles after the collision can be determined, if they are known before the collision.

Depending on the magnitude of U we distinguish between three cases:

- **$U = 0$, elastic collisions**. The total kinetic energy is preserved, while the kinetic energy of the individual particles generally changes.
- **$U < 0$, inelastic collisions**. The total kinetic energy after the collision is smaller than before. Part of the initial kinetic energy has been converted into internal energy of the collision partners.
- **$U > 0$, superelastic collisions** (sometimes called collisions of the second kind). At least one of the collision partners had internal energy before the collision which was transferred into kinetic energy during the collision.
  The kinetic energy after the collision is larger than before the collision.

During **reactive collisions** (for instance during chemical reactions or in high energy collisions) new particles can be produced and the masses of the collision partners may change. An example is the reaction

$$H_2 + Cl_2 \longrightarrow HCl + HCl\,.$$

These reactive collisions are treated later.

Note:

- While the kinetic energy is only preserved in elastic collisions the total momentum is preserved for all kinds of collisions (Fig. 4.6).
- Inelastic, super-elastic and reactive collisions can only occur, if at least one of the collision partners has an internal structure. This means that it must consist of at least two particles, which are bound together. Examples are atoms (consisting of nuclei and electrons) or nuclei (consisting of protons and neutrons). Part of the kinetic energy of the collision partners then can be transferred into the increase of the internal energy (potential or kinetic energy of the constituents). For collision partners consisting of many particles (for example solids) the increase of kinetic energy of the constituents can be defined as an increase of the temperature (see Sect. 7.3) which is then called *"thermal energy"* (see Sect. 10.1).

4.2.2 Elastic Collisions in the Lab-System

The description of collision processes can be essentially simplified when the appropriate coordinate system is chosen. For many situations one of the collision partners, for instance m_2, is at rest before the collision. We choose its position as the origin of our coordinate system, which is fixed relative to the laboratory system. In this system is therefore $\boldsymbol{p}_2 = \boldsymbol{0}$ (Fig. 4.7). We assume that the masses do not change during the collision ($m_1 = m_1'$, $m_2 = m_2'$). With $U = 0$ for elastic collisions we obtain from (4.16) and (4.17)

$$\boldsymbol{p}_1 = \boldsymbol{p}_1' + \boldsymbol{p}_2' = \boldsymbol{p}' \ , \tag{4.16a}$$

$$\frac{p_1^2}{2m_1} = \frac{p_1'^2}{2m_1} + \frac{p_2'^2}{2m_2} \ . \tag{4.17a}$$

We choose the direction of the initial momentum $\boldsymbol{p}_1$ as the x-direction (Fig. 4.8) $\Rightarrow \boldsymbol{p}_1 = \{p_1, 0, 0\}$. The angular momentum $\boldsymbol{L} = \boldsymbol{r} \times \boldsymbol{p}$ points into the z-direction. Because the angular momentum is constant the motion of the collision partners is restricted to the x-y-plane. The endpoint of the vector $\boldsymbol{p}_2'$ is the point $P(x, y)$. From Fig. 4.8 we derive the relations:

$$x^2 + y^2 = p_2'^2 \ ,$$
$$(p_1 - x)^2 + y^2 = p_1'^2 \ .$$

Inserting into (4.17a) yields

$$\frac{p_1^2}{2m_1} = \frac{(p_1 - x)^2 + y^2}{2m_1} + \frac{x^2 + y^2}{2m_2} \ .$$

Rearranging gives with the reduced mass $\mu = m_1 \cdot m_2/(m_1 + m_2)$ the equation

$$(x - \mu v_1)^2 + y^2 = (\mu v_1)^2 \tag{4.18}$$

of a circle in the x-y-plane with the radius $r = \mu v_1$ and the centre $M = \{\mu v_1, 0\}$. This implies that the endpoints of all possible vectors p_2' which fulfil energy-and momentum conservation have to lie on the circle around M, if they start from the origin $\{0, 0\}$ (Fig. 4.9).

The angles θ_1 and θ_2 are the deflection angles of the two collisions partners. The maximum deflectionangle $\theta_1^{\max}$ of the

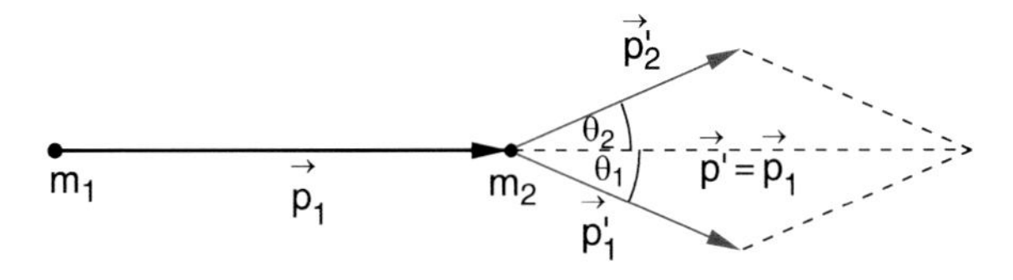

Figure 4.7 Collision of a particle with mass m_1 and momentum p_1 with a mass m_2 at rest, drawn in the Lab-system

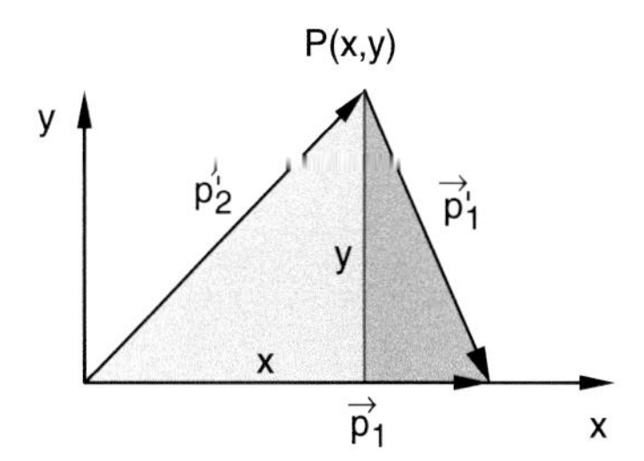

Figure 4.8 Illustration of (4.18)

impinging particle is reached, when $\boldsymbol{p}_1'$ is the tangent to the circle. For $m_1 > m_2 \rightarrow p_1 = m_1 v_1 > 2\mu v_1$, which means that $|p_1| > 2r$. The magnitude of the momentum of the impinging particle is larger than the diameter of the circle. From Fig. 4.9 we can then conclude the relation

$$\sin\theta_1^{\max} = \frac{\mu v_1}{(m_1 - \mu) v_1} = \frac{\mu}{m_1 - \mu} = \frac{m_2}{m_1} \ . \tag{4.19}$$

Examples

1. $m_1 = 1.1 m_2 \Rightarrow \mu = 0.52 m_2 \Rightarrow \sin\theta_1^{\max} = 0.91$
   $\Rightarrow \theta_1^{\max} = 65° \ .$
2. $m_1 = 2m_2 \Rightarrow \mu = 0.67 m_2 \Rightarrow \sin\theta_1^{\max} = 0.5$
   $\Rightarrow \theta_1^{\max} = 30° \ .$
3. $m_1 = 100 m_2 \Rightarrow \mu = 0.99 m_2$
   $\Rightarrow \theta_1^{\max} = 0.6° \ .$ ◀

Special Case: Central Collisions

If $\boldsymbol{p}_2'$ has the same direction as $\boldsymbol{p}_1$ the deflection angle becomes $\theta_2 = 0$ (central or collinear collision). All vectors $\boldsymbol{p}_1$, $\boldsymbol{p}_1'$ and $\boldsymbol{p}_2$ are collinear and coincide in Fig. 4.9 with the x-axis. We obtain

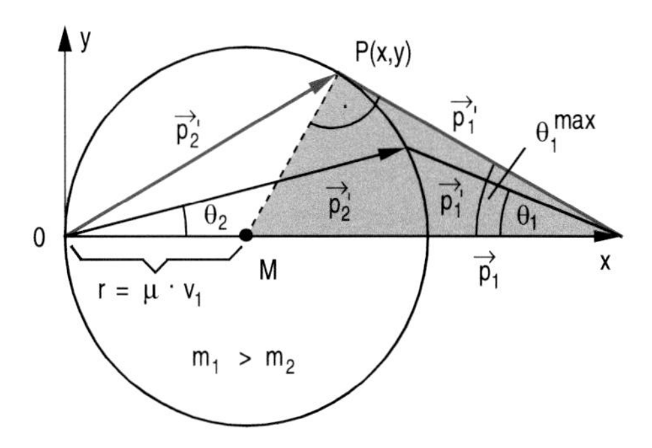

Figure 4.9 Momentum diagram of elastic collisions for $m_1 > m_2$. All possible endpoints of the vector $\boldsymbol{p}_2'$ are located on the circle with radius μv_1 around M

from Fig. 4.9:

$$
\begin{aligned}
p_1 &= 2\mu v_1 + p_1' \\
\Rightarrow m_1 v_1' &= m_1 v_1 - 2\frac{m_1 m_2}{m_1 + m_2} v_1 \\
\Rightarrow v_1' &= \frac{m_1 - m_2}{m_1 + m_2} v_1 \; ; \\
p_2' &= 2\mu v_1 \\
\Rightarrow v_2' &= 2\frac{\mu}{m_2} v_1 = \frac{2m_1}{m_1 + m_2} v_1 \; .
\end{aligned}
\tag{4.20}
$$

The momentum of the pushed particle gets its maximum value $p_2' = 2\mu v_1$ for collinear collisions.

Also the kinetic energy, transferred from m_1 to m_2 during a collinear elastic collision reaches its maximum value

$$
\begin{aligned}
\Delta E_{\mathrm{kin}} &= \frac{p_2'^2}{2m_2} \le \Delta E_{\mathrm{kin}}^{\max} = \frac{2m_1^2 m_2}{(m_1 + m_2)^2} v_1^2 \\
\Delta E_{\mathrm{kin}}^{\max} &= 4\frac{m_1 m_2}{M^2} E_1 = \frac{4\mu^2}{m_1 m_2} E_1 \; ,
\end{aligned}
\tag{4.21}
$$

which equals the fraction $4\mu^2/(m_1 m_2)$ of the initial energy E_1 of the impinging mass m_1. In Fig. 4.10 this maximum transferred fraction is shown as a function of the mass ratio m_1/m_2.

For $m_1 = m_2$ it is $v_1' = 0$ and $v_2' = v_1$. The two masses exchange their momentum during the collision, i. e. after the collision m_1 is at rest and m_2 moves with the momentum $\boldsymbol{p}_2' = \boldsymbol{p}_1$.

For equal masses $m_1 = m_2$ the energy of the incident particle is completely transferred to the resting mass m_2 during a collinear collision.

$$
\frac{\Delta E_{\mathrm{kin}}^{\max}}{E_1} = \frac{4m_1 m_2}{(m_1+m_2)^2} = \frac{4}{(m_1/m_2)+2+1/(m_1/m_2)}
$$

Figure 4.10 Maximum energy transfer $\Delta E = E - E_1'$ for a collinear elastic collison of a particle with mass m_1 onto a mass m_2 at rest for different ratios m_1/m_2

Special cases of non-collinear collisions

We will now discuss the general case of non-collinear collisions and illustrate it by some important special cases of the mass ratio m_1/m_2.

- $\boldsymbol{m_1 = m_2 = m \quad \Rightarrow \quad \mu = \frac{1}{2}m}$.

Equation 4.18 gives for the radius of the circle in Fig. 4.9 $r = \frac{1}{2}mv_1$, which implies that the momentum $p_1 = mv_1$ of the incident particle is the diameter of the circle (Fig. 4.11). For non-collinear collisions the momenta $\boldsymbol{p}_1'$ and $\boldsymbol{p}_2'$ **after the collision** are perpendicular to each other according to the theorem of Thales. For the deflection angles it follows $\theta_1 + \theta_2 = \pi/2$.

The paths of the two particles are perpendicular to each other after the non-collinear collision, i. e. $\boldsymbol{p}_1' \perp \boldsymbol{p}_2'$.

Example

For the deceleration of neutrons in nuclear reactors a material with many hydrogen atoms is the best choice. Because the protons have nearly the same mass as neutrons. ◀

- $\boldsymbol{m_1 \ll m_2 \quad \Rightarrow \quad \mu \approx m_1}$

The radius of the circle in Fig. 4.9 becomes for the limiting case $m_1/m_2 \to 0$ equal to the momentum $p_1 = m_1 v_1$ of the incident particle (Fig. 4.12a). The magnitude of $\boldsymbol{p}_1$ does not change during the collision ($|p_1| = |p_1'|$) but all directions of $\boldsymbol{p}_1'$ are possible. The scattering angle θ_1 can take all values in the range $-\pi \le \theta_1 \le +\pi$.

The maximum momentum transfer onto m_2 is

$$
|p_2'|_{\max} = 2r = 2p_1 \; .
$$

The maximum transferred energy is

$$
\Delta E_{\mathrm{kin}}^{\max} = \frac{(2p_1)^2}{2m_2} = \frac{4p_1^2}{2m_1}\frac{m_1}{m_2} = 4\frac{m_1}{m_2}E_1 \; . \tag{4.22}
$$

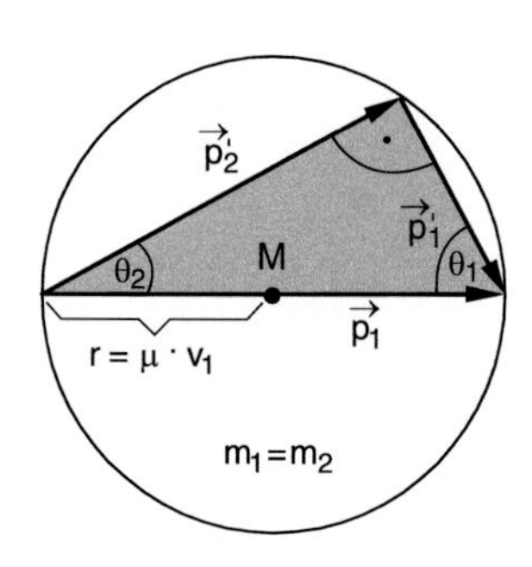

Figure 4.11 Elastic collision between particles of equal mass

Chapter 4

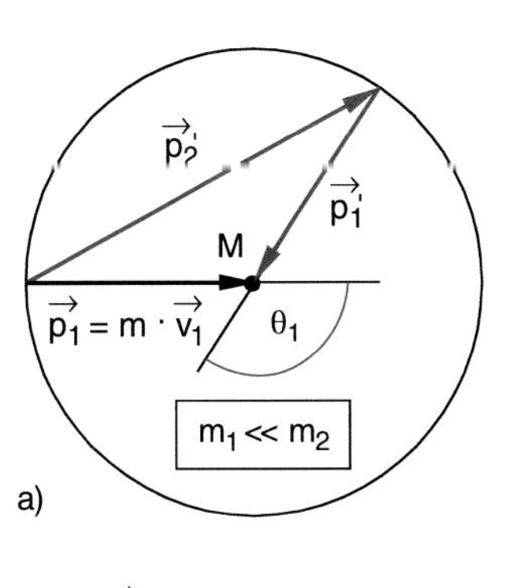

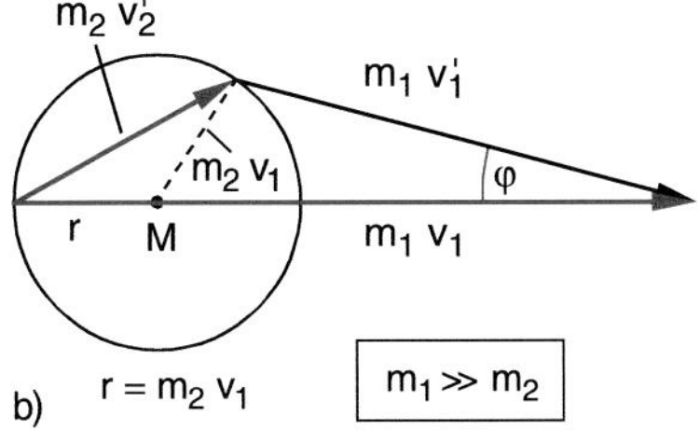

Figure 4.12 Elastic collision for $m_1 \ll m_2$ (**a**) and $m_1 \gg m_2$ (**b**)

In collisions of a small mass m_1 against a large mass m_2 the maximal transferred fraction of the initial kinetic energy is $4(m_1/m_2)$.

Examples

1. *Impact of a particle onto a solid wall.*

$$m_2 = \infty \Rightarrow \Delta E_{\text{kin}}^{\max} = 0 \quad \text{but: } \boldsymbol{p}_2' = -2\boldsymbol{p}_1 .$$

During the elastic collision of a particle with a solid wall the particle is elastically reflected and $\boldsymbol{p}_1' = -\boldsymbol{p}_1$. Therefore twice the initial momentum is transferred to the wall but no energy!

2. *Collision of an electron with a proton at rest.* $m_1 = m_2/1836$. The maximal transferred energy occurs in central collisions and is then $\Delta E_{\text{kin}}^{\max} = 4(m_1/m_2)E_1 = 0.00218\,E_1$. ◀

- $\boldsymbol{m_1 \gg m_2 \quad \Rightarrow \quad \mu \approx m_2}$.

In this case the radius of the circle in Fig. 4.9 is $r = m_2 v_1$ (Fig. 4.12b). For central collisions is

$$m_2 v_2' = 2r = 2m_2 v_1 \Rightarrow v_2' = 2v_1 ,$$

and the transferred energy is

$$\Delta E_{\text{kin}} = \frac{m_2}{2} v_2'^2 = 4\frac{m_2}{m_1} E_1 . \tag{4.23}$$

For non-collinear collisions the energy transferred to m_2 is smaller. The maximum deflection angle $\varphi = \theta_1^{\max}$ of the incident mass m_1 is according to (4.19)

$$\sin\varphi = \frac{m_2}{m_1} .$$

Example

In collisions of α-particles (helium nuclei) with electrons at most the fraction $\Delta E_1 = 0.00054 E_1$ of the initial energy E_1 can be transferred to the electron. The maximum deflection angle of the α-particles is $\varphi \approx \sin\varphi = 1.36 \cdot 10^{-4}\text{rad} = 0.48'$. When α-particles pass through matter the electron shell of the atoms contributes to the deflection only a tiny part. Most of the deflection is caused by the atomic nuclei (see Rutherford scattering in Vol. 3). ◀

4.2.3 Elastic Collisions in the Centre-of Mass system

When none of the collision partners is resting, the description of the collision process is often simpler in the CM-system than in the lab-system. Since, however, the observation of the process always occurs in the lab-system the measured results must be transformed into the CM-system in order to compare them with the predictions calculated in the CM-system. The relations between position vectors and velocities in the two systems is illustrated in Fig. 4.13b and the results are compiled in Tab. 4.1.

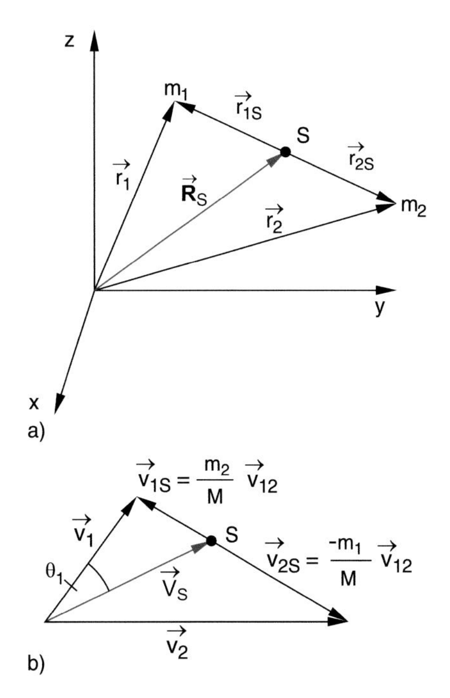

Figure 4.13 Graphical representation of the relations between **a** Lab- and CM-coordinates and **b** Lab- and CM-velocities

Table 4.1 Compilation of quantities relevant for collisions in the lab-system and the CM-system

$M = m_1 + m_2 =$ total mass
$\mu = \frac{m_1 \cdot m_2}{m_1 + m_2} =$ reduced mass
$\boldsymbol{R}_S = \frac{1}{M}(m_1\boldsymbol{r}_1 + m_2\boldsymbol{r}_2) =$ position vector of CM
$\boldsymbol{V}_S = \frac{1}{M}(m_1\boldsymbol{v}_1 + m_2\boldsymbol{v}_2) =$ velocity of CM
$\boldsymbol{r}_{12} = \boldsymbol{r}_1 - \boldsymbol{r}_2 =$ relative distance
$\boldsymbol{v}_{12} = \boldsymbol{v}_1 - \boldsymbol{v}_2 =$ relative velocity
$\boldsymbol{r}_{iS} = \boldsymbol{r}_i - \boldsymbol{R}_S =$ position vector of i-th particle in the CM-system
$\boldsymbol{v}_{iS} = \boldsymbol{v}_i - \boldsymbol{V}_S =$ velocity of i-th particle in the CM-system
$\boldsymbol{p}_{iS} = m_i\boldsymbol{v}_{iS} =$ momentum of i-th particle in the CM-system
$\sum \boldsymbol{p}_{iS} = 0$
$\Theta_i =$ deflection angle of i-th particle in the lab-system
$\vartheta_i =$ deflection angle of i-th particle in the CM-system

Note: We will denote the center of mass with the index S.

Since the total momentum in the CM-system is always zero, we can write for two particles 1 and 2

$$\boldsymbol{p}_{1S} = -\boldsymbol{p}_{2S} \quad \text{and} \quad \boldsymbol{p}'_{1S} = -\boldsymbol{p}'_{2S} \,.$$

The sum of the momenta of the collision partners before the collision and after the collision is in the CM-system always zero.

From the energy conservation (4.17) it therefore follows:

$$\frac{1}{2}\left(\frac{1}{m_1} + \frac{1}{m_2}\right)p'^2_{1S} = \frac{1}{2}\left(\frac{1}{m_1} + \frac{1}{m_2}\right)p^2_{1S} + U \,,$$

which can be written when using the reduced mass μ

energy conservation in the S-system

$$\frac{p'^2_{1S}}{2\mu} = \frac{p^2_{1S}}{2\mu} + U \,. \tag{4.24}$$

For elastic collisions ($U = 0$) in the CM-system is $p^2_{1S} = p'^2_{1S}$ and $p^2_{2S} = p'^2_{2S}$. This means:

In the CM-system each collision partner retains in elastic collisions its kinetic energy.

In the CM-system the result of an elastic collision is merely a turn of the momentum vectors which are always pointing into the opposite direction (Fig. 4.14).

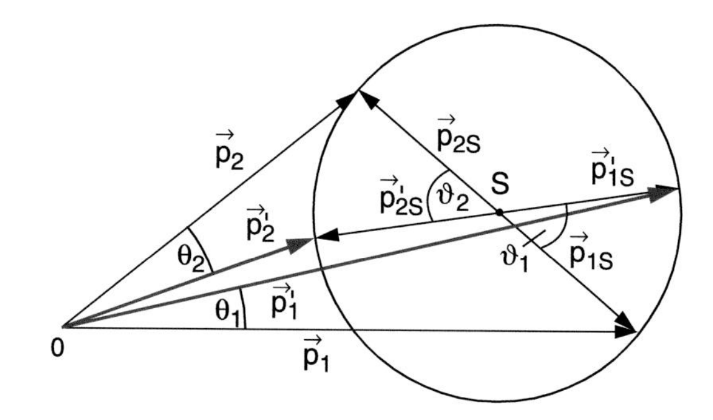

Figure 4.14 In the CM-system an elastic collision is represented by a turn of the momentum vectors without changing their length

Note: in order to distinguish the deflection angles θ in the lab-system written as capital letters, from those in the CM we will label all deflection angles in the CM-system by lower case letters ϑ.

Example

Deceleration of neutrons (mass m_1, velocity v_1) in elastic collisions by atomic nuclei (mass m_2) at rest. The CM-velocity is

$$\boldsymbol{V}_S = \boldsymbol{V}'_S = \frac{m_1\boldsymbol{v}_1}{m_1 + m_2} = \frac{\boldsymbol{v}_1}{1 + A} \quad \text{with} \quad A = m_2/m_1 \,.$$

◀

The velocity of the two particles in the CM-system is according to Fig. 4.15

$$\boldsymbol{v}_{1S} = \boldsymbol{v}_1 - \boldsymbol{V}_S = \frac{A\boldsymbol{v}_1}{1 + A} \,;$$
$$\boldsymbol{v}_{2S} = \boldsymbol{0} - \boldsymbol{V}_S = -\frac{\boldsymbol{v}_1}{1 + A} \,;$$
$$\boldsymbol{v}'_{1S} = \boldsymbol{v}'_1 + \boldsymbol{v}_{2S} = \boldsymbol{v}'_1 - \boldsymbol{V}_S \,.$$

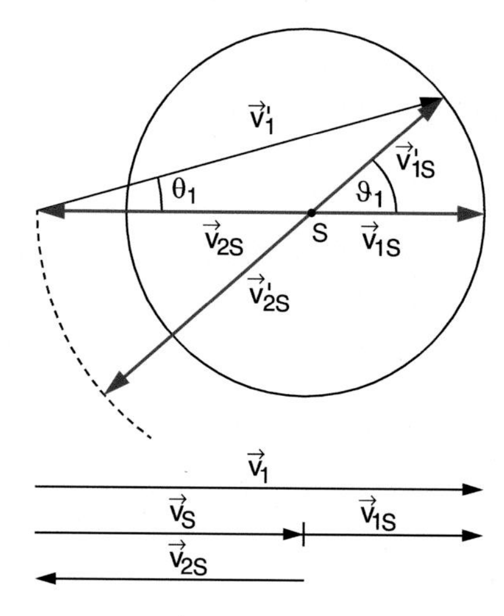

Figure 4.15 Determination of energy transfer at elastic collisions

Squaring gives with $\boldsymbol{v}'_{1S} \cdot \boldsymbol{V}_S = v'_{1S} \cdot V_S \cdot \cos\vartheta_1$

$$v_1'^2 = v_{1S}'^2 + V_S^2 + 2v'_{1S} v_S \cos\vartheta_1 ,$$

where ϑ_1 is the angle between $\boldsymbol{v}'_{1S}$ and $\boldsymbol{V}_S$. Because $\boldsymbol{V}_S \parallel \boldsymbol{v}_{1S}$ it follows that ϑ_1 is also the angle between v'_{1S} and v_{1S}, i.e. the deflection angle of m_1 in the CM-system. Inserting the relations above for V_S we obtain

$$v_{1S}'^2 = v_1^2 \frac{A^2 + 2A\cos\vartheta_1 + 1}{(1+A)^2} .$$

The ratio of the kinetic energies of the neutron after and before the collision is then

$$\left(\frac{E'_{\text{kin}}}{E_{\text{kin}}}\right) = \frac{v_1'^2}{v_1^2} = \frac{A^2 + 2A\cos\vartheta_1 + 1}{(1+A)^2} .$$

For central collisions is $\vartheta_1 = \pi$ and the ratio becomes

$$\left(\frac{E'_{\text{kin}}}{E_{\text{kin}}}\right)^{\text{central}} = \left(\frac{A-1}{A+1}\right)^2 .$$

For the transferred energy $\Delta E = E'_{\text{kin}} - E_{\text{kin}}$ we then obtain

$$\frac{\Delta E}{E_{\text{kin}}} = \frac{4A}{(A+1)^2} = \frac{4m_1 m_2}{(m_1 + m_2)^2} .$$

For $m_1 = m_2$ the transferred energy $\Delta E/E$ takes on its maximum value $\Delta E/E = 1$. This means that the neutron can transfer its total kinetic energy if it suffers a central collision with a proton.

4.2.4 Inelastic Collisions

For inelastic collisions part of the initial kinetic energy is transferred into internal energy U of the collision partners. Such collisions are only possible, if at least one of the partners has a variable internal sub-structure, which means that it has to be composed of two or more particles. For point masses there are no inelastic collisions!

For inelastic collisions momentum conservation remains valid (4.16) and also energy conservation with $U < 0$ (4.17). In the limiting case of *maximal inelastic collisions* the two collision partners stick together after the collision and move with the CM-velocity.

$$\boldsymbol{V}_S = \frac{m_1 \boldsymbol{v}_1 + m_2 \boldsymbol{v}_2}{m_1 + m_2} . \tag{4.25}$$

From (4.17) and (4.25) we obtain for the maximum fraction of the kinetic energy, which is transferred into internal energy

$$\begin{aligned} U &= \frac{1}{2}(m_1 + m_2) V_S^2 - \frac{1}{2}\left(m_1 v_1^2 + m_2 v_2^2\right) \\ &= -\frac{1}{2}\frac{m_1 m_2}{m_1 + m_2}(\boldsymbol{v}_1 - \boldsymbol{v}_2)^2 = -\frac{1}{2}\mu v_{12}^2 , \end{aligned} \tag{4.26a}$$

which is identical to the kinetic energy of the two particles in the CM-system (see (4.11b)).

In a completely inelastic collision, where the two particles stick together after the collision, just the kinetic energy of the two particles in the CM-system is converted into internal energy of one or both collision partners.

From (4.26) it follows that only in cases where the two collision partners have equal but opposite momenta ($m_1 \boldsymbol{v}_1 = -m_2 \boldsymbol{v}_2 \Rightarrow \boldsymbol{V}_S = \boldsymbol{0}$) the total kinetic energy can be converted into internal energy. The two particles then stick together and are at rest, their total momentum is zero before and after the collision. For all other collisions $|U| < |U_{\max}|$. Therefore the general rule is:

For all inelastic collisions not more than $\Delta E = \frac{1}{2}\mu \cdot v_{12}^2$ of the initial kinetic energy can be converted into internal energy. At least the proportion

$$\tfrac{1}{2}(m_1 + m_2)V_S^2 = \tfrac{1}{2}MV_S^2 \tag{4.26b}$$

of the CM-motion remains as kinetic energy of the collision partners.

Examples

1. A glider with mass m_1 on an air-track hits a second glider with mass m_2 at rest ($v_2 = 0$). The two colliding ends are covered with plasticine, which causes the two gliders to stick together after the collision and they move with the CM-velocity

$$\boldsymbol{V}_S = \frac{m_1}{m_1 + m_2}\boldsymbol{v}_1 .$$

The kinetic energy after the collision is

$$E'_{\text{kin}} = \frac{m_1 + m_2}{2} V_S^2 = \frac{m_1^2}{2(m_1 + m_2)} v_1^2 ,$$

and the energy converted into the plasticine energy is

$$U = E'_{\text{kin}} - E_{\text{kin}} = -\frac{m_2}{m_1 + m_2} E_{\text{kin}} .$$

For $m_1 = m_2$ this gives

$$U = -\tfrac{1}{2}E_{\text{kin}} .$$

2. A neutron n with velocity v_1 impinges on a proton p at rest. This produces a deuteron d = np.

$$\text{n} + \text{p} \longrightarrow \text{d} .$$

Because of $m_1 = m_2$ the deuteron moves with the CM-velocity $V_S = \frac{1}{2}v_1$ and has therefore half of the initial kinetic energy $E'_{\text{kin}} = E_{\text{kin}}$ of the incident neutron. The other half is converted into internal energy of the deuteron, which is excited into a higher energy state, that can decay by emission of γ-radiation. ◀

Summarizing the results: In inelastic collisions of particles with equal masses where one collision partner is at rest at most half of the kinetic energy of the incident particle can be converted into internal energy

$$|U| \le |U_{\max}| = \frac{1}{2}\frac{m}{2}v_1^2 \,. \tag{4.27a}$$

The amount $U_{\max} - U$ remains as kinetic energy of the collision partners in addition to the kinetic energy $\frac{1}{2}MV_S^2$ of the CM-motion.

Special Cases

- If a particle with mass m_1 suffers a totally inelastic collision with a wall ($m_2 \gg m_1 \to \mu \approx m_1$) it remains adsorbed at the wall and transfers its kinetic energy completely to the wall, which heats up. ($U = -E_{\text{kin}}, E'_{\text{kin}} = 0$).
- If two equal masses collide head-on with $\boldsymbol{p}_1 = -\boldsymbol{p}_2$ the total momentum after the collision must be zero. With $v_1^2 = v_2^2 = v^2$ the increase of internal energy is

$$U = -\frac{1}{2}(m_1 + m_2)v^2 \,,$$

as in the first case the total kinetic energy is converted into internal energy. These two special cases are illustrated in Fig. 4.16 and compared with the corresponding elastic collisions.

Examples

1. Collisional excitation of mercury atoms by electron impact (Franck-Hertz-Experiment). Because of $m_{\text{Hg}} \gg m_e$ the reduced mass $m \approx m_e$. From (4.26) we can conclude that nearly the total kinetic energy of the electron can be converted into excitation energy of the Hg-atoms.
2. A heavy particle with mass $m_1 = 100m_2$ collides with a particle of mass m_2. Now $\mu = 0.99m_2$ and $U = (0.99/100)m_2v_1^2/2$. This implies that only about 1% of the kinetic energy is converted into internal energy U. ◄

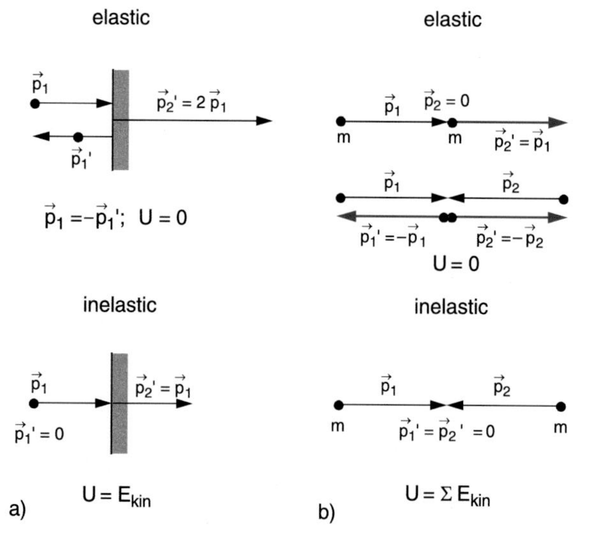

Figure 4.16 Comparison of elastic and completely inelastic collisions: **a** particle against a wall, **b** collision between two equal masses

4.2.5 Newton-Diagrams

The measurements of deflection angles at collisions between atoms or molecules is performed in the laboratory-system. The determination of the interaction potential derived from these deflection angles is, however, much easier in the CM-system. The relations between the relevant parameters in the two systems (velocities, deflection angles, energy transfer) for arbitrary elastic or inelastic collisions can be visualized with the help of **Newton diagrams**, which connects the velocities in the lab-system with those in the CM-system (Fig. 4.17). The parameters used in the following are listed in Tab. 4.1.

With the relations

$$\begin{aligned} \boldsymbol{r}_1 &= \boldsymbol{R}_S + (m_2/M)\,\boldsymbol{r}_{12} \quad \text{and} \\ \boldsymbol{r}_2 &= \boldsymbol{R}_S - (m_1/M)\,\boldsymbol{r}_{12} \,, \end{aligned} \tag{4.28}$$

$$\begin{aligned} \boldsymbol{v}_1 &= \boldsymbol{V}_S + (m_2/M)\,\boldsymbol{v}_{12} \quad \text{and} \\ \boldsymbol{v}_2 &= \boldsymbol{V}_S - (m_1/M)\,\boldsymbol{v}_{12} \,, \end{aligned} \tag{4.29}$$

which can be derived from Fig. 4.13, the kinetic energy can be separated into the two parts

$$E_{\text{kin}} = \underbrace{\tfrac{1}{2}m_1v_1^2 + \tfrac{1}{2}m_2v_2^2}_{E_{\text{kin}}\text{ in lab frame}} = \underbrace{\tfrac{1}{2}MV_S^2}_{\substack{E_{\text{kin}}\text{ of}\\ \text{CM-motion}}} + \underbrace{\tfrac{1}{2}\mu v_{12}^2}_{\substack{E_{\text{kin}}\text{ of relative motion}\\ \text{in the CM-system}}} \,. \tag{4.30}$$

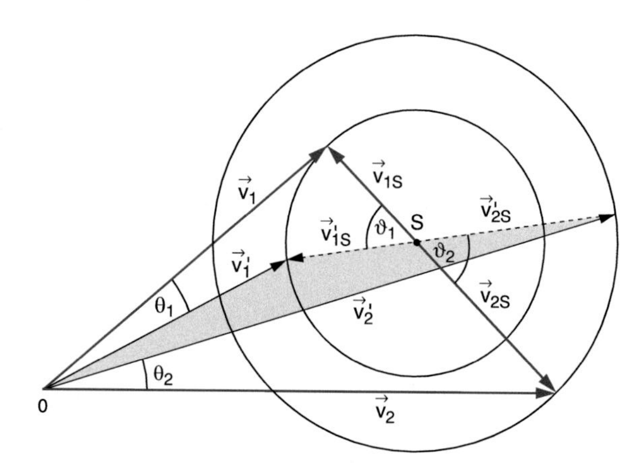

Figure 4.17 Newton diagram of elastic collision between two particles

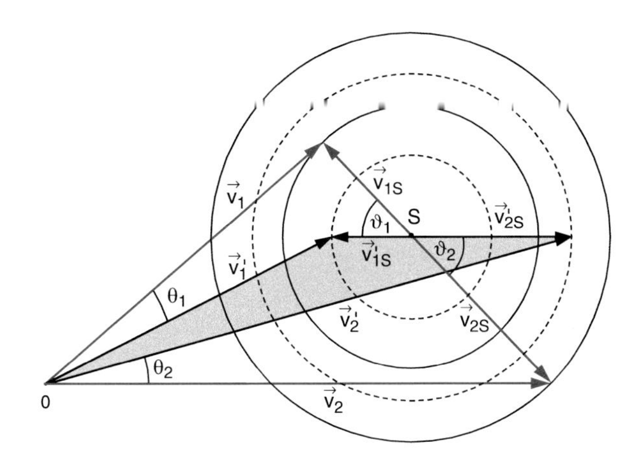

Figure 4.18 Newton-diagram of inelastic collisions between two particles

Since for elastic collisions the kinetic energy of each partner in the CM-system is preserved, the vector of the relative velocity v_{12} retains its magnitude but turns around the centre of mass S where the end of the vector describes a circle with the radii $v_{1S} = (m_2/M)v_{12}$ and $v_{2S} = (m_1/M)v_{12}$. The deflection angles ϑ_1 in the CM-system can be determined graphically from the deflection angles θ_1 in the lab-system.

In particular the maximum deflection angle $\theta_1^{\max}$ can be determined readily. It appears when v'_1 is the tangent to the Newton circle.

For inelastic collisions (Fig. 4.18) part of the kinetic energy $\frac{1}{2}\mu v_{12}^2$ is converted into excitation energy, which means that v'_{12} becomes smaller. However, still the centre-of-mass S divides the connecting line between the endpoints of the vectors $\boldsymbol{v}_1$ and $\boldsymbol{v}_2$ in the ratio m_1/m_2 of the two masses. The endpoints of v'_{12} are now located on a circle with smaller radius (dashed circles in Fig. 4.18).

For both elastic and inelastic collisions the range of possible deflection angles and the maximum deflection angles can be determined from the Newton diagrams. Therefore such diagrams are very useful for the planning of experiments, because they tell us, in which deflection ranges one must look for scattered particles for given initial conditions [4.2].

4.3 What Do We Learn from the Investigation of Collisions?

The deflection of a particle A during the collision with another particle B is due to the momentum transfer

$$\Delta p = \int_{-\infty}^{+\infty} \boldsymbol{F}\,\mathrm{d}t\,, \tag{4.31}$$

which is caused by the force F acting between A and B while passing by each other. The momentum change Δp experienced

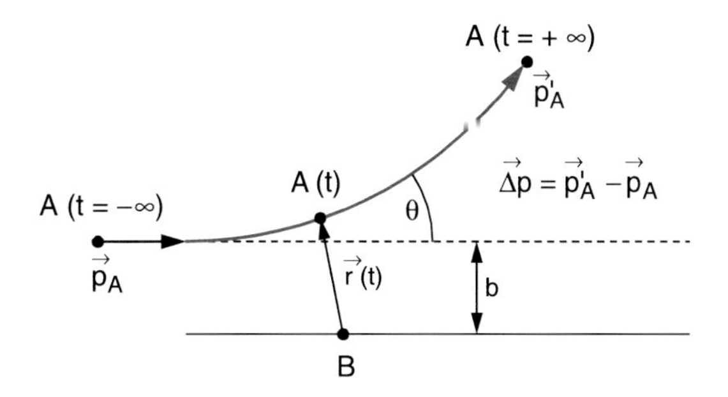

Figure 4.19 Illustration of impact parameter

by A is, of course, compensated by the change $-\Delta p$ experienced by B, because the total momentum has to be conserved.

Since $\boldsymbol{F}(\boldsymbol{r}) = -\nabla E_\mathrm{p}$ the force $\boldsymbol{F}$ is a measure for the potential energy $E_\mathrm{p}(r)$ of the interaction between A and B which depends on the distance r between A and B. The deflection of A therefore depends on the impact parameter b in Fig. 4.19, which is a measure of the closest approach between A and B. It is defined in the following way:

For large distances between A and B the force F is negligible and the incident particle A will follow a straight line. If there would be no interaction between A and B the incident particle A would follow this straight line and pass B at the closest distance b. This line is parallel to the straight line through B where B is resting in the origin of the coordinate system. To each impact parameter b belongs a certain deflection angle θ in the lab-system resp. ϑ in the CM-system., which depends on the interaction potential $V(r)$ between A and B.

4.3.1 Scattering in a Spherical Symmetric Potential

In Sect. 4.1.3 it was shown, that the relative motion of two particles around the CM caused by the mutual interaction force $\boldsymbol{F}(r)$ can be reduced to the motion of a single particle with the reduced mass μ in the spherical symmetric potential with its origin at the position of one of the two particles (usually the one with the larger mass). If the force $F(r)$ is known, the deflection angle ϑ in the CM-system can be determined from Eq. 4.3 and the re-

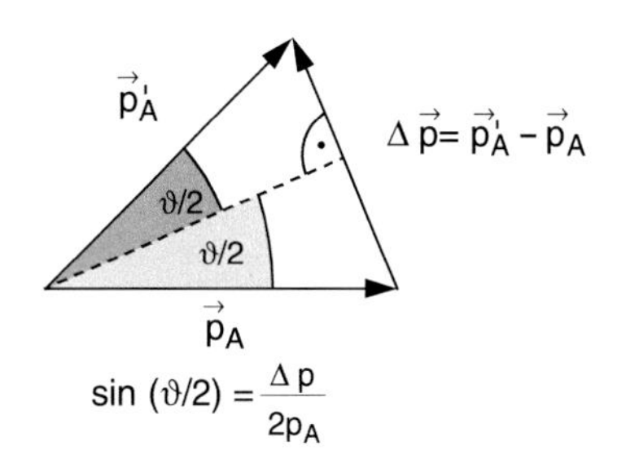

Figure 4.20 Relation between momentum change Δp and deflection angle ϑ in the CM-system

Chapter 4

lation $\sin(\vartheta/2) = \frac{1}{2}\frac{\Delta p}{p_A}$ (Fig. 4.20). In Sect. 4.2.5 it was shown how the angle θ measured in the lab-system can be transformed into the angle ϑ in the CM-system.

The deflection of particles in a potential is called *potential scattering*. We will illustrate this potential scattering and its treatment by some examples.

Examples

1. *Collision of two hard spheres with radii r_1 and r_2 (Fig. 4.21).*
If the impact parameter b is larger the sum $r_1 + r_2$ no collision takes place. The hard sphere A moves on a straight line and passes B without deflection. For $b \leq r_1 + r_2$ the colliding partner A is reflected at the surface of B (Fig. 4.21a). In order to determine the deflection angle ϑ of A we decompose the momentum $\boldsymbol{p}_1$ of A into a component p_r parallel to the connecting line M_1M_2 at the touch of the two spheres and a component p_t in the tangential direction perpendicular to p_r (Fig. 4.21b). We assume the surface of the spheres as frictionless. Then no rotation of the spheres can be excited and the component p_t does not change during the collision. For the component p_r we can conclude from (4.20) for central collisions

$$p_r' = \frac{m_1 - m_2}{m_1 + m_2} p_r \,. \tag{4.32a}$$

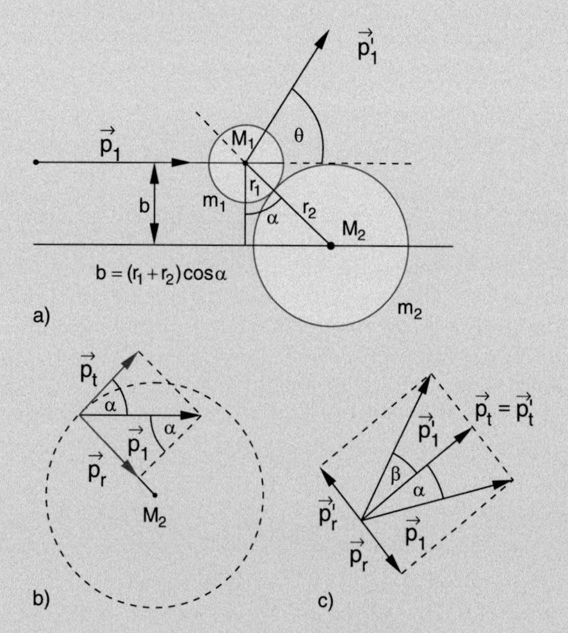

Figure 4.21 Determination of the deflection function for collisions between two hard spheres. **a** Definition of impact parameter, **b** decomposition of impact momentum, **c** momentum vector addition

For $m_2 \gg m_1$ this gives $p_r' = -p_r$. In this case is in Fig. 4.21c $\beta = \alpha$ and the deflection angle becomes $\theta = 2\alpha$.

From Fig. 4.21b one can deduce from $b = r_1 + r_2$ for $\theta = 2\alpha$ the dependence of the deflection angle on the impact parameter b

$$\theta(b) = 2 \arccos \frac{b}{r_1 + r_2} \,. \tag{4.32b}$$

For $b = 0$ is $\theta = \pi$ i.e. A is reflected back., for $b > r_1 + r_2$ is $\theta = 0$. The function $\theta(b)$ is called *deflection function*. Its curve depends on the interaction potential $(V(r))$. For the collision of two hard spheres the potential is a step function (Fig. 4.22a) and the deflection function is the monotonic curved shown in Fig. 4.22b for $m_2 \gg m_1$.
For the general case of arbitrary ratios m_1/m_2 we obtain from Fig. 4.21c:

$$\Theta = \alpha + \beta$$

with $p_r/p_t = \tan\alpha$ it follows

$$\tan\beta = p_r'/p_r = -(m_1 - m_2)/(m_1 + m_2) \,.$$

While (4.32b) is only strictly valid for $m_2 = \infty$, the deflection function

$$\vartheta(b) = \arccos \frac{b}{r_1 + r_2} \tag{4.32c}$$

in the CM-system is correct for arbitrary ratios m_1/m_2.

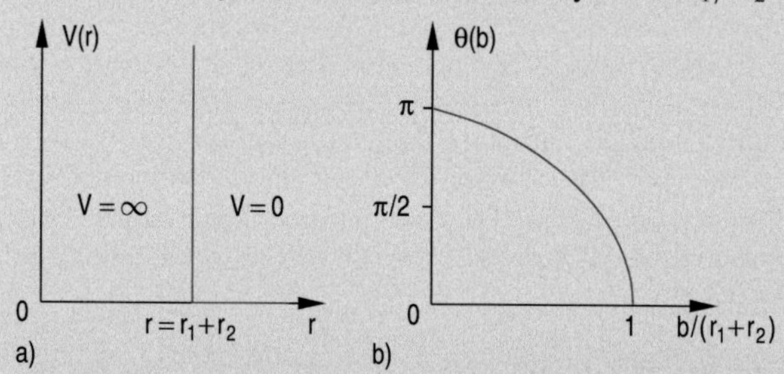

Figure 4.22 **a** Potential $V(r)$ for hard spheres; **b** deflection function for hard sphere collisions

2. *Scattering of a particle in a potential $V(r) \propto 1/r$.*
This important case applies for instance for the Coulomb-scattering of charged particles (electrons or α-particles) on atomic nuclei (see Vol. 3) or the Kepler-orbit of comets in the gravitational field of the sun.
For a potential $V(r)$ with the potential energy $E_p = a/r$ the force between the interacting particles A and B with masses m_1 and m_2 is

$$F = -\operatorname{grad} E_p = \frac{a}{r^2}\hat{\boldsymbol{r}} \,. \tag{4.33}$$

For $a > 0$ a repulsion between the particles occur, for $a < 0$ an attraction. The angular momentum in the CM-system is according to (4.15)

$$\boldsymbol{L} = \boldsymbol{r} \times \mu \cdot \boldsymbol{v} \quad \text{with} \quad \mu = \frac{m_1 m_2}{m_1 + m_2} \,,$$

Chapter 4

where r is the distance between A and B and v is the relative velocity.
Since $\boldsymbol{L}$ is in the central potential temporally constant the orbit of the particle remains in the plane $\perp \boldsymbol{L}$, which we choose as the x-y-plane (Fig. 4.23). The particle A is incident parallel to the x-axis with an impact parameter b and the initial velocity v_0. It is convenient to use polar coordinates for the description of its trajectory. The magnitude of $\boldsymbol{L}$ is then

$$L = |\boldsymbol{r} \times \mu \boldsymbol{v}| = \mu r^2 \frac{\mathrm{d}\varphi}{\mathrm{d}t} = \mu v_0 b \,, \tag{4.34}$$

where the last term describes the angular momentum of A for large distances $r \to \infty$ referred to the particle B which sits at the origin $r = 0$. It should be emphasized that we use the CM-system for our description. In the lab-system B does not stay at $r = 0$ but moves around the common centre of mass. We compose the force $F(r)$ of the components F_x and F_y. For the deflection of A only the component F_y is responsible. From Fig. 4.23 we see that

$$F_y = \frac{a \sin\varphi}{r^2} = \mu \frac{\mathrm{d}v_y}{\mathrm{d}t} \,. \tag{4.35}$$

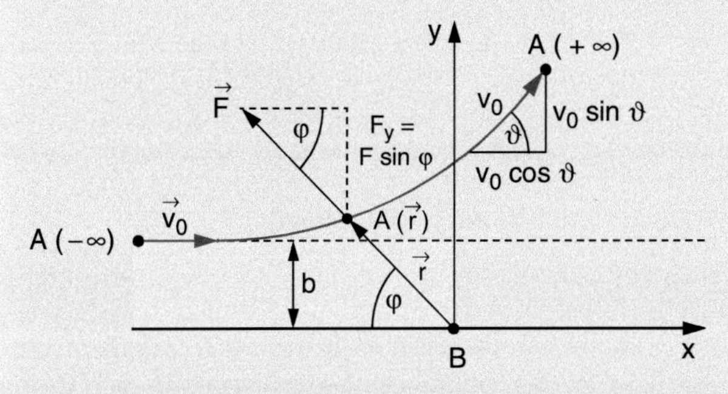

Figure 4.23 Scattering of a particle in the potential $V(r)$ with r = distance AB

From (4.34) and (4.35) we obtain

$$\frac{\mathrm{d}v_y}{\mathrm{d}t} = \frac{a \sin\varphi}{\mu v_0 b} \frac{\mathrm{d}\varphi}{\mathrm{d}t} \,. \tag{4.36}$$

The total deflection of A during its path through the potential $V(r)$ is obtained by integration over the whole range from $r = -\infty$ to $r = \infty$.
For $A(-\infty)$ we have $v_y = 0$ and $\varphi = 0$, for $A(+\infty)$ is $v_y = v_0 \cdot \sin\vartheta$ with $\vartheta = \pi - \varphi_{\max}$. For the elastic potential scattering the magnitude v_0 of the velocity remains constant. Therefore the integration of (4.36) yields

$$\int \mathrm{d}v_y = \frac{a}{\mu v_0 b} \int\limits_0^{\pi-\vartheta} \sin\varphi \, \mathrm{d}\varphi$$

$$\to v_0 \sin\vartheta = \frac{a}{\mu v_0 b}(1 + \cos\vartheta) \,.$$

With the equation $(1 + \cos\vartheta)/\sin\vartheta = \cot(\vartheta/2)$ the relation between deflection angle ϑ and impact parameter b for scattering in the potential with energy $E_\mathrm{p} = a/r$ becomes

$$\cot\left(\frac{\vartheta}{2}\right) = \frac{\mu v_0^2}{a} b = \frac{2E_\mathrm{kin}}{a} b \,. \tag{4.37a}$$

The ratio a/b gives the potential energy of the interaction between the particles A and B at a distance $r = b$. Inserting this into (4.37a) gives the result

$$\cot\left(\frac{\vartheta}{2}\right) = \frac{2E_\mathrm{kin}}{E_\mathrm{p}(b)} \,. \tag{4.37b}$$

The deflection angle ϑ in the cm-system is determined by the ratio of twice the kinetic energy and the potential energy at a distance $r = b$ between the interacting particles. The deflection function $\vartheta(b)$ is shown in Fig. 4.24. For $b = 0$ is $\cot(\vartheta/2) = \infty \to \vartheta = \pi$. The particle A is scattered back into the $-x$-direction. The turning point which is the closest approach r_0 can be obtained from $E_\mathrm{kin} = E_\mathrm{p} \to \mu v_0^2/2 = a/r_0$. This gives $r_0 = 2a/(\mu v_0^2)$.

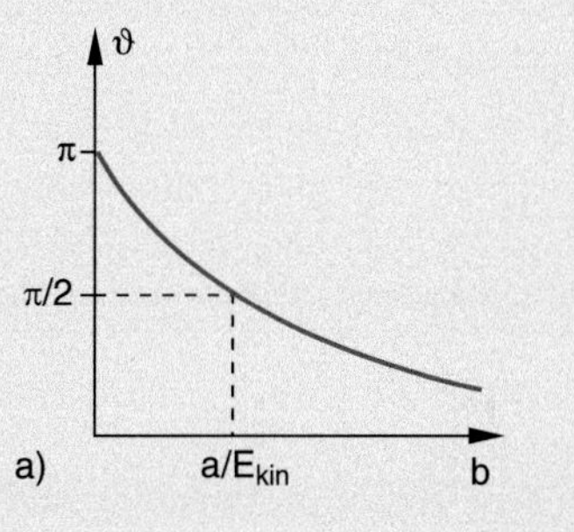

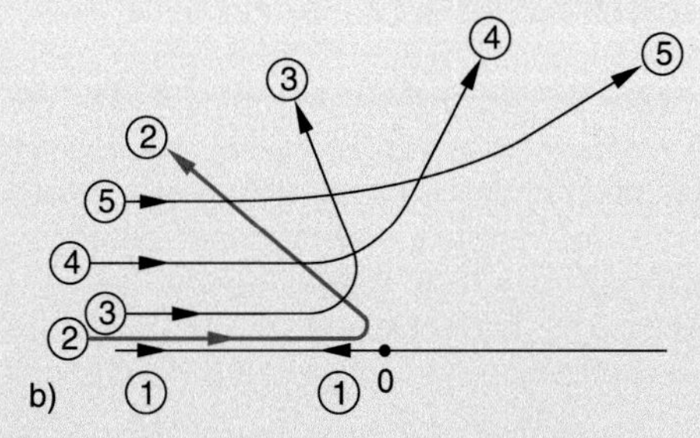

Figure 4.24 Deflection function $\vartheta(b)$ (**a**) and trajectories of a particle in a potential $V(r) \propto 1/r$ for different impact parameters but constant initial energy. Each deflection angle corresponds to a different ratio $f = 2E_\mathrm{kin}/E_\mathrm{pot}(b)$. ①: $\vartheta = \pi$; ②: $\vartheta = \frac{3}{4}\pi$; ③: $\vartheta = 105°$, $f = 0.76$; ④: $\vartheta = 60°$, $f = 1.7$; ⑤: $\vartheta = 30°$, $f = 3.7$

For the gravitational potential is $a = -Gm_1m_2$ (see (2.52) and we get from (4.37b) with $M = m_1 + m_2$ the result

$$\cot\left(\frac{\vartheta}{2}\right) = -\frac{v_0^2}{GM} b \,. \tag{4.37c}$$

The deflection angle depends only on the masses, the initial velocity v_0 and the impact parameter b. For a comet is $m_1 \ll m_2 = M_\odot$. The total mass M is then with a very good approximation $M = M_\odot$. The mass of the comet does not affect the deflection angle. According to (2.60) the trajectories of the particle m_1 for $E = E_{kin} + E_p > 0$ are hyperbolas. In Fig. 4.24b some of these hyperbolas are shown for a repulsive potential ($a > 0$) and different impact parameters b. For the interaction between two positively charged particles with charges q_1 and q_2 is $a = (1/4\pi\varepsilon_0) \cdot q_1 \cdot q_2$ (see Vol. 2). ◀

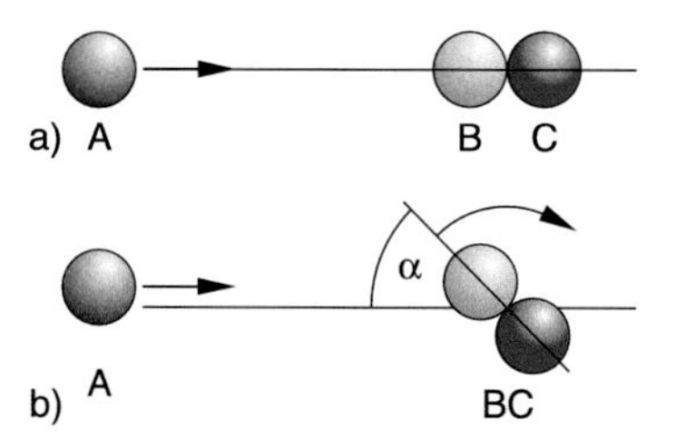

Figure 4.26 Collinear (**a**) and noncollinear (**b**) collision, where angular momentum of the relative motion is transferred to rotational angular momentum of the molecule BC

4.3.2 Reactive Collisions

Reactive collisions provide the basis of all chemical reactions. A simple example is the reaction

$$A + BC \to ABC \to AB + C\ , \tag{4.38}$$

where an atom A with the velocity v_A collides with a molecule BC (velocity v_{BC}), forms a complex ABC, which can decay into the fragments AB + C (Fig. 4.25).

Momentum conservation is also valid for reactive collisions. The momentum of the right side in equation (4.38) must be therefore the same as on the left side. The kinetic energy is, however, in general not conserved because part of this energy may be converted into internal energy ($U < 0$). In cases where the reactants on the left side are already excited, this internal energy may be also transferred to kinetic energy ($U > 0$, superelastic collisions). The measurement of velocities and deflection angles after the collision gives information about the energy balance of the reaction and the interaction potential between the reactants, if the initial conditions are known. The potential is in general no longer spherical symmetric but depends on the spatial orientation of the molecule BC against the momentum direction of A. The reaction probability can differ considerably for collinear collisions, (Fig. 4.26a) from that for non-collinear collisions where the internuclear axis of BC is inclined by the angle α against the momentum direction of A (Fig. 4.26b).

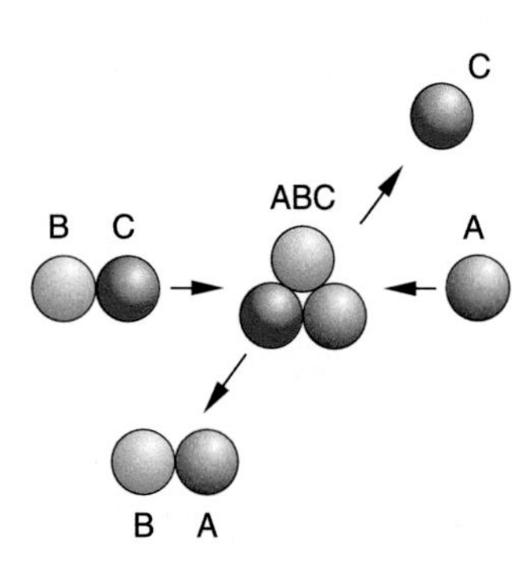

Figure 4.25 Schematic representation of a reactive collision, where a collision complex is formed that decays again

Often the reactants A and BC fly in two perpendicular collimated molecular beams. In the intersection volume of the two beams the reactants collide. For this arrangement the directions of the reactants are known and their velocities can be selected by velocity selectors which interrupt the beams (see Sect. 7.4.1). The initial conditions are then well known (apart from the often unknown internal energies).

Note, that the masses are generally not constant for reactive collisions, because the reduced mass $\mu(A + BC)$ before the collision differs in general from $\mu(AB + C)$ after the collision.

If the kinetic energy E_2 of the reaction products is smaller than the kinetic energy E_1 of the reactants, the reaction is called *endotherm*. One has to put energy into the system in order to make the reaction possible. If energy is released in the reaction it is called *exotherm*. In this case the kinetic energy of the reaction products is larger than that of the reactants. Measurements of the velocities of reactants and reaction products can decide which type applies to the investigated reaction.

The energy balance is illustrated by the potential diagram of Fig. 4.27. Often the reactants have to overcome a potential barriers in order to start the reaction. In this case a minimum initial energy is necessary even for exothermic reactions.

The heights of the potential barrier and with it the reaction probability depends on the internal energy (vibrational- rota-

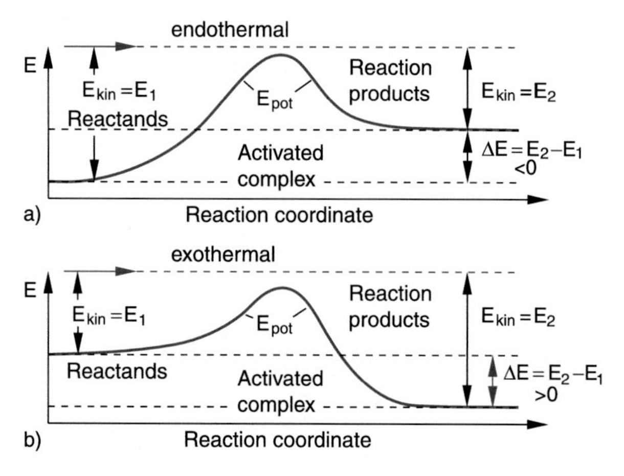

Figure 4.27 Diagram of potential energy as function of the reaction coordinate for **a** endothermic and **b** exothermic reactive collisions

Chapter 4

tional or even electronic energy) of the reaction partners. For the measurement of these internal energy several spectroscopic techniques have been developed which allow to determine the excitation state of the partners involved in the reaction.

An ideal experiments should allow to measure all relevant parameters of a collision process such as the internal energies, the deflection angles and the velocities of all particles. Such modern techniques are discussed in Vol. 3.

4.4 Collisions at Relativistic Energies

Up to now we have used the Newtonian laws (energy- and momentum conservation) for the description of collision processes and we have assumed that the masses of the reaction partners are constant (besides in reactive collisions). This is justified as long as the velocities of the collision partners are small compared with the velocity c of light (see Chap. 3).

For the investigation of interactions between elementary particles and atomic nuclei, higher energies of the collision partners are required. Such energies, where the velocity of particles comes close to the velocity of light can be realized in particle accelerators and storage rings (see Vol. 4). We will now discuss, how the rules governing collisions at relativistic energies (the domain of high energy physics) must be formulated.

4.4.1 Relativistic Mass Increase

We regard two particles A and B which have equal masses $m_1 = m_2 = m$, if they are at rest. We assume that A and B move in a system S with velocities

$$\boldsymbol{v}_1 = \{v_{x1}, -v_{y1}\} \quad \text{and} \quad \boldsymbol{v}_2 = \{0, v_{y2}\}$$

against each other where $v_{y1} = v_{y2}$ (Fig. 4.28a).

The particle B should suffer an elastic striking collision with A such that during the collision the x-component v_{x1} of A remains constant but the y-component is reversed. The velocity of A after the collision is then $\boldsymbol{v}_1' = \{v_{x1}, v_{y1}\}$. The magnitude of its velocity $|\boldsymbol{v}_1| = \sqrt{v_{x1}^2 + v_{y1}^2}$ is therefore also preserved. Because the momentum must be constant the velocity of B after the collision must be

$$\boldsymbol{v}_2' = \{0, -v_{y2}\} .$$

We assume that $v_y \ll v_x$. For the magnitudes of the velocities this implies

$$v_1 = (v_{x1}^2 + v_{y1}^2)^{(1/2)} \approx v_{x1} \quad \text{and} \quad v_2 \ll v_1 .$$

Now we describe this collision between A and B in a system S^* which moves against S with the velocity $v = v_{x1}$ into the x-direction (Fig. 4.28b). In this system we get for the velocities of A and B:

$$v_{x1}^* = 0 \quad \text{but} \quad v_{x2}^* = -v_{x1} ,$$

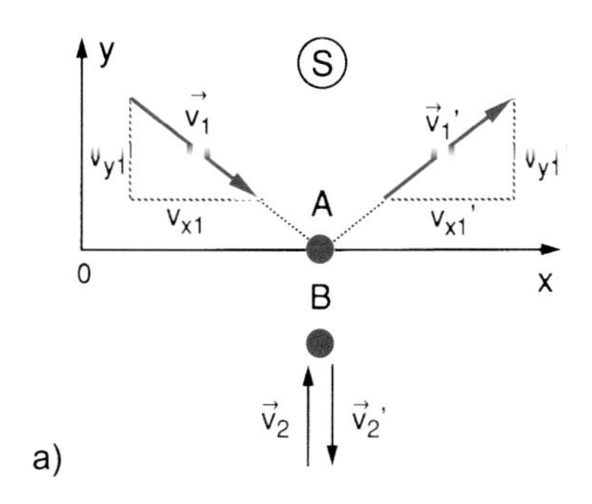

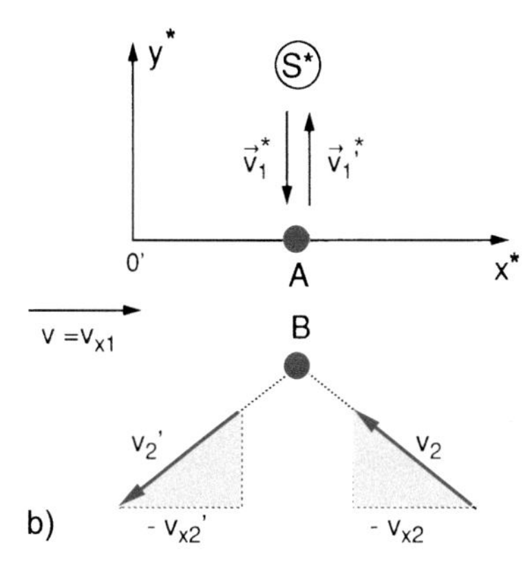

Figure 4.28 Grazing elastic collision between A and B at relativistic velocities. **a** In the system S has A a large and B a small velocity since $V_y \ll V_x$. **b** In the sytem S^*, that moves relativ to S the situation is reversed

which implies that the roles of A and B are just interchanged.

According to Eq. 3.28 for the transformation of velocities when changing from system S to S^* the observer O^* in S^* measures the velocity component

$$v_y^* = \frac{v_y/\gamma}{1 - v_x v/c^2} . \tag{4.39}$$

Since in the system S for the particle A holds: $v_{x1} = v_{x1}' \neq 0$, for B, however, $v_{x2} = v_{x2}' = 0$ the observer O^* measures for the two particles different y-components of the velocities

$$v_{y1}^* = \frac{v_{y1}/\gamma}{1 - v_{x1} v/c^2} = \frac{v_{y1}/\gamma}{1 - v^2/c^2} = \gamma v_{y1} , \quad \text{since } v \approx v_{x1} , \tag{4.40a}$$

$$v_{y2}^* = \frac{v_{y2}/\gamma}{1 - v_{x2} v/c^2} = v_{y2}/\gamma , \tag{4.40b}$$

while for the observer O in S the velocity component of A is v_{y1} and for B it is v_{y2}.

In both inertial systems S and S^* the conservation of total momentum holds, since the physical laws are independent of the chosen inertial system (see Sect. 3.2). This yields for the y-component of the total momentum

$$m_A v_{y1} - m_B v_{y2} = m_A^* v_{y1}^* - m_B^* v_{y2}^* = 0 . \tag{4.41}$$

Chapter 4

For $m_A = m_A^*$ and $m_B = m_B^*$ this condition cannot be fulfilled, i. e. the conservation of momentum would fail, because according to (4.40) v_{y1}/v_{y1}^* is different from v_{y2}/v_{y2}^*. We are therefore forced (if we will not give up the well proved conservation of momentum) to assume that the mass of a particle is changing with its velocity. For the limiting case $v_{x1} \gg v_{y1} \approx 0$ we can write:

$$v_A \approx v_{x1} = v\,; \quad v_A^* \approx 0\,,$$
$$v_B \approx 0\,; \quad v_B^* \approx v_{x1} \approx v\,.$$

We therefore get with $m(v = 0) = m_0$ for (4.41)

$$m(v)v_{y1} + m_0 v_{y2} = 0\,, \tag{4.42a}$$
$$m_0 v_{y1}^* + m(v)v_{y2}^* = 0\,, \tag{4.42b}$$

with (4.40) this gives

$$\frac{(m(v))^2}{m_0^2} = \frac{v_{y2}}{v_{y2}^*} \cdot \frac{v_{y1}^*}{v_{y1}} = \gamma^2$$

$$\Rightarrow m(v) = \gamma m_0 = \frac{m_0}{\sqrt{1 - v^2/c^2}}\,. \tag{4.43}$$

The mass $m(v)$ of a moving particle increases with its velocity v. the mass $m_0 = m(v = 0)$ is called its **rest-mass**. This mass increase is noticeable only for large velocities [4.3].

Examples

1. For $v = 0.01c \Rightarrow m = 1.00005m_0$. The relative mass increase $\Delta m/m = (m - m_0)/m_0 \approx 5 \cdot 10^{-5}$.
2. For $v = 0.9c \Rightarrow m = 2.29m_0$.
3. For $v = 0.99c \Rightarrow m = 7m_0$. ◀

In Fig. 4.29 the increase of the mass $m(v)$ is plotted against the normalized velocity v/c. This illustrates also, that the maximum velocity of a particle with $m_0 \neq 0$ is always smaller than the velocity of light because for $v \to c$ it follows from (4.43) $m(v) \to \infty$.

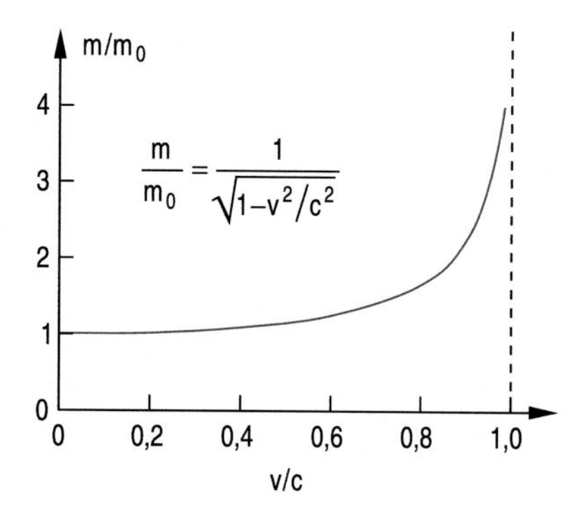

Figure 4.29 Dependence of a mass m on the ratio v/c

4.4.2 Force and Relativistic Momentum

The work, which has to be spent for the acceleration of a mass is used with increasing velocity more and more for the increase of the mass and less for the increase of the velocity.

The Newton-equation (2.18) between force and momentum with the inclusion of the relativistic mass increase (4.43) is

$$\begin{aligned} F &= \frac{\mathrm{d}\boldsymbol{p}}{\mathrm{d}t} = \frac{\mathrm{d}}{\mathrm{d}t}(m\boldsymbol{v}) = \frac{\mathrm{d}}{\mathrm{d}t}\left(\frac{\boldsymbol{v}m_0}{\sqrt{1 - v^2/c^2}}\right) \\ &= \left(\frac{\mathrm{d}}{\mathrm{d}t}\frac{m_0}{\sqrt{1 - v^2/c^2}}\right)\boldsymbol{v} + m\boldsymbol{a}\,. \end{aligned} \tag{4.44a}$$

This gives with $\mathrm{d}/\mathrm{d}t = (\mathrm{d}v/\mathrm{d}t) \cdot (\mathrm{d}/\mathrm{d}v)$

$$\begin{aligned} F &= \frac{m_0\left(v/c^2\right)a}{\left(1 - v^2/c^2\right)^{3/2}}\boldsymbol{v} + m\boldsymbol{a} \\ &= \gamma^3 m_0 a\left[\frac{v^2}{c^2}\hat{\boldsymbol{e}}_v + \left(1 - \frac{v^2}{c^2}\right)\hat{\boldsymbol{e}}_a\right]\,, \end{aligned} \tag{4.44b}$$

where $\hat{\boldsymbol{e}}_v$ and $\hat{\boldsymbol{e}}_a$ are unit vectors in the direction of $\boldsymbol{v}$ and $\boldsymbol{a}$.

These equations show that for large velocities v the force $\boldsymbol{F}$ is no longer parallel to the acceleration $\boldsymbol{a}$ but has a component in the direction of $\boldsymbol{v}$. For $v \ll c$ the first term in (4.44b) can be neglected and we obtain the classical Newton equation $\boldsymbol{F} = m\cdot\boldsymbol{a}$.

In order to keep the Newton equation $F = \mathrm{d}p/\mathrm{d}t$ the relativistic momentum

$$\boldsymbol{p}(v) = m(v) \cdot \boldsymbol{v} = \gamma \cdot m_0 \cdot \boldsymbol{v} \tag{4.45a}$$

has to be used which has the magnitude

$$p = \beta\gamma m_0 c \qquad (\beta = v/c)\,. \tag{4.45b}$$

For the relativistic momentum the conservation law (4.41) is fulfilled for all velocities.

We will now discuss, how the components of the force are transformed for a transition from a system S where a particle has a velocity v and a mass $m = \gamma m_0$ into a system S^* which moves with the velocity $U = +v$ against S. In S^* is therefore $v^* = 0$ and $m^* = m_0$. We choose the axes of the coordinate system such that $\boldsymbol{v} = \{v_x, 0, 0\}$. It follows then from (4.44) with $\hat{\boldsymbol{e}}_v = \hat{\boldsymbol{e}}_a$

$$F_x = \frac{\mathrm{d}p_x}{\mathrm{d}t} = \gamma^3 m_0 a_x\,. \tag{4.45c}$$

In S^* the x-component of the acceleration a becomes

$$a_x^* = \gamma^3 a_x$$

as can be seen from (3.26) and (3.28). Therefore the component of the force in the system S^* becomes

$$F_x^* = m_0 \cdot a_x^* = \gamma^3 m_0 \cdot a_x \equiv F_x\,. \tag{4.45d}$$

We obtain the remarkable result that the x-components in the two systems which move against each other in the x-direction, are equal!

This no longer true for the components perpendicular to the relative motion of the two systems, because we get for $v_y \ll v_x$ the result

$$F_y^* = m_0 \cdot a_y^* = \gamma^2 m_0 a_y = \gamma \cdot F_y \,, \tag{4.45e}$$

and therefore obtain for the ratio

$$\frac{F_x}{F_y} = \gamma^2 \cdot \frac{a_x}{a_y} \,. \tag{4.45f}$$

This shows again that for $\gamma \neq 1$ the force $\boldsymbol{F}$ is no longer parallel to the acceleration $\boldsymbol{a}$ as in the nonrelativistic case.

4.4.3 The Relativistic Energy

In classical mechanics the kinetic energy of a particle

$$E_{\text{kin}} = \tfrac{1}{2} m v^2$$

is different in diverse inertial systems which move against each other, because the velocity v is different.

In order to obey energy conservation when changing from one system to the other the total energy of a particle has to be defined in such a way, that it is Lorentz-invariant. i. e. that it does not change for transformations into different inertial systems (see Sect. 3.3).

We will at first present an intuitively accessible description, which is based on a "Gedanken-experiment" of Einstein. We regard in Fig. 4.30 a box with length L and mass M. We assume, that at the time $t_1 = 0$ a light pulse with energy E is emitted from the left side of the box which travels with the velocity of light c to the right. According to results of classical physics the momentum of the light pulse is $\boldsymbol{p} = (E/c)\hat{e}$ (see Vol. 2). Because of the conservation of momentum the left wall and with it the total box suffers a recoil $-\boldsymbol{p}$ into the left direction. This results in a velocity

$$\boldsymbol{v} = -\boldsymbol{p}/M = (E/Mc)\hat{e}$$

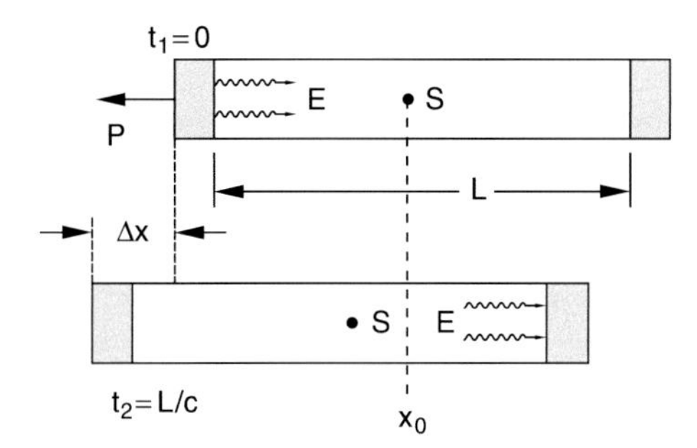

Figure 4.30 Einstein's "Gedanken-experiment" illustrating $E = mc^2$

of the box. For $v \ll c$ the light pulse reaches the right wall of the box at a time $t_2 = L/c$ and is absorbed by the wall. This transfers the momentum $\boldsymbol{p} = +(E/c)\hat{e}$ to the right wall. The total momentum transferred to the box is therefore zero and the box is again at rest. However, during the time t_2 the box has moved to the right by a distance

$$\Delta x = -v \cdot t_2 = -E \cdot \frac{L}{Mc^2} \,. \tag{4.46}$$

Since the box plus light pulse represent a closed system where no external forces act onto the system, the centre of mass cannot have moved. The CM of the box certainly has moved by Δx into the $-x$-direction. Therefore the light pulse must have transported mass into the $+x$-direction in order to leave the CM of the total system (box + light pulse) at rest. If we attribute a mass m to the light pulse with energy E the CM of the total system stays at rest, if

$$m \cdot L - M \cdot \Delta x = 0 \,. \tag{4.47}$$

This gives with (4.46) the result

$$m = E/c^2 \Rightarrow E = mc^2 \,. \tag{4.48a}$$

According to this consideration each mass m is correlated to the energy $E = m \cdot c^2$. Mass and energy are proportional to each other.

When we insert the rest mass m_0 from (4.43) we obtain from (4.48a) the energy of a mass m that moves with the velocity v

$$E = \frac{m_0 c^2}{\sqrt{1 - v^2/c^2}} = m_0 c^2 + (m - m_0)c^2 \,. \tag{4.48b}$$

This energy E can be composed of two parts:

The rest energy $m_0 c^2$ which a particle at rest must have, and the kinetic energy

$$E_{\text{kin}} = (m(v) - m_0)c^2 \,, \tag{4.49a}$$

which is here described as the increase of its mass $m(v)$. If we expand the square root in (4.48b) according to

$$\frac{1}{\sqrt{1 - v^2/c^2}} = 1 + \frac{1}{2}\frac{v^2}{c^2} + \frac{3}{8}\frac{v^4}{c^4} + \cdots \,,$$

the kinetic energy becomes

$$E_{\text{kin}} = \frac{1}{2} m_0 v^2 + \frac{3}{8} m_0 \frac{v^4}{c^2} + \cdots \,. \tag{4.49b}$$

In the limiting case $v \ll c$ we can neglect the higher order terms in (4.49b) and obtain the classical result

$$E_{\text{kin}} = \tfrac{1}{2} m v^2 \,.$$

This shows that the classical expression for the kinetic energy is an approximation for $v \ll c$. Since in daily life the condition

$v \ll c$ is always fulfilled, the relativistic expression is important only for cases where v approaches c as in high energy physics or astrophysics.

Squaring (4.48b) and multiplying both sides with c^2 gives

$$E^2 = \frac{m_0^2 c^6}{c^2 - v^2} = m_0^2 c^4 + m^2 c^2 v^2 . \quad (4.50)$$

Inserting (4.45) for the relativistic momentum yields

$$E^2 = m_0^2 c^4 + p^2 c^2 .$$

This gives the relativistic relation between total energy E and momentum p

$$E = c\sqrt{m_0^2 c^2 + p^2} . \quad (4.51)$$

For $v \ll c$ the square root can be expanded and gives the result

$$E_{\text{kin}} = E - m_0 c^2 \approx \frac{p^2}{2m} = \frac{1}{2} m_0 v^2$$

with the classical momentum $p = m_0 \cdot v$.

4.4.4 Inelastic Collisions at relativistic Energies

The relativistic energy and its conservation can be illustrated by the example of a collinear completely inelastic collision (Fig. 4.31). We regard two particles A and B with equal masses m which fly against each other width velocities $v_1 = \{v_1, 0, 0\}$ and $v_2 = \{-v_1, 0, 0\}$, measured in the system S. In a completely inelastic collision their total kinetic energy is converted into internal energy of the collision partners (see Sect. 4.2.4). After the collision they form a compound particle AB with the velocity $v = 0$ (Fig. 4.31 upper part).

In a system S^*, which moves with the velocity $v = v_1$ against S the particle A has the velocity $v_1^* = 0$, the compound particle AB which rests in S has in S^* the velocity $u = -v_1$. The particle

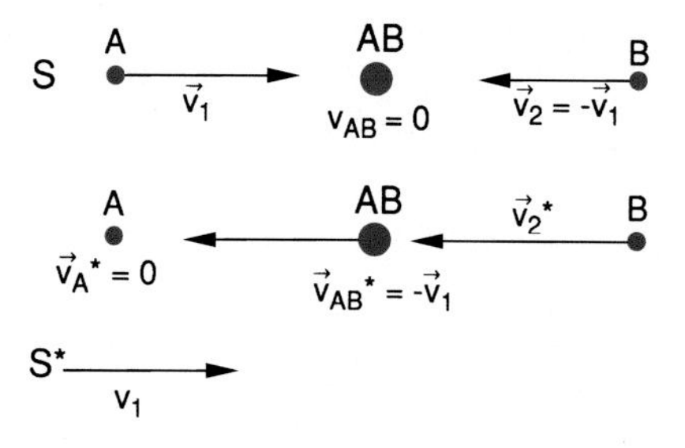

Figure 4.31 Description of a collinear completely inelastic collision in two different inertial systems S and S^*

B has in S^* according to the relativistic addition of velocities (3.28) the velocity

$$v_2^* = \frac{v_2 - v}{1 - v_2 v/c^2} = \frac{-2v}{1 + v^2/c^2} , \quad (4.52)$$

where $v_2 = -v_1$ is the velocity of B in S and $v = v_1$ is the velocity of S^* against S.

The conservation of momentum demands for the collision described in S^*

$$m\left(v_2^*\right) \cdot v_2^* = Mu = -Mv_1 , \quad (4.53)$$

where M is the mass of the compound AB with the velocity $u = -v_1$ measured in S^*.

Conservation of energy requires, when dividing by c^2

$$m\left(v_2^*\right) + m_0 = M . \quad (4.54)$$

Inserting from (4.53) the relation $M = -m(v_2^*) \cdot v_2^*/v$ into (4.54) we obtain

$$\frac{m\left(v_2^*\right)}{m_0} = -\frac{v}{v_2^* + v} . \quad (4.55)$$

Equation 4.52 gives the relation between v and v_2^*

$$v = -\frac{c^2}{v_2^*}\left[1 + \left(1 - \frac{v_2^{*2}}{c^2}\right)^{1/2}\right] ; \quad (4.56)$$

inserting this into (4.55) gives the mass ratio

$$\frac{m\left(v_2^*\right)}{m_0} = \left(1 - \frac{v_2^{*2}}{c^2}\right)^{-1/2} = \gamma\left(v_2^*\right) , \quad (4.57)$$

and therefore again the general relation

$$m(v) = \frac{m_0}{\sqrt{1 - v^2/c^2}} = \gamma(v) m_0 , \quad (4.58)$$

which has been already derived in Sect. 4.4.1.

4.4.5 Relativistic Formulation of Energy Conservation

In order to show, that the relativistic energy $E = m \cdot c^2$ is conserved we must discuss the relativistic formulation of the Newton equation $\boldsymbol{F} = \mathrm{d}\boldsymbol{p}/\mathrm{d}t$. Thereto we replace the classical position vector r by the Lorentz four-vector

$$\mathcal{R} = x\hat{\boldsymbol{e}}_x + y\hat{\boldsymbol{e}}_y + z\hat{\boldsymbol{e}}_z + ict\hat{\boldsymbol{e}}_t = \boldsymbol{r} + ict\hat{\boldsymbol{e}}_t , \quad (4.59)$$

defined in the four-dimensional Minkowski space (x, y, z, ict) (see Sect. 3.6.2), where the unit vector $\hat{e}_t$ is perpendicular to the three spatial axes. From (4.59) one can derive that $\mathcal{R}^2 = r^2 - c^2 t^2$.

This gives the total differential

$$\left(\mathrm{d}\mathcal{R}^2\right) - \mathrm{d}r^2 + \mathrm{d}y^2 + \mathrm{d}z^2 - c^2\mathrm{d}t^2 = -c^2\mathrm{d}\tau^2 \tag{4.60}$$

where we have used as abbreviation the differential

$$\begin{aligned}\mathrm{d}\tau &= \sqrt{\mathrm{d}t^2 - \frac{1}{c^2}\left(\mathrm{d}x^2 + \mathrm{d}y^2 + \mathrm{d}z^2\right)}\\ &= \mathrm{d}t\sqrt{1 - \frac{v^2}{c^2}} = \mathrm{d}t/\gamma\end{aligned} \tag{4.61}$$

of the "*eigen-time*" τ, which approaches the classical time differential $\mathrm{d}t$ for $v \ll c$.

The differentiation of (4.59) gives the four-vector of the velocity

$$\begin{aligned}\frac{\mathrm{d}\mathcal{R}}{\mathrm{d}\tau} &= \frac{\mathrm{d}x}{\mathrm{d}\tau}\hat{e}_x + \frac{\mathrm{d}y}{\mathrm{d}\tau}\hat{e}_y + \frac{\mathrm{d}z}{\mathrm{d}\tau}\hat{e}_z + ic\frac{\mathrm{d}t}{\mathrm{d}\tau}\hat{e}_t\\ &= \frac{\boldsymbol{v} + ic\hat{e}_t}{\sqrt{1 - v^2/c^2}} .\end{aligned} \tag{4.62}$$

The four-momentum is defined as

$$\mathcal{P} = m_0\frac{\mathrm{d}\mathcal{R}}{\mathrm{d}\tau} = m_0\frac{\boldsymbol{v} + ic\hat{e}_t}{\sqrt{1 - v^2/c^2}} . \tag{4.63}$$

In analogy to the Newton equation $\boldsymbol{F} = \mathrm{d}\boldsymbol{p}/\mathrm{d}t$ we define the four-force (also called the Minkowski-force)

$$\begin{aligned}\mathcal{F} &= \frac{\mathrm{d}\mathcal{P}}{\mathrm{d}\tau} = m_0\frac{\mathrm{d}^2\mathcal{R}}{\mathrm{d}\tau^2}\\ &= \gamma\left[\frac{\mathrm{d}}{\mathrm{d}t}(m\cdot\boldsymbol{v}) + ic\frac{\mathrm{d}}{\mathrm{d}t}(m\hat{e}_t)\right]\\ &= \gamma\left[\boldsymbol{F} + i\frac{\mathrm{d}}{\mathrm{d}t}(mc\hat{e}_t)\right] .\end{aligned} \tag{4.64}$$

Using these definitions we can derive the relativistic energy conservation law. We multiply (4.64) with $\mathrm{d}\mathcal{R}/\mathrm{d}\tau$

$$\begin{aligned}\left(\mathcal{F}\frac{\mathrm{d}\mathcal{R}}{\mathrm{d}\tau}\right) &= m_0\left(\frac{\mathrm{d}^2\mathcal{R}}{\mathrm{d}\tau^2}\right)\cdot\frac{\mathrm{d}\mathcal{R}}{\mathrm{d}\tau}\\ &= \frac{m_0}{2}\frac{\mathrm{d}}{\mathrm{d}\tau}\left(\frac{\mathrm{d}\mathcal{R}}{\mathrm{d}\tau}\right)^2 .\end{aligned} \tag{4.65}$$

According to (4.60) is $(\mathrm{d}\mathcal{R}/\mathrm{d}\tau)^2 = -c^2 = \text{const}$. Therefore the right side of (4.65) is zero!

$$\mathcal{F}\frac{\mathrm{d}\mathcal{R}}{\mathrm{d}\tau} = 0 . \tag{4.66}$$

This can be separated in a spatial and a temporal part, which gives

$$\begin{aligned}&\frac{1}{1 - v^2/c^2}\left(\boldsymbol{F}\cdot\frac{\mathrm{d}\boldsymbol{r}}{\mathrm{d}t} - \frac{\mathrm{d}}{\mathrm{d}t}\left(mc^2\right)\right) = 0\\ &\Rightarrow \mathrm{d}\left(mc^2\right) = \boldsymbol{F}\cdot\mathrm{d}\boldsymbol{r} = \mathrm{d}W .\end{aligned} \tag{4.67}$$

The quantity $\mathrm{d}W$ represents the work, performed on the particle with mass m. For conservative forces $F = -\operatorname{grad} V$, which have a potential, is $\mathrm{d}W$ equal to the change of the potential energy E_p. Integration of (4.67) over the time yields

$$E_\mathrm{p} + mc^2 = \text{const} = E , \tag{4.68a}$$

which corresponds to the classical energy conservation (2.38) if E_kin is replaced by mc^2.

For a particle with the velocity v we can write

$$mc^2 = m_0\gamma(v)\cdot c^2 = \frac{m_0c^2}{\sqrt{1 - (v/c)^2}} .$$

The equation (4.68a) of energy conservation then becomes

$$E_\mathrm{p} + \frac{m_0c^2}{\sqrt{1 - v^2/c^2}} = E . \tag{4.68b}$$

4.5 Conservation Laws

In the foregoing sections we have discussed, that there are physical quantities which are conserved in closed systems, i. e. they do not change in the course of time.

As a reminder, please note,

that a closed system is a system which has no interaction with the outside, i. e. there are no external forces acting on the particles of the system, although the particles may interact with each other.

Such conserved quantities are the total momentum $\boldsymbol{p}$, the total energy E and the angular momentum $\boldsymbol{L}$ of a closed system. The conservation of these quantities is, because of its great importance, formulated in special conservation laws, which shall be summarized and generalized in the following sections.

4.5.1 Conservation of Momentum

For a single free particle (no forces acting on it) the momentum conservation reads:

The momentum $\boldsymbol{p} = m\cdot\boldsymbol{v}$ of a free particle is constant in time.

This is identical with the Newton postulate (Sect. 2.6).

Generalized for a system of particles this reads:

The total momentum of a closed system of particles which may interact with each other, does not change with time.

This can be also formulated as: If the vector sum of all external forces acting on a system of particles is zero, the total momentum of the system does not change with time. According to the 3. Newton's axiom *actio* = *reactio* the vector sum of all internal forces is anyway zero.

Note that the momentum of the individual particles can indeed change!

4.5.2 Energy Conservation

We have seen in Sect. 2.7 that in conservative force fields the sum of kinetic anf potential energy is constant. This energy conservation can be generalized to a system of particles and also further types of energy (internal energy, thermal energy, mass energy $E = mc^2$) can be included. The law of energy conservation in the general form is then:

The total energy of a closed system is constant in time, where the different forms of energy can be completely or partially converted into each other

For instance the kinetic energy of a particle can be converted into thermal energy at a collision with the wall, or the mass energy of electron and positron can be converted into radiation energy if the two particles collide.

4.5.3 Conservation of Angular Momentum

If the vector sum of all torques D_i which act on a system of particles is zero, the total angular momentum $\boldsymbol{L}$ of the system remains constant. This follows from the relation $\mathrm{d}\boldsymbol{L}/\mathrm{d}t = \sum \boldsymbol{D}_i$.

Note: For the definition of the angular momentum

$$\boldsymbol{L} = \sum(\boldsymbol{r}_i \times \boldsymbol{p}_i) ,$$

the reference point (generally the origin of the coordinate system from which the position vectors $\boldsymbol{r}_i$ start) has to be defined.

Since for a closed system $\sum \boldsymbol{D}_i = \boldsymbol{0}$ the conservation of angular momentum can be also formulated as

In a closed system the total angular momentum $\boldsymbol{L}$ remains constant in time.

4.5.4 Conservation Laws and Symmetries

A more detailed investigation of the real causes of the conservation laws reveals that these laws are based on symmetry properties of space and time [4.9]. In order to prove this, we introduce the Lagrange function $\mathcal{L}$

$$\begin{aligned}\mathcal{L}(\boldsymbol{r}_i, \boldsymbol{v}_i) &= \sum_i^N \frac{m_i}{2} v_i^2 - E_{\text{pot}}(\boldsymbol{r}_1, \boldsymbol{r}_2 \ldots \boldsymbol{r}_N) \\ &= E_{\text{kin}} - E_{\text{p}}\end{aligned} \tag{4.69}$$

of a closed system with N particles, which represents the difference of kinetic and potential energy. From (4.69) the relations

$$\frac{\partial \mathcal{L}}{\partial \boldsymbol{v}_i} = m_i \boldsymbol{v}_i = \boldsymbol{p}_i \tag{4.70a}$$

$$\frac{\partial \mathcal{L}}{\partial \boldsymbol{r}_i} = -\frac{\partial E_{\text{p}}}{\partial \boldsymbol{r}_i} = \boldsymbol{F}_i \tag{4.70b}$$

follow immediately. This gives the equation of motion $F_i = m_i \cdot \mathrm{d}v_i/\mathrm{d}t$ in the general form

$$\frac{\mathrm{d}}{\mathrm{d}t}\left(\frac{\partial \mathcal{L}}{\partial \boldsymbol{v}_i}\right) = \frac{\partial \mathcal{L}}{\partial \boldsymbol{r}_i} . \tag{4.71}$$

Note: The Lagrange equation (4.71) can be derived quite general from a fundamental variation principle, called the *principle of minimum* action [4.8].

This principle also gives the definite justification for the following statements and their explanation.

1. The conservation of momentum is due to the homogeneity of space.

This homogeneity of space guarantees that the properties of a closed system do not change when all particles are shifted by an amount ε, which means that their position vectors changes from $\boldsymbol{r}$ to $\boldsymbol{r} + \boldsymbol{\varepsilon}$. Because of the homogeneity all masses and velocities remain unchanged.

The Lagrange function in a homogeneous space does not depend on the position vectors $\boldsymbol{r}_i$ i. e.

$$\sum \partial \mathcal{L}/\partial r_i = 0 .$$

From (4.71) we can conclude

$$\begin{aligned}&\sum_i \frac{\mathrm{d}}{\mathrm{d}t}\frac{\partial \mathcal{L}}{\partial \boldsymbol{v}_i} = \frac{\mathrm{d}}{\mathrm{d}t}\sum_i \frac{\partial \mathcal{L}}{\partial \boldsymbol{v}_i} = 0 \\ &\Rightarrow \sum \frac{\partial \mathcal{L}}{\partial \boldsymbol{v}_i} = \sum \boldsymbol{p}_i = \boldsymbol{p} = \text{const} .\end{aligned} \tag{4.72}$$

Chapter 4

2. The conservation of energy follows from the homogeneity of time.

The homogeneity of time implies that the Lagrange function $\mathcal{L}$ does not explicitly depend on time. Which means that $\partial\mathcal{L}/\partial t = 0$.

The total derivation of $\mathcal{L}$ is

$$\frac{\mathrm{d}\mathcal{L}}{\mathrm{d}t} = \sum_{i=1}^{3N} \frac{\partial\mathcal{L}}{\partial x_i}\dot{x}_i + \sum \frac{\partial\mathcal{L}}{\partial \dot{x}_i}\ddot{x}_i \,.$$

If we replace according to (4.71) $\partial\mathcal{L}/\partial x_i$ by $\mathrm{d}/\mathrm{d}t(\partial\mathcal{L}/\partial\dot{e})$ we obtain

$$\begin{aligned}
\frac{\mathrm{d}\mathcal{L}}{\mathrm{d}t} &= \sum \dot{x}_i \frac{\mathrm{d}}{\mathrm{d}t}\frac{\partial\mathcal{L}}{\partial\dot{x}_i} + \sum \frac{\partial\mathcal{L}}{\partial\dot{x}_i}\ddot{x}_i = \sum \frac{\mathrm{d}}{\mathrm{d}t}\left(\frac{\partial\mathcal{L}}{\partial\dot{x}_i}\dot{x}_i\right) \\
\Rightarrow &\frac{\mathrm{d}}{\mathrm{d}t}\left(\sum \dot{x}_i \frac{\partial\mathcal{L}}{\partial\dot{x}_i} - \mathcal{L}\right) = 0 \qquad (4.73)\\
\Rightarrow &\sum \dot{x}_i \frac{\partial\mathcal{L}}{\partial\dot{x}_i} - \mathcal{L} = E = \text{const} \,,
\end{aligned}$$

which means that E is constant in time.

Finally the conservation of angular momentum follows from the isotropy of space, which means that no specific direction in space is preferred.

This isotropy implies that an arbitrary rotation of a closed system does not change the mechanical properties of the system. In particular the Lagrange function should not change when the system rotates by an angle $\partial\varphi$.

We introduce the vector $\boldsymbol{\delta\varphi}$ with the magnitude $\delta\varphi$ and the direction of the rotation axis. The change of the position vector r_i of the point P is (Fig. 4.32)

$$\boldsymbol{\delta r}_i = \boldsymbol{\delta\varphi} \times \boldsymbol{r}_i \,. \qquad (4.74a)$$

The velocity of P is then changing by

$$\delta v_i = \delta\boldsymbol{\varphi} \times \boldsymbol{v}_i \,. \qquad (4.74b)$$

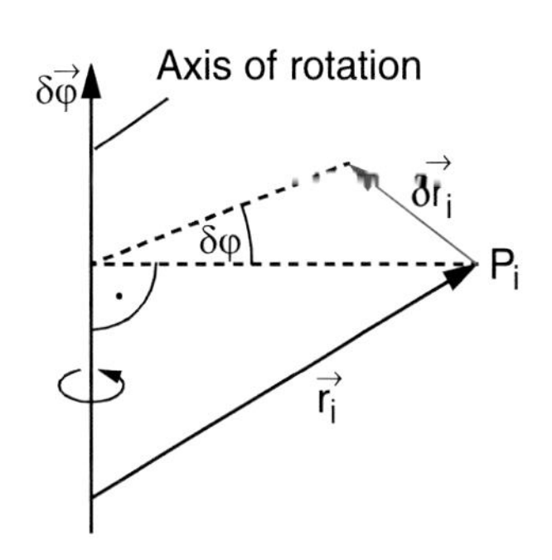

Figure 4.32 Definition of $\delta\varphi$ and δr

For the change $\delta\mathcal{L} = 0$ of the Lagrange function $\mathcal{L}$ we obtain

$$\delta\mathcal{L} = \sum_i \frac{\partial\mathcal{L}}{\partial\boldsymbol{r}_i}\delta\boldsymbol{r}_i + \frac{\partial\mathcal{L}}{\partial\boldsymbol{v}_i}\delta\boldsymbol{v} = 0 \,. \qquad (4.75)$$

With the relations

$$\partial\mathcal{L}/\partial\boldsymbol{v}_i = \boldsymbol{p}_i \quad \text{and} \quad \partial\mathcal{L}/\partial\boldsymbol{r}_i = \boldsymbol{F}_i = \frac{\mathrm{d}\boldsymbol{p}_i}{\mathrm{d}t}$$

we can write (4.75) in the form

$$\begin{aligned}
&\sum_i \dot{\boldsymbol{p}}_i(\delta\boldsymbol{\varphi}\times\boldsymbol{r}_i) + \boldsymbol{p}_i(\delta\boldsymbol{\varphi}\times\boldsymbol{v}_i) = 0 \\
&\Rightarrow \delta\boldsymbol{\varphi}\left[\sum_i((\boldsymbol{r}_i\times\dot{\boldsymbol{p}}_i) + (\boldsymbol{v}_i\times\boldsymbol{p}_i))\right] \qquad (4.76)\\
&= \delta\boldsymbol{\varphi}\frac{\mathrm{d}}{\mathrm{d}t}\sum(\boldsymbol{r}_i\times\boldsymbol{p}_i) = 0 \,.
\end{aligned}$$

Since this must hold for arbitrary values of $\delta\varphi$ it follows

$$\sum(\boldsymbol{r}_i\times\boldsymbol{p}_i) = \boldsymbol{L} = \textbf{const} \,. \qquad (4.77)$$

Summary

- The centre of mass of a system of N point masses m_i with the position vectors $\boldsymbol{r}_i$ has the position vector

$$\boldsymbol{r}_\mathrm{S} = \frac{1}{\sum m_i}\sum m_i\boldsymbol{r}_i = \frac{1}{M}\sum m_i\boldsymbol{r}_i \,.$$

- The coordinate system with the CM as origin is called the centre-of-mass system
- The vector sum of all momenta $m_i\boldsymbol{v}_i$ of the masses m_i in the CM-system is always zero.
- The reduced mass μ of two masses m_1 and m_2 is defined as

$$\mu = \frac{m_1 \cdot m_2}{m_1 + m_2} \,.$$

- The relative motion of two particles with the mutual interaction forces $\boldsymbol{F}_{12} = -\boldsymbol{F}_{21}$ can be reduced to the motion of a single particle with the reduced mass μ which moves with the velocity $\boldsymbol{v}_{12} = \boldsymbol{v}_1 - \boldsymbol{v}_2$ around the centre of m_1.

- A system of particles with masses m_i, where no external forces are present is called a **closed system**. The total momentum and the total angular momentum of a closed system are always constant, i. e. they do not change with time (conservation laws for momentum and angular momentum).
- In elastic collisions between two particles the total kinetic energy and the total momentum are conserved. For inelastic collisions the total momentum is also conserved but part of the initial kinetic energy is transferred into internal energy (e. g. potential energy or kinetic energy of the building blocks of composed collision partners). Inelastic collisions can only occur if at least one of the collision partners has a substructure, i. e. is compound of smaller entities.
- While for elastic collisions in the lab-system the kinetic energies E_i of the individual partners change (although the total energy is conserved), in the CM-system also the E_i are conserved.
- In inelastic collisions only the kinetic energy $\frac{1}{2}\mu v_{12}^2$ of the relative motion can be transferred into internal energy. At least the part $\frac{1}{2}Mv_S^2$ of the CM-motion must be preserved as kinetic energy of the collision partners.
- The collision between two particles with masses m_1 and m_2 can be reduced in the CM-system to the scattering of a single particle with reduced mass

$$\mu = \frac{m_1 \cdot m_2}{m_1 + m_2}$$

by a particle with mass $m\infty$ fixed in the CM. This can be also described by the scattering in a potential depending on the interaction force between the two particles.
- The deflection angle φ of the particle in the CM-system depends on the impact parameter b, the reduced mass μ, the initial kinetic energy $\frac{1}{2}\mu v_0^2$ and the radial dependence of the interaction potential.
- The evaluation of collisions at relativistic velocities v demands the consideration of the relativistic mass increase. Then also energy and momentum conservation remain valid.
- The conservation laws for energy, momentum and angular momentum can be ascribed to general symmetry principles, as the homogeneity of space and time and the isotropy of space.

Problems

4.1 Two particles with masses $m_1 = m$ and $m_2 = 3m$ suffer a central collision. What are their velocities v_1' and v_2' after the collision if the two particles had equal but opposite velocities $v_1 = -v_2$ before the collision

a) For a completely elastic collision
b) For a completely inelastic collision?

4.2 A wooden block with mass $m_1 = 1$ kg hangs on a wire with length $L = 1$ m. A bullet with mass $m_2 = 20$ g is shot with the velocity $v = 10^3$ m/s into the block and sticks there. What is the maximum deflection angle of the block?

4.3 A proton with the velocity v_1 collides elastically with a deuteron (nucleus consisting of proton and neutron) at rest. After the collision the deuteron flies under the angle of 45° against v_1. Determine

a) the deflection angle θ_1 of the proton
b) the CM-velocity
c) the velocities v_1' and v_2' of proton and deuteron after the collision.

4.4 A particle with mass $m_1 = 2$ kg has the velocity $\boldsymbol{v}_1 = \{3\hat{e}_x + 2\hat{e}_y - \hat{e}_z\}$ m/s. It collides completely inelastic with a particle of mass $m_2 = 3$ kg and velocity $\boldsymbol{v}_2 = \{-2\hat{e}_x + 2\hat{e}_y + 4\hat{e}_z\}$. Determine

a) The kinetic energies of the two particle before the collision in the lab-system and the CM-system.
b) Velocity and kinetic energy of the compound particle after the collision.
c) Which fraction of the initial kinetic energy has been converted into internal energy? Calculate this fraction in the lab-system and the CM-system.

4.5 A mass $m_1 = 1$ kg with a velocity $v_1 = 4$ m/s collides with a mass $m_2 = 2$ kg. After the collision m_1 moves with $v_1' = \sqrt{8}$ m/s under an angle of 45° against $\boldsymbol{v}_1$ and m_2 with $v_2' = \sqrt{2}$ m/s under an angle of −45°

a) What was the velocity v_2?
b) Which fraction of the initial kinetic energy has been converted into internal energy in the lab-system and the CM-system?
c) How large are the deflection angles ϑ_1 and ϑ_2 in the CM-system?

4.6 Two cuboids with masses $m_1 = 1$ kg and $m_2 < m_1$ slide frictionless on an air-track, which is blocked on both sides by a vertical barrier (Fig. 4.33). Initially m_1 is at rest and m_2 moves with constant velocity $v_2 = 0.5$ m/s to the left. After the collision with m_1 the mass m_2 is reflected to the right, collides with the barrier ($m = \infty$) and slides again to the left. We assume that all collisions are completely elastic.

a) What is the ratio m_1/m_2 if the two masses finally move to the left with equal velocities?
b) How large should m_2 be in order to catch m_1 before it reaches the left barrier?
c) Where collide the two masses at the second collision for $m_2 = 0.5$ kg?

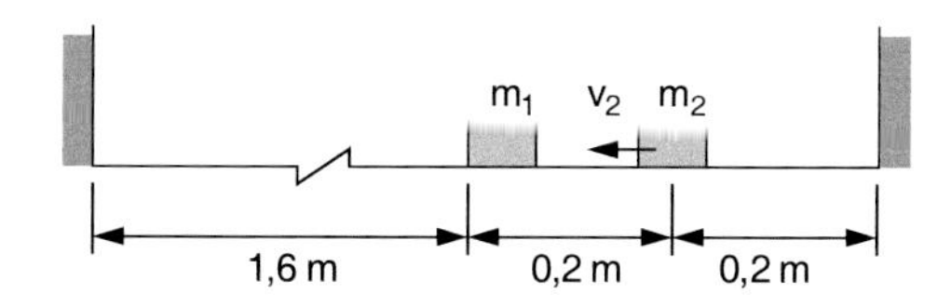

Figure 4.33 Illustrating Probl. 4.6

4.7 A steel ball with mass $m_1 = 1\,\text{kg}$ hangs on a wire with $L = 1\,\text{m}$, vertically above the left edge of a resting mass $m_2 = 5\,\text{kg}$ which can slide without friction on a horizontal air-track. (Fig. 4.34). The steel ball with the wire is lifted by an angle $\varphi = 90°$ from the vertical into the horizontal position and then released. It collides elastically with the glider. What is the maximum angle φ of m_1 after the collision?

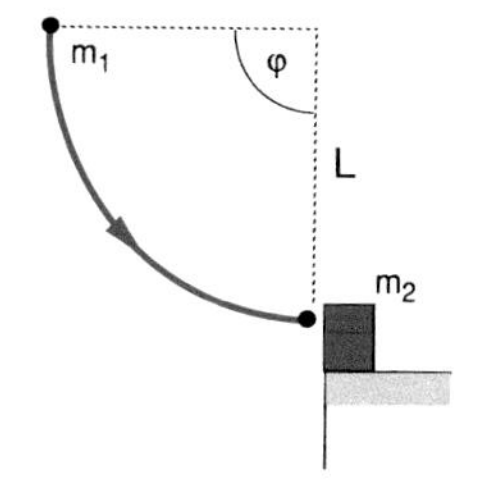

Figure 4.34 Illustration of Probl. 4.7

4.8 An elevator ascends with constant velocity $v = 2\,\text{m/s}$. When its ceiling is still 30 m below the upper point A of the lift shaft a ball is released from A which falls freely down and hits elastically the ceiling of the elevator, from where it is elastically reflected upwards.

a) Where does it hit the elevator ceiling?
b) What is its maximum height after the reflection?
c) Where does it hit the elevator ceiling a second time?

4.9 An α-particle (nucleus of the He-atom) hits with the velocity $\boldsymbol{v}_1$ an oxygen nucleus at rest ($m_2 = 4m_1$). The α-particle is deflected by 64°, the oxygen nucleus by −51° against $\boldsymbol{v}_1$. The collision is completely elastic.

a) What is the ratio v_1'/v_2' of the velocities after the collision?
b) What is the ratio of the kinetic energies after the collision?

4.10 A particle has in a system S a kinetic energy of 6 GeV. and the momentum $P = 6\,\text{GeV/c}$. What is its energy in a system S', where its momentum is measured as 5 GeV/c? What is the relative velocity of S' against S?

References

4.1a. http://berkeleyscience.com/pm.htm
4.1b. A. Tan, *Theoryof Orbital Motion.* (World Scientific Publ., 2008)
4.2. R. D. Levine, *Moleclar Reaction Dynamics.* (Cambridge Univ. Press, 2005)
4.3. A. Einstein, N. Calder, *Relativity: The special and the general Theory.*
4.4a. E.F. Taylor, J.A. Wheeler, *Spacetime Physics: Introduction to Special Relativity,* 2nd ed., (W. H. Freeman & Company, 1992)
4.4b. W. Rindler, *Introduction to Special Relativity,* 2nd ed., (Oxford University Press, 1991)
4.5. R.M. Dreizler, C.S. Lüdde, *Theoretical Mechanics.* (Springer, Berlin, Heidelberg, 2010)
4.6. A.P. French, *Special Relativity.* (W.W. Norton & Co, 1968)
4.7. A. Das, *The Special Theory of Relativity.* (Springer, Berlin, Heidelberg, 1996)
4.8. Goldstein, *Classical Mechanics.* (Addison Wesley, 2001)
4.9. J. Schwichtenberg, *Physics from Symmetry.* (Springer, Heidelberg, 2015)

Dynamics of rigid Bodies

5

© Springer International Publishing Switzerland 2017
W. Demtröder, *Mechanics and Thermodynamics*, Undergraduate Lecture Notes in Physics, DOI 10.1007/978-3-319-27877-3_5

Up to now we have discussed idealized bodies where their spatial extension could be neglected and they were therefore adequately described by the model of a point mass. We have investigated their motion under the influence of forces and have presented besides Newton's laws fundamental conservation laws for linear momentum, energy and angular momentum.

All phenomena found in nature which are due to the spatial extension of bodies demand for their explanation an extension of our model. Besides the translation of point masses, discussed so far, we have to take into account the fact that extended bodies can also rotate around fixed or free axes.

At first, we will restrict ourselves to the motion of free extended bodies under the influence of forces. The motion of single volume elements of an extended body against each other, that results in a deformation of the body will be discussed in the next chapter. Such still idealized extended bodies that do not change their form, are called **rigid bodies**.

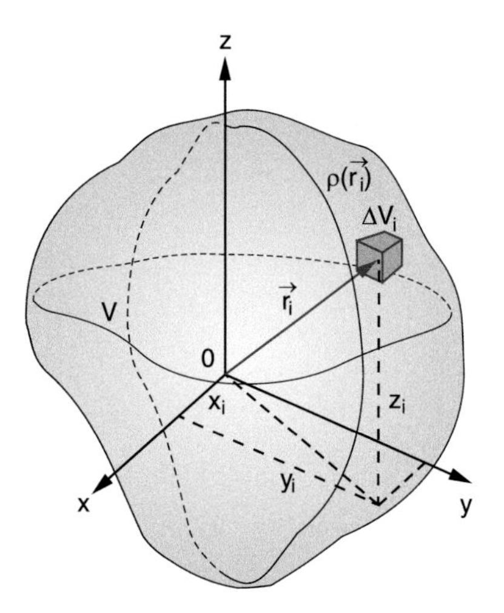

Figure 5.1 Partition of a spatially extended body into small volume elements ΔV_i

5.1 The Model of a Rigid Body

We can partition an extended rigid body with volume V and total mass M into many small volume elements ΔV_i with masses Δm_i which are rigidly bound together (Fig. 5.1). We can regard theses mass elements Δm_i as point masses and treat them according to the rules discussed in Chap. 2.

The total body can then be composed of these volume elements:

$$V = \sum_{i=1}^{N} \Delta V_i \,, \qquad M = \sum_{i=1}^{N} \Delta m_i \,.$$

We name the ratio

$$\varrho = \Delta m / \Delta V \,; \qquad [\varrho] = \mathrm{kg/m^3} \tag{5.1}$$

the mass density of the volume element ΔV. The total mass can then be expressed as

$$M = \sum_{i=1}^{N} \varrho_i \Delta V_i \,. \tag{5.2}$$

If the volume elements ΔV become smaller and smaller, their number N correspondingly larger, the sums converge for the limiting case $\Delta V \to 0$ to volume integrals [5.1]

$$\begin{aligned} V &= \lim_{\substack{\Delta V_i \to 0 \\ N \to \infty}} \sum_{i=1}^{N} \Delta V_i = \int_V \mathrm{d}V \,; \\ M &= \int_V \varrho \mathrm{d}V \,, \end{aligned} \tag{5.3}$$

where the volume integral stands for the three-dimensional integral

$$V = \int_{z_1}^{z_2} \left[\int_{y_1}^{y_2} \left(\int_{x_1}^{x_2} \mathrm{d}x \right) \mathrm{d}y \right] \mathrm{d}z \tag{5.4}$$

for the example of a cuboid, while for a spherical volume with radius R and a volume element $\mathrm{d}V = r^2 \cdot \sin\vartheta \cdot \mathrm{d}r \cdot \mathrm{d}\vartheta \cdot \mathrm{d}\varphi$ (see Sect. 13.2) the integral can be written as

$$V = \int_{r=0}^{R} \int_{\vartheta=0}^{\pi} \int_{\varphi=0}^{2\pi} r^2 \sin\vartheta \, \mathrm{d}r \, \mathrm{d}\vartheta \, \mathrm{d}\varphi \,. \tag{5.5}$$

The mass density $\varrho(x, y, z)$ can generally depend on the location (x, y, z). For homogeneous bodies ϱ is constant for all points of the body and we can extract ϱ out of the integral. The mass M of the body can then be expressed as

$$M = \varrho \int_V \mathrm{d}V = \varrho V \,. \tag{5.6}$$

5.2 Center of Mass

As has been shown in the previous chapter the position vector $\boldsymbol{r}_S$ of the CM of a system with N particles at the positions $\boldsymbol{r}_i$ (Fig. 5.2) is

$$\begin{aligned} \boldsymbol{r}_S &= \frac{\sum_{i=1}^{N} \boldsymbol{r}_i \Delta m_i}{\sum_{i=1}^{N} \Delta m_i} \\ &= \frac{1}{M} \sum_{i=1}^{N} \boldsymbol{r}_i \varrho(\boldsymbol{r}_i) \Delta V_i \,. \end{aligned} \tag{5.7}$$

For the limiting case $\Delta V \to 0$ and $N \to \infty$ this becomes

$$\begin{aligned} \boldsymbol{r}_S &= \frac{1}{M} \int_V \boldsymbol{r} \, \mathrm{d}m \\ &= \frac{1}{M} \int_V \boldsymbol{r} \varrho(\boldsymbol{r}) \, \mathrm{d}V \,. \end{aligned} \tag{5.8a}$$

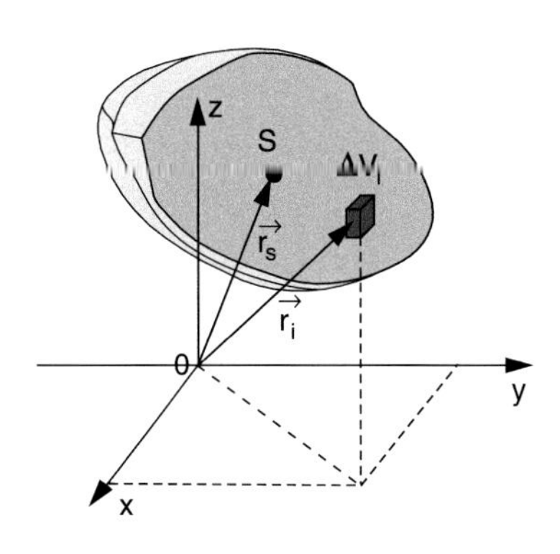

Figure 5.2 Definition of the Center of Mass of an extended body

This corresponds to the three equations for the components

$$x_S = \frac{1}{M}\int_V x\varrho(x,y,z)\,dV\,,$$

$$y_S = \frac{1}{M}\int_V y\varrho(x,y,z)\,dV\,,$$

$$z_S = \frac{1}{M}\int_V z\varrho(x,y,z)\,dV\,.$$

For homogeneous bodies ($\varrho = \text{const}$) (5.8a) simplifies to

$$\boldsymbol{r}_S = \frac{1}{V}\int_V \boldsymbol{r}\,dV\,. \tag{5.8b}$$

Example

Center of Mass of a homogeneous hemisphere.

If the center of the sphere is at the origin ($x = y = z = 0$) (Fig. 5.3) symmetry arguments require $x_S = y_S = 0$. For $\varrho = \text{const}$ we obtain from (5.8b)

$$z_S = \frac{1}{M}\int_V z\varrho\,dV = \frac{1}{V}\int_V z\,dV\,.$$

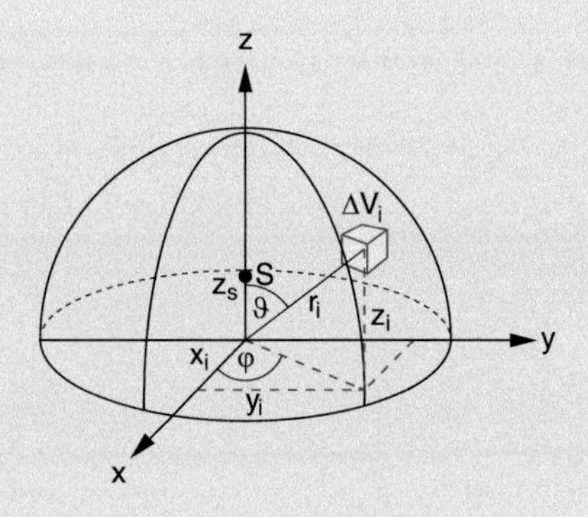

Figure 5.3 Center of Mass of a hemisphere

This becomes with $z = r{\cdot}\cos\vartheta$ and $dV = r^2 dr{\cdot}\sin\vartheta\,d\vartheta\,d\varphi$

$$\begin{aligned} z_S &= \frac{1}{V}\int_{r=0}^{R}\int_{\vartheta=0}^{\pi/2}\int_{\varphi=0}^{2\pi} r^3\cos\vartheta\,\sin\vartheta\,dr\,d\vartheta\,d\varphi \\ &= \frac{3}{8}R\,. \end{aligned} \tag{5.9}$$

◄

5.3 Motion of a Rigid Body

The center points P_i of the volume elements dV_i are defined by their position vectors $\boldsymbol{r}_i$, the CM by r_S. The vector

$$r_{iS} = \boldsymbol{r}_i - \boldsymbol{r}_S$$

points from the center of mass S to the point P_i (Fig. 5.4). The vector

$$dr_{iS}/dt = \boldsymbol{v}_{iS} = \boldsymbol{v}_i - \boldsymbol{v}_S \tag{5.10}$$

gives the relative velocity of P_i with respect to the CM.

In a rigid body all distances are fixed, i.e. $|\boldsymbol{r}_{iS}| = \text{const}$. Differentiation of $r_{iS}^2 = \text{const}$ gives

$$2\boldsymbol{r}_{iS}\cdot\boldsymbol{v}_S = 0\,,$$

which implies that the vector of the relative velocity is perpendicular to the position vector. This can be written as (see Sect. 2.4)

$$\boldsymbol{v}_{iS} = (\boldsymbol{\omega}\times\boldsymbol{r}_{iS})\,, \tag{5.11}$$

where ω is the angular velocity of P_i rotating about an axis through the CM perpendicular to the velocity vector $\boldsymbol{v}_{iS}$.

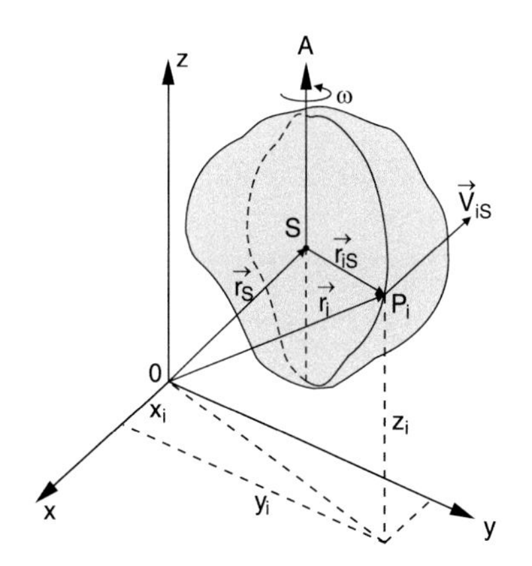

Figure 5.4 The motion of a rigid spatially extended body

For the general motion of the rigid body the velocity of the point P_i

$$\boldsymbol{v}_i = \boldsymbol{v}_S + (\boldsymbol{\omega} \times \boldsymbol{r}_{iS}) \tag{5.12}$$

can be composed of two contributions: The translational velocity v_S of the CM and the rotation $(\boldsymbol{\omega} \times \boldsymbol{r}_{iS})$ of P_i around the CM. Since the consideration discussed above is valid for an arbitrary point P_i we can make the general statement:

The motion of an extended rigid body can always be composed of the translation of its CM and a rotation of the body about its CM.

Note: The rotational axis is not necessarily constant but can change its direction in the course of time, even when no external forces act onto the body (see Sect. 5.7).

The Eq. 5.10 and 5.11 are based on the condition $r_{ik}^2 = \text{const}$ for a rigid body. They are no longer valid, if deformations of the body occur, because then vibrational motions of P_i against the CM can be present as additional movements.

The complete characterization of the motion of a free rigid body demands 6 time-dependent parameters: The position coordinates

$$\boldsymbol{r}_S(t) = \{x_S(t), y_S(t), z_S(t)\}$$

for the description of the CM-motion and three angular coordinates for the description of the rotation of the rigid body about its CM.

The free rigid body has six degrees of freedom for its motion.

If one point of the body (for example the CM) is kept fixed the body can still rotate about this point but cannot perform a translation. The number of degrees of freedom then reduces to three, namely the three rotational degrees of freedom. If the body rotates around a fixed axis, only one degree of freedom is left, namely the one-dimensional rotation described by the angle φ.

5.4 Forces and Couple of Forces

While a force $\boldsymbol{F}$ acting on a point mass is unambiguously defined when its magnitude and direction is given, for forces acting on an extended body the point of origin P has to be added (Fig. 5.5).

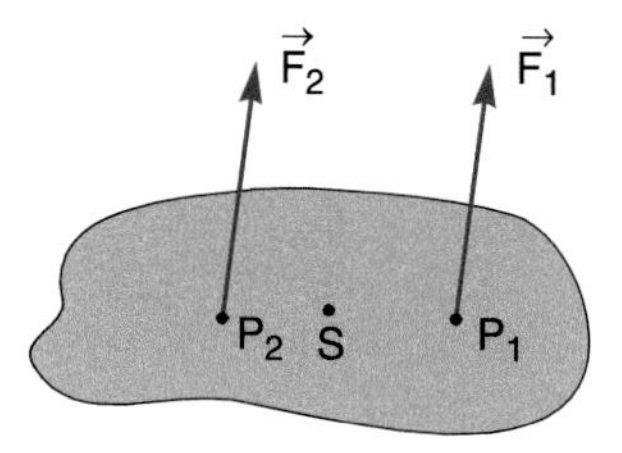

Figure 5.5 The two forces $\boldsymbol{F}_1$ and $\boldsymbol{F}_2$ have equal magnitudes but different points of application P_1 and P_2. They effect different motions of the body

We will investigate the change of motion which an extended body suffers under the action of a force $\boldsymbol{F}(P_i)$ with its origin in an arbitrary point P_i. We can simplify the treatment when we add two antiparallel forces $\boldsymbol{F}_2(S)$ and $\boldsymbol{F}_3(S) = -\boldsymbol{F}_2(S)$ with equal magnitude which both act on the center of mass S and therefore do not affect the motion of the body, because they act on the same point S and since $\boldsymbol{F}_2(S) + \boldsymbol{F}_3(S) = \boldsymbol{0}$ they cancel each other.

Now we combine the two antiparallel forces $\boldsymbol{F}_1$ and $\boldsymbol{F}_3$ with equal magnitude (Fig. 5.6) which form a **couple of forces**, but regard at first the remaining single force $\boldsymbol{F}_2$, which acts on the center of mass S. This force causes a translation of the CM. The couple of forces brings about a torque

$$\boldsymbol{D}_S = (\boldsymbol{r}_{iS} \times \boldsymbol{F}_1) , \tag{5.13a}$$

referred to the center of mass S. Since $\boldsymbol{F}_1 + \boldsymbol{F}_3 = \boldsymbol{0}$ this couple of forces does not cause an acceleration of the CM. It induces, however, a rotation of the body around S. Summarizing we can make the general statement:

A force F acting on an arbitrary point $P \neq S$ of an extended body causes an acceleration of the CM and a rotation of the body about the center of mass S.

An extended rigid body initially at rest suffers an accelerated translation of its center of mass S and a rotation around S when a force F acts on a point $P \neq S$.

In this chapter we will investigate such motions in more detail.

At first we will restrict the treatment to the special case where the body rotates around a space-fixed axis. The motion is then restricted to a rotation which has only one degree of freedom.

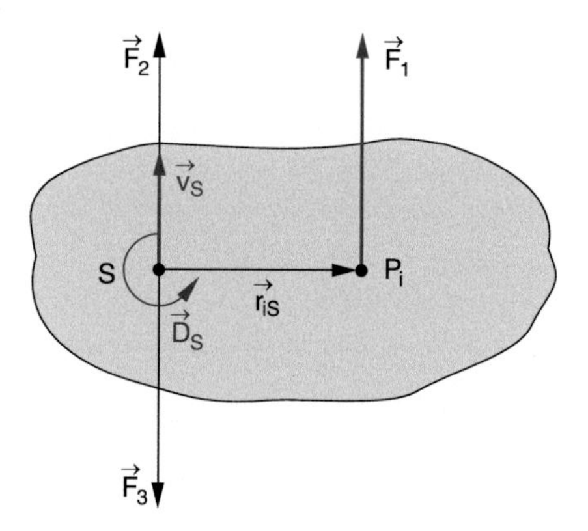

Figure 5.6 Decomposition of a force $\boldsymbol{F}_1$ into a couple of forces $\boldsymbol{F}_1\boldsymbol{F}_3$ and a force $\boldsymbol{F}_2$ that attacks at the Center of Mass

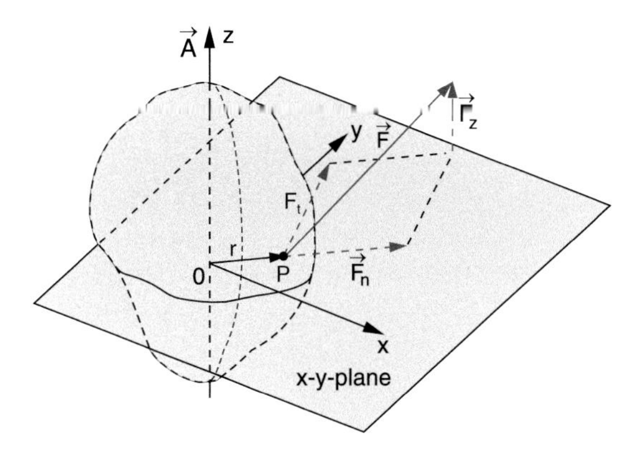

Figure 5.7 Rotation of a rigid body about a fixed axis A, induced by a force F attacking at the point P. Decomposition of the projection of F onto the x-y-plane into a normal and a tangential component

We choose the direction of the rotation axis $\boldsymbol{A}$ as the z-direction and the x-y-plane through the point $P(r)$ where the force $\boldsymbol{F}$ acts on the body, which has the distance r from the rotation axis (Fig. 5.7). We can then decompose the force into the three components $F_z \parallel \boldsymbol{A}$, the radial component $F_r \parallel r$ and the tangential component $F_t \perp \boldsymbol{r}$ and $\perp \boldsymbol{A}$. F_z is perpendicular to the x-y-plane and the other two components are in the x-y-plane.

The torque exerted by the force $\boldsymbol{F}$ onto the body is

$$\boldsymbol{D} = (\boldsymbol{r} \times \boldsymbol{F}) = (\boldsymbol{r} \times \boldsymbol{F}_t) + (\boldsymbol{r} \times \boldsymbol{F}_z) ,$$

since $(\boldsymbol{r} \times \boldsymbol{F}_n) = \boldsymbol{0}$.

The first term causes a torque about the z-axis, and therefore an acceleration of the rotation about the z-axis. The second term would change the direction of the rotation axis. If this axis is fixed by axle bearings the torque only acts on the bearings and does not lead to a change of motion.

If the rotation axis intersects the center of mass S (Fig. 5.8) which we choose as the origin of our coordinate system, the torque exerted by the weight $F_w = M \cdot g$ of the body is zero, as can be seen by the following derivation:

The torque with respect to the rotation axis caused by the gravitational force $\Delta m_i \cdot g$ on the mass element Δm_i is $\boldsymbol{D}_i = (\boldsymbol{r}_{iS} \times \Delta m_i \cdot \boldsymbol{g})$. The torque exerted by the weight of the whole body is then

$$\begin{aligned} D &= \int_V (\boldsymbol{r} \times \boldsymbol{g})\,\mathrm{d}m = -\boldsymbol{g} \times \int_V \boldsymbol{r}\,\mathrm{d}m \\ &= -(\boldsymbol{g} \times M\boldsymbol{r}_S) = \boldsymbol{0} , \end{aligned} \tag{5.13b}$$

because the center of mass S is the origin and therefore is $r_S = 0$.

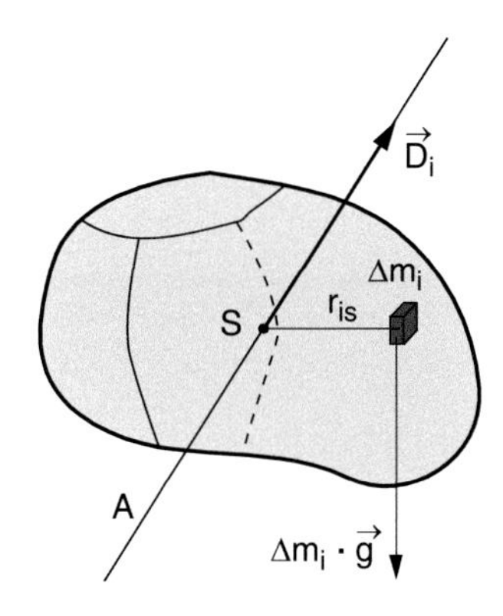

Figure 5.8 Contribution of the mass element Δm to the torque about an axis through the Center of Mass, due to the weight of Δm

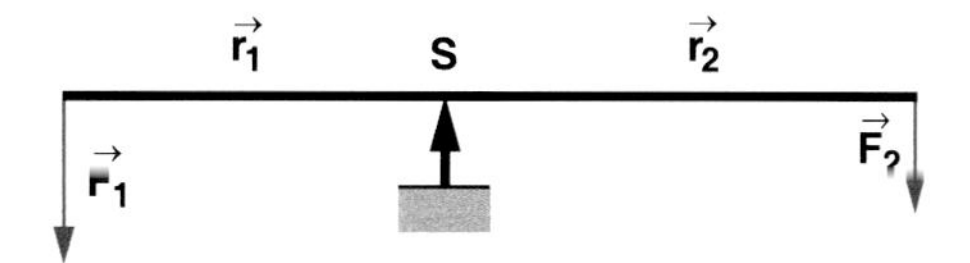

Figure 5.9 Principle of beam balance

> If a body can rotate around an axis through the CM it is always at equilibrium, independent on the space orientation of the axis because the torque exerted by its weight is always zero.

All beam balances are based on this principle (Fig. 5.9). The balance is at equilibrium if $\sum D_i = 0$, which means

$$\boldsymbol{r}_1 \times \boldsymbol{F}_1 + \boldsymbol{r}_2 \times \boldsymbol{F}_2 = \boldsymbol{0} .$$

This is the equilibrium condition for a balance as two-armed lever.

5.5 Rotational Inertia and Rotational Energy

We consider an extended body which rotates about a fixed axis A with the angular velocity ω (Fig. 5.10). The mass element Δm_i with the distance $r_{i\perp} = |\boldsymbol{r}_i|$ from the axis A has the velocity $v_i = r_i \cdot \omega$. Its kinetic energy is then

$$E_{\mathrm{kin}}(\Delta m_i) = \tfrac{1}{2}\Delta m_i v_i^2 = \tfrac{1}{2}\Delta m_i r_i^2 \omega^2 .$$

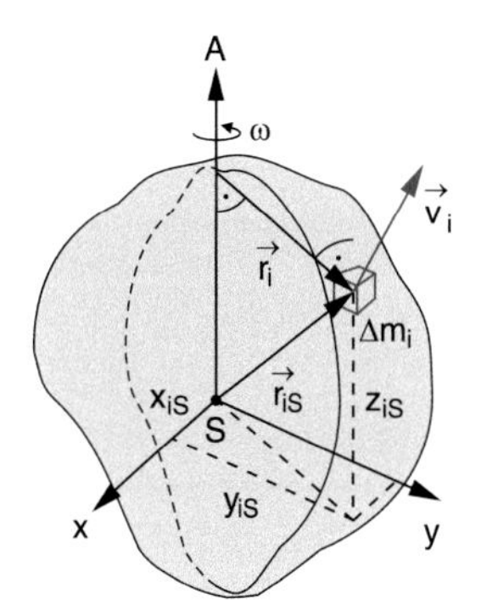

Figure 5.10 Definition of moment of inertia

The summation over all mass elements gives the total rotational energy of the body

$$E_{\text{rot}} = \lim_{\substack{N\to\infty \\ \Delta m_i \to 0}} \left(\frac{1}{2} \sum_{i=1}^{N} \Delta m_i r_{i\perp}^2 \omega^2 \right) = \frac{1}{2}\omega^2 \int r_\perp^2 \, \mathrm{d}m \, . \quad (5.14)$$

The expression

$$I \overset{\text{Def}}{=} \int_V r_\perp^2 \, \mathrm{d}m = \int_V r_\perp^2 \varrho \, \mathrm{d}V \quad (5.15)$$

is called the *rotational inertia* (often also *moment of inertia*) of the rotating body referred to the axis A. With this definition we obtain for the rotational energy

$$E_{\text{rot}} = \tfrac{1}{2} I \omega^2 \, . \quad (5.16a)$$

The angular momentum of Δm_i with respect to the axis A is

$$\boldsymbol{L}_i(\Delta m_i) = \boldsymbol{r}_{i\perp} \times (\Delta m_i \boldsymbol{v}_i) = r_{i\perp}^2 \Delta m_i \boldsymbol{\omega} \, , \quad (5.17a)$$

which gives the total angular momentum of the body as

$$L = I \cdot \omega \, . \quad (5.17b)$$

Replacing in (5.16a) ω^2 by L^2/I^2 we obtain for the rotational energy

$$E_{\text{rot}} = \frac{1}{2} I \omega^2 = \frac{L^2}{2I} \, . \quad (5.16b)$$

The rotational inertia I is a measure for the mass distribution in an extended body relative to the rotational axis. For geometrically simple bodies with homogeneous mass distribution $\varrho = \text{const}$ the rotational inertia I can be readily calculated, as is illustrated in the following examples. For bodies with a complex geometrical structure I has to be measured (see below).

If the rotational axis A intersects the center of mass $S(r = 0)$, the rotational inertia can be written as

$$I_S = \varrho \cdot \int r^2 \mathrm{d}V \, .$$

The rotational inertia is always defined with respect to a definite rotational axis and depends on the location of this axis with relative to the CM.

5.5.1 The Parallel Axis Theorem (Steiner's Theorem)

If a body rotates about an axis B which is parallel to the axis A through the CM, the rotational inertia with respect to the axis B can be readily calculated, if it is known with respect to the axis A. If the distance between the two axes is a (Fig. 5.11) we can write

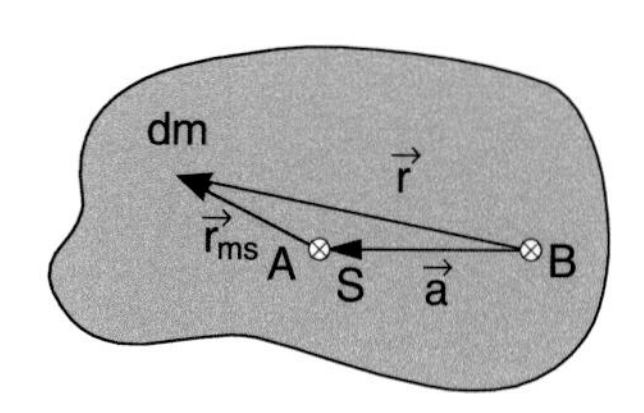

Figure 5.11 Steiner's rule: The drawing shows the plane through dm, perpendicular to the axis A

$$\begin{aligned} I_B &= \int r^2 \mathrm{d}m = \int (r_{mS} + a)^2 \mathrm{d}m \\ &= \int r_{mS}^2 \mathrm{d}m + 2a \int r_{mS} \mathrm{d}m + a^2 \int \mathrm{d}m \, . \end{aligned}$$

According to (5.8) is $\int r_{mS} \mathrm{d}m = R_S \cdot M = 0$, because the center of mass S is the origin of the coordinate system and therefore is $R_{\text{CM}} = R_S = 0$.

We then obtain

$$\boldsymbol{I}_B = \boldsymbol{I}_S + \boldsymbol{a}^2 \boldsymbol{M} \, . \quad (5.18)$$

Equation 5.18 is called the *parallel axis theorem* or *Steiner's theorem*. It states:

The inertial moment of a body rotating around an axis B is equal to the sum of the inertial moment with respect to an axis A through the center of mass S with a distance a from the axis B plus the moment of inertia of the total mass M concentrated in S with respect to B.

This illustrates that it is sufficient to determine the rotational inertia with respect to an axis A through S. With (5.18) we can then obtain the rotational inertia with respect to any axis parallel to A.

In the following we will give examples for the calculation of I for homogeneous bodies with different geometrical structures.

Example

1. *Thin disc.* We assume the height h in the z-direction is small compared to the extension of the body in the x- and y-directions.
   a) Rotational axis in the z-direction:
   $$I_z = \varrho \cdot \int (x^2 + y^2) \mathrm{d}V \, .$$
   b) Rotational axis in the x-direction:
   $$I_x = \varrho \cdot \int (y^2 + z^2) \mathrm{d}V \approx \varrho \cdot \int y^2 \mathrm{d}V \, ,$$
   because $|z| \leq h/2 \ll y_{\max}$.

Chapter 5

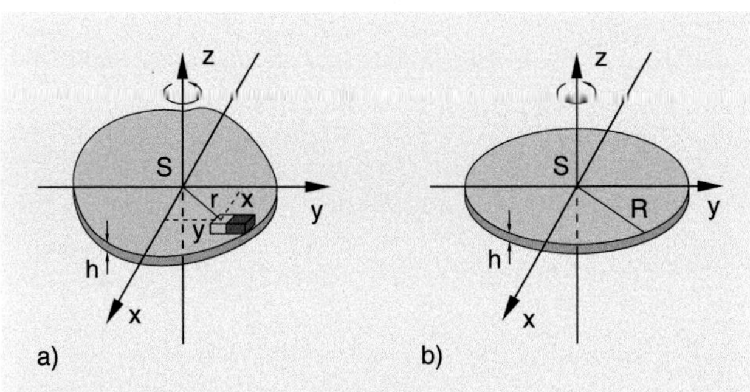

Figure 5.12 Moment of inertia of a thin disc with arbitrary shape (**a**), circular disc (**b**)

c) Rotational axis in the y-direction:

$$I_y = \varrho \cdot \int (x^2 + z^2) \mathrm{d}V \approx \varrho \cdot \int x^2 \mathrm{d}V \,.$$

This shows that with the approximation $z \ll x, y$ one obtains

$$I_z \approx I_x + I_y \,. \tag{5.19}$$

For plane bodies (for example a triatomic molecule) the rotational inertia for the rotation around an axis through the CM perpendicular to the plane is equal to the sum of the two other moments of inertia.
For the case of a thin circular disc we obaqin from (5.19) because of the rotational symmetry (Fig. 5.12b)

$$I_x = I_y = \tfrac{1}{2} I_z \,.$$

For the homogeneous circular disc with radius R it is not difficult to calculate I_z:

$$I_z = \varrho \cdot \int r^2 \mathrm{d}V = 2\pi h \varrho \int r^3 \mathrm{d}r = \varrho \cdot h \cdot \pi \cdot R^4/2 \,,$$

because $\mathrm{d}V = 2\pi r \cdot h \cdot \mathrm{d}r$. With $M = \varrho \cdot \pi \cdot R^2 \cdot h$ this gives

$$I_z = \tfrac{1}{2} M R^2 \,. \tag{5.20a}$$

2. *Hollow cylinder* with height h, outer radius R and wall thickness $d \ll R$ (Fig. 5.13). Rotation about the z-axis as symmetry axis:

$$I_z = \varrho \int_V r^2 \mathrm{d}V = 2\pi h \varrho \int_{R-d}^{R} r^3 \,\mathrm{d}r \,,$$

with $\mathrm{d}V = 2\pi \cdot R \cdot h \cdot \mathrm{d}r$ and $\mathrm{d} \ll R$ one obtains

$$I_z = h \cdot \varrho \cdot \pi \left[R^4 - (R-d)^4\right] \approx 2\pi \varrho h R^3 d \,.$$

This gives with $M = 2\pi \cdot r \cdot \varrho \cdot d \cdot h$

$$I_z = M \cdot R^2 \,. \tag{5.20b}$$

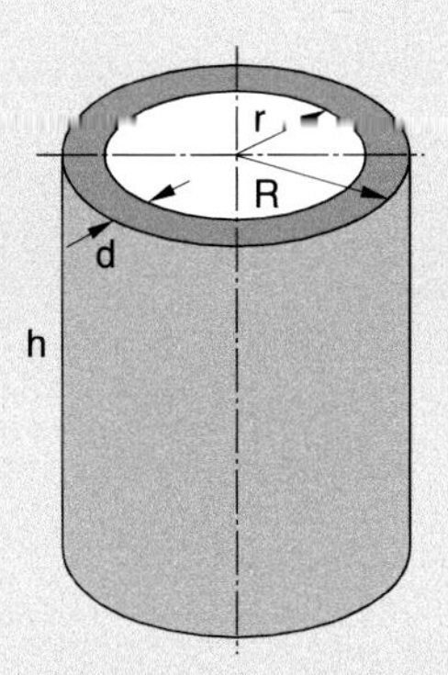

Figure 5.13 Moment of inertia of a hollow cylinder

Remark. A simpler derivation starts with the relation for the total mass of the hollow cylinder with outer radius r_2 and inner radius r_1:

$$M = \pi \cdot \varrho \cdot h \left(r_2^2 - r_1^2\right) \,.$$

$$\begin{aligned} I_z &= \int r^2 \,\mathrm{d}m = 2\pi \varrho \cdot h \int r^3 \mathrm{d}r = \tfrac{1}{2}\pi \varrho \cdot h \left(r_2^4 - r_1^4\right) \\ &= \tfrac{1}{2}\pi \varrho \cdot h \left(r_2^2 - r_1^2\right) \cdot \left(r_2^2 + r_1^2\right) \\ &\approx \tfrac{1}{2} M \cdot 2R^2 = MR^2 \,. \end{aligned}$$

3. *Full cylinder* with radius R and height h.

$$\begin{aligned} I_z &= 2\pi \varrho h \int_0^R r^3 \,\mathrm{d}r = \frac{\pi}{2} h \varrho R^4 \\ &= \frac{M}{2} R^2 \,, \end{aligned} \tag{5.20c}$$

which, of course, concurs with (5.20a).

4. *Thin rod* (length $L \gg$ diameter (Fig. 5.14))
   a) Rotation about the vertical axis ***a*** through the center of mass S.

$$\begin{aligned} I_S &= \varrho \int x^2 \,\mathrm{d}V = \varrho A \int_{-L/2}^{+L/2} x^2 \,\mathrm{d}x \\ &= \frac{1}{12} \varrho A L^3 = \frac{1}{12} M L^2 \,. \end{aligned} \tag{5.21a}$$

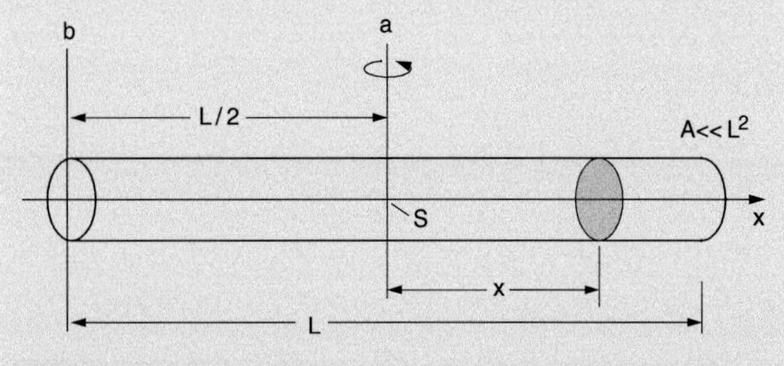

Figure 5.14 Rotation of a thin rod with arbitrary cross section about an axis *a* through the Center of Mass and about an axis *b* at one end of the rod

b) Rotation about the vertical axis $\boldsymbol{b}$ through an end point of the rod. According to the parallel axis theorem (5.18) the moment of inertia is

$$\begin{aligned} I_b &= I_S + M\left(\frac{L}{2}\right)^2 \\ &= \frac{1}{12}ML^2 + \frac{1}{4}ML^2 = \frac{1}{3}ML^2 . \end{aligned} \tag{5.21b}$$

This result could have been obtained also directly from

$$I_b = \varrho A \int_0^L x^2\,\mathrm{d}x = \frac{1}{3}\varrho A L^3 = \frac{1}{3}ML^2 .$$

5. *Diatomic molecule.* Because of the small electron mass ($m_e \approx 1/1836 m_\mathrm{p}$) the electrons do not contribute essentially to the moment of inertia when the molecule rotates around an axis A through the CM perpendicular to the inter-nuclear axis (Fig. 5.15). Because the diameter of the nuclei ($d \approx 10^{-14}$ m) is very small compared with the inter-nuclear distance R ($\approx 10^{-10}$ m) we can treat the nuclei as point masses and obtain

$$I_{SA} = m_1 r_1^2 + m_2 r_2^2 . \tag{5.22a}$$

With the inter-nuclear distance $R = r_1 + r_2$ and the reduced mass $\mu = m_1 \cdot m_2/(m_1 + m_2)$ (5.22a) becomes with $r_1/r_2 = m_2/m_1$

$$I_{SA} = \mu \cdot R^2 . \tag{5.22b}$$

When the molecule rotates around its inter-nuclear axis B the nuclei lie on the rotational axis and do not contribute to the moment of inertia. Now the electrons provide the major contribution. Because of the small electron mass the moment of inertia is now very small and the rotational energy

$$E_\mathrm{rot} = L^2/2I_B$$

for a given angular momentum L becomes much larger than for the rotation around A (see Chap. 11 and Vol. 3).

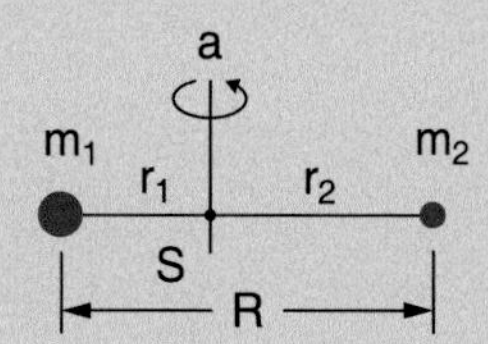

Figure 5.15 Moment of inertia of a diatomic molecule

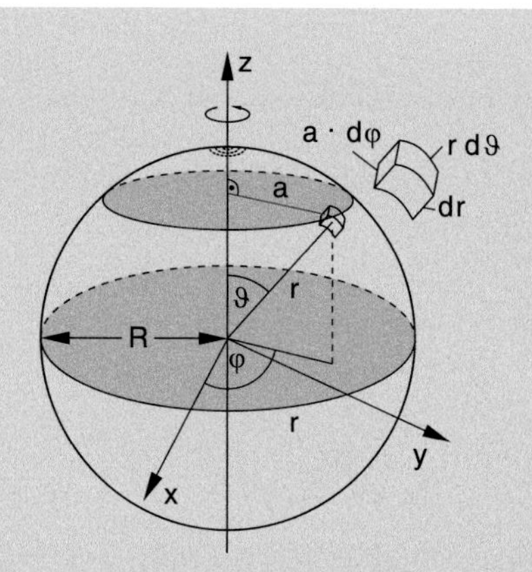

Figure 5.16 Derivation of the moment of inertia of a sphere

6. *Moment of inertia for a homogeneous sphere.* Because of the spherical symmetry the moment of inertia is independent of the direction of the rotational axis through the center of the sphere. The moment of inertia of a mass element Δm with a distance a from the rotation axis (Fig. 5.16) is $\mathrm{d}I = a^2\mathrm{d}m$. For the whole sphere we obtain

$$I_S = \varrho \int a^2 \mathrm{d}V$$

with $\mathrm{d}V = r^2 \sin\vartheta\,\mathrm{d}r\mathrm{d}\vartheta\,\mathrm{d}\varphi$ and $a = r\sin\vartheta$.

$$\begin{aligned} I_S &= \varrho \int_{r=0}^{R} \int_{\vartheta=0}^{\pi} \int_{\varphi=0}^{2\pi} r^4 \sin^3\vartheta\,\mathrm{d}r\,\mathrm{d}\vartheta\,\mathrm{d}\varphi \\ &= \frac{1}{5}\varrho R^5 2\pi \int_{\vartheta=0}^{\pi} \sin^3\vartheta\,\mathrm{d}\vartheta \\ &= \frac{2}{5}\varrho R^2 \frac{4}{3}\pi R^3 = \frac{2}{5}MR^2 . \end{aligned} \tag{5.23}$$

◄

These examples with their rotational inertia are compiled in Tab. 5.1.

5.6 Equation of Motion for the Rotation of a Rigid Body

For the rotation of a rigid body around a space-fixed axis the angular momentum of a mass element Δm_i is:

$$L_i = (\boldsymbol{r}_{i\perp} \times \boldsymbol{p}_i) = \Delta m_i\,(\boldsymbol{r}_{i\perp} \times \boldsymbol{v}_i) = \Delta m_i r_{i\perp}^2 \boldsymbol{\omega} , \tag{5.24}$$

where the velocity $\boldsymbol{v}_i$ is perpendicular to the rotation axis (pointing into the direction of $\boldsymbol{\omega}$ which is the z-direction) and to the radius r.

Table 5.1 Moments of inertia of some symmetric bodies that rotate about a symmetry axis

Geometrical figure	Realization	Moment of inertia
	Thin disc	$\frac{1}{2}MR^2$
	Hollow cylinder with thin wall	MR^2
	Full cylinder	$\frac{1}{2}MR^2$
	Thin rod $L \gg r$	$\frac{1}{12}ML^2$
	Homogeneous sphere	$\frac{2}{5}MR^2$
	Hollow sphere with thin wall	$\frac{2}{3}MR^2$
	Cuboid	$I_x = \frac{1}{12}M(b^2 + c^2)$ $I_y = \frac{1}{12}M(a^2 + c^2)$ $I_z = \frac{1}{12}M(a^2 + b^2)$
	Diatomic molecule	$I = \frac{m_1 m_2}{m_1 + m_2} \cdot R^2$

The time-derivative of (5.24) is

$$\frac{\mathrm{d}L_i}{\mathrm{d}t} = \Delta m_i \left(\boldsymbol{r}_{i\perp} \times \frac{\mathrm{d}\boldsymbol{v}_i}{\mathrm{d}t} \right) = (\boldsymbol{r}_{i\perp} \times \boldsymbol{F}_t) = \boldsymbol{D}_{i\parallel} \,, \tag{5.25}$$

where $D_{i\parallel} = r_{i\perp} \cdot F_i$ is the component of the torque $\boldsymbol{D}_i$ parallel to the rotation axis A. The other components F_z and F_r of the force F are compensated by elastic forces of the mounting of the fixed axis A (Fig. 5.17).

For the magnitude $D_i = |D_{i\parallel}|$ of the not compensated torque we obtain from (5.24) and (5.25)

$$D_i = \Delta m_i r_{i\perp}^2 \frac{\mathrm{d}\omega}{\mathrm{d}t} \,. \tag{5.26}$$

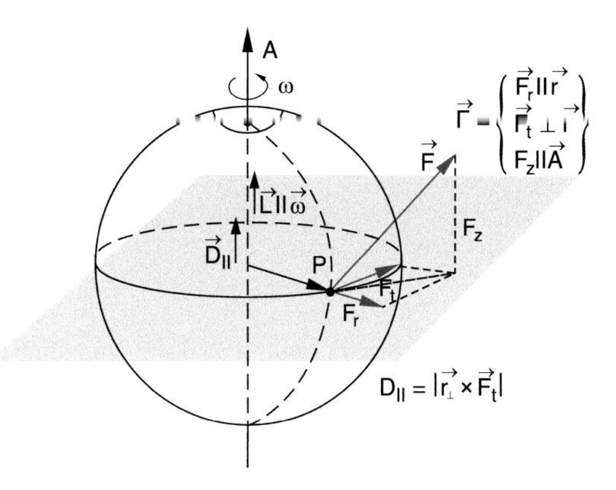

Figure 5.17 Torque acting on the rotation about a fixed axis induced by a force F attacking at the point P

The integration overall mass elements yields

$$D = I \cdot \frac{\mathrm{d}\omega}{\mathrm{d}t} = I \cdot \frac{\mathrm{d}^2\varphi}{\mathrm{d}t^2} \,, \tag{5.27}$$

where φ is the angle of $\boldsymbol{r}$ against the x-axis.

5.6.1 Rotation About an Axis for a Constant Torque

For a constant torque which does not change with time the integration of (5.27) yields the equation of rotation analogous to (2.6) for the translation of a body

$$\varphi = \frac{D}{2I} t^2 + At + B \,. \tag{5.28a}$$

The integration constants A and B are specified by the initial conditions $\varphi(0) = \varphi_0$ and $\mathrm{d}\varphi(0)/\mathrm{d}t = \omega_0$. This gives for (5.28a)

$$\varphi = \frac{D}{2I} t^2 + \omega_0 t + \varphi_0 \,. \tag{5.28b}$$

Example

1. A full cylinder, a hollow cylinder and a ball with equal masses M and equal radii r role down an inclined plane. All three bodies start at the same time. Who will win the race? The question can be answered experimentally as demonstration experiment during the lecture and arises always astonishment.
Solution: The rotation takes place around the momentary rotation axis which is the contact line between the body and the plane (Fig. 5.18). The torque acting on the body is $D = M \cdot g \cdot \sin\alpha$ where α is the inclination angel of the plane. The rotational inertia is

according to the parallel axis theorem $I = I_S + Mr^2$. Equation 5.27 then becomes

$$Mgr\sin\alpha = \left(I_S + Mr^2\right)\dot{\omega} . \quad (5.29)$$

The translational acceleration $a = \mathrm{d}^2 s/\mathrm{d}t^2$ of the center of mass S is equal to the acceleration $r \cdot \mathrm{d}\omega/\mathrm{d}t$ of the perimeter which rolls on the inclined plane. This gives the relation

$$\begin{aligned} \frac{\mathrm{d}^2 s}{\mathrm{d}t^2} &= r\dot{\omega} = r\frac{Mgr\sin\alpha}{I_S + Mr^2} \\ &= \frac{g\sin\alpha}{1 + I_S/Mr^2} = a . \end{aligned} \quad (5.30)$$

Compare this with the acceleration of a body which slides frictionless down the plane without rolling. In this case the acceleration is $a_t = g \cdot \sin\alpha$.

For the rolling body part of the potential energy is converted into rotational energy and only the rest is available for translational energy. The translational acceleration is reduced by the factor $b = (1 + I_S/Mr^2)$, which depends on the moment of inertia I_S of the rolling body. The race is therefore won by the body with the smallest moment of inertia. According to Sect. 5.5.1 these are:

Ball: $b = 7/5 \rightarrow a = 5/7 \cdot g \cdot \sin\alpha$,

Full cylinder: $b = 3/2 \rightarrow a = 2/3 \cdot g \cdot \sin\alpha$,

Hollow cylinder: $b = 2 \rightarrow a = 1/2 \cdot g \cdot \sin\alpha$.

Therefore the ball wins the race barely before the full cylinder, while the hollow cylinder arrives last. It is instructive to consider the situation from another point of view: When the body has travelled the distance s from the starting point on the inclined plane the loss of potential energy is $\Delta E_{\mathrm{pot}} = M \cdot g \cdot h = M \cdot g \cdot s \cdot \sin\alpha$ which is converted into kinetic energy $E_{\mathrm{kin}} = E_{\mathrm{trans}} + E_{\mathrm{rot}} = \frac{1}{2}(Mv^2 + \omega^2 I_S) = \frac{1}{2}Mv^2(1 + I_S/Mv^2)$. This gives for the translational velocity

$$v^2 = \frac{2gs\sin\alpha}{1 + I_S/Mr^2} .$$

Differentiation yields with $\mathrm{d}(v^2)/\mathrm{d}t = 2 \cdot v \cdot a$ the result (5.30) for the acceleration a.

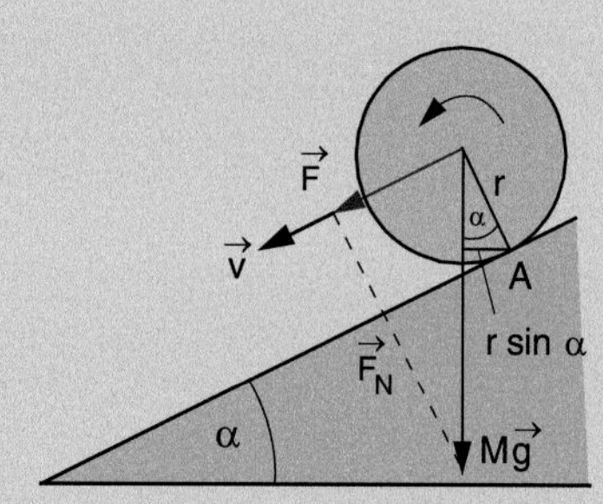

Figure 5.18 Rolling cylinder on an inclined plane

2. *Maxwell's Wheel.* A cylindrical disc with radius R, mass M and rotational inertia $I_S = \frac{1}{2}MR^2$ is centered on a thin axis through S with radius $r \ll R$. The disc hangs on a strand which is wrapped around the axis (Fig. 5.19). The mass of the axis should be negligible compared with the mass M of the disc. When the wheel is released it will roll down on the strand under the influence of the torque $\boldsymbol{D} = \boldsymbol{r} \times M\boldsymbol{g}$ and will move down with the acceleration

$$\begin{aligned} a &= r\frac{\mathrm{d}^2\varphi}{\mathrm{d}t^2} = \frac{rD}{I} \\ &= \frac{r^2 Mg}{\frac{1}{2}MR^2 + Mr^2} \\ &= \frac{g}{1 + R^2/2r^2} . \end{aligned}$$

The acceleration g is therefore reduced by the factor $(1 + \frac{1}{2}R^2/r^2)$. This allows to observe this small acceleration when performing the experiment. After the CM of the wheel has travelled the distance h the total potential energy Mgh has been converted into kinetic energy $Mgh = E_{\mathrm{kin}} = E_{\mathrm{trans}} + E_{\mathrm{rot}}$ where

$$\begin{aligned} E_{\mathrm{trans}} &= \frac{1}{2}Mv_{\mathrm{trans}}^2 = \frac{1}{2}Mr^2\omega^2 \\ &= Mgh\frac{2r^2}{R^2 + 2r^2} \end{aligned}$$

and

$$\begin{aligned} E_{\mathrm{rot}} &= \frac{1}{2}I\omega^2 \\ &= Mgh\frac{R^2}{R^2 + 2r^2} . \end{aligned}$$

Hint: The result is obtained from the relations $r^2\omega^2 = v_t^2$ and $v_t^2 = g^2T^2/(1 + \frac{1}{2}R^2/r^2)^2$ with the fall time $T = [(2h/g)(1 + \frac{1}{2}R^2/r^2)]^{1/2}$. By far the larger fraction $1/(1 + 2r^2/R^2)$ of the total energy is converted into rotational energy.

At the lowest point of its path where the strand is completely unwound, the wheel continues to rotate in the same sense (why?) and the strand winds up again, which causes the wheel to rise nearly up to the starting point. Because of frictional losses it does not completely reach its original starting heights.

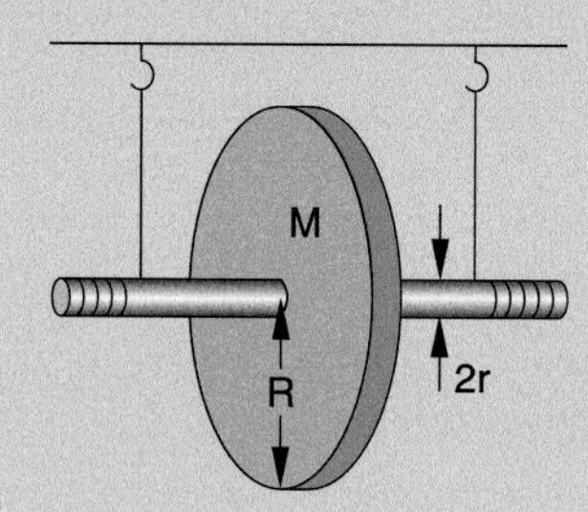

Figure 5.19 Maxwell's wheel ◄

5.6.2 Measurements of rotational inertia; Rotary Oscillations About a Fixed Axis

The experimental determination of inertial moments for bodies with arbitrary form uses a rotary table consisting of a circular disc with a concentric axis which can turn in fixed ball bearings (Fig. 5.20). A coil spring with one end attached to the axis and the other end to the mounting is bent by the turn of the table and causes by its tension a restoring torque which is proportional to the displacement angle φ from the equilibrium position $\varphi = 0$ (see Sect. 6.2)

$$D = -D_r \cdot \varphi \,. \tag{5.31}$$

The proportionality factor D_r is called *torsional rigidity*. Its value depends on the rigidity of the spring. The equation of motion (5.27) for this case is

$$I_0 \ddot{\varphi} = -D_r \varphi \,, \tag{5.32a}$$

where I_0 is the inertial moment of the rotary table. We have neglected any friction. The solution of the differential equation (5.32a) is with the initial condition $\varphi(0) = 0$

$$\varphi = a \sin\left(\sqrt{D_r/I_0}\, t\right) \,. \tag{5.32b}$$

Once deflected from its equilibrium position the rotary table performs a harmonic oscillation with the oscillation period

$$T_0 = 2\pi \sqrt{I_0/D_r} \,. \tag{5.32c}$$

If a circular disc with known mass M and radius R is placed concentrically on the table, the moment of inertia increases to $I = I_0 + \frac{1}{2}MR^2$ and the oscillation period becomes

$$T_1 = 2\pi \sqrt{\left(I_0 + \tfrac{1}{2}MR^2\right)/D_r} \,. \tag{5.32d}$$

From the difference $T_1^2 - T_0^2 = 2\pi^2 \mu R^2 / D_r$ the torsional rigidity D_r can be determined. Now a body A with arbitrary form can be placed on the table. The total moment of inertia I_A now depends on the location of A with respect to the center of the table. The measured oscillation period

$$T = 2\pi \sqrt{(I_0 + I_A)/D_r} \tag{5.32e}$$

allows the determination of I_A. With the parallel axis theorem (5.18) the moment of inertia I_S of A with respect to its center of mass S is $I_S = I_A - Ma^2$, where a is the distance between the center of the rotary table and the CM of A.

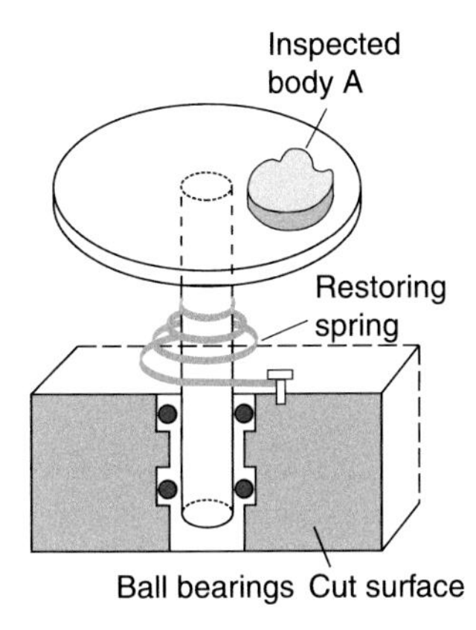

Figure 5.20 Turntable with cut through the ball bearing

5.6.3 Comparison Between Translation and Rotation

Table 5.2 shows a comparison between corresponding quantities for the description of translation of a point mass and rotation of an extended body. Note the analogous notation for momentum, angular momentum, energy and power, if the mass m is replaced by the moment of inertia I.

5.7 Rotation About Free Axes; Spinning Top

Up to now we have discussed only rotations of rigid bodies about space-fixed axes. Even for the example of the cylinder rolling down the inclined plane the direction of the rotational axis remained constant although it performed a translation.

In the present section we will deal with the more complex situation that a body can rotate about a free axis, which might change its direction in space. We will treat at first the case that no external forces act on the body and then discuss the cases where external torques are present.

Such rigid bodies rotating about free axes are called *spinning tops* or *gyroscopes*.

Table 5.2 Comparison of corresponding quantities for rotation and translation

Translation	Rotation
Length L	Angle φ
Mass m	Moment of inertia I
Velocity $\boldsymbol{v}$	Angular velocity $\boldsymbol{\omega}$
Momentum $\boldsymbol{p} = m \cdot \boldsymbol{v}$	Angular momentum $\boldsymbol{L} = I \cdot \boldsymbol{\omega}$
Force $\boldsymbol{F}$	Torque $\boldsymbol{D} = \boldsymbol{r} \times \boldsymbol{F}$
$\boldsymbol{F} = \dfrac{d\boldsymbol{p}}{dt}$	$\boldsymbol{D} = \dfrac{d\boldsymbol{L}}{dt}$
$E_{kin} = \dfrac{m}{2} v^2$	$E_{rot} = \dfrac{I}{2} \omega^2$
Restoring force	Restoring torque
$\boldsymbol{F} = -D \cdot \boldsymbol{x}$	$D = -D_r \cdot \varphi$
Period of a linear oszillation	Period of torsional oszillation
$T = 2\pi \sqrt{m/D}$	$T = 2\pi \sqrt{I/D_r}$

Chapter 5

For the general motion one must take into account the translation of the CM (which can be always treated separately) and the rotation around the CM. If the motion is discussed in the CM-system where the CM is at rest, one has to regard only the rotation about the CM. We will see, that the space-orientation of free axes generally changes with time and the motion of an arbitrary point of the rigid body might perform a complicated trajectory.

In order to calculate the motion about free axes we have to determine the dependence of the moment of inertia on the direction of the rotation axis, which, however, always should intersect the CM.

5.7.1 Inertial Tensor and Inertial Ellipsoid

When a rigid body rotates with the angular velocity ω around an axis through the center of mass S with arbitrary space orientation (Fig. 5.21) the mass element Δm_i moving with the velocity $v_i = \omega \times r_i$ has the angular momentum

$$L_i = \Delta m_i (\boldsymbol{r}_i \times \boldsymbol{v}_i) = \Delta m_i (\boldsymbol{r}_i \times (\boldsymbol{\omega} \times \boldsymbol{r}_i)) , \tag{5.33a}$$

using the vector relation (see Sect. 13.1.5.4)

$$\boldsymbol{A} \times (\boldsymbol{B} \times \boldsymbol{C}) = (\boldsymbol{A} \cdot \boldsymbol{C})\boldsymbol{B} - (\boldsymbol{A} \cdot \boldsymbol{B})\boldsymbol{C} ,$$

this can be transformed into

$$\boldsymbol{L}_i = \Delta m_i \left[(\boldsymbol{r}_i^2 \cdot \boldsymbol{\omega}) - (\boldsymbol{r}_i \cdot \boldsymbol{\omega}) \boldsymbol{r}_i \right] . \tag{5.33b}$$

The total angular momentum of the rigid body is then obtained by integration over all mass elements. This gives

$$\boldsymbol{L} = \int \left(r^2 \boldsymbol{\omega} - (\boldsymbol{r} \cdot \boldsymbol{\omega}) \boldsymbol{r} \right) \mathrm{d}m . \tag{5.34a}$$

This vector equation corresponds to the three equations for the components

$$\begin{aligned} L_x &= I_{xx}\omega_x + I_{xy}\omega_y + I_{xz}\omega_z \\ L_y &= I_{yx}\omega_x + I_{yy}\omega_y + I_{yz}\omega_z \\ L_z &= I_{zx}\omega_x + I_{zy}\omega_y + I_{zz}\omega_z , \end{aligned} \tag{5.34b}$$

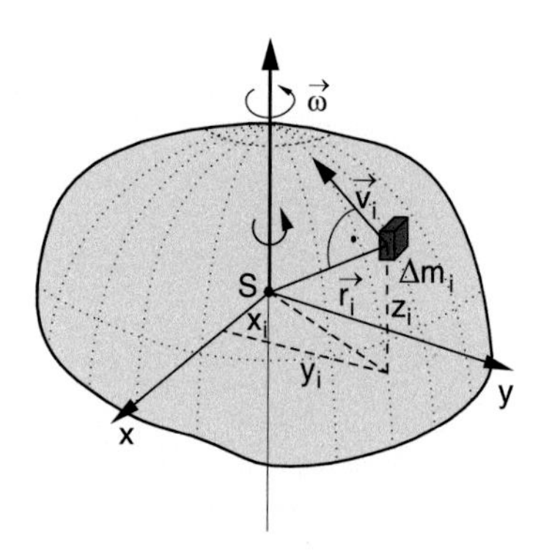

Figure 5.21 Rotation of a body about an arbitrary axis through the Center of Mass

where the coefficients I_{ik} are abbreviations for the expressions

$$\begin{aligned} I_{xx} &= \int (r^2 - x^2)\, \mathrm{d}m \\ I_{xy} &= I_{yx} = -\int xy\, \mathrm{d}m \\ I_{yy} &= \int (r^2 - y^2)\, \mathrm{d}m \\ I_{yz} &= I_{zy} = -\int yz\, \mathrm{d}m \\ I_{zz} &= \int (r^2 - z^2)\, \mathrm{d}m \\ I_{xz} &= I_{zx} = -\int xz\, \mathrm{d}m . \end{aligned} \tag{5.35a}$$

Equation 5.34b can be readily checked when inserting the relations $r^2 = x^2 + y^2 + z^2$ and $\boldsymbol{r} \cdot \boldsymbol{\omega} = x\omega_x + y\omega_y + z\omega_z$ into (5.34a) and using (5.35a). The components I_{ik} can be written in form of the matrix

$$\widetilde{I} = \begin{pmatrix} I_{xx} & I_{xy} & I_{xz} \\ I_{yx} & I_{yy} & I_{yz} \\ I_{zx} & I_{zy} & I_{zz} \end{pmatrix} , \tag{5.35b}$$

which allows to write Eq. 5.34b in the vector form

$$\begin{pmatrix} L_x \\ L_y \\ L_z \end{pmatrix} = \begin{pmatrix} I_{xx} & I_{xy} & I_{xz} \\ I_{yx} & I_{yy} & I_{yz} \\ I_{zx} & I_{zy} & I_{zz} \end{pmatrix} \begin{pmatrix} \omega_x \\ \omega_y \\ \omega_z \end{pmatrix} . \tag{5.34c}$$

This can be shortened to

$$\boldsymbol{L} = \widetilde{I} \cdot \boldsymbol{\omega} . \tag{5.34d}$$

The mathematical term for $\boldsymbol{I}$ is a *tensor of rank two*, which is called **inertial tensor**. The diagonal elements of I give the moments of inertia for rotation axes in the direction of the coordinate axes x, y, z.

To illustrate the advantage of introducing this inertial tensor we will at first calculate the rotational energy of the body for a rotation about an arbitrary axis $\boldsymbol{\omega}$. For a mass element Δm_i (Fig. 5.21) the rotational energy is

$$\begin{aligned} \tfrac{1}{2}\Delta m_i v_i^2 &= \tfrac{1}{2}\Delta m_i (\boldsymbol{\omega} \times \boldsymbol{r}_i)(\boldsymbol{\omega} \times \boldsymbol{r}_i) \\ &= \tfrac{1}{2}\Delta m_i \left[\omega^2 r_i^2 - (\boldsymbol{\omega} \cdot \boldsymbol{r}_i)^2 \right] , \end{aligned}$$

where the right hand side follows from the vector relation $(\boldsymbol{A} \times \boldsymbol{B}) \cdot (\boldsymbol{A} \times \boldsymbol{B}) = \boldsymbol{A}^2\boldsymbol{B}^2 - (\boldsymbol{A} \cdot \boldsymbol{B})^2$.

Chapter 5

The spatial integration over all mass elements gives the rotational energy of the whole rigid body

$$
\begin{aligned}
E_{\text{rot}} &= \frac{\omega^2}{2}\int r^2 \mathrm{d}m - \frac{1}{2}\int (\boldsymbol{\omega}\cdot\boldsymbol{r})^2\,\mathrm{d}m \\
&= \frac{\omega_x^2+\omega_y^2+\omega_z^2}{2}\int \left(x^2+y^2+z^2\right)\mathrm{d}m \\
&\quad - \frac{1}{2}\int \left[\omega_x x + \omega_y y + \omega_z z\right]^2 \mathrm{d}m \\
&= \frac{1}{2}\left[\omega_x^2 I_{xx} + \omega_y^2 I_{yy} + \omega_z^2 I_{zz}\right] \\
&\quad + \omega_x\omega_y I_{xy} + \omega_x\omega_z I_{xz} + \omega_y\omega_z I_{yz}\,,
\end{aligned}
\tag{5.36}
$$

where the definitions (5.35b) have been used. Within the tensor notation (5.36) can be written as

$$E_{\text{rot}} = \tfrac{1}{2}\boldsymbol{\omega}^{\mathrm{T}}\cdot\tilde{\boldsymbol{I}}\cdot\boldsymbol{\omega}\,,$$

which explicitly means

$$E_{\text{rot}} = \frac{1}{2}(\omega_x\omega_y\omega_z)\begin{pmatrix} I_{xx} & I_{xy} & I_{xz} \\ I_{yx} & I_{yy} & I_{yz} \\ I_{zx} & I_{zy} & I_{zz}\end{pmatrix}\begin{pmatrix}\omega_x\\ \omega_y\\ \omega_z\end{pmatrix}.$$

This shows that for arbitrary orientations of the rotation axis all elements of the inertial tensor can contribute to the rotational energy.

When the rotation axis $\boldsymbol{\omega}$ forms the angles α, β, γ with the coordinate axes the components of $\boldsymbol{\omega}$ are

$$\omega_x = \omega\cdot\cos\alpha\,,\quad \omega_y = \omega\cdot\cos\beta\,,\quad \omega_z = \omega\cdot\cos\gamma\,.$$

When the rotational energy is written in the form of Eq. 5.16 as

$$E_{\text{rot}} = \tfrac{1}{2}I\omega^2\,,$$

the comparison with (5.36) yields for the scalar moment of inertia

$$
\begin{aligned}
I &= \cos^2\alpha\, I_{xx} + \cos^2\beta\, I_{yy} + \cos^2\gamma\, I_{zz} \\
&\quad + 2\cos\alpha\cos\beta\, I_{xy} + 2\cos\alpha\cos\gamma\, I_{xz} \\
&\quad + 2\cos\beta\cos\gamma\, I_{yz}\,.
\end{aligned}
\tag{5.37a}
$$

When we introduce a vector R in the direction of the rotation axis with the components $x = R\cdot\cos\alpha$; $y = R\cdot\cos\beta$; $z = R\cdot\cos\gamma$ Eq. 5.37a can be written as

$$
\begin{aligned}
R^2 I &= x^2 I_{xx} + y^2 I_{yy} + z^2 I_{zz} \\
&\quad + 2xyI_{xy} + 2xzI_{xz} + 2yzI_{yz}\,.
\end{aligned}
\tag{5.37b}
$$

This is a quadratic equation in x, y, and z with constant coefficients I_{ik}. All points (x, y, z) for which $R^2\cdot I = \text{const}$ are located on an ellipsoid because (5.37) describes for $R^2\cdot I = k = \text{const}$ an ellipsoid, with axes which depend on the coefficients I_{ik}. Since $I \propto M\cdot R_m^2$ the constant $k = M\cdot R_m^4$ has the dimension $[k] = \text{kg}\cdot\text{m}^4$. Its value depends on the mass M of the rigid body and the mass distribution relative to the center of mass S which is expressed by a mean distance R_m.

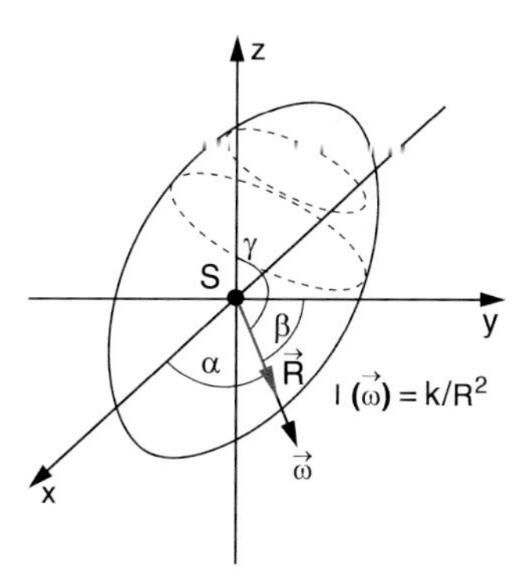

Figure 5.22 Inertial ellipsoid

The moment of inertia $I_\omega = k/R^2$ for a rotation about an arbitrary axis $\boldsymbol{\omega} = \{\omega_x, \omega_y, \omega_z\}$ is proportional to $1/R^2$ where R is the distance from the center of the ellipsoid to its surface (Fig. 5.22). With this notation one can say that the scalar value I of the moment of inertia as a function of the spatial orientation (α, β, γ) of the rotation axis represents the **inertial ellipsoid**.

5.7.2 Principal Moments of Inertia

We introduce a coordinate system (ξ, η, ζ) which is generated by three orthogonal vectors $\boldsymbol{\xi}$, $\boldsymbol{\eta}$ and $\boldsymbol{\zeta}$ with axes which point into the directions of the principal axes $\boldsymbol{a}$, $\boldsymbol{b}$ and $\boldsymbol{c}$ of the inertial ellipsoid (Fig. 5.23). Their magnitude is normalized when dividing by $\sqrt{k}$. In this coordinate system the ellipsoid equation (5.37) becomes with $R^2\cdot I = 1$

$$\xi^2 I_a + \eta^2 I_b + \zeta^2 I_c = 1\,. \tag{5.38}$$

In this *principal axis coordinate system* all off-diagonal elements I_{ik} with $i \neq k$ of the inertial tensor I are zero and the

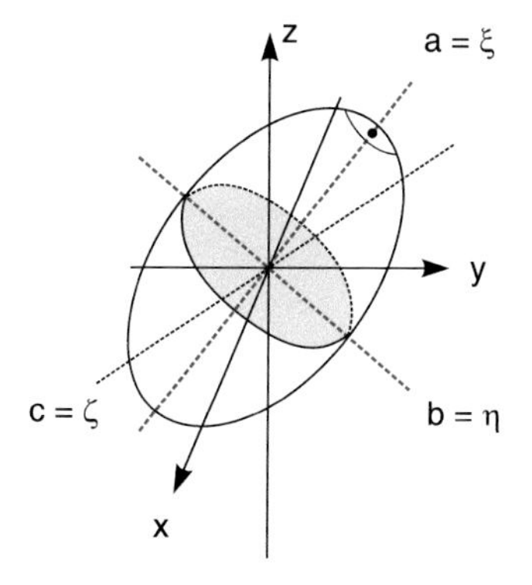

Figure 5.23 Definition of principal axes of inertia

tensor becomes a diagonal tensor

$$\widetilde{I} = \begin{bmatrix} I_a & 0 & 0 \\ 0 & I_b & 0 \\ 0 & 0 & I_c \end{bmatrix} . \tag{5.39}$$

Mathematically such a principal axes transformation can be performed by a diagonalization of the corresponding matrix [5.2]. The principal inertia moments I_a, I_b, I_c (i.e. the moments of inertia for rotations about the principal axes a, b, c) are the solutions of the determinant equation

$$\begin{vmatrix} I_{xx} - I & I_{xy} & I_{xz} \\ I_{yx} & I_{yy} - I & I_{yz} \\ I_{zx} & I_{zy} & I_{zz} - I \end{vmatrix} = 0 . \tag{5.40}$$

Note, that generally the principal moments of inertia do not concur with the elements I_{xx}, I_{yy}, I_{zz}, because all elements of the tensor can change under the principal axes transformation.

According to international agreements [5.3] the assignment of the principal moments follows the definition:

$$I_a \le I_b \le I_c .$$

The moment of inertia for a rotation about an arbitrary axis with direction angels α, β, φ, against the x, y, z, axis is (Fig. 5.24)

$$I = I_a \cos^2 \alpha + I_b \cos^2 \beta + I_c \cos^2 \gamma . \tag{5.41}$$

This equation corresponds to (5.37a) since all off-diagonal elements are zero. The principal axes transformation has made the expression for the general moment of inertia I simpler.

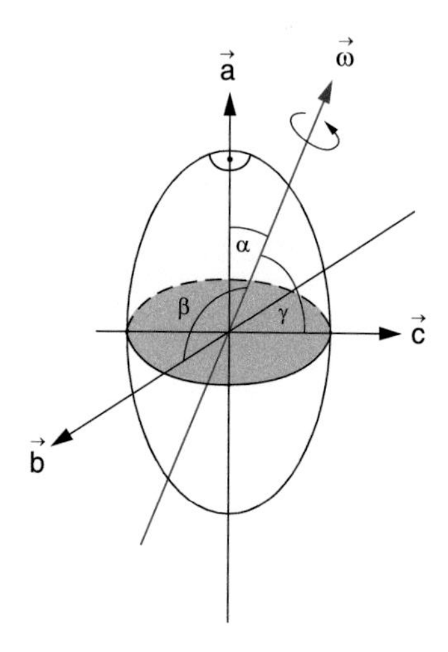

Figure 5.24 Inertial moment about an arbitrary axis

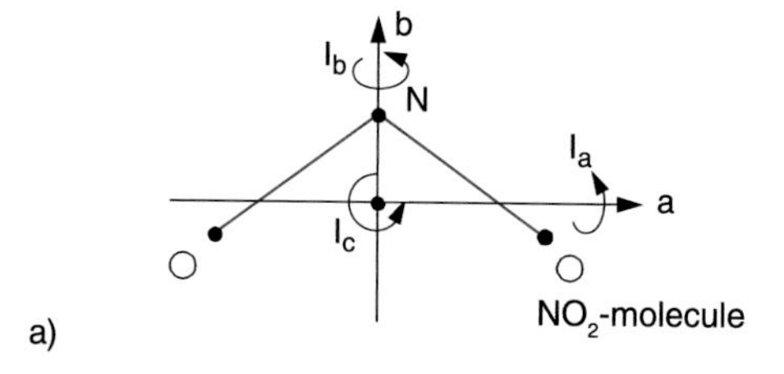

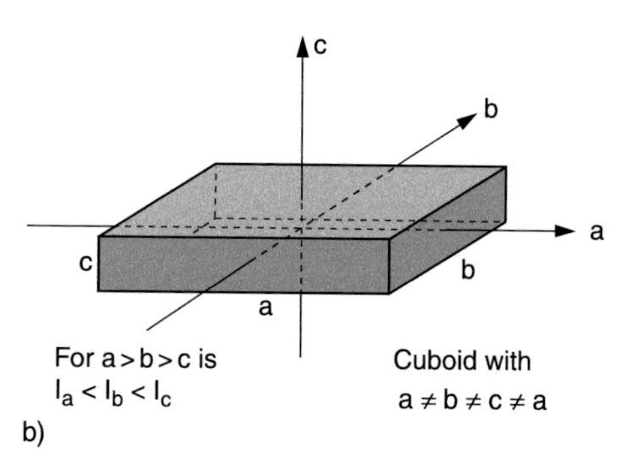

Figure 5.25 Examples of asymmetric tops

With the principal moments of inertia the angular momentum and the rotational energy can be written as

$$L = \{L_a, L_b, L_c\} = \{\omega_a I_a, \omega_b I_b, \omega_c I_c\} , \tag{5.42}$$

$$\begin{aligned} E_{\text{rot}} &= \frac{1}{2}\left(\omega_a^2 I_a + \omega_b^2 I_b + \omega_c^2 I_c\right) \\ &= \frac{L_a^2}{2I_a} + \frac{L_b^2}{2I_b} + \frac{L_c^2}{2I_c} . \end{aligned} \tag{5.43}$$

If all three principal moments are different ($I_a \neq I_b \neq I_c \neq I_a$) the body is called an **asymmetric top**.

Example: A cuboid with three different side lengths a, b, c (Fig. 5.25b) or the NO_2 molecule (Fig. 5.25a).

If two principal moments of inertia are equal the body is called a **symmetric top**.

Example: All bodies with rotational symmetry (circular cylinder linear molecules but also quadratic cuboids).

Every body with rotational symmetry is a symmetric top, but a symmetric top has not necessarily a rotational symmetry (for example a quadratic post). The inertial ellipsoid of a symmetric top is, however, always rotationally symmetric.

We distinguish between

- *Prolate symmetric tops* (Fig. 5.26a) with $I_a < I_b = I_c$. The inertial ellipsoid is a stretched rotational ellipsoid where the diameter along the symmetry axis z is larger than the diameter in the x-y-plane (Fig. 5.27a).
- *Oblate symmetric tops* (Fig. 5.26b) with $I_a = I_b < I_c$. The inertial ellipsoid is a squeezed rotational ellipsoid (disc, Fig. 5.27b).

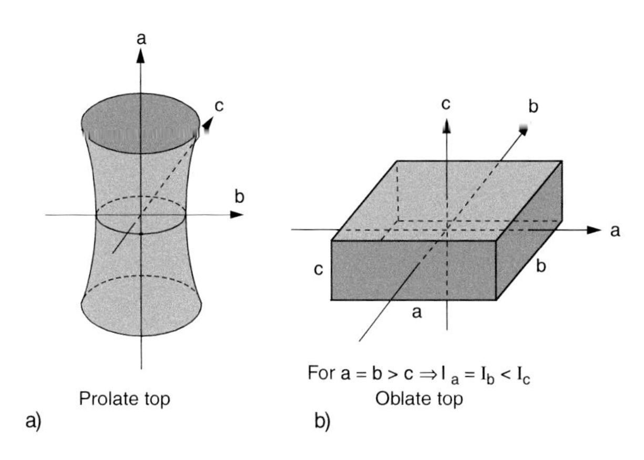

Figure 5.26 Examples of symmetric tops: **a** prolate and **b** oblate symmetric top

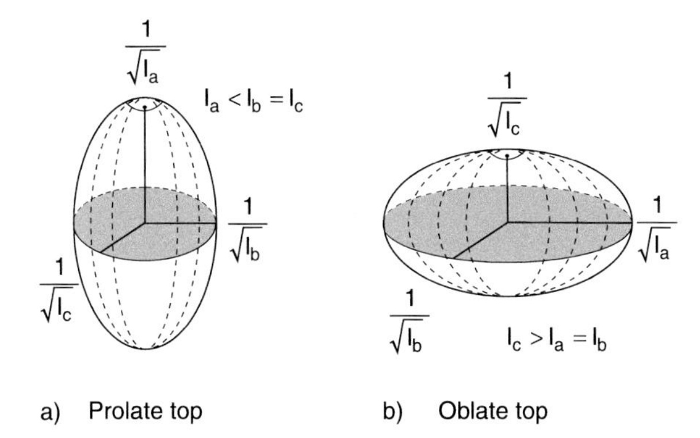

Figure 5.27 Inertial ellipsoids of **a** prolate and **b** oblate symmetric top

For an asymmetric top the angular momentum $\boldsymbol{L}$ and the rotational axis $\boldsymbol{\omega} = \{\omega_x, \omega_y, \omega_z\}$ generally point into different directions, because the components I_x, I_y, I_z, in Eq. 5.42 are different, except if the body rotates aboud one of its principal axes.

When all three principal moments of inertia are equal, the body is a spherical top, because in this case its inertial ellipsoid is a sphere.

Examples: A ball or a cube.

5.7.3 Free Rotational axes

The Eq. 5.42 and Fig. 5.28 give the following important information: Angular momentum $\boldsymbol{L}$ and rotational axis $\boldsymbol{\omega}$ point for all bodies with free axes (where the rotation axis is not fixed by mountings) *only then* into the same direction if at least one of the following conditions is fulfilled.

- $I_a = I_b = I_c$ (spherical top) or

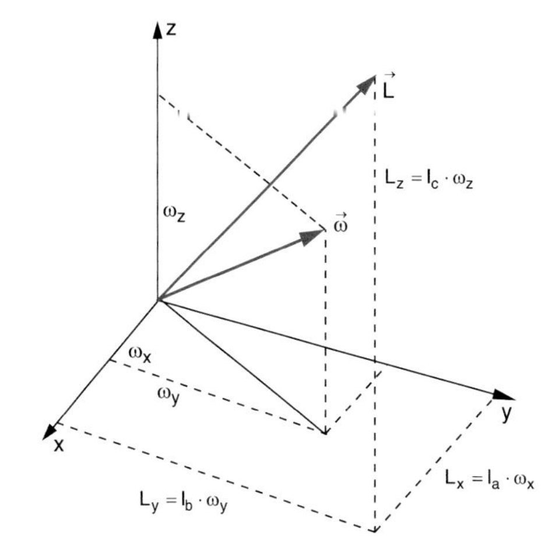

Figure 5.28 Angular momentum L and rotational axis are generally not parallel. This is illustrated in a (x, y, z) coordinate system that coincides with the principal axes (a, b, c)

- the body rotates about one of its principal inertial axes, which implies that only one of the three components ω_x, ω_y, ω_z is not zero.
- For a symmetric top is $\boldsymbol{L} \parallel \boldsymbol{\omega}$ if the body rotates around an arbitrary axis perpendicular to the symmetry axis.

Since without external torque the angular momentum $\boldsymbol{L}$ is constant and has a constant orientation in space the body has in these three cases a space fixed rotational axis and rotates around this constant axis with constant angular velocity $\boldsymbol{\omega}$. Its rotational motion is then identical to the rotation about axes with fixed mountings (see Sect. 5.6)

The principal axes of a body are therefore also called *free axes* because the body can freely rotate about them even if they are not fixed by mountings.

The experiment shows, however, that a stable rotation is only possible about the axes of the smallest and the largest moment of inertia. For the rotation about a free axis of the intermediate moment of inertia any tiny perturbation makes the motion instable and the body finally flips into a rotation about one of the other two principal axes.

Examples

1. A cuboid with $I_a < I_b < I_c$ is suspended by a thread (Fig. 5.29) and can be induced to rotations about the thread by a small motor which twists the thread. The cuboid rotates stable if the thread direction coincides with the axis of the inertial moments I_a or I_c. If it is suspended in a way that the thread direction coincides with the axis of I_b the cuboid flips for faster rotations into a rotation around the axis b, as shown in Fig. 5.29c), it rotates then no longer about the thread but around the dashed line in Fig. 5.29c), which is a free axis.

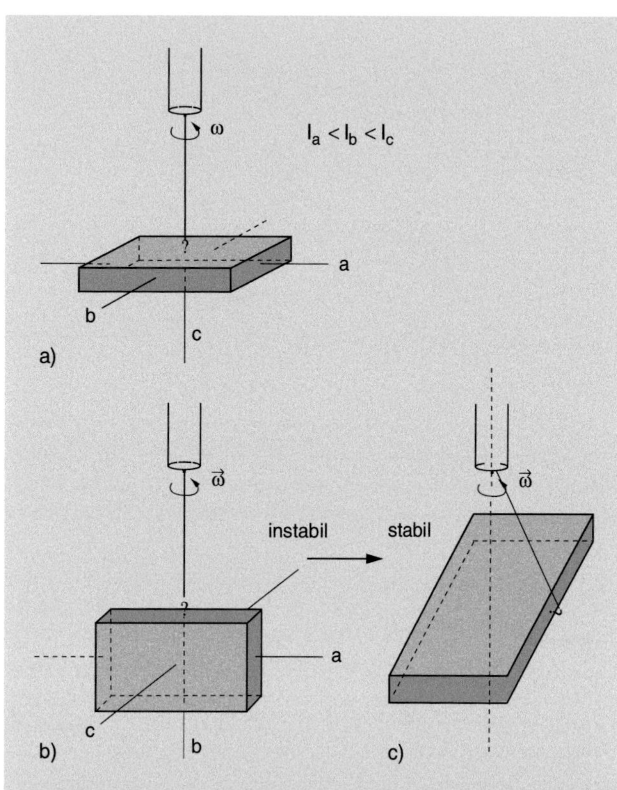

Figure 5.29 Rotation of a cuboid about free axes: **a** stable rotation about the axis of maximum moment of inertia; **b** instable rotation about the axis of median moment of inertia, which jumps into a rotation about the axis *c* of maximum moment of inertia (**c**)

2. A closed chain hangs on a thread and is induced to rotations by a motor (Fig. 5.30). Due to the centrifugal force the chain widens to a circle which orientates itself in a horizontal plane, because in this position the rotation takes place about the axis of the maximum inertial moment and therefore the rotational energy $E_{\text{rot}}L^2/2I$ becomes a minimum. In this stable rotational mode the rotation axis does not coincide with the direction of the thread.

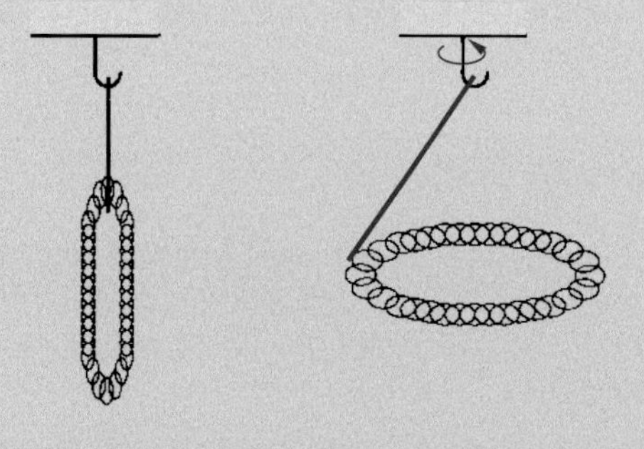

Figure 5.30 Rotation of a chain about the axis of maximum moment of inertia

3. A thrown discus flies stable as long as the rotation proceeds about the symmetry axis (axis of the maximum moment of inertia) (Fig. 5.31).

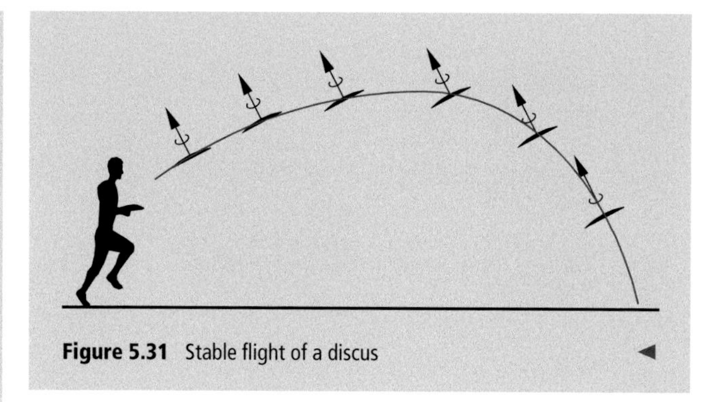

Figure 5.31 Stable flight of a discus

5.7.4 Euler's Equations

For an arbitrary orientation of the rotation axis angular momentum L and rotation axis $\boldsymbol{\omega}$ are no longer parallel. The motion of the body is now more complicated. In order to investigate this motion as seen by an observer sitting in a space-fixed inertial coordinate system S, we have to describe it in this system S.

The time derivative $\mathrm{d}\boldsymbol{L}/\mathrm{d}t$ of the angular momentum is equal to the torque $\boldsymbol{D}$ acting on the body.

$$\left(\frac{\mathrm{d}\boldsymbol{L}}{\mathrm{d}t}\right)_S = \boldsymbol{D} . \tag{5.44}$$

A coordinate system K where the axes are the principal axes of the body, which is therefore rigidly connected to the rotating body rotates with the angular velocity ω against the system S. In this system the time derivative of $\boldsymbol{L}$ is (see Sect. 3.3.2)

$$\left(\frac{\mathrm{d}\boldsymbol{L}}{\mathrm{d}t}\right)_K = \left(\frac{\mathrm{d}\boldsymbol{L}}{\mathrm{d}t}\right)_S - (\boldsymbol{\omega} \times \boldsymbol{L}) , \tag{5.45}$$

which gives the vector equation for the torque $\boldsymbol{D}$

$$\boldsymbol{D} = \left(\frac{\mathrm{d}\boldsymbol{L}}{\mathrm{d}t}\right)_K + (\boldsymbol{\omega} \times \boldsymbol{L}) . \tag{5.46}$$

This equation corresponds formally to (3.14) if we replace $\boldsymbol{L}$ by $\boldsymbol{r}$. **Note**, that in (5.46) $d\boldsymbol{L}/\mathrm{d}t$ is the derivative of $\boldsymbol{L}$ in the body fixed principal axes system K, while $\boldsymbol{\omega}$ is the angular velocity in the space-fixed system S. If (5.46) is written for the three components in the direction of the principal axes one obtains for example for the axis a the relation

$$\begin{aligned} D_a &= \left(\frac{\mathrm{d}\boldsymbol{L}}{\mathrm{d}t}\right)_a + (\boldsymbol{\omega} \times \boldsymbol{L})_a \\ &= \frac{\mathrm{d}}{\mathrm{d}t}(I_a\omega_a) + (\omega_b L_c - \omega_c L_b) \\ &= I_a \frac{\mathrm{d}\omega_a}{\mathrm{d}t} + \omega_b I_c \omega_c - \omega_c I_b \omega_b , \end{aligned}$$

where D_a is the component of the torque in the direction of the principal axis a.

Similar equations can be derived for the other two components. This leads to the **Euler-equations**

$$
\begin{aligned}
I_a\frac{\mathrm{d}\omega_a}{\mathrm{d}t} + (I_c - I_b)\,\omega_c\omega_b &= D_a \\
I_b\frac{\mathrm{d}\omega_b}{\mathrm{d}t} + (I_a - I_c)\,\omega_a\omega_c &= D_b \qquad (5.47) \\
I_c\frac{\mathrm{d}\omega_c}{\mathrm{d}t} + (I_b - I_a)\,\omega_b\omega_a &= D_c\ .
\end{aligned}
$$

5.7.5 The Torque-free Symmetric Top

A symmetric top has two equal principal moments of inertia. If the symmetry axis of its inertial ellipsoid is the axis c we have $I_a = I_b \neq I_c$. For rotational symmetric bodies the symmetry axis is also called the *figure axis*. For a bicycle wheel as symmetric top this is the visible wheel axis (Fig. 5.32). Without any external torque ($\boldsymbol{D} = \boldsymbol{0}$) the magnitude and the direction of the angular momentum $\boldsymbol{L}$ is constant. Such a top with $\boldsymbol{D} = \boldsymbol{0}$ is called force-free top although it should be called more correctly torque-free top.

When the top rotates about its figure axis, $\boldsymbol{L}$ and $\boldsymbol{\omega}$ coincide with this axis. The top rotates as if its axis would be hold by a stable mounting (see Sect. 5.6). If, however, $\boldsymbol{\omega}$ points into an arbitrary direction which does not coincide with the figure axis the motion becomes complicated.

For the description of this motion one has to distinguish between three axes (Fig. 5.33a):

- The space-fixed angular momentum axis $\boldsymbol{L}$
- The momentary (not space-fixed) rotation axis $\boldsymbol{\omega}$
- The figure axis of the symmetric top, which is only space-fixed, if L coincides with this axis.

We can win a qualitative picture for the motion of the figure axis by the following consideration: For $\boldsymbol{D} = \boldsymbol{0}$ the angular momentum $\boldsymbol{L}$ and the rotational energy E_{rot} are both constant.

Figure 5.32 Bicycle wheel as symmetric top

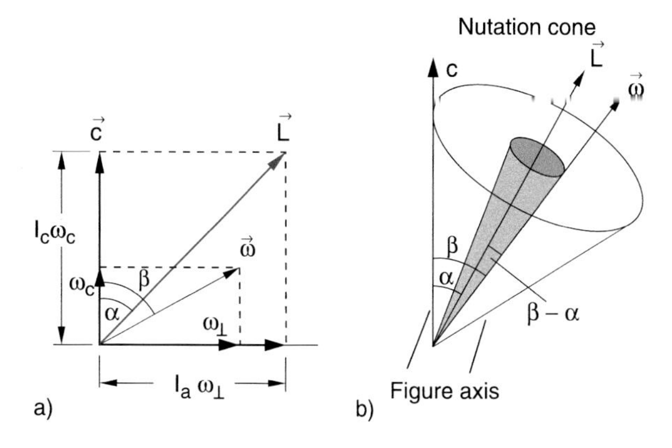

Figure 5.33 Figure axis $\boldsymbol{c}$, angular momentum $\boldsymbol{L}$ and momentary rotation axis $\boldsymbol{\omega}$: **a** Decomposition of $\boldsymbol{\omega}$ and $\boldsymbol{L}$ into the components parallel and perpendicular to the figure axis of the symmetric top. **b** Nutation cone of $\vec{L}$ and $\vec{\omega}$

Then we obtain from Eq. 5.43

$$L_x^2 + L_y^2 + L_z^2 = \text{const} = C_1\ , \qquad (5.48a)$$

$$\frac{L_a^2}{I_a} + \frac{L_b^2}{I_b} + \frac{L_c^2}{I_c} = \text{const} = C_2\ . \qquad (5.48b)$$

In a space-fixed coordinate system with the axes L_x, L_y, L_z (5.48a) represents the equation of a sphere. Equation 5.48b describes an ellipsoid in the principal axes system. Since the components of the space-fixed vector L must obey both equations, the endpoint of $\boldsymbol{L}$ can only move on the curve of intersection between sphere and ellipsoid (Fig. 5.34). Since the ellipsoid is determined by the principal axes system of the top, i.e. rotates with the top, while $\boldsymbol{L}$ is space-fixed, the top and therefore also its inertial ellipsoid move in such a way, that the endpoint of L always remains on the curve of intersection. This causes a nutation of the figure axis and the momentary rotation axis $\boldsymbol{\omega}$ about the space-fixed axis $\boldsymbol{L}$ (Fig. 5.33b).

While the figure axis can be seen straight forward, the momentary rotation axis can be made visible by an experimental trick: A circular disc with red, black and white circular segments is

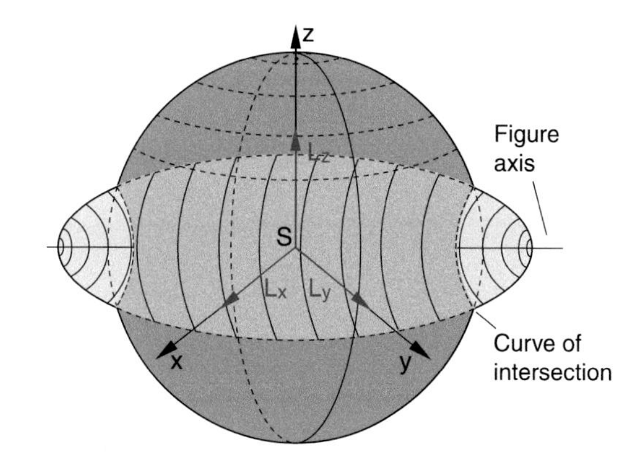

Figure 5.34 The top of the angular momentum vector moves on the intersection curve of angular momentum sphere and energy ellipsoid

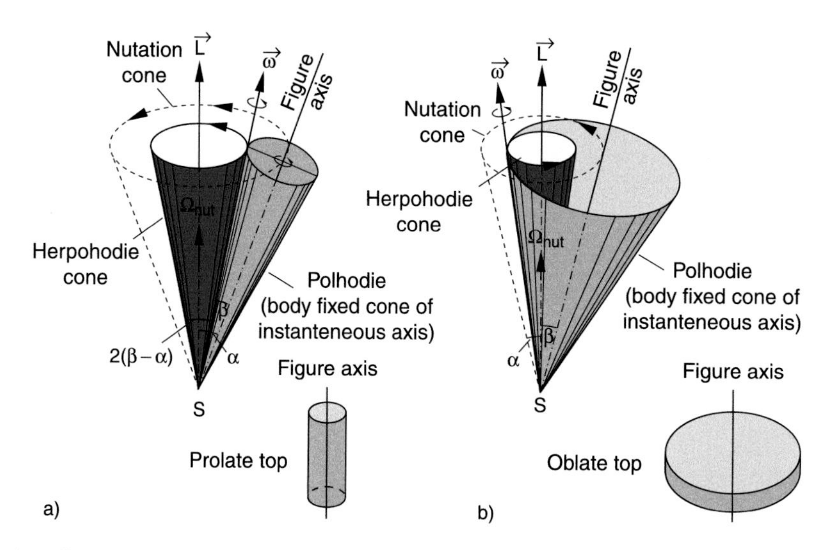

Figure 5.36 Nutation cone, herpolhode cone and polhode cone for **a** the prolate, **b** the oblate top

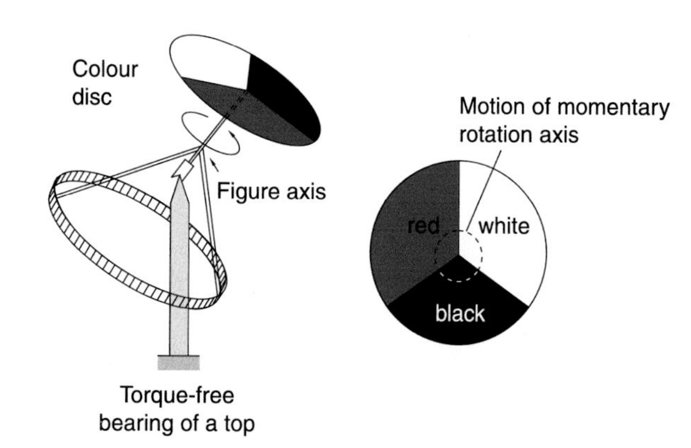

Figure 5.35 Visualization of the momentary rotational axis

centered on the peak of the figure axis (Fig. 5.35). When the top rotates the three colors blur to an olive-brown mixed color. Only at the intersection point with the momentary rotation axis one can see the color of the specific segment which wanders slowly from red over black to white which indicates the motion of the momentary rotation axis.

In order to calculate the motion of the figure axis and the momentary rotation axis more quantitatively we apply the Euler equations (5.47) to the special case $\boldsymbol{D} = \boldsymbol{0}$ and $I_a = I_b$. They simplify to

$$\begin{aligned} &\dot{\omega}_a + \Omega \omega_b = 0\,, \\ &\dot{\omega}_b - \Omega \omega_a = 0\,, \\ &\dot{\omega}_c = 0\,, \end{aligned} \tag{5.49}$$

with the abbreviation $\Omega = ((I_c - I_a)/I_a)\omega_c$. The solutions of this system of equations are

$$\begin{aligned} &\Omega_a = A\cos\Omega t\,, \quad \Omega_b = A\sin\Omega t\,, \\ &\Omega_c = C \quad \text{with} \quad A, C = \text{const}\,, \end{aligned} \tag{5.50}$$

as can be proved by inserting (5.50) into (5.49). While ω is the angular frequency of the spinning top, Ω is the frequency of the nutation. The solutions show, that the magnitude $\omega = |\boldsymbol{\omega}|$ is constant in the body-fixed system as well as in the lab-system, because $\omega^2 = \omega_a^2 + \omega_b^2 + \omega_c^2 = A^2 + C^2 = \text{const}$. However, the individual components ω_a and ω_b can change and therefore the direction of $\boldsymbol{\omega}$.

We separate ω into a component ω_c parallel to the figure axis and a component $\omega_\perp = \sqrt{(\omega_a^2 + \omega_b^2)} = A$ perpendicular to the figure axis (Fig. 5.33a) . Squaring (5.42) yields then

$$\boldsymbol{L} = I_a \boldsymbol{\omega}_\perp + I_c \boldsymbol{\omega}_c\,. \tag{5.51}$$

The figure axis forms a constant angle α against the space-fixed angular momentum axis with

$$\tan\alpha = \frac{I_a \omega_\perp}{I_c \omega_c} = \frac{I_a}{I_c}\frac{\sqrt{\omega_a^2 + \omega_b^2}}{\omega_c} = \frac{I_a}{I_c} \cdot \frac{A}{\omega_c}\,.$$

This means that the figure axis migrates on a cone with the full aperture angle 2α around the space-fixed axis $\boldsymbol{L}$ (Fig. 5.33b and 5.36). This cone is called **nutation-cone**.

The vector $\boldsymbol{\omega}$ with its constant magnitude forms the constant angle β with the figure axis where $\sin\beta = \omega_\perp/\omega = A/\omega_c$. The momentary rotation axis $\boldsymbol{\omega}$ is also wandering on a cone with the opening angel $2(\beta - \alpha)$, called **herpolhode cone** around the space fixed axis of $\boldsymbol{L}$. This common motion of figure axis and momentary rotation axis without external torque is called **nutation**.

The common motion of figure axis and momentary rotation axis can be illustrated by two cones (nutation cone and herpolhode cone) with different opening angles centered around the space-fixed axis $\boldsymbol{L}$. A third cone (*polhode cone*) centered around the nutating figure axis touches the space fixed *herpolhode cone* along the momentary rotation axis and rolls on the outer surface (prolate top Fig. 5.36a) or the inner surface (oblate top Fig. 5.36b) of the herpolhode cone. The contact line shows the momentary rotation line $\boldsymbol{\omega}$. The apex of the three cones lies in the center of mass of the nutating body.

5.7.6 Precession of the Symmetric Top

If an external torque $\boldsymbol{D}$ acts on the body the angular momentum is no longer space-fixed, because of $\boldsymbol{D} = \mathrm{d}\boldsymbol{L}/\mathrm{d}t$. Depending on the direction of $\boldsymbol{D}$ relative to the figure axis the direction and also the magnitude of $\boldsymbol{L}$ changes with time. At first we will discuss the simplest case where the body rotates with the angular velocity ω around its figure axis c and all three axis $\boldsymbol{L}$, $\boldsymbol{\omega}$ and $\boldsymbol{c}$ coincide. In this case there is no nutation and for $\boldsymbol{D} = \boldsymbol{0}$ the body would rotate with $\omega = \text{const}$ about the space-fixed figure axis.

If the top is not supported in its CM, the gravitational force generates a torque

$$\boldsymbol{D} = \boldsymbol{r} \times m \cdot \boldsymbol{g} ,$$

where $\boldsymbol{r}$ is the vector from the support point to the CM. If the symmetric top spins with the angular momentum $\boldsymbol{L}$ the torque is perpendicular to $\boldsymbol{L}$ and therefore changes only its direction but not its magnitude (Fig. 5.37). During the time interval $\mathrm{d}t$ the direction of L changes by the angle $\mathrm{d}\varphi$ and we can derive from Fig. 5.37

$$|\mathrm{d}\boldsymbol{L}| = |\boldsymbol{L}| \cdot \mathrm{d}\varphi \rightarrow D = \frac{\mathrm{d}L}{\mathrm{d}t} = |\boldsymbol{L}|\frac{\mathrm{d}\varphi}{\mathrm{d}t} .$$

The angular momentum axis and with it the coincidental figure axis rotate with the angular velocity

$$\omega_\mathrm{p} = \frac{\mathrm{d}\varphi}{\mathrm{d}t} = \frac{D}{L} = \frac{D}{I\omega} \tag{5.52}$$

about an axis perpendicular to the plane of $\boldsymbol{D}$ and $\boldsymbol{L}$ where we have assumed that $\omega_\mathrm{p} \ll \omega$. This motion is called **precession**.

If the figure axis forms the angle α against the vertical direction the magnitude of the torque is $D = m \cdot g \cdot \sin\alpha$. The change $\mathrm{d}L$ of the angular momentum L is now for $\mathrm{d}L \ll L$ (Fig. 5.38)

$$\mathrm{d}L = |L| \sin\alpha \cdot \mathrm{d}\varphi .$$

For $\omega_\mathrm{p} \ll \omega$ therefore the equation for the precession frequency becomes

$$\omega_\mathrm{p} = \frac{mgr \sin\alpha}{I\omega \sin\alpha} = \frac{mgr}{I\omega} , \tag{5.53}$$

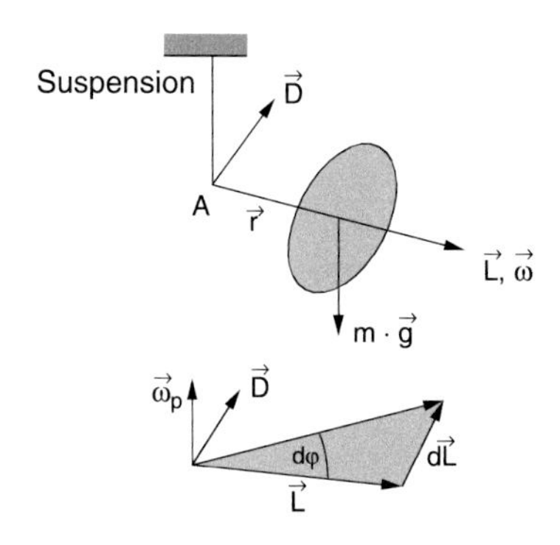

Figure 5.37 The gravitational force causes a torque acting on a top, that is not supported in the Center of Mass

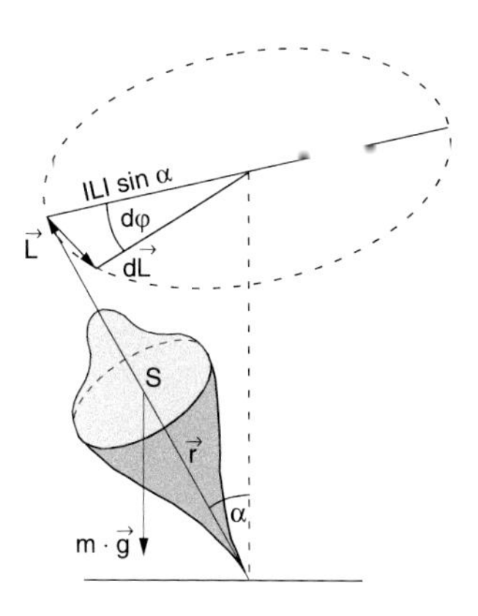

Figure 5.38 Precession of a spinning top

which shows that ω_p is independent of the space orientation of the figure axis and depends only on the angular momentum $\boldsymbol{L}$ and the torque $\boldsymbol{D}$.

The general treatment of precession has to take into account the three vectors (Fig. 5.39)

1. The angular velocity $\boldsymbol{\omega}_\mathrm{F}$ about the figure axis
2. The angular velocity $\boldsymbol{\omega}_\mathrm{p}$ of precession around the vertical z-axis
3. The total angular velocity $\boldsymbol{\omega} = \omega_\mathrm{F} + \omega_\mathrm{p}$

According to Fig. 5.39 we get the relations:

$$\begin{aligned} \boldsymbol{\omega}_\mathrm{F} &= \omega \cdot \boldsymbol{e} \quad \text{with } \boldsymbol{e} = \{\sin\theta\cos\varphi, \sin\theta\sin\varphi, \cos\theta\} \\ \boldsymbol{\omega}_\mathrm{p} &= \dot{\varphi} \cdot \{0, 0, 1\} \\ \boldsymbol{\omega} &= \{\omega \cdot \sin\theta\cos\varphi, \omega \cdot \sin\theta\sin\varphi, \omega \cdot \cos\theta + \dot{\varphi}\} . \end{aligned} \tag{5.53a}$$

We separate $\boldsymbol{\omega}$ into a component $\boldsymbol{\omega}_\parallel$ parallel and $\boldsymbol{\omega}_\perp$ perpendicular to the figure axis $\boldsymbol{e}$.

$$\begin{aligned} \boldsymbol{\omega}_\parallel &= \boldsymbol{e} \cdot (\omega + \dot{\varphi}\cos\theta) \\ \boldsymbol{\omega}_\perp &= \boldsymbol{e} \times (\boldsymbol{\omega} \times \boldsymbol{e}) \\ &= \dot{\varphi}\sin\theta \cdot \{-\cos\theta\cos\varphi, -\cos\theta\sin\varphi, \sin\theta\} . \end{aligned} \tag{5.53b}$$

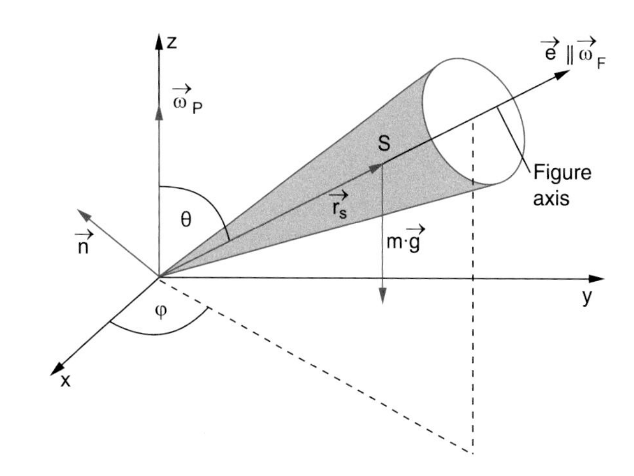

Figure 5.39 Illustration of the equation of motion for a spinning top

The total angular momentum $\boldsymbol{L}$ is

$$\begin{aligned}\boldsymbol{L} &= I_{\|}\boldsymbol{\omega}_{\|} + \left(mr_s^2 + I_{\perp}\right)\boldsymbol{\omega}_{\perp} \\ &= I_{\|}\cdot \boldsymbol{e}(\omega + \dot{\varphi}\cos\theta) \\ &\quad + \left(I_{\perp} + mr_s^2\right) \\ &\quad \cdot \dot{\varphi}\sin\theta\{-\cos\theta\cos\varphi, -\cos\theta\sin\varphi, \sin\theta\}\,,\end{aligned} \tag{5.53c}$$

where $\boldsymbol{I}_{\|}$ is the moment of inertia for a rotation about the figure axis and $\boldsymbol{I}_{\perp}$ about an axis perpendicular to the figure axis.

Because ω, $\mathrm{d}\varphi/\mathrm{d}t$ and θ do not change with time the time derivative of (5.53c) is

$$\begin{aligned}\frac{\mathrm{d}\boldsymbol{L}}{\mathrm{d}t} &= I_{\|}(\omega + \dot{\varphi}\cos\theta)\cdot\dot{\boldsymbol{e}} \\ &\quad - \left(I_{\perp} + mr_s^2\right)\dot{\varphi}^2\cos\theta\sin\theta\{-\sin\varphi, \cos\varphi, 0\} \\ &= \left[I_{\|}\cdot\sin\theta(\omega + \dot{\varphi}\cos\theta)\dot{\varphi}\right. \\ &\quad \left. - \left(I_{\perp} + mr_s^2\right)\dot{\varphi}^2\sin\theta\cos\theta\right]\cdot\hat{\boldsymbol{n}}\,.\end{aligned} \tag{5.53d}$$

where $\boldsymbol{n} = \{-\sin\varphi, \cos\varphi, 0\}$ is the unit vector in the direction of the torque $\boldsymbol{D}$. With $\mathrm{d}L/\mathrm{d}t = \boldsymbol{D} = m\cdot g\cdot r_s\cdot\sin\theta\cdot\boldsymbol{n}$ we obtain the equation

$$\omega_p\cdot I_{\|}\cdot\omega + \omega_p^2\cos\theta(I_{\|} - I_{\perp}) = mgr_s\,. \tag{5.53e}$$

which has two solutions for the precession frequency ω_p. The difference between the two solutions depends on the difference $I_{\|} - I_{\perp}$ of the two moments of inertia [5.4].

5.7.7 Superposition of Nutation and Precession

In the general case the top does not rotate about its figure axis. Without external torque the top would perform a nutation around the space-fixed angular momentum axis $\boldsymbol{L}$. With an external torque the angular momentum axis is no longer constant but precesses with the angular velocity ω_p around an axis through the underpinning point A parallel to the external force (Fig. 5.39) while the figure axis performs a nutation around the precessing axis $\boldsymbol{L}$. With this combination of precession and nutation the end of the figure axis describes a complicated path (Fig. 5.40). The exact form of this trajectory depends on the ratio of nutation frequency Ω to the precession frequency ω_p.

For the demonstration of nutation and precession a special bearing of the top is useful called a *gimbal mounting* where the figure axis can be turned into arbitrary directions and the top is always "torque-free" (Fig. 5.41). This can be realized if the figure axis is mounted by ball bearings in a frame which can freely rotate around an axis perpendicular to the figure axis. The mounting of this axis can again rotate about a vertical axis. If the system is turned around the vertical axis, the figure axis of the top diverts from its horizontal direction. Reversal of the turning direction also reverses the direction of this diversion. If a short push is applied perpendicular to the top axis, the angular momentum axis is forced into another direction and the top starts to nutate. When a mass m is attached to the first frame, a torque acts on the top which starts to precess around a vertical axis.

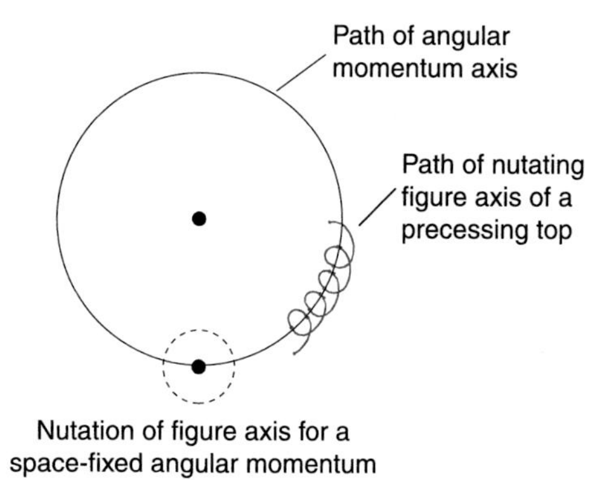

Figure 5.40 Path of figure axis when precession and mutation are superimposed

Before the invention of the GPS system the precession of the top was used for navigation purposes (gyro compass). Its function is explained in Fig. 5.42, where a rotating disc is suspended in a mounting B which can turn around a vertical axis a through the suspension point A. The top axis KA can freely turn only in a horizontal plane. The center of mass lies below the point A. Different from a torque-free top rotating around its figure axis where the figure axis and the angular momentum axis coincide, for the gyro compass the suspension axis through the point A is rigidly connected with the earth and participates in the earth rotation with the angular velocity ω. Therefore a torque is acting on the gyro perpendicular to the drawing plane. The gyro turns around the axis a until the figure axis is parallel to the rotation axis of the earth and points into the south-north direction. Now angular momentum axis L and the forced rotation axis ω_E are parallel (Fig. 5.42b), and the torque forcing the turn of the figure axis becomes zero.

This property can be experimentally studied with the gimbal mount by simulating the earth rotation by the rotation of the outer mounting in Fig. 5.41. The figure axis then turns into the vertical position.

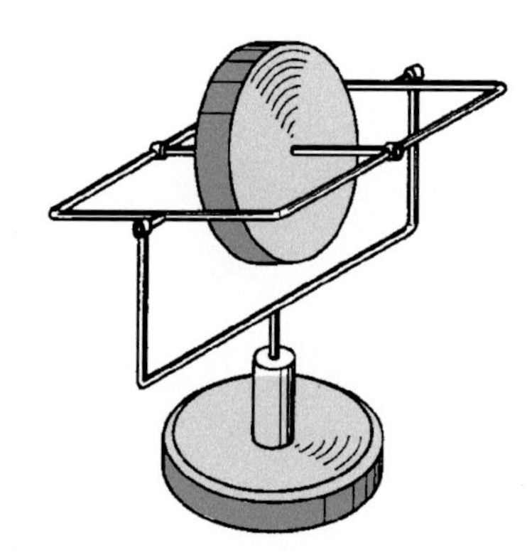

Figure 5.41 Gimbal mount of a symmetric top

Chapter 5

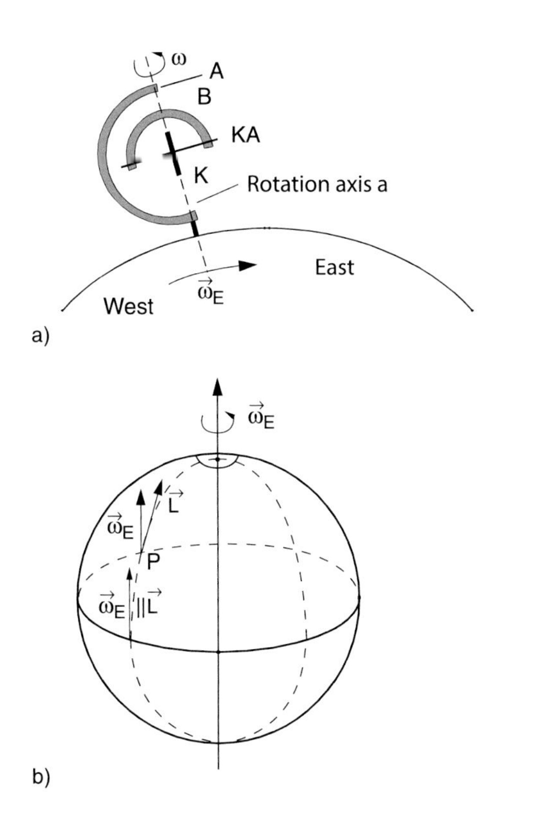

Figure 5.42 Gyro-compass: **a** mount, **b** direction of L and ω_E at the equator and for higher lattitudes

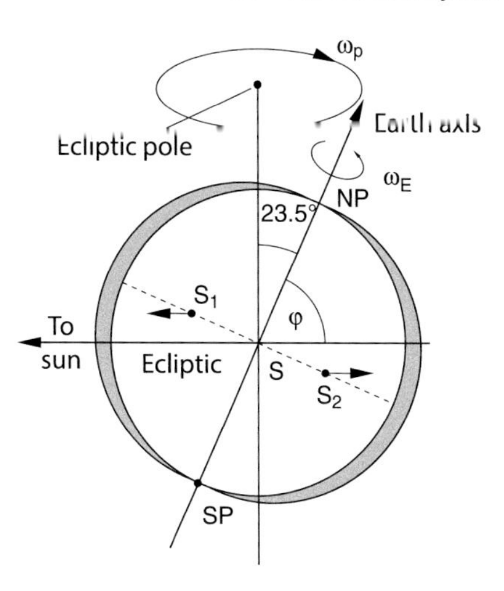

Figure 5.43 The earth as symmetric top. The *arrows* indicate the difference forces $F_1 - F_2$ ath the Centers of Mass S_1 and S_2 of the two opposite sides of the bulge

At the earth equator $\boldsymbol{\omega}_E$ and $\boldsymbol{L}$ are parallel. This is not the case for a point P on another circle of latitude because L has to lie in a horizontal plane. However, also in this case the gyro adjusts itself in such a way that the component L parallel to $\boldsymbol{\omega}_E$ becomes maximum. The vector $\boldsymbol{L}$ becomes the tangent to the circle of longitude, points therefore again to the north. Only at the two poles of the earth the gyro fails, because here $\boldsymbol{L}$ is always perpendicular to $\boldsymbol{\omega}_E$.

5.8 The Earth as Symmetric Top

In a good approximation the earth can be described by a clinched rotational ellipsoid, i.e. an oblate symmetric top with $I_a = I_b < I_c$. The equator diameter is with 12 756 km by about 43 km larger than the pole diameter with 12 713 km. This clinch is caused by the centrifugal force due to the rotation of the earth (see Sect. 6.6). For the following considerations we will compose this oblate ellipsoid by a sphere plus additional bulges which have their maximum thickness at the equator (red area in Fig. 5.43).

Because of the inclination of the earth axis ($\varphi = 90° - 23.5° = 66.5°$) against the ecliptic (orbital plane of the earth's motion around the sun) the two centers of mass S_1 of the bulge towards the sun and S_2 of the bulge opposite to the sun are located above and below the ecliptic (Fig. 5.43) in contrast to the center of mass S of the sphere which lies in the ecliptic. While for the mass m_E of the sphere concentrated in S the centripetal force $F_1 = Gm_E M_{\odot}/r^2$ due to the gravitational attraction between earth and sun is just compensated by the equal but opposite centrifugal force $F_2 = m_E v_E^2/r$ this is no longer true for the centers of mass S_1 and S_2 of the bulges. Since S_1 is closer to the sun the centripetal force predominates while for S_2 the centrifugal force prevails. Since the net forces for S_1 and S_2 are antiparallel they form a couple of forces which act as a torque on the earth and cause the earth axis to precess (solar precession).

Besides the gravitational force between earth and sun the attraction between moon and earth must be taken into account. The calculation is here more complicated because the orbital plane of the moon around the earth is inclined by an angle of 5.1° against the ecliptic. The calculation shows that the influence on the earth is of the same order of magnitude than that of the sun.

Altogether both torques cause the lunar-solar precession where the earth axis propagates on a cone with an opening angle of $2 \times 23.5°$ by an angle $\varphi \approx 50''$ per year which gives a precession period of about 25 750 years for $\varphi = 2\pi$ (*Platonic year*). Within a Platonic year the cone is once circulated. The elongation of the earth axis describes a circle on the celestial sphere around the ecliptic pole (Fig. 5.44).

Remark. This precession causes a turn of the intersecting line between ecliptic and equatorial plane by 360° within 25 850 years. This shifts the vernal equinox (where day and night both last 12 hours) by about $50''$ per year. It causes furthermore a shift of the signs of the zodiac between their naming 2000 years ago and today by about one month. For example the real constellation of the *Gemini* (twins) coincides in our times with the sign of the zodiac *Cancer*. This is unknown to many astrological oriented people who come into trouble if they should explain whether the real stars or the signs of the zodiac are responsible for the fate of a person.

The precession of the earth axis described above is not uniform because of the following reasons:

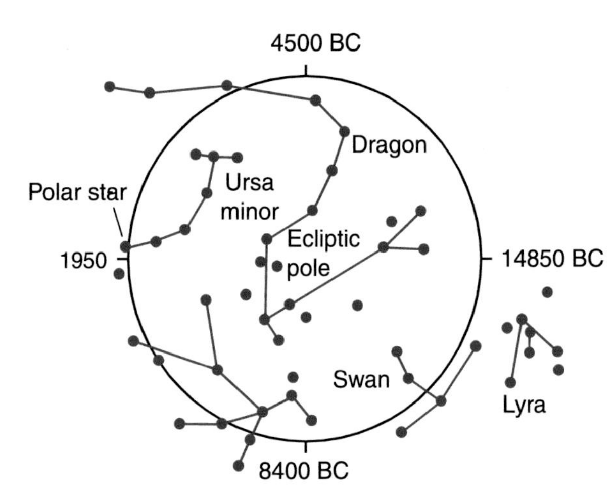

Figure 5.44 Due to the precession of the earth its axis traverses a circle on the celestial sphere around the ecliptic pole. In 1950 it pointed towards the pole star

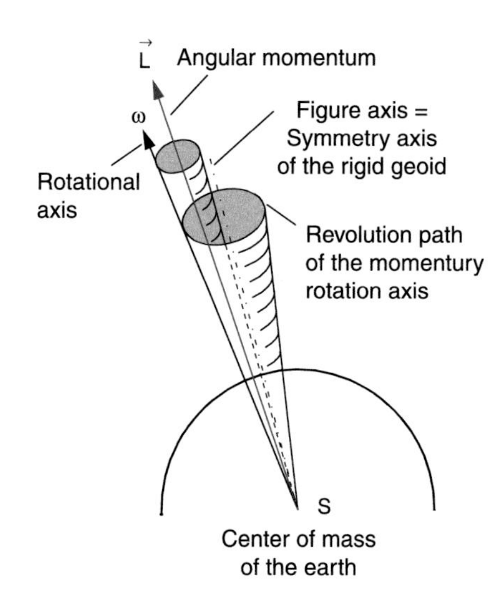

Figure 5.46 Nutation of the earth axis

- Because of the inclination of the earth axis the torque exerted by the sun changes periodically during the year (Fig. 5.45). It is maximal on December 22nd and June 21st and minimal at March 21st and September 23rd
- The torque exerted by the moon changes because the inclination of the moon's orbital plane changes with a period of 9.3 years.
- Also the other planets cause a small torque acting on the earth. Because the relative distances to the earth change in time, this causes a tiny variation of the precession.
- The motion of the earth around the sun proceeds on an elliptical path and therefore the distance r between earth and sun changes periodically. It is minimum in December and maximum in June (Fig. 5.45). Therefore the gravitational force acting on the earth changes correspondingly.

Astronomers call these short-period fluctuations of the precession *nutations* although they are strictly speaking no nutations but perturbations of the precession.

There are real torque-free nutations superimposed on the complicated precession. They are caused by the fact, that the figure axis of the earth and the rotation axis do not exactly coincide (Fig. 5.46). The figure axes (south-north pole intersection) therefore nutates around the precessing angular momentum axis with a measure period of about 303 days. On the other hand the nutation period is

$$T_{\text{nut}} = \frac{2\pi}{\omega}\frac{I_a}{I_c - I_a} \,. \tag{5.54}$$

From the measured nutation period one can therefore determine the difference $I_c - I_a$ of the inertial moments [5.5a, 5.5b].

Since the earth is no rigid body the mass distribution and therefore the inertial moments can change, for instance by volcanic eruptions or by convective currents in the liquid interior of the earth [5.6a, 5.6b]. This causes small fluctuations of the nutation. In Fig. 5.47 the wandering of the north-pole of the rotational axis during the year 1957 is shown.

The above discussion has shown, that the more precise measurements have been made, the more different influences on the motion of the earth axis have to be taken into account. Even today there are discussions about the best model for the earth motion [5.7, 5.8].

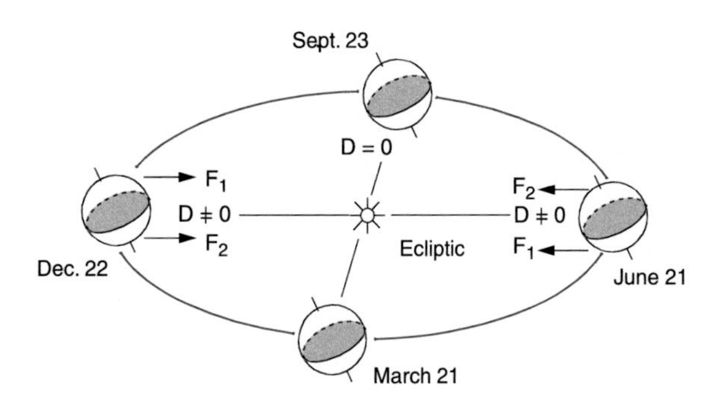

Figure 5.45 Position of the earth axis during the revolution of the earth about the sun. Note, that the direction of the angular momentum does not change during the year

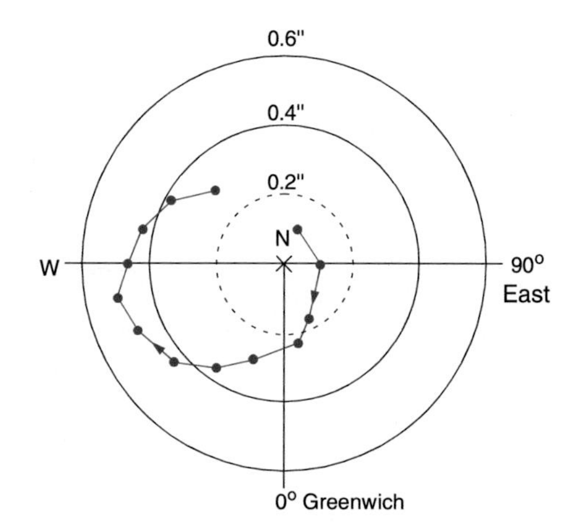

Figure 5.47 Migration of the north pole of the earth's rotation axis during the year 1957 about the average position of the period 1900–1905. One second of arc corresponds to about 30 m

Summary

- The model of the extended rigid body neglects all internal motions (Deformations and vibrations). The center of mass S has the coordinates

$$\boldsymbol{r}_S = \frac{1}{M}\int_V \boldsymbol{r}\varrho(\boldsymbol{r})\mathrm{d}V = \frac{1}{V}\int \boldsymbol{r}\mathrm{d}V \quad \text{for} \quad \varrho = \text{const} .$$

- The motion of a free rigid body can be always composed of a translation of the center of mass S with the velocity v_s and a rotation of the body around S with the angular velocity ω. The motion of the extended body has therefore 6 degrees of freedom.
- For the motion of an extended body not only magnitude and direction of the force acting on the body are important but also the point of action on the body.
- An arbitrary force acting on an extended body can always be composed of a force acting on the center of mass S (translational acceleration) and a couple of forces causing an accelerated rotation.
- The moment of inertia (rotational inertia) for a rotation about an axis through the center of mass S is $I_s = \int r_\perp^2 \varrho \mathrm{d}V$ where $r_\perp$ is the distance of the volume element $\mathrm{d}V$ from the rotation axis. The moment of inertia for a rotation around an arbitrary axis with a distance a from the parallel axis through S is $I = I_s + Ma^2$ (parallel axis theorem or Steiner's theorem).
- The kinetic energy of the rotational motion is $E_{\text{rot}} = \frac{1}{2}I\omega^2$.
- The equation of motion for a body rotating about a space-fixed axis is $D_\parallel = I \cdot \mathrm{d}\omega/\mathrm{d}t$, where $D_\parallel$ is the component of the torque parallel to the rotation axis.
- The moment of inertia I_s depends on the direction of the rotation axis relative to a selected axis of the body. It can be described by a tensor. The directions of the axes with the maximum and the minimum inertial moment determine the principal axes system. In this system the tensor is diagonal. The diagonal elements are the principal moments of inertia.
- If two of the principal moments are equal, the body is a symmetric top. If all three are equal the body is a spherical top.
- Angular momentum $\boldsymbol{L}$ and angular velocity $\boldsymbol{\omega}$ are related by $\boldsymbol{L} = I \cdot \boldsymbol{\omega}$, where I is the inertial tensor, which is diagonal in the principal axes system. In the general case $\boldsymbol{L}$ and $\boldsymbol{\omega}$ are not parallel.
- If the body rotates about a principal axis, $\boldsymbol{L}$ and $\boldsymbol{\omega}$ are parallel and without external torque their directions are space-fixed.
- For an arbitrary direction of ω the momentary rotation axis $\boldsymbol{\omega}$ nutates around the angular momentum axis which is space-fixed without external torque.
- Under the action of an external torque the angular momentum axis $\boldsymbol{L}$ precedes around the external force and in addition the momentary rotation axis nutates around $\boldsymbol{L}$. The relation between $\boldsymbol{L}$ and $\boldsymbol{D}$ is $\boldsymbol{D} = \mathrm{d}\boldsymbol{L}/\mathrm{d}t$.
- The general motion of a top is completely described by the Euler-equations.
- The earth can be approximately described by a symmetric top, which rotates about the axis of its maximum moment of inertia. The vector sum of the gravity forces exerted by the sun, the moon and the planets results in a torque which causes a periodic precession of the earth axis with a period of 25 850 years. In addition changes of the mass distribution in the earth cause a small difference between symmetry axis and momentary rotation axis. Therefor the earth axis performs an irregular nutation around the symmetry axis.

Problems

5.1 Determine the center of mass of a homogeneous sector of a sphere with radius R and opening angle α.

5.2 What are moment of inertia, angular momentum and rotational energy of our earth

a) If the density ϱ is constant for the whole earth
b) If for $r \leq R/2$ the homogeneous density ϱ_1 is twice the density ϱ_2 for $r > R/2$?
c) By how much would the angular velocity of the earth change, if all people on earth ($n = 5 \cdot 10^9$ with $m = 70\,\text{kg}$ each) would gather at the equator and would start at the same time to run into the east direction with an acceleration $a = 2\,\text{m/s}^2$?

5.3 A cylindrical disc with radius R and mass M rotates with $\omega = 2\pi \cdot 10\,\text{s}^{-1}$ about the symmetry axis ($R = 10\,\text{cm}$, $M = 0.1\,\text{kg}$).

a) Calculate the angular momentum L and the rotational energy E_{rot}.
b) a bug with $m = 10\,\text{g}$ falls vertical down onto the edge of the disc and holds itself tight. What is the change of L and E_{rot}?
c) The bug now creeps slowly in radial direction to the center of the disc. How large are now $\omega(r)$, $I(r)$ and $E_{\text{rot}}(r)$ as a function of the distance r from the center $r = 0$?

5.4 The mass density ϱ of a circular cylinder (radius R, height H) increases in the radial direction as $\varrho(r) = \varrho_0(1 + (r/R)^2)$.

a) How large is its inertial moment for the rotation about the symmetry axis for $R = 10\,\text{cm}$ and $\varrho_0 = 2\,\text{kg/dm}^3$?
b) How long does it take for the cylinder to roll down an inclined plane with $\alpha = 10°$ from $h = 1\,\text{m}$ to $h = 0$?

5.5 Calculate the rotational energy of the Na_3-molecule composed of 3 Na atoms ($m = 23\,\text{AMU}$) which form an isosceles triangle with the apex angle $\alpha = 79°$ and a side length of $d = 0.32\,\text{nm}$ when it rotates around the three principal axes with the angular momentum $L = \sqrt{l(l+1)} \cdot \hbar$. Determine at first the three axis and the center of mass.

5.6 A wooden rod with mass $M = 1\,\text{kg}$ and a length $l = 0.4\,\text{m}$, which is initially at rest, can freely rotate about a vertical axis through the center of mass. The end of the rod is hit by a bullet ($m = 0.01\,\text{kg}$) with the velocity $v = 200\,\text{m/s}$, which moves in the horizontal plane perpendicular to the rod and to the rotation axis and which gets stuck in the wood.
What are the angular velocity ω and the rotational energy E_{rot} of the rod after the collision? Which fraction of the kinetic energy of the bullet has been converted to heat?

5.7 A homogeneous circular disc with mass m and radius R rotates with constant velocity ω around a fixed axis through the center of mass S perpendicular to the disc plane. At the time $t = 0$ a torque $D = D_0 \cdot e^{-at}$ starts to act on the disc. What is the time dependence $\omega(t)$ of the angular velocity? Numerical example: $\omega_0 = 10\,\text{s}^{-1}$, $m = 2\,\text{kg}$, $R = 10\,\text{cm}$, $a = 0.1\,\text{s}^{-1}$, $D_0 = 0.2\,\text{Nm}$.

5.8 A full cylinder and a hollow cylinder with a thin wall and equal outer diameters roll with equal angular velocity ω_0 on a horizontal plane and then role up an inclined plane. At which height h do they return? (Friction should be neglected), numerical example: $R = 0.1\,\text{m}$, $\omega_0 = 15\,\text{s}^{-1}$.

References

5.1. Ph. Dennery, A. Krzwicki, *Mathematics for Physicists.* (Dover Publications, 1996)
5.2. J.C. Kolecki, *An Introduction to Tensors for Students of Physics and Engineering.*
5.3. https://en.wikipedia.org/wiki/Moment_of_inertia#Principal_moments_of_inertia
5.4. W. Winn, *Introduction to Understandable Physics.* (Authort House, 2010)
5.5a. K. Lambrecht, *The earth's variable Rotation. Geophysical Causes and Consequences.* (Cambridge University Press, Cambridge, 1980)
5.5b. W.H. Munk, G.J. MacDonald, *The Rotation of the Earth. A Geophysical Discussion.* (Cambridge Monographs, 2010)
5.6a. St. Marshak, *Earth: Portrait of a Planet,* 4th ed. (W.W. Norton & Company, 2011)
5.6b. I. Jackson, *The Earth's Mantle.* (Cambridge University Press, 2000)
5.7. V. Dehant, P.M. Mathews, *Precession, Nutation and Wobble of the Earth.* (Cambridge University Press, 2015)
5.8. A. Fothergill, D. Attenborough, *Planet Earth: As You've Never Seen it Before.* (BBC Books, 2006)

Real Solid and Liquid Bodies

6

© Springer International Publishing Switzerland 2017
W. Demtröder, *Mechanics and Thermodynamics*, Undergraduate Lecture Notes in Physics, DOI 10.1007/978-3-319-27877-3_6

In this chapter, we will proceed with the stepwise refinement of our "*model of reality*". We will take into account the experimental fact that extended bodies can change their shape under the influence of external forces. We will also *discuss* the important question why and under which conditions real bodies can exist in different aggregation states as solids, liquids or gases. We will see, that an atomic model, which considers the different interactions between the atoms, can at least qualitatively explain all observed phenomena. For a quantitative description, a more profound knowledge about the atomic structure is demanded. The quantitative calculation of the detailed characteristics of solids or liquids is still not trivial, even with fast computers and sophisticated programs, because of the enormous number (10^{23} /kg) of atoms involved. Here symmetry considerations are helpful to facilitate the description.

If all physical characteristics of an extended body (density, elasticity, hardness etc.) are constant within the body, we call it a **homogeneous body**. Are they also independent of the direction the body is **isotropic**. A liquid metal is an example of an isotropic and homogeneous body while a NaCl-crystal (table salt) is homogeneous but not isotropic.

6.1 Atomic Model of the Different Aggregate States

Many experiments have proved that all macroscopic bodies are composed of atoms or molecules (see Vol. 3). Between two atoms, which consist of a positively charged small nucleus and a negatively charged extended electron cloud, attractive as well as repulsive interactions can occur. The superposition of all these interactions results in a force $\boldsymbol{F}(r)$ and a potential energy $E_p(r)$ which depend on the distance r between the interacting atoms and which are qualitatively depicted in Fig. 6.1. At the equilibrium distance r_0, the potential energy $E_p(r)$ shows a minimum and the force $F(r) = -\operatorname{grad} E_p$ becomes zero. For shorter distances $r < r_0$ the repulsive forces dominate and for larger distances $r > r_0$ the attractive forces. For both cases, the potential energy increases. When an atom A is surrounded by many other atoms A_i at distances r_i the total force $\boldsymbol{F}$ acting on A is the vector sum of all individual forces $\boldsymbol{F}_i$:

$$\boldsymbol{F} = \sum \boldsymbol{F}_i(r_i) .$$

The resulting potential energy E_p of atom A depends on the spatial distribution of the surrounding atoms A_i and is related to the force F by $\boldsymbol{F} = -\mathbf{grad}\, E_p$.

In *crystalline solids* the atoms are arranged in regular lattices (Fig. 6.2) while in *amorphous solids* they sit on more or less statistically distributed sites. Examples for the first cases are NaCl-crystals, or noble gas crystals at low temperatures, while examples for amorphous solids are glasses or amorphous silicon, which is used for solar cells.

When we place the atom A in a crystalline solid at the origin $\boldsymbol{r} = \boldsymbol{0}$ of our coordinate system the atoms A_i have the position

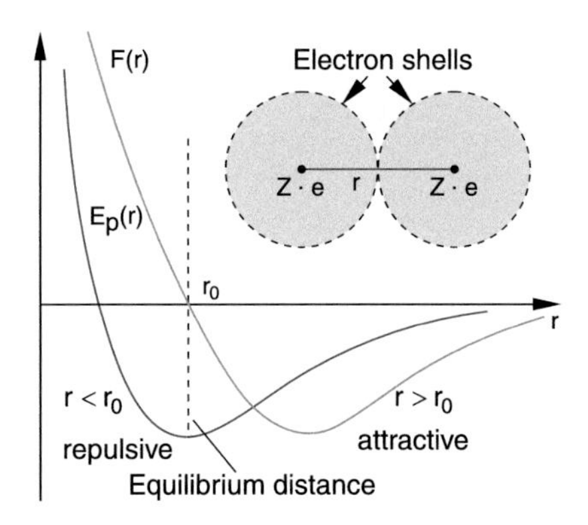

Figure 6.1 Qualitative dependence of potential energy $E_p(r)$ and force $F(r)$ between two atoms as a function of distance r between the nuclei of two adjacent atoms

vectors

$$\boldsymbol{r}_i = n_{1i}\boldsymbol{a} + n_{2i}\boldsymbol{b} + n_{3i}\boldsymbol{c} , \tag{6.1}$$

where the $n_{\alpha i}$ are integers and the basis vectors $\boldsymbol{a}, \boldsymbol{b}, \boldsymbol{c}$ define the unit cell (or elementary cell) in the crystalline solid. They are marked as red vectors in Fig. 6.2. Their magnitudes and directions define the crystal structure of the solid. The forces between the atoms can be modelled by elastic springs (Fig. 6.3) where the restoring force constants k_i can be different in the different directions. At the absolute temperature T the atoms vibrate about their equilibrium position r_0. Their mean kinetic energy is $\langle E_{kin}\rangle = (1/2)kT$ per degree of freedom (see Sect. 7.3) where the equilibrium positions correspond to the minimum of the potential energy E_p in Fig. 6.1. For temperatures far below the melting temperature, $\langle E_{kin}\rangle$ is small compared to the magnitude $|E_p(r_0)|$ of the potential energy at the equilibrium position which means that the atoms cannot leave their equilibrium positions.

If the temperature rises above the melting temperature the mean kinetic energy $\langle E_{kin}\rangle$ becomes larger than the binding energy $E_B = -E_p(r_0)$. The atoms cannot be kept any

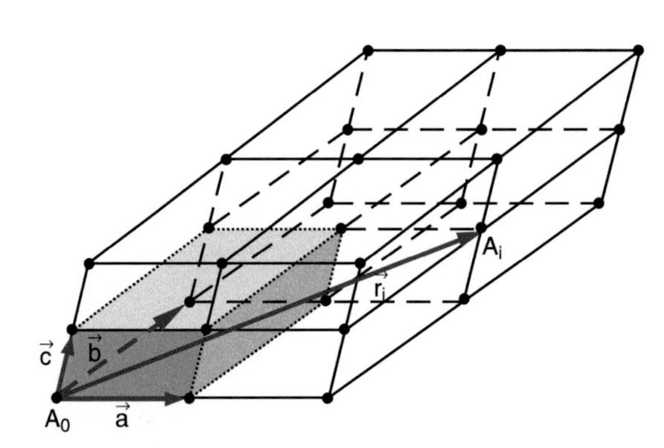

Figure 6.2 Regular structure of a solid crystal. The basis vectors $\boldsymbol{a}$, $\boldsymbol{b}$, $\boldsymbol{c}$ form the elementary cell with volume $V_E = (\boldsymbol{a} \times \boldsymbol{b}) \cdot \boldsymbol{c}$. The position vector of the point A is $\boldsymbol{r}_A = 2\boldsymbol{a} + \boldsymbol{b} + \boldsymbol{c}$

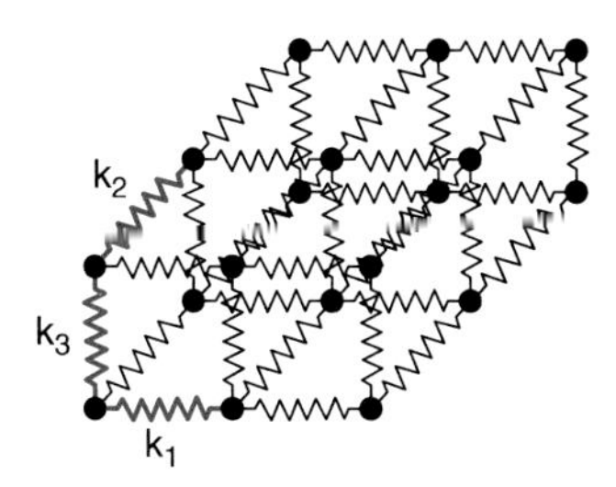

Figure 6.3 Spring model of a solid crystal. The restoring force is for an isotropic crystal equal for all three directions $\boldsymbol{a}$, $\boldsymbol{b}$, $\boldsymbol{c}$, for an unisotropic crystal they differ

longer on their positions r_i but can diffuse around. The crystalline solid melts and converts to the liquid state.

Also in the liquid state the minimum of the potential energy remains at the mean distance $\langle r_0 \rangle$ between the atoms. This means that the density in the liquid state is not very different from that in the solid state. However, now a single atom is no longer bound to a fixed position but can move freely within the liquid. Nevertheless, there is still a certain order. If one plots the probability $W(r)$ that an atom A occupies a position with the distance r from its neighbors (Fig. 6.4) a pronounced maximum is found at $r = r_0$ which is close to the minimum distance r_0 in the crystalline solid. Similar to the amorphous solid the liquid has a *short range order*, while a crystalline solid has a *long range order*, because it is possible to assign a definite position $r = n_1 a + n_2 b + n_3 c$ (with integers n_i) to each atom regardless how far away it is (Fig. 6.2). While the crystalline solid can be described by the spring model of Fig. 6.3, many features of liquids can be modelled by the string model of Fig. 6.5. Here the atoms are connected by strings of constant lengths where the directions can be arbitrarily altered. The balls in this mechanical model can move similar to the atoms in a liquid.

A further increase of the temperature above the boiling point makes the mean kinetic energy large compared to the magnitude of the potential energy. The potential energy is then negligible

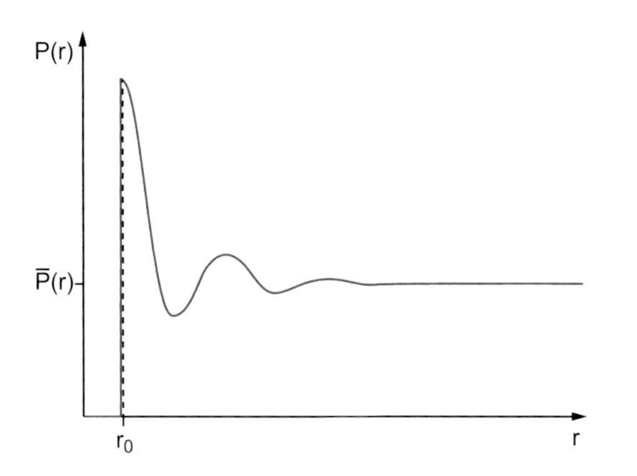

Figure 6.4 Probability $P(r)$ that an atom A_1 in a liquid has the distance r to an arbitrary other atom A_2

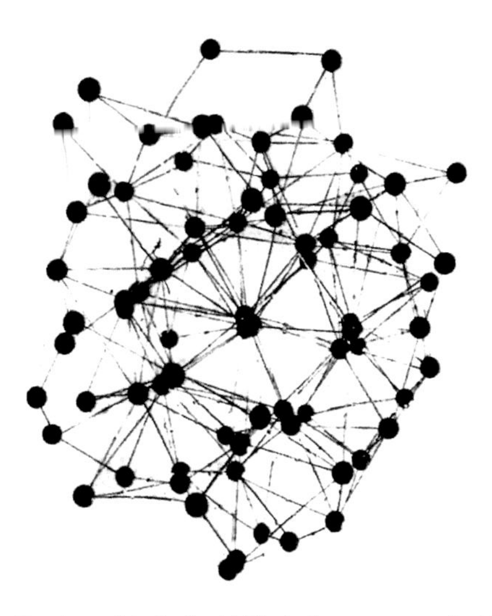

Figure 6.5 Atomic model of a liquid. The balls are connected with each other by strings. The model illustrates the free mobility of each atom

and the atoms can move freely. They form a gas that fills the total accessible volume. The interaction energy is only noticeable during collisions of the atoms with each other.

The mean distance $\langle r \rangle$ between the atoms or molecules and therefore also the density $\varrho = M/V$ of the gas with total mass M depends on the volume V which is accessible to the $N = M/m$ molecules with mass m. At normal pressure $p = 1$ bar the density of the gas is about three orders of magnitude smaller than that of solids or liquids.

Examples

The density of air at $p = 1$ bar and $T = 300\,\mathrm{K}(\approx 20\,°\mathrm{C})$ is $\varrho = 1.24\,\mathrm{kg/m^3}$, while the density of water is about $10^3\,\mathrm{kg/m^3}$ and that of lead is $11.3 \cdot 10^3\,\mathrm{kg/m^3}$. ◄

The considerations above show that the aggregation state of material depends on the ratio $\langle E_{\mathrm{kin}} \rangle / E_{\mathrm{p}}$ and therefore on the temperature and on the binding energy of the atoms or molecules of the body.

We will now discuss the most important characteristic features of the different aggregation states in a phenomenological manner. The more detailed treatment is given in Vol. 3.

6.2 Deformable Solid Bodies

External forces can change the shape of solid bodies. If the body returns to its original shape after the exposure to the external forces we call it elastic. For a plastic body the deformation remains.

6.2.1 Hooke's Law

If a pulling force acts onto the end face of an elastic rod with length L and cross section A, which is hold tight at the other end $x = 0$ (Fig. 6.6) the length L is prolonged by ΔL. The linear relation between the magnitude $F = |\boldsymbol{F}|$ of the force and the prolongation ΔL

$$F = E \cdot A \cdot \Delta L / L \tag{6.2}$$

is called *Hooke's law*, which is valid for sufficiently small relative length changes $\Delta L/L$. The proportional factor E is the *elastic modulus* with the dimension $\mathrm{N/m^2}$. For technical applications often the dimension $\mathrm{kN/mm^2} = 10^9\,\mathrm{N/m^2}$ is used. Table 6.1 gives numerical values for some materials.

For materials with a large elastic modulus E one needs a large force to achieve a given relative change of length $\Delta L/L$. With other words: Materials with a large value of E show for a given force a small relative length change.

Introducing the tensile stress (= pulling force/cross section A)

$$\sigma = F/A$$

and the relative stretch or strain $\varepsilon = \Delta L/L$ Hooke's law can be written in the clearer form

$$\sigma = E \cdot \varepsilon \,. \tag{6.2a}$$

For sufficiently small relative stretches ε, tensile stress and strain are proportional. In this proportional range the distances between neighboring atoms vary only within a small range around r_0 (Fig. 6.1) where the interatomic force $F(r) = a \cdot r$ is approximately a linear function of the distance r and the potential energy $E_\mathrm{p}(r)$ can be approximated by a parabola.

Note: This linear relation is only an approximation for small values of ε. For larger ε nonlinear forces appear that cannot be neglected.

Expanding $E_\mathrm{p}(r)$ into a Taylor series around the equilibrium position r_0

$$E_\mathrm{p}(r) = \sum_{n=0}^{\infty} \frac{(r-r_0)^n}{n!} \left(\frac{\partial^n E_\mathrm{p}}{\partial r^n}\right)_{r=r_0} \tag{6.3a}$$

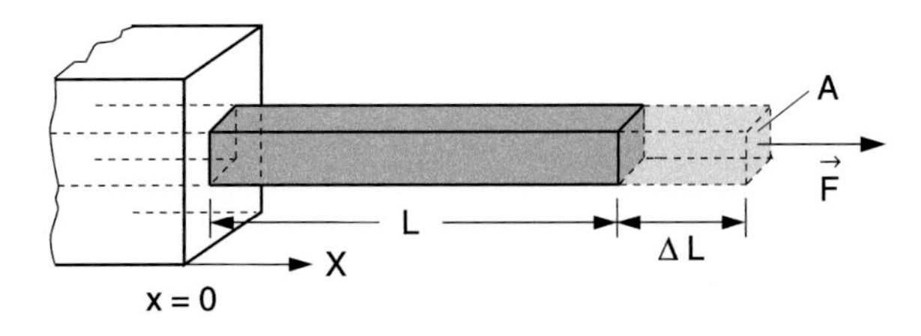

Figure 6.6 A rod fixed at $x = 0$ expands under the action of a force $\vec{F}$ by $\Delta L = F \cdot L/(E \cdot A)$

Table 6.1 Elastic constants of some solid materials. E = elastic modulus; G = modulus of shear; K = compressibility modulus; μ = inverse contraction number = Poisson number; [6.1]

Material	E [$10^9\,\mathrm{N/m^2}$]	G [$10^9\,\mathrm{N/m^2}$]	K [$10^9\,\mathrm{N/m^2}$]	μ
Aluminium	71	26	74	0.34
Cast iron	64–181	25–71	48–137	0.28
Ferrite steel	108–212	42–83	82–161	0.28
Stainless steel	200	80	167	0.3
Copper	125	46	139	0.35
Tungsten	407	158	323	0.29
Lead	19	7	53	0.44
Fused silica	75	32	38	0.17
Water ice (−4 °C)	10	3.6	9	0.33

and choosing the minimum of $E_\mathrm{p}(r)$ as $E_\mathrm{p}(r_0) = 0$, the first two members of the Taylor series (6.3a) vanish because also $\partial E_\mathrm{p}/\partial r|_{r=0} = 0$. This reduces (6.3a) to

$$\begin{aligned} E_\mathrm{p}(r) &= \frac{1}{2}(r-r_0)^2 \left(\frac{\partial^2 E_\mathrm{p}}{\partial r^2}\right)_{r=r_0} \\ &\quad + \frac{1}{6}(r-r_0)^3 \left(\frac{\partial^3 E_\mathrm{p}}{\partial r^3}\right)_{r=r_0} + \dots \,. \end{aligned} \tag{6.3b}$$

For small elongations $(r - r_0)$ all higher order terms with powers $n \geq 3$ can be neglected and (6.3b) gives for the force $\boldsymbol{F} = -\mathbf{grad}\, E_\mathrm{p}$ the linear relation of Hooke's law. For larger elongations, however, the higher order terms can no longer be neglected and must be taken into account.

Surpassing the linear range at the point P in Fig. 6.7 the relative stretch ε increases faster than the tensile stress σ (Fig. 6.7). The material is still elastic until the point F, i.e. it returns nearly to its initial length after the stress is released. Above the yield point F, internal shifts of atomic layers (lattice planes) occur (Fig. 6.8). The body becomes malleable and the plastic flow starts. Permanent changes of shape remain after the termination of the external force. While for the elastic stretch the distances r between the atoms increases linearly by $\Delta r \approx (\Delta L/L) r_0$ the

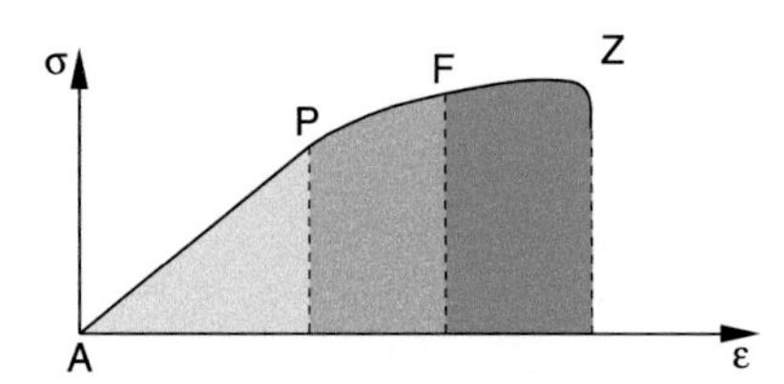

Figure 6.7 Relative length change ε of a body caused by an external tensile force σ. Beyond the point P the linear elongation changes to a nonlinear one. The point F marks the yield point, the point Z the tear point

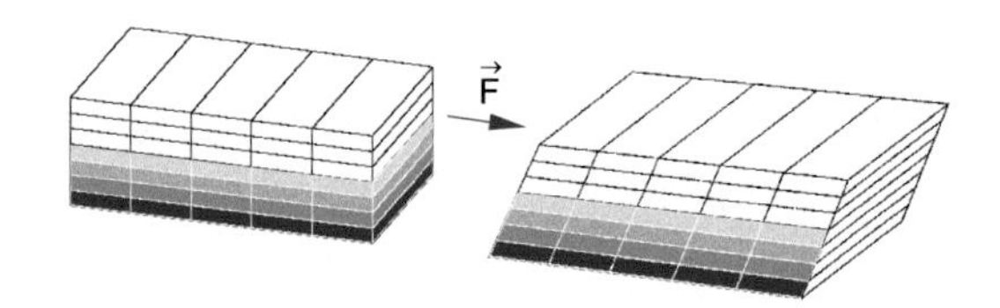

Figure 6.8 Model of plastic flow of a solid explained by shift of atomic planes

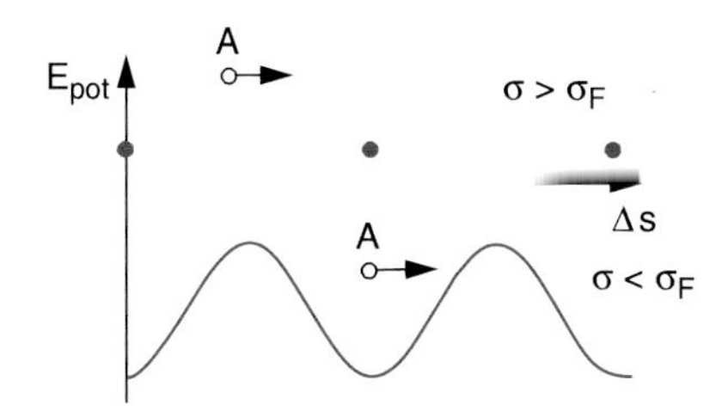

Figure 6.9 Atomic model of elastic expansion of a solid. The atom *A* moves in the potential well but does not leave it

flow process can be achieved by a shift of the atomic planes against each other, as illustrated in Fig. 6.8.

This can be made clear by Fig. 6.9 which shows the potential energy of an atom A in an atomic plane. For small length changes all atoms remain within their potential well. For larger changes the atomic plane is shifted against the adjacent plane. The atom can move from one minimum into the next only if the external pulling force is sufficiently large to lift all atoms of this plane over the potential hills. Since such a shift changes the distance between atoms only slightly, the minima in Fig. 6.9 are much shallower than the minimum of the potential energy between two atoms in Fig. 6.1. The barrier height and the modulation period of $E_p(\Delta L)$ depends for an anisotropic crystal on the direction of the pulling force relative to the crystal axes.

In a real crystal lattice, defects and dislocations are present which influence the flow process and can increase or decrease the renitence against stretches and shifts of crystal planes against each other.

6.2.2 Transverse Contraction

When a rod is stretched by an external pulling force, not only the length L in the direction of the force is prolonged but also the cross section decreases (Fig. 6.10). For a rod with length L and quadratic cross section d^2 the change ΔV of its volume V under a length stretch $\Delta L > 0$ and $\Delta d < 0$ is

$$\begin{aligned}\Delta V &= (d+\Delta d)^2\cdot(L+\Delta L)-d^2L\\ &= d^2\Delta L + 2L\cdot d\Delta d\\ &\quad + \left(L\Delta d^2 + 2d\Delta d\Delta L + \Delta L\Delta d^2\right) .\end{aligned}$$

For small deformations ($\Delta L \ll L$ and $\Delta d \ll d$), the terms in the bracket can be neglected, because they converge quadratic or even cubic towards zero for $\Delta L \to 0$. This reduces the above equation to

$$\frac{\Delta V}{V} \approx \frac{\Delta L}{L} + 2\frac{\Delta d}{d} . \tag{6.4}$$

The quantity

$$\mu \stackrel{\text{Def}}{=} -\frac{\Delta d}{d}\Big/\frac{\Delta L}{L} \tag{6.5}$$

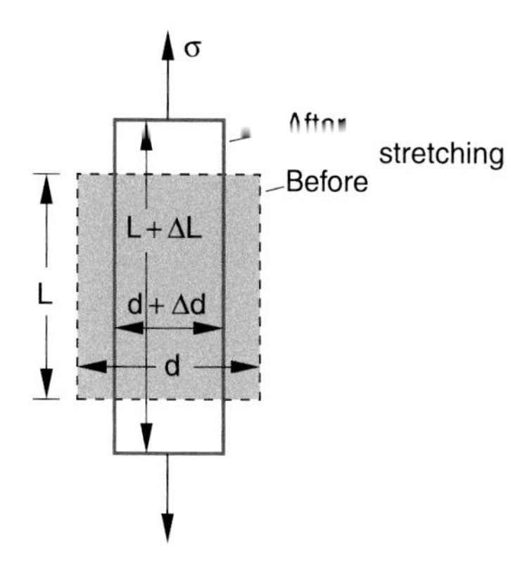

Figure 6.10 Transverse contraction under the influence of longitudinal tensile stress

is called the *transverse contraction ratio* because it is the ratio of transverse contraction to longitudinal elongation. The relative volume change is then expressed as

$$\frac{\Delta V}{V} = \frac{\Delta L}{L}\left(1+\frac{2\Delta d/d}{\Delta L/L}\right) = \varepsilon\,(1-2\mu) . \tag{6.6a}$$

Since a pulling force increases the volume ($\Delta V > 0$), we obtain for μ the condition $\mu < 0.5$. According to Hooke's law (6.2a) is $\Delta L/L = \sigma/E$. Inserting this into (6.6a) gives

$$\frac{\Delta V}{V} = \frac{\sigma}{E}(1-2\mu) . \tag{6.6b}$$

If a pressure instead of a tensile stress is exerted onto the end faces of a rod, ΔL and ΔV become negative but Δd positive because the rod is compressed in the length direction which causes an increase of its cross section. The resulting relative volume change can be obtained from (6.6b) when σ is replaced by the pressure p.

In both cases is $\mu > 0$ because for the pulling force is $\Delta L > 0$ and $\Delta d < 0$ while in case of a pressure is $\Delta L < 0$ and $\Delta d > 0$, which means that the ratio in the bracket in (6.6a) does not change its sign.

If the body is exerted to an isotropic pressure $p = -\sigma$, which acts onto all sides of the body, the resulting volume change can be obtained by the following consideration.

The pressure acting on the end faces d^2 decreases the length L by $\Delta L = -L\cdot p/E$, the pressure acting on the sides decreases the transverse edge length by $\Delta d = -d\cdot p/E$. However, because of the transverse action on the elongation this transverse contraction increases the length by $\Delta L = +\mu\cdot L\cdot p/E$. Taking both effects in account the length L under the action of a uniform pressure p changes by

$$\Delta L = -(L\cdot p/E)(1-2\mu) . \tag{6.7}$$

In a similar way the transverse dimension d is changed by

$$\Delta d = -(d\cdot p/E)(1-2\mu) .$$

Since $\Delta L \ll L$ and $\Delta d \ll d$ the higher order terms in the expansion of $\Delta V/V = \Delta L/L + 2\Delta d/d$ can be neglected and we obtain

$$\frac{\Delta V}{V} = \frac{\Delta L}{L} + \frac{2\Delta d}{d} = -\frac{3p}{E}(1-2\mu) . \tag{6.8}$$

Introducing the compressibility modulus K by the equation

$$p = -K \cdot \frac{\Delta V}{V} \tag{6.9}$$

and the coefficient of compressibility $\kappa = 1/K$, Eq. 6.8 can be written as

$$\kappa = \frac{1}{K} = \frac{3}{E}(1-2\mu) . \tag{6.10}$$

This gives the relations between compressibility modulus K, coefficient of compressibility κ, elastic modulus E and transverse contraction number (Poisson number) μ.

6.2.3 Shearing and Torsion Module

A shear force $\boldsymbol{F}$ is a force, which acts on a body parallel to a plane surface A (Fig. 6.11). The shearing stress

$$\tau = \boldsymbol{F}/A$$

is the tangential shearing force $\boldsymbol{F}$ per unit surface area A. The result of the action of a shearing stress is a tilt of the axis of the cuboid in Fig. 6.11 by an angle α. For sufficiently small tilting angles α, the experiments prove that the tilting angle α is proportional to the applied shearing stress.

$$T = G \cdot \alpha . \tag{6.11}$$

The constant G is called *modulus of shear* (or modulus of torsion).

Since the restoring forces under deformations of an elastic body are due to interatomic forces, all elastic constants E, μ, K and G must be related to each other.

As can be proved [6.2] for isotropic bodies the following relation holds:

$$E/2G = 1 + \mu . \tag{6.12a}$$

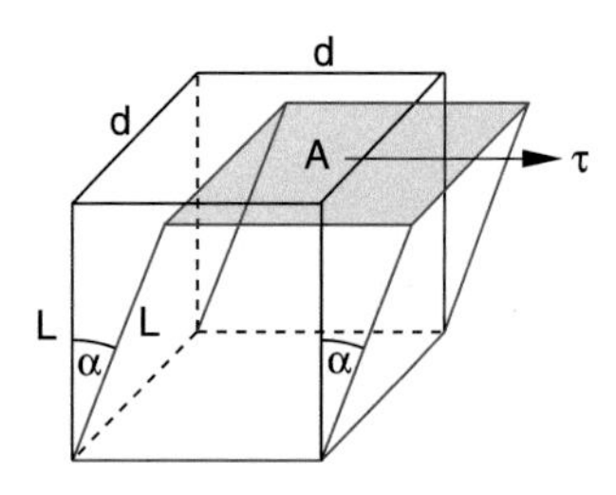

Figure 6.11 Shearing of a cube under the action of shearing stress τ

Rearrangement of (6.10) yields

$$E/3K = 1 - 2\mu . \tag{6.12b}$$

The division of (6.12a) by (6.12b) gives

$$2G/3K = \frac{1-2\mu}{1+\mu} . \tag{6.12c}$$

Example

Torsion of a wire: We assume a force $\boldsymbol{F}$ that acts tangential on a cylinder with radius R and length L and which causes a torsion of the cylinder (Fig. 6.12). We subdivide the cylinder in thin radial cylindrical shucks between the radii r and $r+\mathrm{d}r$ and in axial strips with the angular width $\delta\varphi$. If the upper end of the cylinder twists under the action of a torsional force $\boldsymbol{F}$ by the angle φ the prismatic column marked in red in Fig. 6.12 experiences a shear by the angle α. For $r \cdot \varphi \ll L$ one finds $\alpha = r \cdot \varphi/L$.

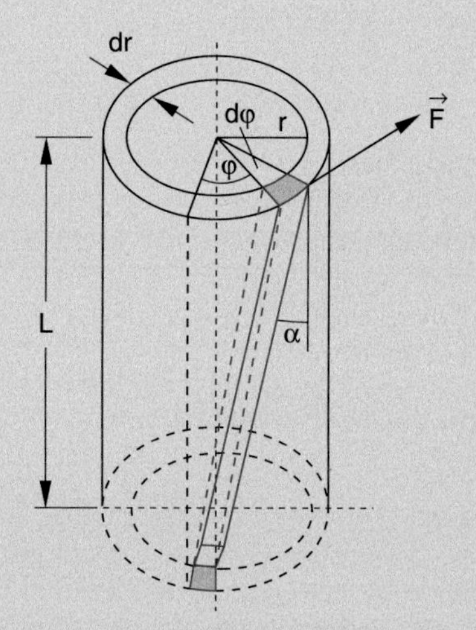

Figure 6.12 Torsion of a circular cylinder

The shearing stress necessary to achieve this torsion is according to (6.11)

$$\tau = G \cdot r \cdot \varphi/L .$$

Since all surface elements of the upper annulus with the area $2\pi r \cdot \mathrm{d}r$ are twisted by the same angle φ against their position for $\tau = 0$ the amount of the force $\mathrm{d}F$ necessary for the shear of the whole cylindrical shuck is

$$\mathrm{d}F = \tau \cdot 2\pi r \cdot \mathrm{d}r = 2\pi r^2 \cdot \mathrm{d}r \cdot \varphi \cdot g/L$$

and the corresponding torque is

$$\mathrm{d}D = r \cdot \mathrm{d}F = 2\pi \cdot r^3 \cdot \mathrm{d}r \cdot G \cdot \varphi/L .$$

The torsion of the whole cylinder with radius R by the angle φ is then accomplished by the torque

$$D = \frac{2\pi G\varphi}{L}\int_0^R r^3\mathrm{d}r = \frac{\pi}{2}G\frac{R^4}{L}\cdot\varphi\,. \qquad (6.13)$$

At equilibrium, the retro-driving torque, due to the elastic twist of the cylinder, must be equal to the external torque. This gives for the retro-driving torque

$$D^* = -D_\mathrm{r}\cdot\varphi \quad \text{with } D_\mathrm{r} = \frac{\pi}{2}G\frac{R^4}{L}\,. \qquad (6.14)$$

The constant D_r, which depends on the shear modulus G and gives the torque per unit angle, is called *restoring torque*.

If a body with the moment of inertia I with respect to the symmetry axis, is fixed to the lower end of a wire this torsional pendulum performs rotary oscillations after the wire has been twisted (see Sect. 5.6.2) with the oscillation period

$$T = 2\pi\sqrt{\frac{I}{D_\mathrm{r}}} = \frac{2\pi}{R^2}\sqrt{\frac{2L\cdot I}{\pi\cdot G}}\,. \qquad (6.15)$$

Such a torsional pendulum is a very sensitive device for measurements of small torques. Examples are the Eötvös's torsional pendulum for the measurement of Newton's gravitational constant (see Sect. 2.9.6), Coulomb's torsional pendulum for the measurement of the electric force between charges (Vol. 2, Chap. 1) and many modifications of Galvanometers for the measurement of small electric currents (Vol. 2, Chap. 2). ◄

6.2.4 Bending of a Balk

For technical constructions (buildings, bridges, etc.) the bending of balks under the influence of suspended weights represents an important problem and can decide about the stability of the construction. We will illustrate the problem with a simple example, where a rod with a rectangular cross section $A = d\cdot b$ is clamped at one end while a force acts on the other free end (Fig. 6.13). The calculation of such bending for arbitrary bodies is very complicated and can be accomplished only numerically.

If a rectangular rod with cross section $A = d\cdot b$ is clamped at $x = 0$ and a force F_0 is acting in the $-z$-direction on the other end at $x = L$, the bending of the rod can be approximately described by approximating a short curved section of the rod by a circle. When the central dashed curve in Fig. 6.14 has the radius of curvature r, the length of the upper edge of the rod section is $(r + d/2)\varphi$, that of the lower edge is $(r - d/2)\varphi$.

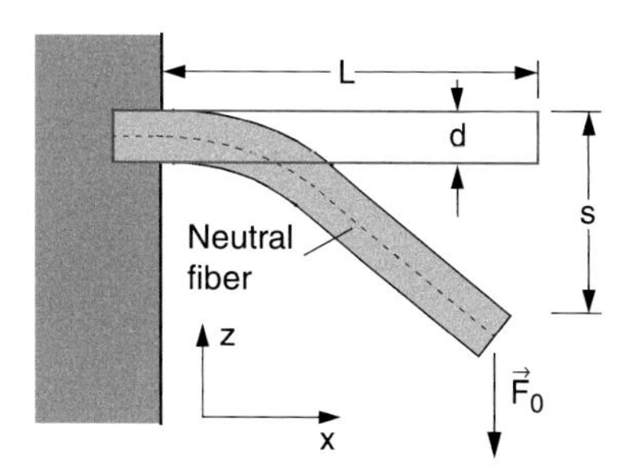

Figure 6.13 Bending of a rod which is clamped at one end

While the length of the central curve does not change by the bending (neutral filament) the length $\ell(z)$ of a layer in the upper half of the rod ($z > 0$) increases by the amount

$$\Delta\ell(z) = z\cdot\varphi = z\cdot\ell/r\,.$$

A corresponding layer in the lower half ($z < 0$) is shortened by this amount. In order to achieve such an increase of the length ℓ a pulling force per unit cross section (tensile stress)

$$\sigma = E\cdot\Delta\ell/\ell = z\cdot E/r$$

has to be applied, while for the layer in the lower half ($z < 0$) a corresponding pressure

$$p = -\sigma = -|z|\cdot E/r$$

is necessary. The force on a rectangular element with width b, heights $\mathrm{d}z$ and distance z from the neutral filament at $z = 0$ is then

$$\mathrm{d}F = \sigma b\mathrm{d}z = \frac{bE}{r}z\mathrm{d}z\,. \qquad (6.16a)$$

The force causes a torque

$$\mathrm{d}D_y = \frac{bE}{r}z^2\mathrm{d}z \qquad (6.16b)$$

in the y-direction.

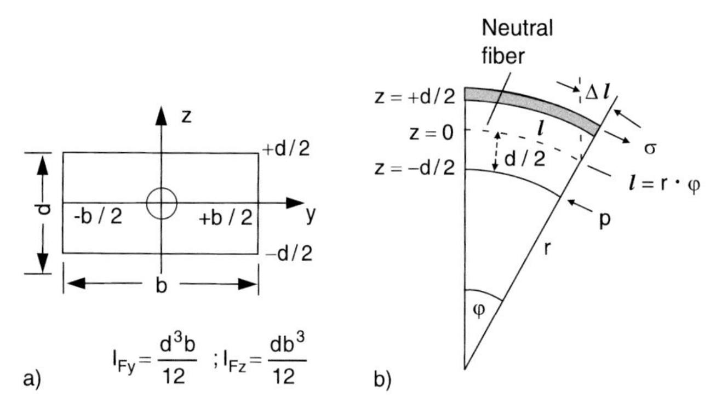

Figure 6.14 **a** Definition of the neutral filament at $z = 0$. **b** Illustration of (6.17)

Integration of this infinitesimal torque dD_y over the total height d of the rod gives

$$D_y = \frac{bE}{r} \int_{-d/2}^{+d/2} z^2 \mathrm{d}z = \frac{Ed^3 b}{12r} . \tag{6.17}$$

This torque is caused by the vertical force F_0 at the end $x = L$ of the rod. On the other hand the torque induced by the force F_0 on a selected part of the rod at the position x is

$$D_y = F_0(L - x) \quad \text{with} \quad F_0 = |\boldsymbol{F}_0| . \tag{6.18}$$

The equilibrium position of the bent rod is determined by the condition that the restoring torque of the elastic material (6.17) must just compensate the torque (6.18). This yields the curvature $1/r$ of the rod at the distance x from the fixed end at $x = 0$:

$$1/r = -\frac{12F_0}{Ed^3 b} \cdot (L - x) . \tag{6.19}$$

The neutral filament which is without a force the horizontal straight line $z = 0$, becomes the bent curve $z = z(x)$. As is shown in books on differential geometry the relation between the curvature $1/r$ and the function $z = z(x)$ is

$$1/r = \frac{z''(x)}{[1 + z'(x)^2]^{3/2}} ,$$

where $z'(x) = \mathrm{d}z/\mathrm{d}x$ and $z''(x) = \mathrm{d}^2z/\mathrm{d}x^2$. For small curvatures is $z'(x) \ll 1$ and therefore $1/r$ can be approximated by $1/r \approx z''(x)$. Integration of the equation

$$z''(x) = a \cdot (L - x) \quad \text{with } a = -12F_0/Ed^3 b$$

derived from (6.17) and (6.18), gives with the boundary conditions $z(0) = 0$ and $z'(0) = 0$ the function of the neutral filament of the strained rod

$$z(x) = \frac{a}{2} Lx^2 - \frac{a}{6} x^3 \quad \text{with } a < 0 .$$

The free end of the rod at $x = L$ bends by

$$s_{\max} = z(L) = -4 \frac{L^3}{E \cdot d^3 b} F_0 \tag{6.20}$$

compared to $z(L) = 0$ for the straight rod. The bend of the rod $s = z(L)$ is also called *pitch of deflection sag*.

The bend of a rectangular rod with length L and thickness d is proportional to L^3 and to $1/d^3$.

For $x = 0$ (at the clamped end of the rod) the curvature $1/r = z''(0) = a \cdot L$ becomes maximum. The tensile stress at the upper edge of the rod ($z = +d/2$) is

$$\sigma_{\max} = \frac{E \cdot d}{2r} = \frac{12F_0 \cdot L}{2d^2 b} . \tag{6.21}$$

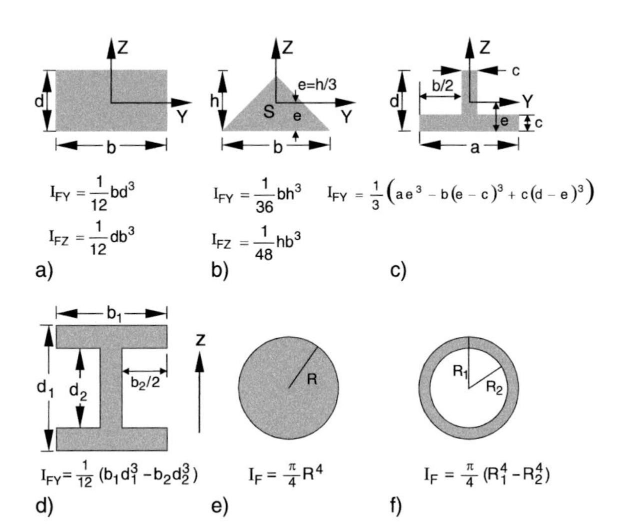

Figure 6.15 Geometrical moments of inertia for some selected cross sections. I_{Fy}: bending about the y-axis; I_{Fz}: about the z-axis

As soon as $\sigma_{\max}$ exceeds the fracture stress of the rod material, the rod starts to notch at the upper edge at $z = +d/2$ and $x = 0$ and the rod cracks.

Remark. The bend of rods with arbitrary cross section $A = \int \mathrm{d}y\mathrm{d}z$ can be treated in a similar way if one introduces the *geometrical moment of inertia* (*second moment of area*)

$$I_F \overset{\text{Def}}{=} \iint z^2 \mathrm{d}y\mathrm{d}z , \tag{6.22a}$$

where z is the direction of the acting force $\boldsymbol{F}$. For the rod with rectangular cross section $A = d \cdot b$ (Fig. 6.15b) we get

$$I_F = \int_{z=-d/2}^{+d/2} \int_{y=-b/2}^{+b/2} z^2 \mathrm{d}y\mathrm{d}z = \frac{1}{12} d^3 b . \tag{6.22b}$$

The maximum deflection $s_{\max}$ (pitch of deflection sag) is, in accordance with (6.20),

$$s_{\max} = -\frac{L^3}{3E \cdot I_F} F . \tag{6.23}$$

For a rod with circular cross section (radius R) (Fig. 6.15e) we get

$$I_F = \frac{1}{4} \pi R^4$$

and therefore

$$s_{\max} = -\frac{4L^3}{3\pi E R^4} F . \tag{6.24}$$

For a double T-beam (Fig. 6.15d) is

$$I_F = \frac{1}{12} \left(b_1 d_1^3 - b_2 d_2^3\right) . \tag{6.25}$$

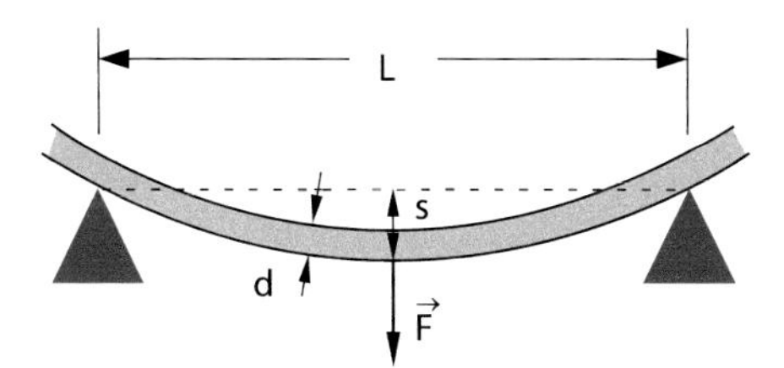

Figure 6.16 Bending of a rod, which is clamped at both ends

A beam with length L, supported on both ends by fixed bearings, suffers by a force F acting at the midpoint $x = L/2$ (Fig. 6.16) the maximum sag

$$s_{\max} = -\frac{1}{4E}\frac{L^3}{d^3 b} \cdot F\,. \tag{6.26}$$

Note, that here the sag is smaller by a factor 16! compared to the rod which is fixed only at one end (because of the L^3 dependence). The force is now distributed onto the two halves of the rod with $L/2$ each.

6.2.5 Elastic Hysteresis; Energy of Deformation

When a rod without deformation is exposed to an external tensile stress σ between the end faces the relative stretch $\varepsilon = \Delta L/L$ follows the curve OA in Fig. 6.17. For small values of ε the curve $\sigma(\varepsilon)$ is linear until a point is reached where the deformation is no longer reversible and $\sigma(\varepsilon)$ rises slower than linear. The point A in Fig. 6.17 is already in the irreversible region. This means that the curve $\sigma(\varepsilon)$ does not return on the same curve when the stress is released but arrives at the point B for $\sigma = 0$. This phenomenon is called *elastic hysteresis*, because the stress-free state of the body depends on its past history (the Greek word *hysteresis* means: lagging behind i.e. the length change lags behind the applied stress).

When the body in the state B is exposed to an external pressure $p = -\sigma$ onto the two end faces the curve $\sigma(\varepsilon)$ reaches the point C where it is also nonlinear. Releasing the pressure ε does not become zero for $\sigma = 0$ but arrives at the point D in Fig. 6.17, which corresponds in the atomic model of Fig. 6.1 to an interatomic distance $r < r_0$.

Under a periodic change between stretch and compression the function $\sigma(\varepsilon)$ passes through the closed loop ABCDA, which is called the elastic hysteresis loop. During a roundtrip one has to expend work against the interatomic forces because the interatomic distances r are periodically increased (stress) and decreased (compression). When the length L of a quadratic rod

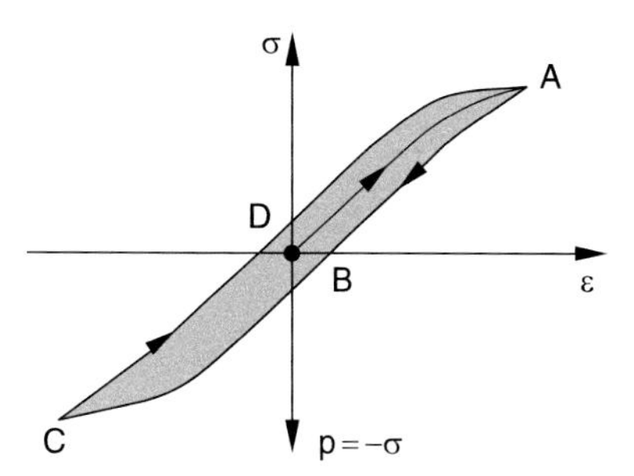

Figure 6.17 Mechanical hysteresis curve

with cross section A increases by ΔL the necessary work is

$$\begin{aligned} W &= \int_0^{\Delta L} F \mathrm{d}L = \int_0^{\Delta L} A \cdot \sigma \mathrm{d}L \\ &= \int_0^{\varepsilon} A\sigma \cdot L \mathrm{d}\varepsilon = V \cdot \int_0^{\varepsilon} \sigma \mathrm{d}\varepsilon\,. \end{aligned} \tag{6.27}$$

The integral $\int \sigma \cdot d\varepsilon$ represents the work per unit volume, necessary for the relative length change ε.

In the region where Hooke's law is valid (linear region of $\sigma(\varepsilon)$ is $\sigma = E \cdot \varepsilon$ and the work for the elastic length change ΔL

$$W_{\text{elast}} = \tfrac{1}{2} E \cdot V \cdot \varepsilon^2\,. \tag{6.28}$$

Returning to the original stress-free state this energy is again released. The hysteresis curve simplifies to a straight line through the origin $\sigma = \varepsilon = 0$.

Example

Elongation and compression of an elastic spiral spring during the oscillation of a mass m that hangs on the spring. During the oscillation with small amplitude with the linear region of Hooke's law, the potential energy of the spring and the kinetic energy of the mass m are periodically converted into each other (see Example 2 in Sect. 2.7.4 and Sect. 11.6). The total energy, however, is always conserved. ◀

This is no longer true for the nonlinear part of the curve $\sigma(\varepsilon)$ in Fig. 6.17. Here the work $\int \sigma \cdot \mathrm{d}\varepsilon$ has to be put into the system in order to proceed from the point O to the point A. This work is equal to the area under the curve OA. However, after releasing the tensile stress, only the work $\int \sigma \cdot \mathrm{d}\varepsilon$ that equals the area under the curve AB can be regained. The rest is converted into thermal energy, due to the non-elastic deformation of the body.

Altogether the net work per unit volume, put into the system during a roundtrip along the curve ABCDA, is given by the area enclosed by this hysteresis curve in Fig. 6.17.

Table 6.2 Hardness scale according to Mohs

Selected materials as measurement standards		Examples	
		Aluminium	2.3–2.9
1. Tallow	6. Feldspar	Lead	1.5
2. Gypsum	7. Quartz	Chromium	8
3. Calcite	8. Topaz	Iron	3.5–4.5
4. Fluorite	9. Corundum	Graphite	1
5. Apatite	10. Diamond	Tungsten	7

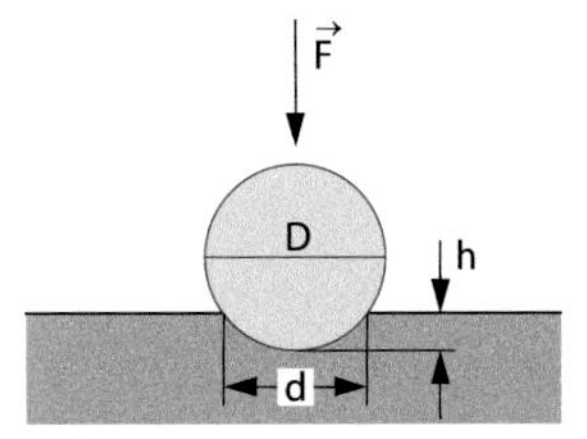

Figure 6.18 Hardness test according to *Brinell*

6.2.6 The Hardness of a Solid Body

The hardness of a body is a measure for the resistance, which the body sets against a penetration of another body. Depending on the measuring technique, there are some slightly different hardness values. The scratch-method introduced 1820 by Mohs, defines a body A as harder than a body B if it is possible to scratch B with A. The *hardness scale of Mohr* is based on this definition. Here the hardness scale is divided into 10 degrees of hardness based on 10 selected minerals, listed in Tab. 6.2.

The scratch method measures in fact mainly the hardness of the surface. This surface hardness is of particular importance for technical applications, because the attrition of tools or of axes and bearings depends on the surface hardness. Therefore, several techniques have been developed for the enhancement of the surface hardness. One example is the transformation of the surface layers of a crystalline solid into an amorphous state by irradiation with a powerful laser. Another example is the cover of solid tools, e.g. drills or steel mills, with a thin layer of a hard material such as carbon-nitride NC or titanium Ti.

For measuring the hardness of a body, often a technique is used which had been proposed by *Brinell* in 1900. Here a hardened steel ball with diameter D is pressed vertically with a constant force $F = a \cdot D^2$ into the sample (Fig. 6.18). The diameter d of the resulting circular notch in the sample gives the penetration depth, which is a measure for the *Brinell-hardness*.

6.3 Static Liquids; Hydrostatics

In order to achieve a change of the shape of solid bodies substantial forces are required, even if the volume of the body does not change (for example for a shear or a torsion). Although similarly large forces are necessary for a compression of liquids, a mere deformation of liquids at constant volume requires only very small forces and are merely caused by friction or surface effects.

At first we will discuss the simplified model of an ideal liquid, where surface effects and friction are neglected. For the static case of a liquid at rest, friction does not occur anyway. Surface effects will be treated in Sect. 6.4 and the influence of friction for streaming liquids is discussed in Chap. 8.

6.3.1 Free Displacement and Surfaces of Liquids

For ideal liquids without friction, there is no force necessary to deform a given liquid volume. In the atomic model this means: While in solid bodies the atoms can vibrate around fixed equilibrium positions, which do not change much under moderate external forces, the atoms or molecules in liquids can freely move around within the given liquid volume, determined by the solid container (Fig. 6.5). In the macroscopic model this free movement can be expressed by the statement:

The shear modulus of an ideal liquid is zero.

This implies that at the surface of an ideal liquid no tangential forces can be present, because they would immediately deform the liquid until the forces disappear and a minimum energy is achieved. This force-free condition represents a stable state of the liquid.

The surface of an ideal liquid is always perpendicular to the total external force.

Examples

1. If only gravity acts onto a liquid, the surface of the liquid forms always a horizontal plane (Fig. 6.19a)
2. The surface of a liquid in a cylinder which rotates about a vertical axis (Fig. 6.19b) forms a surface where the total force composed of gravity $m \cdot g$ in the $-z$-direction and centrifugal force $m\omega^2 \cdot r$ in the radial direction points perpendicular to the surface. The slope of the intersection curve $z(r)$ of the surface at the point A in Fig. 6.19b is

$$\tan\alpha = \frac{m\omega^2 r}{m \cdot g} = \frac{\omega^2 r}{g} .$$

On the other hand the slope of the curve $z(r)$ is given by $\tan\alpha = \mathrm{d}z/\mathrm{d}r$. Integration yields

$$z(r) = \frac{\omega^2}{g} \int r \mathrm{d}r = \frac{\omega^2}{2g} r^2 + C .$$

Chapter 6

For $z(0) = z_0$ is $C = z_0$ and we get

$$z(r) = \frac{\omega^2}{2g} r^2 + z_0 . \quad (6.29)$$

The surface forms a rotational paraboloid with its axis coincident with the rotation axis.

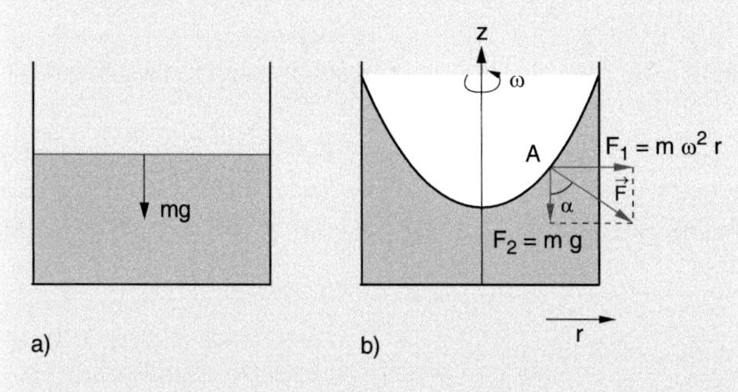

Figure 6.19 **a** Horizontal liquid surface in a container at rest. **b** Surface as rotational paraboloid in a rotating container

6.3.2 Static Pressure in a Liquid

Any external force acts only vertically on the surface of a liquid. If a container with a liquid is closed by a movable piston with surface A, onto which a vertical force F acts (Fig. 6.20) we define the pressure onto the liquid as

$$p = F/A, \quad \text{with} \quad F = |\boldsymbol{F}| .$$

6.3.2.1 Forces onto a Liquid Volume Element

We consider an arbitrary cuboid volume element $\mathrm{d}V = \mathrm{d}x{\cdot}\mathrm{d}y{\cdot}\mathrm{d}z$ inside the liquid (Fig. 6.21). We assume that a pressure p acts in x-direction onto the left side $\mathrm{d}y \cdot \mathrm{d}z$ of the cuboid. Then a pressure

$$p + \partial p/\partial x \cdot \mathrm{d}x$$

acts onto the opposite side. The resulting force on the volume element is then

$$F_x = p \cdot \mathrm{d}y\mathrm{d}z - \left(p + \frac{\partial p}{\partial x}\mathrm{d}x\right)\mathrm{d}y\mathrm{d}z = -\frac{\partial p}{\partial x}\mathrm{d}V .$$

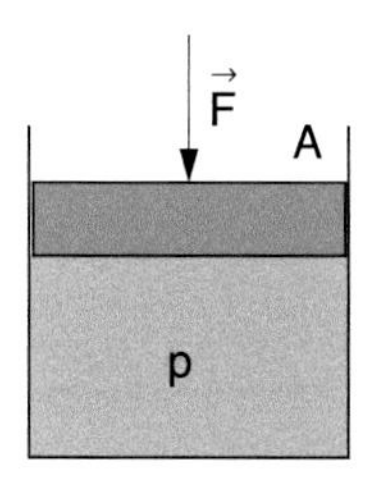

Figure 6.20 The force F, acting on a piston with area A generates a pressure $p = F/A$ in the liquid

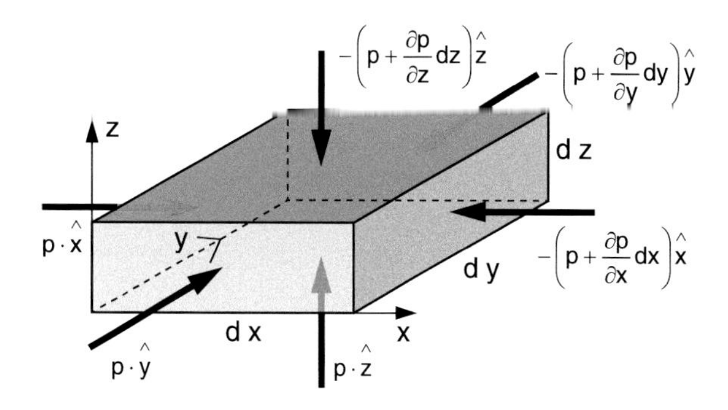

Figure 6.21 Relation between the pressure inside a volume element $\mathrm{d}V$ and the forces acting onto the sides of $\mathrm{d}V$

In an analogous way we obtain the force components in the other directions

$$F_y = -\frac{\partial p}{\partial y}\mathrm{d}V \quad \text{and} \quad F_z = -\frac{\partial p}{\partial z}\mathrm{d}V .$$

We can condense these three equations into the vector equation

$$\boldsymbol{F} = -\mathbf{grad}\, p \cdot \mathrm{d}V . \quad (6.30)$$

Because of the free mobility of any volume element inside the liquid the total force onto a volume element at rest must be zero. This implies that $\mathbf{grad}\, p = \mathbf{0}$.

The pressure inside the whole liquid is constant as long as no unisotropic forces act onto the liquid.

For a static liquid the same pressure acts onto all surface elements of the container!

This can be experimentally demonstrated by the simple device shown in Fig. 6.22. A spherical container with small holes in several directions of the x-y-plane is filled with dyed water and placed above a blotting paper in the plane $z = 0$. When a piston

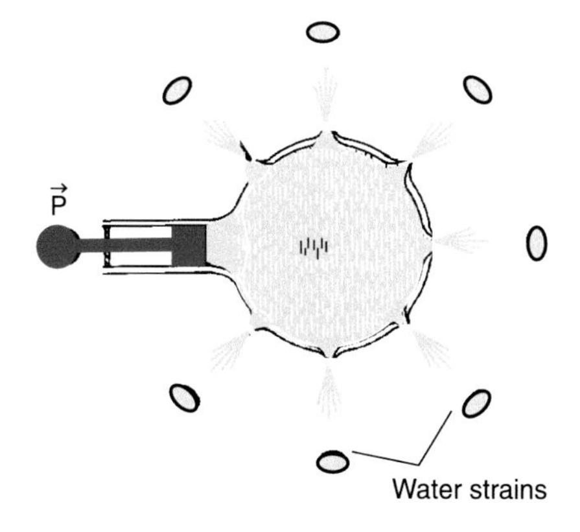

Figure 6.22 Demonstration of the isotropic pressure in a liquid. When the piston is moved the pressure p increases and the dyed water splashes through holes onto a white paper below the device, where the spots form a circle around the center, indication equal pressure

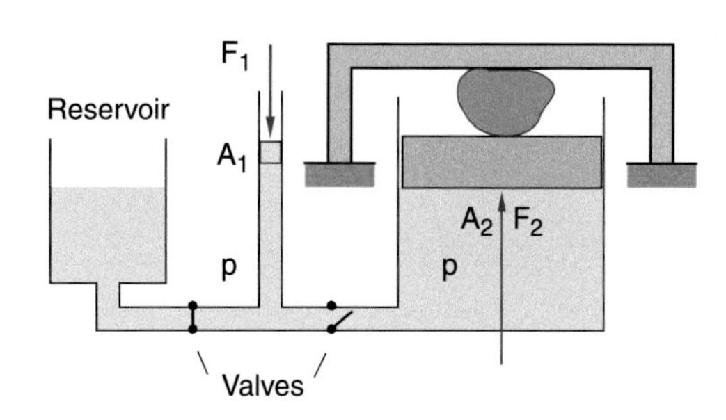

Figure 6.23 Hydraulic press (forces are drawn not to scale)

is pressed to increase the pressure the water streams out of the holes and all water filaments follow projectile trajectories. Their points of impact on the blotting paper in the plane $z = 0$ form a circle which proves that they all had the same initial velocity, i.e. they stream out driven by the same pressure.

Application: Hydraulic press (Fig. 6.23)
Two cylinders with cross sections A_1 and $A_2 \gg A_1$ that are connected with each other, and are therefore at the same pressure p, are filled with a liquid. Applying the force $F_1 = p \cdot A_1$ on a piston in the narrow cylinder causes a force $F_2 = pA_2 = (A_2/A_1) \cdot F_1 \gg F_1$ acting on a piston in the large cylinder which presses a sample against a fixed mounting. For demonstration experiments, large stones can be cracked by this device. For a displacement Δx_2 of the large piston, the small piston has to move by the much larger amount $\Delta x_1 = (A_2/A_1) \cdot \Delta x_1$, because the volume $\Delta V_2 = A_2 \cdot \Delta x_2 = \Delta V_1 = A_1 \cdot \Delta x_1$ transferred from the small to the large cylinder must be of course equal.

6.3.2.2 Hydrostatic Pressure

Taking into account that every volume element ΔV of a liquid has a weight $\varrho \cdot g \cdot \Delta V$ in the gravity field of the earth, even without external force a pressure onto the bottom of the container is present due to the weight of the liquid above the bottom. For a height $z = H$ of the liquid the hydrostatic pressure at the bottom with area A is with $\mathrm{d}V = A \cdot \mathrm{d}z$

$$p(z = 0) = \int_0^H \frac{\varrho \cdot g \cdot A}{A} \mathrm{d}z = \varrho \cdot g \cdot H \,, \qquad (6.31)$$

if we assume that the density ϱ is independent of the pressure p.

For real liquids, there is a small change of ϱ with the pressure p. A measure for this dependence is the compressibility

$$\kappa \stackrel{\text{Def}}{=} -\frac{1}{V}\frac{\partial V}{\partial p} \,, \qquad (6.32)$$

which describes the relative volume change $\Delta V/V$ for a change Δp of the pressure.

For liquids κ is very small (for example for water is $\kappa = 5 \cdot 10^{-10}\,\mathrm{m^2/N}$). This shows that the density ϱ of a liquid changes only by a tiny amount with pressure and in most cases the density $\varrho(p) = \varrho_0$ can be assumed to be constant.

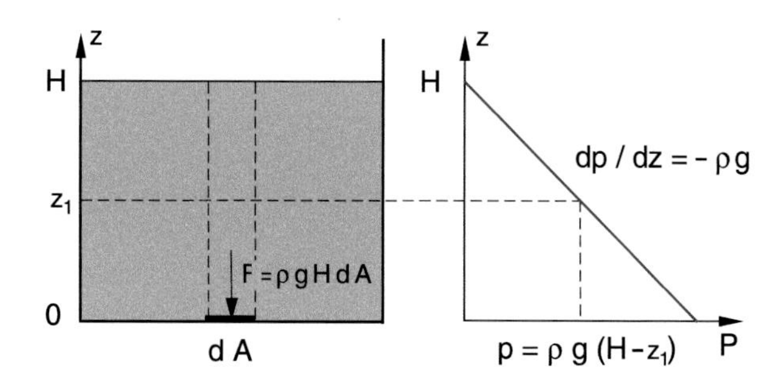

Figure 6.24 Pressure $p(z)$ in an incompressible liquid in the gravity field of the earth, as a function of height z above ground

Then it follows from (6.31) for the pressure $p(z)$ in a liquid with total heights H (Fig. 6.24)

$$p(z) = \varrho \cdot g \cdot (H - z) \,.$$

The SI unit for the pressure is 1 Pascal = 1 Pa = 1 N/m², which corresponds to 10^{-5} bar.

Examples

1. A water column of 10 m heights causes a hydrostatic pressure of $p = \varrho \cdot g \cdot h = 9.81 \cdot 10^4\,\mathrm{Pa} = 0.981\,\mathrm{bar} =$ 1 atmosphere. At an ocean depth of 10.000 m (Philippine rift) the hydrostatic pressure is $\approx 10^8\,\mathrm{Pa}$ (about 1000 atm). The total force onto the outer surface of a hollow steel sphere of an aquanaut with 3 m diameter is at this depth $F = 2.8 \cdot 10^9\,\mathrm{N}$.

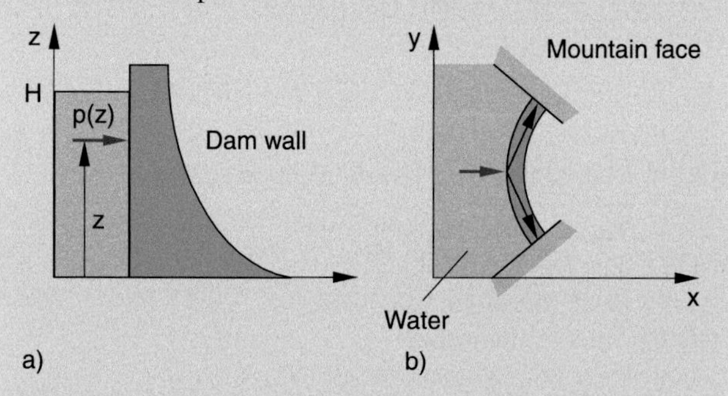

Figure 6.25 **a** Water pressure acting onto a dam wall; **b** Additional support by the mountain walls for a curved dam wall

2. The total force F onto the river dam with length L caused by the water with heights H can be obtained by integration over all contributions $F(z)\mathrm{d}z$ onto the surface elements $L \cdot \mathrm{d}z$ of the dam.

$$F = L \int p(z)\mathrm{d}z = \varrho \cdot g \cdot L \int (H - z)\mathrm{d}z$$
$$= \tfrac{1}{2}\varrho \cdot g \cdot L \cdot H^2$$

This force can be partly supported by choosing a curved dam where part of the force are balanced by the mountain walls (Fig. 6.25b). The thickness of the dam decreases with z in order to take into account the decreasing hydrostatic pressure (Fig. 6.25a). ◄

Figure 6.26 River dam of the river Eder, Germany. The bending of the dam towards the water side conducts part of the waterpressure against the mountain sides (see Sect. 6.3). With kind permission of Cramers Kunstverlag, Dortmund

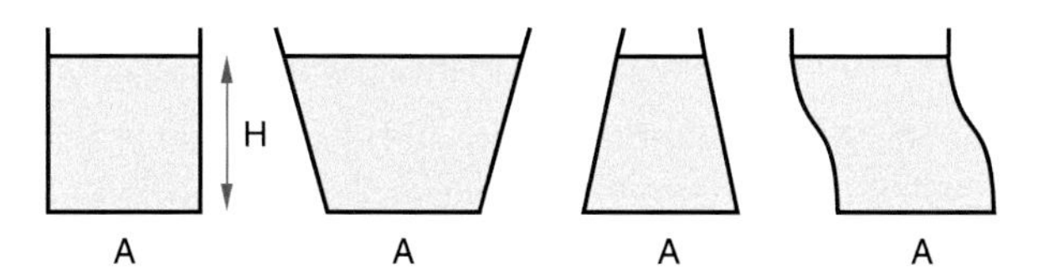

Figure 6.27 Hydrostatic paradoxon. The pressure onto the bottom is equal for all containers filled up to the same height H

Equation 6.31 tells us that the pressure p at the upper surface of a liquid volume element $\Delta V = A \cdot \Delta h$ with height Δh is smaller than at the bottom of this element by the amount $\varrho \cdot g \cdot \Delta h$. This results in an upwards force $F = A \cdot \varrho \cdot g \cdot \Delta h$, which is just compensated by the weight $G = M \cdot g = \varrho \cdot g \cdot \Delta V = \varrho \cdot g \cdot A \cdot \Delta h$ of the volume element. The total force on an arbitrary volume element ΔV inside a homogeneous liquid in a homogeneous gravity force field is therefore zero.

Since the hydrostatic pressure at the bottom of a liquid container depends only on the height H of the liquid but not on the shape of the container, the pressure at the bottom is identical for all four containers shown in Fig. 6.27, although the total mass of the liquid and therefore also its weight is different. This *hydrostatic paradox* leads to the following astonishing but true fact: When a hollow cube with a volume 1 m^3 is filled completely with water, the hydrostatic pressure at the bottom is 0.1 bar. If now a thin tube with 1 cm^2 cross section but 10 m height is put through a small hole in the top wall of the cube and filled with water the pressure in the cube rises to 1 bar. Although the additional mass of water is only 10^{-3} of the water in the cube the pressure rises by a factor of 10.

6.3.3 Buoyancy and Floatage

If we immerse a cuboid with basic area A and volume $V = A \cdot \Delta h$ into a liquid with density ϱ_L the pressure difference between bottom and top surface is (Fig. 6.28)

$$\Delta p = \varrho_L \cdot g \cdot \Delta h\ .$$

This results in an upwards directed buoyancy force

$$F_B = \varrho_L \cdot g \cdot A \cdot \Delta h = -G_L\ ,$$

which is equal to the weight G_L of the liquid displaced by the body, but has the opposite direction.

This can be formulated as *Archimedes' Principle*:

> A body immersed in a liquid looses seemingly as much of its weight as the weight of the displaced liquid.

This principle illustrated for the example of a cuboid, is valid for any body with arbitrary shape as can be seen from the following consideration:

Due to the hydrostatic pressure $p = \varrho_{fl} \cdot g \cdot (H - z)$ at the height z in a liquid with total height H the force on a volume element $\mathrm{d}V$ is

$$\begin{aligned} \boldsymbol{F} &= -\mathbf{grad}\, p \cdot \mathrm{d}V = -(\partial p/\partial z)\hat{\boldsymbol{e}}_z \mathrm{d}V = \varrho_L \cdot g \cdot \mathrm{d}V \cdot \hat{\boldsymbol{e}}_z \\ &= -\varrho_L \cdot \boldsymbol{g} \cdot \mathrm{d}V\ . \end{aligned}$$

The buoyancy force on the whole body immersed in the liquid is then

$$\boldsymbol{F}_B = -\boldsymbol{g} \int \varrho_L \cdot \mathrm{d}V = -\boldsymbol{G}_L\ . \qquad (6.33)$$

Chapter 6

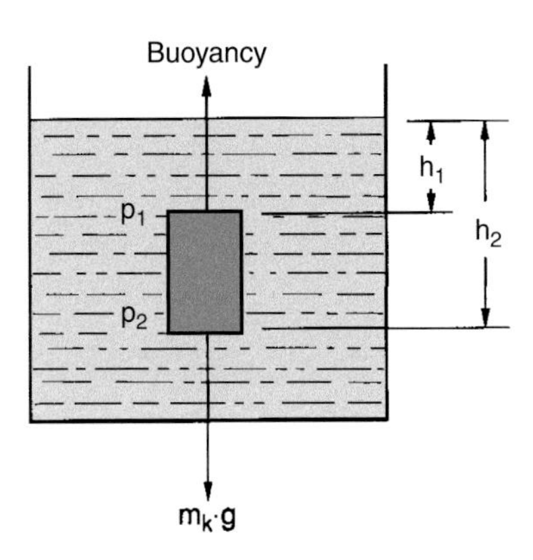

Figure 6.28 Axiom of Archimedes and buoyancy

If the density ϱ_b of a body is smaller than the density ϱ_L of the liquid, the buoyancy force becomes larger than the weight G_b of the body and the body floats on the surface of the liquid. Only part of the body immerses while the other part is above the liquid surface. Equilibrium is reached if the buoyancy (it is the weight G_L of the displaced liquid) just cancels the weight G_b of the total body.

Example

The density of ice is $\varrho_i = 0.95\,\text{kg/dm}^3$, the density of salty sea-water at 0 °C is $\varrho_L = 1.05\,\text{kg/dm}^3$. Therefore, about 10% of the volume of an iceberg stick out of the ocean surface, 90% are under water. ◄

Remark. Of course, the buoyancy is also present in gases. However, because of the much smaller density of gases the buoyancy force is correspondingly smaller. A body in a gas atmosphere loses (seemingly) as much of its weight as the weight of the displaced gas. This is the basis for balloon flights (see Sect. 7.2 and Fig. 7.6.

For the stability of a floating ship it is important that in case of heeling induced by waves there is always a restoring torque which brings the ship back into its vertical position. This stability criterion can be quantitatively formulated in the following way:

We consider the torque generated by the gravity force G_g and the buoyancy F_B for a ship in an oblique position (Fig. 6.29). The two forces form a couple of forces (Sect. 5.4) which cause a torque about the center of mass S_K. The point of origin for the gravity force G_g is the center of mass S_K of the ship, while the point of origin for the buoyancy $F_B = -G_g$ is the center of mass S_B of the displaced water. The symmetry plane of the ship, indicated in Fig. 6.29b by the dashed line, intersects the vertical direction of the buoyancy in the point M, called the *meta-center*. The vector $\boldsymbol{r}$ gives the distance between M and S_K. As long as M lies above S_K the resulting torque

$$\boldsymbol{D} = (\boldsymbol{r} \times \boldsymbol{G}_K) = -(\boldsymbol{r} \times \boldsymbol{F}_B) \,,$$

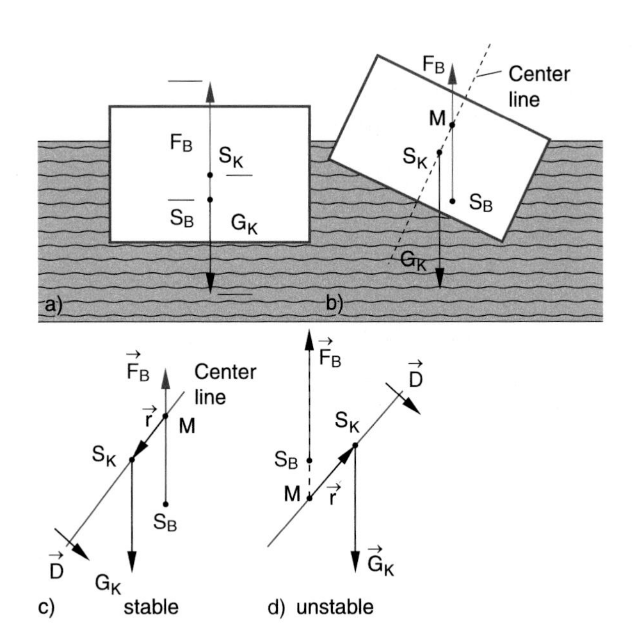

Figure 6.29 Stability of a floating body. **a** equilibrium position, **b** tilting below the critical angle, **c** vector diagram of stable and unstable heeling

which has in Fig. 6.29c the counter-clockwise direction, brings the ship back into the vertical stable position. If the slope becomes so large that M comes below S_K (Fig. 6.29d) the resulting torque acts into the clockwise direction and it brings the ship into a larger slope. It overturns and sinks. It is therefore advantageous for the stability to have the center of mass S_K as low as possible. This can be achieved by putting heavy masses at the bottom of the ship. In case of container ships, the cargo is loaded on top of the ship which decreases the stability. These ships have therefore a double mantel at the bottom where the interspace is filled with water, in order to bring the center of mass down.

6.4 Phenomena at Liquid Surfaces

We will now upgrade our simple model of the ideal liquid in order to introduce effects which occur at surfaces of real liquids and which are not present in ideal liquids. While inside a liquid the resulting time-averaged force on an arbitrary molecule, exerted by all other molecules, is zero, (this allows the free relocatability of each molecule), this is no longer true for molecules at the surface of liquids (Fig. 6.30) which are only attracted by molecules in a half sphere inside the liquid. Therefore a residual force $\boldsymbol{F}_R$ remains, which attracts the molecules towards the interior of the liquid.

6.4.1 Surface Tension

If a molecule is brought from the inside of a liquid to the surface, energy has to be supplied to move the molecule against the residual force F_R. A molecule at the surface has therefore a higher

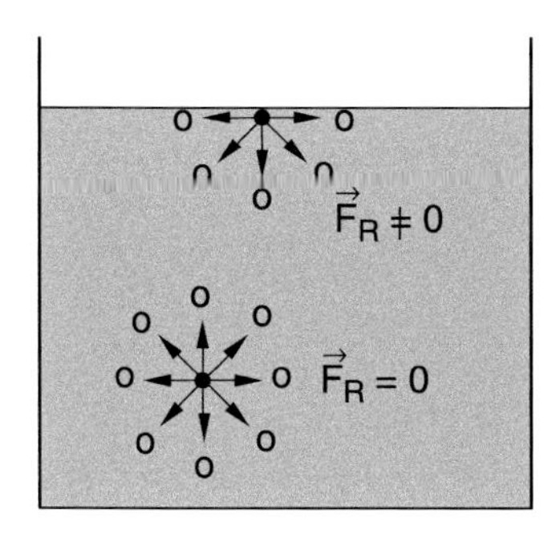

Figure 6.30 Resulting force on a molecule by all other surrounding molecules inside a liquid and at the surface of a liquid

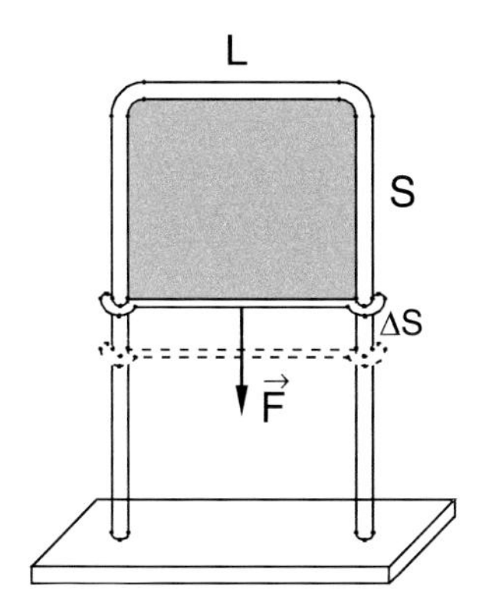

Figure 6.31 Determination of surface tension by measuring the force on a sliding straight wire, that extends a liquid skin

energy than a molecule inside the liquid. In order to enlarge the surface by an amount ΔA molecules have to be transferred to the surface which needs the energy ΔW. The ratio

$$\varepsilon = \frac{\Delta W}{\Delta A}\,; \quad [\varepsilon] = \frac{\mathrm{J}}{\mathrm{m}^2} \tag{6.34}$$

is the *specific surface energy*. The value of ε depends on the binding forces between the molecules of the liquid. It can be measured with the equipment shown in Fig. 6.31. Between the two sides of a U-shaped frame a horizontal wire with length L can be shifted vertically. When the system is dipped into a liquid, a liquid lamella is formed with the surface area (on both sides) $A = 2L \cdot s$. For moving the horizontal wire by Δs, the force F is necessary. One has to supply the energy

$$\Delta W = \boldsymbol{F} \cdot \Delta s = \varepsilon \cdot \Delta A = \varepsilon \cdot 2 \cdot L \cdot \Delta s\,. \tag{6.35}$$

The restoring force $\boldsymbol{F}$, which is directed tangential to the surface of the lamella, produces a tensile strain $\sigma = F/2L$ per length unit which is called *surface tension*. According to (6.35) is

$$\sigma = \varepsilon\,.$$

Surface tension σ and specific surface energy ε are identical.

The surface tension can be impressively demonstrated by the apparatus shown in Fig. 6.32. A metal strip bent into a circle hangs

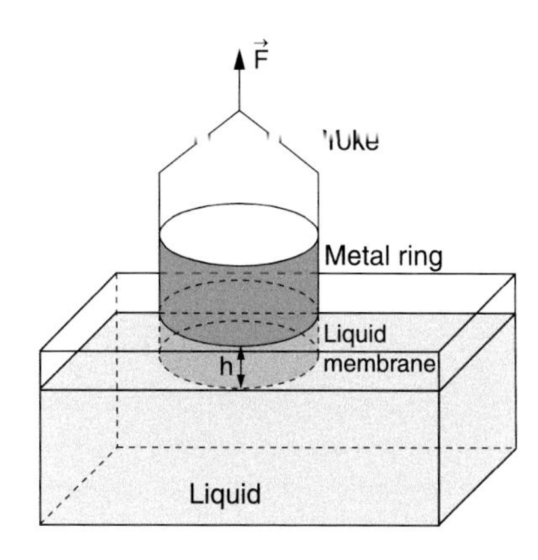

Figure 6.32 Measurement of surface tension by liftig an immersed metal ring

on a spring balance. It is immersed into a glass container filled with a liquid. When the container is lowered or the metal ring is uplifted, the lower rim of the ring emerges more and more out of the liquid, carrying a cylindrical liquid lamella. With soapy water more than 10 cm heights of the lamella can be reached. The spring balance measures the force

$$F = 4\pi \cdot r \cdot \sigma\,,$$

because the lamella has two surfaces, inside and outside of the cylindrical membrane. The work necessary to lift the lamella up to the height h is

$$W = 4\pi \cdot r \cdot \sigma \cdot h\,.$$

Example

Surface tension and pressure in a soap bubble (Fig. 6.33). Because of its surface tension, the bubble tries to reduce its surface. This increases the pressure inside the bubble. Equilibrium is reached, if the work against the increasing pressure during the decrease Δr of the bubble radius is equal to the work gained by the reduction ΔA of the surface area A

$$\varepsilon \cdot 2 \cdot 4\pi(r^2 - (r - \Delta r)^2) = 4\pi \cdot r^2 \cdot \Delta p\,.$$

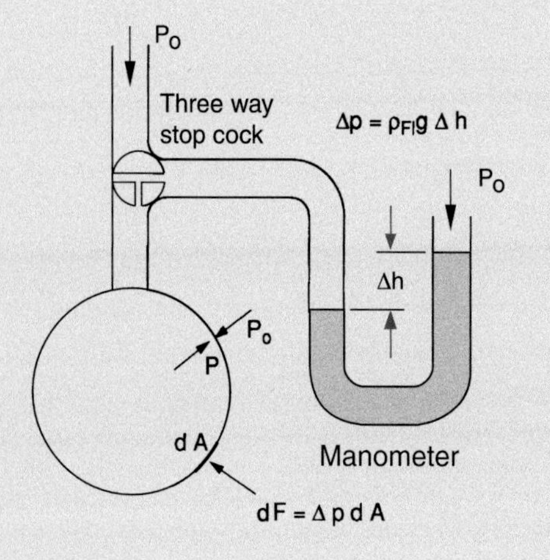

Figure 6.33 Measurement of overpressure Δp in a soap bubble, caused by surface tension

Neglecting the term with $(\Delta r)^2$ gives the excess pressure

$$\Delta p = 4\varepsilon/r , \qquad (6.36)$$

which shows that Δp decreases with increasing radius r. This can be demonstrated by the equipment in Fig. 6.34. The lower ends of the tubes 1 and 2 are immersed into soapy water and then lifted again. With open valves 1 and 2 but closed valve 3, two bubbles with different sizes can be produced by blowing air into the corresponding filling tubes. Now valves 1 and 2 are closed and valve 3 is opened. The smaller bubble starts to shrink and the larger one inflates. This continues until the smaller bubble completely disappears. It's like in daily life. The powerful people (larger ones) increase their power at the cost of the little guys.

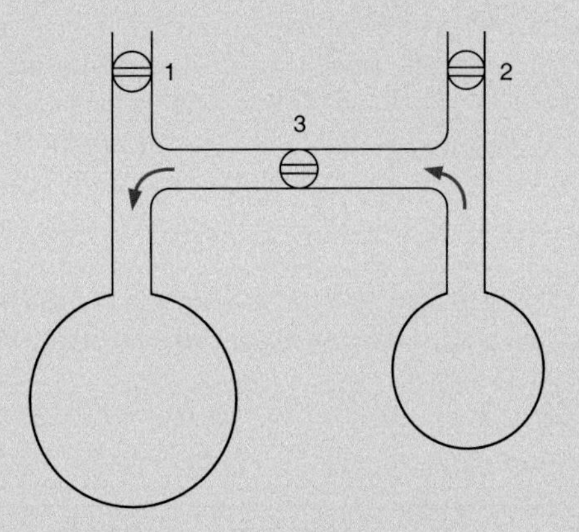

Figure 6.34 Demonstration of overpressure $\Delta p(r)$ which increases with decreasing radius r of a soap bubble

For liquids with positive surface energy each liquid with a given volume tries to minimize its surface area.

This can be demonstrated by adding drop wise mercury through a pipette into a bowl filled with diluted sulfur acid. At first many small mercury droplets are formed which, however, soon merge into a single larger drop. ◄

6.4.2 Interfaces and Adhesion Tension

Up to now, we have only discussed surfaces of liquids as boundaries between liquid and gaseous phases. Often interfaces between different liquids or between liquid and solid bodies can occur. Analogue to the surface tension we define the boundary tension σ_{ik} (identical with the specific interface energy ε_{ik}) as the energy that has to be spend (or is gained) when the interface between the phases i and k is increased by 1 m^2.

The sign of ε_{ik} can be obtained by the following considerations:

- For stable interfaces between liquid and gas ε_{ik} has to be positive. Otherwise the liquid phase would be transferred into the gas phase because energy would be gained, i.e. the liquid would vaporize.
- Also for stable interfaces between two different liquids ε_{ik} must be positive. Otherwise the two liquids would intermix and the interface would disappear.
- For the interface between liquid and solid phases the sign of ε_{ik} depends on the materials of the two phases. If the molecules M_L in the liquid are attracted more strongly by the molecules M_s in the solid, than by neighboring molecules in the liquid, is $\varepsilon_{ik} < 0$. If the attracting forces between molecules M_L are stronger than between M_L and M_s is $\varepsilon_{ik} > 0$.
- Also between a solid surface and a gas an interface energy can occur, because the gas molecules can be attracted by the solid surface (adhesion) or they can be repelled, depending on the gas and the solid material.

We will illustrate these points by some examples: In Fig. 6.35 is the surface of a liquid 2 against the gas phase 3 close to a vertical solid wall 1 depicted. Here the surface tensions $\sigma_{1,2}$; $\sigma_{1,3}$ and $\sigma_{2,3}$ tangential to the corresponding surfaces have to be considered. We regard a line element dl perpendicular to the plane of the drawing through the point A, where all three phases are in contact with each other. The force parallel to the solid surface is $F_{\|s} = (\sigma_{1,2} - \sigma_{1,3})\mathrm{d}l$ and $F_{\|l} = \sigma_{2,3}\mathrm{d}l$ is parallel to the liquid surface. The resulting force causes a change of the liquid

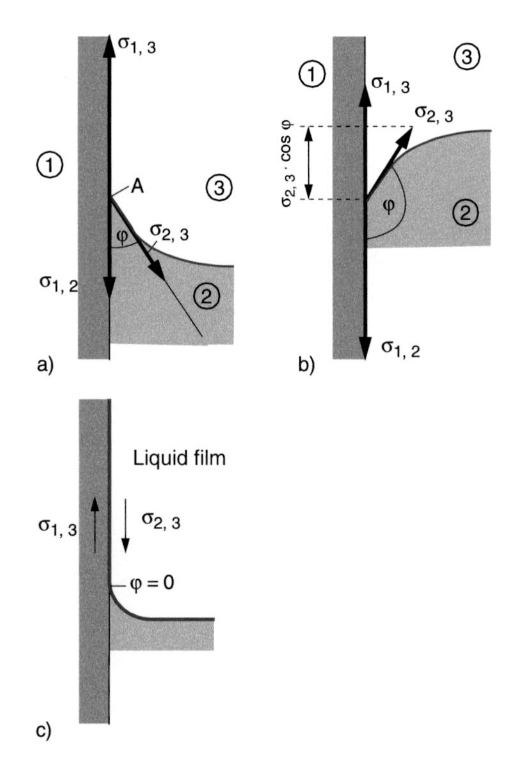

Figure 6.35 Formation of a contact angle of a liquid surface with a vertical solid wall. **a** Concave liquid surface for water-glass ($\sigma_{1,3} > \sigma_{1,2}$); **b** convex surface of Hg-glass ($\sigma_{1,3} < \sigma_{1,2}$); **c** complete wetting for $\sigma_{1,3} - \sigma_{1,2} > \sigma_{2,3}$

surface, which would be a horizontal plane under the action of gravity without surface tension.

If we neglect the small change of the gravitational force due to the change of the surface, which is very small compared to the forces caused by surface tension, we have the equilibrium condition that in the point A the vector sum of all forces must be zero. For the vertical component parallel to the solid wall this implies:

$$\sigma_{1,2} + \sigma_{2,3}\cos\varphi - \sigma_{1,3} = 0\,. \tag{6.37}$$

The horizontal component $\sigma_{2,3} \cdot \sin\varphi$ causes an imperceptibly small deformation of the solid wall. This induces a restoring deformation force which is opposite to the force $\sigma_{2,3}$ and has the same magnitude and therefore compensates it. The wetting angle φ can be obtained from the condition

$$\cos\varphi = \frac{\sigma_{1,3} - \sigma_{1,2}}{\sigma_{2,3}}\,. \tag{6.37a}$$

It has a definite value only for $|\sigma_{1,3} - \sigma_{1,2}| \leq \sigma_{2,3}$. We distinguish the following cases:

- $\sigma_{1,3} > \sigma_{1,2} \rightarrow \cos\varphi > 0 \rightarrow \varphi < 90°$.
  The liquid forms close to the solid wall a concave surface, which forms an acute angle φ with the wall (Fig. 6.35a). It is energetically favorable to increase the interface liquid-solid at the cost of the interface solid–gas.
  Example: Interfaces water–glas–air.
- $\sigma_{1,3} < \sigma_{1,2} \rightarrow \cos\varphi < 0 \rightarrow \varphi > 90°$.
  The liquid forms close to the solid wall a convex surface (Fig. 6.35b).
  Example: interfaces mercury–glas–air.
- For $|\sigma_{1,3} - \sigma_{1,2}| > \sigma_{2,3}$ Eq. 6.37 cannot been fulfilled for any angle φ. In this case a force component parallel to the solid surface is uncompensated. It pulls the liquid along the solid surface until the whole surface is covered by a liquid film (Fig. 6.35c). The interface solid–gas disappears completely.

If external forces are present, such as gravitational or inertial forces in accelerated systems, the vector sum of all forces is in general not zero. However, the liquid surface reacts always in such a way, that the resultant force is perpendicular to the liquid surface, i.e. its tangential component is always zero. This is illustrated in Fig. 6.36 for the cases of a concave and a convex curvature of the liquid surface close to the solid wall where besides the gravitational force also the attractive force F_4 between liquid and solid surfaces is taken into account.

For a liquid in a container the total force is compensated by the restoring elastic force of the container wall.

For two non-mixable liquids 1 and 2 (for example a fat drop on water) the angles φ_1 and φ_2 in Fig. 6.37 adjust in such a way that the equilibrium condition

$$\sigma_{1,3} = \sigma_{2,3}\cos\varphi_2 + \sigma_{1,2}\cos\varphi_1 \tag{6.38}$$

is fulfilled. This shows that a droplet of the liquid 2 can be only formed, if $\sigma_{1,3} < \sigma_{2,3} + \sigma_{1,2}$. Otherwise the droplet would be spread out by the surface tension $\sigma_{1,3}$ until it forms a thin film, which covers the surface of liquid 1.

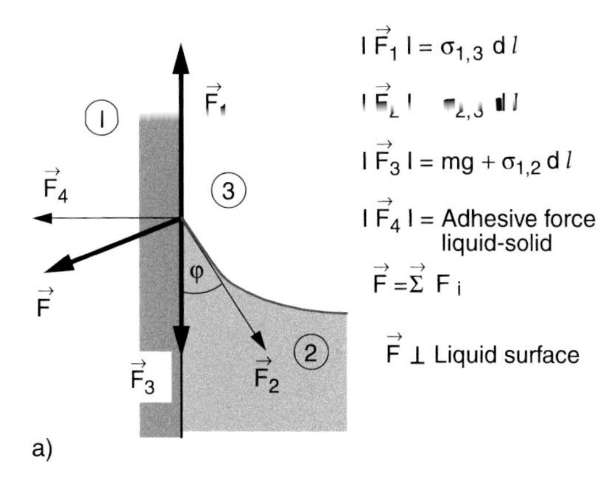

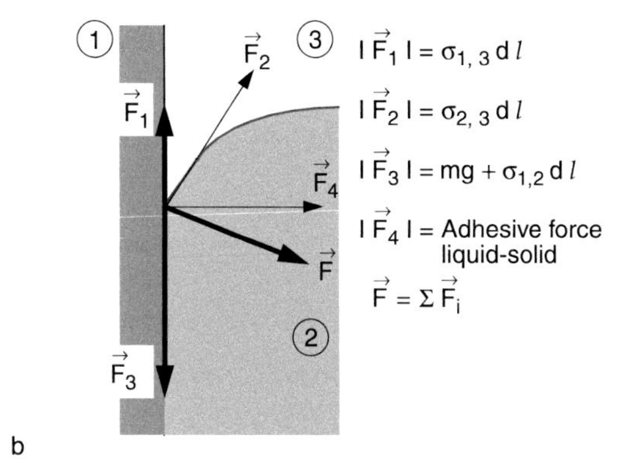

Figure 6.36 The vector sum of all forces acting onto a liquid sureface must be always vertical to the surface, for non-wetting liquids. **a** Concave, **b** convex curved surface

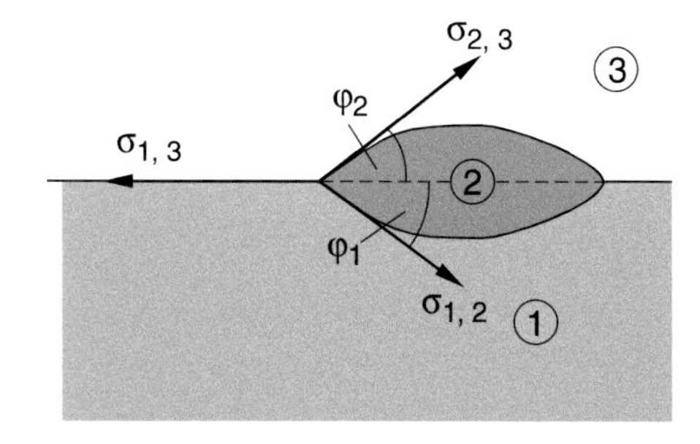

Figure 6.37 Formation of a liquid drop on the surface of another liquid

Example

For the interfaces water–oil–air the numerical values of the surface tensions are:

$$\sigma_{1,3}(\text{water-air}) = 72.5 \cdot 10^{-3}\,\text{J/m}^2$$
$$\sigma_{1,2}(\text{water-oil}) = 46 \cdot 10^{-3}\,\text{J/m}^2$$
$$\sigma_{2,3}(\text{oil-air}) = 32 \cdot 10^{-3}\,\text{J/m}^2\,.$$

This shows that $\sigma_{1,3} > \sigma_{2,3} + \sigma_{1,2}$. Therefore, oil cannot form droplets on a water surface. ◄

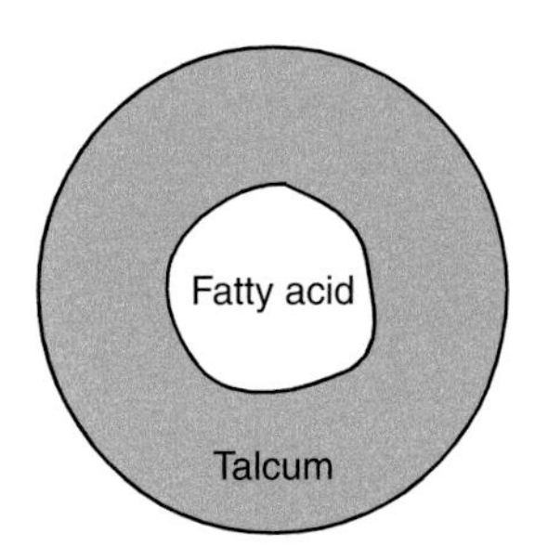

Figure 6.38 Formation of a mono-molecular layer of a fatty acid on a liquid surface covered with a talcum powder layer

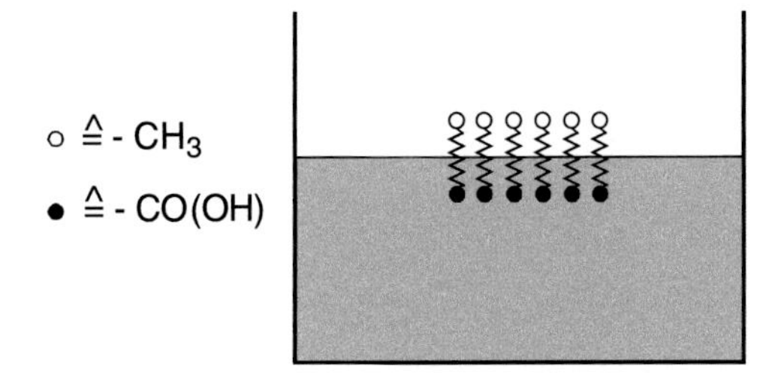

Figure 6.39 Fatty acid molecules forming a mono-molecular layer on a water surface are oriented due to the attractive for one end of the interaction with the water molecules and a repulsive interaction for the other end

If an oil drop is brought onto a water surface, it will spread out to form a mono-molecular layer of oil which covers the whole water surface if sufficient oil is contained in the drop. Otherwise, the oil film forms a cohesive insula of this mono-molecular film. This can be demonstrated by the following experiment (Fig. 6.38): Onto a water surface, powdered with talc, a droplet of fatty acid is supplied through a pipette. The droplet immediately spreads out and displaces the talc layer. The fatty acid molecules are oriented in such a way, that the attractive force with the water molecules becomes maximum (Fig. 6.39). The atomic groups COOH, which are directed against the water surface, are called *hydrophilic* while the groups on the opposite side of the molecule, which are repelled by the water molecules, are called *hydrophobic*. The interaction with the water molecules causes a displacement of the charges in the fatty acid molecules while the water molecules, which are electric dipoles, are orientated in such a way, that their positive pole is directed toward the negative pole of the induced dipole molecules of the fatty acid (see Vol. 2, Chap. 2).

6.4.3 Capillarity

When a capillary tube is dipped into a wetting liquid ($\sigma_{1,3} > \sigma_{1,2}$), the wetting liquid rises in the capillary tube up to the height h above the liquid surface (Fig. 6.40). This observation can be explained as follows: If a liquid column in the capillary with radius r is lifted up to the height h ($h \gg r$) the potential energy is increased by

$$\mathrm{d}E_\mathrm{p} = m \cdot g \cdot \mathrm{d}h = \pi \cdot r^2 g \cdot \varrho \cdot h \cdot \mathrm{d}h\,. \tag{6.39a}$$

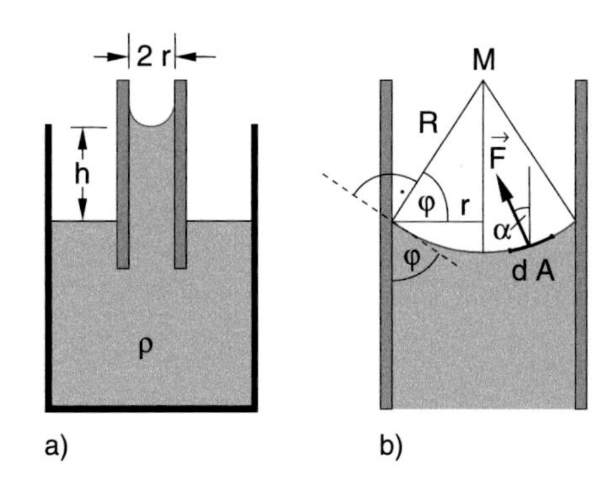

Figure 6.40 **a** Capillary rise of a wetting liquid, **b** derivation of the rise height

On the other hand, the surface energy changes by (see Fig. 6.36)

$$\begin{aligned}\mathrm{d}E_\mathrm{surface} &= -2\pi r \cdot \mathrm{d}h(\sigma_{13} - \sigma_{12}) \\ &= 2\pi r \cdot \mathrm{d}h \cdot \sigma_{23} \cdot \cos\varphi\,,\end{aligned} \tag{6.39b}$$

where Eq. 6.37a has been used. At equilibrium is $\mathrm{d}E_\mathrm{p} + \mathrm{d}E_\mathrm{surface} = 0$. This gives the resulting height

$$\begin{aligned}h &= 2\sigma_{23} \cdot \cos\varphi/(r \cdot g \cdot \varrho) \\ &= 2\sigma \cdot \cos\varphi/(r \cdot g \cdot \varrho)\,.\end{aligned} \tag{6.40}$$

The wetting angle φ is determined by Eq. 6.37). The surface tension $\sigma_{2,3} = \sigma$ is the surface tension of the liquid against air, introduced in Sect. 6.4.1.

For completely wetting liquids ($\sigma_{1,3} > \sigma_{1,2} + \sigma_{2,3}$) is $\varphi = 0$. The complete inner surface of the capillary tube is covered by a thin liquid film and the capillary rise becomes according to (6.40)

$$h = \frac{2\sigma}{rg\varrho}\,. \tag{6.40a}$$

For non-wetting liquids ($\sigma_{1,3} < \sigma_{1,2}$) the liquid surface inside the capillary is convex. This convex curvature causes a force, which is directed downwards and leads to a capillary depression (Fig. 6.41). The depression height $-h$ is again given by (6.40), where now $\cos\varphi = (\sigma_{1,3} - \sigma_{1,2})/\sigma_{2,3} < 0$.

The capillary rise offers an experimental method for the measurement of absolute values of surface tensions. Instead of

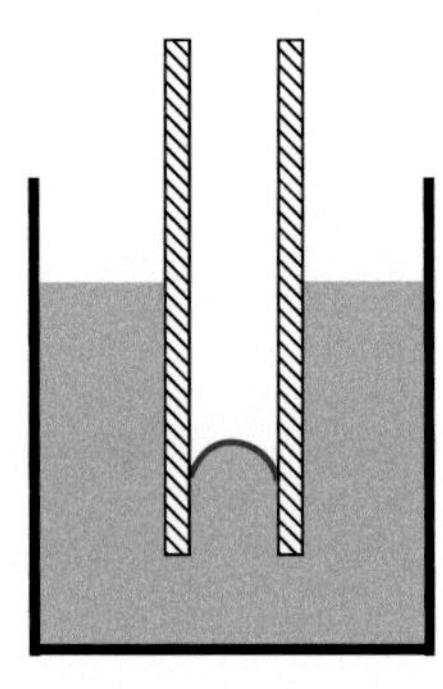

Figure 6.41 Capillary depression

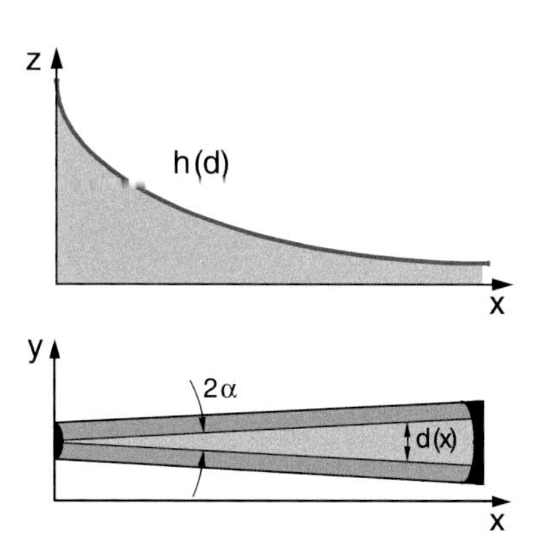

Figure 6.42 Demonstration of capillary rise $h(d) \propto 1/d$ of a liquid confined between two wedged plane walls with the wegde angle 2α

capillary tubes one can also use two parallel plates with the distance d. A liquid between these plates has the capillary rise

$$h = \frac{2\sigma}{\varrho g} \cdot \frac{1}{d} \,. \tag{6.41}$$

The dependence $h(d)$ can be demonstrated by two nearly parallel plates, which are slightly inclined against each other by a small angle α (Fig. 6.42). Since the distance $d(x) = 2x \cdot \tan\alpha$ increases linearly with x the height $h(x) \propto 1/x$ is a hyperbola.

6.4.4 Summary of Section 6.4

The many different phenomena at the boundaries of liquids can be all quantitatively explained by the magnitude of the surface tensions or surface energies. We can make the following statements:

- At each point of a stable liquid surface the total force is always perpendicular to the surface, its tangential component is zero.
- The boundary of a liquid with a given volume always approaches that shape that has the minimum surface area.
- A bent convex liquid surface with radius of curvature r produces an inward pressure, that is proportional to $1/r$ and to the surface tension.

6.5 Friction Between Solid Bodies

If two moving extended bodies touch each other, additional forces occur which depend on the properties of the two surfaces. Examples are a metal block sliding on a plane base, or a wheel rotating around an axis. These forces are due to the interaction between the atoms or molecules in the outer layers of the two bodies. This interaction is reinforced by surface irregularities and deformations, caused by the contact between the two bodies. These forces are called *friction forces*. For point masses they can be completely neglected because their surface area is zero. In daily life and for technical problems they play a very important role. Without friction we would not be able to walk nor cars could run. Also most technical processes of machine work on material, such as drilling, milling or cutting would not be possible without friction. On the other hand, often friction needs to be minimized in order to avoid energy dissipation.

We will therefore discuss the basic principles of friction phenomena in more detail.

6.5.1 Static Friction

A body with a plane base (for example a cuboid) rests on a horizontal plane table. In order to move it across the table we must apply a force in the horizontal direction, which can be measured with a spring balance (Fig. 6.43a). The experiment shows that in spite of the applied force the body with mass m does not move until the force exceeds a definite value F_s. When the body is turned over (Fig. 6.43b) so that now another surface with a different area touches the table, this critical force F_s does not change in spite of the different surface area in contact with the table. However, if the body is pressed by an additional force against the table, the critical pulling force F_s increases. The experiments show, that F_s is proportional to the total vertical force F_N exerted by the body on the table and on the roughness of the two surfaces in contact.

The amount of this static friction force is

$$F_s = \mu_s \cdot F_N \,. \tag{6.42}$$

The static friction coefficient μ_s depends on the materials of the bodies in contact and on the texture of the two surfaces.

The static friction can be explained in a simple model (Fig. 6.44) by the roughness of the two surfaces in contact. Even a polished plane surface is not an ideal plane but shows microscopic deviations from the ideal plane, which may be caused by lattice defects, shifts of atomic planes etc. The envelope of this microroughness gives the macroscopic deviations caused by imperfect polishing or grinding. A measure for these deviations is the mean quadratic deviation $\langle z^2(x, y)\rangle$ from the ideal plane $z = 0$. Since one measures generally not single points but surface elements $dx \cdot dy$, the function $z(x, y)$ is averaged over the surface

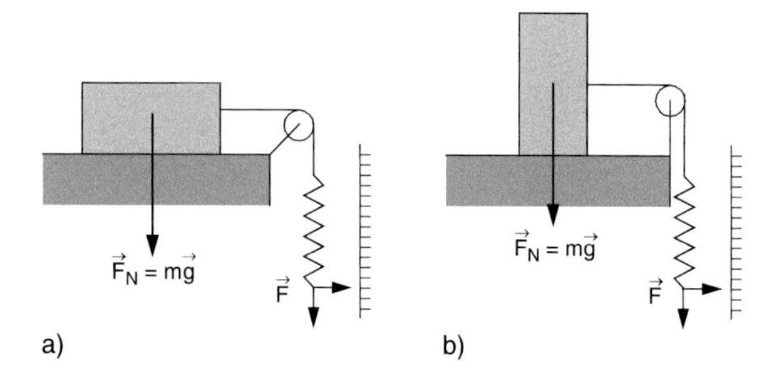

Figure 6.43 Measurement of static friction with a spring balance

Chapter 6

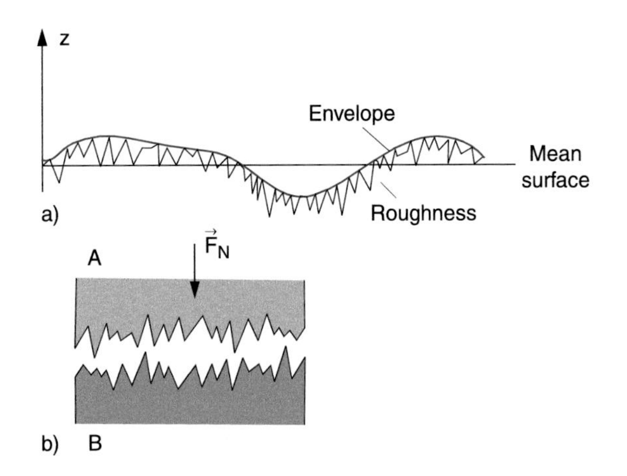

Figure 6.44 Schematic model of the surface roughness as the cause of friction. **a** micro roughness (exagerated) and macroscopic coarseness; **b** static friction caused by interlocking of two rough surfaces

elements $dx \cdot dy$ and this average depends on the spatial resolution of the analyzing instrument i.e. on the size of the resolved elements $dx \cdot dy$. With modern surface analysis, using tunnel-microscopes (see Vol. 3) even the roughness on an atomic scale can be spatially resolved.

The two surfaces in contact interlock each other due to the force that presses them together (Fig. 6.44b) and the force F_s is necessary to release this interlocking. This can be achieved by breaking away the "hills" of the rough surface, or by lifting the body over these hills.

A possible way to determine experimentally the coefficient of static friction uses the inclined plane with a variable inclination angle α in Fig. 6.45. The angle α is continuously increased until the body B with mass m starts to slide down for $\alpha = \alpha_{max}$.

The weight force $\boldsymbol{G} = m \cdot \boldsymbol{g}$ can be decomposed into two components:

1. A component $F_{\parallel} = m \cdot g \cdot \sin\alpha$ parallel to the inclined plane

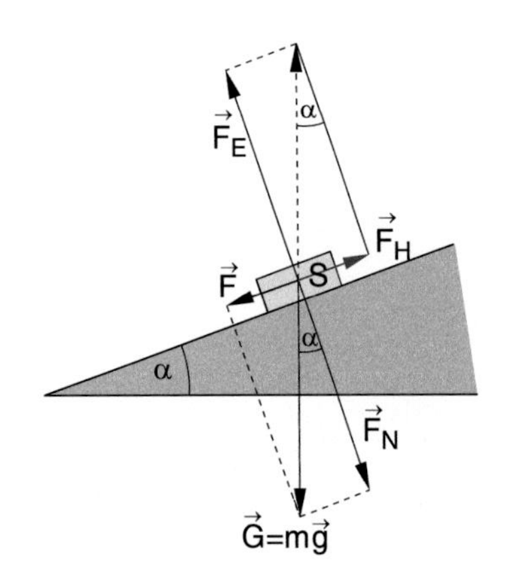

Figure 6.45 Measurement of coefficient of static friction with the inclined plane

2. A component $F_{\perp} = m \cdot g \cdot \cos\alpha$ perpendicular to the inclined plane, which is compensated by the opposite restoring force of the elastic deformation of the plane.

The body starts to slide downwards as soon as $F_{\parallel}$ becomes larger than the static friction force $F_s = \mu_s \cdot F_{\perp} = \mu_s \cdot m \cdot g \cdot \cos\alpha$. This gives the condition for the coefficient μ_s

$$\mu_s = \frac{F_{\parallel}(\alpha_{max})}{F_{\perp}(\alpha_{max})} = \tan\alpha_{max} \; . \tag{6.43}$$

If α is increased beyond α_{max} the body performs an accelerated sliding motion. This indicates that the sliding friction force is smaller than the static friction force.

6.5.2 Sliding Friction

When the body in Fig. 6.43 is moved by a force $|\boldsymbol{F}| > |\boldsymbol{F}_s|$ the sliding motion is accelerated. In order to reach a uniform motion of a sliding body with constant velocity, where the total force is zero, one needs only the smaller force $|\boldsymbol{F}_{sl}| < |\boldsymbol{F}_s|$. Analogue to the static friction force, the sliding friction force F_{sl} is proportional to the force F_N normal to the surface of the table on which it slides.

$$\boldsymbol{F}_{sl} = \mu_{sl} \cdot \boldsymbol{F}_N \; . \tag{6.44}$$

The coefficient of sliding friction μ_{sl} depends again on the material of body and basis, but also on the relative velocity. It is, however, always smaller than the coefficient of static friction. This can be explained by the simplified model of the two surfaces in contact, shown schematically in Fig. 6.44, where the roughness of the surfaces has been exaggerated. If the two bodies are at rest the peaks and the valleys of the micro-mountains interlock. This allows a minimum distance between the two attracting surfaces resulting in a minimum energy. At the sliding motion the two surfaces move above the peaks and the mean distance between the surfaces is larger. During the sliding motion, parts of the peaks are ablated. This results in an attrition of the surfaces.

The sliding motion dissipates energy, even for a horizontal motion. If the body is moved by the distance Δx, the necessary work is $W = F_{sl} \cdot \Delta x$, which is converted into heat.

Experiments show that the sling friction force increases with the relative velocity. The reason is that with increasing velocity more material of the two surfaces is ablated. The power $P = dW/dt$, necessary to maintain the velocity v of a sliding motion, increases with v^n where $n > 1$.

Note: The friction between a moving body and the surrounding air has different reasons. If the body moves through air at rest, a thin layer of air close to the surface of the body sticks at the surface and is therefore accelerated by the moving body. This requires the energy $1/2m_L \cdot v^2$ where m_L is the mass of the air layer that also increases with the velocity v.

6.5.3 Rolling Friction

When a round body rolls over a surface, also friction forces F_R occur which are caused by the interaction between the atoms of the bodies at the line of contact. Furthermore, the base is deformed by the weight of the round body (Fig. 6.46), which leads to deformation forces. For the rolling of a round body with constant angular velocity, a torque around the contact line is necessary that just compensates the opposite torque of the rolling friction. Around the depression of the base at the line of contact, bulges are formed, which have to be overcome when the body rolls.

The experiments tell us that the torque, necessary for keeping a constant angular velocity, is proportional to the force F_N normal to the surface of the base

$$D_R = \mu_R \cdot F_N \,, \tag{6.45}$$

where the coefficient μ_R of rolling friction has the dimension of a length in contrast to the dimensionless coefficients μ_s and μ_{sl}.

Similar to the measurement of μ_s the coefficient μ_R can be measured with an inclined plane (Fig. 6.47). A circular cylinder with mass m and radius r does not roll down the inclined plane, if the inclination angle α is smaller than a critical angle α_R, which is, however, smaller than the angle α_{max} measured for the static friction in Fig. 6.45.

For this critical angle α_R is the counter-clockwise torque $D_G = m \cdot g \cdot r \cdot \sin\alpha_R$ around the contact line just equal to the clockwise torque $D_R = \mu_R \cdot F_N = \mu_R \cdot m \cdot g \cdot \cos\alpha_R$. This yields

$$\mu_R = r \cdot \tan\alpha_R \,. \tag{6.46}$$

The rolling friction is proportional to the radius of the round body. The rolling friction is much smaller than the sliding friction, because the surface irregularities, shown in Fig. 6.44, are partly overrun. Therefore, the invention of the wheel was a great progress for humankind. The comparison of the frictional forces for sliding and rolling gives with (6.44) and (6.45) the ratio

$$\frac{F_s}{F_R} = \frac{F_s}{F_R/r} = r \cdot \frac{\mu_s}{\mu_R} \,. \tag{6.47}$$

The much smaller rolling friction is utilized by ball bearings, which reduce the friction of rotating axes compared to the sliding friction without these ball bearings. In Fig. 6.48, some technical realizations of different ball bearings and axial bearings are shown. In Tab. 6.3, the friction coefficients for some materials are listed.

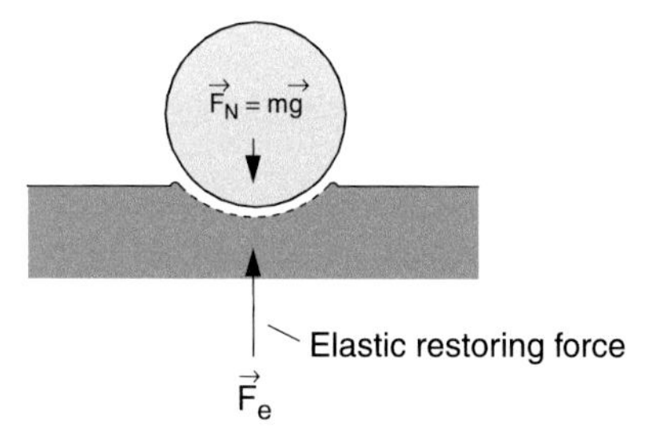

Figure 6.46 Deformation of a surface around the contact line

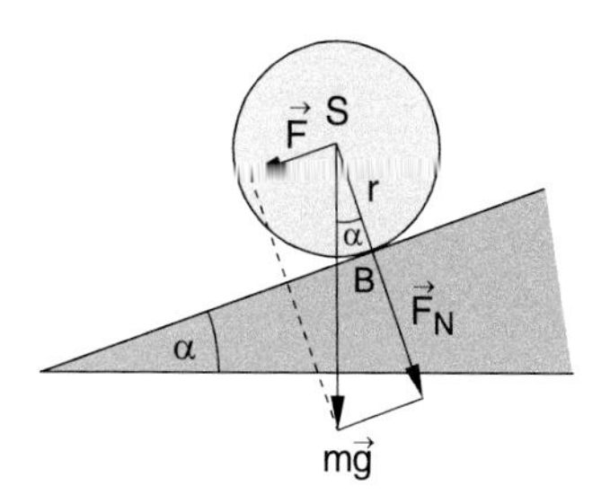

Figure 6.47 Measurement of rolling friction with the inclined plane

Remark. For skating or tobogganing the snow melts under the runners because of heat conduction from the warmer skates and due to the heat produced by friction. The water film under the vats reduces the friction considerably. Often one finds the explanation that the pressure exerted by the weight of the skater is the main reason for melting. This effect plays, however, only a minor part, as can be calculated from the known decrease of the melting point with increasing pressure. (see Sect. 10.4.2.4). The much smaller sliding friction between solid surface and liquids is also utilized by applying lubricants between the two surfaces, for instance between a rotating axis and its fixed support or between the moving pistons of a car engine and the cylinders. The oil film reduces the friction by about two orders of magnitude.

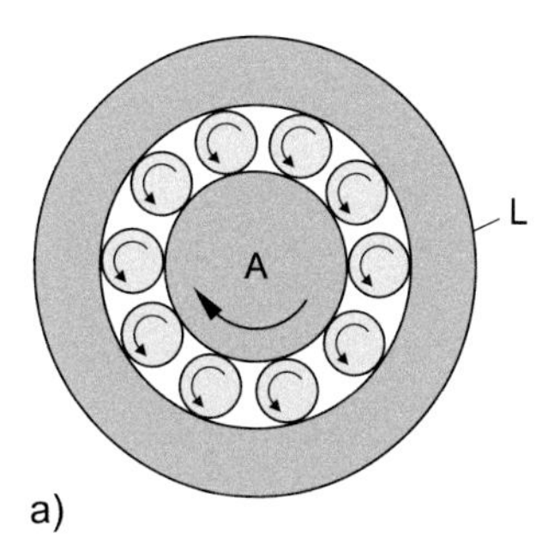

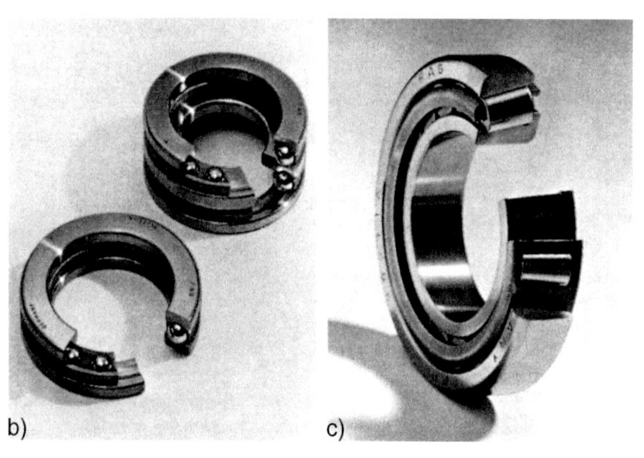

Figure 6.48 Ball bearings. **a** scheme of a radial grooves bearing; **b** realization; **c** axial groove bearing

Chapter 6

Table 6.3 Coefficients of static, sliding and rolling friction of some materials in contact with each other. The values strongly depend of the characteristics of the surfaces. They therefore differ for different authors

Interacting materials	μ_H	μ_G	μ_R/r
Steel–Steel	0.5–0.8	0.4	0.05
Steel with oil film	0.08	0.06	0.03–0.1
Al–Al	1.1	0.8–1.0	
Steel–Wood	0.5	0.2–0.5	
Wood–Wood	0.6	0.3	0.5
Diamond–Diamond	0.1	0.08	
Glass–Glass	0.9–1.0	0.4	
Rubber-tar seal			
– dry	1.2	1.05	
– wet without waterfilm	0.6	0.4	

6.5.4 Significance of Friction for Technology

Friction plays an outstanding role for many technical problems. In some cases it should be as large as possible (for example for clutches in cars or other machinery). The rolling friction for car tires should be as small as possible, but the static friction and the sliding friction should be as large as possible.

For many sliding or rotating parts of machinery, friction is damaging. It causes increased energy consumption and a destruction of the sliding surfaces (attrition). For such cases, it is therefore necessary to minimize friction. This can be achieved either by reducing the sliding friction by liquid films or air buffers or by using ball bearings. Because of its importance, meanwhile a whole branch of science called *tribology* works on problems of friction [6.4].

Example

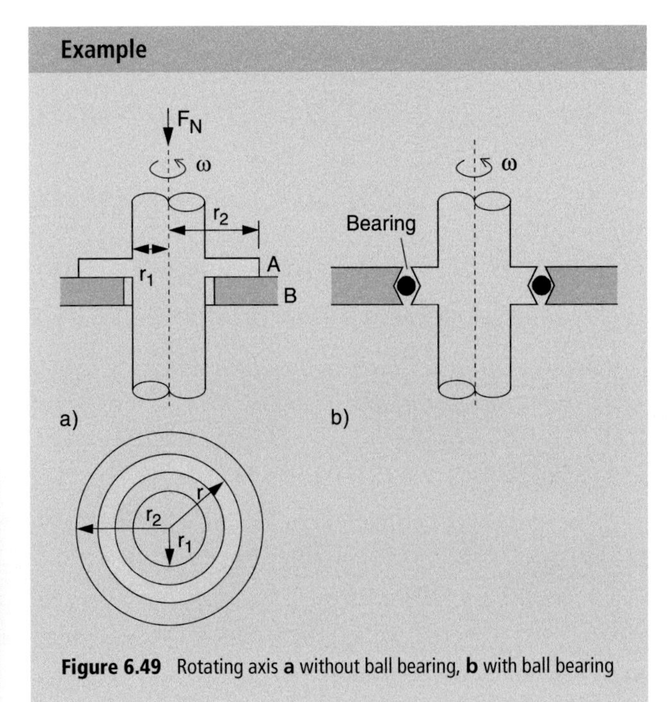

Figure 6.49 Rotating axis **a** without ball bearing, **b** with ball bearing

In Fig. 6.49a is an axis shown that rotates with the angular velocity ω. A circular ring with area $A = \pi(r_2^2 - r_1^2)$ is welded to the axis and exerts a force F_N and a pressure $p = F_N/A$ onto the support base. The sliding friction causes a torque D on the rotating axis, which has to be compensated by an opposite torque supplied by an external force.

On the red annulus in the lower part of Fig. 6.49a acts the force $dF_N = 2\pi \cdot r \cdot dr \cdot p$, which causes the torque $dD = r \cdot dF_s = \mu_s \cdot p \cdot 2\pi \cdot r^2 \cdot dr$. Integrating over all annuli gives the total torque

$$D = \int_{r_1}^{r_2} dD = \frac{2\pi}{3}\mu_{sl} \cdot p \cdot \left(r_2^3 - r_1^3\right) . \qquad (6.48)$$

The friction consumes the power $P = D \cdot \omega$, which is converted into heat. This dissipated power is proportional to the coefficient μ_s of sliding friction, to the contact pressure p and the angular velocity ω. If the annulus with area A is supported by ball bearings (Fig. 6.49b), the torque caused by friction decreases by some orders of magnitude. ◀

Another solution uses the mounting in Fig. 6.49a but now with a liquid film between the contacting surfaces. Often air is blown with high pressure between the two surfaces and an air buffer supports the rotating annulus.. This allows one to realize an extremely low friction. Examples are very fast rotating turbomolecular vacuum pumps (see Sect. 9.2.1.3), where the rotating blades are supported by the air blow.

6.6 The Earth as Deformable Body

At the end of this chapter, we will apply the results of the foregoing sections to the interesting example of our earth, which can be deformed by several forces acting on it. In addition friction plays an important role for phenomena such as the tides or the differential rotation of the inner parts of the earth. Since the earth is composed of solid material as well as of liquid phases, it gives a good example of a realistic and more complicated deformable body.

Our earth is not a rigid homogeneous sphere. It shows an inhomogeneous radial density profile $r(r)$ (Fig. 6.50), which is determined by the pressure profile $p(r)$, but also by the chemical composition, which changes with the radius r. Furthermore, the different solid and liquid phases in the interior of the earth contribute to the inhomogeneous profile. The central region with $r < 1000$ km is a solid kernel of heavy elements (iron, nickel), while for $r > 1000$ km hot liquid phases of metals are predominant, covered by a relatively thin solid crust, consisting of large plates, which float on the liquid material. The earth is therefore not a rigid body but can be deformed by centrifugal forces,

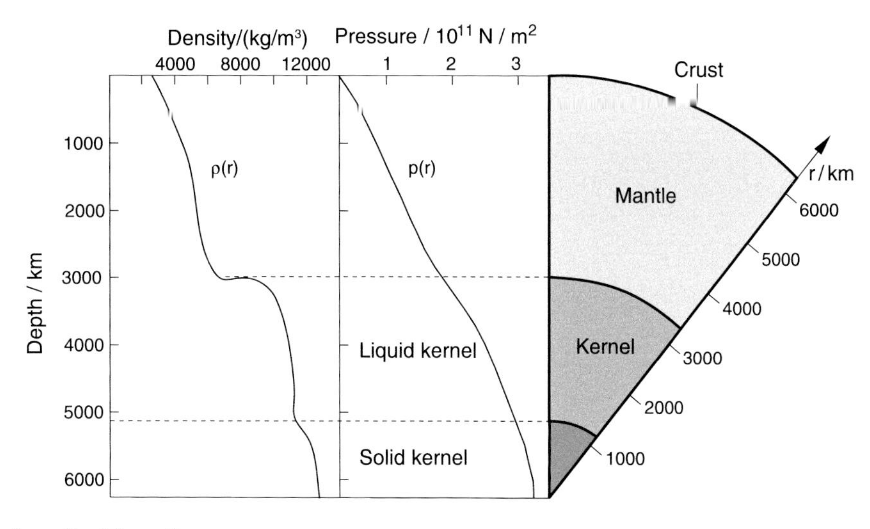

Figure 6.50 Radial density profile of the earth

caused by the earth rotation, and by gravitational, forces due to the attraction by the sun and the moon. These deformations are partly elastic (tides of the earth crust) or plastic (= inelastic). In the latter case, the deformed material does not come back to its original location after the force ends and a permanent change of the shape remains. The shift of the continental plates or the eruption of volcanos with the formation of new islands or mountains are examples of non-elastic deformations.

6.6.1 Ellipticity of the Rotating Earth

The rotation of the earth with the angular velocity $\omega = 2\pi/\text{day} = 7.3 \cdot 10^{-5}\,\text{s}^{-1}$ causes a centrifugal force on a mass element Δm with the distance a from the rotation axis

$$\boldsymbol{F}_{\text{cf}} = \Delta m \cdot a \cdot \omega^2 \cdot \hat{\boldsymbol{e}}_{\text{cf}} \tag{6.49a}$$

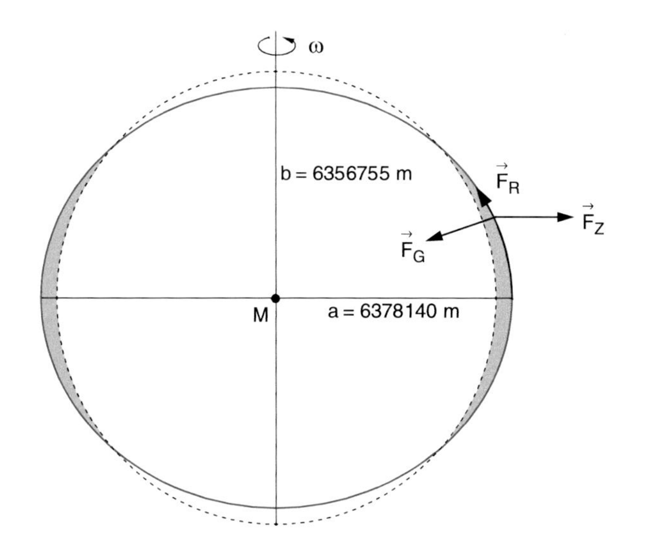

Figure 6.51 Deformation of the rotating earth due to centrifugal force

with the unit vector $\hat{\boldsymbol{e}}_{\text{cf}}$ perpendicular to $\boldsymbol{\omega}$. This force acts in addition to the gravitational force

$$F_{\text{G}} = -G \cdot \frac{\Delta m \cdot M(r)}{r^2} \hat{\boldsymbol{r}} \,, \tag{6.49b}$$

where $M(r)$ is the mass of that part of the earth inside the radius r. Because of the plastic deformation the mass element Δm shifts until the total force $\boldsymbol{F}$ acting on it, is zero.

$$\boldsymbol{F} = \boldsymbol{F}_{\text{G}} + \boldsymbol{F}_{\text{cf}} + \boldsymbol{F}_{\text{R}}$$

is the sum of gravity force $\boldsymbol{F}_{\text{G}}$, centrifugal force $\boldsymbol{F}_{\text{cf}}$ and restoring force $\boldsymbol{F}_{\text{R}}$. For a homogeneous earth this would result in a rotational ellipsoid with the major diameter in the equatorial plane

$$2a = 12\,756.3\,\text{km} \,,$$

and with a minor diameter in the direction of the rotational axis of

$$2b = 12\,713.5\,\text{km} \,.$$

The ellipticity $\varepsilon = (a - b)/a$ of this rotational ellipsoid is $\varepsilon = 3.353 \cdot 10^{-3}$.

Because of the inhomogeneous mass distribution the shape of the rotating earth deviates slightly from this rotational ellipsoid but forms a nearly pear-shaped pattern called *geoid* (Fig. 2.56). The surface of this geoid is the zero-surface for all geodetic measurements. This means: all measurements of elevations z are related to this zero surface $z = 0$ [6.5].

6.6.2 Tidal Deformations

Induced by the additional forces of the gravitational attraction by the sun and the moon the earth surface deforms in a characteristic time-dependent way. This deformation is maximum for

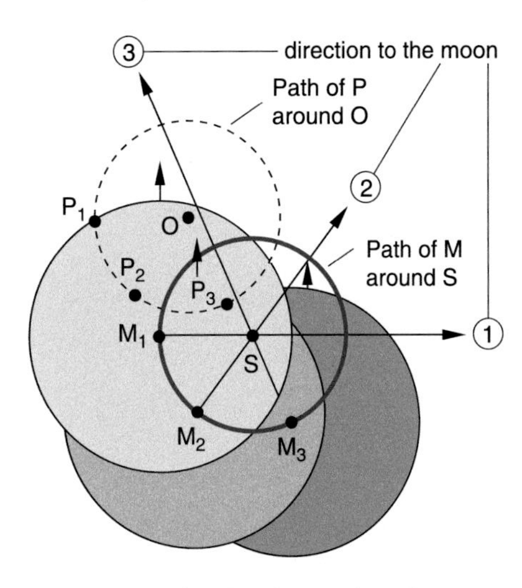

Figure 6.52 The rotation of earth and moon about their common center of mass S causes all points of the earth to rotate about the center S, that moves with the revolution of the moon. This is shown, without the daily rotation of the earth about its axis, for three different positions of the moon

the oceans (low tides and high tides) since for liquids the restoring elastic force is zero. However, it also appears with smaller elongation in the solid crust of the earth.

The deformation of the earth and the resulting tides have three causes:

a) The centrifugal distortion due to the motion of earth and moon about their common center of mass.
b) The gravitational force, effected by the masses of moon and sun.
c) The centrifugal distortion due to the rotation of the earth around its axis.

In order to understand this tidal deformation we discuss at first the simplified model of the deformation of the non-rotating earth and neglect the gravitational attraction by the sun and the revolution of the earth around the sun. We restrict the discussion therefore to the influence of the moon on the non-rotating earth. Under the mutual gravitational attraction

$$\boldsymbol{F}_{\text{G}} = -G \cdot \frac{M_{\text{E}} \cdot M_{\text{Mo}}}{r_0^2} \hat{\boldsymbol{r}}_0 , \tag{6.50}$$

earth and moon move around their common center of mass S (also called bari-center) which lies still inside the earth (about 0.75 of the earth radius from the center). The distance between the centers of earth and moon is r_0. During a moon-period of 27.3 days the center M of the earth moves on a circle with radius $0.75R$ around the baricenter S, which always lies on the line M_{E}-M_{Mo}. All arbitrary points P_i in the earth move around S on circles with radii $P_i - S$. However, the center of mass S has no fixed position inside the earth but moves during one moon period inside the earth on a circle with radius $0.75R_{\text{E}}$ around the center M of the earth, because the space-fixed center of mass S lies always on the line between earth-center and moon center.

The motion of the non-rotating earth as extended body, described in the coordinate system of the earth, is therefore **not** a rotation about a fixed axis but rather a shift since the space-fixed point S. has not a fixed location inside the earth. The revolution of the moon and the earth about S with the angular velocity Ω causes therefore for all points of the non-rotating earth the same centrifugal force

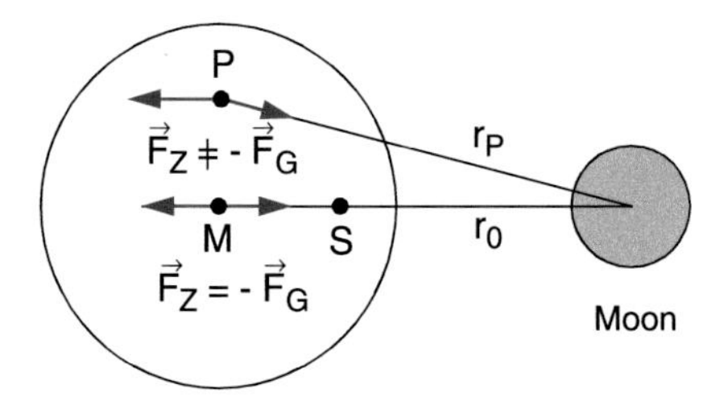

Figure 6.53 Only for the center M of the earth are gravitational attraction by the moon and centrifugal force of the earth–moon rotation about S equal but opposite and cancel each other

$$F_{\text{cf}} = m\Omega^2 \cdot R_{\text{S}} = m\Omega^2 \cdot 0.75R . \tag{6.51}$$

On the other hand, the gravitational attraction between earth and moon is different for the different points of the earth because of their different distance from the moon center. For the earth center M it is

$$\boldsymbol{F}_{\text{G}} = -G\frac{M_{\text{E}} \cdot M_{\text{Mo}}}{r^2}\hat{\boldsymbol{r}}_0 \tag{6.52}$$

with $r = r_0$. Here gravitational force and centrifugal force just compensate each other.

$$F_{\text{cf}} = M_{\text{E}}\Omega^2 \cdot 0.75R \cdot \hat{r}_0 = -F_{\text{G}}(r_0) .$$

The total force in M is zero (Fig. 6.53). This is no longer true for other points P because the distances to the moon are different and therefore the gravitational force differs while for the non-rotting earth the centrifugal force is the same for all points P. For example the gravitational force in the points A and B in Fig. 6.54 is

$$\begin{aligned} \boldsymbol{F}_{\text{G}}(r_{\text{A}}) &= -G\frac{m \cdot M_{\text{Mo}}}{(r_0 + R)^2}\hat{\boldsymbol{r}}_0 , \\ \boldsymbol{F}_{\text{G}}(r_{\text{B}}) &= -G\frac{m \cdot M_{\text{Mo}}}{(r_0 - R)^2}\hat{\boldsymbol{r}}_0 . \end{aligned} \tag{6.53}$$

Compared with the gravitational force $F_{\text{G}}(r_0)$ in M the force differences are $\Delta F(r_{\text{A}}) = F_{\text{G}}(r_{\text{A}}) - F_{\text{G}}(r_0)$ and $\Delta F(r_{\text{B}}) = F_{\text{G}}(r_{\text{B}}) - F_{\text{G}}(r_0)$ which point in the direction of the connecting line earth–moon. The magnitude of these differences can be obtained from (6.52) and (6.53). Because $R \ll r_0$, we can approximate $(1 + R/r_0)^{-2} \approx 1 - 2R/r_0$ and we get:

$$\begin{aligned} \Delta\boldsymbol{F}(r_{\text{A}}) &= -G \cdot \frac{m \cdot M_{\text{Mo}}}{r_0^2} \cdot \left(\frac{1}{(1 + R/r_0)^2} - 1\right)\hat{\boldsymbol{r}}_0 \\ &\approx G \cdot \frac{2m \cdot M_{\text{Mo}}}{r_0^3} R \cdot \hat{\boldsymbol{r}}_0 \\ &= -2\boldsymbol{F}_{\text{G}}(r_0) \cdot \frac{R}{r_0} \cdot \hat{\boldsymbol{r}}_0 , \\ \Delta\boldsymbol{F}(r_{\text{B}}) &= +2\boldsymbol{F}_{\text{G}}(r_0) \cdot \frac{R}{r_0} \cdot \hat{\boldsymbol{r}}_0 , \end{aligned} \tag{6.54}$$

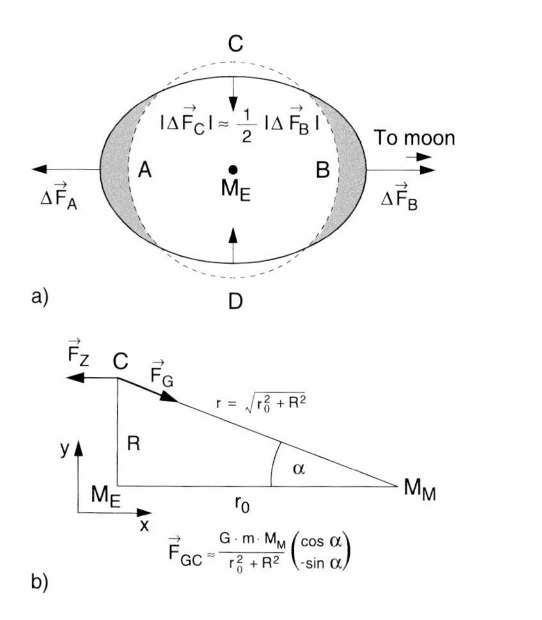

Figure 6.54 Deformation of the earth by the tides (exagerated). The *arrows* give magnitude and direction of the tidal forces

The difference $\Delta F(r_B)$ is directed from M to the center of the moon, while $\Delta \boldsymbol{F}(r_A)$ has the opposite direction. Both differences result in a convex curved deformation of the earth surface, as shown exaggerated in Fig. 6.54. For a mass in the points C or D the gravitational force caused by the moon

$$\begin{aligned} \boldsymbol{F}_G(r_C) &= -G\frac{m \cdot M_{Mo}}{r_0^2 + R^2}\hat{\boldsymbol{r}} = \{F_x, F_y\} \\ &= F_G(r_0)\frac{r_0^2}{r_0^2 + R^2}\begin{pmatrix} \cos\alpha \\ -\sin\alpha \end{pmatrix} \end{aligned} \tag{6.55}$$

points to the center of the moon (Fig. 6.54b), while the centrifugal force is directed as for all points of the earth in the direction of r_0 and is anti-collinear to F_G, while the magnitude of both forces are equal, i.e. $\boldsymbol{F}_G(r_0) = -\boldsymbol{F}_{cf}$. (Note, that we regard a non-rotating earth and $\boldsymbol{F}_{cf}$ is only due to the revolution of earth and moon around the common center of mass S). With $\cos\alpha = r_0/\sqrt{(r_0^2 + R^2)}$ and $\sin\alpha = -R/\sqrt{(r_0^2 + R^2)}$ the resulting residual force is

$$\begin{aligned} \Delta \boldsymbol{F}(r_C) &= \boldsymbol{F}_{cf} + \boldsymbol{F}_G = F_G(r_0)\begin{pmatrix} \dfrac{r_0^3}{(r_0^2 + R^2)^{3/2}} - 1 \\ -\dfrac{r_0^2 R}{(r_0^2 + R^2)^{3/2}} \end{pmatrix} \\ &\approx F_G(r_0)\frac{R}{r_0}\begin{pmatrix} \frac{3}{2}(R/r_0) \\ -1 \end{pmatrix}, \end{aligned} \tag{6.56}$$

because $R \ll r_0 \Delta F(r_C)$ points nearly into the $-y$-directions to the center of the earth. it therefore decreases the curvature of the earth surface (Fig. 6.57b) which causes low tide. Its amount

$$\begin{aligned} \Delta F(r_C) &= |\boldsymbol{F}_G(r_C) - \boldsymbol{F}_G(r_0)| \approx G\frac{m \cdot M_{Mo}}{r_0^3}R \\ &= F_G(r_0) \cdot \frac{R}{r_0} = \frac{1}{2}\Delta F(r_A) \end{aligned} \tag{6.57}$$

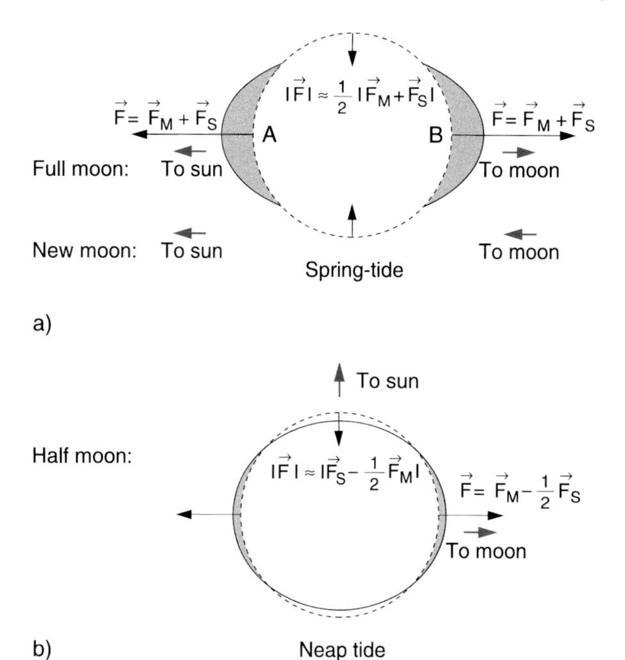

Figure 6.55 Spring tide and neap tide caused by addition or subtraction of the gravitational forces by moon and sun

is smaller by the factor 1/2 than in the points A and B. For all other points of the earth surface the resulting forces ΔF have a radial as well as a tangential component. The tangential component causes an acceleration of the ocean water towards the points A or B. The borderline between the different tangential directions lies in Fig. 6.54a left of the line CD. where the x-component of F_G is

$$F_{Gr} = +\tfrac{3}{2}F_G(r_0)(R/r_0) \,. \tag{6.58}$$

From (6.54) and (6.56) one can infer, that the maximum tide force depends on the ratio M_{Mo}/r^3. If the numerical values for r and M_{Mo} are inserted one obtains $M_{Mo}/r^3 = 1.34 \cdot 10^{-3}\,\mathrm{kg/m^3}$ and a tide acceleration of $a_1 = \Delta F/m = 1.1 \cdot 10^{-6}\,\mathrm{m/s^2}$. This leads to a deformation of the solid earth crust of up to 0.5 m. Since $M_{sun}/r_{sun}^3 = 6.6 \cdot 10^{-4}\,\mathrm{kg/m^3}$ the effect of the sun on the tides is only about half of that of the moon and one obtains for the contribution to the tide-acceleration $a_2 = 5.6 \cdot 10^{-7}\,\mathrm{m/s^2}$. If sun and moon stand both on a line through the center of the earth (this is the case for full moon and for new moon) the actions of moon and sun add (spring-tide). If sun and moon are in quadrature (the connecting lines sun–earth and moon–earth intersect in the earth center under 90° (Fig. 6.55)) the effects subtract (neap tide).

Up to now, we have neglected the daily rotation of the earth. It brings about two effects:

- An additional centrifugal force, which causes the deformation of the earth into a oblate symmetric top (see Sect. 6.6.1) The deformation, which amounts to about 21 km at the equator, is very much larger than that caused by the moon but it is equal for all points on the same latitude and is not time dependent in contrast to the tides caused by the moon.

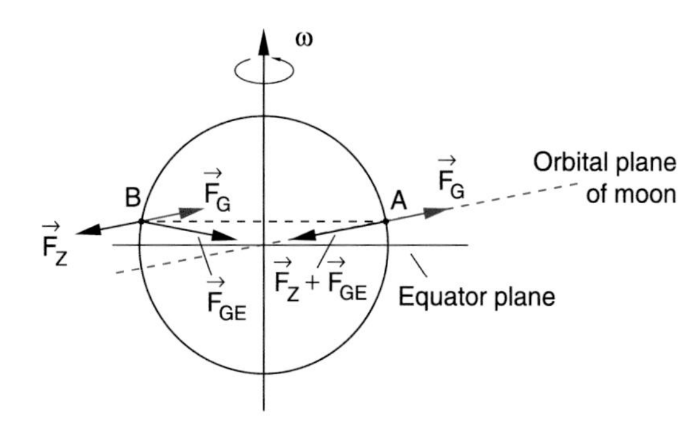

Figure 6.56 Influence of the inclination of the orbital plane of the moon on the periodical variation of the tidal elevation

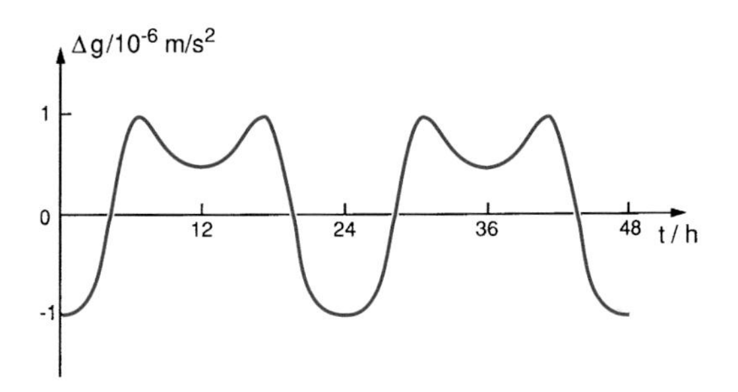

Figure 6.57 Time dependent course of the tidal elevation at a fixed point on the earth surface, measured as the corresponding variation Δg of the earth acceleration g

- When the revolution of the moon around the earth is ignored, the two tide maxima at the points A and B in Fig. 6.54 and the low tides in the point C and D would travel around the earth in 24 hours. At a fixed point one would experience every 12 hours a high tide and a low tide. The deformation of the solid crust is about 0.5 m, that of the ocean away from the coast about 1 m. Tide amplitudes up to 15 m are observed at the coast and in particular in narrow bays. They are generated by nonlinear effects during the propagation of tidal waves.

For a more accurate description of the tides the revolution of the moon has to be taken into account. It demands the following corrections of our simple model:

- The moon moves around the earth–moon-center of mass S in 27.5 days with the same direction as the rotation of the earth. Therefore, the round-trip time of the tides is 24.87 h instead of 24 h.
- The plane of the moon's revolution is inclined against the equator plane (Fig. 6.56). An observer in the point A experiences a higher tide amplitude than an observer in B 12.4 h later. This can be seen as follows: The centrifugal force (6.51) caused by the revolution of the earth–moon system around S is in A parallel to the gravitational force F_{GE} caused by the mass of the earth. The resulting force $F = F_{GE} + F_{GM} + F_{cf}$ is perpendicular to the earth surface. The total force has to include the centrifugal force F_{cE} caused by the rotation of the earth, which is perpendicular to the rotation axis of the earth. In the point B the centrifugal force F_{cf} has the same direction than in A, but the gravitational force F_{GE} has a nearly opposite direction and therefore the vector sum of the two forces is in B smaller than in A. The force F_{cE} has for both points the same direction because they are located on the same circle of latitude (Fig. 6.56). The tide amplitudes show an amplitude modulation with a period of about 12.4 h. The modulation index depends on the geographical latitude.
- The motion of the moon changes the relative positions of the interacting sun, moon and earth. Therefore, the vector sum of the tide forces show also a periodic modulation.

These considerations illustrate that the total tide amplitude is determined by the superposition of many effects and is therefore a complicated function of time (Fig. 6.57). It can be measured with various techniques. One of them uses the time variation of the gravitational acceleration g which depends on the geographic location and is affected by the tides. Another very sensitive interferometric technique measure the local deformation of the earth crust (see Sect. 6.6.4).

6.6.3 Consequences of the Tides

With the tides of the oceans as well as with the periodic deformations of the earth crust, friction occurs which causes a partial transfer of kinetic energy into heat. This lost kinetic energy slows down the rotation of the earth and causes an increase of the rotation period by 90 ns per day. Within 10^6 years this prolongs the duration of the day by 0.5 min (see Probl. 1.4).

The gravitational force between earth and moon causes of course also deformations on the moon. Accurate measurements have proved that the shape of the moon is an ellipsoid with the major axis pointing towards the earth. The general opinion is that in former times the moon also rotated around its axis. This rotation was, however, in the course of many million years slowed down by friction until the moon no longer rotates and shows always the same side to the earth.

The tidal friction of earth and moon has the following interesting effect: The total angular momentum of the earth–moon system is constant in time because the system moves in the central force field of the sun (the additional non-central forces due to interactions with the other planets are negligible). Since the rotation of the earth around its axis slows down and its angular momentum $I \cdot \omega$ decreases, the orbital angular momentum of the earth–moon system

$$|\boldsymbol{L}_{EM}| = r \cdot \mu \cdot v_{rel} \cdot = I_{EM} \cdot \Omega$$

(r = distance earth–moon, v_{rel} = relative velocity of the moon against the earth, I_{EM} = inertial moment of the earth–moon system and Ω = angular velocity of the rotating earth–moon system) has to increase. The moon is accelerated by the tidal wave running around the earth. This can be seen as follows (Fig. 6.58): The earth rotating with the angular velocity $\omega \gg \Omega$ accelerates the tidal waves due to the friction forces: This acceleration brings the tidal maximum slightly ahead of the connecting line between the centers of earth and moon. Due to the slightly increased gravitational force, the moon is accelerated while the earth rotation decreases. The larger kinetic energy

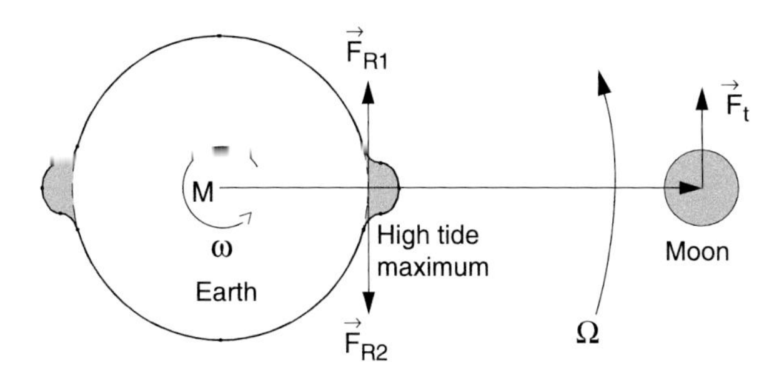

Figure 6.58 Deceleration of the earth rotation and acceleration of the orbital velocity of the moon by the tidal friction

of the moon increases its total energy ($E_{\text{kin}} + E_{\text{pot}}$) and therefore also its distance to the earth. In former times the moon was closer to the earth. The nowadays generally accepted theory [6.7a, 6.7b] assumes that the moon was part of the earth but has been catapulted out of the earth by the impact of a heavy asteroid some billion years ago (see Vol. 4).

6.6.4 Measurements of the Earth Deformation

The deformation of the earth by tidal effects can be measured with different techniques. We will shortly discuss three of them:

6.6.4.1 Changes of the Gravitational Force

According to (6.54) the additional gravitational force caused by the moon in the points A and B in Fig. 6.54 is

$$\Delta F_{\text{G}} \approx \frac{2mM_{\text{Mo}}}{r_{\text{Mo}}^3} R . \tag{6.59}$$

In the gravity meter shown in Fig. 6.59 a mass m is suspended by a spring in such a way, that a small change $\Delta F_{\text{G}} = m \cdot \Delta g$ due to the corresponding change of g causes a large vertical deflection of the arrow on the scale. This is achieved by a sloped mounting of the spring with length L and a restoring force $F_r = -D \cdot \Delta L$. With the slope angle α, the vertical deflection Δz causes a length increase of the spring $\Delta L = \Delta z \cdot \sin\alpha$ (Fig. 6.59b) and a change of the restoring force $\Delta F_r = m \cdot \Delta g \cdot \sin\alpha$.

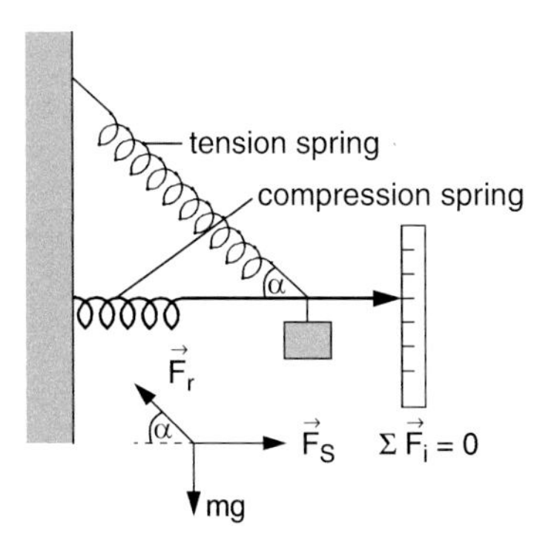

Figure 6.59 Measurement of the gravitational force with a special spring balance

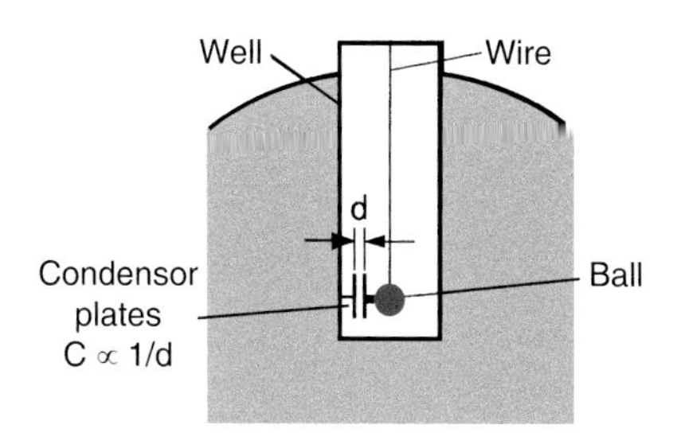

Figure 6.60 Measurement of the deviation of g from the vertical direction

The device measures the periodic changes of F_{G} with a period of 24.87 h from which the tidal amplitudes can be inferred (Fig. 6.57). Because of the different contributing effects, $\Delta g(t)$ follows a complicated curve.

The experimental arrangement of Fig. 6.60 allows to measure the deviation from the vertical direction of the earth acceleration g. Without external perturbation, g would point nearly to the earth center (only for a spherical mass distribution it would point exactly to the center). The additional gravitational force exerted by the moon causes a slight deviation from this direction. The maximum angular deviation, which depends on the latitude, amounts only to about $2.1 \cdot 10^{-6}$ rad($= 0.4''$), the measurement must be sufficiently accurate. This required accuracy can be reached with a pendulum [6.9]. A metal ball suspended on a long wire in a well is connected with one plate of a charged capacitor, while the other plate is fixed on the wall of the well. Any deviation of the pendulum from the vertical direction changes the distance between the two plates and therefore the voltage of the capacitor (see Vol. 2, Sect. 5).

6.6.4.2 Measurements of the Earth Deformation

Here the change ΔL of the length L between two points A_1 and A_2 connected with the earth ground is measured. Figure 6.61 illustrates the method. A very sensitive Laser interferometer is located in a gold mine deep in the ground in order to eliminate acoustic noise from the surroundings. The two mirrors of the laser resonator are mounted on the ground base at the points A_1 and A_2 separated by the distance L. The optical frequency of the laser $\nu_L = m \cdot c/(2L)$ is determined by the length L of the resonator and the large integer $m \gg 1$. If the length L changes due to the deformation of the earth crust, the laser frequency changes accordingly. This frequency change can be measured very accurately, when the laser beam is superimposed on a detector with the output beam of a reference laser with stabilized frequency ν_r. The difference frequency $\nu_L - \nu_r$ in the radio-frequency range can be counted by a digital frequency counter. Existing devices have resonator lengths of 100 m up to several km. They can measure deformations of the earth crust of less than 10^{-9} m (see Vol. 2, Sect. 10.4). This sensitivity is sufficient to measure the deformation of the ground base in the Rocky Mountains caused by the tidal waves of the Pacific Ocean [6.10].

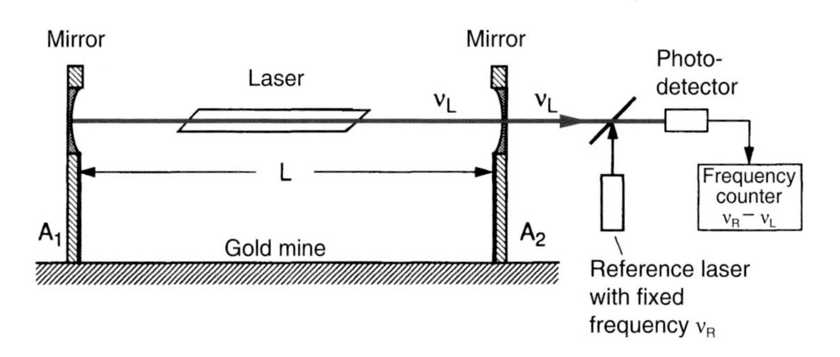

Figure 6.61 Laser interferometer for the measurement of the deformation of the earth crust

Summary

- Elastic bodies show restoring forces for any deformation of their shape. For sufficiently small deformations, these forces are proportional to the elongation from the equilibrium position.
- For a relative length increase $\varepsilon = \Delta L/L$ of a body with length L, constant cross section and elastic modulus E one needs a tensile stress $\sigma = E \cdot \varepsilon$ (Hooke's law).
- A length change ΔL of a rod with length L and quadratic cross section $A = d^2$ caused by the tensile stress σ is accompanied by a change ΔA of its cross section. The relative change of the volume V
$$\frac{\Delta V}{V} = \frac{\sigma}{E}(1 - 2\mu)$$
is determined by the elastic modulus E and the transvers contraction ratio $\mu = -(\Delta d/d)/(\Delta L/L)$.
- Exposed to isotropic pressure p the relative volume change $\Delta V/V = -\kappa \cdot p$ of a body is determined by the compressibility $\kappa = (3/E) \cdot (1 - 2\mu)$.
- A force F acting tangentially on a wall of a body causes a shear of the body. For a cuboid with the side area d^2 the shear angle α is related to the shear stress $\tau = F/d^2$ by $\tau = G \cdot \alpha$ where G is the modulus of shear.
- A rod with length L and cross section $d \cdot b$ is fixed at one end. The vertical force $\boldsymbol{F}$ acting on the other end causes a bending
$$s = (4L^3 \cdot F)/(E \cdot d^3 \cdot b) ,$$
which is proportional to the third power of the length L and the vertical width d.
- Beyond the linear range of Hooke's law plastic deformations occur. If a periodical tensile stress σ acts on a rod with length L, a closed hysteresis curve $\sigma(\varepsilon)$ is traversed. The area enclosed by this curve represents the energy that is transformed into heat for every cycle.
- Inside a liquid the same pressure is present for all volume elements with the same distance Δh from the surface. The hydrostatic pressure $p(z) = p_0 + \varrho \cdot g \cdot (h - z)$ at this height increases linearly with the height $(h - z)$ of the liquid with density ϱ above the layer at z. At the upper surface $z = H$ of a liquid with total height H the pressure is p_0 (for example the barometric pressure of the air above the surface).
- Each solid body with mass m and density ϱ_s experience in a liquid a buoyant force $\boldsymbol{F}_B$ which is equal but opposite to the weight $\boldsymbol{F}_G$ of the liquid volume displaced by the solid body. If $|\boldsymbol{F}|_B > m \cdot g$ the body floats at the liquid surface, for $|\boldsymbol{F}_B| = m \cdot g$ the body can float at any height in the liquid.
- Because of the attractive forces between the molecules of a liquid, energy is required to bring molecules from the interior to the surface. The energy, necessary to increase the surface by 1 m^2, is the specific surface energy. It is equal to the specific surface tension.
- The shape of the surface of a liquid in a container depends on the different surface tensions for the boundaries between container wall and liquid, liquid and air, container wall and air and on the gravity force. It always takes that form, for which the energy is minimum.
- Because of the surface tension a liquid can rise in a capillary (wetting liquid) or descend (non-wetting liquid).
- When two bodies come into touch, friction forces appear, which are different for a relative velocity zero (static friction) or for a relative motion (sliding friction). The smallest friction is found, when a circular body rolls on a plane base. The quantitative description uses friction coefficients μ, which depend on the materials of the two bodies. Generally it is $|\mu_s| > |\mu_{sl}| > |\mu_R|$, where μ_s is the coefficient for static friction, μ_{sl} for sliding friction and μ_R for rolling friction. A liquid film between the two solid bodies reduces the friction considerably.
- The earth is a deformable ellipsoid which is permanently deformed by its rotation and periodically by the gravitational forces exerted by moon and sun, which cause tidal effects. The periodic deformations are partly non-elastic and the friction transfers part of the rotational energy into heat. This causes a slowdown of the earth rotation and a prolongation of the day. Conservation of the total angular momentum leads to an increase of the distance earth–moon.

Chapter 6

Problems

6.1 What is the change ΔL of a steel rope with $L = 9\,\text{km}$,
a) which hangs in a vertical well?
b) What is the maximum length of the rope before its rupture?
c) How large is ΔL when the rope is lowered from a ship into the ocean? ($E = 2 \cdot 10^{11}\,\text{N/m}^2$; $\varrho_{\text{steel}} = 7.7 \cdot 10^3\,\text{kg/m}^3$, $\varrho_{\text{ocean}} = 1.03 \cdot 10^3\,\text{kg/m}^3$)

6.2 A steel beam with $L = 10\,\text{m}$ is clamped at one end. A force $F = 10^3\,\text{N}$ acts on the other end in vertical z-direction. How large is the bending of this end
a) for a rectangular cross section $d \cdot b$ with $d = \Delta z = 0.1\,\text{m}$; $b = \Delta y = d/2$?
b) for a double T-profile (Fig. 6.15d) with $b_1 = d_1 = 0.1\,\text{m}$, $b_2 = d_2 = 0.05\,\text{m}$?

6.3 The deep ocean aquanaut Picard reached in his spherical steel submarine a depth of 10.000 m in the Philippine trench. How large are pressure and total force exerted on the sphere? What is the volume change $\Delta V/V$ caused by the pressure
a) for a hollow sphere with wall thickness of 0.2 m?
b) for a full sphere?

6.4 A turbine drives a generator connected to a steel shaft with length L and diameter D. By which angle α are the two ends of the shaft twisted if the power $P = 300\,\text{kW}$ is transferred at a frequency $\omega = 2\pi \cdot 25\,\text{s}^{-1}$
a) for a steel shaft as full cylinder with $D = 0.1\,\text{m}$; $L = 20\,\text{m}$?
b) for a hollow cylinder with $D_1 = 5\,\text{cm}$ and $D_2 = 10\,\text{cm}$?

6.5 What is the density of water with a compressibility $\kappa = 4.8 \cdot 10^{-10}\,\text{m}^2/\text{N}$ at a depth of 10.000 m?

6.6 A hollow steel cube ($\varrho = 7.8 \cdot 10^3\,\text{kg/m}^3$) with edge length $a = 1\,\text{m}$ and a wall thickness of $d = 0.02\,\text{m}$ and with an open upper side floats on water.
a) How deep does it immerse?
b) What is the location of center of mass and metacenter?
c) What is the maximum angle of its symmetry axis against the vertical direction before it becomes unstable?

6.7 Which energy has to be spent in order to lift a full cube of steel from the bottom of a swimming pool with the water depth of 4 m to a position where the lower side of the cube is at the surface of the water?

6.8 Which force was necessary to separate the two hemispheres in the demonstration experiment by Guericke in Magdeburg with a diameter of 0.6 m, when the pressure difference between inside and outside was $\Delta p = 90\,\text{kPa}$? Guericke had used 16 horses. What should have been done in order to separate the hemi-spheres already with 8 horses?

6.9 In order to verify that a gold bar is really made of gold ($\varrho_{\text{gold}} = 19.3\,\text{kg/dm}^3$) a goldsmith measures its weight in air and when totally immersed in water. Which ratio of the two values is obtained
a) for a 100% gold bar?
b) for a 20% admixture of copper ($\varrho = 8.9\,\text{kg/dm}^3$)?
c) What is the minimum required accuracy of the measurements for unambiguously distinguishing between the two cases? What is the accuracy if an admixture of 1% of copper should be detected?

6.10 A round cylinder of wood ($L = 1\,\text{m}$, $d = 0.2\,\text{m}$, $\varrho = 525\,\text{kg/m}^3$) is floating in water. How deep does it immerse
a) in a horizontal position?
b) if a steel ball with $m = 1\,\text{kg}$ is attached to one end in order to bring it into a vertical floating position?

References

6.1. W.D. Callister, D.G. Rethwisch, *Material Science and Engineering. An Introduction.* (Wiley, 2013)
6.2. *Deformation,* ed. by M. Hazewinkel *Encyclopedia of Mathematics.* (Springer, 2001)
6.3. S. Lipschitz, *Schaum's Outline of mathematical Handbook of Formulas.* (McGraw Hill, 2012)
6.4. M. Abramowitz, I. Stergun, *Handbook of Mathematical Functions.* (Martino Fine Books)
6.5. R.H. Rapp, F. Sanso (ed.), *Determination of the Geoid.* (Springer, Berlin, Heidelberg, 1991)
6.6a. V.M. Lyathker, *Tidal Power. Harnessing Energy from Water Currents.* (Wiley Scrivener, 2014)
6.6b. L. Peppas, *Ocean, Tidal and Wave energy.* (Crabtree Publ. Comp., 2008)
6.6c. https://en.wikipedia.org/wiki/List_of_tidal_power_stations
6.6d. https://en.wikipedia.org/wiki/Tidal_power
6.7a. http://novan.com/earth.htm
6.7b. P. Brosche, H. Schuh, Surveys in Geophysics **19**, 417 (1998)

6.8a. D. Flannagan, *The Dynamic Earth.* (Freeman, 1983)
6.8b. G.M.R. Fowler, *The Solid Earth. An Introduction to Global Geophysics.* (Cambridge Univ. Press, 2004)
6.9. D. Wolf, M. Santoyo, J. Fernadez (ed.), *Deformation and Gravity Change.* (Birkhäuser, 2013)
6.10. J. Levine, J.L. Hall, J. Geophys. Res. **77**, 2595 (1972)
6.11. A.E. Musset, M.A. Khan, *Looking into the Earth: An Introduction to Geological Geophysics.* (Cambridge Univ. Press, 2000)
6.12. M. Goldsmith, M.A. Garlick, *Earth: The Life on our Planet.* (Kingfisher, 2011)

Gases

7

Chapter 7

© Springer International Publishing Switzerland 2017
W. Demtröder, *Mechanics and Thermodynamics*, Undergraduate Lecture Notes in Physics, DOI 10.1007/978-3-319-27877-3_7

Different from solid or liquid bodies, which change their volume only slightly under the action of external forces, gases can be expanded readily. They occupy any volume that is offered to them. Under the action of external pressure their volume can be reduced by orders of magnitude up to a certain limit. The reason for these differences is their much smaller density. At atmospheric pressure, the gas density is smaller than that of solids or liquids by about three orders of magnitude. The mean distance between the atoms or molecules is therefore about ten times larger. This has the consequence that their mean kinetic energy is much larger than the mean potential energy of the mutual attraction or repulsion, while for liquid and solid bodies both energies are nearly equal at room temperature (see Sect. 6.1).

In this chapter we present at first the macroscopic properties of gases before we will discuss in more detail the atomic explanation of the observed macroscopic phenomena. The atomic fundamentals, developed already in the 19th century as kinetic gas theory, was one of the most powerful supports for the existence of atoms and their relevance as constituents of matter.

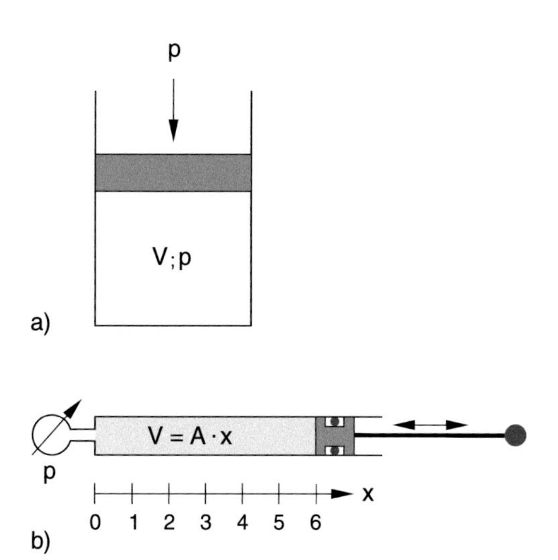

Figure 7.1 A movable piston changes volume V and pressure p of an enclosed gas volume. **a** Principle, **b** demonstration of the Boyle–Marriotte law

7.1 Macroscopic Model

The volume V of an enclosed gas can be changed by a movable piston due to a variable pressure p (Fig. 7.1). For the relation between V and p at a constant temperature T the experiment gives the result

$$p \cdot V = \text{const} \quad \text{(Boyle–Mariotte's law)}\,. \tag{7.1}$$

From $V = \text{const}/p$ we obtain by differentiation

$$\frac{\mathrm{d}V}{\mathrm{d}p} = -\frac{\text{const}}{p^2} = -\frac{V}{p}\,. \tag{7.2}$$

As measure for the compressibility of a gas we define the quantity

$$\kappa = -\frac{1}{V}\frac{\partial V}{\partial p}\,, \quad [\kappa] = \frac{\mathrm{m}^2}{\mathrm{N}}\,. \tag{7.3a}$$

For a constant temperature it follows

$$\frac{\partial V}{\partial p} = \frac{\mathrm{d}V}{\mathrm{d}p} = -\frac{V}{p} \quad \Rightarrow \quad \kappa = \frac{1}{p}\,. \tag{7.3b}$$

A gas is easier to compress for smaller pressures. For a total mass M of a gas in a volume V the density is $\varrho = M/V$. For an enclosed gas its total mass M is constant. Its density ϱ is then inversely proportional to its volume V. Inserting $V = M/\varrho$ into (7.1) gives

$$p = \frac{\text{const}}{M} \cdot \varrho \quad \text{i.e.}\ p \propto \varrho\,. \tag{7.4}$$

For a constant temperature the density ϱ of a gas is proportional to its pressure.

Remark. This is also valid for non-enclosed gases, as for example in the free atmosphere.

This can be seen from the general gas equation $p \cdot V = N \cdot k \cdot T$ (see Sect. 10.3), where N is the total number of molecules in the volume V, which is independent of the boundaries of the volume V. Since $n = N/V$ is the density of molecules with mass m, which is proportional to the mass density $\varrho = n \cdot m/V$ we obtain $p \propto \varrho$.

The gas pressure p with the unit Newton per square meter

$$[p] = \frac{\mathrm{N}}{\mathrm{m}^2} = \text{Pascal} = \text{Pa}$$

can be measured with different techniques (see Sect. 9.3). A simple method uses a mercury manometer (Fig. 7.2). In the container with volume V is a gas under the pressure p. The left branch of a U-shaped tube filled partly with liquid mercury is closed at the upper end, while the right branch is connected with the container. If the upper valve in Fig. 7.2 is open, the pressure p causes a difference Δh of the mercury heights in the two branches. Equilibrium is reached if the pressure caused by the gravity just compensates the pressure p in the container. This gives the relation

$$\varrho \cdot g \cdot \Delta h = p - p_0\,,$$

where p_0 is the vapour pressure of mercury above the mercury surface in the left branch.

Historical the pressure difference corresponding to $\Delta h = 1$ mm between the two mercury columns is named 1 torr, in honour of the Italian physicist Torricelli. The unit used nowadays in the SI-system is 1 Pascal. The following relations hold:

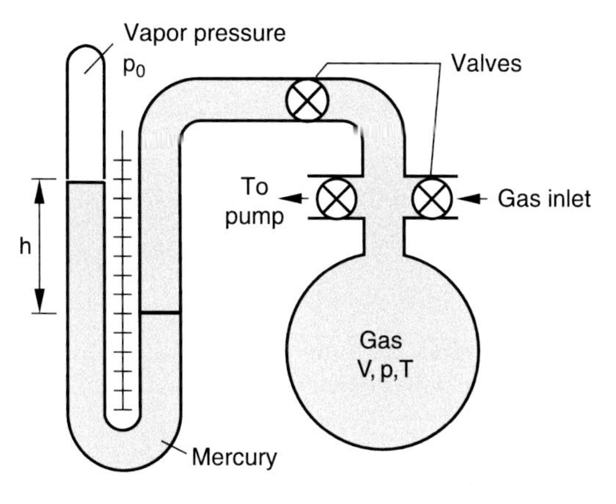

Figure 7.2 Measurement of gas pressure with a mercury manometer

$$1\,\text{Pa} = 1\,\text{N/m}^2$$
$$1\text{ standard atmosphere} = 1\,\text{atm} = 101\,325\,\text{Pa}$$
$$1\text{ torr} = (1/760)\text{ standard atmosphere} = 133.3\,\text{Pa}\,.$$

7.2 Atmospheric Pressure and Barometric Formula

Similar to liquids also in gases a static pressure is present due to the weight of the air. It can be measured with the Torricelli U-tube in Fig. 7.3, filled partly with mercury. The left branch is closed at the upper end, while the right branch is open. In the left branch above the liquid mercury surface is the small vapour pressure p_0 of mercury (about 10^{-3} torr $= 0.13$ Pa at room temperature) which can be neglected. Due to the atmospheric pressure p the mercury in the right branch is depressed by $\Delta h = p/(\varrho \cdot g)$.

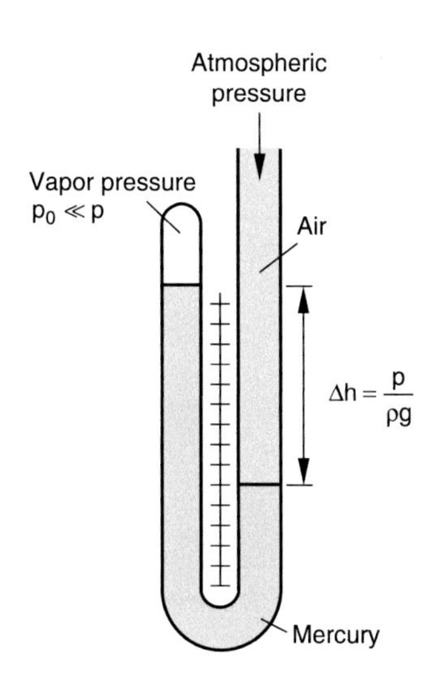

Figure 7.3 Torricelli manometer for measuring the barometric air pressure

Table 7.1 Units of pressure

Unit	Athmosphere abbreviation	Definition	Conversion
1 Pascal	1 Pa	$1\,\text{N/m}^2$	–
1 Hektopascal	1 hPa	$10^2\,\text{N/m}^2$	10^2 Pa
1 Bar	1 bar	$10^5\,\text{N/m}^2$	10^5 Pa
1 Millibar	1 mbar	10^{-3} bar	10^2 Pa
1 Torricelli	1 Torr	1 mm hg	133.32 Pa
1 physical atmosphere	1 atm	760 Torr	101 325 Pa
1 technical atmosphere	1 at	$1\,\text{kp/cm}^2$	$9.8 \cdot 10^4$ Pa

The pressure of the earth atmosphere at sea level $h = 0$ at normal weather conditions is 101 325 Pa and is called *normal pressure* or *standard pressure* which is often given in the unit 1 standard atmosphere (1 atm). This pressure causes in the mercury manometer of Fig. 7.3 a height difference of 760 mm.

In meteorology the hecto-pascal (1 hPa $=$ 100 Pa) is a commonly used unit. In Tab. 7.1 the conversion factors for some pressure units are listed [7.1].

The weight of the air column above an area A in the height h decreases with increasing h (Fig. 7.4). Changing the position of A from h to $h + \mathrm{d}h$ decreases the weight by $\varrho \cdot g \cdot A \cdot \mathrm{d}h$ and therefore the pressure p decreases as

$$\mathrm{d}p = -\varrho \cdot g \cdot \mathrm{d}h\,. \tag{7.5}$$

In the case of liquids the density is independent of the height because of the small compressibility. The solution of (7.5) shows the linear dependence $p = -\varrho \cdot g \cdot h$ of the pressure on the height h (Fig. 7.5b). This is no longer true for gases, where the density is proportional to the pressure and therefore depends on the heights in the atmosphere. From (7.4) we obtain for a constant temperature T

$$\frac{p}{\varrho} = \frac{p_0}{\varrho_0} = \text{const} \Rightarrow \varrho = \frac{\varrho_0}{p_0} \cdot p\,.$$

Inserting this into (7.5) gives

$$\mathrm{d}p = -\frac{\varrho_0}{p_0} g p\,\mathrm{d}h\,. \tag{7.5a}$$

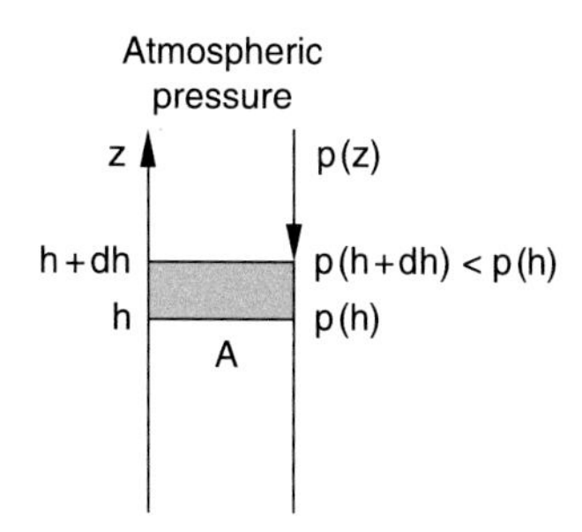

Figure 7.4 Derivation of barometric Eq. 7.6

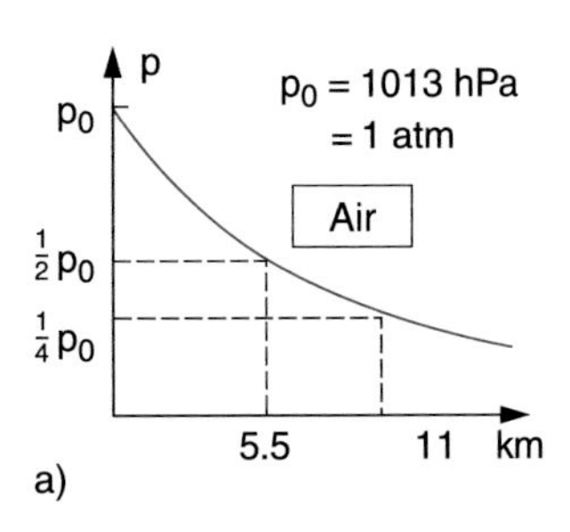

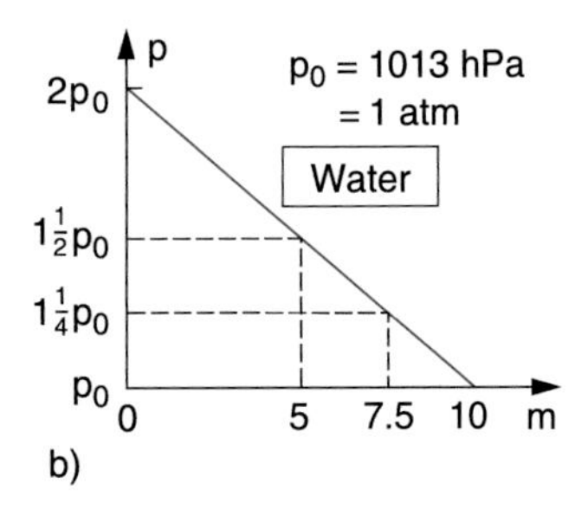

Figure 7.5 Comparison of pressure dependence $p(s)$ in the earth atmosphere and in a water column of 10 m height

Integration yields

$$\ln p = -\frac{\varrho_0}{p_0} gh + C \,. \tag{7.5b}$$

With $p(h = 0) = p_0$ the integration constant C becomes $C = \ln p_0$. Solving for p gives the barometric formula

$$p = p_0 \cdot e^{-\varrho_0 g \cdot h / p_0} \,. \tag{7.6a}$$

Note, that the ratio ϱ_0/p_0 depends on the temperature T.

The atmospheric pressure of an isothermal atmosphere (T = constant, independent of h) decreases exponential with the height h (Fig. 7.5a).

Because $\varrho = (\varrho_0/p_0) \cdot p$ the density ϱ follows the same formula

$$\varrho = \varrho_0 e^{-\varrho_0 g h / p_0} \,. \tag{7.6b}$$

Inserting the numerical values of the atmosphere ($\varrho_0 = 1.24\,\text{kg/m}^3$ and $p_0 = 1013\,\text{hPa}$) into (7.6a) yields

$$p = p_0 \cdot e^{-h/8.33\,\text{km}} \,. \tag{7.6c}$$

For $h = 8.33\,\text{km}$ the pressure p has decreased to $p_0/e \approx 373\,\text{hPa}$.

The height $h_{1/2}$ where the pressure has dropped to $(1/2)p_0$ is obtained from $\exp(g \cdot h_{1/2}\varrho_0/p_0) = 2$ which gives $h_{1/2} = 5.77\,\text{km}$.

On a mountain with an elevation of 5.77 km the barometric pressure sinks to half of its value at $h = 0$.

The pressure decrease of an isothermal atmosphere follows the exponential law (7.6a) contrary to the linear decrease in liquids. The atmosphere has therefore no sharp boundary!

Note: The real earth atmosphere is not isothermal! The temperature decreases with increasing height (see Sect. 7.6). The pressure $p(h)$ is therefore slightly different from (7.6). Nevertheless is (7.6) a useful approximation, which is sufficiently accurate for many applications.

Analogue to the situation in liquids the Archimedes' principle of buoyancy is valid for bodies in air.

A body in air experiences a buoyancy force, that has the opposite direction but the equal amount as the weight of the displaced air.

This principle is the basis for balloon flights. A balloon can only rise in air, if its weight (balloon + car + passengers) is smaller than the buoyancy force. For a balloon with total mass M and total volume V this gives the condition

$$M \cdot g < V \cdot \varrho_{\text{air}} g \,.$$

The balloon must therefore contain a gas with a smaller density than the surrounding air. Generally helium is used, since hydrogen is too dangerous due to its possible explosion.

Another solution is the hot-air balloon, where a burner blows hot air into the balloon shell. The density $\varrho = p/kT$ is inversely proportional to the temperature T (Fig. 7.6).

Example

The density ϱ at a temperature $T = 80\,°\text{C} = 350\,\text{K}$ is smaller than air at room temperature $T = 20\,°\text{C} = 290\,\text{K}$ by the factor $290/350 = 0.83$. For a balloon volume of $3000\,\text{m}^3$ the buoyancy force is then $F_\text{B} = g \cdot V \cdot \Delta\varrho = 8.270\,\text{N}$. The maximum mass of balloon + passengers is then 843 kg. The mass of the balloon is about 100 kg, that of the burner with propane supply about 200 kg. This leaves a maximum weight for the passengers of 543 kg. ◀

Since the density of the earth atmosphere decreases exponentially with the height h, also the buoyancy decreases with h. This limits the maximum altitude of weather- and research-balloons, which should rise up to very high altitudes in order to investigate the upper part of our atmosphere. One uses extremely large balloon volumes which are at the ground only partly filled with helium but blow up with increasing altitude because of the decreasing air pressure (Probl. 7.18) (Fig. 7.7).

Figure 7.6 Lift of the first manned Montgolfiere. At the 21st of November 1783 the hot-air-balloon, named after its inventor Montgolfiere, started in the garden of the castle Muette near Paris with two ballonists and landed safely after 25 min at a distance of 10 km from the starting point. The balloon was constructed with thin branches of a willow tree, that stabilized the envelope of thin fabric, covered with coulorfully painted paper. The air inside the ballon was heated by a coal fire in the center of the lower opening. With kind permission of Deutsches Museum München

Figure 7.7 Start of a helium-filled research balloon for the investigation of the higher stratosphere. The balloon is filled only with a low He-pressure. With increasing height the external pressure decreases and the balloon inflates, increasing its volume and the buoancy (SSC/DLR) (http://www.eskp.de/turmhohe-forschungsballons-messen-ozonschicht/)

7.3 Kinetic Gas Theory

The kinetic gas theory, which was developed by Boltzmann, Maxwell and Clausius in the second half of the 19th century, attributes all observed macroscopic properties of gases to the motion of atoms and their collisions with each other and the wall. Its success has essentially contributed to the acceptance of the atomic hypothesis (see Vol. 3, Chap. 2). The exact theoretical description requires a more detailed and advanced mathematical model. We will therefore restrict the treatment to a simplified model, which, however, describes the essential basic ideas and the experimental findings correctly.

7.3.1 The Model of the Ideal Gas

The most simple gas model is based on the following assumptions: The gas consists of atoms or molecules which can be described by rigid balls with radius r_0. They move with statistically distributed velocities inside the gas container. Collisions with each other or with the walls are governed by the laws of energy-and momentum conservation. The collisions are completely elastic. Any interaction between the balls only occurs during collisions (direct touch of two balls). For distances $d > 2r_0$ the atoms do not interact. The interaction potential in this model is therefore (Fig. 7.8)

$$\begin{aligned} E_{\mathrm{pot}}(r) &= 0 \qquad \text{for } |r| > 2r_0 \\ E_{\mathrm{pot}}(r) &= \infty \qquad \text{for } |r| \le 2r_0\,. \end{aligned}$$

Such a gas model is called **ideal gas**, if r_0 is very small against the mean distance $\langle d \rangle$ between the atoms. This means that the atomic volume is negligible compared to the volume V of the gas container. In this model the atoms can be treated as point masses (see Sect. 2.1).

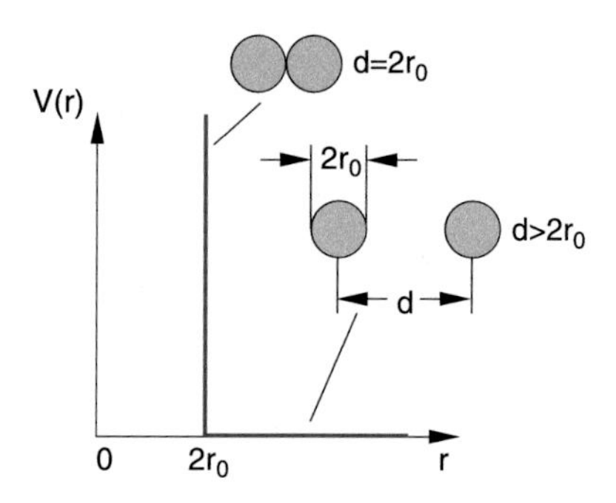

Figure 7.8 Interaction potential between two rigid balls with radius r_0

Example

At a pressure of 1 bar and room temperature ($T = 300\,\mathrm{K}$) $1\,\mathrm{cm}^3$ of a gas contains about $3 \cdot 10^{19}$ molecules. Their mean relative distance is $\langle d \rangle = 3\,\mathrm{nm}$. For helium atoms is $r_0 \approx 0.05\,\mathrm{nm}$, which implies $r_0/\langle d \rangle = 0.017 \ll 1$. Helium at a pressure of 1 bar can be therefore regarded as ideal gas. ◀

The pressure that the gas exerts onto the wall is caused by the momentum transfer of the atoms to the wall. The force F acting on the area A of the wall during collision of the atoms with the wall is equal to the momentum transfer per second to the area A. The pressure $p = F/A$ is then

$$p = \frac{\mathrm{d}}{\mathrm{d}t}\left(\frac{\text{momentum transfer to area } A}{\text{area } A}\right) . \tag{7.7}$$

If for instance $N \cdot \mathrm{d}t$ atoms with mass m hit the wall within the time interval $\mathrm{d}t$ with a velocity v in the direction of the surface normal, the momentum transfer per second for elastic collisions is $2N \cdot m \cdot v$ and the pressure onto the wall is $p = 2N \cdot m \cdot v/A$.

7.3.2 Basic Equations of the Kinetic Gas Theory

We will at first regard the atoms as point masses and only take into account their translational energy. The discussion of rotationally or vibrationally excited molecules demands a farther reaching discussion which will be postponed to Sect. 10.2

For N molecules in a volume V the number density is $n = N/V$. At first we regard only that part n_x of all molecules per cm^3 in a cubical volume, which move with the velocity v_x into the x-direction (Fig. 7.9). Within the time interval Δt

$$Z = n_x \cdot v_x \cdot A \cdot \Delta t$$

molecules hit the surface area A. These are just those molecules in the cuboid with length $v_x \Delta t$ and cross section A, illustrated in Fig. 7.9. Each of these molecules transfers the momentum

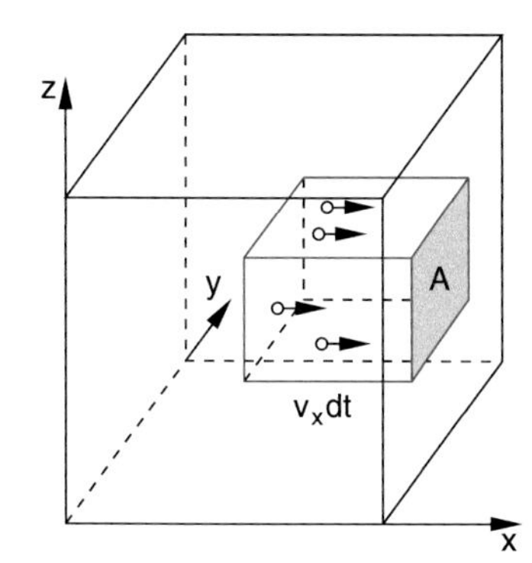

Figure 7.9 Derivation of Eq. 7.8

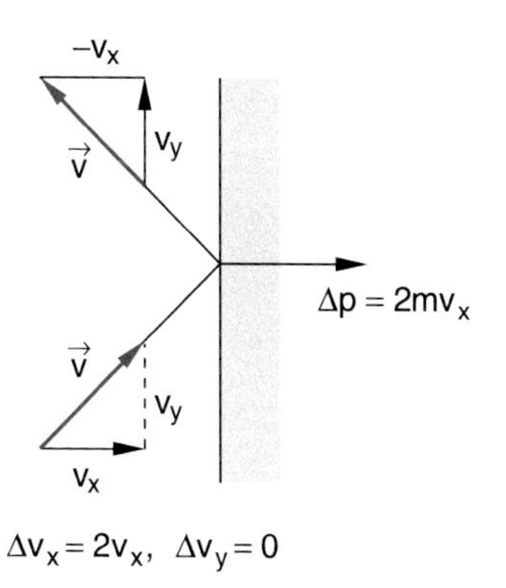

Figure 7.10 Momentum transfer for elastic collisions with a wall

$\Delta p_x = 2\,\mathrm{m} \cdot v_x$. The force onto the area A is then $F = Z \cdot \Delta p_x/\Delta t = 2Z \cdot m \cdot v_x/\Delta t$ and the pressure onto the wall is

$$p = F/A = 2m \cdot n_x v_x^2 . \tag{7.8}$$

When molecules with the velocity $\boldsymbol{v} = \{v_x, v_y, v_z\}$ do not move vertically to the wall, still only the momentum $2m \cdot v_x$ is transferred, because the tangential components do not transfer momentum to the wall (Fig. 7.10).

Not all molecules have the same velocity. Under stationary conditions the velocities of the molecules are isotropically distributed, which means that each direction has the same probability. Since the pressure of a gas is isotropic, the mean momentum transfer must be equal into all directions. The mean square value of the velocities is*

$$\langle v_x^2 \rangle = \overline{v_x^2} = \frac{1}{N}\int N(v_x)\, v_x^2 \,\mathrm{d}v_x = \overline{v_y^2} = \overline{v_z^2} , \tag{7.9}$$

where $N(v_x)\mathrm{d}v_x$ is the number of molecules in the volume V with the velocity components v_x in the interval from v_x to $v_x + \mathrm{d}v_x$. Because on the average half of the molecules move into the $+x$-direction and the other half into the $-x$-direction, the pressure, exerted by all molecules with a number density $n = N/V$ in the x-direction on the wall in the y-z-plane is given by

$$p = \tfrac{1}{2}n \cdot 2m\overline{v_x^2} = n \cdot m \cdot \overline{v_x^2} . \tag{7.10}$$

With $v^2 = v_x^2 + v_y^2 + v_z^2$ it follows from (7.9)

$$\overline{v_x^2} = \overline{v_y^2} = \overline{v_z^2} = \tfrac{1}{3}\overline{v^2} . \tag{7.11}$$

This gives with (7.10)

$$p = \frac{1}{3}m \cdot n\overline{v^2} = \frac{2}{3}n \cdot \frac{m}{2}\overline{v^2} = \frac{2}{3}n \cdot \overline{E_{\text{kin}}} , \tag{7.12a}$$

where $\overline{E_{\text{kin}}} = (m/2) \cdot n \cdot \overline{v^2}$ is the mean kinetic energy per molecule. With $n = N/V$ this can be written as

$$p \cdot V = \tfrac{2}{3}N \cdot \tfrac{1}{2}m\overline{v^2} , \tag{7.12b}$$

where $N = n \cdot V$ is the total number of molecules in the volume V.

* **Remark.** We use for the average values of a quantity A the notations $\langle A \rangle$ as well as $\overline{A}$.

7.3.3 Mean Kinetic Energy and Absolute Temperature

All experiments give the result that the product $p \cdot V$ depends solely on the temperature and is for constant temperature constant (*Boyle–Mariotte's law*). This implies also that the mean kinetic energy $\overline{E_{\text{kin}}} = (m/2)\overline{v^2}$ depends on the temperature. It turns out that it is convenient to define an absolute temperature which is proportional to $\overline{E_{\text{kin}}}$.

The **absolute temperature** T (with the unit 1 Kelvin = 1 K) is defined by the relation

$$\frac{m}{2}\overline{v^2} = \frac{3}{2}kT , \tag{7.13}$$

where $k = 1.38054 \cdot 10^{-23}\,\text{J/K}$ is the **Boltzmann constant**.

With this definition Eq. 7.12 becomes the *general gas-equation*

$$p \cdot V = N \cdot k \cdot T , \tag{7.14}$$

which represents a generalization of Boyle–Mariotte's law (7.1) and which reduces to (7.1) for $T = \text{const}$.

Each molecule can move into three directions x, y, and z. This means it has three **degrees of freedom** for its translation. Collisions with other molecules change direction and magnitude of its velocity. In the time-average all directions are equally probable. We therefore obtain for the mean square velocities the relations

$$\left\langle v_x^2 \right\rangle_t = \left\langle v_y^2 \right\rangle_t = \left\langle v_z^2 \right\rangle_t = \tfrac{1}{3}\left\langle v^2 \right\rangle_t = \tfrac{1}{3}\overline{v^2} .$$

The mean kinetic energy of a molecule at the temperature T is then

$$\overline{E_{\text{kin}}} = \tfrac{1}{2}kT \text{ per degree of freedom} .$$

Remark. In statistical physics it is proved [7.2] that in a closed system of many mutually interacting particles at thermal equilibrium the time average $\langle A \rangle_t$ of a physical quantity A is equal to the ensemble average

$$\overline{A} = \frac{1}{N}\sum A_i ,$$

averaged over all particles of the ensemble and determined at a fixed time (*ergoden hypothesis*). This is however, only true under certain conditions, which have to be proved for each case. The "*ergoden-theory*" is a current field of research in mathematics and statistical physics.

Real molecules can rotate and vibrate. The energy of these degrees of freedom have to be taken into account in addition to the translational energy. The number of degrees of freedom therefore becomes larger. For example diatomic molecules can rotate around two axes perpendicular to the molecular axis. This gives the additional energy $E_{\text{rot}} = L^2/2I$, where $\boldsymbol{L}$ is the angular momentum of the rotation and I the inertial moment (see Sect. 5.5).

At sufficiently high temperatures also vibrations of molecules can be excited (see Sect. 10.3) which contribute to the total energy.

Equipartition Law

In a gas that is kept sufficiently long at a constant temperature T the energy of the atoms or molecules is uniformly distributed by collisions over all degrees of freedom. Therefore each molecule has on the average the energy $E_{\text{kin}} = f \cdot (1/2)kT$ where f is the number of degrees of freedom, accessible to the molecule.

7.3.4 Distribution Function

After the descriptive discussion of the relation between mean kinetic energy and the pressure of a gas for the special case of a cuboid container we will now give a more quantitative representation for the general case of an arbitrary volume. For this purpose the velocity distribution has to be defined in a quantitative way. This can be achieved with the distribution function $f(u)$ (see Sect. 1.8), which describes how the quantity u is distributed among the different molecules. For $u = v_x$ we obtain

$$f(v_x)\mathrm{d}v_x = \frac{N(v_x)\mathrm{d}v_x}{N}$$
$$\text{with } N = \int_{-\infty}^{+\infty} N(v_x)\mathrm{d}v_x . \tag{7.15}$$

The quantity $f(v_x)\mathrm{d}v_x q$ gives the fraction of all molecules with a velocity component v_x in the interval v_x to $v_x + \mathrm{d}v_x$. The number of all particles within the interval v_x to $v_x + \mathrm{d}v_x$ is then

$$N(v_x)\mathrm{d}v_x = N \cdot f(v_x)\mathrm{d}v_x . \tag{7.15a}$$

The number of molecules with $v_x \geq u$ is then

$$N(v_x \geq u) = N \int_{v_x=u}^{\infty} f(v_x)\mathrm{d}v_x . \tag{7.15b}$$

From (7.15) we obtain the normalization condition

$$\int_{-\infty}^{+\infty} f(v_x)\mathrm{d}v_x = \frac{1}{N}\int_{-\infty}^{+\infty} N(v_x)\mathrm{d}v_x = 1 . \tag{7.16}$$

For $u = |\boldsymbol{v}| = v$ the quantity $f(v)\mathrm{d}v$ gives the fraction of all molecules with velocity amounts between v and $v + \mathrm{d}v$. The normalization is now

$$\int_{0}^{\infty} f(v)\mathrm{d}v = 1 . \tag{7.16a}$$

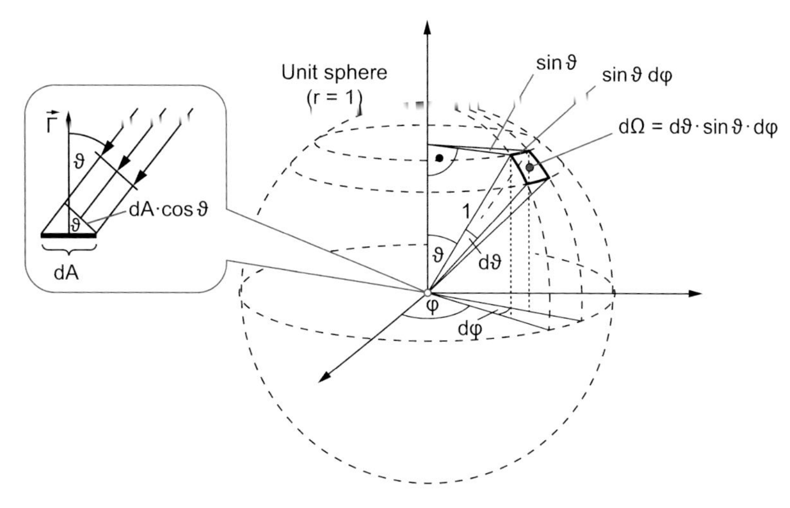

Figure 7.11 Illustration of Eq. 7.17 and 7.18. Because of the small area dA and the solid angle $d\Omega$ the velocity vectors of particles hitting dA within the solid angle $d\Omega$ are approximately parallel

We consider a surface element dA, which is hit by molecules from the upper half space (Fig. 7.11). Within the time interval Δt the number Z of molecules within the velocity interval from v to $v + dv$ coming from the angular range $d\Omega$ around the angle ϑ against the surface normal and impinging on dA is

$$Z = n \cdot f(v)dv \cdot dA \cos\vartheta \cdot v\Delta t \cdot \frac{d\Omega}{4\pi} . \quad (7.17)$$

This can be seen as follows: The product $n \cdot f(v)dv$ gives the particle density within the velocity interval dv. Within the time interval Δt all molecules up to a distance $v \cdot \Delta t$ from the surface element dA can reach dA. From all molecules with isotropic velocity distribution only the fraction $d\Omega/4\pi$ reaches the effective area $dA \cdot \cos\vartheta$ within the solid angle $d\Omega$. The momentum change $|\Delta p|$ of a molecule at the impact on dA is

$$|\Delta p| = 2\,\mathrm{m} \cdot v \cdot \cos\vartheta .$$

The momentum transfer of Z molecules per sec is then

$$\Delta p_{\text{total}}/dt = Z \cdot |\Delta p|/dt . \quad (7.17a)$$

Integration over all velocities v and over all impact angles ϑ yields the total momentum transfer per sec which is equal to the product $p \cdot dA$ of pressure p acting onto dA and the area dA.

Remark. The bold vector $\boldsymbol{p}$ denotes the momentum while the scalar quantity p indicates the pressure. Although both quantities are labelled with the same letter (this is in agreement with the general convention), there should be no confusion, because it is clear from the text, which of the two quantities is meant. With the solid angle

$$d\Omega = \frac{r \cdot d\vartheta \cdot r \cdot \sin\vartheta \cdot d\varphi}{r^2} = d\vartheta \cdot \sin\vartheta \cdot d\varphi , \quad (7.18)$$

the pressure p can be obtained from (7.17)–(7.18) as

$$p = \frac{|\Delta \boldsymbol{p}|_{\text{total}}}{dA \cdot \Delta t} = \frac{2n \cdot m}{4\pi} \int\limits_{v=0}^{\infty} v^2 f(v) dv \times \int\limits_{\varphi=0}^{2\pi} \int\limits_{\vartheta=0}^{\pi/2} \cos^2\vartheta \cdot \sin\vartheta \, d\vartheta \, d\varphi . \quad (7.19)$$

The first integral gives the quadratic means $\overline{v^2}$. The second double integral can be analytically solved and has the solution $2\pi/3$.

This gives finally the pressure p onto the area dA

$$p = \tfrac{1}{3} n \cdot m \cdot \overline{v^2} ,$$

in accordance with (7.12a). In order to calculate the mean square $\overline{v^2}$ we must determine the distribution function $f(v)$. This will be the task of the next section.

7.3.5 Maxwell–Boltzmann Velocity Distribution

The decrease of the air density with increasing height in our atmosphere (barometric formula) discussed in the previous section can be explained by the velocity distribution $f(v)$ of the air molecules.

If we extend the exponent in (7.6b) with the volume V_0 of a gas with mass $M = \varrho_0 \cdot V_0$ and insert for $p_0 \cdot V_0$ the general gas-equation 7.14 Eq. 7.6b becomes

$$\varrho = \varrho_0 \cdot e^{-(Mgh)/(NkT)} . \quad (7.20a)$$

For the number density $n = \varrho/m$ of gas molecules with mass m we obtain with $m = M/N$

$$n(h) = n_0 \cdot e^{-(mgh)/(kT)} = n_0 \cdot e^{-E_p/kT} . \quad (7.20b)$$

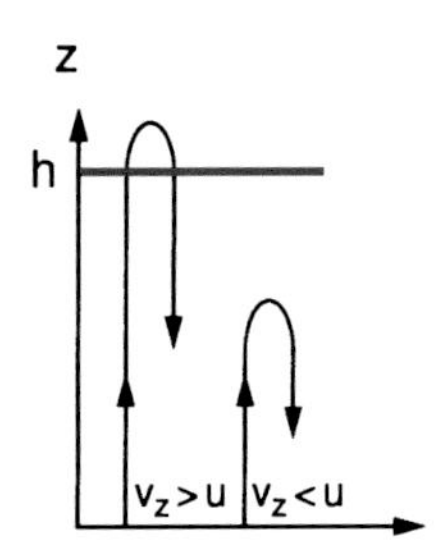

Figure 7.12 Only molecules with initial velocites $V_z(h = 0) > u$ reach the height $z = h$

The exponent in (7.20b) represents the ratio of potential energy $E_p = mgh$ at the height h above ground and twice the mean kinetic energy $E_{kin} = (1/2)kT$ per degree of freedom of a molecule, due to its thermal motion at the temperature T. We regard here at first only a one-dimensional motion in the z-direction, which is described by the distribution of the z-components v_z of the velocity $\boldsymbol{v}$.

The barometric formula gives for an isothermal atmosphere the ratio of particle densities in layers Δz at different heights h. If the molecules had no kinetic energy, they would all rest on the earth surface and form a solid layer, i.e. the earth atmosphere would disappear.

Let us first assume the atmosphere had been built up by molecules that start from the ground $z = 0$ upwards with the velocity $v_z = u$. They reach the height h, given by

$$\frac{1}{2}m \cdot u^2 = m \cdot g \cdot h$$

before they fall down in the earth gravitational field (Fig. 7.12). In fact the primary earth atmosphere had been formed by molecules outgassing through volcanos.

Since we have assumed an isothermal atmosphere, the velocity distribution at $z = h$ must be the same as for $z = 0$. This means, there are at $z = h$ also molecules with $v_z = u$ which move upwards until their kinetic energy is just cancelled by the increase of the potential energy. The density of these molecules is, however, smaller at $z = h$ than at $z = 0$ because the number density $n(h)$ decreases exponentially with h according to the barometric formula (7.20b). The specific choice of the group of molecules starting from $z = 0$ does not constrict the following argumentation.

Remark.

1. Some readers have argued that the decrease of $v(z)$ with increasing z contradicts the assumption of an isothermal atmosphere. This is, however, not true, because the temperature is determined by the velocity distribution of **all** molecules but not only by that of an arbitrarily selected subgroup.
2. In the real atmosphere the temperature decreases with increasing z. The main reason for that is the decreasing heat flow from the ground into the atmosphere and the decreasing absorption of the infrared radiation emitted by the earth surface.

We will assume in the following an isothermal atmosphere where collisions can be neglected, although collisions are responsible for the equipartition of the total energy onto all molecules and therefore for establishing a temperature. However, this does not influence the validity of the following derivation where we select a subgroup of all molecules with a velocity component v_z in the $+z$-direction which fly upwards.

The number $N_{>u}(z = 0)$ of molecules that start from a unit area at $z = 0$ with velocity components $v_z > u$ is equal to the number $N_{>0}(z = h)q$ of molecules that fly through a unit area in the plane $z = h$ with velocities $v_z > 0$.

$$N_{>u}(z = 0) = N_{>0}(z = h) . \tag{7.20c}$$

The number $N(v_z)$ of molecules that pass per unit time with the velocity v_z through a unit surface (flux density) is given by the product

$$N(v_z) = n(v_z) \cdot v_z$$

of number density $n(v_z)$ and velocity v_z (Fig. 7.13).

For an isothermal atmosphere with the constant temperature T the mean square velocity $\overline{v^2}$ and the distribution function $f(v_z)$ must be independent of h, since they depend only on the temperature.

The flux density $N_{>0}(z = h)$ can be expressed as

$$N_{\geq 0}(z = h) = n(h) \cdot \int_{v_z=0}^{\infty} v_z f(v_z) \mathrm{d}v_z . \tag{7.21a}$$

It is smaller than the flux density

$$N_{>0}(z = 0) = n(0) \int_{0}^{\infty} v_z f(v_z) \mathrm{d}v_z . \tag{7.21b}$$

Since the two integrals in the two equations are equal, it follows with (7.20c)

$$\frac{N_{v_z>u}(0)}{N_{v_z>0}(0)} = \frac{N_{v_z>0}(z = h)}{N_{v_z>0}(0)} = \frac{n(h)}{n(0)} . \tag{7.21c}$$

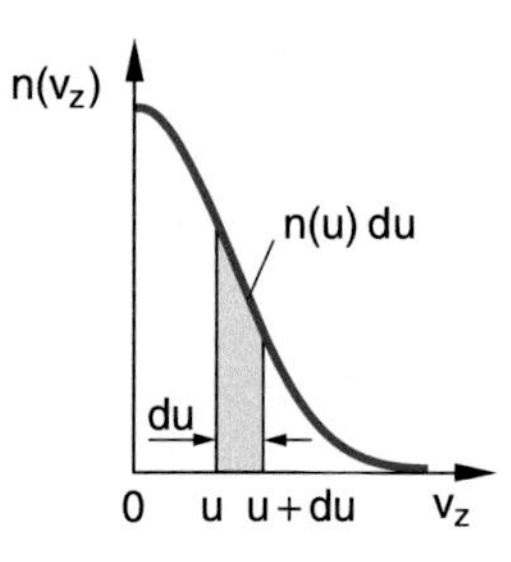

Figure 7.13 Number of particles per unit volume $n(v_z)\mathrm{d}v_z$ within the intervall $u \leq v_z \leq u + du$

From the definition of the distribution function $f(v_z)$ in (7.15b) and (7.15) it follows for the flux densities

$$N_{v_z \geq 0}(z=0) = n_0 \int_{v_z=0}^{\infty} v_z f(v_z) \mathrm{d}v_z$$
$$N_{v_z \geq u}(z=0) = n_0 \int_{v_z=u}^{\infty} v_z f(v_z) \mathrm{d}v_z \ , \tag{7.22}$$

with (7.20b) and the relation $m \cdot g \cdot h = (1/2)m \cdot v^2$ we obtain

$$\int_u^{\infty} v_z f(v_z) \mathrm{d}v_z = C_1 \cdot e^{-\frac{m}{2}u^2/kT} \ , \tag{7.23}$$

where C_1 is a constant which depends on the temperature T.

Differentiating both sides with respect to the lower limit u yields on the left side the negative integrand for $v_z = u$:

$$-u \cdot f(u) = -\frac{m \cdot u}{kT} \cdot C_1 \cdot e^{-\frac{m}{2}u^2/kT}$$
$$\Rightarrow f(u) = C_2 \cdot e^{-\frac{m}{2}u^2/kT}$$
$$\text{with } C_2 = C_1 \cdot \frac{m}{kT} = \text{const} \ .$$

The constant C_2 can be obtained from the normalization condition

$$\int f(u) \mathrm{d}u = 1$$

and the integration $\int e^{-x^2} \mathrm{d}x = \sqrt{\pi}$. This gives $C_2 = (m/2\pi kT)^{(1/2)}$.

Replacing u by v_z, finally yields the distribution function

$$f(v_z) = \sqrt{\frac{m}{2\pi kT}} \cdot e^{-\frac{m}{2}v_z^2/kT} \ . \tag{7.24}$$

This is a symmetric Gauss-distribution illustrated in Fig. 7.14.

For a gas in a closed volume V, where the mean kinetic energy is large compared to the difference of potential energies inside the volume, no direction for the motion of the molecules is preferred, all directions are equally probable as has been discussed in Sect. 7.3.2. The distributions of the velocity components are equal for all three components v_x, v_y, v_z and are described by (7.24).

The probability to find a molecule with the velocity $\boldsymbol{v} = \{v_x, v_y, v_z\}$ is equal to the product of the probabilities for v_x, v_y and v_z. One therefore obtains for the distribution function

$$f(v_x, v_y, v_z) = \left(\frac{m}{2\pi kT}\right)^{(3/2)} e^{-(mv^2)/(2kT)} \ . \tag{7.25}$$

In many cases only the magnitude $|\boldsymbol{v}|$ of the velocity is of interest, where the direction can be arbitrary. The heads of all velocity arrows with a length between v and $v + \mathrm{d}v$ are located within a spherical shell with the volume $4\pi v^2 \mathrm{d}v$. Therefore the integration

$$\int_{v_x, v_y, v_z} f(v_x, v_y, v_z) \, \mathrm{d}v_x \, \mathrm{d}v_y \, \mathrm{d}v_z$$

over all values of $\boldsymbol{v} = \{v_x, v_y, v_z\}$ within this spherical shell gives for the number density $n(v)\mathrm{d}v$ of all molecules per unit volume with velocities between v and $v + \mathrm{d}v$ the result

$$n(v)\mathrm{d}v = n \cdot \left(\frac{m}{2\pi kT}\right)^{(3/2)} \cdot 4\pi v^2 \cdot e^{-mv^2/2kT} \mathrm{d}v \ . \tag{7.26}$$

This is the **Maxwell–Boltzmann velocity distribution** (Fig. 7.15). The normalized distribution function is then $f(v) = n(v)/n$, where n is the total number density of molecules with any velocity.

Note

Contrary to the symmetric distribution for the velocity components, which extends from $-\infty$ to ∞, the distribution for the velocity magnitude is restricted to the range $v \geq 0$, because there are no negative velocities. The distribution is therefore asymmetric. Because of the factor v^2 it is also not symmetric around a mean value $\overline{v}$.

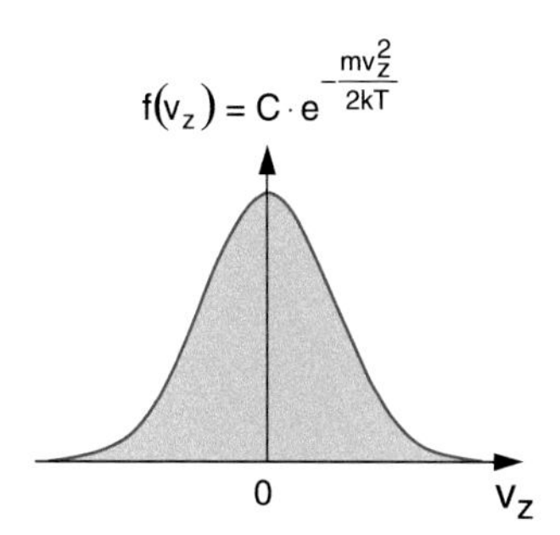

Figure 7.14 Distribution function $f(v_z)$ of the velocity component v_z

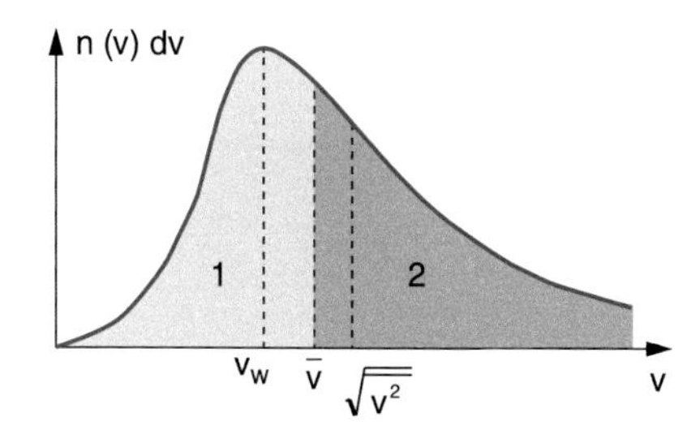

Figure 7.15 Maxwell–Boltzmann velocity distribution $n(v)\mathrm{d}v$ with most probable velocity v_w, mean velocity $\langle v \rangle$ and the square root of the mean velocity square $\sqrt{\overline{v^2}}$

Remark. In unidirectional molecular beams generally not the molecular density $n(v)$ is measured but rather the flux density $N = n(v)\cdot v$, for instance by a detector which measures the number of particles reaching the detector per unit time. The velocity distribution of the flux density $N(v) = n\cdot v\cdot f(v)$ in a collimated molecular beam where all molecules fly within a small angular cone around the x-direction differs from $n(v)$ in (7.26) by the additional factor v. The prefactor of $f(v)$ is therefore v^3 instead of v^2.

The maximum of the distribution (7.26) appears at the most probable velocity v_w. With the condition $\mathrm{d}n(v)/\mathrm{d}v|_{mp} = 0$ one obtains from (7.26) the value

$$v_\mathrm{w} = \sqrt{\frac{2kT}{m}} \,. \tag{7.27}$$

The mean velocity $\overline{v}$ is defined by

$$\begin{aligned}\overline{v} &= \int\limits_0^\infty v\cdot f(v)\mathrm{d}v \\ &= 4\pi\cdot\left(\frac{m}{2\pi kT}\right)^{(3/2)}\int\limits_0^\infty v^3\cdot e^{-mv^2/(2kT)}\mathrm{d}v \,.\end{aligned}$$

Integration yields

$$\overline{v} = \sqrt{\frac{8kT}{\pi\cdot m}} = \frac{2v_\mathrm{w}}{\sqrt{\pi}} \,. \tag{7.28}$$

Finally we get for the mean square $\overline{v^2}$

$$\overline{v^2} = \int\limits_0^\infty v^2 f(v)\mathrm{d}v = \frac{3kT}{m} \,. \tag{7.29}$$

This gives for the mean energy of a particle with three translational degrees of freedom the result

$$\frac{m}{2}\overline{v^2} = \frac{3}{2}kT = f\cdot\frac{1}{2}kT \,,$$

which has been already used in Sect. 7.3.2.

The sequence of magnitudes for the three special velocities is

$$v_w < \overline{v} < \sqrt{\langle v^2\rangle} \,.$$

With the most probable velocity v_w (7.27) Eq. 7.26 can be written as

$$\begin{aligned}n(v)\mathrm{d}v &= n\cdot\frac{4v^2}{v_\mathrm{w}^3\sqrt{\pi}}e^{-mv^2/2kT}\mathrm{d}v \\ &= n\cdot\frac{4v^2}{v_\mathrm{w}^3\cdot\sqrt{\pi}}e^{-v^2/v_\mathrm{w}^2}\mathrm{d}v \,.\end{aligned} \tag{7.30}$$

The velocity distribution depends strongly on the temperature T. In Fig. 7.16 two distributions are shown for two different temperatures T_1 and T_2 which are related by $T_1/T_2 = 1:4$. The numerical values of the velocities $\overline{v}$, v_w and $(\overline{v^2})^{(1/2)}$ change by a factor of 2 because they are proportional to the square root of the temperature (see also Tab. 7.2).

Table 7.2 Mean values of thermal velocities

Quantity	Symbol	Mathematical expression
Most probable velocity	v_w	$\sqrt{\frac{2kT}{m}}$
Mean velocity	$\overline{v}$	$\sqrt{\frac{8kT}{\pi m}} = \frac{2}{\sqrt{\pi}}v_\mathrm{w}$
Square root of mean velocity square	$\sqrt{\langle v^2\rangle}$	$\sqrt{\frac{3kT}{m}} = \sqrt{\frac{3}{2}}v_\mathrm{w}$

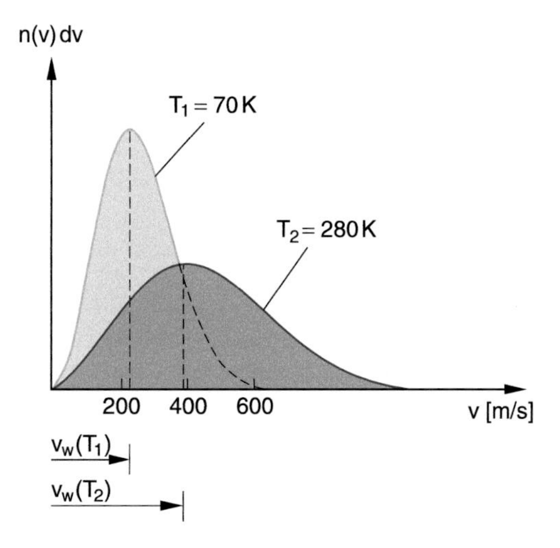

Figure 7.16 Velocity distribution of N_2-molecules at two different temperatures. The area under the two curves represent the total number of particles per unit volume. For a closed gas system the two areas are equal

Example

The density of nitrogen gas N_2 at room temperature $T = 300\,\mathrm{K}$ and at a pressure of 1 bar is

$$\varrho(\mathrm{N_2}) = 1.12\,\mathrm{kg/m^3}, m(n_2) = 4.67\cdot 10^{-26}\,\mathrm{kg} \,.$$

This gives

$$\begin{aligned}n = \varrho/m &= 2.4\cdot 10^{25}\,\mathrm{N_2\ molecules/m^3} \\ &= 2.4\cdot 10^{19}\,\mathrm{N_2\ molecules/cm^3} \,.\end{aligned}$$

The numerical values for the velocities can be calculated from (7.27) and (7.29) as

$$v_\mathrm{w} = 422\,\mathrm{m/s}; \quad \overline{v} = 476\,\mathrm{m/s}; \quad \sqrt{\langle v^2\rangle} = 517\,\mathrm{m/s} \,.$$

The mean kinetic energy of a molecule is $\overline{E_\mathrm{kin}} = (3/2)kT = 6.21\cdot 10^{-21}\,\mathrm{J}$ the energy density of all molecules per cm³ is $n\cdot\overline{E_\mathrm{kin}} = n\cdot(3/2)kT = 0.15\,\mathrm{J/cm^3}$. ◀

7.3.6 Collision Cross Section and Mean Free Path Length

The model of the ideal gas describes the gas particles by small rigid spheres with a radius r_i that is small compared to the average distance $\overline{d}$ between the spheres. A collision takes place if the spheres touch each other, i.e. if $d \leq (r_1 + r_2)$.

We define the impact parameter b for the collision between two particles A_1 and A_2 as the distance between two straight lines (Fig. 7.17):

1. The path of the centre of A_1 without any interaction.
2. The straight line through the centre of A_2 parallel to line 1. (see also Sect. 4.3).

In this model a collision takes place if $b \leq (r_1 + r_2)$. At the collision the closest distance between the centres of A_1 and A_2 is $d = r_1 + r_2$. All particles A_1 for which their centre passes through the circular area

$$\sigma = \pi \cdot (r_1 + r_2)^2 \tag{7.31a}$$

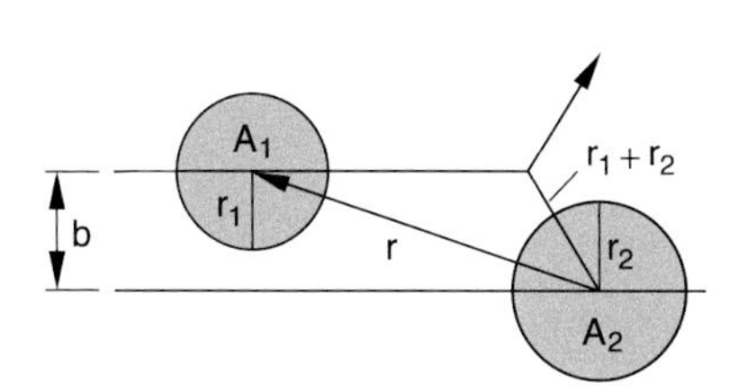

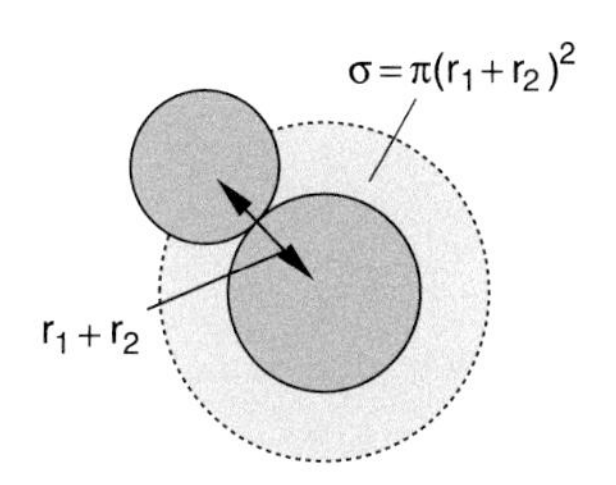

Figure 7.17 Impact parameter b and collision cross section σ for collisions of hard spheres with radiii r_1 and r_2

around the centre of A_2 are deflected from their straight path. This area σ is called the **collisional cross section** (see Sect. 4.3)

If a beam of particles A_1 passes in the x-direction through a gas with n particles A_2 per cm^3 at a sufficiently small number density n (the mean distance d should be large compared to $(r_1 + r_2)$ the probability that a particle A_1 suffers a collision during the path length Δx is given by the quotient

$$\frac{\sum \sigma}{A} = \frac{n \cdot \sigma \cdot \Delta x \cdot A}{A} = n \cdot \sigma \cdot \Delta x \tag{7.31b}$$

where $\sum \sigma$ is the sum of the cross sections of all atoms A_2 in the volume $V = A \cdot \Delta x$ and A is the total cross section of the incident beam (Fig. 7.18).

If N particles impinge per sec onto the area A of the volume $V = A \cdot \Delta x$ the fraction $\Delta N/N$ that suffers a collision after a path length Δx is

$$\frac{\Delta N}{N} = n \cdot \sigma \cdot \Delta x. \tag{7.32a}$$

In its differential form this reads

$$\frac{\mathrm{d}N}{N} = -n \cdot \sigma \cdot \mathrm{d}x\,. \tag{7.32b}$$

The negative sign should indicate that the particle flux N decreases because the collisions deflect the particles out of the x-direction.

Integration of (7.32b) gives the particle flux

$$N(x) = N_0 \cdot e^{-n\sigma x}\,, \tag{7.33}$$

after a path length x through the collision volume.

The path length Λ which a particle A_1 passes on the average without a collision is

$$\begin{aligned} \Lambda &= \frac{1}{N_0} \int_0^\infty x \left| \frac{\mathrm{d}N(x)}{\mathrm{d}x} \right| \mathrm{d}x \\ &= n \cdot \sigma \int_0^\infty x \cdot e^{-n\sigma x} \mathrm{d}x = \frac{1}{n\sigma}\,, \end{aligned} \tag{7.34a}$$

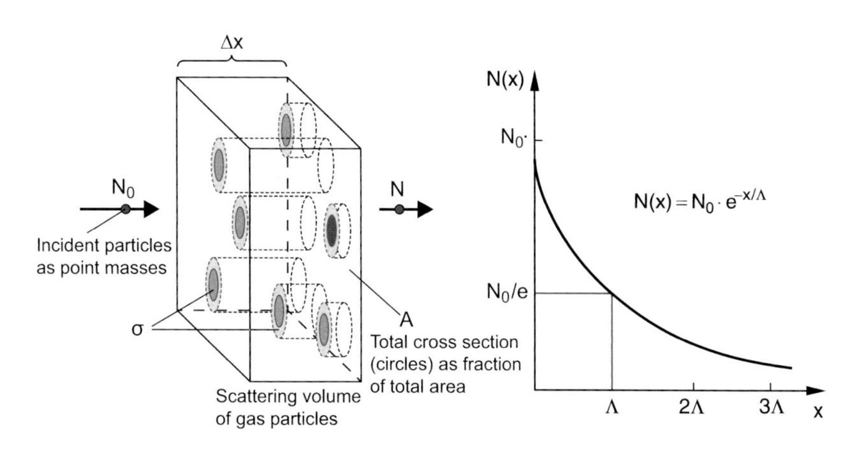

Figure 7.18 Illustration of collision cross section and mean free path length Λ [7.10]

where $|dN(x)/dx| \cdot dx$ gives the number of collisions on the path interval dx. The probability of a collision in the interval dx is then $|dN(x)/N_0|$.

The quantity $\Lambda = 1/(n \cdot \sigma)$ is called *mean free path*.

The mean free path Λ represents that path length after which the number of particles in the incident beam has decreased to $1/e$ of its initial value.

The average time interval τ between two successive collisions can then be defined as

$$\tau = \frac{\Lambda}{\langle v \rangle} = \frac{1}{n\sigma \langle v \rangle} . \tag{7.34b}$$

If both particles A_1 and A_2 move with velocities v_1 and v_2 the mean velocity $\langle v \rangle$ in (7.34b) has to be replaced by the mean relative velocity $\overline{\Delta v} = \overline{v_1 - v_2} = \sqrt{2\overline{v}^2}$. For collisions in a gas at the temperature Tone then obtains instead of (7.34b) the mean free collision time

$$\tau = \frac{1}{n \cdot \sigma \cdot \sqrt{2\overline{v^2}}} . \tag{7.34c}$$

Examples

1. At atmospheric pressure $p = 10^5$ Pa the number density of molecules in the atmosphere is $n \approx 3 \cdot 10^{19}\,\text{cm}^{-3}$. For the elastic collision cross section $\sigma = 45 \cdot 10^{-16}\,\text{cm}^2$ the mean free path is

$$\Lambda = \frac{1}{n \cdot \sigma} \approx 7 \cdot 10^{-6}\,\text{cm} = 70\,\text{nm} .$$

With the mean velocity $\langle v \rangle = 475\,\text{m/s}$ at $T = 300\,\text{K}$ the mean flight time between two collisions becomes

$$\tau = \frac{\Lambda}{\sqrt{2\langle v^2 \rangle}} = \Lambda \sqrt{\frac{m}{6kT}} = 1.1 \cdot 10^{-10}\,\text{s} .$$

This means that nitrogen molecules in a gas under normal conditions ($p = 10^5$ Pa, $T = 300$ K) suffer $1.34 \cdot 10^{10}$ collisions per second!

2. In an evacuated container with a residual pressure of $p = 10^{-4}$ Pa (10^{-9} bar) the density is $n = 3 \cdot 10^{10}\,\text{cm}^{-3}$. Now the mean free path is 70 m and therefore large compared with the dimensions of the container. Collisions between molecules are seldom and the molecules fly on straight lines until they hit the walls of the vacuum container. ◄

7.4 Experimental Proof of the Kinetic Gas Theory

There are many experimental methods to prove the statements of kinetic gas theory and to measure important quantities such as velocity distribution, collision cross sections, mean free path at different gas pressures and temperatures. We will here discuss only a few of them that are based on molecular beams or on transport phenomena in gases such as diffusion, viscosity and heat conduction in gases.

7.4.1 Molecular Beams

When atoms or molecules effuse out of a reservoir (pressure p, volume V, temperature T) through a small hole A into a vacuum chamber they fly on straight paths until they hit the wall, if their mean free path Λ is longer than the dimensions D of the chamber (Fig. 7.19). The pressure in the vacuum chamber must be low enough which can be reached with diffusion pumps (see Sect. 9.2).

Placing a slit B with width b at a distance d from A, only molecules can be transmitted by the slit, that fly into the angular range $|\vartheta| < \varepsilon$ with $\tan\varepsilon = b/2d$ around the x-axis. After passing the slit they form a collimated molecular beam. For not too high pressures p in the reservoir the molecules in the beam follow a modified Maxwell–Boltzmann-distribution $N(v) = n \cdot v \cdot f(v)$ and the angular distribution $N(\vartheta) \propto N_0 \cdot \cos\vartheta$ shows a cosine dependence. The angular distribution can be measured with a detector slewing over the angular range ϑ.

With a velocity selector (Fig. 7.20) subgroups of molecules with velocities within the range $v_s - (1/2)\Delta v < v < v_s + (1/2)\Delta v$ can be selected.

Such a selector consist in principal of two metallic circular discs with radius R at a distance a, each having a slit with width $\Delta S = R \cdot \Delta\varphi$. The two slits are twisted against each other by $S = R \cdot \varphi$. When the discs rotate with the angular velocity ω only

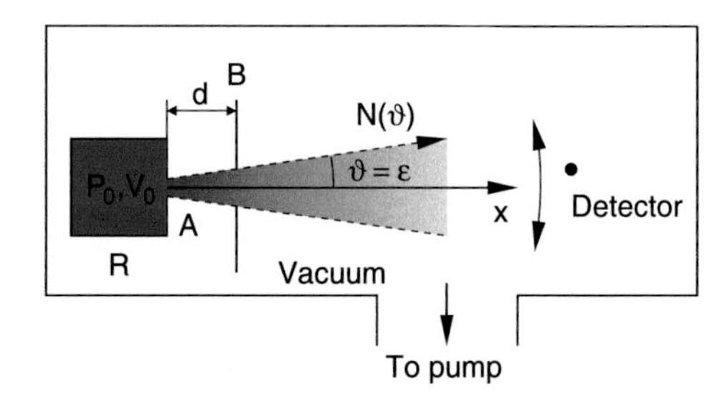

Figure 7.19 Schematic depiction of a molecular beam apparatus

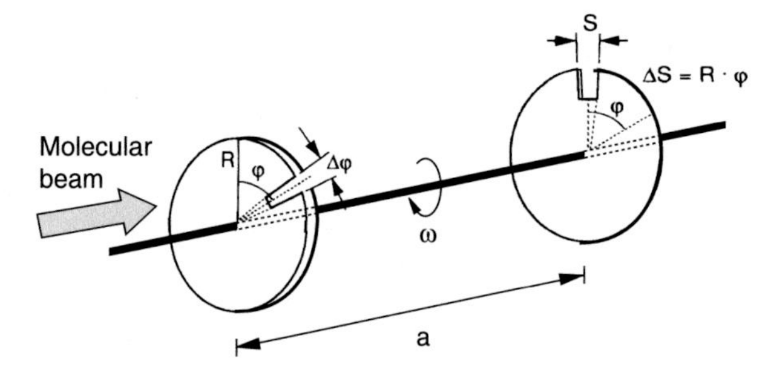

Figure 7.20 Principle of mechanical velocity selector

molecules with velocity v can pass both slits, if their flight time $T = a/v$ between the two slits equals the time $T_2 = R \cdot \varphi/(R \cdot \omega) = \varphi/\omega$. Their velocity is then

$$v = \frac{\omega \cdot a}{\varphi} \tag{7.35}$$

For a slit width $\Delta S = R \cdot \Delta\varphi$ with $\Delta\varphi \ll \varphi$ the transmitted velocity interval is

$$\Delta v = v \cdot \frac{\Delta\varphi}{\varphi} . \tag{7.36}$$

Varying the angular velocity ω of the discs one can select any velocity subgroup with a velocities up to $v_{max} = \omega_{max} \cdot a/\varphi$, where ω_{max} is the maximum value that can be technically realized. This allows one to measure the velocity distribution of the molecules in the beam.

Example

$\varphi = 20° = 0.35$ rad; $a = 10$ cm. If molecules with $v = 400$ m/s should be selected, the velocity selector has to rotate with $\omega = 1.4 \cdot 10^3\,s^{-1}$ which corresponds to 13 370 rpm. ◄

When the density $n(v)$ of molecules in the molecular beam follows the Maxwell–Boltzmann-distribution (7.30) the flux $N(v) = v \cdot n(v)$ is

$$N(v) = n(v) \cdot v = n \cdot \frac{4v^3}{v_w^3\sqrt{\pi}} \cdot e^{-mv^2/2kT} . \tag{7.37}$$

The number of molecules with velocities in the interval from v to $v + dv$, passing per sec through the area of 1 cm^2 is then $N(v) \cdot dv$.

Remark. For the velocity selector in the above example also molecules with $v = 21$ m/s would be transmitted for $\varphi = 20° + 360°$ i.e. at the next full turn. In order to prevent this ambiguity one has to use at least a third disc in the middle between the two discs with a slit tilted by $\varphi/2$. Generally many discs are used with many slits (Fig. 7.21) in order to transmit more molecules per sec. For q discs at a distance a/q the tilt of the slits between two successive discs should be φ/q.

For the detection of the transmitted molecules several different detectors have been developed.

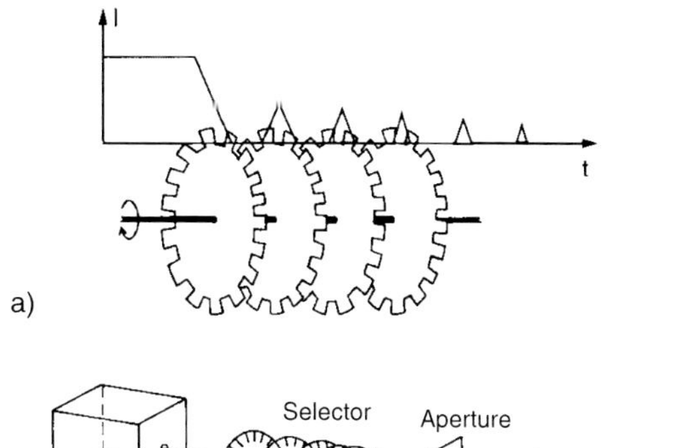

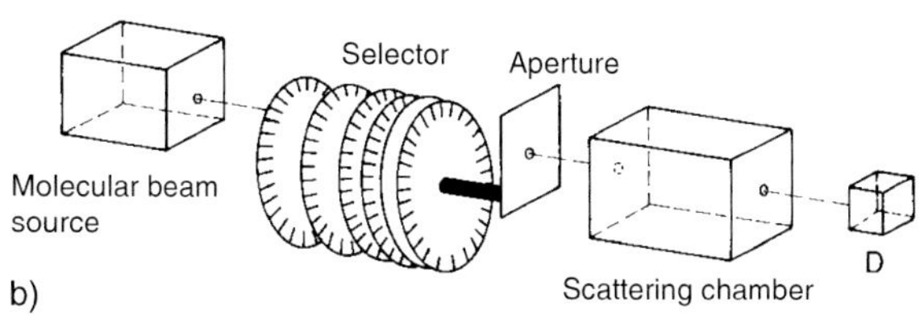

Figure 7.21 Velocity selector with 6 rotating discs. **a** Principle of selecting a velocity class with particle flux $N(v)dv$, **b** arrangement for measuring collision cross sections $\sigma(v)$ by detecting the number $N(v) = N_0 \cdot e^{-n \cdot \sigma(v) \cdot L}$ that have passed the scattering chamber with length L

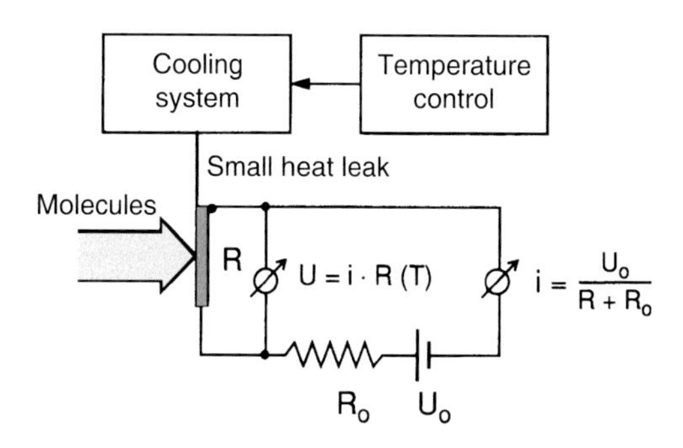

Figure 7.22 Schematic design of a bolometer for measuring the flux $N(v)$ of neutral molecules

- ***Bolometer*** (Fig. 7.22). This is a small semiconductor plate cooled down to very low temperatures with a small heat capacity C and a small heat conductivity G to its surrounding. The molecules impinging onto the cooled surface transfer their kinetic energy $E_{kin} = (1/2)mv^2$ to the semiconductor. This increases its temperature by $\Delta T = N(v) \cdot E_{kin}/G$ (see Sect. 10.2.2). The temperature increase results in a change $\Delta R = (\partial R/\partial T)\Delta T$ of the electric resistance R of the semiconductor. This can be measured by the corresponding change of the electric current $I = U_0/(R + R_0)$ that flows through the circuit of semiconductor and external resistor R_0 in series with R, if a constant voltage U_0 is applied (see Vol. 2). With such a device, it is possible to measure a transferred power as small as 10^{-14} W! This corresponds to a minimum rate of $N(v)dv = 2.8 \cdot 10^6$/s molecules with $v = 400$ m/s.
- ***Ionization detector*** (Fig. 7.23). The neutral molecules are ionized by electron impact (see Vol. 3). The ions with charge $q = +e$ are collected on an electrode at negative voltage. For a flux N of neutral particles the electric output current of the detector is $I = \eta \cdot N \cdot e$ where $\eta \ll 1$ is the ionization probability of each neutral molecule.
- ***Langmuir–Taylor detector***. This is a heated wire where all neutral particles with ionization energies smaller than the work function of the hot wire are ionized if they hit the wire. In this case, energy is gained if the electron is transferred from the molecule to the metal wire. The ions are extracted by an electric field and collected on a detector, for instance a Faraday cup.
- Modern methods for the measurement of the velocity distribution are based on laser-spectroscopic techniques (see Vol. 3).

When the pressure p in the reservoir is increased, the mean free path of the molecules in the hole A of the reservoir becomes

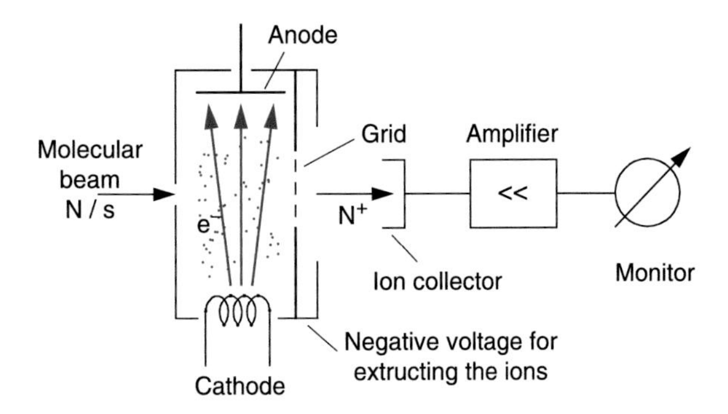

Figure 7.23 Principle of ionization detector

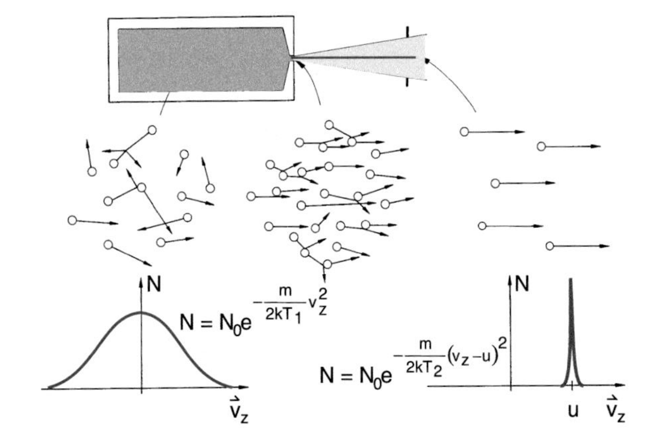

Figure 7.24 Narrowing of the velocity distribution $N(v)$ in a supersonic beam

smaller than the dimensions of the hole. In this case the particles suffer collisions during their expansion into the vacuum. Because the faster particles hit the slower particles ahead of them and they transfer part of their kinetic energy, they become slower and the slow particles faster. This implies that the velocity distribution becomes narrower (Fig. 7.24). If the mean velocity exceeds the local velocity of sound, a supersonic beam is formed. Its velocity distribution is described by

$$N(v) = Cv^3 \cdot e^{-m(u-v)^2/2kT_t} . \tag{7.38}$$

The width of the distribution around the mean velocity $u = \langle v \rangle$ can be characterized by a translational temperature T_t which is a measure for the relative velocities of the particles in the beam. With increasing pressure p the temperature T_t decreases, which means that the relative velocities decrease. Translational temperatures below 1 K have been realized, where all particles have nearly the same velocity u.

7.5 Transport Phenomena in Gases

Because molecules in a gas can freely move around within the gas container many transport processes can occur. When the molecules collide with each other or with the wall of the container energy and momentum can be transferred. In a gas flow also mass is transported. If molecules A in a sub-volume V_1 can move into a volume V_2 where molecules B are present, the two species mix with each other until both sorts are uniformly distributed over the whole volume (diffusion).

There are mainly 3 such transport phenomena:

- Diffusion (mass transport),
- heat conduction (energy transport),
- gas flow with viscosity (momentum transport).

They occur always when local differences (gradients) of density, temperature or flow velocities are present. The important point is that all these transport phenomena can be explained by the kinetic gas theory. The experimental investigation of these macroscopic processes gives information about the size of the gas molecules and their mutual interactions (see Vol. 3).

7.5.1 Diffusion

When a bottle of an intensively smelling substance (for example a pleasantly smelling perfume or the badly smelling hydrogen sulphide H_2S) is opened the odour can be soon sensed in the whole room. The molecules escaping out of the bottle must travel in a short time over several meters through the air at atmospheric pressure in spite of the very small free mean path of $\Lambda = 10$–100 nm! This migration of molecules A through a gas of molecules B resulting in a uniform spatial distribution of both sorts A and B is called *diffusion*.

The diffusion is illustrated by Fig. 7.25, where a volume V is divided by a thin wall into two parts, each containing only molecules of one type. After removing the separating wall the two types A and B of the molecules mix and fill the whole volume with spatially uniform concentration.

> Diffusion is a net transport of particles from a region with higher concentration to a region with lower concentration.

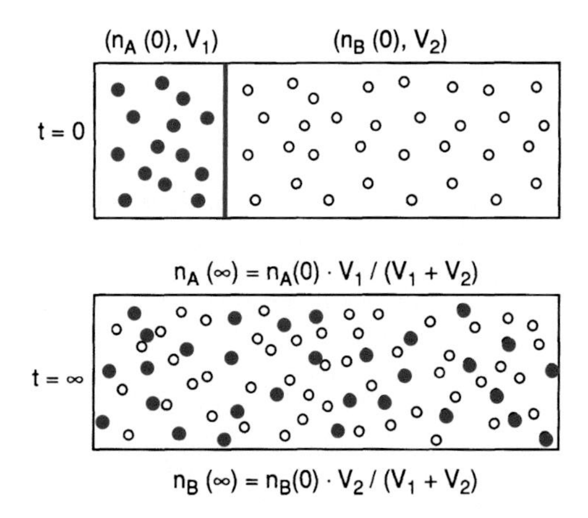

Figure 7.25 Diffusion of two differrent particles with densities $n_A(t)$ and $n_B(t)$ after opening a hole in the dividing wall at $t = 0$

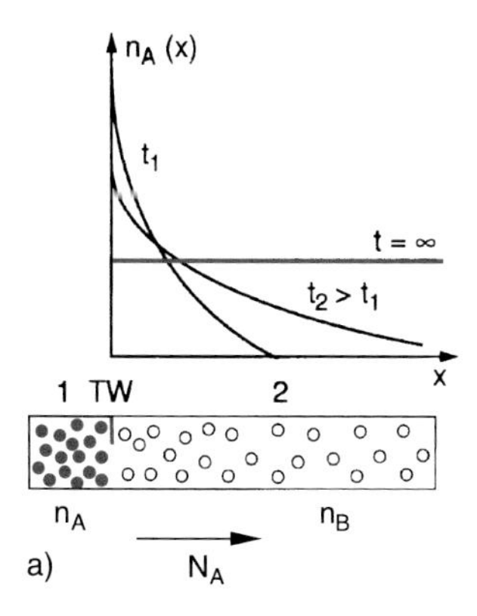

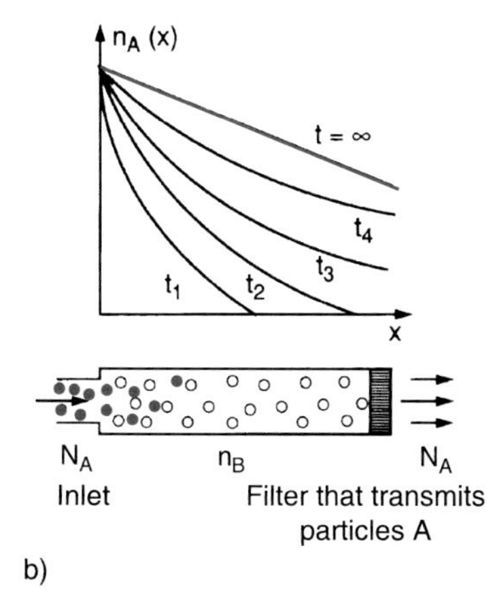

Figure 7.26 **a** Density gradient driving the diffusion of particles A through a gas of particles B. At time $t = 0$ a hole is opened in the dividing wall and the particles A diffuse into the volume V_2. **b** On the left side N_A particles A per sec are continuously supplied, which are pumped away on the right side. The *curves* show the distribution $n_A(x)$ at different times t_i

The examples above illustrate that diffusion always takes place if density gradients are present. Diffusion reduces these gradients until a uniform spatial distribution is reached, where the density gradients are zero (Fig. 7.26a) unless external conditions maintain a stationary density gradient. This can, for example, be realized if particles A are continuously supplied to the left volume in Fig. 7.25 and particles A and B are simultaneously removed on the right side, thus maintaining a concentration gradient (Fig. 7.26b).

We will now discuss diffusion in a more quantitative way. We assume the density $n_A(x)$ of particles A to be constant in the y- and z-direction but to vary in the x-direction (Fig. 7.27). The thermal velocities of the particles A are isotropic. This means that the following two probabilities are equal:

1. The probability P_- that A flies after its last collision at $x_- = x_0 - \Lambda \cdot \cos\vartheta$ (Λ = mean free path) with a velocity v under an angle ϑ against the $+x$-direction and passes the plane $x = x_0$ from left to right.

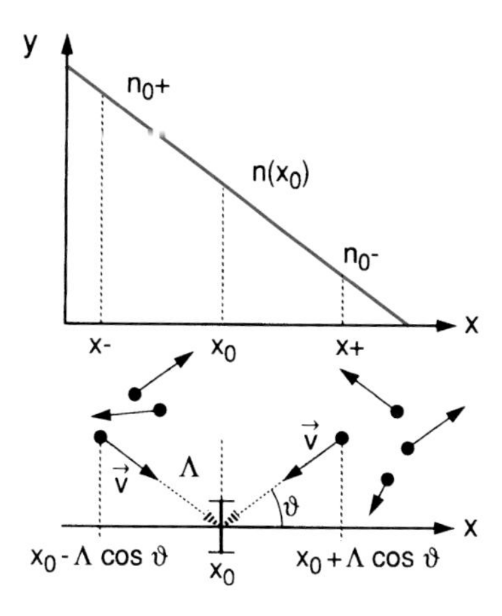

Figure 7.27 Illustration of the derivation of the diffusion coefficient

2. The probability P_+ that A flies after its last collision at $x_+ = x_0 + \Lambda \cdot \cos\vartheta$ with the velocity v under an angle ϑ against the $-x$-direction through the plane $x = x_0$ from right to left.

We have to take into account that the velocities are not equal for all molecules but follow a Maxwell–Boltzmann distribution with the distribution function $f(v)$. The directions of their velocities are randomly distributed. With the density $n_+(x)$ left of the plane $x = x_0$ the flux $\mathrm{d}N_+$ of particles within the velocity interval from v to $v + \mathrm{d}v$ that pass in the time interval $\mathrm{d}t$ from left to right under the angle ϑ within the solid angle $\mathrm{d}\Omega$ the plane $\mathrm{d}A$ at $x = x_0$ is

$$\mathrm{d}N_+(v) = n_+ f(v)\mathrm{d}v\, v\, \mathrm{d}t \cdot \mathrm{d}A \cdot \cos\vartheta \cdot \mathrm{d}\Omega/4\pi\,, \tag{7.39}$$

because the volume from where they come, is $\mathrm{d}V = \mathrm{d}A \cdot \cos\vartheta \cdot v \cdot \mathrm{d}t$ (Fig. 7.28).

A corresponding equation is obtained for $\mathrm{d}N_-(v)$. The question is now which densities n_+ and n_- have to be used?

The particles start from the position $x = x_0 \pm \Delta x$ ($\Delta x \approx \Lambda \cdot \cos\vartheta$) of their last collision. There are the densities

$$\begin{aligned} n_+ &= n_0 + \Delta x \frac{\mathrm{d}n}{\mathrm{d}x}\,, \\ n_- &= n_0 - \Delta x \frac{\mathrm{d}n}{\mathrm{d}x}\,. \end{aligned} \tag{7.40}$$

On the average is $\overline{\Delta x} = \Lambda \cdot \cos\vartheta$. We define the vertical particle flux density by the vector

$$\boldsymbol{j} = \frac{\mathrm{d}N}{\mathrm{d}A \cdot \mathrm{d}t}\hat{\boldsymbol{e}}$$

where $\mathrm{d}N$ is the number of particles which pass the area $\mathrm{d}A$ in the plane $x = x_0$ during the time interval $\mathrm{d}t$. For that part of the net flux of particles with the velocity v in the interval $\mathrm{d}v$ within the solid angle $\mathrm{d}\Omega$

$$\mathrm{d}j(v)\mathrm{d}v = -\frac{1}{\mathrm{d}A}\left(\frac{\mathrm{d}N_+(v)}{\mathrm{d}t} - \frac{\mathrm{d}N_-(v)}{\mathrm{d}t}\right)\mathrm{d}v \tag{7.41}$$

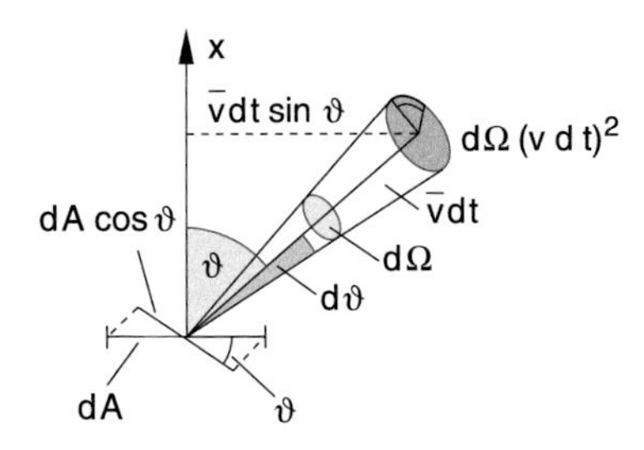

Figure 7.28 The number of particles passing during the time interval dt through the area dA inclined under the angle ϑ against the x-direction is $dn = n \cdot f(v) \cdot v\,dv\,dt\,dA \cdot \cos\vartheta\,d\vartheta/4\pi$

we obtain from (7.39) with $\mathrm{d}\Omega = \sin\vartheta \cdot \mathrm{d}\vartheta \cdot \mathrm{d}\varphi$

$$\mathrm{d}j_x(v)\mathrm{d}v = -2\Lambda f(v)v\,\mathrm{d}v\frac{\cos^2\vartheta\ \sin\vartheta\ \mathrm{d}\vartheta\mathrm{d}\varphi}{4\pi}\frac{\mathrm{d}n}{\mathrm{d}x}\ . \qquad (7.41a)$$

Integration over φ gives the factor 2π, over ϑ $(0 \le \vartheta \le \pi/2)$ the factor $1/3$, while the integration over all velocities gives the mean velocity

$$\overline{v} = \int v \cdot f(v)\mathrm{d}v\ . \qquad (7.42)$$

One obtains finally for the total mean particle flux in the x-direction

Fick's Law:

$$j_x = -\frac{\Lambda \cdot \overline{v}}{3} \cdot \frac{\mathrm{d}n}{\mathrm{d}x} = -D \cdot \frac{\mathrm{d}n}{\mathrm{d}x}\ . \qquad (7.43)$$

For the general three-dimensional case this modifies to the vector equation

$$\boldsymbol{j} = -D \cdot \mathbf{grad}\,n.$$

The particle flux $\boldsymbol{j}$ due to diffusion is equal to the product of diffusion coefficient D and concentration gradient $\mathbf{grad}\,n$.

The diffusion coefficient

$$D = \tfrac{1}{3}\Lambda \cdot \overline{v} \qquad (7.44)$$

is proportional to the product of mean free path Λ and mean velocity v. Using (7.34) and (7.28) we can express the diffusion coefficient

$$D = \frac{\Lambda \cdot \overline{v}}{3} = \frac{1}{n \cdot \sigma}\sqrt{\frac{8kT}{9\pi m}} \qquad (7.45)$$

for the diffusion of particles A through a gas of particles B with density n by the collision cross section σ and the mass m of particles A. This shows that heavy particles diffuse slower than light ones.

This can be demonstrated by the experiment shown in Fig. 7.29. A porous cylinder of clay shows in air the same pressure inside and outside. At time t_1 a baker is put over the cylinder and helium is blown into the baker. The pressure gauge shows at first a higher pressure inside the cylinder, which gradually decreases until it becomes equal to the pressure outside. At time t_2 the baker is removed. Now the pressure inside the cylinder drops below the external pressure until it finally approaches the external pressure.

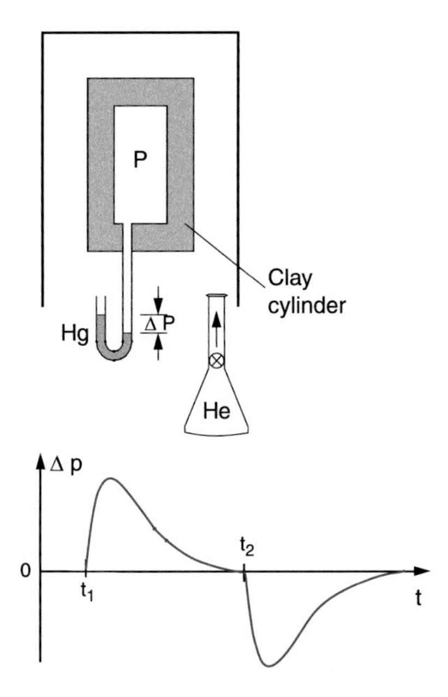

Figure 7.29 Demonstration of the fact that light particles diffuse faster than heavy particles

The explanation is the larger diffusion velocity of the lighter helium atoms. It diffuses faster into the cylinder than the air in the cylinder diffuses out. The total pressure therefore increases, until the air has diffused out and the pressure difference between inside and outside becomes zero. After the baker has been removed there is no longer helium outside. Therefore, the helium diffuses from the inside of the cylinder to the outside faster than the air diffuses the opposite way. Therefore the pressure inside drops at first below the outside pressure.

7.5.2 Brownian Motion

The diffusion of a microscopic particle A suffering collisions with molecules B in a gas at atmospheric pressure can be observed through a microscope if the mass of A is much larger than that of the molecules B. This can be realized with cigarette smoke particles that have diameters of about 0.1 µm. Illuminating the diffusion chamber, makes the particles A visible (even if their diameter is small compared to the wavelength of the illuminating light) because they scatter the incident light and appear in the microscope as bright spots (see Vol. 2). This particle A which contains still about 10^5 molecules, performs a random walk through the gas with particles B (Fig. 7.30a). When observing the motion of A with sufficient local and time resolution the motion appears on short straight lines between successive collisions that abruptly change the direction of the motion. The length of these straight lines is the mean free path length of the particle A. The directions of the straight lines are statistically

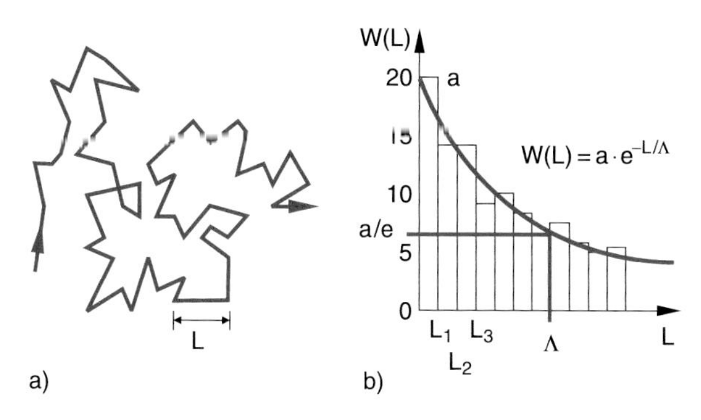

Figure 7.30 **a** Random path of a particle (Brownian Motion) induced by collisions with air molecules. **b** Histogram of the distribution of the mean free path between successive collisions

distributed. Such a random walk can be mathematically simulated, if the lengths and the directions are generated by a random generator. Figure 7.30b shows a histogram $N(L_i)$ for the lengths L_i of the straight lines. Each bar indicates how often the length L_i is observed within the interval L to $L + \Delta L$. The analytical curve as envelope of the different bars gives the probability of finding the length L. It has the form

$$W(L) = a \cdot e^{-L/\Lambda} . \tag{7.46}$$

The random motion of microscopic particles in liquid solutions was first discovered 1827 by the English botanist *Robert Brown* who believed initially that he observed small living microbes until he realized that the motion was completely irregular and should be attributed to lifeless particles. Magnified by a microscope and put on a large screen it allows a fascinating view into the micro-world, accessible to a large auditorium.

The mathematical description first given by Albert Einstein will be treated in Vol. 3.

The Brownian motion can be simulated by moving pucks on an air bearing stage, where a large number of small discs are kept in motion by vibrating wires at the 4 edges of the stage. A larger puck with a small light bulb moves in between the small discs and its random motion is detected by a video camera. In fact, the random path in Fig. 7.30a has been obtained in this way.

7.5.3 Heat Conduction in Gases

Heat conduction is also based on the motion of molecules which transfer during collisions part of their kinetic energy to the collision partner. This results in a transport of energy from regions with higher temperature to those of lower temperature. The mechanism of heat conduction is different for gases, liquids and solid bodies (see Sect. 10.2). In solids the atoms are fixed to definite positions while they can freely move in gases.

We start the discussion of heat conduction in gases with a gas between two parallel plates at a distance d (Fig. 7.31) that are kept at different temperatures T_1 and T_2. The transport of heat

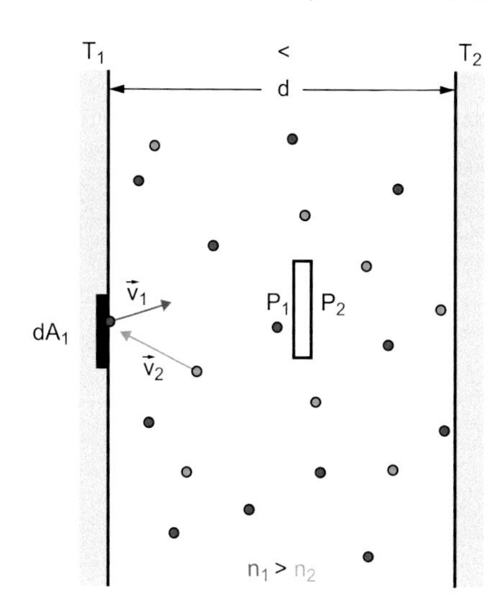

Figure 7.31 Heat conduction in gases. The distance d is small compared to the extensions of the plates. A pressure sensor measures the different pressures p_1 and p_2 at different sides of the sensor

energy from the hotter to the cooler plate depends on the ratio Λ/d of free mean path Λ to plate separation d.

At low gas pressures is $\Lambda > d$ and the molecules can fly between the two plates without suffering collisions in the gas volume. Molecules that leave the plate 1 have the mean kinetic energy

$$E_{k_1} = \tfrac{1}{2} m \cdot v^2 = \tfrac{3}{2} k T_1$$

For an isotropic distribution of the velocity directions of the molecules with a density n

$$\Delta Z = n \cos\vartheta \, \Delta A \left(\int v f(v) \mathrm{d}v \right) \cdot \Delta t \cdot \mathrm{d}\Omega / 4\pi$$

molecules leaving plate 1 with velocities v within the solid angle $\mathrm{d}\Omega = \sin\vartheta \cdot \mathrm{d}\vartheta \cdot \mathrm{d}\varphi$ under the angle ϑ against the surface normal reach within the time interval Δt the surface element ΔA on plate 2 (see Fig. 7.28).

Integration over all possible velocities yields

$$\begin{aligned} Z &= \frac{n \overline{v} \Delta A \, \Delta t}{4\pi} \int_0^{\pi/2} \sin\vartheta \, \cos\vartheta \, \mathrm{d}\vartheta \int_{\varphi=0}^{2\pi} \mathrm{d}\varphi \\ &= \frac{n}{4} \overline{v} \Delta A \quad \text{with} \quad \overline{v} = \int_0^\infty v f(v) \, \mathrm{d}v . \end{aligned} \tag{7.47}$$

We will assume that every molecule impinging on plate 2 will stay there for a short time, adapt the temperature of plate 2 and desorbs again. The surface element ΔA of plate 1 loses energy because of the desorbing molecules

$$\frac{\mathrm{d}W_1}{\mathrm{d}t} \Delta A = -\frac{Z_1}{\Delta t} \Delta A \cdot U_1 , \tag{7.48a}$$

where $\mathrm{d}W/\mathrm{d}t$ is the energy loss per unit surface and unit time and $U_1 = (f/2)kT_1$ is the energy of a molecule with f degrees of freedom (kinetic, rotational and vibrational energy, see Sect. 10.3). On the other hand the surface element wins the energy

$$\frac{\mathrm{d}W_2}{\mathrm{d}t}\Delta A = \frac{Z_2}{\Delta t}\Delta A \cdot U_2 \quad \text{with} \quad U_2 = \frac{1}{2} f\, kT_2 \tag{7.48b}$$

by molecules coming from plate 2 and impinging on plate 1. Under stationary conditions is $Z_1 = Z_2$.

Therefore the net energy flow is

$$\begin{aligned}\frac{\mathrm{d}W}{\mathrm{d}t}\cdot A_1 &= \kappa \cdot A_1(T_2 - T_1)\\ \text{with} \quad \kappa &= \frac{n \cdot \overline{v} \cdot k \cdot f}{8}\end{aligned} \tag{7.49a}$$

The constant κ is called *heat transfer coefficient*. It has the unit $[\kappa] = 1\,\mathrm{J\,s^{-1}\,m^{-2}\,K^{-1}}$.

The heat conduction (energy flux per unit time) $j_W = \mathrm{d}W/\mathrm{d}t$ in gases at low pressures ($\Lambda \gg d$) is proportional to the temperature difference between the walls and to the density n of the gas molecules.

Since $v \propto m^{-1/2}$ heavy molecules have a lower heat conduction than light molecules. Because of their larger number f of degrees of freedom molecules transport more energy than atoms. Since the pressure $p = n \cdot k \cdot T$ is the same within the whole volume between the plates the molecular densities n_1 and n_2 in front of the plates differ according to

$$\frac{n_1}{n_2} = \frac{T_2}{T_1}\,. \tag{7.49b}$$

The gas density is therefore lower in regions with higher temperatures.

In order to decrease the heat conduction, the gas density has to be low, i.e. the space between the plates should be evacuated. The thermos bottle is an example, where low heat conduction is realized in order to keep a liquid for a longer time on a nearly constant temperature.

If the mean free path Λ is low compared to the plate separation ($\Lambda \ll d$) the molecules often collide on their way between the plates. The heat energy of the hotter plate is no longer directly transported by molecules to the cooler plate but transferred within a path $x \approx \Lambda$ during collisions to other molecules. Therefore a temperature gradient $\mathrm{d}T/\mathrm{d}x$ appears in the gas volume. The energy flow per unit time through the unit surface element of the plane $x = x_0$ between the plates is similar to the discussion in Sect. 7.5.1

$$\begin{aligned}\frac{\mathrm{d}W}{\mathrm{d}t} &= \frac{1}{3}\Lambda \cdot \frac{\mathrm{d}}{\mathrm{d}x}(\overline{v} \cdot n \cdot U)\\ &= \frac{1}{3}\Lambda \cdot n \cdot \frac{f}{2}kT \cdot \frac{\mathrm{d}\overline{v}}{\mathrm{d}x}\,.\end{aligned} \tag{7.50a}$$

Here the relation $n \cdot U = (1/2)n \cdot f \cdot kT$ of Eq. 7.49 has been used.

With

$$\mathrm{d}\overline{v}/\mathrm{d}x = \frac{\mathrm{d}\overline{v}}{\mathrm{d}T}\cdot\frac{\mathrm{d}T}{\mathrm{d}x} = \frac{\sqrt{8k/\pi m}}{2\cdot\sqrt{T}}\cdot\frac{\mathrm{d}T}{\mathrm{d}x}$$

this can be written as

$$\frac{\mathrm{d}W}{\mathrm{d}t} = \lambda \cdot \frac{\mathrm{d}T}{\mathrm{d}x}\,. \tag{7.50b}$$

The constant

$$\lambda = \frac{1}{12} f \cdot n \cdot k\overline{v}\Lambda$$

is the **heat conductivity**, with the unit

$$[\lambda] = 1\,\frac{\mathrm{J}}{\mathrm{s \cdot m \cdot K}}\,.$$

The heat conductivity is for $\Lambda \ll d$ independent of the gas density n because, according to (7.34), is $\Lambda = 1/(n \cdot \sigma)$ and therefore

$$\lambda = \frac{1}{12}\frac{f \cdot k \cdot \overline{v}}{\sigma}\,. \tag{7.50c}$$

7.5.4 Viscosity of Gases

As has been discussed in the previous sections diffusion and heat conduction can be ascribed to mass- and energy transport by molecules. They are accomplished by the thermal motion of molecules at local variations of density (diffusion) and temperature (heat conduction). Diffusion and heat conduction also occur if the gas as a whole is at rest, i.e. if the macroscopic momentum $\boldsymbol{P} = \sum \boldsymbol{p}_i = \boldsymbol{0}$.

When in addition to the thermal motion of the molecules a macroscopic flow of the whole gas volume occurs (gas current), further phenomena appear as for instance friction (viscosity), if the flow velocity varies locally (Sect. 8.3). Also the viscosity is related to the thermal motion of the molecules as can be illustrated by the following example:

We consider a gas, which flows into the y-direction with a flow velocity $u(x)$ that varies in the x-direction (Fig. 7.32). An example is the air flow over a lake with the water surface at the plane $x = 0$. Layers of the streaming air close to the water surface are retarded by friction with the water surface and have therefore a smaller flow velocity than higher layers.

We select a layer between the planes $x = x_0 \pm \Delta x/2$ (Fig. 7.32b). The velocity of the gas molecules is a superposition of their thermal velocities with the flow velocity. Because of their thermal velocities, the gas molecules pass from their layer $x = x_0 \pm \Delta x/2$ into adjacent layers and collide there with other molecules. Since the y-component of their velocity is higher than that of molecules in layers $x > x_0$ they transfer part of

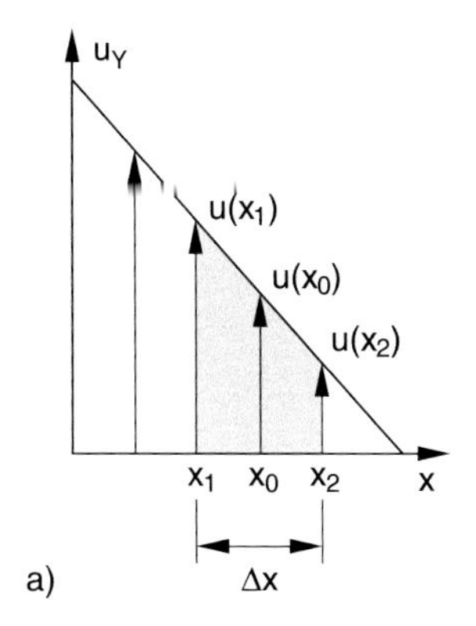

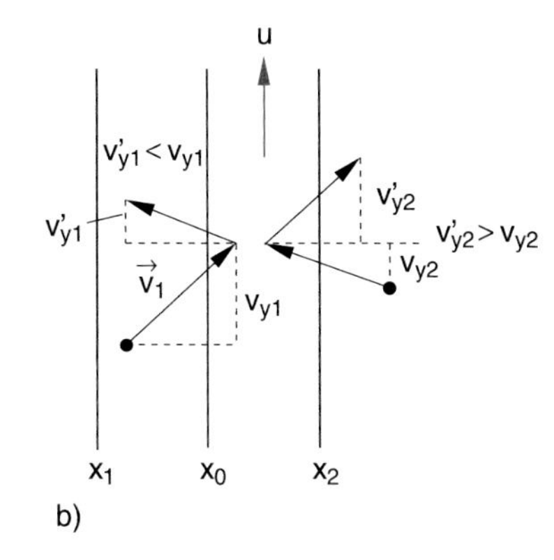

Figure 7.32 Viscosity of gases. The gas streams into the y-direction with a flow velocity $u(x)$, that decreases with increasing x. **a** velocity $u_y(x)$; **b** molecular model of viscosity

their momentum to these molecules and increase their mean y-component. The amount of the transferred momentum depends on the difference between the components $v_y(x)$, i.e. on the gradient $\mathrm{d}u/\mathrm{d}x$ of the flow velocities.

The transport of momentum occurs in the x-direction of decreasing $u(x)$. When we define the momentum flux as the momentum transfer through the unit area in the plane $x = x_0$, the viscosity law for gases with locally varying flow velocities can be written as

$$j_p = \eta \cdot \frac{\mathrm{d}u}{\mathrm{d}x} \,. \tag{7.51}$$

The factor η is the coefficient of internal friction or **coefficient of viscosity**. A consideration similar to the derivation of the diffusion coefficient (see Sect. 7.5.1) yields with (7.34)

$$\eta = \tfrac{1}{3} n \cdot m \cdot \overline{v} \cdot \Lambda \,. \tag{7.52}$$

The momentum transfer in gas flows with a velocity gradient $\mathrm{d}u/\mathrm{d}x$ is proportional to the particle density n and the mean thermal velocity $\overline{v} = (8kT/n \cdot m)^{1/2}$, to the mean free path Λ and the flow velocity gradient $\mathrm{d}u/\mathrm{d}x$.

It decreases with decreasing Λ, i.e. with increasing collision cross section. Heavy particles cause a higher viscosity, because of their higher momentum transfer.

7.5.5 Summary of Transport Phenomena

Diffusion, heat conduction and internal friction can be explained by the thermal motion of the molecules and the exchange of energy and momentum during collisions. All these phenomena can be described by the kinetic gas theory. Therefore from measurement of the macroscopic quantities diffusion coefficient D, heat conductivity λ and viscosity coefficient η, which describe averages over the quantities of the individual molecules, information can be obtained about the microscopic quantities mean free path Λ, collision cross section $\sigma = 1/n \cdot \Lambda$ and mean velocity $\overline{v}$ of the molecules.

In order to recall and clarify the relations between the three transport coefficients they are here again arranged in a compact form.

- ***At high gas pressures ($\Lambda \ll d$):*** Diffusion coefficient

$$D = \tfrac{1}{3}\overline{v} \cdot \Lambda = \tfrac{1}{3}\overline{v}/(n \cdot \sigma) \,. \tag{7.53a}$$

Energy transport through heat conduction between two plates at a distance d and temperature difference ΔT:

$$\frac{\mathrm{d}W}{\mathrm{d}t} = \frac{\lambda}{d} \cdot \Delta T \tag{7.53b}$$

with the heat conductivity

$$\lambda = \frac{1}{12}\frac{f \cdot k \cdot \overline{v}}{\sigma} = \frac{1}{4} D \cdot f \cdot k \cdot n \tag{7.53c}$$

which is independent of the gas pressure, because $D \propto (1/n)$. Viscosity coefficient:

$$\eta = \tfrac{1}{3} nm\overline{v} \cdot \Lambda = n \cdot m \cdot D \tag{7.53d}$$

- ***At low pressures ($\Lambda \gg d$):*** The energy transport becomes proportional to the gas density

$$\begin{aligned} \frac{\mathrm{d}W}{\mathrm{d}t} &= \kappa \cdot \Delta T \\ \kappa &= \frac{n \cdot \overline{v} \cdot k \cdot f}{8} = \frac{3}{2}\frac{\lambda}{\Lambda} = \frac{3}{2}\lambda \cdot n \cdot \sigma \end{aligned} \tag{7.53e}$$

In Tab. 7.3 the transport coefficients of some gases are compiled.

Table 7.3 Selfdiffusion coefficient D, heat conductivity λ, and coefficient of viscosity η of some gases at $p = 10^5$ Pa and $T = 20\,°\mathrm{C}$

Gas	$D/\mathrm{m^2/s}$	$\lambda/\mathrm{J\,m^{-1}s^{-1}K^{-1}}$	$\eta/\mathrm{Pa \cdot s}$
He	$1.0 \cdot 10^{-4}$	$1.5 \cdot 10^{-2}$	$1.5 \cdot 10^{-5}$
Ne	$4.5 \cdot 10^{-5}$	$4.6 \cdot 10^{-2}$	$3.0 \cdot 10^{-5}$
Ar	$1.6 \cdot 10^{-5}$	$1.7 \cdot 10^{-2}$	$2.0 \cdot 10^{-5}$
Xe	$6.0 \cdot 10^{-6}$	$0.5 \cdot 10^{-2}$	$2.1 \cdot 10^{-5}$
H_2	$1.3 \cdot 10^{-4}$	$1.7 \cdot 10^{-1}$	$8.0 \cdot 10^{-6}$
N_2	$1.81 \cdot 10^{-5}$	$2.6 \cdot 10^{-2}$	$1.7 \cdot 10^{-5}$
O_2	$2.4 \cdot 10^{-5}$	$2.0 \cdot 10^{-2}$	$2.0 \cdot 10^{-5}$

7.6 The Atmosphere of the Earth

Our atmosphere consists of a mixture of molecular and atomic gases. In the lower part it contains also water vapour, aerosols and dust particles. Its composition is listed in Tab. 7.4. The density gradient $\mathrm{d}n/\mathrm{d}z$, described by Eq. 7.6 causes diffusion which tries to establish a uniform density. However, the gravity acts against this tendency and results in the exponential density function $n(z)$. Stationary equilibrium is reached, when for all values of z the upwards directed diffusion current j_D is just compensated by the downwards directed current j_g of particles in the gravity field of the earth

$$\boldsymbol{j}_\mathrm{D}(z) + \boldsymbol{j}_g(z) = \mathbf{0}\,. \tag{7.54}$$

The diffusion current is according to (7.43)

$$j_\mathrm{D} = -D\cdot\frac{\mathrm{d}n}{\mathrm{d}z}\,.$$

With $n = n_0\cdot e^{-(mgz/kT)} \Rightarrow \mathrm{d}n/\mathrm{d}z = -(mg/kT)\cdot n$ one obtains

$$j_\mathrm{D} = +D\cdot\frac{m\cdot g}{k\cdot T}\cdot n\,. \tag{7.54a}$$

Opposite to the gravitational force is the friction force acting on the falling molecules. This results in a constant fall velocity v_g and therefore a constant particle current

$$\boldsymbol{j}_g = n\cdot\boldsymbol{v}_g\,. \tag{7.55}$$

Since $j_g = -j_\mathrm{D}$ we obtain for the constant sink velocity

$$v_g = -j_\mathrm{D}/n = \frac{m\cdot g}{kT}\cdot D\,. \tag{7.56}$$

It depends on the diffusion constant D and therefore according to (7.44) on the mean free path Λ.

From (7.54) and (7.56) we obtain

$$-D\frac{\mathrm{d}n}{\mathrm{d}z} = \frac{n\cdot m\cdot g}{kT}\cdot D\,. \tag{7.57}$$

Table 7.4 Gas composition of the earth atmosphere

Component	Volume %
Nitrogen N_2	78.084
Oxygen O_2	20.947
Argon Ar	0.934
Carbon-Dioxyd CO_2	0.032
Neon Ne	0.0018
Helium He	$5.2\cdot10^{-4}$
Methane CH_4	$2\cdot10^{-4}$
Krypton Kr	$1.1\cdot10^{-4}$
Hydrogen H_2	$5\cdot10^{-5}$
trace gases (e.g. SO_2, O_3, NO_2)	$< 5\cdot10^{-4}$

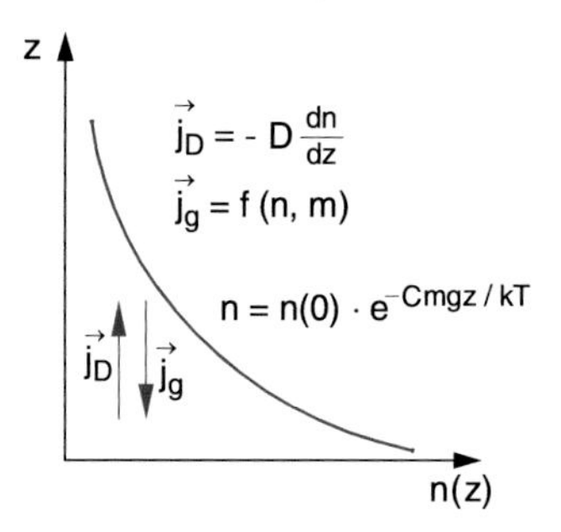

Figure 7.33 Stationary density distribution $n(z)$ in the isothermal atmosphere, caused by the superposition of upwards diffusion and downward particle flux in the gravitational field of the earth

Integration yields again the barometric formula

$$n = n_0\cdot e^{-mgz/(kT)}\,. \tag{7.58}$$

The exponential decrease of the density in the isothermal atmosphere is due to the common action of diffusion and gravitational force, namely the compensation of the upwards diffusion due to the density gradient and the downwards motion of the molecules due to the gravitation (Fig. 7.33).

The concentration $n_i(z)$ of molecules with mass m_i therefore depends on their mass m_i. For the different molecular components in the earth atmosphere a different z-dependence appears (Fig. 7.34). The density of the heavier components should decrease more rapidly with increasing height than the lighter component. However, the measurements show that the composition of the atmosphere up to altitudes of about 30 km does not change much with the altitude. This has the following reason: The atmosphere is not isothermal but the temperature changes with increasing altitude (Fig. 7.35). These temperature differences causes pressure differences and strong upwards and downwards air currents which mix the different layers of the

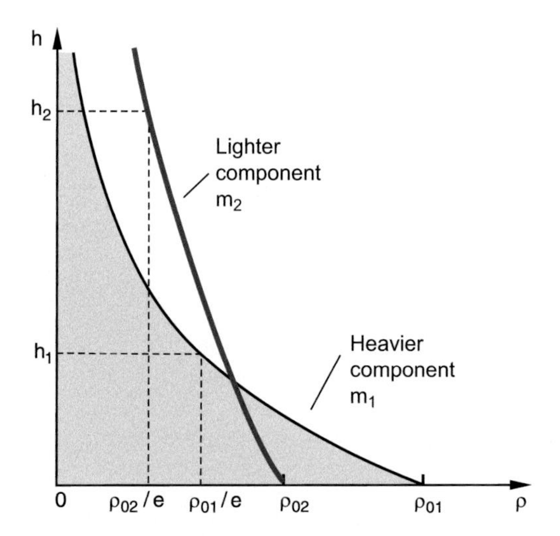

Figure 7.34 Density distribution of two molecular components with different masses in an atmosphere in the gravity field of the earth, if the atmosphere is governed by diffusion

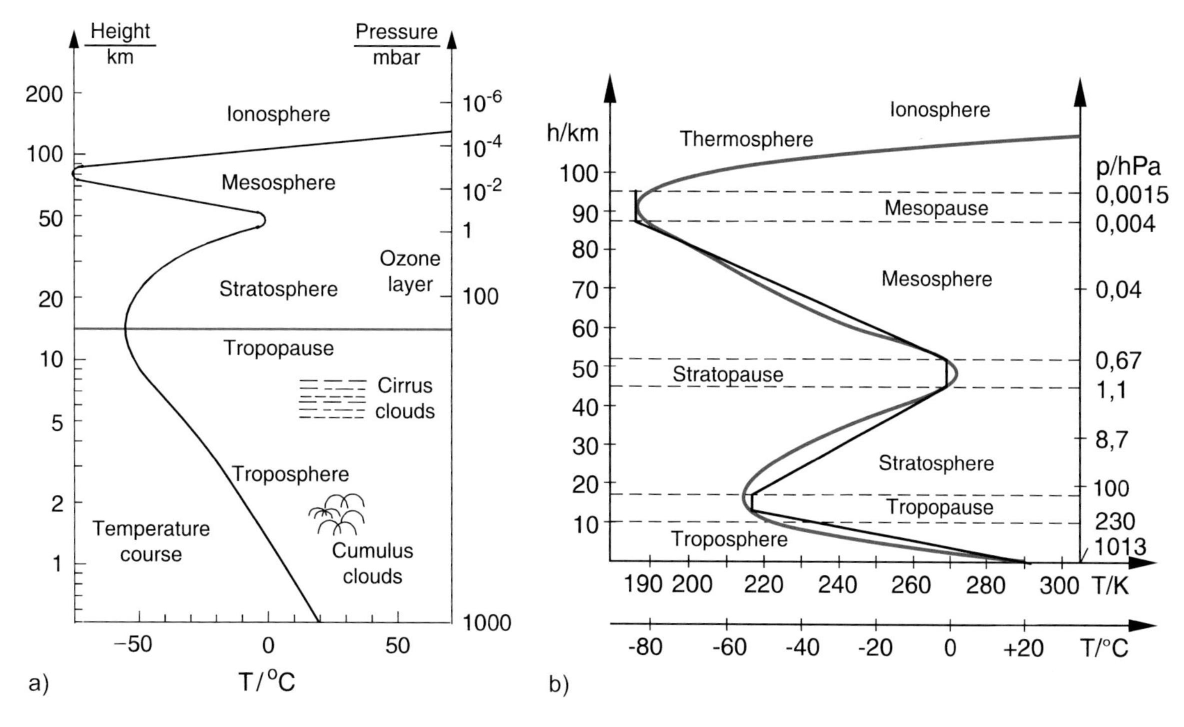

Figure 7.35 Temperature trend $T(h)$ in the earth atmosphere. **a** Measured dependence on a logarithmic scale; **b** Temperature trend on a linear scale. The *black curve* gives the dependence $T(h)$ of the standard atmosphere [7.10]

atmosphere. The temperature and its dependence on z is determined by the locally varying net energy flux into the different layers.

There are at first the absorption of the sun radiation which heats up the atmosphere. Furthermore the absorption of the infrared radiation emitted by the earth surface contributes to the energy flux. The radiation scattered back by the atmosphere diminishes the energy of the different parts of the atmosphere in a different way [7.5–7.8].

In order to standardize the description of the atmosphere, a model atmosphere was defined which serves as the standard atmosphere [7.7]. It is divided into different layers (Fig. 7.35). Within these layers the temperature $T(h)$ is a defined function which should be close to the measured values. Between these layers small regions appear where the temperature is nearly constant. They are called "*pauses*".

The **troposphere** extends from the ground to altitudes of about 8–12 km, where the upper limit depends on the season of the year. In the troposphere our weather takes place. The temperature decreases with increasing height with a slope $\mathrm{d}T/\mathrm{d}z \approx 6°/\mathrm{km}$ from a mean value $T(z = 0) = 17\,°\mathrm{C}$ to $T(z = 12\,\mathrm{km}) = -52\,°\mathrm{C}$. The nearly linear temperature decrease is caused by the heat transport from the earth surface into the atmosphere (convection, heat conduction and infrared radiation) which decreases nearly linear with increasing z.

Above the troposphere lies a thin layer, the **tropopause**, where the temperature stays nearly constant. The range between 10 to 50 km altitudes is the **stratosphere**. In the lower part of the stratosphere the temperature is nearly constant. With increasing altitude it increases up to 0 °C. The reason for the temperature increase is the ozone layer between 30–50 km, which contains O_3-molecules that absorb the UV-radiation from the sun and are excited into higher energy states. The excited ozone molecules collide with other atmospheric molecules and transfer their excitation energy into kinetic energy of the collision partners thus raising the temperature.

Above the stratosphere lies the stratopause, followed by the mesosphere between 50–80 km altitude. Here the temperature decreases with increasing altitude down to −93 °C and the molecular composition changes. With increasing altitude the lighter elements prevail as shown in Fig. 7.34. Because of the lower density the collision rate is much lower than in the troposphere and the mean free path is several kilometres. Therefore air currents are less effective in mixing the different layers.

The **mesopause** separates the mesosphere form the higher **thermosphere**, (85 km until 500–800 km) where the temperature rises up to 1700 °C. The temperature rise is due to collisions with high energy particles (electrons and protons) from the sun (sun wind) which also cause the polar light phenomena (aurora polaris). The density decreases down to 10^{-6} Pa = 10^{-11} bar. In spite of the low density, friction effects are-non negligible. The international space station ISS in about 350 km above ground flies through the thermosphere and it has to be lifted from time to time, because it loses kinetic energy and therefore altitude due to friction. The extreme ultraviolet radiation and the solar wind cause dissociation and ionization of the atmospheric molecules. Part of the components in the thermosphere are therefore ions. The thermosphere is part of a larger range in the atmosphere called the **ionosphere** which extends far into the space around

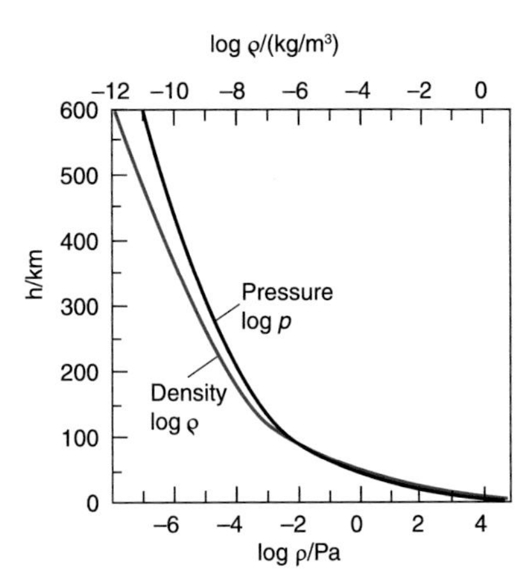

Figure 7.36 Dependence of pressure $p(h)$ and density $n(h)$ on a logarithmic scale [7.11]

the earth. It is no longer spherical symmetric because the magnetic field of the earth influences the path of charged particles (see Vol. 2) which move on spiral paths around the magnetic field lines. The variation of the temperature $T(h)$ with increasing height h implicates that density $n(h)$ and pressure $p(h)$ are no longer strictly proportional (Fig. 7.36).

The evaporation and condensation of water plays an important role in the lower atmosphere. Also the spurious concentrations of molecules with dipole moment such as OH, CO, CO_2, NH_4 etc. and dust particles and other aerosols have a pronounced influence on the weather conditions in the troposphere. The Chemistry of the atmosphere based on reaction of the different species by collisions with each other, is a subject of intense research and there are still many open questions [7.3a–7.5].

Summary

- For a constant temperature the pressure p of a gas in a closed but variable volume V obeys the Boyle–Mariotte law: $p \cdot V = \text{const}$
- The air pressure in an isothermal atmosphere decreases exponential with the altitude h above ground, due to the gravitational force.
- For the particle density $n(h)$ holds:

$$N(h) = n(0) \cdot \exp[-mgh/kT] \, .$$

  Without mixing effects in the atmosphere the concentration of particles with larger mass m therefore decreases faster with h than for those with lighter mass.
- The real earth atmosphere in not isothermal. Due to upwards and downwards air currents the different layers of the atmosphere are mixed which leads to an equilibrium of the concentrations of different masses.
- The kinetic gas theory explains the macroscopic features of gases such as pressure and temperature by the average momentum and the kinetic energy of the gas molecules. With the Boltzmann constant k the mean kinetic energy of molecules with mass m is related to the temperature T by $(1/2)m\overline{v^2} = (3/2)kT$.
- The velocity distribution $n(v)$ of gas molecules at thermal equilibrium is the Maxwell-Boltzmann distribution $n(v)\mathrm{d}v = v^2 \cdot \exp[-(1/2)mv^2/kT]\mathrm{d}v$ for the magnitude $v = |\boldsymbol{v}|$ of the velocity. The distribution $n(v_i)$, $(i = x, y, z)$ of the velocity components v_i is a Gaussian function, symmetric to $v_i = 0$.
- These distributions can be experimentally determined in molecular beams using mechanical velocity selectors. The molecular beams are formed by expanding a gas from a reservoir through a small hole into the vacuum, where the mean free path is longer than the dimensions of the vacuum chamber. The beam can be collimated by small apertures which transmit only molecules with small transverse velocities.
- Always when gradients of concentrations in a gas exist, diffusion processes occur which try to equalize the concentrations. The mean diffusion particle flux $j_\mathrm{D} = -D \cdot \text{grad}\, n$ is proportional to the gradient of the particle density n. The diffusion constant D depends on the kind of particles. Diffusion causes a mass transport from regions of high particle density n to those of low density.
- If velocity gradient in a gas flow appear, viscosity causes momentum transfer from particles with higher flow velocity to those with lower velocity.
- If temperature gradients appear in a gas, energy is transported by diffusing molecules from regions of higher temperature to those of lower temperature. For a one-dimensional temperature gradient $\mathrm{d}T/\mathrm{d}x$ the transferred heat power is $\mathrm{d}W/\mathrm{d}t = \lambda \cdot \mathrm{d}T/\mathrm{d}x$. The heat conductivity λ depends on the particle density n, the mean velocity $\overline{v}$, and the mean free path Λ.
- The density distribution $n(h)$ in the atmosphere is determined by the common action of gravitational attraction of the air molecules by the earth and the diffusion current from regions with higher density to those with lower density. In the real earth atmosphere furthermore vertical and horizontal air currents occur caused by local heat sources due to absorption of sun radiation and infrared radiation from the earth surface. This convection leads to a mixing of different layers in the lower atmosphere.

Problems

7.1 What would be the density distribution in the atmosphere, if the dependence of the gravitational force on the altitude is taken into account?

7.2 At which altitude exists, according to (7.6), a pressure of 1 mbar, if the constant value $T = 300\,\mathrm{K}$ is assumed for the temperature $T(h)$?

7.3 Calculate from (7.6) the pressure at $h = 100\,\mathrm{km}$ and the density n for $T = 250\,\mathrm{K}$.

7.4 A balloon with $V = 3000\,\mathrm{m}^3$ floats at $h = 1000\,\mathrm{m}$ and a temperature of $20\,°\mathrm{C}$. What is the maximum weight of balloon with ballast mass and passengers (without the weight of the filling gas) if one uses as filling gas
a) helium
b) hydrogen gas H_2
at a pressure equal to the external air pressure.
($\varrho_{\mathrm{air}} = 1.293\,\mathrm{kg/m^3}$, $\varrho_{\mathrm{He}} = 0.1785\,\mathrm{kg/m^3}$, $\varrho_{\mathrm{H_2}} = 0.09\,\mathrm{kg/m^3}$ at $T = 20\,°\mathrm{C}$ and $p = 10^5\,\mathrm{Pa}$).

7.5 A shop for diving equipment offers for measuring the diving depth a glass tube with movable piston that compresses a gas volume $V = A \cdot x$. Down to which depth is the uncertainty of the device $\Delta z \leq 1\,\mathrm{m}$ if the piston edge can be read with an accuracy of 1 mm and $x(p_0) = 0.2\,\mathrm{m}$.

7.6 Which fraction of all gas molecules has a free path that is larger than
a) the mean free path Λ
b) 2Λ?

7.7 Calculate the probability that N_2-molecules in a gas at $T = 300\,\mathrm{K}$ have velocities within the interval $900\,\mathrm{m/s} \leq v \leq 1000\,\mathrm{m/s}$. What is the total number $N(v)$ of molecules with velocities within this interval in a volume $V = 1\,\mathrm{m}^3$ at $T = 300\,\mathrm{K}$ and $p = 10^5\,\mathrm{Pa}$?

7.8 What is the thickness Δz of an isothermal atmospheric layer at $T = 280\,\mathrm{K}$ between the altitudes z_1 and z_2 with $p(z_1) = 1000\,\mathrm{hPa}$ and $p(z_2) = 900\,\mathrm{hPa}$?

7.9 What is the square root of the mean square relative velocities between two gas molecules
a) for a Maxwell distribution
b) if the magnitudes of all velocities are equal but the directions uniformly distributed?

7.10 The mean free path Λ in a gas at $p = 10^5\,\mathrm{Pa}$ and $T = 20\,°\mathrm{C}$ is for argon atoms $\Lambda_{\mathrm{Ar}} = 1 \cdot 10^{-7}\,\mathrm{m}$ and for N_2 molecules $\Lambda_{\mathrm{N_2}} = 2.7 \cdot 10^{-7}\,\mathrm{m}$.
a) What are the collision cross sections σ_{Ar} and $\sigma_{\mathrm{N_2}}$?
b) How large are the mean times between two successive collisions?

7.11 In a container is 0.1 kg helium at $p = 10^5\,\mathrm{Pa}$ and $T = 300\,\mathrm{K}$. Calculate
a) the number of He-atoms,
b) the mean free path Λ,
c) the sum $\sum S_i$ of all path lengths S_i which is passed by all molecules in 1 s. Give this sum in the units m and light years.

7.12 The rotating disc of a velocity selector with a slit allows N_2 molecules with a Maxwellian distribution at $T = 500\,\mathrm{K}$ to pass for a time interval $\Delta t = 10^{-3}\,\mathrm{s}$. A detector at 1 m distance from the disc measures the time distribution of the molecules. What is the half width of this distribution?

7.13 What is the minimum velocity of a helium atom at 100 km above ground for leaving the earth into space? At which temperature would half of the N_2-molecules above 100 km altitude escape into space?

7.14 The exhaust gases of a factory escaping out of a 50 m high smokestack have the density $\varrho = 0.85\,\mathrm{kg/m^3}$. How large is the pressure difference at the base of the smokestack to that of the surrounding air with $\varrho_{\mathrm{air}} = 1.29\,\mathrm{kg/m^3}$?

7.15 Up to which volume a children's balloon (m = 10g) has to be blown and filled with helium at a pressure of 1.5 bar, in order to let it float in air?

7.16 In the centre of the sun the density of protons and electrons is estimated as $n = 5 \cdot 10^{29}/\mathrm{m}^3$ at a temperature of $1.5 \cdot 10^7\,\mathrm{K}$.
a) What is the mean kinetic energy of electrons and protons? Compare this with the ionization energy of the hydrogen atom ($E_{\mathrm{ion}} = 13.6\,\mathrm{eV}$).
b) What are the mean velocities?
c) How large is the pressure?

7.17 Determine the total mass of the earth atmosphere from the pressure $p = 1\,\mathrm{atm} = 1013\,\mathrm{hPa}$ the atmosphere exerts onto the earth surface.

7.18 A research balloon has without filling a mass $m = 300\,\mathrm{kg}$. How large must be the volume of helium inside the balloon to let it rise up if the helium pressure at any height is always 0.1 bar higher than that of the surrounding air? ($T(h = 0) = 300\,\mathrm{K}$, $T(h = 20\,\mathrm{km}) = 217\,\mathrm{K}$)

7.19 What would be the height of the earth atmosphere
a) if the atmosphere is compressed with a pressure at the upper edge (assumed to be sharp) of 10 atm at a temperature of 300 K?
b) at $T = 0\,\mathrm{K}$ where all gases are solidified?

References

7.1. L.F. Reichl, *A Modern Course in Statistical Physics.* (Wiley VCH, Weinheim, 2016)
7.2. M. Scott Shell, *Thermodynamics and Statistical Mechanics.* (Cambridge Series in Chemical Engineering, 2015)
7.3a. R.P. Wayne, *Chemistry of the Atmosphere.* (Oxford Science Publ., Oxford Univ. Press, 1991)
7.3b. S. Kshudiram, *The Earth Atmosphere.* (Springer, Berlin, Heidelberg, 2008)
7.4. St.E. Manahan, *Enviromental Chemistry.* (CRC Press, 2009)
7.5. T.E. Graedel, P. Crutzen, *Atmospheric Change: An Earth System Perspective.* (Freeman and Co., 1993)
7.6. J.M. Wallace, P.V. Hobbs, *Atmospheric Science.* (Academic Press, 2006)
7.7. *US Standard Atmosphere.* (Government Printing Office, Washington, 1976) https://en.wikipedia.org/wiki/U.S._Standard_Atmosphere
7.8. C.D. Ahrens, *Essentials of Metereology.* (Thomson Broodge, 2011)
7.9a. St.A. Ackerman, J.A. Knox, *Metereology: Understanding the Atmosphere.* (Jones & Bartlett Learning, 2011)
7.9b. Ch. Taylor-Butler, *Metereology: The Study of Weather.* (Scholastic, 2012)
7.10. P. Staub, *Physics.* (TU Wien)
7.11. https://en.wikipedia.org/wiki/Atmosphere_of_Earth

Liquids and Gases in Motion; Fluid Dynamics

8

© Springer International Publishing Switzerland 2017
W. Demtröder, *Mechanics and Thermodynamics*, Undergraduate Lecture Notes in Physics, DOI 10.1007/978-3-319-27877-3_8

Up to now we have only considered liquids and gases at rest where the total momentum $\boldsymbol{P} = \sum \boldsymbol{p}_i = \boldsymbol{0}$, although the momenta p_i of the individual molecules, because of their thermal motion, are not zero but show a Maxwellian distribution with directions uniformly spread over all directions.

In this chapter, we will discuss phenomena that occur for streaming liquids and gases. Their detailed investigation has led to a special research area, the **hydrodynamics** resp. **aerodynamics** which are treated in more detail in special textbooks [8.1a–8.3b].

The macroscopic treatment of fluids in motion generally neglects the thermal motion of the individual molecules but considers only the average motion of a volume element ΔV, which can depend on the position $\boldsymbol{r} = \{x, y, z\}$. Since even for very small volume elements with dimensions in the mm-range ΔV still contains about 10^{15} molecules, the averaging is justified. The main difference between streaming fluids and gases is the density, which differs by about 3 orders of magnitude. This is closely related to the incompressibility of liquids while gases can be readily compressed. For streaming liquids, the density ϱ is constant in time, while the gas density can vary with time and position.

A complete description of the macroscopic motion of liquids and gases demands the knowledge of all forces acting on a volume element ΔV with the mass $\Delta m = \varrho \cdot \Delta V$. These forces have different underlying causes:

- pressure differences between different local positions induce forces $\boldsymbol{F}_\mathrm{p} = -\mathbf{grad}\, p \cdot \Delta V$ on a volume element ΔV.
- The gravity force $F_g = \Delta m \cdot g = \varrho \cdot g \cdot \Delta V$ leads for fluid flows with a vertical component to acceleration of ΔV.
- If the flow velocity u depends on the position $\boldsymbol{r}$ this results in friction forces F_f between layers of the fluid flow with different values of u.
- Charged particles in streaming fluids experience additional forces by external electric or magnetic fields (Lorentz force see Vol. 2). Such forces play an important role in stars and in laboratory plasmas. They are therefore extensively investigated in Plasmaphysics and Astrophysics. We will discuss them here, however, no longer, because their treatment is the subject of magneto-hydrodynamics and it would exceed the frame of the present textbook.

The Newtonian equation for the motion of a mass element $\Delta m = \varrho \cdot \Delta V$ in motion is then

$$\begin{aligned} \boldsymbol{F} &= \boldsymbol{F}_p + \boldsymbol{F}_g + \boldsymbol{F}_\mathrm{f} = \Delta m \ddot{\boldsymbol{r}} \\ &= \varrho \cdot \Delta V \cdot \frac{\mathrm{d}\boldsymbol{u}}{\mathrm{d}t}\,, \end{aligned} \tag{8.1}$$

where $u = \mathrm{d}r/\mathrm{d}t$ is the flow velocity of the volume element ΔV.

Before we try to solve this equation we will discuss at first some basic definitions and features of fluids in motion.

8.1 Basic Definitions and Types of Fluid Flow

The motion of the whole liquid is known, if it is possible to define the flow velocity $\boldsymbol{u}(r,t)$ of an arbitrary volume element $\mathrm{d}V$ at every location r and at any time t (Fig. 8.1). All values $u(r, t_0)$ for a given time t_0 form the *velocity field* (also named *flow field*) which can change with time. If $u(r)$ does not depend on time, the velocity field is stationary.

For a stationary flow the velocity $u(r)$ is at any position $\boldsymbol{r}$ temporally constant. It can, however, differ for different locations $\boldsymbol{r}$ (Fig. 8.1b).

The location curve $\boldsymbol{r}(t)$, which is traversed by a volume element ΔV (e.g. visualized by a small piece of cork) is called its *streamline* or *stream filament* (Fig. 8.1). The density of streamlines is the number of streamlines passing per second through an area of 1 m^2. All streamlines passing through the area A form a **stream tube**. Since the liquid is always moving along the stream lines no liquid leaks out of the walls of a stream tube.

For a stationary flow the path $\boldsymbol{r}(t)$ of a volume element $\mathrm{d}V$ follows the curve $\boldsymbol{u}(\boldsymbol{r})$ of the flow field. For non-stationary flows ($\partial u/\partial t \neq 0$), this is generally not the case as is illustrated in Fig. 8.2, where the curve $u(r, t_1)$ of the velocity field at time t_1 extends from P_1 via P_2 to P_3. However, when the volume element $\mathrm{d}V$ has arrived in P_2 at the time $t_1 + \Delta t$, the velocity field has changed meanwhile and the volume element follows now the curve $u(r, t_1 + \Delta t)$ from P_2 to P_4.

Since the different forces in (8.1) generally have different directions and furthermore the friction force depends on the velocity gradient the motion of $\mathrm{d}V$ depends on the relative contribution

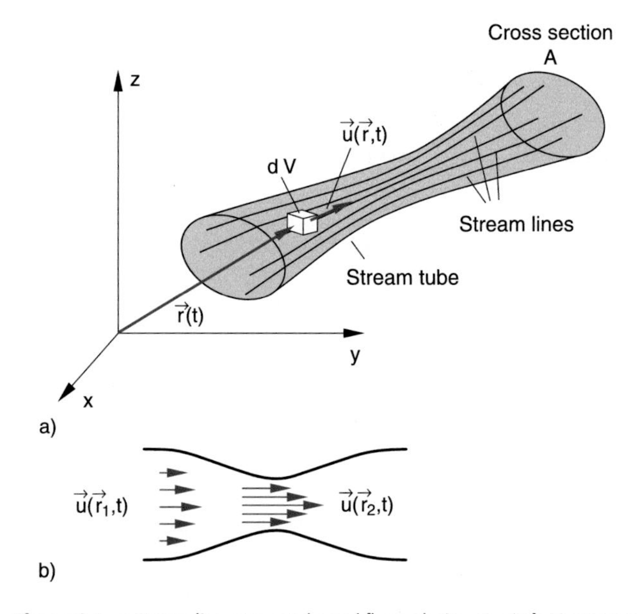

Figure 8.1 **a** Stream line, stream tube and flow velocity $u(\boldsymbol{r}, t)$; **b** Momentary condition of a flow field (velocity field)

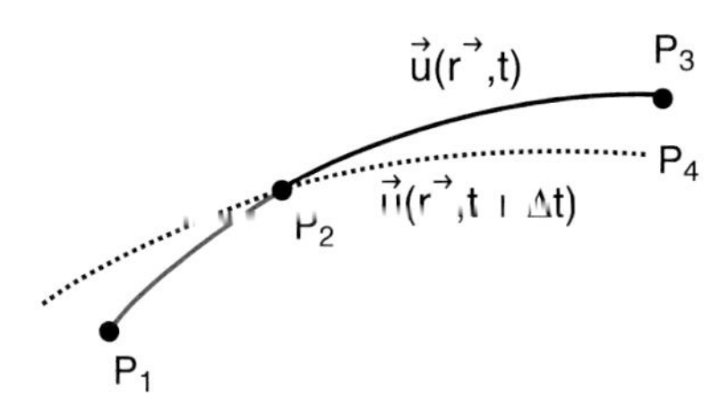

Figure 8.2 In a nonstationary flow the path of a particle does not necessarily follow a streamline $u(r,t)$

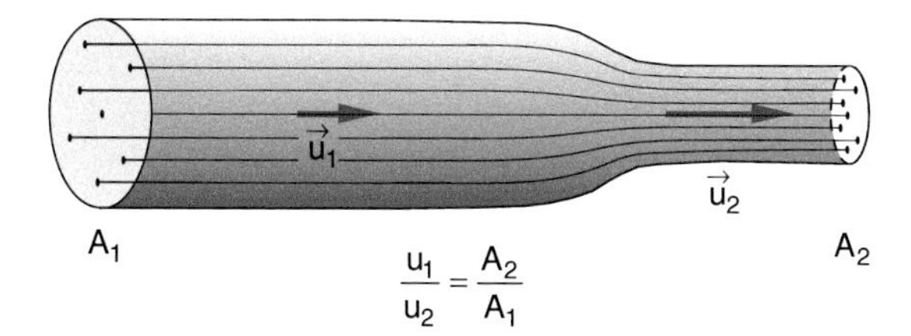

Figure 8.3 Example of a laminar flow

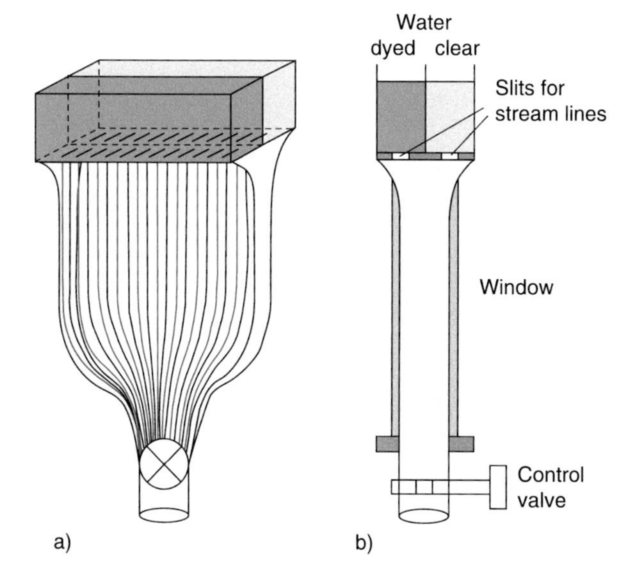

Figure 8.4 Streamline apparatus. **a** Angle view, **b** Side view

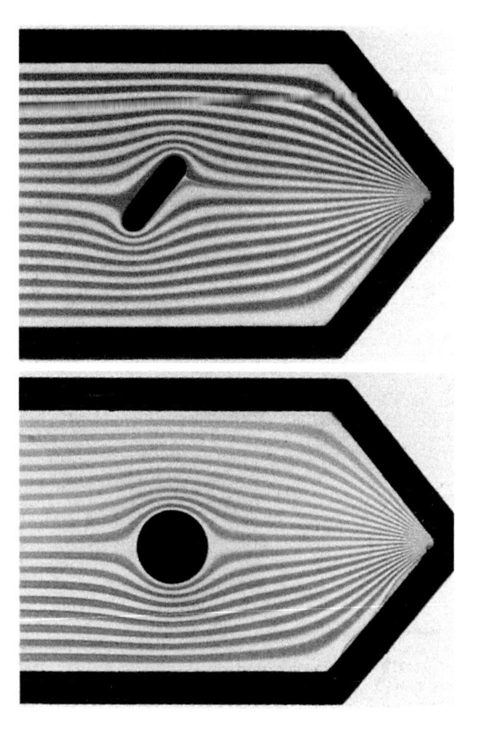

Figure 8.5 Laminar flow from left to right around different obstacles, photographed with the streamline device of Fig. 8.4

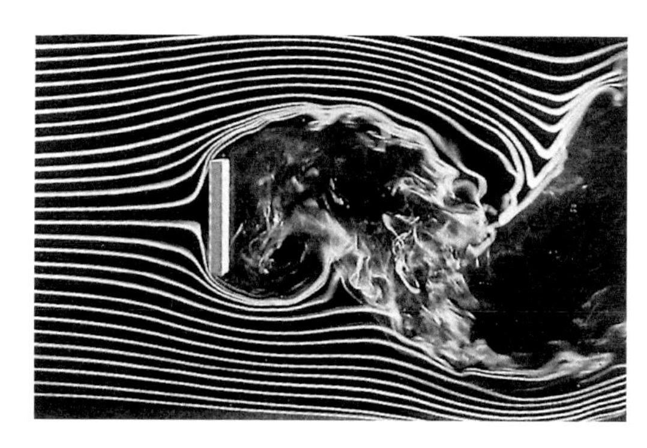

Figure 8.6 A laminar flow coming from left becomes turbulent after impinging on a plate

of the different forces. Liquids where the friction forces are negligible compared to the other forces are called **ideal liquids**. If the frictional forces are large compared to all other forces we have the limiting case of viscous liquids.

Examples for the first case are the flow of liquid helium through a pipe or of air along the smooth wing of an airplane, while the second case is realized by the flow of honey or molasses out of a sloped glass container onto a slice of bread or the slow flow of heavy oil through pipelines. The real liquids and gases are located between these two limiting cases. A flow where the stream lines stay side by side without mixing is called a **laminar flow** (Fig. 8.3). Laminar flows are always realized if the frictional forces are dominant. They can be demonstrated with the streamline generator. This is an apparatus where the bottom of two containers has narrow slits arranged in such a way, that the liquid from each container streams alternately through every second slit downwards between two parallel glass plates (Fig. 8.4). When one container is filled with red dyed water and the other with black tinted water the stream lines are alternately black and red. With such a demonstration apparatus the stream line conditions with different obstacles in the flow can be readily shown to a large auditorium, if projected onto a large screen (Fig. 8.5).

Turbulent flows are generated by friction between the wall and the peripheral layer of the flow if the internal friction of the flow is smaller than the accelerating forces. Vortices are formed which can intermixe the stream lines completely (Fig. 8.6).

8.2 Euler Equation for Ideal Liquids

A volume element $\mathrm{d}V$ with the flow velocity $\boldsymbol{u}(\boldsymbol{r}, t)$ passes during the time interval $\mathrm{d}t$ a path length $\mathrm{d}\boldsymbol{r} = \boldsymbol{u}\mathrm{d}t$. Starting from the position r it reaches the position $\boldsymbol{r} + \boldsymbol{u} \cdot \mathrm{d}t$ at time $t + \mathrm{d}t$ and has there the velocity

$$\boldsymbol{u} + \mathrm{d}\boldsymbol{u} = \boldsymbol{u}(\boldsymbol{r} + \boldsymbol{u}\,\mathrm{d}t, t + \mathrm{d}t) \,. \tag{8.2}$$

Even for stationary flows, the velocity can change with position. For example, a liquid flowing through a pipe increases its velocity when the pipe cross section decreases (Fig. 8.3). The stream line density is there increased. For nonstationary flows the velocity changes also with time even at the same location, because $\partial \boldsymbol{u}/\mathrm{d}t \neq 0$.

We define the substantial acceleration of a volume element $\mathrm{d}V$ as the total change of its velocity $\boldsymbol{u} = \{u_x, u_y, u_z\}$ when $\mathrm{d}V$ passes during the time interval $\mathrm{d}t$ from the position $\boldsymbol{r}$ to $\boldsymbol{r} + \mathrm{d}\boldsymbol{r}$. This total acceleration has two contributions:

1. the temporal change $\partial \boldsymbol{u}/\partial t$ at the same position
2. the change of u when $\mathrm{d}V$ passes from $\boldsymbol{r}$ to $\boldsymbol{r}+\mathrm{d}\boldsymbol{r}$. This change is per second $(\partial \boldsymbol{u}/\partial \boldsymbol{r}) \cdot (\partial \boldsymbol{r}/\partial t)$.

This can be written in components as

$$\frac{\mathrm{d}u_x}{\mathrm{d}t} = \frac{\partial u_x}{\partial t} + \frac{\partial u_x}{\partial x}\frac{\mathrm{d}x}{\mathrm{d}t} + \frac{\partial u_x}{\partial y}\frac{\mathrm{d}y}{\mathrm{d}t} + \frac{\partial u_x}{\partial z}\frac{\mathrm{d}z}{\mathrm{d}t} \tag{8.3a}$$

with corresponding equations for the other components u_y and u_z.

In vector form this reads with $u_x = \mathrm{d}x/\mathrm{d}t$, $u_y = \mathrm{d}y/\mathrm{d}t$, $u_z = \mathrm{d}z/\mathrm{d}t$

$$\frac{\mathrm{d}\boldsymbol{u}}{\mathrm{d}t} = \frac{\partial \boldsymbol{u}}{\partial t} + (\boldsymbol{u} \cdot \nabla)\,\boldsymbol{u} \,. \tag{8.3b}$$

Here $\boldsymbol{u} \cdot \nabla \boldsymbol{u}$ is the scalar product of the vector $\boldsymbol{u}$ and the tensor

$$\nabla \boldsymbol{u} = \begin{pmatrix} \frac{\partial u_x}{\partial x} & \frac{\partial u_x}{\partial y} & \frac{\partial u_x}{\partial z} \\ \frac{\partial u_y}{\partial x} & \frac{\partial u_y}{\partial y} & \frac{\partial u_y}{\partial z} \\ \frac{\partial u_z}{\partial x} & \frac{\partial u_z}{\partial y} & \frac{\partial u_z}{\partial z} \end{pmatrix} .$$

The substantial acceleration is composed of the time derivative $\partial \boldsymbol{u}/\partial t$ of the velocity at a fixed position r and the convection acceleration $(\boldsymbol{u} \cdot \nabla)\boldsymbol{u}$. The first contribution is only nonzero for nonstationary flows, the second only if the velocity changes with the position r.

The equation of motion for an ideal liquid (frictional forces are negligible) which experiences the accelerating forces of gravity $F_g = m \cdot g$ and pressure gradient $\boldsymbol{F}_\mathrm{p} = -\mathbf{grad}\, p \cdot \mathrm{d}V$ is the **Euler equation**

$$\frac{\mathrm{d}\boldsymbol{u}}{\mathrm{d}t} = \frac{\partial \boldsymbol{u}}{\partial t} + (\boldsymbol{u} \cdot \nabla)\,\boldsymbol{u} = \boldsymbol{g} - \frac{1}{\varrho}\mathbf{grad}\, p \,. \tag{8.4}$$

This is the basic equation for the motion of ideal liquids, which was already postulated by L. Euler in 1755.

8.3 Continuity Equation

We consider a liquid volume $\mathrm{d}V = A \cdot \mathrm{d}x$, which flows in x-direction through a pipe with variable cross section $A(x)$ (Fig. 8.7a). Its mass is $\mathrm{d}M = \varrho \cdot \mathrm{d}V = \varrho \cdot A \cdot \mathrm{d}x$. Through the cross section A_1 flows per time unit the mass

$$\frac{\mathrm{d}M}{\mathrm{d}t} = \varrho A_1 \frac{\mathrm{d}x}{\mathrm{d}t} = \varrho A_1 u_{x_1} \,. \tag{8.5}$$

We assume, that at the position $x = x_0$ the cross section A changes to A_2. For incompressible liquids ϱ remains constant. Since the liquid cannot escape through the side walls the mass flowing per time unit through A_2 must be equal to that flowing through A_1. This gives the equation

$$\varrho A_1 u_{x_1} = \varrho A_2 u_{x_2} \Rightarrow \frac{u_{x_1}}{u_{x_2}} = \frac{A_2}{A_1} \,. \tag{8.6}$$

Through the narrow part of the pipe the liquid flows faster than through a wide part. The product

$$\boldsymbol{j} = \varrho \cdot \boldsymbol{u} \tag{8.7}$$

is called **mass flow density**. The product $I = \boldsymbol{j} \cdot \boldsymbol{A}$ is the **total mass flow** and gives the mass flowing per unit time through the cross section $\boldsymbol{A}$.

Equation 8.6 can then be written as $I = \text{const}$. The total mass flow through a pipe is the same at every position in the pipe.

This statement about the conservation of the total mass flow can be formulated in a more general way: The volume V contains at

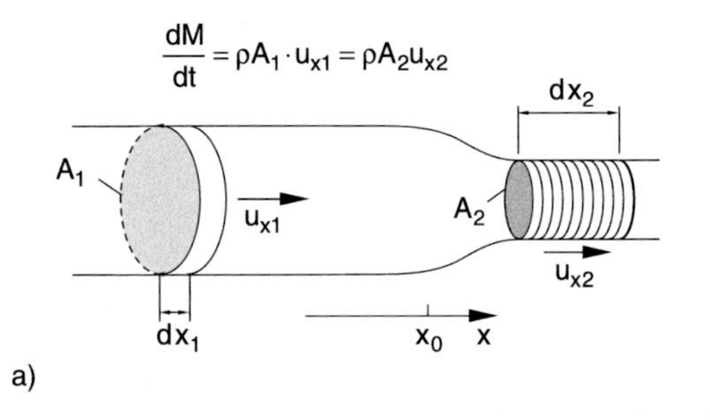

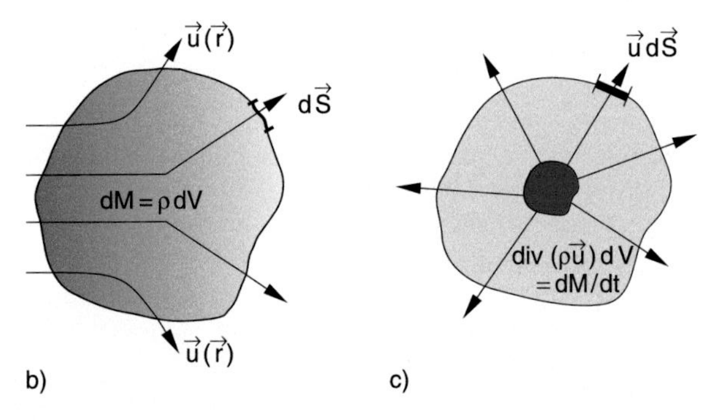

Figure 8.7 Illustration of the continuity equation: **a** in a pipe with changing diameter; **b** in a volume V with surface S with a mass flow $\mathrm{d}M/\mathrm{d}t$ through V; **c** with a source inside a volume V

time t the liquid mass

$$M = \int_V \varrho \, \mathrm{d}V \,. \qquad (8.8)$$

The mass per volume changes with time if mass flows out of the volume or into the volume. The mass flowing per second through its surface S is

$$-\frac{\partial M}{\partial t} = \int_S \varrho \cdot \boldsymbol{u} \, \mathrm{d}\boldsymbol{S} = \int_S \boldsymbol{j} \, \mathrm{d}\boldsymbol{S} \,, \qquad (8.9)$$

where the normal vector $\mathrm{d}\boldsymbol{S}$ is perpendicular to the surface element $\mathrm{d}A$.

According to Gauß' law (see textbooks on vector analysis e.g. [8.8]) the surface integral can be converted into a volume integral over the volume V enclosed by the surface S.

$$\int_S \varrho \cdot \boldsymbol{u} \cdot \mathrm{d}\boldsymbol{S} = \int_V \operatorname{div}(\varrho \cdot \boldsymbol{u}) \mathrm{d}V \,, \qquad (8.10)$$

and we obtain from (8.8)–(8.10) for a constant volume V the relation

$$-\frac{\partial}{\partial t} \int_V \varrho \, \mathrm{d}V = -\int \frac{\partial \varrho}{\partial t} \mathrm{d}V = \int \operatorname{div}(\varrho \boldsymbol{u}) \mathrm{d}V \,. \qquad (8.11)$$

Since this must be valid for arbitrary volumes this gives the continuity equation

$$\frac{\partial \varrho}{\partial t} + \operatorname{div}(\varrho \boldsymbol{u}) = 0 \,, \qquad (8.12)$$

which states that for any mass flow the total mass is conserved, i.e. **mass is neither produced nor annihilated**.

For a constant volume element $\mathrm{d}V$ (8.12) can be written as

$$\operatorname{div}(\varrho \cdot \boldsymbol{u}) \mathrm{d}V = -\frac{\partial \varrho}{\partial t} \mathrm{d}V = -\frac{\partial}{\partial t} (\mathrm{d}M) \,. \qquad (8.12a)$$

The expression $\operatorname{div}(\varrho \cdot u) \cdot \mathrm{d}V$ gives the mass that escapes per second out of the volume element $\mathrm{d}V$. Therefore $\operatorname{div}(\varrho \cdot \boldsymbol{u})$ is called the **source strength** per unit volume. A source which delivers the mass $\mathrm{d}M/\mathrm{d}t$ per sec leads to a mass flow $\operatorname{div}(\varrho \cdot u)$ per sec through the surface surrounding the source (Fig. 8.7c).

The continuity equation (8.12) is valid for liquids as well as for gases. For incompressible liquids is $\partial \varrho / \partial t = 0$ and ϱ is furthermore spatially constant. The equation of continuity simplifies then to

$$\operatorname{div}(\boldsymbol{u}) = 0 \quad \textbf{(continuity equation for incompressible liquids)} \,. \qquad (8.13a)$$

For the three components this equation reads

$$\frac{\partial u_x}{\partial x} + \frac{\partial u_y}{\partial y} + \frac{\partial u_z}{\partial z} = 0 \,. \qquad (8.13b)$$

In pipes with constant cross section A the liquid flows only into one direction which we choose as the x-direction. Then $u_y = u_z = 0$ and (8.13b) becomes $\partial u_x / \partial x = 0 \Rightarrow u_x = \text{const}$.

8.4 Bernoulli Equation

If a liquid or a gas flows in x-direction through a pipe with variable cross section $A(x)$ the flow velocity is larger at locations with smaller cross section (continuity equation). The volume elements therefore have to be accelerated and have a higher kinetic energy than at places with larger cross section. This results in a decrease of the pressure p. This can be seen as follows:

In order to transport the volume element $\mathrm{d}V_1 = A_1 \cdot \Delta x_1$ in the wider part of the pipe through the cross section A_1 it has to be shifted by Δx_1 against the pressure p_1 (Fig. 8.8). This demands the work

$$\begin{aligned} \Delta W_1 &= F_1 \Delta x_1 = p_1 A_1 \cdot \Delta x_1 \\ &= p_1 \Delta V_1 \,. \end{aligned} \qquad (8.14a)$$

In the narrow part of the pipe is $\Delta V_2 = A_2 \cdot \Delta x_2$ and the work necessary to shift ΔV_2 by Δx_2 against the pressure p_2 is

$$\begin{aligned} \Delta W_2 &= p_2 A_2 \Delta x_2 \\ &= p_2 \Delta V_2 \,. \end{aligned} \qquad (8.14b)$$

The kinetic energy of the volume elements is

$$E_{\text{kin}} = \tfrac{1}{2} \Delta M \cdot u^2 = \tfrac{1}{2} \varrho \cdot u^2 \cdot \Delta V \,.$$

For ideal liquids (frictional forces are negligible) the sum of potential and kinetic energy has to be constant (energy conservation). This gives the equation

$$p_1 \Delta V_1 + \tfrac{1}{2} \varrho u_1^2 \Delta V_1 = p_2 \Delta V_2 + \tfrac{1}{2} \varrho u_2^2 \Delta V_2 \,. \qquad (8.15)$$

For incompressible liquid is $\varrho = \text{constant}$ and therefore $\Delta V_1 = \Delta V_2 = \Delta V$. This simplifies (8.15) to

$$p_1 + \tfrac{1}{2} \varrho u_1^2 = p_2 + \tfrac{1}{2} \varrho u_2^2 \,. \qquad (8.16)$$

For a frictionless incompressible liquid flowing through a horizontal pipe with variable cross section (Fig. 8.9) we obtain for a stationary flow from (8.16) the **Bernoulli Equation**

$$p + \tfrac{1}{2} \varrho u^2 = p_0 = \text{const} \,. \qquad (8.17)$$

The constant p_0 is the total pressure which is reached at locations with $u = 0$. the quantity $p_s = (\varrho/2) u^2 = p_0 - p$ is the **dynamic stagnation pressure (ram pressure)**, while $p = p_0 - p_s$ is the **static pressure** of the flowing liquid.

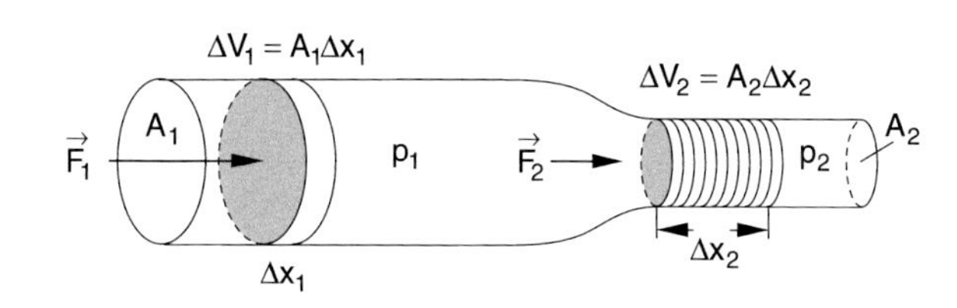

Figure 8.8 Illustration of Bernoulli-equation

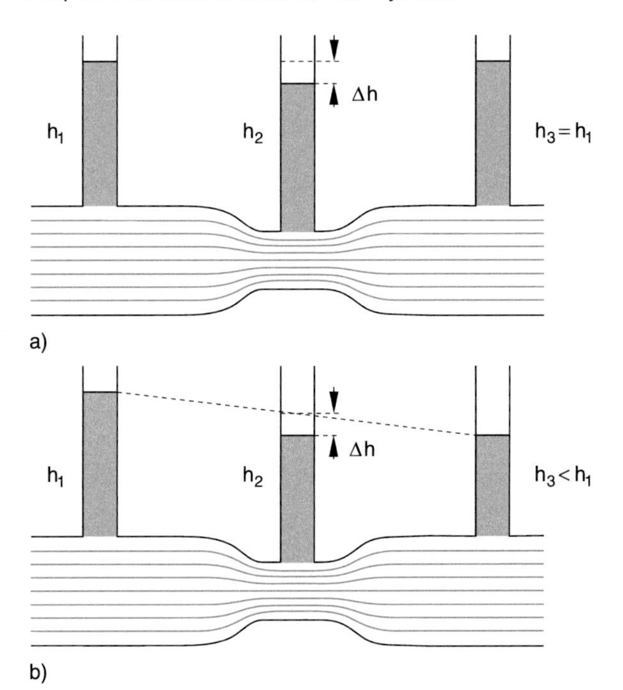

Figure 8.9 Demonstration of Bernoulli equation by pressure measurements in stand pipes. The pressure difference is $\Delta p = \varrho \cdot g \cdot \Delta h$. **a** For ideal liquids without friction; **b** for real liquids with friction. The liquid streams from *left to right*

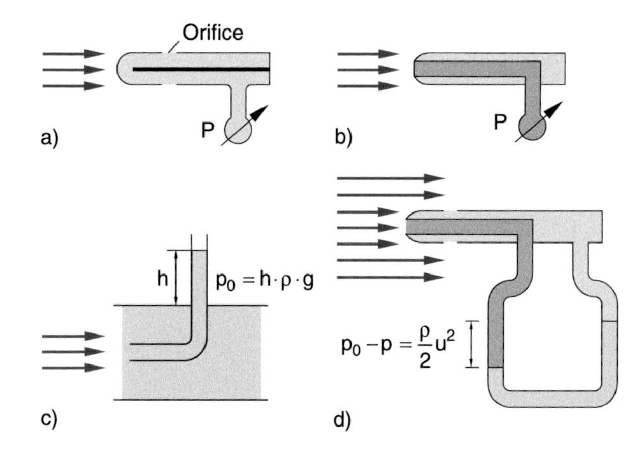

Figure 8.10 Measurement of pressure conditions in flows. **a** Measurement of static pressure; **b** measurement of total pressure p_0 with Pitot tube and pressure manometer; **c** measurement of p_0 with a stand pipe; **d** measurement of stagnation pressure $p_s = p_0 - p$ as difference of total pressure and static pressure

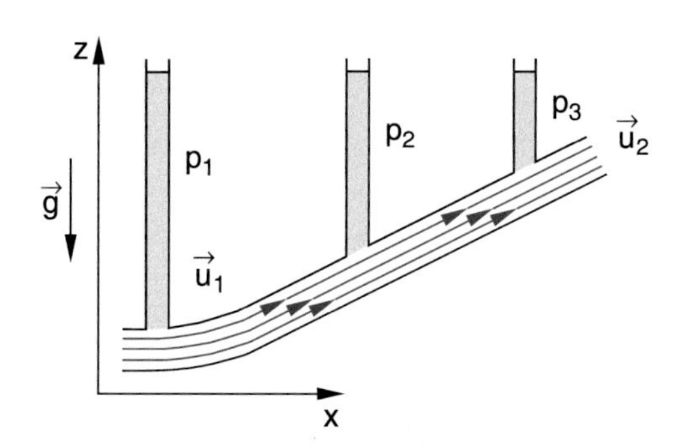

Figure 8.11 Flow of a liquid through an inclined pipe

The Bernoulli equation can be demonstrated with the arrangement shown in Fig. 8.9, where dyed water flows through a horizontal glass tube with variable cross section and vertical stand pipes. The rise h in the vertical stand tube gives the static pressure $p = \varrho \cdot g \cdot h$. At the narrow parts of the horizontal pipe the flow velocity is larger and therefore pressure and height h are smaller. In Fig. 8.9a the situation for an ideal frictionless liquid with $p(x) =$ constant for constant cross section is shown, while Fig. 8.9b illustrates the influence of friction on the pressure $p(x)$. For tubes with constant cross section a linear decrease of $p(x)$ is observed.

The three quantities p, p_0 and p_s can be measured at arbitrary locations in the flow with the devices shown in Fig. 8.10a–d. With a pressure gauge, shown in Fig. 8.10a which has a small hole in the sidewall of a tube, the liquid flow, streaming around the tube creates a static pressure inside the tube, which is monitored by a pressure manometer. The pitot-tube (Fig. 8.10b and c) has a hole at the end of the tube. If the tube is aligned parallel to the stream lines the flow velocity at the head of the tube is $u = 0$, i.e. the measured pressure is the total pressure p_0. It can be measured either with a manometer (Fig. 8.10b), or with a vertical stand pipe (Fig. 8.10c). With a combination of pressure gauge and Pitot tube (Fig. 8.10d) the pressure p_0 is measured at the head of the horizontal tube while a hole in the sidewalls monitors the pressure p. The difference $p_s = p_0 - p$ is shown as the difference of the heights of mercury in the U-shaped lower part of the device.

For liquid flows in inclined pipes the difference of potential energies $\Delta E_{pot} = \varrho \cdot g \cdot \Delta h \cdot \Delta V$ of a volume element ΔV at different heights h has to be taken into account. If the flow, for instance, is directed in the x-z-plane (Fig. 8.11) the height is $h = z(x)$ and we obtain from (8.17) the general equation

$$p + \varrho g z(x) + \tfrac{1}{2}\varrho u^2(x) = \text{const} = p_0 \,. \tag{8.18}$$

For an ideal incompressible liquid ϱ is constant within the whole pipe. If the cross section of the pipe is constant also the flow velocity u is constant throughout the whole pipe. If $p + \varrho \cdot g \cdot z \geq p_0$ the flow ceases and $u = 0$ in the whole pipe.

Note: Although the Bernoulli equation (8.17) has been derived for incompressible liquids the equation allows to obtain also the pressure change of gases for laminar flows at not too high flow velocities. For example, inserting for air flows the numerical values $p_0 = 1$ bar, $u = 100\,\text{m/s}$, $\varrho = 1.293\,\text{kg/m}^3$ into the equation

$$p_0 - p = \tfrac{1}{2}\varrho \cdot u^2$$

one obtains $p = 0.935 p_0$, i.e. a pressure decrease of 6.5% and therefore also a corresponding decrease of the density ϱ. However, if the flow velocity approaches the velocity of sound ($c = 340\,\text{m/s}$) the change of the density becomes so large that the condition of incompressibility is even approximately not fulfilled.

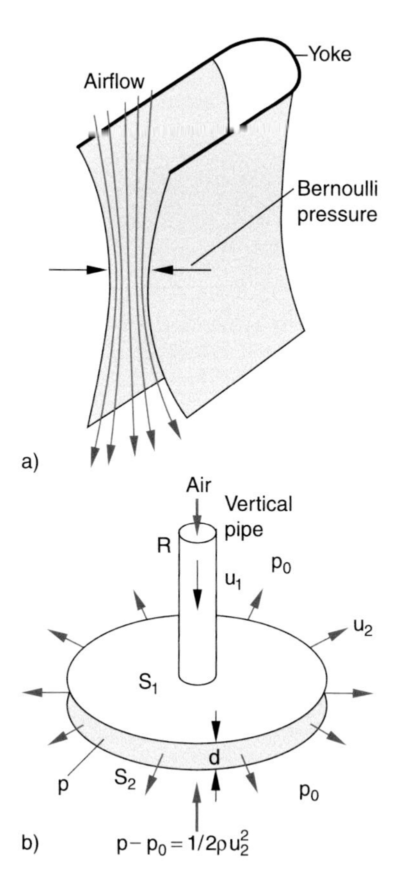

Figure 8.12 Hydrodynamic paradox: **a** Two curved aluminium plates, which can swing around a yoke, are pressed together when blowing air between them; **b** the lower circular plate is attracted to the upper plate when air is blown through the pipe

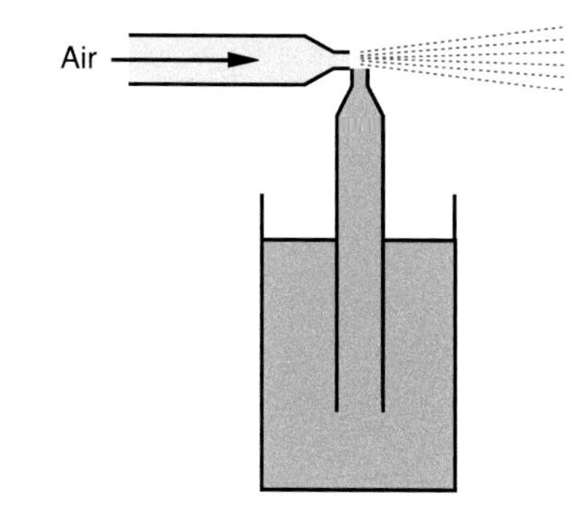

Figure 8.13 Vaporizer

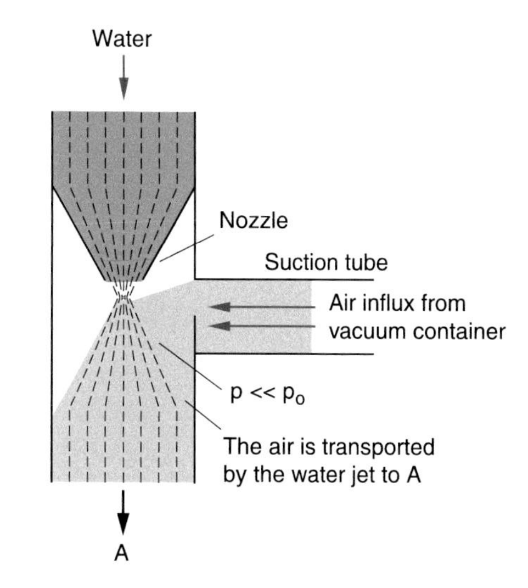

Figure 8.14 Water jet pump

The Bernoulli equation can be demonstrated by many simple experiments which often astound the auditorium. One example is the **hydrodynamic paradox**. Two curved aluminum plates are hanging on a U-shaped wire bar (Fig. 8.12a). If one blows air between the two plates they move towards each other, contrary to the expectation that they will be pushed away from each other. When air is blown through a vertical pipe fixed on one end to a circular disc S_1 with a hole (Fig. 8.12b) a second disc S_2 below the fixed disc is lifted to the upper disc by the air streaming between the two discs. The distance d between the two discs with area A must be below a critical value where the flow velocity u of the air is sufficiently large to cause an attractive force

$$F = A(p_0 - p) = \tfrac{1}{2}\varrho \cdot u^2 \cdot A \geq m \cdot g$$

between the two discs which can balance the weight $m \cdot g$ of the lower disc.

The Bernoulli theorem is used for many practical applications. Examples are the vaporizer or the spray bottle (Fig. 8.13) where air streams out of a narrow nozzle and generates a reduced pressure, which sucks the liquid out of the bottle into the air stream. Here it is nebulized. Another example is the water jet vacuum pump (Fig. 8.14). Here water streams with a large velocity through a narrow nozzle where it generates a reduced pressure. The air from the surrounding diffuses into the region with reduced pressure where it penetrates into the water jet and is transported out of the container into the outer space A, thus evacuating the container. With such a device reduced pressures down to 30 mbar can be achieved.

Undesirable effects of the Bernoulli theorem are the unroofing of houses under the action of typhoons (Fig. 8.15). When wind

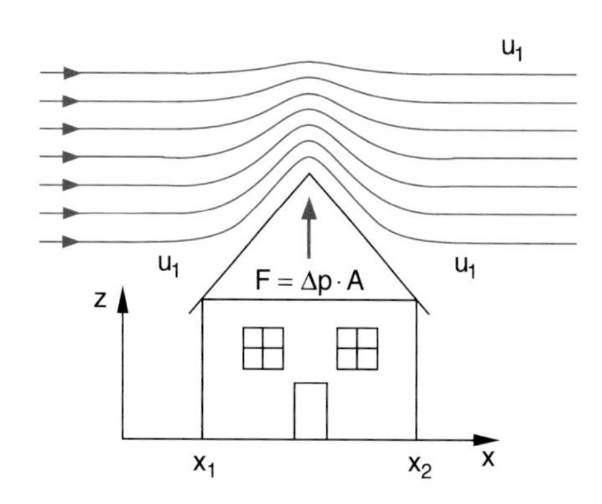

Figure 8.15 A strong wind can unroof a house due to the reduced pressure above the roof

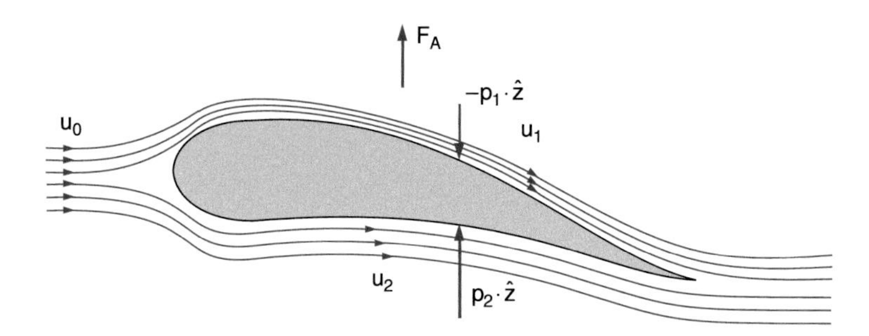

Figure 8.16 Aerodynamic lift at a wing profile due to the higher velocity around the upper side of the profile

blows with the flow velocity $u(x)$ over the roof of a house, the pressure difference $\Delta p = p_0 - p$ results in an upwards directed force

$$F = L_y \cdot \int \Delta p(x)\mathrm{d}x = L_y \cdot \int \tfrac{1}{2}\varrho u^2(x)\mathrm{d}x$$

on the roof, where L_y is the length of the roof in y-direction. The pressure difference depends on the flow velocity $u(x)$ which is maximum at the top of the roof, where the stream lines have the highest density.

The Bernoulli equation is the basis of the aerodynamic lift force and therefore important for the whole aviation. In Fig. 8.16 the profile of an airplane wing is shown with the stream lines of air flowing below and above the wing. For the asymmetric profile the air flows faster above than below the wing. This cause, according to (8.17) for a wing area A and the air density ϱ_a a lift force

$$F = (p_2 - p_1) \cdot A = \tfrac{1}{2}\varrho_L \left(u_2^2 - u_1^2\right) \cdot A \, .$$

Remark. Since air at high flow velocities is compressible and therefore cannot be treated as ideal liquid, the situation for a plane is more complex because the flow velocity of the air relative to the flying plane is very large. Besides friction forces turbulence and density changes play an important role for the calculation of the upwards lift (see Sect. 8.6).

8.5 Laminar Flow

Laminar flows (Fig. 8.3) are always realized when the frictional forces exceed the accelerating forces. Therefore, we will at first discuss the internal friction in liquids and gases and then illustrate the importance of laminar flow by several practical examples.

8.5.1 Internal Friction

Assume a plane sheet with area A in the y-z-plane is pulled through a liquid with the velocity u_0 into the horizontal direction (which we choose in Fig. 8.17 as the z-direction). The liquid layers at $x = x_0 \pm \mathrm{d}x$ adjacent to the two plate surfaces at $x = x_0$

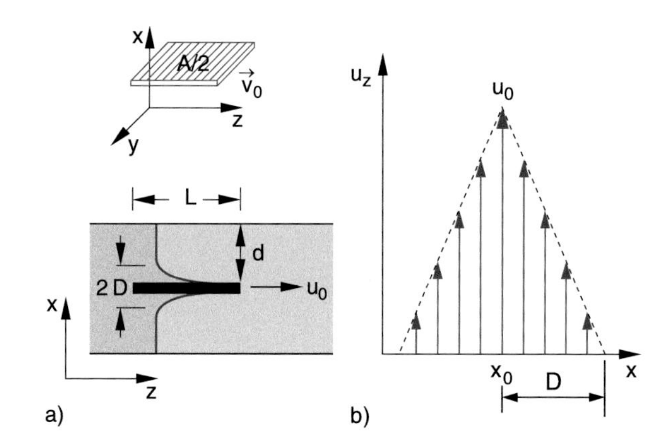

Figure 8.17 Internal friction of liquids. **a** A slab is pulled with the velocity u_x through a viscose liquid. It takes along boundary layers of the liquid. **b** Velocity profile and thickness D of the boundary layer

will be dragged with the moving plate due to the static friction between liquid and plate surfaces. These layers transfer part of their momentum $\varrho_L \cdot u_z \mathrm{d}V$ to the neighbouring liquid layers. This can be demonstrated by the experiment shown in Fig. 8.17a: In a glass trough is a viscous liquid, for example glycerine. The left part of the liquid is dyed. When an immersed plate is slowly pulled through the liquid with the velocity u_0 into the z-direction one can see that the liquid layers adjacent to the plate surfaces stick to the surfaces and are dragged with the velocity u_0. Perpendicular to the plate surfaces a velocity gradient is present (Fig. 8.17b). As has been discussed in Sect. 7.5, this gradient is due to the thermal motion of the liquid molecules, which penetrate by about a mean free path Λ into the neighbouring layers and transfer part of their momentum. This causes a velocity gradient $\mathrm{d}u/\mathrm{d}x$ perpendicular to the velocity of the plate.

In Sect. 7.5.4 it was shown, that the momentum transferred per second and unit area between neighbouring layers is $j_p = \eta \cdot \mathrm{d}u_z/\mathrm{d}x$. Since the time derivative of the momentum is equal to the acting force we obtain for the force between adjacent layers

$$F = \eta \cdot A \cdot \frac{\mathrm{d}u}{\mathrm{d}x} \, , \tag{8.19}$$

which appears when the plate is pulled with the constant velocity u_0 through the liquid, where A is the total surface of the plate (both sides!). This force must just compensate the friction force

Table 8.1 Dynamical viscosities of some liquids and gases at a temperature $T = 20\,°\mathrm{C}$

Substance	$\eta/(\mathrm{mPa}\cdot\mathrm{s})$
Water	1.002
Benzene	0.65
Ethanol	1.20
Glycerine	1480.0
Heavy fuel oil	660
Mercure	1.55
Air (10^5 Pa)	$1.8\cdot10^{-2}$
Helium (10^5 Pa)	$1.9\cdot10^{-2}$

opposite to the direction of u_0

$$F_\mathrm{f} = -\eta\cdot A\cdot\frac{\mathrm{d}u}{\mathrm{d}x}\,. \tag{8.20}$$

The factor η is the *dynamic viscosity*. It has the dimension $[\eta] = \mathrm{N\cdot s/m^2} = \mathrm{Pa\cdot s}$. In the older literature often the unit Poise $= \mathrm{P} = \mathrm{g\cdot cm^{-1}\cdot s^{-1}}$ is used. The conversion is $1\,\mathrm{P} = 0.1\,\mathrm{Pa\cdot s}$; 1 centipoise $= 1\,\mathrm{cP} = 10^{-3}\,\mathrm{Pa\cdot s}$.

In Tab. 8.1 the numerical values of η for some liquids are compiled. They should be compared with the data for gases in Tab. 7.3.

The dynamic viscosity η depends strongly on the temperature, as can be seen from Tab. 8.2. For liquid helium a superfluid phase exists at temperatures below 2.17 K, where $\eta = 0\,\mathrm{Pa\cdot s}$ [8.5].

The distance D where the liquid is dragged by the moving plate is called **fluid dynamic boundary layer**. Its value can be obtained by the following consideration: In order to move the plate by its length L against the frictional force F_f one has to accomplish the work

$$W_\mathrm{f} = -F_\mathrm{f}\cdot L = \eta AL\cdot\left|\frac{\mathrm{d}u}{\mathrm{d}x}\right| = \eta AL\cdot\frac{u_0}{D}\,, \tag{8.21}$$

where we have assumed that a linear velocity gradient $\mathrm{d}u/\mathrm{d}x = u_0/D$ is valid (Fig. 8.17b). The liquid layer with a mass $\mathrm{d}m$ and a velocity u has the kinetic energy $\mathrm{d}E_\mathrm{kin} = (1/2)\mathrm{d}m\cdot u^2$. With the constraint $u(x = \pm D) = 0$ the velocity of the layer is $u = u_0(1-|x|/D)$. Altogether the kinetic energy of all dragged layers is

$$\begin{aligned} E_\mathrm{kin} &= \frac{1}{2}\int u^2\mathrm{d}m = \frac{\varrho}{2}\int_0^D 2u_0^2\,(1-|x|/D)^2\,A\,\mathrm{d}x \\ &= \frac{1}{3}A\varrho Du_0^2\,. \end{aligned} \tag{8.22}$$

Due to friction part of the spent work is converted into heat. Therefore we get $E_\mathrm{kin} < W_\mathrm{f}$. This yields with (8.21) the relation

$$D < \left(\frac{3\eta L}{\varrho u_0}\right)^{1/2}. \tag{8.23}$$

The boundary layer has therefore the order of magnitude

$$D \approx \sqrt{\frac{\eta L}{\varrho u_0}}\,. \tag{8.24}$$

The boundary layer can only develop if the distance d to the container walls is larger than D. For $d < D$ the static friction between the liquid and the wall forces the velocity $u(d) = 0$, the dragged boundary layer becomes smaller and the velocity gradient larger.

For the derivation of the general friction force on a volume element $\mathrm{d}V = \mathrm{d}x\cdot\mathrm{d}y\cdot\mathrm{d}z$ we choose a liquid flowing into the z-direction with an arbitrary velocity gradient

$$\operatorname{grad}u_z = \left\{\frac{\partial u_z}{\partial x},\frac{\partial u_z}{\partial y},\frac{\partial u_z}{\partial z}\right\}.$$

We regard in Fig. 8.18 at first a flow that has only a gradient $\partial u_z/\partial x$ in x-direction ($\partial u_z/\partial y = \partial u_z/\partial z = 0$). The flow velocity $u_z(x)$ can be expanded into a Taylor series

$$u_z(x_0+\mathrm{d}x) = u_z(x_0) + \frac{\partial u_z}{\partial x}\mathrm{d}x + \ldots\,, \tag{8.25}$$

which we truncate after the linear term.

The liquid layer between $x = x_0$ and $x = x_0 + \mathrm{d}x$ experiences a friction force $\mathrm{d}F_\mathrm{f}$ per surface element $\mathrm{d}A = \mathrm{d}y\cdot\mathrm{d}z$. If $\partial u_z/\partial x > 0$ this force is decelerating for the surface layer at $x = x_0$ because here the neighbouring layer at $x = x_0 - \mathrm{d}x$ is slower but it is accelerating for the surface layer at $x = x_0 + \mathrm{d}x$, because here the adjacent layer is faster (Fig. 8.18b). The net tangential force is therefore

$$\begin{aligned} (\delta F_\mathrm{f})_z &= \mathrm{d}F_\mathrm{f}(x_0+\mathrm{d}x) - \mathrm{d}F_\mathrm{f}(x_0) \\ &= \eta\cdot\mathrm{d}y\mathrm{d}z\left[\left(\frac{\partial u_z}{\partial x}\right)_{x=x_0+\mathrm{d}x} - \left(\frac{\partial u_z}{\partial x}\right)_{x=x_0}\right]. \end{aligned}$$

Inserting the derivatives from (8.25) yields for the bracket the expression $(\partial^2u/\partial x^2)\cdot\mathrm{d}x$ and therefore for the net force onto the volume element $\mathrm{d}V$ due to the velocity gradient $\partial u_z/\partial x$

$$(\delta F_\mathrm{f})_z = \eta\cdot\mathrm{d}x\,\mathrm{d}y\,\mathrm{d}z\cdot\frac{\partial^2u_z}{\partial x^2} = \eta\cdot\mathrm{d}V\cdot\frac{\partial^2u_z}{\partial x^2}\,. \tag{8.26}$$

Table 8.2 Temperature dependence of the dynamical viscosity $\eta(T)$ of water and glycerine

$T/°\mathrm{C}$	Viscosity $\eta(T)/(\mathrm{mPa}\cdot\mathrm{s})$	
	Water	Glycerine
0	1.792	12 100
+20	1.002	1480
+40	0.653	238
+60	0.466	81
+80	0.355	31.8
+100	0.282	14.8

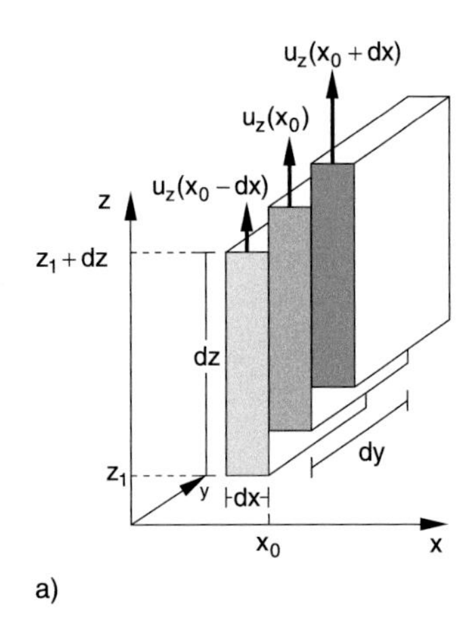

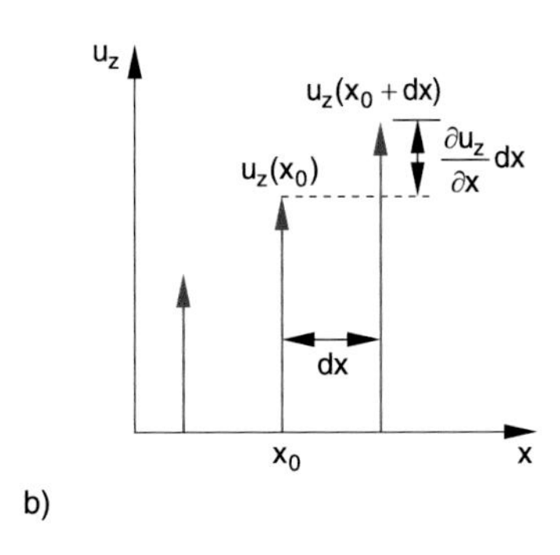

Figure 8.18 Derivation of friction force acting on a volume element dx dy dz in a flow with homogeneous velocity profile

A similar result is obtained for the velocity gradient $\partial u_z/\partial y$ in y-direction.

For compressible media, e.g. for gases, a velocity gradient $\partial u_z/\partial z$ can also appear for a flow into the z-direction if the density changes with z, while for incompressible media $\partial u_z/\partial z \neq 0$ only if the velocity changes, e.g. in tubes with variable cross section.

From (8.25)–(8.26) we finally obtain for the total friction force onto the volume element $\mathrm{d}V$ in case of a laminar flow with the velocity u_z the expression

$$(\mathrm{d}F_\mathrm{f})_z = \eta\,\mathrm{d}V\left(\frac{\partial^2 u_z}{\partial x^2} + \frac{\partial^2 u_z}{\partial y^2} + \frac{\partial^2 u_z}{\partial z^2}\right) . \tag{8.26a}$$

The first two terms cause tangential forces (shear forces, see Sect. 6.2.3), the third term, which is only nonzero for compressible media causes a normal force onto the surface element $\mathrm{d}x{\cdot}\mathrm{d}y$. With the Laplace operator

$$\Delta = \frac{\partial^2}{\partial x^2} + \frac{\partial^2}{\partial y^2} + \frac{\partial^2}{\partial z^2}$$

(see Sect. 13.1.6) the total friction force onto the volume element $\mathrm{d}V$, which moves with the velocity $\boldsymbol{u} = \{0, 0, u_z(x, y, z)\}$ can be written as

$$(\mathrm{d}F_\mathrm{f})_z = \eta \cdot \Delta u_z \mathrm{d}V . \tag{8.26b}$$

For arbitrary flow velocities $\boldsymbol{u} = \{u_x, u_y, u_z\}$ (8.26b) can be generalized to

$$\boldsymbol{F}_\mathrm{f} = \eta \cdot \int_V \Delta \boldsymbol{u}\,\mathrm{d}V , \tag{8.26c}$$

this is equivalent to the three equations $(F_\mathrm{f})_i = \eta \int \Delta u_i \cdot \mathrm{d}V$ for the components $i = x, y, z$.

8.5.2 Laminar Flow Between Two Parallel Walls

In order to maintain a stationary flow with constant velocity into the z-direction between two walls at $x = -d$ and $x = +d$ one has to apply a force opposite to the friction force $\boldsymbol{F}_\mathrm{f}$ which just compensates $\boldsymbol{F}_\mathrm{f}$. This force can be, for instance, caused by a pressure difference between the planes $z = -z_0$ and $z = +z_0$. In the following we assume that the pressure is constant in a plane $z =$ constant, i.e. independent of x and y.

We consider in Fig. 8.19 a volume element $\mathrm{d}V = \mathrm{d}x \cdot \mathrm{d}y \cdot \mathrm{d}z$ with the width $\mathrm{d}y = b$ in y-direction and the height $\mathrm{d}z$. At its end faces $z = z_1$ and $z = z_1 + \mathrm{d}z$ the pressure forces

$$\mathrm{d}F_1 = b \cdot \mathrm{d}x \cdot p(z_1) \quad \text{and} \quad \mathrm{d}F_2 = b \cdot \mathrm{d}x \cdot p(z_1 + \mathrm{d}z)$$

are effective. They result in a total force onto the volume element $\mathrm{d}V$

$$\mathrm{d}F_z = -b\,\mathrm{d}x\frac{\mathrm{d}p}{\mathrm{d}z}\mathrm{d}z . \tag{8.27}$$

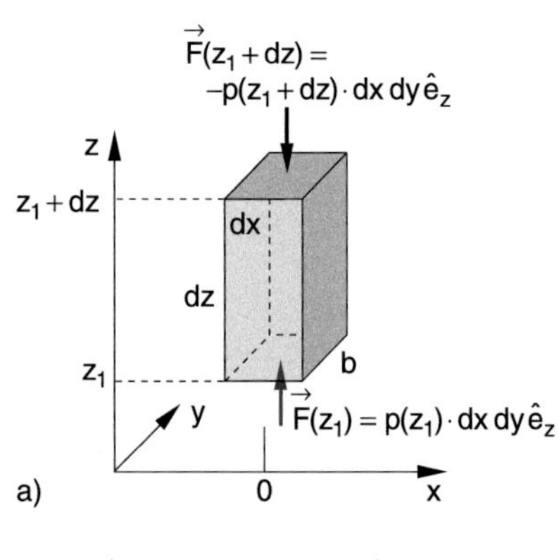

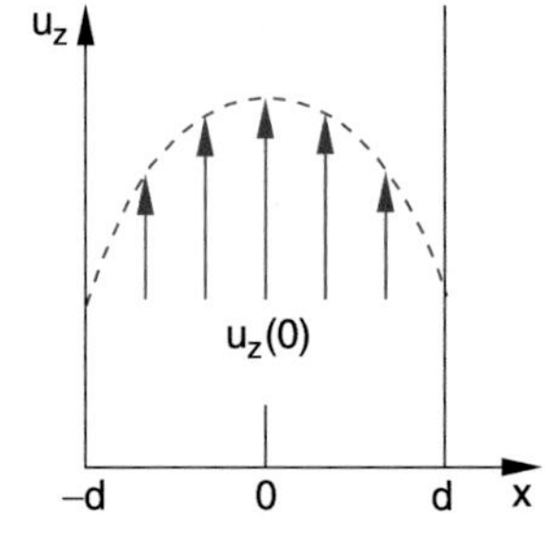

Figure 8.19 Laminar flow between two parallel walls

This pressure force compensates the friction force

$$(\mathrm{d}F_1)_z = \eta\,\mathrm{d}V\,\Delta u_z = \eta\,\mathrm{d}y\,\mathrm{d}x\,\mathrm{d}z\frac{\mathrm{d}^2u}{\mathrm{d}x^2}\,,$$

if the condition

$$\frac{\mathrm{d}^2u_z}{\mathrm{d}x^2} = -\frac{1}{\eta}\frac{\mathrm{d}p}{\mathrm{d}z} \Rightarrow \frac{\mathrm{d}u_z}{\mathrm{d}x} = -\frac{x}{\eta}\cdot\frac{\mathrm{d}p}{\mathrm{d}z} + C_1$$

is fulfilled. The integration constant $C_1 = (\mathrm{d}u_z/\mathrm{d}x)_{x=0}$ gives the slope of the velocity profile $u(x)$ at $x = 0$.

Integration yields

$$u_z = -\frac{x^2}{2\eta}\frac{\mathrm{d}p}{\mathrm{d}z} + C_1x + C_2\,, \qquad (8.28)$$

since p and $\mathrm{d}p/\mathrm{d}z$ do not depend on x.

For a liquid streaming between two parallel walls at $x = -d$ and $x = +d$ symmetry arguments demand $(\mathrm{d}u/\mathrm{d}x)_{x=0} = C_1 = 0$. At $x = \pm d$ the static friction between the liquid and the walls causes $u(x = \pm d) = 0$. This gives for the integration constant C_2

$$C_2 = \frac{d^2}{2\eta}\frac{\mathrm{d}p}{\mathrm{d}z}\,.$$

We then obtain for the velocity profile the parabola

$$u(x) = \frac{1}{2\eta}\frac{\mathrm{d}p}{\mathrm{d}z}(d^2 - x^2)\,, \qquad (8.29a)$$

with the crest at $x = 0$ midway between the two walls. If the friction between the liquid and the walls is not high enough $(u(\pm d) \neq 0)$, we get instead of (8.29a) the more general equation

$$u(x) = \frac{1}{2\eta}\frac{\mathrm{d}p}{\mathrm{d}z}(d^2 - x^2) + u_d\,. \qquad (8.29b)$$

8.5.3 Laminar Flows in Tubes

The flow of liquids in cylindrical tubes plays an important role for many technical applications (water pipes, oil pipelines), and also in medicine (blood flow through veins). It is therefore worthwhile to study this problem in more detail.

We assume, as in the previous example, a pressure difference $p_1 - p_2$ between the planes $z = 0$ and $z = L$ in a cylindrical pipe with radius R (Fig. 8.20) which maintains a stationary flow. Symmetry reasons demand that the flow velocity can only depend on the distance r from the cylinder axis. For a coaxial small cylinder with radius r the same reasoning as in the previous section gives for the condition "friction force must compensate the pressure force"

$$-\eta\cdot 2r\pi\cdot L\frac{\mathrm{d}u}{\mathrm{d}r} = r^2\pi\cdot(p_1 - p_2)\,.$$

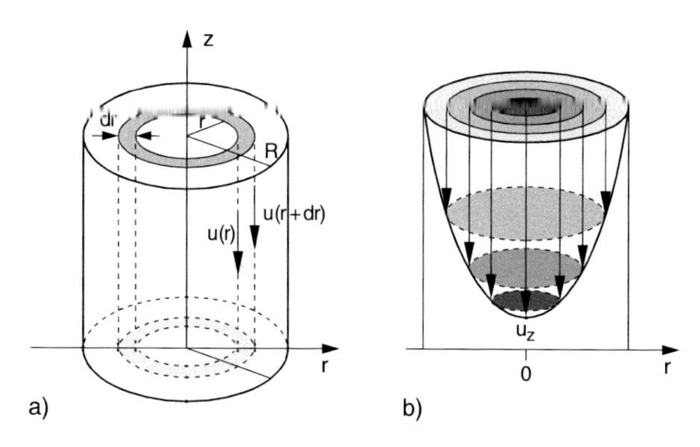

Figure 8.20 **a** Derivation of Hagen–Poiseuille law; **b** velocity profile of a laminar flow in a cylindrical tube

Integration over r yields

$$u(r) = \int_r^R \frac{p_1 - p_2}{2\eta L}r\,\mathrm{d}r = \frac{p_1 - p_2}{4\eta L}\cdot\left(R^2 - r^2\right)\,. \qquad (8.30)$$

This velocity profile is a rotational paraboloid. It can be vividly demonstrated by the flow of coloured glycerine through a vertical pipe (Fig. 8.20b).

The total liquid volume flowing per second through the plane $z =$ constant of the hollow cylinder with radii between r_1 and $r_1 + \mathrm{d}r$ shown in Fig. 8.20a is according to (8.30)

$$\frac{\mathrm{d}}{\mathrm{d}t}(V(r)) = 2\pi r\,\mathrm{d}r\cdot u = \frac{2\pi r\,\mathrm{d}r\cdot(R^2 - r^2)}{4\eta L}(p_1 - p_2)\,.$$

The total volume streaming during the time t through the pipe is

$$\begin{aligned} V &= t\cdot\int_{r=0}^{R} 2\pi r\cdot u\,\mathrm{d}r \\ &= \frac{\pi R^4(p_1 - p_2)}{8\eta L}t = \frac{\pi R^4\Delta p}{8\eta L}t\,. \end{aligned} \qquad (8.31)$$

The ratio $\Delta p/L = \partial p/\partial z$ is the linear pressure gradient along the tube. The total volumetric flowrate (volume per second) through the pipe is then

Hagen–Poiseuille Law

$$\frac{\mathrm{d}V}{\mathrm{d}t} = \frac{\pi R^4}{8\eta L}\Delta p = \frac{\pi R^4}{8\eta}\frac{\partial p}{\partial z} \qquad (8.32)$$

Note the strong dependence of $\mathrm{d}V/\mathrm{d}t$ from the radius R of the pipe $(\sim R^4!)$.

The human body utilizes this dependence for the regulation of the blood flow by adjusting the cross section area of the veins.

8.5.4 Stokes Law, Falling Ball Viscometer

When a ball with radius R is dropped with the initial velocity $u = 0$ into a liquid one observes at first an acceleration of the ball due to the gravity force and after a short falling distance a constant velocity. For this uniform motion the friction force F_f which increases with increasing velocity just cancels the gravity force

$$F_g = m_{\text{eff}} \cdot g = (\varrho_k - \varrho_{\text{Fl}}) \tfrac{4}{3}\pi R^3 \cdot g \quad (8.33)$$

diminished by the buoyancy (Fig. 8.21).

Experiments with different liquids and balls with different radii prove that the friction force is proportional to the viscosity η of the liquid, to the radius R of the ball and to its velocity u. For radii still small compared to the diameter of the container one finds

Stokes Law

$$\boldsymbol{F}_f = -6\pi \cdot \eta \cdot R \cdot \boldsymbol{u}_0 \, . \quad (8.34a)$$

The stationary final velocity u_0 is obtained for $\boldsymbol{F}_f + \boldsymbol{F}_g = 0$

$$u_0 = \frac{2}{9} g \frac{R^2}{\eta} (\varrho_K - \varrho_{\text{Fl}}) \, . \quad (8.35)$$

Measuring u_0 allows the determination of the viscosity η, if the densities of liquid and ball and the ball radius R are known (Falling ball viscometer Fig. 8.22). According to (8.35) the ratio u_0/R^2 should be independent of the ball radius. This is indeed observed for small radii R.

Stokes Law (8.34a) can be derived also theoretically. A more detailed calculation [8.1a, 8.1b, 8.7] shows that (8.34) is only an approximation. The exact expression for the friction force, derived by *Oseen*, is

$$F_f = -6\pi\eta R \cdot u_0 \left(1 + \frac{3\varrho_{\text{Fl}} \cdot R \cdot u_0}{8\eta}\right) \, . \quad (8.34b)$$

The second term in the bracket is for small radii R small compared to 1 and can be neglected.

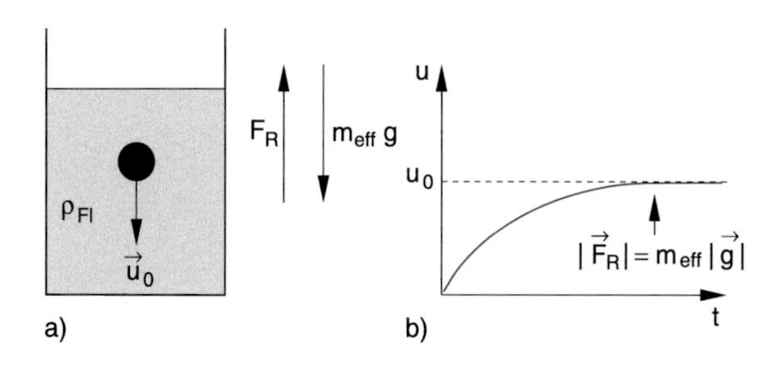

Figure 8.21 Uniform sink speed u_0 of a ball in a viscose liquid

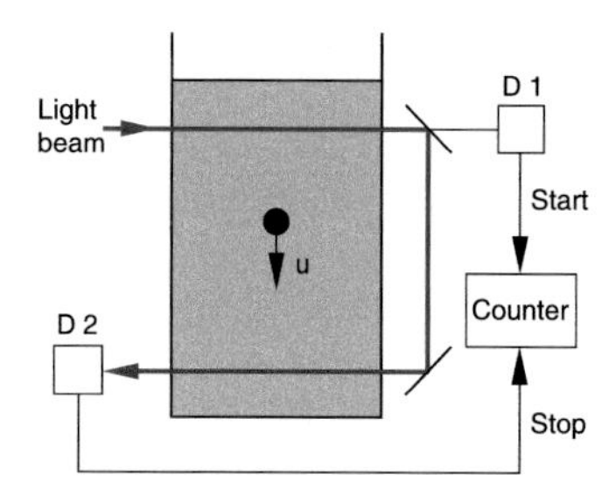

Figure 8.22 Viscosimeter with sinking ball and photoelectric barrier

Example

For steel balls ($\varrho = 7.8\,\text{kg/dm}^3$) with radius $R = 0.1\,\text{cm}$ falling in glycerine ($\varrho = 1260\,\text{kg/m}^3$) and $\eta = 1.48\,\text{Pa}\cdot\text{s}$, the stationary velocity becomes $u_0 = 1\,\text{cm/s}$. In this case the second term is $3.2 \cdot 10^{-3} \ll 1$. For $R = 1\,\text{cm}$, however, $u_0 = 1\,\text{m/s}$ and the second term becomes $2.5 > 1$ and cannot be neglected.

The Stokes Law (8.34a) therefore is correct only for sufficiently small products $R \cdot u_0$ of ball radius and final velocity u_0. ◄

8.6 Navier–Stokes Equation

In the previous sections we have discussed the different forces acting onto a volume element $\text{d}V$ in a streaming liquid. We can now present the general equation of motion for a real viscous streaming liquid. With the different contributions

$$\text{d}\boldsymbol{F}_f = \eta \cdot \Delta\boldsymbol{u} \cdot \text{d}V \qquad \text{(friction force)}$$
$$\text{d}\boldsymbol{F}_p = -\textbf{grad}\, p \cdot \text{d}V \qquad \text{(pressure force)}$$
$$\text{d}\boldsymbol{F}_g = = \varrho \cdot \boldsymbol{g} \cdot \text{d}V \qquad \text{(gravity force)}$$

to the total force and the substantial acceleration (8.3)

$$\frac{\text{d}u}{\text{d}t} = \frac{\partial u}{\partial t} + (u \cdot \nabla)$$

we obtain the Navier–Stokes equation

$$\varrho \left(\frac{\partial}{\partial t} + \boldsymbol{u} \cdot \nabla\right) \boldsymbol{u} = -\,\text{grad}\, p + \varrho \cdot \boldsymbol{g} + \eta \Delta \boldsymbol{u} \, . \quad (8.36a)$$

For ideal liquids with $\eta = 0$ this reduces to the special case of the Euler equation (8.4). The friction term $\eta \cdot \Delta u$ expands the Euler equation, which is a first order differential equation, to a second order differential equation, which is more difficult to solve.

On the right hand side of (8.36a) the forces are listed and on the left hand side the motion induced by these forces, which we will now discuss in more detail.

The first term $\partial u/\partial t$ gives the time derivative of the velocity at a fixed location. The second term describes the change of the velocity of dV while it moves from the position r to $r + dr$. Using the vector relation

$$(u \cdot \nabla)\, u = \tfrac{1}{2}\,\mathrm{grad}\, u^2 - (u \times \mathrm{rot}\; u)\ , \tag{8.36b}$$

that is deduced in textbooks on Vector Analysis [8.8, 8.9] (see also Sect. 13.1.6) we see that this spatial change of u can be composed of two contributions: The first term gives the change of the amount of $\boldsymbol{u}$, the second term the change of the direction of $\boldsymbol{u}$. This second term gives rise to vortices in the liquid, which we will discuss next.

8.6.1 Vortices and Circulation

When a liquid streams around a circular obstacle one observes for small velocities the streamline picture of laminar flow, shown in Fig. 8.5. If the velocity is increased above a critical velocity u_c, which depends on the viscosity η of the liquid, vortices appear behind the obstacle (Fig. 8.23). Such vortices can be made visible by small cork pieces floating on the liquid and moving along the streamlines. One observes that in a region around the centre of the vortex the liquid rotates like a rigid body. The rotational velocity

$$\boldsymbol{u} = \boldsymbol{\omega} \times \boldsymbol{r}$$

increases linear with the distance r from the centre and all particles have the same angular velocity ω. This region $r < r_k$ is called the **vortex kernel** (Fig. 8.24). Inserting small cork pieces with a fixed direction arrow to the surface of the liquid it becomes apparent that they turn once around their own axis while circulating around the vortex. (Fig. 8.25) as expected for a rigid rotation.

Outside of the vortex kernel ($r > r_k$) the angular velocity ω decreases with increasing distance r. The particles do no longer rotate about their axis but keep their spatial orientation (Fig. 8.25). This region of the vortex is called the **circulation**. Here a deformation of the volume elements during the rotation takes place (Fig. 8.26).

We can describe the vortex by the vortex vector

$$\boldsymbol{\Omega} = \tfrac{1}{2}\mathbf{rot}\,\boldsymbol{u}\ . \tag{8.37}$$

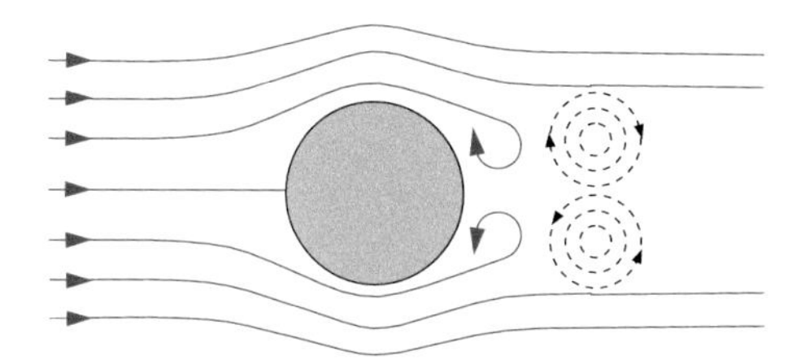

Figure 8.23 Generation of vortices in a turbulent flow around a circular obstacle

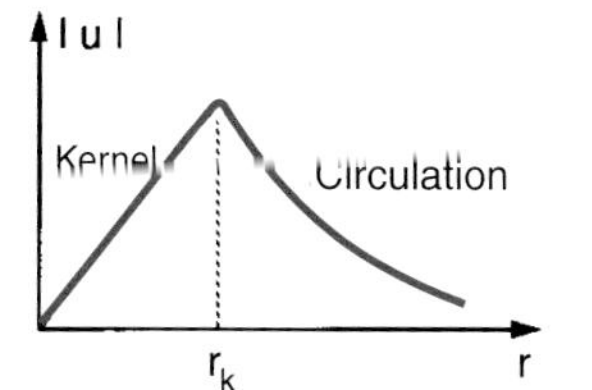

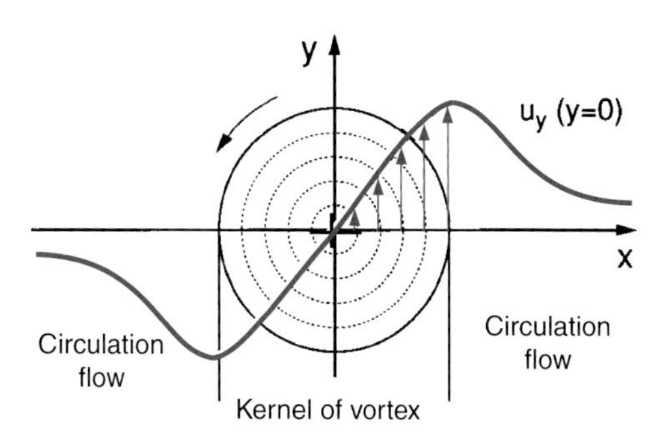

Figure 8.24 Kernel of vortex and circulation region

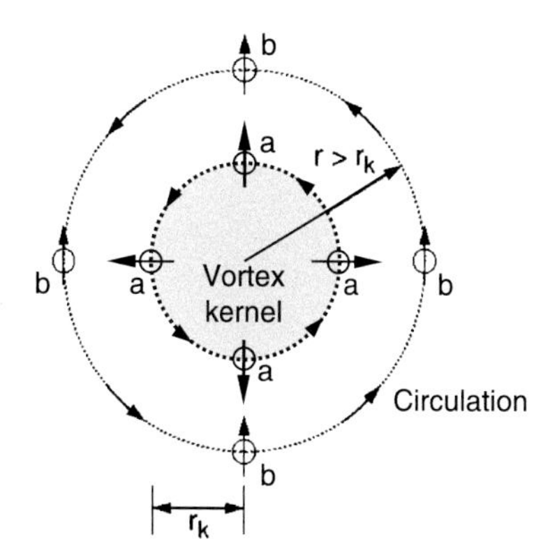

Figure 8.25 Orientation of cork pieces: **a** inside the vortex kernel (circular motion with turning orientation), **b** in the circulation region (non turning orientation)

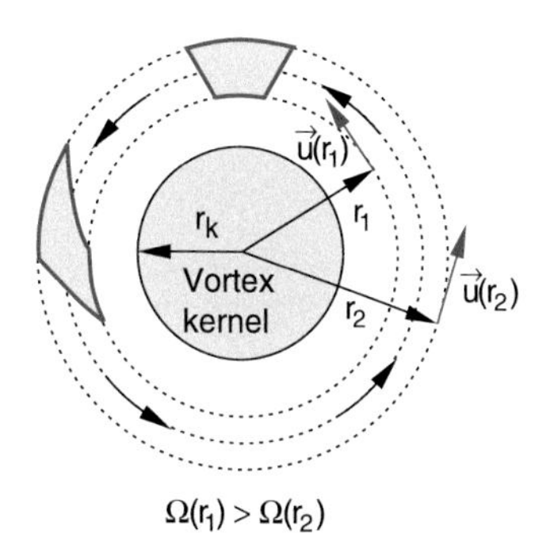

Figure 8.26 Deformation of a plane element in the circulation region outside the vortex kernel

The amount of $\boldsymbol{\Omega}$ gives the angular velocity ω inside the vortex kernel (see below). Magnitude and direction of $\boldsymbol{\Omega}$ in a vortex are generally not constant. They change because the vortex is not necessarily fixed in space but moves with the flowing liquid to

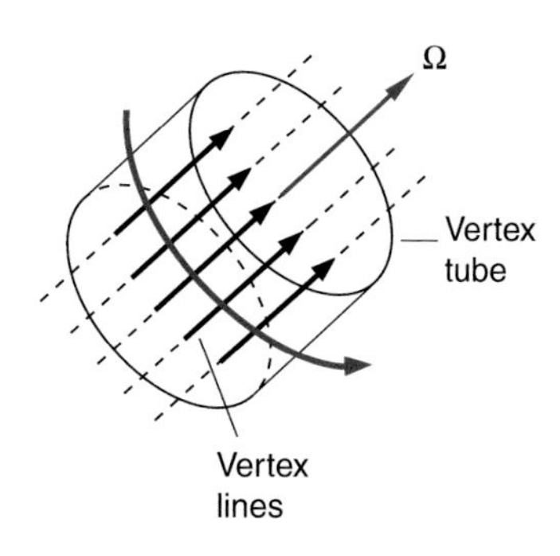

Figure 8.27 Vortex lines and tube

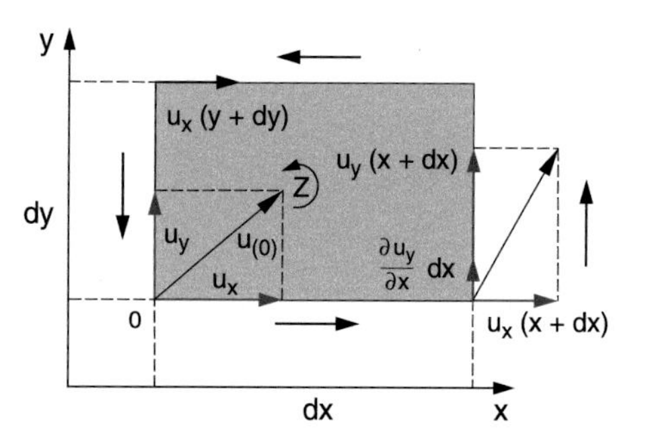

Figure 8.28 Explanation of circulation and its relation with rot u

other locations and furthermore the energy of the vortex changes because of friction and with it the magnitude of $\boldsymbol{\Omega}$ changes. The curves which coincide at every place with the direction of $\boldsymbol{\Omega}$ are called **vortex lines**. When, for instance, the particles move on circles in the x-y-plane the vector $\boldsymbol{\Omega}$ points into the z-direction. All lines parallel to the z-direction inside the kernel with $x^2 + y^2 \le r_k^2$ are vortex lines (Fig. 8.27). All vortex lines through the vortex area A form the **vortex tube**.

For a quantitative description of torques based on the Navier–Stokes equation we have to study the rotational part ($\boldsymbol{u} \times \mathbf{rot}\,\boldsymbol{u}$) in (8.36a,b). At first we must understand, that the term rot $\boldsymbol{u}$ describes the rotation of moving particles. We therefore regard in Fig. 8.28 the tangential velocity components along the edge of the surface element $\mathrm{d}x \cdot \mathrm{d}y$. As a measure of the **torque strength** of the flow through the area A we define the circulation

$$Z = \oint \boldsymbol{u}\,\mathrm{d}\boldsymbol{s} \tag{8.38a}$$

along the edge of the surface in the counterclockwise direction. Our surface element $\mathrm{d}x \cdot \mathrm{d}y$ contributes the share

$$\begin{aligned} \mathrm{d}Z &= u_x\,\mathrm{d}x + \left(u_y + \frac{\partial u_y}{\partial x}\mathrm{d}x\right)\mathrm{d}y \\ &\quad - \left(u_x + \frac{\partial u_x}{\partial y}\mathrm{d}y\right)\mathrm{d}x - u_y\,\mathrm{d}y \\ &= \left(\frac{\partial u_y}{\partial x} - \frac{\partial u_x}{\partial y}\right)\mathrm{d}x\,\mathrm{d}y = (\text{rot}\;\boldsymbol{u})_z \mathrm{d}x\,\mathrm{d}y \end{aligned} \tag{8.38b}$$

to the circulation, because the z-component of $\mathbf{rot}\,\boldsymbol{u} = \nabla \times \boldsymbol{u}$ is defined as $(\nabla \times \boldsymbol{u})_z = (\partial u_y/\partial x - \partial u_x/\partial y)$.

Analogous relations are obtained for the x- and y-components. From these relations one obtains by integration the Stokes' theorem

$$\oint \boldsymbol{u}\,\mathrm{d}\boldsymbol{s} = \int_A \text{rot}\;\boldsymbol{u}\,\mathrm{d}\boldsymbol{A}\;, \tag{8.38c}$$

which states: "The surface integral over $\mathbf{rot}\,\boldsymbol{u}$ equals the path integral along the border of the surface element".

For a circular current of a liquid around a centre the circulation at a distance r from the centre is

$$Z = \oint \boldsymbol{u}\,\mathrm{d}\boldsymbol{s} = 2\pi r u(r)\;. \tag{8.38d}$$

The average amount Ω of the vortex vector $\Omega = \frac{1}{2}\mathbf{rot}\,\boldsymbol{u}$ that points into the direction perpendicular to the surface is, according to Stokes' theorem

$$\begin{aligned} \boldsymbol{\Omega} &= \frac{1}{2A}\int |\text{rot}\;\boldsymbol{u}|\,\mathrm{d}A = \frac{1}{2\pi r^2}\oint \boldsymbol{u}\mathrm{d}\boldsymbol{s} \\ &= \frac{2\pi r u}{2\pi r^2} = \frac{u}{r}\;, \end{aligned} \tag{8.38e}$$

where A is the area of the torque kernel.

Since the torque kernel rotates like a solid body, Ω must be independent of r. As illustrated in Fig. 8.24 the velocity $u = r \cdot \Omega$ increases linear with r.

The average $\Omega = Z/2A$ of the magnitude of the torque vector gives the circulation per surface unit and therefore the torque strength per surface unit.

8.6.2 Helmholtz Vorticity Theorems

For an ideal liquid ($\eta = 0$) the Navier–Stokes equation (8.36a) without external fields (gravity is neglected $\rightarrow g = 0$) can be transformed into an equation that illustrates certain conservation laws. This was first recognized in 1858 by Hermann von Helmholtz.

On both sides of (8.36a) we apply the differential operator **rot**, divide by the density ϱ and obtain from (8.36b) and (8.37) with $\mathbf{rot}\,\mathbf{grad}\,p = \nabla \times \nabla p = \mathbf{0}$ the equation (see Probl. 8.11)

$$\frac{\partial \boldsymbol{\Omega}}{\partial t} + \nabla \times (\boldsymbol{\Omega} \times \boldsymbol{u}) = \mathbf{0}\;. \tag{8.39}$$

Together with the equation of continuity $\text{div}\,\boldsymbol{u} = 0$ (8.13) for incompressible media this equation determines completely and for all times the velocity field of an ideal streaming liquid. This means: If the quantities $\boldsymbol{\Omega}$ and $\boldsymbol{u}$ are given at a certain time (8.39) describes their future development unambiguously.

For example: If for $t = t_0$ the vortex vector $\boldsymbol{\Omega}$ for the total liquid is $\Omega = 0$, it follows from (8.39) $\partial\Omega/\partial t = 0$. This means: If an ideal liquid without vortices is set into motion it will stay vortex-free for all times.

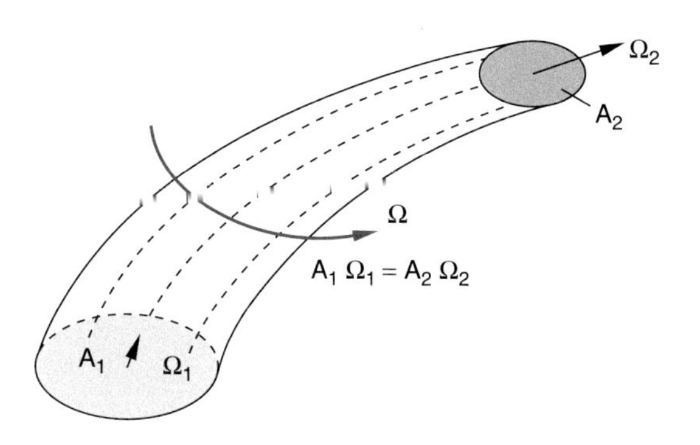

Figure 8.29 Deformation of a circular vortex during its flow with conservation of total mass and angular momentum

If there are vortices in a liquid, it follows

$$\Omega = \frac{1}{2} \operatorname{rot} \boldsymbol{u} \Rightarrow \operatorname{div} \boldsymbol{\Omega} = \nabla \cdot (\nabla \times \boldsymbol{u}) \equiv 0 \,. \tag{8.40}$$

This means: Inside an ideal liquid, there are no sources or sinks for the vortex lines. They are either closed lines or they end at the boundary of the liquid, for instance at the walls of the liquid tube.

Inside an ideal liquid the vortex strength $Z = 2\boldsymbol{\Omega} \cdot \boldsymbol{A}$ is constant in time. Vortices cannot be generated nor vanish.

The constancy of Z in a frictionless liquid is equivalent to the conservation of angular momentum of the mass circulating in a vortex. Because of $\eta = 0$ no tangential forces can act and the pressure forces have only radial components. Therefore, there is no torque and the angular momentum has to be constant.

These conservation laws can be summarized by the following model: Vortices move like solid but strongly deformable bodies through a liquid or a gas. Without friction, their total mass and their angular momentum remain constant although the angular velocity and the radius of a vortex can change. This is illustrated in Fig. 8.29 by a cylindrical vortex. The constancy of the angular momentum $\boldsymbol{L} = I \cdot \boldsymbol{\omega}$ (see Sect. 5.5) with the moment of inertia $I = (1/2)Mr^2$ results in the equation

$$M_1 r_1^2 \Omega_1 = M_2 r_2^2 \Omega_2 \,.$$

Since $M_1 = M_2$ and the vortex area $A = \pi \cdot r^2$ this gives

$$A_1 \cdot \Omega_1 = A_2 \cdot \Omega_2 \,.$$

This means the vortex strength is constant.

8.6.3 The Formation of Vortices

In the previous section we have seen, that friction is essential for the formation of vortices. On the other hand, it was discussed in Sect. 8.5, that liquids with large friction show a laminar flow

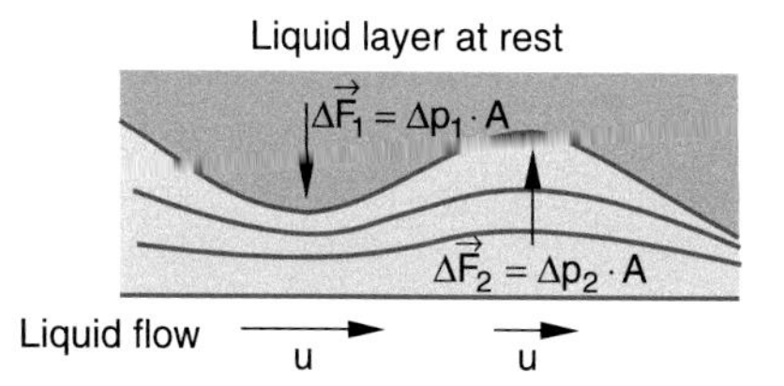

Figure 8.30 Generation of vortices by instabilities at boundaries between liquid layers with different velocities

where no vortices occur. Vortices must be therefore formed in liquids with small viscosity where at certain places, e.g. at the boundaries with walls, the friction has maxima. Here velocity gradients occur between adjacent liquid layers because of the static friction between these liquid layers and the wall. These velocity gradients produce, due to friction, tangential forces, which give rise to vortices.

When such boundary layers show small irregularities as shown exaggerated in Fig. 8.30, the adjacent stream lines are deformed. At the narrow positions the stream lines are compressed and the flow velocity u increases. According to the Bernoulli equation, a pressure gradient Δp develops which further increases the irregularities. Finally, an unstable condition arises which results in the formation of vortices.

We will illustrate this vortex formation for the example of a flow around a circular cylinder (Fig. 8.31). For sufficiently small flow velocities u the influence of friction is small and a laminar flow occurs (Fig. 8.5 and 8.31a). At the stagnation point S_1 on the forefront of the cylinder, the flow velocity is zero and according to (8.17) the pressure equals the total pressure p_0. From S_1 the liquid moves along the upper side of the cylinder and is accelerated until it reaches the point P, where the velocity reaches its maximum and the pressure its minimum. The acceleration is caused by the pressure difference $\Delta p = p_0(S_1) - p(P)$. At the stagnation point S_2 at the backside of the cylinder the velocity becomes zero again, because the opposite pressure difference decelerates the flow and brings the velocity down to zero.

When the flow velocity is increased the velocity gradient between the wall and the adjacent liquid layers also increases. This increases the friction which is proportional to the velocity gradient. The liquid volume elements do not reach their full velocity in the point P and therefore reach the velocity $v = 0$ already in the point W before S_2 (Fig. 8.31b and 8.32). The pressure force caused by the pressure gradient between S_2 and W now accelerates the volume elements into the opposite direction against the flow velocity of the liquid layers farther away from the wall. There are two opposite forces acting on the liquid layers close to the wall (Fig. 8.32):

a the friction force due to the friction between the liquid layers close to the wall and the layers farther away which have different velocities,
b the force due to the pressure gradient.

These two forces exert a torque onto the liquid layers which cause a rotation. On each side of the cylinder a vortex is created. The two vortices have an opposite direction of rotation

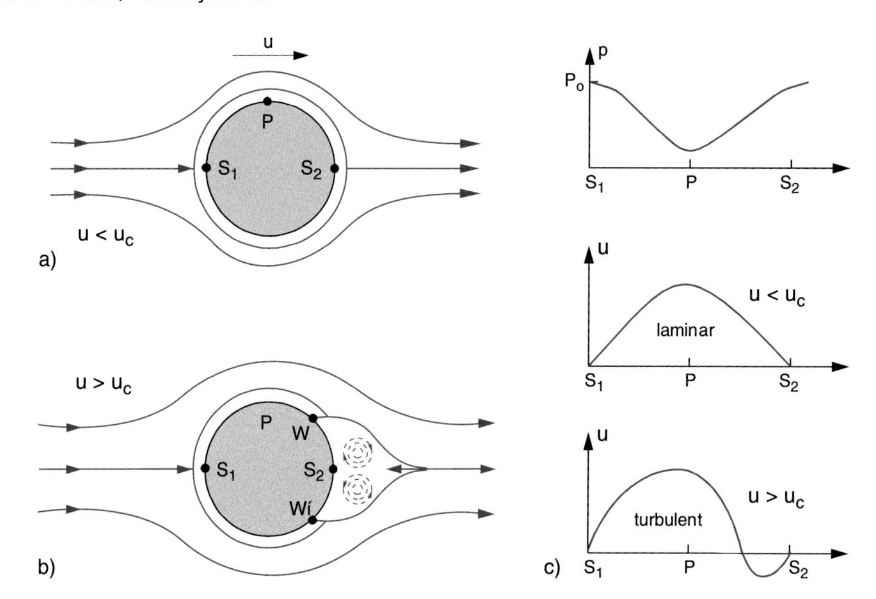

Figure 8.31 **a** Laminar flow for small velocities around a circular cylinder. **b** Generation of vortices behind a circular cylinder for large velocities. **c** Pressure and velocity behaviour for $u < u_c$ and $u > u_c$

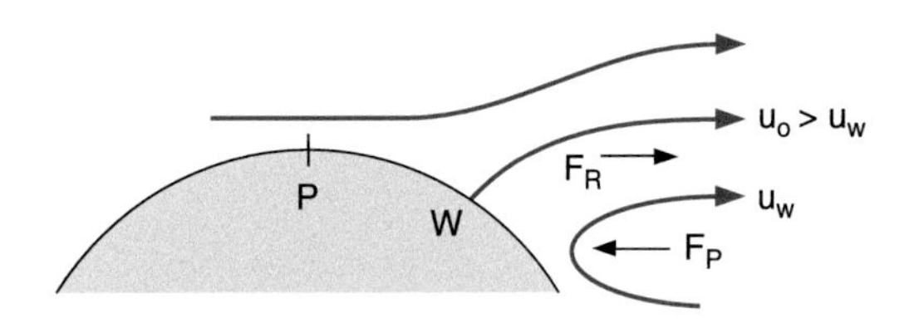

Figure 8.32 Illustration of torque necessary for the generation of vortices

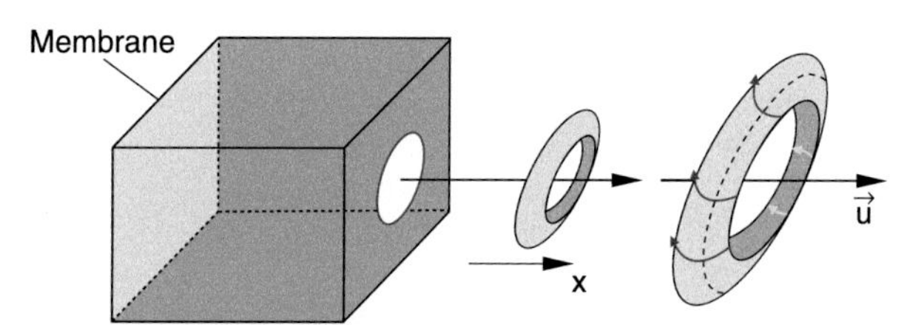

Figure 8.34 Generation of smoke vortex by beating a membrane at the backside of a box filled with smoke

(Fig. 8.31b), i.e. the vortex vector Ω_1 points into the direction into the drawing plane while Ω_2 points out of this plane.

These vortices can be visualized by dyed streamlines produced with the apparatus shown in Fig. 8.4.

Vortices can be also produced at the end of a circular tube, through which a liquid flows with sufficently high velocity (Fig. 8.33).

A nice demonstration experiment where vortices are produced in air mixed with cigarette smoke, is shown in Fig. 8.34. A box with a thin membrane at one side and a hole with 20–30 cm diameter on the opposite side is filled with cigarette smoke. Beating the membrane with a flat hand, produces a sudden pressure increase inside the box and drives the air-smoke mixture through the hole out of the box. At the edges of the hole vortices are produced which travel through the open air and can be readily seen by a large auditorium. These vortices can extinguish a candle flame, several meters away from the box. The vortices in air move nearly like a solid body through the air at atmospheric pressure. Without vortices a pure pressure wave would not be able to extinguish the candle flame because its intensity decreases with the distance d from the box as $1/d^2$ (see Sect. 11.9).

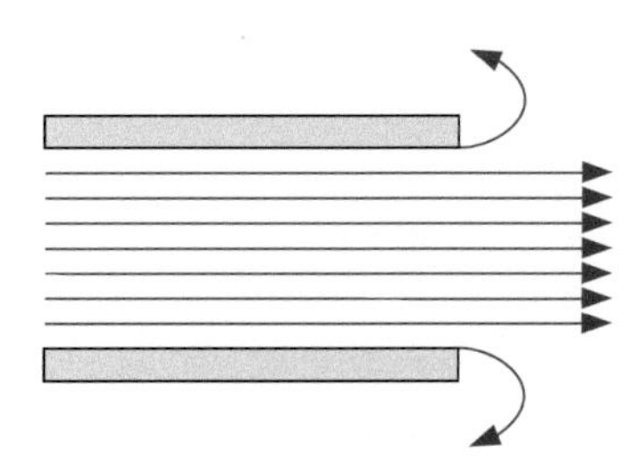

Figure 8.33 Generation of vortices at the end edge of a tube

8.6.4 Turbulent Flows; Flow Resistance

The curls shown in Fig. 8.31 behind an obstacle, do not stay at the location of their generation but move with the streaming liquid due to internal friction. At the original location new vortices can now emerge, which again detach from the surface of the immersed body and follow the liquid flow. This leads to the formation of a "Karman vortex street" (Fig. 8.35). It turns out that the two vortices of a vortex pair do not detach simultaneously but alternatively from the upper and the lower side of the obstacle. In the vortex street therefore the vortices with opposite angular momentum are shifted against each other. Car

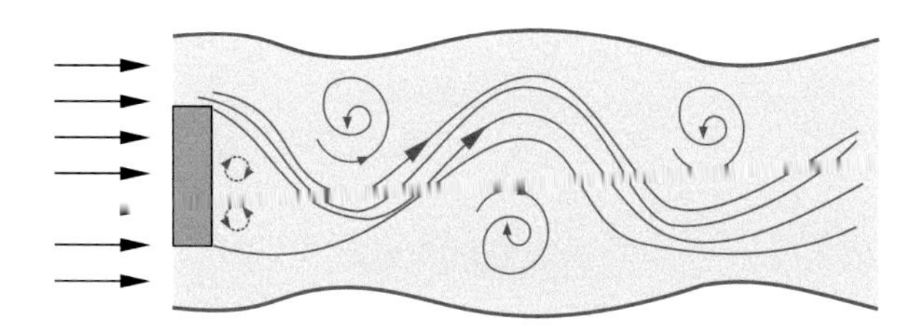

Figure 8.35 Karman vortex street

drivers can experience such a vortex street, when driving behind a fast truck, where they can feel the alternating directions of the transverse gust of winds. Behind a starting jet plane the vortex street can extend over several kilometres. Therefore there must be always a minimum safety distance between starting planes.

The rotational energy $E_{\text{rot}} = (I/2) \cdot \Omega^2$ (I = inertial moment), necessary for the generation of vortices has to come from the kinetic energy of the liquid flow. The flow velocity must therefore decrease when vortices are formed.

In a laminar friction-free flow the flow velocity $\boldsymbol{u}$ in the point S_2 in Fig. 8.31 is zero and in S_2 the same stagnation pressure p_0 appears as in S_1. In a turbulent flow the velocity behind the obstacle is not zero and therefore, according to the Bernoulli theorem the pressure is lower than p_0, causing a pressure difference between the regions before and behind the obstacle. This results in a force $F = \Delta p \cdot A$ on the obstacle with the cross section A in the direction of the flow. In order to keep the body at a fixed place, an opposite force has to be applied in addition to the force against the friction force.

The pressure difference at S_2 is according to Bernoulli's theorem $\Delta p \propto (1/2)\varrho \cdot u^2$. Therefore the force due to the pressure difference can be written as

$$F_{\text{D}} = c_{\text{D}} \cdot \frac{\varrho}{2} u^2 A \,, \tag{8.41a}$$

where the dimensionless constant c_{D} is the *pressure drag coefficient*. It depends on the form of the body (Fig. 8.36). This force adds to the friction force that is also present for laminar flows. According to the Hagen–Poiseuille law (8.31) the friction causes a pressure loss Δp_{f} (see Fig. 8.9b). The Bernoulli equation for a viscose liquid flowing through a horizontal tube, has to be augmented to

$$p_1 + \tfrac{1}{2}\varrho u_1^2 = p_2 + \Delta p_{\text{f}} + \tfrac{1}{2}\varrho u_2^2 \,; \quad \Delta p_{\text{f}} < 0 \,.$$

The pressure difference Δp_{f} depends on the square of the velocity u. We can write the total resistance force

$$F_{\text{total}} = F_{\text{f}} + F_{\text{D}} = \tfrac{1}{2} c_{\text{w}} \cdot \varrho \cdot u^2 \cdot A \,. \tag{8.41b}$$

The proportional factor c_{w} is called flow resistance coefficient. It depends analogue to c_{D} on the form of the body in the flow. In Fig. 8.36 the values of c_{w} for Air flows at atmospheric pressure are compiled for some profiles. This figure illustrates that the streamlined profile has the smallest flow resistance coefficient. Bodies with edges on the side of the incoming flow have larger flow resistance coefficients than spherical profiles.

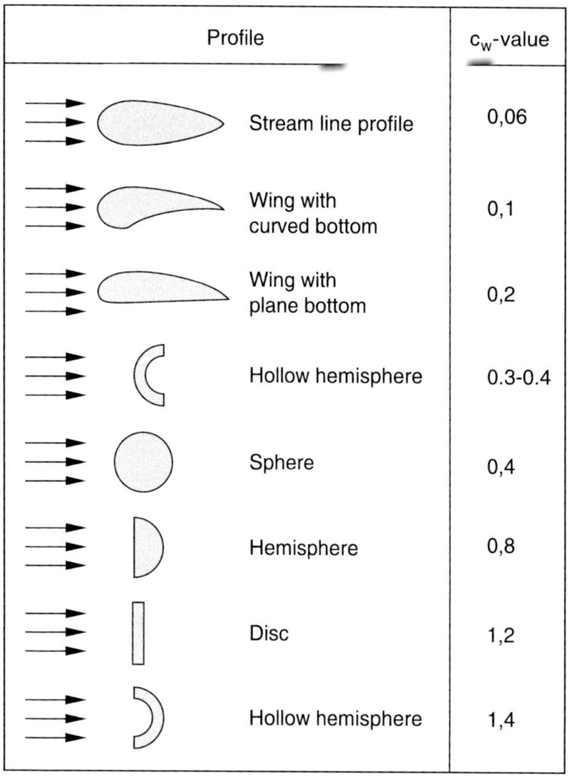

Profile	c_{w}-value
Stream line profile	0,06
Wing with curved bottom	0,1
Wing with plane bottom	0,2
Hollow hemisphere	0.3-0.4
Sphere	0,4
Hemisphere	0,8
Disc	1,2
Hollow hemisphere	1,4

Figure 8.36 Flow resistance coefficents c_{w} for different shapes of obstacles

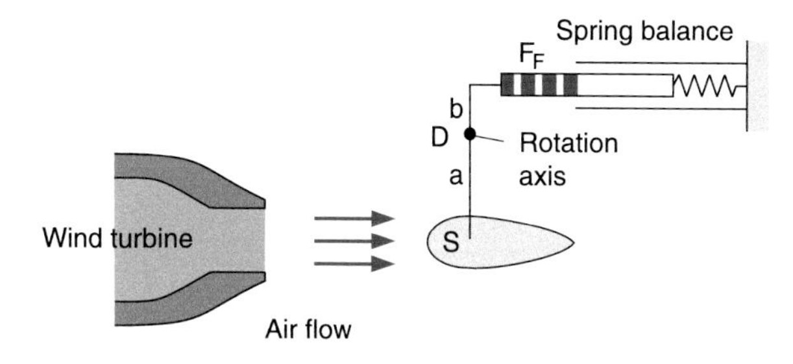

Figure 8.37 Experimental arrangement for the measurement of flow resistance

By means of the stagnation pressure $p_{\text{s}} = (1/2)\varrho \cdot u^2$ Eq. 8.41b can be written as

$$F_{\text{w}} = c_{\text{w}} \cdot p_{\text{s}} \cdot A \,. \tag{8.41c}$$

Experimental values of c_{w} can be measured with the arrangement shown in Fig. 8.37. The body to be measured is suspended by a bar that can turn around a horizontal axis. A fan blows air against the body. Due to its flow resistance the body is pressed to the right, thus expanding a spring balance on the other side of the bar. The torque exerted by the flow resistance of the body acting on the lever arm with length a is $F_{\text{w}} \cdot a$ where

$$F_{\text{w}} = \tfrac{1}{2} \cdot c_{\text{w}} \cdot \varrho \cdot u^2 \cdot A \,,$$

while the opposite torque of the spring balance is $F_{\text{s}} \cdot b$. The force F_{s} measured with the spring balance is a measure of the flow resistance F_{w} and allows the determination of the coefficient c_{w}.

8.7 Aerodynamics

The knowledge of all forces that are present when air streams around bodies with different shapes is very important not only for aviation but also for the utilization of wind energy and the optimization of car profiles. In this section only one aspect will be discussed, namely the aerodynamic buoyancy (lift) and its relation to the flow resistance of different body profiles. For a more extensive treatment, the reader is referred to the special literature [8.11a, 8.11b]

8.7.1 The Aerodynamical Buoyancy

In addition to the force on bodies in streaming media that acts in the direction of the flow also a force perpendicular to the streamlines can occur. We will illustrate this by two examples: In Fig. 8.38a we consider a laminar stream that flows around a circular cylinder. Because of symmetry reasons there could be no net force perpendicular to the current and only a force in the direction of the stream can occur which is caused by the friction between the flowing medium and the surface of the cylinder. However, if the cylinder rotates clockwise the relative velocity between surface and flowing medium is smaller at the upper side than at the lower side. This leads to a different friction on the two sides causing a net force upwards. This can be seen as follows: Due to friction a layer of the flowing medium close to the surface is dragged into the direction of the rotation causing a circulation of the layers close to the surface, which is partly transferred to adjacent layers (Fig. 8.38b).

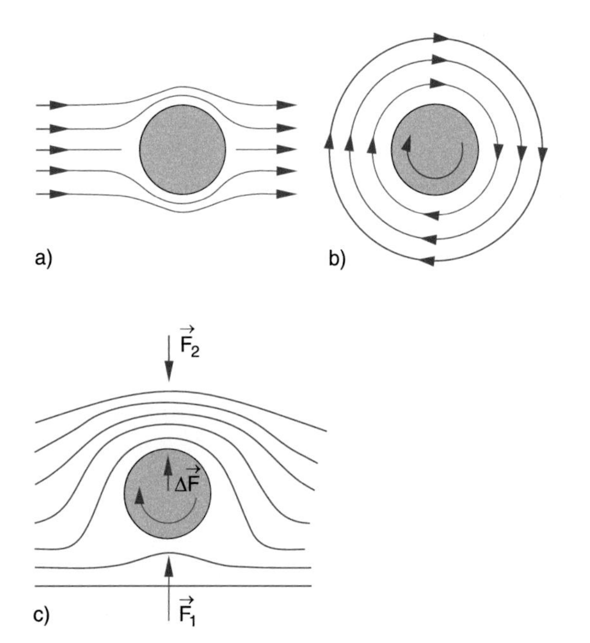

Figure 8.38 Magnus effect: **a** laminar flow around a circular cylinder, **b** circulation around a rotating cylinder in a liquid at rest, **c** streamlines around a rotation cylinder in an airflow as a superposition of **a** and **c**

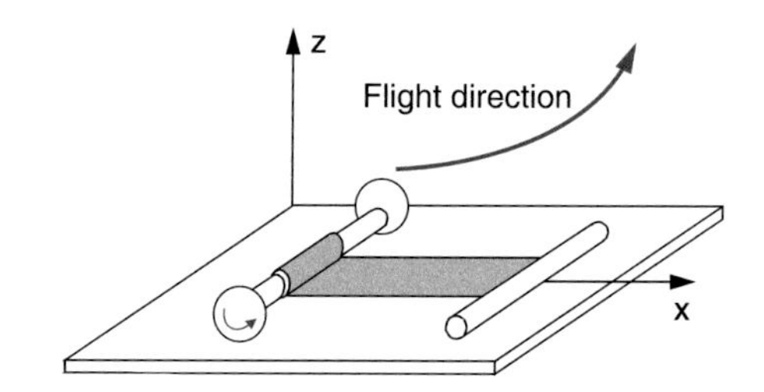

Figure 8.39 Demonstration of Magnus effect in air

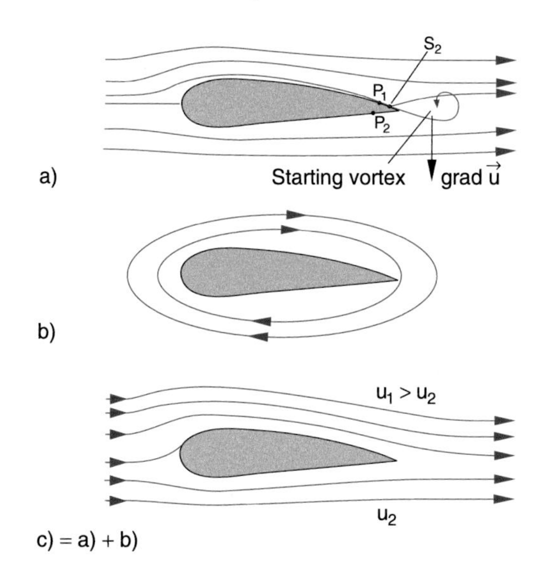

Figure 8.40 Generation of dynamical lift of a wing profile. **a** Without circulation; **b** With sole circulation; **c** Superposition of **a** and **b**

The superposition of this circulation and the laminar flow leads to an increase of the flow velocity on the upper side and a decrease on the lower side, resulting in the streamlines shown in Fig. 8.38c. The Bernoulli equation (8.17) tells us that this difference of the velocities results in a net upwards force $\Delta \boldsymbol{F} = \boldsymbol{F}_1 - \boldsymbol{F}_2$ with $|\boldsymbol{F}_1| > |\boldsymbol{F}_2|$. This effect was first discovered by Magnus and was used for the propulsion of ships. The Magnus Effect can be demonstrated in Physics lectures with a hollow cylinder of cardboard that can be brought into fast rotation by a thin ribbon around the cylinder, which is fast pulled (Fig. 8.39). The cylinder moves then against the pulling direction and rises upwards because of the Magnus effect until its rotation is slowed down due to friction and then slowly sinks down.

For bodies with asymmetric profiles in a flowing medium a perpendicular net force occurs even without rotation of the body (dynamical buoyancy). It is again explained by the superposition of a circulation and the laminar flow. In this case, however, the circulation is not caused by rotation but by the formation of vortices. We will discuss this for the example of a wing profile (Fig. 8.40).

For a laminar flow around the asymmetric wing profile the layers of the flow medium close to the surface of the wing are decelerated due to friction. Because the path along the surface is

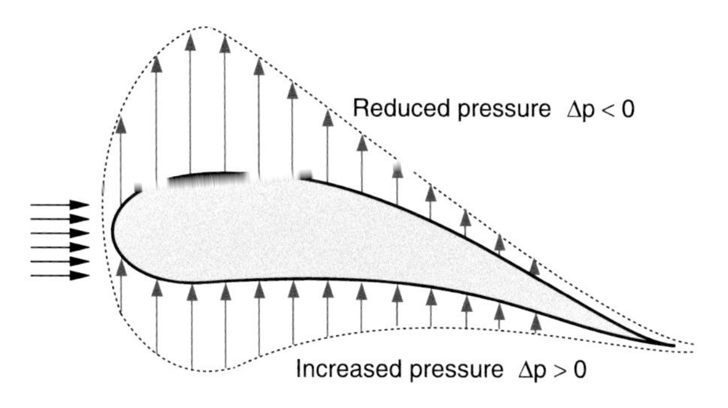

Figure 8.41 Distribution of lift force along lower and upper surface of a wing profile

longer at the upper side than at the lower side, the streaming medium arrives at the point P_1 at the upper side with lower velocity than at P_2 at the lower surface. The stagnation point S_2 at the backside is at the upper side behind P_1. Behind the profile a large velocity gradient grad u occurs between neighbouring layers of the streaming medium. If this gradient surpasses a limiting value, which depends on the velocity u and the viscosity η of the medium, a vortex develops behind the wing profile.

This can be demonstrated, when the profile is moved with increasing velocity through air or a liquid at rest. Above a critical velocity u_c the generation of a vortex is observed (starting vortex). Since the total angular momentum of the streaming medium must be conserved, the angular momentum of this vortex has to be compensated by a circulation around the total profile with opposite direction of rotation. (Fig. 8.40b). The superposition with the laminar flow leads, analogous to Fig. 8.38c, to an increase of the velocity above the wing profile and a decrease below the wing (Fig. 8.40c). According to the Bernoulli equation (8.17) this generates an upward force with the amount

$$F_L = \Delta p \cdot A = c_L \cdot \frac{\varrho}{2} \cdot \left(u_1^2 - u_2^2\right) A \,, \tag{8.42}$$

which is called the **aerodynamical lift**.

The lift coefficient c_L depends on the shape of the profile. With pressure probes, the pressure distribution along the wing profile can be measured. Figure 8.41 shows a typical pressure distribution (difference Δp to the pressure in the surrounding air) along the upper and lower side of a wing profile, where the length of the arrows indicates the magnitude of Δp [8.12].

8.7.2 Relation between Dynamical and Flow Resistance

The Eq. 8.41 and 8.42 show that the flow resistance F_D and the F_L are both proportional to the kinetic energy per unit volume of the medium streaming around the profile, where the proportionality constants c_D and c_L both depend on the shape of the profile and the smoothness of its surface.

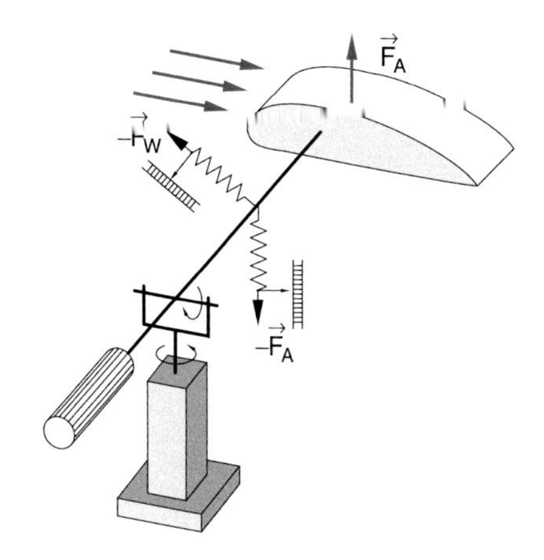

Figure 8.42 Arrangement for simultaneous measurements of flow resistance F_D and lift force F_L

Figure 8.42 shows a device (two-component balance) that allows the simultaneous measurement of the resistance force F_D and the lift force F_L for different model profiles.

It turns out that both forces (lift and drag) depend on the angle α of the profile relative to the laminar flow (Fig. 8.43). Even a flat plank shows for a certain range of α a lift force, which is, however, smaller than for a wing profile. The two curves $c_D(\alpha)$ and $c_L(\alpha)$ can be plotted in a polar diagram (Fig. 8.44) (polar profile) which illustrates the relation between c_D and c_L for all possible angles of attack. The optimum angel α is chosen such that the flow resistance is as small as possible, but the lift force is still high enough. If α is too large, vortices are generated at the upper side of the wing profile which decrease the flow velocity drastically and therefore reduce the force which can even become negative.

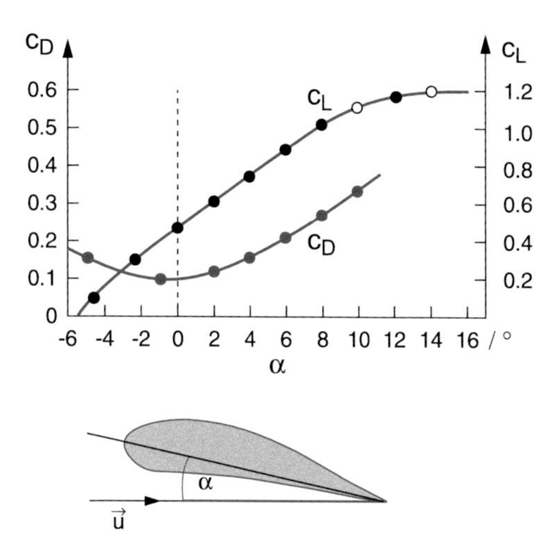

Figure 8.43 Dependence of flow resistance coefficient c_D and lift coefficient c_L on the angle of attack α of a wing profile

Chapter 8

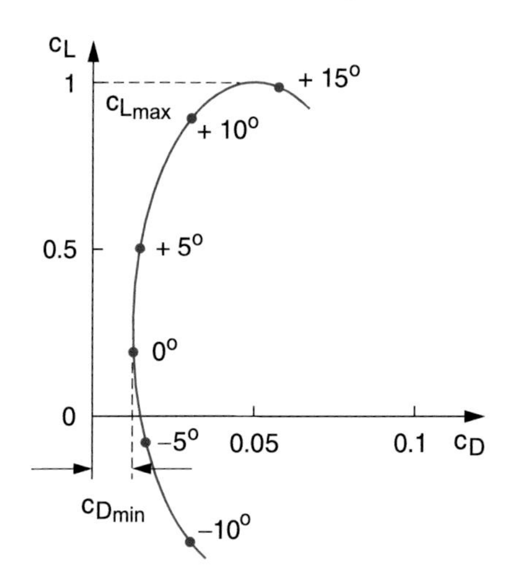

Figure 8.44 Polar diagram of a modern wing profile with small flow resistance coefficient

8.7.3 Forces on a flying Plane

At first we will discuss the flight without motor (glider). For a stationary flight of a glider with constant velocity $\boldsymbol{v}$ the total force on the glider (including gravity) must be zero. The total force is the vector sum of the lift force F_L, the flow resistance force F_D (Fig. 8.45) which depend on the flow velocity $u = -v$ of the air streaming around the glider, and the gravity force $m \cdot g$. A stable flight is only possible, if the glider flies on a declining path with the glide angle γ. From the condition $\boldsymbol{F} = \boldsymbol{0}$ we obtain with $F_D = |\boldsymbol{F}_D|$ and $F_L = |\boldsymbol{F}_L|$

$$\tan \gamma = -\frac{F_D}{F_L} \quad \text{and} \quad \sin \gamma = \frac{F_D}{mg}. \tag{8.43a}$$

The ratio F_D/F_L is called glide ratio. In order to realize a small glide angle, the force F_L should be as large as possible. As can be seen from Fig. 8.44 there is a lower limit for the glide ratio.

Modern gliders reach glide ratios of 1/50. This implies that a glider can reach 100 km flight distance without thermal lift, when it starts from a height of 2 km. If the glide angle γ is made larger by operating the elevation unit, the velocity v of the glider becomes larger, when γ is made smaller, the velocity decreases until the uplift breaks down and the glider becomes unstable. Without an experienced pilot, this might lead to a crash down.

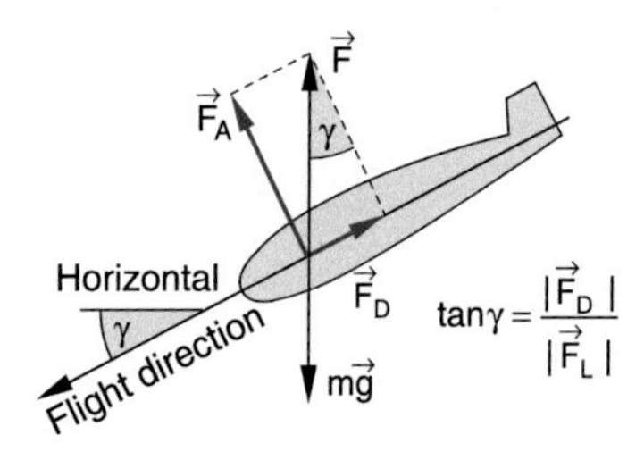

Figure 8.45 Forces at the gliding flight

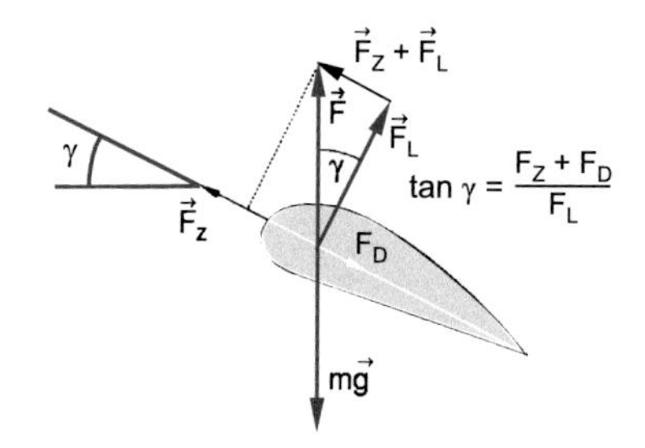

Figure 8.46 Forces at the ascent of a motor plane

When the air locally heats up (for instance above a hot ground or above chimneys of power stations) the air expands, its density decreases and it rises upwards (thermal lift, see Sect. 6.3). This gives an additional vertical component to the flow velocity of the air relative to the glider. In this case the glide angle γ can become negative, i.e. the glider rises upwards.

For planes with an engine (Fig. 8.46) an additional pulling force is produced by the propeller (or a corresponding propulsive force for jet planes). A climb is only possible, if the pulling force F_Z is larger than the magnitude of the opposite drag F_D. For the flight with constant velocity v at constant height the pulling force must just compensate the drag ($F_Z = F_D$). The angle of climb, γ is given by the ratio

$$\tan \gamma = \frac{F_Z - F_D}{F_L}. \tag{8.43b}$$

For $F_Z < F_D$ the angle γ becomes negative and the plane can fly with constant velocity only on a continuous descent.

8.8 Similarity Laws; Reynolds' Number

In Sect. 8.6.3 we have seen that vortices are caused by friction in the layers between liquid flow and walls. Although friction inside the liquid flow is small compared to that at the walls, it essentially influences the behaviour of the liquid flow, because this internal friction acts on the surface layers and starts turbulent flow.

Such boundary conditions are not included in the Navier–Stokes equation, because this equation describes the motion of an infinitesimal volume element and its motion under the influence of the different forces. It does not contain the special geometry of the flow pipe. Its geometry, however, plays an important role for the characteristics of the flow. It can be inserted into (8.36a) as special boundary conditions, but a reliable solution demands the knowledge of all details of such boundary conditions, which is often missing. Therefore generally experimental solutions are preferred which are obtained in the following way:

In hydro-and aero-dynamics the flow conditions for the motion of large objects (ships, airplanes) is studied with small models that have a similar but scaled down geometry. With such model experiments in wind channels or in small liquid flow chambers, the optimum shape of a wing or a hulk can be found. In order to obtain realistic results, not only the shape of the model must be a true scaled down version of the true object, but also the flow conditions must be accordingly scaled down in a correct way. How this can be achieved, will be shortly outlined:

We normalize all length dimensions by a unit length L, all times by a unit time T and all velocities by L/T. We therefore define new values of length, time and velocity:

$$l' = l/L\,; \quad t' = t/T\,; \quad u' = u/(L/T) = u \cdot T/L. \tag{8.44a}$$

This gives for the gradient ∇' and the pressure p'

$$\nabla' = \nabla \cdot L\,; \quad p' = p \cdot (T/L)^2/\varrho\,, \tag{8.44b}$$

where $\nabla' = L \cdot (\partial/\partial x, \partial/\partial y, \partial/\partial z)$, L', t', u' and p' are dimensionless quantities.

With these normalized quantities the Navier–Stokes equation becomes

$$\frac{\partial \boldsymbol{u}'}{\partial t'} + (\boldsymbol{u}' \cdot \nabla')\,\boldsymbol{u}' = -\nabla' p' + \frac{1}{\mathrm{Re}} \Delta' \boldsymbol{u}'\,, \tag{8.45}$$

with the dimensionless Reynolds'-Number

$$\mathrm{Re} = \frac{\varrho \cdot L^2}{\eta \cdot T} = \frac{\varrho \cdot U \cdot L}{\eta}, \quad \text{with} \quad U = \frac{L}{T}\,. \tag{8.46}$$

The quantity $U = L/T$ is the flow velocity averaged over the length L. For ideal liquids is $\eta = 0 \rightarrow \mathrm{Re} = \infty$. Here the following statement can be made:

Flows of ideal liquids in geometrical similar containers for which (8.44) is valid, are described by the same equation (8.45) with the same boundary conditions. This means: At corresponding positions $\boldsymbol{r}'$ and times t' one obtains the same dimensionless quantities p' and u' in (8.45). Even non-stationary flows have the same progression within time intervals that are proportional to the container dimensions and inversely proportional to the flow velocity u.

For viscous liquids with $\eta \neq 0$ this is only valid if the Reynolds number Re has the same value. Flows of viscous liquids are only similar if the Reynolds number Re has the same value and the flow proceeds in containers with similar geometrical dimensions.

We will illustrate the physical meaning of the Reynolds number Re. When we multiply numerator and denuminator in the fraction (8.46) by $L^2 \cdot U$ we obtain

$$\mathrm{Re} = \frac{\varrho \cdot L^3 \cdot U^2}{\eta L^2 \cdot U} = \frac{2E_{\mathrm{kin}}}{W_{\mathrm{f}}}\,. \tag{8.47}$$

The numerator gives twice the kinetic energy of a volume element L^3, which moves with the velocity U, while the denominator is the friction energy W_{f}, which is dissipated when the volume element L^3 moves with the velocity U over a distance L.

For small Reynolds' numbers $E_{\mathrm{kin}} \ll W_{\mathrm{f}}$, which implies that the accelerating forces are small compared to the friction forces. The flow is laminar. Turbulent flows occur above a critical Reynolds' number Re_{c}.

Experimental findings give for water flows in circular pipes with diameter d the critical Reynolds' Number

$$\mathrm{Re}_{\mathrm{c}} = \varrho \cdot d \cdot U_{\mathrm{c}}/\eta = 2300\,.$$

For prevention of turbulent flows the normalized flow velocity must always obey the condition $U < U_{\mathrm{c}} \rightarrow \mathrm{Re} < \mathrm{Re}_{\mathrm{c}}$. If Re is only slightly smaller than Re_{c} vortices are formed, which, however, have diameters that are smaller than the flow pipe diameter. Their rotational energy is small compared to the kinetic energy of the laminar flow and they therefore do not impede the flow very much. Only for $\mathrm{Re} \geq \mathrm{Re}_{\mathrm{c}}$ their rotational energy becomes comparable to the friction energy and macroscopic vortices are generated. The flow becomes completely turbulent.

8.9 Usage of Wind Energy

The kinetic energy of streaming air can be utilized for the generation of electric power by wind energy converters. This had been already realized for many centuries by wind mills for grinding grain or for pumping water.

Modern wind converters generally have three rotor blades (coloured pictures 3 and 4). According to new insight in the flow conditions of air around the rotor blades, the shape of these rotors is formed in a complicated way in order to optimize the conversion efficiency of wind energy into mechanic rotation energy of the rotor, which is then further converted through transmission gears and electric generators into electric power.

Most of the wind energy converters produce alternating current, which is then rectified and again converted by dc-ac converters into alternating current. This is necessary in order to synchronize the phase of the ac-current with that of the countrywide network.

The kinetic energy of a volume element $\mathrm{d}V$ of the airflow moving with the velocity v is

$$E_{\mathrm{kin}} = \tfrac{1}{2} m v^2 = \tfrac{1}{2} \varrho v^2 \mathrm{d}V\,. \tag{8.48}$$

The air volume impinging per second onto the vertical area A is $\mathrm{d}V = v \cdot A$. The maximum power (energy per second) of the air flow hitting the area A is then

$$P_{\mathrm{W}} = \tfrac{1}{2} \varrho \cdot v^3 \cdot A\,. \tag{8.49}$$

In reality, only a fraction of the power can be converted into rotational energy of the wind converter. Firstly the wind is not completely decelerated to $v = 0$, because for $v = 0$ a tailback of air would build up behind the wind converter which would impede the air flow to the converter. Secondly, friction losses diminish the kinetic flow energy and rise the temperature.

If the velocity of the air inflow is v_1, it is decelerated to $v < v_1$ because of the stagnation at the rotor blades, where the pressure

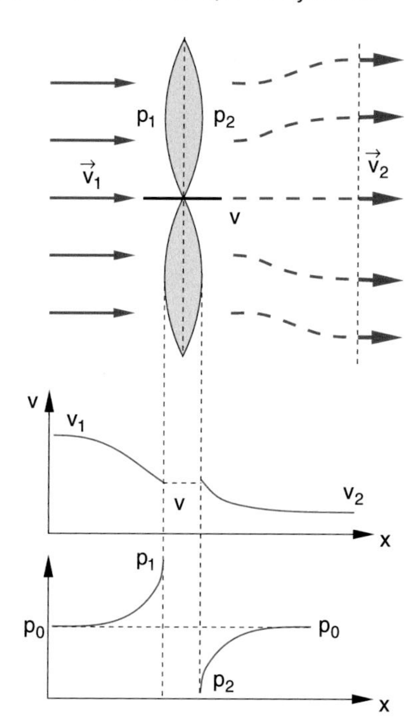

Figure 8.47 Schematic illustration of velocity and pressure conditions for a rotor blade at rest in an air flow [8.13]

increases from the initial value p_0 to $p_1 > p_0$ (Fig. 8.47). At the backside of the rotor the pressure sinks to $p_2 < p_0$. Behind the rotor the pressure increases again to p_0 and the airflow velocity is down to $v_2 < v$. Only after a larger distance behind the rotor the wind velocity rises again to its initial value $v = v_1$.

According to the Bernoulli equation is

$$p_1 - p_2 = \varrho\left(v_1^2 - v_2^2\right)/2\,. \tag{8.50}$$

The force acting on the rotor blades with area A is

$$F = (p_1 - p_2)A = \varrho\left(v_1^2 - v_2^2\right)A/2\,. \tag{8.51}$$

On the other hand this force can be written as

$$F = (v_1 - v_2)\mathrm{d}/\mathrm{d}t(mv) = (v_1 - v_2)\varrho vA\,. \tag{8.52}$$

The comparison between (8.51) and (8.52) shows that $v = (v_1 + v_2)/2$.

The power, transferred to the wind converter is then

$$\Delta P_\mathrm{w} = F\cdot v = \left(v_1^2 - v_2^2\right)\varrho v\frac{A}{2} = a\cdot P_\mathrm{W}\,. \tag{8.53}$$

Inserting P_w from (8.49) gives for the conversion factor a the value $a = (v_1 + v_2)\cdot(v_1^2 - v_2^2)/2v_1^3 < 1$. With a given initial velocity v_1 the maximum transferred power $\Delta P_\mathrm{w}(v_2)$ is reached for $\mathrm{d}(\Delta P_\mathrm{w})/\mathrm{d}v_2 = 0$. This gives with $v = (v_1 + v_2)/2$ the condition

$$-2v_2(v_1 + v_2)\varrho\cdot A/4 + (v_1^2 - v_2^2)\varrho\cdot A/4 = 0\,,$$

which yields $v_2 = \frac{1}{3}v_1$.

For the efficiency factor a one obtains $a = 0.59$. This means that without any other losses at most 59% of the initial wind energy can be converted into rotational energy of the wind converter!

Example

$v_1 = 10\,\mathrm{m/s}$, $v_2 = 4\,\mathrm{m/s} \rightarrow v = 7\,\mathrm{m/s}$ and $a = 0.588$. A typical wind converter has rotor blades with $L = 50\,\mathrm{m}$ length and deliver several Megawatt electric power. At the rotational frequency $f = 1/\mathrm{s}$ the velocity of the rotor ends is already $300\,\mathrm{m/s} = 1080\,\mathrm{km/h}$, which is close to the limit of tensile strength of the blade material. ◀

Note, that the power transferred to the wind converter is proportional to the third power of the initial wind velocity. This means that already small changes of the wind velocity will cause large changes of the power available from wind converters. Modern wind converters can operate at wind velocities between $4\,\mathrm{m/s}$ and $25\,\mathrm{m/s}$. For smaller velocities the transferred power is too small for a profitable operation. For higher velocities $v > 25\,\mathrm{m/s}$ the converters are shut down because of possible destruction.

The efficiency of the energy conversion is reduced by several losses. Firstly there are friction losses between different air layers with different velocities. They correspond to the friction losses η in the Navier–Stokes equation. Furthermore there are mechanical losses of the rotating blades and the transmission gear. Finally the losses in the electric generator have to be considered.

Figure 8.48 Offshore Windpark in the North Sea

Figure 8.49 Windfarm Krummhörn. Rotor span width is 30 m, the nominal electric power output is 300 kW per wind converter. (With kind permission of EWE corporation, Oldenbourg)

The available electric power is then

$$P_{\text{electric}} = a \cdot \eta_{\text{air}} \cdot \eta_{\text{mech}} \cdot \eta_{\text{electric}} \cdot P_{\text{w}} .$$

For our considerations about the wind velocities before and behind the wind converter we have assumed that the rotor blades are at rest. Because of their rotation the relative velocity between initial wind velocity and rotor velocity is smaller and the wind does no longer impinge perpendicular to the blades. This gives not only a smaller effective area $A_{\text{eff}} < A$ but also a smaller value of the transferred energy.

The power delivered by a wind converter, averaged over one year, is only between 10% and 40% of the installed power depending on the wind conditions at the converter location. The highest efficiency is reached for offshore wind converters (Fig. 8.48), because here the wind blows continuously and has generally a higher velocity than above undulating solid ground. For wind converters on solid ground the height should be as large as technically possible, because the wind velocity at 100 m altitude is much higher than directly above ground (Fig. 8.49).

In Tab. 8.3 the total installed electric power of wind converters is listed for the countries with the highest usage of wind energy and in Fig. 8.50 the impressive increase of worldwide annually new installed electric power from wind converters is illustrated.

Now we will discuss the energy conversion of wind converters in more detail: The forces driving the rotor blades can be composed of the flow resistance force and the Bernoulli-force. Their ratio depends on the shape of the blades and on their angle of attack α. This is similar to the situation for air flowing around a wing profile of an air plane (see Sect. 8.7.2 and Figs. 8.41 and 8.46). The pressure dependence $p(x)$ along a wing profile at rest is shown for the upper and lower side of the profile in Fig. 8.51. The pressure difference generates a lift force and a torque about an axis in x-direction. Depending on the orientation of the pro-

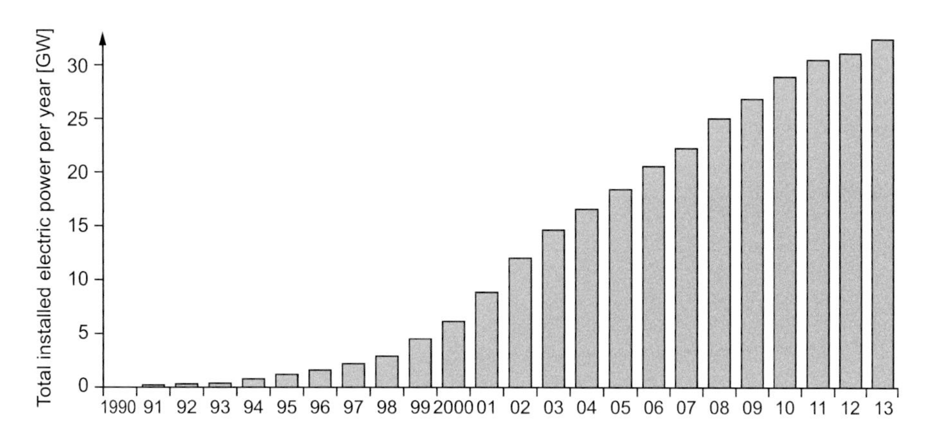

Figure 8.50 Growth of worldwide annually installed eletric power of wind converters in GW

Table 8.3 Installed electric power of wind converters for different countries (2014)

Ranking	Country	Power in GW
1	China	114.763
2	USA	65.879
3	Germany	39.165
4	Spain	22.987
5	India	22.465
6	UK	12.440
7	Canada	9.694
8	France	9.285
9	Italy	8.663
10	Brazil	5.939
11	Sweden	5.425
12	Portugal	4.914
13	Denmark	4.845
14	Poland	3.834
15	Australia	3.806
16	Turkey	3.763
17	Rumania	2.954
18	Netherland	2.805
19	Japan	2.789
20	Mexico	2.381
	Worldwide	369.553
	Europe	133.969

file against the direction of wind flow, the lift force as well as the flow resistance force can be used for driving the rotor blades.

When the rotor blade rotates with the angular velocity ω, the velocity $\boldsymbol{v}_B(r)$ of the section of the blade at a distance r from the rotation axis adds to the wind velocity v to an effective velocity $\boldsymbol{v}_{eff} = \boldsymbol{v} + \boldsymbol{v}_B(\boldsymbol{r})$ (Fig. 8.52). The angle of attack α against the direction of $\boldsymbol{v}_{eff}$ must be chosen in such a way (Fig. 8.43), that the optimum force can be used. Since $\boldsymbol{v}_{eff}$ changes with r the profile of the blade must change with r. With increasing r the blade must become slimmer and the direction of the profile changes. The whole blade is therefore twisted (Fig. 8.53) in order to reach for all sections of the rotating blade the optimum usage of the lift force.

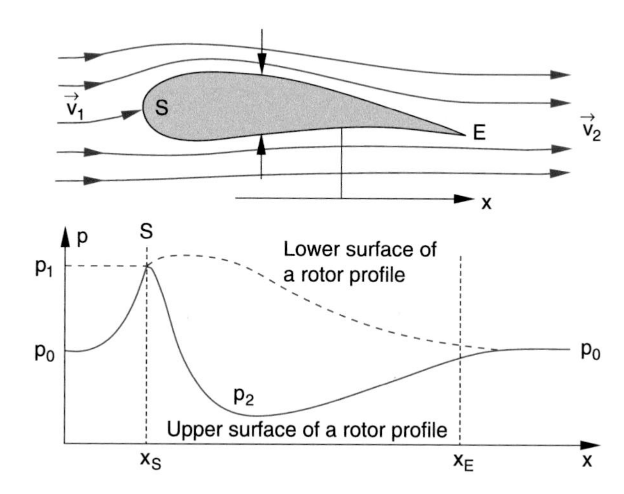

Figure 8.51 Pressure variation along the lower and upper surface of a rotor profile at rest, with air flowing around the profile. The rotation axis lies above the drawing plane [8.14]

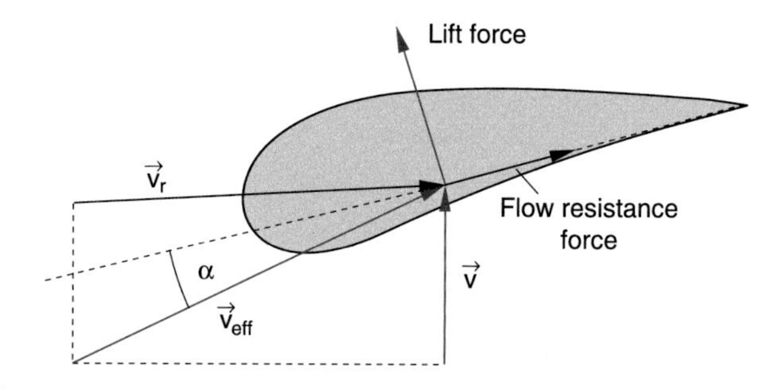

Figure 8.52 Velocities and forces on the rotating rotor. The rotation axis points into the direction of $\boldsymbol{v}$ and is above the drawing plane [8.13]

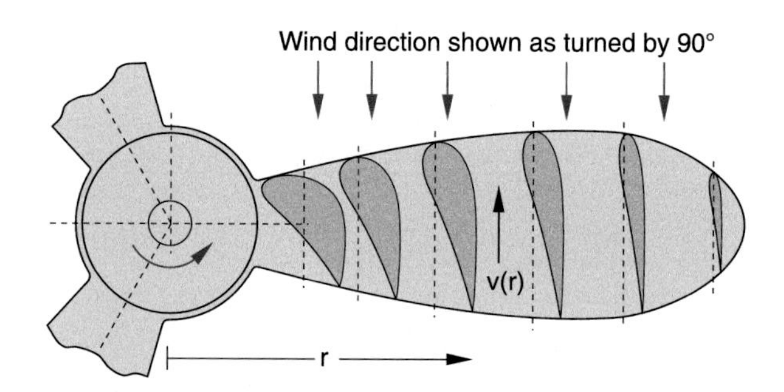

Figure 8.53 Rotor blade of a fast rotating wind converter. The *red areas* represent the rotor profile at different distances r from the rotation axis. In this drawing they are turned by 90° into the drawing plane. Also the wind direction is turned. The wind comes really from above the drawing plane

Summary

- The motion of particles of a flowing medium (liquids or gases) is determined by the total force $\boldsymbol{F} = \boldsymbol{F}_g + \boldsymbol{F}_p + \boldsymbol{F}_f$ which is the vector sum of gravity force, pressure force and friction force. The equation of motion is

$$\boldsymbol{F} = \varrho \cdot \Delta V \cdot \frac{d\boldsymbol{u}}{dt},$$

where $\boldsymbol{u}$ is the flow velocity of the volume element ΔV with the mass density ϱ.
- In a stationary flow $\boldsymbol{u}(\boldsymbol{r})$ is at every position $\boldsymbol{r}$ constant in time but can vary for different positions $\boldsymbol{r}_i$.
- Frictionless liquids ($\boldsymbol{F}_f = \boldsymbol{0}$) are called **ideal liquids**. For them the Euler equation

$$\frac{\partial \boldsymbol{u}}{\partial t} + (\boldsymbol{u} \cdot \nabla)\boldsymbol{u} = \boldsymbol{g} - \frac{1}{\varrho} \operatorname{grad} p$$

describes the motion of the liquid.
- The continuity equation

$$\frac{\partial \varrho}{\partial t} + \operatorname{div}(\varrho \cdot \boldsymbol{u}) = 0$$

describes the conservation of mass for a flowing medium. For incompressible media ($\varrho = \text{const}$) the continuity equation reduces to $\operatorname{div} \boldsymbol{u} = 0$.
- For frictionless incompressible flowing media the Bernoulli-equation

$$p + \tfrac{1}{2}\varrho \cdot u^2 = \text{const}$$

represents the energy conservation $E_p + E_{kin} = E = \text{const}$. The pressure decreases with increasing flow velocity u.
The Bernoulli equation is the basic equation for the explanation of the dynamical buoyancy and therefore also for aviation.
- For flow velocities u below a critical value u_c laminar flows are observed, while for $u > u_c$ turbulent flows occur. This critical value u_c is determined by the Reynolds number $\text{Re} = 2E_{kin}/W_f$ which gives the ratio of kinetic energy to the friction energy of a volume element $\Delta V = L^3$ when ΔV is shifted by L.
- For laminar flows where the inertial forces are small compared to the friction forces no turbulence occurs and the stream lines are not swirled.
- For a laminar flow through a tube with circular cross section πR^2 the volumetric flow rate

$$Q = \frac{\pi R^4}{8\eta} \operatorname{grad} p$$

flowing per second through the tube is proportional to $R^4 \cdot \operatorname{grad} p$ but inversely proportional to the viscosity η.
- A ball with radius r moving with the velocity u through a medium with viscosity η experiences a friction force

$$\boldsymbol{F}_f = -6\pi\eta r \cdot \boldsymbol{u},$$

that is proportional to its velocity $\boldsymbol{u}$.
- The complete description of a flowing medium is provided by the Navier–Stokes equation (8.36a) which reduces for ideal liquids ($\eta = 0$) to the Euler equation. The Navier–Stokes equation describes also turbulent flows, but for the general case no analytical solutions exist and the equation can be solved only numerically.
- For the generation and the decay of vortices friction is necessary. Vortices are generally generated at boundaries (walls and solid obstacles in the liquid flow).
- The flow resistance of a body in a streaming medium is described by the resisting force $F_D = c_D \cdot \varrho \cdot \frac{1}{2}u^2 \cdot A$. It depends on the cross section A of the body and its drag coefficient c_D which is determined by the geometrical shape of the body. The force is proportional to the kinetic energy per volume element ΔV of the streaming medium. In laminar flows, F_D is much smaller than in turbulent flows.
- The aero-dynamical buoyancy is caused by the difference of the flow velocities above and below the body. This difference is influenced by the geometrical shape of the body and can be explained by the superposition of a laminar flow and turbulent effects (circulation).

Problems

8.1 Estimate the force that a horizontal wind with a velocity of 100 km/h (density of air = 1.225 kg/m^3) exerts (ϱ = 1.225 kg/m^3; $c_D = 1.2$)
a) on a vertical square wall of 100 m^2 area
b) on a saddle roof with 100 m^2 area and length $L = 6$ m and a cross section that forms an isosceles triangle with $\alpha = 150°$.

8.2 Why can an airplane fly "on the head" during flight shows, although it should experience according to Fig. 8.41 a negative buoyancy?

8.3 Why do the streamlines not intermix in a laminar flow although the molecules could penetrate a mean free path Λ into the adjacent layers?
Hint: Estimate the magnitude of Λ in a liquid.

8.4 Prove the relation (8.36b) using the component representation.

8.5 A cylinder is filled with a liquid up to the height H. The liquid can flow out through a pipe at height h (Fig. 8.54)

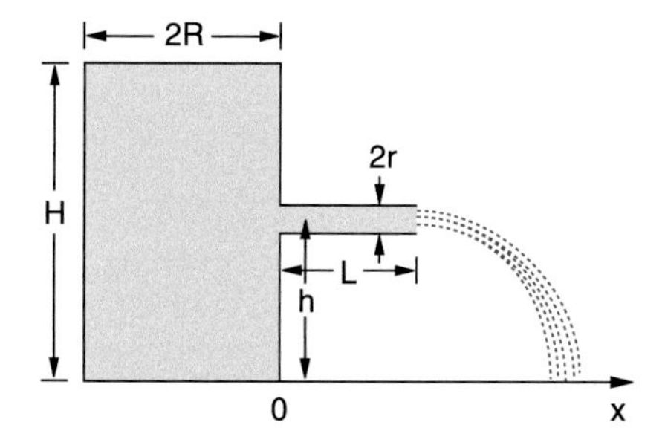

Figure 8.54 To Probl. 8.5

a) Calculate for an ideal liquid (no friction) the position $x(H)$ where the outflowing liquid hits the ground and the velocity $v_x(H)$ and $v_z(H)$ for $z = 0$. Compare this result with the velocity of a free falling body starting from $z = H$.
b) What is the function $z(t)$ of the liquid surface in the cylinder with radius R for a liquid with the viscosity η streaming through a pipe with length L and radius $r \ll R$ at the height $z = 0$?

8.6 A pressure gauge as shown in Fig. 8.10c is placed into flowing water. The water in the stand pipe rises by 15 cm. The measurement with the device of Fig. 8.10a shows a pressure of $p = 10$ mbar. How large is the flow velocity?

8.7 A funnel with the opening angle $\alpha = 60°$ is filled with water up to the height H. The water can flow into a storage vessel with volume V through a horizontal pipe at the bottom of the funnel with length L and inner diameter d.
a) What is the height $H(t)$ in the funnel as a function of time?
b) What is the total flow mass $M(t)$?
c) After which time is the funnel empty for $H = 30$ cm, $d = 0.5$ cm, $L = 20$ cm, and $\eta = 1.002$ mPa · s?
d) After which time is the storage vessel with a volume V = 4 litre full, if the water in the funnel is always kept at the height H by supplying continuously water?

8.8 A water reservoir has at Δh below the water surface a drain pipe with inner diameter $d = 0.5$ cm and length $L = 1$ m which is inclined by the angel α below the horizontal.
a) How much water flows per second through the pipe for a laminar flow with $\eta = 10^{-3}$ Pa · s and $\Delta h = 0.1$ m?
b) Above which angle α the flow becomes turbulent if the critical Reynolds number is 2300?

8.9 What is the minimum diameter of a horizontal tube with $L = 100$ m to allow a laminar flow of water of 1 l · s^{-1} from a vessel with a water level 20 m above the horizontal tube?

8.10 What is the vertical path $z(t)$ of a ball with radius r falling through glycerine ($\eta = 1480$ mPa · s) if it immerses at $t = 0$ and $z = 0$ into the glycerine with the initial velocity $v_0 = 2$ m/s
a) for $r = 2$ mm,
b) for $r = 10$ mm?

8.11 Derive the Helmholtz equation (8.39) starting from (8.36a).

References

8.1a. G. Birkhoff, *Hydrodynamics.* (Princeton Univ. Press, 2015)

8.1b. E. Guyon, J.P. Hulin, *Physical Hydrodynamics.* (Oxford Univ. Press, 2012)

8.2. J.E.A. John, Th.G. Keith, *Gasdynamics.* (Prentice Hall, 2006)

8.3a. R.K. Bansal, *A Textbook on Fluid Mechanics.* (Firewall Media, 2005)

8.3b. G.K. Batchelor, *Introduction to Fluid Mechanics.* (Cambridge Univ. Press, 2000)

8.4a. J.P. Freidberg, *Ideal Magnetohydrodynamics.* (Plenum Press, 1987)

8.4b. P.A. Davidson, *An Introduction to Magnetohydrodynamics.* (Cambridge University Press, Cambridge, England, 2001)

8.5. F. Pobell, *Matter and Methods at Low Termperatures.* (Springer, Berlin, Heidelberg, 1992)

8.6. D. Even, N. Shurter, E. Gundersen, *Applied Physics.* (Pentice Hall, 2010)

8.7. A.E. Dunstan, *The Viscosity of Liquids.* (Hard Press Publ., 2013)

8.8. L. Brand, *Vector Analysis.* (Dover Publications, 2006)

8.9. K. Weltner, S. John, *Mathematics for Physicists and Engineers.* (Springer Berlin, Heidelberg, 2014)

8.10. G.B. Arfken, H.J. Weber, *Mathematical Methods for Physicists. A Comprehensive Guide.* (Elsevier Ltd., Oxford, 2012)

8.11a. L.J. Clancy, *Aerodynamics.* (Pitman, London, 1978)

8.11b. J.D. Anderson, *Fundamentals of Aerodynamics.* (McGrawHill, 2012)

8.12. D. Pigott, *Understanding Gliding.* (A&C Black, London)

8.13. M. Diesendorf, *GreenhouseSolutions with sustainable Energy.* (University of New South Wales, 2007)

8.14. J.F. Manwell, J.G. McGowan, A.L. Rogers, *Wind Energy Explained: Theory, Design and Applications.* (Wiley, 2010)

8.15. R. Ferry, E. Monoian, *A Field Guide to Renewable Energy Technologies.* (Society for Cultural Exchange, 2012)

8.16a. T. Burton et al., *Wind Energy Handbook.* (Wiley, 2001)

8.16b. N. Walker, *Generating Wind Power.* (Crabtree Publ., 2007)

8.17. The European Wind Energy Association, *Wind Energy: The Facts. A Guide to the Technology, Economics and Future of Wind Power.* (Routledge, 2009)

Vacuum Physics

9

© Springer International Publishing Switzerland 2017

W. Demtröder, *Mechanics and Thermodynamics*, Undergraduate Lecture Notes in Physics, DOI 10.1007/978-3-319-27877-3_9

The importance of vacuum physics for the development of modern physics and technology can be hardly overestimated. Only after the realization of a sufficiently low vacuum, many experiments in atomic, molecular and nuclear physics became possible. These experiments have essentially contributed to the understanding of the micro-structure of matter, of electrons and nuclei as the building blocks and of the internal structure of atoms and nuclei. Based on the results of these experiments the quantum theory of matter could be successfully developed (see Vol. 3).

Without vacuum technology, the manufacturing of semiconductor elements and integrated circuits would have been impossible and therefore we would be still without computers.

Besides for basic research vacuum technology is nowadays used as indispensable tool in many technical applications, which reach from vacuum melting of special metal alloys over the production of thin optical films to the dry freezing of food. It is therefore essential for every physics student to study at least some basic facts of vacuum physics and technology.

In this chapter we will discuss, after a summary of the most important fundamentals, some techniques for the generation of vacuum and the measurement of low pressures. More detailed presentations can be found in [9.1, 9.2, 9.3].

9.1 Fundamentals and Basic Concepts

Vacuum is produced in a container, when most of the gases or vapours have been removed and the pressure p in the volume V becomes small compared to the atmospheric pressure $p_0 \approx 1$ bar. Devices, that can achieve such a reduction of the pressure, are called **vacuum pumps**, because they pump part of the gases or vapours in the volume V into other containers or into the open air (Fig. 9.1). The achieved pressure, given in the unit Pascal ($1\,\text{Pa} = 1\,\text{N/m}^2 = 10^{-2}\,\text{hPa}$) or often quoted in millibar ($1\,\text{mbar} = 1\,\text{hPa} = 100\,\text{Pa}$) (see Tab. 7.1) depends essentially on the type of vacuum pumps used for the evacuation. At low pressures ($p < 10^{-4}\,\text{hPa}$) the walls of the vacuum container and the gas molecules attached to the walls play an important role for further evacuation because their outgasing essentially influence the achievable vacuum pressure.

9.1.1 The Different Vacuum Ranges

We distinguish four different vacuum ranges, depending on the lower pressure limit of the achievable vacuum.

- Low *vacuum*
  $1\,\text{hPa} < p < 1000\,\text{hPa} = 1\,\text{bar}$
- Medium *vacuum*
  $10^{-3}\,\text{hPa} < p < 1\,\text{hPa}$
- *High vacuum*
  $10^{-7}\,\text{hPa} < p < 10^{-3}\,\text{hPa}$
- *Ultrahigh Vacuum*
  $p < 10^{-7}\,\text{hPa}$.

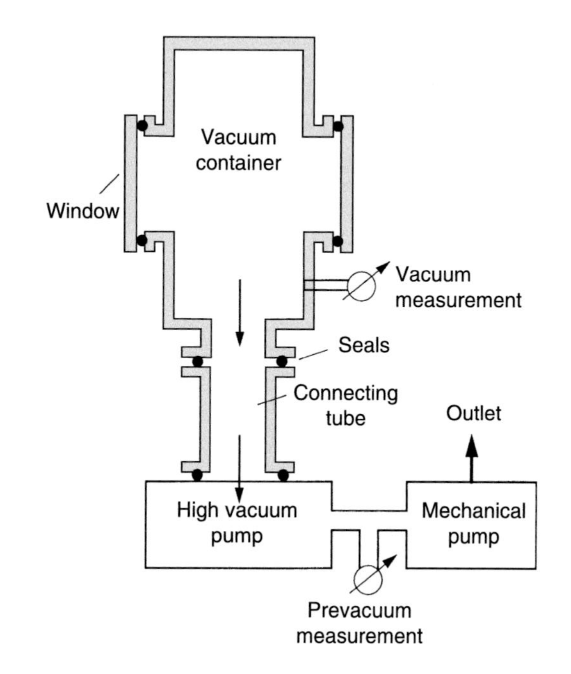

Figure 9.1 Schematic drawing of a vacuum apparatus

The best vacuum, achievable today is about $10^{-13}\,\text{hPa}$.

In order to give an impression how empty an evacuated container really is, Tab. 9.1 compiles the number of gas molecules per m^3 for different pressures. It is illustrative to compare these numbers with the number n_w of molecules sitting per m^2 in a monomolecular layer on the surface of the container walls. With a mean distance of 0.3 nm between the molecules of a monomolecular layer we get $n_w = 10^{19}/\text{m}^2$. A cubic vacuum container with $V = 1\,\text{m}^3$ has a wall surface of $6\,\text{m}^2$. At a pressure of $2 \cdot 10^{-5}\,\text{hPa}$ the number of molecules on the wall therefore equals the number of all molecules in the volume of the container. For pressures below $10^{-5}\,\text{hPa}$ the number of molecules on the wall therefore exceeds the number in the volume and in order to reach a much lower vacuum the walls have to be outgassed by heating.

Generally there are different gases (N_2, O_2, He, Ar with pressures p_i) and vapours (water, oil or other liquids with saturation

Table 9.1 Particle density n of air molecules, mean free path Λ and particle flux density ϕ onto the container surface for different pressures at room temperature

p/hPa	n/m^{-3}	Λ/m	ϕ/m^{-2}s^{-1}
10^3	$2.5 \cdot 10^{25}$	$6 \cdot 10^{-8}$	$3 \cdot 10^{27}$
1	$2.5 \cdot 10^{22}$	$6 \cdot 10^{-5}$	$3 \cdot 10^{24}$
10^{-3}	$2.5 \cdot 10^{19}$	$6 \cdot 10^{-2}$	$3 \cdot 10^{21}$
10^{-6}	$2.5 \cdot 10^{16}$	60	$3 \cdot 10^{18}$
10^{-9}	$2.5 \cdot 10^{13}$	$6 \cdot 10^{4}$	$3 \cdot 10^{15}$

pressures p_{si}) in the vacuum container. The total pressure

$$p = \sum (p_i + p_{si}) \qquad (9.1)$$

is then the sum of all partial pressures. The saturation pressure which adjusts itself at the equilibrium between liquid and vapour depends on the temperature (see Sect. 10.4.2).

For planning an experiment in the vacuum chamber the mean free path Λ of the molecules is of great importance. It determines the collision probability between the molecules in the chamber (see Sect. 7.3). Table 9.1 shows that in the fine vacuum range Λ is small compared with the dimensions of commonly used vacuum chambers. Collisions can be therefore not neglected. On the other hand, in the high vacuum range for $p < 10^{-6}$ hPa, Λ becomes large compared with the dimensions of the chamber and the molecules fly freely through the chamber without suffering collisions until they hit a wall.

9.1.2 Influence of the Molecules at the Walls

The number of molecules, hitting per sec an area of 1 m^2 of the walls of a vacuum container (particle flux density Φ, last column of Tab. 9.1) depends on the particle density n in the evacuated volume V and on the mean thermal velocity $\overline{v}$ (see Sect. 7.3). A molecule with the velocity $\boldsymbol{v} = \{v_x, v_y v_z\}$ with the distance z from the surface can reach the surface within the time $\Delta t \geq z/v_z$ as long as z is smaller than the mean free path Λ (Fig. 9.2). For a mean particle density n the number of wall collisions per second of molecules in the upper half volume is

$$Z = \frac{nA\overline{v}}{4\pi} \int_0^{\pi/2} \sin\vartheta \, \cos\vartheta \, \mathrm{d}\vartheta \int_0^{2\pi} \mathrm{d}\varphi \, . \qquad (9.2a)$$

The first integral gives the value 1/2, the second gives 2π. The particle flux density $\Phi = Z/A$ onto the unit area of the wall surface is then

$$\Phi = (1/4)n \cdot v \, . \qquad (9.2b)$$

The numerical values in Tab. 9.1 show, that at a pressure of $p = 3 \cdot 10^{-6}$ hPa and a mean velocity $v = 500$ m/s nearly as many molecules hit the surface per second as are contained in a mono-molecular layer on the surface. If all impinging molecules would stick at the surface a clean surface would be covered within 1 s with a monomolecular layer. This illustrates that a really clean surface can be only realized at very low pressures (ultrahigh vacuum) and if the molecules do not stick on the surface. This can be achieved, when the surface is heated, which causes all impinging molecules to leave the surface immediately.

With decreasing temperature the evaporation decreases and the inner wall of a vacuum chamber is therefore at low temperatures always covered by a layer of adsorbed molecules. An equilibrium adjusts itself which depends on the temperature of the surface, on the density n in the chamber and on the molecular

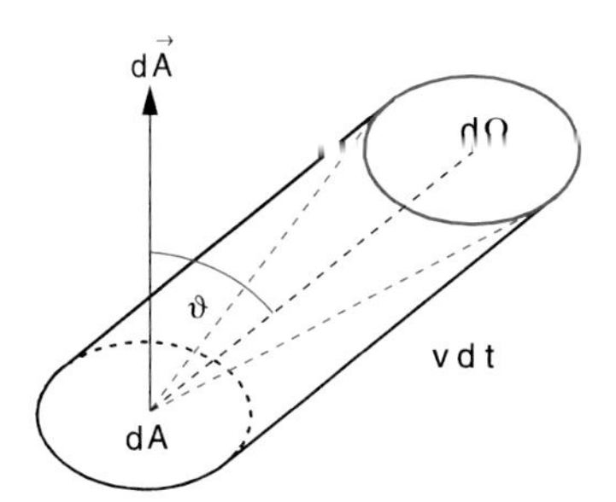

Figure 9.2 Illustration of the collision rate with the wall

species, where the rates of adsorbing and desorbing molecules become equal.

Our example above has shown that at pressures $p \leq 10^{-3}$ hPa the number of molecules adsorbed on the wall becomes larger than the number in the evacuated volume. When a vacuum chamber is evacuated, the pressure in the chamber below 10^{-3} hPa will be at first essentially determined by the rate of molecule desorbing from the wall, until the desorbing rate becomes smaller than the pumping rate that removes the molecules out of the vacuum chamber.

9.1.3 Pumping Speed and Suction Capacity of Vacuum Pumps

When a vacuum chamber is evacuated the gas in the chamber has to pass through an opening and through pipes in order to reach the vacuum pump. The **volume flow rate** of the pipe (often given in the unit litre per second = l/s or cubic meter per hour = m^3/h) is the gas volume that flows per sec through a cross section of the pipe at a given pressure p and temperature T.

Note, that the molecular density $n = N/V$ decreases with the pressure according to

$$p \cdot V = NkT \to N = \frac{pV}{kT} \to n = \frac{p}{kT} \, . \qquad (9.3)$$

Therefore even for a constant volume flow rate the number of molecules passing per second through the cross section decreases with p! This means for equal volume flow rates $\mathrm{d}V/\mathrm{d}t$, the number $\mathrm{d}N/\mathrm{d}t$ of molecules pumped out of the vacuum chamber depends on pressure p and temperature T.

The **suction capacity**

$$S_V = \frac{\mathrm{d}V}{\mathrm{d}t} \text{ given in [l/s] or in [m}^3\text{/h]} \qquad (9.4)$$

of a vacuum pump is defined as the volume flow rate $\mathrm{d}V/\mathrm{d}t$ at the suction intake of the pump.

The total mass flow of molecules with mass m

$$\frac{\mathrm{d}M}{\mathrm{d}t} = \varrho \cdot \frac{\mathrm{d}V}{\mathrm{d}t} = \frac{m}{kT} p \cdot \frac{\mathrm{d}V}{\mathrm{d}t} \, , \qquad (9.5)$$

Chapter 9

that is pumped per second out of the vacuum chamber is the mass suction capacity. It depends on the pressure and the volume flow rate through the chamber opening and the pipes.

Manufacturers of pumps generally give the suction capacity or pumping speed of a pump in the unit

$$S_L = p \cdot \frac{dV}{dt}\,, \qquad [S_L] = \text{hPa} \cdot \text{l/s} \tag{9.6}$$

as the product of pressure p and volume flow rate dV/dt.

Example

With a suction capacity $S_V = 500$ l/s about 10^{22} molecules per sec are pumped out of a vacuum chamber at room temperature and $p = 1$ hPa. At the lower pressure $p = 10^{-6}$ hPa these are only 10^{16} molecules per second at the same value of S_V.

The suction capacity is for the first case $S_L = 500\,\text{hPa} \cdot \text{l/s}$ (corresponding to 50 Watt) while for the second case S_L is only $5 \cdot 10^{-4}\,\text{hPa} \cdot \text{l/s}$ (50 μW). ◀

9.1.4 Flow Conductance of Vacuum Pipes

The dimensions of vacuum pipes play an important role for the design of a vacuum apparatus. The mass flow

$$\frac{dM}{dt} = L_m \cdot (p_2 - p_1) \tag{9.7a}$$

through a vacuum pipe is proportional to the pressure difference $(p_2 - p_1)$ between entrance and exit of the pipe. The proportionality factor L_m is the coefficient of mass flow conductance given in the unit [1 m · s]. Generally the pumping speed

$$p \cdot \frac{dV}{dt} = L_S \cdot (p_2 - p_1) \tag{9.7b}$$

is used with the unit $[\text{hPa} \cdot \text{m}^3/\text{h}]$. Because of $p \cdot V = N \cdot kT \to p = (\varrho/m)kT$ with $m = M/N$ = mass of one molecule the coefficient of volume flow conductance L_S can be related to the mass flow conductance by

$$L_S = \frac{kT}{m} \cdot L_m\,. \tag{9.7c}$$

L_S depends on the mass m of the molecules, on the mean free path Λ (because $\Lambda \propto 1/p$) and on the geometry of the vacuum pipes. For simple geometries it can be calculated. For complex geometries it must be determined experimentally. The values are compiled in special tables [9.1].

The gas flow through pipes

$$\frac{dV}{dt} = C_S \frac{\Delta p}{p} \tag{9.7d}$$

strongly depends on the pressure. Here $C_S = L_S$ is named the volume flow conductance. The different pressure ranges are characterized by the Knudsen number

$$\text{Kn} = \frac{\Lambda}{d}\,, \tag{9.8}$$

which gives the ratio of mean free path Λ and the diameter d of openings or pipes. According to the magnitude of Kn we distinguish between three ranges:

- Range of laminar gas flow (for Reynolds numbers Re < 2200) or turbulent flow (for Re > 2200) which occurs for Kn $\ll 1$. Here is $\Lambda \ll d$.
- Range of Knudsen flow (also called transition range) where Kn ≈ 1 and $\Lambda \approx d$.
- Range of free molecular flow where Kn $\gg 1$ and $\Lambda \gg d$.

In the range Kn $\ll 1$ the gas flow is essentially governed by collisions between the gas molecules, which means that the viscosity plays an important role. The flow can be described by hydro-dynamical models (see Chap. 8). Depending on the magnitude of the Reynolds number Re and the viscosity η the flow is laminar for Re < 2200 or turbulent for Re > 2200. Under the conditions relevant for most vacuum systems the Reynolds number is generally smaller than 2200 and the flow is therefore laminar.

In the range Kn $\gg 1$ collisions between the molecules can be neglected. The viscosity η does no longer influence the gas flow and collisions with the wall determine the suction capacity. The flow conductance becomes independent of the pressure.

We will illustrate these conditions by some examples:

Examples

1. Volume flow conductance C_S of a circular opening with diameter d in the range of molecular flow ($\Lambda \gg d$). According to Eq. 9.2 the number of molecules passing per sec through the hole with area $A = \pi d^2/4$ is
$$Z = \tfrac{1}{4} A \cdot n \cdot \overline{v}\,,$$
with $p \cdot V = N \cdot kT$ and $Z = dN/dt$ we obtain for the volume gas flow through the hole
$$\frac{dV}{dt} = \frac{1}{4} A \cdot \frac{n}{p} kT\,\overline{v} = \frac{1}{4} A \cdot \overline{v},$$
since $n = N/V = p/kT$.
Since $n \sim p$ the volume flow dV/dt becomes independent of pressure. Inserting numerical values of v for air at $T = 300$ K gives $dV/dt = 11.6 \cdot A$ in l/s if A is given in cm². A circular opening with $d = 10$ cm has therefore at low pressures ($\Lambda \gg d$) the volume flow conductance $C_S = 900$ l/s.
2. Flow through a pipe with length L and diameter d in the range of laminar flow ($\Lambda \ll d$). The pressures at

the two ends of the pipe are p_1 and p_2. According to the Hagen-Poisseuille-law (8.32)

$$p \cdot \frac{\mathrm{d}V}{\mathrm{d}t} = \frac{\pi \cdot d^4}{128\,\eta L} \cdot \frac{p_1 + p_2}{2}(p_1 - p_2)\,, \tag{9.9}$$

we obtain for $d = 5\,\mathrm{cm}$, $L = 1\,\mathrm{m}$, $p_1 = 2\,\mathrm{hPa}$, $p_2 = 0$, $\eta_{air} = 0.018\,\mathrm{mPa \cdot s}$, the numerical value $p \cdot \mathrm{d}V/\mathrm{d}t = 170\,\mathrm{Pa \cdot m^2/s}$. According to (9.7) the volume flow conductance then becomes $C_\mathrm{S} = 0.85\,\mathrm{m^3/s} = 850\,\mathrm{l/s}$. At a lower pressure of $10^{-1}\,\mathrm{hPa}$, where $\Lambda = 0.06\,\mathrm{cm}$ (Tab. 9.1) which is still smaller than d the flow conductance decreases according to (9.7) and (9.9) to $C_\mathrm{S} = 42\,\mathrm{l/s}$. Equation 9.9 is in this range, however, only approximately valid and the accurate value is $C_\mathrm{S} = 80\,\mathrm{l/s}$.
In the range of molecular flow ($\Lambda > d$) C_S converges with decreasing pressure towards the value $C_\mathrm{S} = 16\,\mathrm{l/s}$. ◀

The reciprocal of the flow conductance

$$R_\mathrm{S} = 1/C_\mathrm{S} \tag{9.10}$$

is the flow resistance. Completely analogue to the electrical resistance in electricity the *flow resistance* of consecutive flow pipes is the sum of the individual resistances, while for flow pipes in parallel arrangements the individual *flow conductances* add to the total conductance, as can be immediately seen from (9.7).

9.1.5 Accessible Final Pressure

Every vacuum chamber has openings that allow access to the experimental setup in its inside. They are closed by flange seals. However, there are always leaks which are often difficult to find and to close. Through these leaks molecules can penetrate from the outside into the vacuum chamber. We define the gas rate $\mathrm{d}G_\mathrm{L}/\mathrm{d}t = p_0 \cdot \mathrm{d}V_\mathrm{L}/\mathrm{d}t$ (p_0 = atmospheric pressure) which penetrates through all leaks into the vacuum chamber as the **leak rate**. It is given in the same units hPa · l/s as the pumping speed defined in (9.6).

As has been previously discussed molecules can also be desorbed from the inner walls and delivered into the volume of the vacuum chamber. This leads without pumping to a pressure increase Δp.

For the rate $\mathrm{d}N_\mathrm{d}/\mathrm{d}t$ of desorbing molecules we obtain with $p \cdot V = N \cdot kT$ the pressure increase per second

$$\frac{\mathrm{d}p}{\mathrm{d}t} = \frac{kT}{V}\frac{\mathrm{d}N_\mathrm{d}}{\mathrm{d}t}\,. \tag{9.11}$$

The total rate of desorbed gas is

$$\frac{\mathrm{d}G_\mathrm{d}}{\mathrm{d}t} = V \cdot \frac{\mathrm{d}p}{\mathrm{d}t} = kT\frac{\mathrm{d}N_\mathrm{d}}{\mathrm{d}t}\,. \tag{9.12}$$

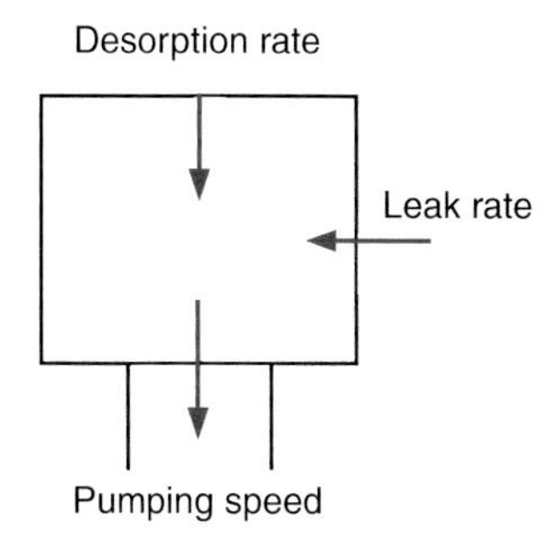

Figure 9.3 The achievable final pressure is determined by the compensation of pumping speed by leak rate + desorption rate

The final pressure achieved in the vacuum chamber is determined by the pumping speed, the leak rate and the total rate of desorbing molecules (Fig. 9.3). At the final pressure the pumping speed just equals the sum of leak rate and desorbing gas rate.

$$S_\mathrm{L}^\mathrm{eff}(p) = \frac{\mathrm{d}G_\mathrm{L}}{\mathrm{d}t} + \frac{\mathrm{d}G_\mathrm{d}}{\mathrm{d}t}\,, \tag{9.13}$$

where $S_\mathrm{L}^\mathrm{eff}$ is the effective pumping speed at the outlet opening of the chamber to the pumping pipes. It is equal to the pumping speed of the pump minus the flow conductance of the vacuum lines between chamber and pump.

The attainable final pressure p_f results then from (9.13) with (9.6) and (9.12):

$$p_\mathrm{f} = \frac{\mathrm{d}G_\mathrm{d}/\mathrm{d}t + \mathrm{d}G_\mathrm{L}/\mathrm{d}t}{S_\mathrm{V}}\,, \tag{9.14}$$

where $S_\mathrm{V} = \mathrm{d}V_\mathrm{p}/\mathrm{d}t$ is the effective suction capacity at the exit of the vacuum chamber.

Example

For a suction capacity $S_\mathrm{V} = 10^3\,\mathrm{l/s}$, a leak rate of $10^{-4}\,\mathrm{hPa \cdot l/s}$ and a desorbing gas rate of $10^{-3}\,\mathrm{hPa \cdot l/s}$ a final pressure of $p_\mathrm{f} = 1.1 \cdot 10^{-6}\,\mathrm{hPa}$ can be reached. After heating the walls the desorbing rate sinks below the leak rate and a final pressure of $p_\mathrm{f} = 10^{-7}\,\mathrm{hPa}$ can be achieved. ◀

9.2 Generation of Vacuum

In order to remove gas particles out of the vacuum chamber vacuum pumps are used. The different types can be divided into three classes (Tab. 9.2):

- Mechanical pumps,
- Diffusion pumps (fluid acceleration vacuum pump),
- Cryo pumps and sorption pumps.

We will briefly discuss these three classes. In Fig. 9.4 the pressure ranges are compiled where the different pumps can be used.

Table 9.2 Classification of the most important types of vacuum pumps

Mechanical pumps	Fuel acceleration pumps	Condensation pumps, sorption pumps
Rotary vane pumps	Liquid jet pumps	Cool traps
Roots pumps	Vapor jet pumps	Kryo pumps, sorption pumps
Turbopumps	Diffusion pumps	ion getterpumps

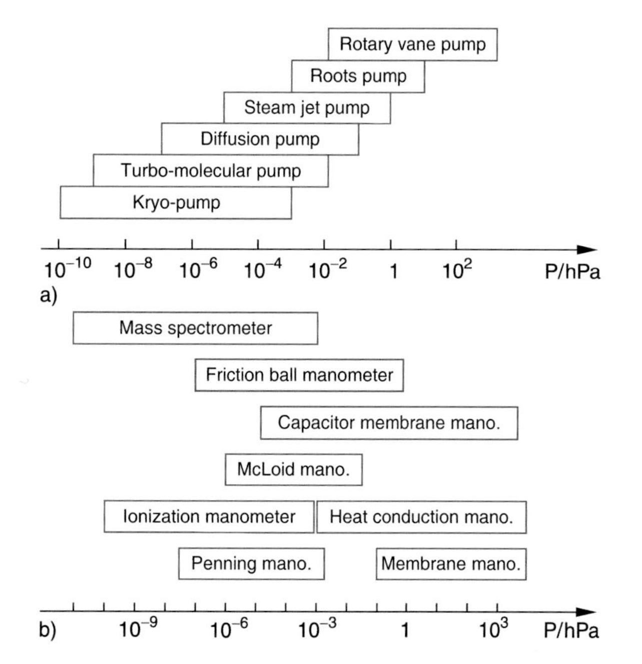

Figure 9.4 Pressure ranges **a** of the differrent types of vacuum pumps, **b** of pressure detectors

9.2.1 Mechanical Pumps

Already around 1600 Galileo Galilei has produced a low vacuum in a container by using a movable piston. More detailed experiments were performed in 1643 by Evangelista Torricelli, who was Galileo' successor in Florence. In honour of Torricelli the unit of pressure has been named torr (1 torr is the pressure of 1 mm mercury column = 133.3 Pa). The unit torr has been used for several centuries before the SI unit 1 Pa was introduced.

Spectacular experiments with evacuated spheres were performed 1645 by Otto von Guericke, the major of the German city Magdeburg. He put two hemi-spheres together, sealed them up with leather gaskets and evacuated the interior. This pressed the two hemi-spheres tightly together. In order to demonstrate the force on the hemi-spheres due to the external pressure he roped 8 horses in on each side who tried unsuccessfully to separate the hemi-spheres. The large auditorium was very much astonished that 16 horses could not separate the hemi-spheres although they could be readily separated after the evacuated sphere was filled again with air at the external pressure. Superstitious people believed in a ghost inside the sphere. An engraving of Caspar Schott illustrates this spectacular experiment (Fig. 9.5). At that time the evacuation with piston air pumps (Fig. 9.6) was tedious, because the seals were imperfect.

Figure 9.5 The demonstration experiment by Otto von Guericke. Engraving by Caspar Schott

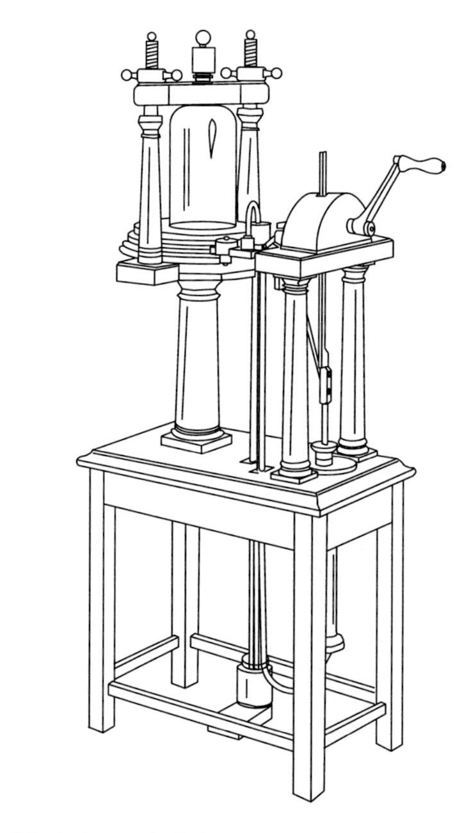

Figure 9.6 Ancient mechanical vacuum pump

Nowadays the mechanical pumps are mainly rotary vane pumps, Roots pumps or turbo-molecular pumps, which are all driven by electro motors.

9.2.1.1 Rotary Vane Pumps

The basic principle of rotary vane pumps is schematically illustrated in Fig. 9.7. An eccentrically mounted rotor R_1 rotates in a cylindrical bore with an inlet S_1 from the vacuum chamber and an outlet A_1 into the open air at atmospheric pressure. The rotor has a slit in which two sliders are pushed by a coil spring against the wall of the bore. When the rotor rotates in the direction of the arrow the right side of the sliders sucks the gas from S_1 and drives it during half of a rotation period towards A_1. This is repeated every half turn thus continuously evacuating the vacuum chamber behind S_1.

For single-stage pumps the outlet A_1 is connected to the open air (or an exhaust gas line) and the pressure p in A_1 equals the atmospheric pressure. Due to the pressure difference between A_1 and S_1 always some gas can flow back from A_1 to S_1 because the slider in the rotor does not completely seal the connection between A_1 and S_1. This limits the attainable final pressure in S_1. In order to keep the leak rate as small as possible the pump is filled with oil which forms a film between slider and wall and not only gives a better seal but also acts as lubricant that prevents jamming of the rotor. With such single-stage pumps final pressures of 10^{-1}–10^{-2} hPa are reached.

In order to obtain lower final pressures the outlet A_1 can be connected to a second pump (Fig. 9.8), which produces in A_1 already a pressure of 10^{-1} hPa, thus reducing the back-streaming considerably. This leads to a final pressure of permanent gases in S_1 of about 10^{-3} to 10^{-4} hPa. However, now the saturation vapour pressure of the pump oil ($p_s \approx 10^{-3}$ hPa at $T = 350$ K) is the limiting factor for the final pressure. Using a cool trap between S_1 and the vacuum chamber can reduce the saturation pressure and realizes an oil-free vacuum in the chamber.

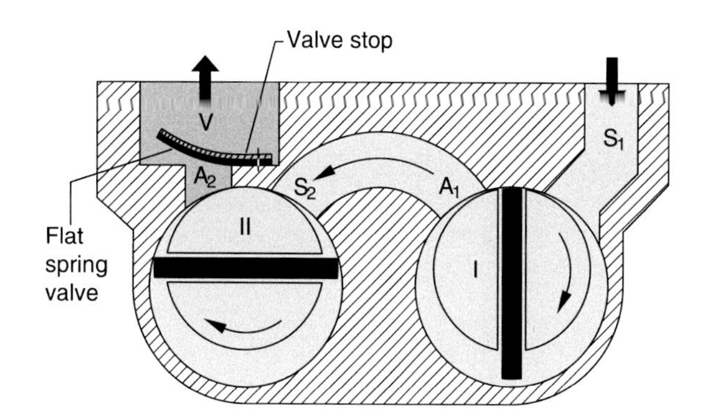

Figure 9.8 Two stage rotary vane pump [9.1]. With kind permission of Leybold GmbH

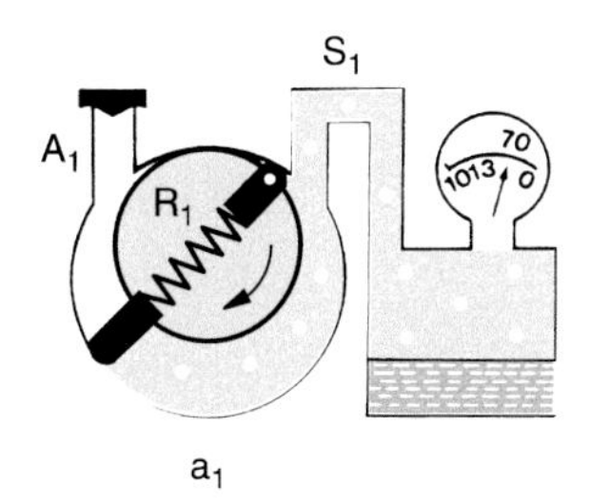

Figure 9.7 Principle operation of a rotating vane pump [9.1]. With kind permission of Leybold GmbH

Typical pumping speeds of such rotary van pumps reach from 1 m^3/h for small pumps to 60 m^3/h for larger ones. To prevent back streaming of atmospheric pressure from A to S in case of an accidental standstill of the pump a blocking valve V is built in at A_2.

9.2.1.2 Roots-Pump

The principle of a roots pump is shown in Fig. 9.9. Two symmetrically shaped rotors R_1 and R_2 rotate with opposite directions about two axes. They are arranged in such a way that their surfaces nearly touch each other. The gap width between the two rotors and between the rotors and the wall are only a few tenth of a millimetre. For the momentary situation shown in Fig. 9.9 the gas volume V_1 enclosed by the left rotor R_2 is compressed and pushed to the outlet A when the rotor rotates counterclockwise. A quarter of a full turn later the oppositely turning rotor R_1 pushes gas from S to a similar enclosed volume on the right side and presses it to A. Since the rotors do not touch, there is no

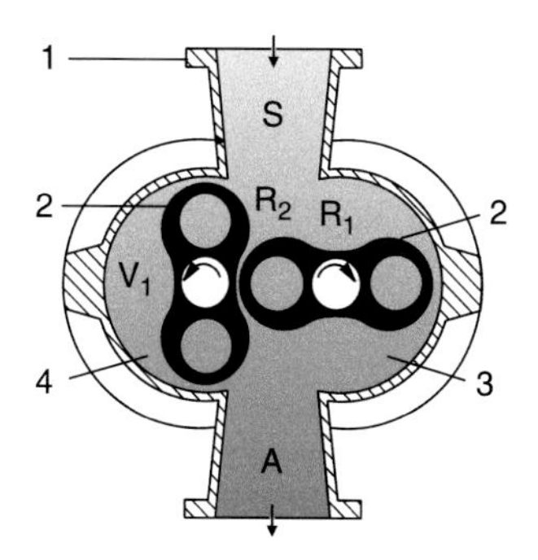

Figure 9.9 Principle operation of roots pump [9.1]. With kind permission of Leybold GmbH

Chapter 9

material abrasion and roots pumps can rotate with high angular velocities, thus increasing the pumping speed. The disadvantage of the gap between the rotors is the backstream of gas from A to S. With decreasing pressure in A the flow resistance of the gaps with width d becomes larger as soon as $\Lambda \gg d$. Therefore the pressure in A should be lowered by a one-stage rotary vane pump. Roots pumps need a forepump. Large roots pumps reach pumping speeds of up to $10^5\,\mathrm{m}^3/\mathrm{h}$.

9.2.1.3 Turbo-Molecular Pumps

The turbo-molecular pump, developed 1958 by W. Becker [9.5] is based on the principle that molecules hitting a fast moving surface, gain momentum in the direction of the surface motion (Fig. 9.10).

The turbo pump consists of a staple of fast rotating rotors with many blades (Fig. 9.11). Assume a gas molecule M with the thermal velocity $\boldsymbol{v}$ impinges on a blade of the rotor which has the same temperature T as the gas. For a resting rotor the molecule M would desorb from the blade after a short time with the velocity $\boldsymbol{v}'$ which has about the same magnitude as v, ($|\boldsymbol{v}| \approx |\boldsymbol{v}'|$) while its directions are distributed around the surface normal.

If the rotor blade moves with the velocity u the total velocity of the desorbing molecules is the vector sum $\boldsymbol{v}^* = \boldsymbol{v}' + \boldsymbol{u}$. Due to the direction of $\boldsymbol{u}$ to the left in Fig. 9.10, the number of impinging molecules from the left half space is larger than that of molecules from the right one. Because of the inclined blades the velocities v' are preferentially directed into the downward direction. The rotating blades therefore transport molecules from the upper space (inlet of the pump) to the lower space (outlet of the pump).

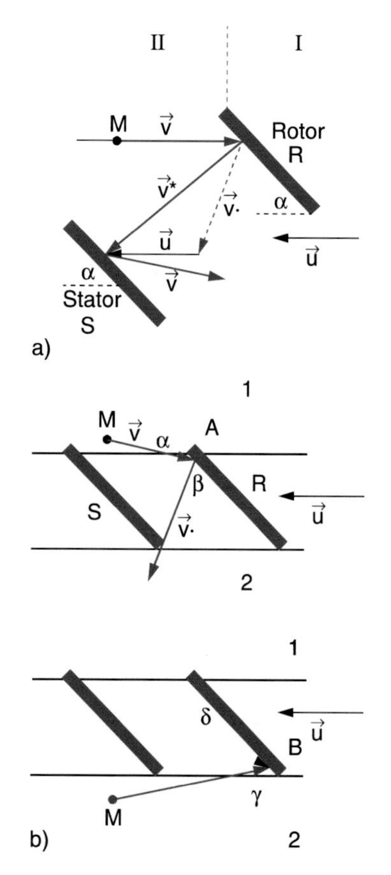

Figure 9.10 **a** Momentum transfer at the reflection of molecules M at a fast moving surface; **b** basic principle of turbo pump

Figure 9.11 Rotor of a turbo pump. With kind permission of CIT Vacuum Technique

This is illustrated in Fig. 9.10b. If the rotor blade moves with the velocity $\boldsymbol{u}$ to the left, molecules moving with the velocity $v' < u$ from the upper space 1 can hit the blade only on the left side.

9.2.2 Diffusion Pumps

Diffusion pumps are used for the generation of high- and ultrahigh-vacuum. Their principle is shown in Fig. 9.12. A pumping fluid 2 (oil or mercury) is evaporated by the heater at the bottom of the pump. The vapour rises in the inner part of the pump and leaves it at the upper end through nozzles where it gains supersonic speed forming fast vapour jets, which are directed downwards. Molecules from the vacuum chamber diffuse into the vapour jets and are pushed downwards by collisions

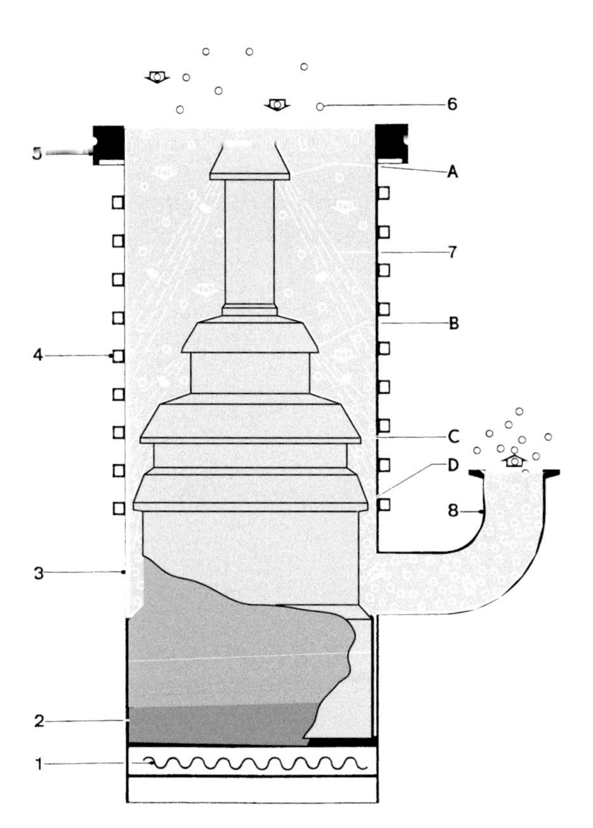

Figure 9.12 Operation principle of a diffusion pump. *1* = Heating, *2* = boiling region, *3* = pump body, *4* = water cooling, *5* = high vacuum side, *6* = particles from high vacuum side, *7* = vapor jet, *8* = pre vacuum tube, *A–D* = nozzles for vapor jets. With kind permission of Leybold GmbH [9.1]

with the vapour molecules. Since the vapour jets are initially free of gas, the diffusion rate into the jet is higher than out of the jet.

After diffusion into the vapor jet the gas molecules experience by collisions with the jet molecules an additional momentum downwards into the direction of the jet. They come into lower regions where they diffuse into the lower jets where they experience more collisions and are transported farther downwards. Finally they reach the outlet of the diffusion pump where they are pumped away by a mechanical pump. A pressure ratio $p_i/p_0 \approx 10^{-7}$ between the pressure p_i at the input and p_0 at the outlet of the diffusion pump can be reached. When the forepump maintains a pressure $p_0 = 10^{-2}$ hPa a pressure as low as 10^{-9} hPa can be realized at the high vacuum side.

The hot vapor jets hit the cooled wall of the pump where they condense and flow as liquid film down to the heater. Here they are again vaporized. In order to form oil vapor jets the free mean path Λ must be sufficiently large, i.e. the pressure sufficiently low. Diffusion pumps therefore can operate only at pressures below 10^{-2}–10^{-3} hPa. They do need a forepump, which generates the necessary minimum starting pressure.

The total pressure at the high vacuum side of the diffusion pump is the sum of all partial pressures, including the saturation pressure of the pump fuel. For mercury as fuel the saturation pressure is 10^{-3} hPa at room temperature. For mercury pumps

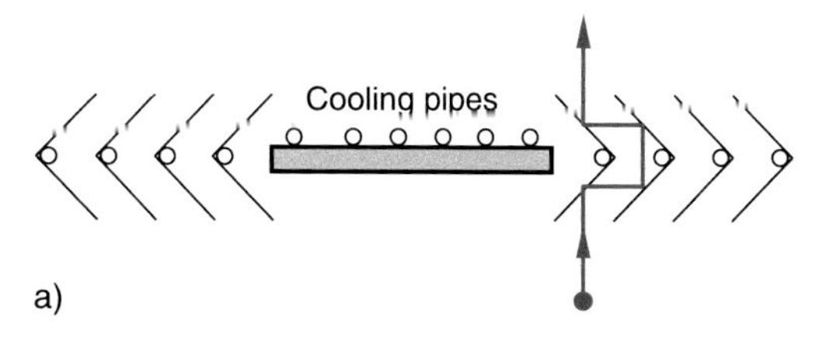

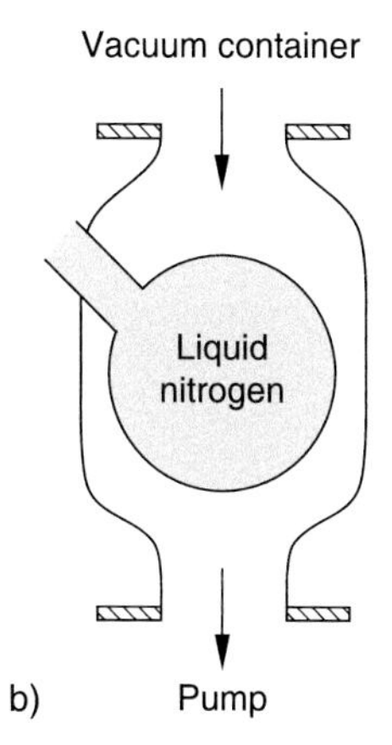

Figure 9.13 a Cooled baffle for reduction of oil return flow. b Liquid nitrogen condensation trap

therefore liquid nitrogen cool traps above the diffusion pump are necessary in order to obtain a better high vacuum. Oil-diffusion pumps operate with special oils that have saturation pressures below 10^{-7} hPa. Therefore nowadays mainly oil diffusion pumps are used.

In order to prevent oil molecules from reaching the vacuum container a cooled baffle is mounted above the pump (Fig. 9.13a) which blocks the direct way of the molecules. Another solution is a liquid nitrogen trap (Fig.9.13b) where every oil molecule on its way to the vacuum container hits at least one cooled wall where the molecules are adsorbed.

The pumping speed of modern vacuum diffusion pumps ranges from 60 l/s (for a small pump with 20 cm heights) to 50 000 l/s (about 4–5 m high). Diffusion pumps are the favorite types of high vacuum pumps. In Fig. 9.14 the pumping speed of medium sized diffusion pumps as a function of the pressure on the high vacuum side is compared with the performance of a turbo pump. Important for the optimum performance of a pumping system is

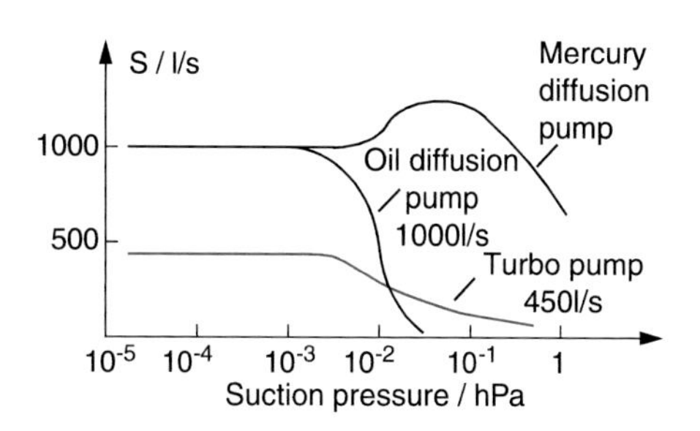

Figure 9.14 Pumping speed $S(p)$ of different types of pumps

the choice of the best forepump, which should be always able to maintain a pressure below 10^{-2} hPa on the high pressure side of the diffusion pump.

Example

A diffusion pump with a pumping speed of 2000 l/s should maintain a pressure of 10^{-5} hPa in a container into which continuously a gas streams. A gas volume of 2000 l at $p = 10^{-5}$ hPa corresponds to a volume of 2 l at a pressure of 10^{-2} hPa. Therefore the forepump must have at least a pumping speed of $2\,\text{l/s} = 7.2\,\text{m}^3/\text{h}$. Since the vacuum pipe between forepump and diffusion pump reduces the pumping speed, a forepump with a pumping speed of $12\,\text{m}^3/\text{h}$ should be used. ◀

9.2.3 Cryo- and Sorption-Pumps; Ion-Getter Pumps

A cryopump consists essentially of one or several cooled surfaces inside the vacuum container. All gases or vapors with condensation temperatures above the temperature of the surfaces condense and are adsorbed as liquids or solids on the surfaces. Liquid nitrogen cooltraps therefore can condense all gases and vapors except hydrogen and helium which need liquid helium traps. Most cryo-pumps use closed cycle lquid helium cooling systems (Fig. 9.15), which reach temperatures down to about $T = 10\,\text{K}$.

The achievable final pressure is determined by the equilibrium of the rate of molecules impinging onto the cold surface and the rate of evaporating molecules. The latter is determined by the vapor pressure of the component with the lowest evaporation temperature. The impinging molecules have a mean velocity $\overline{v} \sim \sqrt{T_\text{w}}$ which depends on the temperature T_w of the walls of the vacuum chamber, while the mean velocity of the evaporating molecules $\overline{v} \sim \sqrt{T_\text{c}}$ depends on the lower temperature T_c of the cold surface.

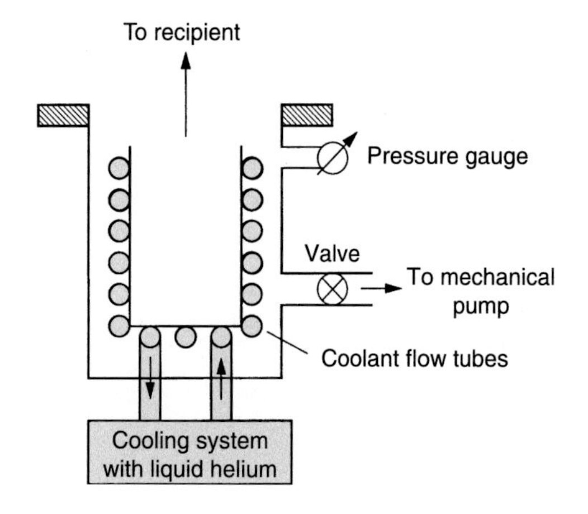

Figure 9.15 Principle of cryo pump with closed cooling cycle

The rate of molecules impinging onto the area A is

$$Z = \frac{1}{4} n \cdot \overline{v_\text{w}} \cdot A\,, \tag{9.15}$$

which equals the desorbing rate under equilibrium conditions. With $p = nkT$ and $\overline{v} \sim \sqrt{T}$ we obtain the partial pressure of the i-th vapor component in the container

$$p_\text{e}(i) = p_\text{s}(i) \cdot \sqrt{T_\text{w}/T_\text{c}}\,, \tag{9.16a}$$

where p_s is the saturation pressure. The attainable final pressure is then

$$p_\text{total} = \sum p_\text{e}(i) \cdot \sqrt{T_\text{w}/T_\text{c}}\,. \tag{9.16b}$$

Cryo-pumps need a forepump which lowers the pressure in the container down to about 10^{-3} hPa, because for $p > 10^{-3}$ hPa the mean free path Λ is smaller than the dimensions of the container and the heat conduction from the cold surfaces to the wall of the container becomes too large (see Sect. 7.5). Furthermore at low pressures the layer of condensed gases becomes too thick which lowers the heat conduction from the surface of this layer to the cooling body and increases the temperature of the surface.

The pumping speed of a cold surface A_c at a pressure p in the vacuum container is according to (9.2) and (9.15)

$$\begin{aligned} L_s &= \frac{1}{4} A_\text{c} \overline{v} \left(1 - \frac{p_\text{s}}{p} \sqrt{T_\text{w}/T_\text{c}}\right) \\ &= \frac{1}{4} A_\text{c} \overline{v} \cdot \alpha \cdot \frac{1 - p_\text{e}}{p}\,, \end{aligned} \tag{9.17}$$

where $\alpha \leq 1$ is the sticking propability of an impinging molecule on the cold surface, and $p_\text{s} = \sum p_\text{s}(i)$ is the sum of the saturation pressures of all vapor components at the temperature T_c of the cold surfaces.

Example

$\overline{v} = 400\,\text{m/s}$, $\alpha = 1$, $P_\text{e} \ll p$, $A = 1\,\text{cm}^2 \rightarrow L_\text{s} = 10\,\text{l/s}$, i.e. the cold surface has a maximum pumping speed of 10 l/s per cm^2. ◀

The growth rate $\text{d}\Delta/\text{d}t$ of the adsorbed layer with thickness $\Delta(t)$ on the cold surface depends on the density $n = N/V$ of molecules with mass m in the container and on their mean velocity $\overline{v} = (8kT/m\pi)^{(1/2)}$ at the temperature T.

According to Fig. 7.28 the number of molecules with mass m hitting per sec the area $\text{d}A$ is

$$\text{d}Z = \frac{1}{4} n \cdot \overline{v} \text{d}A\,. \tag{9.18a}$$

With $\overline{v} = \sqrt{8kT/m\pi}$ we get the mass increase of the layer per sec

$$\begin{aligned} \frac{\text{d}M}{\text{d}t} &= \text{d}Z \cdot m = \frac{1}{4} n \cdot m \cdot \overline{v} \cdot \text{d}A \\ &= \frac{1}{4} n \cdot \sqrt{8kTm/\pi}\,. \end{aligned} \tag{9.18b}$$

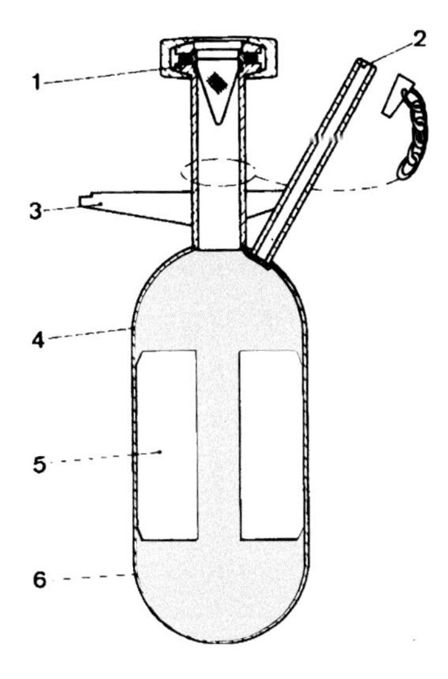

Figure 9.16 Setup of a sorption pump. *1:* inlet connector; *2:* degasing connector; *3:* mechanical support; *4:* pump body; *5:* heat conduction sheets; *6:* adsorption material [9.1]

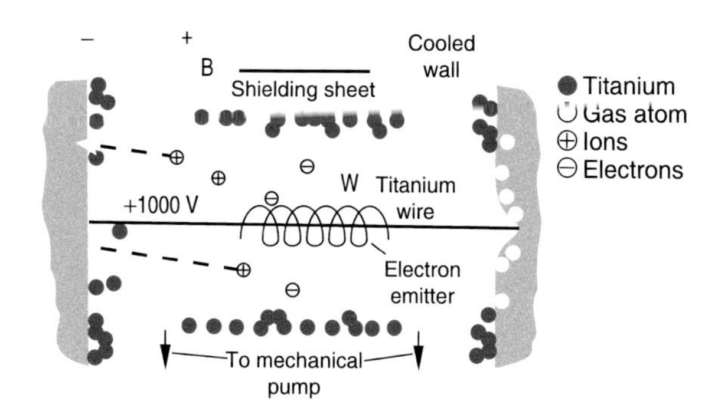

Figure 9.17 Principle of getter ion pump [9.1]

The mass M of the layer with thickness Δ and density ϱ is $M = \varrho \cdot \Delta \cdot dA$.

This gives finally the growth rate of the layer

$$\frac{d\Delta}{dt} = \frac{n}{\varrho}\sqrt{kTm/2\pi} \quad (9.18c)$$

Example

For N_2-molecules at a pressure of 10^{-5} hPa the growth rate is 5 µm/h. ◄

The adsorbed layer should not be too thick, because the heat conduction becomes worse with increasing thickness Δ and the surface temperature of the layer increases. This increases the evaporation rate of the adsorbed molecules.

The adsorbing surface can be greatly increased by using molecular sieves (zeolites = alkali-aluminum silicate)). They consist of small balls with many fine pores, into which the molecules can diffuse and are then adsorbed. The effective surface of Zeolith is about $10^3\,m^2$ per gramm. For Zeolith the diameter of the pores is about 1.3 nm. For typical sizes of 0.5 nm for molecules 1 g Zeolith can adsorb about $2.5 \cdot 10^{21}$ molecules in a monolayer. This corresponds to a gas volume of 10 000 l at a pressure of 10^{-2} hPa.

The adsorption of the molecular sieves depends strongly on the temperature. They can be therefore used at low temperatures (liquid nitrogen temperature = 78 K) as a cryopump and later on they can be degassed at higher temperatures and used again as pumps. Such a sorption pump is shown schematically in Fig. 9.16.

Another solution for high vacuum pumps are ion-getter pumps. In a gas discharge ions are produced which are accelerated onto the cathode. Here they sputter the cathode material (e.g. Titanium) which is adsorbed on cold surfaces, where already a layer of condensed gases has been formed. The titanium atoms form a film, that covers the layer of adsorbed atoms and burries it completely. A new fresh metal surface is formed where further gas molecules can be adsorbed. Since the vapor pressure of titanium is very low, even at room temperature very low pressures can be obtained.

Such ion-getter pumps (chemical getter pumps) are useful for the generation of oil-free ultrahigh vacuum ($p < 10^{-6}$ hPa). In Fig. 9.17 a possible realization is shown. A titanium wire is heated by direct electric current or by electron bombardment. The sputtered titanium atoms are ionized by collisions with electrons and are accelerated onto the cooled walls, which are kept at ground potential. Here they push the adsorbed molecules deeper into the wall and burry them under a metallic film of neutral titanium atoms. A sputter rate of 5 mg/min represents at $p = 10^{-6}$ hPa a pumping speed of 3000 l/s.

9.3 Measurement of Low Pressures

For the measurement of pressures a variety of different measuring techniques and instruments have been developed. We will present only a small selection. Table 9.3 compiles some of these devices suitable for the different pressure ranges.

Table 9.3 Pressure ranges of different pressure measuring devices

Device	Pressure range/mbar
Liquid manometer	0.1–10^3
Mechanical spring vacuum meter	1–10^3
Membrane manometer	1–10^3
Capacity manometer	10^{-4}–10^3
Heat conduction manometer	10^{-3}–1
Heat conduction manometer with control feedback	10^{-3}–100
McLeod manometer	10^{-6}–10^{-1}
Penning ionization manometer	10^{-7}–10^{-3}
Ionization manometer	10^{-12}–10^{-3}
Friction manometer	10^{-7}–10^{-1}

9.3.1 Liquid Manometers

Liquid manometers (Fig. 7.3) are simple devices for pressure measurements, that have been already used in 1643 by Torricelli. Here the height difference Δh of a liquid with density ϱ in the two legs of a U-shaped tube are measured. The pressure difference $\Delta p = p_2 - p_1$ between the two ends of the U-tube is then

$$\Delta p = \varrho \cdot g \cdot \Delta h \,. \tag{9.19}$$

Example

With oil ($\varrho = 900\,\mathrm{kg/m^3}$) a pressure difference $\Delta p = 1\,\mathrm{hPa}$ causes a height difference $\Delta h = 11.3\,\mathrm{mm}$. For mercury ($\varrho = 13{,}546\,\mathrm{kg/m^3}$) one obtains $\Delta h = 1\,\mathrm{mm}$ for $\Delta p = 1.33\,\mathrm{hPa} = 1\,\mathrm{torr}$. ◀

When one leg of the U-tube is closed and evacuated (Fig. 7.2) the volume above the liquid is filed with the vapour of the liquid with the vapour pressure $p_s(T)$, that depends on the temperature T. The height difference is then

$$\Delta h = \frac{1}{\varrho \cdot g}(p - p_s) \approx \frac{1}{\varrho g} p \quad \text{for} \quad p_s \ll p \,, \tag{9.19a}$$

which gives directly the pressure p above the open leg of the U-tube.

The accuracy and sensitivity of liquid manometers can be considerably increased with a device, developed by McLeod (Fig. 9.18), which is based on the Boyle–Mariotte Law (see Sect. 7.1). At the beginning of the measurement, the container B is lowered until the liquid level in the left leg is at h_1. The pressure p_1 above h_1 is the pressure in the vacuum chamber. Now B is lifted again until the liquid level rises above the point z. The volume $V = V_0 + V_c$ of G and the capillary above G is now separated from the vacuum chamber. The container is lifted up to a height where the liquid level in the very left tube (pressure p_1) is by Δh higher than in the capillary above G where the higher pressure p_2 is present due to the compression of the closed volume V to the much smaller volume $V_c = \pi r^2 \cdot x$. According to the Boyle–Marriot law we obtain

$$p_1 \cdot (V_0 + V_c) = p_2 \cdot \pi \cdot r^2 \cdot x \,.$$

The measured height difference of the liquid between the left tube and the capillary

$$\Delta h = \frac{p_2 - p_1}{\varrho \cdot g} = \frac{p_1}{\varrho \cdot g}\left(\frac{V_0}{\pi r^2 x} + \frac{L}{x} - 1\right) \tag{9.20}$$

yields the pressure p_1 in the vacuum chamber, after the volumes V_0, $V_c = \pi \cdot r^2 \cdot L$ and the length x of the gas-filled part of the capillary have been determined. in case of mercury one has to take into account the capillary depression to mercury (see Sect. 6.4).

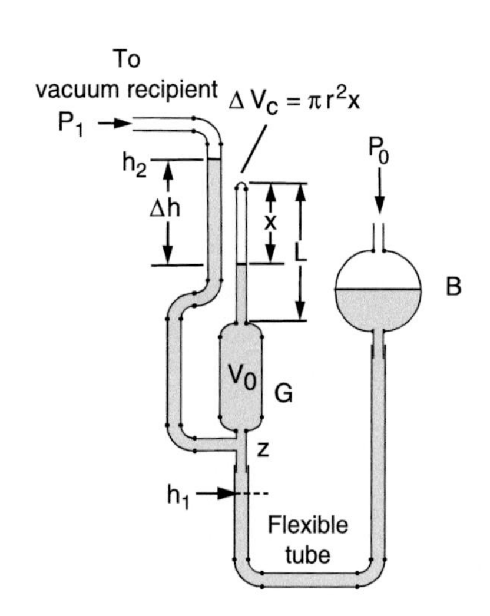

Figure 9.18 Principle of McLeod vacuum meter

9.3.2 Membrane Manometer

To measure the pressure in the low vacuum range robust and simple membrane manometers are available (Fig. 9.19b). A thin membrane separates the vacuum from the upper part at atmospheric pressure. A wire is connected at one end with the centre of the membrane and at the other end with a hand that can rotate around a fixed axis. Due to the pressure difference, the membrane sags and turns the hand by an angle that is proportional to the pressure difference, which can be read on a calibrated scale.

Another realization (Fig. 9.19a) uses a bent thin hollow tube that is connected to the vacuum chamber. The bending radius is

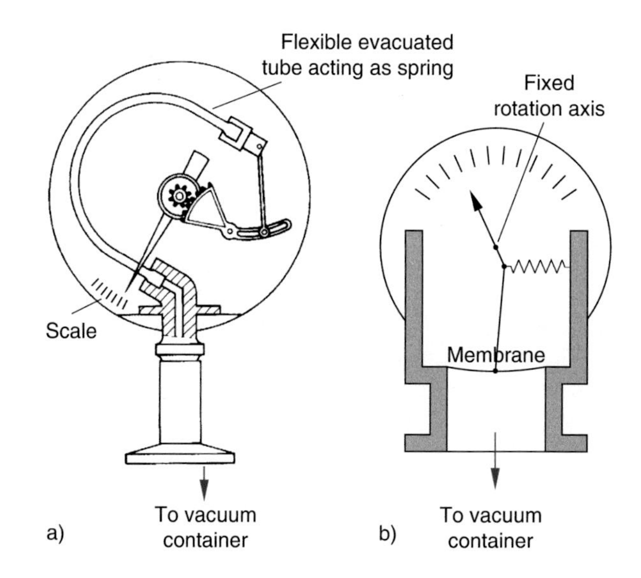

Figure 9.19 Two designs of robust and compact pressure detectors. **a** Spring pressure gauge; **b** membrane pressure gauge

Chapter 9

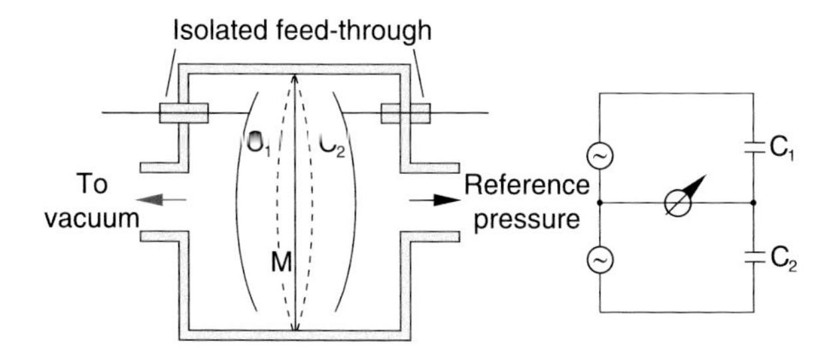

Figure 9.20 Membrane capacitor vacuum gauge. A thin membrane *M* provides together with two curved fixed plates two capacities C_1 and C_2, which are arranged in a bridge circuit (see Vol. 2, Chap. 2). They are fed by two identical ac-voltage sources

dependent on the pressure. When it changes the upper end of the tube moves a hand which indicates the pressure on a calibrated scale.

For lower pressures in the high vacuum range ($p < 10^{-5}$ hPa) capacitance membrane manometers (Fig. 9.20) can be used. Here a thin membrane which separates the vacuum chamber from a chamber with a fixed reference pressure. It forms one electrode of two electric capacitors C_1 and C_2, When the pressure in the vacuum chamber decreases the membrane bends to the left side and decreases the electrode separation of C_1 but increases that of C_2, thus increasing the capacitance of C_1 and decreases that of C_2. This changes their AC resistance in an opposite direction, which can be measured with an electric bridge arrangement where the two capacitors are charged by two identical AC voltage supplies (see Vol. 2. Chap. 1).

9.3.3 Heat Conduction Manometers

As has been shown in Sect. 7.5 the heat conduction of a gas in the pressure range where the mean free path Λ is larger than the dimensions of the vacuum chamber, is proportional to the pressure p. This fact is used in the heat conduction manometer (Fig. 9.21) for measuring pressures. A filament of length L, heated by an electric current I, is clamped between to yokes along the axis of a small cylindrical tube. Its temperature T_d is determined by the supplied electric power $I^2 \cdot R$ and the power loss due to heat conduction.

$$\frac{\mathrm{d}W}{\mathrm{d}t} = 2\pi r \cdot L \cdot \kappa (T_\mathrm{d} - T_\mathrm{w}) , \qquad (9.21)$$

which is given by the surface $2\pi r \cdot L$ of the filament, the heat conduction κ of the gas and the temperature difference $\Delta T = (T_\mathrm{d} - T_\mathrm{w})$ between filament and wall (see Sect. 7.5.3). Stationary conditions are established, when the supplied power equals the power loss. This yields

$$I^2 \cdot R = 2\pi r \cdot L \cdot \kappa \cdot \Delta T . \qquad (9.21a)$$

The coefficient of heat transfer

$$\kappa = n \cdot v \cdot k \cdot f/8 = v \cdot p \cdot f/8T , \qquad (9.21b)$$

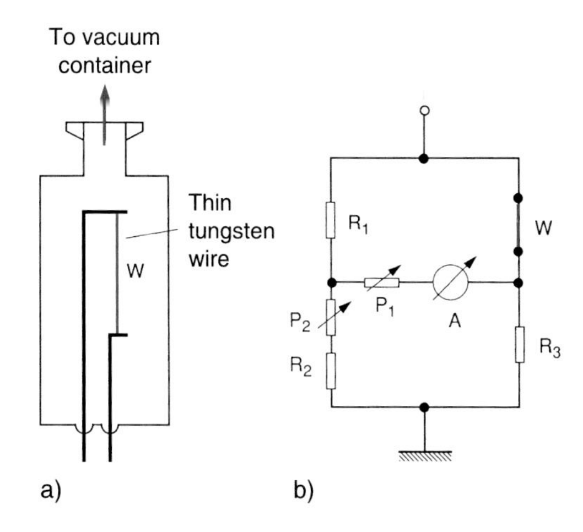

Figure 9.21 Heat conduction vacuum gauge. **a** Mechanical design; **b** electric circuit

(7.49a) is proportional to the gas density $n = p/kT$ and therefore to the pressure p and to the degrees of freedom f of the gas molecules. The electric resistance then becomes

$$R(T_\mathrm{d}) = \frac{2\pi r \cdot L \cdot p \cdot v \cdot f \cdot (T_\mathrm{d} - T_\mathrm{w})}{4(T_\mathrm{d} + T_\mathrm{w})} . \qquad (9.21c)$$

It depends on the temperature T_d. It can be measured with an electric bridge (Fig. 9.21b) (see Vol. 2, Sect. 2.4.3) and yields the wanted pressure measurement.

Since the heat conduction in the low vacuum range ($\Lambda \ll d$) is independent of the pressure (see Tab. 9.1) heat conduction manometers can be used only in the medium vacuum range ($1 - 10^{-3}$ hPa), for instance between diffusion pump and backing pump. For pressures below 10^{-3} hPa the heat conduction through the gas becomes smaller than other heat leaks (for example through the yokes of the filament). Therefore the accuracy of pressure measurements decreases strongly below $p = 10^{-3}$ hPa.

9.3.4 Ionization Gauge and Penning Vacuum Meter

The vacuum meters that are used most often in the high vacuum range are the ionization gauge (Fig. 9.22) and the Penning vacuum meter (Fig. 9.23). The ionization gauge consists of a heated filament as cathode K emitting electrons that are accelerated onto the anode *A*. On their way from *K* to *A*, they collide with gas molecules and ionize them (see Vol. 3). When the free mean path of the electrons is larger than the distance *K*–*A* the number N_ion of produced ions is proportional to the density n of the gas molecules in the manometer and therefore also to the pressure $p = n \cdot kT$. It is

$$N_\mathrm{ion} = N_\mathrm{el} \cdot \sum_i n_i \cdot \alpha_i(E_\mathrm{el}) , \qquad (9.22)$$

Chapter 9

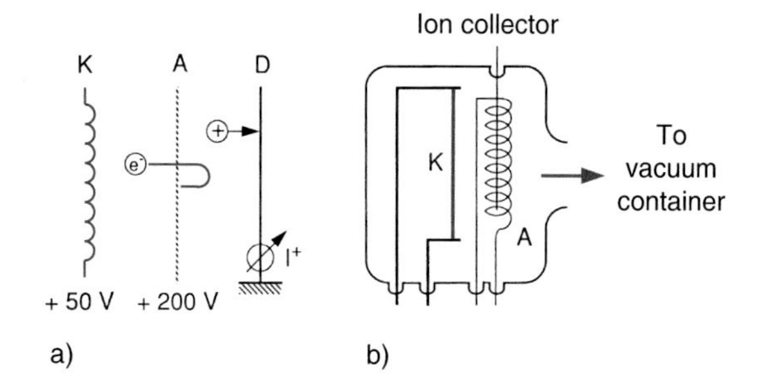

Figure 9.22 Ionization vacuum meter. **a** Schematic principle; **b** design of Bayard–Alpert tube

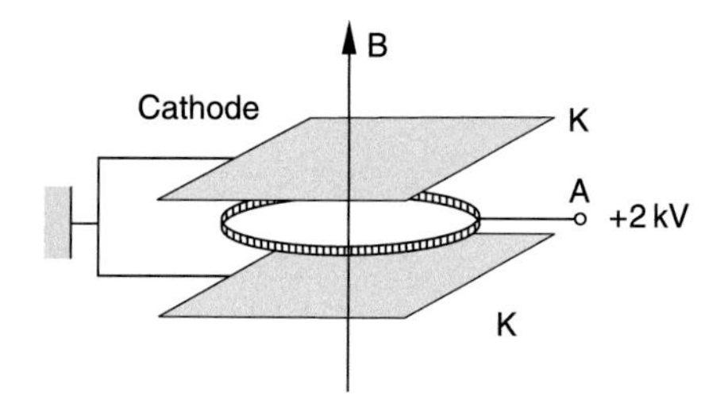

Figure 9.23 Penning vacuum gauge

where n_i is the partial density of molecules of type i and $\alpha_i(E_{el})$ the ionization probability, which depends on the energy E_{el} of the electrons. The positive ions are collected on a wire D at a negative potential against the anode.

The minimum of still detectable pressure is limited by several factors: Firstly, the ion current decreases with decreasing pressure, which demands good current amplifiers. Secondly, the electrons impinging onto the anode generate X-rays that can release electrons from the ion collector. Their number is independent of the pressure and form an underground current that overlaps the wanted signal current.

Typical pressure ranges where the ionization gauge can be used are 10^{-3} hPa $\geq p \geq 10^{-12}$ hPa, where for the lower pressures special designs have been developed which minimize the underground current (Bayard–Alpert tube Fig. 9.22b).

Instead of the thermionic emission of electrons from a heated filament at higher voltages ($\approx$ 1000 V) also a cold electron emission between two metal plates can be realized. Since the ionization probability is small at such high electron energies, the ionization path must be enlarged. This is achieved by a permanent magnet with a magnetic field B that forces the electrons on spiral paths until they reach the anode (Fig. 9.23).

These Penning manometers are robust but not as accurate as the ionization gauges. They can be used in the vacuum range from 10^{-3} to 10^{-7} hPa.

9.3.5 Rotating Ball Vacuum Gauge

The principle of this vacuum gauge is based on the deceleration of a rotating ball due to friction with the rest gas molecules. A small steel hollow sphere is contact-free hold in its position by a magnetic field (Fig. 9.24). A rotating magnetic field produced by special coils is superimposed onto the static magnetic field. It causes the ball to rotate with an angular velocity of about $\omega = 2\pi \cdot 400\,\mathrm{s}^{-1}$. After shut off the rotating field, the ball rotates freely and is only decelerated by friction due to collisions with the gas molecules. The slowing down time depends on the rate of collisions and therefore on the gas pressure.

The angular momentum of the rotating ball is

$$L = I\omega = \frac{2}{5}MR^2\omega = \frac{8}{15}\pi\varrho R^5\omega\,. \tag{9.23}$$

The retarding collisions produce a mean torque

$$D = \frac{\mathrm{d}L}{\mathrm{d}t} = I \cdot \dot{\omega}\,, \tag{9.24}$$

onto the ball which is proportional to the gas pressure. The decrease rate of the angular velocity ω is then

$$\frac{\mathrm{d}\omega}{\mathrm{d}t} = \frac{D}{I} = a \cdot \omega \cdot p\,. \tag{9.25}$$

The proportionality factor a depends on the radius R of the ball, on the density ϱ and on the mean molecular velocity v. After calibrating the system the factor a can be measured. The pressure p, which is proportional to the density ϱ can then be determined from the relative deceleration $\mathrm{d}\omega/\mathrm{d}t/\omega$. The accuracy of the measurement is about $\Delta p/p = 3\%$. Therefore the rotating ball gauge is the most accurate vacuum meter in the vacuum range 0.1–10^{-7} hPa [9.6].

For more detailed and recent information on modern techniques of vacuum physics, the reader is referred to the literature [9.7].

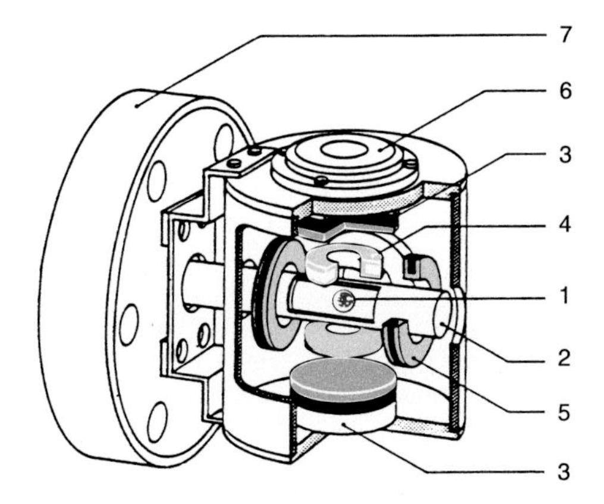

Figure 9.24 Section through the gauge head of a friction vacuum gauge. *1* = Steel ball, *2* = gauge tube with one open end, which is welded to the flange *7*, *3* = permanent magnet, *4* = stabilization coils, *5* = four driving coils, *6* = horizontal position detector. With kind permission of Leybold GmbH

Summary

- A volume V is called evacuated if the total gas pressure in V is small compared to the atmospheric pressure.
- The different vacuum ranges are:
  Low vacuum $(1\,\text{hPa} \leq p \leq 10^3\,\text{hPa})$
  Medium vacuum $(10^{-3}\,\text{hPa} \leq p \leq 1\,\text{hPa})$
  High vacuum $(10^{-7}\,\text{hPa} \leq p \leq 10^{-3}\,\text{hPa})$
  Ultrahigh vacuum $(p \leq 10^{-7}\,\text{hPa})$
- Vacuum is generated with vacuum pumps. The most important types are mechanical pumps (rotary vane pumps) and roots-pumps, (which are used as fore pumps for the generation of fine vacuum), turbo-molecular pumps for the generation of oil-free high- and ultrahigh vacuum, oil- and mercury diffusion pumps, cryo-pumps and ion getter pumps for the generation of ultrahigh vacuum.
- The gas pressure in a vacuum chamber can be measured with one of the following devices:
  liquid barometer $(0.1\,\text{hPa} \leq p \leq 10^3\,\text{hPa})$
  membrane manometer $(p \geq 1\,\text{hPa})$
  heat conduction manometer $(p \geq 10^{-3}\,\text{hPa})$
  capacitance manometer $(p \geq 10^{-5}\,\text{hPa})$
  friction vacuum manometer $(p \geq 10^{-7}\,\text{hPa})$
  ionization gauge $(p \geq 10^{-12}\,\text{hPa})$
- The suction capability $S_\text{V} = \text{d}V/\text{d}t$ is the gas volume flow through the suction connection of a pump. Often the product $S_\text{L} = p \cdot S_\text{V}$ of pressure and suction capability is called the pumping speed.
- The vacuum lines (tubes and pump connectors) between vacuum chamber and pump reduce the total suction capability. Their flow conductance $L_\text{S} = p \cdot \text{d}V/(p_2 - p_1)$ should be as high as possible, in order to make the pressure difference between entrance and exit of the vacuum line small.
- The achievable final pressure in the vacuum chamber is determined by the pumping speed of the pump, by the leak rate and the desorption rate of molecules from the inner walls of the chamber.

Problems

9.1 A vacuum chamber is connected to the outside at atmospheric pressure through a capillary tube with length $L = 10\,\text{cm}$ and 0.5 mm inner diameter. What should be the effective suction capacity of the vacuum pump in order to maintain a pressure of $10^{-3}\,\text{hPa}$?

9.2 Which force was necessary to separate the two hemispheres of Guericke's demonstration experiment, when the diameter of the spheres was 60 cm and the inner pressure 100 hPa?

9.3 In a cubic vacuum chamber with a volume $V = 0.4\,\text{m}^3$ a pressure of $p = 10^{-5}\,\text{hPa}$ is maintained. What are the particle density n, the mean free path Λ and the mean time τ between two successive collisions between particles at room temperature? How large is the ratio Z_1/Z_2 of the rate Z_1 for mutual collision between particles to the rate Z_2 for collision of particles with the walls? How large is the total mean path length that a particle traverses within 1 s, and what is the sum of the path lengths of all particle in the chamber?

9.4 Assume, the vacuum chamber of Probl. 9.3 is operated under ultrahigh vacuum and the inner walls are free from all adsorbed molecules. At $t = 0$ oxygen is let in until the pressure rises to $10^{-7}\,\text{hPa}$. How long does it take until the walls are covered by a monomolecular layer, if each oxygen molecule covers an area of $0.15 \times 0.2\,\text{nm}^2$ and its sticking probability is 1?

9.5 A vacuum chamber should be evacuated down to a pressure of $10^{-6}\,\text{hPa}$ using a diffusion pump with the effective pumping speed of 3000 l/s. What is the minimum effective pumping speed of the mechanical fore pump in order to maintain a vacuum of 0.1 hPa at the outlet of the diffusion pump?

9.6 The ionization cross section of nitrogen molecules N_2 for collisions with electrons of 100 eV energy is $\sigma = 1 \cdot 10^{-18}\,\text{cm}^2$. How large is for an electron current of 10 mA the ion current at a pressure of $10^{-7}\,\text{hPa}$ in the ionization gauge when the path length of the electrons is 2 cm?

9.7 Through the heated filament of a thermal conductivity gauge flows the electric current $I = U/R(T)$ at a constant voltage U. The heating power under vacuum conditions is $P_\text{el} = U^2/R_0$. What is the dissipation power due to heat conduction in a cylindrical chamber with diameter of 2 cm at a gas pressure of $p = 10^{-2}\,\text{hPa}$ when the temperature of the filament is $T_1 = 450\,\text{K}$ and that of the wall is $T_2 = 300\,\text{K}$? (The length of the filament is 5 cm, its diameter 0.5 mm, the distance filament-wall is 1 cm). Which fraction of the electric energy $E_\text{el} = U \cdot I$ is dissipated by heat conduction if $U = 0.5\,\text{V}$ and $I = 2\,\text{A}$?

9.8 The total angular momentum transfer onto a ball at rest in a gas at thermal equilibrium is zero. Why is the rotating ball in a Langmuir friction gauge slowed down? Estimate the torque that the gas molecules transfer to a ball with a radius of 1 cm rotating with the angular velocity $\omega = 2\pi \cdot 400\,\mathrm{s}^{-1}$ at a temperature of $T = 300\,\mathrm{K}$ and a pressure of $p = 10^{-3}\,\mathrm{hPa}$. How long does it take until ω has decreased by 1%?

References

9.1. D.M. Hoffman, B. Singh, J.H. Thomas, *Handbook of Vacuum Science and Technology.* (Academic Press, 1998)
9.2. N.S. Harris, *Modern Vacuum Practice,* reprint 1997 (MacGraw Huill Publ., 1989)
9.3. J.F. Lafferty, *Foundations of Vacuum Science and Technology.* (John Wiley and Sons, 1998)
9.4. Ph. Dnielson, *A Users Guide to Vacuum Technology.* (John Wiley and Sons, 1989)
9.5. https://en.wikipedia.org/wiki/Turbomolecular_pump
9.6. J.F. O'Hanion, *A User's Guide to Vacuum Technology.* (John Wiley & Sons, 2005), p. 385
9.7. K. Jouston (ed.), *Handbook of Vacuum Technology.* (Wiley VCH, Weinheim, 2008)
9.8. J.F. O'Hanion, *A User's Guide to Vacuum Technology.* 3rd ed. (John Wiley and Sons, 2003)
9.9. N. Yoshimura, *Vacuum Technology.* (Springer, 2014)

10 Thermodynamics

© Springer International Publishing Switzerland 2017
W. Demtröder, *Mechanics and Thermodynamics*, Undergraduate Lecture Notes in Physics, DOI 10.1007/978-3-319-27877-3_10

The insight, that heat is just one of several forms of energy and can be explained by a mechanical model, is today common knowledge, but it is only about 170 years old. The physician *Julius Robert Mayer* (1814–1878) formulated in 1842 his ideas about the energy conservation for the conversion of mechanical energy into heat, and he could already give a numerical value for the thermal energy equivalent (Sect. 10.1.5). However, only after the development of the kinetic gas theory (see Sect. 7.3) the microscopic explanation of heat of a macroscopic body as the total energy (kinetic plus potential energy) of all molecules of the body was possible. As has been explained in Sect. 7.3, a measure of the mean kinetic energy of all particles with mass m in a gas volume, which have three degrees of freedom for their motion is the absolute temperature

$$T = \frac{1}{k} \cdot \frac{2}{3} \cdot \frac{m}{2} \cdot \overline{v^2} . \tag{10.1}$$

With this definition of the temperature all macroscopic phenomena and the general laws derived from them (Boyle.-Marriott, general gas law) could be reduced to microscopic models describing matter as composed of atoms and molecules.

In this chapter, we will discuss in more detail the measurement of temperature, the definition of temperature scales, the experimental findings of energy transport and conversion, of material changes with temperature such as thermal expansion and phase transitions. An important subject is the formulation of basic laws of thermodynamics which can be regarded as a summary of many experimental results. We try to explain all macroscopic phenomena as far as possible by microscopic models, where, however, some explanations need a deeper knowledge of atomic physics, which will be imparted in volume 3 of this textbook series.

At the end of this chapter a short excursion to the thermodynamics of real gases and liquids is presented, which might be helpful for the explanation of many phenomena observed in nature.

10.1 Temperature and Amount of Heat

The definition of the absolute temperature, given in (10.1), is for most practical applications of temperature measurements not very helpful. One has to use measuring techniques that are reliable, accurate and easy to handle.

Qualitative information about the temperature can be already obtained with the heat sensibility of our body. Our skin has sensors that inform us whether a body is cold or hot. This sensing is, however, not very accurate and depends on the previous experience, as the following experiment illustrates: Three containers with (1) hot water, (2) lukewarm water and (3) cold water are placed side by side. Dipping a finger at first into (1) and then into (2) the lukewarm water seem to be cold, but dipping at first into (3) and then into (2) the same lukewarm water seems to be warm.

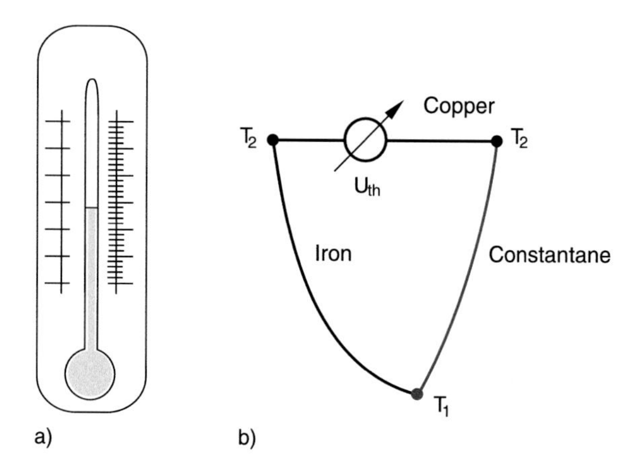

Figure 10.1 **a** Liquid thermometer; **b** thermo couple

This demonstrates that the sensing of our body is not reliable and cannot be used for quantitative measurements. In order to measure temperatures and the amount of heat, measuring instruments and techniques have to be developed that are for practical applications sufficiently easy to handle and which give reliable and reproducible results.

10.1.1 Temperature Measurements, Thermometer, and Temperature Scales

For the measurement of temperatures in principal all physical effects can be used that depend on the temperature. These are for instance:

- The geometrical dimensions of solid, liquid or gaseous bodies, which generally increase with the temperature. Metal wires become longer, liquid or gas volumes expand with increasing temperature at constant pressure.
- The electrical resistance of a body changes with the temperature T. For metals it increases with T for semiconductors it decreases (see Vol. 2, Chap. 2)
- The electric contact potential difference between two different metals in contact changes with temperature (Thermovoltage see Vol. 2, Sect. 2.9)
- The radiation power emitted by a hot body increases with T^4 and can be used for the measurement of the temperature of remote bodies such as stars (Radiation Pyrometer see Vol. 2 Chap. 12).

Devices for the measurement of temperatures are called **thermometers** (Tab. 10.1). For daily practice the expansion of liquids are generally used (liquid thermometer, Fig. 10.1a) or the change of the contact voltage (thermo-elements, Fig. 10.1b).

For the quantitative specification of a temperature, numerical values for fixed temperatures have to be defined that can be accurately reproduced under readily realizable external conditions (temperature fix points). Furthermore, a temperature scale has to be defined. This has been historically realized in different ways.

Table 10.1 The mostly used thermometers

Thermometer type	Temperature range / °C	Measuring principle	Error limits
Liquid thermometers:			
Mercury	−38 to +800	Thermal expansion of liquid in glas capillary	depending on scale division 0.1–1 °C
Alcohol	−110 to +210		
Pentane mixture	−200 to ≈+30		
Solid state thermometers:			
Metal rod	−150 to +1000 dependent on specific metal	Thermal expansion of metals	1–2% of scale range
Bimetal	−150 to +500	Length expansion difference	Dependent on model
Resistance thermometers:			
Metal wire	−250 to +1000	Temperature dependence of electric resistance	0.1–1 °C
Semiconductor	−273 to +400		
Thermo couple:			
Fe-CuNi (iron-constantan)	−200 to +760	Temperature dependence of thermovoltage	0.1–1 °C
Ni-CrNi	−270 to +1000		
Ni-CrNi	−200 to +1370		
Pt-PtRh	−50 to +1700		
W-WMo	−200 to +3000		
Pyrometer	+800 to +3000	Heat radiation	2–10 °C

10.1.1.1 The Celsius Scale

The astronomer *Anders Celsius* (1701–1744) proposed 1742 to use the expansion of a mercury column for the measurement of temperatures (mercury thermometer). Two fix points were defined for the temperature scale: The melting point of ice ($T_C = 0\,°C$) and the boiling point of water ($T_C = 100\,°C$ at a pressure of 1 atm = 1013.25 hPa). The range between these two fix points is divided into 100 equal units, where each unit corresponds to 1 °C.

Note: We will label the Celsius temperature with T_C in order to distinguish it from the Fahrenheit temperature (T_F) and Kelvin temperature T.

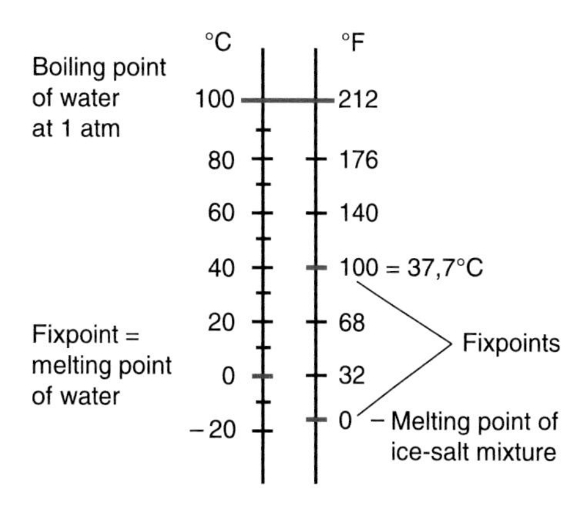

Figure 10.2 Comparison of Celsius and Fahrenheit scale. For the Kelvin scale there is only one fixpoint (triple point of water at 0.01 °C)

10.1.1.2 The Fahrenheit Scale

The Fahrenheit scale, which was proposed by Daniel Gabriel Fahrenheit (1686–1736) is still used in the USA. It defines as fix points the melting point of a defined ice-water-ammonia chloride solution at $T_F = 0\,°F$ ($-17.8\,°C$) and the normal body temperature at $T_F = 100\,°F$ ($+37.7\,°C$). The range between the two fix points is equally divided into 100 units where 1 unit corresponds to 1 °F. From this definition it follows that $0\,°C = 32\,°F$ and $100\,°C = 212\,°F$. The conversion between the two scales is as follows (Fig. 10.2):

$$\begin{aligned} T_C/°C &= \tfrac{5}{9}\,(T_F/°F - 32) \\ T_F/°F &= \tfrac{9}{5}\,(T_C/°C + 32)\ . \end{aligned} \tag{10.2}$$

10.1.1.3 The Absolute Temperature Scale

The absolute temperature scale needs only one fix point, which is the triple point of water (see below). It is measured with gas thermometers.

Its definition is: The Kelvin is the unit of the thermodynamic temperature scale. 1 K is the 273.16th part of the temperature T_p of the triple point of water. The zero point of the Kelvin scale is the lower absolute limit of possible temperatures and is defined by general laws of thermodynamics (see Sect. 10.3).

10.1.1.4 Accuracy of Thermometers

The temperature scale of liquid thermometers depends on the choice of the liquid and also of the glass of the thermometer capillary, because not only the liquid but also the glass expands with rising temperature. The thermal expansion of liquids and solids is generally not constant over the temperature range measured by thermometers and is not necessarily linear (see next Section). For mercury, the deviation from linearity is small. The comparison with an alcohol thermometer shows that its scale differs from that for the mercury thermometer and is not equidistant (Fig. 10.3).

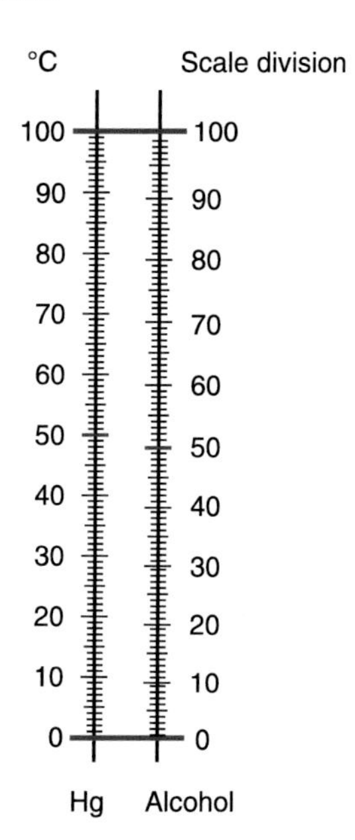

Figure 10.3 Demonstration of non-uniform expansion of liquids by comparing mercury and alcohol themometers

If higher accuracy is demanded, other thermometers have to be found which have a better linear temperature scale. A possible solution is the thermal expansion of gas volumes at constant pressure or the pressure increase at constant volume that are utilized in the gas thermometer (see Sect. 10.1.3). They are used for the definition of the absolute temperature scale (Kelvin scale, Sect. 10.1.4).

10.1.2 Thermal Expansion of Liquids and Solids

The length L of a rod changes with temperature. Experiments show that the relative length change $\Delta L/L$ within a restricted temperature range is approximately proportional to the temperature change ΔT:

$$L(T_C) = L(0) \cdot (1 + \alpha T_C) . \tag{10.3}$$

The expansion coefficient

$$\alpha = (\mathrm{d}L/\mathrm{d}T)/L \tag{10.4}$$

gives the relative length change for a temperature change $\Delta T = 1\,°\mathrm{C}$.

Integration of (10.4) gives

$$L(T) = L(0) \cdot e^{\alpha \Delta T} \quad \text{with} \quad \Delta T = T - T_0 . \tag{10.4a}$$

Table 10.2 compiles numerical values of α for some materials. One can see that for most materials α is positive, i. e. the length L increases with T. The coefficients α can be measured with the device shown in Fig. 10.4. A tube made of the material to be inspected, is clamped on one end A, but can freely slide on

Table 10.2 Thermal expansion of solids and liquids at $T = 293\,\mathrm{K} = 20\,°\mathrm{C}$

Solids	Linear expansion coefficient $\alpha/(10^{-6}\,\mathrm{K}^{-1})$	Liquids	Volume expansion coefficient $\gamma/(10^{-4}\,\mathrm{K}^{-1})$
Aluminium	23.8	Water	2.07
Iron	12	Ethanol	11
V2A Steel	16	Acetone	14.3
Copper	16.8	Benzene	10.6
Sodium	71	Mercury	1.8
Tungsten	4.3	Glycerin	5.0
Invar	1.5	N-Pentane	15
Cerodur	<0.1	Water at	
Hard rubber	75–100	$T = 0\,°\mathrm{C}$	−0.7
		$T = 20\,°\mathrm{C}$	+2.07

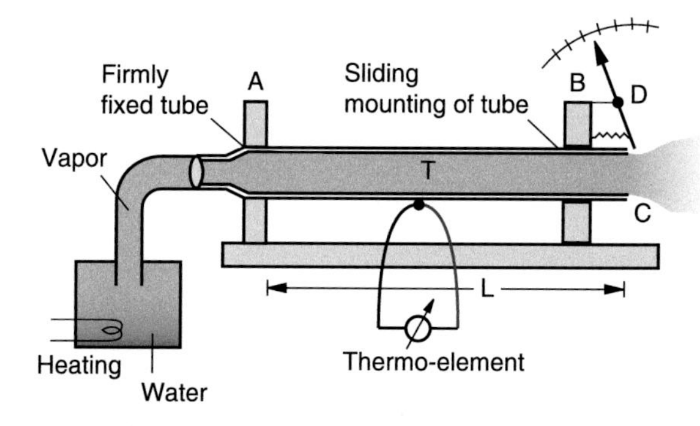

Figure 10.4 Demonstration of thermal expansion of metal tubes

the other end B where a turnable tongue is connected to the rod that shows the length change on a calibrated scale. When hot water vapour streams through the tube, it expands and turns the tongue. The scale is set to zero at $T_C = 20\,°\mathrm{C}$. The temperature of the tube is measured with a calibrated thermo-element. Metal tubes can be also heated by an electric current through the tube.

The reason for the thermal expansion is the asymmetric potential of the interaction between neighbouring atoms (Fig. 10.5). The atoms of a solid are not fixed at a constant value r_0 of the distance between neighbouring atoms but oscillate around r_0 (see Vol. 3). The length L of a rod is determined by the mean distance $\langle r \rangle$ of this oscillation. Increasing the temperature causes an increase of the vibrational energy and of the oscillation amplitude $r(t)$. Because of the asymmetric potential the mean value $\langle r \rangle$ increases with increasing amplitude $r(t)$ thus causing an increase of the length L.

More detailed measurements prove that the thermal expansion is not strictly linear. Expansion of (10.4a) gives

$$L(T) = L(0)(1 + \alpha \cdot \Delta T + \tfrac{1}{2}(\alpha \Delta T)^2 + \ldots) . \tag{10.4b}$$

This nonlinear expansion can be also expressed by a temperature-dependent expansion coefficient α

$$\alpha(T_C) = \alpha(T_C = 0) + \beta \cdot T_C = \alpha_0 + \beta \cdot T_C . \tag{10.4c}$$

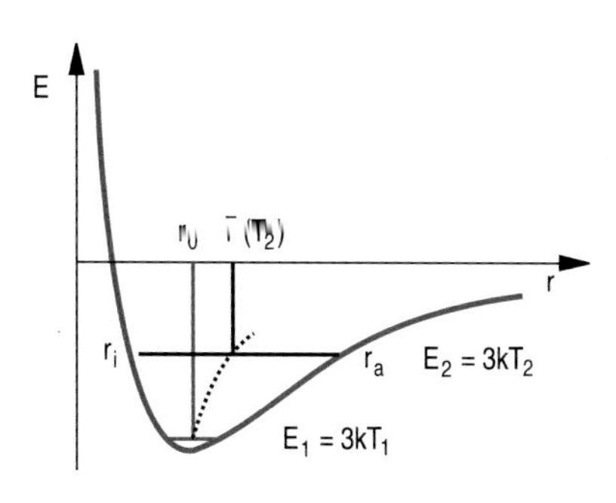

Figure 10.5 Atomic model of thermal expansion due to the anharmonic interaction potential

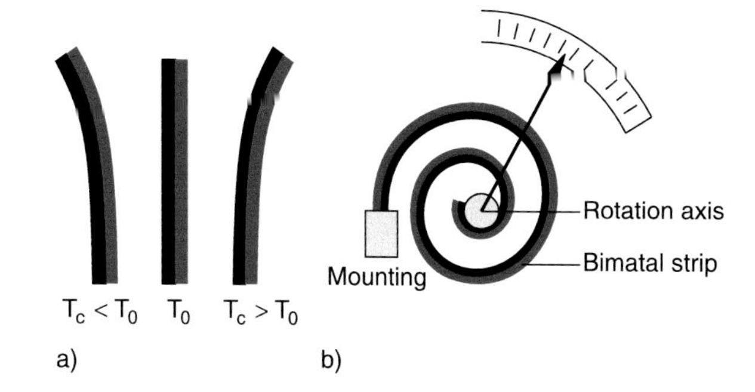

Figure 10.6 Bimetal thermometer. **a** Principle, **b** technical design

Instead of (10.3) the more accurate equation is then

$$L(T_C) = L_0\left(1 + \alpha_0 T_C + \beta T_C^2\right) . \tag{10.5}$$

However, within the temperature range between 0 °C and 100 °C the deviation from linearity is very small, i. e. $\beta \cdot T_C \ll \alpha_0$. For small temperature intervals ΔT the length $L(T_C)$ can be approximated by a straight line with a slope $dL/dT = \alpha_0 + \beta \cdot T_C$ that slightly depends on T_C.

Example

At room temperature the expansion coefficients of aluminium are $\alpha(T_C = 20\,°C) = 23.8 \cdot 10^{-6}/°C$, $\beta = 1.8 \cdot 10^{-8}/(°C)^2$.

The relative proportion of the nonlinear expansion is therefore $\beta \cdot T_C/\alpha = 7.5 \cdot 10^{-4}$. This implies that the coefficient α changes within a temperature range $\Delta T = 100\,°C$ only by 7.5%. ◄

For some alloys the expansion coefficient is very small. Examples are INVAR (64% iron and 36% nickel) or the glass ceramics CERODUR (see Tab. 10.2 and 10.3) [10.1].

Since all length dimensions of a three-dimensional body vary with the temperature, also the volume of the body must change.

Table 10.3 Dependence of mean thermal expansion coefficient $\alpha/10^{-6}\,K^{-1}$ on temperature (given in K)

T/K	Al	Cu	Fe	Al_2O_3	SiO_2
50	3.5	3.8	1.3	0.0	−0.86
100	12.0	10.5	5.7	0.2	−0.80
150	17.1	13.6	8.4	1.0	−0.45
200	20.2	15.2	10.1	2.8	−0.1
250	22.4	16.1	11.1	4.0	+0.2
300	23.8	16.8	12.0	5.0	+0.4
350	24.1	17.3	12.6	6.0	+0.5
400	24.9	17.6	13.2	6.4	+0.55
500	26.5	18.3	14.3	7.2	+0.58

For homogenous and isotropic bodies applies

$$\begin{aligned} V(T_C) &= V_0(1 + \alpha T_C)^3 \quad &&\text{with} \quad V_0 = V(T_C = 0\,°C) \\ &\approx V_0(1 + 3\alpha T_C) \quad &&\text{for} \quad \alpha T_C \ll 1 \\ &= V_0(1 + \gamma T_C) \quad &&\text{with} \quad \gamma = 3\alpha . \end{aligned} \tag{10.6}$$

For non-isotropic bodies the expansion may differ for the different directions and one obtains, instead of (10.6), the equation

$$\begin{aligned} V(T_C) &= V_0(1 + \alpha_1 T_C) \cdot (1 + \alpha_2 T_C) \cdot (1 + \alpha_3 T_C) \\ &\approx V_0\left[1 + (\alpha_1 + \alpha_2 + \alpha_3)T_C\right] \\ &= V_0(1 + 3\overline{\alpha} T_C) \quad \text{with} \quad \overline{\alpha} = \tfrac{1}{3}(\alpha_1 + \alpha_2 + \alpha_3) . \end{aligned} \tag{10.6a}$$

The difference of expansion coefficients of different metals is utilized for bimetal thermometers (Fig. 10.6). When two metal strips of different materials are bonded (e. g. by welding or soldering) the double strip will bend when the temperature changes. A special device converts the bending, which is proportional to the temperature change ΔT, into the turn of a hand with a scale (Fig. 10.6b) where after calibration the temperature can be read on the scale.

If the thermal expansion should be prevented by an external force very large forces are necessary, as the following experiment demonstrates (Fig. 10.7). A thick rod S made of wrought iron, is clamped between two stable mountings L_1 and L_2. On one end a bolt B with 5 mm diameter fixes the rod S to the mounting L_1. Now the rod is heated with a Bunsen burner until it is red glowing. The resulting thermal expansion loosens the screw M on the right side at L_2, which is tightened again at the highest temperature of the rod. Now the rod cools down and contracts. The contraction force is so large that the bold at the left side cracks.

A quantitative calculation of the forces necessary to prevent thermal expansion or contraction proceeds as follows:

The force necessary to achieve an elongation of a rod with length L and cross section $A \ll L^2$ and with an elastic modulus E is according to (6.2) and (10.4)

$$F = E \cdot A \cdot \Delta L/L = E \cdot A \cdot \alpha \cdot \Delta T . \tag{10.7}$$

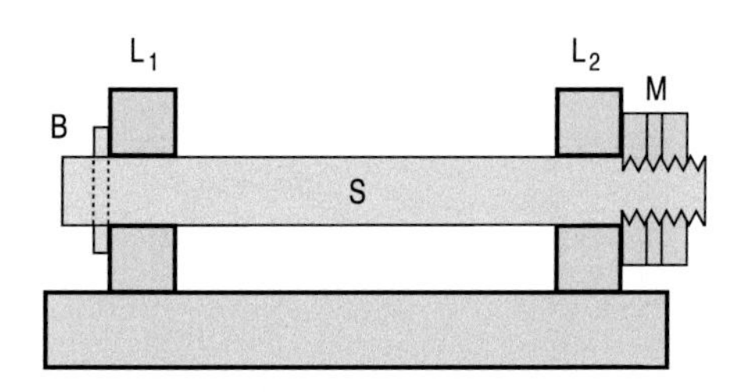

Figure 10.7 Bolt cracker. Demonstration of large forces when the thermal expansion is hindered. (B = bolt, S = hot rod, L_1, L_2 = mounts, M = screw nut)

If the thermal expansion should be prevented by application of external pressure, we obtain from (6.7) and (10.7) the required pressure

$$p = \frac{\alpha \cdot E}{1 - 2\mu} \cdot \Delta T\,, \tag{10.8}$$

where μ is the transverse contraction ratio (Poisson number).

Examples

1. A steel rod ($E = 120\,\mathrm{GN/m^2}$, $\alpha = 16 \cdot 10^{-6}/\,°\mathrm{C}$) with the cross section $A = 100\,\mathrm{cm}^2$ suffers a temperature change $\Delta T = 30\,°\mathrm{C}$. In order to prevent its expansion a force $F = 5.76 \cdot 10^5\,\mathrm{N}$ is necessary.
2. A section of a railroad track of steel with $L = 20\,\mathrm{m}$ and $\alpha = 16 \cdot 10^{-6}/\,°\mathrm{C}$ expands for a temperature difference $\Delta T = 40\,°\mathrm{C}$ by $\Delta L = \alpha \cdot L \cdot \Delta T = 1.3\,\mathrm{cm}$. For modern railroad tracks all sections are welded together at $T = 20\,°\mathrm{C}$ without gap. Without strong mountings each section between the welding spots would bend in such a way, that the length expansion $\Delta L = \alpha \cdot L\Delta T$ could be realized. This would give for $\Delta T = 60\,°\mathrm{C}$ a maximum deviation from the straight line of about 30 cm. This bending is prevented by strong supports where at every meter the rails are mounted. The force on the welding surfaces with a cross section $A = 0.02\,\mathrm{m}^2$ ($d = 10\,\mathrm{cm}$, $b = 20\,\mathrm{cm}$) is then (see Probl. 10.2) $F = 1.5 \cdot 10^6\,\mathrm{N}$. When the rail track cools down to $T = -20\,°\mathrm{C}$ a tensile force of the same magnitude acts onto the welding surfaces which corresponds to a tensile stress of $8 \cdot 10^7\,\mathrm{N/m^2}$. This is still sufficiently far below the break stress of $7 \cdot 10^8\,\mathrm{N/m^2}$. ◄

For the thermal expansion of liquids only the volume expansion can be given. When measuring this volume expansion one has to take into account that the solid container also expands. For the measurement of thermal expansion of liquids a device proposed by *Dulong* and *Petit* (Fig. 10.8) has been developed. The liquid is contained in a U-shaped tube where one side is encased in a jacket containing melting ice, while the other side is heated to 100 °C by water vapour. Of course the inspected liquid should not boil at 100 °C and should not freeze at 0 °C. Since the total mass $M = \varrho \cdot V$ of the liquid is constant, independent of the temperature, the density

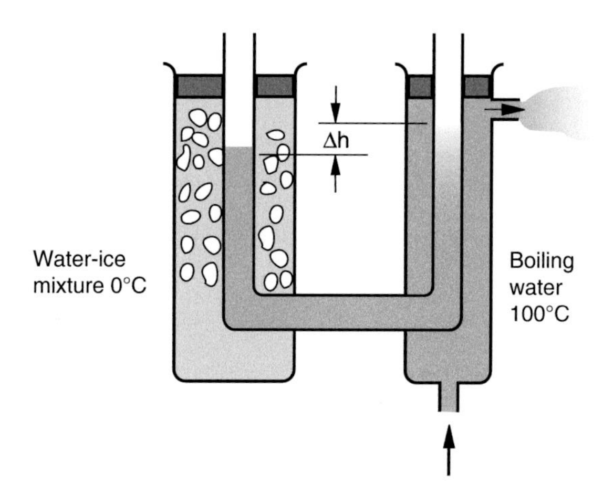

Figure 10.8 Design of Dulong–Petit for the measurement of thermal expansion of liquids

$$\varrho(T_\mathrm{C}) = \frac{\varrho_0}{1 + \gamma T_\mathrm{C}}$$

does depend on T: It is smaller in the hot side and larger in the cold side. The height h_0 of the liquid surface will be therefore lower in the cold side by Δh. Rearrangement of the equation yields

$$\gamma = \frac{1}{T_\mathrm{C}}\left(\frac{\varrho_0}{\varrho(T_\mathrm{C})} - 1\right) = \frac{1}{T_\mathrm{C}} \cdot \frac{\Delta\varrho}{\varrho(T_\mathrm{C})}\,.$$

From the equilibrium condition

$$h_0 \cdot \varrho_0 \cdot g = h(T_\mathrm{C}) \cdot \varrho(T_\mathrm{C}) \cdot g$$

gives

$$\varrho_0/\varrho = h/h_0\,.$$

The thermal volume expansion coefficient γ is then

$$\gamma = \frac{1}{T_\mathrm{C}} \cdot \frac{\Delta h}{h_0}\,. \tag{10.9}$$

Table 10.2 compiles some values of γ.

Note, that they are much larger than the volume expansion coefficients 3α of solids. This justifies the neglect of the glass tube expansion for liquid thermometers.

10.1.3 Thermal Expansion of Gases; Gas Thermometer

Experiments show that the volume of ideal gases (see Sect. 7.3) increases at constant pressure proportional to the temperature.

$$V(T_\mathrm{C}) = V_0(1 + \gamma_\mathrm{V} \cdot T_\mathrm{C})\,, \tag{10.10}$$

where the temperature is measured in °C and $V_0 = V(T_\mathrm{C} = 0\,°\mathrm{C})$.

Table 10.4 Thermal expansion coefficient of some gases

Gas	$\gamma/(10^{-3}/\text{K})$
Ideal Gas	3.661
He	3.660
Ar	3.671
O_2	3.674
CO_2	3.726

The expansion coefficient

$$\gamma_V = \frac{V(T_C) - V_0}{V_0 \cdot T_C} \tag{10.11}$$

gives the relative change $\Delta V/V_0$ per 1 °C. The experimentally obtained numerical values of γ_V are compiled in Tab. 10.4.

For Helium, which comes closest to an ideal gas, one finds

$$\gamma_V = \frac{1}{273.15}\,{}^\circ\text{C}^{-1} = 3.661 \cdot 10^{-3}\,{}^\circ\text{C}^{-1} .$$

Accurate experiments performed at a constant gas volume give for the temperature dependence of the pressure the completely analogue relation

$$p = p_0(1 + \gamma_p \cdot T_C)$$
$$\text{with} \quad \gamma_p = \gamma_V = \gamma = \frac{1}{273.15}\,{}^\circ\text{C}^{-1} . \tag{10.12}$$

(**Law of Gay-Lussac**).

The gas thermometer (Fig. 10.9) utilizes this pressure dependence for the measurement of temperatures. The volume V is connected with a U-shaped tube filled with mercury. The height of Hg in the left side of the U-tube can be changed by up- and down lifting of the right side, which is connected with the left side by a flexible tube. When the gas volume is heated, the pressure rises. In order to keep the gas volume constant, the level of the Hg in the left side is always kept at the same height. The pressure is then indicated by the difference Δh between the left and the right side. It is $p = \varrho_{Hg} \cdot g \cdot \Delta h$. The temperature, obtained from (10.12)

$$T_C = \frac{1}{\gamma} \cdot \frac{p - p_0}{p_0} = 273.15 \frac{\Delta p}{p_0}\,{}^\circ\text{C} \tag{10.13}$$

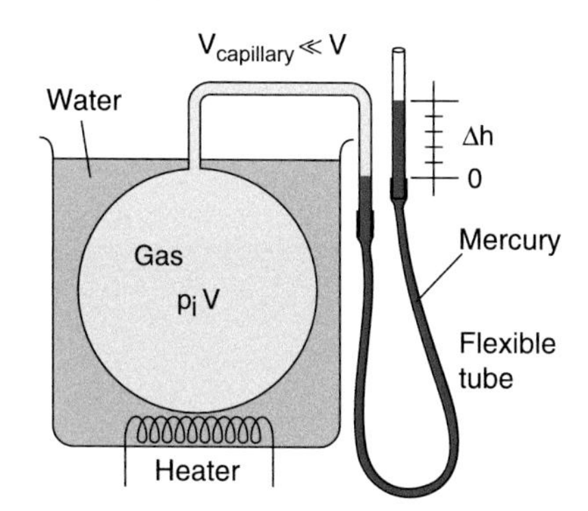

Figure 10.9 Gas thermometer

is therefore determined by a pressure measurement. At the temperature $T_C = 0$ which is realized by immersing the gas volume into a melting ice-water mixture, the height difference is adjusted to $\Delta h = 0$. The pressure in the gas volume V is then $p = p_0$.

10.1.4 Absolute Temperature Scale

We will now discuss the relation between the Celsius scale and the absolute temperature scale, which is also called (after its creator Lord Kelvin) **Kelvin scale** or **thermo-dynamical temperature scale**. In Sect. 7.3 the absolute temperature T was used in the general gas equation

$$p \cdot V = N \cdot k \cdot T , \tag{10.14a}$$

relating pressure p, particle number N in the gas volume V with T. The absolute temperature T was defined by Eq. 10.1, which is based on the results of the kinetic gas theory (see Sect. 7.3). The general gas equation (10.14a) states that for constant pressure and temperature the gas volume has a definite value, which is the same for all ideal gases independent of the specific kind.

At a temperature T_0 (at 0 °C) and a pressure p_0 = 1 bar = 10^3 hPa (normal conditions) Eq. 10.14a becomes

$$p_0 \cdot V_0 = N \cdot k \cdot T_0 . \tag{10.14b}$$

From (10.14a) and (10.14b) we can conclude

$$p = p_0 \cdot \frac{V_0}{V} \cdot \frac{T}{T_0} . \tag{10.14c}$$

In the gas thermometer the volume $V = V_0$ is kept constant. The comparison of (10.14c) with (10.12) yields

$$p = p_0 \cdot \frac{T}{T_0} = p_0(1 + \gamma \cdot T_C) . \tag{10.15}$$

This gives with the experimental value $\gamma = (273.15)^{-1}$ the relation

$$T = T_0 \cdot (1 + \gamma T_C) = T_0 + \frac{T_0}{273.15} \cdot T_C \tag{10.16}$$

between the absolute temperature T and the Celsius scale T_C.

Note: ***The unit oft the absolute temperature scale is the Kelvin. It is the 273.16th part of the thermodynamic temperature of the triple point of water.***

A definition of the absolute temperature that is independent of the specific substance, can be given with their help of the Carnot-Cycle (see Sect. 10.3.5)

10.1.5 Amount of Heat and Specific Heat Capacity

When a defined energy ΔW is transferred to a body, its temperature rises by $\Delta T \sim \Delta W$. A simple demonstration (Fig. 10.10) uses an immersion heater which is immersed into water in a thermally isolated Dewar flask and heated for a time Δt. The electric energy $\Delta W = I \cdot U \cdot \Delta t$ (I = electric current, U = voltage, see Vol. 2 Chap. 2) causes a temperature rise ΔT that depends on the mass of the water. The increase ΔQ of the heat Q (often also called the amount of heat) is given by

$$\Delta Q = \Delta W = c \cdot M \cdot \Delta T \,. \tag{10.17}$$

The proportional constant c is the *specific heat*. It depends on the specific material of the heated body. It gives the amount of heat that increases the temperature of a body with $M = 1$ kg by $\Delta T = 1$ K. The product $C = c \cdot M$ is the *heat capacity* of a body with mass M.

In former times the unit was the large calorie (1 kcal). This is the amount of heat that increases the temperature of 1 kg water from 14.5 to 15.5 °C. Nowadays the unit is 1 Joule ($1\,\mathrm{J} = 1\,\mathrm{W \cdot s} = 1\,\mathrm{N \cdot m}$). It has the great advantage, that for the conversion of heat into electrical or mechanical energy the same units are used and therefore the conversion factor is 1. This is not the case, if the unit calorie is used. Here measurements give the electrical heat equivalent

$$\mathrm{WE_{el}} = \frac{\Delta Q\,[\mathrm{cal}]}{\Delta W_{\mathrm{el}}\,[\mathrm{W\,s}]} = 0.23885\,[\mathrm{cal/W\,s}] \,. \tag{10.18}$$

This equation means: If ΔW_{el} is measured in Joule, but ΔQ in calories, the ratio $\Delta Q/\Delta W_{\mathrm{el}}$ has the numerical value 0.23885 i.e. $1\,\mathrm{W\,s} = 0.2389\,\mathrm{cal}$ or $1\,\mathrm{cal} = 4.1868\,\mathrm{W\,s}$.

The temperature rise in the experiment shown in Fig. 10.10 does not occur abruptly but continuously over the time interval Δt of the heating (Fig. 10.10b). During this time interval, a steady heat flux takes place between the hot water and its surrounding which decreases the temperature difference ΔT. In order to consider this, a temperature progression for a sudden change of ΔT is simulated, indicated by the vertical dashed line in Fig. 10.10b. The time t_1 is chosen such that the areas A_1 and A_2 are equal in order to maintain the same value for the integral $\int T \mathrm{d}t \sim \int \mathrm{d}Q = \Delta Q$.

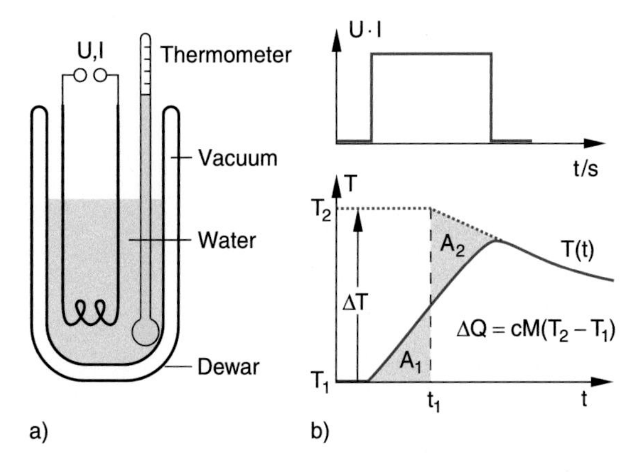

Figure 10.10 Measurement of the electric heat equivalent with immersion heater and Dewar flask. **a** Experimental setup; **b** time progression of electric power and temperature

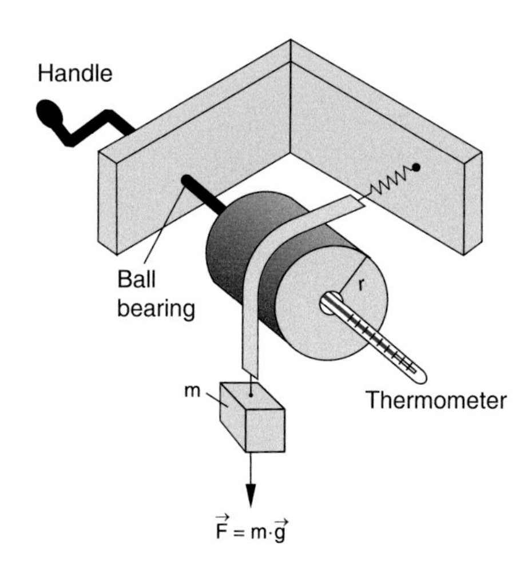

Figure 10.11 Measurement of mechanical heat equivalent

Instead of heat generation by electrical power, heat can be also produced by mechanical work due to friction. This is demonstrated by the arrangement in Fig. 10.11. Here a metal tape is wrapped around a copper cylinder, filled with water. A weight with mass m presses the tape tightly onto the cylinder. Now the cylinder with radius r is turned by a handle during a time interval Δt with such a frequency f that the weight $G = m \cdot g$ is just compensated by the friction force between tape and cylinder. The work performed against the friction force by turning the cylinder N-times during the time interval Δt is

$$\begin{aligned}\Delta W &= m \cdot g \cdot 2\pi r \cdot N \\ &= (c_{\mathrm{W}} \cdot M_{\mathrm{W}} + c_{\mathrm{Co}} \cdot M_{\mathrm{Co}})\Delta T_1 \,,\end{aligned} \tag{10.19a}$$

where M_{W} is the mass of the water and M_{Co} that of the copper cylinder. Repeating the experiment without water filling a larger temperature difference ΔT_2 is measured. From these two measurements we obtain from the relation

$$(c_{\mathrm{W}} m_{\mathrm{W}} + c_{\mathrm{Co}} \cdot m_{\mathrm{Co}})\Delta T_1 = c_{\mathrm{Co}} \cdot m_{\mathrm{Co}} \cdot \Delta T_2 = \Delta W$$

for the heat, put into the water the relation

$$\Delta Q = c_{\mathrm{W}} M_{\mathrm{W}} \Delta T_1 = \left(1 - \frac{\Delta T_1}{\Delta T_2}\right) \Delta W_{\mathrm{mech}} \,. \tag{10.19b}$$

The **mechanical heat equivalent**, determined with such experiments is

$$\mathrm{WE_{mech}} = \frac{\Delta Q/\mathrm{cal}}{\Delta W_{\mathrm{mech}}/\mathrm{Nm}} = 4.186 \,, \tag{10.19c}$$

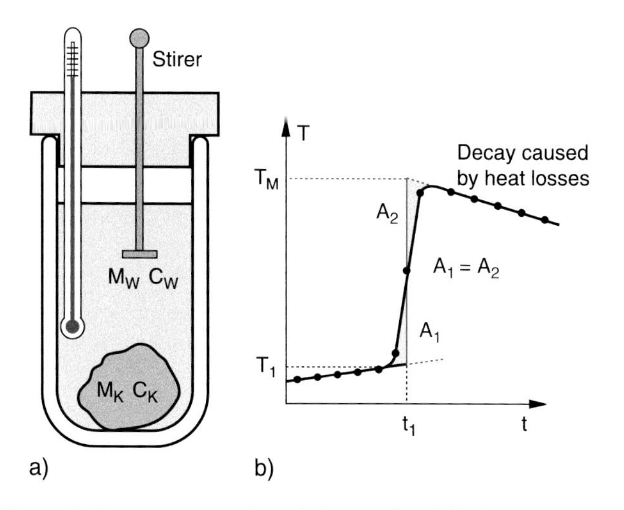

Figure 10.12 Measurement of specific heat c_K of a solid body using a mixture calorimeter. **a** Experimental setup; **b** measurement of time-dependent temperature (t_1 = immersion of solid body, T_1 = initial temperature, T_M = temperature of mixture)

should of course have the same mechanical value as the electrical heat equivalent, because of the definition $1\,\text{W}\cdot\text{s} = 1\,\text{N}\cdot\text{m}$ in the SI system.

The specific heat c_K of a body can be measured with the mixture calorimeter shown in Fig. 10.12. In a well-isolated Dewar container is water with the mass M_W at the temperature T_1. Now a solid body with the mass M_K that had been heated up to the temperature $T_2 > T_1$ is inserted into the water. The temperature $T_M(t)$ of the mixed system (body + Dewar + water) is measured as a function of time (Fig. 10.12b).

The heat $(c_K M_K(T_1 - T_M)$ transferred from the body to water plus Dewar is equal to the heat change $c_W \cdot M_W + c_D \cdot M_D$ of water plus Dewar. This gives the specific heat of the body

$$c_K = \frac{(M_W \cdot c_W + C_D)(T_M - T_1)}{M_K(T_2 - T_M)}\,, \tag{10.20}$$

where $C_D = c_D \cdot M_D$ is the heat capacity of the dewar. The temperature T_M of the mixture is determined in the same way as in Fig. 10.10. The measured curve $T(t)$ in Fig. 10.12b is replaced by the simulated red curve where the vertical line is placed at the time t_1 where the areas $A_1 = A_2$. This takes into account the heat loss during the heat transfer form body to water. The intersection points of the vertical line with the two horizontal red curves give the correct temperatures T_1 and T_M.

The heat capacity $C_D = c_D \cdot M_D$ of the Dewar can be measured when two portions of water with masses M_1 and M_2 at temperatures T_1 and T_2 are mixed in the Dewar and the mixing temperature T_M is measured [10.1].

10.1.6 Molar Volume and Avogadro Constant

One mole is according to the definition given in Sect. 1.6 the amount of a substance that contains as many atoms or molecules as 12 g carbon ^{12}C. The molar mass of a substance ^{y}X with the atomic mass number y is then equal to $(y/12) \cdot m(^{12}\text{C})$g.

The molar volume V_M contains 1 mol of the gas.

Examples

1 mol helium gas He are 4 g He,
1 mol hydrogen gas H_2 are 2 g H_2,
1 mol nitrogen N_2 are 28 g N_2. ◀

The number of atoms or molecules per mole is the *Avogadro constant* N_A. This number is independent of the specific substance.

It can be measured with different methods (see Vol. 3, Chap. 2).

The average value of many measurements is

$$N_A = 6.022 \cdot 10^{23}/\text{mol}\,.$$

One mole of atoms or molecules always fills the same volume under equal external conditions, independent of their specific kind. One finds under normal conditions

$$V_M(p = 1\,\text{atm} = 101.3\,\text{kPa}, T_C = 0\,°\text{C}) = 22.4\,\text{dm}^3$$
$$V_M(p = 1\,\text{bar} = 100\,\text{kPa}, T_C = 0\,°\text{C}) = 22.7\,\text{dm}^3\,.$$

The general gas equation (10.14a) can be written for 1 mol with $V = V_M$ and $N = N_A$

$$p \cdot V_M = N_A \cdot kT = R \cdot T\,, \tag{10.21}$$

where the **general gas constant**

$$R = N_A \cdot k = 8.31\,\text{J}/(\text{K}\cdot\text{mol}) \tag{10.22}$$

is the product of Avogadro number N_A and Boltzmann constant k. All gases that obey this equation are called **ideal gases**.

For an arbitrary volume $V = \nu \cdot V_M$ Eq. 10.14a can be written as

$$p \cdot V = \nu \cdot R \cdot T\,, \tag{10.21a}$$

where the number ν quotes how many moles are contained in V.

10.1.7 Internal Energy and Molar Heat Capacity of Ideal Gases

The amount of heat ΔQ supplied to one mole of a gas with molar mass M (kg/mol) leads to a temperature rise ΔT:

$$\Delta Q = c \cdot M_M \cdot \Delta T = C \cdot \Delta T\,.$$

The product $C = c \cdot M_M$ of specific heat and molar mass is the **molar heat capacity** with the unit $[C] = [\text{J}/(\text{mol} \cdot \text{K})]$. It is the heat energy that increases the temperature of 1 Mole by $\Delta T = 1\,\text{K}$. For an arbitrary mass $M = \nu \cdot M_M$ is

$$\Delta Q = \nu \cdot C \cdot \Delta T \,.$$

The quotient $\Delta Q / \Delta T = \nu \cdot C$ [J/K] is the **heat capacity** of the body with mass M.

The molar specific heat of a gas depends on whether the gas is heated at constant volume or at constant pressure. We will at first discuss the situation for a constant volume.

We define the internal energy of a gas with volume V as the total energy of its N molecules. It is composed of translational energy plus possible rotational and vibrational energy. For non-ideal gases also the potential energy of their mutual interaction has to be taken into account (Fig. 10.5). The internal energy of a gas depends on the number f of degrees of freedom of the molecules. In Sect. 7.3 it was shown that the mean energy of a molecule is $\langle E \rangle = f \cdot \frac{1}{2} kT$. The internal energy of a gas volume with N molecules is then

$$U = \tfrac{1}{2} f \cdot N \cdot kT$$

and for 1 mol with $N = N_A$ it is

$$U(V_M) = \tfrac{1}{2} f \cdot N_A \cdot kT = \tfrac{1}{2} f \cdot R \cdot T \,. \tag{10.23}$$

Under thermal equilibrium the energy U is uniformly distributed among all degrees of freedom.

This equipartition is accomplished by collisions between the molecules (see Sect. 4.2 and Vol. 3, Chap. 8).

When the heat ΔQ is supplied, the internal energy U increases by $\Delta U = \Delta Q$, if the volume V of the gas stays constant. We therefore obtain the equation

$$\Delta Q = \Delta U = \nu C_V \cdot \Delta T \,, \tag{10.24}$$

and with $\Delta U = \frac{1}{2} f \cdot \nu \cdot R \cdot \Delta T$ **the molar heat capacity at constant volume**

$$C_V = \tfrac{1}{2} f \cdot R \,. \tag{10.25}$$

10.1.8 Specific Heat of a Gas at Constant Pressure

When a gas is heated at constant volume the pressure increases according to the general gas equation (10.14a). In order to achieve a temperature increase at constant pressure, the gas volume must expand (Fig. 10.13b). Such an expansion can be

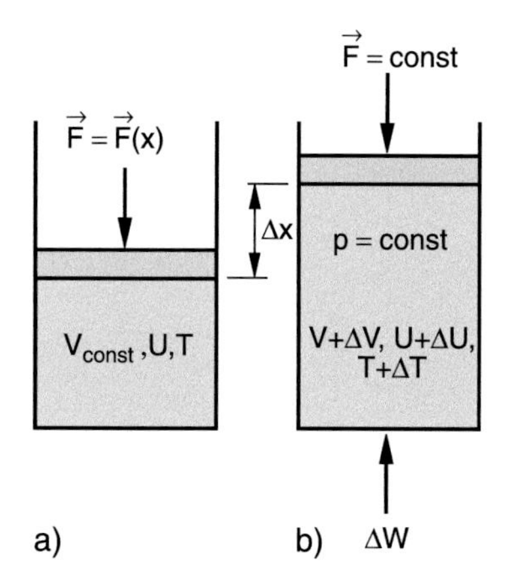

Figure 10.13 Determination of C_p. Heating of gas **a** at constant volume, **b** at constant pressure

realized when the piston with area A moves by the distance Δx against the external pressure p. This requires the work

$$\Delta W = F \cdot \Delta x = p \cdot A \cdot \Delta x = p \cdot \Delta V \,. \tag{10.26}$$

This energy ΔW must be supplied additionally. The heat ΔQ is therefore

$$\Delta Q = C_V \cdot \Delta T + p \cdot \Delta V \,. \tag{10.27}$$

The general gas equation for 1 mol of the gas before and after the expansion is

$$p \cdot V = R \cdot T,$$
$$p(V + \Delta V) = R \cdot (T + \Delta T) \,.$$

Subtraction yields

$$p \cdot \Delta V = R \cdot \Delta T \,.$$

Inserting this into (10.27) we obtain

$$\Delta Q = (C_V + R)\Delta T = C_p \cdot \Delta T \,. \tag{10.28}$$

The factor C_p is the molar specific heat at constant pressure

$$C_p = C_V + R \,. \tag{10.29a}$$

With

$$C_V = \tfrac{1}{2} f \cdot R \rightarrow C_p = \tfrac{1}{2}(f + 2)R \,, \tag{10.29b}$$

the quotient C_p/C_V is called **adiabatic index** or **specific heat ratio**

$$\kappa = \frac{C_p}{C_V} = \frac{f+2}{f} \,. \tag{10.29c}$$

10.1.9 Molecular Explanation of the Specific Heat

Since atoms or molecules can move into three directions, they have three degrees of freedom of translation. Their mean translational energy is therefore

$$E_{\text{trans}} = 3 \cdot \tfrac{1}{2} kT ,$$

the molar specific heat of atomic gases is therefore

$$C_V = (3/2)R .$$

For molecules the supplied energy can be also converted into rotational or vibrational energy. Nonlinear molecules can rotate around three orthogonal axis. They have therefore three degrees of freedom for the rotation. Linear molecules have only 2 rotational degrees of freedom, because of the following reason:

The rotational energy

$$E_{\text{rot}} = L^2/2I$$

is determined by the angular momentum L and the moment of inertia I (see Sect. 5.5). As shown in Quantum physics (see Vol. 3 Chap. 4) the angular momentum has the amount $L = (l \cdot (l+1))^{(1/2)} \cdot \hbar$ with $l = 1; 2; 3; \ldots$. The smallest angular momentum is then $L_{\min} = \sqrt{2} \cdot \hbar$, where $\hbar = h/2\pi$ is Planck's quantum constant, divided by 2π. The moment of inertia for a rotation around the axis of a linear molecule is very small because the heavy nuclei are located on the axis and the light electrons do not contribute much to I. Therefore the rotational energy is very large, generally much larger than the translational energy at accessible temperatures. Collisions cannot excited this rotation and it therefore cannot contribute to the accessible energy.

The vibration of diatomic molecules is one-dimensional and has therefore only one degree of freedom. However, the vibrational energy has two contributions: The kinetic and the potential energy (see Sect. 11.6). The mean value of both contributions is equal to $\frac{1}{2}kT$ and the thermal energy of the vibration is kT. Therefore two degrees of freedom ($f = 2$) are formally attributed to the vibration. A diatomic molecule has then $f = 3 + 2 + 2 = 7$ degrees of freedom, if the temperature is sufficiently high to excite the vibrations.

Note: Quantum Theory shows (see Vol. 3) that the classical model of a vibrating oscillator with regard to the total energy $E = E_{\text{kin}} + E_{\text{pot}}$ is correct, but that the energy can be only absorbed in discrete quanta $h \cdot \nu$. This does, however, not influence our argumentation above.

For polyatomic molecules with j atoms each atom has three degrees of freedom. If we subtract 3 degrees of freedom for the translational motion of the whole molecule and 3 degrees of freedom for the rotation (2 degrees for a linear molecule) we end up with $f_{\text{vib}} = 3j - 6$ ($3j - 5$ for linear molecules) vibrational degrees of freedom.

The total internal energy U of a molecule with j atoms is then

$$U = \tfrac{1}{2} \cdot f \cdot N_A \cdot kT \quad \text{with} \quad f = f_{\text{trans}} + f_{\text{rot}} + f_{\text{vib}} . \qquad (10.30)$$

a

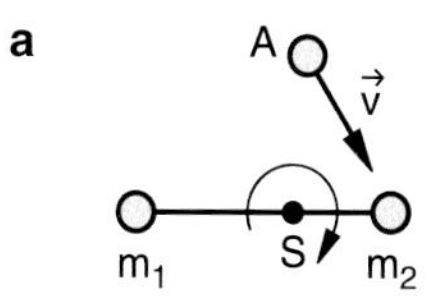

Rotational excitation

b

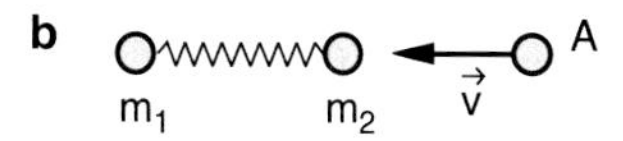

Vibrational excitation

Figure 10.14 Excitation of rotational degrees of freedom of a diatomic molecule induced by collisions. The collision with an atom *A* causes a rotation of the molecule about an axis perpendicular to the drawing plane (**a**), or causes the excitation of a molecular vibration (**b**)

Since collisions can transfer translational energy to rotations or vibrations only if the thermal energy of the collision partners is sufficiently high, at low temperatures only translational degrees of freedom are available and $f_{\text{eff}} = 3$. With increasing temperature at first the rotation can be excited ($f_{\text{eff}} = 6$ resp. 5 for linear molecules) and at still higher temperature also the vibrations contribute to the specific heat, because their energy is higher than that of the rotations ($f_{\text{eff}} = 3 + 3 + 2 \cdot (3j - 6)$ resp. $3 + 2 + 2(3j - 5)$ for linear molecules). This gives for diatomic molecules $f_{\text{eff}} = 3 + 2 + 2 = 7$.

The molar specific heat is

$$C_V = \left(\frac{\partial U}{\partial T}\right)_V = \frac{1}{2} f_{\text{eff}} \cdot R . \qquad (10.31)$$

Here the partial derivative is used, because U can depend on several variables (p, V, T). The index V indicates that the energy supply occurs at constant volume.

Examples

1. For the atomic gas Helium is $f = 3$. Since the translational energy is not quantized all three degrees of freedom are excited even at low temperatures. Therefore the specific heat of Helium is independent of the temperature (Fig. 10.15).

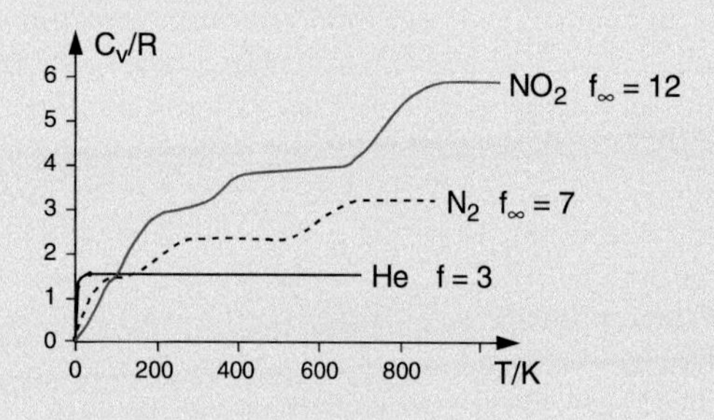

Figure 10.15 Temperature dependence of molar heat capacity of helium, nitrogen N_2 and nitrogen dioxyde NO_2 (see also Sect. 10.1.10)

2. For nitrogen gas consisting of the diatomic molecules N_2 kT is for very low temperatures smaller than the energy of the lowest rotational level (see Probl. 10.3). The rotational energy cannot be excited. Therefore $f_{\text{eff}} = 3$. With increasing temperature when $kT \approx E_{\text{rot}} f_{\text{eff}}$ approaches the value $f_{\text{eff}} = 5$. For still higher temperatures $kT \approx E_{\text{vib}} f_{\text{eff}}$ becomes $f_{\text{eff}} = 7$ because the vibrational degrees of freedom are counted twice (see Sect. 10.1.9). The specific heat of a molecular gas is therefore dependent on the temperature and reaches its maximum value only if kT is sufficiently high to excite all degrees of freedom.
3. Polyatomic gas (e. g. NO_2 at $T > 200\,\text{K}$ where NO_2 has a sufficiently high gas pressure). At this temperature already all three rotational modes can be excited. We then obtain $f = 3 + 3 = 6$.
   Above $T = 300\,\text{K}$ the bending vibration can be excited, rising f to $f = 8$. Only above $T = 800\,\text{K}$ all three vibrational mods can be excited and we have $f = 12$. The molar specific heat is then $C_V = 6R$. ◂

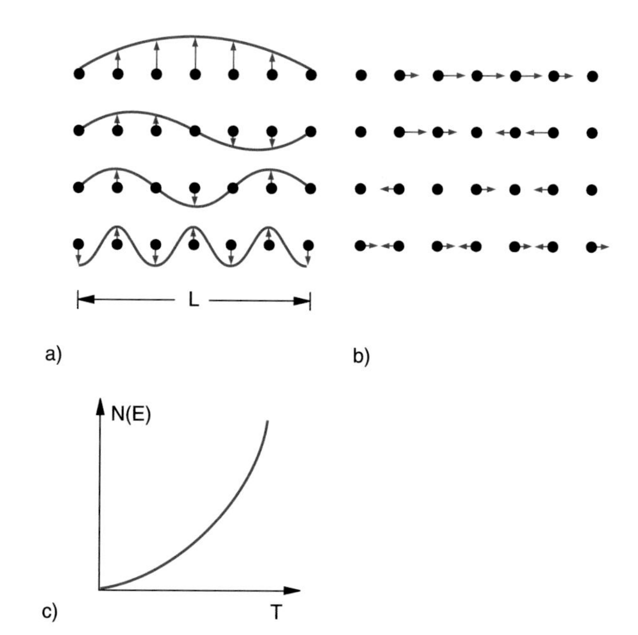

Figure 10.16 Stationary vibrational modes of a linear chain. **a** Transversal; **b** longitudinal standing waves. **c** Number of vibrational modes per energy interval dE of a solid body as function of temperature T

10.1.10 Specific Heat Capacity of Solids

With decreasing temperature all gases become liquids and pass finally into the solid state (except helium, which solidifies only under high pressures). Considerations about the specific heat of solids require a knowledge about the possible degrees of freedom for atoms and molecules in solids. Since the atoms in a solid can only oscillate in three directions around their fixed equilibrium positions but have no degrees of freedom for translation or rotation we would expect $f = 2 \cdot 3 = 6$. However, it turns out that the oscillation frequencies of all atoms are not the same but spread over a large frequency range. In order to get an idea about the frequency distribution, we regard a one-dimensional arrangement of atoms in an ideal crystal where all atoms are regularly placed at a distance d (Fig. 10.16a). When an atom oscillates around its equilibrium positions, the oscillation can be transferred to the neighbouring atoms, due to the coupling force between the atoms. This results in elastic waves travelling through the crystal (see Sect. 11.8). The waves are reflected at the end faces of the crystal, superimpose with the incoming waves and form stationary standing waves. Longitudinal as well as transversal standing waves can develop, depending on whether the oscillation occurs in the direction of wave propagation or perpendicular to it. The standing wave with the smallest possible wavelength λ (i. e. the highest frequency $\nu = c/\lambda$) is realized, when the neighbouring atoms oscillates against each other (Fig. 10.16a, b lowest line). The oscillation with the largest possible wavelength ($\lambda = L$ with $L =$ length of the crystal) has the lowest energy $h \cdot \nu$. At low temperatures only those vibrations with the lowest energy can be excited. With increasing temperature more and more vibrations can be excited. The number of possible vibrations $Z \sim N^3$ is proportional to the third power of the number N of atoms in the crystal. This means that the specific heat rises continuously with the temperature (Fig. 10.16c) until at $kT \geq E_{\text{vib}}^{\max}$ all vibrations are excited and the specific heat takes its maximum value. Since the interaction between neighbouring atoms depends on the specific kind of atoms the progression $C(T)$ differs for the different materials (Fig. 10.17). However, all curves $C(T)$ approach for high temperatures the same value of the molar specific heat

$$C_V = 6 \cdot \tfrac{1}{2} N_A \cdot k = 3R \qquad \text{(Dulong–Petit law)}\,. \tag{10.32}$$

Measurements of the temperature-dependent progression of $C_V(T)$ gives information about the distribution of the vibrational frequencies and therefore about the coupling forces between the atoms of the solid. They are furthermore a convincing experimental proof of quantum theory (see Vol. 3).

Table 10.5 gives numerical values of C_V for some materials.

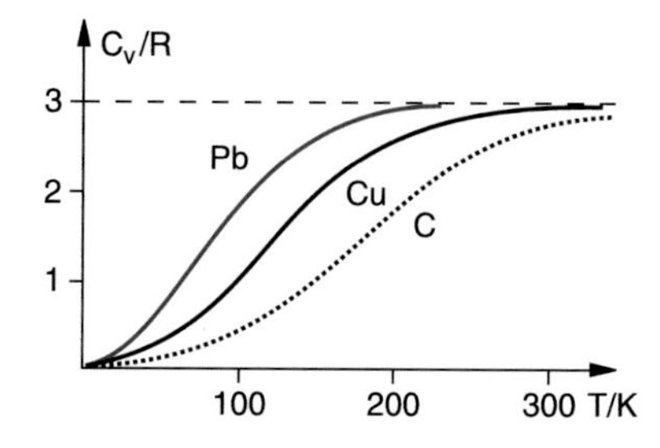

Figure 10.17 Qualitative temperature dependence of molar heat capacity of different solids

Chapter 10

Table 10.5 Specific heat c of some materials at 20 °C and 1013.25 hPa, specific heat of fusion λ_f and heat of evaporation λ_e

Substance	c/kJ kg^{-1}K^{-1}	λ_f/kJ kg^{-1}	λ_e/kJ kg^{-1}
Water	4.182	333.5	2256
Ethanol	2.43	105	840
Mercury	0.14	12.4	285
Aluminium	0.896	397	10,900
Iron	0.45	277	6340
Gold	0.13	65	16,500
Copper	0.383	205	4790
Ice at 0 °C	2.1	332.8	–

10.1.11 Fusion Heat and Heat of Evaporation

When a constant heat power dQ/dt is supplied to a container with 1 kg ice (specific heat c_i) at a temperature $T_C < 0$ °C the temperature

$$T(t) = T_i + a \cdot t \quad \text{with} \quad a = (dQ/dt)/c_i \tag{10.33}$$

rises linearly with the slope $a = (dQ/dt) \cdot c_i$, (c_i = specific heat of ice) up to $T_m = 0$ °C at $t = t_1$. Here the temperature stays constant until t_2, when the ice is completely molten, in spite of a constant power supply dQ/dt (Fig. 10.18). Then the temperature rises again but with a different slope $b = (dQ/dt)/c_W$ (c_W = specific heat of water) up to $T = 100$ °C at $t = t_3$, where the water starts to boil (at p = bar 1). Again the temperature remains constant until $t = t_4$ when the whole water is evaporated. Then the temperature rises further with the slope $(dQ/dt)/c_{vap}$.

The energy $(dQ/dt) \cdot (t_2 - t_1)$ supplied during the melting process is called **fusion heat**, the energy $dQ/dt) \cdot (t_4 - t_3)$ is the **heat of evaporation**.

The energy $\lambda_f = (dQ/dt)/m$ [J/kg] necessary to melt 1 kg of a substance is the **specific fusion heat** while the molar fusion heat is labelled by Λ_f [J/mol]. Analogue label λ_e and Λ_e the specific and the molar heat of evaporation.

Since the temperature has not changed during the melting process, also the kinetic energy must have stayed constant.

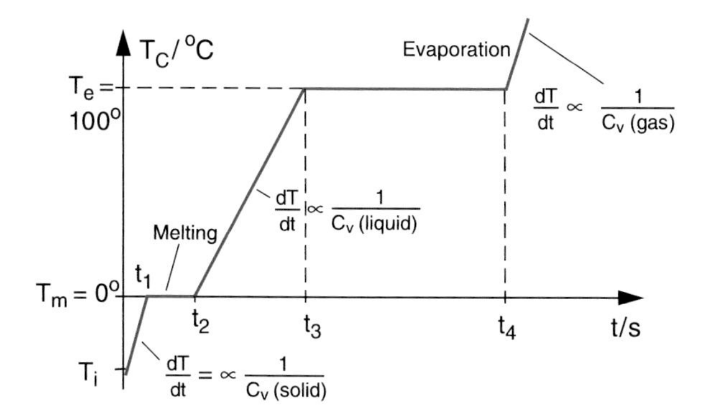

Figure 10.18 Temperature $T(t)$ of water under constant energy supply within the temperature range from below the melting temperature up to above the evaporation temperature (ice–water–water vapor)

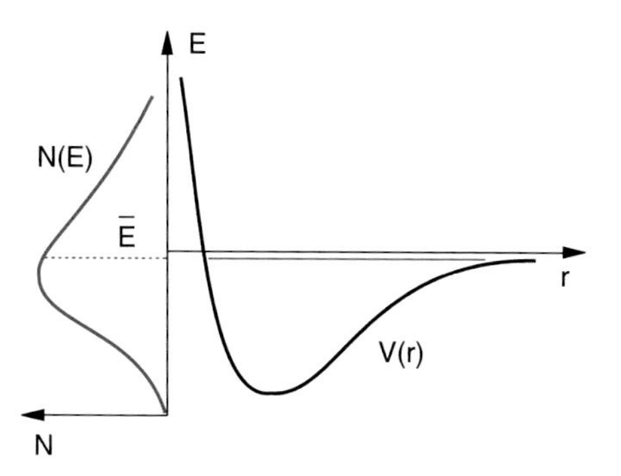

Figure 10.19 Energy distribution $N(E)$ and mean total energy E of molecules just below the melting temperature, illustrated by the interaction potential $V(r)$

Therefore the supplied energy $(dQ/dt) \cdot (t_2 - t_1)$ must have increased the potential energy of the atoms or molecules. This can be explained by the molecular model as follows:

The molecules in a solid body are bound to the equilibrium positions by attractive forces. This means that all molecules vibrate in a potential $V(r)$ that is determined by the vector sum $\boldsymbol{F} = \sum \boldsymbol{F}_i = -\nabla V(r)$ of all forces. At the melting temperature the mean total energy $\overline{E} = \overline{E}_{kin} + \overline{E}_{pot}$ is illustrated in Fig. 10.19 by the horizontal line close to the dissociation energy E_D of the interaction potential between the atoms or molecules. The energy distribution $N(E)$ of the molecules follows a Maxwell–Boltzmann distribution, as depicted in Fig. 10.19. Those molecules with $E > E_D$ can leave their fixed equilibrium position without changing their kinetic energy.

Continuous energy supply increases the number of molecules that leave their fixed position until all molecules can freely move: The solid body has dissolved and has become a liquid.

The energy supplied during the melting process keeps the kinetic energy constant but increases the potential energy.

An analogous process occurs during the evaporation process. The molecules from the higher energy part of the Maxwell–Boltzmann-distribution have sufficient energy to leave the liquid against the attractive forces and enter the vapour phase. Since the density of the vapour at atmospheric pressure is about 3 orders of magnitude smaller than that of the liquid, the mean distance between the molecules if about 10 times larger. The negative potential energy of the mutual attraction is therefore in the gas phase negligible against their kinetic energy. Similar to the melting process the supplied energy increases the potential energy but not the kinetic energy because the temperature remains constant. The potential energy increases from a negative value (work function, surface tension, see Sect. 6.4) to nearly zero.

The numerical values of fusion energy and evaporation energy depend on the substance. In Tab. 10.5 the values for some materials are listed.

10.2 Heat Transport

Always when a temperature difference exists between two different locations, heat is transported from the warmer to the colder region (see Sect. 7.5.3) Such a heat transport is very important for many technical problems and also for different measuring methods. In many cases one tries to maximize heat transport (for example for cooling heat generating systems) in other cases it is minimized (for heat isolating devices such as Dewars or refrigerators).

There are essentially three mechanisms of heat transport: **Convection**, **heat conduction** and **thermal radiation**.

10.2.1 Convection

When the bottom of a container with water is heated, (Fig. 10.20) the lowest liquid layer is heated first. Its temperature increases and its density therefore decreases. This causes a rise of this lower layer through the layers across, which sink down. This process is called **convection**. It results in a heat transport from the warmer to the colder region. This convection of liquids can be demonstrated by colouring the lower layer and observing how this coloured layer moves upwards when the bottom of the container is heated.

Convection occurs also for gases. It plays an essential role in the earth atmosphere and is responsible for the generation and the equalization of pressure differences (Fig. 10.21). Heated air rises from the bottom just above the earth surface, creating a local low pressure region. Air from the surrounding with higher pressure streams into this region. The wind transports not only mass but also heat [10.2a, 10.2b]. This mass- and heat transport depends on the wind velocity and the temperature difference between high and low pressure region. The wind flow can be either laminar or turbulent, depending on the boundary conditions.

Although the total energy received by the earth is due to radiation from the sun, the local distribution of this energy is essentially determined by convection. This is illustrated by sudden local temperature changes when the wind direction changes, although the intensity of the sun radiation has not changed.

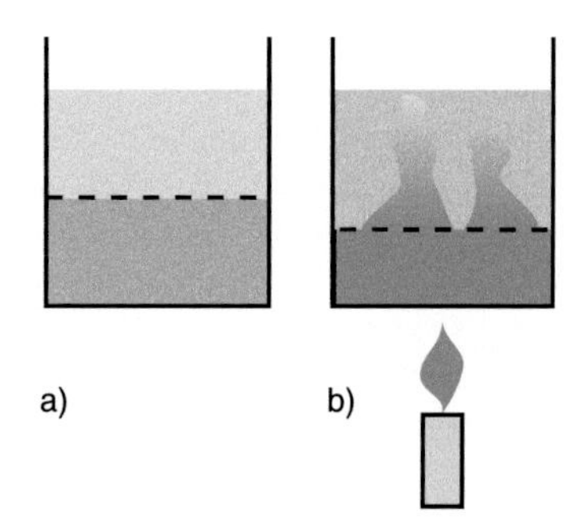

Figure 10.20 Convection in a liquid. **a** Lamination of dyed and pure water at equal temperatures; **b** mixing of the layers by convection due to heating at the bottom

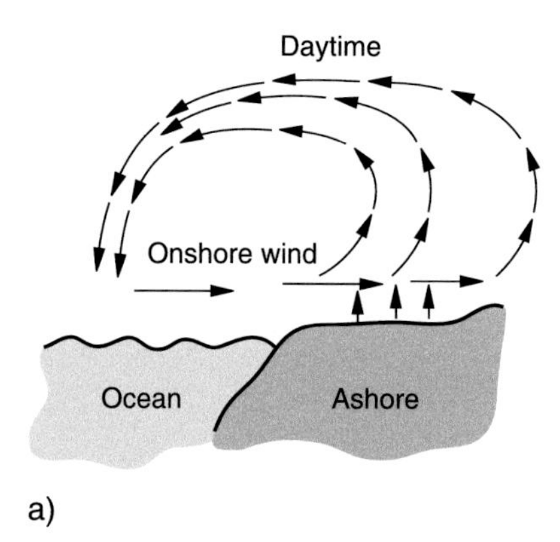

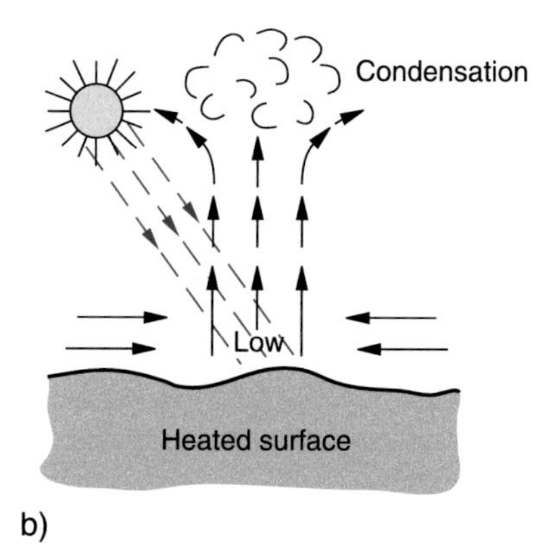

Figure 10.21 Convection in the earth atmosphere. **a** Onshore wind during the daytime, when the ocean surface is colder than the land surface; **b** wind flow into a low pressure region which is created by uprising air (thermal lift)

Also the temperature distribution in the oceans is mainly determined by convection. Examples are the gulf-stream, which influences the climate in the northern part of Europe, or the Humboldt current along the west coast of south America, causing the Atacama desert, because the cold water induces the moisture of the west wind to rain down before it reaches the dry areas.

When the temperature gradient of a liquid, heated at the bottom, exceeds a certain value that depends on the viscosity of the liquid, ordered macroscopic structures of the velocity field can develop. Current roles are created and the liquid moves along cylindrical stream lines (Fig. 10.22b). This sudden start of a

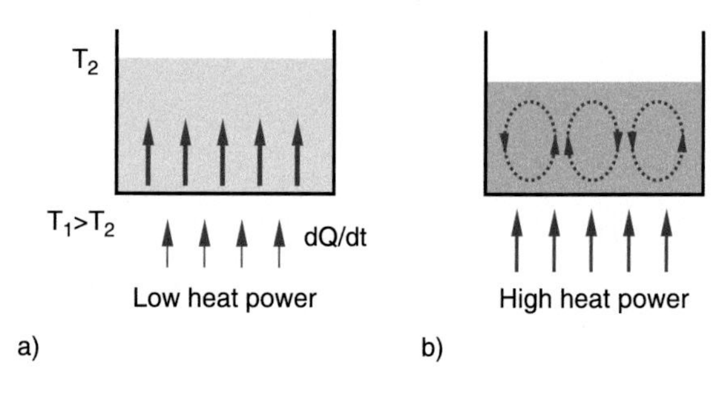

Figure 10.22 **a** Linear vertical temperature gradient; **b** Bénard instability

"self-organization" is called *Bénard-instability*. When the liquid is further heated, the rolls begin a wave-like motion along the cylinder axis. Such organized motions that develop from random conditions play an important role for the creation of organized structures from disordered systems. These processes are investigated in the rapidly developing field of *Synergetics*, which represents a frontier area between Physics, Chemistry, Biology and Computer Science [10.3, 10.4].

10.2.2 Heat Conduction

Contrary to the situation for convection, for heat conduction only energy transport takes place, but generally no mass transport. Heat conduction can only occur in matter, i. e. in vacuum no heat conduction is possible while thermal radiation also takes place in vacuum (otherwise we would not receive the sun radiation).

We will at first discuss heat conduction in solids, where the atoms or molecules are bound to fixed equilibrium positions and no convection can happen.

10.2.2.1 Heat Conduction in Solids

A rod with length L and cross section A is connected at both ends with thermal reservoirs that keep the two ends always at the fixed temperatures T_1 and T_2 with $T_1 < T_2$ (Fig. 10.23). After a sufficient long time a stationary state appears, where a temperature gradient $\mathrm{d}T/\mathrm{d}x$ is established that depends on the temperature difference $\Delta T = T_1 - T_2$, and on the length L. If we neglect heat losses through the side wall of the rod, a constant heat energy

$$\frac{\mathrm{d}Q}{\mathrm{d}t} = -\lambda \cdot A \cdot \frac{\mathrm{d}T}{\mathrm{d}x} \qquad (10.34a)$$

flows per sec through the cross section A of the rod . The constant λ ($[\lambda] = [\mathrm{W \cdot m^{-1} \cdot K^{-1}}]$) depends on the substance of the rod and is called **heat conductivity**. In Tab. 10.6 the heat conductivities of some substances are listed.

For a homogeneous rod with constant cross section A, the stationary temperature T(x) is a linear function of x, as can be seen by integrating (10.34a), which yields

$$T(x) = -\frac{\mathrm{d}Q/\mathrm{d}t}{\lambda \cdot A} x + C\,. \qquad (10.34b)$$

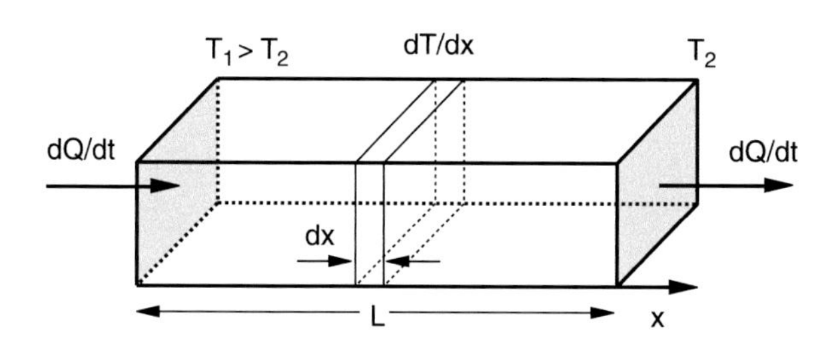

Figure 10.23 Heat conduction in a stab

Table 10.6 Heat conduction coefficient of some materials at 20 °C

Substance	$\lambda/(\mathrm{W\,m^{-1}K^{-1}})$
Aluminium	221
Iron	67
Gold	314
Copper	393
Zinc	112
Lead	35
Normal concrete	2.1
Foamed concrete	0.22
Glas	0.8
Glas wool	0.04
Wood	0.13
Ice	2.2
Water	0.6
Air (p = 1 atm)	0.026
CO_2 (p = 1 atm)	0.015
Helium (p = 1 atm)	0.14

The integration constant C is determined by the boundary condition $T(x = 0) = T_1 = C$. The energy supply necessary to maintain the given temperature gradient $\mathrm{d}T/\mathrm{d}x$, is obtained from

$$\mathrm{d}Q/\mathrm{d}t = \lambda \cdot A \cdot (T_1 - T_2)/L\,.$$

For the general nonstationary heat conduction through inhomogeneous bodies with variable cross section the temperature function $T(x,t)$ is more complicated. For its derivation we regard a volume element $\mathrm{d}V$ between the planes $x = x_1$ and x_2 (Fig. 10.24).

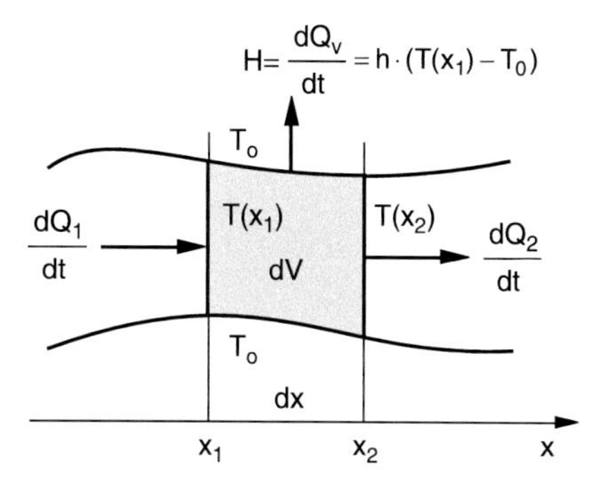

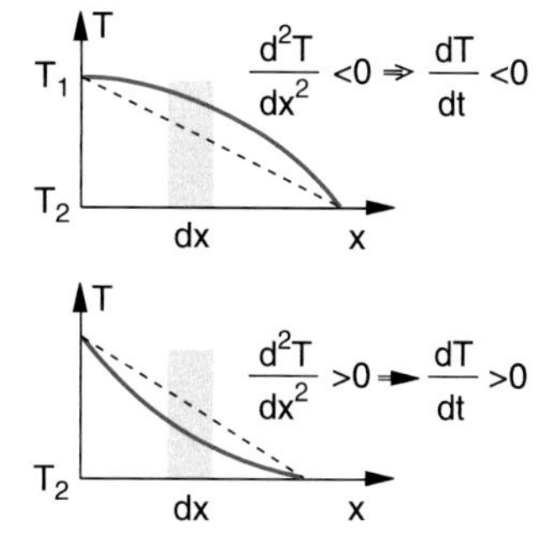

Figure 10.24 Derivation of the heat conduction equation

Chapter 10

For the one-dimensional case (for instance a thin homogeneous rod) the temperature changes only in one direction and we obtain for the heat power, transpórted through the cross section A at the position x_1

$$\frac{dQ_1}{dt} = -\lambda \cdot A \cdot \frac{\partial T}{\partial x} . \tag{10.35}$$

The partial derivative is used here, because the temperature $T(x, t)$ depends on the two variables x and t. At the plane $x_2 = x_1 + dx$ the temperature has changed to

$$T(x_2) = T(x_1) + (\partial T/\partial x) \cdot dx .$$

The heat passing per sec through the plane at $x = x_2$ is

$$\frac{dQ_2}{dt} = -\lambda \cdot A \cdot \frac{\partial}{\partial x}\left(T + \frac{\partial T}{\partial x} \cdot dx\right) . \tag{10.36}$$

When the temperature is higher at x_1 than at x_2, the heat dQ_1/dt flows per sec from the left side in Fig. 10.24 into the volume $dV = A \cdot dx$, and the heat dQ_2/dt leaves it per sec to the right side. The change dQ/dt of the heat per second in the volume dV is then

$$\begin{aligned}\frac{dQ}{dt} &= \frac{dQ_1}{dt} - \frac{dQ_2}{dt} = \lambda \cdot \frac{\partial^2 T}{\partial x^2} \cdot A \cdot dx \\ &= \lambda \cdot \frac{\partial^2 T}{\partial x^2} \cdot dV .\end{aligned} \tag{10.37}$$

Because $dQ = c \cdot m \cdot dT$ and $m = \varrho \cdot dV$ this net supply of heat power dQ changes the temperature T according to (10.37) by

$$\frac{\partial T}{\partial t} = \frac{\lambda}{\varrho \cdot c} \cdot \frac{\partial^2 T}{\partial x^2} . \tag{10.38a}$$

If the rod has heat losses $H = dQ_V/dt$ through the side walls (for example through cooling by the surrounding air) a loss term $H = h \cdot (T - T_0)$ has to be added to (10.37) which is proportional to the temperature difference between the rod temperature at the position x and the surrounding temperature T_0. The factor h has the unit $[W \cdot K^{-1}]$. Equation 10.38a can then be generalized with $h* = h/(c \cdot m)$ as

$$\frac{\partial T}{\partial t} = \frac{\lambda}{\varrho \cdot c} \cdot \frac{\partial^2 T}{\partial x^2} - h^* \cdot (T - T_0) . \tag{10.38b}$$

If T depends also on y and z all net heat power contributions supplied from all directions to the volume element dV add to the total energy increase of dV. One obtains for this three-dimensional case the general equation for the heat conduction

$$\begin{aligned}\frac{\partial T}{\partial t} &= \frac{\lambda}{c \cdot \varrho}\left(\frac{\partial^2 T}{\partial x^2} + \frac{\partial^2 T}{\partial y^2} + \frac{\partial^2 T}{\partial z^2}\right) \\ &= \frac{\lambda}{c \cdot \varrho} \cdot \Delta T = \lambda_T \cdot \Delta T ,\end{aligned} \tag{10.39}$$

with the Laplace operator Δ (see Sect. 13.1.6). The factor $\lambda_T = (\lambda/c \cdot \varrho)$ is the **thermal diffusivity**.

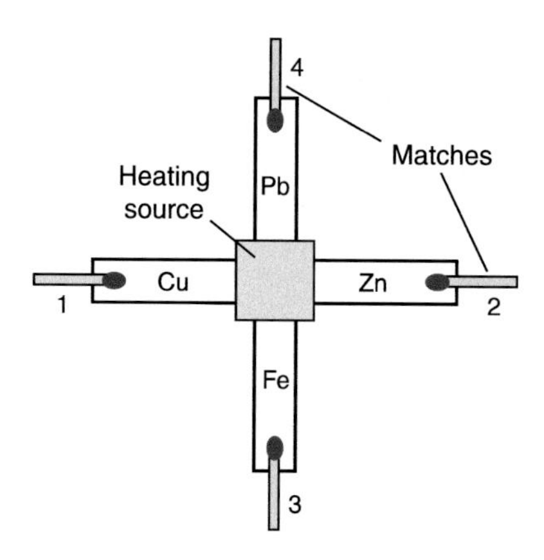

Figure 10.25 Demonstration of the different heat conduction of some metals

The heat conduction in solids is accomplished by the coupling between adjacent atoms, which causes the transport of the vibrational energy of atoms at the plane x to the neighbouring position $x + dx$ without a transport of the atoms themselves.

In metals the freely moving electrons contribute essentially to heat conduction by collisions with each other and with the atoms. Because of their small mass, their thermal velocities and in particular their Fermi-velocities (which can be only explained by quantum theory, see Vol. 3) are very high. They can therefore transfer their large kinetic energy much faster by collisions. The heat conductivity in metals is therefore mainly due to the electrons. Experiments confirm in deed that for metals the thermal conductivity (λ) is proportional to the electrical conductivity (σ), which is solely caused by electron transport.

This is expressed by the *Wiedemann–Franz law*

$$\begin{aligned}\lambda/\sigma &= a \cdot T \quad \text{with } a = \pi^2 k^2/3e^2 \\ &= 2.45 \cdot 10^{-8} \mathrm{V}^2/\mathrm{K}^2 ,\end{aligned} \tag{10.40}$$

where the constant a is determined by the Boltzmann constant k and the elementary charge e.

This can be readily demonstrated by a simple experiment (Fig. 10.25). The red centre plate of a cross with four arms of different metals is heated by a small burner. At the ends of the arms four matches are placed. After the heating starts it takes different times t_i until the ends of the arms reach the ignition temperature. The matches are ignited at times $t_1 < t_2 < t_3 < t_4$. This time sequence reflects the electrical conductivity of the metals, where the arm 1 (Cu) has the highest electrical and thermal conductivities.

In solids (even in non-metals) the thermal conductivity is much larger than in gases, because of the much larger density and the resulting larger coupling strength between neighbouring atoms (see Tab. 10.6). However, the coefficient of heat conductivity $\lambda_T = \lambda(c \cdot \varrho)$, which gives the time constant of reaching a stationary temperature, is for solids and gases nearly the same because of the much smaller density ϱ of gases.

Chapter 10

In gases temperature differences are equated in times comparable to those in solids.

One of the reasons is the much smaller heat energy to reach a temperature rise ΔT for a given volume of a gas than for the same volume of a solid.

For the measurement of heat conductivities stationary as well as time resolving techniques have been developed [10.5].

For the stationary methods a constant heat power $\mathrm{d}Q/\mathrm{d}t$ is supplied to one end of the body (for instance a rod), which is extracted on the other side by cooling. According to (10.35) this results for a rod with constant cross section A in a constant temperature gradient

$$\frac{\partial T}{\partial x} = \text{const} = \frac{T_1 - T_2}{L} = \frac{1}{\lambda \cdot A} \cdot \frac{\mathrm{d}Q}{\mathrm{d}t} , \tag{10.41}$$

which can be determined by measuring the temperatures T_1 and T_2 and the length L.

The dynamical methods for the measurement of the heat conduction under non-stationary conditions are based on a time-dependent supply of the heat power. The heat power $\mathrm{d}Q/\mathrm{d}t$ is either periodically modulated or supplied in short pulses. If for example the heat power supplied at $x = 0$ is

$$\mathrm{d}Q/\mathrm{d}t = \mathrm{d}Q_0/\mathrm{d}t + a \cdot \cos(\omega t) ,$$

the temperature at $x = 0$ is

$$T(0, t) = T_1 + \Delta T \cdot \cos(\omega t) ,$$

and one obtains from the heat conduction equation (10.38b) for a thin cylindrical rod with heat losses $h \cdot (T - T_0)$ through the side walls (Fig. 10.26) the solution

$$\begin{aligned} T(x, t) = {} & T_0 + (T_1 - T_0)\mathrm{e}^{-\alpha_1 x} \\ & + \Delta T \mathrm{e}^{-\alpha_2 x} \cdot \cos(\omega t - kx) . \end{aligned} \tag{10.42}$$

Inserting this into (10.38b) yields for the coefficients

$$\begin{aligned} \alpha_1 &= \sqrt{\frac{\varrho c h^*}{\lambda}} = \sqrt{\frac{h^*}{\lambda_T}} , \\ \alpha_2 &= \left[\frac{(h^{*2} + \omega^2)^{1/2} + h^*}{2\lambda_T}\right]^{1/2} ; \\ k &= \left[\frac{(h^{*2} + \omega^2)^{1/2} - h^*}{2\lambda_T}\right]^{1/2} . \end{aligned}$$

The temperature $T(x)$ along the rod is a superposition of a constant time-independent contribution that decays exponentially with x due to the heat losses through the sidewalls, and a damped temperature wave with an exponentially decreasing amplitude. The phase of this wave is determined by the loss coefficient h,

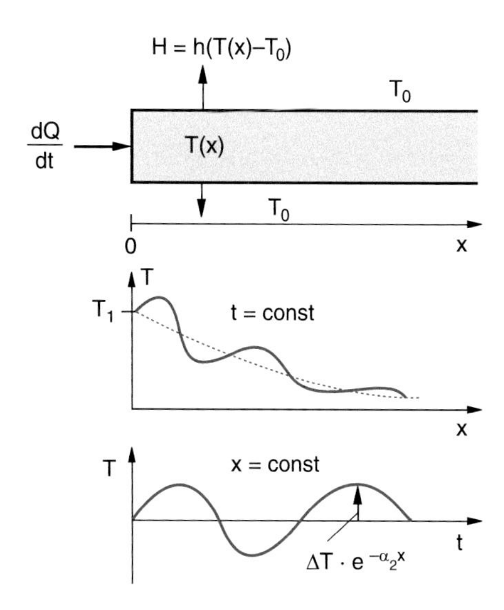

Figure 10.26 Damped temperature waves in a stab under periodic energy supply

the frequency ω and the wavelength λ_T of the temperature wave. The phase velocity of the wave

$$v_{\mathrm{Ph}} = \frac{\omega}{k} = \left[\frac{2\omega^2 \cdot \lambda_T}{(\omega^2 + h^{*2})^{1/2} + h^*}\right]^{1/2} \tag{10.42a}$$

depends on the frequency ω.

Temperature waves show dispersion!

When amplitude and phase of temperature waves are measured at selected points x for different frequencies ω, the quantities h and $\lambda_T = \lambda/\varrho \cdot c$ can be obtained.

Without heat losses ($h = 0$) (10.42) reduces to

$$\begin{aligned} T(x, t) &= T_1 + \Delta T \mathrm{e}^{-\alpha x} \cos(\omega t - kx) \\ \text{with} \quad \alpha &= k = (\omega/2\lambda_T)^{1/2} . \end{aligned} \tag{10.42b}$$

10.2.2.2 Heat Conduction in Liquids

In liquids, there are no shear forces (see Sect. 6.2). Therefore, the coupling between neighbouring atoms is much weaker than in solids and the heat transport is slower. The heat conduction in liquids that have no electrical conductivity, is therefore smaller than in solids (see Tab. 10.6). However, in liquids the freely moving molecules can transfer energy by collisions. The effective energy transfer depends on the mean velocity of the molecules, the time between two collisions and the cross section for energy transferring collisions.

In electrically conducting liquids (for example mercury or melted metals) the free electrons make the major contribute to

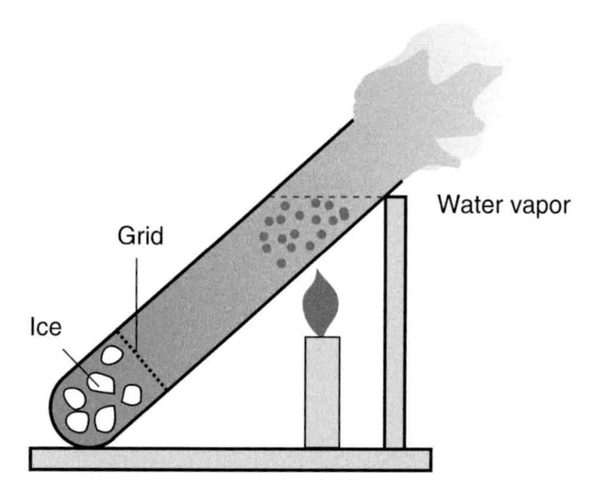

Figure 10.27 Demonstration of the small heat conductivity of water

the energy transfer, similar to the situation in solids. Their heat conductivity is therefore much larger than for non-conductive liquids, because of their much larger mass, the ions give only a minor contribution.

Because of the free mobility of the molecules in liquids, generally convection occurs besides heat conduction. This can be prevented when the liquid is heated from above, because then the hot liquid layer has a smaller density and therefore remains at the top.

The small heat conductivity of water can be demonstrated by the experiment shown in Fig. 10.27. At the bottom of a glass tube filled with water, are small ice cubes which are prevented from uprising by a mesh. One can heat the upper part of the water until it boils and emits water vapour. Nonetheless, the ice cubes do not melt in spite of the temperature difference of $\Delta T = 100\,°\mathrm{C}$ between the lower and upper part of the tube due to the poor heat conductivity of glass and water and the absence of convection.

10.2.2.3 Heat Conduction in Gases

In Sect. 7.5 it was shown, that heat conduction in gases is caused by collisional energy transfer between the molecules which move with thermal velocities. According to Eq. 7.49 the heat energy transferred per m^2 between two parallel walls at temperatures T_1 and T_2 is

$$J_w = \kappa \cdot (T_1 - T_2) \,.$$

According to (7.49a) the heat conduction coefficient is

$$\alpha = n \cdot v \cdot k \cdot f/8 \sim n \cdot \sqrt{T/m} \,.$$

Because of the much smaller density n of gases compared to liquids the heat conduction in gases is generally much smaller, except for ionized gases where the electrons contribute essentially to heat conduction. For neutral gases it is maximum for hydrogen because of the small mass m of hydrogen molecules.

When the mean free path Λ is larger than the dimensions of the gas container, the heat conduction becomes independent of the gas pressure.

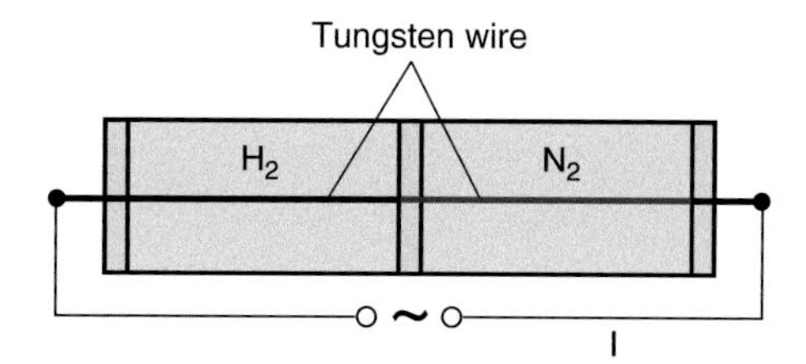

Figure 10.28 Demonstration of heat conduction in gases and its dependence on the molecular mass

The influence of the molecular mass m on the heat conduction can be demonstrated by the device shown in Fig. 10.28. A tungsten wire runs coaxially through two separated parts of a glass tube, which are filled with hydrogen gas in the left part and nitrogen gas in the right part. When the wire is heated by an electric current the right part has a higher temperature and is glowing red while the left part remains much colder due to the different heat conduction of the two gases. When removing the two gases both parts of the wire glow equally strong. This effect is intensified by two causes:

1. The electrical resistance R of tungsten decreases with decreasing temperature T. Therefore the electrical power $\mathrm{d}W/\mathrm{d}t = I^2 \cdot R$ supplied to the wire is smaller in the cold part.
2. The visible radiation power of the glowing wire is proportional to T^4. Even a small change of the temperature T results in a large change of the radiation power.

A modification of the demonstration experiment (Fig. 10.29) uses a vertical glass tube with the coaxial wire, which is filled with a gas mixture of H_2 and N_2. At first the heated wire glows equally bright along the whole tube. After some minutes, the lighter H_2-gas diffuses to the upper part while the heavier N_2-gas sinks to the bottom (see Sect. 7.6 and Fig. 7.34). This effect is even amplified by convection where the hot gas around the wire rises up while the colder gas close to the inner wall of the glass tube sinks down. Now the lower part of the wire is brighter than the upper part.

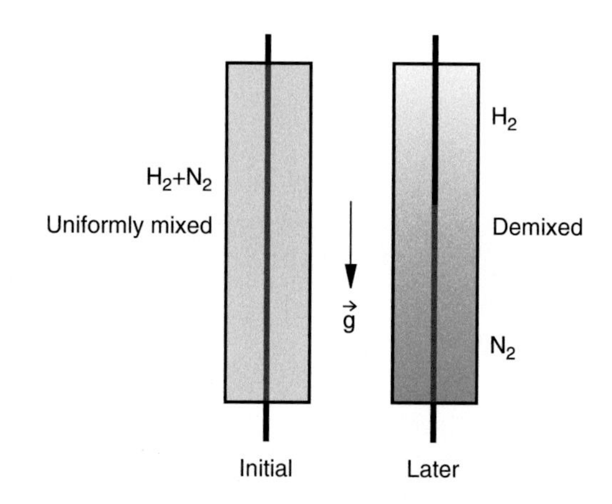

Figure 10.29 Demixing of a gas mixture by convection and diffusion in the gravitation field of the earth

For most situations the convection in gases at atmospheric pressure gives a larger contribution to the energy transport than heat conduction.

10.2.3 The Heat Pipe

Often the problem arises that heat produced in a volume V should be extracted as effectively as possible, in order to reach a sufficient cooling power. For the solution of this problem a special device was developed, which uses evaporation of a liquid on the hot side and condensation of the vapour on the cold side. The heat transport occurs by convection. This heat pipe allows a heat transport through the unit area that is larger by two orders of magnitude than can be achieved with metals. Its basic principle is illustrated in Fig. 10.30.

A tube of metal or another material is connected at the hot side (left) with the volume at the temperature T_1 that should be cooled and on the cold side with a cooling bath at $T = T_2 < T_1$. The evacuated tube is filled with a substance that has an evaporation temperature $T_e < T_1$ and a melting temperature $T_m < T_2$. For instance, if water is used, the temperatures should be $T_1 > 100\,°C$ and $T_2 > 0\,°C$.

At the hot side the substance boils which extracts the evaporation heat from the volume to be cooled. The vapour streams to the cold end where it condenses and delivers its heat of fusion to the cooling bath. Along the tube a gradient of the vapour density develops and an opposite gradient of the liquid density. An essential part of the heat pipe is a mesh that is wrapped around the inner part of the tube close to the wall. For the correct choice of the materials for tube and mesh the liquid substance wets both the mesh and the inner wall of the tube. Due to capillary action the liquid then flows between mesh and wall from the fusion zone back to the evaporation zone where it can be again evaporated and extract heat. The heat transport of this cyclic process depends on the vapour density and its flow velocity from the hot to the cold zone, but mainly on the magnitude of evaporation and fusion energy. For cooling media with a large evaporation energy (for example water) and a large convection velocity a very large heat transport per sec can be achieved.

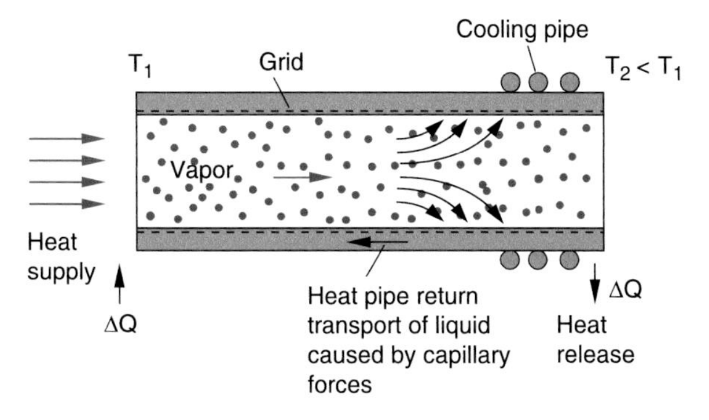

Figure 10.30 Heat pipe

With the mass dm/dt evaporated per sec the energy extracted per sec from the hot volume is

$$dW/dt = \lambda_e \cdot dm/dt\ ,$$

where λ_e is the specific evaporation energy. This is much larger than the heat $c_p(dm/dt)\Delta T$ transported through the heat pipe with a temperature difference $\Delta T = T_1 - T_2$ (see Tab. 10.5). At the cold end the heat

$$dW_2/dt = (\lambda_f \cdot + c_p \Delta T) \cdot dm/dt$$

has to be transferred to the cooling water.

More details about the technical design and the applications of heat pipes can be found in [10.6].

10.2.4 Methods of Thermal Insulation

While in Sect. 10.2.3 the realization of devices with a maximum heat transport was discussed, in this section we will treat methods to prevent heat transport out of a volume or to make it at least as small as possible. In order to reach this goal one has to take into account the contributions of all three heat transport mechanisms and minimize them. We will illustrate this by consider the thermal isolation of a residential house.

The heat transport between the inside and outside is mainly governed by heat conduction through walls and windows and to a minor part by air convection through leaky joints and during airing of a room. Depending on the size and the technical features of the windows also heat radiation can be important for heat exchange.

The heat flux through the area A of walls or windows with thickness d and a temperature difference $\Delta T = T_i - T_o$ between inside and outside is

$$dQ/dt = -(\lambda/d) \cdot A \cdot \Delta T\ , \qquad (10.43)$$

where λ [W/(m · K)] is the heat conductivity, which depends on the material. It is generally characterized by the constant $k = \lambda/d$, which gives the energy flux through the unit area $A = 1\,m^2$ at a temperature difference $\Delta T = 1\,K$. For most estimations of the heat isolation of houses the k value [W/(m^2 · K)] of walls and windows is given. For good heat insulation it should be as small as possible. In Fig. 10.31 the k-values and the temperature rise from an outside temperature $T_0 = -15\,°C$ to the room temperature $T_i = +20\,°C$ are depicted for different wall compositions. These figures illustrate, that even a thin layer of Styrofoam considerably improves the thermal insulation. The largest heat losses are caused by the windows, where the heat transport process is more complex. We regard at first a single-layer window (Fig. 10.32). In spite of the small heat conductivity of glass ($\lambda = 0.9\,W/(mK)$) the k-value $k = 200\,W/(m^2 \cdot K)$ is much larger than that of the thick walls, due to the small thickness ($d = 4\,mm$) of the window.

Because of the temperature gradient in the air layers close to the inside and outside of the glass a convective air current develops,

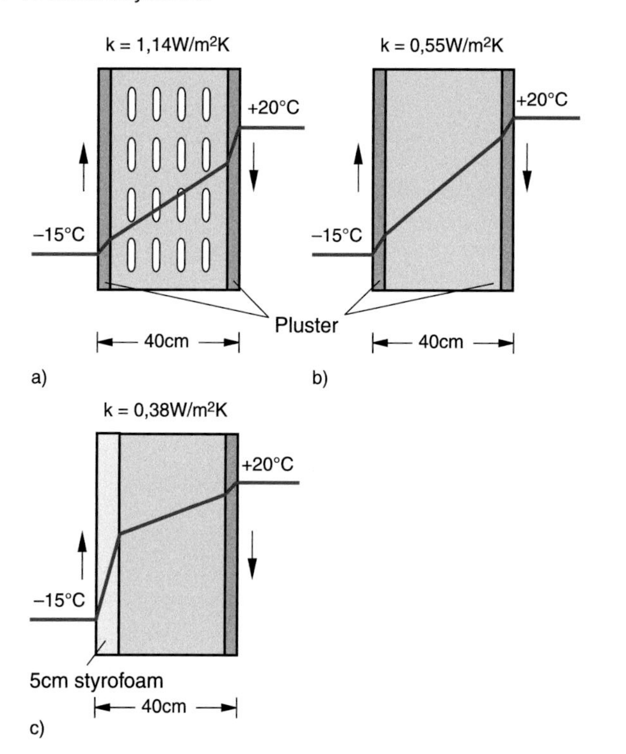

Figure 10.31 Thermal insulation. **a** Temperature behaviour across a plastered claybrick wall; **b** plastered wall of pumice stone; **c** pumice stone wall with styrofoam layer. The arrows give the direction of the convection current

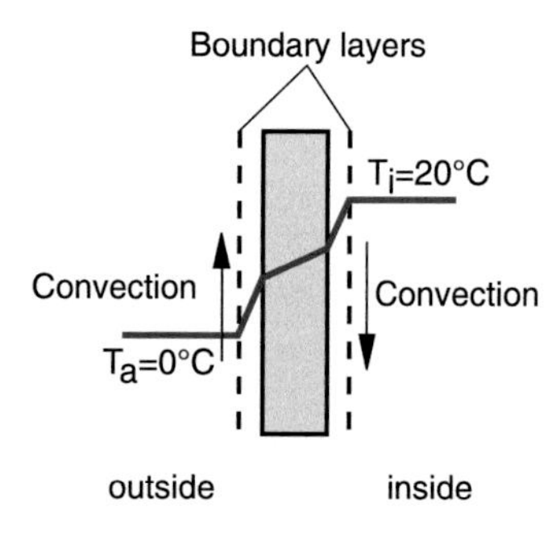

Figure 10.32 Heat transport across a single pane window

which is for $T_o < T_i$ downwards at the inside and upwards at the outside (Fig. 10.32). Due to friction between glass and air a thin boundary layer of air adheres on both sides of the glass (see Sect. 8.4). The heat passes through these layers to the convective air layers. Since the heat conduction is smaller for gas at atmospheric pressure than for glass, the k-value is smaller for these adhered air layers then for the pane of glass.

From Eq. 8.24 one obtains a thickness of 5 mm for the boundary air layer with a k-value $k = 3.4\,\mathrm{W/(m^2 \cdot K)}$ for heat conduction, compared to $k = 200\,\mathrm{W/(m^2 \cdot K)}$ for the pane of glass.

Another heat transport mechanism is heat radiation (see Sect. 10.2.5). The room temperature in the inner part of the house causes infrared radiation, which can escape through the windows. The heat loss can be estimated as $4.6\,\mathrm{W/(m^2 \cdot K)}$. This gives a total k-value of the inner air layer $k = 8\,\mathrm{W/(m^2 \cdot K)}$.

For the outer convective air layer the k-value is different because the air flows upwards against the gravitation. Detailed calculations give a value $k = 20\,\mathrm{W/(m^2 \cdot K)}$ including radiation losses. For successive layers the reciprocal k-values add (analogous to electrostatics where the reciprocal electric conductivities add) and we obtain from

$$\frac{1}{k} = \frac{1}{k_i} + \frac{1}{k_g} + \frac{1}{k_o} \tag{10.44}$$

the total k-value $k = 5.5\,\mathrm{W/(m^2 \cdot K)}$. The comparison with the k-value of the walls $k < 1\,\mathrm{W/(m^2 \cdot K)}$ shows that windows with a single pane of glass constitute a major heat loss.

A much better heat insulation can be achieved with windows of two panes of glass and an inert gas enclosed between the panes (Fig. 10.33a).

The k-value of the gas depends on the thickness d of the gas between the glass panes. For $d \leq 1$ cm the heat conduction is dominant, while for larger values of d convection undertakes the major part of heat transfer. Fig. 10.33b shows, that for $d = 1$ cm the minimum k-value is reached because the boundary layers that adhere to the glass walls, prevent convection.

For such a double glass window, the k-value for heat conduction is substantially smaller than for a single pane window. In order

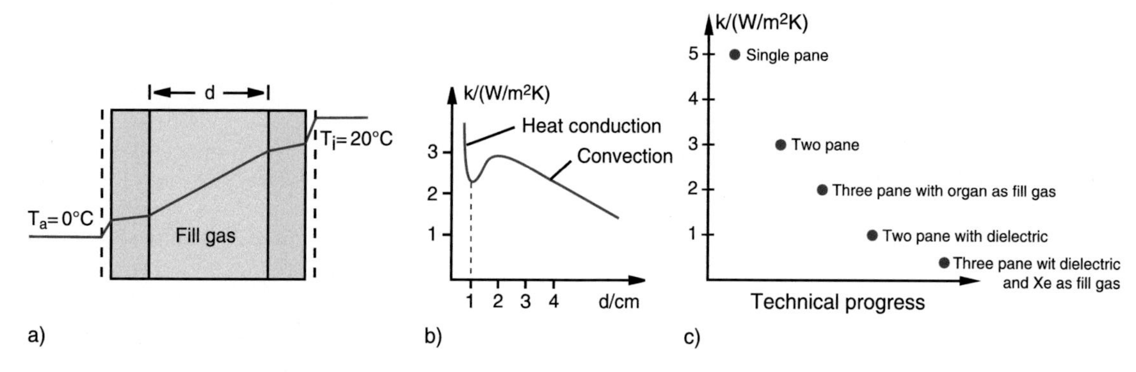

Figure 10.33 Double pane glas window. **a** Composition and temperature change across the window; **b** k-values as a function of the thickness d of the gas layer between the two glas panes; **c** decrease of k-values through technical progress

to minimize also the radiation losses, the glass panes are covered by a thin dielectric layer [10.8], which reflects the infrared radiation from the inside of the house (see Vol. 2, Sect. 10.4). Without a reflecting layer, k-values of $k \leq 3\,\mathrm{W/(m^2 \cdot K)}$ can be realized, while with reflecting layer the k-value decreases down to $k \leq 0.6\,\mathrm{W/(m^2 \cdot K)}$. The k-values are then comparable to those of the walls [10.8, 10.9].

The considerations above illustrate that all three heat-transfer processes as heat conduction, convection and radiation have to be taken into account in order to optimize the heat insulation of a house. In Fig. 10.33c the technical progress of minimizing the total k-value is illustrated.

A more quantitative representation of heat insulation can be found in [10.7] and the references gives there and also in many books on energy saving new house construction [10.9].

10.2.5 Thermal Radiation

Every body at a temperature T_K exchanges energy with its surrounding. If T_K is higher than the temperature T_S of the surrounding, the energy emitted by the body is larger than the energy received from the surrounding. If no energy is supplied to the system body +surrounding, the system approaches thermal equilibrium, where the temperature of the body is equal to that of the surrounding (Fig. 10.34). This energy balance can be reached by heat conduction, convection or radiation. If the body is kept in vacuum, (for instance our earth) radiation is the only way to exchange energy with the surrounding, because both heat conduction and convection need matter for the transport of energy.

Extensive experiments have proved, that radiation emitted by hot bodies represents electromagnetic waves, which can transport energy through matter and also through vacuum.

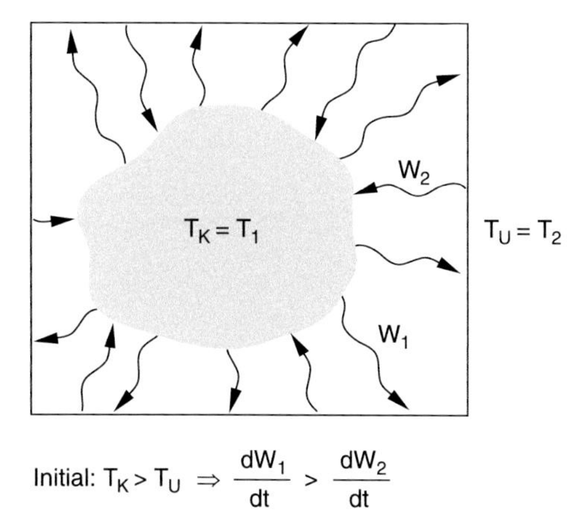

Figure 10.34 Energy exchange by thermal radiation between a body and its surroundings. At thermal equilibrium is $\mathrm{d}W_1/\mathrm{d}t = \mathrm{d}W_2/\mathrm{d}t$ and $T_K = T_2$

Since the intensity and the spectral distribution of the radiation emitted by a body depends essentially on the temperature of the body, this radiation is called **heat radiation** or **thermal radiation**. In this section we will discuss the characteristics of thermal radiation.

10.2.5.1 Emissivity and Absorptivity of a Body

At first we will experimentally study, how the intensity of thermal radiation depends on the surface conditions of the body. We use a metal hollow cube filled with hot water, where the four side walls have a different surface structure (black, white, shiny and rough). All side walls have the same temperature. Four equal radiation detectors, which measure the total radiation (integrated over all wavelengths) are placed at the same distance d from the four walls (Fig. 10.35). They all show different radiation powers. When the cube is turned by $n \cdot 90°$ ($n = 1, 2, 3, \ldots$) about a vertical axis, it can be proved that the difference is not due to differences of the detectors but that the different sidewalls really emit different radiation powers. The experiment shows surprisingly that the black side wall emits the maximum power and the shiny white surface the minimum power. The radiation power emitted from the surface area $\mathrm{d}A$ into the solid angle $\mathrm{d}\Omega$ can be quantitatively described by

$$\frac{\mathrm{d}W}{\mathrm{d}t} = E^* \cdot \mathrm{d}A \cdot \mathrm{d}\Omega\,.$$

The constant E^* is the emissivity of the surface. It gives the radiation power $\mathrm{d}W/\mathrm{d}t$, integrated over all wavelengths that is emitted from a surface element $\mathrm{d}A = 1\,\mathrm{m}^2$ into the solid angle $\mathrm{d}\Omega = 1\,\mathrm{sr}$ around the surface normal (Fig. 10.36). According to the experiment the emissivity E^* of a black surface is larger than that of a white surface at the same temperature.

The integral absorptivity A^* is defined as the mean value of the quotient $A^* =$ absorbed radiation power / incident radiation power, averaged over all wavelengths.

The ratio

$$K(T) = \frac{E^*(T)}{A^*(T)} \tag{10.45}$$

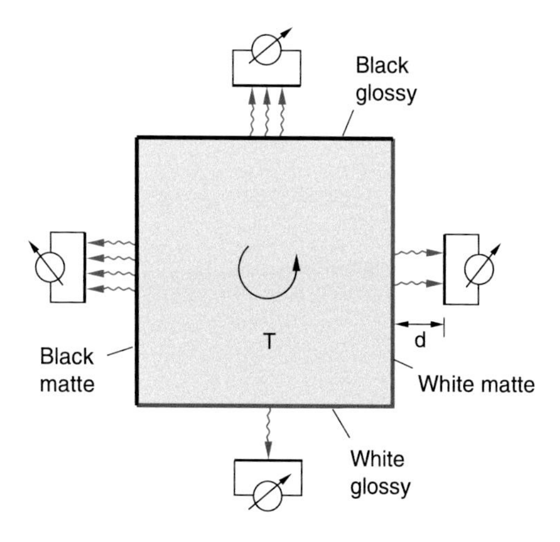

Figure 10.35 Experimental setup for the measurement of emission

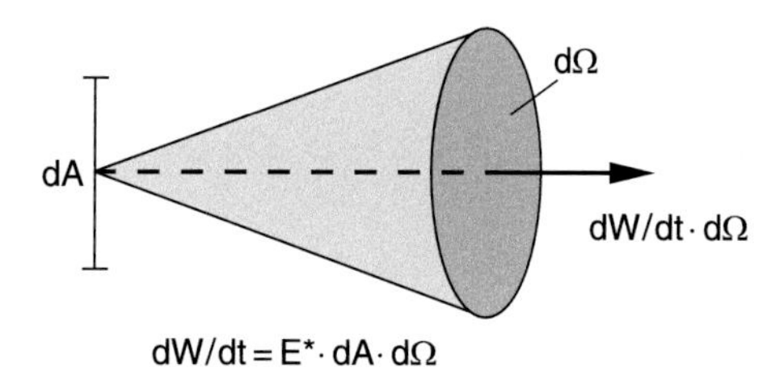

Figure 10.36 Illustration of the emissivity E^* of a surface element dA

depends solely on the temperature T and not on the material of the body, as can be demonstrated by the following experiment:

Experiment

We place in Fig. 10.37a in front of the black surface A_1 of the hot cube an equivalent surface A'_1 of the detector at a distance d and in front of the shiny surface A_2 a shiny detector surface A'_2 at the same distance d. Measuring the temperatures T_1 of A'_1 and T_2 of A'_2 one finds that $T_1 > T_2$.

Since the surface structure of A'_1 is equal to that of A_1, and that of A'_2 is equal to that of A_2, the absorptivity A^*_1 must be equal to A'^*_1 and $A^*_2 = A'^*_2$.

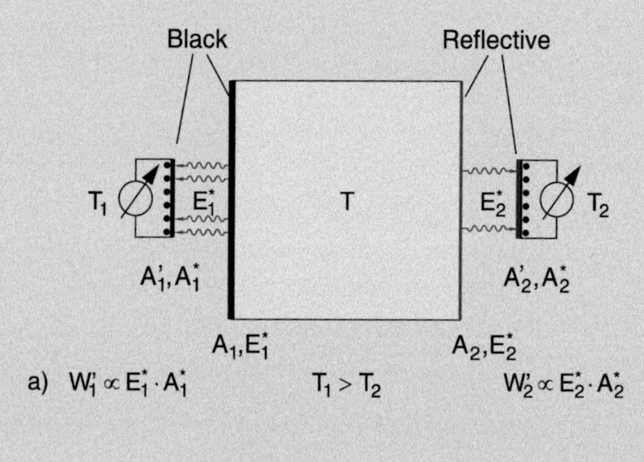

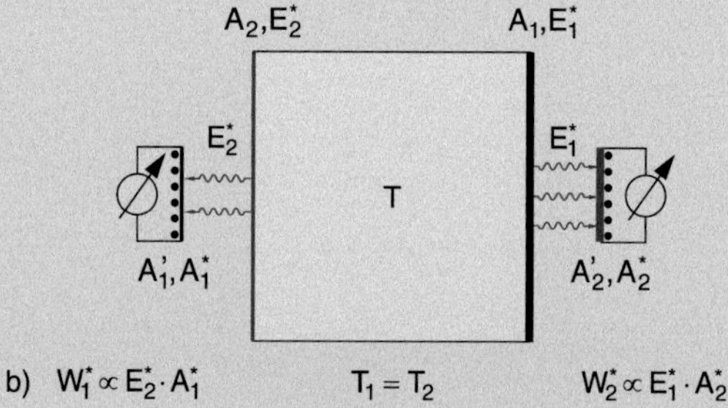

Figure 10.37 Derivation of (10.47)

The power absorbed by the two detector surfaces is

$$\mathrm{d}W'_1/\mathrm{d}t \sim E^*_1 \cdot A'^*_1 \quad \text{and} \quad \mathrm{d}W'_2/\mathrm{d}t \sim E^*_2 \cdot A'^*_2 \, .$$

Since $E^*_1 > E^*_2$ (according to the foregoing experiment, a black surface emits a larger power than a shiny one) and $A'^*_1 > A'^*_2$ (a black surface has a higher absorptivity than a shiny one) it follows that $\mathrm{d}W'_1/\mathrm{d}t > \mathrm{d}W'_2/\mathrm{d}t \to T_1 > T_2$.

Now the cube is turned about a vertical axis by 180° and the surface A_1 now faces A'_2 and A_2 faces A'_1 (Fig. 10.37b). The absorbed powers are

$$\mathrm{d}W_1/\mathrm{d}t \sim E^*_2 \cdot A'^*_1 \quad \text{and} \quad \mathrm{d}W_2/\mathrm{d}t \sim E^*_1 \cdot A'^*_2 \, .$$

The experimental result is now $T_1 = T_2 \to \mathrm{d}W_1/\mathrm{d}t = \mathrm{d}W_2/\mathrm{d}t$.

$$\Rightarrow \frac{E^*_1(T)}{A^*_1} = \frac{E^*_2(T)}{A^*_2} \, . \tag{10.46}$$

◀

A separate experiment proves that the absorptivity of the surfaces does not depend on the temperature at least within the temperature range from 0–100 °C, which is covered in the experiment above. Therefore it follows from (10.46) for an arbitrary body

$$\frac{E^*_1(T)}{A^*_1} = \frac{E^*_2(T)}{A^*_2} = K(T) \, . \tag{10.47}$$

The ratio of emissivity to absorptivity can be described for any body by a function $K(T)$ that depends solely on the temperature T.

A body with $A^* = 1$ is called a ***black body***.

It completely absorbs any incident radiation. According to (10.47) a black body must also have the maximum emissivity compared to all other bodies with equal temperature.

Note: Bodies with a large absorption coefficient α but a sudden increase of α at the glossy surface are not a black body, because their reflectivity also increases (Fig. 10.38a). Therefore the major part of the incident radiation is reflected and only the minor part, that penetrates into the body is absorbed (see Vol. 2, Chap. 8). In order to realize a black body, the absorption coefficient should not increase suddenly at the surface but must continuously increase over a distance $\Delta z > \lambda$ (λ = wavelength of the incident radiation) from zero to its maximum value (Fig. 10.38b). This can be for instance realized by a roughened surface (black velvet, soot or graphite with a rough surface) where the optical density rises slowly from the outside to the inner part of the body. The sun is an example of a nearly perfect black body, because the gas density and with it the absorptivity increases slowly from the diffuse outer edge of the photosphere to the interior.

Often the problem arises to keep a body at a constant temperature T_K, that differs from the temperature T_S of its surrounding by supporting ($T_K > T_S$) or extracting ($T_K < T_S$) energy. This energy can be minimized when heat conduction, convection and radiation are minimized. The experimental realization uses materials with low heat conductance and radiation shields.

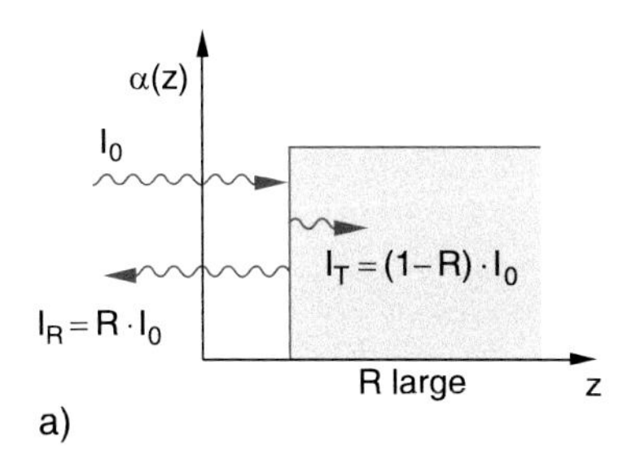

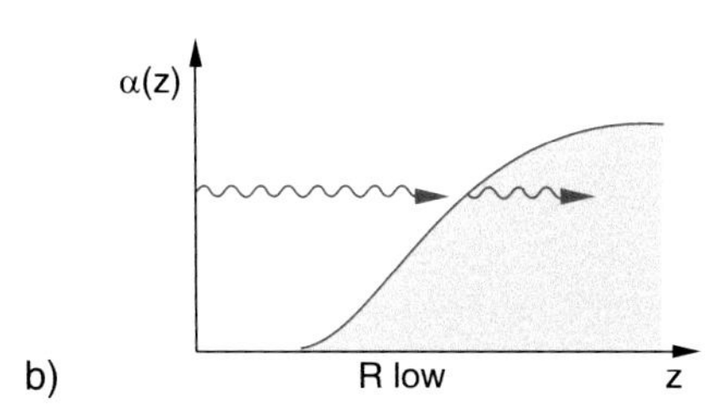

Figure 10.38 **a** For bodies with a large gradient $d\alpha/dz$ at the surface incident radiation is mainly reflected and only partly absorbed inspite of the large absorption coefficient α. **b** Most of the incident radiation is absoorbed, if $d\alpha/dz$ is small, i. e. $\alpha(z)$ rises slowly from $\alpha = 0$ to $\alpha = \alpha_{max}$

Examples

Figure 10.39 **a** Thermos bottle; **b** Dewar gasket

1. A *thermos bottle* (Fig. 10.39a) consists of a double wall glass flask. The space between the two walls is evacuated and the two inner sides of the walls are mirrored. The vacuum prevents heat conduction and convection. The reflective walls minimize the escape of thermal radiation to the outside. Therefore, the heat losses from the inner volume are very small and the coffee stays hot for a long time or cold drinks remain cold.
2. For the storage of liquid nitrogen a *Dewar* is used (Fig. 10.39b), which is based on the same principle as the thermos bottle. Here the heat transfer from the outside is minimized in order to keep the evaporation of the cold liquid nitrogen ($T = 77$ K) as low as possible. The small portion of evaporating nitrogen extracts the heat of evaporation and keeps the temperature in the Dewar low.
   If liquid air is used (78% N_2 and 21% O_2), the nitrogen evaporates faster because of its higher vapour pressure and the concentration of the reactive oxygen increases until an explosive concentration is reached. Therefore, generally liquid air is dangerous and is only used for special purposes. ◄

10.2.5.2 Characteristic Features of Thermal Radiation

The energy that is emitted by the surface element dA into the solid angle $d\Omega$ around the direction Θ against the surface normal can be measured with a radiation detector (for example a thermo-couple connected to a black surface). The detector area dA_2 at a distance r from the radiation source receives the radiation within the solid angel

$$d\Omega = \frac{dA_2}{r^2} \,.$$

Experiments prove that for many radiation sources the angular distribution of the measured radiation power is

$$dW(\Theta)/dt = S^* \cos\Theta \cdot dA \cdot d\Omega \,. \qquad (10.48)$$

The quantity S^* is the *emittance* or *radiation density* of the source. It describes the radiation power per m^2 of the radiation source, emitted into the solid angle $d\Omega = 1$ sr around the surface normal (Fig. 10.40a).

The radiant intensity

$$J(\Theta) = \int_F S^* \cos\theta dA \,, \quad [J] = 1\,\frac{\text{W}}{\text{sr}} \qquad (10.49)$$

is the total radiation power emitted by the radiation source into the solid angle $d\Omega = 1$ sr around the direction Θ against the surface normal.

Note: The relation between the *radiation density* S^* and the *emissivity* E^* is outlined in Sect. 10.2.5.3.

Chapter 10

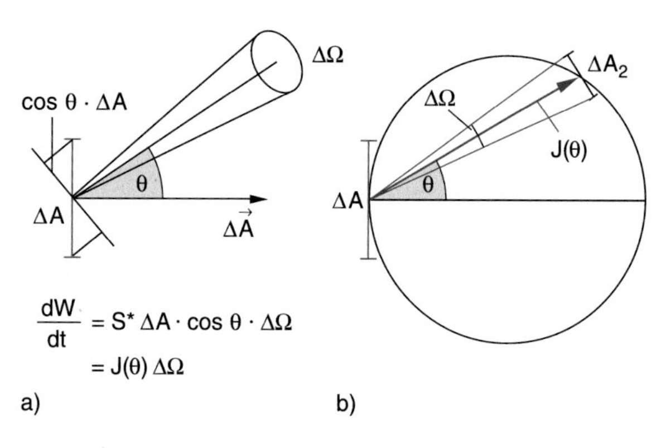

Figure 10.40 **a** Illustration of radiant intensity $J(\theta)$. **b** The length of the arrow is proportional to the radiant intensity $J(\theta)$

The emitted radiation power generally depends on the wavelength λ resp. the frequency $\nu = c/\lambda$ of the electromagnetic wave. The spectral radiation intensity S^*_ν is defined by the equation

$$S^* = \int_{\nu=0}^{\infty} S^*_\nu \mathrm{d}\nu \,. \tag{10.50}$$

The radiation of the source results in an electromagnetic field with the energy density w [J/m^3] and the intensity I [W/m^2]. The *spectral energy density* w_ν is the energy per m^3 within the spectral frequency interval $\Delta\nu = 1\,\mathrm{s}^{-1}$. It is related to the total energy density w by

$$w = \int w_\nu \cdot d\nu \,. \tag{10.51}$$

For a radiation source with isotropic radiation (for instance the sun) the relation between $I = |S|$ and w is

$$I = (c/4\pi) \cdot w \,, \tag{10.52a}$$

and similar for the spectral quantities

$$I_\nu = (c/4\pi) \cdot w_\nu \,, \tag{10.52b}$$

where c is the velocity of light. For plane waves the relations are $I = c \cdot w$ and $I_\nu = c \cdot w_\nu$.

The detector element ΔA_2 at a distance r from the isotropic source receives from the source element ΔA_1 the radiation power

$$\begin{aligned}\frac{\mathrm{d}W_1}{\mathrm{d}t} &= S^*_1 \cos\theta_1 \Delta A_1 \Delta\Omega \\ &= \left(S^*_1 \cos\theta_1 \Delta A_1 \Delta A_2 \cdot \cos\theta_2\right)/r^2 \,,\end{aligned} \tag{10.53}$$

where $\Delta\Omega = \Delta A_2 \cdot \cos\theta_2 / r^2$ is the solid angle under which the tilted surface element ΔA_2 appears from the source (Fig. 10.41).

The Equation 10.53 is symmetric. Replacing S^*_1 by the radiation intensity of the surface element ΔA_2 the equation describes the radiation power $\mathrm{d}W_2/\mathrm{d}t$ received by ΔA_1 from ΔA_2.

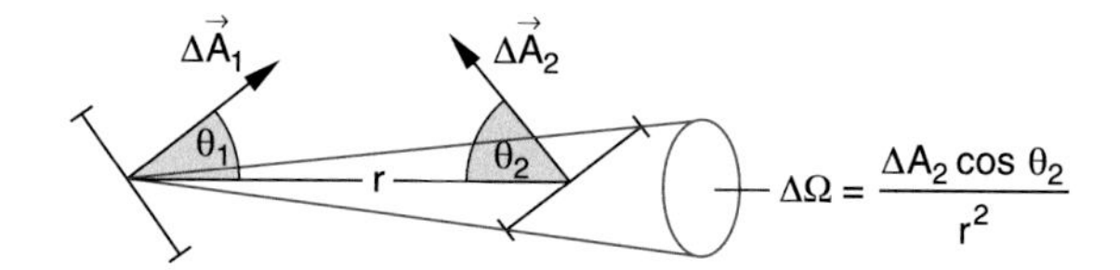

Figure 10.41 The surface element ΔA_2 receives from ΔA_1 the radiation power $\mathrm{d}W/\mathrm{d}t = (S^* \Delta A_1 \cdot \Delta A_2 \cdot \cos\theta_1 \cdot \cos\theta_2)/r^2$

The ratio

$$\frac{\mathrm{d}W_2/\mathrm{d}t}{\Delta A_2} = \int_{A_1} \left(\mathrm{d}A_1 S^*_1 \cos\theta_1 \cos\theta_2\right)/r^2 \tag{10.54}$$

is the **irradiance** or **intensity** at the detector [W/m^2].

Note: The radiation power, absorbed by the detector with the normalized absorptivity A^*, the reflectivity R and the transmission T ($A^* + R + T = 1$) is

$$\frac{\mathrm{d}W_{\text{abs}}}{\mathrm{d}t} = A^* \cdot \frac{\mathrm{d}W_1}{\mathrm{d}t} = (1 - R - T) \cdot \frac{\mathrm{d}W_1}{\mathrm{d}t} \,,$$

because the fraction $(R+T)$ of the incident radiation is reflected and transmitted.

10.2.5.3 Black Body Radiation

A black body with the absorptivity A^* can be experimentally realized by a cavity with absorbing walls and a small hole with an area ΔA that is small compared to the total inner wall area A of the cavity (Fig. 10.42). Radiation that penetrates through the hole into the cavity, suffers many reflections at the absorbing walls before it can eventually escape with a very small probability through the hole. The absorptivity of the hole area ΔA is therefore $A^* \approx 1$.

When the cavity is heated up to a temperature T, the hole area ΔA acts as radiation source with an emissivity E^* that is, according to (10.47) larger than that of all other bodies with $A^* < 1$ at the same temperature T (the black body radiation is therefore also called *cavity radiation*). This can be demonstrated by the following experiment (Fig. 10.43):

The letter H is milled deeply into a graphite cube. At room temperature, the letter H appears darker than the surface of the cube (left picture of Fig. 10.43). When the cube is heated, up to

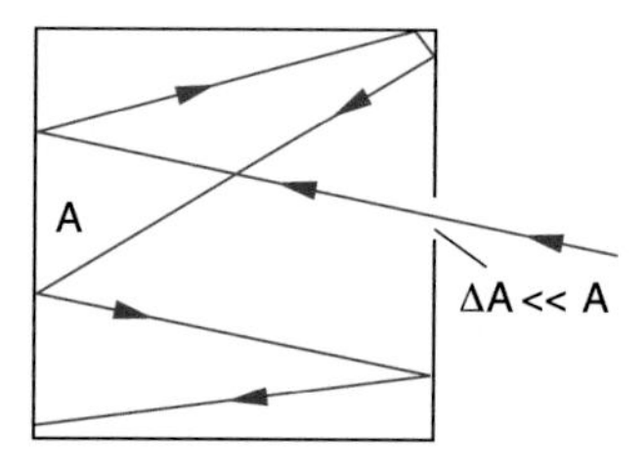

Figure 10.42 A cavity with a small hole ΔA absorbs nearly all of the radiation incident onto ΔA

Chapter 10

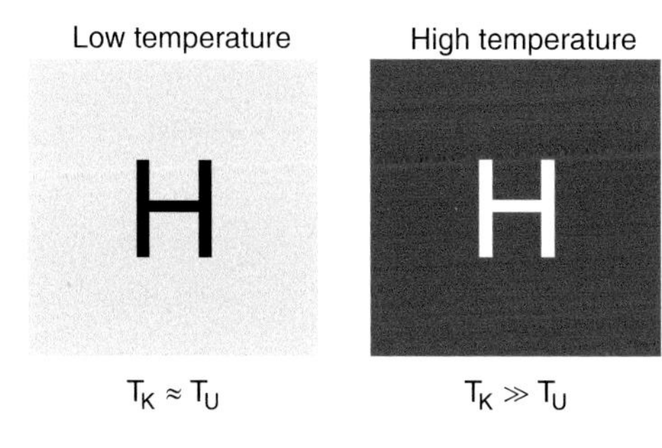

Figure 10.43 The letter H milled into a graphite block appears darker as its surrounding at low temperatures but brighter at high temperatures

$T = 1000$ K the letter H appears much brighter than the other surface elements (right figure).

Some simple considerations allow one to postulate some basic laws of the black body radiation:

- Under stationary conditions (T = const) emission and absorption of the cavity walls must be balanced. This implies for all frequencies ν of the radiation that the absorbed power of an arbitrary surface element ΔA of the walls must be equal to the emitted power:

$$\mathrm{d}W_\mathrm{a}(\nu)/\mathrm{d}t = \mathrm{d}W_\mathrm{e}(\nu)/\mathrm{d}t\,.$$

  At this equilibrium we define the temperature T of the black body radiation as the temperature of the walls.
- The black body radiation is isotropic. The spectral irradiance I_ν [W/(m$^2 \cdot$ s$^{-1} \cdot$ sr] is for any point in the cavity independent of the direction in the cavity and also of the material or structure of the walls. If the radiation were not isotropic, one could place a black disc into the cavity and orientate it in such a way, that its surface normal points into the direction of maximum radiation intensity S^*. The disc would then absorb more radiation power and would heat up to higher temperatures than the walls. This contradicts the second law of thermodynamics (see Sect. 10.3).
- The black body radiation is homogeneous, i. e. its energy density is independent of the specific location inside the cavity. Otherwise one could construct a perpetuum mobile of the second kind (see Sect. 10.3).

When we place a body in the radiation field of the cavity, the spectral radiation power $S_\nu^* \cdot d\nu \cdot \mathrm{d}A \cdot \mathrm{d}\Omega$, falls within the solid angle $\mathrm{d}\Omega$ onto the body. The spectral power absorbed by the surface element $\mathrm{d}A$ is

$$\frac{\mathrm{d}W_\mathrm{a}}{\mathrm{d}t} = A_\nu^* S_\nu^* \mathrm{d}A \cdot \mathrm{d}\Omega \cdot \mathrm{d}\nu\,, \tag{10.55a}$$

while the emitted power is

$$\frac{\mathrm{d}W_\mathrm{e}}{\mathrm{d}t} = E_\nu^* \mathrm{d}A \cdot \mathrm{d}\Omega \cdot \mathrm{d}\nu\,. \tag{10.55b}$$

At thermal equilibrium the absorbed power must be equal to the emitted power. Since the cavity radiation is isotropic, this must be valid for all directions. Therefore it follows from (10.55a,b) the Kirchhoff-Law

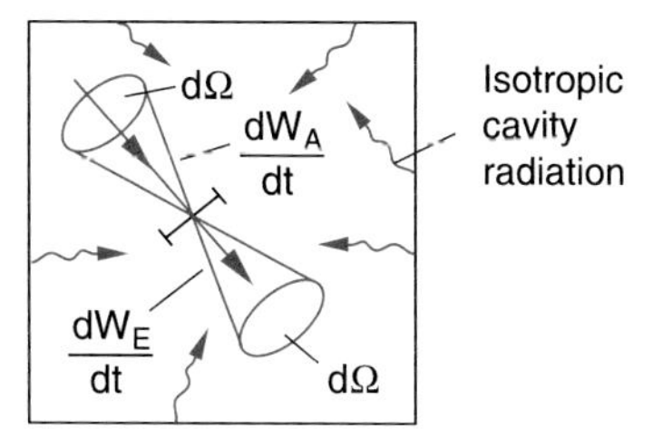

Figure 10.44 A body inside a closed cavity at thermal equilibrium with the radiation field

$$E_\nu^*/A_\nu^* = S_\nu^*(T)\,. \tag{10.56}$$

For all bodies in the radiation field of the cavity the ratio of spectral emissivity and absorptivity equals the spectral radiation density S_ν^* of the radiation field.

For a black body is $A^* = 1$ for all frequencies ν. We can therefore conclude:

The spectral emissivity E_ν^* of a black body is equal to the spectral radiation density S_ν^* of the cavity radiation.

10.2.5.4 The Emitted Radiation Power of a Hot Body

The surface S of a black body at the temperature T emits, according to the **Stefan–Boltzmann Law**, (see Vol. 2, Sect. 12.3) the radiation power

$$\frac{\mathrm{d}W}{\mathrm{d}t} = \sigma \cdot S \cdot T^4\,. \tag{10.57}$$

For a black surface with $A^* = 1$ the Stefan–Boltzmann constant σ has the numerical value $\sigma = 5.67051 \cdot 10^{-8}\,\mathrm{W/(m^2 \cdot K)}$. For bodies with $A^* < 1$ the emissivity is smaller and therefore also the emitted radiation power at the same temperature is smaller than for a black body. The Stefan–Boltzmann Law can be derived from Planck's radiation law (see Vol. 3, Chap. 3). The deviation of the experimental results for small wavelengths from those predicted by the Stefan–Boltzmann law,. gave the impetus for the development of quantum theory.

Note:

- The radiation power of a hot body is proportional to the fourth power of the surface temperature. With increasing temperature it therefore represents an increasing fraction of the total energy loss of a body.
- The thermal radiation is an electromagnetic wave and therefore propagates also through vacuum. The energy transport by radiation is not bound to matter. We own our existence to the heat radiation from the sun because this is the only energy transport mechanism from the sun to the earth (except the negligible contribution of particles such as electrons and protons emitted by the sun).

A more detailed and quantitative treatment of heat radiation will be postponed to Vol. 3, because it demands some basic knowledge of quantum theory.

10.2.5.5 Practical Use of Solar Energy

The radiation energy of the sun, received on earth, can be either directly converted to heat by solar energy collectors or transformed into electrical power by photovoltaic semiconductor elements. While the second technique is treated in Vol. 3, the first will be shortly discussed here [10.10, 10.11].

The radiation power of the sun, incident on $1\,\text{m}^2$ of a surface black surface element outside of the earth atmosphere has an annual average $P_\odot = 1.4\,\text{kW/m}^2$ (*solar spectral irradiance*). However, even at a clear day without clouds only a smaller power P_E reaches the earth surface because of absorption and light scattering in the atmosphere. For geographical latitudes $\varphi = 40°–50°$ one measures $P_\text{E} \approx 0.5 P_\odot$. For an inclination angle α of the incident radiation to the surface normal the received power at a clear sky is $P_\text{E} \approx 730 \cdot \cos\alpha\,\text{W/m}^2$.

With the absorptivity A^* of the surface, the power absorbed within the time interval Δt by a plane surface with area ΔA is

$$P_\text{a} = A^* \cdot \Delta A \cdot P_\text{E} \cdot \cos\alpha \cdot \Delta t\,.$$

This results in a temperature increase ΔT of a sun collector with mass m and specific heat c

$$\Delta T = A^* \cdot \Delta A \cdot P_\text{E} \cdot \cos\alpha \cdot \Delta t/(c \cdot m)\,, \tag{10.58}$$

if no heat losses occur.

The temperature increases with irradiation time if the heat is not dissipated. This dissipation can be achieved, when on the backside of the sun collector tubes are welded with a good heat contact to the sun collector and a liquid is pumped through the tubes, which takes away the heat. In order to keep the temperature of the sun collector constant, the pumping speed is chosen such that the heat transport just balances the received radiation power.

With a mass flow $\text{d}m_\text{l}/\text{d}t$ of the heat transporting liquid with the specific heat c_l and the temperature increase ΔT the energy balance is given by the equation

$$A^* \cdot P_\text{E}\,\Delta A\,\cos\alpha = (\text{d}m_\text{l}/\text{d}t)\,c_\text{l}\,\Delta T + (\text{d}W/\text{d}t)_\text{v}\,. \tag{10.59}$$

The angle α depends on the inclination of the energy collecting plane, on the latitude φ and on the daytime. In Fig. 10.45 the daytime dependence of the sun energy received by a collector with $\alpha = 45°$ in Kaiserslautern ($\varphi = 49°$) is illustrated for three different dates. Two effects cause this variation with the daytime: 1) The variation of the angle α due to the apparent motion of the sun and 2) The variation of the path length of the sun radiation through the atmosphere during the day, where absorption and scattering attenuates the radiation energy.

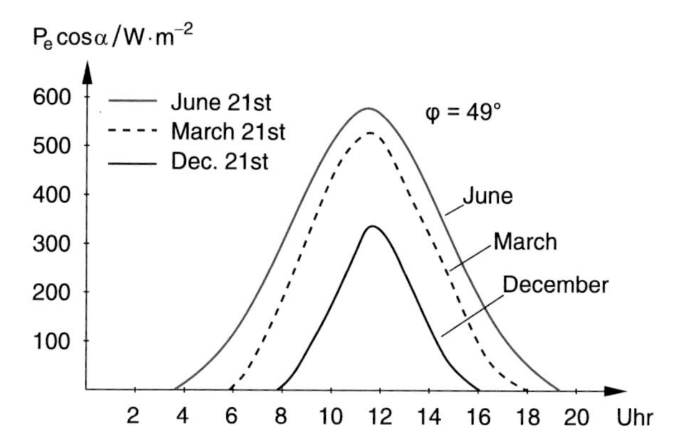

Figure 10.45 Variation of the sun radiation incident onto a sun radiation collector at the lattitude $\varphi = 49°$ as a function of daytime for three different times of the year

The area under the curves gives the integrated energy ($\text{J/m}^2\cdot$ sun hours) received during a whole day between sunrise t_1 and sun set t_2.

$$W_\text{E} = \int_{t_1}^{t_2} P_\text{E} \cos\alpha\,\text{d}t \tag{10.60}$$

The average radiation power per day is then $\langle P_\text{E} \cdot \cos\alpha \rangle = W_\text{E}/(t_2 - t_1)$.

Example

$A^* = 0.8$; $\langle P_\text{E} \cdot \cos\alpha \rangle = 250\,\text{W/m}^2$ during a clear day in August at $\varphi = 45°$; $\Delta A = 8\,\text{m}^2$. With water as the heat transporting liquid ($c_\text{W} = 4186\,\text{Ws/(kg}\cdot\text{K)}$) which is heated from 20 to 60 °C. With a good heat insulation the heat losses $\text{d}W_\text{l}/\text{d}t$ can be kept down to $50\,\text{W/m}^2$ for a temperature difference of $\Delta T = 40\,°\text{C}$. The amount of water heated per sec is then given by $\text{d}m_\text{W}/\text{d}t = (A^* \cdot \langle P_\text{E} \cdot \cos\alpha \rangle - \text{d}W_\text{l}/\text{d}t) \cdot A/(c_\text{W} \cdot \Delta T) = 0.0072\,\text{kg/s}$. Within one hour 26 l water are heated from 20 to 60 °C. ◀

Figure 10.46 shows a possible realization of a sun power collector for the heating of houses. It consists of a blackened

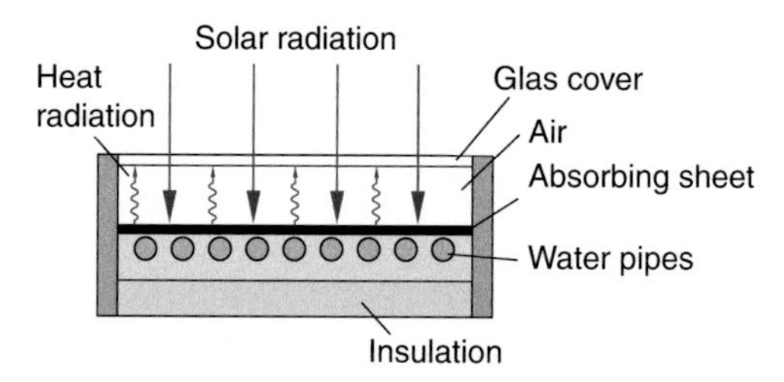

Figure 10.46 Cross section of a flat solar radiation collector which is mounted on house roofs

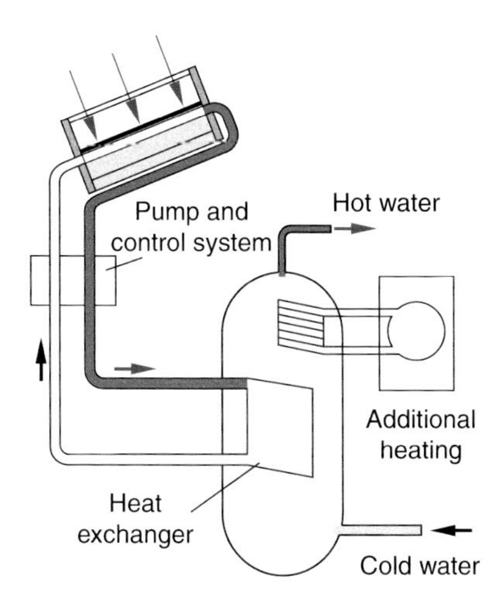

Figure 10.47 Thermal solar radiation collector for heating water with heat exchanger

absorber plate with pipes connected to the backside with good heat contact. Through the pipes a water-glycol mixture (to avoid freezing at low temperatures) is pumped. In cases where temperatures above 100 °C are reached, low viscosity oil is used. The absorber plate is placed inside a heat insulating housing with a glass plate in front. Heat losses are due to reflection of the sun radiation by the glass plate, by heat radiation of the black absorber plate and by heat conduction from the hot part of the sun radiation collector (including the pipes for the transport of the hot liquid) and convective cooling by the wind that blows along the device.

The heated liquid transfers its heat through a heat exchanger to a thermal storage system inside the house that generates hot service water (Fig. 10.47). A temperature sensor and a feedback system controls the temperature of the service water and takes care that it always has the wanted temperature. In case the sun energy is not sufficient, a conventional heating system is connected which only operates if the temperature sinks below the wanted value. When the hot water is used for room heating, a floor heating system is advantageous, because here the water temperature can be lower than that for radiator heating [10.9–10.11].

In large facilities for thermal solar energy conversion, it is more effective to heat the liquid above its boiling point. The generated vapour drives turbines which can produce electric current through electric generators. The technical realization uses large spherical mirrors that focus the sun radiation onto a black surface connected to a pipe system that transports the hot vapour. Temperatures above 1000 °C can be achieved and an electric output power of many kW has been demonstrated. The installation costs for such systems are up to now very high and therefore only a few pilot plants have been built. One example is the system in Almeria in Spain.

10.3 The Three Laws of Thermodynamics

We will define a thermodynamic system as a system of atoms or molecules that interacts with its surroundings by exchange of energy in form of heat or mechanical work. The system can be described by physical quantities such as temperature, pressure, volume, particle density etc. In this section we will discuss, how the state of such a system changes by the exchange of energy with its surroundings. The results of all investigations can be condensed in three laws of thermodynamics, which have a comparable importance for Physics as the conservation laws of mechanics for momentum, angular momentum and energy. These three laws are solely based on experimental data and cannot be derived mathematically from first principles contrary to a widespread false opinion.

At first we must discuss, which quantities are necessary to describe the state of a thermodynamic system.

10.3.1 Thermodynamic Variables

The state of a system is defined by all characteristic properties, which are determined by the external conditions. A thermodynamic system is completely determined if the chemical composition is known and the quantities pressure p, volume V and temperature T are given. If these quantities do not change with time, the system is in an equilibrium state and it is called a *stationary system*. Most of the thermodynamic considerations deal with stationary systems. Often a system changes so slowly, that it can be described by a succession of equilibrium states.

Systems far away from equilibrium play an important role for all chemical and biological reactions and they are intensively discussed in modern physics. They are therefore shortly treated at the end of this chapter. *In this section, we will restrict the discussion to ideal gases.* The thermodynamics of real bodies will be discussed later.

An equilibrium state of a system is unambiguously determined, if the three quantities pressure p, volume V and temperature T are fixed. These quantities are therefore called *thermodynamic variables*.

Definition

A thermodynamic variable is a variable in the equation of state of a thermodynamic system. It describes the momentary state of the system and is independent of the way on which the system has reached its momentary state. Besides V, p and T also the total energy, the entropy and the enthalpy are thermodynamic variables.

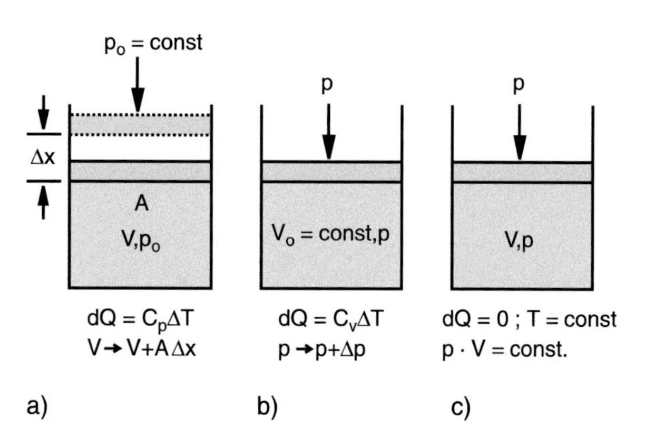

Figure 10.48 **a** Heating at constant pressure; **b** heating at constant volume; **c** no heat supply

The thermodynamic variables are related to each other by the equation of state for a gas volume V of an ideal gas with N molecules

$$p \cdot V = \nu \cdot R \cdot T\,, \tag{10.61}$$

where $\nu = N/N_A$ is the number of moles and N_A the Avogadro number. Also for real gases a corresponding equation can be derived (see Sect. 10.4). For a given volume V and a pressure p the temperature T determines the internal energy

$$U = \tfrac{1}{2} \cdot \nu \cdot f \cdot R \cdot T \tag{10.62}$$

of molecules with f degrees of freedom. For ideal gases (for instance helium) is $f = 3$. If the volume decreases ($dV < 0$) at a constant pressure p the necessary work is

$$dW = -p \cdot dV\,. \tag{10.63}$$

The sign is chosen in such a way that the applied work is positive, if the energy of the system increases. Work performed by the system means a decrease of its internal energy and is therefore defined as negative.

When a gas is heated at constant pressure p its volume increases (Fig. 10.48a). The quantity

$$\gamma_p = \frac{1}{V} \cdot \left(\frac{\partial V}{\partial T}\right)_p\,, \tag{10.64}$$

that describes the relative volume change per Kelvin temperature rise, is the *isobaric expansion coefficient*.

In an analogous way the heating of a gas at a constant volume (Fig. 10.48b, where the pressure increases, is described by the isochoric pressure coefficient

$$\gamma_V = \frac{1}{p} \cdot \left(\frac{\partial p}{\partial T}\right)_V\,, \tag{10.65}$$

which describes the relative pressure increase $\Delta p/p$ for a temperature rise of 1 K.

The isothermal compressibility

$$\gamma_T = \kappa = -\frac{1}{V} \cdot \left(\frac{\partial V}{\partial p}\right)_T \tag{10.66}$$

gives the relative volume change $\Delta V/V$ for a pressure change Δp at a constant temperature T.

As recollection keep in mind:

isothermal: $T = \text{const}$
isobaric: $p = \text{const}$
isochoric: $V = \text{const}$.

The total change dV of the volume $V(p, T)$, when both quantities p and T are changing is

$$\begin{aligned} dV &= \left(\frac{\partial V}{\partial p}\right)_T dp + \left(\frac{\partial V}{\partial T}\right)_p dT \\ &= -\kappa \cdot V \cdot dp + \gamma_p \cdot V \cdot dT\,. \end{aligned} \tag{10.67}$$

For isochoric processes the volume V stays constant, i. e. $dV = 0$. Then (10.67) reduces to

$$\begin{aligned} 0 &= -\kappa \cdot V \cdot (dp)_V + \gamma_p \cdot V \cdot (dT)_V \\ &\Rightarrow \kappa \cdot dp = \gamma_p \cdot dT\,. \end{aligned} \tag{10.68}$$

Division by dT yields with $(dp/dT)_V = \gamma_V \cdot p$ the relation

$$\gamma_p = \kappa \cdot \gamma_V \cdot p \tag{10.69}$$

between isobaric expansion coefficient γ_p, isothermic compressibility κ, isochoric expansion coefficient γ_V and pressure p.

10.3.2 The First Law of Thermodynamics

The heat ΔQ applied to a system can be either used for rising the temperature T at a constant volume V, or for the expansion of the volume V against the external pressure p where the system has to perform the work ΔW. Energy conservation demands

$$\Delta Q = \Delta U - \Delta W\,, \tag{10.70a}$$

where, as defined before, $\Delta W < 0$ if the system performs work (which decreases its own energy). This sign definition is in agreement with the definition (2.35) for the work. If the system, for instance, performs work against an external force $\boldsymbol{F} = -p \cdot \boldsymbol{A}$ when a piston with area A is moved along the distance Δx against the external pressure p, the work is

$$\Delta W = \boldsymbol{F} \cdot \boldsymbol{\Delta x} = -p \cdot A \cdot \Delta x = -p \cdot \Delta V \quad \text{with } \Delta V > 0\,.$$

Equation 10.70 is the **first law of thermodynamics**. It is a special case of the general law of energy conservation. It can be formulated as:

The sum of the external heat ΔQ, applied to a thermodynamic system, and the supplied mechanical energy ΔW is equal to the increase ΔU of the total internal energy U.

$$\Delta U = \Delta Q + \Delta W \qquad (10.70b)$$

When the system performs work against an external force, is $\Delta W < 0$ and therefore $\Delta U < 0$. Many inventors have tried to construct machines that deliver more energy than they consume. Such a machine could use part of the delivered energy for its own operation. It could run continuously delivering energy without external energy input. Therefore this hypothetical machine is called a **perpetuum mobile**. Because it contradicts the first law of thermodynamics it is also called *perpetuum mobile of the first kind*.

Equation 10.70 can be also formulated in a more floppy way as:

A perpetuum mobile of the first kind is impossible.

Note: This statement cannot be proved mathematically. It is solely based on empirical knowledge.

For ideal gases the work performed during the expansion of the volume V by infinitesimal amount $\mathrm{d}V$ against the external pressure p is

$$\mathrm{d}W = p \cdot \mathrm{d}V \,.$$

The first law of thermodynamics for ideal gases can therefore be written in a differential form as

$$\mathrm{d}U = \mathrm{d}Q - p \cdot \mathrm{d}V \,. \qquad (10.71)$$

For $\mathrm{d}V > 0$ the system releases energy and according to (10.71) $\mathrm{d}U < \mathrm{d}Q$, i. e. the loss of internal energy cannot be compensated by the supplied heat $\mathrm{d}Q$. For $\mathrm{d}V < 0$ the volume is compressed and the system gains the energy $p \cdot \mathrm{d}V$. Now $\mathrm{d}U > \mathrm{d}Q$, the gain of internal energy is larger than the supplied heat.

The relation between the thermodynamic variables p, V, T can be derived from (10.71) for special processes where in each case one of the variables p, V, T or the quantity Q is kept constant.

Note, that the quantity Q is not a thermodynamic variable! The state of a system does change with the supply of heat $\mathrm{d}Q$, but one cannot unambiguously determine the final state of the system, because either U or V or both variables can change. In a mathematical language this means: $\mathrm{d}Q$ is not a complete differential.

10.3.3 Special Processes as Examples of the First Law of Thermodynamics

Note: We will discuss the following processes for one mole of a gas where the number of moles is $\nu = V/V_\mathrm{M} = 1$.

10.3.3.1 Isochoric Processes (V = const)

With $\mathrm{d}V = 0$ it follows from (10.71)

$$\mathrm{d}Q = \mathrm{d}U = C_V \cdot \mathrm{d}T \,. \qquad (10.72)$$

The heat supplied to the system is used solely for the increase of the internal energy U. We can therefore relate the specific heat to the internal energy U by

$$C_V = \left(\frac{\partial U}{\partial T}\right)_V \,. \qquad (10.73)$$

10.3.3.2 Isobaric Processes (p = const)

The first law of thermodynamics has now the form

$$\mathrm{d}Q = \mathrm{d}U + p \cdot \mathrm{d}V = C_p \cdot \mathrm{d}T \,, \qquad (10.74)$$

where we have used (10.28). When we introduce the *enthalpy*

$$H = U + p \cdot V \qquad (10.75)$$

as new thermodynamic variable with

$$\mathrm{d}H = \mathrm{d}U + p \cdot \mathrm{d}V + V \cdot \mathrm{d}p = \mathrm{d}Q + V \cdot \mathrm{d}p \,, \qquad (10.76)$$

we can write the first law of thermodynamics as

$$\mathrm{d}H = \mathrm{d}U + p \cdot \mathrm{d}V = \mathrm{d}Q \,. \qquad (10.77)$$

For isobaric processes the increase $\mathrm{d}H$ of the enthalpy H is equal to the supplied heat $\mathrm{d}Q$.

The specific heat at constant pressure is then

$$C_p = \left(\frac{\partial H}{\partial T}\right)_p \,. \qquad (10.78)$$

The variable H is often used for phase changes, chemical reactions or other processes that take place at constant pressure, but where the volume can change. A further example is the expansion of a gas from a reservoir with constant pressure into the vacuum where the pressure $p = 0$ is maintained by pumps.

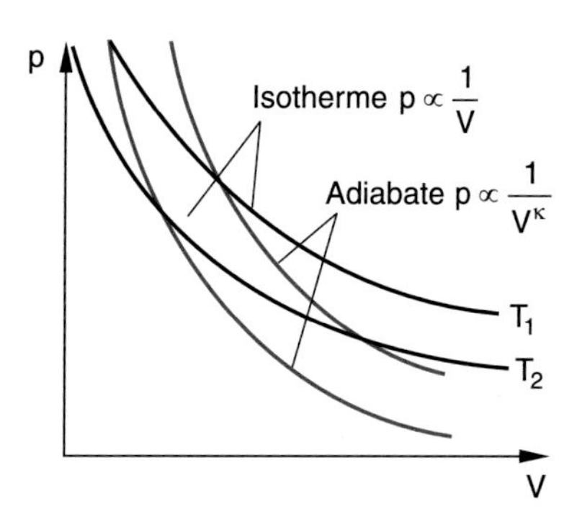

Figure 10.49 Isothermal and adiabatic curves in a p-V-diagram

10.3.3.3 Isothermal Processes (T = const)

Since the internal energy per mole of a gas depends solely on the temperature T but not on the pressure p or the volume V, for isothermal processes U must be constant, i. e. $\mathrm{d}U = 0$. From (10.71) it follows

$$\mathrm{d}Q = p \cdot \mathrm{d}V\,. \tag{10.79}$$

The external heat energy $\mathrm{d}Q$ supplied to the system is completely transferred to the work $p \cdot \mathrm{d}V$ that the system releases to the outside. Its internal energy does not change.

The equation of state $p \cdot V = R \cdot T$ then reduces to the Boyle–Marriott law (see Sect. 7.1)

$$p \cdot V = \text{const}\,. \tag{10.80}$$

The state $p(V)$ of the system can be plotted in a p-V-diagram (Fig. 10.49) for isothermal and adiabatic processes at different temperatures T. This gives for isothermal processes the hyperbolas

$$p = \frac{R \cdot T_\mathrm{K}}{V} = \frac{\text{const}}{V}\,,$$

which are called *isotherms* (black curves in Fig. 10.49).

We will now discuss how large the work is that a system has to perform for an isothermal expansion from a volume V_1 to $V_2 > V_1$ at constant temperature T.

$$\begin{aligned} W &= -\int_{V_1}^{V_2} p \cdot \mathrm{d}V = -R \cdot T \cdot \int_{V_1}^{V_2} \frac{\mathrm{d}V}{V} \\ &= -R \cdot T \cdot \ln \frac{V_2}{V_1} = R \cdot T \cdot \ln \frac{V_1}{V_2}\,. \end{aligned} \tag{10.81}$$

10.3.3.4 Adiabatic Processes

During adiabatic processes no heat is exchanged between the system and its surroundings. Adiabatic processes occur in nature, when changes of volume or pressure are so fast, that the energy exchange during this short time period can be neglected. An example is the propagation of acoustic waves at high frequencies ν through a medium (see Sect. 11.9). During one oscillation period $\Delta T = 1/\nu$ nearly no energy exchange between maxima and minima of the wave can occur.

The first law of thermodynamics (10.71) can be written with (10.73) for adiabatic processes

$$\mathrm{d}U = C_V \cdot \mathrm{d}T = -p \cdot \mathrm{d}V\,. \tag{10.82}$$

From the equation of state (10.21) $p \cdot V = R \cdot T$ we obtain $p = R \cdot T/V$. Inserting this into (10.82) yields

$$C_V \cdot \mathrm{d}T/T = -R \cdot \mathrm{d}V/V\,.$$

Integration gives

$$\begin{aligned} C_V \cdot \ln T &= -R \cdot \ln V + \text{const} \\ \Rightarrow \ln\left(T^{C_V} \cdot V^R\right) &= \text{const}\,. \end{aligned}$$

With $R = C_p - C_V$ this can be written as

$$T^{C_V} \cdot V^{(C_p - C_V)} = \text{const}\,. \tag{10.83a}$$

The $1/C_V$-th power of (10.83a) yields with the adiabatic index $\kappa = C_\mathrm{p}/C_V$ the equation

$$T \cdot V^{\kappa-1} = \text{const}\,, \tag{10.83b}$$

because $T = p \cdot V/R$, this can be also written as

$$p \cdot V^\kappa = \text{const}\,. \tag{10.83c}$$

The Eq. 10.83a–c describe the relations between the thermodynamic variables T, p, V for adiabatic processes. They are called Poisson-adiabatic equations.

In a p–V-diagram (Fig. 10.49) the red adiabatic curves $p(V) \propto 1/V^\kappa$ ($\kappa > 1$) are steeper than the isothermal curves $p(V) \propto 1/V$.

For an ideal gas is $f = 3$ and $\kappa = (f+2)/f = 5/3$. For molecular nitrogen N_2 is $f = 5 \rightarrow \kappa = 7/5$.

Example

In the pneumatic cigarette lighter, the volume V filled with an air–benzene-mixture is suddenly compressed to 0.1 V. According to (10.83b) the temperature T rises from room temperature ($T_1 = 293\,\mathrm{K}$) to $T_2 = T_1(V_1/V_2)^{\kappa-1}$. For air is $\kappa = 7/5$ which gives $T_2 = 736\,\mathrm{K} = 463\,°\mathrm{C}$. This is above the ignition temperature of the air–benzene-mixture. ◀

10.3.4 The Second Law of Thermodynamics

While the first law of thermodynamics represents the energy conservation when thermal energy is converted into mechanical energy, the second law of thermodynamics gives the maximum fraction of thermal energy that can be really transferred into mechanical energy.

As we will see, this is connected with the question, into which direction the transfer of one form of energy into another form proceeds by its own, i. e. without external action. All of our experience tells us, that heat flows by its own only from the hotter region into the colder one, not vice versa. Furthermore all experiments show, that mechanical energy can be completely converted into heat, but that for the opposite process only part of the heat can be converted into mechanical energy.

This fact, that is based solely on experimental experience, is formulated in the second law of thermodynamics:

> Heat flows *by its own* only from the warmer body to the colder one, never into the opposite direction.

We will now discuss more quantitatively the transformation of heat into mechanical work. This will be illustrated by considering thermodynamic cyclic processes, which leads us to a quantitative formulation of the second law of thermodynamics.

10.3.5 The Carnot Cycle

A thermodynamic cycle is a series of processes where a thermodynamic system passes through several different states until it finally reaches again its initial state. At the end of this cycle, the system shows again the same thermodynamic variables as in the initial state, although it has passed during the cycle through different states with different variables. A simple example is a system that is heated and then cooled again until it has reached the initial temperature.

If the cyclic process can traverse into both directions the cycle is called *reversible* (Fig. 10.50) otherwise it is called *irreversible*. Although such reversible processes can occur in micro-physics if only a few particles are involved, they represent in the real world of many-particle systems only idealized "*Gedanken-experiments*", which represent limiting cases of real processes that are always irreversible.

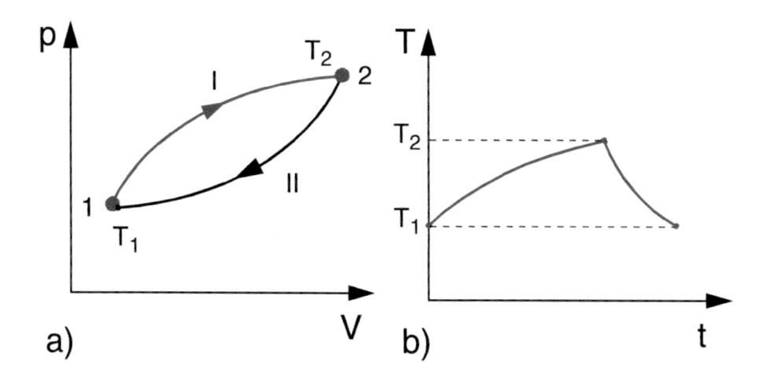

Figure 10.50 Thermodynamic cycle from the state 1 (T_1, p_1, V_1) via the state 2 (T_2, p_2, V_2) back to the state 1. **a** In a p–V-diagram; **b** in the temperature-time diagram. Note: The cycle shown here, can only proceed, if energy is fed into the system during the first step and energy is taken away from the system during the second step

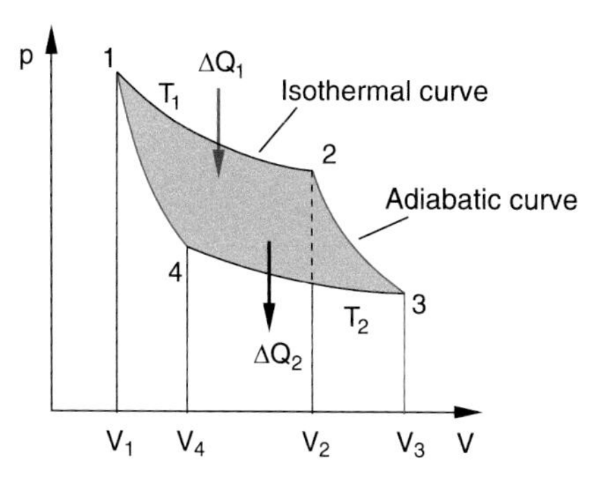

Figure 10.51 Carnot's cycle

All periodically operating machines, such as steam engines or car motors traverse irreversible cyclic processes. Although they arrive at the end of the cycle again at the initial state, if they are regarded as isolated systems, but they have lost energy during the cycle (for instance friction losses) which has to be replaced for each cycle.

The most famous reversible cyclic process is the Carnot-cycle that represents an idealized loss-free cycle. It was published in 1824 by *Nicolas Leonard Sadi Carnot*. This cyclic process will enable us to calculate the maximum fraction of heat that can be transformed into mechanical energy and therefore allows the quantitative formulation of the second law of thermodynamics. Furthermore, it illustrates nicely the difference between reversible and irreversible processes.

The Carnot Cycle is a "*Gedanken-Experiment*", where a thermodynamic system passes through two isothermal and two adiabatic processes during two expansion and compression events, until it finally reaches its initial state again (Fig. 10.51).

Note: The following considerations are valid for 1 mol of an ideal gas, where in Eq. 10.61 $V = V_M$ and $\nu = 1$.

The state of the system at the starting point 1 is defined by the thermodynamic variables (V_1, p_1, T_1). The isothermal expansion brings the system to the state $2 = (V_2, p_2, T_1)$. During this process, the heat ΔQ_1 has to be supplied to the system in order to keep the temperature constant. Now an adiabatic expansion follows and the system gets to the state $3 = (V_3, p_3, T_2 < T_1)$. In the next step the system is isothermally compressed and reaches the state 4 with the conditions $(V_4,\ p_4,\ T_2)$. Here the heat ΔQ_2 has to be removed from the system. Finally an adiabatic compression brings the system back to its initial state $1 = (V_1, p_1, T_1)$. Such a virtual thermodynamic system that passes through a Carnot cycle is called a *Carnot Machine*.

We will now calculate the heat energies ΔQ_1 and ΔQ_2 which are exchanged between the system and a heat reservoir during the isothermal processes.

1st process: Isothermal expansion from the state 1 to the state 2. According to the first law of thermodynamics we obtain for an isothermal expansion

$$dQ = p \cdot dV .$$

The heat supplied to the system is equal to the mechanical work the system performs during the expansion.

With (10.81) it follows:

$$\Delta Q_1 = -\Delta W_{12} = \int_{V_1}^{V_2} p\,dV = R \cdot T_1 \cdot \ln(V_2/V_1) . \quad (10.84a)$$

2nd process: Adiabatic expansion from state 2 to state 3. For adiabatic processes the heat exchange is zero. We therefore obtain:

$$dQ = 0 \rightarrow dU = -p \cdot dV = \Delta W_{23} . \quad (10.84b)$$

The work performed during the expansion is negative, because it is delivered from the system to the surrounding. This results in a decrease $\Delta U = U(T_2) - U(T_1)$ of the internal energy U because $T_2 < T_1$.

3rd process: Isothermal compression from state 3 to state 4. Similar to step 1 is the heat ΔQ_2 delivered at the lower temperature T_2 to the heat reservoir equal to the work ΔW_{34} necessary to compress the volume V

$$\Delta W_{34} = R \cdot T_2 \cdot \ln(V_3/V_4) = -\Delta Q_2 > 0 . \quad (10.84c)$$

4th process: Adiabatic compression from state 4 to the starting conditions in state 1. Similar to step 2 is here no heat exchange with the surrounding and the work performed during the compression is converted to the increase ΔU of the internal energy

$$\Delta U = U(T_1) - U(T_2) . \quad (10.84d)$$

Total energy balance: The work delivered to the surrounding during the 2nd process is equal to the work supplied to the system during the 4th process. Therefore, only during the isothermal processes a net energy is transferred. The net mechanical work during the Carnot cycle (Fig. 10.52) is

$$\Delta W = \Delta W_{12} + \Delta W_{34} = R \cdot T_1 \cdot \ln(V_1/V_2) + R \cdot T_2 \cdot \ln(V_3/V_4) .$$

For the adiabatic processes 2 → 3 and 4 → 1 the relations hold

$$T_1 \cdot V_2^{\kappa-1} = T_2 \cdot V_3^{\kappa-1} \quad \text{and}$$
$$T_1 \cdot V_1^{\kappa-1} = T_2 \cdot V_4^{\kappa-1} .$$

Division of the two equations yields

$$V_2/V_1 = V_3/V_4 \Rightarrow \ln(V_3/V_4) = -\ln(V_1/V_2) .$$

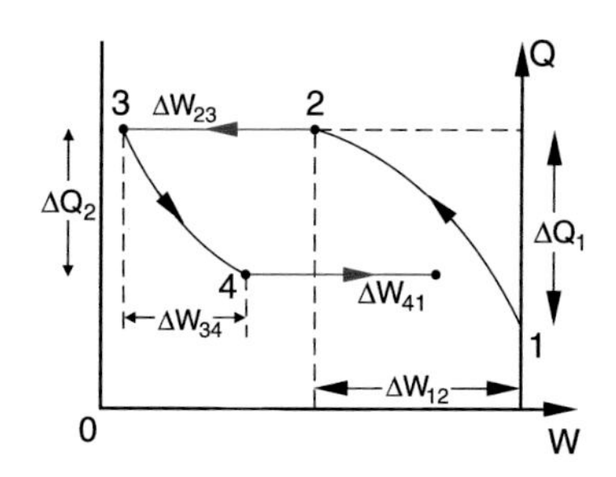

Figure 10.52 Heat exchange and net mechanical energy $\Delta W = \Delta W_{34} - \Delta W_{12}$ of Carnot's cycle

The net work is then

$$\Delta W = R \cdot (T_1 - T_2) \cdot \ln(V_1/V_2) . \quad (10.85)$$

The Carnot Engine has received the heat ΔQ_1 and has supplied the mechanical work $\Delta W < 0$ to the outside.

Such a machine that transfers heat into mechanical energy is called **heat engine**.

The heat ΔQ_2 supplied to the surrounding, is generally lost. Therefore the efficiency of the engine is defined as the mechanical work supplied by the engine divided by the heat ΔQ_1 put into the engine.

The efficiency of the Carnot Engine is then

$$\eta = \left|\frac{\Delta W}{\Delta Q_1}\right| = \frac{R(T_1 - T_2) \cdot \ln(V_2/V_1)}{R \cdot T_1 \cdot \ln(V_2/V_1)} = \frac{T_1 - T_2}{T_1}$$

$$\eta = \frac{T_1 - T_2}{T_1} . \quad (10.86)$$

This is a remarkable result: During the cycle the total received heat cannot be transformed into mechanical work, but only the fraction $\eta = (T_1 - T_2)/T_1 < 1$. This fraction is called *exergy*. The remaining part $(1 - \eta)$ of the input energy (*Anergy*) is exchanged as heat ΔQ_2 to the surrounding at the lower temperature T_2. The conservation of total energy can be written as

Energy = Exergy + Anergy .

The efficiency of the Carnot Engine increases with increasing temperature difference $T_1 - T_2$. It is therefore advantageous to choose T_1 as high as possible and T_2 as low as possible. We will see in Sect. 10.3.10 that it is impossible to reach $T_2 = 0\,\text{K}$.

This implies that η is always smaller than 1.

When the Carnot cycle is traversed into the opposite direction, heat is transported from the lower temperature T_2 to the higher temperature T_1. This requires the work

$$\Delta W = R \cdot (T_2 - T_1) \cdot \ln(V_1/V_2) .$$

This represents the ideal limiting case of a heat pump, which is also used as refrigerating machine (see Sect. 10.3.14). Its coefficient of performance (also called figure of merit) is defined as the ratio of delivered heat ΔQ to the input work ΔW.

$$\varepsilon_{hp} = \frac{\Delta Q_1}{\Delta W} = \frac{T_1}{T_1 - T_2} = \frac{1}{\eta} .$$

Note, that $\varepsilon_{hp} > 1$.

Example

$T_1 = 30\,°C = 303\,K$,
$T_2 = 10\,°C = 283\,K \Rightarrow \varepsilon_{hp} = 15.2$. ◀

Note:

1. The heat pump does not contradict the second law of thermodynamics, because here the heat does not flow *by its own* from the colder to the hotter place but needs mechanical work to drive this heat transport.
2. The Carnot Engine works with an ideal gas and all energy losses are neglected. The Carnot Cycle is reversible. Real engines have always losses that cannot be avoided. They are due to friction of the moving pistons, friction in the non-ideal gas, heat conduction from the system to the surroundings etc. These losses decrease the efficiency of the engine. We will now indeed prove, that:

There is no periodically working machine with a higher efficiency than that of the Carnot engine.

Proof

Assume, there is a machine M_x with a higher efficiency than the Carnot Engine. This "magic machine" needs for a given mechanical energy output a smaller heat input than the Carnot Engine, i. e. $\Delta Q_x < \Delta Q_1$. We now combine M_x with a Carnot engine that passes the cycle in opposite direction, i. e. it works as a heat pump (Fig. 10.53). We adapt the size of M_x in such a way that it delivers just the mechanical work ΔW, which the Carnot engine needs as heat pump. The Carnot engine then transports the heat

$$|\Delta Q_1| = |\Delta Q_2| + |\Delta W|$$

from the colder to the warmer reservoir. Since we have assumed that the magic machine M_x has a higher efficiency than the Carnot Engine, it needs less heat from the reservoir at the temperature T_1 for its operation than the Carnot Engine transports to this reservoir. It furthermore delivers less heat to the cold reservoir at T_2 than the Carnot engine needs for its operation as heat pump.

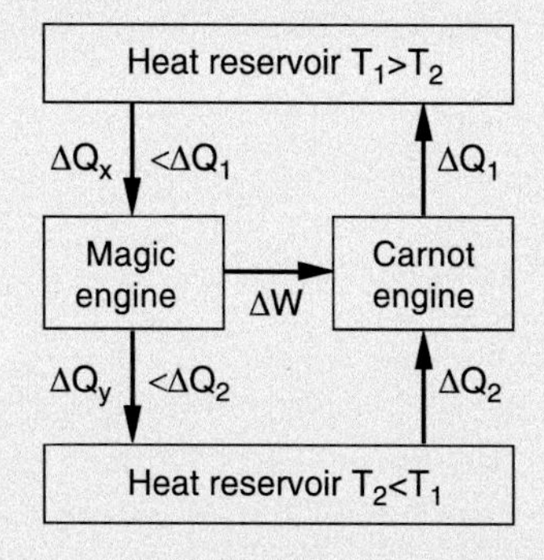

Figure 10.53 Proof of the impossibility of the perpetuum mobile of the second kind

The combined system therefore transports heat from the colder to the hotter reservoir without mechanical energy input. This contradicts the second law of thermodynamics which has been proved by numerous experiments. Therefore a heat engine with a higher efficiency than that of the Carnot engine is not possible! ◀

Remark. These considerations can be also applied to a heat pump, where the cycle is traversed into the opposite direction. We replace the Carnot Engine in Fig. 10.53 by a "magic heat pump" and the magic machine M_x by the Carnot Engine (Fig. 10.54) and assume that the coefficient of performance ε_x is larger than that of a Carnot heat pump. An analogous consideration shows that $\varepsilon_x < \varepsilon_C = 1/\eta_C$. This can be seen as follows:

The Carnot engine in Fig. 10.54 now runs as heat engine that extracts the heat ΔQ_1 from the hot reservoir at the temperature T_1 and delivers the heat $\Delta Q_2 = \Delta Q_2 - \Delta W$ to the cold reservoir at $T_2 < T_1$. The output energy ΔW is transferred to the magic heat pump, which takes the heat ΔQ_4 from the cold reservoir

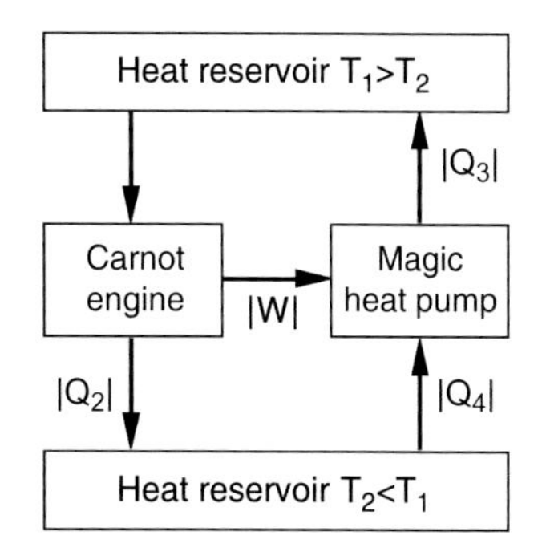

Figure 10.54 Proof, that the energy efficiency ratio of a heat pump is always smaller than that of a Carnot-engine that operates as a heat pump

Chapter 10

and transports the heat $\Delta Q_3 = \Delta Q + \Delta W$ to the hot reservoir. Assume that the coefficient of performance $\varepsilon_x = \Delta Q_3/\Delta W$ of the magic heat pump is larger than $\varepsilon_C = 1/\eta_C = \Delta Q_1/\Delta W$ of the Carnot engine. Then $|\Delta Q_3| > |\Delta Q_1|$ and $|\Delta Q_4| = |\Delta Q_3| - W > |\Delta Q_2| = |\Delta Q_1|$ This implies that the combined system Carnot engine plus magic heat pump can pump heat from the cold to the hot reservoir without mechanical energy input. This again contradicts the second law of thermodynamics.

The coefficient of performance ε_x of an arbitrary heat pump cannot be larger than $\varepsilon_C = 1/\eta_C$ where η_C is the efficiency of the Carnot engine.

With other words: The coefficient of performance ε_x of any heat pump cannot be larger than that of a Carnot heat pump $\varepsilon_C = T_1/(T_1 - T_2)$.

From the considerations above it follows: All reversible cycles have the same efficiency

$$\eta = \Delta W/\Delta Q_1 = (T_1 - T_2)/T_1 ,$$

independent of the working material, which can be different from an ideal gas.

10.3.6 Equivalent Formulations of the Second Law

The considerations above allow the following statements:

- The efficiency $\eta = \Delta W/\Delta Q_1 < 1$ of any heat engine is always smaller than 100%. This means that heat cannot completely converted into mechanical work.
- The Carnot engine has the maximum possible efficiency: $\eta = (T_1 - T_2)/T_1$.

The value $\eta = 1$ would be only possible for $T_2 = 0$. However, we will see that this is excluded by the third law of thermodynamics (see Sect. 10.3.13)

The first and second laws of thermodynamics are purely rules of thumb, based on numerous experimental facts. They cannot be proved mathematically without additional assumptions.

The second law can be formulated in different ways:

- Heat flows by its own only from the hot to the cold region, never into the opposite direction.
- There is no periodically acting machine that can convert heat completely into mechanical work without additional energy supply.

Such a machine is called a *perpetuum mobile of the second kind*. The second law can then be formulated similar to the first law:

The realization of a perpetuum mobile of the second kind is impossible.

Example

A perpetuum mobile of the second kind could be a ship with engines that receive their energy solely from the heat of the sea. Such a ship could move without additional energy and would not need oil or coal. ◄

A perpetuum mobile of the second kind does not contradict the first law of thermodynamics, because it does not violate the energy conservation. Therefore numerous inventors have tried to construct such machines, however unsuccessful!

The Carnot cycle allows a method to measure the Kelvin temperature, which is independent of the thermometer substance and works down to very low temperatures where gas thermometers are no longer useful, because all gases condense at such low temperatures. From (10.84) we can deduce the ratio of the heat energies ΔQ_1 and ΔQ_2 supplied from and released to the heat reservoirs

$$\frac{\Delta Q_1}{\Delta Q_2} = \frac{T_1}{T_2} .$$

The temperatures of the two heat reservoirs can be compared, when the heat energies, exchanged between the system and the reservoirs, are measured. For instance, if one of the reservoirs is kept at the temperature $T_1 = 273.16\,\mathrm{K}$ of the triple point of water the temperature T_2 is obtained from

$$T_2 = 273.16 \cdot \frac{|\Delta Q_2|}{|\Delta Q_1|} = 273.16 \cdot (1 - \eta) .$$

The heat reservoir, kept at the temperature T_1, can be electrically heated, which allows the determination of ΔQ_1. The efficiency η can then be measured as the ratio of mechanical work $\Delta W = p \cdot \Delta A \cdot \Delta x$ when a piston with area A moves by the distance Δx against the external pressure p and the supplied heat ΔQ_1.

The temperature scale obtained by this way is called the *thermodynamic temperature scale*.

One Kelvin (1 K) is 1/273.16 times the temperature of the triple point of water.

10.3.7 Entropy

By introducing the entropy as new thermodynamic variable, the second law of thermodynamics can be mathematically formulated in an elegant way. When the heat $\mathrm{d}Q$ is supplied to a system at the temperature T we define as *reduced heat* the ratio $\mathrm{d}Q/T$.

For the Carnot cycle in Fig. 10.51 we can bring the system from the point 1 to the point 3 on two different ways: $1 \to 2 \to 3$ or $1 \to 4 \to 3$. Only during the isothermal processes, heat is

Chapter 10

exchanged with the surroundings. The absolute values of the reduced heat energies $|\Delta Q_1|/T_1$ and $|\Delta Q_2|/T_2$ on the two ways are equal, as can be seen from (10.84a–c). This means: The reduced energies do not depend on the way but only on starting and final state of the system. This is not only valid for the Carnot cycle but for all reversible processes.

We introduce the thermodynamic variable S called the *entropy* with the dimension $[S] = [\mathrm{J/K}]$, in the following way. We define the change $\mathrm{d}S$ of the entropy as the reduced heat exchanged on an infinitesimal part of a reversible process

$$\mathrm{d}S = \mathrm{d}Q/T\,.$$

Since the change ΔS for a system that is brought from a defined initial state into a defined final state is independent of the way between these two states, and depends solely on initial and final states of the system, the quantity S is a thermodynamic variable which describes together with pressure p, temperature T and volume V the state of a thermodynamic system.

In the Carnot Cycle the reduced heat energies change only during the isothermal processes. According to (10.84) the entropy then changes by

$$\Delta S = \frac{\Delta Q}{T} = \pm R \cdot \ln \frac{V_2}{V_1}\,. \tag{10.87}$$

For the complete reversible cycle we have

$$\frac{\Delta Q_1}{T_1} = -\frac{\Delta Q_2}{T_2}\,,$$

and therefore

$$\Delta S = 0\,.$$

For a reversible cycle the entropy S is constant.

Processes where $S = \text{const}$ are called **isentropic processes**. For these processes is $\Delta S = 0$ and therefore $\Delta Q = 0$ and $T = \text{const}$. During isentropic processes the system must be kept at a constant temperature. This distinguishes isentropic processes from adiabatic processes where also $\Delta Q = 0$ but where the temperature changes.

With the first law of thermodynamics (10.71) the entropy change $\mathrm{d}S$ during reversible processes of an ideal gas can be calculated as

$$\mathrm{d}S = \frac{\mathrm{d}Q_{\mathrm{rev}}}{T} = \frac{\mathrm{d}U + p\,\mathrm{d}V}{T}\,. \tag{10.88}$$

For 1 mol of the gas is $\mathrm{d}U = C_V \cdot \mathrm{d}T$ and $p \cdot V_{\mathrm{M}} = R \cdot T$. This converts (10.88) to

$$\mathrm{d}S = C_V \frac{\mathrm{d}T}{T} + R \cdot \frac{\mathrm{d}V}{V}\,. \tag{10.89}$$

Integration over the temperature range from T_1 to T_2 where the molar heat capacity can be assumed as constant, yields for isobaric processes where V and T can change but $p = \text{const}$.

$$\Delta S_{\mathrm{isobar}} = C_V \ln \frac{T_2}{T_1} + R \ln \frac{V_2}{V_1}\,. \tag{10.90}$$

In a similar way one obtains for isochoric processes ($V = \text{const}$) with $C_V = C_p - R$ and $p_1/T_1 = p_2/T_2$

$$\begin{aligned}\Delta S_{\mathrm{isochor}} &= C_V \ln \frac{T_2}{T_1} = (C_p - R) \ln \frac{T_2}{T_1}\\ &= C_p \ln \frac{T_2}{T_1} - R \ln \frac{T_2}{T_1}\end{aligned}$$

$$\Delta S_{\mathrm{isochor}} = C_p \ln \frac{T_2}{T_1} - R \cdot \ln \frac{p_2}{p_1}\,. \tag{10.91}$$

Since the entropy S is a thermodynamic variable, its change ΔS does not depend on the kind of process but only on initial and final state of the process. We can therefore determine ΔS also for irreversible processes. This can be seen as follows:

We consider a substance at the temperature T_1 (e. g. a solid body) in a gas volume V. The body should be in thermal contact with the gas. Now the gas is slowly expanded in an adiabatic process, which results in a slow decrease of the temperature. If this proceeds sufficiently slowly, the temperature of the body is always equal to that of the gas, because sufficient time is available for reaching temperature equilibrium. Finally the temperature has decreased to T_2. This process is reversible because the initial state can be retrieved by slow adiabatic compression.

When the solid body is regarded as isolated body without the gas, the cooling process is irreversible, because heat is transferred to the surrounding. The entropy change of the body is

$$\begin{aligned}\Delta S &= \frac{\Delta Q_{\mathrm{irr}}}{T}\\ &= C_V \int_{T_1}^{T_2} \frac{\mathrm{d}T}{T}\\ &= C_V \cdot \ln \frac{T_2}{T_1} < 0 \quad \text{for} \quad T_2 < T_1\,,\end{aligned} \tag{10.92}$$

as in the reversible process. However, since heat has been transferred to the surroundings the entropy of the surroundings increases. For the total change of the system body + surroundings $\Delta S_{\mathrm{irrev}} > 0$, i. e. the entropy increases!

Example

1. We regard in Fig. 10.55 two equal bodies with mass m and specific heat c which have been brought by different energy supply to different temperatures T_1 and $T_2 < T_1$. Their heat energies are then $Q_1 = m \cdot c \cdot T_1$ and $Q_2 = m \cdot c \cdot T_2$. When they are brought into

thermal contact heat flows from the hot body 1 to the colder body 2 until the temperatures are equal to the average temperature T_m. If no heat is transferred to the surroundings the body 1 has lost the energy $\Delta Q_1 = m \cdot c(T_m - T_1)$ and the body 2 has received the energy $\Delta Q_2 = -m \cdot c \cdot (T_2 - T_m)$. Because $\Delta Q_1 = \Delta Q_2$ we obtain the average temperature

$$T_m = \frac{T_1 + T_2}{2} . \tag{10.93}$$

The entropy change ΔS of body 1 is

$$\Delta S_1 = \int_{T_1}^{T_m} \frac{dQ}{T} = mc \int_{T_1}^{T_m} \frac{dT}{T} = mc \ln(T_m/T_1) .$$

Since $T_m < T_1 \to \Delta S_1 < 0$.
The change of S_2 is accordingly

$$\Delta S_2 = mc \ln(T_m/T_2) ,$$

where $\Delta S_2 > 0$. The total change of the entropy of the system of bodies is therefore

$$\Delta S = \Delta S_1 + \Delta S_2 = mc \ln \frac{T_m^2}{T_1 \cdot T_2} . \tag{10.94}$$

Since $T_m = \frac{1}{2}(T_1 + T_2)$ we get $T_m^2/(T_1 \cdot T_2) > 1$ because the arithmetic mean is always $\geq$ geometric mean. This gives $\Delta S > 0$. The entropy increases during the irreversible process. The combination of the two bodies at different temperatures to a combined system is an irreversible process, because the cooled body cannot heat up again by cooling the other body without the supply of energy from outside (second law).

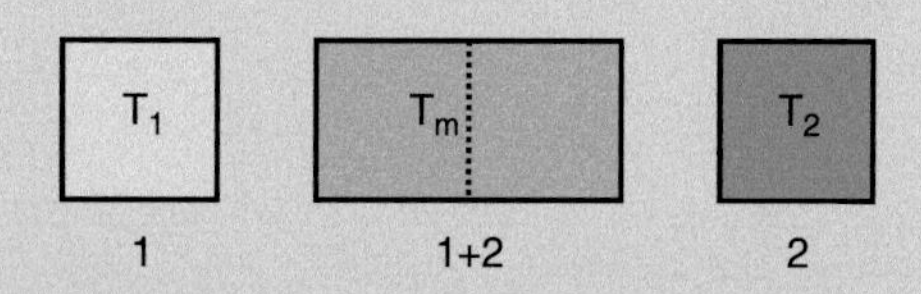

Figure 10.55 Increase of entropy during the equalization of temperatures at the contact of two bodies with different temperatures

2. The second example, which will give us a deeper insight into the meaning of entropy, is related to the diffusion of an ideal gas from a small volume V_1 through a hole into a larger volume V_2. Initially (for times $t \leq 0$) the gas is confined to the small volume V_1. At $t = 0$ the hole in the barrier separating V_1 from V_2 is opened and the gas molecules diffuse into the evacuated volume V_2 (Fig. 10.56). After a sufficiently long time $t > 0$ they are uniformly distributed over the whole volume $V = V_1 + V_2$. The gas temperature remains constant during this isothermal expansion (experiment of Gay-Lussac) because no work is needed for the expansion into the vacuum ($p \cdot dV = 0$).

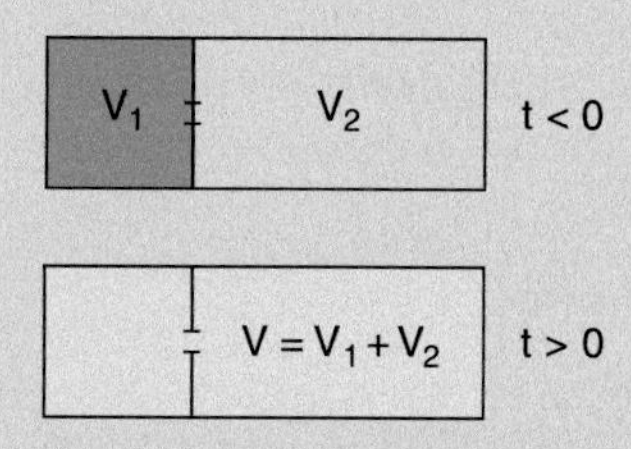

Figure 10.56 Diffusion of molecules from a small volume V_1 through a hole into the large volume V_2. After a sufficiently long time the molecules are uniformly distributed over the total volume $V_1 + V_2$

The diffusion is irreversible, because it is highly improbable that all gas molecules diffuse back through the hole into the small volume V_1 (see below). It is, however, nevertheless possible to calculate the entropy change by using as reversible substitute process the isothermal expansion (against an external pressure) with the same initial and final states as the diffusion. For this process, the supply of heat ΔQ is necessary in order to keep the temperature constant (Sect. 10.3.5). Since the reduced heat does not depend on the way during the expansion but solely on initial and final states the entropy change ΔS

$$\Delta S = R \cdot ln \frac{V}{V_1} \tag{10.95}$$

for the adiabatic expansions must be the same as for the diffusion.
This can be also understood, when we substitute the diffusion by a Gedanken-experiment, where the diffusion is separated into two steps (Fig. 10.57). The gas drives during the isothermal expansion a piston and extracts from a heat reservoir the heat ΔQ_1 as in the Carnot cycle. The work $\Delta W = \Delta Q_1$ performed during the expansion drives a stirrer that releases the heat ΔQ_1 again to the heat reservoir due to frictional losses. For this *Gedankenexperiment* initial and final states are identical to those of the diffusion. Therefore, the entropy change must be the same. Since in (10.95) $V \gg V_1 \to \Delta S > 0$.
Based on this diffusion process a statistical explanation of the entropy can be derived. We regard a molecule in volume V_1. Before the hole in the barrier is opened, the probability of finding the molecule in V_1 is $P_1 = 1$, because it must be for sure in V_1. After opening the hole the probability has decreased to $P_1 = V_1/(V_1 + V_2) = V_1/V$.
For two molecules the probability of finding both molecules in V_1 is equal to the product $P_2 = P_1 \cdot P_1 =$

P_1^2 of the probabilities for each molecule. For N molecules we therefore obtain

$$P_N = \left(\frac{V_1}{V}\right)^N . \tag{10.96}$$

For 1 mol is $N = N_A = R/k$, where k is the Boltzmann constant and N_A the Avogadro number. We then get

$$P_{N_A} = \left(\frac{V_1}{V}\right)^{R/k} . \tag{10.97}$$

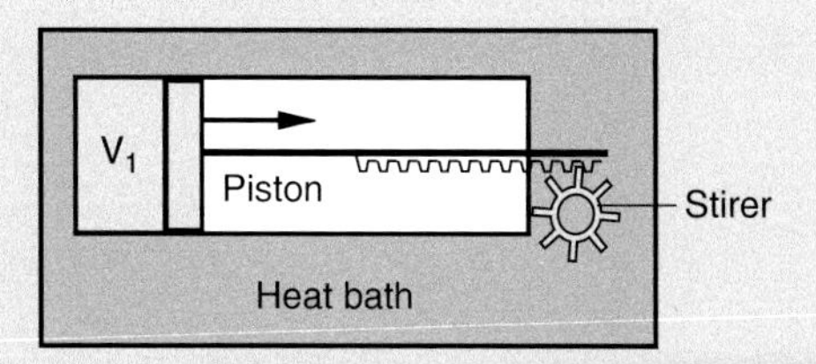

Figure 10.57 Separation of the diffusion process in Fig. 10.56 into two steps: Isothermal expansion and conversion of the mechanical work into heat at the heat reservoir ◀

Example

For $V_1 = (1/2)V, N_A = 6 \cdot 10^{23}$ /mol the probability that all molecules are found in V_1 is

$$\begin{aligned} P_N &= 2^{-6 \cdot 10^{23}} \\ &= 10^{-1.8 \cdot 10^{23}} \\ &\approx 0 \quad \text{(Fig. 10.58)} . \end{aligned}$$

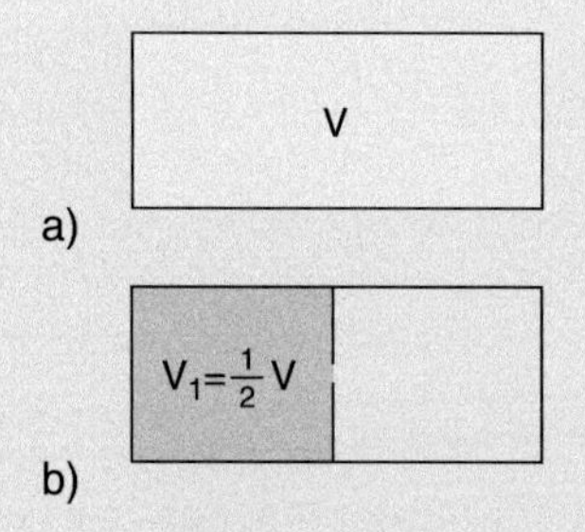

Figure 10.58 The probability P that all N molecules are simultaneously in the volume $V_1 = (1/2)V$ is $P = (1/2)^N$ ◀

Because of the large numbers in the exponent it is more convenient to use the logarithm of the probability P. From (10.97) we obtain

$$k \cdot \ln P = R \cdot \ln \frac{V_1}{V} = -R \ln \frac{V}{V_1} . \tag{10.98}$$

According to (10.95) the right side of (10.98) is equal to the change ΔS of the entropy. We therefore obtain for the change ΔS during the transition from state 1 (all molecules are in volume V_1) to state 2 (all molecules are distributes over the volume $V = V_1 + V_2$ the result

$$\begin{aligned} \Delta S &= S(V) - S(V_1) = k \cdot (\ln P(V) - \ln P(V_1)) \\ &= k \cdot \ln \frac{P_{\text{final}}}{P_{\text{initial}}} . \end{aligned} \tag{10.99a}$$

This illustrates that the entropy change ΔS during the diffusion from state 1 with the probability P_{initial} to the state 2 with the probability P_{final}

$$\Delta S = k \cdot \left(\ln \frac{P_{\text{final}}}{P_{\text{initial}}}\right) \tag{10.99b}$$

is a measure for the probability that a system undergoes a transition from the initial to the final state. This can be formulated in a more general way:

The probability that a system occupies a state i is proportional to the number Z of possible ways that lead to this state.

Example

When N particles, each with the energy $n_i E_0$ that are integer multiples of a minimum energy E_0 occupy a state with the total energy $E = \sum n_i E_0$, the number Z of possible realizations of this state equals the number of possible combinations of the integers n_i that fulfil the condition $\sum n_i = E/E_0$. ◀

The entropy S of a thermodynamic system state that can be realized by $Z = P$ possible ways is

$$S = k \cdot \ln P . \tag{10.99c}$$

The entropy S of a thermodynamic state is proportional to the number of possible realizations for this state.

As a third example, we will discuss the increase of entropy for the mixing of two different kinds of molecules X and Y. Initially all N_X molecules X should be in volume V_1 and all N_Y molecules Y in volume V_2. We assume that pressure p and temperature T are equal in both volumes, which demands $N_X/V_1 = N_Y/V_2$. When we open a hole in the barrier between the two volumes the molecules will diffuse through the hole until a uniform distribution of both kinds of molecules is reached. This process is irreversible and the entropy increases because the molecules N_X as well as the molecules N_Y fill now a larger volume $V = V_1 + V_2$, and the number of possible realizations of this situation is larger than that of the initial state. The increase of entropy for the N_X molecules is

$$\begin{aligned}\Delta S_X &= k \cdot \ln \left(\frac{V}{V_1}\right)^{N_X} = k \cdot N_X \ln \frac{V}{V_1} \\ &= k \cdot N_X \ln \frac{N_X + N_Y}{N_X} ,\end{aligned}$$

and for the N_Y molecules

$$\Delta S_Y = k \cdot N_Y \ln \frac{N_X + N_Y}{N_Y} .$$

The total change of entropy (called mixing entropy) is then

$$\begin{aligned}\Delta S_m &= \Delta S_X + \Delta S_Y \\ &= k \left[N_X \ln \frac{N_X + N_Y}{N_X} + N_Y \ln \frac{N_X + N_Y}{N_Y} \right] .\end{aligned} \tag{10.100}$$

These examples illustrate that only *changes* of entropy can be measured. The absolute value of the entropy $S(V, p, T)$ of a thermodynamic state

$$S = S_0 + \Delta S \tag{10.101}$$

is only defined, if the constant term S_0 is known. We will show in Sect. 10.3.10, how S_0 can be determined.

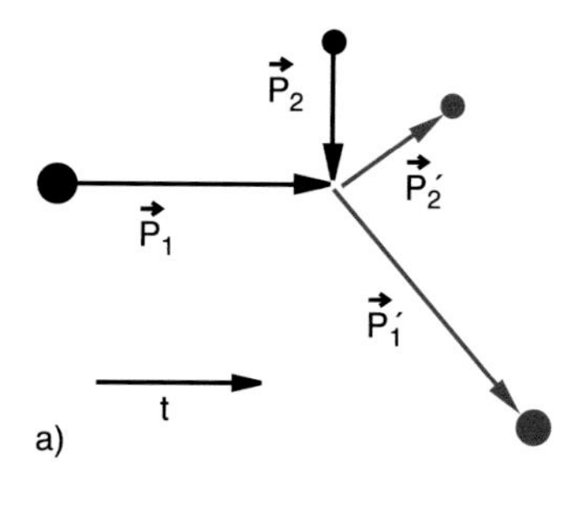

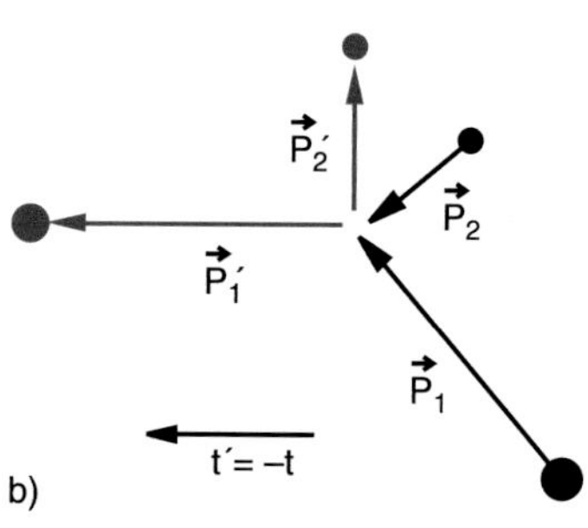

Figure 10.59 Reversible collision process. Reversing the time course of the process inverts the direction of all momentum vectors

10.3.8 Reversible and Irreversible Processes

For a completely elastic collision between two particles, energy and momentum of the two-body system are conserved (Sect. 4.2). A movie of such a collision process could run backwards and the observer would not notice this, because the reverse process is equally probable (Fig. 10.59). The collision process is reversible. One can also say that the process is time-invariant, i. e. one con exchange t with $-t$ without violating any physical law.

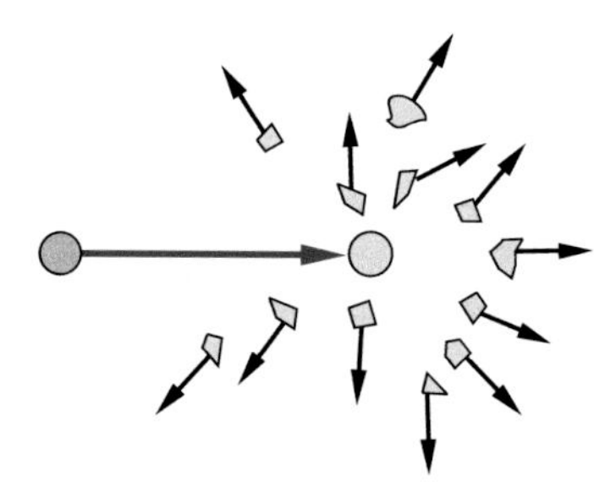

Figure 10.60 Collision of a steel ball with a glas hollow sphere, which bursts into many fragments, as example of an irreversible process

Contrary to this situation, the collision process in Fig. 10.60, where a bullet hits a glass ball that shatters in numerous pieces, is irreversible. The inverse process, where the pieces come again together to form a glass ball, which then emits the bullet, is highly improbable. The following question now arises: Since the glass ball consists of atoms and for each atom a reversible collision process should occur, why is the macroscopic process irreversible?

The answer to this question leads us again to the number of possible realizations of a macroscopic state that depends on a huge number of atoms or molecules. While before the collision the glass ball was at rest and the bullet had a well-defined energy and momentum the final state could have a very large number of possible realizations, because energy and momentum conservation still allows many different flight paths of the fragments as long as the sum of all individual pieces fulfils the conservation laws. The distribution of the fragments, observed for one experiment, represents only one of many possible distributions. At a second experiment under identical conditions, another distribution will be observed although both distributions obey energy and momentum conservation. The entropy increases during this collision process because the process starts from a state with a well-defined realization and ends at a large number of possible realizations. This is the signature of an irreversible process.

We can define an irreversible process as follows:

- The change of a thermodynamic state of a closed system is irreversible, if the reverse process that leads to the initial state, does not proceed by its own but only with additional energy supply from outside.
- The change of a thermodynamic state is irreversible, if the entropy increases during this process.
- The transition from an ordered state to a non-ordered state (for instance the melting of a crystal) always increases the entropy.

Note, that these statements are only valid for closed systems, which do not interact with their surroundings. For a macroscopic subsystem the entropy can indeed decrease if that of the other subsystems which, interact with the selected one, does increase.

Examples

1. When single crystals are formed out of the molten bath, a non-ordered state (the liquid) is transferred into an ordered state (the single crystal). The entropy S of the crystal is lower than that of the liquid, however, the decrease of S for a subsystem is over-compensated by the increase of S for the surroundings of the crystal.
2. All living beings (plants, animals, humans) decrease their entropy S by building up ordered structures, but at the expense of an increase of S of their surroundings (for example the digestion of food increases S). ◀

In all of these cases the entropy of the total system increases!

Ordered structures therefore cannot be formed in closed systems. Their formation needs open systems far away from thermal equilibrium. This nonequilibrium allows the exchange of energy between the open system and its surroundings, which can induce the decrease of the entropy of the open system.

For all macroscopic closed systems strictly speaking reversible processes cannot occur, because always part of the kinetic energy (even if it is very small) is converted into heat by the unavoidable friction. One of many examples is a swinging pendulum, where the amplitude continuously decreases due to friction by the air. The oscillation to the right will not be exactly reproduced by the following oscillation to the left, because of this amplitude decrease.

The interesting question, why the time has only one direction, can be related to the increase of entropy. The time-derivative $\mathrm{d}S/\mathrm{d}t$ can define a time arrow that allows us to distinguish between past and future [10.12]. For completely reversible processes time reversal would not change the validity of physical phenomena.

10.3.9 Free Energy and Enthalpy

The first and second law of thermodynamics contain the essential statements of thermodynamics. For their application to special problems it is useful to introduce as a new parameter the free energy

$$F = U - T \cdot S . \qquad (10.102)$$

With the entropy S we can formulate the first law in a more specific form:

$$\mathrm{d}U = \mathrm{d}Q_{\mathrm{rev}} + \mathrm{d}Q_{\mathrm{irr}} + \mathrm{d}W . \qquad (10.103a)$$

With $\mathrm{d}Q_{\mathrm{rev}} = T \cdot \mathrm{d}S$ and inserting the free energy $F = U - T \cdot S \Rightarrow \mathrm{d}F = \mathrm{d}U - T \cdot \mathrm{d}S - S \cdot \mathrm{d}T$, this becomes

$$\mathrm{d}F = \mathrm{d}Q_{\mathrm{irr}} + \mathrm{d}W - S\,\mathrm{d}T \leq \mathrm{d}W - S\mathrm{d}T , \qquad (10.103b)$$

where the equal sign holds for reversible and the $<$-sign for irreversible processes.

For isothermal processes is $\mathrm{d}T = 0$. This reduces (10.103) to

$$\mathrm{d}F \leq \mathrm{d}W \Rightarrow -\mathrm{d}W \leq -\mathrm{d}F . \qquad (10.104a)$$

This means:

- For isothermal processes, the maximum increase of the free energy is equal to the mechanical work supplied to the system.
- The maximum work that a system can deliver during an isothermal process, is equal to the decrease of its free energy.

The difference

$$U - F = T \cdot S$$

is called bound energy. From the relations

$$\mathrm{d}U = \mathrm{d}Q + \mathrm{d}W \quad \text{and} \quad \mathrm{d}F \leq \mathrm{d}W$$

follows by subtraction

$$d(U - F) \geq \mathrm{d}Q .$$

For isothermal processes the bound energy $U - F$ is completely converted into heat and is therefore not available for mechanical work. This explains the label "bound energy".

The second law of thermodynamics makes the following statement:

For isothermal processes, the change of the bound energy is at least equal to the supplied heat. The increase of the free energy is at most equal to the supplied mechanical energy.

If the isothermal process occurs at a constant volume ($\mathrm{d}V = 0$) no mechanical energy is exchanged, i. e. $\mathrm{d}W = 0$. Then

$$\mathrm{d}F \leq \mathrm{d}W = 0 , \qquad (10.104b)$$

which means that the free energy decreases.

A spontaneous isothermal process without exchange of work always proceeds in the direction where the free energy decreases. The entropy S then increases because of

$$T \cdot S = U - F \quad \text{and} \quad U = \text{const} .$$

Since most of the processes occurring in nature are irreversible, the free energy of the universe decreases and therefore also the capability to perform mechanical work. All irreversible processes always tend to decrease existing temperature differences, because then the entropy increases (see the examples in the previous section).

Pessimists say: "The universe strives towards its *heat death*. This means, that all temperature differences approach zero, where no longer any chemical and biological processes are possible. However, it will take quite a while until this might happen and furthermore it is not clear, which fate the universe after many billion years will suffer, because it is still an open question, whether the universe represents a closed or an open system."

As the last thermodynamic parameter we will introduce besides the ethalpiy $H = U + p \cdot V$ the **free enthalpy** G (also called *Gibb's chemical potential*) defined by the relation

$$G = U + pV - TS = H - TS \,. \tag{10.105}$$

The total differential of G is

$$\mathrm{d}G = \mathrm{d}U + p \cdot \mathrm{d}V + V \cdot \mathrm{d}p - T \cdot \mathrm{d}S - S \cdot \mathrm{d}T \,. \tag{10.106a}$$

With the first law of thermodynamics

$$\mathrm{d}U + p \cdot \mathrm{d}V = T \cdot \mathrm{d}S \,, \tag{10.106b}$$

this converts to

$$\mathrm{d}G = V \cdot \mathrm{d}p - S \cdot \mathrm{d}T \,. \tag{10.106c}$$

Chapter 10

10.3.10 Chemical Reactions

Chemical reactions represent the basis of all living processes. The utilization of food or the decomposition of waste products proceed by chemical reactions. It is therefore of essential interest, under which conditions chemical reactions proceed by themselves and when they need external energy supply for their start. For all reactions that proceed at constant pressure and constant temperature the Gibbs' potential is constant. That is the reason why G is called *chemical potential*. Often several components react with each other. If ν_i moles of the i-th component exist before the reaction, the total free enthalpy is $G = \sum \nu_i \mu_i$ (μ_i = Gibbs potential for one mole of the i-th component).

The mixing of the different components increases the entropy (see (10.100)) by the amount

$$\Delta S_\mathrm{m} = -R \cdot \sum_i \nu_i \ln x_i \,, \tag{10.107}$$

where $x_i = \nu_i / \sum \nu_i$ is the mole fraction of the i-th component.

A chemical reaction between the molecules A which results in the formation of molecules B is then described by

$$\sum_{i=1}^{k} \nu_i \mathrm{A}_i \rightarrow \sum_{j=k+1}^{p} \nu_j \mathrm{B}_j \,. \tag{10.108}$$

The number of moles can change by the reaction. For instance for the reaction

$$2\mathrm{H_2} + \mathrm{O_2} \rightarrow 2\mathrm{H_2O}$$

is $\nu_1 = 2$; $\nu_2 = 1$ and $\nu_3 = 2 \rightarrow \sum \nu_i \neq \sum \nu_j$.

If the number ν_i of moles for the i-th component changes by $\Delta\nu_i$, the free enthalpy G changes for processes with $\Delta p = \Delta T = 0$ according to (10.106a) by

$$\begin{aligned} \Delta G &= \sum \Delta\nu_i \mu_i - T \cdot \Delta S_\mathrm{m} \\ &= \sum \Delta\nu_i \mu_i - RT \cdot \sum (\nu_i + \Delta\nu_i) \ln x_i' \,, \end{aligned} \tag{10.109a}$$

where $x_i' = (\nu_i + \Delta\nu_i) / \sum (\nu_i + \Delta\nu_i)$.

When a reaction proceeds by its own (without external energy supply), its free enthalpy must decrease; i. e. $\Delta G < 0$. Equilibrium is reached if G becomes minimal.

If the number of moles does not change during the reaction, one can define a chemical equilibrium constant K by

$$K = exp\left[\sum \nu_i G_i / RT\right] \,.$$

The change of the free enthalpy can then be written as

$$\Delta G = RT\left[\ln K + \sum \xi_i \ln \nu_i\right] \,, \tag{10.109b}$$

where ξ_i is the fraction of the component i that reacts.

When the quantity of all components is one mole ($\nu_i = 1$) Eq. 10.109b can be reduced to

$$\Delta G(1\,\mathrm{mol}) = RT \ln K \,. \tag{10.109c}$$

The equilibrium constant K is therefore directly related to the change ΔG of the chemical potential G.

10.3.11 Thermodynamic Potentials; Relations Between Thermodynamic Variables

The thermodynamic variables: internal energy U, free energy F, Gibbs' potential G and enthalpy H are also called **thermodynamic potentials**. The advantage of their introduction is based on the fact that all thermodynamic variables can be written as partial derivatives of these potentials. The total differentials of the potentials are

$$dF = \left(\frac{\partial F}{\partial V}\right)_T dV + \left(\frac{\partial F}{\partial T}\right)_V dT$$

$$dU = \left(\frac{\partial U}{\partial V}\right)_S dV + \left(\frac{\partial U}{\partial S}\right)_V dS$$

$$dG = \left(\frac{\partial G}{\partial p}\right)_T dp + \left(\frac{\partial G}{\partial T}\right)_p dT$$

$$dH = \left(\frac{\partial H}{\partial S}\right)_p dS + \left(\frac{\partial H}{\partial p}\right)_S dp ,$$

where the lower index at the brackets denotes the quantity that is kept constant. The comparison with the equations derived in the previous sections

$$dF = -p dV - S dT , \quad (10.103)$$

$$dU = -p dV + T dS , \quad (10.88)$$

$$dG = V dp - S dT , \quad (10.106c)$$

$$dH = dU + p dV + V dp = dQ + V dp \quad (10.76)$$

gives the following relations between the thermodynamic variables and the potentials.

For the entropy we obtain

$$S = -\left(\frac{\partial G}{\partial T}\right)_p = -\left(\frac{\partial F}{\partial T}\right)_V , \quad (10.110a)$$

and for the pressure

$$p = -\left(\frac{\partial F}{\partial V}\right)_T = -\left(\frac{\partial U}{\partial V}\right)_S , \quad (10.110b)$$

while the relation for the volume V is

$$V = \left(\frac{\partial G}{\partial p}\right)_T = \left(\frac{\partial H}{\partial p}\right)_S . \quad (10.110c)$$

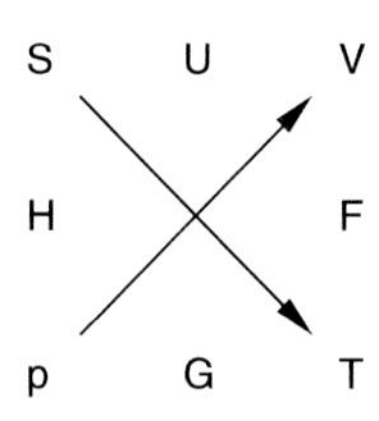

In the "Guggenheim Square", each thermodynamic potential can be placed in such a way, that the results of their derivatives can be immediately seen by the following procedure: One goes in the scheme from the potential symbol to the derivative variable and from there to the opposite corner on the diagonal. If this way on the diagonal is in the direction of the arrow, the result is positive, if it is opposite to the arrow, it is negative.

Examples

$$\left.\frac{\partial U}{\partial S}\right|_V = T ; \quad \left.\frac{\partial G}{\partial T}\right|_p = -S ;$$

$$\left.\frac{\partial F}{\partial V}\right|_T = -p ; \quad \left.\frac{\partial H}{\partial p}\right|_V = V .$$

◄

An example for the application of thermodynamic potentials is given in Sect. 10.4.2.

10.3.12 Equilibrium States

The thermodynamic potentials play a comparable role in thermodynamics as the mechanical potential E_p that determines the forces $\boldsymbol{F} = -\mathbf{grad}\, E_p$ which governs the motion of particles. In a similar way the gradient of the thermodynamic potentials keeps the chemical processes running until the minimum of the potentials is reached.

A system is at equilibrium, if without the action of external influences the state of the system does not change. If the state of a system changes due to external action, but returns to its original conditions after the external action ends, the equilibrium is stable. If, however, the system further removes from equilibrium, even after the termination of the external influence, the equilibrium is unstable. A mechanical example is a mass m which is fixed to a rigid rod that can rotate around a horizontal axis. At the minimum of the potential energy, where the mass is just below the horizontal axis the mass is in a stable equilibrium. When the mass is at its maximum height vertically above the horizontal axis, the equilibrium is unstable. Every slight perturbation brings m downwards. In a thermodynamic system, the thermodynamic potentials take the role of the potential energy in our example. We will illustrate this for several specific processes.

We assume a thermodynamic system with the internal energy U and the volume V at the temperature T and the pressure p. An arbitrary change of the conditions of the system is described by the differentials dU, dV, dT and dp. If the change is reversible, the work $dW = -p \cdot dV$ performed by the system during an adiabatic expansion, causes a decrease $dU = dW$ of the internal energy.

For irreversible processes, the system loses heat which causes a decrease of the total energy. Equilibrium is reached, if no further irreversible process is possible. Since for all irreversible processes with constant volume the entropy increases, the equilibrium condition can be formulated as

$$dS \leq 0 . \quad (10.111)$$

For all possible processes, which bring a thermodynamic system away from equilibrium the entropy must decrease.

Chapter 10

With other words:

A closed system with constant volume is in an equilibrium state if its entropy is maximal.

The thermodynamic potentials of an equilibrium state have their minimum value. This can be seen as follows:

With the change of entropy $\mathrm{d}S = \mathrm{d}Q/T$ we obtain from the first law of thermodynamics (10.71)

$$\mathrm{d}U + p \cdot \mathrm{d}V - T \cdot \mathrm{d}S = 0 . \qquad (10.112)$$

For isothermal-isochoric processes is $\mathrm{d}V = 0$ and $T =$ const. From (10.111) it follows: $\mathrm{d}S \leq 0$. Inserting this into (10.112) gives

$$\mathrm{d}(U - T\mathrm{d}S) \geq 0 \rightarrow \mathrm{d}F \geq 0 . \qquad (10.113)$$

Under isothermal and isochoric conditions a system has reached its equilibrium state, if the free energy F has its minimum value.

Under isothermal-isobaric conditions ($\mathrm{d}T = 0$ and $\mathrm{d}p = 0$) equilibrium is reached if

$$\mathrm{d}U + p\mathrm{d}V - T\mathrm{d}S = d(U + p \cdot V - T \cdot S) = \mathrm{d}G = 0 ,$$

because for all processes that drive the system away from equilibrium $\mathrm{d}G > 0$, as can be seen in an analogous way as the arguments above for $\mathrm{d}F > 0$.

Under isothermal-isobaric conditions a system is in an equilibrium state, if the Gibbs' potential is minimal.

In a similar way it can be proved, that for adiabatic-isobaric processes ($\mathrm{d}Q = 0$ and $\mathrm{d}p = 0$) the system is in an equilibrium state if the enthalpy $H = U + p \cdot V$ is minimal.

For adiabatic-isochoric processes ($\mathrm{d}Q = 0$ and $\mathrm{d}V = 0$) the internal energy $< U$ must be minimal at equilibrium.

All reactions that are possible without external interaction must start from states far away from equilibrium.

Therefore the thermodynamic treatment of chemical reactions and biological processes is based on the description of systems that are not at thermodynamic equilibrium.

10.3.13 The Third Law of Thermodynamics

We have seen in Sect. 10.3.7 that the entropy S is only determined apart from an additive constant S_0. We will now show, that for $T \rightarrow 0$

$$\lim S(T) = 0 .$$

This fixes the constant $S_0 = S(T = 0) = 0$.

For the proof we start with the free energy $F = U - T \cdot S$. Because of (10.110a) this can be also written as

$$F = U + T\left(\frac{\partial F}{\partial T}\right)_V . \qquad (10.114a)$$

We regard an isothermal chemical reaction where the system starts with the free energy F_1 and ends at F_2. The change $\Delta F = F_1 - F_2$ is

$$\Delta F = \Delta U + T\left(\frac{\partial}{\partial T}\Delta F\right)_V . \qquad (10.114b)$$

For $T > 0$ the changes ΔF and ΔU differ, but for $T \rightarrow 0$ the difference approaches zero

$$\lim(\Delta F - \Delta U) = 0 . \qquad (10.114c)$$

Nernst observed that with decreasing temperature the derivatives $\mathrm{d}(\Delta F)/\mathrm{d}T$ and $\mathrm{d}(\Delta U/\mathrm{d}T)$ decreased and that they approached zero for $T \rightarrow 0$. This means that the curves $\Delta F(T)$ and $\Delta U(T)$ come towards each other with horizontal slopes (Fig. 10.61).

Nernst therefore postulated that also for the general case

$$\lim_{T\rightarrow 0}\left(\frac{\partial \Delta F}{\partial T}\right)_V = 0 \quad \text{and} \qquad (10.115a)$$

$$\lim_{T\rightarrow 0}\left(\frac{\partial \Delta U}{\partial T}\right)_V = 0 . \qquad (10.115b)$$

because of (10.114a) it follows then

$$\lim_{T\rightarrow 0}\left(\frac{\partial U}{\partial T} - \frac{\partial F}{\partial T}\right) = 0 . \qquad (10.115c)$$

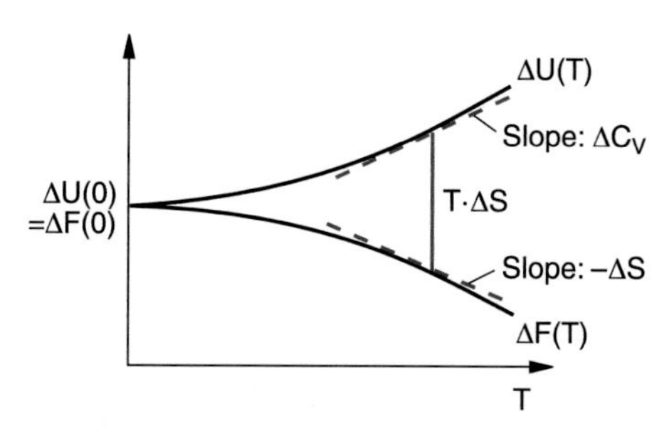

Figure 10.61 Nernst's theorem: Temperature course of $\Delta U(T)$, $\Delta F(T)$ near the absolute zero temperature

With

$$\left(\frac{\partial}{\partial T}\Delta F\right)_V = -\Delta S$$

$$\Rightarrow \lim_{T\to 0} \Delta S(T) = 0 .$$

This means that at sufficiently low temperatures reactions in pure condensed substances proceed always without changes of entropy i. e. $\Delta S = 0$. These reactions therefore proceed reversible.

Investigating the dependence $S(T)$ of condensed substances (liquified or solid gases) at very low temperatures, one finds indeed that the entropy does not depend on the crystal modification or on the specific substance as long as it is a pure substance, i. e. not a mixture of different substances. This suggests that the entropy of all pure substances approaches for $T \to 0$ the same value. Quantum theoretical considerations show (see Vol. 3) that for all pure substances the entropy S approaches zero for $T \to 0$.

$$\lim S(T) = 0 . \qquad (10.116)$$

The relations (10.115)–(10.116) are named the **3rd law of thermodynamics** or **Nernst's Theorem**.

Regarding the statistical interpretation of the entropy $S = k \cdot \ln W$ the 3rd law can be also formulated as:

The thermodynamic equilibrium state at $T = 0$ is a state with maximum order, which has only one possible realization with $P = 1$. The entropy is then $S = 0$.

Note: The statement $S(T = 0) = 0$ is only valid for pure substances. Mixed substances (for example mixed crystals) have even for $T = 0$ an entropy $S > 0$, called the mixing entropy (see Sect. 10.3.7).

The definition $S_0 = 0$ for the zero point of the entropy allows the determination of the absolute value of $S(T > 0)$.

For one mole one obtains

$$S(T) = \int_0^T \frac{\mathrm{d}Q_{\text{rev}}}{T'} = \int_0^T \frac{C(T')}{T'}\mathrm{d}T' . \qquad (10.117)$$

In order to fulfil the condition $\lim S(T \to 0) = 0$ the specific heat $C(T)$ must converge sufficiently fast towards zero for $T \to 0$ This is indeed observed experimentally (see Sect. 10.1.10). More detailed measurements show that for solids at very low temperatures $C(T) \propto T^3$ (see Vol. 3). This is indeed observed experimentally. The entropy $S(T)$ is then, according to (10.117), also proportional to T^3.

Remark. The first and second law of thermodynamics could be formulated as the impossibility to realize a perpetuum mobile of the first resp. the second kind. Also the third law can be formulated as an impossibility statement:

It is impossible to reach the absolute zero $T = 0$ of the thermodynamic = absolute temperature scale.

This can be seen as follows by an experimental argument.

If one tries to reach experimentally the absolute zero $T = 0$ this could be only realized by an adiabatic process, because every cooling process where heat is exchanged, requires a system that is colder than the system to be cooled.

During an adiabatic process no entropy change occurs because $\mathrm{d}Q = S \cdot \mathrm{d}T = 0$. For an adiabatic isobaric process is

$$\mathrm{d}S = \frac{\partial S}{\partial V}\mathrm{d}V + \frac{\partial S}{\partial T}\mathrm{d}T = 0 .$$

This gives

$$\mathrm{d}T = -\frac{(\partial S/\partial V)_T}{(\partial S/\partial T)_p}\mathrm{d}V . \qquad (10.118a)$$

For the partial derivative applies

$$\frac{\partial S}{\partial T} = \lim_{\Delta T\to 0}\left(\frac{\Delta S}{\Delta T}\right)_p = \lim_{\Delta T\to 0}\left(\frac{1}{T}\frac{\Delta Q}{\Delta T}\right)_p = \frac{C_p}{T} ,$$

and with (10.110a) and (10.114a) it follows

$$\left(\frac{\partial S}{\partial V}\right)_T = \frac{\partial}{\partial V}\left(\frac{\partial F}{\partial T}\right) .$$

For $T \to 0$ is with (10.115c)

$$\frac{\partial F}{\partial T} \to \frac{\partial U}{\partial T} = C_V ,$$

and we obtain from (10.118a)

$$\mathrm{d}T = -T \cdot \frac{C_V}{C_p} = -T \cdot \frac{C_V}{C_V + R} . \qquad (10.118b)$$

This shows that for $T \to 0$ also $\mathrm{d}T \to 0$. The absolute zero $T = 0$ for the temperature can be therefore not reached.

10.3.14 Thermodynamic Engines

When the Carnot cycle in Fig. 10.51 is traversed into the opposite direction, i. e. counterclockwise, the corresponding engine uses mechanical work to transport heat from the cold to the warmer part of a system (Fig. 10.62). This has technical applications in refrigerators and heat pumps.

10.3.14.1 Refrigerators

In refrigerators the heat Q_2 is extracted at a temperature T_2 from the volume V_2 that should be cooled and a larger heat energy $Q_1 = Q_2 + W$ is transported to a warmer environment. This

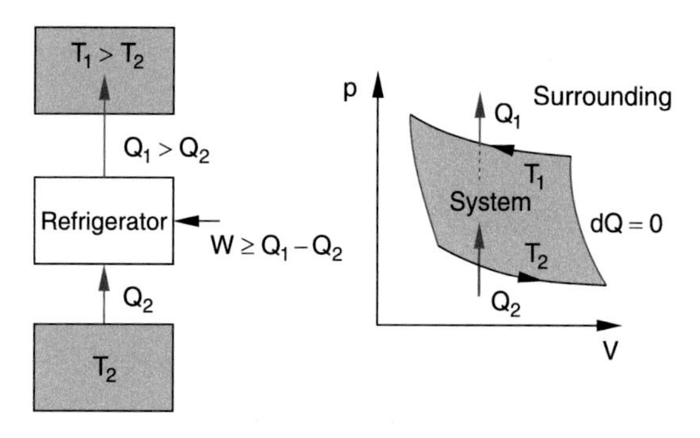

Figure 10.62 Principle of a refrigeration machine and heat pump based on the inverse Carnot's cycle

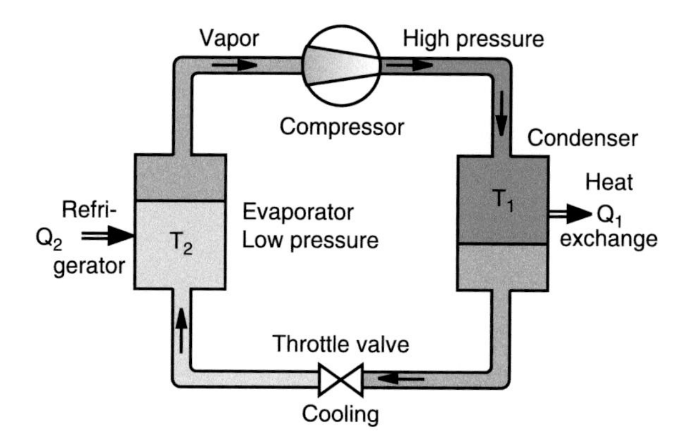

Figure 10.63 Technical realization of a refrigerator

demands the supply of mechanical or electrical work W to the system. We have neglected all energy losses by friction or heat conduction. The coefficient of performance

$$K_{\text{ref}} = \frac{Q_2}{W} = \frac{\text{d}Q_2/\text{d}t}{\text{d}W/\text{d}t} \tag{10.119a}$$

gives the ratio of cooling rate $\text{d}Q_2/\text{d}t$ and power input $\text{d}W/\text{d}t$.

From the efficiency η of the Carnot engine we obtain for the inverse Carnot cycle the coefficient of performance for the refrigerator the relation

$$K_{\text{ref}} = 1/\eta = \frac{T_2}{T_1 - T_2} . \tag{10.119b}$$

This shows that a refrigerator works more efficiently for small temperature differences $(T_1 - T_2)$ between the cooled volume and the warmer environment.

10.3.14.2 Heat Pumps

Heat pumps use the heat reservoir of the environment (air, ground) for heating water for floor heating of rooms or for swimming pools. The basic principle is the same as that of refrigerators. Heat is transported from a cold to a warmer volume. This demands the supply of mechanical or electrical energy. The useful energy is the heat transported to the warmer volume. Therefore the coefficient of performance is defined as in (10.119b)

$$K_{\text{hp}} = \frac{Q_1}{W} = \frac{\text{d}Q_1/\text{d}t}{\text{d}W/\text{d}t} = \frac{T_1}{T_1 - T_2} . \tag{10.120}$$

Contrary to the efficiency $\eta < 1$ of the Carnot engine the coefficient $K_{\text{hp}} = 1/\eta$ is larger than 1! It increases with decreasing temperature difference $\Delta T = T_1 - T_2$.

Example

A heat pump used for heating a swimming pool takes the heat from a river with a water temperature of $10\,°\text{C} = 283\,\text{K}$ and heats the swimming pool to a temperature of $T = 27\,°\text{C} = 300\,\text{K}$. The maximum coefficient of performance is then $K_{\text{hp}} = 17.6$. One therefore saves a factor of 17.6 of heating costs compared with the direct heating of the swimming pool. In this idealized example all other losses of the heat pump system have been neglected. Realistic values, taking into account all losses, are $K_{\text{hp}} = 5$–10. ◄

For practical applications heat pumps and refrigerators operate with special cooling liquids, which are not permanent gases but evaporate and condense during one cycle. This is illustrated schematically in Fig. 10.63. The heat Q_2 is transported from the room to be cooled to the liquid cooling agent at the low temperature T_2 in the evaporator. The resulting temperature increase of the cooling liquid results in the evaporation of the liquid . In the condenser the heat Q_1 is extracted by a heat exchanger from the vapour at high pressure. This causes the condensation of the vapour. The liquid under high pressure expands through a throttle valve, which decreases its temperature and is again used for heat extraction from the volume to be cooled.

10.3.14.3 Stirling Engine (Hot Air Engine)

The Stirling engine uses air as working agent, which is periodically expanded and compressed in a cycle of two isotherms and two isochors (Fig. 10.64a). The red arrows indicate the heat exchange between the environment (white) and the system (red). During the isothermal expansion $1 \to 2$ the heat Q_1 is supplied at the temperature T_1 to the system. During the isochoric cooling $2 \to 3$ the temperature drops to $T_2 < T_1$. Now isothermal compression $3 \to 4$ occurs where the heat Q_2 is transported to the environment. Finally the isochoric compression $4 \to 1$ with heat supply Q_4 brings the system back to its original state 1. The heat Q_4 is necessary to increase the temperature from T_2 to $T_1 > T_2$. Since no work is performed during the isochoric processes, the energy balance demands

$$Q_2 = -Q_4 = C_v \Delta T .$$

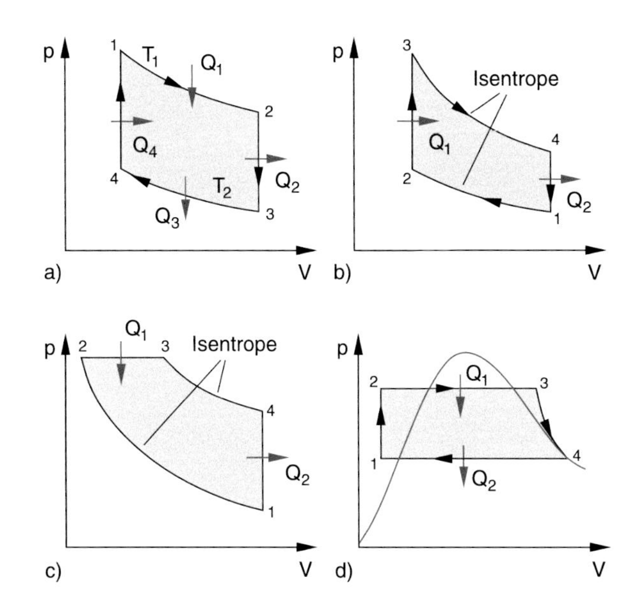

Figure 10.64 Cycle of **a** Stirling engine, **b** Otto engine (gasoline engine), **c** Diesel engine, **d** steam engine. The *red curve* gives the vapor pressure $p(V)$ of water vapor

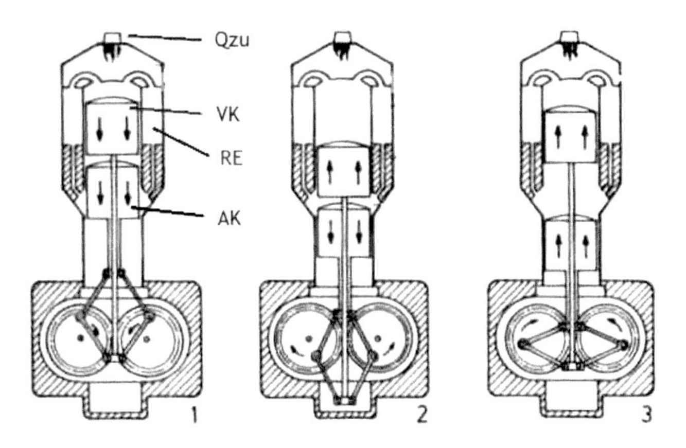

Figure 10.67 Stirling motor with two pistons in one cylinder

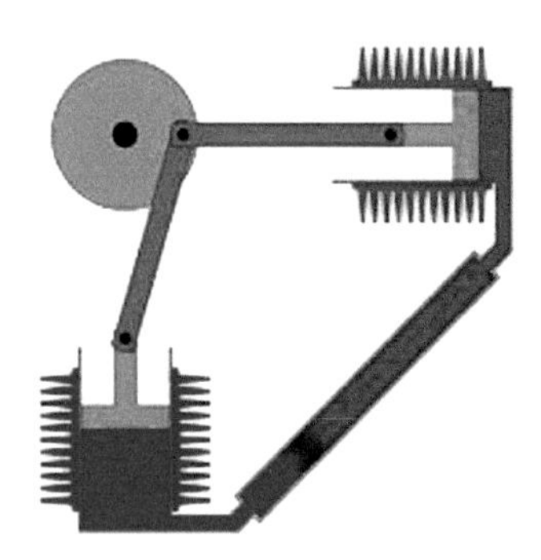

Figure 10.65 Stirling engine with two pistons and two cylinders

When the extracted heat Q_2 can be stored and resubmitted to the system during the process $4 \rightarrow 1$, the system does not loose energy during a cycle and the efficiency of the Stirling process would be comparable to that of the Carnot cycle.

This can be technically realized, at least approximately, by using two pistons, the working piston and the displacer piston in two different cylinders: a hot cylinder and a cold one (Fig. 10.65). The two pistons are driven by the same crankshaft with a 90° phase shift against each other. The two cylinders are connected by a pipe filled with an energy storage material (regenerator). When the piston compresses the gas in the hot cylinder the hot gas flows from the hot to the cold cylinder through the connecting pipe and heats up the storage material. In the next step the cold cylinder is compressed and the cold gas flowing through the pipe is heated by the storage material. About 80% of the energy exchanged during one cycle can be stored in the regenerator. In the diagrams of Fig. 10.66 the time sequence of the total volume, the hot and the cold volume are depicted for the Stirling engine, used as heat engine and as heat-pump.

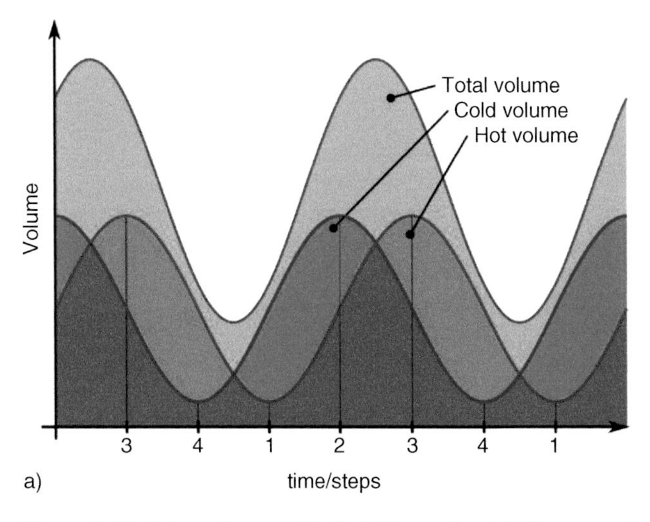

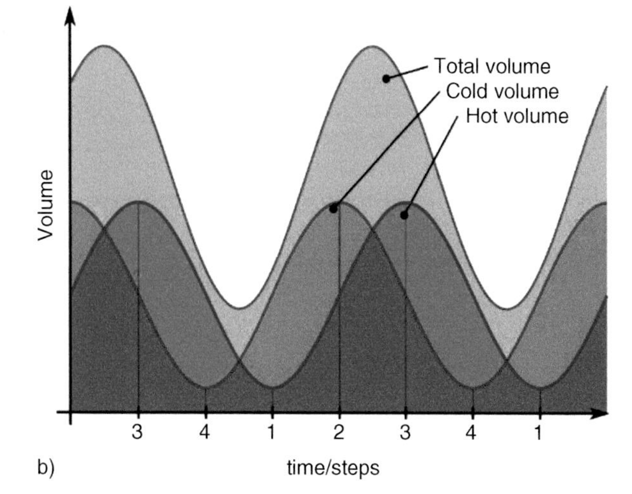

Figure 10.66 Volume-diagram $V(t)$ of a Stirling engine: *Left diagram:* used as heat engine, *right diagram:* used as heat pump

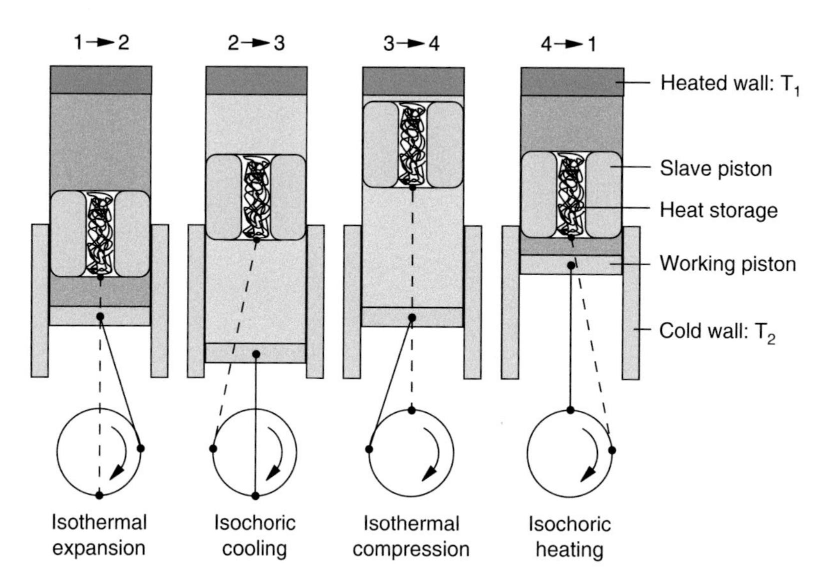

Figure 10.68 Positions of working piston and slave piston during the four sections of a Stirling cycle. The energy necessary for the operation of the engine is supplied by heating the upper wall

Another version of the Stirling engine uses only one cylinder but still two pistons. The displacer piston presses the air periodically into the upper hot volume and in the next step into the lower cold volume. During these processes the air streams through a hole in the piston which is filled with metal cuttings. They are heated during the passage $2 \rightarrow 3$ of the hot air and they transfer their heat during the passage $4 \rightarrow 1$ to the cold air.

In Fig. 10.67 and 10.68 the different steps durng a working cycle are illustrated for a Stirling engine with one cylinder and two pistons.

10.3.14.4 The Otto-Engine

The Otto-engine is used in many cars as effective drive. In the p-V-diagram it passes a cycle consisting of two isentropic and two isochoric processes (Fig. 10.64b). In the state 1 the gasoline-air mixture is sucked in and compressed. At point 2 the ignition oocurs where the mixture explodes so fast that the volume does not change essentially. The heat Q_1 released at the explosion is fed into the system and increases the pressure very fast up to point 3. Now an isentropic expansion (no further heat supply) follows until point 4 is reached. Here the exhaust valve is opened and the exhaust gas streams into the exhaust pipe. This causes a decrease of the pressure, a release of the heat Q_2 and the restitution of state 1.

The efficiency η depends on the compression ratio V_1/V_2. One obtains (see Probl. 10.12)

$$\eta = 1 - \frac{1}{(V_1/V_2)^{\kappa-1}}\,, \qquad (10.121)$$

where $\kappa = C_p/C_v$ is the specific heat ratio.

Example

$V_1/V_2 = 9$ and $\kappa = 9/7 \rightarrow \eta = 0.44$. Note, that the real efficiency is only about 0.3–0.35 due to energy losses by friction and heat conduction. ◄

10.3.14.5 Diesel Engine

For the Diesel engine the cycle in the p-V-diagram (Fig. 10.64c) consists of two isentropic, one isobaric and one isochoric process. In the state 1 air is sucked in and the volume is compressed until point 2 is reached. The compression ratio is much larger (up to 1:20) as in the Otto-engine. During this compression the temperature rises to 700–900 °C, which is above the ignition termperature of Diesel-fuel. Now Diesel fuel is injected, which does not explode as in the Otto-engine but burns more slowly (there is no electrical ignition). This causes the air-fuel-mixture to expand isobaric until point 3 is reached, where the combustion stops. The volume now further expands isentropically to the point 4, where the exhaust valve opens and the pressure suddenly drops to the atmospheric pressure outside. Here the intial point 1 is reached again.

The efficiency of the Diesel engine is higher thatn that of the Otto-engine because of the higher compression ratio. Its theoretical value is about 0.55 but due to unavoidable losses the real enegines only reach about 0.45. The disadvantage of the Diesel engine is the higher output of NO_x gases and soot particles.

10.3.14.6 Steam Engine

In a Steam Engine the cyclic process (Clausius–Rankine Process) consists of two isentropic and two isobaric parts (Fig. 10.64d). In the initial state 1 the system contains water. A pump increases the pressure at a constant volume isentropic from p_1 to p_2. From point 2 to 3 heat is supplied at constant pressure, which causes the volume to expand and increases the temperature above the boiling point of water. The hot vapour drives a piston during the isentropic expansion and the system reaches point 4 where the temperature is cooled down, the vapour condenses and the heat Q_2 is transferred to the surrounding. Now the initial point 1 is reached again. Mechanical work is performed on the part $3 \to 4$.

The red curve in Fig. 10.64d gives part of the Van der Waals curve $p(V)$ for water vapour (see Sect. 10.4.1). Inside this curve water and vapour can exist simultaneously, in the region left of the curve only the liquid phase exists, to the right hand of the curve only the vapour phase.

10.3.14.7 Thermal Power Plants

In thermal power plants, heat is produced by burning fossil fuels, such as coal, oil, wood or gas, or by fission of atomic nuclei. For fossil fuels the heat comes from the reaction heat that is released during the oxidation of atoms or molecules and is due to the different chemical binding energies of reaction partners and reaction products. The essential part of this energy stems from the oxidation of carbon atoms C to CO_2. The produced heat is 8 kcal = 33 kJ for 1 g C. The fission of 1 g Uranium produces an energy of $2.5 \cdot 10^7$ kJ. This is $7.5 \cdot 10^5$ times more!

The heat produced in thermal power plants is converted into the generation of hot water vapour under high pressure, driving turbines that propel electric generators for the production of electric energy. The maximum efficiency depends, according to the second law, on the initial temperature T_1 and the final temperature T_2.

The initial temperature is limited by technical conditions (heat and pressure resistance of the hot vapour tank. Typical values are between 600 and 700 °C. Only for the high temperature reactors, temperatures above 800 °C are realized.

For the choice of the final temperature T_2 two options exist:

1. One chooses $T_2 = 100\,°\mathrm{C}$ (condensation temperature of water) and uses the rest energy of the hot water for heating of houses. The efficiency for the conversion of heat into mechanical (or electric) energy is then for an initial temperature of $T_1 = 600\,°\mathrm{C} = 873\,\mathrm{K}$: $\eta = 500/873 = 0.57$. In addition the heat ΔQ of the hot water can be delivered to houses nearby the power station.
2. The final temperature of the water vapour is chosen as $T_2 = 30\,°\mathrm{C}$. In order to avoid condensation one has to lower the pressure below the atmospheric pressure by pumping the expanding volume. This increases the efficiency to $\eta = 570/873 = 0.65$. The work needed for evacuating the expanding volume against the external pressure is smaller than the additional energy gain due to the lower final temperature.

In case 1 one does not win the total energy of the hot water ΔQ compared to case 2 because here one could use the extra energy due to the higher efficiency $\eta_2 = (600 - 30)/873$ to transport electric energy for heating. The increase of the efficiency for case 2 compared to case 1 is

$$\begin{aligned}\eta_2 - \eta_1 &= (70/873 - \varepsilon)\Delta Q\\ &= (0.19 - \varepsilon)\Delta Q\,,\end{aligned}$$

where $\varepsilon \cdot \Delta Q$ is the mechanical work of the pump, necessary to evacuate the volume down to a pressure that is equal to the vapour pressure of water at $T = 30\,°\mathrm{C}$.

10.4 Thermodynamics of Real Gases and Liquids

Up to now, we have discussed the thermodynamics of ideal gases, where the interaction between the atoms of the gas has been neglected.

We will now discuss, which rules have to be generalized and which are still valid without restrictions, when we treat the thermodynamics of real atoms and molecules including their size and their mutual interactions.

While ideal gases remain gaseous at any temperature, real gases condense below their boiling temperature and they can even become solids below the melting temperature. In this section, we will investigate what are the conditions for transitions between the different phases solid, liquid and gaseous and what are the equilibrium conditions of the different phases.

10.4.1 Van der Waals Equation of State

At very high pressures, the density of atoms or molecules becomes so high, that the internal volume of the molecules (also called covolume) cannot be neglected compared with the free volume V that is available for the molecules.

When we describe the atoms as rigid balls with radius r, two atoms cannot come closer to each other than at a minimum distance $d = 2r$. If one atom is in the volume V the other atoms cannot penetrate into the volume $V_{\text{forbidden}} = (4/3)\pi d^3 = 8V_a$, where V_a is the volume of one atom in the model of rigid balls (Fig. 10.69a). Furthermore the centres of all balls must have the minimum distance $d = r$ from the walls of the container.

Assume there were only two atoms in the cubic volume $V = l_3$. The volume available for the second atom is then

$$V_2 = (L - 2r)^3 - 8V_a \quad \text{(Fig. 10.69b)}\,.$$

A third atom could only be found in the volume

$$V_3 = (L - 2r)^3 - 2 \cdot 8V_a\,,$$

and the rest volume available for the n-th atom is then

$$V_n = (L - 2r)^3 - (n - 1) \cdot 8V_a\,.$$

Chapter 10

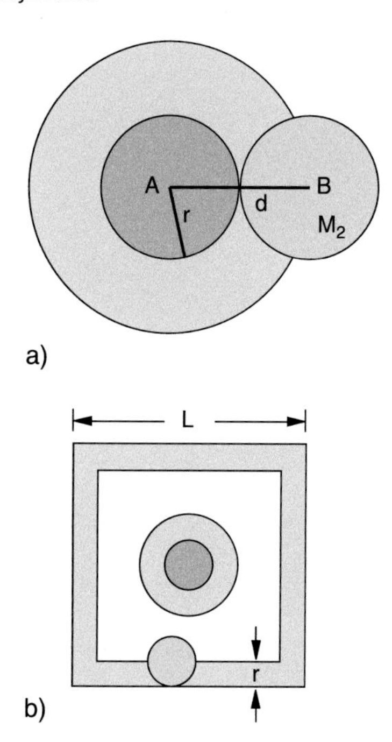

Figure 10.69 Illustration of the co-volume. **a** The center of B cannot be in the *bright red circle*. **b** Forbidden volume (*bright red*) of molecule B in the volume L^3 with one atom A

An estimation of the real sizes shows, however, that, for example, for $N = 10^{20}$ atoms in a volume with $L = 0.1$ m the forbidden volume at the wall is with $r \approx 10^{-10}$ m completely negligible compared to the internal volume $N \cdot V_a$ of the N atoms in the volume V. The average over all N atoms gives then for the mean volume available to each atom

$$\begin{aligned} V &= (L-2r)^3 - \frac{1}{N}\sum_{n=1}^{N}(n-1)\cdot 8V_a \\ &= L^3 - 6r\cdot L^2 - 6r^2\cdot L + 8r^3 - 4NV_a \\ &\approx L^3 - 4NV_a \quad \text{for} \quad N \gg 1\,. \end{aligned} \tag{10.122}$$

because the 2nd, 3rd and 4th term are neglible compared to the 1st and last term.

We therefore have to replace in the general gas-equation (10.21) the volume V by the reduced volume

$$V - b = L^3 - 4\cdot N\cdot V_a \quad \text{with} \quad b = 4\mathrm{N}\cdot V_a\,.$$

For the situation $N \gg 1$ the volume $V_{\text{available}}$ available to the N atoms in a volume V is reduced by 4 times the total atomic volume $N \cdot V_a = N \cdot (4/3)\pi \cdot r^3$.

The next question concerns possible corrections for the pressure p due to the attractive or repulsive forces between the atoms. At low temperatures or for high densities the interaction between the atoms can be no longer neglected. The total force on a selected atom resulting from the interaction with all other atoms

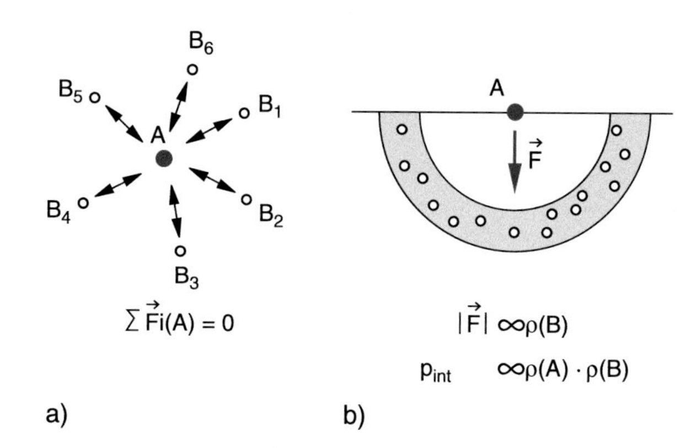

Figure 10.70 Illustration of the internal pressure. Forces on an atom A **a** inside a gas, **b** at the boundary between wall and gas

cancels for atoms inside a liquid or a gas volume, because the interaction forces are in the average uniformly distributed over all directions (Fig. 10.70a) (see the similar discussion about the surface tension in Sect. 6.4.1). At the boundaries between liquid and gas or between gas and wall, the interaction forces are no longer uniformly distributed but are directed only into the half space of the medium. They do not cancel and the total force F_a on one atom is not zero but is proportional to the number density n_a of atoms in the half-sphere shown in Fig. 10.70b, which means to the density $\varrho = M/V$, where M is the total mass of the gas in the volume V.

The amount of the total force $F = |\sum \boldsymbol{F}_i| \propto n_a \cdot F_a$ onto all n_a atoms is therefore proportional to $n_a^2 \propto \varrho^2$. The force is directed towards the interior of the gas and causes an intrinsic pressure

$$p_b = a\cdot\varrho^2 \propto a/V^2\,,$$

which acts onto the atoms in addition to the external pressure p.

Taking into account this intrinsic pressure and the co-volume $b = 4N \cdot V_a$ the general gas equation for one mole of an ideal gas

$$p\cdot V_M = R\cdot T$$

has to be modified to the **van der Waals-equation** of real gases

$$\left(p + \frac{a}{V^2}\right)\cdot(V_M - b) = R\cdot T\,, \tag{10.123}$$

where the constant $b = 4 \cdot N_a \cdot V_a$ gives 4-times the internal volume of the N_A atoms in the mole volume V_M.

The progression of the function $p(V)$ at constant T for a real gas, described by (10.123), depends on the constants a and b. In Fig. 10.71 the isotherms of CO_2 are shown for different temperatures. They confirm, that for high temperatures ($E_{kin} \gg |E_{pot}|$) the curves are similar to those of an ideal gas, but for low temperatures closely above the condensation temperature they deviate strongly.

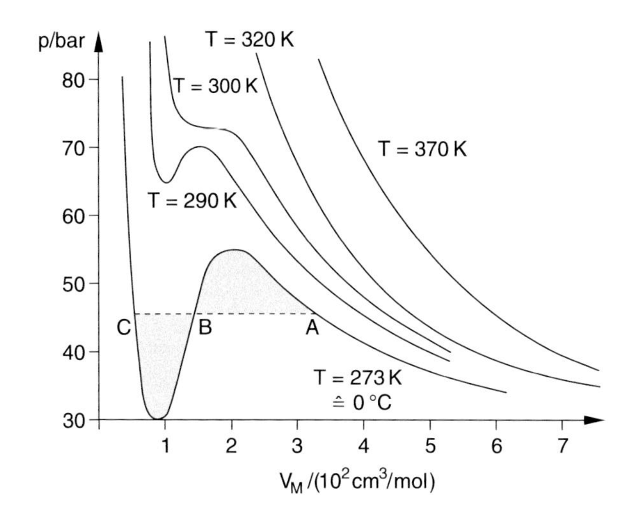

Figure 10.71 Van-der-Waals-isotherms of CO_2 for different temperatures

Solving (10.123) for p at constant T gives a polynomial $p(V)$ of third order, which shows for low temperatures a maximum and a minimum (Fig. 10.71). How looks the comparison of this theoretical curve with experimental results? Let's see this for the example of CO_2.

If one mole of CO_2 is compressed at the temperature $T = 0\,°C$ starting at low pressures one finds in deed that the curve $p(V)$ follows quite nicely the theoretical curve until the point A in Fig. 10.71. Further compression does **not** increase the pressure p, which stays constant until the point C is reached where the pressure shows a steep increase and follows again the van der Waals curve.

The reason for this strange behaviour is the condensation of the CO_2 vapour that starts at the point A. On the way from A to C the fraction of the liquid phase continuously increases until in C the vapour is completely liquefied. On the way from C to smaller volumes, the pressure increases steeply because of the small compressibility of the liquid. Between A and C gas and liquid can both exist (co-existence range).

For a quantitative description of the condensation process we must discuss the different phases (aggregation states) in more detail.

10.4.2 Matter in Different Aggregation States

The different aggregation states of matter (solid, liquid gas) are called its phases. In this section we will discuss, under which conditions a phase transition solid → liquid, liquid → gas or solid → gas can occur and when two or three phases can exist side by side.

10.4.2.1 Vapour Pressure and Liquid–Gas Equilibrium

When a liquid is enclosed in a container which it fills only partly, one finds that part of the liquid is vaporized and in the volume

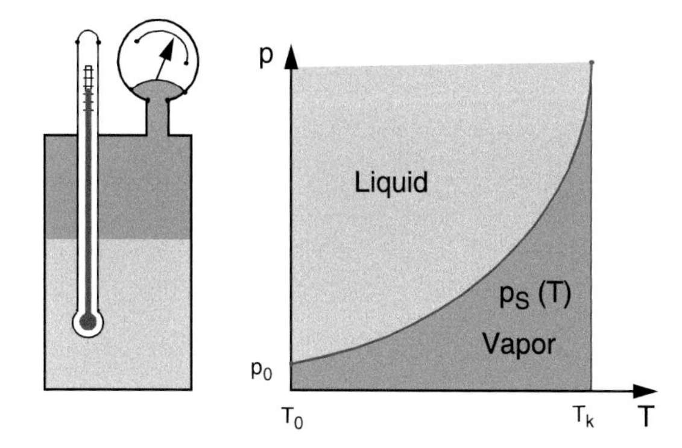

Figure 10.72 Measurement of vapor pressure curve $p_s(T)$

above the liquid surface a vapour phase has established at a vapour pressure $p_s(T)$ that acts upon the walls and the liquid surface. The dependence of the vapour pressure $p_s(T)$ on the temperature can be measured with the pressure tank shown in Fig. 10.72 that is equipped with a thermometer and a manometer.

At a constant temperature T a constant saturation vapour pressure $p_s(T)$ is present where the liquid and the gaseous phase can exist simultaneously under stable conditions.

The explanation given by molecular physics is based on the kinetic gas theory (Sect. 7.4). Similar to the situation in a gas also the molecules in a liquid show a velocity distribution with kinetic energies that follow the Maxwell–Boltzmann distribution. The fastest molecules in the high energy tail of the velocity distribution can leave the liquid, if their energy is larger than the surface tension of the liquid (See Sect. 6.4). On the other hand, when molecules in the gas phase hit the liquid surface, they can enter into the liquid.

At the saturation vapour pressure $p_s(T)$ the liquid and the gas-phase are at equilibrium, which means that the rate of molecules leaving the liquid is equal to the rate of molecules that reenter the liquid from the gas phase.

The higher the temperature the more molecules have sufficient energy to leave the liquid, i. e. the vapour pressure rises with increasing temperature (Fig. 10.72). The quantitative form of the vapour pressure curve $p_s(T)$ can be calculated in the following way:

In Fig. 10.73 we regard for 1 mol of the evaporating liquid a cyclic process in the p, V-diagram of Fig. 10.71. In the state $C'(T + dT, p_s + dp_s)$ the vapour should be completely condensed and the liquid occupies the volume V_l. Now the volume is isothermally expanded at the temperature $T + dT$, while the pressure is kept constant. Here the heat $dQ_1 = \Lambda$, which is equal to the evaporation energy of 1 mol, has to be supplied in order to keep the temperature constant. At A' the total liquid is evaporated. During the next step, the adiabatic expansion $A' \to A$, pressure and temperature are lowered by an infinitesimal small amount. The system remains in the vapour phase and reaches the point $A(p_s, T)$. Now the vapour is isothermally compressed while the pressure remains constant, because condensation progresses during the path from A to the point C. The

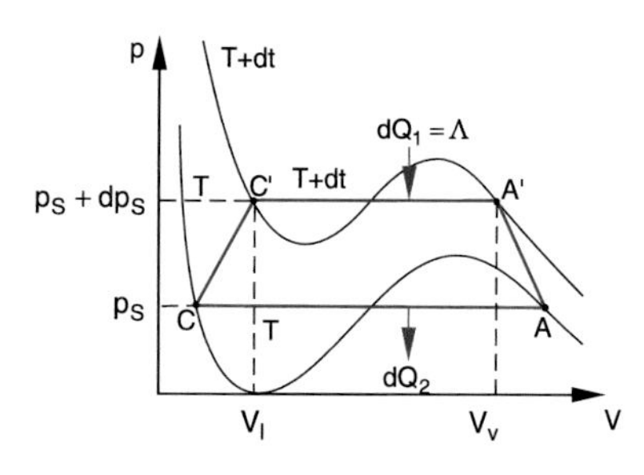

Figure 10.73 Carnot cycle C′A′ACC′ in a p-V diagram of Fig. 10.71 illustrating the derivation of the Clausius–Clapeyron-equation

condensation heat dQ_2 is released to the surrounding. The step $A \to C$ corresponds to the curve ABC in Fig. 10.71. The liquid state in point C is then transferred by an infinitesimal step to the initial point $C'(p + dp, T + dT)$.

The temperature of the system changes only on the short paths $A' \to A$ and $C \to C'$. During the isothermal expansion $C' \to A'$ the system has delivered the work $dW_1 = (p_s + dp_s) \cdot (V_l - V_v)$, while during the compression $A \to C$ the work $dW_2 = p_s \cdot (V_v - V_l)$ has to be supplied to the system. The net work is therefore $dW = dW_1 + dW_2 = (V_l - V_v) \cdot dp_s$.

In Sect. 10.3.5 it was shown that the efficiency η of the Carnot engine for an arbitrary working material is

$$\eta = \frac{|\Delta W|}{\Delta Q_1} = \frac{(V_v - V_l)dp_s}{\Lambda}$$
$$= \frac{T + dT - T}{T + dT} \approx \frac{dT}{T},$$

because here is $dT \ll T$. This gives for the evaporation energy for 1 mol evaporated liquid the **Clausius–Clapeyron equation**

$$\Lambda = T\frac{dp_s}{dT}(V_v - V_l). \qquad (10.124)$$

The evaporation heat is proportional to the difference of the mole-volumes of the liquid and gaseous phases and to the slope dp_s/dT of the vapour pressure curve.

Note: Often the specific evaporation energy λ [kJ/kg] is given instead of the molar evaporation energy [kJ/mol]. The conversion factor is

$$1\,\text{kJ/mol} = (10^{-3}M)\,\text{kJ/kg},$$

where M is the molar mass in g/mol.

The evaporation energy pro molecule is $w = \Lambda/N_A$ with N_A = Avogadro number.

The heat of evaporation has two causes: The first cause is the energy necessary to enlarge the volume V_l of the liquid to the larger volume V_v of the vapour against the external pressure p.

The second cause is the energy spend to enlarge the distance between the molecules against their mutual attraction. The second contribution is by far the largest one. It is therefore nearly equal to the heat of evaporation.

Example

The volume of 1 kg water expands during the evaporation from $V_l = 1\,\text{dm}^3$ to $V_v = 1700\,\text{dm}^3$. The work performed during the expansion against the external pressure of 1 bar is $W = p \cdot dV = 10^5\,\text{Nm} \cdot 1.7\,\text{m}^3 = 170\,\text{kJ}$. The measured specific evaporation heat is $\lambda = 2080\,\text{kJ/kg}$. Therefore the first contribution only amounts to 8%. ◀

As one of many applications of the thermodynamic potentials we will derive the Clausius–Clapeyron equation (10.124) with the help of the thermodynamic potentials, where here the Gibbs'-potential $G(p, T)$ of (10.105) is used.

Differentiation of (10.105) gives

$$dG = \left.\frac{\partial G}{\partial p}\right|_T dp + \left.\frac{\partial G}{\partial T}\right|_p dT.$$

The compilation scheme of the potentials in Sect. 10.3.11 shows

$$\left.\frac{\partial G}{\partial p}\right|_T = V \quad \text{and} \quad \left.\frac{\partial G}{\partial T}\right|_p = -S.$$

At the phase equilibrium is $dG_1 = dG_2$

$$\to dG_1 = V_1 dp - S_1 dT = V_2 dp - S_2 dT = dG_2$$
$$(S_2 - S_1)dT = (V_2 - V_1)dp$$
$$\frac{dp}{dT} = \frac{S_2 - S_1}{V_2 - V_1}.$$

From the definition of the entropy we conclude

$$S_2 - S_1 = \int_1^2 \frac{dQ_{rev}}{T} = \frac{\Lambda}{T},$$

which finally gives

$$\frac{dp}{dT} = \frac{\Lambda}{T(V_2 - V_1)}.$$

The heat supply does not increase the kinetic energy of the molecules, (because the temperature stays constant), but only the potential energy. Therefore in Fig. 10.18 the long horizontal line T(t) appears during the evaporation process.

Since in (10.124) $V_v \gg V_l$ we can neglect V_l in (10.124) and we can also approximate in the general gas equation $p \cdot V = RT$ the volume $V \approx V_v \to V_v = R \cdot T/p_s$. Inserting this into (10.124) we get

$$\frac{1}{p_s}\frac{dp_s}{dT} = \frac{\Lambda}{RT^2}.$$

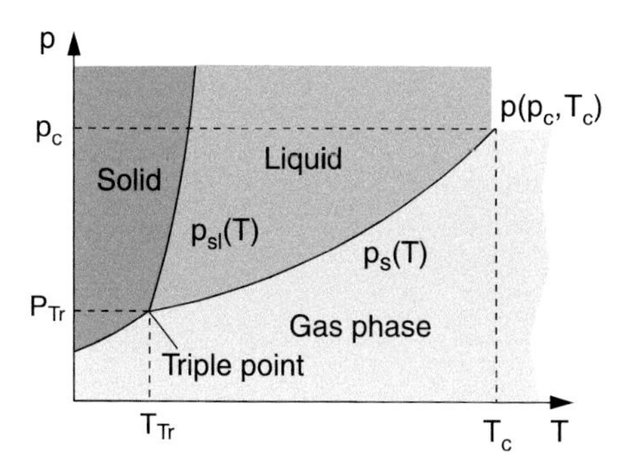

Figure 10.74 Phase diagram with vapor pressure curve $p_s(T)$ representing the separating line between liquid and vapor phase from the triple point up to the critical point $P(p_c, T_c)$ and melting curve $P_{sl}(T)$ as separation line betweeen solid and liquid phase

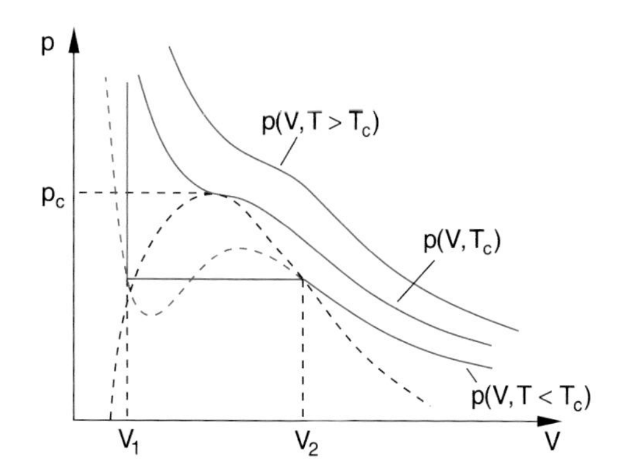

Figure 10.75 Behaviour of Van-der-Waals isotherms $p(V)$ around the critical point (p_c, T_c)

Integration yields

$$\ln p_s = -\frac{\Lambda}{RT} + C$$

with the integration constant C. This gives with the boundary condition $p_s(T_0) = p_0$

$$p_s = p_0 \cdot A \cdot e^{-\Lambda/(RT)} \quad \text{with} \quad A = e^{\Lambda/RT_0} . \tag{10.125}$$

This **van't-Hoff equation** shows that the vapour pressure rises proportional to $\exp(-1/T)$.

Along the vapour pressure curve $p_s(T)$ the vapour phase and the liquid phase are at equilibrium, i. e. at each temperature there exists a vapour pressure $p_s(T)$ where the two phases exist simultaneously and are both stable.

The vapour pressure curve divides the area in the p-T-diagram into two sections (Fig. 10.74). For $p(T) < p_s(T)$ only the vapour phase exists under equilibrium conditions, for $p(T) > p_s(T)$ only the liquid phase.

The vapour pressure curve terminates at the critical temperature $T = T_c$. The corresponding vapour pressure $p_c = p_s(T_c)$ is the critical pressure. Above the critical temperature T_c no distinction between liquid and vapour phase is possible. The densities of both phases become equal. The slope of the vapour pressure curve is there

$$\left(\frac{dp_s}{dT}\right)_{T_c} = \frac{p_c \cdot \Lambda}{RT_c^2} . \tag{10.126}$$

The evaporation heat decreases with increasing temperature and becomes zero at the critical temperature T_c. Just below T_c part of the liquid changes statistically into the vapour phase and back. This causes striations in the optical density which can be seen in the transmitted light.

The critical temperature is related to the interaction potential between the molecules. Above T_c the mean kinetic energy of the molecules is larger than the amount of the mean potential energy. In the p-V-diagram of Fig. 10.71 the isotherms have for T_c three intersection points with the horizontal line $p = \text{const} < p_c$.

When the volume V is compressed, the real pressure curve (Fig. 10.75) shows a kink at $V = V_2$ and follows until V_1 not the van der Waals curve but the horizontal line $p = \text{const}$ because here condensation takes place. The dashed black curve in Fig. 10.75 gives the volume V_2 where condensation starts and V_1 where the whole gas is liquefied. At the critical temperature T_c the curve $p(V)$ has no longer minima and maxima but only an inflection point, which indicates that there are no longer phase transitions but only a unique phase is present, which is called the supercritical phase. The tangent to the curve p(V) in the critical point $p(T_c, V_c)$ is horizontal. The critical point can be calculated from the van der Waals equation (10.123) with the conditions

$$\left(\frac{\partial p}{\partial V}\right)_{T_c,V_c} = 0 \quad \text{and} \quad \left(\frac{\partial^2 p}{\partial V^2}\right)_{T_c,V_c} = 0 .$$

This gives for p_c and T_c the results

$$p_c = \frac{1}{27}\frac{a}{b^2} ; \quad V_c = 3b ; \quad T_c = \frac{8}{27}\frac{a}{Rb} , \tag{10.127a}$$

and for the van der Waals constants a and b

$$a = 3p_c V_c^2 ; \quad b = \frac{1}{3} V_c . \tag{10.127b}$$

It is therefore possible to gain information about the attractive interaction between the molecules and their internal volume from measurements of the critical parameters p_c and T_c.

10.4.2.2 Boiling and Condensation

If the vapour pressure p_s becomes larger than the external pressure p acting on the liquid surface, vapour bubbles can form in the inside of a liquid. They rise, due to buoyancy, to the liquid surface: The liquid boils. The boiling temperature T_b depends on the external pressure p. From (10.125) one obtains

$$T_b(p) = T_b(p_0) \cdot \frac{1}{1 - \frac{RT_b(p_0)}{\Lambda} \ln(p/p_0)} . \tag{10.128}$$

Chapter 10

Example

Water boils under a pressure $p = 1$ bar at $T_b = 373$ K $= 100\,°$C. For $p = 400$ mbar $T_b = 77\,°$C. Since the cooking time of food strongly depends on the temperature, cooking at high altitudes becomes tedious. Therefore one uses a pressure cooker, which operates at about 1.5–2 bar and reduces the cooking time considerably. ◄

If the vapour pressure becomes smaller than the external pressure the vapour starts to condensate.

In our atmosphere the air mixed with water vapour generally does not reach an equilibrium state (p, T), because the conditions in the atmosphere change faster than the time necessary to establish an equilibrium. The water vapour pressure is therefore in general lower than the saturation pressure.

The concentration of water vapour in our atmosphere, measured in $\mathrm{g/m^3}$, is called the *absolute humidity* φ_a. The maximum possible concentration of water vapour is reached, when the water vapour pressure p_w is equal to the saturation pressure p_s. The humidity φ at this pressure is the *saturation humidity* φ_s.

The relative humidity is the quotient

$$\varphi_{rel} = \frac{\varphi_a}{\varphi_s} = \frac{p_w}{p_s}\,. \tag{10.129}$$

Example

A relative humidity of 40% is reached, when the vapour pressure of water is $p_w = 0.4 p_s\,(H_2O)$. ◄

For a given absolute humidity the relative humidity increases with decreasing temperature because the vapour pressure of water decreases with T. (Fig. 10.76). When $\varphi_{rel} = 1$ it starts to rain. The temperature T_d where $\varphi_{rel} = 1$ is the *dew point* or saturation temperature.

For the operation of air conditioning systems, this has to be taken into account. If the air is cooled below the dew point, the water vapour will condense and increase the humidity in the cooled room. The air has therefore to be dried before it is cooled down.

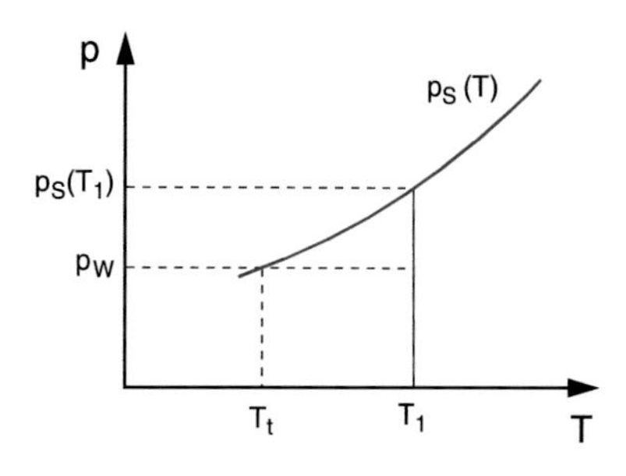

Figure 10.76 Illustration of relative and absolute humidity of air and of dew point

10.4.2.3 Liquefaction of Gases; Joule–Thomson Effect

In order to liquefy gases at the pressure p one has to lower their temperature below the pressure-dependent boiling temperature $T_s(p)$. There are several experimental realizations:

Adiabatic Cooling with Energy Output

Here the internal energy U of the gas at the pressure p_1 decreases because the expanding gas delivers the work $\mathrm{d}W = p{\cdot}\mathrm{d}V$ against the lower external pressure p, while no heat is exchanged ($\mathrm{d}Q = 0$). From the first law (10.82) we obtain for 1 mol

$$\mathrm{d}U = C_v \cdot \mathrm{d}T = -p \cdot \mathrm{d}V\,.$$

This yields the temperature decrease

$$\mathrm{d}T = -\frac{p_p}{C_v}\mathrm{d}V\,.$$

Example

10 mol of a gas at room temperature $T = 300\kappa$ are expanded against an external pressure of 10 bar $= 10^6$ Pa by $\Delta V = 10^{-2}\,\mathrm{m^3}$ (this corresponds to 5 mol volumes. With $C_v = 20.7\,\mathrm{J/(mol\cdot K)}$ we obtain $\Delta T = -4.8$ K. ◄

This adiabatic cooling can be realizes for ideal and also for real gases. It comes from the decrease ΔU of the internal energy due to the partial transfer into mechanical work.

Joule–Thomson Effect

For real gases cooling can be also achieved without the transfer into mechanical work. The expansion of the volume V increases the mean distance between the molecules. This requires work against the attractive forces between the molecules, which means that the potential energy of the system increases at the expense of the kinetic energy and the temperature decreases.

When a real gas expands adiabatically through a nozzle at a pressure p_1 that is kept constant, from the volume V_1 into the volume V_2 (Fig. 10.77) with the pressure $p_2 < p_1$ there is no heat exchange with the surrounding ($\mathrm{d}Q = 0$) and the enthalpy $H = U + p \cdot V$, is constant because the cooling is due to the work against the attractive forces between the molecules during the expansion.

The internal energy U of a real gas is the sum of the kinetic energy $E_{kin} = (f/2) \cdot R \cdot T$ and a the potential energy

$$E_p = \int_{\infty}^{V_1} \frac{a}{V^2}\mathrm{d}V = -\frac{a}{V_1}\,,$$

which is due to the attractive forces between the molecules and causes the internal pressure (cohesion pressure).

Chapter 10

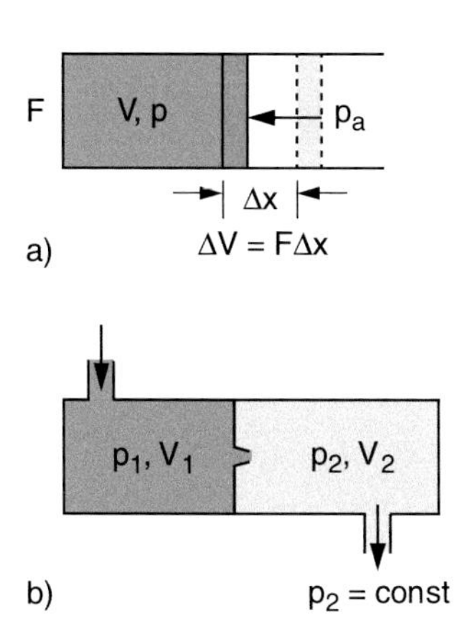

Figure 10.77 Comparison between adiabatic cooling and Joule–Thomson effect. **a** Adiabatic expansion with work delivery $\Delta W = p_a \cdot \Delta V$; **b** adiabatic expansion through a nozzle without work output

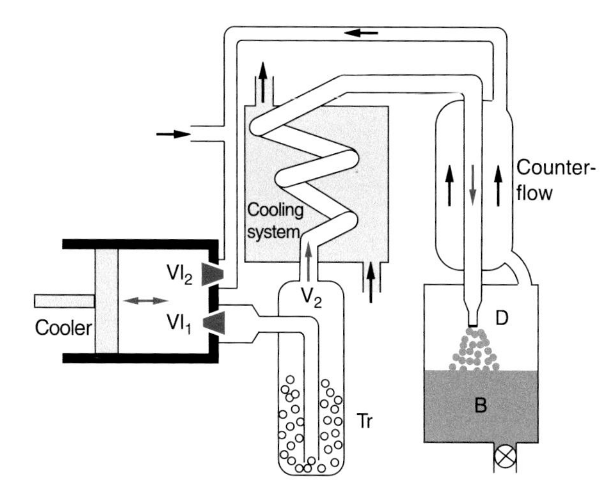

Figure 10.78 Schematic illustration of the Linde-process for liquefaction of air

Solving the van der Waals equation (10.123) for p, we obtain

$$p = \frac{R \cdot T}{V - b} - \frac{a}{V^2} .$$

The enthalpy then becomes

$$\begin{aligned} H = U + p \cdot V &= \frac{f}{2} RT - \frac{a}{V} + \left(\frac{RT}{V - b} - \frac{a}{V^2} \right) \cdot V \\ &= RT \left(\frac{f}{2} + \frac{V}{V - b} \right) - \frac{2a}{V} . \end{aligned} \tag{10.130}$$

Since H is constant during the adiabatic expansion through the nozzle, we get

$$\mathrm{d}H = \frac{\partial H}{\partial V} \mathrm{d}V + \frac{\partial H}{\partial T} \mathrm{d}T = 0$$

$$\begin{aligned} \Rightarrow \mathrm{d}T = -\frac{\frac{\partial H}{\partial V}\mathrm{d}V}{\frac{\partial H}{\partial T}} &= \frac{\frac{bT}{(V-b)^2} - \frac{2a}{RV^2}}{\frac{f}{2} + \frac{V}{V-b}} \mathrm{d}V \\ &\approx \frac{bRT - 2a}{(\frac{1}{2}f + 1)RV^2} \mathrm{d}V . \end{aligned} \tag{10.131}$$

For temperatures below the inversion temperature

$$T_\mathrm{i} = \frac{2a}{bR} , \tag{10.132}$$

we get $\mathrm{d}T < 0$. The gas cools down although no heat exchange with the surrounding takes place. The value of the inversion temperature depends on the ratio of the amount of the attractive forces (described by the constant a) and the covolume $b = 4N \cdot V_\mathrm{a}$ of the molecules. For ideal gases is $a = b = 0 \Rightarrow \mathrm{d}T = 0$ and no cooling occurs. The cooling of real gases through adiabatic expansion through a nozzle is the **Joule–Thomson effect**. It is only realized for real gases, not for ideal gases. This can be seen as follows: When the gas flows through the nozzle, driven by the pressure p_1, the energy $p_1 \cdot V_1$ is released. The gas, streaming into V_2, builds up the pressure p which requires the energy $p_2 \cdot V_2$. Energy conservation demands $U_1 + p_1 \cdot V_1 = U_2 + p_2 \cdot V_2 \Rightarrow H_1 = H_2$ (10.76). The expansion therefore proceeds at constant enthalpy.

For temperatures above the inversion temperature is $\mathrm{d}T > 0$, i.e. the gas heats up. In order to use the Joule–Thomson effect for cooling, the gas has at first to be precooled below the inversion temperature. For higher pressures the density of molecules increases and with it the relative share of the covolume $b = 4N \cdot V_\mathrm{a}$ and the inversion temperature T_i becomes pressure dependent. More detailed information on the curves $T_\mathrm{i}(p)$ can be found in [10.14].

In Tab. 10.7 the maximum values of T_i are compiled for some gases. The numbers show that for air the inversion temperature lies above room temperature. Therefore precooling is not necessary. The gases N_2 and O_2 can be cooled below their condensation temperatures solely with the Joule–Thomson effect. This is realized with the Linde-gas liquefying system, which uses the counter-current principle (Fig. 10.78). The gas is compressed by the piston K and streams through the valve Vl_1 into the volume V_2 where it is dehumidified. It then passes through a cooling system where it is precooled, before it streams through the counter current cooler and finally through a nozzle into the container D at low pressure. During this last step, it further cools down. The cooling rate is for air $\Delta T/\Delta p = 0.25\,\mathrm{K/bar}$. For a pressure difference $\Delta p = 100\,\mathrm{bar}$ one reaches a cooling rate of 25 K per step. The cold vapour is guided through the counter-current cooler and helps to precool the incoming gas. Finally, it is sucked in through the valve Vl_2 into the initial chamber during the expansion phase of the piston. The next step starts then already with a colder gas and reaches therefore a lower final temperature. After several steps the cooling during the expansion through the nozzle reaches the condensation temperature and the gas is liquefied.

During the cooling of air, which is composed of N_2 and O_2, at first the higher condensation temperature of oxygen is reached

Table 10.7 Critical temperatures T_c, critical pressure p_c, maximum inversion temperature T_i and boiling temperature for some gases

Gas	T_c/K	P_c/bar	$a/\mathrm{N\cdot m^4/mol^2}$	$b/10^6\,\mathrm{m^3/mol}$	T_i/K	T_s/K at $p_0 = 1.013$ bar
Helium	5.19	2.26	0.0033	24	30	4.2
Hydrogen	33.2	13	0.025	27	200	20.4
Nitrogen	126	35	0.136	38.5	620	77.4
Oxygen	154.6	50.8	0.137	31.6	765	90.2
Air	132.5	37.2	–	–	650	80.2
CO_2	304.2	72.9	0.365	42.5	>1000	194.7
NH_3	405.5	108.9	0.424	37.2	>1000	–
Water vapor	647.15	217.0	–	–	–	373.2

before N_2 liquefies. Therefore the two gases can be readily separated.

Nowadays liquid nitrogen rather than liquid air is used for many applications, because liquid oxygen contains the explosive ozone O_3. Liquid air that is kept in a Dewar increases its O_2 and O_3 concentration in the course of time since N_2 evaporates faster due to its higher vapour pressure and therefore after some time liquid air reaches a critical concentration of O_3 which explodes above a critical temperature.

The gases H_2, He of Ne can be liquefied by precooling them with liquid nitrogen below the inversion temperature before they can be further cooled by the Joule–Thomson effect.

10.4.2.4 Equilibrium Between Solid and Liquid Phase; Melting Curve

If the temperature of a solid material is increased above a certain temperature that depends on the material, the solid phase starts to convert into the liquid phase. Only at the melting temperature T_m, both phases can coexist under equilibrium conditions. The pressure dependence $\mathrm{d}T_m/\mathrm{d}p$ of the melting temperature is much smaller than that of the evaporation temperature, i. e. the slope of the curve $p(T)$ in the p-T-diagram of Fig. 10.79 is much larger than that of the evaporation curve. One of the reasons is the much smaller change of the volume during the melting process, compared with the much larger change during the evaporation process. A similar consideration as that resulting in Eq. 10.124 for the heat of evaporation gives the heat of fusion

$$\Lambda_m = T \cdot \frac{\mathrm{d}p}{\mathrm{d}T}(V_{\text{liquid}} - V_{\text{solid}})\,. \tag{10.133}$$

For most materials the density decreases during the melting process, i. e. $V_{\text{liquid}} > V_{\text{solid}}$. This gives $\mathrm{d}p/\mathrm{d}T > 0$, because $\Lambda_m > 0$. There are some substances (e. g. water) where $V_{\text{liquid}} < V_{\text{solid}}$. For these substances is $\mathrm{d}p/\mathrm{d}T < 0$, the melting curve has a negative slope (anomaly of water) (Fig. 10.79b).

Note: The fact that for water $V_{\text{liquid}} < V_{\text{solid}} \rightarrow \varrho_{\text{liquid}} > \varrho_{\text{solid}}$ is essential for many processes in nature. Lakes freeze up from the top to the bottom. Since the heat conductivity of ice is small, this gives an isolating layer at the top, preventing the complete freezing of the water, thus protecting fishes and other sensitive creatures.

The fact, that water has its maximum density at $T = 4\,°\mathrm{C}$ is called its *anomaly*. It is due to the temperature dependent molecular structure of water. Liquid water does not solely consist of H_2O molecules but also contains multimers $(H_2O)_n$ in a concentration that depends on the temperature and on the distance from the surface of water. In the multimers the different H_2O molecules are connected by hydrogen bonds. At higher temperatures theses weak bonds break and a structural change results in a change of the mean distance and therefore also a change of the density. In the solid phase the H_2O-molecules form a regular lattice with empty space between the molecules. Therefore the density of the solid is smaller than that of the liquid phase.

Example

At $T = 0\,°\mathrm{C}$ the density of solid ice is $\varrho = 0.917\,\mathrm{kg/dm^3}$, that of sea water is $\varrho = 1.04\,\mathrm{kg/dm^3}$. Therefore, only about 12% of an iceberg are above the seawater surface, but 88% are below. ◀

Application of external pressure decreases the mean distance between the molecules and therefore the ice can melt, according to the principle of minimum constraint. This is utilized by skaters,

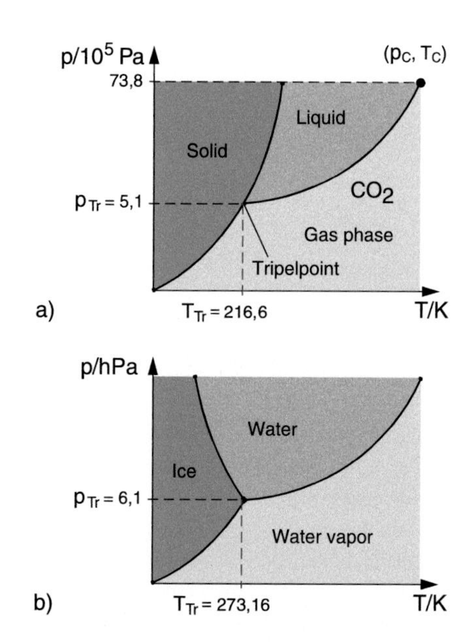

Figure 10.79 Melting curve, vapor pressure curve and triple point for **a** a positive, **b** a negative slope of the melting curve. **a** represents the phase diagram of CO_2, **b** that of water

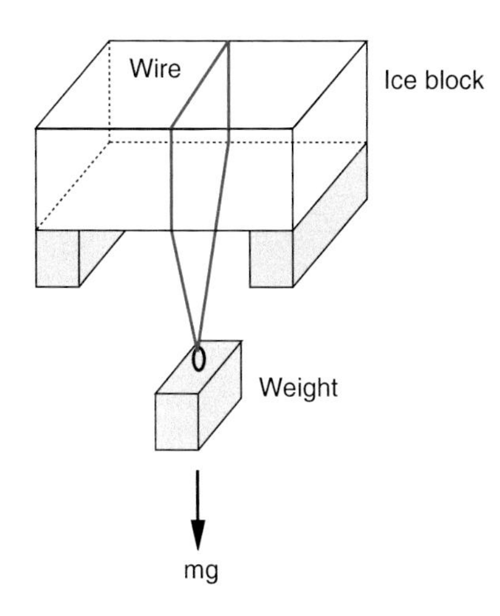

Figure 10.80 Apparently convincing demonstration of the lowering of the melting temperature by pressure

where the high pressure below the sharp ice skates forms a thin liquid layer with low friction (see, however, the remarks below).

The lowering of the melting temperature $\Delta T_m = (dT_m/dp) \cdot \Delta p$ is often demonstrated by a wire that is pulled through an ice block by a heavy weight (Fig. 10.80).

Remark. More detailed calculations show, however, that the major part of the necessary melting energy comes from heat conduction from the higher temperature of the wire to the ice surface (see Probl. 10.11).

Even without external pressure a thin liquid layer is formed at the surface of ice above $T = -33\,°C$. The necessary energy for melting this layer is provided by the gain in surface energy. The boundary ice-air needs more energy for evaporating molecules than the boundary ice-liquid. A liquid layer therefore has a lower potential energy and the loss of potential energy is larger than the melting energy.

10.4.2.5 Coexistence of different phases; Triple Point

Since the melting curve in the p-V-diagram of Fig. 10.76 has a larger slope than the vapor curve the two curves must intersect in a point (p_{tr}, T_{tr}), called the **triple point**. Here the three phases solid, liquid and gase can coexist.

For $T < T_{tr}$ there is one boundary curve (*sublimation curve*) that separates the solid and the gaseous phases. It has in the (p, V)-diagram generally a positive slope. Solid materials can directly pass into the gaseous phase without becoming liquid. This process is called **sublimation**. Because of the small vapor pressure of solids this process is, however, very slow.

If there are more than one phase of a material in a container, pressure p and temperature T are no longer independent of each other. For example, the coexistence of the liquid and the vapor phase is only possible on the vapor curve $p_S(T)$. This implies that p and T are related by the evaporation coefficient Λ in (10.125). It is possible to change T but then $p_S(T)$ is fixed. At the triple point (p_{tr}, T_{tr}) p and T are connected by two conditions: the vapor curve and the sublimation curve. This means that none of the two variables p and T can be changed without leaving the triple point.

This can be generalized by **Gibb's phase rule**, which relates the number f of the degrees of freedom in the choice of the variables p and T with the number q of coexisting phases. It states:

$$f = 3 - q \qquad (10.134)$$

At the triple point is $q = 3 \rightarrow f = 0$, i.e. no degree of freedom in choosing the variables p and T. If only one phase is present ($q = 1$) we obtain $f = 2$. The pressure p as well as the temperature T can be chosen independently (within certain limits). On the vapor curve is $q = 2$ and therefore $f = 1$. We can choose one variable and the other is then fixed.

For a mixture of different chemical components, which can be present in different phases the generalized Gibbs phase rule states:

$$f = k + 2 - q \qquad (10.135)$$

where k is the number of components.

10.4.3 Solutions and Mixed States

Up to now we have discussed only pure substances, which are composed of only one component and do not contain any impurities. We have explained the different phases of solid, liquid and gaseous states and possible transitions between these phases. In nature, however, often mixed substances are present where molecules of different species are mixed together. Examples are NaCl-molecules or sugar molecules, which are dissolved in water and dissociate into their atomic components. Other examples are metal alloys

For the complete characterization of such mixed states pressure and temperature are not sufficient, but also the concentration of the different components have to be defined.

The concentration of a substance dissolved in a liquid is generally given in g/litre or in mole/litre. Often not the complete substance has dissolved but a rest remains as solid sediment (if $\varrho_{solid} > \varrho_{liquid}$) or as layer on the liquid surface (if $\varrho_{solid} < \varrho_{liquid}$).

The solution of substances can alter the characteristic features of the liquid considerably. In this section we will shortly discuss the most important features of solutions.

10.4.3.1 Osmosis and Osmotic Pressure

Assume a container with a semipermeable membrane including a solution with the concentration c of the dissolved substance is submerged into a reservoir with the pure liquid (Fig. 10.81). One observes that the level of the solution in a standing pipe rises above the level of the pure solution, if the molecules of the solvent can penetrate through the semipermeable membrane but not the molecules of the dissolved substance. Such permeable

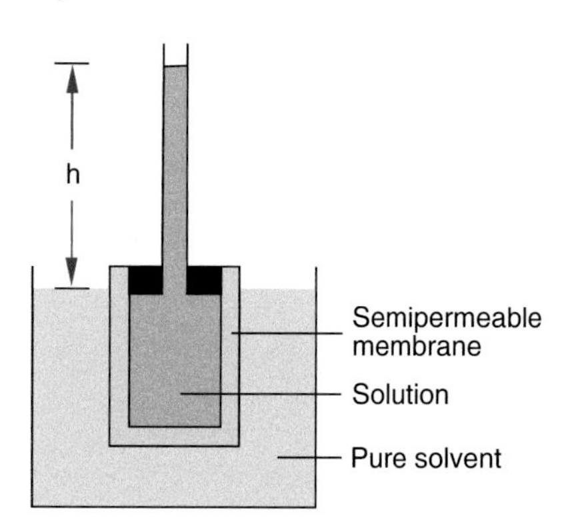

Figure 10.81 Demonstration of osmosis in a Pfeffer cell

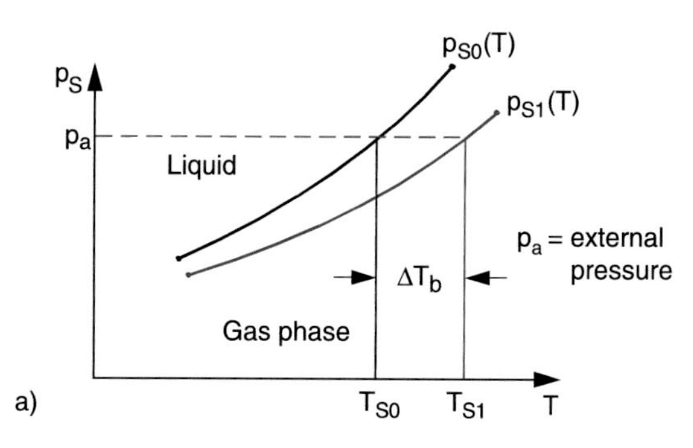

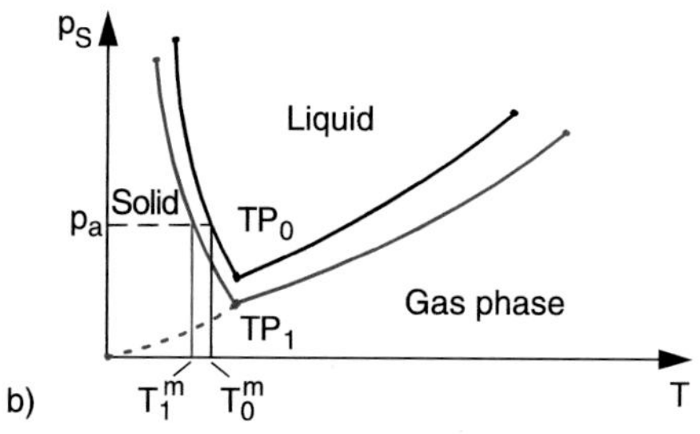

Figure 10.82 **a** Decrease of vapor pressure and increase of boiling temperature ΔT_b of a solution compared to that of a pure solvent; **b** decrease of melting temperature ΔT_m

membranes with substance-specific transmission play an important role in biological cells.

In the example of Fig. 10.81 the concentration difference of the dissolved substance between outside and inside of the container results in a diffusion of the solvent molecules into the solution through the permeable membrane. This builds up a pressure difference, indicated by the height h in the standing pipe.

$$\Delta p = \varrho \cdot g \cdot h ,$$

which stops the net diffusion, because now an equal number of molecules diffuses into and out of the container.

The net diffusion caused by the concentration difference is called *Osmosis* and the pressure difference Δp is the *osmotic pressure*.

The osmotic pressure p_{osm} is proportional to the concentration of the dissolved molecules and to the temperature.

Experiments show that

$$p_{osm} \cdot V = \nu \cdot R \cdot T , \qquad (10.136)$$

where ν is the number of moles dissolved in the volume V of the solvent.

This **van't Hoff's Law** is the analogue to the general gas equation

$$p \cdot V = \nu \cdot R \cdot T .$$

The osmotic pressure of a solution exerted onto the walls of the container equals the pressure that would be present, if the dissolved molecules were in the gas phase at the temperature T.

10.4.3.2 Reduction of Vapour Pressure

Due to the additional attractive forces between the dissolved molecules and the solvent molecules the work function of the solvent molecules increases. This means that at identical temperatures less molecules evaporate than in pure liquids. The vapour pressure is therefore lower than in a pure solvent.

The vapour pressure reduction Δp is proportional to the concentration of the dissolved molecules (if their vapour pressure is negligible).

Francois Marie Raoult formulated in 1882 the law

$$\frac{\Delta p_s}{p_{s0}} = -\frac{\nu_1}{\nu_0 + \nu_1} , \qquad (10.137a)$$

here p_{s0} is the vapour pressure of the pure solvent, ν_0 is the number of moles of the solvent and ν_1 that of the dissolved substance. For diluted solutions is $\nu_1 \ll \nu_0$ and (10.137a) reduces to

$$\Delta p_s = -p_{s0} \cdot \frac{\nu_1}{\nu_0} . \qquad (10.137b)$$

The lowering of the vapour pressure causes an increase ΔT_b of the boiling temperature as shown in Fig. 10.82a. The vapour pressure has to rise by Δp_s to reach the external pressure p_a. From the vapour pressure curve $p_s(T)$ in Eq. 10.125 we can derive the relation between Δp_s and ΔT_b. Differentiation of (10.125) gives

$$\frac{dp_s}{dT} = \frac{\Lambda}{RT^2} p_s \Rightarrow \Delta T = \frac{RT^2}{\Lambda} \frac{\Delta p_s}{p_s} . \qquad (10.137c)$$

Together with (10.137) this yields the Raoult' Law

$$\Delta T_b = \frac{RT^2}{\Lambda}\frac{\nu_1}{\nu_0}\,. \qquad (10.138a)$$

When several substances with the molar concentrations ν_i are dissolved, this generalizes to

$$\Delta T_b = \frac{RT^2}{\Lambda\nu_0}\sum_i \nu_i\,. \qquad (10.138b)$$

Since ΔT_b depends on the molar evaporation heat Λ, it is dependent not only on the dissolved substances but also on the specific solvent.

For dissolved substances that partly dissociate (for instance dissociates NaCl into $Na^+ + Cl^-$) the sum in (10.138b) extends over all dissociated and non-dissociated components dissolved in the solvent.

The lowering of the vapour pressure also results in a lowering of the melting temperature T_m (Fig. 10.82b). Similar to (10.138a) one gets

$$\Delta T_m = -\frac{RT^2}{\Lambda_m}\frac{\nu_1}{\nu_0}\,, \qquad (10.139)$$

where Λ_m is the molar melting heat.

Example

For water with the concentration of ν_1 moles of a dissolved substance is the lowering of the melting temperature

$$\Delta T_m = -1.85\,\text{K}\cdot\nu_1\,.$$

When 50 g NaCl are dissolved in 1 litre water, (1 mol NaCl are 58 g), the lowering of the melting temperature is with $\sum\nu_i = 2\cdot 50/58 = 1.72\,\text{mol}$: $\Delta T_m = -3.2\,\text{K}$. ◀

Seawater has a melting temperature that lies several degrees below 0 °C depending on the salt concentration.

The lowering of the melting temperature is used to clear icy roads from ice and snow by salting the roads.

The zero point of the Fahrenheit temperature scale is defined by the melting temperature of a specific salt-water solution. From (10.2) and (10.139) the zero point can be obtained as

$$0\,°\text{F} = -17.8\,°\text{C}\,.$$

Solutions with dissolved substances have generally a larger temperature range of the liquid phase than pure solvents, because the boiling point rises and the melting point is lowered.

10.5 Comparison of the Different Changes of State

Here we will summarize all possible changes of thermodynamic states and the corresponding equations.

1. Isochoric processes: $V = \text{const}$

$$dQ = C_V\cdot dT \qquad (10.140a)$$

2. Isobaric processes: $p = \text{const}$

$$dQ = C_p\cdot dT = dU + p\cdot dV \qquad (10.140b)$$

3. Isothermal processes: $T = \text{const}$

$$dU = 0,\quad dQ = p\cdot dV,\quad p\cdot V = \text{const} \qquad (10.140c)$$

4. Adiabatic processes: $dQ = 0$

$$p\cdot V^\kappa = \text{const};\quad \kappa = C_p/C_V \qquad (10.140d)$$

5. Isentropic processes: $S = const$

$$dS = C_V\cdot dT/T + R\cdot dV/V = 0$$
$$\Rightarrow T\cdot V^{\kappa-1} = \text{const} \qquad (10.140e)$$

A reversible adiabatic process is always isentropic, but not every isentropic process is also adiabatic.

6. Isoenthalpic processes

$$H = U + p\cdot V = \text{const}$$
$$dH = (\partial H/\partial p)_{T=\text{const}} + (\partial H/\partial T)_{p=\text{const}} \qquad (10.140f)$$

10.6 Energy Sources and Energy Conversion

The supply of sufficient energy that can replace to a large extent manual work, has changed our life considerably. It is fair to say that only the provision of sufficient and affordable energy has essentially improved our standard of life. This is the reason why in developing countries the desire for more energy will cause a drastic increase of worldwide energy consumption.

The first law of thermodynamics teaches us, however, that energy can be neither generated nor annihilated. The phrase "energy generation" (for example in power stations) means correctly speaking the conversion of energy from a specific form into another (for instance from thermal energy into electric energy).

In fossil power stations the potential energy of CO- and CO_2-molecules is transferred into heat (kinetic energy of the molecules and atoms), which is further converted via turbines into mechanic energy of the rotating turbine, which drives an electric generator that produces electric energy.

Figure 10.83 Cooling towers of the coal power plant Staudinger. The plant delivers 500 MW electric power and 300 MW heat power. It reaches an efficiency of 42.5% (With kind permission of Preußen Elektra AG, Hannover)

In car engines this molecular potential energy is converted into mechanical energy that drives the car. In nuclear power stations the potential energy of uranium nuclei (which exceeds that of molecular bindings by 6 orders of magnitude) is converted by nuclear fission into kinetic energy of the fission products and then into heat of circulating cooling water.

Wind energy converters convert the kinetic energy of airflow into rotation energy of the converter rotor blades, which drive an electric generator. The wind energy has its origin in the solar radiation energy, which in turn stems from nuclear fusion energy in the interior of the sun.

In order to realize an energy conversion efficiency as high as possible, one has to understand the basic physical processes of the different conversion processes. We have learned in Sect. 10.3.3 that the maximum possible conversion factor for the conversion of heat into mechanical energy is given by the efficiency of the Carnot engine which depends on initial and final temperature during the conversion. The maximum initial temperature is generally limited by the material of the container walls which enclose the working gas. The lowest final temperature is often limited by the temperature of the surrounding. By using the rest energy of the cooled gas for heating (combined heat and power) the energy efficiency can be improved. This reduces the waste of energy which would otherwise heat up the environment. The non-usable rest heat energy is taken away by cooling towers (Fig. 10.83).

The increasing concern about the warming of our atmosphere (global warming) by man-made emission of molecular gases such as CO_2, CH_4, NO_2 etc., which absorb the infrared emission of the earth surface thus heating up the atmosphere, has led to the proposal and partly realization of several different "energy sources", i. e. energy conversion processes. In particular regenerative energy sources, where the working material is available in unlimited quantities, or where the consumption of the working material is replaced by nature over time intervals of many centuries, are favourable candidates. Such energy conversion processes should not contribute to global warming. Examples are nuclear energy conversion, wind energy, solar energy and energy conversion based on the tides of the ocean. The most important renewable energy conversion processes include:

- Hydro-electric power plants (based on the potential or kinetic energy of water)
- Wind-energy converters
- Geothermic plants
- Solar-thermal power plants
- Solar-electric conversion (photo-voltaic devices)
- Bio-energy (burning of regrowing biological material such as wood, plants)

Some examples shall illustrate these different "energy sources".

At first we will clarify some often used definitions.

The **primary energy** is the energy directly obtained from the different sources (coal, oil, gas, water, wind, sun radiation, nuclear fission) while the **secondary energy** is won by conversion of the primary energy into other energy forms (mechanical energy, electric energy, etc.). The conversion of primary into secondary energy has an efficiency $\eta < 1$. This means a fraction $(1 - \eta)$ is lost and is delivered as heat into the surrounding. If the consumption of primary energy in a country is larger than the production of energy sources, the country has to import coal, oil or gas.

In Tab. 10.8 the increase of the worldwide primary energy consumption is summarized from 1990 to 2012. Note the large increase of the electric power consumption. In Fig. 10.84 the contributions of the different energy sources to the total worldwide energy consumption are illustrated and Tab. 10.9 lists some countries with the highest energy consumption. It illustrates the enormous increase during the last 40 years.

Table 10.8 Worldwide total energy consumption (in 10^3 TWh) and electric energy

Year	Total energy	Electric energy
1990	71	6
2000	117	15
2012	155	23

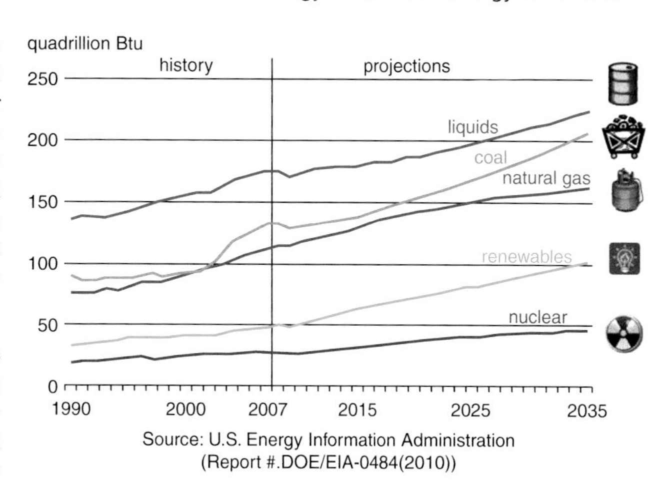

Figure 10.84 Worldwide energy consumption of different energy sources in units of 10^{15} Btu $\approx 10^{18}$ J $\approx$ 300 TWh

The units for energy and their abbreviations are given below:

$$\begin{aligned}
&1\ \text{Kilojoule} &&= 1\,\text{kJ} &&= 10^3\,\text{J}\\
&1\ \text{Megajoule} &&= 1\,\text{MJ} &&= 10^6\,\text{J}\\
&1\ \text{Gigajoule} &&= 1\,\text{GJ} &&= 10^9\,\text{J}\\
&1\ \text{Terrajoule} &&= 1\,\text{TJ} &&= 10^{12}\,\text{J}\\
&1\ \text{Petajoule} &&= 1\,\text{PJ} &&= 10^{15}\,\text{J}\\
&1\ \text{Exajoule} &&= 1\,\text{EJ} &&= 10^{18}\,\text{J}\\
&1\ \text{Kilowatt hour} &&= 1\,\text{kWh} &&= 3.6\,\text{MJ}
\end{aligned}$$

It is interesting to compare the total energy consumption of Germany (13 400 PJ per year) with the energy that it receives per year from the sun. The energy of solar radiation per sec and m^2 outside the atmosphere (solar constant) is $1.367\,\text{kW}/(\text{m}^2 \cdot \text{s})$. During its transit through the atmosphere, the radiation power decreases through backscattering (30%) and absorption (20%)

Table 10.9 The countries with the highest consumption of primary energy (in Megatons Oil-Units) [10.31]

Pos.	Country	1970	2000	2010	2013	%
1	China	202.1	980.3	2339.6	2852.4	22.4
2	USA	1627.7	2313.7	2284.9	2265.8	17.8
3	Russia	483.0	619.4	674.1	699.0	5.5
4	India	64.8	295.8	510.2	595.0	4.7
5	Japan	279.9	518.0	506.8	474.0	3.7
6	Canada	156.4	303.0	315.6	332.9	2.6
7	Germany	309.6	333.0	322.5	325.0	2.6
8	Brasilia	36.8	185.8	257.4	284.0	2.2
9	South Korea	14.3	193.9	254.6	271.3	2.1
10	France	155.8	258.7	253.3	248.4	2.0

and therefore only about 50% reach the earth surface. Since the sun radiation generally does not incide vertically onto a surface element but under an angle α that depends on the daytime, the latitude and the yearly season, the annual average $P = P_0 \cdot \cos\alpha$ of the incident radiation intensity is for a latitude of 45° about $300\,\mathrm{W/(m^2 \cdot s)}$. With an annual sunshine duration of 1000 h/year ($3.6 \cdot 10^6$ s/year) we receive the annual average of solar radiation energy of about 1 GJ per m^2 and year. In order to match the energy consumption one needs in Germany an area of $3 \cdot 10^4\,\mathrm{km}^2$ for solar energy collectors with an efficiency of 50%.

For the whole earth (the hemi-sphere with the area of $2.55 \cdot 10^8\,\mathrm{km}^2$) the total incident sun radiation power is $10^{15} \cdot 2.55 \cdot 10^8 = 2.55 \cdot 10^{23}$ J/year. The total energy consumption in the year 2011 was, however, only $5 \cdot 10^{20}$ J/year which is about 0.2% of the incident sun energy.

10.6.1 Hydro-Electric Power Plants

Most of the hydro-electric power plants use water reservoirs where the water outlet streams through pipes and drives turbines that generate electric power. Here the potential energy of the damned up water is converted into kinetic energy of the water flowing through the pipe.

For a storage height h an area A of the reservoir and a density ϱ of the water the total potential energy is

$$E_{\mathrm{pot}} = \varrho \cdot g \cdot h \cdot A \cdot \Delta h\,,$$

when the storage height is lowered by $\Delta h \ll h$.

Example

$A = 1\,\mathrm{km}^2 = 10^6\,\mathrm{m}^2$, $h = 30\,\mathrm{m}$, $\Delta h = 5\,\mathrm{m} \to E_{\mathrm{pot}} = 1.5 \cdot 10^{12}\,\mathrm{J} = 1.5\,\mathrm{TJ}$. ◄

Some hydro-electric power stations use the flow energy of rivers, where in most cases, however, the river has several barrages where again the potential energy of the dammed river is used to drive turbines. This method was often used in earlier times to drive corn mills and hammer mills which only need moderate powers.

Example

When a channel with a width of 5 m and a depth of 3 m is branched off a river the water with a velocity of $v = 6\,\mathrm{km/h} = 1.67\,\mathrm{m/s}$ drives a turbine, the maximum available power is

$$P = \tfrac{1}{2} M v^2 = \varrho \cdot B \cdot h \cdot v^2/2\,.$$

With the numerical values given above this yields $P = 21\,\mathrm{kW}$. ◄

10.6.2 Tidal Power Stations

Tidal power stations use the tidal range between low and high tide for power generation. This range is in particular large in the mouth of rivers, where it can reach up to 16 m. The water passes through turbines built into logs in the river. At low tides, the water streams seawards and at high tides against the river. This streaming water drives the turbines at low tides as well as at high tides (Fig. 10.85), which activates generators for producing electric energy. At a water level difference Δh between the dammed river and the sea level the energy that can be converted is

$$W = \int (\mathrm{d}M/\mathrm{d}t) \cdot g \cdot \Delta h(t)\mathrm{d}t\,,$$

where $\mathrm{d}M/\mathrm{d}t$ is the mass of water passing pro second through the turbines, Δh is the time dependent level difference and T (about 5 h) the time duration of low resp. high tide.

Here the gravitational energy of earth-moon attraction and the decrease of the rotation energy of the earth (due to friction by the tides) are the primary energy sources. During the time intervals where $\Delta h = 0$ the tidal power station cannot deliver energy.

The first tidal power station was built in France in the mouth of the river *Rance* (Fig. 10.86) where a tidal range of 16 m is obtained. The river dam is 750 m long and has 24 passages where the turbines are located. The total power station delivers an electric power of 240 MW and per year an electric energy of 600 GWh. This equals the energy delivered by 240 wind converters with 1 MW power each and 3000 hours of full operation per year.

The disadvantage of such tidal power stations is the separation of the bay at the mouth of the river. This can change the biological conditions for plants and fishes and it can furthermore influence the tidal range in neighbouring bays with the danger of flooding.

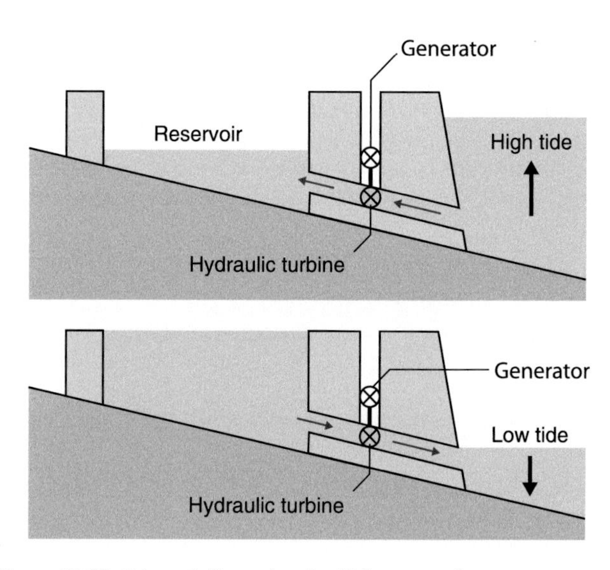

Figure 10.85 Schematic illustration of a tidal power station

Figure 10.86 Tidal Power Station St.Malo. Aerial view of the power station in the estuary of the river Rance in France. Here the tidal amplitude is about 8 m (see Sect. 6.6). With kind permission © Foto dpa

10.6.3 Wave Power Stations

Wave power stations use the kinetic and potential energy of sea waves for the generation of electric energy. Their basic principle is illustrated in Fig. 10.87. A pneumatic chamber is filled with air in its upper part while the lower part has a connection to the sea. The incoming waves induce a periodic change of the water level in the lower part of the chamber. This causes a periodic change of the air pressure in the upper part and an air flow through the pipe at the top of the chamber that periodically changes its direction. In the upper part of the pipe a *Wells-turbine* is installed, that always rotates in the same direction independent of the direction of the air flow. This turbine has symmetric blade profiles in contrast to normal turbines that have asymmetric blade profiles, optimized for one direction of the airflow. The efficiency of the Wells-turbine is smaller than that of normal turbines. It has, however, the advantage that it rotates continuously for both directions of the air flow.

Wave power stations do not use the tide difference between high and low tide but the wave energy, which is in turn driven mainly by the wind energy and only to a minor part by the tides [10.32, 10.33].

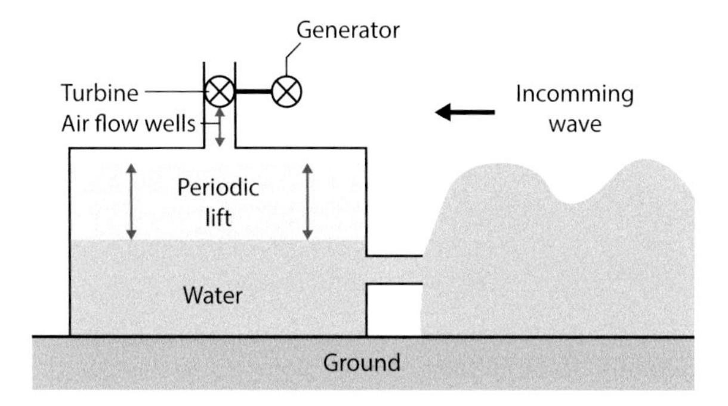

Figure 10.87 Concept of a wave-power station

10.6.4 Geothermal Power Plants

The temperature of the earth's interior increases with increasing depth by about 3–5 °C/100 m, because heat flows from the hot kernel to the outer parts of the earth. The heat in the kernel was mainly generated in the formation period of the earth (about 4 billion years ago) where heavier elements dropped down to the kernel due to gravitational forces. This increased the temperature of the kernel. Another cause for the production of heat is the radioactive decay of elements such as Uranium, Thorium and Potassium that are contained in the kernel as well as in the earth mantle.

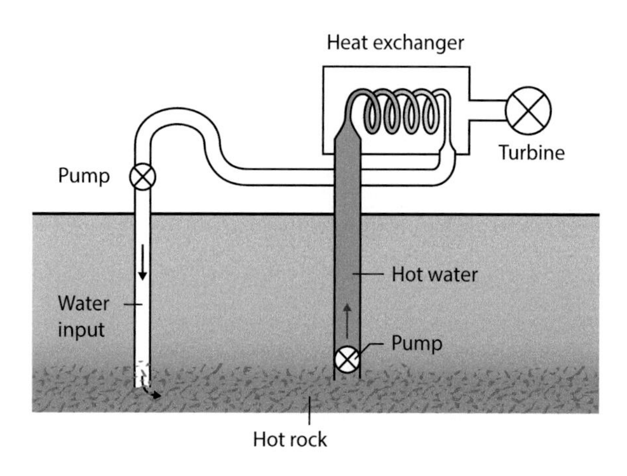

Figure 10.88 Schematic illustration of a geothermic power station

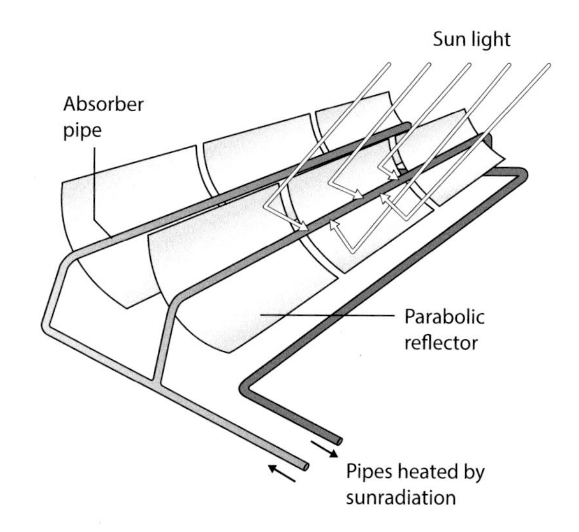

Figure 10.89 Thermal solar power station using parabolic reflectors

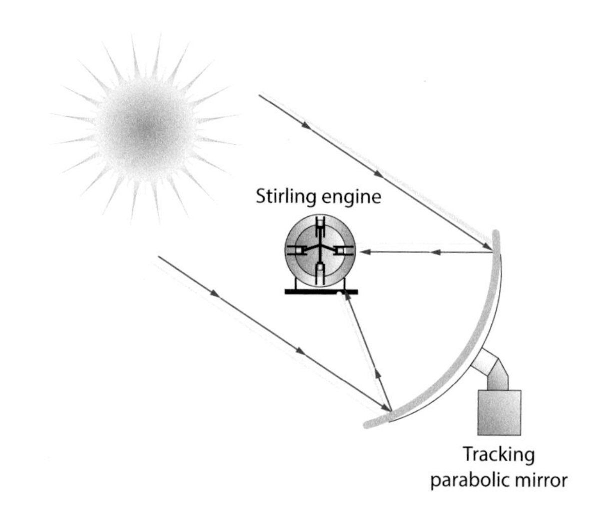

Figure 10.90 Solar power station with parabolic mirrors, that follow up the sun position and focus the sun radiation onto a Stirling motor [10.26]

In regions with volcanic activities, water rising from the interior to the surface of the earth has a sufficiently high temperature to be useful as energy source. For example in *Iceland* residences, green houses and swimming pools are heated by hot water from the earth interior. This energy streaming out of the earth interior is called **geo-thermal energy**.

Iceland can cover about 80% of its primary energy consumption ($40\,\text{PJ} = 40 \cdot 10^{15}\,\text{J}$) by this geothermal sources.

In regions without such active volcanism, one can drill deep bore holes (3000–5000 m) in order to utilize geothermal energy. The temperature at a depth of 5000 m is about 200–300 °C. For energy production water is pumped into the bore holes which interacts with the hot rock and is heated up. The hot water is pumped back to the surface and can be used for heating purposes or, if its temperature is above 100 °C it can produce through heat exchangers hot steam that drives turbines (Fig. 10.88).

The limitations of the geothermal energy usage in non-volcanic regions is the slow transport of heat from the surrounding of a bore hole, which is mainly due to heat conduction. When the heat extraction becomes larger than the supply of energy from the surroundings the temperature drops and the efficiency of the plant decreases accordingly [10.24]. A much more serious problem are possible geological dislocations. The water pumped under high pressure into the bore hole can modify the rock in the surroundings of the bore hole and can increase the volume of such chemically altered porous rocks. This will cause local uplifts at the earth surface which can damage buildings. Such geothermal plants should be therefore operated far away from inhabited areas.

10.6.5 Solar-Thermal Power Stations

These power stations use the heating of material that absorbs the sun radiation and transfers the heat to a liquid transport medium, such as water or oil. In order to reach sufficiently high temperatures the sun radiation is focused by parabolic or spherical mirrors onto the heated devices.

In the *parabolic gullies construction* the water or oil is pumped through pipes that are located in the focal line of cylindrical mirrors with parabolic profile (Fig. 10.89), which concentrates the sun radiation onto the pipes [10.25].

Another modification consists of several hundred parabolic mirrors (heliostats) that follow up the changing sun position during the day (Fig. 10.90) and concentrate the sun radiation, nearly independent of the position of the sun, onto a small volume at the top of a high tower (Fig. 10.91).

The achievable radiation density of this device is much higher than in the parabolic gully construction and temperatures of about 1000 °C can be reached. This increases the efficiency for the conversion into electric energy. The generated hot steam drives turbines as in fossil power stations.

An example of such a solar-thermal power station is the plant "Plataforma solar de Almeria" in Spain (Fig. 10.92). Here 300 heliostats with $40\,\text{m}^2$ parabolic mirror surface each concentrate the sun radiation onto the radiation collector at the top of an

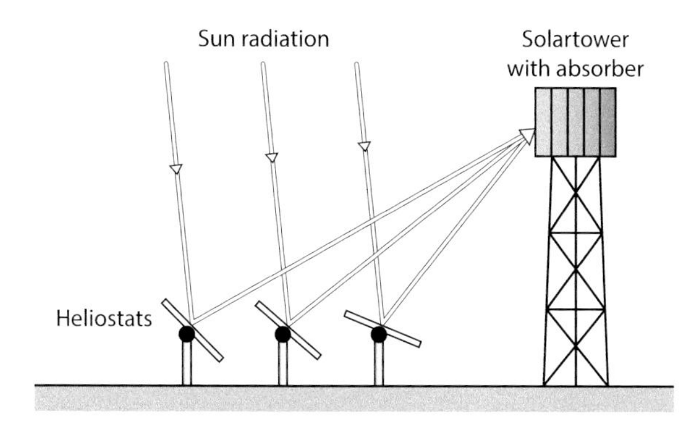

Figure 10.91 Solar tower power station

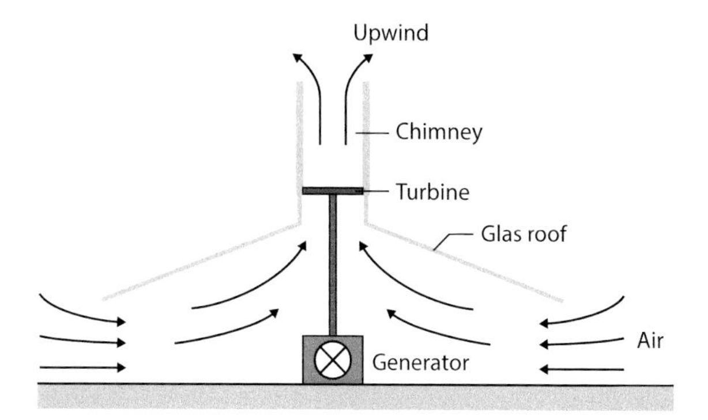

Figure 10.93 Upwind solar power station

80 m high tower. This plant produces an average electric power of 40 MW.

The solar-thermal power plants discussed so far only work efficiently for direct sun radiation, i. e. under a clear sky, because the concentration by the heliostats onto a small volume does not work efficiently for the diffuse radiation at cloudy days.

Here another type of solar power stations, the upwind plant, is favourable (Fig. 10.93). The air above a large area under a sloped glass roof that is formed like a cone, is heated by direct or diffuse sun radiation. The hot air streams to the centre of the area where it rises up into a chimney, driving a turbine. Although the efficiency of such a plant is only about 5% it still pays off because of the low construction and operation costs.

An example is the plant "Fuente el Fresno" in Spain, where an area of 2.5 km^2 is covered by the glass roof. The chimney is 750 m high and the delivered electric power amounts to 40 MW. The large area is not lost for agriculture but can be used for growing plants and fruits even during colder periods, since it operates like a green house.

10.6.6 Photovoltaic Power Stations

Here the sun radiation energy is directly converted into electric energy by photovoltaic semiconductors. The efficiency amounts to 5–20% depending on the semiconductor material. The basic physics of these devices is explained in Vol. 3.

The price per kWh was very high in the beginning but decreases now rapidly due to mass production of solar cells or thin film photovoltaic devices. In view of the rising prices for fossil energy sources it will soon be able to compete with conventional power plants.

The large disadvantage of all solar power plants is the dependence on the unreliable sunshine duration. It is therefore necessary to realize energy storage devices which can bridge the time periods where the sun does not shine.

Figure 10.92 Gemasolar power station close to Sevilla. 2650 mirrors reflect the sun light onto a tower where a salt solution is heated. It will deliver electric energy of 110 GWh per year. (Torresol Energy Investment S.A.)

10.6.7 Bio-Energy

The burning of farming refuse, such as waste wood, stray, garbage or biogas, which remains in agriculture can produce useful energy, named **bio-energy**, because the burned material is of biological origin. The advantage of this energy source is that it is renewable as long as the consumption does not exceed the natural production. Its disadvantage is the emission of CO_2 which generally exceeds the consumption of CO_2 by the growing plants, although the net emission balance is more favourable than for conventional fossil power plants. Furthermore other species such as SO_2, phosphor and heavy metal compounds are emitted. If only substances as burning material are used, which cannot be utilized for other purposes, the bio-energy can be judged positively. However, if food is used for the production of gasoline, this is contra-productive and should be rejected. Also the burning of wood pellets only makes sense, if they are produced from wood waste, but this technique is nonsense if the pellets are pressed from trees that could have been used elsewhere.

10.6.8 Energy Storage

The increasing production of energy from renewable resources that are not continuously available, demands the realization of sufficient energy storage systems in order to bridge time periods where these sources cannot deliver sufficient energy. There are several proposals for such storage systems, where some of them have been already realized.

The oldest energy storage systems are pumped hydro storage plants. Here water is pumped from a lower storage reservoir into a higher one during times, where sufficient energy is available. During periods where more energy is needed, the water runs back from the higher into the lower reservoir and drives turbines, which activate electric generators. The generators are used during the up-pumping period as electric motors that drive the pumps. This method is up to now the most efficient, but it needs sufficient space on the top of mountains for the upper reservoir. One of many examples is the *Walchensee plant* in Bavaria, Germany, where the water is pumped from the lower Kochelsee into the 200 m higher Walchensee.

For bridging the night periods, where solar plants cannot work, salt storage systems have been developed. Here the surplus energy produced during daytime is used to heat up and melt a salt solution. During night time the heat of the hot solution and the heat of fusion that is released when the solution solidifies, can be used to bridge the energy gap. With multi-component salt solutions, there are several melting temperatures and the heat of fusion is more uniformly delivered during the cooling of the solution. Examples of such salt solutions are $Mg(NO_3)_2 \cdot 6H_2O$, or $CaCl_2 \cdot 6H_2O$.

For small energy demands during night-time compact lithium batteries have been developed which have a storage capacity of 0.2 kWh per kilogram mass. For a volume of 0.5 m^3 of the battery system one can reach a storage energy of 20 kWh. This is sufficient for most private households, which may have solar collectors on the roof and can provide with such a combined system their energy demands during day and night.

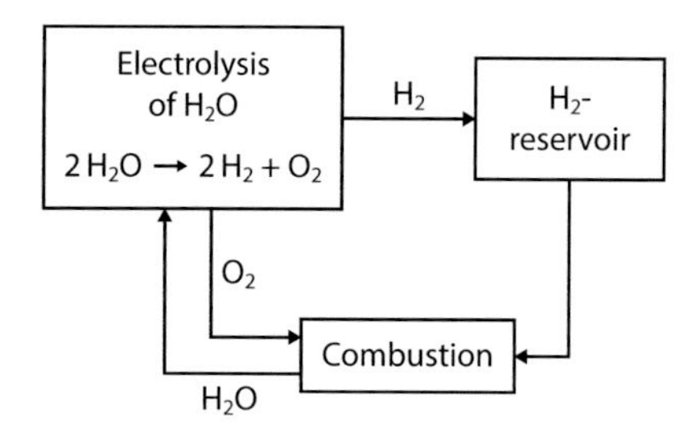

Figure 10.94 Hydrogen gas as energy reservior

A promising storage medium is hydrogen gas H_2, which can be produced by electrolysis during times where surplus solar or wind energy is available. According to the endothermic reaction

$$2H_2O \rightarrow 2H_2 + O_2 \,, \tag{a}$$

(Fig. 10.94) hydrogen gas is produced by electrolysis of water. In the reverse exothermic reaction

$$2H_2 + O_2 \rightarrow 2H_2O \tag{b}$$

energy is released. The advantage is, that no environmentally dangerous gases such as CO_2 or NH_3 are emitted. The electrolytic systems can be placed directly inside the tower of wind converters and the produced hydrogen gas can be stored in high pressure bottles. The systems can be controlled in such a way, that reaction (a) operates during the time period of wind energy surplus and reaction (b) during times of wind energy shortage. For large plants the hydrogen gas is stored in huge underground caverns, for instance in no longer used salt mines and is transported by underground pipes to special power stations which can burn hydrogen gas. Meanwhile long-time experience is present for the storage of hydrogen gas in caverns.

Example

The cavern *Clemens Dome*, close to Lake Jackson USA, has a volume of 580 000 m^3. The stored gas at a pressure of 10 MPa can deliver an energy of 90 GWh. It is operated since 1986. ◀

While for H_2 storage the reaction energy of the reaction (b) is used, for air storage at high pressures the potential energy $p \cdot V$ of the gas volume is utilized. When the gas flows from the storage tank through a pipe, the potential energy is converted into kinetic energy, which is used to drive a turbine (Fig. 10.95).

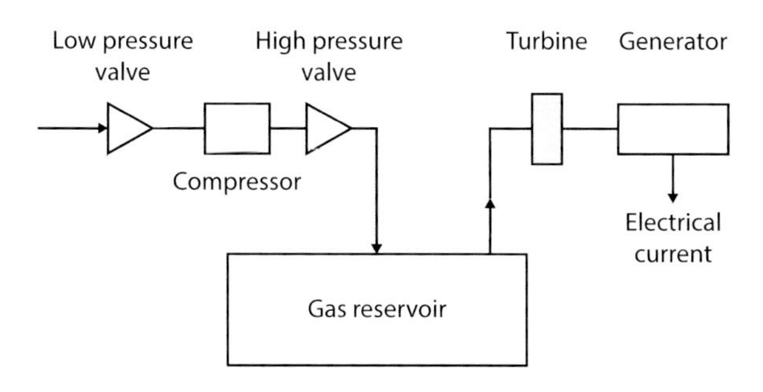

Figure 10.95 Scheme of a high pressure gas reservior

Example

$p = 100\,\text{atm} = 10\,\text{MPa}$, $V = 500\,000\,\text{m}^3 \to E_\text{pot} = p \cdot V = 5 \cdot 10^{12}\,\text{N} \cdot \text{m} = 5 \cdot 10^{12}\,\text{Ws} = 1.39\,\text{GWh}$. ◄

Summary

- The temperature of a body is given either as absolute temperature T in Kelvin or as Celsius temperature $T_\text{C}/\,°\text{C}$ or in the US as Fahrenheit temperature. The relations are

$$\begin{aligned} T/\text{K} &= T_\text{C}/\,°\text{C} + 273.15, \\ T/\text{F} &= (9/5)T_\text{C}\,°\text{C} + 32 \\ &= (9/5)[T/\text{K} - 273.15) + 32 \\ &= (9/5)T/\text{K} - 459.67\,. \end{aligned}$$

  For temperature measurements all quantities can be used, that depend on the temperature (expansion of a liquid volume, electric resistance, thermo-voltage, conductivity of semiconductors).
- The thermal expansion of bodies is caused by the non-harmonic interaction potential between neighbouring atoms.
- The absolute temperature is determined with the gas thermometer, where the increase of the gas pressure with temperature in a constant volume is proportional to the temperature increase.
- The thermal energy of a body is determined by the kinetic and potential energy of the atoms or molecules. The temperature increase ΔT of the system is proportional to the supplied heat energy $\Delta Q = C \cdot \Delta T$.
- The molar heat capacity for a constant volume of a gas $C_V = R \cdot f/2$, is equal to the product of gas constant $R = k \cdot N_\text{A}$ times one half of the number f of degrees of freedom of the atoms or molecules in the gas.
- The molar heat capacity at constant pressure is $C_p = C_V + R$
- The transition from the solid to the liquid phase requires the molar melting energy $W = \Lambda_\text{m}$ per mole. During the melting the potential energy of the atoms or molecules increases while the kinetic energy stays constant. Similar the transition from the liquid to the gaseous phase needs the energy per mole $W = \Lambda_\text{e}$ (heat of evaporation).
- Thermal energy can be transported from one area to another
  - by heat conduction
  - by convection
  - by thermal radiation
- The amount of heat transported per second by heat conduction in the direction $\boldsymbol{r}$ through the area A is $\text{d}Q/\text{d}t = -\lambda \cdot A \cdot (\text{grad}\,T)_{\boldsymbol{r}}$, i.e. the product of heat conductivity λ, area A and temperature gradient in the direction of $\boldsymbol{r}$.
- For metals the heat conductivity is proportional to the electrical conductivity, which indicates that the electrons are mainly responsible for both conductivities.
- The thermodynamic state of a system is unambiguously determined by the state variables pressure p, volume V and temperature T. For ν moles of an ideal gas in the volume V the general gas equation is

$$p \cdot V = \nu \cdot R \cdot T\,.$$

  The number of internal state variables in real gases is given by Gibbs' phase rule (10.134).
- The entropy S of a system is a measure for the number of possible ways the state of the system can be realized. The change of the entropy is $\text{d}S = \text{d}Q/T$ where $\text{d}Q$ is the heat energy supplied to or by the system.
- The first law of thermodynamics $\Delta U = \Delta Q + \Delta W$ describes the energy conservation. The change ΔU of internal energy $U = N \cdot (f/2)kT$ of a system with N atoms or molecules equals the sum of supplied heat ΔQ and mechanical work ΔW performed on or by the system. For real gases is $U = E_\text{kin} + E_\text{pot}$, because the interaction energy between the atoms has to be taken into account.
- Special processes in a system of an ideal gas are:
  isochoric processes ($V = \text{const}) \Rightarrow \text{d}U = C_V \cdot \text{d}T$,
  isobaric processes ($p = \text{const}) \Rightarrow \text{d}U = \text{d}Q - p\text{d}V$,
  isothermal processes ($T = \text{const}) \Rightarrow p \cdot V = \text{constant}$,
  adiabatic processes ($\text{d}Q = 0) \Rightarrow \text{d}U = \text{d}W$ and $p \cdot V^\kappa = \text{constant}$ with $\kappa = C_p/C_V = $ adiabatic index.
- The second law of thermodynamics states that at the conversion of heat into mechanical energy at most the fraction $\eta = (T_1 - T_2)/T_1$ can be converted when the heat reservoir is cooled from the temperature T_1 to T_2.
- The entropy $S = k \cdot \ln P$ is a measure for the number P of realization possibilities for a system with a given temperature T and total energy E.

- Reversible processes are ideal processes where a system passes a cycle of processes and reaches its initial state without any losses. An example is the Carnot Cycle where the system passes through two isothermal and two adiabatic processes.
- For reversible processes the entropy remains constant. For all irreversible processes the entropy increases and the free energy $F = U - T \cdot S$ decreases.
- The entropy S approaches zero for $T \to 0$ (third law of thermodynamics).
- For real gases the Eigen-volume of the atoms and the interaction between the atoms cannot be neglected as for ideal gases. The equation of state $p \cdot V = \nu \cdot R \cdot T$ of ideal gases is modified to the van der Waals equation $(p+a/V^2) \cdot (V-b) = R \cdot T$, where a/V^2 denotes the internal pressure and $b/4$ the Eigen-volume of the N_A molecules per mole.
- The heat of evaporation of a liquid $\Lambda = T \cdot dp_s/dT \cdot (V_v - V_l)$ is due to the mechanical work necessary to enlarge the volume V_l of the liquid to the much larger volume V_v of the vapour against the external pressure and against the internal attractive forces between the molecules. The second contribution is much larger than the first one.
- In a $p(T)$ phase diagram the liquid and gaseous phases are separated by the vapour phase curve and the liquid and solid phase by the melting curve. The two curves intersect in the triple point (T_{tr}, p_{tr}) where all three phases can coexist.
- The vapour pressure of a liquid is lowered by addition of solvable substances, which increases the evaporation temperature. Also the melting temperature can be lowered.

Problems

10.1 Give a physically intuitive explanation, why the thermal expansion coefficient for liquids is larger than that of solids.

10.2 Prove example 2 in Sect. 10.1.2.

10.3 A container with 1 mol helium and a container of equal size with 1 mol nitrogen are heated with the same heat power of 10 W. Calculate after which time the temperature of the gas in the containers has risen from 20 to 100 °C. The heat capacity of the containers is 10 Ws/K. How long does it take, until $T = 1000\,°C$ is reached, when we assume that the vibrational degrees of freedom of N_2 can be excited already at $T = 500\,°C$? All heat losses should be neglected.

10.4 Give a vivid and a mathematical justification for the time dependent temperature function $T(t)$ during the mixing experiment of Fig. 10.12.

10.5 A container ($m = 0.1$ kg) with 10 mol air at room temperature rests on the ground. What is the probability that it lifts by itself 10 cm above ground? Such an event would cause a cooling (conversion of thermal into potential energy). How large is the decrease of the temperature? (Specific heat of the gas is $(5/2)R$, that of the container is 1 kJ/(kg · K).)

10.6 A volume of 1 dm³ of helium under standard conditions ($p_0 = 1$ bar, $T_0 = 0\,°C$) is heated up to the temperature $T = 500$ K. What is the entropy increase for isochoric and for isobaric heating?

10.7 The critical temperature for CO_2 ($M = 44$ g/mol) is $T_c = 304.2$ K and the critical pressure $p_c = 7.6 \cdot 10^6$ Pa, its density at the critical point is $\varrho = 46\,\text{kg/m}^3$. What are the van der Waals constants a and b?

10.8 What is the entropy increase ΔS_1 when 1 kg water is heated from 0 to 50 °C? Compare ΔS_1 with the entropy increase ΔS_2 when 0.5 kg water of 0 °C is mixed with 0.5 kg of 100 °C.

10.9 A power station delivers the mechanical work W_1 when water vapour of 600 °C drives a turbine and cools down to 100 °C.

a) What is the Carnot efficiency?
b) How many % of the output energy can one win, when the water of 100 °C is used for heating and cools down to 30 °C?

10.10 A hot solid body ($m = 1$ kg, $c = 470$ J/(kg · K), $T = 300\,°C$) is immersed into 10 kg of water at 20 °C.

a) What is the final temperature?
b) What is the entropy increase?

10.11 Calculate the pressure that a wire with 1 mm diameter exerts onto an ice block with a width of 10 cm (according to Fig. 10.80) when both ends are connected with a mass $m = 5$ kg. What is the increase of the melting temperature? What is the heat supplied to the ice block by the wire, if the outside temperature and the wire temperature are 300 K? How much ice can be melted per second by the wire?

10.12 Calculate from the diagram of Fig. 10.64b the theoretical efficiency of the Otto-motor.

10.13 Show that for a periodically supplied heat at $x = 0$ Eq. 10.42 is a solution of the Eq. 10.38b for one-dimensional heat conduction.

10.14 What is the maximum power an upwind power plant can deliver (area 5 km², temperature below the glass roof $T = 50\,°C$, height of the tower 100 m, outside temperature 20 °C at the top of the chimney).

References

10.1. F.W.G. Kohlrausch, *An Introduction to Physical Measurements,* 2nd ed. (Univ. Toronto Libraries, Toronto, 2011)
10.2a. J.V. Iribane, H.R. Cho, *Atmospheric Physics.* (D. Reidel, Dordrecht, 1980)
10.2b. D.G. Andrews, *An Introduction to Atmospheric Physics.* (Cambridge Univ. Press, Cambridge, 2010)
10.3. H. Haken, *Synergetics. An Introduction.* (Springer, Berlin, Heidelberg, 2014)
10.4. H. Haken, *Synergetics. Introduction and Advanced Topics.* (Springer, Heidelberg, 2004)
10.5. J.E. Parrot, A.D. Stuckes, *Thermal Conductivity of Solids.* (Pion Ltd, London, 1975)
10.6. P. Dunn, D.A. Reay, *Heatpipes,* 2nd ed. (Pergamon, Oxford, 1978)
10.7. N. Rice, *Thermal Insulation. A Building Guide.* (NY Research Press, New York, 2015)
10.8. R.T. Bynum, *Insulation Handbook.* (McGrawHill, New York, 2000)
10.9. J. Fricke and W.L. Borst, *Essentials of Energy Technology: Sources, Transport, Storage Conservation.* (Wiley VCH, Weinheim, 2014)
10.10. Ph. Warburg, *Harvest the Sun. America's Quest for a Solar Powered Future.* (Beacon Press, Boston, 2015)
10.11. M. Green, *Third Generation Photovoltaics. Advanced Solar Energy Conversion. Springer Series in Photonics, Vol. 12.* (Springer, Berlin, Heidelberg, 2005)
10.12. S.A. Goudsmit, R. Clayborne, *Time.* (Time-Life Amsterdam, 1970)
10.13. S.C. Colbeck, Am. J. Phys. **63**, 888 (1995)
10.14. CRC handbook of Chemistry and Physics, 96th ed. (CRC Press, Boca Raton, Florida, USA, 2015)
10.15. http://cdn.intechopen.com/pdfs/20377/InTech-Practical_application_of_electrical_energy_storage_system_in_industry.pdf
10.16. http://www2.hesston.edu/physics/201112/regenerativeenergy_cw/paper.html
10.17. https://en.wikipedia.org/wiki/Climate_change
10.18. http://www.bounceenergy.com/blog/2013/05/wind-energy-grid-part-3-future/, http://www.pasolar.org/index.asp?Type=B_BASIC&SEC=%7B9D644D34-AF8E-475C-A330-4396D09F454B%7D
10.19. https://en.wikipedia.org/wiki/World_energy_consumption
10.20. https://en.wikipedia.org/wiki/Electric_energy_consumption
10.21. https://en.wikipedia.org/wiki/Renewable_energy
10.22. https://en.wikipedia.org/wiki/Tidal_power, https://en.wikipedia.org/wiki/Rance_Tidal_Power_Station, http://www.darvill.clara.net/altenerg/tidal.htm
10.23. https://en.wikipedia.org/wiki/Wave_power, https://en.wikipedia.org/wiki/List_of_wave_power_stations
10.24. https://en.wikipedia.org/wiki/Geothermal_energy, https://en.wikipedia.org/wiki/Geothermal_electricity
10.25. https://en.wikipedia.org/wiki/Solar_thermal_energy
10.26. http://www.volker-quaschning.de/articles/fundamentals2/index_e.php, https://en.wikipedia.org/wiki/List_of_solar_thermal_power_stations
10.27. http://de.total.com/en-us/making-energy-better/worldwide-%20%20%20projects/sunpower-puts-total-cutting-edge-solar?gclid=CLKnluvxjckCFVZAGwod6mQFh
10.28. http://energy.gov/eere/fuelcells/hydrogen-storage, https://en.wikipedia.org/wiki/Hydrogen_storage
10.29. https://en.wikipedia.org/wiki/Energy_storage
10.30. https://en.wikipedia.org/wiki/Renewable_energy
10.31. BP, *Workbook of historical data.* Microsoft Excel document
10.32. http://www.darvill.clara.net/altenerg/wave.htm
10.33. http://thinkglobalgreen.org/WAVEPOWER.html

Mechanical Oscillations and Waves

11

© Springer International Publishing Switzerland 2017
W. Demtröder, *Mechanics and Thermodynamics*, Undergraduate Lecture Notes in Physics, DOI 10.1007/978-3-319-27877-3_11

Mechanical oscillations play an important role in basic sciences as well as for technical applications. Their significance as sources of acoustic waves and for the realization of musical performances, in sensors for hearing is obvious. Often the prevention of unwanted acoustic resonances of buildings and bridges represents a technical challenge. All these points justify a more detailed study of the basic physics of oscillations and waves.

Their mathematical treatment is in many aspects very similar to that of electric oscillations and waves (see Vol. 2, Chap. 6). The investigation of common features and differences between mechanical and electro-magnetic oscillations and waves not only intensifies our knowledge of macroscopic oscillation phenomena but also gives a deeper insight into the microscopic structure of matter. (atomic and molecular vibrations in solids).

In this chapter we will discuss mechanical oscillations, where matter is moved, and mechanical waves where this motion is transported by couplings between neighbouring layers of gases, liquids or solids. At the end of this chapter some interesting applications of ultrasonics in medicine and of acoustics in music are presented.

11.1 The Free Undamped Oscillator

In Chap. 2 the basic equations of motion for the simplified model of point masses were derived. In a similar way the basic facts of mechanical oscillations can be best understood when we start with the idealized model of point masses before we proceed to oscillations of extended bodies.

A point mass m suspended by a spring has its equilibrium position at $x = 0$ where the gravity force is just compensated by the opposite restoring force of the spring. When the mass m is removed from its equilibrium position by a small displacement x (Fig. 11.1) a restoring force occurs, which is, according to Hooke's Law (Sect. 6.2) proportional to x:

$$\boldsymbol{F} = -D \cdot \boldsymbol{x} ,$$

where D is the spring constant that depends on the strength of the spring. This force drives the mass m back to its equilibrium position $x = 0$. The one-dimensional equation of motion is then

$$m \cdot \frac{\mathrm{d}^2 x}{\mathrm{d}t^2} = -Dx . \tag{11.1a}$$

With the abbreviation $\omega_0^2 = D/m$ this becomes

$$\frac{\mathrm{d}^2 x}{\mathrm{d}t^2} + \omega_0^2 x = 0 . \tag{11.1b}$$

This is the equation for the harmonic oscillator (which is called "harmonic" because its oscillation generates a "pure" sinusoidal tone at the frequency ω_0. Together with its overtones $n \cdot \omega_0$ it forms a superposition of tones that are felt by human ears as harmony).

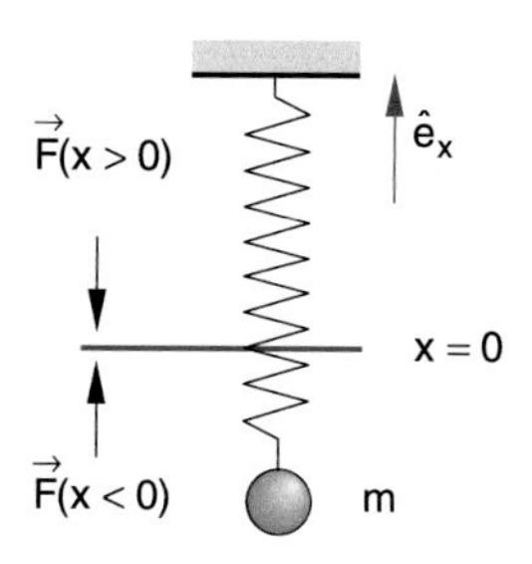

Figure 11.1 Undamped harmonic oscillator

In Sect. 2.9.7 we had already derived the oscillation equation for the simple pendulum, where we had found for small elongations the same equation $\ddot{\varphi} + (g/L)\varphi = 0$ for the angle φ (Eq. 2.79b).

The equation (11.1) has the solution

$$x = c \cdot \mathrm{e}^{\lambda t} , \tag{11.2}$$

where c is an arbitrary constant. Inserting (11.2) into (11.1) gives the quadratic equation for the parameter λ:

$$\lambda^2 + \omega_0^2 = 0 ,$$

which has the two solutions

$$\lambda_1 = +\mathrm{i} \cdot \omega_0 \quad \text{and} \quad \lambda_2 = -\mathrm{i} \cdot \omega_0 .$$

We therefore obtain the two solutions of (11.1)

$$x_1(t) = c_1 \mathrm{e}^{\mathrm{i}\omega_0 t} \quad \text{and} \quad x_2(t) = c_2 \mathrm{e}^{-\mathrm{i}\omega_0 t} ,$$

which are linearly independent for $\omega_0 \neq 0$. The general solution of the linear differential equation (11.1) is then the linear combination of the two solutions

$$x(t) = c_1 \mathrm{e}^{\mathrm{i}\omega_0 t} + c_2 \mathrm{e}^{-\mathrm{i}\omega_0 t} . \tag{11.3}$$

Since $x(t)$ must be a real function (not complex) it follows for the complex constants $c_1 = c_2^* = c$. The solution for the oscillation amplitude is then

$$x(t) = c\mathrm{e}^{\mathrm{i}\omega_0 t} + c^* \mathrm{e}^{-\mathrm{i}\omega_0 t} \quad \text{with} \quad c = a + \mathrm{i}b . \tag{11.4a}$$

The real constants a and b can be determined from the initial conditions for the special oscillation problem.

Example

When the mass m in Fig. 11.1 passes at $t = 0$ with the velocity v_0 through the equilibrium position $x = 0$, we obtain from (11.4a): $c^* + c = 0 \Rightarrow a = 0$ and $v_0 = \mathrm{i} \cdot \omega_0 (c - c^*) = \mathrm{i}\omega_0 \cdot 2\mathrm{i}b => b = v_0/2\omega_0$. Therefore is

$$x(t) = \frac{v_0}{\omega_0} \sin \omega_0 t .$$

◀

Remark. The oscillating mass on a spring is only one example for a harmonic oscillator. Other examples are a mass that oscillates on a parabolic air track, or the simple pendulum suspended by a string, or an electron in the lowest energy level of the hydrogen atom.

11.2 Mathematical Notations of Oscillations

When we write the complex amplitudes c and c^* in (11.4) as polar representation

$$c = |c| \cdot e^{i\varphi} , \quad c^* = |c| \cdot e^{-i\varphi} .$$

We obtain the representation

$$x(t) = |c| \left[e^{i(\omega_0 t+\varphi)} + e^{-i(\omega_0 t+\varphi)} \right] , \qquad (11.4b)$$

that is equivalent to (11.4a).

According to Euler's formula for complex numbers

$$e^{ix} = \cos x + i \cdot \sin x .$$

We can write (11.4a) also in the form

$$x(t) = C_1 \cos \omega_0 t + C_2 \sin \omega_0 t$$
$$\text{with} \quad \left\{ \begin{array}{l} C_1 = c + c^* \\ C_2 = i(c - c^*) \end{array} \right\} . \qquad (11.4c)$$

A forth equivalent representation is

$$x(t) = A \cdot \cos(\omega_0 t + \varphi) . \qquad (11.4d)$$

The comparison with (11.4c) gives

$$C_1 = A \cdot \cos \varphi , \quad C_2 = -A \cdot \sin \varphi$$
$$\Rightarrow -\tan \varphi = \frac{C_2}{C_1} \quad \text{and} \quad A = \sqrt{C_1^2 + C_2^2} .$$

All 4 representations (11.4a–d) for the solution of (11.1) are equivalent (Fig. 11.2). They represent a harmonic oscillation with the frequency ω_0 and the amplitude $A = 2|c|$ (Fig. 11.3).

For our example above with the initial conditions $x(0) = 0$ and $(dx/dt)_0 = v_0$ all forms (11.4a–d) give the solution

$$x(t) = \frac{v_0}{\omega_0} \sin(\omega_0 t) ,$$

as can be immediately proved by inserting $x(t)$ into (11.4a–d).

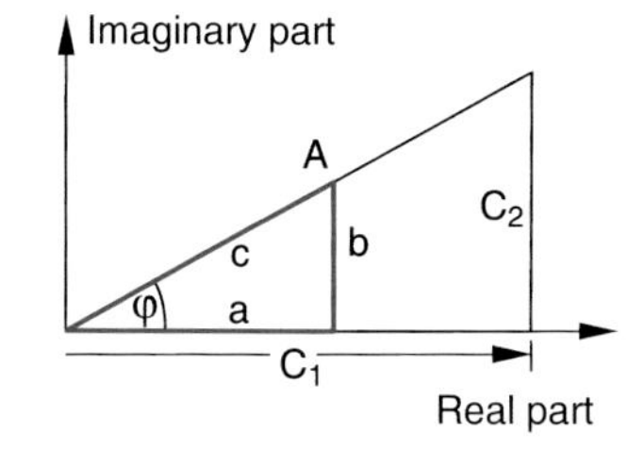

Figure 11.2 Relations betwenn different equivalent representations of harmonic oscillations

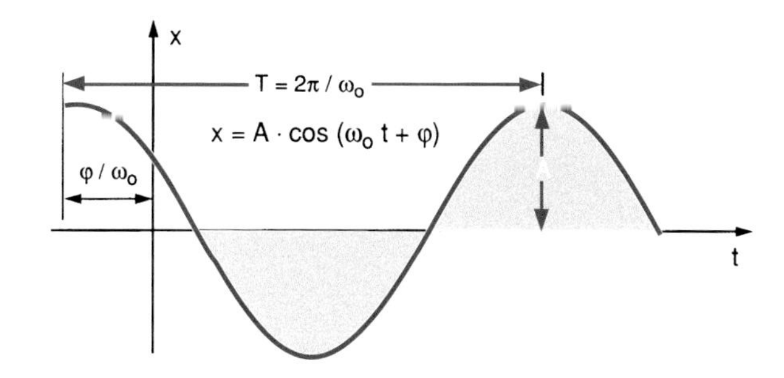

Figure 11.3 Period T, amplitude A and phase shift φ of a harmonic oscillation

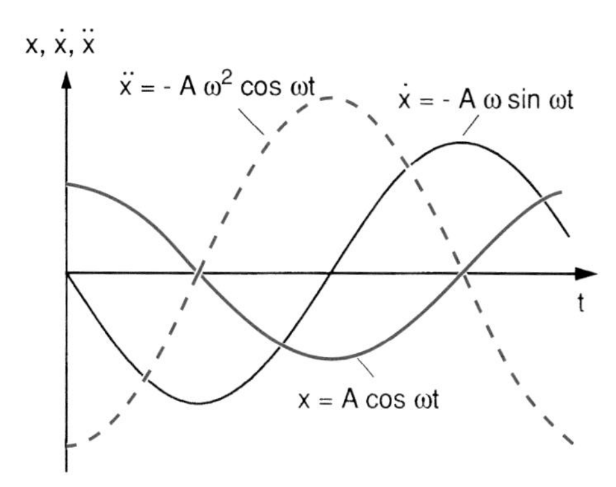

Figure 11.4 Elongation $x(t)$, velocity $\dot{x}(t)$ and acceleration $\ddot{x}(t)$ of a harmonic oscillation

The argument $(\omega_0 t + \varphi)$ in the cosine function (11.4d), which determines the momentary value of the elongation $x(t)$, is called the *phase* of the oscillation. The time origin $x = 0$ can be chosen in such a way that $\varphi = 0$. This reduces (11.4d) to

$$x(t) = A \cdot \cos \omega_0 t . \qquad (11.4e)$$

After a time $t = 2\pi/\omega_0 = T$ always the same value of $x(t)$ is reached. This means

$$x(t + T) = x(t) .$$

The time interval T is called the **oscillation period**, while the reciprocal $\nu = 1/T$ is the **oscillation frequency** and $\omega = 2\pi\nu$ is the circular frequency. The mass that experiences a restoring force proportional to the displacement $(x - x_0)$ from the equilibrium position x_0 is called a **harmonic oscillator**.

In Fig. 11.4 the elongation $x(t) = A \cdot \cos(\omega t)$, the velocity dx/dt and the acceleration d^2x/dt^2 are shown. The figure illustrates that the acceleration always has the opposite phase as the elongation, i. e. $x(t)$ shows a phase shift of π against d^2x/dt^2.

Example

$x_1(t) = A \cdot \cos(\omega_0 t)$ and $x_2(t) = A \cdot \cos(\omega_0 + \varphi)$ are two harmonic oscillators with the same frequency and the same amplitude but with a phase shift φ against each other. The maxima of the two oscillations are shifted against each other by the time $\Delta t = \varphi/\omega_0$ (Fig. 11.5).

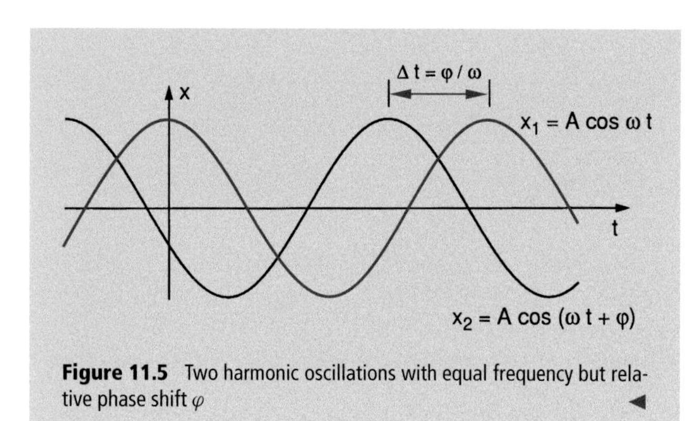

Figure 11.5 Two harmonic oscillations with equal frequency but relative phase shift φ

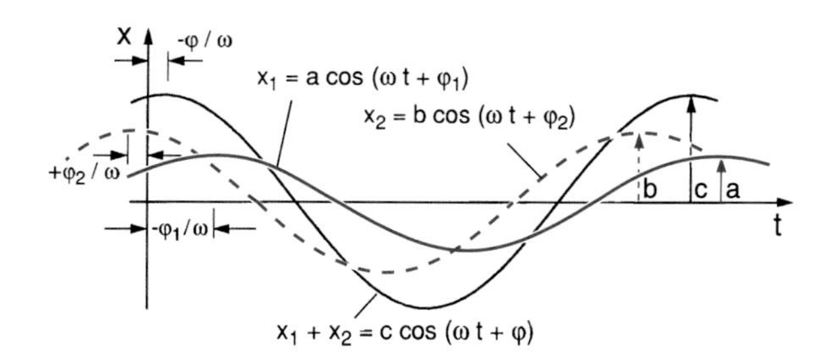

Figure 11.6 One-dimensional superposition of two oscillations with equal frequencies but different phases φ_1 and φ_2

11.3 Superposition of Oscillations

In nature, pure harmonic sine oscillations are rare. Generally more or less complex forms of oscillations occur. It turns out, however, that even complex non-harmonic oscillations can be represented by a superposition of pure harmonic oscillations with different amplitudes, frequencies and phases. We will discuss in this Section such superpositions. If the elongations of all pure harmonic oscillations point into the same direction (for instance into the x-direction), we have a one-dimensional superposition. In the general case of two- or three-dimensional superpositions the elongations of the different oscillators can point into arbitrary directions.

11.3.1 One-Dimensional Superposition

The sum of the different oscillations

$$x(t) = \sum_n x_n(t) = \sum_n a_n \cos(\omega_n t + \varphi_n) \tag{11.5}$$

depends on the amplitudes a_n, the frequencies ω_n and the phases φ_n of the different summands.

11.3.1.1 Two Oscillations of Equal Frequencies

If the two oscillations

$$x_1(t) = a \cdot \cos(\omega t + \varphi_1)$$
$$x_2(t) = b \cdot \cos(\omega t + \varphi_2)$$

with equal frequencies ω, but different amplitudes and phases are superimposed, one obtains according to the addition theorem of trigonometric functions

$$\begin{aligned} x(t) = x_1(t) + x_2(t) &= A \cdot \cos \omega t + B \cdot \sin \omega t \\ &= C \cdot \cos(\omega t + \varphi) \end{aligned} \tag{11.6}$$

with the relations

$$A = a \cdot \cos \varphi_1 + b \cdot \cos \varphi_2 ,$$
$$B = -a \cdot \sin \varphi_1 - b \cdot \sin \varphi_2 ,$$
$$C = \sqrt{A^2 + B^2} \quad \text{and} \quad \tan \varphi = -\frac{B}{A} .$$

The superposition is therefore again a harmonic oscillation with the same frequency but amplitude and phase differ from that of the partial oscillations (Fig. 11.6).

Special Cases:

1. $a = b$ and $\varphi_1 = \varphi_2 = \varphi$

$$\Rightarrow x = x_1 + x_2 = 2a \cdot \cos(\omega t + \varphi)$$

Both oscillations add in phase and the resulting oscillation has twice the amplitude of the two summands.

2. $a = b$ but $\varphi_1 \neq \varphi_2$

$$\begin{aligned} x(t) &= a[\cos(\omega t + \varphi_1) + \cos(\omega t + \varphi_2)] \\ &= a[\cos \omega t(\cos \varphi_1 + \cos \varphi_2) - \sin \omega t(\sin \varphi_1 + \sin \varphi_2)] \end{aligned}$$

Ansatz:

$$\begin{aligned} x(t) &= b \cdot \cos(\omega t + \varphi) \\ &= b \cdot [\cos \omega t \cos \varphi - \sin \omega t \sin \varphi] \end{aligned}$$

$$\left.\begin{aligned} \Rightarrow a(\cos \varphi_1 + \cos \varphi_2) &= b \cdot \cos \varphi \\ a(\sin \varphi_1 + \sin \varphi_2) &= b \cdot \sin \varphi \end{aligned}\right\} \Rightarrow$$

$$\tan \varphi = \frac{\sin \varphi_1 + \sin \varphi_2}{\cos \varphi_1 + \cos \varphi_2} = \tan \frac{\varphi_1 + \varphi_2}{2}$$

$$\Rightarrow \varphi = \frac{\varphi_1 + \varphi_2}{2}$$

$$\Rightarrow b = a \cdot \sqrt{2 + 2\cos(\varphi_1 - \varphi_2)}$$

$$\Rightarrow x(t) = a \cdot \sqrt{2 + 2\cos(\varphi_1 - \varphi_2)} \cos(\omega t + \varphi)$$

The resultant amplitude is smaller than $2a$ and the phase differs from φ_1 and φ_2. For $\varphi_1 = \varphi_2 + \pi$ the two oscillations have opposite phases. The two oscillations cancel each other and $x(t) \equiv 0$, i. e. the total amplitude is zero.

11.3.1.2 Different Frequencies, Beats

A different situation arises when two oscillations with different frequencies are superimposed (Fig. 11.7). For equal amplitudes $a = b$ the sum of the two oscillations

$$x_1(t) = a \cdot \cos \omega_1 t\,; \quad x_2(t) = a \cdot \cos \omega_2 t$$

gives with the trigonometric theorem

$$\cos\alpha + \cos\beta = 2\cos\frac{\alpha-\beta}{2} \cdot \cos\frac{\alpha+\beta}{2}\,,$$

the superposition

$$x(t) = 2a \cdot \cos\left(\frac{\omega_1 - \omega_2}{2} t\right) \cdot \cos\left(\frac{\omega_1 + \omega_2}{2} t\right)\,. \tag{11.7}$$

If the two frequencies do not differ much, i. e. $(\omega_1 - \omega_2) \ll \omega = \frac{1}{2}(\omega_1 + \omega_2)$ Eq. 11.7 can be interpreted as an oscillation

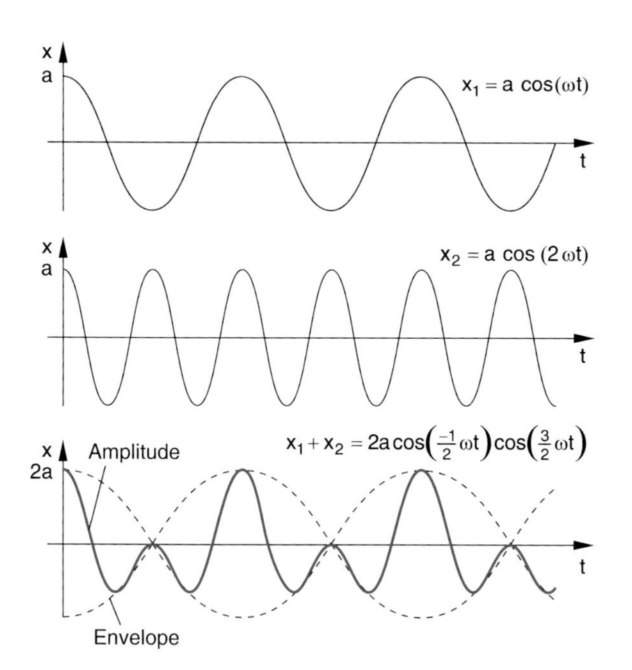

Figure 11.7 One-dimensional superposition of two oscillations with different frequencies

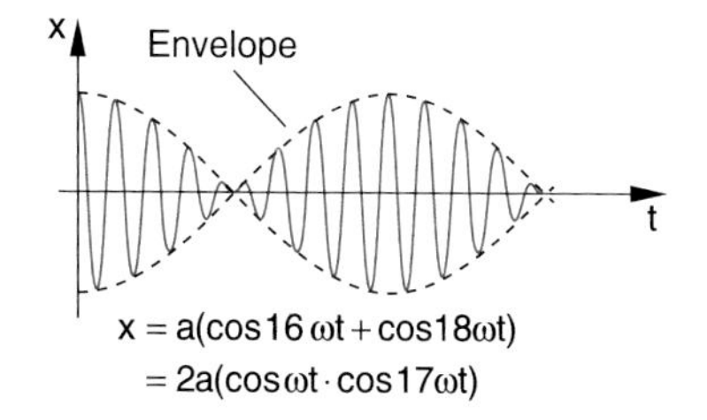

Figure 11.8 Beat pattern of the superposition of two oscillations with frequencies ω_1 and ω_2 and $\delta\omega \ll \omega = \frac{1}{2}(\omega_1 + \omega_2)$

with the frequency ω and an amplitude $A(t) = 2a \cdot \cos[\frac{1}{2}(\omega_1 - \omega_2)t]$ that oscillates slowly with a period $\tau = 2\pi/(\omega_1 - \omega_2)$, which is long compared with the mean oscillation period $T = 2\pi/\omega$ (Fig. 11.8). This oscillation $x(t)$ is called a **beat** and the period τ is the **beat period** or beat cycle.

Acoustic beats can be realized by two vibrating tuning forks, which are slightly detuned against each other. With a microphone they can be made audible to a large auditorium and they can be also made visible on an oscilloscope (Fig. 11.9) (see also Sect. 11.10).

11.3.1.3 Superposition of Several Oscillations; Fourier-Analysis

When N oscillations with frequencies ω_n are superimposed, the resultant oscillation

$$x(t) = \sum_{n=1}^{N} a_n \cos(\omega_n t + \varphi_n) \tag{11.8}$$

is generally complex (Fig. 11.10). It is, however, still periodic with a period $T = 2\pi/\omega_m$ where ω_m is the maximum common factor of all involved frequencies ω_n. For the special case $\omega_n = n \cdot \omega_1$ ($n = 1; 2; 3; \ldots$) the period of the superposition is $T = 2\pi/\omega_1$.

Conversely every periodic function $f(t)$ with $f(t) = f(t + T)$, can be expressed by a sum of sin- and cos-functions with frequencies $\omega_n = n \cdot \omega_1$ (n = integer). It is

$$f(t) = a_0 + \sum_{n=1}^{\infty} a_n \cos(n \cdot \omega_1 \cdot t + \varphi_n)\,. \tag{11.9}$$

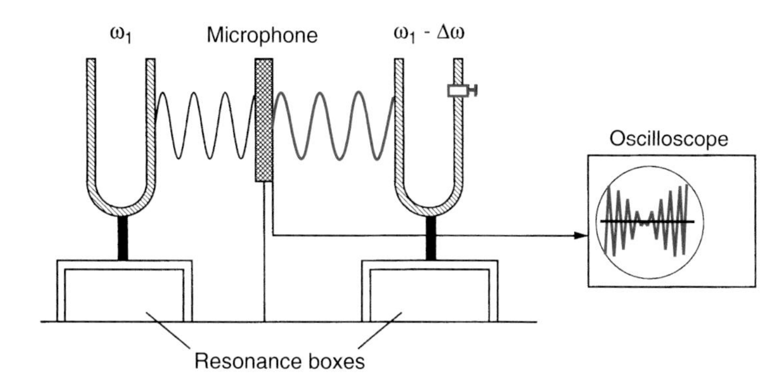

Figure 11.9 Experimental setup for the acoustic and visible demonstration of beats

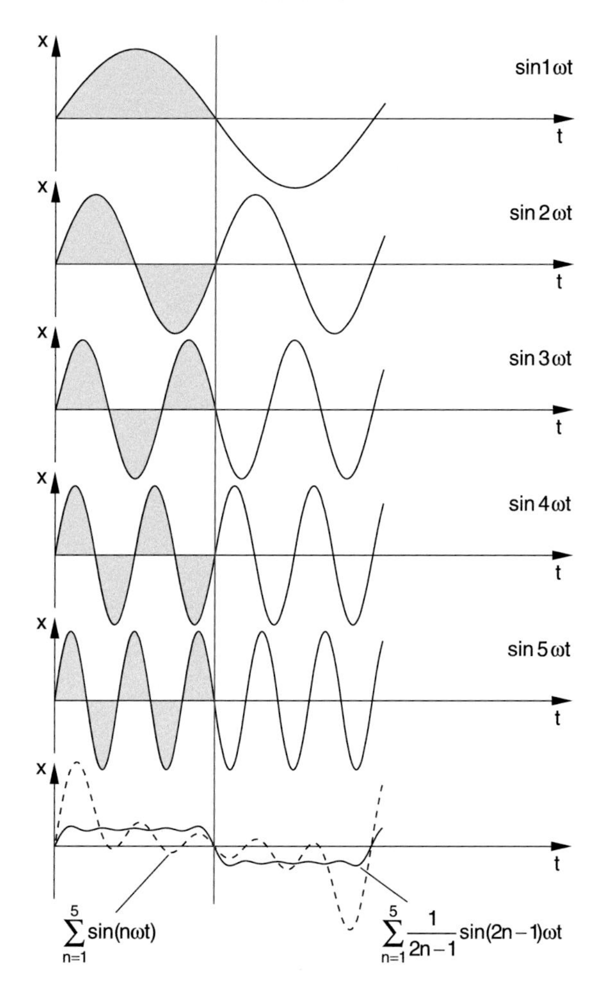

Figure 11.10 Superposition of five oscillations $x_n = a \cdot \sin(n\omega t)$ with equal amplitudes A_n and frequencies $\omega_n = n \cdot \omega_1$ (dashed curve) and $\omega_n = (2n-1)\omega_1$ (solid curve). The amplitude of the superposition has been reduced by a factor 3.66

> The oscillation $a_1 \cdot \cos \omega_1 t$ is the **fundamental oscillation**, the members with higher frequencies $n \cdot \omega_1$ are **higher harmonics**. In acoustics, they are called **fundamental tone** and **overtones**.

The partition of a periodic function into harmonics i. e. the representation of $f(t)$ as a Fourier-series is called **Fourier analysis**. Its general from is discussed in Sect. 13.4.

The experimental Fourier analysis can be performed with the vibrating reed frequency meter, shown in Fig. 11.11. It consists of a series of flat springs with different lengths and resonance frequencies $\omega = n \cdot \omega_1$ $(n = 1, 2, 3, \ldots)$. When they are excited by a mechanical vibration their oscillation amplitude becomes maximum, when the exciting frequency matches the resonance frequency determined by the length of the spring.

The human inner ear embodies such vibrating springs in the form of thin hairs with different resonance frequencies. Their vibrations are transformed into electric signals that are conducted to the brain (see Sect. 11.5 and 11.14). If the hairs are exposed to excessive sound intensities, they can break and the ear can no longer hear the corresponding frequencies.

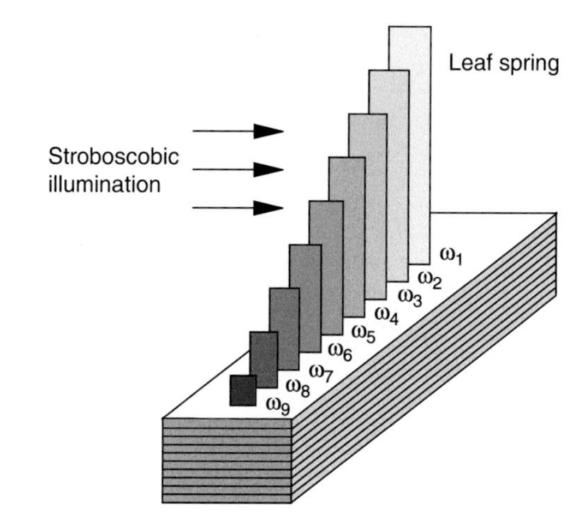

Figure 11.11 Reed frequency meter

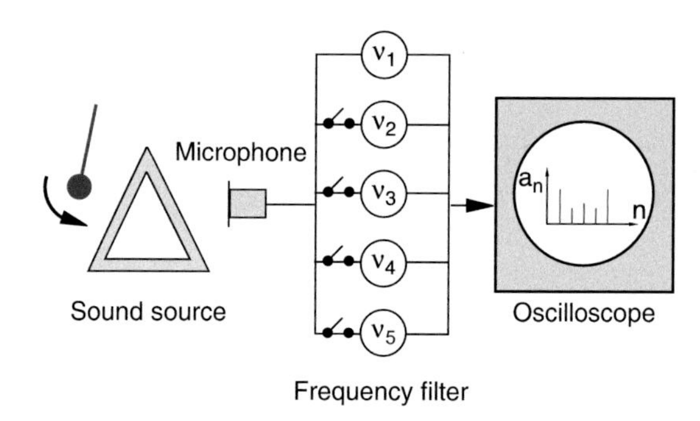

Figure 11.12 Electrical detection of mechanical oscillation and Fourier-analysis with frequency filters

In a more elegant way mechanical oscillations can be detected and measured by microphones, which transform them into electrical signals that can be viewed for instance, on the screen of an oscilloscope. If the source emits several frequencies, they can be selected by parallel-connected electrical filters (Fig. 11.12). that act as Fourier-analysers.

In Fig. 11.13 the Fourier-analysis of a periodic rectangular function

$$f(t) = \begin{cases} A & \text{for} \quad 0 < t < \frac{T}{2} \\ 0 & \text{for} \quad \frac{T}{2} \le t \le T \end{cases} \quad \text{and} \quad f(t+T) = f(t) \, ;$$

$$f(t) = a_0 + \sum_{n=1}^{\infty} a_n \sin(n\omega t) \, ; \quad \omega = \frac{2\pi}{T}$$

$$a_0 = \frac{A}{2} \, ; \quad a_{2n-1} = \frac{2A}{(2n-1)\pi} \, ; \quad a_{2n} = 0$$

is shown with the corresponding magnitude distribution of the amplitudes a_n.

Remark. Even an arbitrary not necessary periodic function can be represented as a superposition of periodic functions. Instead

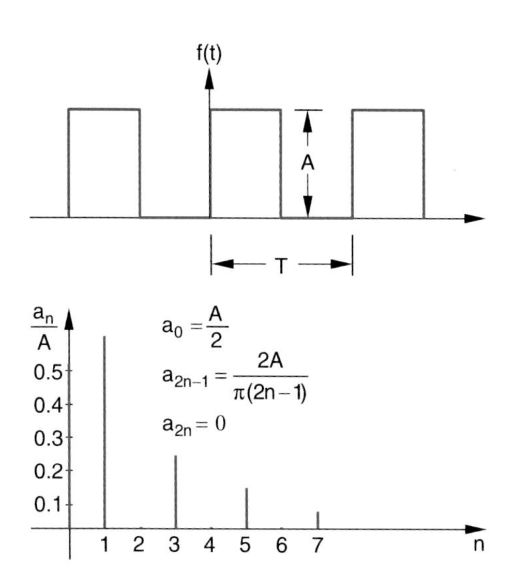

Figure 11.13 Fourier-analysis of an equidistant sequence of rectangular profiles

of the sum one obtains now the integral

$$f(t) = \int_0^\infty [a(\omega)\cos\omega t + b(\omega)\sin\omega t]\,d\omega$$

with the coefficients

$$a(\omega) = \frac{1}{\pi}\int_{-\infty}^{+\infty} f(t)\cos\omega t dt\,, \quad b(\omega) = \frac{1}{\pi}\int_{-\infty}^{+\infty} f(t)\sin\omega t dt\,.$$

11.3.2 Two-dimensional Superposition; Lissajous-Figures

When two oscillations with the same frequency ω and a phase shift φ,

$$x = a\cdot\cos(\omega t)\,, \quad y = b\cdot\cos(\omega t+\varphi)\,, \tag{11.10}$$

which oscillate into directions that are orthogonal to each other, are superimposed, one obtains the two-dimensional representation

$$\frac{x}{a} = \cos\omega t\,; \qquad \begin{aligned}\frac{y}{b} &= \cos\omega t\cos\varphi - \sin\omega t\sin\varphi\\ &= \frac{x}{a}\cos\varphi - \sin\omega t\sin\varphi\,.\end{aligned}$$

Rearrangement gives

$$\sin\omega t\cdot\sin\varphi = \frac{x}{a}\cos\varphi - \frac{y}{b}$$

$$\cos\omega t\cdot\sin\varphi = \frac{x}{a}\sin\varphi\,.$$

Elimination of time t by squaring and adding the two equations yields

$$\sin^2\varphi = \frac{x^2}{a^2} + \frac{y^2}{b^2} - \frac{2xy}{ab}\cos\varphi\,.$$

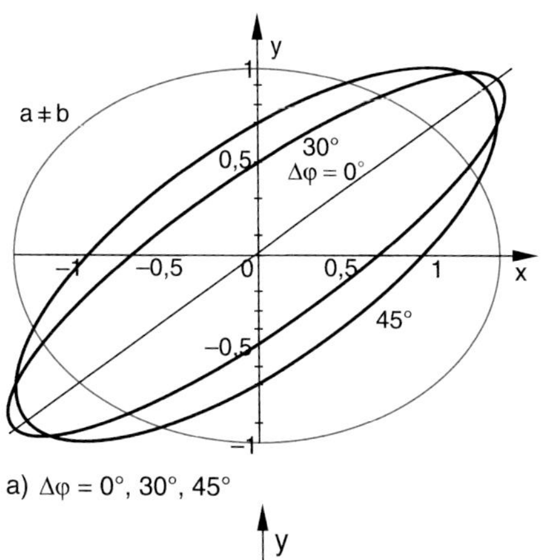

a) Δφ = 0°, 30°, 45°

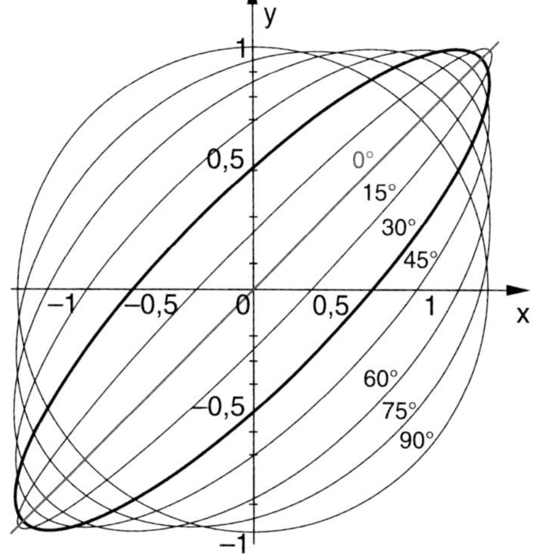

b) a = b; Δφ = 0°, 15°, 30°, 45°, 60°, 75°, 90°

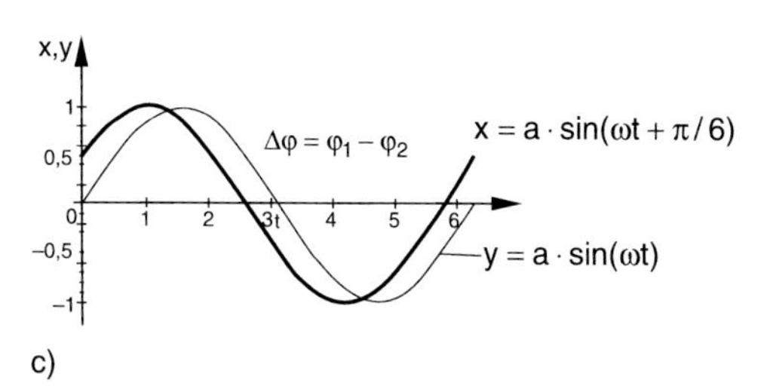

c)

Figure 11.14 Trajectories $y(x)$ of the superposition $x(t) + y(t)$ with $x(t) = b\cdot\sin(\omega t + \Delta\varphi)$ and $y(t) = a\cdot\sin\omega t$ for different values of the phase shift $\Delta\varphi$. **a** for $a \neq b$ and **b** for $a = b$. **c** Illustration of the functions $y(t)$ and $x(t)$ for $\Delta\varphi = \pi/6$

This can be rearranged into the ellipse equation for the motion of the oscillating point mass in the x–y-plane.

$$\frac{x^2}{a^{*2}} + \frac{y^2}{b^{*2}} - \frac{2xy\cos\varphi}{a^* b^*} = 1\,, \tag{11.11}$$

which describes an ellipse with axes tilted by the angle φ against the x- and y-axes. The length of the half axes $a^* = a\cdot\sin\varphi$ and $b^* = b\cdot\sin\varphi$ depends on the amplitudes a and b of the oscillations and on the phase shift φ (Fig. 11.14).

For the special case $\varphi = 0$ the ellipse reduces to the straight line $y = (b/a)x$.

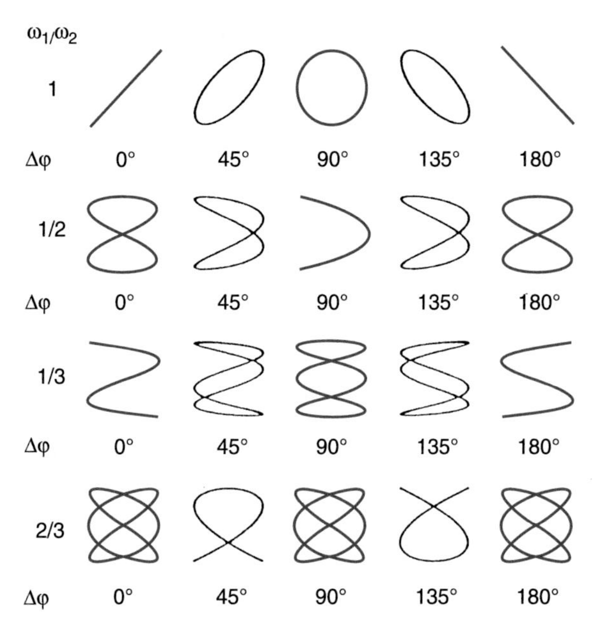

Figure 11.15 Lissajous-figures $y = f(x)$ of oscillations (11.12) for different values of the frequency-ratio ω_1/ω_2 and phase differences $\Delta\varphi = \varphi_1 - \varphi_2$ and $\varphi_1 = 0$

For $\varphi = \pi/2$ one obtains an ellipse with the axes $a^* = a$ and $b^* = b$ in the direction of the x- and y-axes. For $\varphi = \pi/2$ and $a = b$ the trajectory is a circle.

For demonstration experiments these two-dimensional oscillations can be realized when a pendulum with a hollow sphere of magnetic material, filled with white sand is excited in the x- and the y-directions by electromagnets. When the sand runs through a small hole in the hollow sphere onto a plane of black velvet it writes the resulting trajectories for different values of a, b and φ. In a simpler demonstration the x- and y-input of an oscilloscope are fed by periodic voltages with different amplitudes and phase shifts. In Fig. 11.14 the trajectories in the x–y-plane are illustrated for some ratios a/b and different phase shifts φ.

If two oscillations in x- and y-directions

$$\begin{aligned} x &= a \cdot \cos(\omega_1 t + \varphi_1) \\ y &= b \cdot \cos(\omega_2 t + \varphi_2) , \end{aligned} \tag{11.12}$$

with different frequencies ω_1 and ω_2 are superimposed the resultant trajectory is generally more complex. It describes a closed curve only if the ratio ω_1/ω_2 is a rational number. Such a curve is called a Lissajous-figure. If ω_1/ω_2 is irrational the trajectory fills in the course of time the whole area of the rectangle

$$-a \le x \le +a\,; \quad -b \le y \le +b$$

in the x–y-plane [11.5]. In Fig. 11.15 some Lissajous figures are shown for different values of ω_1/ω_2 and φ.

11.4 The Free Damped Oscillator

If the mass of the oscillator in Fig. 11.1 moves through a viscous liquid (Fig. 11.16) the friction can no longer be neglected. In addition to the restoring spring force $\boldsymbol{F} = -D \cdot \boldsymbol{x}$ the Stokes friction force (8.34)

$$\boldsymbol{F}_{\mathrm{f}} = -6\pi\eta \cdot r \cdot \boldsymbol{v} \tag{11.13}$$

influences the oscillation.

For the general case where the friction force is opposite to the velocity $\boldsymbol{v}$ and proportional to the magnitude $v = |\boldsymbol{v}|$ we can describe oscillations by the differential equation

$$m \cdot \ddot{x} = -D \cdot x - b \cdot \dot{x}\,. \tag{11.14}$$

With the abbreviations

$$\omega_0^2 = \frac{D}{m} \quad \text{and} \quad 2\gamma = \frac{b}{m}$$

we obtain the general equation of motion

$$\ddot{x} + 2\gamma\dot{x} + \omega_0^2 x = 0 \tag{11.15}$$

of the damped oscillation with the damping constant γ. Similar to the problem in Sect. 11.1 we make the ansatz for the solution

$$x(t) = c \cdot \mathrm{e}^{\lambda t}\,.$$

Inserting this into (11.15) one obtains the equation for the parameter λ

$$\lambda^2 + 2\gamma\lambda + \omega_0^2 = 0$$

with the solutions

$$\lambda_{1,2} = -\gamma \pm \sqrt{\gamma^2 - \omega_0^2}\,. \tag{11.16}$$

This gives analogous to (11.3) the general solution for the amplitude

$$x(t) = \mathrm{e}^{-\gamma t}\left[c_1 \mathrm{e}^{\sqrt{\gamma^2-\omega_0^2}\cdot t} + c_2 \mathrm{e}^{-\sqrt{\gamma^2-\omega_0^2}\cdot t}\right]. \tag{11.17a}$$

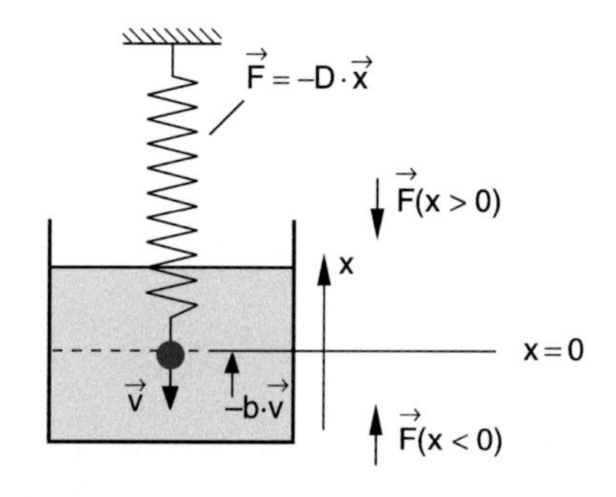

Figure 11.16 Damped oscillator

The time behaviour of $x(t)$ essentially depends on the ratio of

$$\frac{\text{mean restoring force}}{\text{mean friction force}} = \frac{\langle|D\cdot x|\rangle}{\langle|b\cdot\dot{x}|\rangle} = \frac{m\omega_0^2\sqrt{\overline{x^2}}}{2\gamma m\sqrt{\overline{\dot{x}^2}}} = \frac{\omega_0}{2\gamma}\,.$$

This means that it depends on the relative magnitude of ω_0 and γ. We distinguish between the following three cases:

11.4.1 $\gamma < \omega_0$, i. e. weak damping

With the abbreviation $\omega^2 = \omega_0^2 - \gamma^2$ is

$$\lambda_{1,2} = -\gamma \pm \sqrt{-\omega^2} = -\gamma \pm \mathrm{i}\omega\,.$$

The general solution (11.17a) is then

$$\begin{aligned} x(t) &= \mathrm{e}^{-\gamma t}\left(c\mathrm{e}^{\mathrm{i}\omega t} + c^*\mathrm{e}^{-\mathrm{i}\omega t}\right) \\ &= A\mathrm{e}^{-\gamma t}\cos(\omega t + \varphi)\,, \end{aligned} \tag{11.17b}$$

where, as in Sect. 11.1

$$A = 2|c| \quad \text{and} \quad \tan\varphi = -\frac{\mathrm{i}(c - c^*)}{c + c^*} = \frac{\mathrm{Im}\{c\}}{\mathrm{Re}\{c\}}\,.$$

Equation 11.17b describes a damped oscillation, where the amplitude $A\cdot\mathrm{e}^{-\gamma t}$ decays exponentially. It is illustrated in Fig. 11.17 for the initial conditions $x(0) = A$ and $(\mathrm{d}x/\mathrm{d}t)(0) = v_0$, for which the solution (11.17b) is

$$x(t) = A\mathrm{e}^{-\gamma t}\cos\omega t\,. \tag{11.17c}$$

With $v_0 = -A\cdot\gamma$ we obtain

$$v(t) = v_0\mathrm{e}^{-\gamma t}\left[\cos\omega t + \frac{\omega}{\gamma}\sin\omega t\right]\,. \tag{11.17d}$$

Two successive maxima of the damped oscillation have the amplitude ratio

$$\frac{x(t+T)}{x(t)} = \mathrm{e}^{-\gamma T} \tag{11.18}$$

with the period $T = 2\pi/\omega$.

The natural logarithm of the inverse ratio

$$\ln\left[\frac{x(t)}{x(t+T)}\right] = \gamma\cdot T = \delta$$

is called the **logarithmic decrement**.

After the time $t = \tau = 1/\gamma$ the envelope $f(t) = A\cdot\mathrm{e}^{-\gamma t}$ of the oscillation has decreased to $1/\mathrm{e}$ of its initial value $f(0) = A$.

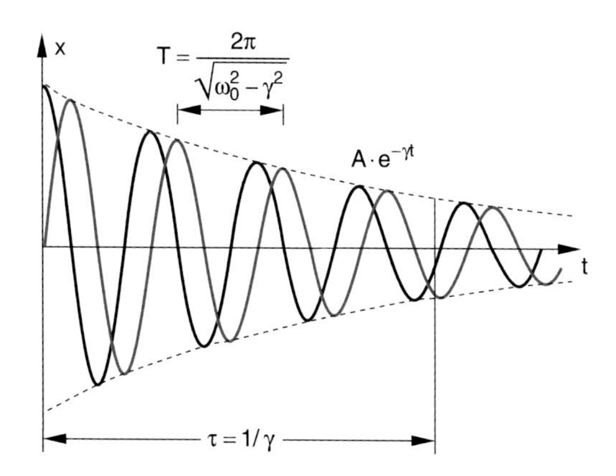

Figure 11.17 Damped oscillation with damping constant γ and oscillation period T for the initial conditions $x(0) = A$ (*black curve*) and $x(0) = 0$ (*red curve*)

The frequency $\omega = \sqrt{\omega_0^2 - \gamma^2}$ of the damped oscillator is for the same spring constant D slightly smaller than that of the undamped oscillator. The frequency shift $\Delta\omega$ increases with increasing damping γ.

Examples

1. $\gamma/\omega_0 = 0.01 \Rightarrow (\omega_0 - \omega) = 5\cdot 10^{-5}\cdot\omega_0$; $\delta = 0.06$, i. e. after about 16 oscillation periods the amplitude has dropped to $1/\mathrm{e}$ of its initial value, i. e. $\tau = 16T$.
2. $\gamma/\omega_0 = 0.1 \Rightarrow (\omega_0 - \omega) = 5\cdot 10^{-3}\omega_0$; $\delta = 0.6$, $\Rightarrow \tau = 1.6T$, i. e. after 1.6 oscillation periods the amplitude has already decreased to $1/\mathrm{e}$. ◀

11.4.2 $\gamma > \omega_0$, i. e. strong Damping

The coefficients λ (11.16) are now real.

$$\lambda_{1,2} = -\gamma \pm \sqrt{\gamma^2 - \omega_0^2} = -\gamma \pm \alpha$$
$$\text{with} \quad \alpha = \sqrt{\gamma^2 - \omega_0^2}$$

The general solution (11.17) is therefore

$$x(t) = \mathrm{e}^{-\gamma t}\left[c_1\mathrm{e}^{\alpha t} + c_2\mathrm{e}^{-\alpha t}\right]\,. \tag{11.19a}$$

With the initial conditions $x(0) = 0$ and $\mathrm{d}x/\mathrm{d}t(0) = v_0$ one obtains $c_1 + c_2 = 0$ and $c_1 - c_2 = v_0$. This gives the special solution

$$x(t) = \frac{v_0}{2\alpha}\mathrm{e}^{-\gamma t}\left[\mathrm{e}^{\alpha t} - \mathrm{e}^{-\alpha t}\right]\,. \tag{11.19b}$$

With the hyperbolic sine-function $\sinh(\alpha t) = 1/2(\mathrm{e}^{\alpha t} - \mathrm{e}^{-\alpha t})$ this can be written as

$$x(t) = \frac{v_0}{\alpha}\mathrm{e}^{-\gamma t}\sinh(\alpha t)\,. \tag{11.19c}$$

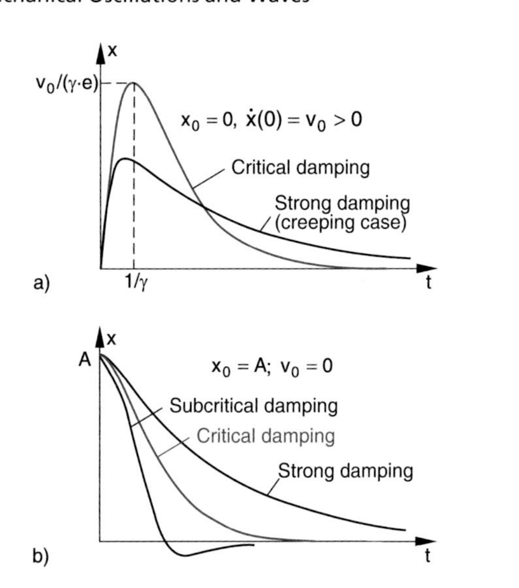

Figure 11.18 Elongation $x(t)$ of the damped oscillation for critical damping (*red curve*) and super-critical damping (*black curve*, creeping case). Initial conditions: **a** $x(0) = 0, \dot{x}(0) = v_0$; **b** $x(0) = A, \dot{x}(0) = 0$

The "oscillation" (11.19c) consist of a single elongation, which converges slowly to zero for $t \to \infty$ (Fig. 11.18a). This situation is also called the "creep-case" because $x(t)$ creeps after its maximum very slowly towards zero.

For different initial conditions $x(0) = A$ and $\mathrm{d}x/\mathrm{d}t(0) = 0$ the coefficients c_i become

$$c_1 = \frac{\alpha + \gamma}{2\alpha} A ; \quad c_2 = \frac{\alpha - \gamma}{2\alpha} A ,$$

which gives

$$x(t) = \frac{A}{\alpha} \mathrm{e}^{-\gamma t} [\alpha \cosh(\alpha t) + \gamma \sinh(\alpha t)] . \qquad (11.19\mathrm{d})$$

11.4.3 $\gamma = \omega_0$ (aperiodic limiting case)

The two parameters λ (11.16) now become

$$\lambda_1 = \lambda_2 = \lambda = -\gamma .$$

The general solution of the differential equation (11.15) must have two independent integration constants. In order to find these constants we try the ansatz

$$x(t) = C(t) \cdot \mathrm{e}^{\lambda t} \qquad (11.20)$$

with the time-dependent factor $C(t)$. Inserting this into (11.15) we obtain for the function $C(t)$ the equation

$$\ddot{C} + (2\lambda + 2\gamma)\dot{C} + (\lambda^2 + 2\gamma\lambda + \omega_0^2)C = 0 .$$

For the solution $\lambda = -\gamma = -\omega_0$ the two expressions in the brackets become zero and our equation reduces to

$$\frac{\mathrm{d}^2 C}{\mathrm{d}t^2} = 0 \to C = c_1 t + c_2 .$$

The general solution (11.20) then becomes

$$x(t) = (c_1 t + c_2)\mathrm{e}^{-\gamma t} . \qquad (11.21)$$

With the initial conditions $x(0) = 0$, $\mathrm{d}x/\mathrm{d}t(0) = v_0$ this reduces to

$$x(t) = v_0 t \mathrm{e}^{-\gamma t} , \qquad (11.21\mathrm{a})$$

which is illustrated in Fig. 11.18a). The oscillation is degenerated to a single elongation, like in the creeping case (Sect. 11.4.2), it starts, however, with a linear rise and, after the maximum, it reaches zero faster than for the case in Sect. 11.4.2. The maximum is reached at $t = 1/\gamma$. For $t = 5/\gamma$ the amplitude has already decreased to $x = 0.1 x_{\max}$.

For different initial conditions $x(0) = A$ and $\mathrm{d}x/\mathrm{d}t(0) = 0$ the solution is

$$x(t) = A(1 + \gamma t)\mathrm{e}^{-\gamma t} . \qquad (11.21\mathrm{b})$$

The amplitude starts at $x = A$ and proceeds initially with horizontal slope until it decreases exponentially to zero (Fig. 11.18b).

11.5 Forced Oscillations

If the upper end of the spring in Fig. 11.16 is not kept at a fixed position, but is moved up and down by a periodic external force (Fig. 11.19), this force is transferred through the spring onto the mass m. The equation of motion is now

$$m \cdot \ddot{x} = -Dx - b\dot{x} + F_0 \cos \omega t . \qquad (11.22\mathrm{a})$$

With the abbreviations

$$\omega_0^2 = \frac{D}{m} ; \quad \gamma = \frac{b}{2m} ; \quad K = \frac{F_0}{m} ,$$

this changes into the inhomogeneous differential equation

$$\ddot{x} + 2\gamma\dot{x} + \omega_0^2 x = K \cdot \cos \omega t , \qquad (11.22\mathrm{b})$$

which differs from the homogeneous equation (11.15) of the damped free oscillator by the expression $K \cdot \cos(\omega t)$ of the external force, which is independent of x.

The general solution of this inhomogeneous equation is the sum of the general solution (11.17a) of the homogeneous equation (11.15) with $F = 0$, and a special solution of the inhomogeneous equation (11.22b). It therefore must have the form

$$x(t) = A_1 \mathrm{e}^{-\gamma t} \cdot \cos(\omega_1 t + \varphi_1) + A_2 \cos(\omega t + \varphi) , \qquad (11.23\mathrm{a})$$

where $\omega_1 = \sqrt{\omega_0^2 - \gamma^2}$ is the frequency of the free damped oscillator (11.15) with weak damping.

After a sufficiently long time $t \gg 1/\gamma$ the amplitude $A_1 \cdot e^{-\gamma t}$ becomes so small that we can neglect it against the second term in (11.23a). This second term depends on the frequency ω of the periodic driving force, which enforces its frequency ω onto the system (forced oscillation).

The second term in (11.23a) therefore represents the **stationary oscillation state** while the first term gives the **transient response** valid for times $t < 1/\gamma$.

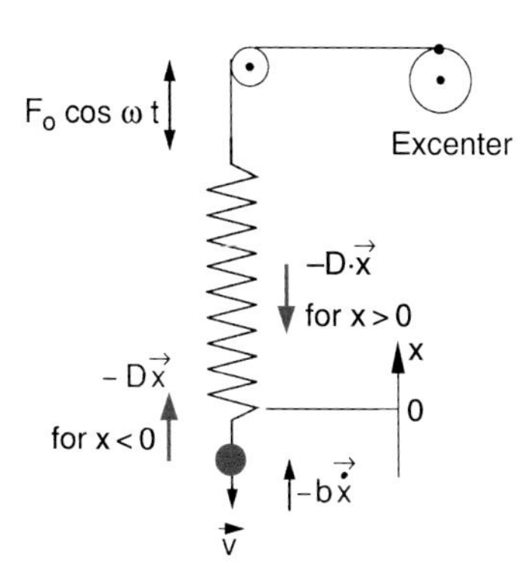

Figure 11.19 Forced oscillation

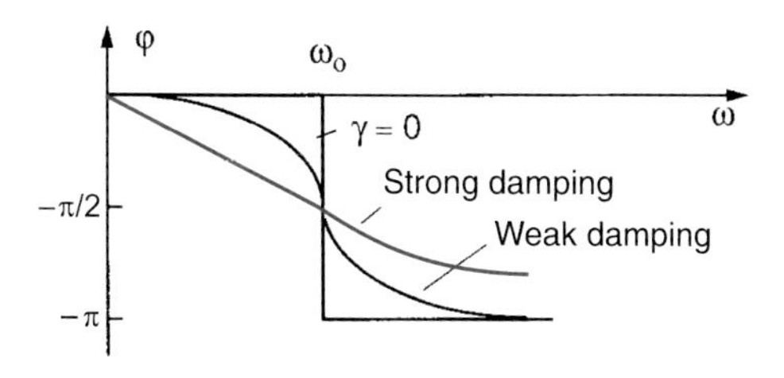

Figure 11.20 Phase shift φ between forced oscillation and exciting oscillation for different damping coefficients

11.5.1 Stationary State

We will at first discuss the stationary state of the forced oscillation, where the transient response of the damped oscillation of the free oscillator has already decayed. We make the ansatz

$$x(t) = A_2 \cdot \cos(\omega t + \varphi) , \qquad (11.23b)$$

which has the two free parameters amplitude A_2 and phase φ of the forced oscillation. Inserting (11.23b) into (11.22b) gives with the trigonometric addition theorem and after rearrangement of the summands the result

$$\begin{aligned}&\left[\left(\omega_0^2 - \omega^2\right) A_2 \cos\varphi - 2\gamma A_2 \omega \sin\varphi - K\right] \cos\omega t\\ &\quad - \left[\left(\omega_0^2 - \omega^2\right) A_2 \sin\varphi + 2\gamma A_2 \omega \cos\varphi\right] \sin\omega t = 0 .\end{aligned}$$

Since this equation must be valid for arbitrary times the two prefactors of the time dependent terms $\cos\omega t$ and $\sin\omega t$ in the cornered brackets must be identical zero. This gives the two equations

$$A_2 \left(\omega_0^2 - \omega^2\right) \cos\varphi - 2A_2 \gamma\omega \sin\varphi - K = 0 . \qquad (11.24a)$$

$$\left(\omega_0^2 - \omega^2\right) \sin\varphi + 2\gamma\omega \cos\varphi = 0 . \qquad (11.24b)$$

From (11.24b) we conclude

$$\tan\varphi = -\frac{2\gamma\omega}{\omega_0^2 - \omega^2} . \qquad (11.25)$$

The phase shift $\varphi(\omega)$ of the forced oscillation with $\gamma > 0$ against the enforcing oscillation increases for $\omega \le \omega_0$ from 0 to $-\pi/2$. For $\omega > \omega_0$ from $-\pi/2$ to $-\pi$ (Fig. 11.20). It is negative, i. e. the forced oscillation lags behind the enforcing oscillation.

The phase shift $\varphi(\omega)$ depends on the damping constant γ and on the frequency difference $\omega_0 - \omega$ between the eigen-frequency ω_0 of the free oscillator and the frequency ω of the driving force (Fig. 11.20).

For $\omega = 0$ is $\varphi = 0$. For $\omega = \omega_0$ the phase shift φ reaches the value $\varphi = -\pi/2$, where $d\varphi/d\omega$ has its maximum value and converges for $\omega \to \infty$ towards $\varphi = -\pi$.

When we solve (11.24a) for $A_2 \cos\varphi$ and $A_2 \sin\varphi$ and insert (11.24b), we obtain

$$A_2 \sin\varphi = -\frac{2\gamma\omega K}{(\omega_0^2 - \omega^2)^2 + (2\gamma\omega)^2} ,$$

$$A_2 \cos\varphi = \frac{(\omega_0^2 - \omega^2) K}{(\omega_0^2 - \omega^2)^2 + (2\gamma\omega)^2} .$$

Squaring both equations and adding the results gives

$$A_2(\omega) = \frac{F_0/m}{\sqrt{(\omega_0^2 - \omega^2)^2 + (2\gamma\omega)^2}} . \qquad (11.26)$$

Note: With a complex Ansatz for the driving force we get instead of the real equation (11.22) the complex equation for $z = x + iy$

$$\ddot{z} + 2\gamma\dot{z} + \omega_0^2 z = K \cdot e^{i\omega t} \quad \text{with} \quad z = x + iy . \qquad (11.27)$$

This equation allows a faster and more elegant way to the solution than the derivation (11.24)–(11.26). Inserting the Ansatz $z = A \cdot e^{i\omega t}$ into (11.27) gives immediately the amplitude

$$A = \frac{K \cdot (\omega_0^2 - \omega^2 - 2i\gamma\omega)}{(\omega_0^2 - \omega^2)^2 + (2\gamma\omega)^2} = a + ib = |A| \cdot e^{i\varphi} , \qquad (11.27a)$$

with the real part

$$a = \frac{K(\omega_0^2 - \omega^2)}{(\omega_0^2 - \omega^2)^2 + (2\gamma\omega)^2} = A_2 \cos\varphi , \qquad (11.27b)$$

and the imaginary part

$$b = -\frac{2K\gamma\omega}{(\omega_0^2 - \omega^2)^2 + (2\gamma\omega)^2} = A_2 \sin\varphi . \qquad (11.27c)$$

With $\tan\varphi = b/a$ this allows the direct determination of the phase φ and the real amplitude $|A| = \sqrt{a^2 + b^2}$ (Fig. 11.21).

This derivation uses the fact that for complex solutions of linear differential equations the real part as well as the imaginary part are solutions. It can be shown (see Probl. 11.3) that only the imaginary part b describes the consumption of energy, which is supported by the driving force. The real part describes the periodic acceptance and delivery of energy during the oscillation cycles.

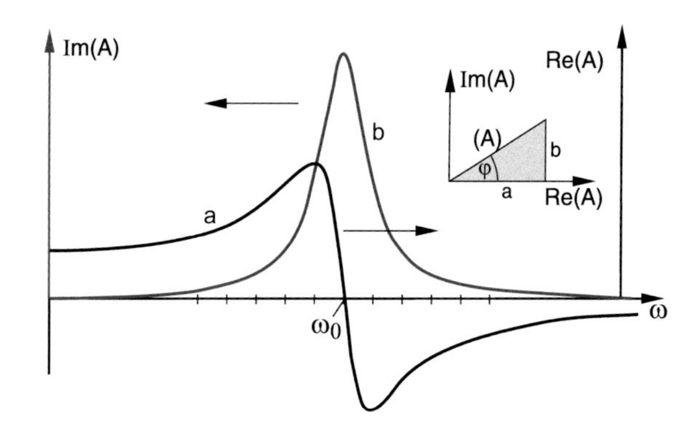

Figure 11.21 Complex representation of a forced oscillation with $a = \text{Re}(A)$ and $b = \text{Im}(A)$

The amplitude of the forced oscillation depends on

- the amplitude $K = F_0/m$ of the driving force
- the damping constant γ
- the frequency difference $(\omega_0 - \omega)$ between the eigen-frequency ω_0 of the free oscillator and the frequency ω of the driving force.

If we differentiate the radicand in (11.26) with respect to ω and set the derivative equal to zero, we obtain the minimum of the denominator, i.e., the maximum of the amplitude $A_2(\omega)$. This yields for the resonance frequency of the forced oscillator

$$\omega_R = \sqrt{\omega_0^2 - 2\gamma^2}\,. \tag{11.27d}$$

This is not exactly the resonance frequency $\omega_1 = \sqrt{\omega_0^2 - \gamma^2}$ of the free damped oscillator. The deviation is, however, small for $\gamma \ll \omega_0$.

The resonance curve $A_2(\omega)$ of the forced oscillation is shown in Fig. 11.22 for different values of the ratio γ/ω_0 of damping constant and eigen-frequency ω_0. Note, that the curves are not symmetric with respect to ω_0. and are also not centered around the resonance frequency $\omega_R = \sqrt{\omega_0^2 - \gamma^2}$ of the free damped oscillator. The asymmetry increases with increasing damping constant.

We will now determine the half width (FWHM = full width at half maximum) of the resonances. The amplitude $A(\omega)$ in (11.26) becomes maximum for the resonance frequency ω_R in (11.27d), where the denominator in (11.26) has the value $2\gamma(\omega_R^2 + \gamma^2)^{1/2}$. The amplitude decreases to 1/2 of its maximum value at the frequencies $\omega_{1,2}$, when the radicand in (11.26) takes four times the value of the radicand for the resonance frequency ω_R.

Taking into account (11.27d) we obtain the condition

$$\begin{aligned}(\omega_0^2 - \omega_{1,2}^2)^2 + (2\gamma\omega_{1,2})^2 &= 4\cdot[(\omega_0^2 - \omega_R^2)^2 + (2\gamma\omega_{1,2}^2)^2] \\ &= 16\gamma^2(\omega_R^2 + \gamma^2)\,.\end{aligned}$$

Solving for $\omega_{1,2}$ gives

$$\omega_{1,2} = \omega_R^2 \pm \sqrt{3\omega_R^2 + 3\gamma^2}\,.$$

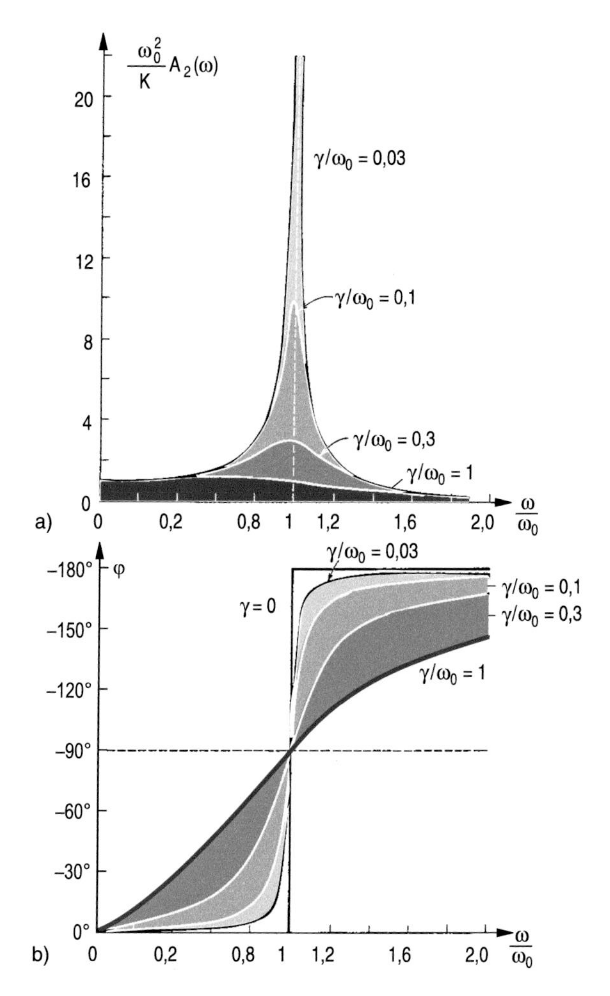

Figure 11.22 **a** Resonance curve of forced oscillation for different damping. Note the shift of the maximum with increasing damping. **b** Quantitative behaviour of $\varphi(\omega/\omega_0)$

The full half width $\Delta\omega = \omega_1 - \omega_2$ is then

$$\begin{aligned}\Delta\omega = &\left[\omega_R^2 + 2\gamma\sqrt{3\omega_R^2 + 3\gamma^2}\right]^{1/2} \\ &- \left[\omega_R^2 - 2\gamma\cdot\sqrt{3\omega_R^2 + 3\gamma^2}\right]^{1/2}\,.\end{aligned} \tag{11.27e}$$

The asymmetry of the curve $A_2(\omega)$ and the shift of the maximum becomes more obvious in Fig. 11.23, where the frequency scale is compressed in order to show a larger frequency range.

For $\gamma \ll \omega_R$ is $\omega_{1,2}^2 - \omega_R^2 = (\omega_{1,2} + \omega_R)\cdot(\omega_{1,2} - \omega_R) \approx 2\cdot\omega_R\cdot 1/2\Delta\omega$ and we can approximate the half width by

$$\Delta\omega = \frac{2\gamma}{\omega_R}\sqrt{3\omega_R^2 + 3\gamma^2} \approx 2\gamma\cdot\sqrt{3}\,. \tag{11.27f}$$

For $\gamma = 0$ is $\Delta\omega = 0$ and the amplitude is $A_2(\omega_R) = \infty$ (Resonance disaster). The damping restricts the amplitude to such finite value, where the energy supplied per second by the external force just compensates the friction losses (see Sect. 11.15).

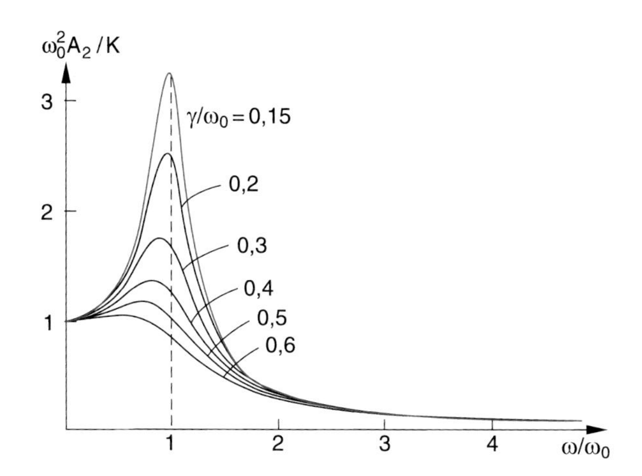

Figure 11.23 Normalized amplitude $A_2(\omega)$ of a forced oscillation for different values of γ/ω_0

11.5.2 Transient State

In the stationary phase of the forced oscillator the eigen-frequency ω_0 appears only indirectly: Although the system oscillates with the frequency ω of the external periodic force, amplitude and phase of the system depend on the difference $\omega_0 - \omega$ between eigen-frequency ω_0 and enforced frequency ω.

The situation is different during the initial transient time where the eigen-frequency has not yet decayed. The characteristic tone quality of musical instruments is mainly caused by this initial transients (see Sect. 11.15). If these transients are suppressed it will be very difficult to distinguish between the sound of the different instruments.

11.6 Energy Balance for the Oscillation of a Point Mass

The kinetic energy of the harmonic oscillator is according to (11.4e)

$$E_{\text{kin}} = \tfrac{1}{2} m\dot{x}^2 = \tfrac{1}{2} m\omega_0^2 A^2 \sin^2 \omega_0 t\,. \tag{11.28a}$$

The average over one oscillation period is

$$\langle E_{\text{kin}}\rangle = \frac{1}{T}\int_0^T \frac{1}{2} m\dot{x}^2 \mathrm{d}t = \frac{1}{4} mA^2\omega_0^2\,. \tag{11.28b}$$

For the potential energy we obtain

$$\begin{aligned} E_{\text{pot}} &= \int_0^x F\mathrm{d}x = \frac{1}{2}Dx^2 = \frac{1}{2}DA^2\cdot\cos^2\omega_0 t \\ &= \frac{1}{2}m\omega_0^2 A^2\cos^2\omega_0 t \end{aligned} \tag{11.29a}$$

with the average

$$\langle E_{\text{pot}}\rangle = \frac{1}{T}\int_0^T \frac{1}{2}Dx^2\mathrm{d}t = \frac{1}{4}mA^2\omega_0^2\,. \tag{11.29b}$$

The sum of kinetic plus potential energy

$$\begin{aligned} E_{\text{kin}}(t) + E_{\text{pot}}(t) &= \tfrac{1}{2}m\omega_0^2 A^2(\cos^2\omega_0 t + \sin^2\omega_0 t) \\ &= \tfrac{1}{2}m\omega_0^2 A^2 = E = \text{const} \end{aligned} \tag{11.29c}$$

remains for all times constant and is equal to the constant total energy.

For the harmonic oscillation the time averaged values of kinetic and potential energy are equal. They are proportional to the squares of amplitude A and frequency ω.

Example

A mass with $m = 1$ g, which oscillates with the frequency $\omega_0 = 2\pi\cdot 10^3\,\text{s}^{-1}$ and the amplitude $A = 1$ cm has a total energy $E = 1/2m\cdot\omega_0^2\cdot A^2 \approx 2\,\text{J}$. ◀

For the damped oscillation, part of the total energy $E = E_{\text{kin}} + E_{\text{pot}}$ is converted by friction into heat. This can be quantitatively derived when we multiply (11.14) on both sides by $\mathrm{d}x/\mathrm{d}t$. This gives

$$m\ddot{x}\dot{x} + Dx\dot{x} = -b\dot{x}^2\,, \tag{11.30}$$

which is equivalent to

$$\frac{\mathrm{d}}{\mathrm{d}t}\left(\frac{m}{2}\dot{x}^2 + \frac{1}{2}Dx^2\right) = -b\dot{x}^2 = -2\gamma m\dot{x}^2\,. \tag{11.31}$$

The term in the bracket is the sum of kinetic and potential energy and the right side of (11.31) gives the loss of energy per second due to friction.

Integration over one oscillation period yields the energy loss per oscillation period.

$$\begin{aligned} W &= -2\gamma m\int_0^T \dot{x}^2\mathrm{d}t \\ &= -2\gamma m\int_0^T A^2\mathrm{e}^{-2\gamma t}(\gamma\cos\omega t + \omega\sin\omega t)^2\mathrm{d}t \\ &= \frac{m}{2}A^2\left(\omega_0^2 + \gamma^2\right)\left(\mathrm{e}^{-2\gamma T} - 1\right)\,, \end{aligned} \tag{11.32}$$

where A is the initial amplitude of the first maximum at $t = 0$.

For weak damping ($\gamma \ll \omega_0$) is $\gamma \cdot T \ll 1$ and the expansion of the exponential $e^{-x} \approx 1 - x$ gives the approximate energy loss per oscillation period T

$$W \approx -mA^2 \left(\omega_0^2 + \gamma^2\right) \cdot \gamma T . \qquad (11.33)$$

The friction losses for weak damping increase linearly with γ. They are proportional to the square A^2 of the amplitude and the square ω_0^2 of the frequency. For strong damping ($\gamma \geq \omega_0$) they are proportional to γ^3. Here there is no longer a real oscillation (see Sect. 11.4.3) and therefore ω_0 is not really defined.

We get a better insight into the energy balance of the *forced oscillation*, when we multiply (11.22) with $\dot{x}$. This gives, similar to (11.31)

$$\begin{aligned} &m\ddot{x}\dot{x} + Dx\dot{x} = -b\dot{x}^2 + F(t)\dot{x} \\ \Rightarrow &\frac{\mathrm{d}}{\mathrm{d}t}\left(\frac{m}{2}\dot{x}^2 + \frac{D}{2}x^2\right) = -b\dot{x}^2 + F(t) \cdot \dot{x} . \end{aligned} \qquad (11.34)$$

The left side describes the change of the total energy (kinetic + potential energy) with time. It is reduced by the amount $-b \cdot \dot{x}^2$ due to friction and enlarged by $F(t) \cdot \dot{x}$ supplied by the external force. In the stationary state the total energy is constant, i. e. the energy loss by friction is just compensated by the energy supplied by the external force. The energy supplied by the external force is completely transferred into friction heat.

The energy, supplied per oscillation period to the system is

$$\begin{aligned} W &= \int_t^{t+T} b\dot{x}^2 \mathrm{d}t = b\omega^2 A_2^2 \int_t^{t+T} \sin^2(\omega t + \varphi)\mathrm{d}t \\ &= \frac{b}{2}\omega^2 A_2^2 T , \end{aligned} \qquad (11.35)$$

where A_2 is the amplitude of the forced oscillation (11.26) in the stationary state.

The power received by the system is with $b = 2\gamma \cdot m$

$$P = \frac{W}{T} = m\gamma\omega^2 A_2^2 . \qquad (11.36)$$

It is proportional to the damping constant γ and the squares of amplitude A and frequency ω.

Inserting for A_2 the expression (11.26) one can derive that $\mathrm{d}P/\mathrm{d}\omega = 0$ for $\omega = \omega_0$. This means that the energy fed into the oscillating system becomes maximum for the resonance case $\omega = \omega_0$.

$$P_{\max}(\omega_0) = \frac{F_0^2}{4m \cdot \gamma} \qquad (11.37)$$

The curve $P(\omega)$ of the received external power has a maximum at $\omega = \omega_0$, (Fig. 11.24) different from the oscillation amplitude, which becomes maximum for $\omega = \omega_R$ (11.27d).

Resonance phenomena play an important role in daily life. All machines and cars that are subjected to periodic forces must avoid resonance frequencies within the frequency range of the external forces. Such resonances can be extremely dangerous for bridges which are exposed to turbulent wind (see the impressive film on the breakdown of the Tacoma suspension bridge (http://en.wikipedia.org/wiki/Tacoma_Narrows_Bridge).

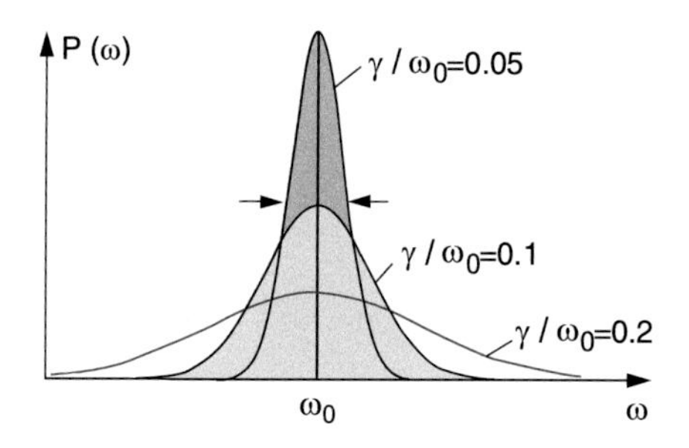

Figure 11.24 Accepted power $P(\omega)$ of the damped oscillator

11.7 Parametric Oscillator

In the equation (11.1b) of the harmonic oscillator the frequency ω_0 can be regarded as a parameter that is defined by the mass m and the spring constant $D = m \cdot \omega_0^2$.

If ω_0 is not constant but changes periodically with time, the other characteristic features (amplitude, phase) of the oscillator must also change. Such a system with periodical changes of its parameters is called a **parametric oscillator**. Its equation of motion is

$$\ddot{x} + \omega^2(t)x = 0 . \qquad (11.38)$$

A simple example is a string pendulum with a periodically changing string length $L(t)$. It can be realized when the upper end of the string is suspended by an eccentric wheel (Fig. 11.25). If frequency and phase of this periodic length change are correctly chosen, the amplitude of the oscillation can increase. If, for instance, the string length L is shortened by ΔL at the phase $\varphi = 0$ the potential energy of the pendulum increases by

$$\Delta E_1 = m \cdot g \cdot \Delta L .$$

If L is again lengthened by ΔL at the turning points $\varphi = \varphi_0$, the potential energy decreases by $\Delta E_2 = m \cdot g \cdot \Delta L \cdot \cos\varphi_0$. Since $\Delta E_2 < \Delta E_1$ the energy of the system increases. Energy is pumped into the system. Every child on a swing does that intuitively when it tries to increase the swing amplitude by periodic changes of its posture at the right phases. This causes a periodic change of the position of the centre of mass.

The periodic change of the string length L (= distance between suspension point A and centre of mass results in a corresponding change of the oscillation frequency $\omega = \sqrt{g/L}$ (see Sect. 2.9). For a quantitative description we define

$$\omega^2(t) = \omega_0^2(1 + h\cos\Omega t) . \qquad (11.39)$$

We assume, that the maximum relative frequency swing $h = (\omega^2 - \omega_0^2)/\omega_0^2 \approx 2(\omega - \omega_0)/\omega_0 \ll 1$ is small compared to 1. The equation of motion (11.38) then reduces to

$$\ddot{x} + \omega_0^2\left[1 + h\cos\Omega t\right] \cdot x = 0 . \qquad (11.40)$$

This is a Mathieu's differential equation [11.7]. The example of the swing (Fig. 11.25) illustrates that the optimum frequency of

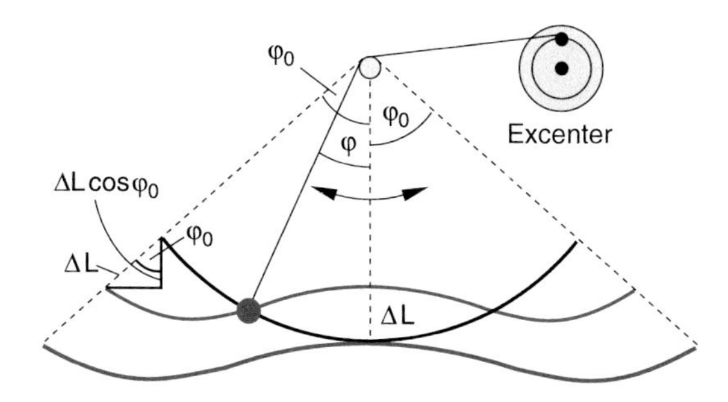

Figure 11.25 Parametric oscillator as string pendulum with periodically changing string length. The *lower red curve* represents the trajectory of the center of mass for the childrens swing, the upper red curve corresponds to the here discussed case of the string pendulum with periodic change of the string length

the energy supply should occur with twice the eigen-frequency of the swing. This means that the excitation frequency Ω should be

$$\Omega = 2\omega_0 + \varepsilon \quad \text{with} \quad |\varepsilon| \ll \omega_0 . \qquad (11.41)$$

The system then performs a forced oscillation with the frequency $\Omega/2$. For the solution of (11.40) we try, similar to (11.4c), the ansatz:

$$x = c_1(t)\cos\left(\frac{\Omega}{2}t\right) + c_2(t)\sin\left(\frac{\Omega}{2}t\right) , \qquad (11.42)$$

where now the coefficients c_i are, compared with $1/\Omega$, slowly varying functions of time. This implies that the second derivatives d^2c_i/dt^2 can be neglected.

Inserting (11.42) into (11.40) one obtains, when using the relation $\cos\alpha \cdot \cos 2\alpha = 1/2(\cos\alpha + \cos 3\alpha)$ and neglecting all terms that contain $\cos(3\Omega/2)$ or ε/ω_0

$$-\left[2\dot{c}_1 + \left(\varepsilon + \frac{h\omega_0}{2}\right)c_2\right]\sin\left(\frac{\Omega}{2}t\right) + \left[2\dot{c}_2 - \left(\varepsilon - \frac{h\omega_0}{2}\right)c_1\right]\cos\left(\frac{\Omega}{2}t\right) = 0 .$$

Since this equation must be fulfilled for all times, the expressions in the cornered brackets have to be zero (see the similar argumentation in (11.24)). This gives the result:

$$\dot{c}_1 = -\frac{1}{2}\left(\varepsilon + \frac{h\omega_0}{2}\right)c_2 ,$$
$$\dot{c}_2 = +\frac{1}{2}\left(\varepsilon - \frac{h\omega_0}{2}\right)c_1 .$$

Differentiating the first equation with respect to t and inserting for c_2 we obtain

$$\ddot{c}_1 = -\frac{1}{4}\left[\varepsilon^2 - \left(\frac{h\omega_0}{2}\right)^2\right]c_1 = -\beta^2 c_1$$

with the solution

$$c_1(t) = A \cdot e^{-i\beta t} \quad \text{with} \quad \beta = \frac{1}{2}\sqrt{\varepsilon^2 - \left(\frac{h\omega_0}{2}\right)^2} .$$

For $\beta^2 < 0 \rightarrow \varepsilon^2 < (\frac{h\cdot\omega_0}{2})^2$ the function

$$c_1(t) = A \cdot \exp\left\{\frac{1}{2}\left[\left(\frac{h\omega_0}{2}\right)^2 - \varepsilon^2\right]^{1/2} t\right\} \qquad (11.43)$$

grows exponentially with time t. This means that only within the narrow frequency range

$$2\omega_0 - \varepsilon \le \Omega \le 2\omega_0 + \varepsilon$$

the amplitude of the undamped parametric oscillator rises exponentially towards $c = \infty$. For all other frequencies, the amplitude remains finite.

For a damped oscillator with the damping constant γ the range of possible ε-values is more restricted to

$$-\sqrt{\left(\frac{h\omega_0}{2}\right)^2 - \gamma^2} < \varepsilon < +\sqrt{\left(\frac{h\omega_0}{2}\right)^2 - \gamma^2} . \qquad (11.44)$$

An amplitude increase is only possible for $h \ge 2\gamma/\omega_0$. This means that the frequency swing must have a minimum value for compensating the friction losses through the energy provided by the excitation force.

Parametric oscillators are not only significant for mechanical oscillation problems but also in the quadrupole mass spectrometer (see Vol. 3) and for optical parametric oscillators. These devices, which have been developed during the last decades, represent wavelength-tunable coherent radiation sources, which are pumped by lasers but often surpass most lasers regarding tunability over extended frequency ranges (see Vol. 3).

11.8 Coupled Oscillators

Coupled oscillating systems play an important role in physics and for technical applications. The coupling causes an energy transfer between different oscillating subsystems. If many locally separated oscillating systems are coupled, the oscillation energy can travel as mechanical wave through the total system. Without such coupling no mechanical waves can develop. In this section we will at first deal with the coupling of point masses before we discuss the more complex case of coupling between extended bodies.

11.8.1 Coupled Spring Pendulums

If two point masses m_1 and m_2 that are bound to their equilibrium positions $x_1 = 0$ and $x_2 = 0$ by springs with spring

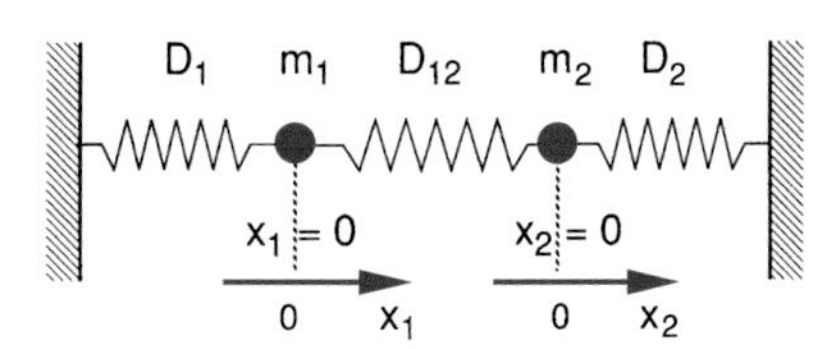

Figure 11.26 Two coupled spring pendulums

constants D_1 and D_2, are coupled to each other by a spring with D_{12} (Fig. 11.26), the elongation of the coupling spring depends on the momentary positions of the two masses. Therefore the force acting on one mass depends not only on its own position but also on the position of the other mass. The two oscillating masses are coupled to each other.

The equations of motion are

$$m_1\ddot{x}_1 = -D_1x_1 - D_{12}(x_1 - x_2) \tag{11.45a}$$

$$m_2\ddot{x}_2 = -D_2x_2 - D_{12}(x_2 - x_1)\,, \tag{11.45b}$$

where x_i is the deviations from the equilibrium position of mass m_i. The Eq. 11.45 represent a system of coupled differential equations, because each equation contains the time-dependent positions x_1 and x_2 of both masses.

A suitable transformation of the variables allows the separation of the two coupled equations. For example, for equal masses $m_1 = m_2 = m$ and equal spring constants $D_1 = D_2 = D$ addition and substraction of the two equations yields the decoupled equations

$$m(\ddot{x}_1 + \ddot{x}_2) = -D(x_1 + x_2)$$
$$m(\ddot{x}_1 - \ddot{x}_2) = -D(x_1 - x_2) - 2D_{12}(x_1 - x_2)\,.$$

With the new coordinates

$$\xi^+ = 1/2(x_1 + x_2)\,; \quad \xi^- = 1/2(x_1 - x_2)\,,$$

this gives the simple decoupled equations

$$\begin{aligned} m\cdot\ddot{\xi}^+ &= -D\cdot\xi^+ \\ m\cdot\ddot{\xi}^- &= -(D + 2D_{12})\cdot\xi^-\,. \end{aligned} \tag{11.46}$$

The general solutions are the harmonic oscillations

$$\xi^+(t) = A_1\cdot\cos(\omega_1 t + \varphi_1)$$
$$\text{with}\quad \omega_1^2 = D/m \tag{11.47a}$$
$$\xi^-(t) = A_2\cdot\cos(\omega_2 t + \varphi_2)$$
$$\text{with}\quad \omega_2^2 = (D + 2D_{12})/m\,. \tag{11.47b}$$

These harmonic oscillations of the coupled system are called **normal vibrations** and the coordinates ξ the **normal coordinates**.

For this simple example the normal coordinates $\xi^+ = 1/2(x_1 + x_2)$ and $\xi^- = 1/2(x_1 - x_2)$ give the arithmetic mean and half the difference of the local coordinates. The transformation to the normal coordinates allows the description of the coupled system as superposition of two harmonic oscillations with the frequencies ω_1 and ω_2.

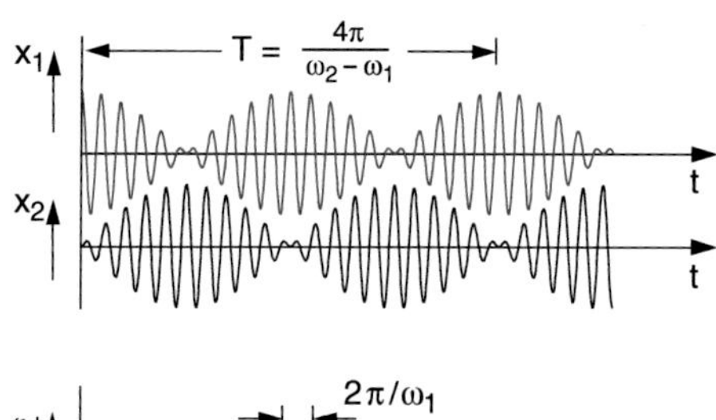

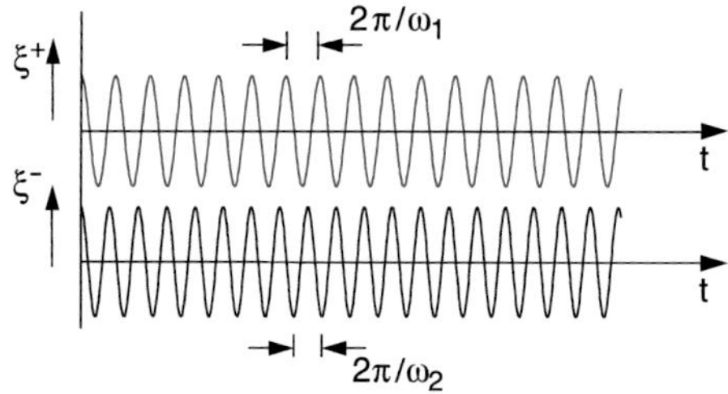

Figure 11.27 Oscillation amplitudes $x_1(t)$ and $x_2(t)$ of coupled oscillators, showing the beat period T and the two normal oscillations $\xi^+(t)$ and $\xi^-(t)$

For equal amplitudes $A_1 = A_2 = A$ the oscillations of the two masses can be described in local coordinates by back-transformation to x_i. This gives

$$\begin{aligned} x_1 &= (\xi^+ + \xi^-) = A\left[\cos(\omega_1 t + \varphi_1) + \cos(\omega_2 t + \varphi_2)\right] \\ &= 2A\cdot\cos\left(\frac{\omega_1 - \omega_2}{2}t + \frac{\varphi_1 - \varphi_2}{2}\right) \\ &\quad\cdot\cos\left(\frac{\omega_1 + \omega_2}{2}t + \frac{\varphi_1 + \varphi_2}{2}\right)\,, \end{aligned} \tag{11.48a}$$

$$\begin{aligned} x_2 &= (\xi^+ - \xi^-) = A\left[\cos(\omega_1 t + \varphi_1) - \cos(\omega_2 t + \varphi_2)\right] \\ &= -2A\cdot\sin\left(\frac{\omega_1 - \omega_2}{2}t + \frac{\varphi_1 - \varphi_2}{2}\right) \\ &\quad\cdot\sin\left(\frac{\omega_1 + \omega_2}{2}t + \frac{\varphi_1 + \varphi_2}{2}\right)\,. \end{aligned} \tag{11.48b}$$

These are the beats with the period $T = 4\pi/(\omega_1 - \omega_2)$, shown in Fig. 11.27. The maxima of the beats for x_1 are shifted against those for x_2 where the shift depends on the phases φ_1 and φ_2.

After a half cycle of the beat

$$\begin{aligned} \tau &= \frac{T}{2} = \frac{2\pi}{\omega_2 - \omega_1} \\ &= 2\pi\Big/\left(\sqrt{\frac{D + 2D_{12}}{m}} - \sqrt{\frac{D}{m}}\right)\,, \end{aligned} \tag{11.49a}$$

the oscillation energy has been transferred from m_1 to m_2. This is the time interval between two successive standstills of each oscillator i. e. when the energy of one of the masses is zero. During the full beat period T the oscillation energy is transferred from m_1 to m_2 and back to m_1.

The first square root in (11.49a) can be written as $\sqrt{D/m} \cdot \sqrt{1 + 2D_{12}/D}$. For $D_{12} \ll D$ this can be expanded according to $\sqrt{1+x} \approx 1 + x/2$, which reduces (11.49a) to

$$\tau = 2\pi \sqrt{m \cdot \frac{D}{D_{12}^2}} \,. \tag{11.49b}$$

Example

For $D_{12} = 0.1D$ the half beat period τ is 10 times the period of the free uncoupled oscillator, i. e. after 10 oscillation periods the energy of one oscillator has been completely transferred to the other (Fig. 11.27).

For special initial conditions the normal oscillations can be directly excited and pure harmonic oscillations are obtained. In order to excite $\xi^+(t)$, both oscillators must start in phase. In this case is in (11.47)

$$\xi^- = 0\,; \quad \varphi_1 = \varphi_2 = \varphi\,; \quad \text{and} \quad A_1 = A_2 \Rightarrow$$
$$x_1 = x_2 = \xi^+ = A_1 \cdot \cos(\omega_1 t + \varphi)\,. \tag{11.50a}$$

If both masses oscillate with opposite phases, i. e. $x_1 = -x_2$, we have

$$\xi^+ = 0\,; \quad \varphi_1 = \varphi_2 = \varphi\,; \quad \text{and} \quad A_1 = -A_2 \Rightarrow$$
$$x_1(t) = -x_2(t) = \xi^-(t) = A_2 \cdot \cos(\omega_2 t + \varphi) \tag{11.50b}$$

with

$$\omega_1 = \sqrt{\frac{D}{m}} \quad \text{and} \quad \omega_2 = \sqrt{\frac{D + D_{12}}{m}}\,.$$

Coupled oscillators can be demonstrated by two simple pendulums, coupled by a spring (Fig. 11.28) where two masses $m_1 = m_2 = m$ are suspended by strings with length L. The coupling spring is connected to the strings at the distance L_1 below the suspension points. The frequencies of the normal oscillations are now

$$\omega_1 = \sqrt{\frac{g}{L}}\,; \quad \omega_2 = \sqrt{\frac{g}{L} + \left(\frac{L_1}{L}\right) \cdot \frac{2D_{12}}{m}}\,. \tag{11.50c}$$

When the two string pendula are coupled by a bar (Fig. 11.29a) the frequencies are

$$\omega_1 = \sqrt{\frac{g}{L}}\,; \quad \omega_2 = \sqrt{\frac{g}{L_2}}\,. \tag{11.50d}$$

A special case of a spring-coupled pendulum is the torsion pendulum (Fig. 11.29c). A bar with two end masses at the lower end of a spring is induced to torsional oscillation. The torsion of the spring changes the length of the spring and causes up- and down oscillations. After N torsional periods the total torsional energy is transferred to the vertical oscillation. Now the transfer begins in the opposite direction until the energy of the vertical oscillation is completely transferred into torsional motion. The number N depends on the coupling strength of the spring.

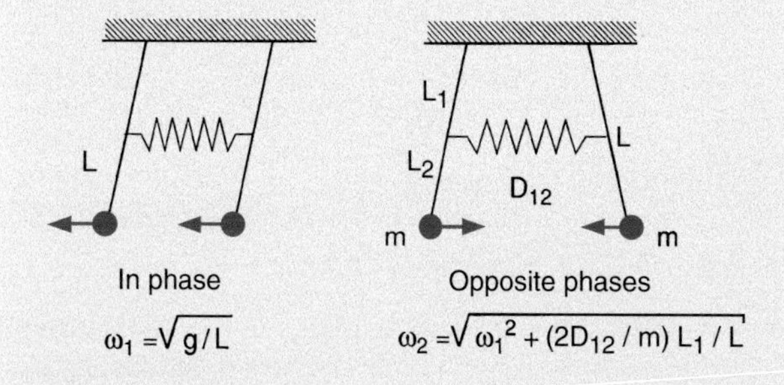

Figure 11.28 Initial conditions for the excitation of the two normal oscillations $\xi^+(t)$ and $\xi^-(t)$

Coupled oscillators can be demonstrated with several experimental setups. In Fig. 11.29 some examples are shown. The coupling between two pendulums can be realized by a spring or by a bar. In case of the spring the spring constant D_{12} determines the strength of the coupling, in case of the bar the coupling is proportional to the ratio L_1/L_2.

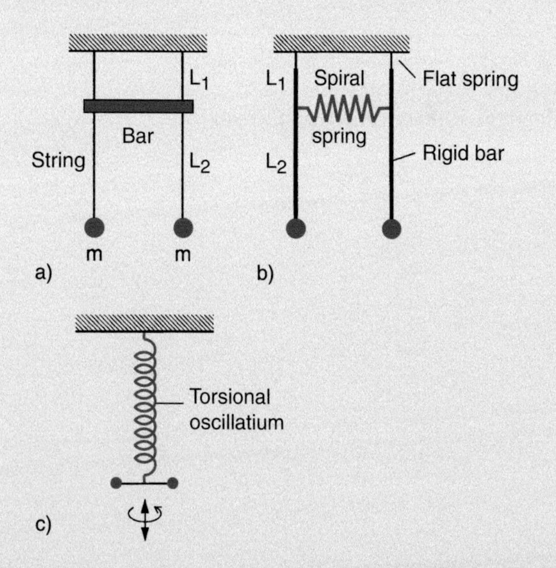

Figure 11.29 Some experimental setups for demonstrating coupled oscillators. **a** The coupling strength is given by the ratio L_1/L_2. **b** The restoring force of the spring determines the coupling strength. **c** The torsional oscillation is induced by the torsion of the oscillating spring

◀

11.8.2 Forced Oscillations of Two Coupled Oscillators

When the coupled oscillators in Fig. 11.26 are exposed to an external periodic force $F = F_0 \cos \omega t$ that acts on m_1 (Fig. 11.30a) both masses are excited to forced oscillations, due to the coupling between m_1 and m_2. The equations of motion (11.45) then become with $m_1 = m_2 = m$

$$\begin{aligned} m\ddot{x}_1 &= -D_1 x_1 - D_{12}(x_1 - x_2) \\ &\quad - 2m\gamma\dot{x}_1 + F_0 \cdot \cos \omega t \end{aligned} \tag{11.51a}$$

$$\begin{aligned} m\ddot{x}_2 &= -D_2 x_2 + D_{12}(x_1 - x_2) \\ &\quad - 2m\gamma\dot{x}_2 \,, \end{aligned} \tag{11.51b}$$

where $\gamma = b/2m$ is the damping constant.

For $\gamma = 0$ and $F_0 = 0$ we obtain Eq. 11.45 of the undamped coupled oscillators. When we introduce again the normal coordinates

$$\xi^+ = \frac{x_1 + x_2}{2} \quad \text{and} \quad \xi^- = \frac{x_1 - x_2}{2} \,.$$

We obtain after adding and subtracting the two equations for the case $D_1 = D_2 = D$ the decoupled equations

$$m\ddot{\xi}^+ = -D\xi^+ - 2m\gamma\dot{\xi}^+ + \tfrac{1}{2}F_0 \cos \omega t \,, \tag{11.52a}$$

$$m\ddot{\xi}^- = -(D + 2D_{12})\xi^- - 2m\gamma\dot{\xi}^- + \tfrac{1}{2}F_0 \cos \omega t \,. \tag{11.52b}$$

Each of these equations represents a forced oscillation. This means that the two normal oscillations can be regarded as forced oscillations with the eigen-frequencies

$$\omega_1 = \sqrt{(D/m - \gamma^2)} \,; \quad \omega_2 = \sqrt{(D + D_{12}/m - \gamma^2)} \,.$$

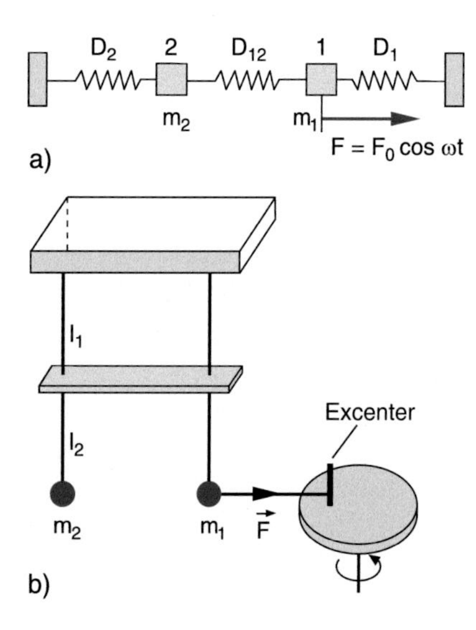

Figure 11.30 Forced oscillation of coupled oscillators. **a** Coupling by springs; **b** coupling by a bar

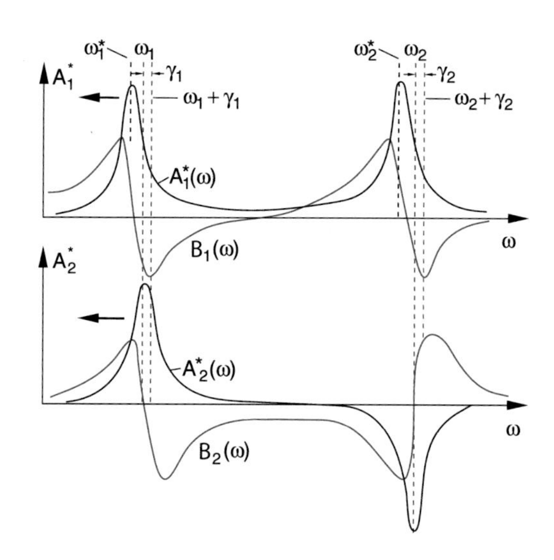

Figure 11.31 Amplitudes A_1^* (*black curves*), B_1 (*red curves*) and A_2^*, B_2 of the amplitudes of coupled forced oscillators, that are driven by the force $F_0 \cdot \cos \omega t$, acting on pendulum 1

Their amplitudes can be determined from (11.52), where for ξ^+ the eigen-frequency is $\omega = \omega_1$ and for ξ^- it is $\omega = \omega_2$.

When the exciting frequency ω is tuned one observes resonances at the frequencies $\omega_1^* = \sqrt{D/m - 2\gamma^2}$ and $\omega_2^* = \sqrt{(D + 2D_{12})/m - 2\gamma^2}$ (see (11.27d).

For the oscillations of the two pendulums in Fig. 11.30b with

$$\begin{aligned} x_1 &= \xi^+ + \xi^- = A_1 \cos \omega t + B_1 \cdot \sin \omega t \\ x_2 &= \xi^+ - \xi^- = A_2 \cos \,, \end{aligned}$$

we obtain for the representation

$$x_i = A_i \cdot \sin \varphi + B_i \cdot \cos \varphi$$

the amplitudes A_1, B_1 and A_2, B_2. Art the frequency ω_1^* only the oscillation ξ^+ contributes essentially, for ω_2^* only ξ^-.

In Fig. 11.31 the amplitudes $A_1^* = 2(\omega_1^2/K) \cdot A_1$ and $A_2^* = 2(\omega_2^2/K) \cdot A_2$, are plotted, normalized to the driving force

$$F = K \cdot e^{i\omega t} = K(\cos \omega t + i \cdot \sin \omega t) \,,$$

with $K = F_0/m$ (black curves) and $B_i^* = 2(\omega_i^2/K)B_i$ (red curves).

Such coupled forced oscillations can be demonstrated by different arrangements. One example is shown in Fig. 11.30b, where two spring pendulums are coupled by a rigid bar while m_1 is coupled to the external force driven by an excenter. The coupling strength between the two pendulums can be varied with the height of the bar and the coupling to the external force by the modulation amplitude of the excenter.

11.8.3 Normal Vibrations

In nature one finds many examples of coupled vibrations. The number N of coupled oscillations is not restricted to $N = 2$ but can be large integers. For instance in a solid crystal all N atoms are coupled by electromagnetic forces interacting between the atoms. For a crystal with a volume of $1\,\mathrm{cm}^3$ is $N = 10^{23}$. If one atom is excited to vibrations, it transfers its excitation energy to many atoms in the crystal.

We will at first consider the one-dimensional case of N atoms on a line, which are all coupled by springs with the same coupling constant as illustrated for $N = 5$ in Fig. 11.32. The possible 5 coupled oscillations are also shown.

Analogue to the Eq. 11.45 one obtains a system of 5 equations. They can be arranged in form of a matrix, where equal masses are assumed.

$$m \cdot \begin{pmatrix} \ddot{x}_1 \\ \ddot{x}_2 \\ \ddot{x}_3 \\ \ddot{x}_4 \\ \ddot{x}_5 \end{pmatrix} = \begin{pmatrix} -2D & D & 0 & 0 & 0 \\ D & -2D & D & 0 & 0 \\ 0 & D & -2D & D & 0 \\ 0 & 0 & D & -2D & D \\ 0 & 0 & 0 & D & -2D \end{pmatrix} \begin{pmatrix} x_1 \\ x_2 \\ x_3 \\ x_4 \\ x_5 \end{pmatrix} . \tag{11.53a}$$

The solutions of these equations give the frequencies ω_N ($N = 0, 1, 2, 3, 4$). The result is

$$\omega_0 = \sqrt{(2-\sqrt{3})D/m}\,; \quad \omega_1 = \sqrt{D/m}\,;$$
$$\omega_2 = \sqrt{2D/m}\,; \quad \omega_3 = \sqrt{3D/m}\,; \tag{11.53b}$$
$$\omega_4 = \sqrt{(2+\sqrt{3})D/m}\,.$$

Each of these 5 normal vibrations describes a state where all masses perform harmonic oscillations with the same frequency ω_N.

Every arbitrary oscillation of the system can be always described as a superposition of normal vibrations.

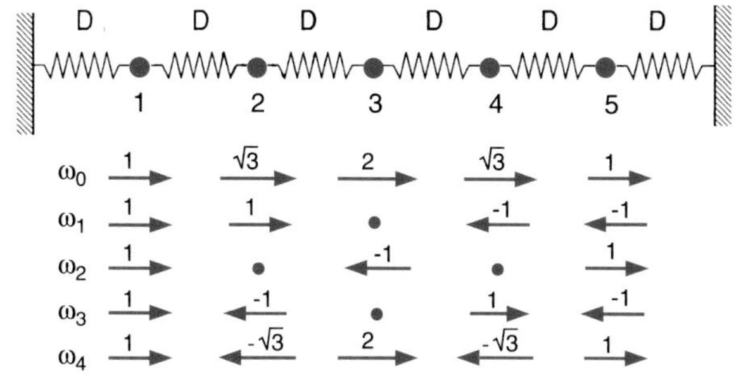

Figure 11.32 Longitudinal oscillation of a linear chain of five equal masses

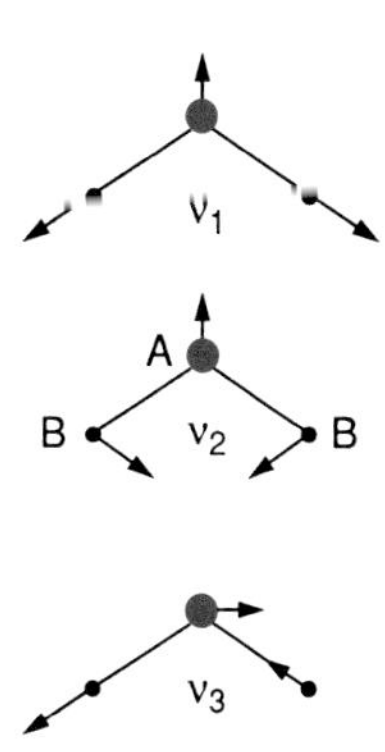

Figure 11.33 The three normal vibrations of a nonlinear triatomic molecule AB_2, where the center of mass remains at rest

Examples of oscillations where the masses do not move along a line, are the vibrations of polyatomic molecules (see Vol. 3). They can be also described as superposition of normal vibrations, which can be deduced by the following considerations.

Each of the N atoms of a molecule has 3 degrees of freedom for its motion. A molecule with N atoms therefore has $3N$ degrees of freedom. If the molecule would be a rigid body (i. e. the atoms could not vibrate) its motion could be described as a superposition of the translation of its centre of mass (3 degrees of freedom) and a rotation around this centre (3 degrees of freedom, see Sect. 5.1). For the vibrations therefore $3N - 6$ degrees of freedom are left. This is valid for nonlinear molecules. Linear molecules cannot rotate around their axis (because the moment of inertia would be extremely small and therefore its rotation energy extremely high (see Eq. 5.16b)).

In summary: Linear molecules with N atoms have $3N - 5$ degrees of freedom for their vibrations and therefore $3N-5$ normal vibrations. Nonlinear molecules have $3N-6$ vibrational degrees of freedom and $3N - 6$ normal vibrations. Since translation and rotation are already subtracted normal vibrations are those motions of the nuclear frame of the molecule, where the centre of mass does not move and the rotational angular momentum is zero.

In Fig. 11.33 the three normal vibrations of a triatomic bend molecule are shown. For all three normal vibrations the centre of mass is locally fixed and the angular momentum is zero.

11.9 Mechanical Waves

When an oscillating point mass m_1 is coupled with other neighbouring masses the oscillation energy is transported from m_1 to the neighbours. (Fig. 11.34). For a transport velocity v and a distance to the neighbouring masses Δz the energy transport needs the time $\Delta t = \Delta z/v$.

This spatial transport of oscillation energy is called a **wave**. Examples for mechanical waves are water waves, acoustic waves or pressure waves in solids, liquids or gases.

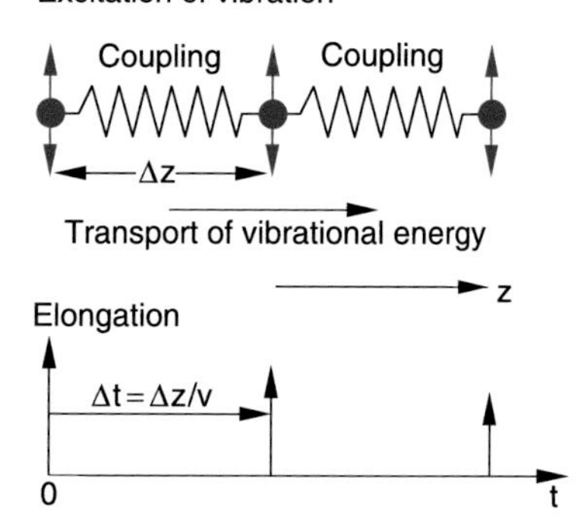

Figure 11.34 Propagation of a mechanical wave with velocity v as local transport of oscillation energy due to the coupling between neighbouring oscillators with distance Δz

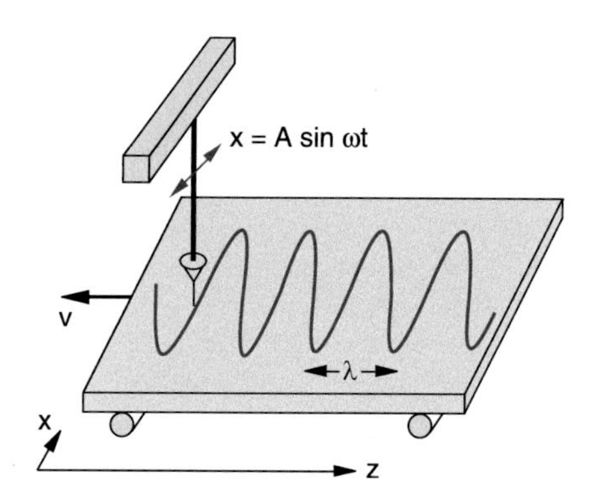

Figure 11.35 Demonstration of the relation between oscillation and wave with the help of a sand pendulum

A wave is a process, where oscillation energy is transported from the location of its generation to adjacent places where matter can oscillate. The transport is due to couplings between the excitation place and its surroundings. The wave velocity depends on the coupling strength and on the oscillating masses.

In this section we will at first discuss those waves that travel only into one direction before we deal with the more general case of wave propagation into all directions.

Remark. We will denote the displacement amplitude of the wave by the letter ξ. In the general case ξ is not necessarily the displacement of matter, but can also describe a local pressure change or for electromagnetic waves a local change of the electric field strength. All such waves can, however, be described by the Eq. 11.54 to 11.58 discussed below.

11.9.1 Different Representations of Harmonic Plane Waves

We start with a simple experiment: We pull into the $-z$-direction with constant velocity v a plate with black velvet below a sand pendulum, oscillating in the x-direction. The white sand that pours out of the pendulum writes a sine wave onto the velvet (Fig. 11.35). For an observer O sitting on the plate and moving with the velocity $-v$ into the $-z$-direction, the linear harmonic oscillation $\xi = A \cdot \sin(\omega t)$ has spread with the velocity $+v$ into the $+z$-direction. For O the elongation $\xi(t, z)$ becomes a function of time t and location z:

$$\xi(z,t) = A \cdot \sin\left[\omega \frac{t-z}{v}\right] . \tag{11.54a}$$

We call all waves that describe the propagation of harmonic oscillations as *harmonic waves*.

With the wavenumber $k = 2\pi/\lambda$ we can write (11.54a) as

$$\xi(z,t) = A \cdot \sin(\omega t - kz) . \tag{11.54b}$$

At a given time $t = t_0$ the phase $\varphi = \omega \cdot t - k \cdot z$ of the wave is equal for all locations on the plane $z = z_0$. The plane $z = z_0$ is called the *phase plane* and all waves (11.54) are *plane waves*. For plane waves propagating into the z-direction the displacement ξ does not depend on x or y, but only on z and t.

The wavelength λ of a wave is defined as the distance $\Delta z = z_1 - z_2$ between two planes $z = z_1$ and $z = z_2$ where the phases $\varphi(t_0, z_1)$ and $\varphi(t_0, z_2)$ at the same time $t = t_0$ differ by 2π. The displacements ξ are the same for both planes. For instance, λ is equal to the distance between two maxima of the wave (Fig. 11.36): The velocity of the propagation of constant phase is the *phase velocity* v_{Ph}. The relations

$$\begin{aligned} &\omega \cdot z_1/v_{\mathrm{Ph}} + 2\pi = \omega(z_1 + \lambda)/v_{\mathrm{Ph}} \\ &\Rightarrow \lambda = 2\pi \cdot v_{\mathrm{Ph}}/\omega = v_{\mathrm{Ph}}/\nu = v_{\mathrm{Ph}} \cdot T \qquad (11.55a) \\ &\text{with } \nu = \omega/2\pi = 1/T . \end{aligned}$$

The wavelength λ is the distance that the wave propagates during one oscillation period T.

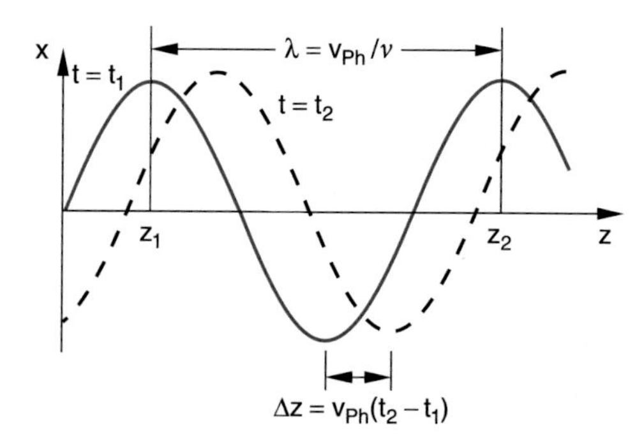

Figure 11.36 Relations between oscillation period $T = 1/\nu$, wavelength λ and phase velocity v_{Ph}

For an observer who moves with the phase velocity of the wave (he is, for example, always sitting on a maximum of the wave), the phase of the wave stays constant. For him is

$$\frac{\mathrm{d}(\omega t - kz)}{\mathrm{d}t} = 0 \Rightarrow \omega - k \cdot \frac{\mathrm{d}z}{\mathrm{d}t} = 0 \Rightarrow \frac{\mathrm{d}z}{\mathrm{d}t} = \frac{\omega}{k} .$$

The phase velocity in (11.55a) is then with $\nu = \omega/2\pi$

$$v_{\mathrm{Ph}} = \frac{\omega}{k} = \nu \cdot \lambda . \qquad (11.55\mathrm{b})$$

The phase velocity v_{Ph} is equal to the ratio of angular velocity ω and wave number k.

It depends on the coupling strength between neighbouring atoms. their density and on the mass of the oscillating atoms.

The following descriptions of plane waves propagating into the z-direction are equivalent:

$$\begin{aligned} \xi(z,t) &= A \cdot \sin(\omega t - kz) \\ &= A \cdot \sin\left[\frac{2\pi}{\lambda}(v_{\mathrm{Ph}} \cdot t - z)\right] \\ &= A \cdot \sin\left[2\pi \frac{v_{\mathrm{Ph}} \cdot t - z}{\lambda}\right] . \end{aligned} \qquad (11.56)$$

In Sect. 11.1 it was shown, that oscillations can be also described by complex functions. This means that the real part as well as the imaginary part of the complex function

$$e^{\mathrm{i}\omega t} = \cos \omega t + \mathrm{i} \cdot \sin \omega t$$

are solutions of the oscillation equation. In a similar way harmonic waves can be written in a complex form as

$$\xi(z,t) = C \cdot \mathrm{e}^{\mathrm{i}(\omega t - kz)} + C^* \cdot \mathrm{e}^{-\mathrm{i}(\omega t - kz)} . \qquad (11.57\mathrm{a})$$

This is equivalent to

$$\begin{aligned} &\xi(z,t) = A \cdot \cos(\omega t - kz) + B \cdot \sin(\omega t - kz) \\ &\text{with} \quad A = C + C^* \quad \text{and} \quad B = \mathrm{i}(C - C^*) . \end{aligned} \qquad (11.57\mathrm{b})$$

Often only the complex amplitude is written as an abbreviated version:

$$\xi(z,t) = C \cdot \mathrm{e}^{\mathrm{i}(\omega t - kz)} . \qquad (11.57\mathrm{c})$$

The real and the imaginary part are then the representations of the real wave.

11.9.2 Summary

Each of the representations (11.56), (11.57) of a harmonic wave describes:

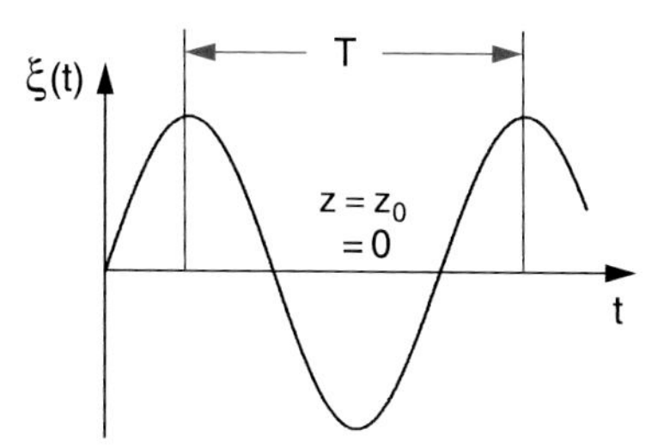

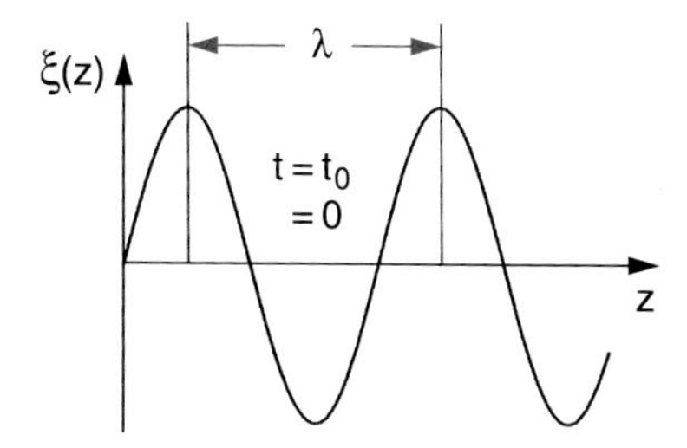

Figure 11.37 Illustration of a harmonic wave **a** as stationary oscillator $\xi(z_0, t)$ at a fixed position z_0, **b** as spatially periodic function $\xi(z, t_0)$ at a fixed time t_0

- at a fixed position $z = z_0$ a periodic harmonic oscillation

$$\xi(t) = A \cdot \sin(\omega t - kz_0) = A \cdot \sin(\omega t - \varphi) \qquad (11.58\mathrm{a})$$

  with the period $T = 2\pi/\omega$ and the phase $\varphi(t = 0) = k \cdot z_0$ (Fig. 11.37a),
- at a fixed time $t = t_0$ a spatially period function

$$\xi(z) = A \cdot \cos(\omega t_0 - kz) \qquad (11.58\mathrm{b})$$

  with the wavelength $\lambda = 2\pi/k$ and the initial phase $\varphi(z = 0) = \omega \cdot t_0$ (Fig. 11.37b).

11.9.3 General Description of Arbitrary Waves; Wave-Equation

The harmonic sine wave described so far, is only a special type of a great variety of different wave forms. We will therefore discuss a more general description of waves propagating into the z-direction. The following consideration will lead us to such a general description:

A physical quantity (for example pressure, temperature, mechanical displacement, electrical field strength etc.) should experience a local perturbation ξ as deviation from the equilibrium conditions. Due to the coupling with the neighbouring particles this perturbation will propagate in space in the course of time. We call this propagation process a wave. We will study its characteristics at first for one-dimensional waves that propagate into the z-direction. A simple example is the deflection of a string stretched in the z-direction at $t = 0$ (Fig. 11.38). In the course of time this deflection will propagate into the z-direction.

If the perturbation ξ occurs at the time $t = 0$ at the position $z = z_0$ it will propagate with the velocity v and reach at a later

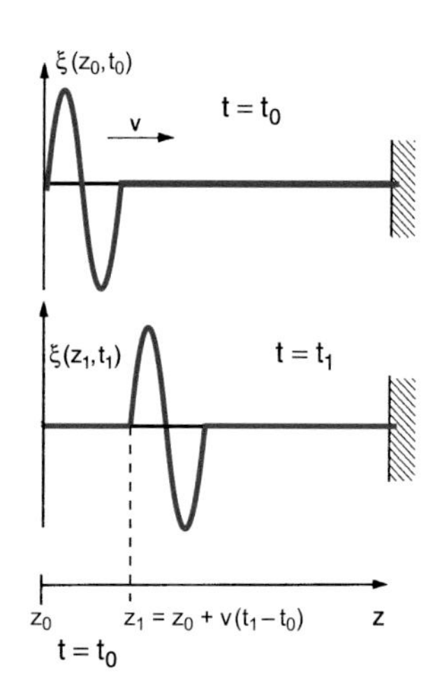

Figure 11.38 Propagation of a short pulse on a stretched string

time t_1 the position $z_1 = z_0 + v \cdot t_1$. If we assume that the form of the deviation does not change during its propagation into the z-direction, we obtain the equation for the time- and position-dependent perturbation ξ as

$$\xi(z_1, t_1) = \xi(z_1 - vt_1, 0) = \xi(z_0, 0) \,. \tag{11.59}$$

This shows that the function $\xi(z,t)$ remains constant for a constant argument $(z - v \cdot t)$, i. e. for all positions $z = v \cdot t + z_0$. We therefore can write the wave function $\xi(z,t)$ as functions of the argument $(z - v \cdot t)$ in the general form

$$\xi(z,t) = f(z - vt) \,. \tag{11.60}$$

If the function $f(z - v \cdot t)$ does not depend on x or y, the amplitude ξ is constant on the plane $z = \text{const}$ for a given time $t = t_0$.

Such a wave is called a plane wave that propagates in the $+z$-direction with the phase velocity $v = v_{\text{Ph}}$. The planes $z = \text{const}$ where the phase $(z - v \cdot t)$ of the wave function is the same for all points of this plane, is called the *equiphase surface* of the plane wave.

A plane wave propagating into the $-z$-direction can be described by the general function $f(z + v \cdot t)$.

The second derivative of Eq. 11.60 with respect to z and to t gives with the abbreviation $u = z - v \cdot t$, $\xi(z,t) = f(u)$ and $f'(u) = df/du$ the equations

$$\begin{aligned} \frac{\partial \xi}{\partial z} &= \frac{df}{du} \cdot \frac{du}{dz} = f'(u) \cdot 1 \\ \frac{\partial^2 \xi}{\partial z^2} &= \frac{d^2 f}{du^2} = f''(u) \end{aligned} \tag{11.61a}$$

$$\begin{aligned} \frac{\partial \xi}{\partial t} &= \frac{df}{du} \cdot \frac{\partial u}{\partial t} = -v \cdot f'(u) \\ \frac{\partial^2 \xi}{\partial t^2} &= \frac{d^2 f}{du^2} \cdot v^2 = f''(u) \cdot v^2 \,. \end{aligned} \tag{11.61b}$$

The comparison of (11.61a and b) gives the wave equation

$$\frac{\partial^2 \xi}{\partial z^2} = \frac{1}{v^2} \frac{\partial^2 \xi}{\partial t^2} \tag{11.62}$$

of a wave $\xi(z,t)$ that propagates with the phase velocity v into the $+z$-direction. All solutions of this equation represent possible waves. By imposing certain initial conditions, special waves are selected from the infinite number of possible solutions.

Note: The solutions of (11.62) are not necessary harmonic waves and even not periodic waves. Also short pulses $\xi(z,t)$ that propagate into the z-direction (Fig. 11.38) can fulfil the wave equation (11.62) and are therefore also called "waves".

The quantity ξ can be also the electric or magnetic field strength. Therefore (11.62) not only describes mechanical waves but also electro-magnetic waves. The phase velocity $v = c$ then becomes the velocity of light c (see Vol. 2).

11.9.4 Different Types of Waves

The different types of waves depend on the physical meaning of the quantity ξ and on the time behaviour $\xi(z_0, t)$ of the oscillation at the location z_0 where the wave is generated.

11.9.4.1 Plane Waves in z-Direction

For the harmonic wave $\xi = A \cdot \cos(\omega t - kz)$ the wave equation (11.62) becomes

$$\frac{\partial^2 \xi}{\partial z^2} = -k^2 \xi \quad \text{and} \quad \frac{\partial^2 \xi}{\partial t^2} = -\omega^2 \cdot \xi \,. \tag{11.63}$$

The phase velocity is then $v_{\text{Ph}} = \omega/k = v \cdot \lambda$, which is equal to the Eq. 11.55b.

If the quantity x describes a mechanical deviation of particles in a medium, the displacement can occur either in the z-direction of wave propagation or perpendicular to this direction. In the first case the wave is called **longitudinal**, in the second case **transversal**.

The Fig. 11.39 and 11.40 illustrate both cases by the examples of a transverse sine wave

$$\Delta \boldsymbol{x}(z,t) = A\hat{\boldsymbol{x}} \sin(kz - \omega t) \,, \tag{11.64a}$$

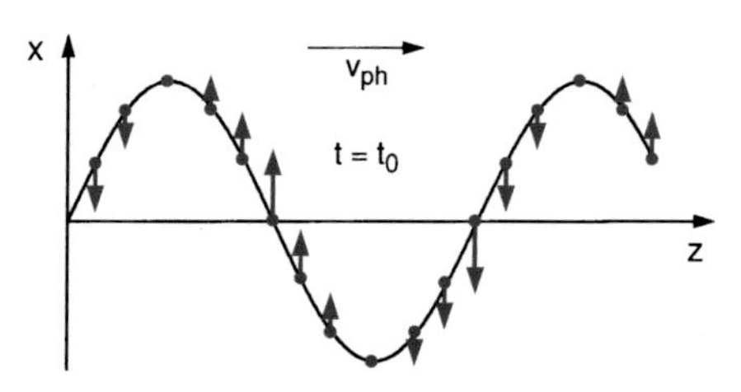

Figure 11.39 Momentary state of a transversal wave at time t_0. The *arrows* give the velocity $\dot{x}(t_0)$ of the elongation $x(t_0)$

Chapter 11

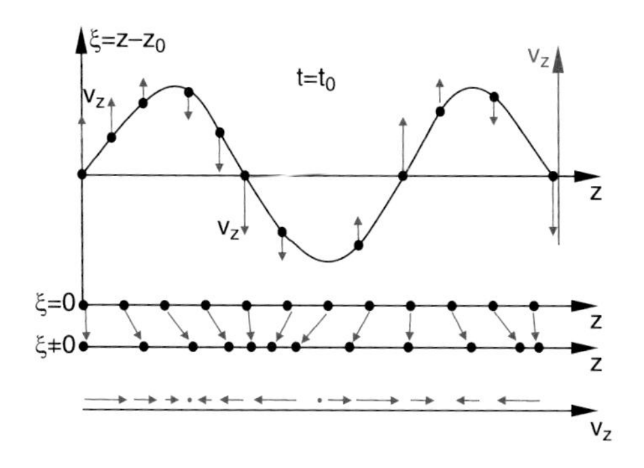

Figure 11.40 Longitudinal wave with elongation ξ of the particles and their velocities v_z (*arrows*) in the propagation direction z. The *lower part* illustrates schematically the shifts of the oscillating particles at $t = t_0$

with the amplitude $\Delta x = x - x_0$, where $\hat{x}$ is the unit vector in x-direction.

For the longitudinal wave we obtain

$$\Delta z(z, t) = B\hat{z}\sin(kz - \omega t) , \qquad (11.64b)$$

where the displacement of the oscillating particles occurs in the propagation direction of the wave.

We will see below, that the phase velocity $v_{\text{Ph}} = \omega/k$ of the different wave types generally will be different because the restoring forces that determine the oscillation frequency ω is different for shear displacements and for compression. The velocity depends, of course, also on the medium, because the coupling strength between neighbouring atoms may be different for different materials.

When the displacement of the particles in a transverse wave occurs in a fixed plane (for instance in the x-direction) the wave is linearly polarized (Fig. 11.41a). The wave (11.64a) is for example for A = const a transverse wave, linearly polarized in x-direction and propagating into the z-direction.

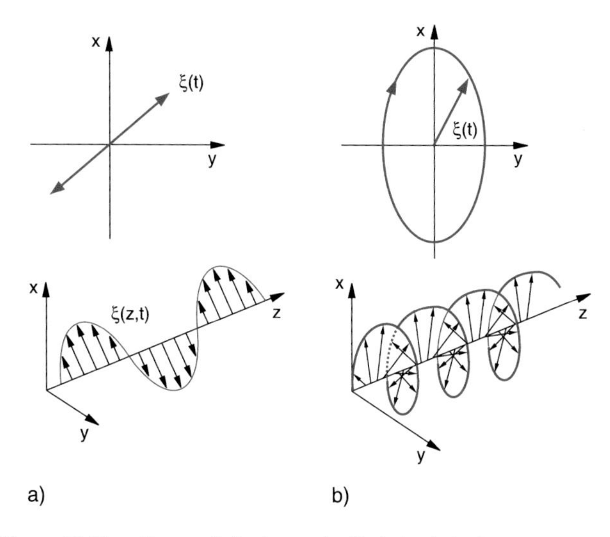

Figure 11.41 **a** Linear polarized wave; **b** elliptical polarized wave

When two oscillations

$$x = A\sin(\omega t + \varphi_1)$$
$$y = B\sin(\omega t + \varphi_2)$$

with $\varphi_1 \neq \varphi_2$ are superimposed, an elliptical oscillation in the x–y-plane is generated (see Sect. 11.3). If such an oscillation propagates into the z-direction, an elliptical polarized transverse wave emerges (Fig. 11.41b) which becomes for $A = B$ and $\varphi_1 = \varphi_2 \pm \pi/2$ a circular polarized wave. Any elliptically polarized wave can be generated by superposition of two linearly polarized waves with orthogonal polarization.

The complex representation of an elliptically polarized wave is for equal phase velocity of the two linearly polarized waves

$$\boldsymbol{\xi} = \boldsymbol{\xi}_1 + \boldsymbol{\xi}_2 = \left(A\hat{\boldsymbol{x}} + B\hat{\boldsymbol{y}}\mathrm{e}^{\mathrm{i}\Delta\varphi}\right)\mathrm{e}^{\mathrm{i}(\omega t - kz)} . \qquad (11.65)$$

The real part as well as the imaginary part give the amplitude ξ of the wave (Fig. 11.41b).

11.9.4.2 Plane Waves with Arbitrary Propagation Direction

When a plane wave propagates into an arbitrary direction we describe the propagation direction by the wave vector

$$\boldsymbol{k} = \{k_x, k_y, k_z\} . \qquad (11.66)$$

The absolute value $k = |\boldsymbol{k}| = 2\pi/\lambda$ is the wavenumber, already introduced in Sect. 11.9.1 (Fig. 11.42).

Since the equiphasic surfaces of a plane wave are planes perpendicular to the propagation direction $\boldsymbol{k}$, the position vector $\boldsymbol{r}$ of a point on this surface must obey the condition $\boldsymbol{k}\cdot\boldsymbol{r}$ = const. This can be seen as follows: For the position vectors $\boldsymbol{r}_1$ and $\boldsymbol{r}_2$ of two points on the surface the difference $r_1 - r_2$ must be a vector in the surface. This implies: $\boldsymbol{k}\cdot(\boldsymbol{r}_1 - \boldsymbol{r}_2) = 0 \rightarrow \boldsymbol{k}\cdot\boldsymbol{r}_1 = \boldsymbol{k}\cdot\boldsymbol{r}_2$ = const.

The representation of the plane harmonic wave is then

$$\boldsymbol{\xi} = \boldsymbol{A}\sin(\omega t - \boldsymbol{k}\cdot\boldsymbol{r}) , \qquad (11.67a)$$

because this ensures that for a fixed time all point on the equiphasic surface $\boldsymbol{k}\cdot\boldsymbol{r} = const$ have the same phase. The amplitude vector $\boldsymbol{A}$ is perpendicular to $\boldsymbol{k}$.

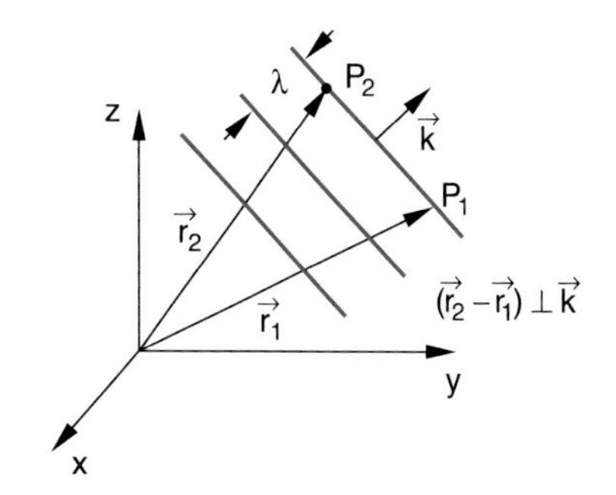

Figure 11.42 Plane wave with arbitrary propagation direction $\boldsymbol{k}$

The shorter complex description is

$$\boldsymbol{\xi} = A\mathrm{e}^{\mathrm{i}(\omega t - \boldsymbol{k}\cdot\boldsymbol{r})} \quad \text{with} \quad \boldsymbol{A} = \boldsymbol{a} + \mathrm{i}\boldsymbol{b} \,, \tag{11.67b}$$

where the amplitude of the wave is $|\boldsymbol{A}|$.

The general description of a wave that propagates into an arbitrary direction is the function

$$\boldsymbol{\xi} = Af(\omega t - \boldsymbol{k}\cdot\boldsymbol{r}) \,.$$

The three partial second derivatives are

$$\frac{\partial^2 \boldsymbol{\xi}}{\partial x^2} = Ak_x^2 \frac{\mathrm{d}^2 f}{\mathrm{d}u^2} \,, \quad \frac{\partial^2 \boldsymbol{\xi}}{\partial y^2} = Ak_y^2 \frac{\mathrm{d}^2 f}{\mathrm{d}u^2} \,,$$

$$\frac{\partial^2 \boldsymbol{\xi}}{\partial z^2} = Ak_z^2 \frac{\mathrm{d}^2 f}{\mathrm{d}u^2} \,, \quad \frac{\partial^2 \boldsymbol{\xi}}{\partial t^2} = A\omega^2 \frac{\mathrm{d}^2 f}{\mathrm{d}u^2} \,,$$

with the abbreviation $u = \omega t - \boldsymbol{k}\cdot\boldsymbol{r}$. The addition of all three derivatives yields the general wave equation

$$\Delta \boldsymbol{\xi} = \frac{1}{v^2}\frac{\partial^2 \boldsymbol{\xi}}{\partial t^2} \tag{11.68}$$

with the Laplace operator $\Delta = \partial^2/(\partial x^2) + \partial^2/(\partial y^2) + \partial^2/(\partial z^2)$ and the phase velocity $v = v_{\mathrm{Ph}} = \omega/k$.

It is easy to prove, that the special waves (11.67a, b) obey this general wave equation.

11.9.4.3 Spherical Waves

When a perturbation proceeds from a pointlike excitation source into all directions, the equiphasic surfaces must be spheres. The radial propagation directions are perpendicular to these spheres (Fig. 11.43). The description of a spherical wave is then

$$\xi(r,t) = f(r)\sin(\omega t - kr) \,, \tag{11.69}$$

where $f(r)$ is a spherical symmetric function. In (11.29) it was shown that the energy of the oscillation is proportional to the square of the amplitude. Since the energy produced in the excitation centre propagates into all directions, the energy flow through the surface $4\pi \cdot r^2$ must be independent of the radius r.

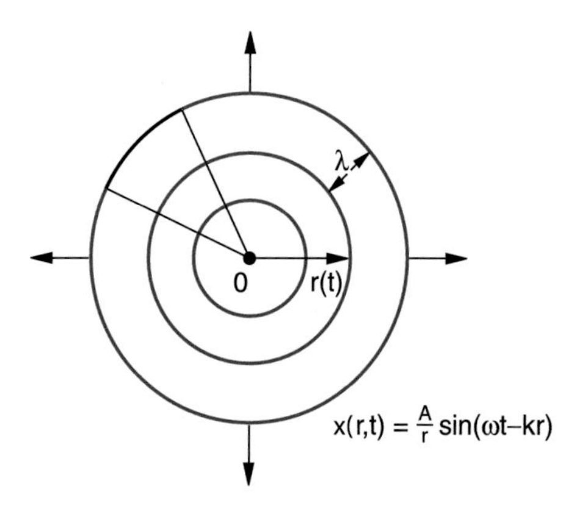

Figure 11.43 Spherical wave

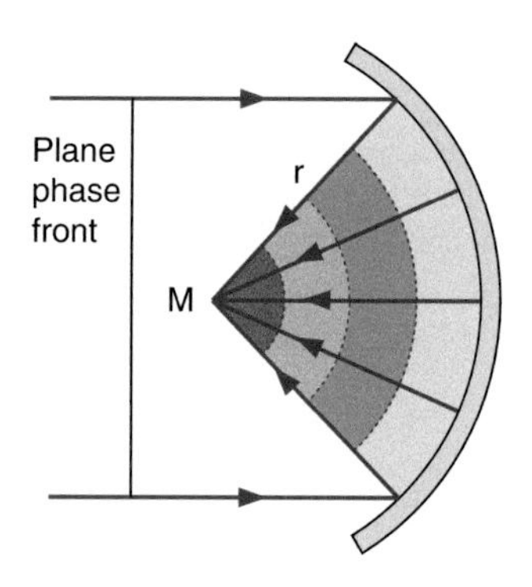

Figure 11.44 Reflection of a plane wave by a spherical mirror with focal point M and radius r

We will prove in Sect. 11.9.6 that the product $v \cdot \xi^2$ of phase velocity v and the square of the amplitude ξ is proportional to the energy flux density (this is the energy that is transported per sec through 1 m^2 of the surface). Therefore the amplitude $f(r)$ of the spherical wave must obey the condition

$$v \cdot f(r)^2 \cdot 4\pi r^2 = \text{const}$$
$$\Rightarrow f(r) \propto A/r \,.$$

The representation of a spherical harmonic wave that is excited in the centre $r = 0$ is then for all $r > 0$

$$\xi(r,t) = \frac{A}{r}\sin(\omega t - kr) \,, \tag{11.70a}$$

or in the complex notation

$$\xi(r,t) = \frac{A}{r}\mathrm{e}^{\mathrm{i}(\omega t - kr)} \,. \tag{11.70b}$$

It differs from the plane wave not only in the decreasing amplitude $A \propto 1/r$ but also in the phase $(\omega t - kr)$, because here the product kr of two scalars that is the same for all directions, replaces the scalar product $\boldsymbol{k}\cdot\boldsymbol{r} = kr \cdot \cos\vartheta$ of two vectors in the plane wave.

A spherical wave can be produced by reflecting a plane wave by a spherical mirror. The wave is then focused into the focal point M of the mirror (Fig. 11.44). The reflected wave

$$\xi(r,t) = \frac{A}{r}\mathrm{e}^{\mathrm{i}(\omega t + kr)} \tag{11.70c}$$

is a spherical wave propagating towards the focus of the mirror into the opposite direction as an outgoing spherical wave.

11.9.5 Propagation of Waves in Different Media

The mathematical description of waves, given in the previous section, is true for all kind of waves, also for electromagnetic waves with a phase velocity c that is higher by 5–6 orders of magnitude than that of mechanical waves. In this section we will show, how the phase velocity of mechanical waves depends on the characteristic properties of media. For illustration, the propagation of mechanical waves through gaseous, liquid and solid materials will be discussed.

11.9.5.1 Elastic Longitudinal Waves in Solids

As an example of a longitudinal wave we regard in Fig. 11.45 a compression wave that runs through a long solid rod with cross section A, density ϱ and elastic modulus E. The wave can be generated, for instance, by a loudspeaker attached to one end of the rod.

The particles of the rod in the layer $z = z_0$ shall have the oscillation amplitude ξ. Particles in a neighbouring layer $z = z_0 + \mathrm{d}z$ have then the amplitude

$$\xi + \mathrm{d}\xi = \xi + \frac{\partial \xi}{\delta z}\mathrm{d}z \,.$$

The longitudinal oscillation of the particles changes the thickness $\mathrm{d}z$ of the layer by the amount $(\partial\xi/\partial z)\mathrm{d}z$. Due to the resulting elastic tension this causes an elastic force $F = \sigma \cdot A$ where σ is the mechanical stress (force per unit area). According to Hooke's law σ is related to the elastic modulus E by

$$\sigma = E \cdot \frac{\partial \xi}{\partial z} \,. \tag{11.71}$$

At the right side of the layer, the elastic tension is

$$\sigma + \mathrm{d}\sigma = \sigma + \frac{\partial \sigma}{\partial z} \cdot \mathrm{d}z = \sigma + E \cdot \frac{\partial^2 \xi}{\partial z^2}\mathrm{d}z \,.$$

The net force on the volume element $\mathrm{d}V = A \cdot \mathrm{d}z$ is then

$$\begin{aligned} \mathrm{d}F &= A \cdot (\sigma + \mathrm{d}\sigma - \sigma) = A \cdot \mathrm{d}\sigma \\ &= A \cdot \frac{\partial \sigma}{\partial z}\mathrm{d}z = A \cdot E \cdot \frac{\partial^2 \xi}{\partial z^2}\mathrm{d}z \,. \end{aligned} \tag{11.72a}$$

This net force results in an acceleration $\partial^2\xi/\partial t^2$ of the mass element $\mathrm{d}m = \varrho \cdot \mathrm{d}V$, which can be obtained from Newton's equation of motion

$$\begin{aligned} \mathrm{d}F &= \mathrm{d}m \cdot \frac{\partial^2 \xi}{\partial t^2} = \varrho \cdot \frac{\partial^2 \xi}{\partial t^2}\mathrm{d}V \\ &= \varrho \cdot A\mathrm{d}z \frac{\partial^2 \xi}{\partial t^2} \,. \end{aligned} \tag{11.72b}$$

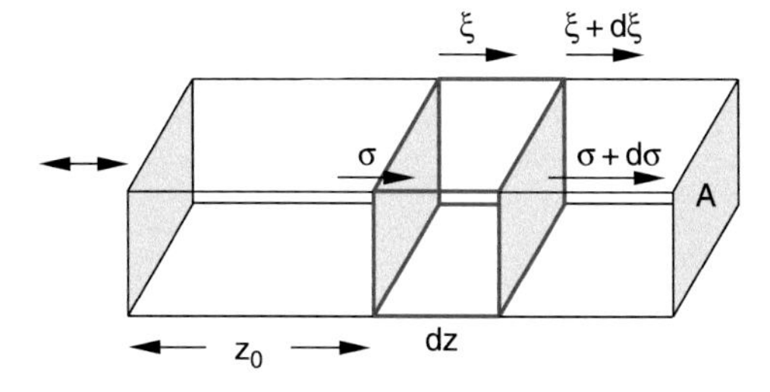

Figure 11.45 Illustration of the derivation of the wave equation (11.73)

Inserting (11.72a) gives the wave equation

$$\frac{\partial^2 \xi}{\partial t^2} = \frac{E}{\varrho}\frac{\partial^2 \xi}{\partial z^2} \,. \tag{11.73}$$

The comparison with (11.62) yields the velocity of a sound wave in a solid isotropic medium with elastic modulus E and density ϱ

$$v_{\mathrm{Ph}} = \sqrt{E/\varrho} \,. \tag{11.74a}$$

Measurements of the velocity of sound therefore allows the determination of the elastic modulus E.

The longitudinal contortions cause transverse contractions and elongations. If this is taken into account, Eq. 11.74a must be refined to

$$v_{\mathrm{Ph}} = \sqrt{\frac{E(1-\mu)}{\varrho(1+\mu)(1-2\mu)}} \,. \tag{11.74b}$$

11.9.5.2 Transverse Waves in Solids

When a transverse wave $\xi(z,t)$ propagates through a medium, neighbouring layers are shifted against each other in the direction of ξ perpendicular to the propagation direction. The coupling between neighbouring layers is caused by shear forces. The wave velocity therefore depends on the shear-modulus G of the medium (see Sect. 6.2.3). Since for ideal liquids $G = 0$ no transversal waves can propagate through the interior of frictionless liquids, while at the surface the surface tension provides transversal forces and transversal waves are therefore possible on the surface of liquids.

The derivation of the wave equation proceeds similar to that in the last section. Instead of the displacement $\mathrm{d}\xi$ in the propagation direction $\boldsymbol{k}$ we now have a displacement $\mathrm{d}\xi$ perpendicular to k against the shear force with a shear stress $\tau = G \cdot \alpha$ (see Eq. 6.11). For small displacements we obtain $\mathrm{d}\xi/\mathrm{d}z = \tan\alpha \approx \alpha$ and the phase velocity becomes

$$v_{\mathrm{Ph}} = \sqrt{G/\varrho} \,. \tag{11.75}$$

Table 11.1 Phase velocities of longitudinal and transverse acoustic waves in some isotropic solids at $T = 20\,°\mathrm{C}$

Material	$v_{\mathrm{long}}/\mathrm{m\,s^{-1}}$	$v_{\mathrm{trans}}/\mathrm{m\,s^{-1}}$
Aluminium	6420	3040
Titanium	6070	3125
Iron	5950	3240
Lead	1960	690
Optical glas	5640	3280
Flint glas	3980	2380
Nylon	2620	1070

Table 11.1 lists for some materials the phase velocity of longitudinal and transversal sound waves in isotropic media at $T = 20\,°\text{C}$. It illustrates that the velocity of longitudinal waves is higher than that of transversal waves, because $E > G$.

11.9.5.3 Sound Waves in Anisotropic Solids

If the solid body is not isotropic (for example a single crystal) the restoring forces depend on the direction (the elastic modulus becomes a tensor). and therefore the phase velocity depends on the propagation direction and for transversal waves also on the direction of the displacement ξ.

Measurements of the sound velocity in anisotropic solids gives information about the restoring forces and their dependence on the direction, i. e. on the components of the elastic tensor $\boldsymbol{E}$ [11.8, 11.9].

When we denote with F_{xy} the shear force component acting in the x-direction on a plane with its normal in the y-direction we can formulate the connection between the tensile forces (F_{xx}, F_{yy}, F_{zz}) or the shear forces (F_{xy}, F_{xz}, $F_{yx}\ldots$) and the deformations of an anisotropic elastic solid body. The deformations are defined as follows:

Before the deformation a Cartesian Coordinate system $\hat{\boldsymbol{r}} = \{\hat{\boldsymbol{x}}, \hat{\boldsymbol{y}}, \hat{\boldsymbol{z}}\}$ with the unit vector $\hat{\boldsymbol{r}}$ is defined with its origin at the point P. The deformation transfers these coordinates into new coordinates

$$x' = (1 + e_{xx})\hat{\boldsymbol{x}} + e_{xy}\hat{\boldsymbol{y}} + e_{xz}\hat{\boldsymbol{z}}$$

$$y' = e_{yx}\hat{\boldsymbol{x}} + (1 + e_{yy})\hat{\boldsymbol{y}} + e_{yz}\hat{\boldsymbol{z}}$$

$$z' = e_{zx}\hat{\boldsymbol{x}} + e_{zy}\hat{\boldsymbol{y}} + (1 + e_{zz})\hat{\boldsymbol{z}}\,,$$

which can be written as linear combinations of the old coordinates.

The diagonal components e_{xx}, e_{yy}, e_{zz} give the relative stretches, the non-diagonal components e_{xy}, e_{xz}, e_{yx}, ... the shear.

With the notations $1 = xx$, $2 = yy$, $3 = zz$, $4 = yz$, $5 = zx$, $6 = xy$ for the double indices the relation between the stretch- resp. shear forces and the deformations can be written as

$$F_k = \sum_{i=1}^{6} C_{ki} \cdot e_i \qquad (k = 1, 2, \ldots 6)\,.$$

The C_k are the components of the symmetric elasticity tensor E, written in reduced form. They can be determined by measuring the velocity of longitudinal acoustic waves in different directions $\boldsymbol{k}$ and of transversal waves for different polarizations ξ and different propagation directions $\boldsymbol{k}$.

In Fig. 11.46 the phase velocities for longitudinal- and transversal waves in a cubic crystal for 3 different propagation directions are shown. In Fig. 11.46a v_{Ph} is parallel to an edge of the cube. Here the two polarizations of the transversal waves T_1 and T_2 give the same velocity, which is, however, different from that of the longitudinal wave. In Fig. 11.46b v_{Ph} is parallel to the surface diagonal. Here T_1 and T_2 have different phase velocities

Table 11.2 Elastic constants C_{ki} in units of $10^{10}\,\text{N/m}^2$ for some single crystals with cubic symmetry

Substance	C_{11}	C_{12}	C_{44}
Aluminium	10.82	6.1	2.8
Iron	23.7	14.1	11.6
NaCl	4.9	1.24	1.26

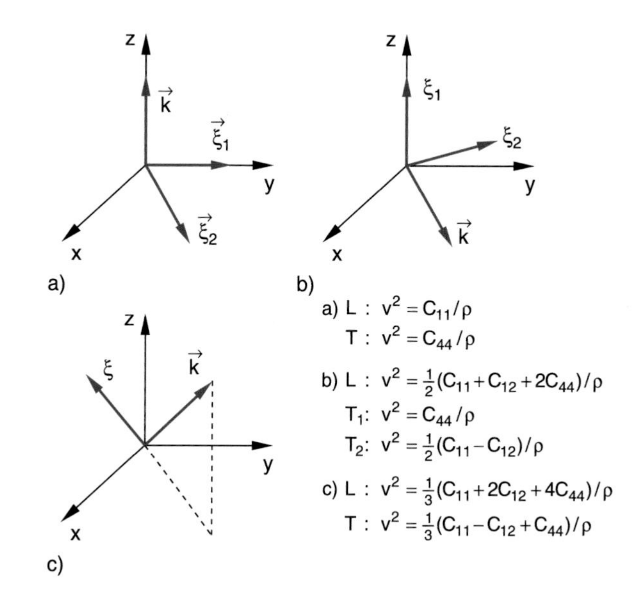

Figure 11.46 Different possible transversal waves in a cubic non-isotropic crystal with propagation in the direction **a** of the edges of the cube, **b** of the surface diagonal and **c** the space diagonal

that also differ from that of the longitudinal wave. In Fig. 11.46c v_{Ph} is parallel to the space diagonal. In Tab. 11.2 the elastic components C_{ik} of the elasticity tensor are listed for some single crystals of different materials.

11.9.5.4 Transversal Waves Along a Stretched String

When a stretched string in z-direction, that is clamped at both ends is pulled into the x-direction, a restoring force in the $-x$-direction acts on the length element $\text{d}s$ (Fig. 11.47) which is

$$\begin{aligned}\text{d}F_x &= (F \cdot \sin\vartheta)_{z+\text{d}z} - (F \cdot \sin\vartheta)_z \\ &= (F \cdot \sin\vartheta)_z + \frac{\partial}{\partial z}(F \cdot \sin\vartheta)\text{d}z - (F \cdot \sin\vartheta)_z \\ &= \frac{\partial}{\partial z}(F \cdot \sin\vartheta)\text{d}z\,,\end{aligned}$$

where F is the force in z-direction that stretches the string. For small displacements $\text{d}x$ the angle ϑ is small and we can approximate $\sin\vartheta \approx \tan\vartheta = \partial x/\partial z$. The force on $\text{d}s$ is then

$$\text{d}F_x = F \cdot \frac{\partial^2 x}{\partial z^2}\text{d}z\,.$$

With the mass density μ of the string (mass per unit length) we obtain for $\text{d}s \approx \text{d}z$ the Newton equation of motion as wave

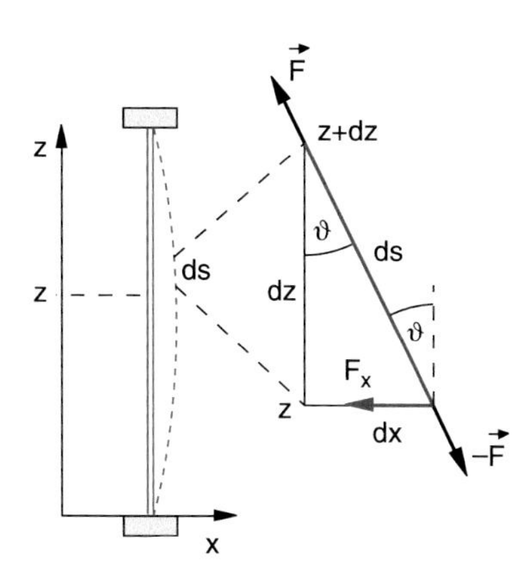

Figure 11.47 Illustration of the propagation of a transversal wave along a stretched string

equation

$$\mu \cdot \mathrm{d}z \cdot \frac{\partial^2 x}{\partial t^2} = F \cdot \frac{\partial^2 x}{\partial z^2}\mathrm{d}z\,. \tag{11.76}$$

The velocity of the transversal wave propagating along the string is then

$$v_{\mathrm{Ph}} = \sqrt{F/\mu}\,. \tag{11.77}$$

The velocity of the transverse wave depends on the force that stretches the string and on the mass density μ, but not on the elastic modulus E as for longitudinal waves or from the torsion modulus G as for transverse waves in extended solid bodies.

11.9.5.5 Sound Waves in Gases

While in solids longitudinal waves (where the elastic modulus E determines the strength of the coupling) as well as transverse waves (where the shear modulus determines the coupling) are possible, in gases only longitudinal waves can occur, because the shear modulus in gases is zero. In gas flows a viscosity (see Sect. 8.5) is indeed present that causes a coupling between neighbouring layers with different velocities. This coupling causes, however, only damping effects and there is no restoring force, because this friction force is proportional to ∇u for a flow velocity u and not to $\mathrm{d}u/\mathrm{d}t$ as for restoring forces.

For longitudinal waves the local compression leads to pressure maxima and minima. We regard in Fig. 11.48 similar to Sect. 11.9.5.1 a volume $\mathrm{d}V = A \cdot \mathrm{d}z$ which is traversed by a plane longitudinal wave in z-direction. The displacement amplitude at the position $z = z_0$ is again denoted by ξ. The displacement at $z = z_0 + \mathrm{d}z$ is then

$$\xi(z_0 + \mathrm{d}z) = \xi(z_0) + \frac{\partial \xi}{\partial z}\mathrm{d}z\,. \tag{11.78a}$$

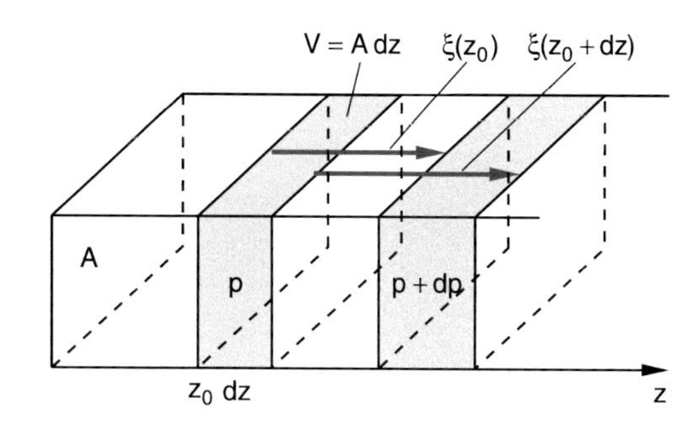

Figure 11.48 Propagation of a plane acoustic wave in gases

The volume $\mathrm{d}V$ changes then by

$$\mathrm{d}V = A \cdot \frac{\partial \xi}{\partial z}\mathrm{d}z\,. \tag{11.78b}$$

In an isothermal gas this volume change causes a pressure change

$$\mathrm{d}p = -p\frac{\mathrm{d}V}{V} = -p\frac{\partial \xi}{\partial z}\,. \tag{11.79a}$$

The force on the volume element $\mathrm{d}V$ is $F = \operatorname{grad} p \cdot \mathrm{d}V$. The net force (force in the direction of the pressure gradient) has the z-component

$$\mathrm{d}F_z = -A \cdot \mathrm{d}z \cdot \frac{\partial}{\partial z}(\mathrm{d}p) = pA \cdot \frac{\partial^2 \xi}{\partial z^2}\mathrm{d}z\,, \tag{11.79b}$$

that acts on the mass $\mathrm{d}m = \varrho \cdot \mathrm{d}V = \varrho \cdot A \cdot \mathrm{d}z$. Newton's equation of motion is therefore

$$p \cdot A \cdot \frac{\partial^2 \xi}{\partial z^2} = \varrho \cdot A \cdot \frac{\partial^2 \xi}{\partial t^2} \quad \Rightarrow \quad \frac{\partial^2 \xi}{\partial t^2} = \frac{p}{\varrho}\frac{\partial^2 \xi}{\partial z^2}\,. \tag{11.80}$$

This is the wave equation for the displacement ξ of particles in a gas with density ϱ and pressure p. The comparison with (11.62) gives the phase velocity of the longitudinal wave

$$v_{\mathrm{Ph}} = \sqrt{p/\varrho}\,. \tag{11.81a}$$

Introducing the compression modulus K the comparison of (6.9) and (11.79a) shows that the pressure p in (11.79a) can be replaced by the compression modulus K. This gives another relation for the phase velocity

$$v_{\mathrm{Ph}} = \sqrt{K/\varrho}\,. \tag{11.81b}$$

From the gas equation (7.14) we get the relation

$$\frac{p}{\varrho} = \frac{p \cdot V}{n \cdot m} = \frac{nkT}{n \cdot m} = \frac{kT}{m}\,,$$

where n is the number of molecules with mass m in the volume V. According to (7.29) the square root of the mean velocity square

$$\sqrt{\langle v^2 \rangle} = \sqrt{\frac{3kT}{m}} = \sqrt{3p/\varrho} = v_{\mathrm{Ph}} \cdot \sqrt{3} \tag{11.81c}$$

Chapter 11

Table 11.3 Velocities of sound waves in gases and liquids at $p = 1$ bar, $T = 0\,°\mathrm{C}$ und $100\,°\mathrm{C}$

Medium	$v_{Ph}/\mathrm{m\,s^{-1}}$ at 0 °C	$v_{Ph}/\mathrm{m\,s^{-1}}$ at 100 °C
Air	331.5	387.5
H_2	1284	1500
O_2	316	369
Helium	965	1127.1
Argon	319	372.6
CO_2-Gas	259	313
Water	1402	1543
Methanol	1189	
Pentane	951	
Mercury	1450	

is larger by a factor $\sqrt{3}$ than the velocity of sound (phase velocity of the wave).

Note: We have assumed that the temperature of the gas remains constant. This is, however, no longer true for sound frequencies above 1kHz. The periodic compression and expansion of the gas causes a general temperature rise. The temperature at the pressure maxima is higher than in the pressure minima. If the settlement of temperature equilibrium takes longer than the period T of the wave the heat flow from pressure maxima to the minima can be neglected (adiabatic approximation, see Sect. 10.3). With the adiabatic equation $p \cdot V^{\kappa} = \text{const}$ one obtains instead of (11.81a) the relation

$$v_{Ph} = \sqrt{\frac{p}{\varrho} \cdot \kappa}\,, \tag{11.81d}$$

where the adiabatic exponent $\kappa = C_p/C_v$ gives the ratio of the molar specific heats at constant pressure or constant volume. For air is $\kappa \approx 1.4$ (see Sect. 10.1).

The velocity of sound in gases depends on the temperature (Tab. 11.3). For constant pressure one obtains with $\varrho = \text{const}/T$ the relation

$$v_{Ph}(T) = v_{Ph}(T_0)\sqrt{T/T_0}\,. \tag{11.81e}$$

The adiabatic exponent $\kappa = (f+2)/f$ depends on the number f of degrees of freedom. For molecular gases the vibrations with higher energy can be only excited by waves with sufficient energy. Since the energy density of a wave is proportional to the square of its frequency (see Sect. 11.9.6), the index κ depends on the frequency ω of the wave. Therefore also the phase velocity depends on the frequency, i. e. **acoustic waves in molecular gases show dispersion**.

11.9.5.6 Waves in Liquids

Inside liquids only longitudinal waves can propagate, because the shear modulus of liquids is zero. The particles in a liquid can freely move without restoring forces (see Sect. 6.3.1).

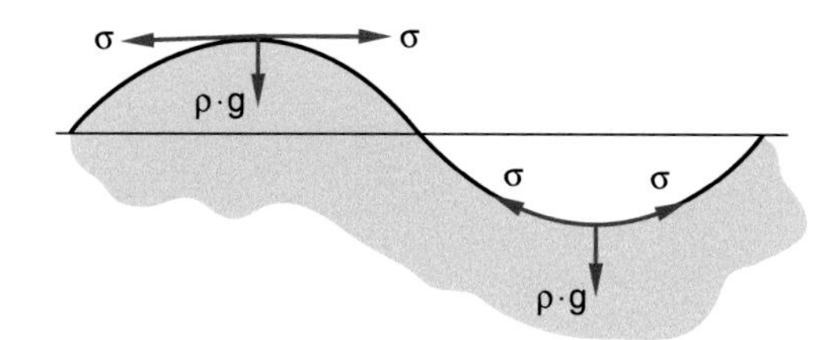

Figure 11.49 Surface wave in liquids

For the wave equation a completely analogous result is obtained as for gases. Because of the much higher compression modulus K the phase velocity

$$v_{Ph} = \sqrt{K/\varrho} \tag{11.81f}$$

is higher than in gases in spite of the higher density ϱ (see Tab. 11.3).

Note: At the surface of liquids, surface tension and gravity can act as restoring forces. Therefore, transversal surface waves are possible (Fig. 11.49).

The detailed description of surface waves and their velocities is rather elaborate. It turns out that each volume element $\mathrm{d}V$ at the surface traverses a curve that can be approximated by a circle around a fixed centre in the middle plane of the wave (Fig. 11.50). The liquid particles themselves are not transported with the wave but stay essentially locally fixed, besides its motion on the circle with a radius that equals about half the wavelength. The wave itself does not transport material but only energy.

Remark. If ocean currents are superimposed on the wave, there is, indeed a material transport, as many swimmers in the ocean have experienced.

The phase velocity of the wave depends on the surface tension, the density ϱ and the ratio h/λ of water depth h and wavelength λ. A detailed calculation proves [11.12] that

$$v_{Ph} = \sqrt{\left(\frac{g \cdot \lambda}{2\pi} + \frac{2\pi\sigma}{\varrho \cdot \lambda}\right) \cdot \tanh\left(\frac{2\pi h}{\lambda}\right)}\,. \tag{11.82}$$

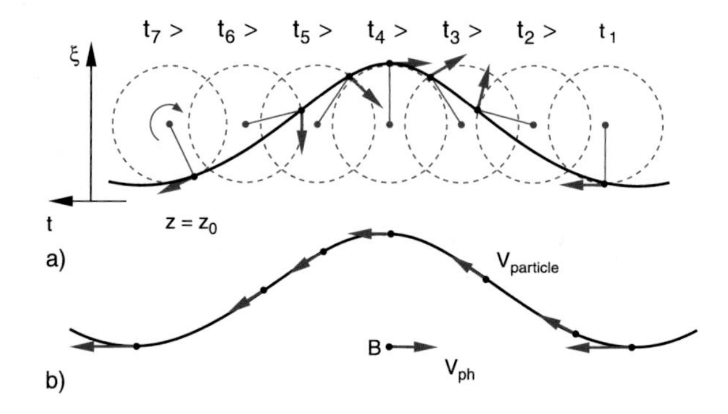

Figure 11.50 Momentary picture of the motion of liquid volume elements during an oscillation period **a** at different times at a fixed position. **b** Velocities of volume elements, measured by an observer who moves with the phase velocity of the wave, always sitting on a wave peak

Chapter 11

Water waves show dispersion, i. e. their phase velocity depends on the wavelength $\boldsymbol{\lambda}$ (Fig. 11.51).

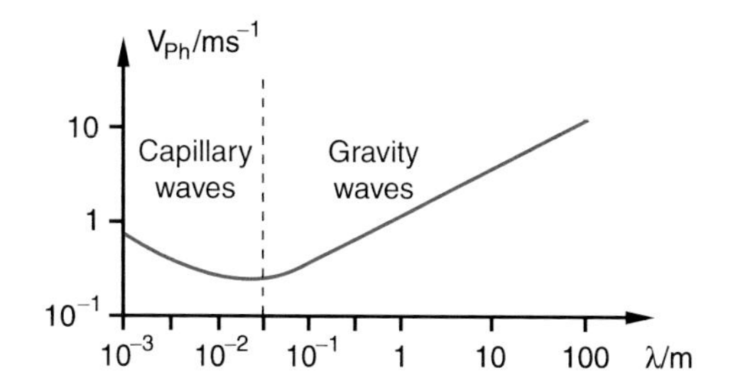

Figure 11.51 Dispersion of surface waves on water

If the first term in (11.82) is dominant, gravity represents the main part of the restoring force and the waves are called *surface gravity waves*. Their wavelength is rather large ($\lambda \gg 2\pi \cdot \sqrt{\sigma/\varrho \cdot g}$). In this case $\mathrm{d}v_{\mathrm{Ph}}/\mathrm{d}\lambda > 0$. The phase velocity increases with increasing wavelength (Fig. 11.51). For $\lambda < 2\pi \cdot \sqrt{\sigma/\varrho g} \approx 5\,\mathrm{cm}$ the second term in (11.82) is dominant and *capillary waves* occur. Because $\mathrm{d}v_{\mathrm{Ph}}/\mathrm{d}\lambda < 0$ the velocity decreases with increasing wavelength. Inserting numerical values for sea water ($\sigma = 7.3 \cdot 10^{-2}\,\mathrm{N/m}$ and $\varrho = 10^3\,\mathrm{kg/m^3}$ the function $v_{\mathrm{Ph}}(\lambda)$ has a minimum at $\lambda_{\mathrm{m}} = 2\pi \cdot \sqrt{\sigma/\varrho g} = 1.5\,\mathrm{cm}$. Capillary waves on water surfaces have wavelengths below 1 cm, while for surface gravity waves $\lambda > 10\,\mathrm{cm}$. The range between these limits is a transition range where both types superimpose.

Figure 11.52 Breaking waves which occur when the water depth becomes smaller than the wavelength. (With kind permission of Elmar Hauck, Creative Studio, Lauda-Königshofen)

11.9.6 Energy Density and Energy Transport in a Wave

In a mechanical wave mass elements $\mathrm{d}m = \varrho \cdot \mathrm{d}V$ oscillate. The kinetic and potential energy of this oscillation propagates with the wave caused by the coupling between neighbouring mass elements. **It should be again stressed that no mass is transported by the wave, but only the energy of the oscillation.**

The kinetic energy of an oscillating mass element Δm in the wave

$$\xi = A \cdot \cos(\omega t - kz)$$

is

$$E_{\text{kin}} = \tfrac{1}{2}\Delta m \cdot \dot{\xi}^2 = \tfrac{1}{2}\varrho \cdot \Delta V \cdot A^2\omega^2 \sin^2(\omega t - kz)\,.$$

The mean energy density, averaged over one oscillation period is then

$$\overline{(E_{\text{kin}}/\Delta V)} = \frac{1}{4}\varrho A^2\omega^2\,. \tag{11.83}$$

The potential energy of a mass element $\mathrm{d}m$, that oscillates with the amplitude ξ against the restoring force $F = -D \cdot \xi$ is with $D = \omega^2 \mathrm{d}m$

$$\begin{aligned} E_{\text{pot}} &= -\int_0^{\xi} F(x)\mathrm{d}x = \tfrac{1}{2}D\xi^2 \\ &= \tfrac{1}{2}DA^2\cos^2(\omega t - kz)\,. \end{aligned} \tag{11.84}$$

The time average of the potential energy per volume element ΔV is then

$$\overline{E_{\text{pot}}/\Delta V} = \tfrac{1}{4}\varrho A^2\omega^2 = \overline{E_{\text{kin}}/\Delta V}\,. \tag{11.85}$$

The total energy density $W/\Delta V = (E_{\text{pot}} + E_{\text{kin}})/\Delta V$ of the wave is therefore

$$\varrho_e = \frac{W}{\Delta V} = \frac{1}{2}\varrho A^2\omega^2\,. \tag{11.86}$$

Definition

The intensity I or energy flux density of a wave is the energy that is transported per second through a unit area perpendicular to the propagation direction of the wave.

Since the energy transport occurs with the wave velocity v_{Ph} the intensity can be written as the product

$$I = v_{\text{Ph}} \cdot \varrho_e = \tfrac{1}{2}v_{\text{Ph}} \cdot \varrho A^2\omega^2\,. \tag{11.87}$$

The intensity of a wave is proportional to the square A^2 of the wave amplitude and the square ω^2 of its frequency.

11.9.7 Dispersion; Phase- and Group-Velocity

We have seen in the previous section that for some wave types the phase velocity v_{Ph} depends on the wavelength λ of the wave (see for example Eq. 11.82). This phenomenon is called *dispersion*. The relation between v_{Ph} and wavelength λ resp. wavenumber $k = 2\pi/\lambda$

$$v_{\text{Ph}} = \frac{\omega}{k} \tag{11.88a}$$

is the dispersion relation.

A well-known example is the dispersion of light waves in glass

$$v_{\text{Ph}} = c = \frac{\omega}{k} = \frac{\omega}{k_0 n(\omega)}\,, \tag{11.88b}$$

where this relation is described by the wavelength-dependent refractive index $n(\lambda)$ of the glass and $c_0 = \omega/k_0$ is the velocity of light waves in vacuum (see Vol. 2). If a parallel ray of white light passes through a glass prism it is diffracted into the different contributions of different wavelength, which are dispersed into different directions. Behind the prism appears a band of light with separated colours (rainbow).

For monochromatic harmonic waves (only one wavelength) there is a unique phase velocity $v_{\text{Ph}} = \omega/k$ with $\omega = 2\pi\nu = 2\pi v_{\text{Ph}}/\lambda$.

The situation is different for waves with many wavelengths or with a broad continuous distribution over the interval $\Delta\lambda$, resp. the frequency interval $\Delta\omega = \omega_m \pm \Delta\omega/2$. A short pulse with time duration Δt contains a continuous frequency spectrum within the interval $\Delta\omega = 2\pi/\Delta t$. If the different frequencies propagate with different velocities the relative phases of the different contributions change and therefore the time profile of the pulse is modified.

We regard in Fig. 11.53a an arbitrary perturbation $\xi(t)$, which propagates into the z-direction. According to the Fourier-theorem (see Sect. 11.3.1) the function $\xi(t,z)$ can be described by a superposition of an infinite number of harmonic waves with frequencies ω

$$\xi(t,z) = \int_0^{\infty} A(\omega) \cdot \mathrm{e}^{\mathrm{i}(\omega t - kz)}\mathrm{d}\omega \tag{11.89a}$$

with amplitudes $A(\omega)$ that follow the frequency distribution shown in Fig 11.53b. The amplitudes $A(\omega)$ can be determined by the inverse Fourier-transformation

$$A(\omega) = \frac{1}{\pi}\int_{-\infty}^{+\infty} \xi(t,z)\mathrm{e}^{-\mathrm{i}(\omega t - kz)}\mathrm{d}t\,. \tag{11.89b}$$

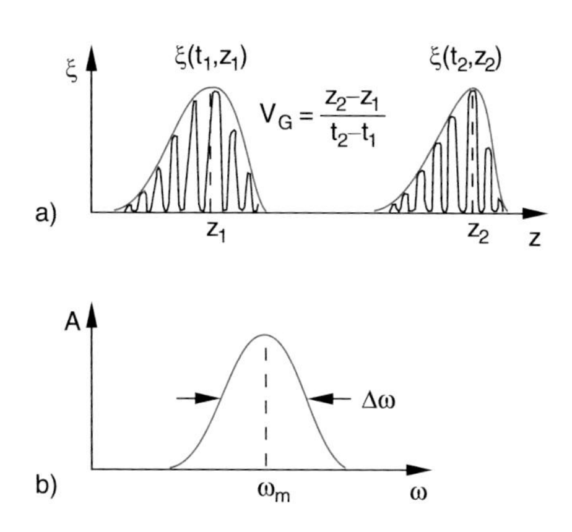

Figure 11.53 Phase velocity and group velocity. **a** Propagation of a perturbation $\xi(t)$ in z-direction; **b** Fourier-transform of the amplitudes $A(\omega)$

This superposition results only in the interval $\Delta z \approx v_{\mathrm{Ph}}/\Delta\omega$ to a significant total amplitude $\xi(t, z \pm \Delta z)$ of the perturbation. In all other spatial points, ξ averages to zero due to destructive interference of the partial waves with different frequencies.

Such a superposition of an infinite number of harmonic waves with frequencies ω within the range $\omega = \omega_{\mathrm{m}} \pm \Delta\omega/2$ is called a wave packet (sometimes also a wave group). The wave packet is characterized by its amplitude distribution $A(\omega)$, its centre frequency ω_{m} and the interval width $\Delta\omega$, which also determines its spatial extension $\Delta z \propto 1/\Delta\omega$.

The velocity of the maximum of the wave packet is the group velocity $v_{\mathrm{G}} = \mathrm{d}\omega/\mathrm{d}k$.

While the phase velocity $v_{\mathrm{Ph}}(\omega)$ can differ for the different waves in the wave group, the group velocity is unambiguously defined . We will illustrate this by a simple example:

We select out of the wave packet two harmonic waves with frequencies ω_1 and ω_2 and amplitudes $A_1 = A_2 = A$.

$$\xi_1 = A \cdot \cos(\omega_1 t - k_1 z) ,$$
$$\xi_2 = A \cdot \cos(\omega_2 t - k_2 z) .$$

Their superposition is

$$\begin{aligned}\xi &= \xi_1 + \xi_2 \\ &= 2A \cdot \cos\left(\frac{\Delta\omega}{2}t - \frac{\Delta k}{2}z\right) \cdot \cos(\omega_{\mathrm{m}} t - k_{\mathrm{m}} z) .\end{aligned} \tag{11.90}$$

This is a beat wave (Fig. 11.54a) which can be described as a wave with the mid frequency $\omega_{\mathrm{m}} = (\omega_1 + \omega_2)/2$, the wave number $k_{\mathrm{m}} = (k_1 + k_2)/2$ and an envelope of the amplitudes which propagates like a wave with the frequency $(\omega_1 - \omega_2)/2$ and a wave number $(k_1 - k_2)2$. While a selected maximum of the wave $\cos(\omega_{\mathrm{m}} t - k_{\mathrm{m}} z)$ (marked by a black dot in Fig. 11.54) propagates with the phase velocity $v_{\mathrm{Ph}} = \omega_{\mathrm{m}}/k_{\mathrm{m}}$, the maximum of the envelope (marked in Fig. 11.54 by the symbol $\oplus$) propagates with the group velocity

$$v_{\mathrm{G}} = \frac{\mathrm{d}\omega}{\mathrm{d}k} . \tag{11.91}$$

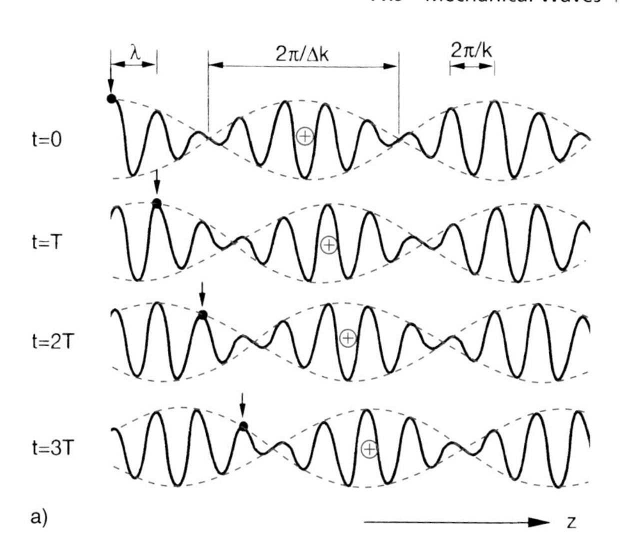

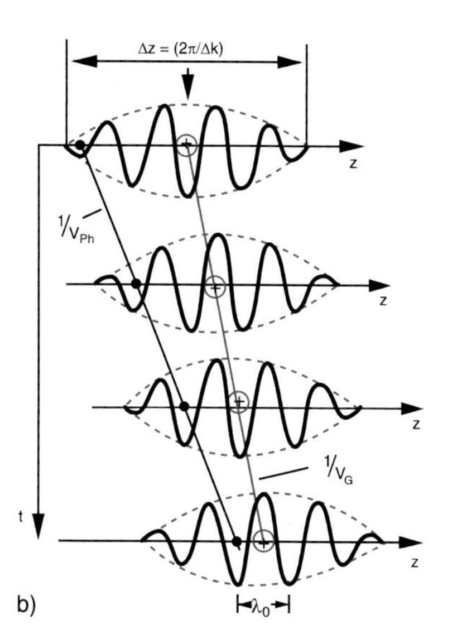

Figure 11.54 **a** Beat wave for the superposition of two monochromatic waves with slightly different frequencies. The *black point with the arrow* gives the locations of equal phase during the wave propagation. The symbol $\oplus$ indicates the maximum of the envelope of the wave packet, that moves with the group velocity $v_{\mathrm{G}} \neq v_{\mathrm{Ph}}$. **b** Different slopes of $(1/v_{\mathrm{Ph}})$ and $(1/v_{\mathrm{G}})$

In Fig. 11.54b the propagation of the phase (indicated by the black dots) is compared with that of the maximum of the wave group (indicated by the symbols $\oplus$). The slopes of the black and red lines are $\propto (1/v_{\mathrm{Ph}})$ resp. $(1/v_{\mathrm{G}})$.

The relation between phase- and group-velocities can be derived as follows:

$$v_G = \frac{d\omega}{dk} = \frac{d}{dk}(v_{Ph} \cdot k) = v_{Ph} \cdot \frac{dk}{dk} + k \cdot \frac{dv_{Ph}}{dk} = v_{Ph} + k \cdot \frac{dv_{Ph}}{dk} \,. \tag{11.92a}$$

With $k = 2\pi/\lambda$ this can be written as

$$v_G = v_{Ph} - \lambda \cdot \frac{dv_{Ph}}{d\lambda} \,. \tag{11.92b}$$

Without dispersion is $dv_{Ph}/d\lambda = 0$ and therefore $v_G = v_{Ph}$. Phase velocity and group velocity are equal. The wave packet does not change its form during the propagation.

The relation between phase- and group-velocities can be illustrated graphically by the functions $\omega(k)$ or $v_{Ph}(\lambda)$. This is demonstrated in Fig. 11.55) for the example of water waves. For small wavelength λ (capillary waves with $\lambda \ll 2\pi \cdot \sqrt{\sigma/(\varrho \cdot g)}$) we obtain from (11.82)

$$\frac{dv_{Ph}}{d\lambda} = -\frac{\pi\sigma}{\varrho \cdot \lambda^2 v_{Ph}} \left(\tanh x + \frac{2\pi h}{\lambda \tanh x}\right) \leq 0 \,.$$

For gravity waves is $\lambda \gg 2\pi \cdot \sqrt{\sigma/(\varrho \cdot g)}$ and the dispersion function is

$$\frac{dv_{Ph}}{d\lambda} = \frac{g}{2v_{Ph}} \left(\frac{1}{2\pi} \tanh x - \frac{2\pi h}{\lambda \tanh x}\right) > 0 \quad \text{for} \quad \lambda > h \quad \text{and} \quad x < 1 \,.$$

For the function $\omega(k)$ it follows from (11.82) with $v_{Ph} = \omega/k$

$$\omega(k) = \sqrt{(g \cdot k + (\sigma/\varrho)k^3)\tanh(h \cdot k)} \,, \tag{11.92c}$$

which reduces for gravity waves to

$$\omega^2 \approx g \cdot k = g \cdot 2\pi/\lambda \,. \tag{11.92d}$$

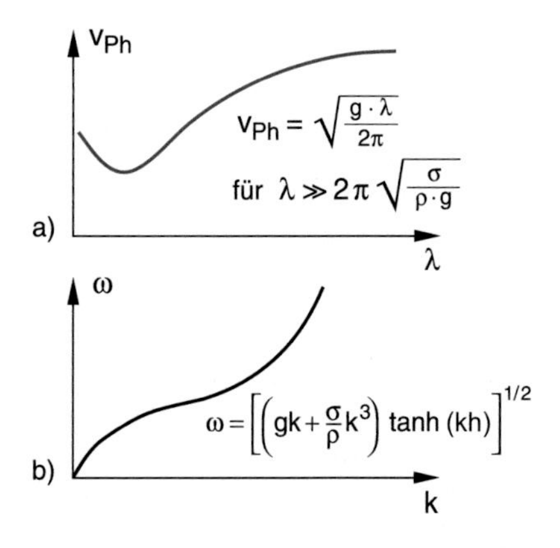

Figure 11.55 Dispersion curves for $\omega(k)$ and $v_{Ph}(\lambda)$ for surface gravity waves on water

The group velocity is

$$v_G = \frac{d\omega}{dk} = \frac{1}{2}g/\sqrt{gk} = \frac{1}{2}g/\sqrt{\tfrac{1}{2}g\sqrt{g} \cdot 2\pi/\lambda} = \frac{1}{2}v_{Ph} \,. \tag{11.92e}$$

The group velocity of water waves with large λ is equal to one half of the phase velocity.

11.10 Superposition of Waves; Interference

For linear differential equations, such as the wave equation (11.68) the following statement holds:

If $\xi_1(r,t)$ and $\xi_2(r,t)$ are solutions of the linear equation (11.68) then also every linear combination of ξ_1 and ξ_2 is a solution, in particular the sum $\xi_1 + \xi_2$. This implies for the superposition of waves: When different waves superimpose each other, their amplitudes at the same location and at the same time add. Such a superposition is called **interference**.

11.10.1 Coherence and Interference

The superposition of different waves results in a stationary wave field with visible interference structures only if some essential conditions are fulfilled:

- All waves must have the same frequency, because otherwise beats would occur that wash out in the time average all interference structures.
- The phase differences between all partial waves at the same position $\boldsymbol{r}$ must be constant in time. It can differ, of course, for different positions $\boldsymbol{r}$. Such waves are called **spatially coherent**. The superposition of coherent partial waves result in a stationary wave field, which generally changes with the position $\boldsymbol{r}$. Stationary interference structures can be only observed for the superposition of coherent waves.
  There are two experimental possibilities to produce coherent partial waves
  a) Two oscillators Q_1 and Q_2 with equal frequencies at two different positions are coupled to each other with constant phase difference (Fig. 11.56a). This results in a temporally constant phase difference between the waves emitted by Q_1 and Q_2 at each position $\boldsymbol{r}$.
  b) The wave emitted by one source Q is divided by reflection or diffraction into two partial waves which are subsequently superimposed after having traversed paths s_1 and s_2 with different lengths (Fig. 11.56b). The phase difference $\Delta\varphi = (2\pi/\lambda) \cdot \delta s$ between the two partial waves depends on the path difference $\delta s(\boldsymbol{r}) = s_1 - s_2$ which is different for different observation points $\boldsymbol{r}_i$.
  For acoustic waves, the case a) can be realized with two phase-locked loud speakers. For optical waves, this is only

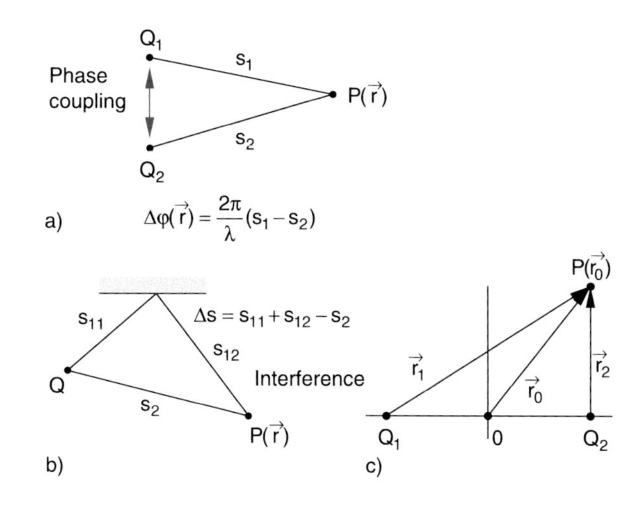

Figure 11.56 Two possibilities for the generation and superposition of coherent waves; **a** phase-coupled sources; **b** splitting of the wave emitted by a single source and superposition of the two partial waves after they have transversed different path length Δs; **c** general case

possible with lasers (see Vol. 3) because a phase-locked coupling is only possible between coherent light sources but not between two incoherent light sources, such as light bulbs or gas discharge lamps where the excited atoms emit the radiation with randomly varying phases. With "classical" incoherent light sources only case b) can be realized for producing stationary interference patterns.

11.10.2 Superposition of Two Harmonic Waves

The superposition of two harmonic waves emitted from two phase-locked sources Q_1 and Q_2 givers at a fixed position $P(r_0)$ the total amplitude (Fig. 11.56c)

$$\begin{aligned}\xi &= \xi_1 + \xi_2\\ &= A_1\cos(\omega t - \boldsymbol{k}_1\cdot\boldsymbol{r}_0 + \varphi_{01})\\ &\quad + A_2\cos(\omega t - \boldsymbol{k}_2\cdot\boldsymbol{r}_0 + \varphi_{02})\,,\end{aligned} \tag{11.93a}$$

where φ_{01} and φ_{02} are the phases of the two partial waves at the location of the sources at the time $t = t_0$.

With $\varphi_i = \boldsymbol{k}_i\cdot\boldsymbol{r}_0 - \varphi_{0i}$ and $\Delta\varphi = \varphi_1 - \varphi_2$ we obtain

$$\begin{aligned}\xi &= A_1\cos(\omega t - \varphi_1) + A_2\cos(\omega t - \varphi_2)\\ &= C\cos(\omega t - \varphi)\\ &= C(\cos\omega t\cos\varphi + \sin\omega t\sin\varphi)\,.\end{aligned} \tag{11.93b}$$

Comparison of the coefficients in a) and b) yields

$$\begin{aligned}C\cdot\cos\varphi &= A_1\cos\varphi_1 + A_2\cos\varphi_2\\ C\cdot\sin\varphi &= A_1\sin\varphi_1 + A_2\sin\varphi_2\,.\end{aligned}$$

Squaring and addition of the two equations gives the equation for the coefficient C and the phase φ

$$\begin{aligned}C &= \left[A_1^2 + A_2^2 + 2A_1A_2\cos\Delta\varphi\right]^{1/2}\,,\\ \tan\varphi &= \frac{A_1\sin\varphi_1 + A_2\sin\varphi_2}{A_1\cos\varphi_1 + A_2\cos\varphi_2}\,.\end{aligned}$$

The superposition of the two waves results again in a harmonic wave with an amplitude C that depends on the amplitudes of the partial waves and their phase difference $\Delta\varphi$. For the phase difference $\Delta\varphi = 2m\cdot\pi(m = 1, 2, 3, \ldots)$ the total amplitude is $C = A_1 + A_2$ (constructive interference). For $\Delta\varphi = (2m+1)\pi$ is $C = A_1 - A_2$ (destructive interference).

The intensity of a wave is proportional to the square of its amplitude:

$$\begin{aligned}I &\propto (\xi_1 + \xi_2)^2 = \xi_1^2 + \xi_2^2 + 2\xi_1\xi_2\\ &= A_1^2\cos^2(\omega t + \varphi_1) + A_2^2\cos^2(\omega t + \varphi_2)\\ &\quad + 2A_1A_2\cos(\omega t + \varphi_1)\cdot\cos(\omega t + \varphi_2)\,.\end{aligned} \tag{11.94}$$

If the period $T = 2\pi/\omega$ is short compared to the detection time the detector measures the time average over the period T. With $\langle\cos^2 x\rangle = 1/2$ and $\cos x\cdot\cos y = 1/2[\cos(x+y) + \cos(x-y)]$ we obtain

$$\begin{aligned}&2\cdot\cos(\omega t + \varphi_1)\cdot\cos(\omega t + \varphi_2)\\ &\quad = \cos(2\omega t + \varphi_1 + \varphi_2) + \cos(\varphi_1 - \varphi_2)\,.\end{aligned}$$

Since the time average $\langle\cos(2\omega t + \varphi_1 + \varphi_2)\rangle = 0$, we obtain the time averaged intensity

$$\langle I\rangle = \frac{1}{2}\left(A_1^2 + A_2^2\right) + A_1A_2\cos\Delta\varphi\,. \tag{11.95}$$

For coherent waves the phase difference $\Delta\varphi = (k_1 - k_2)\cdot r$ has at any location r a constant value for all times. The intensity is therefore only a function of r. it varies for constant $\Delta\varphi$ between $\langle I\rangle = 1/2(A_1 - A_2)^2$ for $\Delta\varphi = (2m+1)\pi$ (interference minima) up to $\langle I\rangle = 1/2(A_1 + A_2)^2$ for $\Delta\varphi = 2\,\mathrm{m}\cdot\pi$ (interference maxima) (Fig. 11.57).

For incoherent waves is $\Delta\varphi(t)$ a randomly varying function of time. Therefore, the time average $\langle r\cdot\cos\Delta\varphi\rangle = 0$ for all locations r. This implies that no stationary interference structures are observed. The average intensity $\langle I\rangle = 1/2(A_1 + A_2)^2$ is equal to the sum of the average intensities of the partial waves.

The average total energy of the wave field must be, of course, equal for both cases, because no energy is lost. In case of the coherent superposition the energy is non-uniformly distributed in space, while for the incoherent case it is equally spread over the whole interference region. For the coherent superposition, the energy density in the interference maxima is larger than for the incoherent case, while for the minima it is smaller.

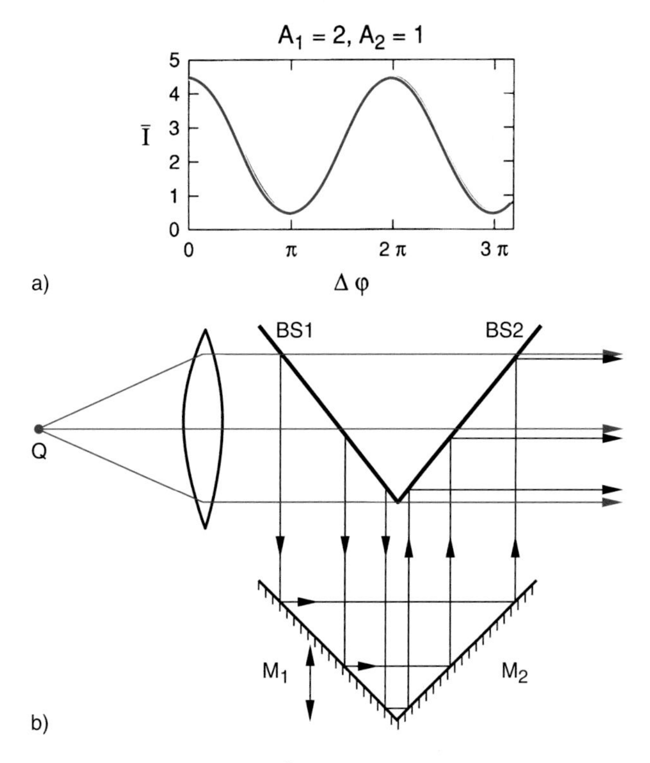

Figure 11.57 **a** Mean intensity $\bar{I}$ of the superpostion of two coherent waves as a function of their phase different $\Delta\varphi$. **b** Experimental arrangement for the interference of copropagation plane waves

Examples

1. *Superposition of plane waves with equal frequencies and equal amplitudes but different phases, propagating both into the z-direction.*

$$\xi_1 = A \cdot \cos(\omega t - kz) ;$$

$$\xi_2 = A \cdot \cos(\omega t - kz + \varphi)$$

Their superposition gives the total amplitude

$$\begin{aligned} \xi &= \xi_1 + \xi_2 = 2A \cdot \cos(\varphi/2) \\ &\cdot \cos(\omega t - kx + \varphi/2) \qquad (11.96) \\ &= B(\varphi) \cdot \cos(\omega t - kx + \varphi/2) . \end{aligned}$$

This is again a harmonic plane wave with a phase that is equal to the mean value $\langle\varphi\rangle = (1/2)\,\varphi$ and an amplitude $B(\varphi)$ that depends on the phase difference $\Delta\varphi$. For $\Delta\varphi = (2m+1)\pi$ the total amplitude ξ is zero in the whole superposition region. *Where has the energy gone?*
Figure 11.57b shows an experimental arrangement that can realize the superposition described above (Mach-Zehnder Interferometer). The wave, emitted by the source Q is divided by the beam splitter $BS\,1$ into a reflected partial wave and a transmitted wave. The reflected wave is again reflected by the mirrors M_1 and M_2 and is then superimposed by beam splitter $BS\,2$ onto the first transmitted partial wave. The phase difference between the two partial waves can be adjusted by moving the mirrors up or down, thus changing the path length difference between the two interfering waves.

2. *Superposition of two spherical waves, emitted by the sources Q_1 and Q_2* (Fig. 11.58).
At the point $P(r)$ the phase difference is $\Delta\varphi = k \cdot (r_1 - r_2)$. The interference maxima are located on the curves for which $\Delta\varphi = k \cdot (r_1 - r_2) = 2m \cdot \pi$. This are hyperbolas (in Fig. 11.58 marked by dashed red curves), where the two sources Q_1 and Q_2 are the focal points.

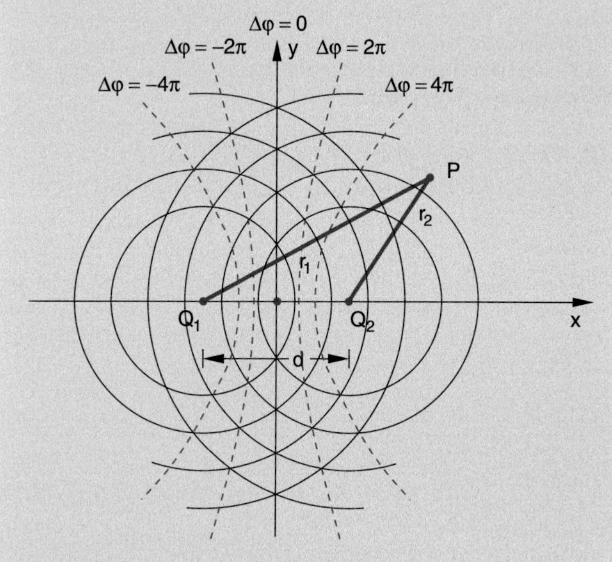

Figure 11.58 Superposition of the spherical waves emitted by two phase-coupled sources Q_1 and Q_2 ◀

When spherical waves, emitted by two phase-coupled sources, are superimposed, one observes hyperbolas as interference structures. The form of the hyperbolas depends on the distance d between Q_1 and Q_2.

11.11 Diffraction, Reflection and Refraction of Waves

The propagation direction and the local amplitude distribution can be altered by reflection on surfaces, by refraction in media with a different refractive index or by diffraction at obstacles such as slits, apertures, or small objects in the path of the wave. All these processes can be described by *Huygen's principle*, which was postulated by the Dutch scientist Christian Huygens around 1600.

11.11.1 Huygens's Principle

The propagation of waves in space can be described when each point of a phase surface is considered as the source of a new spherical wave (secondary wave or elementary wave). These spherical waves propagate into all directions and superimpose (Fig. 11.59a). Assume that all elementary waves on a phase surface $\varphi(r_0, t_0)$ are generated at the same time t_0. They proceed during the time Δt over a distance $r = v_{\mathrm{Ph}} \cdot \Delta t$. The tangent surface as the envelope of all spherical waves at the time $t = t_0 + \Delta t$ forms again a phase surface of the wave which has propagated in space by the distance $r = v_{\mathrm{Ph}} \cdot \Delta t$.

This principle that has been formulated more than 400 years ago, can be explained by modern concepts of atomic physics. In case of sound waves which can propagates only in matter but not in vacuum, all atoms on a phase surface are excited by the acoustic wave to oscillate in phase. They transfer their oscillation energy to neighbouring atoms due to their mutual coupling. The oscillating atoms are the sources for the elementary waves. In isotropic media, the coupling is independent of the direction. This means that the phase velocity of the elementary waves is isotropic. Therefore, spherical waves are generated. In vacuum, there are no atoms, and therefore no sources for acoustic waves are present. This can be experimentally demonstrated by an electric alarm clock in a container that is evacuated. As soon as the air pressure in the container drops below a certain value, the alarm clock can no longer be heard.

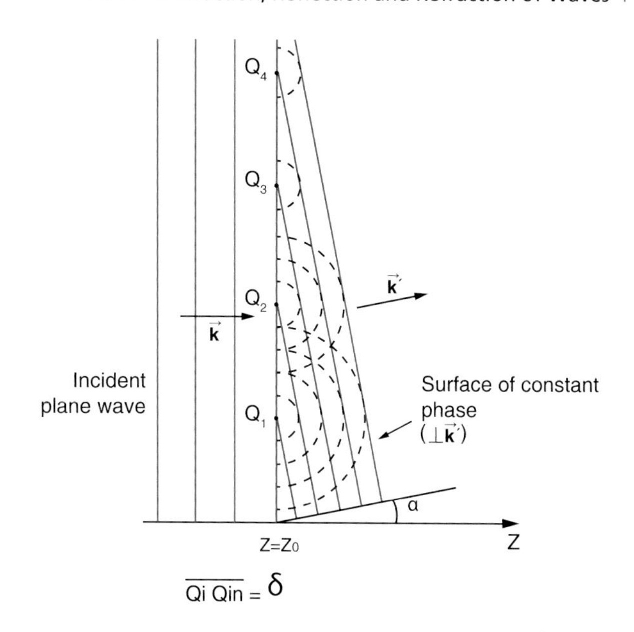

Figure 11.60 Superposition of the spherical waves emitted by N sources located on a plane $z = $ const

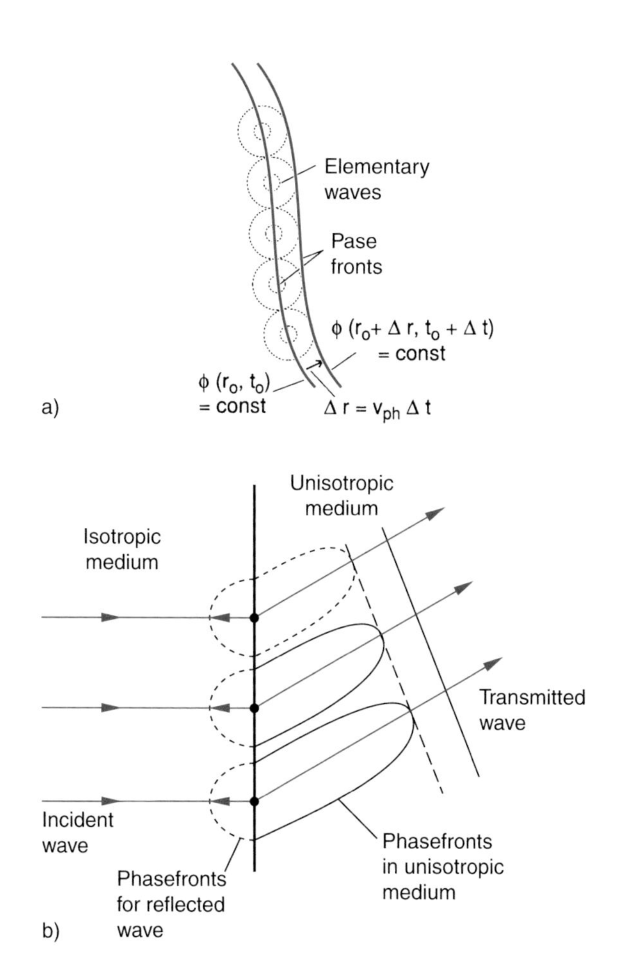

Figure 11.59 **a** Huygen's principle. **b** Transmission of a wave in an unisotropic medium. The phasefront is the tangent to the phasefronts of the elementary waves

In contrast to acoustic waves electromagnetic waves can also propagate through vacuum (otherwise we would not see the sun and the stars). Their propagation also follows Huygens's principle. This is explained in Vol. 2.

For the general case of non-isotropic media, the phase velocity does depend on the direction and the elementary waves have an elliptic envelope (Fig. 11.59b). The envelope to the elliptical waves at a time $t = t_0 + \Delta t$ is a phase-plane that is tilted against the phase-plane of the incoming wave by an angle α.

Up to now we have considered the case of an infinite number of source points for the elementary waves with a continuous distribution on the phase surface. We will now discuss the case of N distinct sources on the phase-plane with a distance δ between adjacent source points Q_i that are phase-locked to each other (Fig. 11.60). The phase difference between the different elementary waves depends on the direction α against the direction k_0 of the incident plane wave.

The path difference of waves from adjacent points in the direction α is $\Delta s = \delta \cdot \sin\alpha$ and the corresponding phase difference

$$\Delta\varphi = \frac{2\pi}{\lambda} \cdot \Delta s = k \cdot \delta \sin\alpha \quad \text{with} \quad k = \frac{2\pi}{\lambda} . \tag{11.97}$$

The superposition of all spherical waves emitted into the direction α from all N sources Q_i that are located on the phase front over a distance $d = N \cdot \delta$ gives the total amplitude at the point P

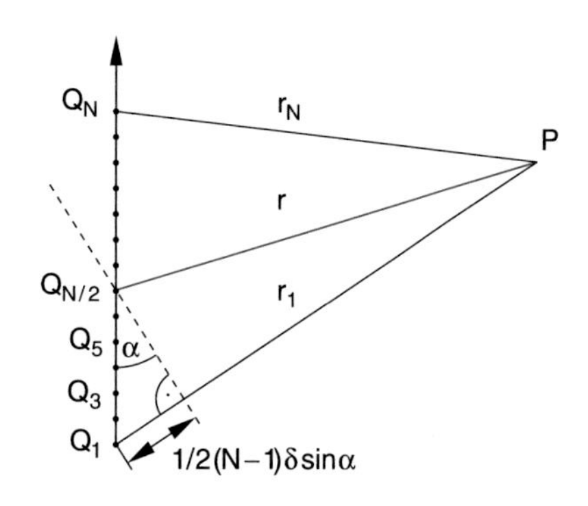

Figure 11.61 Illustration of the calculation of the interference intensity at the point *P*

(Fig. 11.61)

$$\xi(\alpha) = \sum_{n=1}^{N} a \cdot e^{i(kr_n - \omega t)}\,, \tag{11.98}$$

where we have assumed that all amplitudes a_i of the partial waves are equal ($a_i = a$) and that the distance $r \gg d$ of P from the sources is large compared to the total extension d of all N sources.

With $r_n = r + (1/2(N+1) - n) \cdot \delta \cdot \sin\alpha$ (Fig. 11.61) and $\Delta\varphi = k \cdot \delta \cdot \sin\alpha$ we obtain from (11.98)

$$\begin{aligned}\xi(\alpha) &= a \cdot e^{i\frac{(N+1)}{2}\Delta\varphi} \cdot \sum_{n=1}^{N} e^{-in\Delta\varphi} \cdot e^{i(kr-\omega t)}\\ &= A \cdot e^{i(kr-\omega t)}\,.\end{aligned} \tag{11.99}$$

The calculation of the geometrical series yields

$$\begin{aligned}\sum_{n=1}^{N} e^{-in\Delta\varphi} &= e^{-i\Delta\varphi}\frac{e^{-iN\Delta\varphi} - 1}{e^{-i\Delta\varphi} - 1}\\ &= e^{-i\frac{(N+1)}{2}\Delta\varphi} \cdot \frac{e^{i\frac{N}{2}\Delta\varphi} - e^{-i\frac{N}{2}\Delta\varphi}}{e^{i\Delta\varphi/2} - e^{-i\Delta\varphi/2}}\,.\end{aligned}$$

The last factor can be written as

$$\frac{\sin\left(\frac{1}{2}N \cdot \Delta\varphi\right)}{\sin\left(\frac{1}{2}\Delta\varphi\right)}\,.$$

This gives for the amplitude A in (11.99) the result

$$A(\alpha) = a \cdot e^{i\Delta\varphi} \cdot \frac{\sin\left(\frac{N}{2}\Delta\varphi\right)}{\sin\left(\frac{1}{2}\Delta\varphi\right)}\,. \tag{11.100}$$

The intensity of the wave $I(\alpha) \propto |A(\alpha)|^2$ is proportional to the square of the amplitude. This gives the intensity

$$\begin{aligned}I(\alpha) &\propto a^2 \cdot \frac{\sin^2\left(\frac{N}{2}\Delta\varphi\right)}{\sin^2\left(\frac{1}{2}\Delta\varphi\right)}\\ &= a^2\frac{\sin^2\left(\frac{1}{2}Nk\delta\sin\alpha\right)}{\sin^2\left(\frac{1}{2}k\delta\sin\alpha\right)}\,.\end{aligned} \tag{11.101}$$

In Fig. 11.62 $I(\alpha)$ is plotted for $N \gg 1$ and the two cases $\delta < \lambda$ and $\delta > \lambda$ (b) and (c).

This illustrates that for $\lambda > N \cdot \delta = d$ (the wavelength is larger than the extension of all N sources) only a single maximum appears at $\alpha = 0$. The wave propagates only within a small angular range $\Delta\alpha \propto 1/N$ around $\alpha = 0$. The angular width of this maximum is proportional to the inverse number of sources.

For the second case $\lambda < N \cdot \delta$ further intensity maxima appear at angles α_n that obey the condition

$$\sin\alpha_n = n \cdot \frac{\lambda}{N \cdot \delta} \quad \text{with} \quad n = 0, 1, 2, 3, \ldots\,. \tag{11.102}$$

11.11.2 Diffraction at Apertures

We will now increase the number N in (11.101) towards the limit $N = \infty$, while the total width $d = N \cdot \delta$ remains constant, which implies that the distance δ between the sources approaches zero. This can be realized when a parallel plane wave impinges normal ($k \perp$ slit) on a slit with width d. The intensity $I(\alpha)$ is measured behind the slit as a function of the deflection angle α (Fig. 11.63).

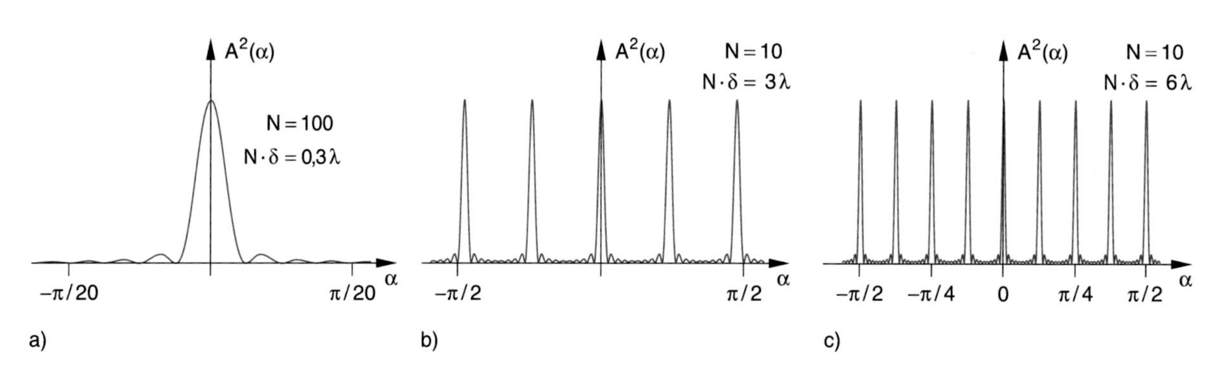

Figure 11.62 Intensity $I(P)$ as a function of the angle α: **a** for $\delta < \lambda$, **b** and **c** $\delta > \lambda$. The total width $d = N \cdot \delta$ is equal for (**a**) and (**b**), in (**c**) twice as large. Note the different abscissa scale in (**a**) compared to that in (**b**) and (**c**)

Chapter 11

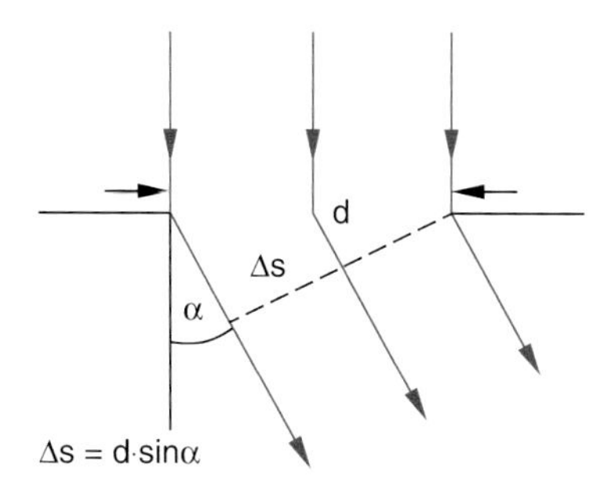

Figure 11.63 Diffraction of a wave when passing through a slot with width d

We can write (11.101) as

$$I(\alpha) \propto a^2 \frac{\sin^2\left(\frac{1}{2}kd\sin\alpha\right)}{\sin^2\left(\frac{kd}{2N}\sin\alpha\right)} .$$

For large values of N the sine function in the denominator can be replaced by its argument and we obtain with $\Delta\varphi = k\cdot d\cdot\sin\alpha = 2\pi d\cdot\sin\alpha/\lambda$.

$$I(\alpha) \propto N^2 a^2 \cdot \frac{\sin^2(\Delta\varphi/2)}{(\Delta\varphi/2)^2} . \qquad (11.103)$$

The function $\sin^2 x/x^2$ is plotted in Fig. 11.64a, where $x = \Delta\varphi/2 = (\pi\cdot d\cdot\sin\alpha)/\lambda$. It is 1 for $x = 0$ and is 0 for $x = \pm\pi$, i.e. $\sin\alpha = \lambda/d$. The two maxima at $x = 3\pi/2$ have only the height $4/(9\pi^2) \approx 0.045$ of the central maximum. In Fig. 11.64b the form of $I(\alpha)$ is illustrated for different ratios d/λ of slit width d and wavelength λ. The angular width of the central maximum at the base (distance between the two adjacent zeros) is for $\lambda/d \ll 1$.

$$\Delta\alpha_0 = 2\cdot\lambda/d$$

For $d \to \infty$ the function (11.103) converges towards a delta function that is always zero except for $x = 0$ where it is 1 (black vertical line in Fig. 11.64b).

The foregoing considerations have brought the following important result:

In spite of the fact, that the different elementary waves propagate into all directions their superposition leads for $d \gg \lambda$ to constructive interference only for $\alpha = 0$ i.e. into the direction of the incident wave. In all other directions the partial waves interfere destructively, they extinguish themselves. This can be understood a follows: We order all elementary waves in the direction $\alpha > 0$ in pairs of waves with a phase shift of π. For each elementary wave within the interval $d \gg \lambda$ there is another elementary wave with a phase shift of π. These two waves therefore interfere completely destructive and extinguish themselves. This is true for all pairs and therefore all waves extinguish themselves for $\alpha > 0$.

This is no longer true for $d/\lambda \leq 1$. If the cross section of the incident wave is limited by an aperture, and the diameter of the wave is of the same order of magnitude or smaller than the wavelength. In this case the maximum path difference $\Delta s = d\cdot\sin\alpha$ between the elementary waves from different points of the aperture is smaller than λ, i.e. the phase difference is smaller than π. Now we do not find pairs with a phase shift of π and therefore destructive interference cannot be complete. In Fig. 11.64b the intensity distribution $I(\alpha)$ is plotted for different values of d/λ.

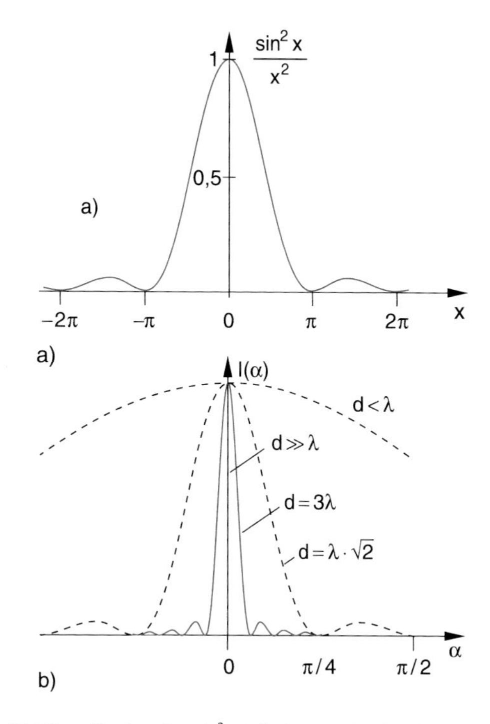

Figure 11.64 **a** The function $\sin^2 x/x^2$. **b** Intensity distribution $I(\alpha)$ of the diffracted wave behind a slit with width d for different values of λ/d

The deflection of waves transmitted through apertures into an angular range $-\infty < \alpha < +\infty$ is called **diffraction**. It is caused by incomplete interference. Not every elementary wave finds a partner with a phase delay $\Delta\varphi = \pi$. Therefore a residual total amplitude remains even for $\alpha \neq 0$. With increasing values of d/λ the residual amplitudes for $\alpha > 0$ decreases.

When P_0 is the total power transmitted through the aperture, the central diffraction maximum around $\alpha = 0$ contains the fraction $\eta = P_1/P_0$. This fraction can be calculated as

$$\eta = \frac{\int_{-\lambda/d}^{+\lambda/d} I(\alpha)\mathrm{d}\alpha}{\int_{-\pi/2}^{+\pi/2} I(\alpha)\mathrm{d}\alpha} = \frac{1}{P_0}\int\limits_{-\lambda/d}^{+\lambda/d} I(\alpha)\mathrm{d}\alpha \approx 0.95 ,$$

which proves that 95% of the transmitted power is included in the central maximum.

11.11.3 Summary

- Without limiting boundaries waves propagate in isotropic media straightforward. Their propagation can be described by Huygens's principle where each phase front of the wave can be regarded as an infinite number of sources that emit elementary waves, which interfere with each other. For $d \gg \lambda$ this interference is always completely destructive for deflection angles $\alpha \neq 0$. Therefore the wave propagates straightforward.
- The intensity distribution $I(\alpha)$ as a function of the deflection angel α against the direction of the incident wave depends on the ratio d/λ of aperture diameter d and wavelength λ.

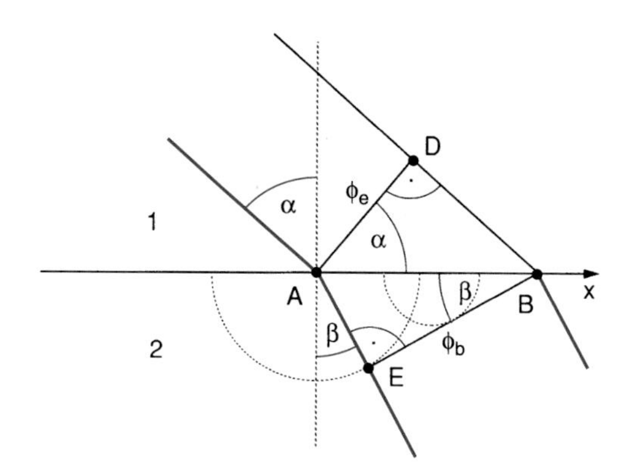

Figure 11.66 Derivation of the refraction law from Huygen's principle

11.11.4 Reflection and Refraction of Waves

When a wave passes through a boundary between two media, where the wave has different phase velocities, part of the wave is reflected and the transmitted part changes its direction. This can be also described by Huygens's principle.

We will show this at first for the reflection of a plane wave at the plane boundary between two different media (Fig. 11.65). When the phase plane φ_e = constant has reached the point A at the time $t = 0$, an elementary spherical wave is emitted from A, which spreads out into the whole upper plane. When the same phase front has reached the point B at the time $t_0 = \overline{DB}/v_{Ph}$ also an elementary spherical wave is emitted from B. Meanwhile the first elementary wave has propagated along the distance $r = v_{Ph} \cdot t_0$.

The elementary waves emitted from points P_n between A and B have been excited at times $t_n = t_0 \cdot (\overline{AP_n})/\overline{AB}$ and until the time t_0 they have travelled the distance $r_n = v_{Ph} \cdot (t_0 - t_n) = v_{Ph}(\overline{P_nB}/\overline{AB})$. The tangent $\overline{BE}$ to all spheres of the elementary waves at time t_0 is the phase plane of the reflected wave. From Fig. 11.65 it can be seen that

$$\overline{AE} = \overline{DB} \Rightarrow \sphericalangle(ABE) = \sphericalangle(BAD) = \alpha\,.$$

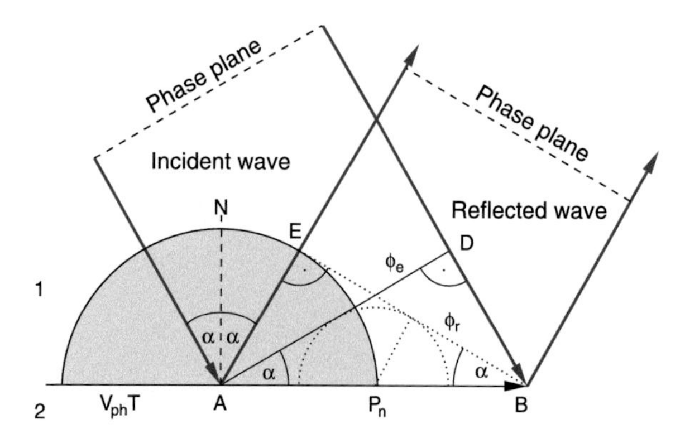

Figure 11.65 Explanation of the reflection law based on Huygen's priciple

We name the angle α between the wave vector $\boldsymbol{k}_e$ of the incident wave and the normal vector $\boldsymbol{N}$ of the reflecting plane the *angle of incidence*. We then can formulate the law of reflection:

> When a plane wave is reflected at a plane boundary between two media, the angle of reflection is equal to the angle of incidence.

For that part of the wave that penetrates into the medium 2 (Fig. 11.66), the velocity $v = v_2$ of the elementary waves is different from the velocity $v = v_1$ in medium 1. During the time $t_0 = \overline{DB}/v_1$, where the wave front in medium 1 travels from D to B the elementary wave starting from A has travelled in medium 2 the distance $\overline{AE} = v_2 \cdot t_0 = \overline{DB} \cdot v_2/v_1$. This implies $\overline{AE}/\overline{DB} = v_2/v_1$. From the triangles ABD and AEB with the same base length we conclude:

$$\frac{\sin\alpha}{\sin\beta} = \frac{v_1}{v_2}\,. \tag{11.104}$$

Equation 11.104 is **Snellius' refraction law** (after the Dutch Astronomer Willebrord Snellius van Royen 1580–1626, who published this law for the first time although it was known before). It holds for arbitrary waves, not only for acoustic but also for light waves (see Vol. 2).

Reflection and refraction of waves can be substantiated by a more general law, called **Fermat's principle**.

A wave always travels that path for which the runtime for a phase front between two points $P_1(x_1, y_1)$ and $P_2(x_2, y_2)$ is minimum. This is illustrated for the refraction of waves in Fig. 11.67). The run-time T for a phase front from $P_1(x_1, y_1)$ to $P_2(x_2, y_2)$ is

$$\begin{aligned} T &= \frac{S_1}{v_1} + \frac{S_2}{v_2} \\ &= \frac{1}{v_1}\sqrt{(x - x_1)^2 + y_1^2} + \frac{1}{v_2}\sqrt{(x_2 - x)^2 + y_2^2}\,. \end{aligned} \tag{11.105}$$

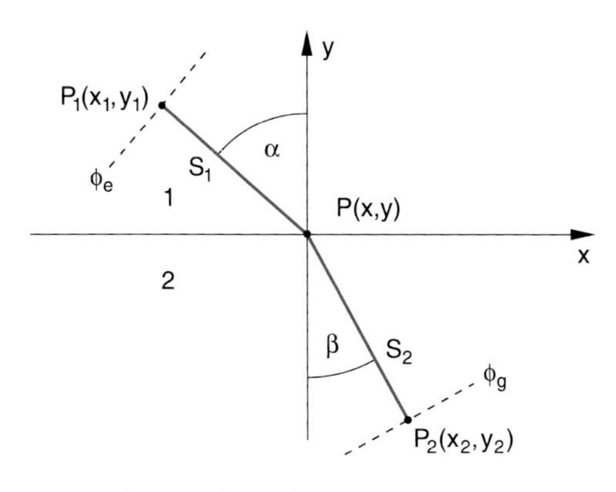

Figure 11.67 Explanation of the refraction law with the Fermat principle

When we shift the point $P(x, y)$ along the boundary surface $y = 0$, the run-time T changes by

$$\frac{\mathrm{d}T}{\mathrm{d}x} = \frac{(x - x_1)/v_1}{\sqrt{(x - x_1)^2 + y_1^2}} - \frac{(x_2 - x)/v_2}{\sqrt{(x_2 - x)^2 + y_2^2}} .$$

If T should be minimum, the derivative $\mathrm{d}T/\mathrm{d}x$ must be zero. This gives with $\sin\alpha = (x - x_1)/s_1$ and $\sin\beta = (x_2 - x)/s_2$ the condition

$$\frac{\sin\alpha}{v_1} = \frac{\sin\beta}{v_2} , \qquad (11.106)$$

which is identical to Snellius' law (11.104).

11.12 Standing Waves

Special superpositions of running waves result in stationary oscillation patterns where certain points, lines or planes in space are always at rest, which means that here the oscillation amplitude is zero (nodes of oscillation). The pattern of nodes depends on the frequency and the boundary conditions. We will illustrate this by several examples.

11.12.1 One-Dimensional Standing Waves

When a plane wave $\xi = A \cdot \cos(\omega t + kz)$ that propagates into the $-z$-direction is reflected by a plane $z = 0$ (Fig. 11.68) the reflected wave $\xi = A \cdot \cos(\omega t - kz)$ propagates into the $+z$-direction. For $z > 0$ the two waves superimpose and the total wave field is

$$\begin{aligned} \xi &= \xi_1 + \xi_2 \\ &= A[\cos(\omega t + kz) + \cos(\omega t - kz + \varphi)] , \end{aligned} \qquad (11.107)$$

where a phase jump φ at the reflection has been taken into account. According to the addition theorem of the cos-functions this can be written as

$$\xi = 2A \cdot \cos\left(kz - \frac{\varphi}{2}\right) \cdot \cos\left(\omega t + \frac{\varphi}{2}\right) . \qquad (11.108)$$

This represents an oscillation $\cos(\omega t + \varphi/2)$ with the amplitude $2A \cdot \cos(k \cdot z - \varphi/2)$ that depends periodically on the location z. It is called a **standing wave**. At the positions $z = (\lambda/4\pi)[(2n+1)\pi + \varphi]$ the amplitude of the standing wave is zero. These zero points are the *oscillation nodes*. For $z = (\lambda/4\pi)(2n\pi + \varphi)$ the amplitude becomes maximum. It changes during one oscillation period $T = 2\pi/\omega$ from $-2A$ to $+2A$. These maxima of the standing wave are called the *oscillation antinodes*.

Note the difference to the running wave $\xi = A \cdot \cos(\omega t - k \cdot z)$. Here the nodes and antinodes propagate with the velocity $v_{\mathrm{Ph}} = \omega/k$ into the $+z$-direction, while for the standing wave they are fixed in place.

The spatial amplitude distribution of the standing wave and the positions of the nodes and antinodes depend on the phase jump at the reflection. We will discuss some special cases:

- *Reflection at a fixed end at* $z = 0$ (Fig. 11.69a).
  This can be realized by a rope with its right end fixed to a wall, while the left end is connected to an oscillator, for example a hand that shakes up and down. Since $\xi(0) = 0$ it follows from (11.108)

$$\varphi/2 = \pm\pi/2 \Rightarrow \varphi = \pm\pi .$$

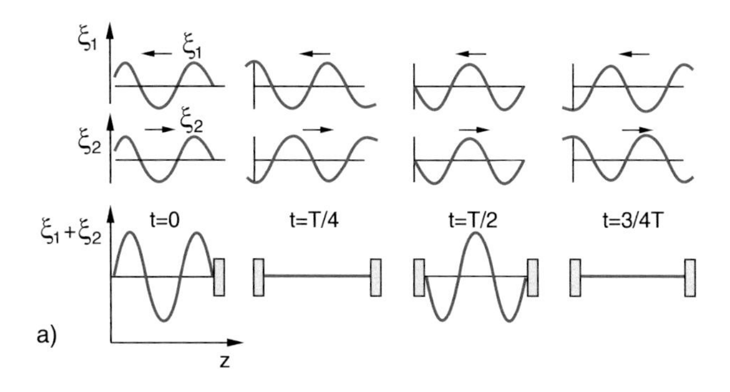

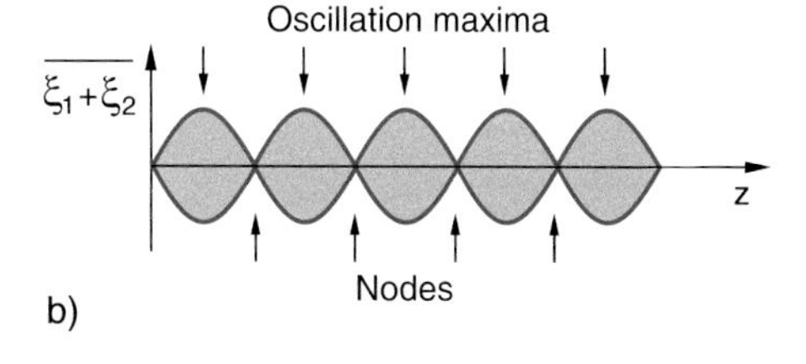

Figure 11.68 **a** Standing wave as superposition of a wave propagating into the $+z$-direction and the reflected wave running into the $-z$-direction. **b** Periodic maxima and nodes of a standing wave

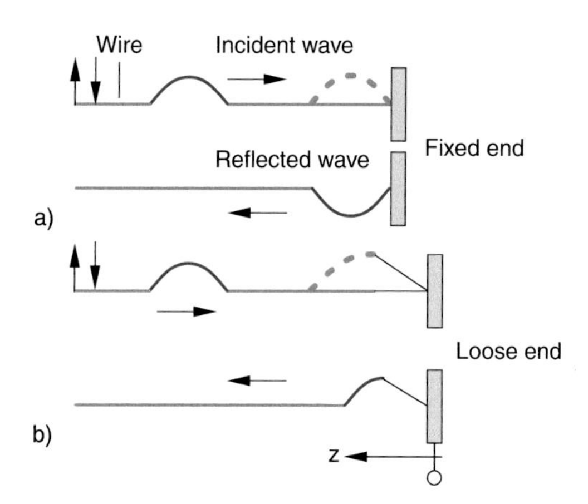

Figure 11.69 Reflection of a wave on a rope: **a** at a fixed end, **b** at a loose end

For the reflection at a fixed end the wave undergoes a phase jump of π. The standing wave is then

$$\xi(z,t) = -2A \cdot \sin(k \cdot z) \cdot \sin(\omega t) . \tag{11.108a}$$

- *Reflection at a loose end at* $z = 0$ (Fig. 11.69b).
  This can be realized with a rope that is not fixed directly to a wall but hangs freely on a strand that is connected to the wall. Here is $\xi(0) = \xi_0 = 2A \Rightarrow \varphi = 0$. There is no phase jump at the reflection and the superposition of the incident and the reflected wave gives the standing wave

$$\xi(z,t) = 2A \cdot \cos(kz) \cdot \cos(\omega t) . \tag{11.108b}$$

In optics this corresponds to the reflection of a light wave at the boundary from a medium with refractive index n_1 to one with $n_2 < n_1$ (see Vol. 2). The standing wave is shifted in time by 90° and in space by $\lambda/4$ against the standing wave in case a).

The one-dimensional standing waves are general solutions of the one-dimensional wave equation (11.62), just as the running waves (11.54). Special forms are selected by the boundary conditions.

Standing waves can be regarded as resonant oscillations of a one-dimensional medium, for instance a clamped string (see Sect. 11.9.5.4).

With a length L of the string and a tensile force $\boldsymbol{F}$ all frequencies ν_n are possible for which

$$\lambda_n = v_{\text{Ph}}/\nu_n = 2L/n \quad \text{with} \quad n = 1, 2, 3, \ldots . \tag{11.109a}$$

According to (11.77) is the frequency ν_1 of the fundamental oscillation with $n = 1$

$$\nu_1 = v_{\text{Ph}}/2L = \frac{1}{2L} \cdot \sqrt{F/\mu} . \tag{11.109b}$$

The fundamental oscillation and all overtones ν_n with $n > 1$ are resonant oscillations of the string. Their frequencies depend on the length L of the string, on the tensile force F and on the mass μ per unit length.

11.12.2 Experimental Demonstrations of Standing Waves

One-dimensional standing waves can be demonstrated in many different ways. One example is the demonstration of standing waves in a gas visualized with Kundt's cork-powder structures (Fig. 11.70). In a glass tube, cork powder is uniformly distributed. At the end of the tube a loud speaker generates acoustic oscillations which produce standing waves if the resonance condition (11.109) is fulfilled. These standing acoustic waves change the distribution of the cork powder. At the maxima of the standing wave, the powder is slung away by the acoustic oscillations, while at the nodes it remains at rest. From the distance $\Delta L = \lambda/2$ between the nodes the wavelength λ can be determined and with the known frequency ν of the loud speaker the phase velocity $v_{\text{Ph}} = \nu \cdot \lambda$ of acoustic waves in the gas is obtained.

Instead of using a loud speaker a metal rod with length L, which is clamped at its centre, can be excited to oscillations by a piezo crystal at its outer end. (Fig. 11.70). A metal plate is connected to the inner end, which converts the vibrations into acoustic waves in the gas. The ratio of the wavelengths in the gas and in the metal rod are equal to the ratio of the phase velocities, because the frequency is the same in both media.

$$\frac{\lambda_{\text{gas}}}{\lambda_{\text{solid}}} = \frac{v_{\text{Ph}}^{\text{gas}}}{v_{\text{Ph}}^{\text{solid}}} = \frac{\sqrt{p \cdot \kappa/\varrho_{\text{gas}}}}{\sqrt{E/\varrho_{\text{solid}}}} = \frac{\sqrt{K\kappa/\varrho_{\text{gas}}}}{\sqrt{E/\varrho_{\text{solid}}}} \tag{11.110}$$

with $\kappa = C_p/C_v$ and $E =$ elastic modulus. If the densities ϱ are known, the measurement yields the ratio of compression modulus of the gas to the elastic modulus of the rod.

A very impressive demonstration of standing waves in a gas is Rubens's flame tube (Fig. 11.71). This is a tube with many small holes on the upper side along the 1–2 m long tube. When the tube is connected to gas supply and filled with a few millibar of propane gas the gas streams out of the holes. With a match this out streaming gas can be ignited resulting in a row of small

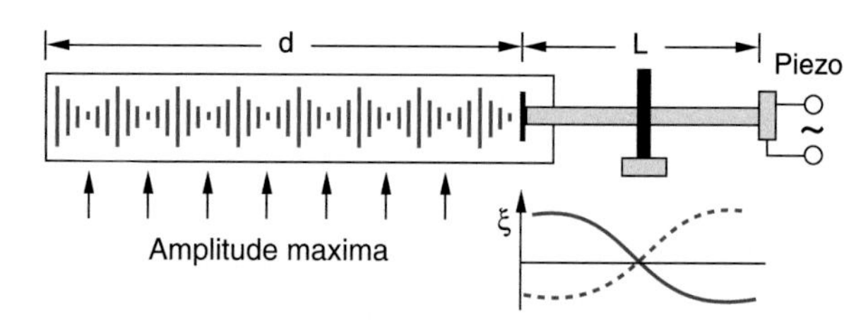

Figure 11.70 Generation of Kundt's cork dust figures

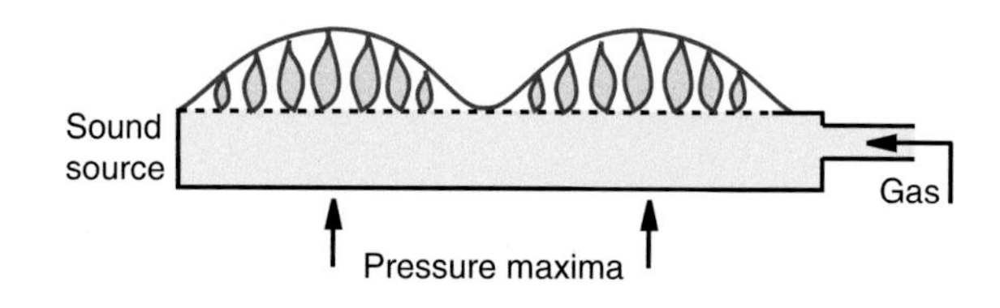

Figure 11.71 Flame tube of Rubens

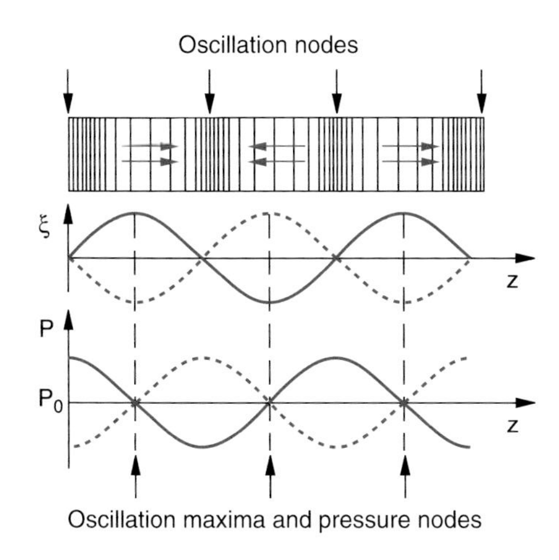

Figure 11.72 Oscillation amplitude and spatial pressure variation of a standing wave in a gas

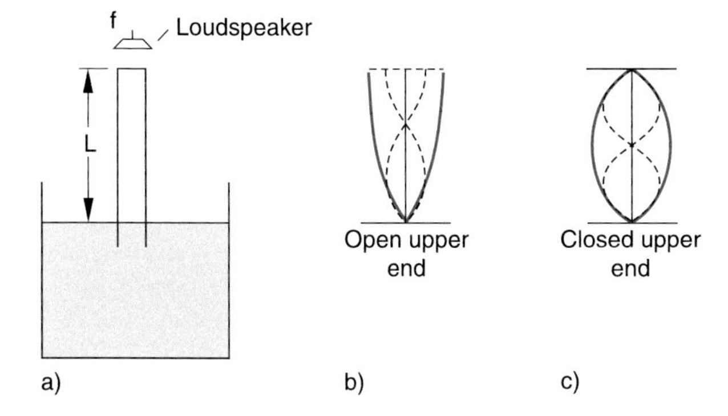

Figure 11.73 Resonance tube of Quinck

flames all with the same height. A loudspeaker at one end of the tube produces acoustic waves in the gas. For certain frequencies resonant standing waves are generated. In the amplitude nodes of the standing waves the gas pressure is maximum because here the gas molecules flow against each other thus increasing the pressure (Fig. 11.72). The standing pressure wave is therefore shifted by $\lambda/4$ against the standing amplitude wave. At the pressure maxima more gas streams out of the holes and the height of the flame becomes larger. Therefore the standing waves in the inside of the tube are visualized as periodic modulation of the flame height.

A very simple demonstration of standing waves in a gas is possible with the resonance tube of Quincke that has two open ends (Fig. 11.73). A loudspeaker is mounted above the upper end of the tube. The lower end is immersed into water. When the tube is lowered or raised the length L of the air column changes. At certain lengths L_n resonances occur which can be heard by a significant increase of the sound level.

Since at the water surface an amplitude node appears but at the upper end of the tube a pressure node (because of the constant pressure of the air above the tube) the resonance condition is

$$L = (2n+1)\cdot\lambda/4 \quad \Rightarrow \nu_n = (2n+1)\frac{v_{\text{Ph}}}{4L}\,.$$

When the upper end of the tube is closed by a cap, then amplitude nodes occur also at this end and the resonance condition is now

$$L = (n+1)\lambda/2 \quad \Rightarrow \nu_n = (n+1)\frac{v_{\text{Ph}}}{2L}\,.$$

The fundamental resonance appears at

$$\lambda_0 = 2L \quad \Rightarrow \nu_0 = \frac{v_{\text{Ph}}}{2L}\,.$$

Tubes that are closed on both ends, or open on both ends have a fundamental resonance frequency that is twice as high as that for tubes with one end open and the other closed. This can be utilized for organ pipes to cover a large frequency range.

11.12.3 Two-dimensional Resonances of Vibrating Membranes

For the investigation of eigen-resonances of two-dimensional surfaces we have to solve the two-dimensional wave equation

$$\frac{\partial^2\xi}{\partial x^2} + \frac{\partial^2\xi}{\partial y^2} = \frac{1}{v_{\text{Ph}}^2}\frac{\partial^2\xi}{\partial t^2} \tag{11.111}$$

with the given boundary conditions. Examples for two-dimensional standing waves are the resonance oscillations of plates, drum membranes or soap bubbles inserted in a frame.

The solutions of (11.111) can be written in the form

$$\xi(x,y,t) = A(x,y)\cdot\cos\omega t\,, \tag{11.112a}$$

where the amplitude function depends on the boundary conditions. For a thin rectangular membrane which is fixed along the boundary lines $x = 0$, $x = a$, $y = 0$ and $y = b$ the solutions are

$$\begin{aligned}\xi_{m,n}(x,y) = A\cdot\sin\frac{(m+1)\pi x}{a}\\ \cdot\sin\frac{(n+1)\pi y}{b}\cdot\cos(\omega_{m,n}t)\,,\end{aligned} \tag{11.112b}$$

as can be readily checked by inserting these functions into the wave equation (11.111) and taking into account the boundary conditions. The integer m gives the number of nodal lines $x = x_m$ vertical to the x-direction, while n gives the number of nodal lines $y = y_n$ vertical to the y-direction (Fig. 11.74).

Such eigen-oscillations can be regarded as superposition $\xi = \xi_1 + \xi_2$ of two waves with wave vectors $\boldsymbol{k} = \{k_1, k_2\}$ and $\boldsymbol{k}_2 = -\boldsymbol{k}_1$ which obey the boundary conditions

$$\xi(x = 0, a; y = 0, b) = 0\,.$$

The relation

$$k^2 = k_x^2 + k_y^2 = 4\pi^2\left(\frac{1}{\lambda_x^2} + \frac{1}{\lambda_y^2}\right) \tag{11.113}$$

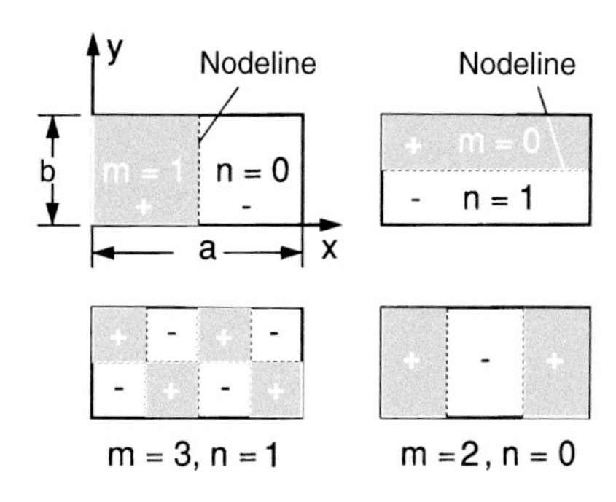

Figure 11.74 Two-dimensional oscillation modes for a rectangular clamp of the membrane

between wave vector and wavelength gives for the eigen-frequencies $\omega = 2\pi \cdot v_{\text{Ph}}/\lambda$ the equation

$$\omega_{m,n} = \pi \sqrt{\frac{\sigma}{\varrho}\left[\left(\frac{m+1}{a}\right)^2 + \left(\frac{n+1}{b}\right)^2\right]} . \tag{11.114}$$

Here we have assumed that the membrane is sufficiently thin that torsion- and bending forces can be neglected. The only restoring forces are then due to tensile stress σ, which acts because of the length change in x- and y-directions.

For a circular frame where the membrane is fixed the node lines have circular symmetry. Therefore, polar coordinates are convenient for the wave equation (11.111)

$$\frac{1}{r}\frac{1}{\partial r}\left(r\frac{\partial \xi}{\partial r}\right) + \frac{1}{r^2}\frac{\partial^2 \xi}{\partial \varphi^2} = \frac{1}{v_{\text{Ph}^2}}\frac{\partial^2 \xi}{\partial t^2} . \tag{11.115}$$

The solutions

$$\xi(r, \varphi, t) = \chi(r) \cdot \phi(\varphi) \cdot \cos \omega t$$

can be written as the product of three one-dimensional functions, analogous to (11.112a, b). For a radius R of the circular membrane one obtains the solutions

$$\begin{aligned} \xi_{n,p}(r, \varphi, t) = J_p\left(r \cdot \frac{r_{n,p}}{R}\right) \cdot [A_1 \cos(p\varphi) \\ + A_2 \sin(p\varphi)] \cdot \cos(\omega_{n,p} t) , \end{aligned} \tag{11.116}$$

where n and p are integers and J_p is the Bessel-function of order p. The frequency

$$\omega_{n,p} = \frac{r_{n,p}}{R}\sqrt{\frac{\sigma}{\mu}} \tag{11.117}$$

is the resonance frequency of the standing wave with n radial circular nodes and p azimuthal nodes along radial lines on a circular membrane with mass density μ (mass per unit area) and the tensile stress σ (Fig. 11.75) The dimensionless number $r_{n,p}$ is the n-th root of the Bessel function J_p.

Such two-dimensional standing waves can be demonstrated by different means:

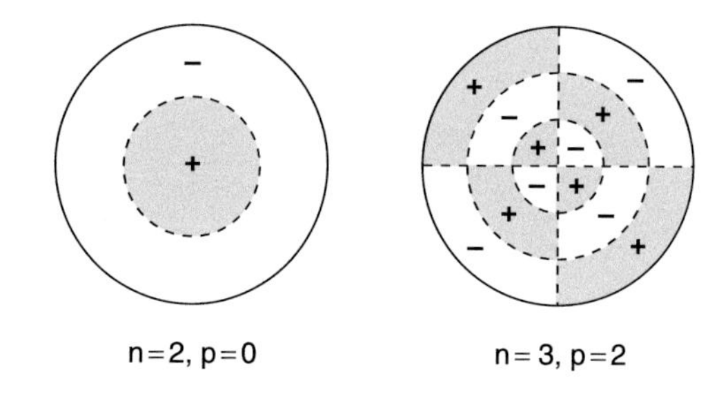

Figure 11.75 Oscillation modes of a circularly clamped membrane

- ***Chladni's Figures.*** A thin rectangular or circular dark metal plate is clamped at one point in the mid of the plate. It is uniformly powdered with colophony and then a violin bow is stroked along an edge of the plate, which excites two-dimensional resonant vibrations of the plate. On the nodal lines the powder remains on the surface, while at all other places it is removed by the vibration. The nodal lines appear as bright lines on a dark background. Depending on the contact pressure of the violin bow and on the velocity of the bow motion different figures appear, which are typical for the sound of the violin (Fig. 11.76).
- ***Oscillations of lamellar soap bubbles.*** When a rectangular or circular wire-frame with a handle is immersed into a soap solution and then carefully lifted, a soap lamella is formed over the whole area within the wire frame, which can be made visible by a proper illumination. by skilful movements of the frame different eigen-vibrations (11.112b) resp. (11.116) of the soap lamella can be excited. For a frame area of $40 \times 100\,\text{cm}^2$ vibrational amplitudes up to 20 cm can be realized. (see the demonstration movie "Standing Waves" by Ealing corporation [11.22]).

Figure 11.76 Oscillation modes of a quadratic metal plate, clamped in the center. The oscillations are excited by a violin bow, sweeping along the edges of the plate (Chladnic's sound patterns)

11.13 Waves Generated by Moving Sources

Up to now we have considered waves emitted from sources at rest and detected by observes at rest. For situations where the source or the observer move against a chosen coordinate frame, new phenomena occur, which will be discussed in this section.

11.13.1 Doppler-Effect

When a source of acoustic waves moves relative to the medium that transmits the acoustic waves, the frequency of the sound changes for an observer. This can be seen as follows:

During the oscillation period $T = 1/\nu_0$ the wave emitted in z-direction travels the distance $\Delta z = v_{\mathrm{Ph}} \cdot T = \lambda_0$. If the source moves with the velocity $u_{\mathrm{S}} = u_z$ in the direction towards the observer, the distance between the phase surfaces with phases φ and $\varphi + 2\pi$ has decreased to

$$\Delta z = \lambda = \lambda_0 - u_{\mathrm{S}} \cdot T = (v_{\mathrm{Ph}} - u_{\mathrm{S}})/\nu_0 .$$

This distance between two phase-fronts with phases differing by 2π is defined as the wavelength λ of the wave. The wavelength measured by an observer at rest is therefore shorter and the frequency

$$\begin{aligned}\nu &= \frac{v_{\mathrm{Ph}}}{\lambda} = \nu_0 \cdot \frac{v_{\mathrm{Ph}}}{v_{\mathrm{Ph}} - u_{\mathrm{S}}} \\ &= \nu_0 \frac{1}{1 - u_{\mathrm{S}}/v_{\mathrm{Ph}}}\end{aligned} \tag{11.118a}$$

higher as for sources at rest.

If the source moves with the velocity $u_{\mathrm{S}} = -u_z$ away from the observer. the frequency measured by the observer

$$\nu = \nu_0 \frac{1}{1 + u_{\mathrm{S}}/v_{\mathrm{Ph}}} \tag{11.118b}$$

is smaller.

This **Doppler-effect**, first described in 1846 by Christian Doppler (1803–1853) is familiar to us when a police car with its siren blaring passes an observer. As long as it approaches the observer the tone is high. As soon as it passes by the tone drops noticeably.

A similar effect occurs when the observes moves with the velocity u_{obs} towards a source at rest or away from it. During the oscillation period $T = 1/\nu_0$ the observer moves along a distance $\Delta z = u_{\mathrm{obs}} \cdot T$. He therefore measures $\Delta n = \Delta z/\lambda_0$ additional oscillation periods. The oscillation frequency measured by him is therefore

$$\begin{aligned}\nu &= \nu_0 + \frac{u_{\mathrm{obs}}}{\lambda_0} = \nu_0 + \frac{u_{\mathrm{obs}}}{v_{\mathrm{Ph}}} \cdot \nu_0 \\ &= \nu_0 \left(1 + \frac{u_{\mathrm{obs}}}{v_{\mathrm{Ph}}}\right) ,\end{aligned} \tag{11.119a}$$

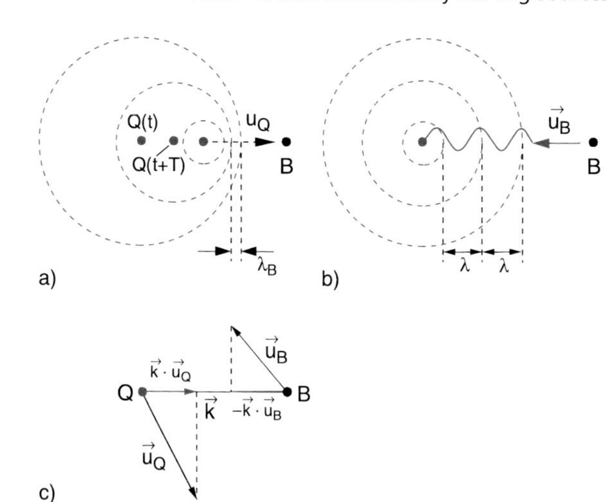

Figure 11.77 Doppler-effect: **a** moving source; **b** moving observer; **c** general case; B = observer, Q = source

when he moves towards the source, and

$$\nu = \nu_0 \left(1 - \frac{u_{\mathrm{obs}}}{v_{\mathrm{Ph}}}\right) , \tag{11.119b}$$

when he moves away from the source.

The Eq. 11.118, 11.119 can be generalized for the case that source as well as observer move. This gives the equation

$$\nu = \nu_0 \frac{(1 \pm u_{\mathrm{obs}}/v_{\mathrm{Ph}})}{(1 \mp u_{\mathrm{S}}/v_{\mathrm{Ph}})} , \tag{11.120}$$

where the upper sign applies when source and observer move towards each other and the lower sign if they move away from each other. The frequency shift $\Delta\nu = \nu - \nu_0$ is the *Doppler-shift*.

For the general case of arbitrary directions of the velocities $\boldsymbol{u}_{\mathrm{S}}$ and $\boldsymbol{u}_{\mathrm{obs}}$ Eq. 11.120 can be written in vector form as

$$\omega = \omega_0 \frac{\omega_0 - \boldsymbol{k} \cdot \boldsymbol{u}_{\mathrm{obs}}}{\omega_0 + \boldsymbol{k} \cdot \boldsymbol{u}_{\mathrm{S}}} \tag{11.120a}$$

with $\omega = 2\pi\nu$ and the wave vector $\boldsymbol{k}$, $(|\boldsymbol{k}| = 2\pi/\lambda = \omega/v_{\mathrm{Ph}})$ (see Fig. 11.77c).

Equation 11.120a can be verified by starting with the general representation of a wave propagating into the direction of $\boldsymbol{k}$

$$\xi = A \cdot \cos(\omega_0 t - \boldsymbol{k} \cdot \boldsymbol{r}) ,$$

where $\boldsymbol{r}$ is the position vector starting from the origin $\boldsymbol{r} = \boldsymbol{0}$. The equation of motion for the moving observer is

$$\boldsymbol{r} = \boldsymbol{u}_{\mathrm{obs}} t + \boldsymbol{r}_0 .$$

The wave for the observer is then

$$\begin{aligned}\xi &= A \cdot \cos\left[\omega_0 t - \boldsymbol{k} \cdot (\boldsymbol{u}_{\mathrm{obs}} \cdot t + \boldsymbol{r}_0)\right] \\ &= A \cdot \cos(\omega t - \boldsymbol{k} \cdot \boldsymbol{r}_0) \\ &\text{with } \omega = \omega_0 - \boldsymbol{k} \cdot \boldsymbol{u}_{\mathrm{obs}} .\end{aligned}$$

A similar expression can be derived when the source is moving.

Note: For light waves the Doppler-shift depends only on the relative velocity between source S and observer. This is different for sound waves, where the Doppler shift differs for the case when the source is moving from that for a moving observer. The reason for this difference is the fact that the propagation of sound waves needs a medium and the motion of the source S against this medium has another effect than the motion of the observer. Only for $u \ll v_{Ph}$ the two Eq. 11.118 and 11.119 converge because of $1/(1-x) \approx 1+x$ for $x \ll 1$.

The Doppler-effect can be already quantitatively demonstrated for small velocities u. One acoustic source is mounted on a glider moving on an air-track while a second source with the same frequency is at rest. Due to the Doppler-shift the superposition of the two sound waves generates a beat signal which can readily be measured, thus giving the Doppler-shift

$$\Delta\nu = \nu_0 \cdot \frac{u}{v_{Ph}} .$$

Numerical Example

$\nu_0 = 5\,\mathrm{kHz}$, $v_{Ph} = 330\,\mathrm{m/s}$, $u = 0.05\,\mathrm{m/s} \Rightarrow \Delta\nu = 0.75\,\mathrm{Hz}$. One hears a modulation of the 5 kHz frequency at 0.75 Hz. ◄

In another demonstration experiment the acoustic source is mounted at the swinging end of a long pendulum with length L. If the pendulum moves in the direction x of the line of sight the frequency shift is

$$\Delta\nu = \nu_0 \cdot \frac{u_x}{v_{Ph}} = \frac{\nu_0}{v_{Ph}} \cdot \sqrt{2g \cdot L(1-\cos\varphi)} .$$

With $L = 10\,\mathrm{m}$ the oscillation period is $T = 6\,\mathrm{s}$. With an elongation $\varphi_{max} = 10°$ from the equilibrium position $\varphi = 0$ the maximum velocity of the acoustic source at $\varphi = 0$ is

$$u_{max} = \sqrt{2g \cdot L(1-\cos\varphi_{max})} = \sqrt{2.9807} = 1.73\,\mathrm{m/s} .$$

(Check this result by using the energy conservation during one oscillation period!). At a frequency $\nu_0 = 5\,\mathrm{kHz}$ the frequency changes during one oscillation period from 4.974 kHz to 5.026 kHz. This can be measured by superimposing a sound wave with fixed frequency ν_0 and determining the beat frequency $\nu_0 - \nu$.

11.13.2 Wave Fronts for Moving Sources

We consider a point-like acoustic source, which moves with the velocity u into the z-direction. During its motion, it emits continuously spherical waves with frequency ν_0 (Fig. 11.78). According to the considerations in the last section the distance between two phase fronts that differ by $\Delta\varphi = 2\pi$

$$\lambda(\alpha) = \frac{1}{\nu_0}(v_{Ph} - u \cdot \cos\alpha) \tag{11.121}$$

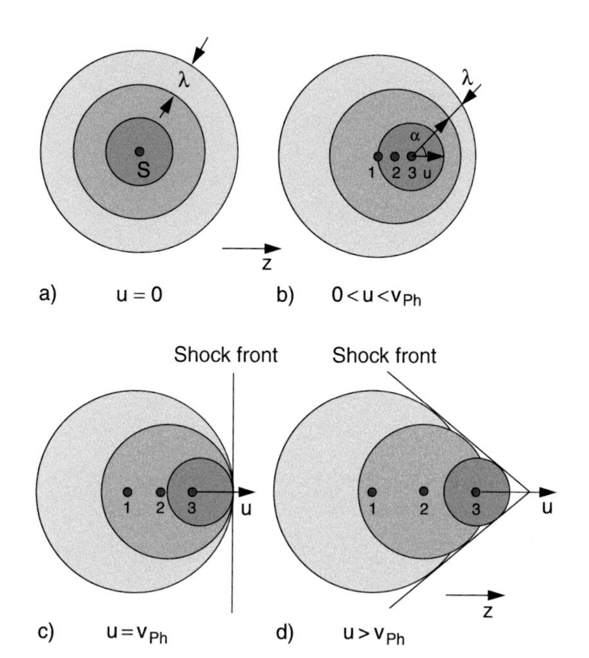

Figure 11.78 Generation of a shock front, if the velocity of the source S approaches the velocity of sound

depends on the angle α against the direction of the velocity $\boldsymbol{u}$ of the source. If the velocity u of the source reaches the velocity v_{Ph} of the acoustic waves, the phase front distance becomes $\lambda(0) = 0$ (Fig. 11.78c). The amplitudes of the waves emitted at different times into the z-direction all superimpose in phase resulting in a shock-wave with exceedingly large amplitude.

For $u > v_{Ph}$ $\lambda(\alpha)$ becomes zero for $\alpha = \arccos(v_{Ph}/u)$. The amplitudes of all waves emitted at different times all add up in phase at a cone with the opening angle $\beta = 90° - \alpha$. According to (11.121) and Fig. 11.79 is $\cos\alpha = v_{Ph}/u$ which gives

$$\sin\beta = \frac{v_{Ph}}{u} = \frac{1}{\mathrm{M}} . \tag{11.122}$$

This shock-wave cone is called *Mach's cone* and the ratio $\mathrm{M} = u/v_{Ph}$ of source velocity to sound velocity is the **Mach number**, named after the Austrian physicist *Ernst Mach* (1838–1916).

Such shock-wave cones can be observed as bow waves along a ship on a lake, when the velocity u of the ship becomes larger than the velocity v_{Ph} of water surface waves [11.16]. The situation is here, however, more complex, because water surface waves show dispersion (see Sect. 11.9.7). The ship generates

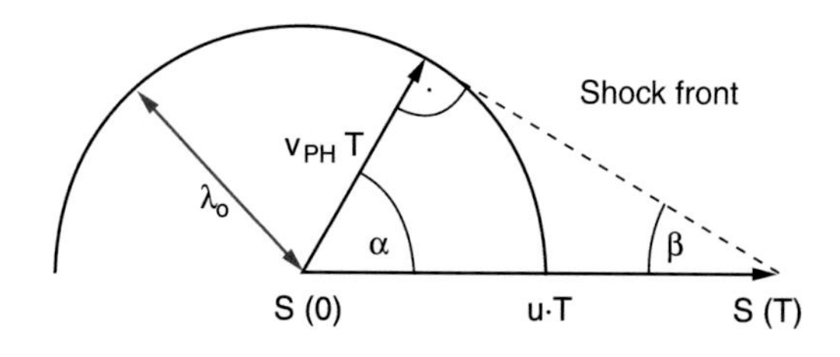

Figure 11.79 Calculation of the aperture angle of the Mach-cone

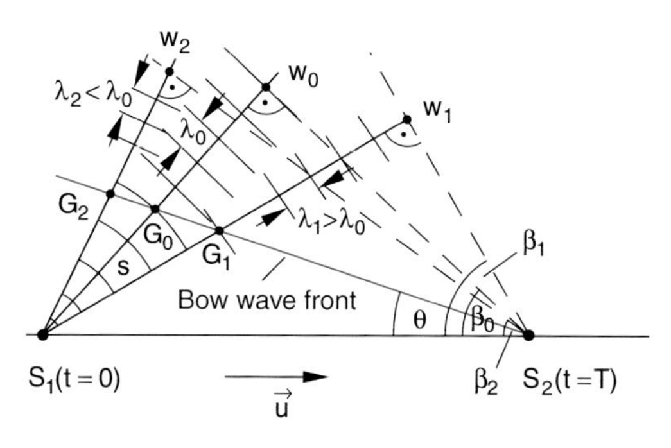

Figure 11.80 Generation of the bow wave by a ship with velocity $|\boldsymbol{u}| > |\boldsymbol{v}|_{\text{Ph}}$

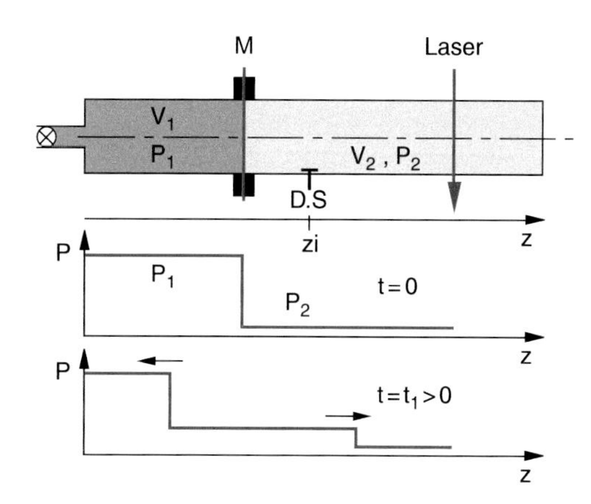

Figure 11.81 Shockwave tube

waves within a broad frequency range. In Fig. 11.80 these waves are generated at time $t = 0$ in the point S_1. At time $t = T$ when the ship has reached the point S_2 the wave with the mid wavelength λ_0 has reached the point W_0. Since the partial waves with other frequencies have different phase velocities their phases are distributed in W_0 between $\varphi = 0$ around $\varphi = 2\pi$. their superposition will therefore have the average zero. No shock wave is generated in W_0. The maximum of the wave group, which propagates with the group velocity $v_G = (1/2)\, v_{\text{Ph}}(\lambda_0)$ has reached at time T only the point G_0, where for the distances holds: $\overline{S_1G_0} = (1/2)\,\overline{S_1W_0}$. The angle β_0 between the straight lines $\overline{S_1S_2}$ and $\overline{W_0S_2}$ is larger than the angle Θ between $\overline{S_1S_2}$ and $\overline{S_2G_0}$. The straight line $\overline{S_2G_0}$ represents the bow wave front because the maximum of the wave group has reached at time T the point G_0. With $\overline{S_2G_0} = d$ and $\overline{S_1G_0} = s = (1/2)\,\overline{S_1W_0}$ we obtain the relations

$$\tan\beta_0 = 2s/d\,; \quad \tan(\beta_0 - \Theta) = s/d$$
$$\to \Theta = \arctan(2s/d) - \arctan(s/d)\,.$$

When the line $\overline{S_1G_0}$ represents the bow wave front, the condition $\mathrm{d}\Theta/\mathrm{d}s = 0$ must be fulfilled. This gives

$$\frac{\mathrm{d}\Theta}{\mathrm{d}s} = \frac{2}{1+4s^2} - \frac{1}{1+s^2} = 0$$
$$\Rightarrow s_{\text{obs}} = d/\sqrt{2} \Rightarrow \Theta_{\text{obs}} = \tanh\sqrt{2} - \tanh\left(1/\sqrt{2}\right)$$
$$= 19.5^\circ\,.$$

While β in fact depends according to (11.122) on the velocity u of the boat, this is not true for Θ, which is independent of u.

11.13.3 Shock Waves

The enhancement of the wave amplitude in case when the source velocity u approaches or surpasses the sound velocity is not the only cause for the generation of shock waves. A very impressive example for the generation of excessive increase of the wave amplitude can be seen, when the phase velocity of ocean water waves approaching the coast decreases because it reaches shallow water. Now following waves can surpass the preceding waves and their amplitudes add up. If the total amplitude becomes larger than the water depth, the wave turns over because the wave maxima still propagate but the minima are slowed down due to friction with the ground. One observes rollers, which can give an impressive spectacle. This phenomenon of rollers can be demonstrated in a wave trough, with a sloped bottom.

Another spectacular example of shock waves is provided by exploding stars. The star material ejected with very high velocities collides with the molecules in the interstellar space. This results in a spherical compression wave with very high temperatures. It can be observed as luminous Ring Nebula. A famous example is the Ring Nebula in the constellation Lyra.

The investigation of shock waves in gases is performed in shock wave tubes (Fig. 11.81). A thin membrane separates the volume V_1 with a high gas pressure p_1 from the volume V_2 with low gas pressure p_2. At time $t = 0$ the membrane is burst, which generates the propagation of a pressure wave into the volume V_2. Pressure sensors measure the pressure $p(z)$ at different locations z. Spectroscopic techniques allow the determination of the temperature $T(z)$ and density $\varrho(z)$.

If the phase velocity of a wave depends on the wave amplitude nonlinear phenomena occur. Under certain conditions special wave fronts are generated which are called **Solitons** and which travel with constant amplitude without damping [11.14]. The have meanwhile found many interesting applications, as for instance in the telecom unication with laser pulses. They cannot

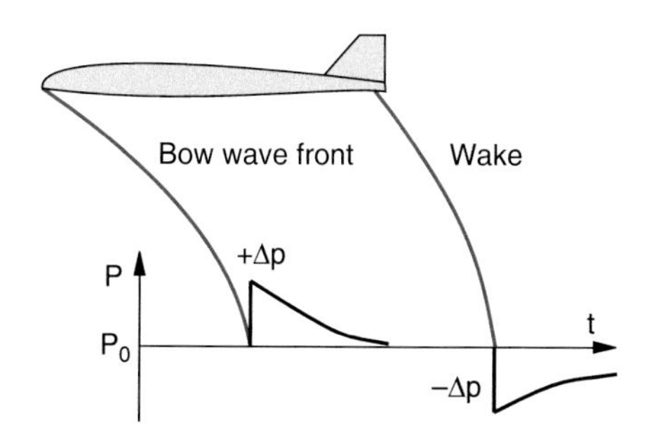

Figure 11.82 Shock waves of a supersonic plane

be described by the linear wave equation (11.62) because they represent essential nonlinear behaviour.

Shock waves are experienced as unpleasant bang, when a plane flies at low altitudes with supersonic velocities. Two bangs are heard generated by the bow wave and the tail wave (Fig. 11.82). The wave fronts of both contributions are curved, because the sound velocity depends on the altitude h since pressure $p(h)$ and temperature $T(h)$ vary with the height.

11.14 Acoustics

The acoustics covers the generation, propagation and detection of mechanical vibrations and acoustic waves. It therefore deals with the various applications of the subjects discussed in the previous sections of this chapter. Periodic pressure changes in air, which cause vibrations of our ear drum, are perceived by the human ear as sound. The frequency scale of sound waves extends, however, far beyond the audible range between 16 Hz and 16 kHz. The frequency range below 16 Hz is named infra-sound, that above 16 kHz ultra-sonic.

11.14.1 Definitions

The whole frequency spectrum of acoustics can be divided into the 4 ranges:

Infra-sound $\nu < 16\,\text{Hz}$,
audible sound $16\,\text{Hz} < \nu < 16\,\text{kHz}$,
ultra-sonics $\nu > 16\,\text{kHz}$,
hyper-sonics $\nu > 10\,\text{MHz}$.

Since the human ear is still the dominant instrument for the detection of sound, the sound waves audible for the ear are classified according to their frequency spectrum and their time-dependent amplitude $A(t)$.

We distinguish between:

- ***tone:*** A pure harmonic oscillation with an amplitude that is either constant or changes slowly compared with the period T of the oscillation. The pitch is determined by the frequency of the oscillation, the intensity of sound by the square A^2 of the amplitude.
- ***clang:*** A periodic but not purely sinusoidal oscillation. The Fourier-analysis gives a sum of sin- or cos oscillations. A clang is therefore a superposition of different tones.
- ***noise:*** Completely inharmonic oscillations. The frequencies and amplitudes of its Fourier components are randomly changing in time.
- ***bang:*** A sound pulse with rapidly increasing and decaying amplitude and a broad frequency spectrum.

The sensitivity of the human ear depends on the frequency. It has its maximum at about 3–4 kHz. The sound level sensed by the brain is proportional to the logarithm of the sound intensity (*Weber–Fechner-law*). The lowest sound intensity still audible by a healthy ear at $\nu = 1\,\text{kHz}$ is the threshold of hearing

$$I_{\min}(\nu = 1\,\text{kHz}) = 10^{-12}\,\text{W/m}^2\,.$$

Since the area of the outer ear is about $10^{-3}\,\text{m}^2$, a healthy ear can still detect a sound power of 10^{-15} W! Nature has optimized this hearing threshold in such a way, that the sound produced by the blood flow is just at this level. Children often press a shell against one ear and hear a noise. They are told that this is the noise of the ocean waves, but in fact it is the noise of the blood flow, which is amplified by the enlargement of the ear detection area.

One defines the subjectively sensed sound pressure level SL of a sound wave with the pressure p and the intensity $I(\nu) = p \cdot u \propto p^2$ (u = velocity of the particles oscillating in the sound wave) as

$$\text{SL} = 20 \cdot \log_{10}\left(\frac{p}{p_0}\right) = 10 \cdot \log_{10}\left(\frac{I(\nu)}{I_{\min}}\right)\,, \tag{11.123}$$

where $I_{\min}(1\,\text{kHz}) = 10^{-12}\,\text{W/m}^2 = p_0 u$ is the hearing threshold and $p_0 = 20\,\mu\text{Pa} = 2{\cdot}10^{-5}$ Pa is the lower threshold pressure of the ear at a frequency of 2 kHz. It is given in the unit 1 phon (although it is a dimensionless number). The threshold of pain of our ear is at 130 phons, this is 10^{13} times higher than the hearing threshold, which is at 0 phons. When the phon number increases by 10, the sound intensity increases by a factor 10, i.e. 10 times.

In a harmonic acoustic wave

$$\xi = A \cdot \cos(\omega t - kz)$$

is the velocity u of the oscillating particles

$$u = \frac{\partial \xi}{\partial t} = -\omega \cdot A \cdot \sin(\omega t - kz)\,.$$

Its maximum value

$$u_0 = \omega \cdot A$$

is the velocity amplitude (maximum sound particle velocity).

Rather than using the subjectively sensed units *phon* an objective measure of the sound power, which is independent of the special person is the **decibel**.

It is defined by the **sound power level**

$$L_P = 10 \cdot \log \frac{P}{P_0}\ \text{decibel}\,,$$

where P is the power emitted by a source and $P_0 = 10^{-12}$ W is the hearing threshold for sound waves with a cross section of $1\,\text{m}^2$ in air.

Table 11.4 Examples of phon numbers for some acoustic sources

Low whisper	10 phon
Clear speech	50 phon
Jet plane at 100 m distance	≈ 120 phon
Discotheque	100–130 phon
Jack hammer at 1 m distance	130 phon

Examples

An acoustic source that emits a power of 1 W has a sound power level of $10 \cdot \log 10^{12} = 120$ decibel. The sound power level at the hearing threshold 10^{-12} W is $L_P = 10 \cdot \log 1 = 0$ decibel. A source S_1 that emits 100 times the power emitted by S_2 differs by 20 decibels by a factor of 100. If the decibel level increases by a factor of ten, the sound power also increases 10 fold (Tab. 11.4). ◄

The sound power is a characteristic property of the sound source. It is independent of the distance, in contrast the sound pressure level and the sound intensity which decrease as $1/r^2$ with the distance r from the source.

11.14.2 Pressure Amplitude and Energy Density of Acoustic Waves

According to Eq. 11.79a and 11.80 the relation between the pressure $p = p_0 + \Delta p$ in an acoustic wave in gases and the amplitude ξ is given by

$$\frac{\partial p}{\partial z} = -\varrho \cdot \frac{\partial^2 \xi}{\partial t^2} . \qquad (11.124)$$

With $\xi = \xi_0 \cdot \cos(\omega t - kz)$ this gives

$$\frac{\partial p}{\partial z} = \varrho \cdot \omega^2 \xi_0 \cos(\omega t - kz) . \qquad (11.125)$$

Integration over z yields with $k = 2\pi/\lambda$

$$p = -\varrho \cdot \omega^2 \cdot \frac{\lambda}{2\pi} \xi_0 \cdot \sin(\omega t - kz) + C .$$

The integration constant C is determined by the condition that without the sound wave ($\xi = 0$) the pressure is $p = p_0$, because $\Delta p = 0$. This gives the equation for the pressure wave

$$p = p_0 + \Delta p_0 \cdot \sin(\omega t - kz) \qquad (11.126)$$

with the pressure amplitude

$$\Delta p_0 = -v_{\text{Ph}} \varrho \omega \xi_0 = -v_{\text{Ph}} \varrho u_0 , \qquad (11.127)$$

where the relations $v_{\text{Ph}} = \omega/k$ and $u_0 = (\partial\xi/\partial t)_0 = \omega \cdot \xi_0$ have been used.

The mean energy density of the wave follows from (11.86) and (11.127) as

$$\frac{dW}{dV} = w = \frac{1}{2} \varrho \omega^2 \xi_0^2 = \frac{1}{2} \frac{\Delta p_0^2}{\varrho \cdot v_{\text{Ph}}^2} . \qquad (11.128)$$

The relation between the energy density and the particle velocity u in a sound wave can be derived as follows:

In a harmonic acoustic wave

$$\xi = \xi_0 \cdot \cos(\omega t - kz) ,$$

the velocity of the oscillating particles is

$$u = \partial \xi / \partial t = -\omega \cdot \xi_0 \cdot \sin(\omega t - kz) .$$

Its maximum value is according to (11.127)

$$u_0 = \omega \cdot A = \Delta p_0 / (\varrho \cdot v_{\text{Ph}}) ,$$

and is named the **acoustic particle velocity**.

Inserting u_0 into (11.128), the energy density can be written as

$$w = \frac{1}{2} \varrho \cdot u_0^2 ,$$

which is equal to the kinetic energy of the particles per unit volume oscillating in the acoustic wave.

The energy flux density of the wave (intensity I with the dimension W/m^2 is then

$$I = v_{\text{Ph}} \cdot \frac{dW}{dV} = \frac{1}{2} \frac{\Delta p_0^2}{\varrho \cdot v_{\text{Ph}}} = \frac{1}{2} v_{\text{Ph}} \varrho u_0^2 . \qquad (11.129)$$

The total power P emitted by an acoustic source is equal to the intensity of the emitted wave, integrated over a closed surface surrounding the source

$$P = \oint I \cdot dA .$$

The sound pressure level is defined as

$$L_{\text{p}} = 10 \log_{10} \left(\frac{\Delta p}{\Delta p_{\text{s}}} \right)^2 = 20 \log_{10} \frac{\Delta p}{\Delta p_{\text{s}}} , \qquad (11.129\text{a})$$

where $\Delta p_{\text{s}} = 2 \cdot 10^{-5}\,\text{Pa} = 2 \cdot 10^{-10}$ bar is the sound pressure at the threshold of hearing.

Both quantities, the sound pressure level L_{p} and the sound power level L_{P} are relative quantities, giving the ratio of sound pressure to a given lower limit Δp_{s}, measured in decibel dB resp. the ratio of sound power to the power at the threshold of hearing, also given in decibel (Tab. 11.5). They are a measure for the logarithm of the ratios $\Delta p/\Delta p_{\text{s}}$ of sound pressures or P/P_{s} of power levels.

Table 11.5 Sound power and sound power levels of some sound sources

Sound source	Sound power	Sound power level
Rocket propulsion engine	10^6 W	180 dB
Jet propulsion engine	10^4 W	160 dB
Sirene	10^3 W	150 dB
Jack hammer	1 W	120 dB
Loud speech	10^{-3} W	90 dB
Normal conversation	10^{-5} W	70 dB
Average apartment in quiet surrounding	10^{-7} W	50 dB
Whisper	10^{-9} W	30 dB
Rustling of leaves	10^{-10} W	20 dB
Hearing threshold	10^{-12} W	0 dB

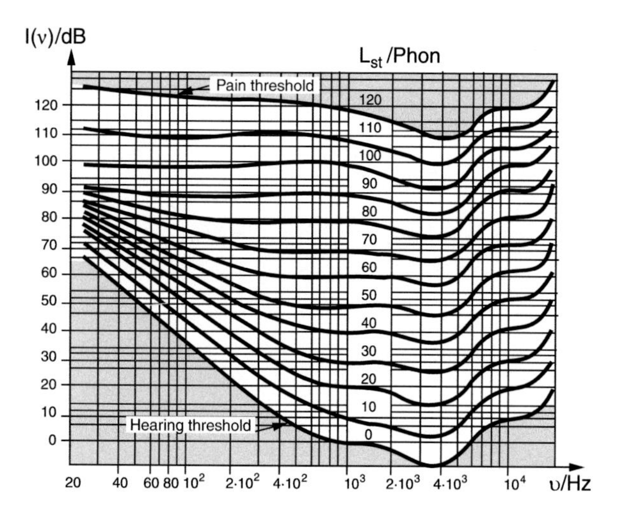

Figure 11.83 Frequency dependence of sound intensity $I(\nu)$ (measured in dB) compared with the phon numbers illustrated by the horizontal lines

Examples

1. The sound pressure level is 60 dB, when $20 \cdot \log(\Delta p/\Delta p_s) = 60 \Rightarrow \log(\Delta p/\Delta p_s) = 3 \Rightarrow \Delta p = 10^3 \cdot \Delta p_s = 2 \cdot 10^{-2}$ Pa.
2. The sound power level is 80 dB if $\log(P/P_s) = 8 \Rightarrow P = 10^8 \cdot P_s = 10^{-4}$ W. ◀

The sensitivity of the ear depends on the frequency. Therefore the curves of equal phon values as a function of frequency are not horizontal lines in a diagram of sound power level against frequency (Fig. 11.83) but have a minimum at the frequency of about 4 kHz where the sensitivity is maximum.

11.14.3 Sound Generators

Sound waves can be generated by free or forced oscillations of solid bodies, which emit their oscillation energy into their surroundings. Examples are loud speakers, vibrating strings and tuning forks or vibrating membranes (Fig. 11.84). Also a gas

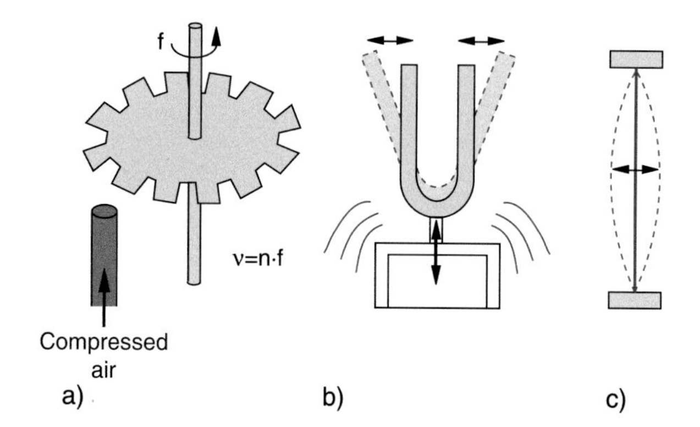

Figure 11.84 Different soundwave sources: **a** Hooter with n teeth of the rotation disc; **b** tuning fork; **c** vibrating string

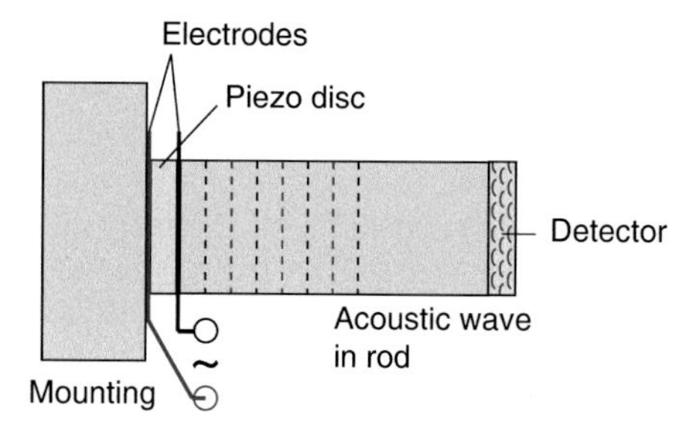

Figure 11.85 Piezo disc as acouctic source mounted on the end of a solid rod

stream, periodically interrupted, produces sound waves. It is used in hooters and in all wind instruments, such as trumpets, horns or flutes.

For the investigation of sound waves in solids and their propagation piezo-crystals can be used. They consist of materials that expand or contract if an external electric voltage is applied. If a disc of piezo material is attached to the end face of a rod (Fig. 11.85), application of an alternating voltage between the two end faces of the piezo disc produces mechanical oscillation of the disc, which propagate as acoustic waves through the rod. A second piezo disc at the other end of the rod serves as detector of the acoustic waves. Measuring the phase difference between the sound generation and the detection yields the sound velocity through the rod.

A commonly used acoustic source is the loudspeaker (Fig. 11.86), where a membrane is attached to a solenoid in a permanent magnet. If an electric current is send through the solenoid, it becomes magnetic and is attracted or repelled in the field of a permanent magnet, depending on its magnetic polarity (see Vol. 2). The attached membrane follows the movement of the solenoid which produces pressure waves that propagate as acoustic waves in the surrounding air.

11.14.4 Sound-Detectors

Besides the human ear, all devices, sensitive to pressure changes, can be used as sound detectors. Examples for demon-

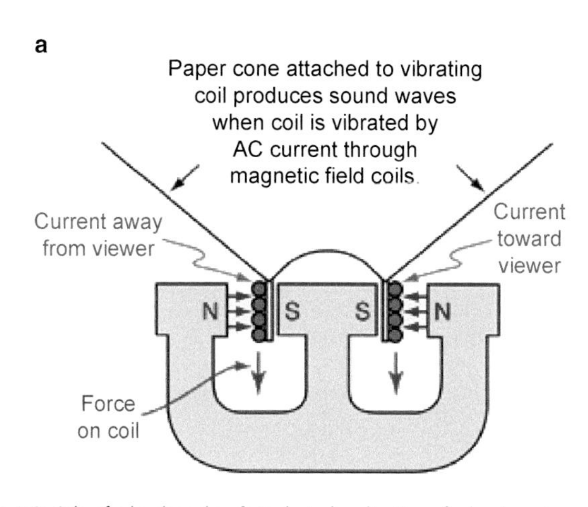

Figure 11.86 **a** Principle of a loudspeaker; **b** Technical realization of a loudspeaker

stration experiments are the Rubens's flame tube (Fig. 11.71) or Kundt's powder figures (Fig. 11.70). In practice mainly microphones are used because of their sensitivity and simple usage. They convert mechanical oscillations into electric alternating voltages. They can be regarded as the reversals of loudspeakers.

For higher frequencies (ultrasonics), the inverse piezo effect in ceramic materials (for instance $BaTiO_3$) can be utilized where pressure changes between the end faces of a piezo crystal induce electric voltages.

Nowadays more and more optical detectors for acoustic waves are used.

For instance, the vibration of a membrane can be made visible by a stroboscopic technique: The membrane is illuminated by a periodic sequence of optical pulses with a variable frequency f. If the difference frequency $f - \nu$ between the pulse frequency f and the vibration frequency ν is sufficiently small, the vibration can be seen at the difference frequency. This allows the observation of details during one cycle of the vibration. If $f = \nu$ the membrane seems to stand still.

Another optical detection technique is based on the Doppler shift. When a vibrating plane $z = z_0 + a \cdot \cos(\Omega t)$ is illuminated with a monochromatic light beam with the optical frequency ω_0 (for example a laser beam) the frequency of the reflected light is Doppler-shifted to

$$\omega' = \omega_0 \left(1 + \frac{2a}{c} \Omega \sin \Omega t\right) . \qquad (11.130)$$

The superposition of the reflected with the incident light beams results in a beat frequency Ω, which is equal to the vibration frequency of the plane.

A standing acoustic wave in transparent media results in a spatially periodic variation of the pressure and therefore also of the refractive index (See Vol. 2). This periodic variation of the refractive index acts as an optical phase grating. When a light beam passes through this grating diffraction phenomena occur which depend on the period of the grating and can be therefore used to measure the acoustic wavelength.

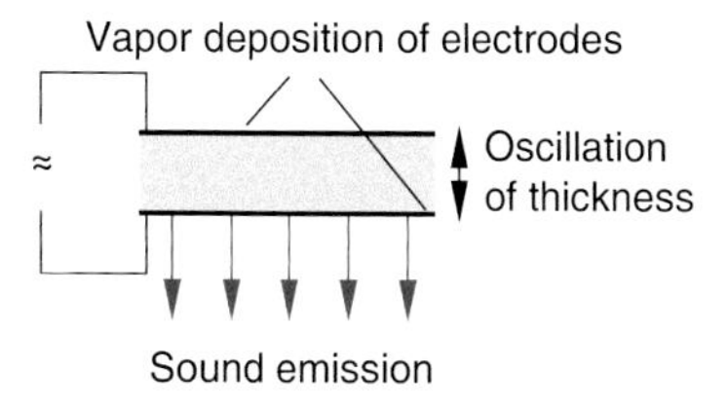

Figure 11.87 Principle of piezo-electric sound generation

11.14.5 Ultrasound

Acoustic waves with frequencies above the audible range ($\nu > 20\,\mathrm{kHz}$) are named supersonic waves. They are generated by electro-acoustic transducers, which convert electrical high frequency voltages into acoustic vibrations. Examples are Piezo-discs (consisting of silica, Barium-Titanate $BaTiO_3$, $PbTiO_3$ or $PbZrO_3$: These are crystals that change the thickness of the disc when an electric voltage is applied (Fig. 11.87). An rf-voltage between the two sides of the disc induces mechanical vibrations. Either single discs can be used or stacks of many discs up to 150.

Sometimes also stacks of nickel sheets are used as magnetostrictive transducers. Here an external magnetic rf-field presses the sheets together or increases the distance between the sheets, thus causing a periodic change of the thickness of the whole stack.

The mechanical oscillation amplitude becomes maximum, when a resonant standing acoustic wave is built up in the oscillating system. For the piezo discs this means that the thickness d of the disc must be an odd integer multiple of the half wavelength λ ($d = (2m+1) \cdot \lambda/2$), because at both surfaces must be oscillation maxima of the standing ultrasonic wave.

Example

From the relation $v_{\mathrm{Ph}} = \nu \cdot \lambda$ is follows that for a frequency $\nu = 1\,\mathrm{MHz}$ and a phase velocity $v_{\mathrm{Ph}} = 5000\,\mathrm{m/s}$

in the piezo crystal the wavelength is $\lambda = 5\,\mathrm{mm}$. For $d = \lambda/2 \Rightarrow d = 2.5\,\mathrm{mm}$. For a higher frequency $\nu = 10\,\mathrm{MHz}$ the thickness d for the fundamental resonance would be $d = 0.25\,\mathrm{mm}$ and therefore too thin for sufficient mechanical stability. In such cases one has to use higher harmonics with $d = (2m+1)\lambda/2$ for the resonances. For $m = 4 \Rightarrow d = 2.25\,\mathrm{mm}$, which ensures a sufficiently stable disc. ◀

The propagation of ultrasonic waves in solids is influenced by absorption, scattering and reflection at surfaces. The absorption strongly depends on the frequency and on the material of the probe. Scattering is mainly caused by inhomogeneities with spatial extensions between 0.1λ to 5λ. The reflection occurs at boundary surfaces between different materials with a different acoustic impedance Z_{acoustic}, which is defined as

$$Z_{\mathrm{acoustic}} = \varrho \cdot v_{\mathrm{s}} ,$$

where ϱ is the density of the material and v_{s} the sound velocity. The reflection coefficient R is

$$R = I_{\mathrm{r}}/I_{\mathrm{inc}} = [(Z_1 - Z_2)/(Z_1 + Z_2)]^2 ,$$

where I_{inc} is the incident intensity and I_{r} the reflected intensity. The reflection of sound waves at boundary surfaces increases with the difference of the two acoustic impedances. This is quite similar to the reflection of light waves where the acoustic impedance is replaced by the index of refraction.

The reflection can be reduced by placing an anti-reflection layer of a material with acoustic thickness $\lambda/4$ between ultrasonic source and sample. This causes nodes of the standing wave at both boundaries, which suppresses any reflection.

For medical ultrasonic inspection a special gel which is applied between the ultrasonic transducer and the skin is used as antireflection layer.

11.14.6 Applications of Ultrasound

The technical development of new sources and detectors for ultrasonic with increasing performance have greatly enlarged the different fields of their applications. For the solution of technical problems as well as in medical diagnostics ultrasonic investigations are routinely applied [11.17–11.18b]. The optimum frequency of the ultrasonic waves depends on the wanted spatial resolution and the penetration depth into the sample.

11.14.6.1 Technical Applications

There are many examples where ultrasonic can be applied for the solution of technical problems: measurements of wall thickness in pipes and containers, investigations of inhomogeneity in solids, e. g. internal fissures in walls or formation of granules in crystals. The spatial resolution is limited by the wavelength λ of the ultrasonic wave. For a frequency of $10\,\mathrm{MHz}$ and a phase velocity $v_{\mathrm{Ph}} = 5000\,\mathrm{m/s}$ in solids the wavelength is $\lambda = v_{\mathrm{Ph}}/\nu = 0.5\,\mathrm{mm}$. For achieving a higher resolution, one has to increase the frequency. Typical frequencies are within the range from $250\,\mathrm{kHz}$ to $100\,\mathrm{MHz}$ resulting in a spatial resolution between $2\,\mathrm{cm}$ and $50\,\mu\mathrm{m}$.

The determination of wall thicknesses uses short ultrasonic pulses. It is based on the measurement of time intervals between the pulse reflected at the front side of the sample and the pulse reflected at the backside.

Example

With a time resolution of $10^{-7}\,\mathrm{s}$ and a sound velocity of $5000\,\mathrm{m/s}$ the thickness can be measured with an accuracy of $0.5\,\mathrm{mm}$. ◀

An important field of applications is the cleaning of surfaces and fabrics. The sample to be cleaned is placed in a tank filled with a liquid. An ultrasonic transducer at the wall of the tank irradiates the sample. The vibrations of the water molecules impinging on the sample leads to a mechanical removal of the dirt particles. Modern developments use already washing machines without detergents, which clean the laundry with ultrasonic, thus avoiding the pollution of the drain water by phosphates.

11.14.6.2 Applications in Medicine

While X-ray diagnostics mainly detects hard substances in the body (bones, cartilages and sclerotic precipitates) ultrasonic inspection is also sensitive for soft tissue (kidney, liver, hall bladder, stomach, sinews or ligaments). Because of the smaller sound velocity in tissue ($v = 1000\,\mathrm{m/s}$) a higher spatial resolution as in technical solids can be obtained at the same ultrasonic frequency.

Example

For frequencies $\nu > 1\,\mathrm{MHz}$ and a sound velocity of $v_{\mathrm{s}} = 1000\,\mathrm{m/s}$ a spatial resolution $\Delta r < 1\,\mathrm{mm}$ can be reached. Modern devices allow a resolution of $0.1\,\mathrm{mm}$ and are able to resolve finer details inside our body. In order to reduce the sound resistance between the transducer and the skin of the body a coupling gel is applied. An important application of ultrasonic is the examination of pregnant women, where the development of the foetus can be followed up during the different stages of the pregnancy. ◀

An advantage of ultrasonic inspection against X-ray diagnosis is not only the absent radiation damage but also the better time resolution which allows the inspection of dynamical processes in the body. For instance, it is meanwhile possible to visualize in detail the contraction and expansion of the heart during one beat cycle [11.18a, 11.18b].

11.14.7 Techniques of Ultrasonic Diagnosis

Different techniques for ultrasonic diagnosis have been developed. Here we will discuss only two of them.

11.14.7.1 The Echo-Pulse Method

Here short ultrasonic pulses with a pulse sequence frequency f (i. e. a time interval $T = 1/f$ between the pulses) are sent into the sample (Fig. 11.88). The pulse reflected by a surface at the distance d from the source is received by the detector at time $\Delta t = 2d/v_s < T$. During the pulse-free time T between two pulses, the ultrasonic source consisting of a piezo-crystal with an applied rf-voltage acts now as detector. The reflected sound waves induce vibrations in the piezo-crystal, which generate electrical oscillations. On the oscilloscope the emitted pulse and the reflected pulses appear. The third pulse comes from the back surface of the sample. The heights of the pulses depends on the reflectivity of the surfaces. The time interval between the pulses allows the determination of the location of the reflecting surface and the pulse height gives the sound wave resistances of the different media in the sample.

This technique gives a one-dimensional image of the inspected sample. Since the amplitude of the pulses are measured, the technique is called A-image technique.

11.14.7.2 The B-Image Technique

This method yields sectional images, where the letter B stands for "brightness-modulation". The reflected pulses control the brightness of the oscilloscope. The ultrasonic transducer is shifted perpendicular to the sound wave propagation or is tilted and the resulting sectional images are recorded, stored and added to a two-dimensional picture on the scope. The electron beam impinging on the oscilloscope screen is synchronously shifted with the motion of the ultrasonic transducer. The intensity of the reflected ultrasonic waves is converted into the grey scale or the colour of the composite oscilloscope picture. This facilitates the recognition of structures in the investigated tissue. If only reflected pulses that arrive within a certain time interval are selected, a specified layer of the tissue is monitored. This tomogram technique gives a three-dimensional picture of the inspected material.

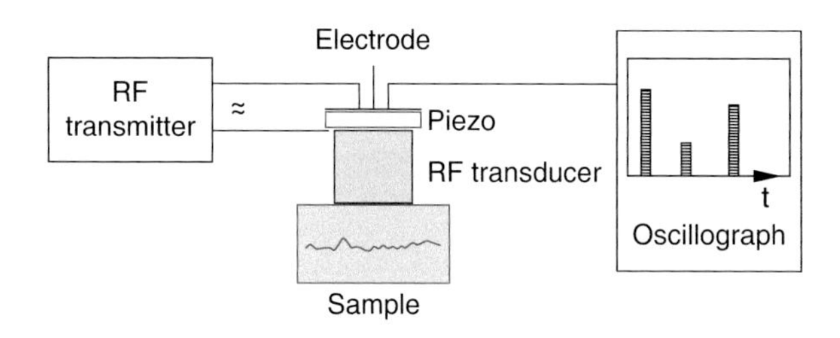

Figure 11.88 Schematic illustration of the design for the ultrasonic echo-pulse technique

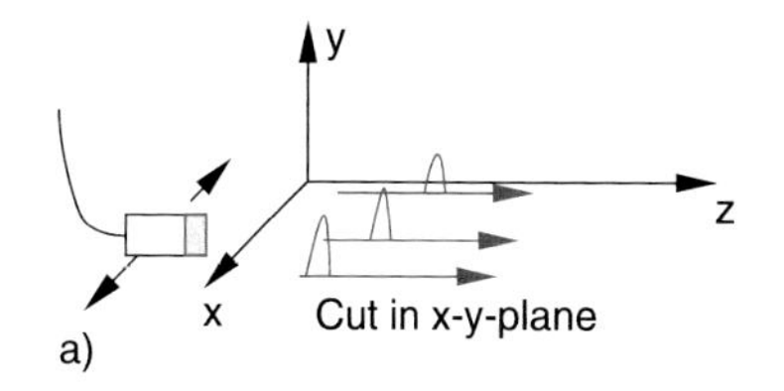

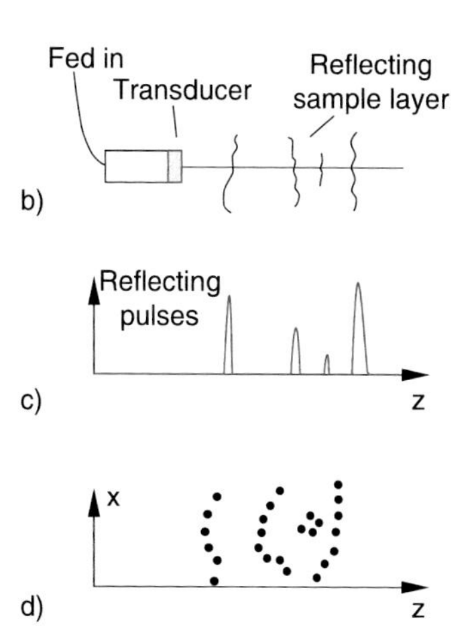

Figure 11.89 **a** Shift of ultrasonic transducer in x-direction for getting different sectional images in the x–z-plane; **b** principle of echo-technique; **c** one-dimensional recording; **d** two-dimensional echo-image obtained with the B-method

In Fig. 11.89 the one-dimensional A-method is schematically compared with the three-dimensional B-method, where the time-measurement gives information on the third dimension.

The advantage of the A-method is its larger recording speed, which allows the visualization of fast processes in the human body in real time. Examples are the detailed visualization of the motion of selected parts of the pulsating heart, or the motion of an embryo in the uterus.

Example

Heart inspection by ultrasonic:
The ultrasonic transducer is places on the skin of the patient. The distance from the skin to the heart is about 15 cm. The time between emitted and received ultrasonic pulse is

$$\Delta t = 2 \cdot \frac{\Delta s}{v_{Ph}} = 2 \cdot \frac{0.15\,\mathrm{m}}{1500\,\mathrm{m/s}} = 100\,\mu\mathrm{s}\,.$$

The maximum pulse repetition frequency is then 5 kHz. ◄

The advantage of the B-method is its capability of producing two-or three-dimensional pictures, which give a better impression of the spatial structure of selected parts of the body. This is further improved by the ultrasonic tomography, where similar to the X-ray tomography many sectional images are composed by the computer to a total three-dimensional image. By combination with the time-resolved echo-technique the spatial resolution and the contrast of the imaged structures can be greatly enhanced. With modern high frequency ultrasonic devices a spatial resolution in the sub-millimetre range can be achieved. This allows the recognition of abnormalities of a foetus in the uterus and opens the possibility of early corrective action.

Blood stream velocities in the artery or veins can be measured with the *Doppler-ultrasonic method*. Here the frequency shift between the emitted and the reflected pulse is utilized to measure with high spatial resolution the blood stream velocity and detect a local narrowing of the artery or widening of the arteries (aneurism). Both phenomena can lead to the death of a patient. Also the motion of the heart valves can be visualized thus allowing the detection of heart valve failures.

The great advantage of ultra-sonic versus X-ray diagnostics is the fact that no damage of the body occurs (if excessive sound intensities are avoided). Therefore, also pregnant women or small children can be examined without the danger of damages.

11.15 Physics of Musical Instruments

The basic physical laws of musical instruments have fascinated many scientists already in the 19th century [11.19]. This had, however, nearly no influence on the construction and optimization of these instruments, which were still based on old traditional recipes, and on techniques that were passed on to many generations within a family. The skilfulness and the sense of hearing of the instrument maker were the essential basis for making a good instrument.

During recent years, the scientific investigation of the quality of musical instruments has found increasing interest, both by physicists as well as by instrument makers. This has led to an intense cooperation. The main reason is the availability of new measuring techniques and detectors such as microphones, storage oscilloscopes, Fourier-analysers and computer simulations, which are able to characterize the tone colour of a musical instrument in more detail and can reproduce it in form of a Fourier-diagram in an objective way, independent of the musical sensibility of the listener. The question why a *Stradivari* violin sounds so much better than an ordinary violin or what distinguishes a *Bechstein* grand piano from an ordinary piano can thus be answered by scientific methods.

The aim of such investigations is the answer to the following questions:

- Which properties of the material influence in which way the tone quality of the instrument?
- How do acoustic resonances determine the tone quality?
- Which relations exist between the Fourier-spectrum of an instrument and the subjective feeling of its quality?
- Why do aging processes of the material affect the tone quality?

The hope is a scientific understanding of the quantitative relations between the Fourier-spectrum and the appraisal of an instrument and the possibility of giving quantitative instructions for making a Stradivari violin or a Bechstein-piano and to select the proper material in order to achieve this goal.

11.15.1 Classification of Musical Instruments

Musical instruments can be sorted in 5 groups according to the kind of sound generation:

- *String instruments*, where tight strings are excited to oscillations by plucking (guitar, harp and harpsichord), by striking (piano) or by bowing (violin, cello, contrabass).
  The different tone colour of the various string instruments is caused by the string tension, the string materials, the form and material of the resonance body and in particular by the strength and duration of the string plucking, i. e. by the musician.
- *Wind instruments*, which use the oscillations of air columns in cylindrical, conical or bent pipes for the generation of tones. To this group belong all wood instruments such as recorders, oboe and bassoon, the brass instruments (Trumpet, Trombone, tuba, natural horn), and the organ with its different wind pipes.
- *Percussion instruments*, where oscillations of membranes (drum, tympanum), or of special forms of solid bodies (triangle, cimbalom, xylophone, button gong, chimes and carillon).
- *Electronic musical instruments*, where electronic oscillations are generated, which are converted into acoustic oscillations by a loud speaker. The frequency spectrum of these instruments and its time variations can be controlled mechanically or by computer programs. Therefore the tone colour of all other instruments can be imitated.
- Of course the *human voice* should be also regarded as musical instrument. Because of its wide variety of tone colour and frequency spectrum and its modulation capability, it is often called the *Queen of all instruments*. Lovers of the organ use this name, however, for the organ. In a physical sense the human voice belongs to the wind instruments.

11.15.2 Chords, Musical Scale and Tuning

A "tone" of an musical instrument is generally no pure sine-oscillation but includes besides the fundamental frequency ν several overtones with frequencies $n \cdot \nu$ $(n = 2, 3, 4, \ldots)$. This overtone spectrum $I(n \cdot \nu)$ is characteristic for the different instruments. The violin, for example, has a completely different overtone spectrum than the piano.

We perceive the superposition of two or more tones as *harmonic*, if the tones have as many common overtones as possible. The superposition of more than two such tones is called a *chord*.

The most common overtones have two fundamental tones with a frequency ratio $\nu_1 : \nu_2 = 2 : 1$. The interval between these

Chapter 11

Table 11.6 Tone interval for a "pure tune" within one octace

Interval	Frequency ratio $\nu_2 : \nu_1$
Octave	2 : 1
Fifth	3 : 2
Fourth	4 : 3
Major third	5 : 4
Minor third	6 : 5
Major sixth	5 : 3
Minor sixth	8 : 5
Minor seventh	9 : 5
Major seventh	15 : 8
Major second	9 : 8
Minor second	16 : 15

two tones is the *octave*. The intervals of tones within one octave are in the European music ratios of small integers, which allow the formation of many harmonic chords. Our polyphonic music is based on these harmonic chords. They are compiled in Tab. 11.6.

The series of tones that is ordered within one octave according to increasing frequencies is called a *musical scale*. In Tab. 11.7 the C-major scale is listed as one example.

The two tables illustrate that the frequency ratios of most intervals are indeed ratios of small integers and have therefore many common overtones. A musical instrument (e. g. a piano) that is tuned according to the frequency intervals of Tab. 11.7 has a pure tuning, because it sounds in this key particularly pure and harmonic. This pure tuning is also called Pythagorean tuning, because Pythagoras found already 2000 years ago that a string, fixed on both sides generates harmonic tones when it is subdivided in two parts with certain lengths ratios.

The numerical values given in Tab. 11.7 illustrate, that the frequency ratios of two successive tones do not always have the same value, but attain the ratios $9/8$, $10/9$, and $16/15$. This has the consequence that an instrument that has a pure tuning for the C-major key, does not sound purely in another key. For example the quint C–G in the C-major key has the frequency ratio $3:2$

Table 11.7 C-major scale

Tone	Relative Frequency	Frequency ratio	Equally tempered scale		Interval
c	1				
		9 : 8		$\sqrt[6]{2}$	
d	9/8		$2^{2/12}$		Major second
		10 : 9		$\sqrt[6]{2}$	
e	5/4		$2^{4/12}$		Major third
		16 : 15		$\sqrt[12]{2}$	
f	4/3		$2^{5/12}$		Fourth
		9 : 8		$\sqrt[6]{2}$	
g	3/2		$2^{7/12}$		Fifth
		10 : 9		$\sqrt[6]{2}$	
a	5/3		$2^{9/12}$		Major sixth
		9 : 8		$\sqrt[6]{2}$	
h	15/8		$2^{11/12}$		Major seventh
		16 : 15		$\sqrt[12]{2}$	
c	2		2		Octave

but the quint D-A for the D-major key the ratio $40/27$. Musical instruments (e. g. the violin) that can be readily tuned to any key by varying the frequency, can overcome this problem and can be played with a pure tuning. The tone "a" has then in the C-major key a slightly different frequency than in the D-major key.

In order to use instruments that cannot be tuned readily to any key (for instance the piano or the organ) the musicians have agreed to the compromise of the *equally tempered scale*. Here the frequency ratios of two subsequent tones within an octave of twelve halftones is always $\sqrt[12]{2} = 1.05946$. The only pure interval is now the octave with a frequency ratio $2:1$ (see Fig. 11.90). In order to show that with this agreement an instrument could be played in all keys, J. S. Bach wrote his famous work "Wohltemperiertes Klavier" = piano with equally tempered scale), where he composed pieces in all possible keys.

In this equally tempered scale there are intervals with $\nu_x/\nu_{x-1} = \sqrt[6]{2}$ called *whole-tones* and intervals with $\nu_x/\nu_{x-1} = \sqrt[12]{2}$, called

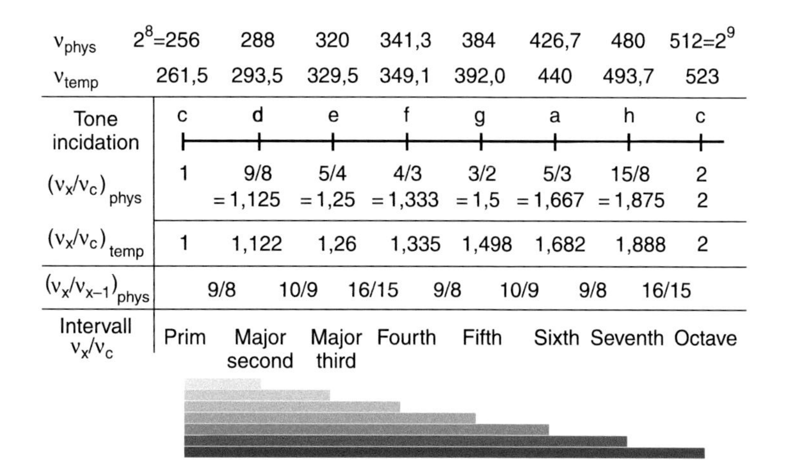

Figure 11.90 Pure tuning ν_{phys}, equally tempered scale ν_{temp} and frequency ratios of the successive tones of the C-major scale

Chapter 11

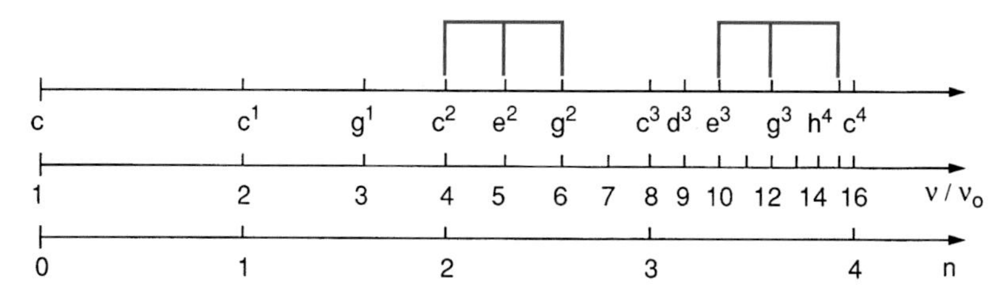

Figure 11.91 Triads and their frequency-ratios

half-tones. In the C-major key the only half-tones are the intervals e–f and h–c, while all other intervals are whole-tones. The only really pure interval in the equally tempered scale is the octave.

Since most instruments span a scale over several octaves, the tones in different octaves must be differently marked. This is done by defining the absolute frequency of a tone. One chooses as standard tone the *concert pitch* "a" with a frequency $f_a = 440\,\mathrm{Hz}$. All tones within the same octave as the concert pitch a are marked by the upper exponent 1 or by an upper apostrophe '. The C-major scale around the concert pitch a is then c', d', .e', f', g', a', h'. All tones in the octave above are marked by the upper exponent 2 or by a double apostrophe ", etc. The octave below the concert pitch are named as *small octave*, *grand octave* and *contra-octave*. The lowest tone c on the piano has the frequency $f = 32\,\mathrm{Hz}$, the next higher c has $f = 65\,\mathrm{Hz}$. The highest c = c"" has $f = 4158\,\mathrm{Hz}$.

Consonant chores are a superposition of overtones with the same fundamental tone. Examples are shown in Fig. 11.91.

11.15.3 Physics of the Violin

The primary sound source of the violin is the string, clamped on both sides, which is bowed by a violin bow. The vibration of the string is transferred by a violin bridge and through the air to the resonance body of the violin. This resonance body has a special form and is made of selected wood, which has been stored for many years in order to have the optimum resonance condition for the sound produced by bowing the strings.

When the violin bow is uniformly bowed over the string, the string will be taken along with the moving bow at the touch point due to the static friction, until the restoring force of the deflected string becomes larger than the static friction force. Now the string jumps back into its equilibrium position, because the sliding friction is smaller than the static friction (see Sect. 5.6.2). The detailed motion of the string depends on the force with which the bow is pressed against the string and on the position of the touch point relative to the fix points of the string (Fig. 11.92).

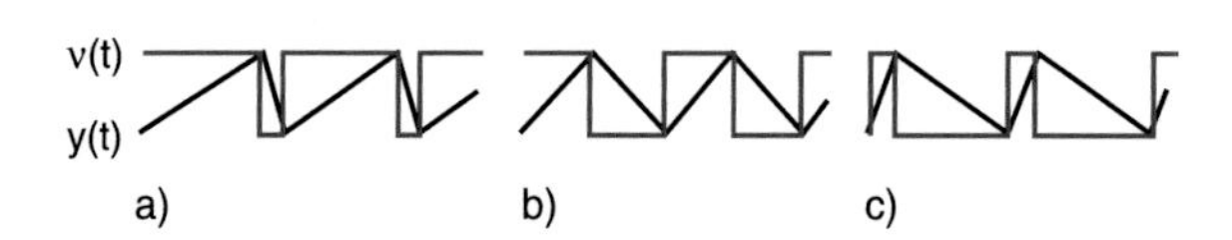

Figure 11.92 Elongation $y(t)$ and velocity $v(t) = \dot{y}(t)$ of the string at the location of the bow under a uniform bow stroke. **a** Strong; **b** medium; **c** weak press-on string [11.20]

The oscillation of the string is therefore by no means a sine function, it has a rich overtone spectrum (see Sect. 11.3.1). The fundamental tone is determined by the length L of the string, its mass μ per unit length and by its tensile force. From equation (11.77) we obtain for the phase velocity

$$\begin{aligned} v_{\mathrm{Ph}} &= \sqrt{F/\mu} = \nu \cdot \lambda = 2\nu_0 \cdot L \\ \Rightarrow \nu_0 &= \frac{1}{2L}\sqrt{F/\mu}\,. \end{aligned} \tag{11.131}$$

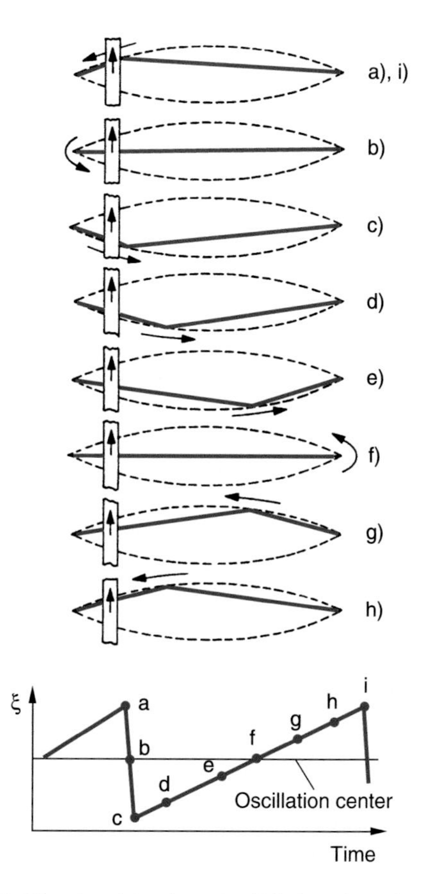

Figure 11.93 Migration of the elongation kink along the violin string during one oscillation period

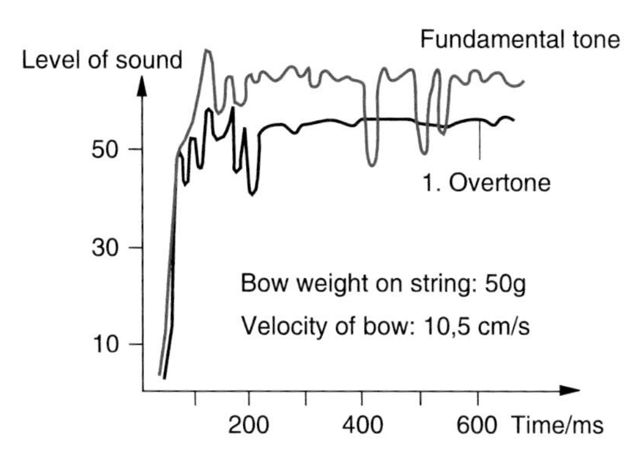

Figure 11.94 Amplitudes $a_n(t)$ of fundamental tune and first overtune for uniform bow

The kink-shaped deflection of the string moves during one oscillation period along the whole string, as shown in Fig. 11.93. The kink passes along the dashed curve, which represents a momentary picture of a standing sine-wave with the wavelength $\lambda = 2L$ at the time of the maximum elongation.

The time-dependent deflection of the string can be described by the Fourier series

$$\xi(x, t) = \sum_n a_n(t) \cdot \sin \frac{n\pi x}{L} \sin(n\omega t) . \qquad (11.132)$$

The amplitudes $a(t)$ of the different overtones are not constant, even for a uniform bow motion but show a behaviour as illustrated in Fig. 11.94.

Of particular importance for the sound of a violin is the resonance body. For a good violin, it has a broad resonance spectrum around the maximum of the ear sensitivity. The wider the resonance spectrum is the more brilliant is the sound of the violin. For comparison, Fig. 11.95 shows the average values of the resonance curves of ten Old Italian violins (upper curve) and that of trivial violins. Significant differences in the range between 1 kHz and 4 kHz are obvious.

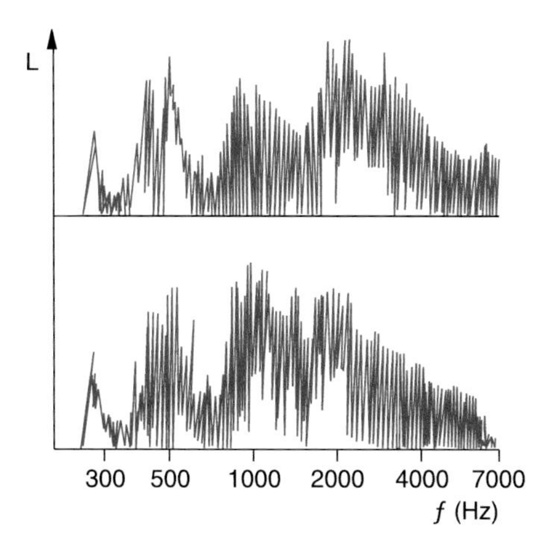

Figure 11.95 Comparison of the frequency spectrum of an old Italian violin (*above*) and a cheap manufactured violin (*below*)

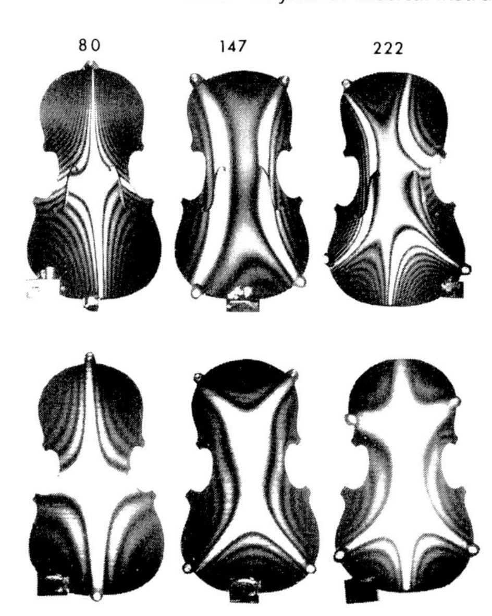

Figure 11.96 Holographic interferograms of the violin top (*above*) and bottom (*below*) for different frequencies. The patterns give mean elongations of the surface, which differ between lines of equal elongation by $\lambda(\approx 0.6\,\mu\text{m})$

With stroboscopic and holographic methods, the elongations of the cap (above) and the bottom (below) of the resonance body can be visualized for different frequencies. Such pictures show which part of the resonance body preferentially emits sound wave (Fig. 11.96).

11.15.4 Physics of the Piano

In the piano, a tone is generated by a key stroke, activating a hammer of felt that strikes a tense string. The sound of the generated tone depends on the material of the hammer and the character of the string (thickness and tensile force) and on the resonance body. In Fig. 11.97 the amplitudes $\xi(t)$ of the transverse elongations are shown for three strings excited at different fundamental frequencies. It shows that the duration t_a of the hammer stroke is for low frequencies small compared with the oscillation period $T = 1/\nu_0$. It furthermore illustrates that the oscillation strongly deviates from a sine function. For higher frequencies, the stroke duration t_a becomes comparable with the oscillation period.

The overtone spectrum of a C_4 string with the fundamental frequency $\nu_1 = 262\,\text{Hz}$, shown in Fig. 11.98, demonstrates that for hard strokes the overtone spectrum is much more pronounced than for soft strokes.

The frequencies of the overtones of the fundamental ν_1 are not exactly at $n \cdot \nu_1$ (Fig. 11.99), because the acoustic velocity v_{Ph} depends on the frequency and since the frequency $\nu(n) = v_{\text{Ph}}/\lambda = v_{\text{Ph}}(\nu)/L/n$ of the n-th harmonic is not equal to $n \cdot \nu_1$. The physical reason for this inharmonicity is the larger rigidity of thick strings (for low tones) compared with that of thin strings for the higher tones. This inharmonicity essentially influences the sound of the piano tone.

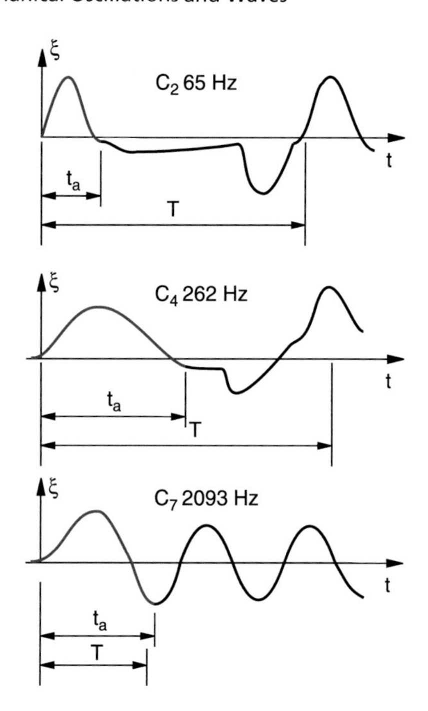

Figure 11.97 Transversal elongations $\xi(t)$ of three piano strings for different tone frequencies. Relation between strike time of hammer t_a and oscillation period T

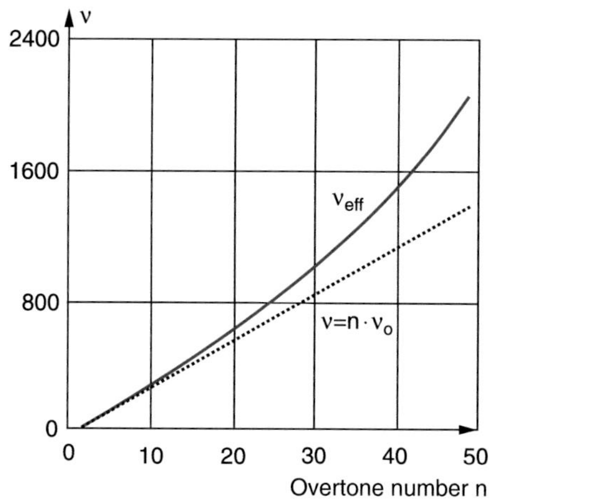

Figure 11.99 Anharmonicity of overtunes $n \cdot \nu_0$ of the lowest piano string A. The real frequenciy of the overtunes is higher than $n \cdot \nu_0$ [11.21]

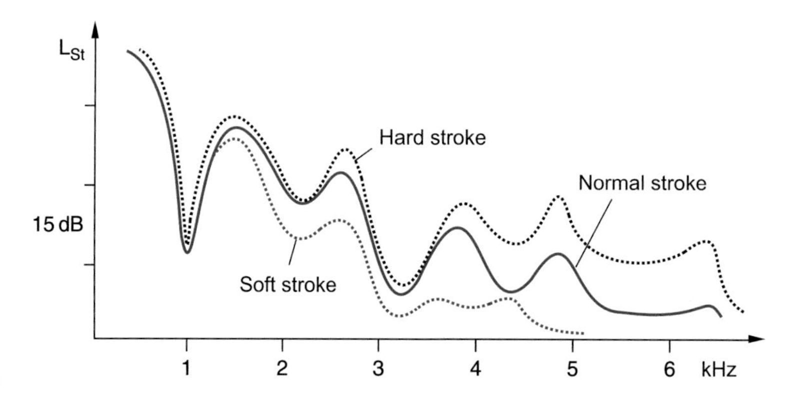

Figure 11.98 Frequency spectrum $L_{st}(\nu)$ of the C_4-string ($\nu_0 = 262$ Hz) of the piano under soft, normal und hard strokes

Summary

- The free undamped one-dimensional oscillator performs a harmonic motion $x = A \cdot \cos(\omega t + \varphi)$ that is determined by its amplitude A, its circlar frequency $\omega = 2\pi \cdot \nu$ and its phase shift φ. The total energy of the oscillation as the sum of kinetic and potential energy is constant in time.
- As long as the deviation from the equilibrium position is sufficiently small, the restoring force is proportional to the elongation and the oscillation amplitude is a linear function of the restoring force. Such linear oscillations are mathematically described by real sine-or cos-functions or by complex functions $e^{i\omega t} + e^{-i\omega t}$ which, however, must be combined in such a way that the total amplitude of the oscillation is always real.

- The superposition of different one-dimensional oscillations with equal frequencies gives again an oscillation with the same frequency but a phase φ, that depends on the phases of the superposed oscillations. For different frequencies the superposition is a more complicated oscillation with time dependent amplitude which is determined by the amplitudes, the phases and the frequencies of the different partial oscillations. The superposition can be reduced to the sum of the partial oscillations by a Fourier-analysis.
- The superposition of oscillations in the x and the y-direction gives two-dimensional curves in the x–y-plane (Lissajous-figures). These are closed curves if the frequency ratio ω_1/ω_2 of the two frequencies are rational numbers.
- For a damped oscillator kinetic energy of the oscillation is transferred into other forms of energy, e. g. into heat of friction. For small damping the oscillation amplitude decreases exponentially. For large friction no oscillation can be realized. The amplitude decreases after a first elongation exponentially against zero.
- For a forced oscillation external energy is periodically fed into the oscillating system. After an initial transient period a stationary oscillation appears with the frequency of the external force. The losses of the damped system are just compensated by the external energy supply. For the resonance case (exciting frequency = resonance frequency of the system) very large amplitudes are reached which might even destroy the system (resonance disaster).
- For a parametric oscillator the oscillation parameters (amplitude, frequency or phase) are periodically changed. This is achieved by a periodic external energy supply. The parametric oscillation characteristics depends on the ratio of exciting frequency to resonance frequency of the system.
- A coupling between two independent oscillators leads to a periodic energy exchange between the two oscillators. The oscillation amplitudes change periodically, where the frequency of the modulation depends on the strength of the coupling. The complicated motion of the system of N coupled oscillators with different frequencies can be always reduced to the superposition of N normal vibrations.
- A wave is the propagation in space of a local perturbation of the equilibrium state. The propagation of a harmonic oscillation gives a harmonic sine-wave. The propagation of a mechanical wave is accomplished by the coupling of oscillating mass particles to its neighbouring particles.
- For transverse waves the elongation occurs perpendicular to the propagation direction, for longitudinal waves in the propagation direction. Transverse wave can occur only in media with a shear modulus $G \neq 0$. In the inside of liquids is $G = 0$ and no transverse waves can appear. At the surface of liquids surface tension and gravity act as restoring forces which can enable the propagation of transvers waves.
- The phase velocity $v_{\mathrm{Ph}} = \omega/k$ of a wave traveling through a medium depends on the properties of this medium. When $v_{\mathrm{Ph}}(\lambda)$ depends on the wavelength, the medium shows dispersion.
- Longitudinal waves in solids have the phase velocity $v_{\mathrm{Ph}} = \sqrt{E/\varrho}$, which depends on the elasticity modulus E and density ϱ, while for transverse waves $v_{\mathrm{Ph}} = \sqrt{G/\varrho}$ depends on the shear modulus G.
- In gases only longitudinal waves are possible. Their phase velocity $v_{\mathrm{Ph}} = \sqrt{p/\varrho}$ depends on pressure and density of the gas.
- The phase velocity of transverse waves at liquid surfaces depends on the surface tension, on the depth of the liquid and on the wavelength. These waves show dispersion.
- The intensity of a wave $I = (1/2)\, v_{\mathrm{Ph}} \cdot \varrho \cdot A^2 \cdot \omega^2$ gives the energy flux density measured in $[\mathrm{W/m^2}]$. It is proportional to the squares of amplitude and frequency.
- A wave packet (= wave group) is generated by the superposition of an infinite number of partial waves with frequencies inside a spectral range $\Delta\omega$ around a mid-frequency ω_0. Its group velocity $v_{\mathrm{G}} = \mathrm{d}\omega/\mathrm{d}k$ gives the velocity of the maximum $A_{\max}(\omega_0)$ of the amplitude distribution $A(\omega)$. In media with dispersion, group velocity and phase velocity are different.
- Huygens' principle states that every point on a wave front is the source of a spherical wave. The resultant wave is the superposition of all such elementary waves. This principle explains all phenomena of the wave propagation, such as reflection, refraction and diffraction.
- Two waves with equal frequencies are called coherent if their superposition gives at every point within the superposition region a constant phase difference between the two waves. The superposition of coherent waves results in interference structures.
- Transverse waves can be realized with linear, circular or elliptical polarization. For longitudinal waves there is only one type because the oscillation direction is always in the propagation direction.
- All linear waves are solutions of the linear wave equation

$$\Delta\xi = (1/v^2)\partial^2\xi/\partial t^2$$

with $v = v_{\mathrm{Ph}}$.
- Standing waves are generated by the superposition of running waves under specific boundary conditions. They represent spatially periodic stationary oscillation patterns.
- Shock waves are non-periodic waves where a short local perturbation propagates as a singular steep pressure change.
- The energy of a sound wave in the volume ΔV

$$\Delta W = (1/2)\,\Delta p^2/(\varrho \cdot v_{\mathrm{Ph}}^2) \cdot \Delta V$$

is proportional to the square of the pressure change Δp. The energy density is $\Delta W/\Delta V$, the energy flux density (intensity) of a sound wave is $I = v_{\mathrm{Ph}} \cdot \Delta W/\Delta V$.
- Sound waves in gases are generated by oscillating solid bodies that transfer their oscillation energy partly to the surrounding gas. Examples are electro-strictive piezo materials or loud speakers.
- Sound waves can be detected by membranes that are excited to oscillations by the acoustic wave. If the oscillations occur in a magnetic field electric signals are generated (microphone).

Problems

11.1 An elastic spring is elongated by the force $F = 1\,\text{N}$ by 5 cm. What is the oscillation period T if a mass of 1 kg is attached to one end of the spring? The mass of the spring can be neglected.

11.2 A homogeneous steel wire with length L and mass M is vertically suspended. At its lower end a mass m is attached. Now the upper suspension is shifted for a short moment in the horizontal direction, causing a transvers wave pulse travelling downwards. At the same time a steel ball is released from the suspension point of the wire and falls down. Where does the ball overtake the wave pulse (air friction can be neglected). What is the minimum value of the ratio m/M in order to realize the overtaking?

11.3 Show, that for a complex representation of a forced oscillation only the imaginary part consumes energy. What is the role of the real part?

11.4 A soap bubble with radius R, wall thickness d and density ϱ performs radial oscillations due to the restoring force of the surface tension σ. Calculate the oscillation period T as a function of R, ϱ, and σ.

11.5 A plane longitudinal acoustic wave with the frequency $\nu = 10\,\text{kHz}$ and the amplitude $A = 10^{-4}\,\text{m}$ propagates through a steel rod ($E = 22 \cdot 10^{10}\,\text{N/m}^2$, density $\varrho = 8 \cdot 10^3\,\text{kg/m}^3$). How large are the maximum tension σ and the phase velocity v_{Ph}?

11.6 How large is the oscillation amplitude of a sound wave in air and the maximum velocity of the oscillating particles for a frequency $\nu = 1\,\text{kHz}$

a) at the hearing threshold (0 dB)
b) at the absolute threshold of pain (130 dB)?

Compare the results with the mean free pathlength Λ in air and the thermal velocity of the molecules at $T = 300\,\text{K}$.

11.7 A U-shaped tube with 2 cm inner diameter contains water with the mass 0.5 kg. When the water column in one branch is shortly pressed down by $\Delta z = 10\,\text{cm}$ and then released, the water column begins to oscillate. How large is the oscillation period and what are the maximum velocity and acceleration of the water? How large is the damping, if we use the values for the viscosity η in Tab. 8.1.

11.8 A sound wave with a frequency of 2 kHz impinges vertically onto a sound-damping wall, which has, however, a vertical free slit with 0.5 m width. A pedestrian walks on the other side of the wall parallel to the wall at a distance of 20 m. Along which path length can he receive more than 50% (5%) of the sound power incident onto the other side of the wall.

11.9 A plane sound wave impinges vertically onto a water surface The sound velocities are: $v_{\text{air}} = 334\,\text{m/s}$; $v_{\text{water}} = 1480\,\text{m/s}$. Which fraction of the incident sound power is reflected, which fraction propagates into the water? Compare also the intensities of the reflected and transmitted sound waves.

11.10 Two plane sound waves $\xi = A \cdot \cos(800t - 2z)$ and $\xi_2 = A \cdot \cos(630t - 1.5z)$ superimpose. How looks the interference pattern? What is the group velocity compared with the phase velocities of the partial waves?

11.11 What is the phase velocity of ocean waves with $\lambda = 500\,\text{m}$ at a large water depth. Compare this result with waves on a lake with $\lambda = 0.5\,\text{m}$, which are generated by throwing a stone into the water.

11.12 A string with length $L = 1\,\text{m}$, density $\varrho = 7.8 \cdot 10^3\,\text{kg/m}^3$ and mass $m = \mu L$ is clamped on both ends. A tensile stress $\sigma = 3 \cdot 1010\,\text{N/m}^2$ acts on the string. The string is deflected in its mid by $\Delta r \ll L$. The form of the deflected string can be represented by a triangle. What are oscillation frequency and oscillation period after ending the deflection.

11.13 A mass $m = 2000\,\text{kg}$ is suspended at the lower end of a thin steel rope with $L = 2\,\text{m}$. What is the period of vertical oscillations of the mass m? Compare the result with the horizontal pendulum oscillation of the mass m.

11.14 At the end of a thin laminated spring of length $L = 10\,\text{cm}$ with the resonance frequency $\omega = 2\pi \cdot 100\,\text{s}^{-1}$ is a mass $m = 100\,\text{g}$ attached. What is the frequency shift by this mass?

11.15 A buoy consisting of a cylindrical pipe with length L floats vertically in water. Without waves, the part $a \cdot L (a < 1)$ immerses in the water. What is the amplitude of the vertical oscillation of the buoy, when sine waves with the total heights $2h$ (from the wave maximum to the minimum) with the period T appear? Numerical example: $a \cdot L = 30\,\text{m}$, $h = 2\,\text{m}$, $T = 5\,\text{s}$. How large must L be that the wave maximum just reaches the upper peak of the buoy?

11.16 Prove that the spherical wave $\xi = (A/r) \cdot e^{i(\omega t - kr)}$ is a solution of the wave equation $\Delta \xi = (1/v_{\text{Ph}}^2)\partial^2 \xi / \partial t^2$.

11.17 The first tone of a police siren has the frequency $\nu_1 = 390\,\text{Hz}$, the second tone has $\nu_2 = 520\,\text{Hz}$ (the tone ratio is a major fourth).

a) At which speed of the approaching police car are both tones higher by a whole tone ($\nu_1' = 1.12246 \cdot \nu_1$, $\nu_2' = 1.12246 \cdot \nu_2$).
b) At which velocity has the observer to approach the police car at rest, in order to hear the same tone shift?

11.18 A cube with mass $m = 2\,\text{kg}$ is at rest in the mid between two springs with the spring constants $D_0 = 100\,\text{N/m}$. It can slide on a horizontal rail (Fig. 11.100). The friction force is $\boldsymbol{F}_\text{f} = f \cdot m \cdot g$. The coefficient of the sliding friction is $f_1 = 0.3$, that of the static friction $f_0 = 0.9$.

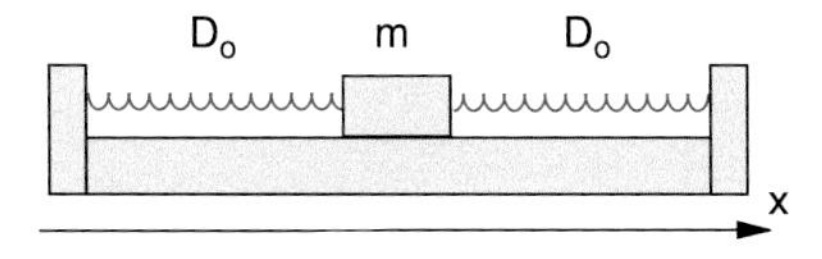

Figure 11.100 To Probl. 11.18

a) According to which physical law do the amplitudes decrease? (Hint: Consider the energy ratio at successive reversal points x_n and x_{n+1} at opposite sides of the oscillation.)
b) At which x-position comes the mass to a rest, if it has been released at a position 22 cm away from the equilibrium position?
c) What is the time difference between two successive reversal points?

References

11.1. R. Bracewell, *The Fourier Transform and its Applications.* (McGraw Hill, New York, 1999)
11.2. Ph. Dennery, A. Krtzywiki, *Mathematics for Physicists.* (Dover Publications, Mineola, New York, USA, 2012)
11.3. B.R. Martin, G.P. Shaw, *Mathematics for Physicists.* (John Wiley and Sons, Hoboken, N.J., USA, 2015)
11.4. K. Weltner et al., *Mathematics for Physicists and Engineers.* (Springer, Berlin, Heidelberg, 2014)
11.5. C.S. Lüdde, R.M. Dreizler, *Theoretical Mechanics.* (Springer, Berlin, Heidelberg 2010)
11.6. M. Tenenbaum, H. Pollard, *Ordinary Differential Equations.* (Dover Publications, Mineola, USA, 1985)
11.7. R. Bronson, G. Costa, *Schaum's Outline of differential Equations,* 4th ed. (Mc GrawHill, New York, 2014)
11.8. N.W. Ashcroft, N.D. Mermin, *Solid State Physics.* (Cengage Learning Inc., Boston, MA, USA, 1976)
11.9. Ch. Kittel, *Introduction to Solid State Physics.* (John Wiley and Sons, Hoboken, N.J., USA, 2004)
11.10. American Institute of Physics, *Handbook,* 3rd ed. (Mcgraw-Hill, Texas, USA, 1972)
11.11. CRC, *Handbook of Chemistry and Physics,* 96th ed. (CRC Press, Cleveland, Ohio, USA, 2015)
11.12. B. Le Mehauté, *An Introduction to Hydrodynamics and Water Waves.* (Springer, Berlin, Heidelberg, 1976)
11.13. G. Joos, I.M. Freeman, *Theoretical Physics.* (Dover Publications, Mineola, USA, 1986)
11.14. Th. Dauxoid, M. Peyrard, *Physics of Solitons.* (Cambridge Univ. Press, Cambridge, M.A., USA, 2006)
11.15. D. Ensminger, L.J. Bond, *Ultrasonics: Fundamentals, Technologies and Applications,* 3rd ed. (CRC Press, Cleveland, Ohio, USA, 2011)
11.16. F.S Crawford, Am. J. Phys. **52**, 782 (1984)
11.17. L.C. Lynnworth, *Industrial Applications of Ultrasound.* (IEEE Transactions of Sonics and Ultrasonic, Vol. 8U 22, March 1975)
11.18a. F.W. Kremkau, *Diagnostic Utrasound: Principles and Instrumentation,* 5th ed. (W.B. Saunders, Philadelphia, 2001)
11.18b. F.W. Kremkau, *Sonography: Principles and Applications.* (Saunders-Elsevier, New York, 2015)
11.19. N.H. Fletcher, Th.D. Rossing, *The Physics of Musical Intruments.* (Springer, Berlin, Heidelberg, 2008)
11.20. B. Parker, *Good Vibrations: The Physics of Music.* (John Hopkins University Press, Baltimore, USA, 2009)
11.21. A. Wood, *The Physics of Music.* (Read Books, 2008)
11.22. https://www.amazon.co.uk/Definitive-Ealing-Studios-Collection-DVD/dp

Nonlinear Dynamics and Chaos

12

© Springer International Publishing Switzerland 2017
W. Demtröder, *Mechanics and Thermodynamics*, Undergraduate Lecture Notes in Physics, DOI 10.1007/978-3-319-27877-3_12

In Chap. 2 we have discussed the motion of point masses under the influence of forces. The equations of motion are linear differential equations. If the complete initial conditions are given, (e. g. location and velocity at time $t = 0$) the solution of the differential equation determines exactly the future fate of the point mass (its location and velocity at future times $t > 0$), as long as the forces and their changes with time are known.

For cases where the equation of motion has no analytical solutions and requires a numerical integration the accuracy of the results is only limited by numerical uncertainties, which can be minimized by using sufficiently fast computers.

In such cases of exact predictions, the motion of a body or the time development of a system are called *strictly deterministic*. For exact initial conditions, exact predictions of the future development are possible.

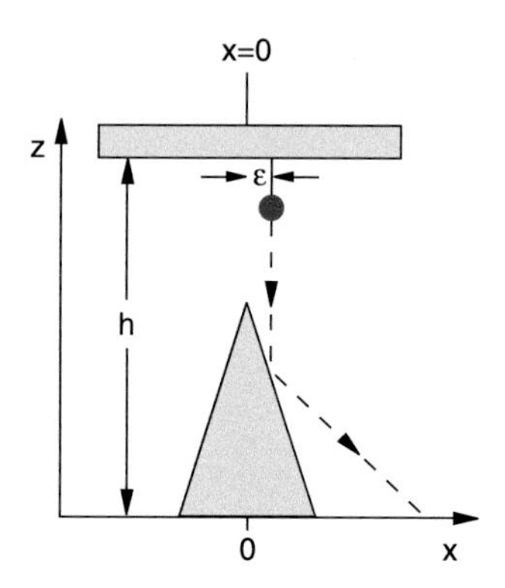

Figure 12.1 The final location of the falling ball depends very sensitive on the initial position around $x = 0$

> If small deviations of the initial conditions cause only small changes of the future development of the system, we call the solutions of the equation of motion stable.

Examples of such stable solutions are the motion of the moon around the earth, or of the earth around the sun. Small perturbation of the central gravitational force field by the influences of the other planets (or in case of the moon by the sun) lead only to small corrections of the trajectories of the earth or the moon, which can be calculated within the framework of perturbation theory. As long as the solutions of the equation of motion are stable, such small corrections do not destroy the predictability of future positions and velocities.

Although we are accustomed in daily life to the normality of stable solutions, there are numerous examples of unstable solutions, where tiny changes of the initial conditions result in a completely different future development of a system, which then lead to completely different final states.

A simple example is a ball, that is released from the point ($x = 0$, $z = h$) and falls down. During its fall it hits a body with two sloped plain surfaces and a sharp edge at the top (Fig. 12.1). If the initial point is only shifted by a tiny amount δx to the right, the ball hits the right slope of the obstacle and is reflected to the point ($x = x_1$, $z = 0$) while for $\delta x < 0$ the ball hits the left side and is deflected to the point ($x = -x_1$, $z = 0$).

A second example is the parametric oscillator (see Sect. 11.7) as an oscillating system that is driven by an external periodic force. It can be realized, for instance, by a simple pendulum with a string length $L = L_0 + \Delta L_0 \cdot \cos(\omega t + \alpha)$ that is periodically changed and a large oscillation amplitude where the restoring force $m \cdot g \cdot \sin\varphi$ can no longer approximated by $m \cdot g \cdot \varphi$ (see Sect. 2.9.6), Eq. 2.79). The equation of motion of such a driven pendulum

$$(L_0 + \Delta L_0 \cos(\omega t + \alpha))\,\ddot{\varphi} + \gamma\dot{\varphi} + g \cdot \sin\varphi = 0 \qquad (12.1)$$

is nonlinear. For certain regimes of the parameters ΔL_0, ω and α (amplitude, frequency and phase of the external force) the amplitude grows until the angel φ exceeds the value π. Then the periodic pendulum motion changes into an irregular circular motion where the function $\varphi(t)$ shows a chaotic behaviour (Fig. 12.2).

Another example is the motion of a planet that moves in the gravitational field of two stars (double star system) that is quite common in the universe. Its trajectory depends critically on the initial conditions. It can be stable, for instance, but for tiny changes of the initial conditions the motion of the planet becomes unstable. It either leaves the system or it collides with one of the two stars.

A more difficile example is the motion of a body around the planet Saturn in the range of the ring system. Here ranges for the distances r to Saturn are found where the superposition of the gravitational forces by Saturn and by its inner moons leads to unstable trajectories of the body while slightly different radii show stable motions. The unstable ranges are those where the gaps in the ring system occurs. In the unstable ranges the ratio T_b/T_m of the circulation periods of the body and of one of the inner moons equals the ratio p/q of small integers p and q. A very small change of the initial radius r can convert a stable trajectory into an unstable one.

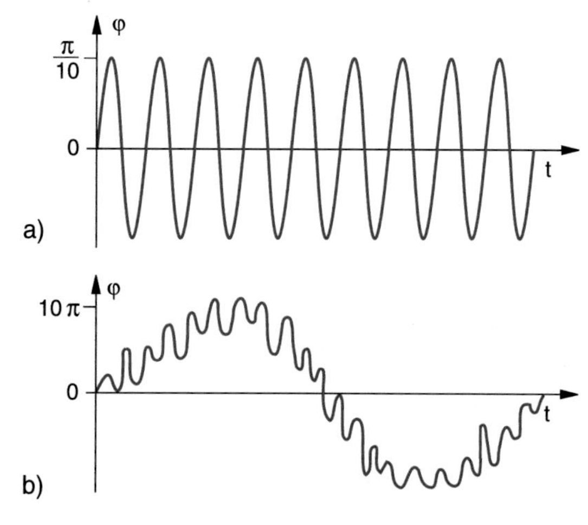

Figure 12.2 Pendulum oscillations driven by an external force. **a** Small amplitude in the stable linear range; **b** in the unstable range with large amplitude. The ordinate scale is in (**b**) 100 times larger than in (**a**)

A similar phenomenon occurs in the asteroid belt between Mars and Jupiter, where the time-dependent gravitational attraction between asteroid and Jupiter leads for certain radii to unstable asteroid trajectories and causes the observed gaps in the asteroid belt.

All of these examples correspond to equations of motion that contain at least one nonlinear term that is responsible for the unstable motion. For the example of the parametric pendulum, these are the nonlinear restoring force $m \cdot g \cdot \sin\varphi$ and the nonlinear driving force. For certain ranges of the parameters a and b (a = ratio of the resonance frequency ω_0 to the driving frequency ω, and b gives the ratio of the amplitudes of driving force to restoring force) the motion becomes unstable.

The solutions of linear equations have the following noteworthy property: They obey the superposition principle, which has been demonstrated in Sect. 11.3 for the example of linear oscillations: If $x_1(t)$ and $x_2(t)$ are solutions of the linear equation, every linear combination $x(t) = a \cdot x_1(t) + b \cdot x_2(t)$ is also a solution.

We will show that this superposition principle no longer holds for the solutions of nonlinear equations It is replaced by another principle: The *scale invariance* or *self-similarity*.

As was illustrated by the forgoing examples, for many parameter ranges there are no stable solutions of nonlinear equations. This means that even for very small changes of the initial conditions the time development of a system leads to completely different final states. Since generally the initial states are not exactly defined (because of measuring uncertainties), the predictability is severely limited for such systems.

Most processes in nature can be described only approximately by linearized equations, although in many cases an admirable accuracy is reached (for example by prediction of moon- or solar eclipses). The exact equation of motion should contain the nonlinear terms. If these terms lead to unstable developments of the system, we called this a **chaotic behaviour**. Examples of such chaotic behaviour can be for instance, found in Meteorology. They demonstrate that the difficulty to predict exactly the forthcoming weather is not an indication of the missing capability of the meteorologists but an inherent feature of the chaotic system.

In spite of these difficulties a lot of surprising statements can be made about the solutions of nonlinear equations and the behaviour of nonlinear systems. The investigation of chaotic systems is the subject of *Chaos Research*, which can be only shortly discussed here. For a more extensive study of this fascinating field the reader is referred to the literature [12.1a–12.6b]

12.1 Stability of Dynamical Systems

A dynamical system changes with time, in contrast to a stationary system, which has reached an equilibrium state that does no longer change in time.

The dynamical system can be described by time-dependent parameters $\xi_i(t)$ $(i = 1, 2, 3, \ldots, N)$. The quantities $\xi(t)$ can be, for example, the coordinates $x_i(t)$ and the velocities $v_i(t)$ of a point mass moving on its trajectory, or they characterize the time dependent state of a system of many particles, for instance pressure $p(t)$ and temperature $T(t)$. They can also describe the number of subjects in a biological system where the population changes with time.

If the state of a dynamical system at time t_2 is unambiguously determined by its state at the earlier time $t_1 < t_2$ we call the dynamics deterministic, in contrast to the stochastic or random dynamics, where for the development of the system only probabilities can be given, no certain and unambiguous predictions.

When the state of a system at time t is characterized by the N quantities

$$X(t) = \{\xi_1(t), \xi_2(t), \ldots, \xi_N(t)\} , \tag{12.2}$$

which we can condense in the vector $X(t)$, the change of the system in time is described by

$$\dot{X}(t) = \left\{ \frac{\mathrm{d}\xi_1(t)}{\mathrm{d}t}, \frac{\mathrm{d}\xi_2(t)}{\mathrm{d}t}, \ldots, \frac{\mathrm{d}\xi_N(t)}{\mathrm{d}t} \right\} . \tag{12.3}$$

If the system converges towards a stationary (time-independent) state and reaches it in a finite time t_f the condition

$$\dot{X}(t_f) = 0$$

must be fulfilled. If the stationary state is only reached at $t = \infty$, the condition is

$$\lim_{t\to\infty} \dot{X}(t) = 0 .$$

An example for the first kind is the function

$$X(t) = X_0 + a \cdot t^2 \quad \text{for} \quad t < 0 \quad \text{and} \quad X = X_0 \quad \text{for} \quad t > 0,$$

$$\to \dot{X}(t) = 2at \quad \text{for} \quad t < 0 \quad \text{and} \quad \dot{X}(t) = 0 \quad \text{for} \quad t \geq 0 .$$

In many cases the system approaches a stationary state only asymptotically and reaches it for $t = \infty$.

Example: The population of radioactive atoms, decaying with a decay constant λ is

$$N = N_0 e^{-\lambda t} ,$$

it approaches $N(t = \infty) = 0$ only after an infinite time.

Often the situation occurs that the state of a system does not change continuously but in finite steps. An example is the number of living species in a biological population, where the birth rate is not constant over the year but births happen only in spring. Such discrete dynamics can be described by finite difference-equations compared to differential equations for continuous dynamics. For example the number N_{n+1} of subjects in the $(n + 1)$th generation is determined by the population in the nth generation and the birth- and death-rate:

$$N_{n+1} = N_n + B_n - D_n , \tag{12.4}$$

where the difference $N_{n+1} - N_n$, which is determined by the birth- and death-rate, is not a continuous but discrete function of time (see Sect. 12.2).

We name the N-dimensional space with the coordinates $\{\xi_1, \xi_2, \ldots, \xi_N\}$ the **phase space** of the system. In this phase space the state of the system at time t_0 is represented by the point $\boldsymbol{X}(t_0)$. The time-development of the system then corresponds to the trajectory $\boldsymbol{X}(t)$ in the phase space. This representation of the time development of the system by a trajectory in the phase space is called the *mapping* of the system. The vector $\dot{\boldsymbol{X}}(t)$, which gives the velocity of the point $\boldsymbol{X}(t)$ in the phase space, maps how fast the system changes its state. The stationary states of the system, given by $\dot{\boldsymbol{X}}(t) = 0$ are called the **fix points**. If the system is deterministic, only one trajectory can pass through each point $X(t)$, which is not a fix point. Only in a fix point many (often an infinite number) of trajectories can concur. Therefore such a fix point is also called an **attractor**. The range of all X-values that converge towards an attractor is called the intake area of the attractor:

In nonlinear systems not only points but also curves or areas can occur as attractors. However, they do not represent fix points (see Example 5).

Examples

1. The undamped harmonic oscillator with linear restoring force $F = -D \cdot x$ has the energy (see Sect. 11.6)

$$E = \frac{m}{2}\dot{x}^2 + \frac{1}{2}Dx^2 , \qquad (12.5)$$

and the oscillation frequency $\omega_0 = \sqrt{D/m}$.
From (12.5) one obtains immediately the two-dimensional phase space trajectory

$$x^2 + \left(\frac{\dot{x}}{\omega_0}\right)^2 = \frac{2E}{D} .$$

In a phase space with the axes x and $\dot{x}/\omega_0$ the trajectory becomes a circle around the origin with the radius $R = \sqrt{2E/D}$ (Fig. 12.3a) For each value of E (initial condition) the system passes with constant frequency a well-defined circle. The motion is stable.

2. For the damped oscillator the energy decreases exponentially (Sect. 11.6). The corresponding trajectory in phase space is obtained from the equation of motion

$$\ddot{x} + 2\gamma\dot{x} + \omega_0^2 x = 0 . \qquad (12.6)$$

Equation 12.6 can be also written as (see Eq. 11.31)

$$\frac{\mathrm{d}}{\mathrm{d}t}\left(x^2 + \left(\frac{\dot{x}}{\omega_0}\right)^2\right) = -4\gamma\left(\frac{\dot{x}}{\omega_0}\right)^2 . \qquad (12.7)$$

Equation 12.7 represents in a phase space with coordinates x and $\dot{x}/\omega$ a spiral, which converges against the origin as a stable attractor (Fig. 12.3b)

3. For a negative damping (the energy loss is overcompensated by an external force) the oscillation amplitude increases with time and the trajectory is a spiral with increasing radius, approaching $r = \infty$.

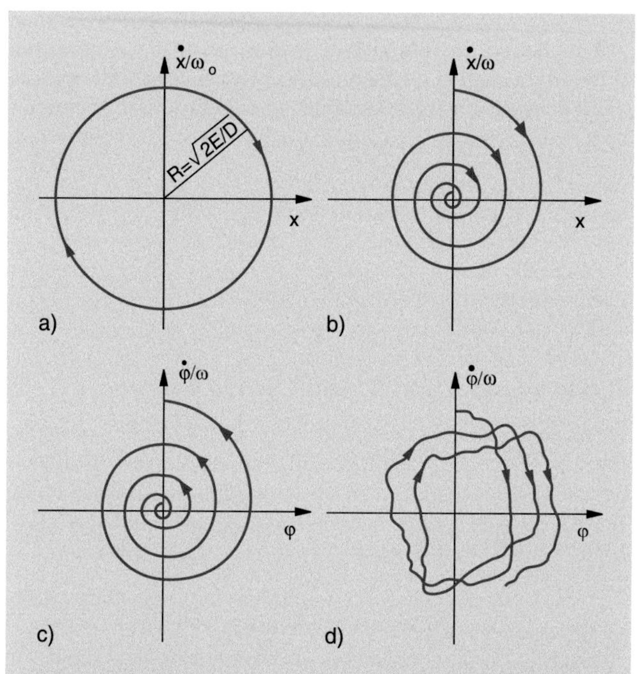

Figure 12.3 Phase space trajectories of the undamped (**a**) and damped (**b**) harmonic oscillators. In (**c**) with negative damping ($\gamma < 0$); (**d**) shows the trajectoriy of the chaotic motion in the unstable range of the nonlinear forced oscillator with large amplitude

For a pendulum with negative damping, the deflection angle φ and its time derivative $\mathrm{d}\varphi/\mathrm{d}t$ can be used as coordinates in phase space. One then obtains the open spiral in Fig. 12.3c.

4. For the parametric oscillator with the equation of motion (12.1) the trajectory in phase space that corresponds to the motion in the chaotic range, is an irregular non-closed curve, which is schematically shown in Fig. 12.3d.
5. A rotating paraboloid $z(r) = a \cdot r^2 = a \cdot (x^2 + y^2)$ contains steel balls that participate in the rotation (Fig. 12.4). The forces that act on the balls are the gravity $m \cdot g$, the centrifugal force $m\omega^2 \cdot r \cdot \hat{\boldsymbol{r}}_0$ and a friction force $F_\mathrm{f} \propto v_\parallel$ parallel to the wall. A stable path for the balls is at the heights $z = z_s = (1/2)(\omega^2/g)r^2$ where the vector sum of gravity force and centrifugal force is perpendicular to the wall (Fig. 6.19b).

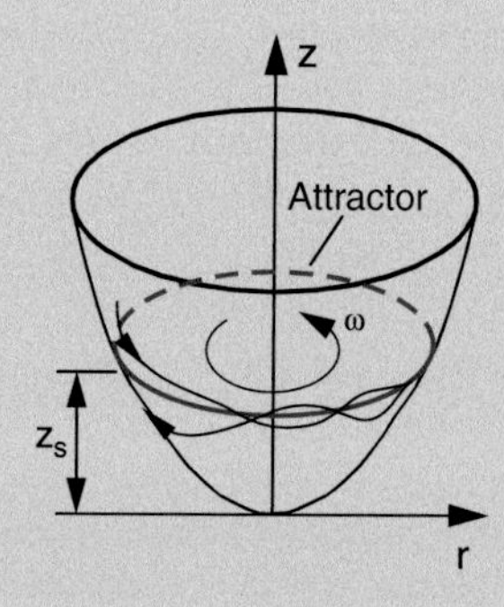

Figure 12.4 Potential surface and attractor curve (*red*) for the example 5

Balls at $z > z_s$ experience a force parallel to the wall, that drives the balls downwards, balls with $z < z_s$ are driven upwards. For a given frequency ω the curve $z = a \cdot r_s^2$ as stable curve is an attractor for all unstable circular trajectories.

6. A further interesting example is the simple pendulum (Eq. 2.79a) with the equation of motion

$$m \cdot L \cdot \ddot{\varphi} + m \cdot g \cdot \sin \varphi = 0 . \tag{12.8}$$

Integration and division by $m \cdot L$ gives with $\omega^2 = g/L$

$$\tfrac{1}{2} \cdot \dot{\varphi}^2 - \omega^2 \cdot \cos \varphi = C . \tag{12.9}$$

Plotting the trajectories in a phase diagram one obtains, depending on the values of φ the curves in Fig. 12.5. They are closed for $\varphi < \pi$ but lead into an unstable region for $\varphi \geq \pi$, here the angle increases continuously. The two regions are separated by the red curve, which is called the *seperatrix*. Multiplying (12.9) with $m \cdot L^2$ and adding $m \cdot g \cdot L$ to both sides yields

$$\begin{aligned} &\tfrac{1}{2} m \cdot L^2 \dot{\varphi}^2 + m \cdot Lg(1 - \cos \varphi) = C_1 \\ &\quad \text{with } C_1 = C \cdot mL^2 + mgL \\ &\quad \Rightarrow E_{\text{kin}} + E_{\text{pot}} = C_1 = E . \end{aligned} \tag{12.10}$$

This shows that the seperatrix is the curve on which $E = 0 \to C = -g/L$. The point A in Fig. 12.5 is the attractor, its intake area are all φ-values with $|\varphi| < \pi$. The point B is a metastable equilibrium point, because it corresponds to the metastable position of the pendulum with $\varphi = \pi$. Every small perturbation can completely change the state.

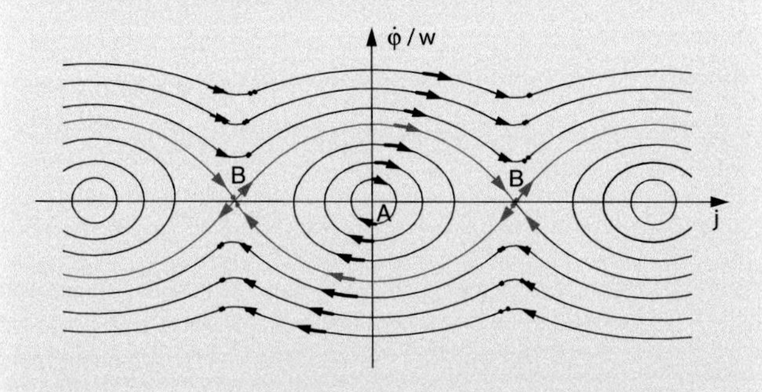

Figure 12.5 Phase space trajectories of the non-linear undamped oscillator. The *red curve* is the seperatrix between the stable ($|\varphi| < \pi$) and unstable ($|\varphi| > \pi$) range

We will now define more quantitatively the stability of fix points of a dynamical system. We consider at first a nonlinear system that depends only on one coordinate x. At discrete times t_n it passes through the coordinates x_n. The value x_{n+1}, which the system takes at the time t_{n+1} depends on the foregoing value x_n:

$$x_{n+1} = f(x_n) , \tag{12.11}$$

where the function f describes the development of the system.

If the system has reached a fixpoint, the development stops and stays stationary. This means

$$x_\text{f} = f(x_\text{f}) . \tag{12.12}$$

If the system converges towards a fixpoint ($x_\text{f} = \lim_{n\to\infty}(x_n)$) the deviation

$$\delta = x_n - x_\text{f} \to 0$$

must converge towards zero for $n \to \infty$. For the difference δ_{n+1} one obtains

$$\begin{aligned} \delta_{n+1} &= x_{n+1} - x_\text{f} = f(x_n) - x_\text{f} \\ &= f(x_\text{f} + \delta_n) - x_\text{f} . \end{aligned} \tag{12.13}$$

Expanding $f(x_\text{f} + \delta_n)$ into a Taylor series around x_f and neglecting for small δ_n the higher order terms, Eq. 12.13 gives

$$\delta_{n+1} = \left.\frac{\mathrm{d}f(x)}{\mathrm{d}x}\right|_{x=x_\text{f}} \cdot \delta_n . \tag{12.14}$$

If the deviations δ_n should converge towards zero for $n \to \infty$ the condition

$$\left|\frac{\mathrm{d}f(x)}{\mathrm{d}x}\right|_{x=x_\text{f}} < 1 \tag{12.15}$$

must be fulfilled.

A system that starts with two slightly different initial values x_0 and $x_0 + \varepsilon_0$ can only reach the same final stationary state (fixpoint), if condition (12.15) hold.

The deviation after the first step is according to (12.11)

$$x_1 + \varepsilon_1 = f(x_0 + \varepsilon_0) \Rightarrow \varepsilon_1 = f(x_0 + \varepsilon_0) - f(x_0) ,$$

and after the second step

$$x_2 + \varepsilon_2 = f(x_1 + \varepsilon_1) = f(f(x_0 + \varepsilon_0)) = f^2(x_0 + \varepsilon_0) ,$$

and therefore after the nth step

$$\varepsilon_n = f^n(x_0 + \varepsilon_0) - f^n(x_0) . \tag{12.16}$$

As a measure of the stability, one defines the *Ljapunov exponent* λ

$$\begin{aligned} \lambda(x_0) &= \lim_{n\to\infty} \lim_{\varepsilon_0 \to 0} \frac{1}{n} \log \left| \frac{f^n(x_0 + \varepsilon_0) - f^n(x_0)}{\delta_0} \right| \\ &= \lim_{n\to\infty} \frac{1}{n} \log \left| \frac{\mathrm{d}f^n(x)}{\mathrm{d}x} \right|_{x=x_0} . \end{aligned} \tag{12.17}$$

The condition (12.15) can then be written for large values of n as

$$\delta_{n+1} = \delta_n \cdot e^{\lambda} . \tag{12.18}$$

For $\lambda < 0$ the system converges against a stable fixpoint. For $\lambda > 0$ the deviations increase exponentially and no fixpoint exists (Fig. 12.6). The case $L = 0$ will be discussed later.

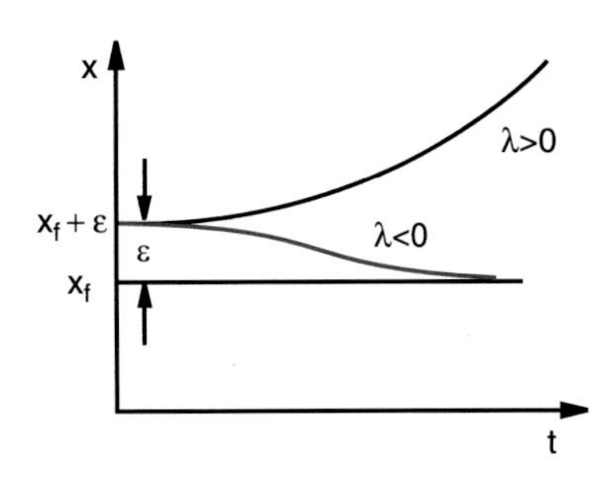

Figure 12.6 Time development of a small deviation ε from the fixpoint x_f for $\lambda > 0$ and $\lambda < 0$

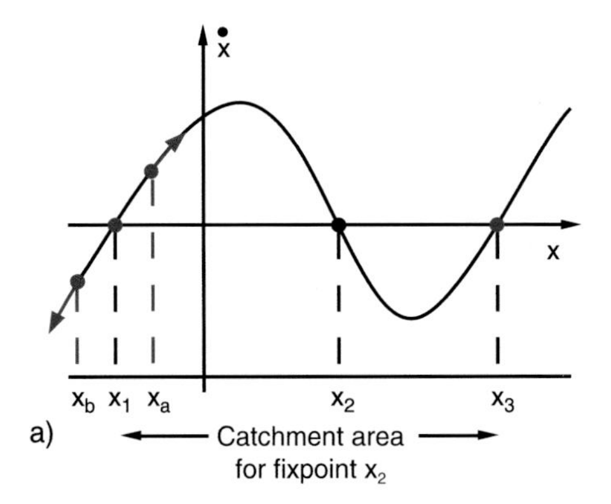

The stability of fixpoints is illustrated by Fig. 12.7a, where the schematically drawn trajectory in the phase space $(x, \dot{x})$ is the representation of a one-dimensional equation of motion. The intersection points of the curve with the horizontal line $\dot{x} = 0$ are the fixpoints. The points x_1 and x_3 are unstable fixpoints, because small negative deviations bring the system to negative velocities, that further remove the system from the fixpoint, while positive deviations cause positive velocities that bring the system further upwards in the diagram. On the other hand is x_2 a stable fixpoint because any deviation brings the system back to x_2. It acts as attractor with a intake range from x_1 to x_3. All states of the system within this range tend to converge to x_2.

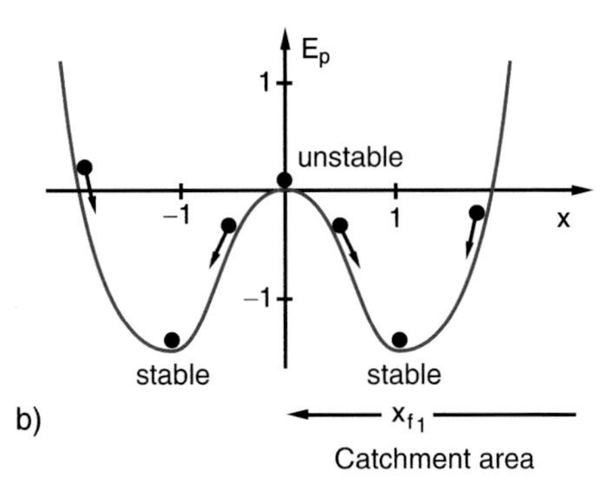

Figure 12.7 a Trajectory in phase space with unstable fixpoints x_1, x_3 and an attractor at x_2 with a catchment area from x_1 to x_3. b Particle in a potential $E_{pot} = -ax^2 + bx^4$ (for $a = 2$, $b = 0.5$) has a maximum at $x = 0$, which is unstable and a stable minimum, where $\dot{x} \to 0$

An example for such a system is a particle with mass m in a double well potential (Fig. 12.7b) with the potential energy

$$E_{pot}(x) = -ax^2 + bx^4 , \tag{12.19}$$

where the nonlinear force

$$F_x(x) = -\frac{dE_{pot}}{dx} = 2ax - 4bx^3 \tag{12.20}$$

acts on the particle. Its equation of motion is

$$m\ddot{x} - 2ax + 4bx^3 = 0 . \tag{12.21}$$

Energy conservation $(1/2)\, m\dot{x}^2 + E_{pot}(x) = E$ yields the velocity of the particle

$$v = \dot{x} = \sqrt{\frac{2}{m}\left(E - E_{pot}\right)} . \tag{12.22}$$

For $E = E_{pot} \to E_{kin} = 0$ the velocity becomes $\dot{x} = 0$ at the position $x = 0$ because there is $E_{pot}(0) = 0$. However, this is no stable fixpoint, because small deviations bring the particle either to the left or the right minimum. If its velocity converges to zero due to frictional losses, it finally rests in one of the two minima at $x = \pm\sqrt{a/2b}$. They are stable fixpoints. The intake range for the fixpoint $x_1 = \sqrt{a/2b}$ includes all x-values $x > 0$, while for the other fixpoint $x_2 = -\sqrt{a/2b}$ all negative x-values $x < 0$ belong to its intake range.

12.2 Logistic Growth Law; Feigenbaum-Diagram

A very instructive example of a nonlinear system is the biological population where the number N_{n+1} of the members in the $(n + 1)$th generation is proportional to the number N_n in the foregoing generation.

$$N_{n+1} = a \cdot N_n , \tag{12.23}$$

where a is the growth factor. Due to food shortage the growth factor a decreases to $a(1 - b \cdot N_n)$, because the food consumption is proportional to the number N_n of consumers. Inserting this into (12.23) gives

$$N_{n+1} = a \cdot N_n(1 - bN_n) . \tag{12.24}$$

A stationary state (fixpoint) is reached for

$$N_{n+1} = N_n = N_{st} \Rightarrow b = \frac{a-1}{a \cdot N_{st}} . \tag{12.25}$$

This is realized for $a = 1 \Rightarrow b = 0$. No food shortage is present.

For $a < 1$ is $N_{n+1} < N_n$ i.e. the population decreases even for $b = 0$ when no food shortage occurs, while for $a > 1$ the

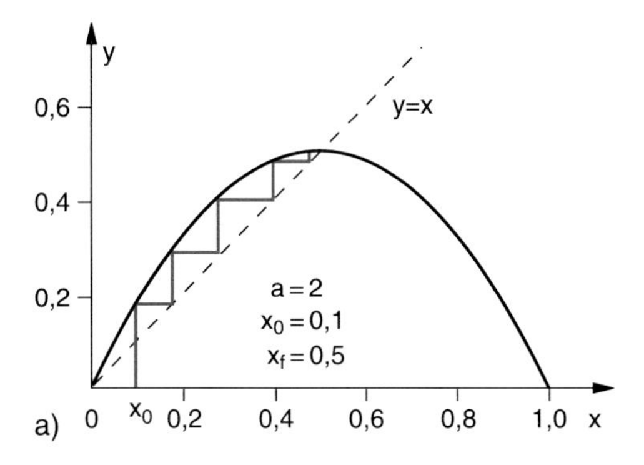

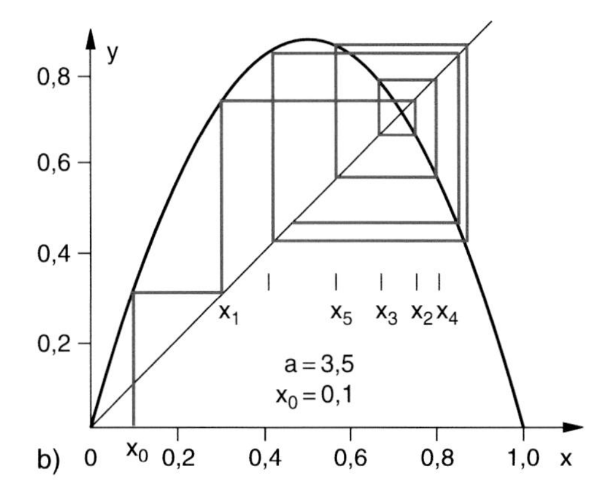

Figure 12.8 Logistic diagram: **a** in the stable range $a = 2$ with stable fixpoint $x_f = 0.5$; **b** in the oscillation range with $a = 3.5$

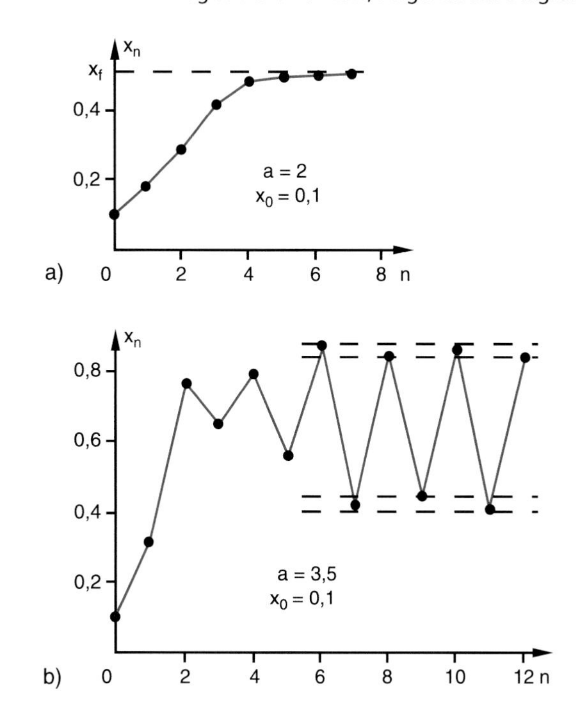

Figure 12.9 The first members of the series x_n: **a** fast convergence for $a = 2$; **b** oscillating behaviour for $a = 3.5$

population increases until the food shortage $b \cdot N_n$ increases and brings the effective growth factor again back to 1.

With the normalization $x = bN \leq 1$ (12.24) converts to the **Verhulst-Equation**

$$x_{n+1} = ax_n(1 - x_n) = ax_n - ax_n^2 \,. \tag{12.26}$$

With the given normalization $x \leq 1$ the possible values of the parameter a are restricted to the interval $0 \leq a \leq 4$.

The solution of Eq. 12.26 for the different generations n and their dependence on the growth parameter a can be illustrated graphically, when the parabola $y = ax - ax^2$ and the linear slope $y = x$ are plotted in a (y, x)-diagram (Fig. 12.8). For each value $x_n < 1$ one finds from (12.26) the corresponding value x_{n+1} as ordinate $y = ax - ax^2 = x_{n+1}$ on the parabola. In order to find the new starting point x_{n+1} one must go from $(x_n, y_n = x_{n+1})$ into the horizontal direction to the intersection with the straight line $y = x$. A vertical line through this intersection point reaches the parabola at the point $(x_{n+1}, y_{n+1} = x_{n+2})$.

In this way one obtains the sequence x_n $(n = 0, 1, 2, \ldots)$ as step function starting from an arbitrarily chosen initial starting point x_0.

This is illustrated in Figs. 12.8a and 12.9a for $x = 0.1$ and $a = 2$. One can see that the sequence x_n converges relatively fast against the fixpoint $x_f = 0.5$.

A completely different situation occurs for the same starting point $x_0 = 0.1$ but another growth factor $a = 3.5$ (Figs. 12.8b and 12.9b). Here the sequence oscillates between 4 limiting values.

It turns out that for $a > 3.57$ the behaviour of the sequence depends critically on the growth factor a, while the values of the sequence memers x_n do not depend on the initial value x_0 as long as $a < 3.57$.

Plotting the limits $\lim_{n\to\infty} x_n$ of the logistic equation (12.26) as a function of the growth parameter a one gets the Feigenbaum-diagram shown in Fig. 12.10, which was first published 1978 by *S. Großmann* [12.6a] and analysed by ***M. Feigenbaum*** [12.6b]. One can see the following surprising results of the logistic growth law:

- For $a \leq 1$ the sequence x_n converges against zero. The closer the value of a comes to $a = 1$ the slower the sequences converges. The stable fixpoint is $x_f = 0$.
- For $1 < a < 3$ a stable fixpoint exists: $\lim_{n\to\infty} x_n = x_f < 1$ but $\neq 0$.
- For $3 < a_\infty$ the values of x_n oscillate between 2^k limiting values, where $a_\infty = 3.57$ (see below). The exponent k is $k = 1$ for $3 < a < 3.449$; $k = 2$ for $3.449 < a < 3.544$. The points (a, x) in the Feigenbaum diagram where k increases by 1 are called **bifurcation points**. At the first bifurcation point in Fig. 12.10 the curve $x_f = 1 - 1/a$ represents the fixpoints x_f as a function of a, until $a = 3$, where the curve $x_f(a)$ splits into two curves. These cures give the limits $x_f(a)$ as a function of the growth parameter a between which the values of x_n oscillate. Each of these curves splits again at the second bifurcation point, etc.

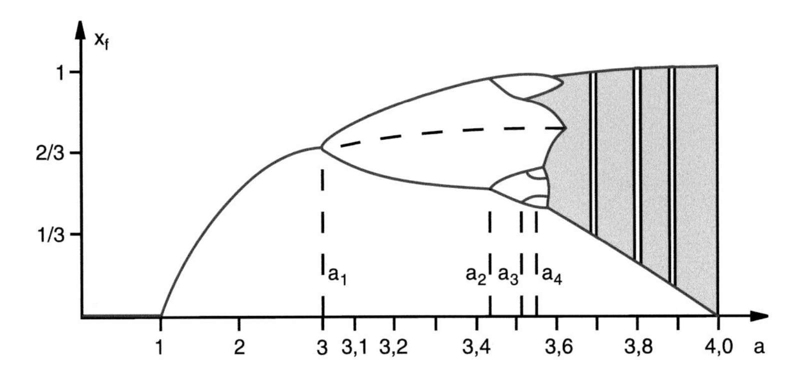

Figure 12.10 Feigenbaum diagram: Value of the fixpoint x_f as a function of the control parameter a. The values a_i give the bifurcation points $a_1 = 3.0$; $a_2 = 3.449\ldots$; $a_3 = 3.544\ldots$; $a_4 = 3.564\ldots$

The system has therefore 2^k attractors for all values of the growth parameter a in the range $a_k \leq a \leq a_{k+1}$ between the bifurcation points a_k and a_{k+1}.
With increasing values of a the interval between two bifurcation points becomes smaller and smaller. The values of the bifurcation points of order k follow a geometrical sequence

$$a_k = a_\infty - c \cdot \delta^{-k} \quad \text{for} \quad k \gg 1 . \tag{12.27}$$

For the distance $\Delta_k = a_k - a_{k-1}$ we obtain

$$\Delta_k = c \cdot \delta^{-k}(\delta - 1) . \tag{12.28}$$

The Feigenbaum-constant $\delta_F = \lim_{n\to\infty}(\Delta_k/\Delta_{k+1})$ has the numerical value

$$\delta_F \approx 4.669201660910\ldots .$$

The sequence of bifurcation points converges against the limit

$$a_\infty = \lim_{k\to\infty} a_k = 3.5699456\ldots .$$

The Ljapunov exponent λ is in the range $3 < a < a_\infty$ always negative, except at the bifurcation points where is $\lambda = 0$.

- In the range $a_\infty < a < 4$ chaotic regions occur where the values of the fixpoints scatter randomly. Here is the Ljapunov exponent $\lambda > 0$ (Fig. 12.11) between these chaotic ranges periodic windows appear where stable fixpoints occur. The sequence x_n oscillates between these fixpoints. The Ljapunov exponent λ is negative in these windows. With increasing values of the growth parameter a the chaotic regions more and more displace the windows of stable regions.
- In the chaotic regions rational start values give fixpoints, while for irrational start values no convergence is possible For $a = 4$ the logistic equation

$$x_{n+1} = 4x_n(1 - x_n) \tag{12.29}$$

can be exactly solved and gives the solution

$$x_n = \sin^2(2^n \pi x_0) , \tag{12.30}$$

where x_0 is the start value.

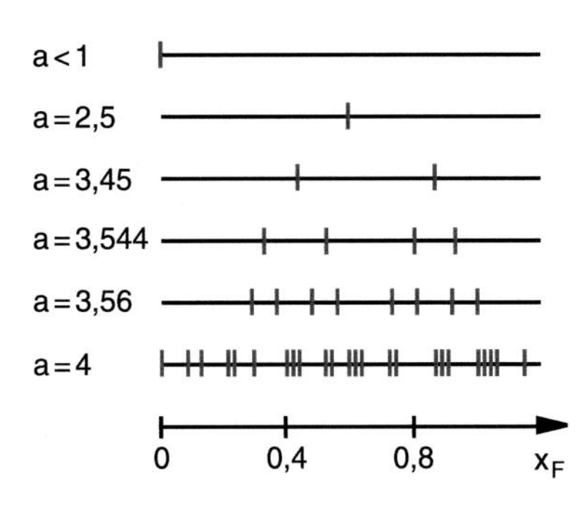

Figure 12.11 The fixpoints of the logistic mapping for different values of the parameter a for illustration of the bifurcation

12.3 Parametric Oscillator

The equation of the undamped parametric oscillator, depicted in Fig. 11.25 with the pendulum length

$$L(t) = L_0 - \Delta L_0 \cos \Omega t \tag{12.31}$$

and $\Delta L_0/L_0 \ll 1$, can be written as

$$\ddot{\varphi} + \omega_0^2 \left[1 + \frac{\Delta L_0}{L_0} \cos \Omega t\right] \sin \varphi = 0 , \tag{12.32}$$

where the approximations

$$\frac{1}{L} = \frac{1}{L_0\left(1 - \frac{\Delta L_0}{L_0} \cos \Omega t\right)} \approx \frac{1}{L_0}\left(1 + \frac{\Delta L_0}{L_0} \cos \Omega t\right)$$

have been used with the frequency

$$\omega^2 = \omega_0^2 \left[1 + \frac{\Delta L_0}{L_0} \cos \Omega t\right] .$$

Introducing the abbreviations $\omega_0^2 = g/L_0$; $\alpha = \omega_0^2/\Omega^2$; $\beta = \Delta L_0/L_0$ and $\tau = \Omega \cdot t$ the Eq. 12.32 converts into the Mathieu-equation

$$\varphi'' + \alpha (1 + \beta \cos \tau) \varphi = 0 , \tag{12.33}$$

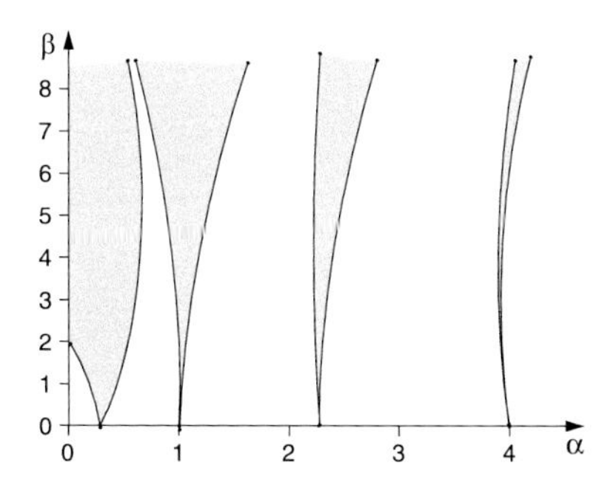

Figure 12.12 Stable (*white*) and unstable (*red*) ranges for the solutions of the Mathieu equation

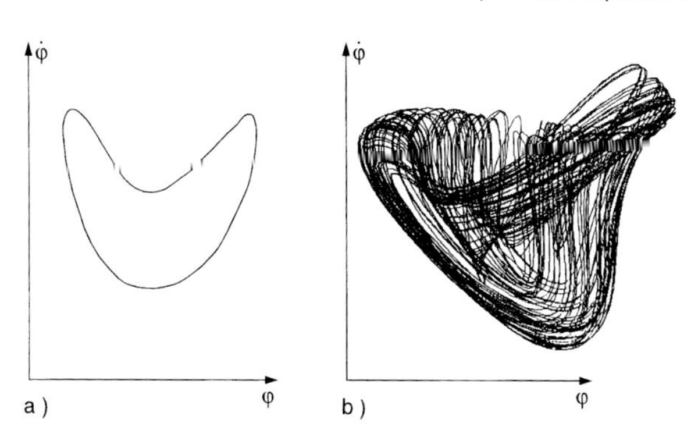

Figure 12.14 Phase diagram of the parametrically driven pendulum **a** in the stable, **b** in the unstable region

where $\varphi'' = \mathrm{d}^2\varphi/\mathrm{d}\tau^2$. The solutions of this linear differential equation depend on the parameters α and β. There exist stable solutions for certain ranges (α, β) which are shown as white areas in Fig. 12.12 while for the red areas (resonance ranges) unstable solutions exist, where φ increases unlimited.

The first instable range appears for $\alpha = (1/4)$, $\rightarrow \Omega = 2\omega_0$. This is the parameter range which a child instinctively uses to enhance the oscillation amplitude of its swing by uplifting and lowering its centre of mass at the right moment twice per oscillation period. Without damping the oscillation amplitude becomes infinite in the red regions and the solutions of (12.33) lead to useless results.

The situation changes if we do not use the approximate linear equation but the exact nonlinear equation (12.32) for the solution of the problem. We start with the trivial case $\varphi = 0$, where the pendulum does not perform angular oscillations $\varphi(t)$ but only vertical periodic changes of the pendulum length $L(t)$. The mass m then executes vertical oscillations with the exciting frequency Ω, where always $\varphi = 0$. This is true for $\alpha < 1/4$ and $\beta \ll 1$. If for $\alpha = 1/4$ the amplitude exceeds a critical amplitude β_c the oscillation becomes unstable and the vertical motion switches into a φ-oscillation (Fig. 12.13) with an amplitude that depends on the parameter β. The frequency ω_0 of this $\varphi(t)$ oscillation is one half of the exciting frequency Ω $(\omega_0 = (1/2)\,\Omega)$. At the bifurcation point B in Fig. 12.13 a doubling of the period length T occurs. When further increasing β, more and more bifurcation points are passed until the chaotic range is reached where the motion of the pendulum becomes random. Here a statistic motion of vertical and angular oscillation takes place. The phase diagram of the regula and chaotic ranges is shown in Fig. 12.14.

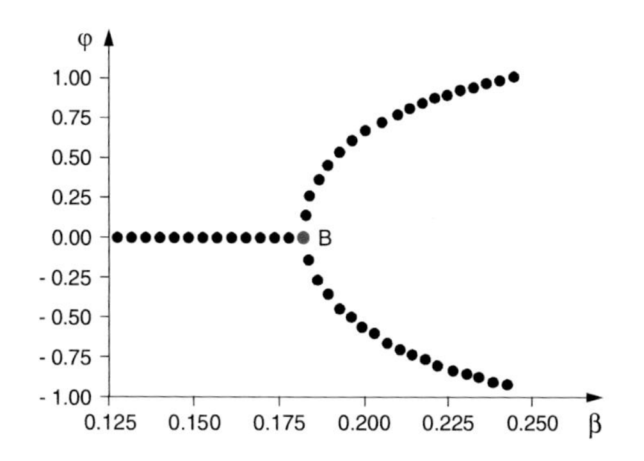

Figure 12.13 The first bifurcation of the pendulum that oscillates initially only vertical. The abscissa $\beta = \Delta L/L_0$ corresponds in the stability diagram of Fig. 12.12 a vertical line

In the chaotic regime the motion is very sensitive to small changes of the initial conditions. Plotting $\varphi(t)$ against t small changes of $\varphi(0)$ or $\mathrm{d}\varphi/\mathrm{d}t(0)$ result in large changes of $\Delta\varphi(t)$ which can grow exponentially with time t [12.4].

12.4 Population Explosion

We will describe the growth of the world population by a simple model that allows in spite of its simplicity, a good insight into the problem [12.5].

We will take $z_f(t)$ as the female and $z_m(t)$ as the male population at time t. We denote the death rate of the females as $a_f \cdot z_f$ and that of the males as $a_m \cdot z_m$. The birth-rate is proportional to the product $z_f \cdot z_m$. For the change of the population per unit time we then obtain

$$\dot{z}_m = -a_m z_m + b_m z_m z_f \,, \tag{12.34a}$$

$$\dot{z}_f = -a_f z_f + b_f z_f z_m \,. \tag{12.34b}$$

The "symbiosis"-terms $b_m z_m \cdot z_f$ and $b_f \cdot z_f \cdot z_m$ cause the nonlinearity of the equation and couple them with each other.

It turns out that birth rates and death rates do not differ much between males and females. We therefore can approximate $a_m = a_f = a$ and $b_m = b_f = b$. Furthermore the population statistics shows that the populations of males and females is approximately equal, i. e. $z_m \approx z_f$.

With these assumptions we obtain for the total population $z = z_m + z_f$ by addition of the two Eqs. 12.34a,b the nonlinear equation

$$\dot{z} = -a \cdot z + \frac{b}{2} z^2 \,. \tag{12.35}$$

For $b = 0$ (zero birth-rate) we get the solution

$$z(t) = z_0 \mathrm{e}^{-at} \,. \tag{12.36}$$

Chapter 12

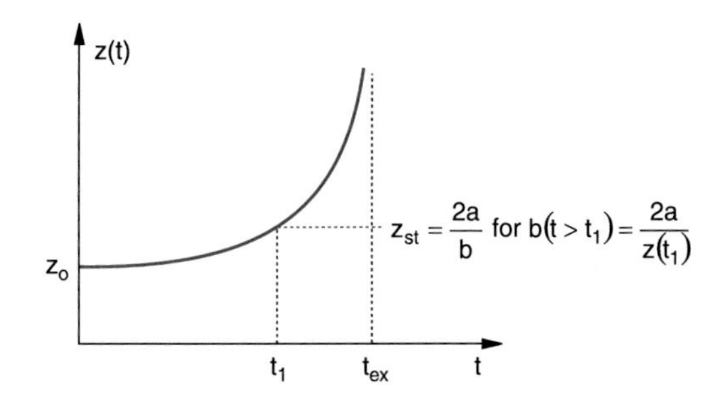

Figure 12.15 Population explosion for fixed values of a and b (*red curve*). If the birth rate b is suddenly reduced to $b = 2a/z(t_1)$ at the time t_1 the population stagnates at $z = z(t_1)$ (*red dashed lines*)

For $a = 0$ and $b \neq 0$ (zero death rate) one obtains an unlimited growth

$$z(t) = z_0 \frac{2}{2 - bz_0 t} , \tag{12.37}$$

as can be verified by inserting (12.37) into(12.36). After a time

$$t_{ex} = \frac{2}{bz_0} , \tag{12.38}$$

the population grows to infinity with our unrealistic assumption $a = 0$.

For $a \neq 0$ and $b \neq 0$ the solution of (12.36) is

$$z(t) = z_0 \frac{2a}{bz_0 - 2 \cdot (bz_0/2 - a)\, e^{+at}} . \tag{12.39}$$

For $a = bz_0/2$ birth rate and death rate just compensate and the population remains stable ($dz/dt = 0$) at its initial value z_0 (Fig. 12.15). **Note**, that the birth-rate depends quadratic on the population, but the death rate only linearly.

For $bz_0 > 2a$ is $dz/dt > 0$ and the population "explodes", At the finite time

$$t = t_{ex} = -\frac{1}{a} \ln\left(1 - \frac{2a}{bz_0}\right) \tag{12.40}$$

the population becomes $z(t_{ex}) = \infty$. **Note**, that in our model this explosion takes place not at $t = \infty$ but at the finite time t_{ex}. Of course, in reality the death rate would increase and the birth rate decrease *before* this time, because of food shortage and conflicts and wars for food supply. In order to avoid such catastrophic situations, the condition $a \leq b \cdot z_{st}/2$ must be reached early enough. Since in our real world the death rate decreases (in particular for children), due to a better medical treatment and a larger food supply, the birth rate $b \cdot z2/2$ has to be drastically reduced in order to avoid this catastrophic case. Comparing in Fig. 12.16 the growth function (12.39) of our model with the real population growth as investigated by the UNESCO, one can see that the population growth proceeds with increasing growth factor $(b \cdot z_0/2 - a)$. While within the time span from 1750 to 1880 the doubling time of the population was 130 years, it dropped from 1950 to 1985 to 35 years! Therefore the actual population curve $z(t)$ increases faster than the growth function (12.39) of our model. This aggravates the problem further.

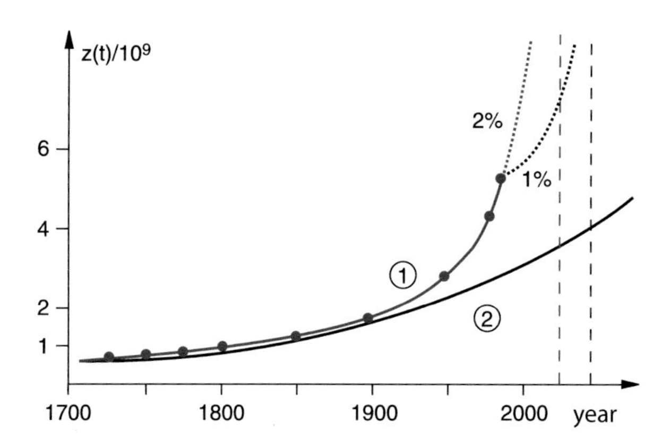

Figure 12.16 Comparison of the population development, empirically determined by the UNO (*curve 1*) with the model of Eq. 12.18 for the numerical values $bz_0/2 - a = 0.003$ for the year 1750 (*curve 2*). The *dashed red curve* is the extrapolation assuming a net growth of 2%. The *black dashed curve* for 1%

Example

For the year 1992 the world population was estimated as $z = 6 \cdot 10^9$. With an average life expectation $\tau = 50$ years the death rate constant becomes $a = 1/\tau = 0.02$ and the death rate $a \cdot z = 120$ Million/year. With an average birth rate of 240 Millions/year the population increases by 2% per year, which means by 120 Millions. Under the assumption of constant birth and death rates the population would double after 50 years. With an increasing growth factor, which is in reality observed, the doubling time would be shorter.

In fact, the real growth factor increases with time.

Inserting the numerical values into (12.40) gives the "explosion time" when the population becomes infinite:

$$t_{ex} = -50 \cdot \ln \frac{0.02}{0.04} = 50 \ln 2 \approx 35 \text{ years} .$$

This means that without decreasing the birth rate the catastrophe will happen in 35 years, which means in the year 2050.

Even when the birth rate is lowered to such a value, that the net population growth decreases to 1%, t_{ex} increases only to 55 years, which means that the catastrophe is not abandoned but only delayed. ◄

12.5 Systems with Delayed Feedback

In many real situations, systems with delayed feedback are found. An example is a microphone that receives not only the direct words of the speaker, but also, with a time delay, the output

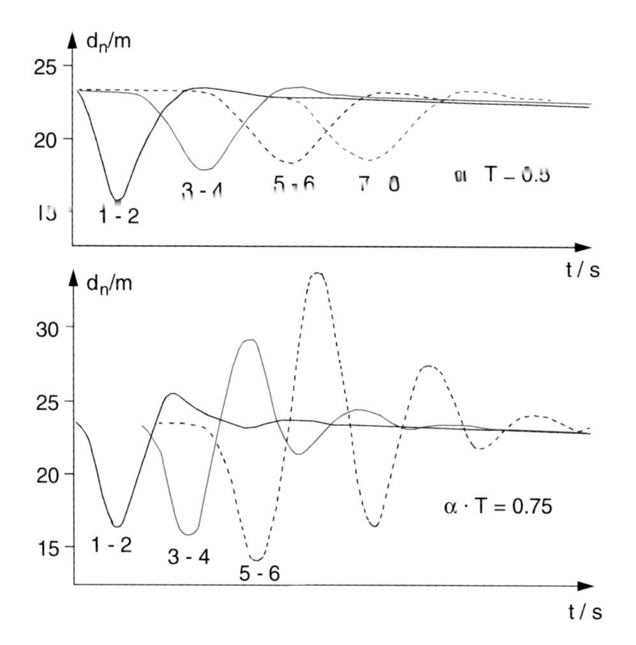

Figure 12.17 Time variation of the distance $d_n = x_n - x_{n+1}$ in a car convoy, when the first car ($n = 1$) brakes and accelerates again, for two different values of $a \cdot T$ [12.7]

of a loud speaker in a large room that amplifies this speech. This time-delayed signal received by the microphone, is again fed into the amplifier and the loud speaker. If the electronic system is not optimized to handle this problem properly, an overload of the loud speaker occurs, resulting in a distortion of the words or even to loud howling noise.

Another example that is a nuisance for every motorist, is the traffic jam arising without obvious reasons when the traffic density surpasses a critical value. We will discuss the reasons for this situation.

In a car sequence on the motorway the driver of the $(n + 1)$ car changes his speed v_{n+1}, if the foregoing nth car diminishes his sped v_n. The more n brakes the more $(n+1)$ does so. because of the finite reaction time T the change of his velocity occurs with a time delay T. We therefore assume, that the braking (deceleration) of $(n + 1)$ at time $t + T$ is proportional to the velocity difference $v_n(t) - v_{n+1}(t)$.

$$\ddot{x}_{n+1}(t + T) = a \cdot [\dot{x}_n(t) - \dot{x}_{n+1}(t)] \; . \tag{12.41}$$

The factor a states how strongly the driver $(n + 1)$ reacts on the change of the relative velocity $v_n(t) - v_{n+1}(t)$. It can depend on his velocity v_{n+1}, on the distance $d_n(t) = x_n - x_{n+1}$ and on the reaction time T.

The most simple case is present for $a =$ constant. Even for this case (12.41) cannot be solved analytically, but only numerical solutions exist. They are plotted in Fig. 12.17 for an initial distance $d(t = 0) = 23\,\text{m}$ and for different values of the product $a \cdot T$.

The curves $d_n(t)$ illustrate, that for $a \cdot T = 0.5$ the distance changes decrease for increasing n. This means, when the first driver brakes, a damped distance wave propagates along the following cars. For the 10th driver it is barely noticeable. For the higher value $a \cdot T = 0.75$, however, the distance changes increase with n. If the minimum of the distance wave reaches $d = 0$ the two sequenced cars collide and cause a traffic jam.

But also without collision a jam can arise.

When driver n brakes because the foregoing car reduces its speed, he will generally over-react and reduces his speed below that of the foregoing car. The following car $(n + 1)$ reduces its speed even more, until the $(n + x)$th car comes to a standstill. Such a traffic jam often occurs when the traffic density is high and the distance between the cars is small.

In order to avoid such unnecessary jams the product $a \cdot T$ ust be sufficiently small. Since the reaction time of most drivers has a lower limit of about 0.1 s (at a speed of 130 km/h this time corresponds to a distance of 36 m) **the best way for safe driving without causing a traffic jam is a sufficiently large distance between successive cars.**

12.6 Self-Similarity

The linear differential equation

$$\dot{x} = -a \cdot x(t) \tag{12.42}$$

has the solution

$$x(t) = x_0 \cdot e^{-at} \; . \tag{12.43}$$

The arbitrary initial value of x and the constant parameter a determine the time dependence of $x(t)$. If we choose two solutions $x_1(t)$ and $x_2(t)$ with different initial conditions, e. g. with different values of $x_1(0)$ and $x_2(0)$ every linear combination $c_1 \cdot x_1(t) + c_2 \cdot x_2(t)$ is again a solution of (12.42), and is again an exponential function.

For a nonlinear equation this is no longer true, as can be exemplified by the equation

$$\dot{x} = -a \cdot x^2 \; . \tag{12.44}$$

The solution of this nonlinear equation is

$$x(t) = \frac{x_0}{1 + a x_0 t} \; . \tag{12.45}$$

For two different solutions

$$x_1(t) = \frac{x_{01}}{1 + a x_{01} t} \quad \text{and} \quad x_2(t) = \frac{x_{02}}{1 + a x_{02} t} \; ,$$

the sum $x_1(t) + x_2(t)$ is not a solution of (12.44).

For long times t, when $a \cdot x_0 \cdot t \gg 1$ the function $x(t)$ can be approximated by

$$x(t) \approx \frac{1}{at} \; . \tag{12.46}$$

Now the solution does not depend on the initial value x_0.

When the time is measured in other units (for example in hours instead of seconds) the time t is replaced by $\lambda \cdot t$. This converts (12.46) to

$$x(\lambda \cdot t) = \frac{1}{\lambda a t} = \frac{x(t)}{\lambda} . \tag{12.47}$$

The solutions are similar even for different time scales! For example, with $\lambda = 10$ one obtains the same time behaviour for the function $x(\lambda t)$ as for $x(t)$ if a tenfold stretched time scale is used.

This scale similarity can be mathematically expressed as

$$x(\lambda t) = \lambda^{\kappa} \cdot x(t) . \tag{12.48}$$

The quantity κ is the scale exponent or similarity exponent. For the nonlinear equation (12.44) with the approximation (12.46) is $\kappa = -1$.

The time dependence of the solution $x(t)$ of (12.48) can be expressed as

$$x(t) \propto t^{\kappa} , \tag{12.49}$$

because this gives $x(\lambda t) \propto \lambda^{\kappa} \cdot x(t)$.

Note: For linear equations such a scale similarity does not exist. This can be seen by replacing in (12.43) t by $t' = \lambda t$. This gives another exponential decay

$$x(t') = x_0 \cdot \mathrm{e}^{-a\lambda t} = \frac{(x(t))^{\lambda}}{x_0^{\lambda-1}} . \tag{12.50}$$

Only if the relaxation constant a is changed to a/λ the same time behaviour is obtained.

This means: The constant "a" fixes a time scale for the solution of the linear equation (12.42) After the time $t = 1/a$ has $x(t)$ decreased to $1/\mathrm{e}$ of its initial value at time $t = 0$. The mean lifetime $\tau = 1/a$ gives a natural time scale for the solution (12.43).

In contrast to this behaviour of the solutions of linear equations the parameter a in the nonlinear equation does not determine such a benchmark. An arbitrary time stretch can be always compensated by a corresponding change of the x-scale.

The self-similar solutions of nonlinear equations do not have a natural benchmark.

This is not only valid for time-dependent problems but also for many other interesting phenomena, which can be only partly presented in the next section. For further examples the reader is referred to the literature [12.1a–12.6b].

12.7 Fractals

The measured length of a real coastline with many bays, juts and mountain ledges depends on the resolution of the measuring gauge. This is illustrated by the famous example of

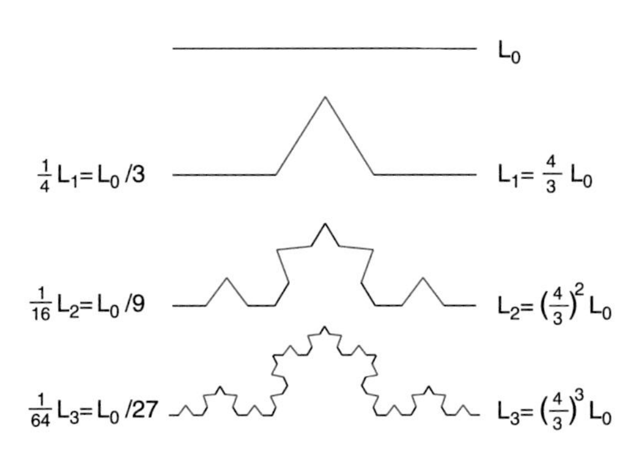

Figure 12.18 Construction of Koch's curve

the *Koch's curve* that is constructed in the following way: A straight line with length L_0 is divided into three sections. The middle section is replaced by the two sides of an equilateral triangle (second line in Fig. 12.18). Each section has the length $L_0/3$, the total length is then $(4/3)\,L_0 = 1.33L_0$. Now each of the 4 sections is again divided into three subsections and the middle subsection is replaced by the two sides of an equilateral triangle (third line in Fig. 12.18). The total length is now $L = (16/9)\,L_0 = (4/3)^2 L_0 = 1.78L_0$.

This procedure is continued. After n steps the total length is

$$L_n = \left(\frac{4}{3}\right)^n L_0 . \tag{12.51}$$

With increasing number n of steps, the total length L_n becomes infinite.

$$\lim_{n\to\infty} L_n = \infty$$

The Koch's curve shows self-similarity, because at the nth iteration the scale length l_n (i. e. the length l_n of each subsection) is $l_n = L_0/3^n$. The total number of subsections is $N_n = 4^n$. We therefore obtain the relation

$$N(l/3) = 4N(l) , \tag{12.52}$$

because at each step the scale length is reduced by a factor 3, but the number of subsections increases by the factor 4. The comparison with the scale law (12.49), which can be written as

$$N(\lambda \cdot l) = \lambda^{\kappa} \cdot N(l) \tag{12.53}$$

yields the value $\lambda = 1/3$ and $\lambda^{\kappa} = 4$. The scale parameter κ is then

$$\kappa = -\frac{\ln 4}{\ln 3} = -1.2618 .$$

The scale law (12.53) can be written as

$$N(l) \propto l^{\kappa} . \tag{12.54}$$

The length of the Koch's curve is then with a scale length l

$$L(l) = l \cdot N(l) \propto l^{\kappa+1} . \tag{12.55}$$

Figure 12.19 Koch's curve in closed form

This shows again that $\lim_{l \to 0} L(l) = \infty$, although the direct distance $\Delta = x_1 - x_2$ between the two ends (start point and endpoint) of the Koch's curve remains finite. The reason for the infinite length of the curve is the increasing refinement of the tooth structure.

Covering the curve by $N(l)$ squares with side length l, which are put together, the total area of all squares is

$$A(l) = l^2 \cdot N(l) \propto l^2 \cdot l^\kappa = l^{0.7382} .$$

The limit of this area is

$$\lim_{l \to 0}(A(l)) = 0 .$$

The Koch's curve is in a certain sense more than a one-dimensional line (because its length tends with $l \to 0$ towards infinity. It is, however, less than a two-dimensional area (because its area, defined by the squares, converges to zero).

If a fictional dimension is attribute to the Koch's curve it should be between 1 and 2.

One can formally define the d-dimensional volume, where d is an arbitrary number (not necessarily an integer)

$$V_d(l) \propto l^d \cdot N(l) \propto l^{d+\kappa} , \tag{12.56}$$

the value of $\lim_{l \to 0}(V_d(l))$ jumps at $d = -\kappa$ from ∞ to 0.

The number

$$d = d_\mathrm{f} = -\kappa \tag{12.57}$$

is defined as the *fractional dimension* of the curve or area, because $d_\mathrm{f} = 1.2618$ is not an integer but by the fraction 0.2618 larger the 1. For this value of d_f the d_f-dimensional volume V_{d_f} has a finite value, that is independent of the scale actor. From (12.47) it follows for $d = -\kappa$

$$V_{d_\mathrm{f}}(\lambda \cdot l) \propto (\lambda \cdot l)^{d_\mathrm{f}+\kappa} = (\lambda \cdot l)^0 = 1 . \tag{12.58}$$

When the Koch's curve is drawn in a closed form (Fig. 12.19) one can see, that it surrounds a finite area, that remains finite even for $l \to 0$ although the length of the surrounding curve tends to infinity.

The fractional dimensions were already introduced by Felix Hausdorff (1868–1942). The fractional dimension d_f of the volume V_d which jumps from ∞ to 0 at $d_\mathrm{f} = -\kappa$ is therefore also called the *Hausdorff dimension*.

There are many more examples for entities with fractional dimensions. One of them is the plane *Sierpinski lattice*, shown in Fig. 12.20. It is constructed by dividing the area of an equilateral triangle into four sub-triangles with equal areas and then remove the middle triangle. Its fractional dimension is

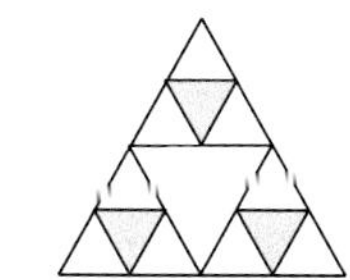

Figure 12.20 Construction of the Sierpinski-grid

$$d = \frac{\ln 3}{\ln 2} = 1.5849\ldots .$$

12.8 Mandelbrot Sets

In Sect. 12.2 we have illustrated for the example of the Verhulst dynamics the path from a stable system over the bifurcation point to the chaotic regime. A more general way to chaos, which leads to very beautiful computer graphics, was shown 1980 by B. Mandelbrot [12.9].

The basic idea relies on a nonlinear feedback algorithm for complex numbers. Instead of the one-dimensional iteration (12.26) of the logistic growth here points in the two-dimensional plane of complex numbers are used, following the iteration rule

$$z_{n+1} = z_n^2 + c , \tag{12.59}$$

where c is a complex number, which determines the pattern of the generated points in the complex plane. For a given initial starting point z_0 the sequence z_n can be calculated by the computer according to the scheme in Fig. 12.21 and plotted in the x–y-plane.

We will illustrate this by some examples:

- $c = 0$, initial start value z_0 with $|z_0| < 1$. With increasing n the z_n decrease more and more and the points in the x–y-plane spiral towards $z_\infty = 0$ which is the attractor for all z-values with $|z| < 1$, i.e. for all points inside the circle with radius $r = 1$ (Fig. 12.22).
  For a starting value z with $|z| > 1$ the sequence z_n diverges. One may formally call $z = \infty$ as the attractor for all points with $|z| > 1$. For start values z_0 with $|z_0| = 1$ all points of the sequence remain on the circle, because $|z_n| = |z_0| = 1$, the circle represents the borderline between the intake areas of the two attractors $z(A_1) = 0$ and $z(A_2) = \infty$.

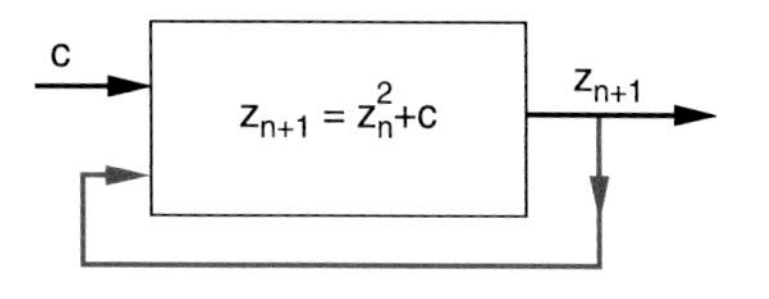

Figure 12.21 Iteration scheme for the generation of Mandelbrot sets

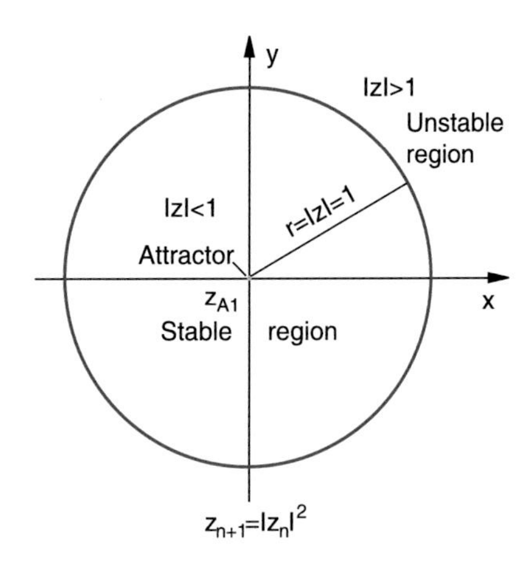

Figure 12.22 Stable range of the progression $z_{n+1} = z_n^2$ is the area inside the circle $|z| = 1$. All initial values with $|z_0| < 1$ converge against $z_\infty = 0$, with $|z_0| = 1$ against points on the circle and for $|z_0| > 1$ they diverge

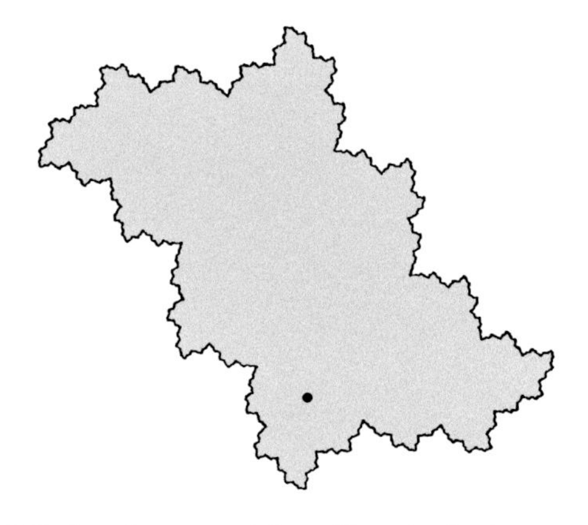

Figure 12.23 Boundary curve of the stable region for the parameter $c = -0.12375 + 0.56508i$. The *black point* is the attractor

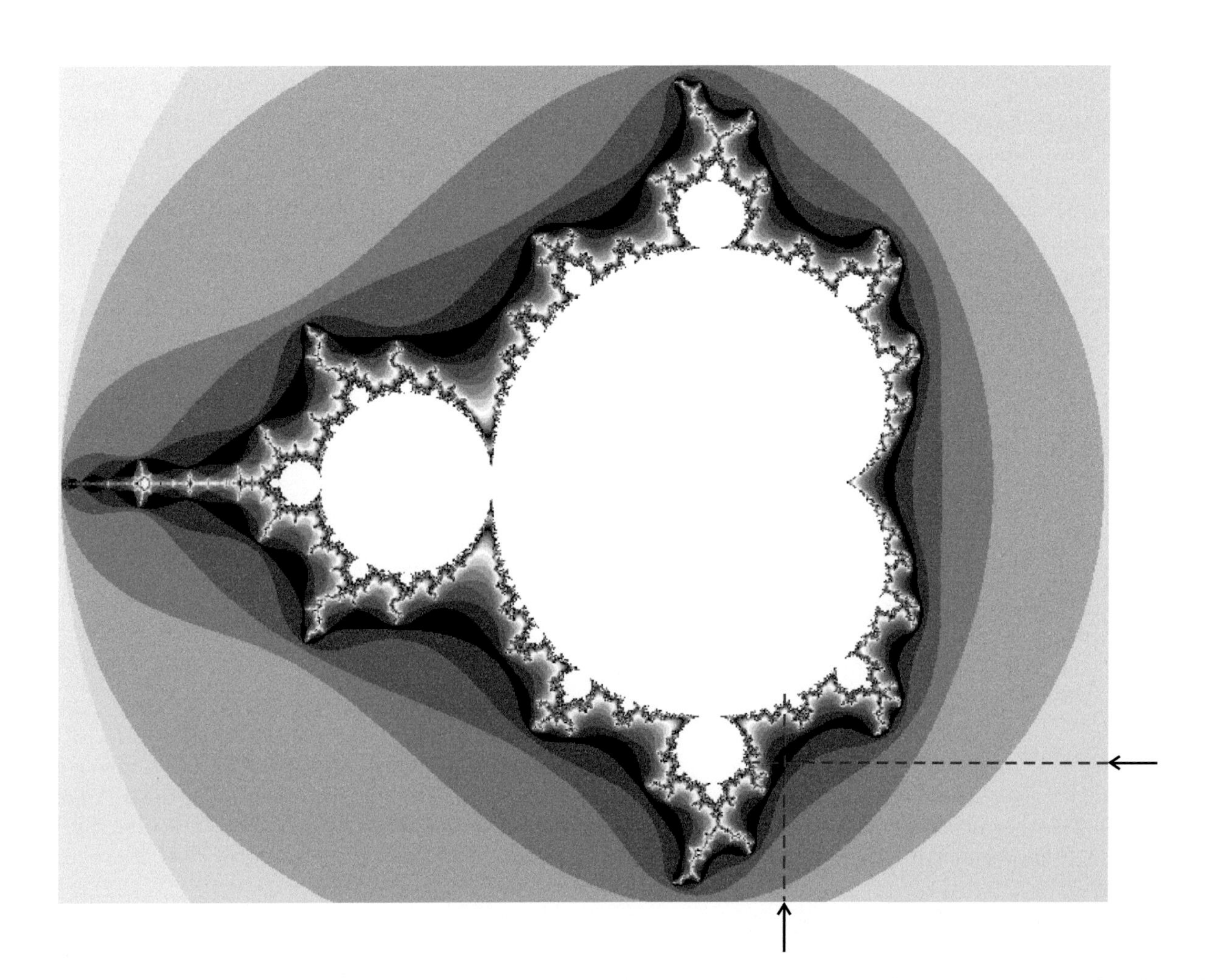

Figure 12.24 Mandelbrot set of all c-values for which the progression (12.59) with $z_0 = 0$ converges (*white area*). All *grey areas* represent c-values, that result in diverging progressions. The *shading* indicates the value of n at which z_n leaves the white area

Figure 12.25 Strongly enlarged section of Fig. 12.24 around the intersection point of the two arrows, with an area of 1 mm^2 in Fig. 12.24

- If we choose $c \neq 0$ we get surprising sequences. For instance, starting with $z_0 = 0$ the sequence (12.59) reads:

$$z_1 = c\,; \qquad z_2 = c^2 + c\,;$$
$$z_3 = (c^2 + c)^2 + c\,; \quad \ldots\,.$$

For $c = 1 + \mathrm{i}$ this gives:

$$z_1 = 1 + \mathrm{i}\,; \quad z_2 = 1 + 3\mathrm{i}\,; \quad z_3 = -7 + 7\mathrm{i}\,;$$
$$z_4 = 1 - 97\mathrm{i}\,; \quad z_5 = -9407 - 193\mathrm{i}\,; \quad \ldots\,.$$

This shows that the points z_n in the x–y-plane perform large jumps from one to the next iteration step and that the sequence of our example diverges.

Also for $c \neq 0$ there is a region in the complex plane where the sequence converges. The attractor is now generally not zero and the borderline between stable and unstable regions is no longer a circle but (Fig. 12.23) a more complicated curve. Similar to the Koch's curve the borderline shows self-similarity i. e. it has a fractional structure. When it is magnified, every magnified segment shows a similar structure as the non-magnified larger section. Such self-similar border curves are called **Julia sets**.

The Mandelbrot set consists of all numbers c of the sequence (12.59) with $z_0 = 0$ that do not diverge. These are for instance all values of $c = a + ib$ with $-2 < a < +1$ and $-1.5 < b < +1.5$. In order to generate the Mandelbrot set, one has to design a computer program that calculates for such sequences the border curves of the stable region.

Choosing, for instance, a rectangle $-A \leq x \leq +A$; $-B \leq y \leq +B$ in the complex plane, one can find out for every value of $c = x + \mathrm{i}y$ within this range, whether the sequence (12.59) converges or diverges. Now a certain colour is assigned to each value of c, depending on the number of iterations before the points c leave the rectangle. In Fig. 12.24 the colours are substituted by different grey shades. The white areas are the Mandelbrot set, imaging those z_n values which lead to stable sequences. Enlarging the tiny region in Fig. 12.24 which lies around the intersection of the two arrows, one gets the magnified Fig. 12.25.

The aesthetic beauty of such Mandelbrot sets becomes obvious with coloured computer graphics (see Fig. 12.26 and [12.10a–12.13]).

Figure 12.26 Coloured picture of the iteration $x \to \left((x^2 + q - 1)/(2x + q - 2)\right)^2$ with $\mathrm{Re}(q) = 1.2882–1.2963$ and $\mathrm{Im}(q) = 0.9695–0.9753$. The *large upper picture* is a magnified section of the central part in the *left lower picture* with different color choice. The *right lower picture* is the enlarged section, marked in the *lower left picture*, with $\mathrm{Re}(q) = 1.290681–1.291136$ and $\mathrm{Im}(q) = 0.97277–0.973098$ (With kind permission of Prof. H.O. Peitgen and Prof. P.H. Richter, Bremen)

12.9 Consequences for Our Comprehension of the Real World

Until the end of the 19th century most physicists were convinced, that all processes in nature proceed strictly deterministic and that it is, at least in principle, possible to determine the initial conditions of a system so accurately, that the future fate of the system can be exactly described for all times. This is substantiated by the famous statement of *Laplace*, published in 1776:

> The momentary state of a system is obviously the consequence of the state at an earlier moment. If we assume an intelligent creature that is able to calculate at one moment all relations between the different parts of the universe, it could predict all motions and all relations at all locations now and forever.

Such a **Laplace demon** would be able to predict the fate of all mankind if he knows the relevant data at a given time.

This strictly deterministic conception of the world arises the question: What can the free will of a human being change, if everything is already determined by initial conditions which we cannot influence? How much is a criminal responsible for his crime, if his future fate is already determined?

This strictly mechanistic conception has been shaken by two developments: The nonlinear dynamics and the quantum theory.

Poicaré, who has performed important spade work in nonlinear dynamics, wrote 1903:

> A very tiny effect, which we even may not notice, causes a large effect, that we cannot overlook. Then we say: This effect is accidental, because we have overlooked the real cause.

We see, that by no means the deterministic character of nature is questioned. This means that the principle of causality is accepted. However, the assumption that the initial conditions can be determined with sufficient accuracy, is no longer valid for instable systems, because here tiny changes of the initial conditions can cause large deviations of the final states. Often the uncertainty of the measurement limits the accuracy of predictions for many cases in nonlinear dynamics.

Quantum mechanics adds another principle argument: The uncertainty principle (see Vol. 3) states that it is impossible to determine exactly both the momentum and the position of a particle simultaneously. The more accurate one quantity is measured, the larger is the uncertainty for the other quantity. this implies that the initial state of a system cannot be determined with arbitrary accuracy, not only because of measurement errors, but because of the principle quantum mechanical restrictions.

For stable systems these small uncertainties have no big effect, but for unstable systems (chaotic systems) they can be disastrous, because they principally limit the predictions of future behaviour.

These few considerations illustrate that investigations of nonlinear phenomena bring about many new and surprising results when leaving the approximations of basic linear equations. This research field has developed only recently and more and more scientists are now interested in its basic physics and possible applications besides the gain of insight into the complexity of our real world [12.17–12.20].

There is in addition a psychological problem: How the spontaneous human interference in natural phenomena changes the predictability of processes and how spontaneous such interference really is, belongs into the field of psychology and cannot be solved in Physics.

Summary

- For phenomena described by nonlinear equations, the time development of the solutions often depends critically on the initial conditions. Unstable solutions are those, where infinitesimal small changes of the initial conditions cause large changes of the final states.
- The dynamical development of a system can be represented by a curve in phase space.
- The development of a system described by the Verhulst equation

  $$x_{n+1} = a \cdot x_n - a \cdot x_n^2$$

  depends on the control parameter a. For certain values of a the solutions $x_f = \lim_{n\to\infty}(x_n)$ split into two possible values (bifurcation). For larger values of a each of these two values split again into two possible values. This splitting continuous with increasing a, until the chaotic regime is reached where no predictions of the final states are possible.
- Examples of applications of this equation are the population explosion, the parametric oscillator and the origin of traffic jams without identifiable causes, often caused by the delayed reaction of the driver.
- While for linear equations the superposition principle holds (i. e. with two independent solutions also their linear combination is a solution), this is no longer true for nonlinear equations. For special nonlinear equations the solutions show self-similarity.

- Self similarity is present, if
$$x(\lambda t) = \lambda^{\kappa} x(t) \Rightarrow x(t) t^{\kappa} ,$$
where κ is a positive or negative number, not necessary an integer.
The number κ is the fractional dimension.
- For self-similar solutions of nonlinear equations a natural scale is missing. An arbitrary stretching of time can be compensated by a corresponding change of the x-scale.
- All complex numbers which are generated by the sequence $z_{n+1} = z_n^2 + c$ form a set in the complex plane. The set of all non-divergent sequences with $z_0 = 0$ generate the Mandelbrot set.
- Such sets can be graphically displayed by simple computer programs.

Problems

12.1 A mass m is hold in its equilibrium position $(0, 0)$ in the x, y-plane by four springs with length L and restoring force constant k. aligned in the $\pm x$- and $\pm y$-directions.

a) What is the equation of motion, when the mass m is displaced in the $+x$-direction?
b) Bring this equation for $x \ll L$ in the form $\mathrm{d}^2x/\mathrm{d}t^2 + ax + bx^3 = 0$. How large are a and b?
c) What is the oscillation frequency for $bx^3 \ll a$ in the linear approximation and how does it change in the nonlinear form of the equation in b)?

12.2 Show, that the nonlinear equation in Probl. 12.1b with the initial conditions $x(0) = x_0$ and $\mathrm{d}x/\mathrm{d}t(0) = 0$ has periodical solutions.

12.3

a) Show that the non linear equation $m\,\mathrm{d}^2x/\mathrm{d}t^2 = -k_1x - k_2x$ with the initial conditions $x(0) = x_0$, $(\mathrm{d}x/\mathrm{d}t)(0) = 0$ can be transformed by the substitutions $\omega_0^2 = k_1/m$, $y = x/x_0$, $L^* = \omega_0 L$ into the dimensionless form $\mathrm{d}^2y/\mathrm{d}L^{*2} + y + \varepsilon y^2$ with $\varepsilon = x_0k_2/k_1$.
b) Calculate for $\varepsilon = 0.1$ the frequency shift against ω_0.

12.4 Determine the fix points for the system of differential equations $\mathrm{d}x_1/\mathrm{d}t = \lambda_1x_1 - \lambda_2x_1x_2$.
For which values of λ_1 and λ_2 are the fix points stable, metastable or unstable?

12.5 The equation of motion for the damped pendulum oscillation is $\ddot{\varphi} + \gamma\dot{\varphi} + \omega_0^2 \sin\varphi = 0$ with $\omega_0^2 = g/L$.
Determine the oscillation period $T(\varphi)$ and calculate the ratios $T(\varphi)/T(0)$ for $\varphi = \pi/4, \pi/2, (3/4)\,\pi$ and π.

12.6

a) What is the solution of the logistic growth function $\dot{z}(t) = az - bz^2$?
b) After which time has the function $z(t)$ doubled for $a = b$?
c) What is the limit for $z(t \to \infty)$?

12.7 Determine the fixpoints x_f and the Ljapunov exponent λ of the logistic equation $x_{n+1} = ax_n(1 - x_n)$ for $a = 3.1$ and $a = 3.3$.

12.8 Show, that the fractional dimension of the Sierpinski grid is $d_1 = 1.5849$.

12.9 Determine fixpoints and attractor for the differential equation in polar coordinates $\mathrm{d}r/\mathrm{d}t = -r(-a + r^2)$ for $a < 0$ and $a > 0$, $\mathrm{d}\varphi/\mathrm{d}t = \omega_0 = \text{const.}$

12.10 A particle with mass m moves in the potential $E_\mathrm{pot} = E_\mathrm{pot}(x_0) + a(x - x_0)^2 + b(x - x_0)^3$.

a) Determine the nonlinear equation of motion.
b) Up to which amplitude x_max is the solution a harmonic oscillation?

References

12.1a. J. Gleick, *Chaos: Making a New Science.* (Penguin Books, New York, 2008)

12.1b. E. Lorenz, *The Essence of Chaos.* (University of Washington Press, Washington, 1995)

12.2a. P. Bak, *How Nature works: The Science of Self-organizes Criticality.* (Copernicus Books, New York, 1996)

12.2b. St.A. Kaufman, *At Home in the Universe: The Search for the Laws of Self-Organization and Complexity.* (Oxford University Press, Oxford, 1996)

12.3. F.D. Peat, J. Briggs, *Timeless Wisdom from the Science of Change.* (Harper Collins Publ., New York, 1999)

12.4. T. Mullin, *The Nature of Chaos.* (Clarendon Press, Watton-under-Edge, UK, 1995)

12.5. P. Bergé, K. Pomeau, Ch. Vidal, *Order within Chaos.* (John Wiley, New York, 1984)

12.6a. G.L. Baker, J.P. Gollub, *Chaotic Dynamics.* (Cambridge University Press, Cambridge, 1996)

12.6b. M.J. Feigenbaum, J. Stat. Phys. **19**, 25–52 (1978)

12.7. W. Leutzbach, *Introduction in to the Theory of Traffic Flow.* (Springer, New York, 1987)

12.8. https://en.wikipedia.org/wiki/Self-similarity

12.9. B.B. Mandelbrot, *The Fractal Geometry of Nature.* (Freeman, San Francisco, 1982)

12.10a. H.-O. Peitgen, P.H. Richter, *The Beauty of Fractals.* (Springer, Berlin, Heidelberg, 1986)

12.10b. H.-O. Peitgen, *Chaos: Bausteine der Ordnung.* (Rowohlt, Hamburg, 1998)

12.11. See many articles in: *Chaos: An Interdisciplinary Journal of Nonlinear Science.* (American Institut of Physics)

12.12. H.J. Korsch, H.J. Jodl, *Chaos, A Program Collection for the PC.* (Springer, Berlin, Heidelberg, 1994)

12.13. H. Jürgens, H.O. Peitgen, D. Saupe (ed.), *Chaos and Fractals.* (Springer, Berlin, Heidelberg, 1993)

12.14. St.H. Strogatz, *Nonlinear Dynamics and Chaos: Applications to Physics, Biology, Chemistry and Engineering.* (Westview Press, Boulder, Colorado, USA, 2014)

12.15. G. Mayer-Kress (ed.), *Dimensions and Entropies in Chaotic Systems.* (Springer, Berlin, Heidelberg, 1986)

12.16. W.H. Steeb, A. Kunick, *Chaos und Quantenchaos in dynamischen Systemen,* 2nd ed. (Bibliographisches Institut, Mannheim, 1994)

12.17. B. Kaye, *Chaos and Complexity.* (VCH, Weinheim, 1993)

12.18. J. Parisi, St. Müller, W. Zimmermann (eds.), *A Perspective Look at Nonlinear Media.* (Springer, Berlin, Heidelberg, 1998)

12.19. L. Lam, *Nonlinear Physics for Beginners.* (World Scientific, Singapore, 1998)

12.20. R. Gilmore, M. Lefrane, *The Topology of Chaos.* (Wiley VCH, Weinheim, 2002)

Supplement

13

Chapter 13

© Springer International Publishing Switzerland 2017
W. Demtröder, *Mechanics and Thermodynamics*, Undergraduate Lecture Notes in Physics, DOI 10.1007/978-3-319-27877-3_13

13.1 Vector Algebra and Analysis

13.1.1 Definition of Vectors

A **vector** is an oriented line segment. Its length is the **magnitude of the vector**. Vectors are denoted in this textbook by bold letters.

Two vectors are equal, if they have the same direction and magnitude. The magnitude of a vector is a pure number (**scalar**), independent of the direction of the vector.

> Since a parallel displacement of a vector in space does not change its direction nor its magnitude, all parallel vectors with the same magnitude are equal, independent of the coordinates of their starting points.

The starting point of a vector is also called **point of origin**.

A vector starting from the origin and ending at a point P is called **position vector**, because it defines the position of P in space.

Multiplying a vector with a scalar number changes its length but not its direction.

13.1.2 Representation of Vectors

Every vector in a three-dimensional space can be represented by three linear independent **basis vectors**. The selection of these basis vectors depends on the chosen coordinate system.

13.1.2.1 Cartesian Coordinates

When we plot a vector $\boldsymbol{r}$ in a Cartesian coordinate system (x, y, z) with the point of origin $(0, 0, 0)$ it ends at the point $P(x, y, z)$ with the coordinates x, y, z (Fig. 13.1). These coordinates are the projection of the vector onto the three coordinate axes. They are called the components of the vector.

The component representation of the vector r is

$$\boldsymbol{r} = \{x, y, z\} . \qquad (13.1\text{a})$$

A vector $\boldsymbol{r}$ is uniquely defined by its components, because its magnitude (written as $|\boldsymbol{r}|$) is

$$r = |\boldsymbol{r}| = \sqrt{x^2 + y^2 + z^2} , \qquad (13.1\text{b})$$

as can be derived by Fig. 13.1 and the theorem of Pythagoras.

The direction of a vector is defined by its components. It can be also represented by the three angles α, β, γ against the coordinate axes. It is

$$\cos\alpha = x/r , \quad \cos\beta = y/r , \quad \cos\gamma = z/r .$$

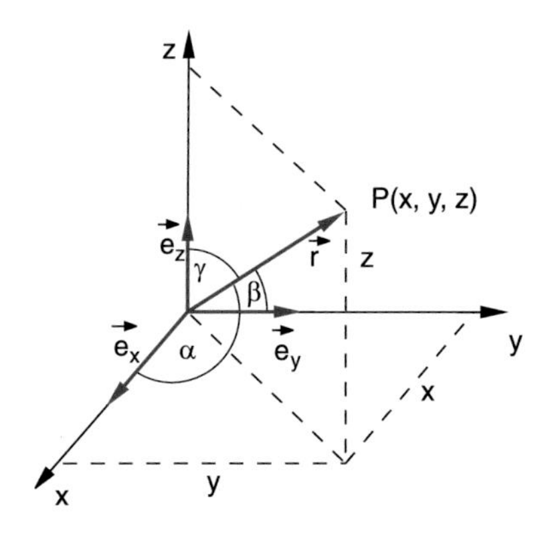

Figure 13.1 Cartesian coordinate system

A vector with the length $L = 1$ ($\sqrt{x^2 + y^2 + z^2} = 1$) is called **unit vector**. It is often represented by

$$\hat{\boldsymbol{r}} = \boldsymbol{r}/|\boldsymbol{r}| = \hat{\boldsymbol{e}} . \qquad (13.1\text{c})$$

Special unit vectors are the three vectors

$$\hat{\boldsymbol{e}}_1 = \{1, 0, 0\} ; \quad \hat{\boldsymbol{e}}_2 = \{0, 1, 0\} ; \quad \hat{\boldsymbol{e}}_3 = \{0, 0, 1\} . \qquad (13.1\text{d})$$

Every vector $\boldsymbol{r} = \{x, y, z\}$ can be written as linear combination of the three basis vectors

$$\boldsymbol{r} = x\hat{\boldsymbol{e}}_1 + y\hat{\boldsymbol{e}}_2 + z\hat{\boldsymbol{e}}_3 . \qquad (13.2)$$

13.1.2.2 Spherical Coordinates

The position vector $\boldsymbol{r}$ pointing from the origin $(0, 0, 0)$ to the point $P(r, \vartheta, \varphi)$ is defined in spherical coordinates (also called polar coordinates) by its length $r = |\boldsymbol{r}|$ and the angles ϑ and φ that define uniquely its direction (Fig. 13.2).

The conversion to Cartesian coordinates is given by

$$\begin{aligned} x &= r \cdot \sin\vartheta \cdot \cos\varphi , \\ y &= r \cdot \sin\vartheta \cdot \sin\varphi , \\ z &= r \cdot \cos\vartheta . \end{aligned}$$

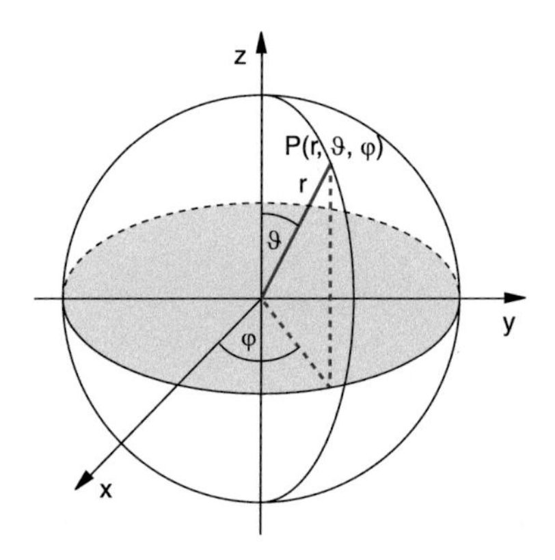

Figure 13.2 Spherical coordinates r, ϑ, φ

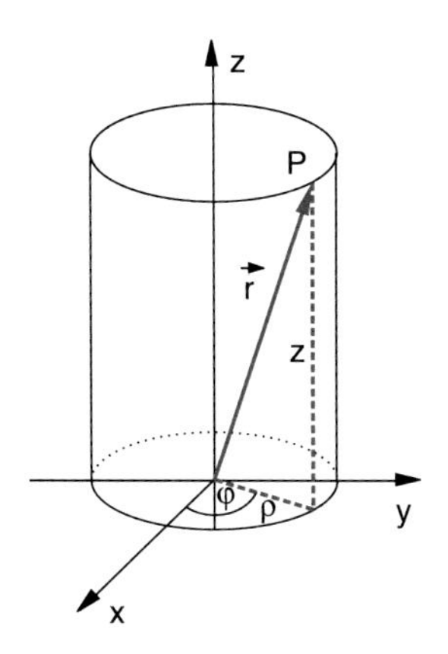

Figure 13.3 Cylindrical coordinates ϱ, φ, z

13.1.2.3 Cylindrical Coordinates

The point $P(\varrho, \varphi, z)$ in Fig. 13.3 is defined in cylindrical coordinates $(\varrho, \varphi . z)$ by the vector $\boldsymbol{r} = \{\varrho, \varphi, z\}$ where ϱ gives the distance from the z-axis, z the distance from the x–y-plane and φ the angle of the projection of $\boldsymbol{r}$ onto the x–y-plane against the x-axis. The conversion to Cartesian coordinates is:

$$x = \varrho \cdot \cos\varphi ,$$
$$y = \varrho \cdot \sin\varphi ,$$
$$z = z .$$

The length of the vector r is

$$|\boldsymbol{r}| = \sqrt{\varrho^2 + z^2} .$$

The direction of r is defined by the angle φ and the ratio z/ϱ.

13.1.3 Polar and Axial Vectors

The transformation $x \to -x$; $y \to -y$; $z \to -z$ (mirror imaging of the coordinate system) transforms the position vector $\boldsymbol{r} \to -\boldsymbol{r}$. Therefore r is called a **polar vector**.

Besides these polar vectors which are defined by their length and their direction, there are also vectors that define apart from direction and magnitude a sense of rotation.

Example

Magnitude and orientation of a surface element can be characterized by the *normal vector* $\boldsymbol{A}$ perpendicular to the surface element (Fig. 13.4). The magnitude of the vector gives the area of the surface element and its direction the orientation of the surface element. In order to define uniquely on which of the two sides of the surface element the vector $\boldsymbol{A}$ starts, its direction is defined such, that it forms a right hand screw (like a corkscrew) when the surface element is counterclockwise circulated.

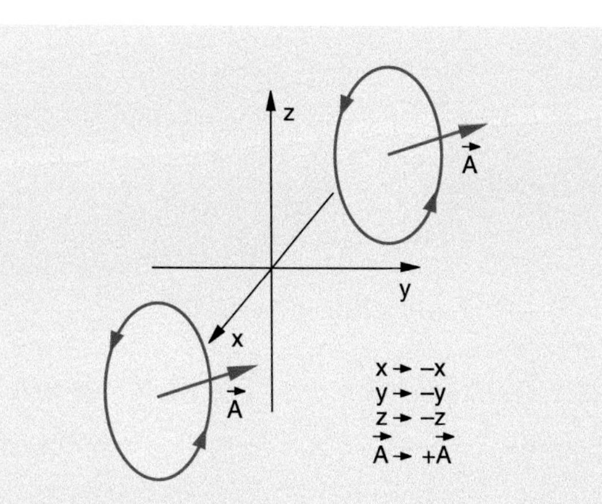

Figure 13.4 Surface normal vector A as axial vector, that is characterized by its direction of orientation. At the coordinate transformation $\boldsymbol{r} \to -\boldsymbol{r}$, $\boldsymbol{A}$ keeps its orientation as a right hand helix ◀

Under the coordinate transformation $x \to -x$, $y \to -y$, $z \to -z$ the sense of rotation of A is preserved, i. e. $\boldsymbol{A} \to +\boldsymbol{A}$ but not $\boldsymbol{A} \to -\boldsymbol{A}$, i. e. $\boldsymbol{A}$ forms again a right-handed screw. Such vectors are called **axial vectors**. Examples are the angular momentum vector $\boldsymbol{L} = \boldsymbol{r} \times \boldsymbol{p}$ (see Sect. 2.8 and 13.1.5.3).

13.1.4 Addition and Subtraction of Vectors

Definition

Vectors are added by adding their components (Fig. 13.5). The sum of the two vectors $\boldsymbol{a} = \{a_1, a_2, a_3\}$ and $\boldsymbol{b} = \{b_1, b_2, b_3\}$ is the vector

$$\boldsymbol{c} = \boldsymbol{a} + \boldsymbol{b} = \{a_1 + b_1, a_2 + b_2, a_3 + b_3\} . \qquad (13.3)$$

According to this rule each vector can be written as the sum of its component vectors

$$\boldsymbol{a} = \{a_1, a_2, a_3\} = a_1\hat{\boldsymbol{e}}_1 + a_2\hat{\boldsymbol{e}}_2 + a_3\hat{\boldsymbol{e}}_3 .$$

The graphical representation of vector addition shown in Fig. 13.5, illustrates that the sum vector $\boldsymbol{c}$ is the diagonal in the parallelogram of the vectors $\boldsymbol{a}$ and $\boldsymbol{b}$.

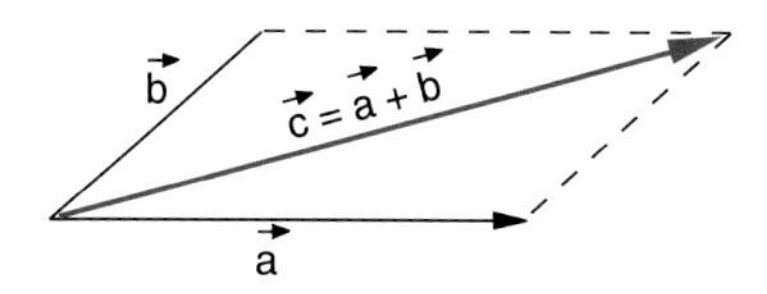

Figure 13.5 Addition of vectors

Chapter 13

Problem

Show that this graphical representation fulfils the rules of vector addition. ◄

13.1.5 Multiplication of Vectors

13.1.5.1 Multiplication of a Vector with a Scalar

The vector $\boldsymbol{a}$ is multiplied by a scalar c by multiplying each component of $\boldsymbol{a}$ by c.

$$c \cdot \boldsymbol{a} = c \cdot \{a_1, a_2, a_3\} = \{c \cdot a_1, c \cdot a_2, c \cdot a_3\} ,$$
$$|c \cdot \boldsymbol{a}| = |c| \cdot |\boldsymbol{a}| .$$

13.1.5.2 The Scalar Product

The scalar product of two vectors

$$\boldsymbol{a} = \{a_1, a_2, a_3\} \quad \text{and} \quad \boldsymbol{b} = \{b_1, b_2, b_3\}$$

with the angle α between them is defined as the scalar

$$c = \boldsymbol{a} \cdot \boldsymbol{b} = |\boldsymbol{a}| \cdot |\boldsymbol{b}| \cdot \cos\alpha . \qquad (13.4a)$$

It is the product of the projection $|\boldsymbol{b}| \cdot \cos\alpha$ of $\boldsymbol{b}$ on $\boldsymbol{a}$ times the amount $|\boldsymbol{a}|$ of $\boldsymbol{a}$. For $\alpha = 90°$ is the scalar product zero (Fig. 13.6).

Two vectors $\neq 0$ are perpendicular to each other only if their scalar product is zero.

For the three unit vectors $\hat{\boldsymbol{e}}_1, \hat{\boldsymbol{e}}_2, \hat{\boldsymbol{e}}_3$ hold the relation

$$\hat{\boldsymbol{e}}_i \cdot \hat{\boldsymbol{e}}_k = \delta_{ik} ,$$

where δ_{ik} is the Kronecker symbol which is defined by

$$\delta_{ik} = \begin{cases} 1 & \text{for } i = k , \\ 0 & \text{for } i \neq k . \end{cases}$$

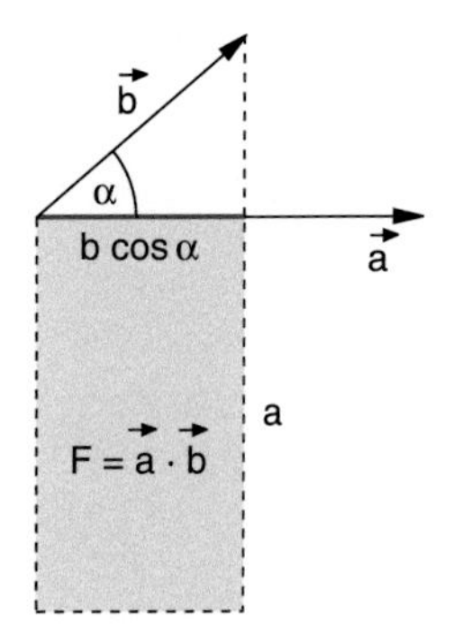

Figure 13.6 The scalar product $\boldsymbol{a} \cdot \boldsymbol{b}$ represents the area $A = \boldsymbol{a} \cdot \boldsymbol{b} = a \cdot b \cdot \cos\alpha$

The scalar product can be also expressed by the vector components. For

$$\boldsymbol{a} = a_1\hat{\boldsymbol{e}}_1 + a_2\hat{\boldsymbol{e}}_2 + a_3\hat{\boldsymbol{e}}_3 ,$$
$$\boldsymbol{b} = b_1\hat{\boldsymbol{e}}_1 + b_2\hat{\boldsymbol{e}}_2 + b_3\hat{\boldsymbol{e}}_3 , \qquad (13.4b)$$

the scalar product becomes

$$\boldsymbol{a} \cdot \boldsymbol{b} = (a_1\hat{\boldsymbol{e}}_1 + a_2\hat{\boldsymbol{e}}_2 + a_3\hat{\boldsymbol{e}}_3) \cdot (b_1\hat{\boldsymbol{e}}_1 + b_2\hat{\boldsymbol{e}}_2 + b_3\hat{\boldsymbol{e}}_3) ,$$
$$= a_1b_1 + a_2b_2 + a_3b_3 \quad \text{since} \quad \hat{\boldsymbol{e}}_i \cdot \hat{\boldsymbol{e}}_k = \delta_{ik} . \qquad (13.5)$$

13.1.5.3 The Vector Product

Definition

The vector product of two vectors $\boldsymbol{a}$ and $\boldsymbol{b}$ is the vector $\boldsymbol{c} = \boldsymbol{a} \times \boldsymbol{b}$

- that is perpendicular to $\boldsymbol{a}$ and $\boldsymbol{b}$,
- that forms a right handed screw when $\boldsymbol{a}$ is rotated toward $\boldsymbol{b}$ on the shortest way,
- that has the magnitude $|\boldsymbol{c}| = |\boldsymbol{a}| \cdot |\boldsymbol{b}| \cdot \sin\alpha$, where α is the angle between $\boldsymbol{a}$ and $\boldsymbol{b}$.

The vector $\boldsymbol{c}$ defines besides magnitude and direction also the orientation. It is therefore an **axial vector**.

Note, that

$$(\boldsymbol{a} \times \boldsymbol{b}) = -(\boldsymbol{b} \times \boldsymbol{a}) .$$

The absolute value $|\boldsymbol{a}| \cdot |\boldsymbol{b}| \cdot \sin\alpha$ of the vector product $\boldsymbol{a} \times \boldsymbol{b}$ is equal to the area of the parallelogram, formed by $\boldsymbol{a}$ and $\boldsymbol{b}$ (Fig. 13.7). The vector product can be therefore regarded as the surface normal of the parallelogram generated by the two vectors $\boldsymbol{a}$ and $\boldsymbol{b}$.

$\boldsymbol{c} = (\boldsymbol{a} \times \boldsymbol{b})$ is an axial vector, because under reflection of all coordinates at the origin we have $\boldsymbol{a} \to -\boldsymbol{a}, \boldsymbol{b} \to -\boldsymbol{b} \Rightarrow \boldsymbol{c} \to \boldsymbol{c}$ (Fig. 13.7b).

For the unit vectors we get the relations:

$$\hat{\boldsymbol{e}}_1 \times \hat{\boldsymbol{e}}_2 = \hat{\boldsymbol{e}}_3 , \quad \hat{\boldsymbol{e}}_2 \times \hat{\boldsymbol{e}}_3 = \hat{\boldsymbol{e}}_1 , \quad \hat{\boldsymbol{e}}_3 \times \hat{\boldsymbol{e}}_1 = \hat{\boldsymbol{e}}_2 .$$

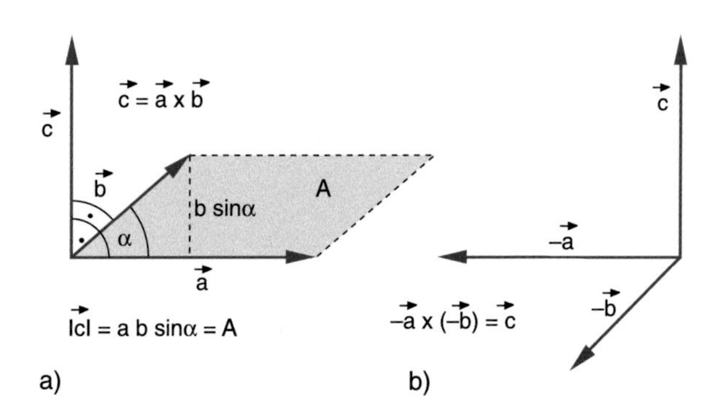

Figure 13.7 The vectorial product as normal vector to the area $|\boldsymbol{a} \times \boldsymbol{b}|$

We therefore get the component representation of the vector product

$$(\boldsymbol{a} \times \boldsymbol{b}) = \{a_2b_3 - a_3b_2, a_3b_1 - a_1b_3, a_1b_2 - a_2b_1\} . \quad (13.6)$$

Check this relation by multiplication of the six vector components.

This component representation can be abbreviated by the symbolic determinant notation

$$\boldsymbol{c} = \begin{vmatrix} \hat{\boldsymbol{e}}_1 & \hat{\boldsymbol{e}}_2 & \hat{\boldsymbol{e}}_3 \\ a_1 & a_2 & a_3 \\ b_1 & b_2 & b_3 \end{vmatrix} \quad (13.7)$$
$$= (a_2b_3 - a_3b_2)\hat{\boldsymbol{e}}_1 + (a_3b_1 - a_1b_3)\hat{\boldsymbol{e}}_2 + (a_1b_2 - a_2b_1)\hat{\boldsymbol{e}}_3 .$$

13.1.5.4 Multiple Products

Scalar Products of a Polar and an Axial Vector

The dot product (scalar product) $d = \boldsymbol{c}(\boldsymbol{a} \times \boldsymbol{b})$ of the polar vector $\boldsymbol{c}$ and the axial vector $(\boldsymbol{a} \times \boldsymbol{b})$ gives a scalar number d that transforms into $-\boldsymbol{c}$ when all coordinates are reflected at the origin, because the axial vector $(\boldsymbol{a} \times \boldsymbol{b})$ does not change its sign, while the scalar number $\boldsymbol{c}$ does. The number d is called a **pseudo-scalar**.

The product $d = |\boldsymbol{c}| \cdot |\boldsymbol{a} \times \boldsymbol{b}| \cdot \cos\beta$ describes the volume of the parallel-epiped (oblique angled cuboid) which is formed by the vectors $\boldsymbol{a}, \boldsymbol{b}$ and $\boldsymbol{c}$ (Fig. 13.8).

This scalar triple product can be written as a determinant

$$\boldsymbol{c} \cdot (\boldsymbol{a} \times \boldsymbol{b}) = \begin{vmatrix} a_1 & a_2 & a_3 \\ b_1 & b_2 & b_3 \\ c_1 & c_2 & c_3 \end{vmatrix} . \quad (13.8)$$

Vector Product of a Polar and an Axial Vector

$$\boldsymbol{d} = \boldsymbol{c} \times (\boldsymbol{a} \times \boldsymbol{b}) .$$

Since the vector $(\boldsymbol{a} \times \boldsymbol{b})$ is perpendicular to $\boldsymbol{a}$ and to $\boldsymbol{b}$ and the vector $\boldsymbol{d}$ is perpendicular to $(\boldsymbol{a} \times \boldsymbol{b})$, $\boldsymbol{d}$ must lie in the plane of $\boldsymbol{a}$ and $\boldsymbol{b}$ which we choose as the x–y-plane. It therefore can be described as a linear combination of $\boldsymbol{a}$ and $\boldsymbol{b}$.

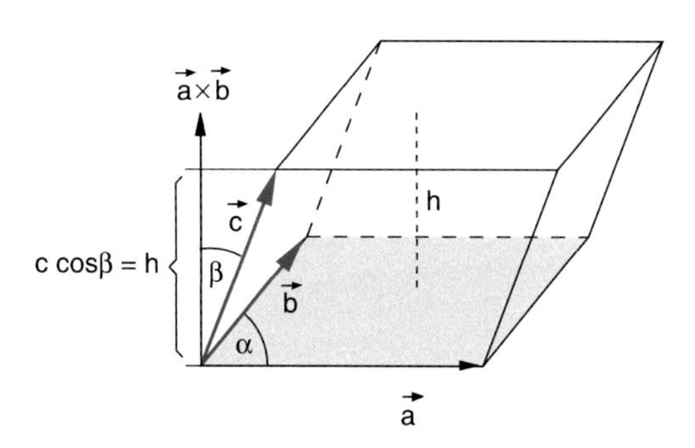

Figure 13.8 Scalar triple products $\boldsymbol{d} = \boldsymbol{c} \cdot (\boldsymbol{a} \times \boldsymbol{b})$ represents the volume of the parallel-epiped generated by vectors $\boldsymbol{a}, \boldsymbol{b}$ and $\boldsymbol{c}$

$$\boldsymbol{d} = x \cdot \boldsymbol{a} + y \cdot \boldsymbol{b} , \quad (x \text{ and } y \text{ are real numbers}) .$$

Inserting the components of $\boldsymbol{a}, \boldsymbol{b}$ and $\boldsymbol{c}$ gives, following the rules given above for the components,

$$x = c_1b_1 + c_2b_2 + c_3b_3 ,$$
$$y = -c_1a_1 - c_2a_2 - c_3a_3 .$$

These relations give the vector equation

$$\boldsymbol{c} \times (\boldsymbol{a} \times \boldsymbol{b}) = (\boldsymbol{c} \cdot \boldsymbol{b})\boldsymbol{a} - (\boldsymbol{c} \cdot \boldsymbol{a})\boldsymbol{b} . \quad (13.9)$$

Since the vector product changes its sign when the sequence of the factors are interchanged, we get the relations

$$(\boldsymbol{a} \times \boldsymbol{b}) \times \boldsymbol{c} = -\boldsymbol{c} \times (\boldsymbol{a} \times \boldsymbol{b}) ,$$
$$(\boldsymbol{a} \times \boldsymbol{b}) \times \boldsymbol{c} \neq \boldsymbol{a} \times (\boldsymbol{b} \times \boldsymbol{c}) \neq (\boldsymbol{a} \times \boldsymbol{c}) \times \boldsymbol{b} .$$

Neither the commutative law nor the associative law are valid for the triple vector products.

From Eq. 13.9 follows

$$\boldsymbol{a} \times (\boldsymbol{b} \times \boldsymbol{c}) + \boldsymbol{b} \times (\boldsymbol{c} \times \boldsymbol{a}) + \boldsymbol{c} \times (\boldsymbol{a} \times \boldsymbol{b}) = \boldsymbol{0} . \quad (13.10)$$

Scalar Product of Two Axial Vectors

From the relations above we can conclude

$$(\boldsymbol{a} \times \boldsymbol{b}) \cdot (\boldsymbol{c} \times \boldsymbol{d}) = (\boldsymbol{a} \cdot \boldsymbol{c})(\boldsymbol{b} \cdot \boldsymbol{d}) - (\boldsymbol{a} \cdot \boldsymbol{d})(\boldsymbol{b} \cdot \boldsymbol{c}) , \quad (13.11a)$$
$$(\boldsymbol{a} \times \boldsymbol{b})^2 = a^2b^2 - (\boldsymbol{a} \cdot \boldsymbol{b})^2 . \quad (13.11b)$$

13.1.6 Differentiation of Vectors

13.1.6.1 Vector-Fields

If it is possible to attribute to each space point $P(x, y, z)$ a vector $\boldsymbol{a} = \{a_x, a_y, a_z\}$ the entity of all vectors $\boldsymbol{a} = \boldsymbol{a}(x, y, z)$ is called a vector field. Each component of $\boldsymbol{a}$ is a function of the coordinates (x, y, z):

$$a_x = f_1(x, y, z) ; \quad a_y = f_2(x, y, z) \quad \text{and} \quad a_z = f_3(x, y, z) .$$

This means that length and direction of the vector a depend on the coordinates (x, y, z). If the components depend additionally on the time t, $\boldsymbol{a}(x, y, z)$ represents a time-dependent vector field. If $\boldsymbol{a}$ does not depend on time, the field is called stationary or static.

Examples

1. The velocity of particles in a fluid, flowing through a pipe with locally variable cross section represents a vector field. If the pressure is time dependent, the vector field $\boldsymbol{v} = \boldsymbol{f}(x, y, z, t)$ is non-stationary.

2. The force on a mass in the gravitation field of the earth depends on the distance from the earth centre. The force field is stationary because the force is independent of time. ◄

13.1.6.2 Scalar Differentiation of a Vector

We assume that the position vector $\boldsymbol{r}(x, y, z, t)$ is a continuous function $f(t)$ of time t, i. e. its components are continuous function of time. The variation of r with time is determined by the corresponding variation of the components. The equation (Fig. 13.9)

$$\frac{\Delta \boldsymbol{r}}{\Delta t} = \frac{\boldsymbol{r}(t + \Delta t) - \boldsymbol{r}(t)}{\Delta t}$$

is the abbreviation for the three equations for the components

$$\frac{\Delta x}{\Delta t} = \frac{x(t + \Delta t) - x(t)}{\Delta t}$$

with corresponding equation for y and z.

For the limes $\Delta t \to 0$ the equation converges towards

$$\lim_{\Delta t \to 0} \frac{\Delta x}{\Delta t} = \frac{\mathrm{d}x}{\mathrm{d}t}, \quad \text{ect. for } y \text{ and } z\,.$$

The time derivative of the vector r then becomes

$$\frac{\mathrm{d}\boldsymbol{r}}{\mathrm{d}t} = \dot{\boldsymbol{r}} \stackrel{\text{Def}}{=} \{\dot{x}, \dot{y}, \dot{z}\}\,. \tag{13.12}$$

The derivative of a vector with respect to a scalar (e. g. the time t) is formed by differentiating all three components.

For the differentiation of products of vectors, the same rules are valid as for scalar quantities:

$$\frac{\mathrm{d}}{\mathrm{d}t}(\boldsymbol{a} \cdot \boldsymbol{b}) = \dot{\boldsymbol{a}} \cdot \boldsymbol{b} + \boldsymbol{a} \cdot \dot{\boldsymbol{b}} \tag{13.13a}$$

$$\frac{\mathrm{d}}{\mathrm{d}t}(\boldsymbol{a} \times \boldsymbol{b}) = (\dot{\boldsymbol{a}} \times \boldsymbol{b}) + (\boldsymbol{a} \times \dot{\boldsymbol{b}})\,. \tag{13.13b}$$

Note, that the succession of the factors in the product is essential because $(\boldsymbol{a} \times \boldsymbol{b}) = -(\boldsymbol{b} \times \boldsymbol{a})$.

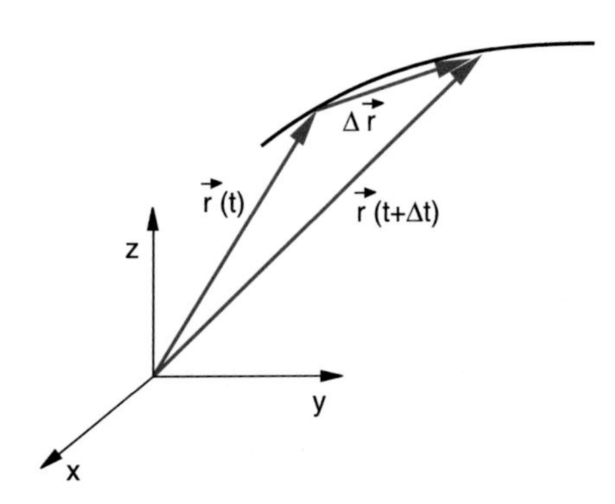

Figure 13.9 Differentiation of a vector with respect to time

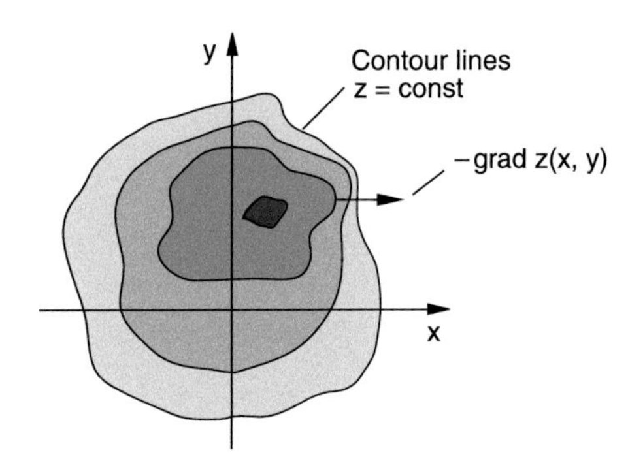

Figure 13.10 The gradient ∇f is a vector, perpendicular to the contour lines $f = z(x, y) = \text{const}$. The *dark red spot* indicates the top of the mountain

13.1.6.3 The Gradient of a Scalar Quantity

The partial derivative $\partial f / \partial x$ of a scalar function $f(x, y, z)$ gives the change of f per unit length in x-direction, while y and z are kept constant. For example, for a surface $z = f(x, y)$ the expression $(\partial f / \partial x)_P$ gives the slope of the surface in the x-direction at the point P. Analogue expressions apply for $\partial f / \partial y$ and $\partial f / \partial z$.

The vector

$$\textbf{grad} f = \nabla f = \left\{ \frac{\partial f}{\partial x}, \frac{\partial f}{\partial y}, \frac{\partial f}{\partial z} \right\}, \tag{13.14}$$

with the partial derivatives as components is called the **gradient** of the function $f(x, y, z)$. It is denoted b the symbol $\boldsymbol{\nabla}$ (nabla, which is the symbol for an Egyptian musical string instrument, similar to our harp).

The differential operator is then expressed by the vector

$$\nabla = \left\{ \frac{\partial}{\partial x}, \frac{\partial}{\partial y}, \frac{\partial}{\partial z} \right\}, \tag{13.15}$$

which gets its significance only when applied to the scalar function $f(x, y, z)$. One can formally write $\textbf{grad} f$ as the product of the differential operator $\boldsymbol{\nabla}$ and the function f.

The total change of the function $f(x, y, z)$ obtained when passing from a point $P(x, y, z) = P(\boldsymbol{r})$ into an arbitrary direction $\mathrm{d}\boldsymbol{n}$ to the neighbouring point $P(\boldsymbol{r} + d\boldsymbol{n})$ is

$$(\mathrm{d}f)_n = \frac{\partial f}{\partial x}\mathrm{d}n_x + \frac{\partial f}{\partial y}\mathrm{d}n_y + \frac{\partial f}{\partial z}\mathrm{d}n_z = \mathrm{d}\boldsymbol{n} \cdot \nabla f\,. \tag{13.16}$$

$(\mathrm{d}f_n)$ becomes maximum, when $\mathrm{d}\boldsymbol{n}$ is parallel to $\textbf{grad} f$.

The gradient ∇f gives the direction of the maximum change of $f(x, y, z)$ (Fig. 13.10).

Example

For the surface $z = f(x, y)$ which gives the height z as a function of the coordinates x and y, ∇z is always perpendicular to the contour lines $z = \text{const}$. The gradient is therefore the tangent vector to the trajectory of maximum slope. ◄

13.1.6.4 The Divergence of a Vector Field

The scalar product of the vector ∇ with a vector function $\boldsymbol{u}(x,y,z)$ (for example the locally varying velocity field of a fluid flow) is called the divergence of the vector field.

$$\operatorname{div} \boldsymbol{u}(x,y,z) = \nabla \cdot \boldsymbol{u} \tag{13.17}$$

According to the definition of the Nabla-operator in (13.15) this is equal to

$$\nabla \cdot \boldsymbol{u} = \frac{\partial u_x}{\partial x} + \frac{\partial u_y}{\partial y} + \frac{\partial u_z}{\partial z} . \tag{13.18}$$

As has been shown in Sect. 8.3 the divergence of a volume element

$$\operatorname{div} \boldsymbol{u} \cdot \mathrm{d}V = \boldsymbol{u} \cdot \mathrm{d}\boldsymbol{S}$$

gives the vector flux passing per sec through the surface $\mathrm{d}S$ that surrounds the volume element $\mathrm{d}V$. It is therefore also called the source function of the vector field $\boldsymbol{u}(x,y,z)$.

13.1.6.5 The Curl of a Vector Field

The vector product

$$\nabla \times \boldsymbol{u} = \mathbf{curl}\, \boldsymbol{u} \tag{13.19}$$

of the vector ∇ with the vector $\boldsymbol{u}(x,y,z)$ is the **curl** of the vector field $\boldsymbol{u}(x,y,z)$.

According to the algorithm for vector products we obtain for the components of $\nabla \times u$

$$\begin{aligned}
(\nabla \times \boldsymbol{u})_x &= \left(\frac{\partial u_z}{\partial y} - \frac{\partial u_y}{\partial z}\right) , \\
(\nabla \times \boldsymbol{u})_y &= \left(\frac{\partial u_x}{\partial z} - \frac{\partial u_z}{\partial x}\right) , \\
(\nabla \times \boldsymbol{u})_z &= \left(\frac{\partial u_y}{\partial x} - \frac{\partial u_x}{\partial y}\right) .
\end{aligned} \tag{13.20}$$

As has been shown in Chap. 8 is $\mathbf{curl}\, \boldsymbol{u}$ (also written as $\mathbf{rot}\, \boldsymbol{u}$) a measure for the rotation of a vortex in a fluid flow with the velocity field $\boldsymbol{u}(x,y,z)$.

13.1.6.6 Second Derivatives

With the nabla operator ∇ higher derivatives of scalar functions $f(x,y,z)$ or of vector fields $\boldsymbol{u}(x,y,z)$ can be written in a clear way:

- $\nabla \cdot (\nabla f) = \operatorname{div} \mathbf{grad} f$

$$\begin{aligned}
&= \frac{\partial}{\partial x}\left(\frac{\partial f}{\partial x}\right) + \frac{\partial}{\partial y}\left(\frac{\partial f}{\partial y}\right) + \frac{\partial}{\partial z}\left(\frac{\partial f}{\partial z}\right) \\
&= \frac{\partial^2 f}{\partial x^2} + \frac{\partial^2 f}{\partial y^2} + \frac{\partial^2 f}{\partial z^2} = \Delta f ,
\end{aligned} \tag{13.21}$$

where the symbol Δ is the **Laplace operator**.

- $\nabla(\nabla \cdot \boldsymbol{u}) = \mathbf{grad} \operatorname{div} \boldsymbol{u}$ is a vector with the three components

$$\begin{aligned}
\nabla(\nabla \cdot \boldsymbol{u})_x &= \frac{\partial}{\partial x}\left(\frac{\partial u_x}{\partial x} + \frac{\partial u_y}{\partial y} + \frac{\partial u_z}{\partial z}\right) \\
&= \frac{\partial}{\partial x}(\operatorname{div} \boldsymbol{u}) ,
\end{aligned} \tag{13.22}$$

and similar equations for the y- and z-components.

- $\nabla \times (\nabla \times \boldsymbol{u}) = \mathbf{curl}\,\mathbf{curl}\, \boldsymbol{u}$.

From the rules in Sect. 13.1.5 we obtain

$$\begin{aligned}
\nabla \times (\nabla \times \boldsymbol{u}) &= \nabla(\nabla \cdot \boldsymbol{u}) - \nabla \cdot (\nabla \boldsymbol{u}) \\
&= \mathbf{grad} \operatorname{div} \boldsymbol{u} - \operatorname{div} \mathbf{grad}\, \boldsymbol{u} .
\end{aligned} \tag{13.23}$$

This is a vector equation because $\nabla \cdot \nabla \boldsymbol{u}$ is the scalar product of the vector ∇ with the tensor $\nabla \boldsymbol{u}$ (see below). The equation for the x-components is

$$\begin{aligned}
&(\nabla \times \nabla \times \boldsymbol{u})_x \\
&= \frac{\partial}{\partial x}\left(\frac{\partial u_x}{\partial x} + \frac{\partial u_y}{\partial y} + \frac{\partial u_z}{\partial z}\right) - \Delta u_x .
\end{aligned} \tag{13.23a}$$

Similar equations hold the y- and z-component.

- Besides the gradient of a scalar field there is also a vector gradient $\nabla \boldsymbol{u}$, which can be written in tensor form as

$$\nabla \boldsymbol{u} = \begin{pmatrix} \dfrac{\partial u_x}{\partial x} & \dfrac{\partial u_x}{\partial y} & \dfrac{\partial u_x}{\partial z} \\ \dfrac{\partial u_y}{\partial x} & \dfrac{\partial u_y}{\partial y} & \dfrac{\partial u_y}{\partial z} \\ \dfrac{\partial u_z}{\partial x} & \dfrac{\partial u_z}{\partial y} & \dfrac{\partial u_z}{\partial z} \end{pmatrix} . \tag{13.24}$$

The product $\nabla \cdot \nabla \boldsymbol{u}$ gives a vector with the components

$$\nabla \cdot \nabla \boldsymbol{u} = \{\Delta u_x, \Delta u_y, \Delta u_z\} , \tag{13.25}$$

where Δ is the Laplace operator.

- $$(\nabla \times \nabla f) = \mathbf{curl}\,\mathbf{grad} f \equiv 0 , \tag{13.26}$$

which can be proved with (13.14) and (13.20) for functions f that have a continuous second derivative.
Finally we consider the product div $\mathbf{rot}\, \boldsymbol{u}$

- $$\nabla \cdot (\nabla \times \boldsymbol{u}) = \operatorname{div} \mathbf{curl}\, \boldsymbol{u} \equiv 0 , \tag{13.27}$$

because

$$\begin{aligned}
\nabla \cdot (\nabla \times \boldsymbol{u}) &= \frac{\partial}{\partial x}\left(\frac{\partial u_z}{\partial y} - \frac{\partial u_y}{\partial z}\right) \\
&+ \frac{\partial}{\partial y}\left(\frac{\partial u_x}{\partial z} - \frac{\partial u_z}{\partial x}\right) + \frac{\partial}{\partial z}\left(\frac{\partial u_y}{\partial x} - \frac{\partial u_x}{\partial y}\right) \equiv 0 .
\end{aligned}$$

13.2 Coordinate Systems

The mathematical description of a physical process can be often essentially simplified when choosing the optimum coordinate system.

13.2.1 Cartesian Coordinates

The coordinate system consists of three coordinate axes (x, y, z) that are perpendicular to each other.

The coordinate planes

$$x = \text{const} \,, \quad y = \text{const} \,, \quad z = \text{const}$$

are planes perpendicular to the x, resp. y or z-axis. The intersection lines between two coordinate planes give the coordinate axes.

The intersection line between the(x–y)-plane ($z = $ const) and the (x–z)-plane ($y = $ const) is the x-axis. The y-axis is the intersection of the (x–y)-plane ($z = $ const) with the (y–z)-plane ($x = $ const), while the z-axis is the intersection of the (x–z) and the (y–z)-planes.

The vector $\boldsymbol{r}$ from the origin to the point $P(x, y, z)$ has the components

$$\boldsymbol{r} = \{x, y, z\} \,.$$

The three orthogonal unit vectors pointing int the three coordinate axes are

$$\hat{\boldsymbol{e}}_1 = \{1, 0, 0\} \,,\ \hat{\boldsymbol{e}}_2 = \{0, 1, 0\} \,,\ \hat{\boldsymbol{e}}_3 = \{0, 0, 1\} \,,$$
$$\to \boldsymbol{r} = x \cdot \hat{\boldsymbol{e}}_1 + y \cdot \hat{\boldsymbol{e}}_2 + z \cdot \hat{\boldsymbol{e}}_3 \,.$$

The line element $\mathrm{d}\boldsymbol{r}$ on a curve between the points $P(\boldsymbol{r})$ and $P(\boldsymbol{r} + \mathrm{d}\boldsymbol{r})$ is

$$\mathrm{d}\boldsymbol{r} = \{\mathrm{d}x, \mathrm{d}y, \mathrm{d}z\} \,. \tag{13.28}$$

The velocity of a point moving along the curve is

$$\boldsymbol{v} = \frac{\mathrm{d}\boldsymbol{r}}{\mathrm{d}t} = \{\dot{x}, \dot{y}, \dot{z}\} \,. \tag{13.29}$$

The acceleration is then

$$a = \frac{\mathrm{d}v}{\mathrm{d}t} = \frac{\mathrm{d}^2 r}{\mathrm{d}t^2} = \dot{\boldsymbol{v}} = \{\ddot{x}, \ddot{y}, \ddot{z}\} \,. \tag{13.30}$$

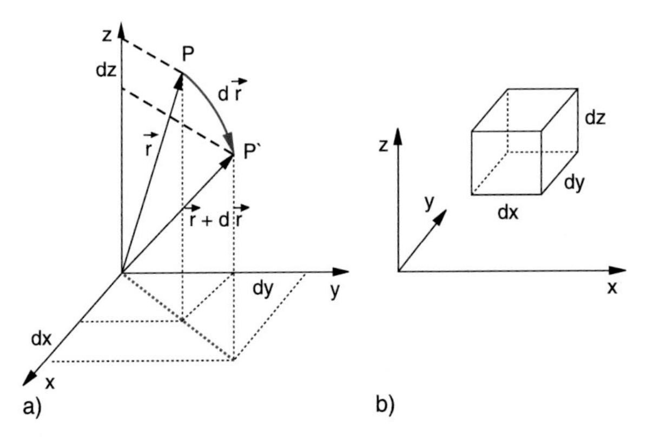

Figure 13.11 **a** Line element $\mathrm{d}\boldsymbol{r}$ and its components $\mathrm{d}x$, $\mathrm{d}y$, $\mathrm{d}z$; **b** volume element $\mathrm{d}V = \mathrm{d}x \cdot \mathrm{d}y \cdot \mathrm{d}z$

A volume element $\mathrm{d}V$ consists of the side edges $\mathrm{d}x$, $\mathrm{d}y$, $\mathrm{d}z$ and has the volume

$$\mathrm{d}V = \mathrm{d}x \cdot \mathrm{d}y \cdot \mathrm{d}z \,. \tag{13.31}$$

13.2.2 Cylindrical Coordinates

When we use in the x–y-plane polar coordinates but keep for the z-direction the Cartesian z-coordinate we get **cylindrical coordinates** (ϱ, φ, z) (Fig. 13.12). They are useful for the description of problems with rotational symmetry (calculation of bodies with cylindrical symmetry, two-atomic molecules, fluid flow through circular pipes etc).

A space point $P(\varrho, \varphi, z)$ is described in cylindrical coordinates by its three coordinates ϱ, φ and z, where ϱ is the distance from the z-axis, φ the angle between the x-axis and the projection of P onto the x–y-plane and z its distance from the x–y-plane.

The conversion to cartesian coordinates is

$$\left.\begin{aligned} x &= \varrho \cos\varphi \\ y &= \varrho \sin\varphi \\ z &= z \end{aligned}\right\} \Rightarrow \begin{aligned} \varrho &= \sqrt{x^2 + y^2} \\ \varphi &= \arctan(y/x) \\ z &= z \,. \end{aligned} \tag{13.32}$$

The coordinate planes are

- $\varrho = $ const
  = rotational cylinder surface around the z-axis,
- $\varphi = $ const
  = planes through the z-axis,
- $z = $ const
  = planes perpendicular to the z-axis.

The coordinate lines are

- **ϱ-lines** ($\varphi = $ const, $z = $ const)
  = straight lines through the z-axis parallel to the x–y-plane,
- **φ-lines** ($\varrho = $ const, $z = $ const)
  = horizontal circles around the z-axis,
- **z-lines** ($\varphi = $ const, $\varrho = $ const)
  = straight lines parallel to the z-axis.

The three unit vectors form for each point $P(\varrho, \varphi, z)$ an orthogonal tripod

$$\begin{aligned} \hat{\boldsymbol{e}}_\varrho &= \{\cos\varphi, \sin\varphi, 0\} \\ \hat{\boldsymbol{e}}_\varphi &= \{-\sin\varphi, \cos\varphi, 0\} \\ \hat{\boldsymbol{e}}_z &= \{0, 0, 1\} \,. \end{aligned} \tag{13.33}$$

The line element $\mathrm{d}\boldsymbol{s}$ (Fig. 13.12) has the three components

$$\mathrm{d}\boldsymbol{s} = \{\mathrm{d}\varrho, \varrho\mathrm{d}\varphi, \mathrm{d}z\} \,. \tag{13.34}$$

The velocity $\boldsymbol{v} = \mathrm{d}\boldsymbol{s}/\mathrm{d}t$ therefore has the components

$$\boldsymbol{v} = \{\dot{\varrho}, \varrho\dot{\varphi}, \dot{z}\} = \dot{\varrho}\hat{\boldsymbol{e}}_\varrho + \varrho\dot{\varphi}\hat{\boldsymbol{e}}_\varphi + \dot{z}\hat{\boldsymbol{e}}_z \,, \tag{13.35}$$

and the acceleration is

$$\begin{aligned} \boldsymbol{a} = \frac{\mathrm{d}\boldsymbol{v}}{\mathrm{d}t} &= \ddot{\varrho}\hat{\boldsymbol{e}}_\varrho + \dot{\varrho}\frac{\mathrm{d}\hat{\boldsymbol{e}}_\varrho}{\mathrm{d}t} + \dot{\varrho}\dot{\varphi}\hat{\boldsymbol{e}}_\varphi + \varrho\ddot{\varphi}\hat{\boldsymbol{e}}_\varphi \\ &\quad + \varrho\dot{\varphi}\frac{\mathrm{d}\hat{\boldsymbol{e}}_\varphi}{\mathrm{d}t} + \ddot{z}\hat{\boldsymbol{e}}_z + \dot{z}\frac{\mathrm{d}\hat{\boldsymbol{e}}_z}{\mathrm{d}t} \,. \end{aligned} \tag{13.36}$$

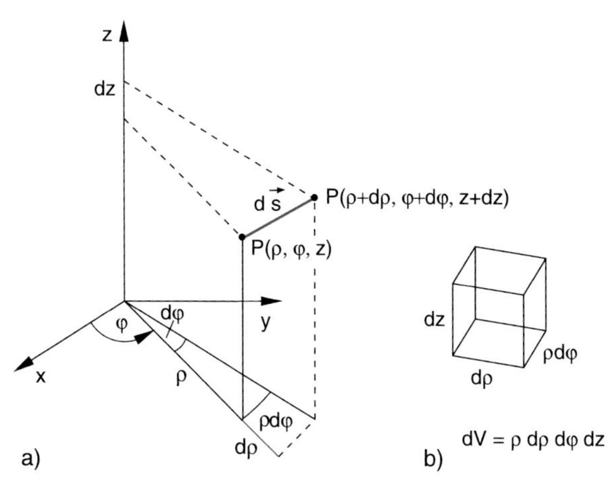

Figure 13.12 **a** Line element $ds = \{d\varrho, \varrho d\varphi, dz\}$ in cylindrical coordinates; **b** volume element $dV = \varrho d\varrho d\varphi dz$

Inserting (13.33) gives

$$\boldsymbol{a} = (\ddot{\varrho} - \varrho\dot{\varphi}^2)\hat{\boldsymbol{e}}_\varrho + (2\dot{\varrho}\dot{\varphi} + \varrho\ddot{\varphi})\hat{\boldsymbol{e}}_\varphi + \ddot{z}\hat{\boldsymbol{e}}_z . \tag{13.37a}$$

The absolute value of the acceleration is then

$$a = |\boldsymbol{a}| \sqrt{(\ddot{\varrho} - \varrho\dot{\varphi}^2)^2 + (2\dot{\varrho}\dot{\varphi} + \varrho\ddot{\varphi})^2 + \ddot{z}^2} . \tag{13.37b}$$

The surface element on the cylinder surface is

$$dS = \varrho \cdot d\varphi \cdot dz , \tag{13.38}$$

and the volume element

$$dV = d\varrho \cdot dS = \varrho \cdot d\varrho \cdot d\varphi \cdot dz . \tag{13.39}$$

13.2.3 Spherical Coordinates

They are useful for all spherical symmetric problems, i. e. if the calculated quantities depend solely on the distance r from the centre.

Example

Motion of particles in a central force field. ◄

The position vector from the origin to the point $P(r; \vartheta, \varphi)$ is defined by its length r and the angles ϑ and φ. (Fig. 13.13).

The conversion relations between spherical and Cartesian coordinates are

$$\left.\begin{aligned} x &= r\sin\vartheta\cos\varphi \\ y &= r\sin\vartheta\sin\varphi \\ z &= r\cos\vartheta \end{aligned}\right\} \qquad \begin{aligned} r &= \sqrt{x^2+y^2+z^2} \\ \vartheta &= \arccos\frac{z}{\sqrt{x^2+y^2+z^2}} \\ \varphi &= \arctan(y/x) . \end{aligned}$$

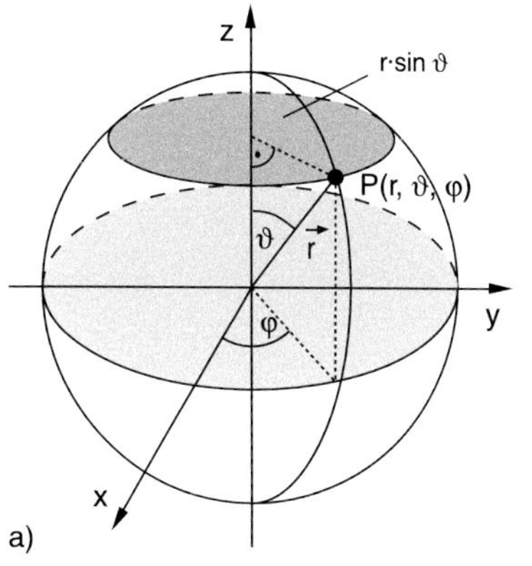

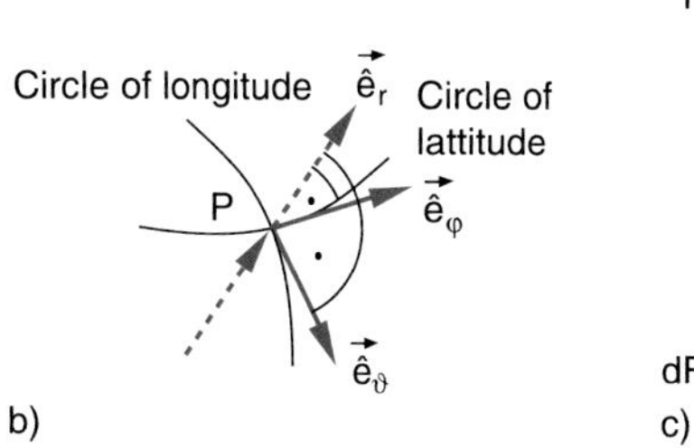

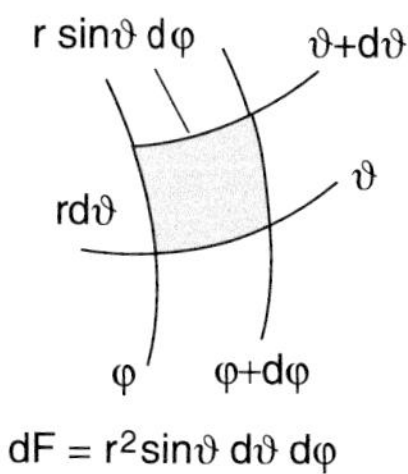

Figure 13.13 **a** Spherical coordinates; **b** orthogonal tripod of the unit vectors $\hat{\boldsymbol{e}}_r, \hat{\boldsymbol{e}}_\vartheta, \hat{\boldsymbol{e}}_\varphi$ at the point P; **c** surface element $dS = r^2 \cdot \sin\vartheta \, d\vartheta \, d\varphi$ on the surface of the sphere

Example

The unit sphere has in spherical coordinates the equation $r = 1$, in Cartesian coordinates $x^2 + y^2 + z^2 = 1$, in cylindrical coordinates $\varrho^2 + z^2 = 1$. ◄

The coordinate surfaces are:

- $r = \text{const}$: concentric spheres around $r = 0$,
- $\vartheta = \text{const}$: rotational cones around the z-axis with the peak at the origin,
- $\varphi = \text{const}$ planes through the z-axis.

The coordinate lines are:

- r-lines ($\varphi = \text{const}$, $\vartheta = \text{const}$): straight lines through the origin,
- ϑ-lines ($r = \text{const}$, $\varphi = \text{const}$): longitudinal circles (meridians),
- φ-lines ($r = \text{const}$, $\vartheta = \text{const}$): parallel circles around the z-axis (circles of lattitude).

The unit vectors in the point $P(r, \vartheta, \varphi)$ are (Fig. 13.13b)

$$\begin{aligned} \hat{\boldsymbol{e}}_r &= \{\sin\vartheta\cos\varphi; \sin\vartheta\sin\varphi; \cos\vartheta\} , \\ \hat{\boldsymbol{e}}_\vartheta &= \{\cos\vartheta\cos\varphi; \cos\vartheta\sin\varphi; -\sin\vartheta\} , \\ \hat{\boldsymbol{e}}_\varphi &= \{-\sin\varphi; \cos\varphi; 0\} . \end{aligned} \tag{13.40}$$

$\hat{\boldsymbol{e}}_r$ points into the $\boldsymbol{r}$-direction, $\hat{\boldsymbol{e}}_\vartheta$ is tangent to the longitudinal circles (meridians) in the point P and $\hat{\boldsymbol{e}}_\varphi$ is tangent to the circles of lattitude in P (Fig. 13.13b).

The line elements of the coordinate lines are

$$\mathrm{d}r, \quad r \cdot \mathrm{d}\vartheta, \quad r \cdot \sin\vartheta\, \mathrm{d}\varphi \,. \tag{13.41a}$$

The line element of an arbitrary curve in the threedimensional space is then

$$\mathrm{d}s = \{\mathrm{d}r, r\mathrm{d}\vartheta, r \cdot \sin\vartheta\, \mathrm{d}\varphi\} \,. \tag{13.41b}$$

A surface element on the surface of the sphere is

$$\mathrm{d}A = r^2 \sin\vartheta\, \mathrm{d}\vartheta\, \mathrm{d}\varphi \,. \tag{13.42}$$

A volume element is

$$\mathrm{d}V = r^2 \sin\vartheta\, \mathrm{d}r\, \mathrm{d}\vartheta\, \mathrm{d}\varphi \,. \tag{13.43}$$

The velocity of a point mass m on the trajectory $s(t)$ is according to (13.41b)

$$\boldsymbol{v} = \frac{\mathrm{d}s}{\mathrm{d}t} = \left\{\dot{r}, r \cdot \dot{\vartheta}, r \cdot \sin\vartheta\dot{\varphi}\right\} \,. \tag{13.44}$$

The acceleration $\boldsymbol{a} = \mathrm{d}\boldsymbol{v}/\mathrm{d}t$ is obtained by differentiation of (13.44). This gives

$$\begin{aligned} \boldsymbol{a} &= \ddot{r}\hat{\boldsymbol{e}}_r + \dot{r}\frac{\mathrm{d}\hat{\boldsymbol{e}}_r}{\mathrm{d}t} + \left(\dot{r}\dot{\vartheta} + r\ddot{\vartheta}\right)\hat{\boldsymbol{e}}_\vartheta + r\dot{\vartheta}\frac{\mathrm{d}\hat{\boldsymbol{e}}_\vartheta}{\mathrm{d}t} \\ &+ \left(\dot{r}\sin\vartheta\dot{\varphi} + r\cos\vartheta\dot{\vartheta}\dot{\varphi} + r\sin\vartheta\ddot{\varphi}\right)\hat{\boldsymbol{e}}_\varphi \\ &+ r\sin\vartheta\dot{\varphi}\frac{\mathrm{d}\hat{\boldsymbol{e}}_\varphi}{\mathrm{d}t} \,. \end{aligned} \tag{13.45}$$

This can be written (using (13.40)) as linear combination of $\hat{\boldsymbol{e}}_r, \hat{\boldsymbol{e}}_\vartheta, \hat{\boldsymbol{e}}_\varphi$.

13.3 Complex Numbers

The solution of the quadratic equation $x^2 + 1 = 0$ gives $x_{1,2} = \pm\sqrt{-1}$, which do not belong to the real numbers, for which the square of a real number x must be always ≥ 0.

Numbers x with $x^2 < 0$ are named imaginary numbers. Their unit element is $\mathrm{i} = +\sqrt{-1}$.

Similar to the assignment of real numbers to the axis of real numbers (generally the x-axis) the imaginary numbers are assigned to the y-axis, called the imaginary axis.

The two axis define a plane, called the complex plane. One attributes to each point $P(x, y)$ in the complex plane a complex number

$$z = x + iy \,, \tag{13.46}$$

where the first number x is the real part and the second number y the imaginary part of the complex number (Fig. 13.14).

Introducing the unit vectors $\hat{\boldsymbol{e}}_x$ and $\hat{\boldsymbol{e}}_y$ in the complex plane, each point $P(x, y)$ in this plane with the position vector $\boldsymbol{r}$ can be characterized by

$$\boldsymbol{r} = x \cdot \hat{\boldsymbol{e}}_x + \mathrm{i}y\hat{\boldsymbol{e}}_y \,. \tag{13.47}$$

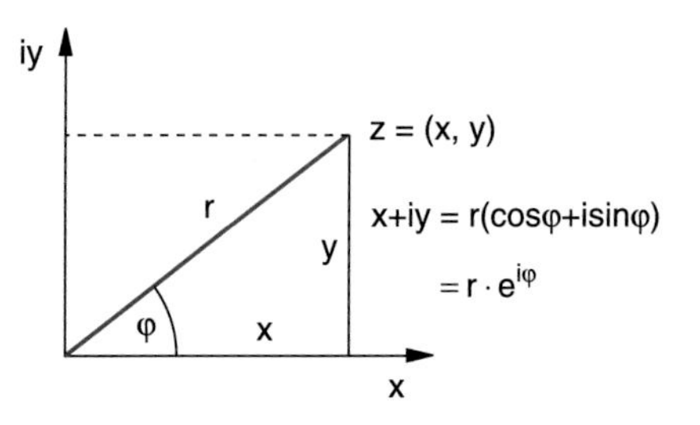

Figure 13.14 Representation of a complex number z as a point in the complex plane (x, iy)

The absolute value of the complex number z is

$$|z| = \sqrt{x^2 + y^2} \,. \tag{13.48}$$

It represents the distance $r = |\boldsymbol{r}|$ of $P(x, y)$ from the origin $(0, 0)$. The number

$$z^* = x - iy \tag{13.49}$$

is the conjugate complex of $z = x + iy$.

13.3.1 Calculation rules of Complex Numbers

The following rules for addition, multiplication and division of complex numbers are analogous to those of two-dimensional vectors $\boldsymbol{r} = \{x, y\}$.

Addition

$$z_1 + z_2 = \{x_1, y_1\} + \{x_2, y_2\} \stackrel{\text{Def}}{=} \{x_1 + x_2, y_1 + y_2\} \,. \tag{13.50}$$

Two complex numbers are added by adding the real parts and the imaginary parts. When z and z^* are added, this gives

$$z + z^* = \{x + \mathrm{i}y\} + \{x - iy\} = 2x \,, \tag{13.51}$$

i.e. twice the real part.

Multiplication

$$\begin{aligned} z_1 \cdot z_2 &= \{x_1, y_1\} \cdot \{x_2, y_2\} = (x_1 + iy_1) \cdot (x_2 + iy_2) \\ &= (x_1 \cdot x_2 - y_1 \cdot y_2) + \mathrm{i} \cdot (x_1 y_2 + x_2 y_1) \end{aligned} \tag{13.52}$$

The product

$$z \cdot z^* = (x + iy) \cdot (x - iy) = (x^2 + y^2) = |z|^2 \tag{13.53}$$

gives the square of the absolute value $|z|$.

Division

$$\frac{z_1}{z_2} = \frac{x_1 + iy_1}{x_2 + iy_2} .$$

Multiplication of numerator and denumerator with $(x_2 - iy_2)$ gives

$$\begin{aligned}\frac{z_1}{z_2} &= \frac{(x_1 + iy_1)\cdot(x_2 - iy_2)}{x_2^2 + y_2^2} \\ &= \frac{(x_1x_2 + y_1y_2) + \mathrm{i}(x_2y_1 - x_1y_2)}{x_2^2 + y_2^2} = a + ib .\end{aligned}$$

This gives again a complex number with the real part

$$a = \frac{(x_1x_2 + y_1y_2)}{x_2^2 + y_2^2}$$

and the imaginary part

$$b = \frac{x_2y_1 - x_1y_2}{x_2^2 + y_2^2} .$$

13.3.2 Polar Representation

Often the representation of a complex number in polar coordinates r and φ is more convenient (Fig. 13.14).

With

$$x = r\cdot\cos\varphi ; \quad y = r\cdot\sin\varphi$$

we obtain

$$\begin{aligned} z &= x + iy = r(\cos\varphi + \mathrm{i}\sin\varphi) = r\cdot\mathrm{e}^{\mathrm{i}\varphi} \\ z^* &= x - iy = r\cdot\mathrm{e}^{-\mathrm{i}\varphi} \\ &\Rightarrow z\cdot z^* = r^2 ; \quad |z| = \sqrt{x^2 + y^2} = r .\end{aligned}$$

From Fig. 13.14 we see that

$$\tan\varphi = \frac{iy}{x} = \frac{\mathrm{Im}(z)}{\mathrm{Re}(z)} \quad \Rightarrow \varphi = \arctan\frac{\mathrm{Im}(z)}{\mathrm{Re}(z)} .$$

Note: The polar representation is not unambiguous, because all angles $\varphi_n = \varphi_0 + n\cdot 2\pi$ $(n = 1, 2, 3, \ldots)$ represent the same complex number z. The representation with $n = 0$ is called the principal value.

From $z = r\cdot\mathrm{e}^{\mathrm{i}\varphi} \Rightarrow \ln z = \ln r + \mathrm{i}(\varphi_0 + 2n\pi)$.

The polar representation facilitates multiplication and division of complex numbers

$$\begin{aligned} z_1\cdot z_2 &= r_1\cdot r_2\cdot\mathrm{e}^{\mathrm{i}(\varphi_1+\varphi_2)} \\ \frac{z_1}{z_2} &= \frac{r_1}{r_2}\cdot\mathrm{e}^{\mathrm{i}(\varphi_1-\varphi_2)} .\end{aligned}$$

For arbitrary n one obtains

$$z^n = (r\cdot\mathrm{e}^{\mathrm{i}\varphi})^n = r^n\cdot\mathrm{e}^{\mathrm{i}n\varphi} .$$

Complex numbers are raised to higher powers n by calculatuing the n-th power of r and multiplying φ by n. In the same way we see the relation

$$\sqrt[n]{z} = z^{1/n} = \sqrt[n]{r}\cdot\mathrm{e}^{\mathrm{i}\varphi/n} .$$

The general rule for complex numbers can be formulated as (see mathematics text books):

> The set of complex numbers $z = (x, y)$ forms a body which includes the real numbers $(x, 0)$ as subset.

13.4 Fourier-Analysis

In mathematical textbooks it is proved, that every continuously differentiable function $f(x)$ can be written as infinite series of basis functions $g(x)$, if the $g(x)$ represent a complete set.

We choose as basis functions the trigonometric functions $\sin(nx)$ and $\cos(nx)$ $(n = 0, 1, 2, \ldots)$.

The Fourier-theorem states:

$$f(x) = \frac{a_0}{2} + \sum_{n=1}^{\infty}\left[a_n\cos(nx) + b_n\sin(nx)\right] . \qquad (13.54)$$

(13.54) is called "Fourier-Series". By multiplication of (13.54) with $\cos(mx)$ or $\sin(mx)$ respectively and integration over x from $x = 0$ to $x = 2\pi$ one obtains because of

$$\begin{aligned} &\int_0^{2\pi}\cos(nx)\cos(mx)\,\mathrm{d}x = \begin{cases}0 & \text{for } m \neq n \\ \pi & \text{for } n = m \neq 0\end{cases} \\ &\int_0^{2\pi}\cos(nx)\sin(mx)\,\mathrm{d}x = 0 \quad \text{for } m \gtrless n \qquad (13.55) \\ &\int_0^{2\pi}\sin(nx)\sin(mx)\,\mathrm{d}x = \begin{cases}0 & \text{for } n \neq m \\ \pi & \text{for } n = m \neq 0\end{cases}\end{aligned}$$

the coefficients a_n and b_n as

$$\begin{aligned} a_0 = b_0 &= \frac{1}{\pi}\int_0^{2\pi} f(x)\,\mathrm{d}x ; \quad a_n = \frac{1}{\pi}\int_0^{2\pi} f(x)\cos(nx)\,\mathrm{d}x ; \\ b_n &= \frac{1}{\pi}\int_0^{2\pi} f(x)\sin(nx)\,\mathrm{d}x . \end{aligned} \qquad (13.56)$$

It is therefore possible to determine the coefficients a_n, b_n in the Fourier-Series (13.54) by integration of $f(x)$.

For $x = \omega \cdot t$ the coefficients give the amplitudes of the contributions $n\omega$ to the total function $f(\omega, t)$ (Fourier-Analysis) with $x = \omega t$ and $T = 2\pi/\omega$ (13.55) transfers to

$$a_n = \frac{2}{T}\int_0^T f(t)\cos(n\omega t)\,\mathrm{d}t\,,$$
$$b_n = \frac{2}{T}\int_0^T f(t)\sin(n\omega t)\,\mathrm{d}t\,. \qquad (13.57)$$

Solutions of the Problems

14

© Springer International Publishing Switzerland 2017
W. Demtröder, *Mechanics and Thermodynamics*, Undergraduate Lecture Notes in Physics, DOI 10.1007/978-3-319-27877-3_14

14.1 Chapter 1

1.1 180 km/h.

1.2 The length measurement can be performed in different ways:
For example the period of the earth's rotation depends on the moment of inertia and is therefore proportional to R^{-2}. Independent measurements of length and time can decide the question.

1.3 This is one, but not the only requirement for a length standard. Similarly important is the accuracy of the comparison between the length to be measured and the standard.

1.4 The length T of a day increases per year by 10^{-4} s. The relative prolongation of a day per year is then

$$\mathrm{d}T/T = a = \frac{10^{-4}}{24 \cdot 3600} = 1.1 \cdot 10^{-9}\,.$$

a) Since the length of a day increases per year by 10^{-4} s, it is after 10^4 years longer by 1 s.
b) How often must a leap second be inserted?
The length of a day increases per day by $\delta t = 10^{-4}\,\mathrm{s}/365\,\mathrm{d} = 2.74 \cdot 10^{-8}\,\mathrm{s/d}$. The total time delay after x days is

$$\Delta t = \delta t \int_0^x n\,\mathrm{d}n = \frac{1}{2}\delta t \cdot n^2\Big|_0^x$$

$$= \tfrac{1}{2}x^2\delta t = 1\,\mathrm{s} \to x^2 = \frac{2}{\delta t}$$

$$\to x = 8600\,\mathrm{d} = 23.5\,\text{years}\,.$$

The distance $x_{n+1} - x_n = (2/aT_0)^{1/2} \cdot (\sqrt{n+1} - \sqrt{n})$ between two leap times becomes shorter and shorter and reaches for $n \gg 1$ the value $1/(aT_0)^{1/2}/\sqrt{2n}$.

1.5 1 light year $= 9.46 \cdot 10^{15}\,\mathrm{m} \Rightarrow T = 4.5$ years. The distance is 1.39 parsec, the angle $1/1.39 = 0.7''$.

1.6 $L = 2 \cdot 10^3 \cdot \tan(\alpha/2) = 17.45\,\mathrm{m}$ for $\alpha = 1°$. For $\alpha = 1° \pm 1' \Rightarrow L = 17.45 \pm 0.29\,\mathrm{m}$.

1.7 Since the orbital speed of the earth varies during one revolution, the time between two culminations of the sun also varies (see Fig. 1.22 and 1.23). Further reasons are variations of the mass distribution inside the earth, which changes the moment of inertia, caused by magma flow, earth quakes and melting of glaciers.

1.8 The mass of a hydrogen atom is $m_\mathrm{H} = 1.673 \cdot 10^{-27}\,\mathrm{kg} \Rightarrow N = 5.98 \cdot 10^{26}\,/\mathrm{kg}$.

1.9 The mass of a H_2O-molecules is $m_{H_2O} = 3.0 \cdot 10^{-26}\,\mathrm{kg}$; $\varrho_{H_2O} = 1\,\mathrm{kg/litre}$; $\Rightarrow N = 3.0 \cdot 10^{25}\,/\mathrm{litre}$.

1.10 The mass of the uranium nucleus is $m(^{238}\mathrm{U}) = 1.661 \cdot 238 \cdot 10^{-27}\,\mathrm{kg}$. Its density is then $\varrho = m/(4 \cdot \frac{1}{3}\pi r^3) = 1.4 \cdot 10^{17}\,\mathrm{kg/m^3}$.

1.11 From $s = 1/2gt^2$ it follows for the falling time $t = \sqrt{2s/g} = 0.45\,\mathrm{s}$.

$$\sigma_\mathrm{m} = \left[\frac{\sum(\bar{x} - x_i)^2}{n(n-1)}\right]^{1/2} = \left[\frac{40 \cdot 0.01}{40 \cdot 39}\right]^{1/2}\mathrm{s}$$

$$= 1.6 \cdot 10^{-2}\,\mathrm{s}$$

$$\Rightarrow \sigma_\mathrm{m}/\bar{x} = \frac{1.6 \cdot 10^{-2}}{0.45} = 3.5\%\,.$$

1.12 a) $\mathrm{e}^{-x^2/2} = 0.5 \Rightarrow x^2 = 2\ln 2 \Rightarrow x = \sqrt{2\ln 2} \approx 1.177$;
b) $\mathrm{e}^{-x^2/2} = 0.1 \Rightarrow x^2 = 2\ln 10 \Rightarrow x = \sqrt{2\ln 10} \approx 2.156$.

1.13 $A = x - y^2 \Rightarrow \partial A/\partial x = 1$ and $\partial A/\partial y = -2y$

$$\sigma_A = \left[(1000 \cdot 10^{-3} \cdot 1)^2 + (30 \cdot 3 \cdot 10^{-3} \cdot 60)^2\right]^{1/2}$$

$$= [1 + 29]^{1/2} \approx 5.5\,.$$

1.14 Quartz clock $\Delta T_\mathrm{max} = 10^{-9} \cdot 3.16 \cdot 10^7\,\mathrm{s} \approx 0.03\,\mathrm{s} = 30\,\mathrm{ms}$.
Atomic Clock: $\Delta T_\mathrm{max} = 0.3\,\mathrm{\mu s}$.

1.15 For the five points we obtain from (1.35) with $n = 5$:

$$a = \frac{5 \cdot \sum x_i y_i - \sum x_i \sum y_i}{5 \cdot (\sum x_i^2) - (\sum x_i)^2}$$

$$= \frac{5 \cdot (3 + 6 + 20 + 25) - 12 \cdot 18}{5 \cdot (1 + 4 + 16 + 25) - 12^2} = 0.628\,,$$

$$b = \frac{(\sum x_i^2) \cdot (\sum y_i) - (\sum x_i) \cdot (\sum x_i y_i)}{5 \cdot (\sum x_i^2) - (\sum x_i)^2}$$

$$= \frac{828 - 684}{230 - 144} = 2.093$$

$$\Rightarrow y = 0.628x + 2.093\,;$$

$$\sigma_y = \sqrt{\frac{0.430}{n-2}} = 0.38\,.$$

Note, that here $(n-2)$ instead of $(n-1)$ has to be used, because already two values are determined by the equation $y = ax + b$.

$$\sigma_a^2 = \frac{5 \cdot \sigma_y^2}{86} = 0.006 \Rightarrow \sigma_a = 0.077\,,$$

$$\sigma_b^2 = \frac{\sigma_y^2 \cdot \sum x_i^2}{86} = \frac{0.102 \cdot (1 + 4 + 16 + 25)}{86} = 0.055$$

$$\Rightarrow \sigma_b = 0.23\,.$$

14.2 Chapter 2

2.1 a) The acceleration time t_1 can be obtained from

$$v_1 = v_0 + at_1$$

$$\Rightarrow t_1 = \frac{v_1 - v_0}{a} = \frac{(100 - 80)\,\mathrm{m/s}}{3.6 \cdot a\,\mathrm{m/s^2}} = 4.27\,\mathrm{s}\,.$$

b) The time t_2 from the end of the acceleration period until the end of overtaking is obtained from the equation for the distance

$$s = v_0 t_1 + \tfrac{1}{2} a t_1^2 + v_1 t_2 .$$

This distance s is

$$s = [v_0(t_1 + t_2) + (40 + 25 + 40)]\,\text{m} .$$

where the first term gives the distance, the truck has passed during the overtaking time. The comparison yields

$$v_0 t_1 + \tfrac{1}{2} a t_1^2 + v_1 t_2 = v_0(t_1 + t_2) + 105 .$$

With $t_1 = 4.27\,\text{s} \Rightarrow t_2 = 16.77\,\text{s}$.
c) The total overtaking time is then $t = t_1 + t_2 \approx 21\,\text{s}$ and the total overtaking distance 570.6 m.
The overtaking would have been therefore not successful but lethal and should not have been tried!

2.2 The driving times are: $t_1 = \frac{x}{2v_1}$; $t_2 = \frac{x}{2v_2}$.
The total driving time is $t = t_1 + t_2$.
The average velocity is

$$\langle v \rangle = \frac{x}{t_1 + t_2} = \frac{2v_1 v_2}{v_1 + v_2} = \frac{2 \cdot 40 \cdot 80}{120} = 53.33\,\text{km/h} .$$

2.3 From $s = v_0 t + \frac{1}{2} a t^2$ one obtains: $a = -1\,\text{cm/s}^2$.

2.4 From $v = v_0 + at \Rightarrow t = (v - v_0)/a$. Inserting into $s = v_0 t + \frac{1}{2} a t^2 = 0.04\,\text{m}$ gives $v_0 = 5 \cdot 10^6\,\text{m/s}$.

2.5 From $s = h + v_0 t - \frac{1}{2} g t^2 = 0$ it follows:
a) $t_1(v_0 = 5\,\text{m/s}) = 2.3\,\text{s}$,
b) $t_2(v_0 = -5\,\text{m/s}) = 1.3\,\text{s}$,
c) derivation of Eq. 2.13 (see figure)

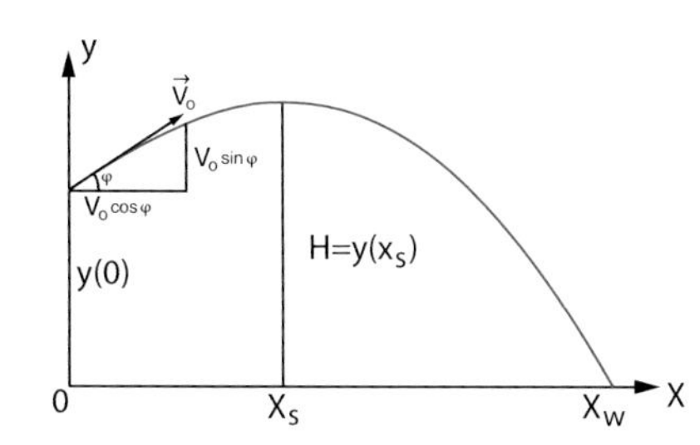

$v_x = v_0 \cdot \cos\varphi$, $v_y = v_0 \cdot \sin\varphi$, t_s = rise time, t_f = fall time, H = rise height

$$y(t) = -\frac{g}{2} t^2 + v_0 \sin\varphi + h ,$$

$$g \cdot t_s = v_0 \sin\varphi \Rightarrow t_s = \frac{v_0 \sin\varphi}{g} ,$$

$$H = y(t_s) .$$

For t_f the equation holds (free fall)

$$H = \frac{1}{2} g t_f^2 \Rightarrow t_f = \sqrt{\frac{2H}{g}} ,$$

$$x_w = v_x \cdot (t_s + t_f)$$

$$= v_0 \cdot \cos\varphi \left[\frac{v_0 \cdot \sin\varphi}{g} + \sqrt{\frac{2H}{g}} \right] ,$$

$$H = -\frac{g}{2} t_s^2 + v_0 \cdot \sin\varphi + h .$$

Inserting H into x_w yields

$$x_w = \frac{v_0^2}{g} \cdot \cos\varphi \left[\sin\varphi + \sqrt{\frac{2gh}{v_0^2} + \sin^2\varphi} \right] .$$

2.6 If the vector of acceleration is not parallel to the tangent on the trajectory the result is a curved trajectory and no straight line.

2.7 $v = gt$;

$$s = \frac{1}{2} g t^2 = \frac{1}{2} v^2/g = \frac{1}{2} \left(\frac{100}{3.6} \right)^2 \frac{1}{9.81}\,\text{m} = 39.3\,\text{m} .$$

2.8 a) If ω is constant the acceleration $a = \omega^2 \cdot R$ is also constant. This demands an additional tangential acceleration, that compensates in each point of the trajectory the tangential component $g \cdot \cos\alpha$ of the gravity acceleration, where α is the angle between the vertical direction and the tangent to the trajectory.
b) Velocity in point C: $v = \sqrt{2gh}$; in the point B: $v = \sqrt{2g(h - 2R)}$.
From the condition $(v^2/R) > g \Rightarrow v(R) > \sqrt{R \cdot g} \Rightarrow h > \frac{5}{2} R$; $v_{\min}(B) = \sqrt{g \cdot R}$.

2.9 a) Potential energy of the moon

$$E_p = -G \frac{M_{\text{Mo}} M_{\text{E}}}{r} .$$

Kinetic energy:

$$E_{\text{kin}} = \tfrac{1}{2} M_{\text{Mo}} r^2 \omega^2 .$$

From $M_{\text{Mo}} \omega^2 \cdot r = G \cdot M_{\text{Mo}} \cdot M_{\text{E}}/r^2 \Rightarrow \omega_{\text{Mo}} = (G \cdot M_{\text{E}}/r^3)^{1/2}$.
The condition $E_p + E_{\text{kin}} = E > 0$ yields

$$\omega^2 > \frac{2G \cdot M_{\text{E}}}{r^3} = 2\omega_{\text{Mo}}^2 .$$

The velocity of the moon at the same radius r must be enlarged by the factor $\sqrt{2}$.
b) $v > (2G \cdot M_{\text{Mo}}/r_{\text{Mo}})^{1/2}$ gives with $M_{\text{Mo}} = 7.36 \cdot 10^{22}\,\text{kg} \Rightarrow v = 2.38\,\text{km/s}$.

2.10 Velocity of the starting point at the equator:

$$v_0 = \pm \frac{4 \cdot 10^7}{24 \cdot 3600}\,\text{m/s} = \pm 4.6 \cdot 10^2\,\text{m/s} .$$

Rocket equation (neglecting the vertical acceleration if $g \cdot T \ll v_0$):

$$v = v_0 + v_e \ln(m_0/m) .$$

With $m_0 = m + m_x$ (m_x = fuel mass) it follows:

$$\ln\left(1 + \frac{m_x}{m}\right) = (7.9 \cdot 10^3 \mp 4.6 \cdot 10^2)/(4.5 \cdot 10^3)$$

$\Rightarrow m_x = 2103\,\mathrm{kg}$ for a launch into the east direction and $m_x = 2705\,\mathrm{kg}$ into the west direction, if the mass of the fuel container is neglected.

2.11 $(mv_0^2/2) > \frac{G \cdot m \cdot M_E}{R} = mgR \Rightarrow v_0 > \sqrt{2gR}$.

2.12 Because the velocity of the earth at the launch point is $v_E = 463\,\mathrm{m/s}$. The initial velocity of the rocket is not zero but $v_0 = v_E \cdot \cos 30° = 400\,\mathrm{m/s}$. One wins the initial kinetic energy $\frac{1}{2}mv_0^2$. For a vertical launch one would need the escape velocity $v_E = \sqrt{2gR} = 11{,}200\,\mathrm{m/s}$ (see (2.30)). Therefore one only needs to accelerate from 400 m/s until 11,200 m/s. This requires the kinetic energy $\Delta E_{kin} = \frac{1}{2}m \cdot (v_E^2 - v_0^2)$. The relative energy saving is $(1 - \Delta E/E) = 0.004$. One saves only 0.4%.

2.13 The total force is $F = F_A - m \cdot g = g \cdot \pi r^2 (z \cdot \varrho_w - h \cdot \varrho_H)$. For $z = \frac{2}{3}h \to F = 0$,

$$\Rightarrow \varrho_H = \tfrac{2}{3}\varrho_w .$$

The work is

$$W = \int_{z=2h/3}^{0} F\,\mathrm{d}z = \int_{2h/3}^{0} g\pi r^2 (z\varrho_w - h\varrho_H)\,\mathrm{d}z$$

$$= \tfrac{2}{9} g\pi r^2 h^2 \varrho_w = 24.7\,\mathrm{J} .$$

Without water twice the work would be necessary.

2.14 $E_{kin}(h_1) = \frac{1}{2}mv^2(h_1) = 200\,\mathrm{N \cdot m}$,

$$mg(h_2 - h_1) = E_{kin}(h) ,$$

$$\Rightarrow h_2 = \frac{E_{kin}(h_1)}{mg} + h_1 = 35.5\,\mathrm{m} .$$

2.15 $F = -Dx_1 \Rightarrow D = \frac{F_1}{x_1} = 400\,\mathrm{N/m}$.

$$W = \int_0^{l_0} Dx\,\mathrm{d}x = \tfrac{1}{2}Dl_0^2 = 128\,\mathrm{N \cdot m} .$$

2.16 The neutral point between earth and moon, where the opposite gravitational forces just compensate, has the distance r_2 from the earth and r_1 from the moon. From $F = 0$ it follows:

$$\frac{G \cdot M_{Mo}}{r_1^2} = \frac{G \cdot M_E}{r_2^2} \quad \text{and} \quad r = r_1 + r_2$$

$$\Rightarrow r_2 = \frac{r}{1 + (M_{Mo}/M_E)^{1/2}} = \frac{3.84 \cdot 10^8\,\mathrm{m}}{1 + 0.11}$$

$$\approx 3.46 \cdot 10^8\,\mathrm{m} .$$

In order to reach the distance r_1 for the earth the initial kinetic energy must be

$$\frac{1}{2}mv_0^2 \geq G \cdot M_E m \int_R^{r_2} \frac{\mathrm{d}r}{r^2} .$$

Since $M_{Mo} = 0.012 M_E$ we can neglect the attraction by the moon at the start of the rocket. It follows:

$$v_0^2 \geq 2G \cdot M_E \left(\frac{1}{R} - \frac{1}{r_2}\right) \approx 0.98\, v_0^2(\infty) .$$

The energy saving is 2% compared with the case where the second escape velocity is required to reach $r = \infty$.

2.17 a) $m\omega^2 r = mG \cdot M_E/r^2 \Rightarrow r^3 = G \cdot M_E/\omega^2$,

$$T = \frac{2\pi}{\omega} = 1\,\text{day} = 24 \cdot 3600\,\mathrm{s}$$

$$\Rightarrow \omega = 7.2 \cdot 10^{-5}\,\mathrm{s}^{-1}$$

$$\Rightarrow r = 4.25 \cdot 10^7\,\mathrm{m} = 42{,}500\,\mathrm{km} .$$

b) The energy of the body in the geostationary orbit compared to a body resting on the earth surface is

$$E = E_{kin} + E_p = \frac{mv^2}{2} + \int_{r=R}^{r_S} \frac{GmM_E}{r^2}\,\mathrm{d}r$$

$$= \frac{m\omega^2 r_S^2}{2} + GmM_E\left(\frac{1}{R} - \frac{1}{r_S}\right)$$

with $r_S = 42{,}500\,\mathrm{km}$. It needs the energy supply

$$E = m\left[\tfrac{1}{2} \cdot \omega^2 r_S + gR(1 - R/r_S)\right] .$$

c) In order to get the accuracy 0.1 km/day of its position the upper limit for the accuracy of the angular velocity is

$$\frac{\Delta\omega}{\text{day}} = \frac{0.1}{42{,}500} = 2.4 \cdot 10^{-6}\ \text{per day} .$$

The minimum relative stability has to be $\Delta\omega/\omega \leq 2.4 \cdot 10^{-6}/2\pi = 3.8 \cdot 10^{-7}$
Since $\omega^2 = G \cdot M_E/r^3 \Rightarrow \Delta r/r = -2/3 \Delta\omega/\omega \Rightarrow \Delta r \leq 10.6\,\mathrm{m}$.

2.18 $E_p = -G \cdot m \cdot \frac{M_E}{r} = -m \cdot g \cdot \frac{R^2}{r}$ with R = radius of earth.

$$E_{kin} = +G \cdot m \cdot \frac{M_E}{2r} = -\frac{1}{2}E_p ,$$

$$E = E_p + E_{kin} = -G \cdot m \cdot \frac{M_E}{2r} = -E_{kin} .$$

2.19 $E_p = mgL(1 - \cos\varphi)$; $E_{kin} = \frac{mv^2}{2} = \frac{m}{2}L^2\dot{\varphi}^2$;

$$E = E_{kin} + E_p .$$

The equation of motion is $m \cdot L \cdot d^2\varphi/dt^2 = -m \cdot g \cdot \sin\varphi$. Multiplication with $L \cdot d\varphi/dt$ gives

$$\frac{d}{dt}\left(\frac{m}{2}L^2\dot{\varphi}^2\right) = \frac{d}{dt}(m \cdot g \cdot L \cdot \cos\varphi) ,$$

$$\frac{d}{dt}E_{kin} = \frac{d}{dt}(E - E_p)$$

$$\Rightarrow E_{kin} + E_p = E = \text{const} .$$

2.20 $T = 2\pi\sqrt{L/g} \Rightarrow g = 2\pi L/T^2$

$$\Delta g = \left[\left(\frac{dg}{dL}\Delta L\right)^2 + \left(\frac{dg}{dT}\Delta T\right)^2\right]^{1/2}$$

$$= g\left[\left(\frac{\Delta L}{L}\right)^2 + \left(\frac{2\Delta T}{T}\right)^2\right]^{1/2} ,$$

$$\frac{\Delta L}{L} = 10^{-5} .$$

The uncertainty of the length measurement results in a relative error of time determination

$$\frac{\Delta T_1}{T} = \frac{1}{2}\frac{\Delta L}{L} = 5 \cdot 10^{-6} .$$

The uncertainty of time measurement $\Delta T = 10^{-2}$ s gives with $T = 6.34$ s a relative error of

$$\frac{\Delta T_2}{T} = 1.5 \cdot 10^{-3} .$$

The uncertainty $2\Delta T_2/nT = 10^{-5}$ which corresponds to the error in the length measurement can be only achieved for $n \cdot T = 2000\,\text{s} \rightarrow n \geq 316$.
With this uncertainty the relative error $\Delta g/g = \{2 \cdot 10^{-10}\}^{1/2} \approx 1.4 \cdot 10^{-5} \Rightarrow \Delta g = 1.37 \cdot 10^{-4}\,\text{m/s}^2$.

2.21 From (2.84) one obtains $\Delta G/G = \Delta\varphi/\varphi = 10^{-4} \Rightarrow \Delta\varphi = 10^{-4} \cdot \varphi \propto \varrho \cdot R_2^3/r^2$. Since $r > R_1 + R_2 \approx R_2$ the maximum elongation angle is $\varphi_{max} \propto R_2$. For a tenfold mass the elongation angle increases only by $10^{1/3} \approx 2.1$. For the angle φ the limitation is $\varphi \leq R_2/L$ (Fig. 2.60). If the measuring uncertainty $\Delta\varphi$, which is due to air turbulence and vibrations of the ground is reduced by a factor of 10, the uncertainty of the value of G is only reduced by a factor $10^{1/3}$.

2.22 According to Kepler's 3rd law the major axis a of the comet trajectory is

$$a = \left[\frac{T^2}{4\pi^2}GM_\odot\right]^{1/3} = 2.68 \cdot 10^{12}\,\text{m} .$$

With $r_{min} = a(1 - \varepsilon) = 0.59\,\text{AU} = 0.88 \cdot 10^{11}\,\text{m} \Rightarrow \varepsilon = 1 - r/a = 0.967$.

2.23 The escape velocity is $v_0 = 23.6\,\text{km/s}$, $g = 11.6\,\text{m/s}^2$.

$$v_0 = \sqrt{2Rg} \Rightarrow R = v_0^2/2g = 2.4 \cdot 10^7\,\text{m} .$$

The centripetal acceleration is $a = \omega^2 \cdot R$

$$\Rightarrow \omega = \sqrt{a/R} = 1.12 \cdot 10^{-4}\,\text{s}^{-1}$$

$$\Rightarrow T = \frac{2\pi}{\omega} = 5.71 \cdot 10^4\,\text{s} = 15.8\,\text{h}$$

$$g = G \cdot M/R^2 \Rightarrow M = g \cdot R^2/G$$

$$= 11.6 \cdot 2.4^2 \times 10^{14}/(6.67 \cdot 10^{-11})\,\text{kg}$$

$$= 1 \cdot 10^{26}\,\text{kg} .$$

The wanted planet is Neptune.

2.24 The gravitational force between the sun and the earth-moon system causes the accelerated motion of the system around the sun. In order to remove the moon from its orbit around the earth, the difference-acceleration $\Delta a = a_1 - a_2$ between a_1 (sun–moon) and a_2 (sun–earth) must be larger than the acceleration a_3 (earth–moon). A fast estimation shows that this is not the case.

2.25 The pendulum period is $T = 2\pi \cdot \sqrt{L/g_{Mo}} = (g_E/g_{Mo})^{1/2}$. T_E. Because $g = G \cdot M/R^2 \Rightarrow (g_E/g_{Mo})^{1/2}$. $T_E = (R_{Mo}/R_E) \cdot (M_E/M_{Mo})^{1/2} = 2.47 \Rightarrow T = 2.47$ s.

2.26 From (2.81) one obtains for the force, that causes the acceleration $F = -a \cdot R$ with $a = G \cdot M_E \cdot m/R_0^3$.
The force is proportional to R and the motion of the body therefore a harmonic oscillation $R = R_0 \cdot \cos(\omega t)$ with $\omega^2 = a/m = GM_E/R_0^3$.
The travel time T_t (half the oscillation period)

$$T_t = \frac{\pi}{\omega} = \pi R_0\sqrt{R_0/G \cdot M_E}$$

is exactly as long as that of the satellite flying around half of the earth at a low distance above the surface.

2.27 From $\omega^2 \cdot r = G \cdot M_E/r^2 = g \cdot R^2/r^2$

$$\Rightarrow r = \left[\frac{g \cdot R^2T^2}{4\pi^2}\right]^{1/3} = 3.8 \cdot 10^8\,\text{m} .$$

2.28 $M = \frac{4}{3}\pi R^3\varrho \Rightarrow R = [3M/4\pi\varrho]^{1/3} = 5.8 \cdot 10^7$ m

$$\Rightarrow g = G \cdot M/R^2 = 11.3\,\text{m/s}^2 .$$

2.29

$$\frac{\Delta g}{g} = \frac{1/R^2 - 1/(R+h)^2}{1/R^2} = 1 - \frac{R^2}{(R+h)^2} \approx \frac{2h}{R}$$

$$= \frac{320}{6380} = 0.05 = 5\% .$$

2.30 The acceleration which the moon causes on the earth due to the gravitational force is $g_{Mo} = G \cdot M_{Mo}/r^2 \approx 3.3 \cdot 10^{-6} g_E$. It causes the accelerated motion of the earth around the common centre of mass of the earth-moon-system. For the sun we get: $g_\odot = G \cdot M_\odot/r^2 \approx 5.4 \cdot 10^{-4} g$. However, measurements on earth detect only the difference between the gravitational attraction by the sun acting on the centre of the earth (which is compensated by the centrifugal force of the earth motion around the sun) and the effect on the earth surface. This difference causes the contribution of the sun to the tides on earth (see Chap. 6).

2.31 The distance between the centres of the balls is
a) $d = 0.2(1 - L/R) = 0.2(1 - 1.57 \cdot 10^{-5})$.
b) Due to the gravitational attraction between the balls the balls do not hang exactly vertical but form an angle $\Delta\varphi = G \cdot m/(d^2 \cdot g) \approx 3.4 \cdot 10^{-9}$ against the vertical direction. The distance between the balls changes therefore by $\Delta d = L \cdot \Delta\varphi = 3.4 \cdot 10^{-7}\,\text{m} = 0.34\,\mu\text{m}$.

2.32 $E_{\text{kin}} = E - E_{\text{p}} = E + G \cdot M_{\text{E}} M_{\odot}/r$ with $E = \text{const}$. In the perihelion is $r = r_{\min} = a(1-\varepsilon)$ with the eccentricity $\varepsilon = 0.0167$, in the aphelion is $r = r_{\max} = a(1+\varepsilon)$. The potential energy changes between perihelion and aphelion by

$$\frac{\Delta E_{\text{p}}}{E_{\text{p}}} = \frac{2\varepsilon}{1-\varepsilon^2} = 0.033 = 3.3\% \,.$$

Since $E_{\text{kin}} \approx -\frac{1}{2}E_{\text{p}}$ is $\Delta E_{\text{kin}}/E_{\text{kin}} = -3.3\%$. Because $\Delta v/v = \frac{1}{2}\Delta E_{\text{kin}}/E_{\text{kin}} \Rightarrow \Delta v/v \approx 1.65\%$. With $\overline{v} = 2\pi a/T \approx 30\,\text{km/s} \Rightarrow v_{\max} = 30.25\,\text{km/s}$ and $v_{\min} = 29.75\,\text{km/s}$.

2.33 Conservation of energy demands:

$$\frac{1}{2}mv^2 = E + G\frac{mM_{\text{E}}}{r} \,.$$

With $GM_{\text{E}} = gR^2$ we get

$$v_{\max}^2 \frac{2E}{m} + \frac{gR^2}{a(1-\varepsilon)}, \quad v_{\min}^2 \frac{2E}{m} + \frac{gR^2}{a(1+\varepsilon)} \,,$$

subtraction yields

$$\overline{v}\Delta v = gR^2 \frac{\varepsilon}{a(1-\varepsilon^2)} \,.$$

The semi-major axis can be obtained from

$$\overline{v}^2/a = gR^2/a^2 \,.$$

The result is $a = 1.1 \cdot 10^7\,\text{m}$. The solution of the quadratic equation for ε gives

$$\varepsilon = 0.268 \Rightarrow r_{\max} = 13{,}950\,\text{km} \,; \quad r_{\min} = 8050\,\text{km} \,.$$

14.3 Chapter 3

3.1 For the motion of the ball relative to the elevator we get
a) $s = 2.50\,\text{m} = \frac{1}{2}(g-a)t^2 = \frac{1}{2} \cdot 8.81t^2 \Rightarrow t = \sqrt{5/8.81}\,\text{s} \approx 0.75\,\text{s}$ after the release at $t = 0$.
b) The fall distance in the lift shaft is

$$s_2 = \left[\tfrac{1}{2}a(t+t_0)^2 + s\right] = 9.35\,\text{m} \quad \text{with } t_0 = 3\,\text{s} \,. \tag{14.1a}$$

In the coordinate system at rest one obtains:

$$s = \tfrac{1}{2}gt^2 + v_0 t = 2.5\,\text{m} + \tfrac{1}{2}at^2 + v_0 t \,, \tag{14.1b}$$

where $v_0 = 3\,\text{m/s}$ is the velocity of the lift at the time of the release $t = 0$. The result is, of course, identical with that obtained in the moving system.
c) In the rest system the ball has the velocity $v = v_0 + g \cdot t = (3a + 9.81 \cdot 0.7)\,\text{m/s} = 9.87\,\text{m/s}$. In the system of the moving lift, the ball has the impact velocity $v_2 = (g - a)t = 6.17\,\text{m/s}$.

3.2 a) When launching into the north direction is $\boldsymbol{v} \parallel \boldsymbol{\omega} \Rightarrow a_{\text{c}} = 0$. The trajectory of the rocket is along a circle of longitude (meridian).
b) When launching into the north-east direction (45° against the equator) the magnitude of the Coriolis acceleration is $|\boldsymbol{a}_{\text{c}}| = 2v' \cdot \omega \cdot \sin 45° = v'\omega \cdot \sqrt{2} = 4.2 \cdot 10^{-2}\,\text{m/s}^2$. The acceleration $\boldsymbol{a}_{\text{c}}$ points into the radial direction away from the centre of the earth. The effective acceleration is the difference between $\boldsymbol{g}$ and $\boldsymbol{a}_{\text{c}}$. The bullet flies on a slightly upwards curved trajectory in the north-east direction.
c) When $\boldsymbol{v}$ points into the north-west direction, a_{c} is pointing radially downwards towards the centre of the earth. The two accelerations $\boldsymbol{g}$ and $\boldsymbol{a}_{\text{c}}$ are parallel and must be added. The trajectory if curved downwards.

3.3 $\tan\alpha = \omega^2 r/g$; $r/L = \sin\alpha = \omega^2 r/\sqrt{\omega^4 r^2 + g^2} \Rightarrow r = \sqrt{L^2 - g^2/\omega^4} = 7.836\,\text{m} \Rightarrow \sin\alpha = 0.7836 \Rightarrow \alpha = 51.6°$; $v = \omega r = 9.85\,\text{m/s}$.

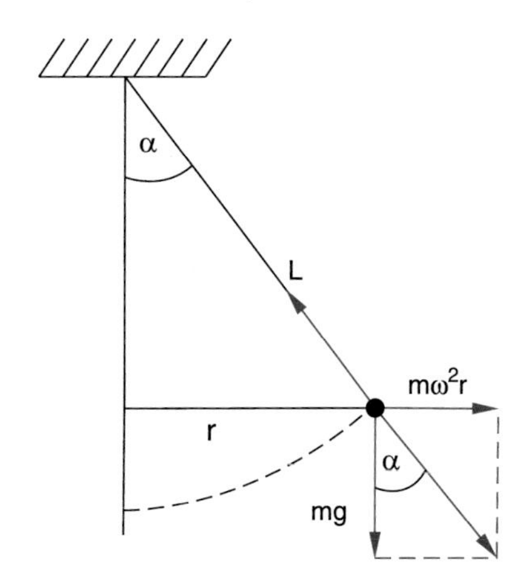

3.4 The vertical component of the angular velocity of the rotating earth is $\omega_s = \omega \cdot \sin\varphi \approx 4.7 \cdot 10^{-5}\,\text{s}^{-1}$. The Coriolis acceleration $a_c = 2\omega \cdot v \cdot \sin\varphi = 9.4 \cdot 10^{-5} \cdot 33.3\,\text{m/s}^2$ points into the horizontal direction. It causes the curvature of the air flow which would stream radially into the centre of the deep pressure region for a non-rotating earth. For the radius r of curvature one obtains from $a_c = v^2/r \Rightarrow r = v^2/a_c = 3.5 \cdot 10^5\,\text{m} = 350\,\text{km}$.

3.5 $F_{\text{c}} = m \cdot a_{\text{c}} = 2m \cdot \omega \cdot v \cdot \sin\varphi = 1.8 \cdot 10^4\,\text{N}$. The Coriolis force is directed toward west.

3.6 The centrifugal force is a) for a horizontal motion

$$F_{\text{cf}} = m\omega^2 r \Rightarrow \omega = (F/mr)^{1/2}$$

$$\omega_{\max} = (1000/5)^{1/2}\,\text{s}^{-1} = 14.14\,\text{s}^{-1}$$

$$\Rightarrow \nu_{\max} = 2.25\,\text{s}^{-1} \,.$$

For constant ω the force on the string is constant.
b) If the body rotates around a horizontal axis in the gravity field of the earth ω is not constant. The tangential acceleration is

$$a_{\text{t}} = g \cdot \sin\varphi(t) \,,$$

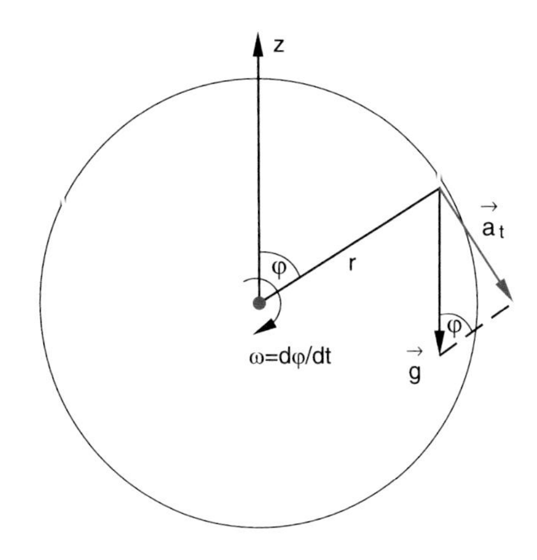

where φ is the angle between the radius vector $\boldsymbol{r}$ and the vertical z-axis. The angular velocity is then

$$\omega = \omega_0 + (g/r)\int_0^{\varphi} \sin\varphi(t)\,\mathrm{d}t\,,$$

where $\omega_0 = \omega(\varphi = 0)$ is the velocity at the upper point of the circle. The maximum value of ω is reached for $\varphi = \pi$ at the lowest point of the circle. The following relations hold:

$$\begin{aligned}\omega_{\max} &= \omega_0 + \left(\frac{g}{r}\right)\int_0^{\pi} \sin\varphi(t)\mathrm{d}t \\ &= \omega_0 + \left(\frac{g}{r}\right)\frac{(\cos\varphi)}{\dot{\varphi}}\bigg|_{\pi}^{0}\,.\end{aligned} \tag{14.2}$$

With $\mathrm{d}\varphi/\mathrm{d}t = \omega$ we obtain

$$\begin{aligned}&\to \omega_{\max} = \omega_0 + \frac{g}{(r\cdot\omega_0)} + \frac{g}{(r\cdot\omega_{\max})}\,,\\ &\to \omega_{\max} = \frac{1}{2r}\left(r\cdot\omega_0 + \frac{g}{\omega_0}\right)\\ &\qquad + \frac{1}{2}\sqrt{\omega_0^2 + \frac{6g}{N} + \frac{g^2}{N^2\omega_0^2}}\,.\end{aligned} \tag{14.3}$$

For $\omega_{\min}$ the plus sign before the square root must be replaced by a minus sign.
The maximum force onto the string occurs at the lower point $\varrho\varphi = \pi$.
The condition that the string does not break gives the relation

$$\begin{aligned}F &= m\omega_{\max}^2\cdot r + m\cdot g \le 1000\,\mathrm{N}\\ \to \omega_{\max} &\le \left[\frac{(1000/5 - g)}{r}\right]^{1/2} = 13.8\,\mathrm{s}^{-1}\,.\end{aligned}$$

The maximum allowed angular velocity ω_0 at the lower point $\varphi = 0$ can be calculated either from (3) or from the law of energy conservation

$$\tfrac{1}{2}mv_0^2 = \tfrac{1}{2}mv_{\max}^2 - 2mg\cdot r \quad \text{with} \quad v = r\cdot\omega\,.$$

Inserting the numerical values gives

$$\omega_0 = 12.3\,\mathrm{s}^{-1}\,.$$

The difference between the velocities in the upper and lower point is

$$\tfrac{1}{2}v_{\min}^2 + 2rg = \tfrac{1}{2}v_{\max}^2 \Rightarrow \omega_{\max}^2 - \omega_{\min}^2 = \frac{4g}{r}\,.$$

3.7 At a radial velocity $v_\mathrm{r} = 10\,\mathrm{m/s}$ the ball needs $10^{-2}\,\mathrm{s}$ to reach the outer edge of the disc. For a non-rotating disc the ball would fly on a straight line with the velocity $v = \{v_\mathrm{r}, v_\varphi\}$ in the lab system as well as in the system of the disc (which are identical for the non-rotating disc). It would reach the edge of the disc at the displacement $R\cdot\varphi$ from the position $\varphi = 0$ which can be obtained from $R\cdot\varphi = \Delta t\cdot v_\varphi = 0.05\,\mathrm{m}$ at the angle $\varphi = 0.05\,\mathrm{m}/0.2\,\mathrm{m} = 0.25\,\mathrm{rad} = 14.8°$. When the disc rotates with $\omega = 2\pi\cdot 10\,\mathrm{s}^{-1}$, its edge turns during the time $T = 0.01\,\mathrm{s}$ by $R\varphi = R\omega T = 0.126\,\mathrm{m} \Rightarrow \varphi = 39°$. For a radial velocity ($\mathrm{d}\varphi/\mathrm{d}t = 0$) the bullet would reach the edge at $\varphi = -39°$. With $v_\varphi = 5\,\mathrm{m/s}$ the bullet reaches the edge at $\varphi = 14.5° - 39° = -24.5°$. From the point of view of the observer at rest the bullet flies on a straight line, but viewed by an observer on the disc it flies along a curved line with the tangential acceleration

$$a_\varphi = 2v_\mathrm{r}\cdot\omega\,.$$

The trajectory on the disc is a parabola. The velocity is $v = \{v_r, v_\varphi - 2v_\mathrm{r}\omega t\}$ where v_r and v_φ are the velocity components in the rest frame while v_r and $v_\varphi - 2v_\mathrm{r}\omega t$ are the components in the system of the rotating disc.

3.8 The centrifugal acceleration is

$$\begin{aligned}a_\mathrm{cf} &= \omega^2 r = \omega^2 R\cos\varphi \quad (R = \text{earth radius})\\ &= 3.7\cdot 10^{-9}\cdot 6.37\cdot 10^6\,\mathrm{m/s}^2\\ &= 2.36\cdot 10^{-2}\,\mathrm{m/s}^2\,.\end{aligned}$$

It is acting radially outwards, perpendicular to the angular velocity $\boldsymbol{\omega}$.
The Coriolis acceleration is parallel to $\boldsymbol{a}_\mathrm{cf}$. Numerical values: $a_\mathrm{cf} = 0.023\,\mathrm{m/s}^2$, $a_\mathrm{C} = 1.02\,\mathrm{m/s}^2$. The Coriolis acceleration is for this case 44 times as large as the centrifugal force.

3.9 a) According to the Galilei-transformations is:

$$\begin{aligned}u_x' &= u_x - v_x = 0.5c - \tfrac{1}{3}c = \tfrac{1}{6}c\\ u_y' &= u_y = 0.1\,\mathrm{c}\\ u_z' &= u_z = 0\,.\end{aligned}$$

b) The Lorentz-transformations give:

$$\begin{aligned}&u_x' = \tfrac{1}{5}c\,;\quad u_y' = 0.113c\,;\quad u_z' = 0\\ &\Rightarrow \boldsymbol{u}' = \{0.2,\,0.113,\,0\}c\,.\end{aligned}$$

The relative errors of the Galilei transformations are with $\Delta u = u_{\text{Lo}} - u_{\text{Ga}}$

$$\frac{\Delta u'_x}{u'_x} = \frac{1/5 - 1/6}{1/5} = \frac{1}{6} \approx 16.7\% ,$$

$$\frac{\Delta u'_y}{u'_u} = \frac{0.013}{0.113} \approx 11.5\% ,$$

$$\Delta u'_z = 0 .$$

3.10 $\gamma = (1 - v^2/c^2)^{-1/2} = 2.785 \Rightarrow L' = L/\gamma = 0.36L$.

3.11 The nearest distance Earth–Neptune is (Tab. 2.1)

$$L = 28.8\,\text{AU} = 4.3 \cdot 10^{12}\,\text{m} .$$

The travel time according to the measurement of the pilot:

$$T' = \frac{2L}{\gamma v} = \frac{2L}{v}\sqrt{1 - v^2/c^2} = 1\,\text{d} \mathrel{\hat{=}} 8.64 \cdot 10^4\,\text{s} .$$

Resolving for v yields

$$v = \frac{2L}{(T^2 + 4L^2/c^2)^{1/2}} = 0.94 \cdot 10^8\,\text{m/s} = 0.3c$$

$$\Rightarrow \gamma = 1.048 .$$

Travel time according to the observers at rest on earth:

$$T = T'\gamma = 9.05 \cdot 10^4\text{s}$$

$$= 1\,\text{d} + 1.41 \cdot 10^4\,\text{s} = 1\,\text{d} + 1.15\,\text{h} .$$

3.12 a) In the rest frame the observer O sits in the middle between A and B. This is also true, when A, B and O move with the same constant velocity v.

b) When O' moves with the velocity v_x against the length $\overline{AB}$, he measures the simultaneous arrival of the light pulses from A and B at the point C, if C is away from A by $(L/2)(1 - v/c)$. C is therefore closer to A than to B.

3.13 $v = 0.8c \Rightarrow \gamma = 5/3$.
The travelling time is according to B: $T = 2L/v = 10$ years. According to A is $T' = 2L/(v\gamma) = 6$ years.
The number of pulses sent by B is

$$N = fT = 1 \cdot 10 = 10 .$$

Number of pulses sent by A:

$$N' = fT' = 6 .$$

Number of pulses received by A during the journey out:

$$N'_1 = (L/v)(1 - \beta) = 5 \cdot 0.2 = 1 .$$

On the journey back:

$$N'_2 = (L/v)(1 + \beta) = 5 \cdot 1.8 = 9 .$$

3.14 For C the two astronauts A and B meet after a distance x (in ly) with $x = 0.8ct + 0.1ct = 0.9ct$, when t is measured in years.
$\Rightarrow t = 8$ years after the departure of B, $\Rightarrow x = 7.2$ ly.
For A the travelling time is $t'_A = 1/\gamma_A$ with $\gamma_A = (1 - 0.8^2)^{-1/2} = 1.67 \Rightarrow t'_A = 4.8$ years.
For B is $t'_B = 1/\gamma_B$ with $\gamma_B = (1 - 0.9^2)^{-1/2} = 2.3 \Rightarrow t'_B = 3.49$ years.

14.4 Chapter 4

4.1 All particles move into the $\pm x$-direction, $\Rightarrow \boldsymbol{v} = \{v_x, 0, 0\}$; $|\boldsymbol{v}| = v$.
The centre of mass velocity is

$$v_{\text{CM}} = \frac{mv - 3mv}{4m} = -\frac{1}{2}v .$$

The particle velocity in the CM-system is

$$v_{1\text{CM}} = v_1 - v_{\text{CM}} = \tfrac{3}{2}v ,$$

$$v_{2\text{CM}} = v_2 - v_{\text{CM}} = \tfrac{1}{2}v .$$

a) Elastic collision:

$$\left.\begin{aligned} v'_{1\text{CM}} &= -v_{1\text{CM}} = -\tfrac{3}{2}v , \\ v'_{2\text{CM}} &= -v_{2\text{CM}} = +\tfrac{1}{2}v \end{aligned}\right\}$$

$$\Rightarrow \left.\begin{aligned} v'_1 &= v'_{1\text{CM}} + v_{\text{CM}} = -2v , \\ v'_2 &= v'_{2\text{CM}} + v_{\text{CM}} = 0 \end{aligned}\right\}$$

$$\Rightarrow \begin{aligned} E'_{\text{kin}}(m_1) &= \tfrac{m}{2}v_1'^2 = 2mv^2 , \\ E'_{\text{kin}}(m_2) &= 0 . \end{aligned}$$

Before the collisions was

$$E_{\text{kin}}(m_1) = \frac{m}{2}v^2 ; \quad E_{\text{kin}}(m_2) = \frac{3}{2}mv^2$$

$$\Rightarrow \sum E_{\text{kin}} = \sum E'_{\text{kin}} .$$

b) Completely inelastic collision: The two particles stay together after the collision. The total mass $M = 4\,\text{m}$ moves with the velocity $v_{\text{CM}} = v'_{\text{CM}} = -\frac{1}{2}v$

$$\Rightarrow E'_{\text{kin}} = \frac{4m}{2}v_{\text{CM}}^2 = \frac{1}{2}mv^2 .$$

The rest $(3/2)mv^2$ of the initial energy $2mv^2$ is transferred into heat energy, $\Rightarrow$ 75% are converted into heat, only 25% remain as kinetic energy.

4.2 The momentum of the bullet: $m_2 v$
$\Rightarrow$ the velocity of wooden block + bullet is

$$v' = \frac{m_2 v}{M} \quad \text{with} \quad M = m_1 + m_2 .$$

$$\Rightarrow \quad E_{\text{kin}} = \tfrac{1}{2}Mv'^2 = \tfrac{1}{2}\tfrac{m_2^2}{M}v^2$$

$$= E_{\text{p}} = MgL(1 - \cos\varphi_0)$$

$$\Rightarrow \quad \cos\varphi_0 = 1 - \frac{1}{2}\frac{m_2^2}{M^2 gL}v^2 = 1 - 0.196 = 0.804$$

$$\Rightarrow \quad \varphi_0 = 36.5^\circ \ .$$

4.3 We assume that the incident proton moves into the $+x$-direction.
a) The momentum conservation for the x- and the y-direction demands:

$$x: \quad mv_1' \cos\theta_1 + 2mv_2' \cos 45^\circ = mv_1 \qquad (14.4a)$$

$$y: \quad mv_1' \sin\theta_1 = 2mv_2' \sin 45^\circ \ . \qquad (14.4b)$$

Division by m gives for (14.4b):

$$v_1' = 2v_2' \frac{\sin 45^\circ}{\sin\theta_1} \ .$$

Inserting in (14.4a) gives:

$$v_2' = \frac{1}{2}\frac{v_1}{\cos 45^\circ + \sin 45^\circ/\tan\theta_1} = \frac{v_1}{\sqrt{2}\,(1+\cot\theta_1)} \ ,$$

$$v_1' = \frac{v_1}{\sin\theta_1 + \cos\theta_1} \ .$$

Energy conservation demands

$$v_1^2 = v_1'^2 + 2v_2'^2$$

$$\Rightarrow 1 = \frac{1}{(\sin\theta_1 + \cos\theta_1)^2} + \frac{1}{(1+\cot\theta_1)^2}$$

$$\Rightarrow \tan\theta_1 = 2 \Rightarrow \theta_1 = 63.435^\circ \ .$$

b) $$v_{CM} = \frac{m_1 v_1}{m_1 + m_2} \quad \text{with} \quad 2m_1 = m_2$$

$$\Rightarrow v_{CM} = \tfrac{1}{3}v_1 = v_{CM}' \ .$$

c) $$v_1'^2 = \frac{v_1^2}{\left(1+\frac{m_2}{m_1}\right)^2}\left[\left(\frac{m_2}{m_1}\right)^2 + 2\frac{m_2}{m_1}\cos\vartheta_1 + 1\right]$$

$$= v_1^2 \frac{4 + 4\cos 63.435^\circ + 1}{9} = 0.75\, v_1^2$$

$$\Rightarrow v_1' = 0.866\, v_1 \ .$$

$$v_2'^2 = \frac{1}{2}(v_1^2 - v_1'^2) = \frac{1}{2}0.25\, v_1^2 = 0.125\, v_1^2$$

$$\Rightarrow v_2' = 0.35\, v_1 \ .$$

4.4 a) Energies in the Lab-system:

$$E_{\text{kin}}(m_1) = \frac{m}{2}\left(v_x^2 + v_y^2 + v_z^2\right) = 1\cdot(9+4+1) = 14\,\text{N m} \ ,$$

$$E_{\text{kin}}(m_2) = 36\,\text{N m} \ .$$

Velocity of the centre of mass:

$$\boldsymbol{v}_{CM} = \frac{1}{M}\sum_i m_i \cdot \boldsymbol{v}_i = \left\{v_{xCM}, v_{yCM}, v_{zCM}\right\} = \{0, 2, 2\}\ \text{m/s} \ .$$

Relative velocities:

$$\left.\begin{aligned} \boldsymbol{v}_{1CM} &= \boldsymbol{v}_1 - \boldsymbol{v}_{CM} = \{3, 0, -3\}\,\text{m/s} \\ \boldsymbol{v}_{2CM} &= \boldsymbol{v}_2 - \boldsymbol{v}_{CM} = \{-2, 0, 2\}\,\text{m/s} \end{aligned}\right\} \Rightarrow$$

$$E_{\text{kin}}^{(CM)}(m_1) = \frac{m_1}{2}v_{1CM}^2 = 18\,\text{N m} \ ,$$

$$E_{\text{kin}}^{(CM)}(m_2) = \frac{m_2}{2}v_{2CM}^2 = 12\,\text{N m} \ .$$

b) The centre of mass momentum equals the momentum of the compound particles after the collision.

$$M\boldsymbol{v}_{CM} = M\,\{0, 2, 2\}\,\text{kg m/s} \ ,$$

$$E_{\text{kin}}'(M) = \frac{M}{2}v_{CM}^2 = 20\,\text{N m} \ .$$

c) The fraction of the converted kinetic energy is

$$\eta = 1 - \frac{E_{\text{kin}}'(M)}{E_{\text{kin}}(m_1) + E_{\text{kin}}(m_2)} = \frac{50-20}{50} = 0.6 \ .$$

In the centre of mass system is $E_{\text{kin}}^{\text{CM}} = 0$. The total kinetic energy is converted into heat.

4.5 We choose the x-axis as the direction of v_1.
a) Conservation of momentum for the x- and y-components yields:

$$m_1 v_{1x} + m_2 v_{2x} = m_1 v_{1x}' + m_2 v_{2x}' \ ,$$

$$\boldsymbol{v}_1 = \{4, 0\}\,\text{m/s} \ ; \quad \boldsymbol{v}_1' = \{2, 2\}\,\text{m/s} \ ;$$

$$\boldsymbol{v}_2' = \{1, -1\}\,\text{m/s} \Rightarrow v_{2x} = 0 \ ,$$

$$m_1 v_{1y} + m_2 v_{2y} = m_1 v_{1y}' + m_2 v_{2y}'$$

$$0 + 2v_{2y} = 2 - 2\cdot 1\,\text{m/s} \ ,$$

$$\Rightarrow v_{2y} = 0\,\text{m/s}$$

$$\Rightarrow \boldsymbol{v}_2 = \{0, 0\}\,\text{m/s} \ ,$$

i. e. m_2 was at rest before the collision.
b) Energy conservation (4.17) gives:

$$Q = E_{\text{kin}}' - E_{\text{kin}} = \tfrac{1}{2}\left(m_1 v_1'^2 + m_2 v_2'^2 - m_1 v_1^2 - m_2 v_2^2\right) = -2\,\text{N m} \ .$$

$E_{\text{kin}} = 8\,\text{N m} \Rightarrow 25\%$ of the initial energy is converted into heat.
The centre of mass velocity is

$$v_{CM} = \frac{1}{M}\left\{m_1 v_{1x} + m_2 v_{2x}; m_1 v_{1y} + m_2 v_{2y}\right\} = \tfrac{1}{3}\{4, 0\}\,\text{m/s} \ .$$

The energy of the centre of mass is

$$E_{\text{kin}}^{(\text{CM})} = \tfrac{1}{2}Mv_{\text{CM}}^2 = 2.66\,\text{N m}\,.$$

For a completely inelastic collision the fraction $Q = E_{\text{kin}} - E_{\text{kin}}^{(\text{CM})}$ is converted into heat. Since the collision of our example is not a central collision, $|Q|$ is smaller. In the C-system 37.5% are converted.

c) Velocities in the CM-system:

$$\boldsymbol{v}_{1\text{CM}} = \boldsymbol{v}_1 - \boldsymbol{v}_{\text{CM}} = \left\{\tfrac{8}{3}, 0\right\}\,\text{m/s}\,,$$

$$\boldsymbol{v}'_{1\text{CM}} = \boldsymbol{v}'_1 - \boldsymbol{v}_{\text{CM}} = \left\{\tfrac{2}{3}, 2\right\}\,\text{m/s}\,;$$

$$\cos\vartheta_1 = \frac{\boldsymbol{v}_{1\text{CM}} \cdot \boldsymbol{v}'_{1\text{CM}}}{|v_{1\text{CM}}||v'_{1\text{CM}}|} = \frac{16/9}{\sqrt{\frac{64}{9} \cdot \frac{40}{9}}} = 0.316$$

$$\Rightarrow \vartheta_1 = 71.578°\,;$$

$$\boldsymbol{v}_{2\text{CM}} = \left\{-\tfrac{4}{3}, \tfrac{2}{3}\right\}\,\text{m/s}\,;\ \boldsymbol{v}'_{2\text{CM}} = \left\{-\tfrac{1}{3}, -\tfrac{7}{3}\right\}\,\text{m/s}$$

$$\Rightarrow \vartheta_2 = 121.6°\,.$$

4.6 Conservation of momentum gives

$$m_1\boldsymbol{v}'_1 + m_2\boldsymbol{v}'_2 = m_2\boldsymbol{v}_2\,.$$

Conservation of energy gives:

$$m_1 v_1'^2 + m_2 v_2'^2 = m_2 v_2^2\,.$$

a) After the collision is $v'_2 = -v'_1$

$$\Rightarrow \boldsymbol{v}'_2\left(1 - \frac{m_1}{m_2}\right) = \boldsymbol{v}_2\,,\quad v_2'^2\left(1 + \frac{m_1}{m_2}\right) = v_2^2$$

$$\Rightarrow m_1/m_2 = 3\,.$$

b) The travel time for m_1 resp. m_2 until the left barrier are

$$t_1 = \frac{1.6\,\text{m}}{v'_1} > t_2 = \frac{2.4\,\text{m}}{v'_2} \Rightarrow \frac{v'_2}{v'_1} > 1.5\,.$$

Energy conservation demands with $x = m_1/m_2$

$$\frac{v_2'^2}{v_1'^2} = \frac{v_2^2}{v_1'^2} - x\,.$$

Using momentum conservation gives:

$$\frac{v'_2}{v'_1} = \frac{1}{2}(x-1) \Rightarrow x > 4\,.$$

c) The velocity of the CM is

$$v_{\text{CM}} = v'_{\text{CM}} = \frac{1}{3}v_2\,.$$

The velocities in the lab-system are after the collision:

$$v'_1 = \frac{2}{3}v_2;\quad v'_2 = -\frac{1}{3}v_2\,.$$

The two masses meet for the first time at $x_0 = 1.6$ m at the time $t_1 = 0$, for the second time t_2 at the location x ($x = 0$ is at the left wall). According to the calculation in a) the masses meet for the second time only after the reflection of m_1 at the left wall. It is:

$$t_2 = \frac{x_0 + x}{v'_1} = \frac{0.8 + x_0 - x}{v'_2}$$

$$\Rightarrow x = 1.07\,\text{m}\,.$$

The two masses meet at $x = 1.07$ m from the left wall after m_1 has suffered a reflection at the left wall and m_2 at the right wall.

4.7 The velocity of the steel ball at the impact is

$$m_1 Lg = \tfrac{1}{2}m_1 v_1^2 \Rightarrow v_1 = \sqrt{2gL} = 4.43\,\text{m/s}\,.$$

The energy transferred to m_2

$$\Delta E = 4\frac{m_1 m_2}{M^2}E_1\,.$$

The steel ball has therefore the energy after the collision

$$E'_{\text{kin}} = \left(1 - 4\frac{m_1 m_2}{M^2}\right)E_1 = \frac{4}{9}E_1\,.$$

It rises up to the height $H = L(1 - \cos\varphi) = \frac{4}{9}L \Rightarrow \cos\varphi = \frac{5}{9} \Rightarrow \varphi = 56.15°$.

4.8 The distance between ball and lift is $\Delta s = 20$ m. The time until the impact onto the ceiling of the lift is obtained from

$$\tfrac{1}{2}gt_1^2 + vt_1 = \Delta s = 20\,\text{m} \Rightarrow t_1 = 1.8\,\text{s}\,.$$

During this time the lift has moved over the distance $vt_1 = 3.6$ m. The impact point is therefore 26.4 m below A.

b) In the lab system the impact velocity of the ball is $v_1 = gt_1 = 17.66$ m/s. The centre of mass moves because $M \gg m$ with the velocity $v_{\text{CM}} = v = 2$ m/s upwards. In the centre of mass system (which is nearly identical with the system of the lift) the ball has the velocity $v_{1\text{CM}} = v_1 + v_{\text{CM}} = 19.66$ m/s downwards. After the completely elastic reflection at the lift ceiling the ball has the upward velocity $v'_{1\text{CM}} = 19.66$ m/s.

In the Lab system is $v'_1 = v'_{1\text{CM}} + v_{\text{CM}} = 21.66$ m/s. The ball has won twice the velocity of the lift by the reflection at the moving lift. It rises now by the distance $\Delta h_1 = v_1'^2/2g$ above the impact point. Inserting the numerical values gives: $\Delta h_1 = 23.9$ m. Its upper return point is then 2.5 m below A.

c) It hits the ceiling of the lift for a second time at the time t_2. During the time $\Delta t = t_2 - t_1$ the lift has moved upwards by $\Delta h_2 = v\Delta t$.

The ball needs a rise time Δt_1 obtained from $v(\Delta h_1) = 0 = v'_1 0 g\Delta t_1 \Rightarrow \Delta t_1 = v'_1/g = 2.2$ s. Its drop time is $\Delta t_2 = 1.9$ s which is obtained from

$$\tfrac{1}{2}g\Delta t_2^2 = \Delta h_1 - v(2.2\,\text{s} + \Delta t_2)\,.$$

This gives the time $\Delta t = \Delta t_1 + \Delta t_2 = 4.1\,\text{s} \Rightarrow \Delta h_2 = 8.2$ m. The second impact occurs 8.2 m above the first impact point, i. e. 18.2 m below A.

4.9 a) The α-particle should fly into the $+x$-direction. For the y-components of the momenta we get

$$0 = m_1 v_1' \sin 64° - m_2 v_2' \sin 51°$$

$$\Rightarrow \frac{v_1'}{v_2'} = 4 \cdot \frac{\sin 51°}{\sin 64°} = 3.46 \quad \text{since} \quad m_2 = 4m_1 \ .$$

b)
$$\frac{E_{\text{kin}}'(m_1)}{E_{\text{kin}}'(m_2)} = \frac{m_1 v_1'^2}{m_2 v_2'^2} = \frac{1}{4} \cdot 3.46^2 \approx 3.0 \ .$$

4.10
$$E = c\sqrt{m_0^2 c^2 + p^2} \ ,$$

with $E = 6\,\text{GeV}$ and $pc = 4\,\text{GeV}$

$$\Rightarrow m_0 c^2 = \sqrt{20}\,\text{GeV} \ ,$$

$$\Rightarrow E' = \sqrt{20 + 25}\,\text{GeV} = 6.71\,\text{GeV} \ .$$

With

$$\left.\begin{aligned} E &= mc^2 \\ E' &= m'c^2 \end{aligned}\right\} \Rightarrow \frac{m'}{m} = \frac{6.7}{6} = \frac{1}{\sqrt{1 - v^2/c^2}}$$

$$\Rightarrow \frac{v}{c} = 0.445 \ .$$

The two systems move with $v = 0.445c$ against each other.

14.5 Chapter 5

5.1 When we cut a cone with full aperture angle 2α out of a sphere we choose the origin of our coordinate system at the peak of the cone. The z-axis is the symmetry axis. Then the coordinates of the centre of mass are

$$x_{\text{CM}} = y_{\text{CM}} = 0 \ ,$$

$$\begin{aligned} z_{\text{CM}} &= \frac{1}{V} \int\limits_{r=0}^{R} \int\limits_{\vartheta=\pi/2-\alpha}^{\pi/2} \int\limits_{\varphi=0}^{2\pi} r^3 \cos\vartheta \sin\vartheta \, \mathrm{d}r \, \mathrm{d}\vartheta \, \mathrm{d}\varphi \\ &= \frac{1}{V} \frac{\pi}{4} R^4 \left[1 - \sin^2\left(\frac{\pi}{2} - \alpha\right)\right] \\ &= \frac{1}{V} \frac{\pi}{4} R^4 \sin^2\alpha \ . \end{aligned}$$

The volume of the cone is $V = \frac{2}{3}\pi R^3(1 - \cos\alpha)$. Then we get

$$z_{\text{CM}} = \frac{3}{8} R \left(\frac{\sin^2\alpha}{1 - \cos\alpha}\right) = \frac{3}{8} R(1 + \cos\alpha) \ .$$

5.2 a)
$$I_{\text{CM}} = \tfrac{2}{5} MR^2 = 9.7 \cdot 10^{37}\,\text{kg}\,\text{m}^2,$$

$$L = I_{\text{CM}}\omega = 7.07 \cdot 10^{33}\,\text{kg}\,\text{m}^2\,\text{s}^{-1}$$

$$\Rightarrow E_{\text{curl}} = \frac{1}{2} I_{\text{CM}}\omega^2 = \frac{1}{5} MR^2\omega^2 = 2.57 \cdot 10^{29}\,\text{J} \ .$$

b) The mass of the earth is for this case

$$M_{\text{E}} = \frac{4}{3}\pi\varrho_1 \frac{R^3}{8} + \frac{4}{3}\pi\varrho_2 \left(R^3 - \frac{1}{8}R^3\right) = \frac{4}{3}\pi\varrho R^3 \ .$$

With $\varrho = M/V$ = mean density

$$\Rightarrow \varrho_1 + 7\varrho_2 = 8\varrho \ .$$

With $\varrho_1 = 2\varrho_2$

$$\Rightarrow \varrho_2 = \frac{8}{9}\varrho \ , \quad \varrho_1 = \frac{16}{9}\varrho \ .$$

The moment of inertia is therefore

$$\begin{aligned} I_{\text{CM}} &= \frac{2}{5} \cdot \frac{4}{3}\pi \left(\varrho_1 \left(\frac{R}{2}\right)^3 \left(\frac{R}{2}\right)^2 + \varrho_2 \left[R^3 R^2 - \left(\frac{R}{2}\right)^3 \left(\frac{R}{2}\right)^2\right]\right) \\ &= \frac{8}{15}\pi \left(\varrho_1 \frac{R^5}{32} + \varrho_2 \frac{31}{32} R^5\right) \\ &= \frac{1}{60}\pi R^5 \left(\frac{16}{9}\varrho + \frac{31 \cdot 8}{9}\varrho\right) \\ &= \frac{22}{45}\pi R^5 \varrho = \frac{11}{30} MR^2 = 0.367\,MR^2 \ . \end{aligned}$$

This should be compared with the moment of inertia $I_{\text{CM}} = (2/5)MR^2 = 0.4MR^2 = 9.72 \cdot 10^{37}\,\text{kg}\,\text{m}^2$ of the homogeneous earth.

c) If all $N = 5 \cdot 10^9$ adults on earth would run simultaneously on the equator eastwards their torque exerted on the earth would be $D = N \cdot m \cdot a \cdot R = 5 \cdot 10^9 \cdot 70 \cdot 2 \cdot 6.37 \cdot 10^6\,\text{N} \cdot \text{m} = 4.46 \cdot 10^{18}\,\text{N}\,\text{m}$. This would lead to a relative decrease $\Delta\omega/\omega = \Delta L/L$ of the earth rotation. Inserting the numerical value for the angular momentum L

$$L = I_{\text{CM}}\omega = 0.71 \cdot 10^{34}\,\text{kg}\,\text{m}^2/\text{s}$$

we get

$$\frac{\Delta\omega/\Delta t}{\omega} = \frac{1}{L}\frac{\Delta L}{\Delta t} = \frac{D}{L} = 6.3 \cdot 10^{-16}\,\text{s}^{-1} \ ,$$

which is so small, that it falls below the detection limit.

5.3 a)
$$I_0 = \tfrac{1}{2} MR^2 = 5 \cdot 10^{-4}\,\text{kg}\,\text{m}^2$$

$$L = I_0\omega_0 = \tfrac{1}{2} MR^2\omega_0 = 3.14 \cdot 10^{-2}\,\text{N}\,\text{m}\,\text{s} \ ,$$

$$E_{\text{curl}}^0 = \tfrac{1}{2} I_0 \omega_0^2 = 0.987\,\text{N}\,\text{m} \ .$$

b) $I = I_0 + mR^2 = (5 + 1) \cdot 10^{-4}\,\text{kg}\,\text{m}^2 = 6 \cdot 10^{-4}\,\text{kg}\,\text{m}^2$. The angular momentum does not change, because the bug falls onto the disc parallel to the rotation axis.

$$\Rightarrow \omega = \frac{L_0}{I} = \frac{5}{6}\omega_0$$

$$E_{\text{curl}} = \frac{1}{2} I\omega^2 = \frac{5}{6} E_{\text{curl}}^0 = 0.823\,\text{N}\,\text{m} \ .$$

The energy difference $\Delta E = 0.164\,\mathrm{N\cdot m}$ is converted by friction into heat energy, which is lost during the equalization of the tangential velocities of bug and disc (which are here assumed to occur instantaneously).

c)
$$\omega(r) = \frac{1}{1 + mr^2/I_0}\omega_0$$

$$L(r) = L_0,\ \text{independent of } r\ ,$$

$$E_{\text{curl}} = \frac{E^0_{\text{curl}}}{1 + mr^2/I_0}\ .$$

5.4 a)
$$I = \int_V r^2 \varrho\, dV = 2\pi H\varrho_0 \int_{r=0}^{R} \left[1 + \left(\frac{r}{R}\right)^2\right] r^3\, dr$$
$$= 2\pi\varrho_0 H\left[\frac{1}{4}R^4 + \frac{1}{6}R^4\right] = \frac{10\pi}{12}\varrho_0 H R^4$$
$$= \frac{5}{6}\varrho_0 R^2 V\ .$$

The mass is $M = \int_V \varrho\, dV = \frac{3}{2}\pi\varrho_0 H R^2$

$$\Rightarrow I = \frac{5}{9}MR^2\ .$$

Numerical values: $M = 18.85\,\mathrm{kg}$, $I = 0.105\,\mathrm{kg\,m^2}$.

b)
$$a = \frac{g\sin\alpha}{1 + I_{\text{CM}}/(MR^2)} = \frac{g\sin 10°}{14/9}$$

$$h = \tfrac{1}{2}at^2 \Rightarrow t = (2h/a)^{1/2} = 1.35\,\mathrm{s}\ .$$

5.5 For the isosceles triangle with height h and side length d the centre of mass $S = (x_{\text{CM}}, y_{\text{CM}})$ has the coordinates $x_{\text{CM}} = 0$, $y_{\text{CM}}(\alpha)$.
The moments of inertia around the principal axes are

$$I_a = 2my^2_{\text{CM}} + m(h - y_{\text{CM}})^2\ ,$$
$$I_b = 2mx^2\ ,$$
$$I_c = m(h - y_{\text{CM}})^2 + 2m(x^2 + y^2_{\text{CM}}) = I_a + I_b\ ,$$
$$x = d\sin(\alpha/2) = 0.204\,\mathrm{nm}\ ,$$
$$h = d\cos(\alpha/2) = 0.247\,\mathrm{nm}\ ,$$
$$y_{\text{CM}} = \tfrac{1}{3}h = 0.082\,\mathrm{nm}\ .$$

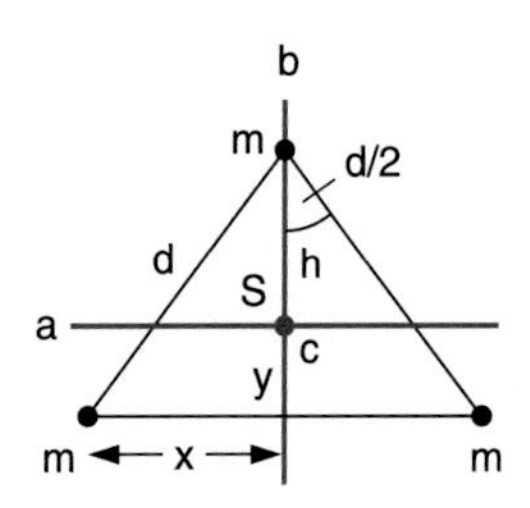

For the moments of inertia we get:

$$I_a = 0.93\,\mathrm{AMU\,nm^2}\ ,$$
$$I_b = 1.91\,\mathrm{AMU\,nm^2}\ ,$$
$$I_c = 2.85\,\mathrm{AMU\,nm^2}\ .$$

$1\,\mathrm{AMU} = 1.67\cdot 10^{-27}\,\mathrm{kg}$.
The rotational energy is then

$$E_{\text{rot}} = \frac{L^2}{2I} \quad \text{with} \quad L^2 = l\cdot(l+1)\hbar^2\ .$$

Where $l = 1, 2, 3, \ldots$ and $\hbar = h/2\pi = 1.06\cdot 10^{-34}\,\mathrm{J\cdot s}$ is the reduced Planck constant and it is the smallest unit of the rotational angular momentum.
This gives with $1\,\mathrm{eV} = 1.6\cdot 10^{-19}\,\mathrm{J}$

$$E_a = 2.2\cdot 10^{-5}\,\mathrm{eV}\ ,$$
$$E_b = 1.1\cdot 10^{-5}\,\mathrm{eV}\ ,$$
$$E_c = 0.73\cdot 10^{-5}\,\mathrm{eV}\ .$$

5.6 The inertial moment of a rod with length L is

$$I_{\text{CM}} = \tfrac{1}{12}ML^2 = 1.33\cdot 10^{-2}\,\mathrm{kg\,m^2}\ .$$

The angular momentum of the bullet referred to the CM of the rod is

$$L_{\text{B}} = |\boldsymbol{r}\times\boldsymbol{p}| = \tfrac{1}{2}Lmv = 0.4\,\mathrm{N\,m\,s}\ .$$

The rotational velocity of the rod is

$$\omega = \frac{L_{\text{B}}}{I} = \frac{L_{\text{B}}}{I_{\text{CM}} + m(L/2)^2}$$
$$= 29.2\,\mathrm{s^{-1}} \Rightarrow \nu = 4.65\,\mathrm{s^{-1}}\ .$$
$$\Rightarrow E_{\text{rot}} = \tfrac{1}{2}I\omega^2 = 5.67\,\mathrm{N\,m}\ ,$$
$$E_{\text{kin}} = \tfrac{1}{2}mv^2 = 200\,\mathrm{N\,m}\ ,$$
$$\Rightarrow E_{\text{rot}}/E_{\text{kin}} = 2.8\cdot 10^{-2} = 2.8\%\ .$$

97.1% of the kinetic energy of the bullet is lot as heat energy. Compare this with the case of a completely inelastic central collision of a bullet with mass m hitting a free mass M. Here the ratio

$$E'_{\text{kin}}/E_{\text{kin}} = \frac{(m+M)v^2_{\text{CM}}}{mv^2} \quad \text{(see Sect. 4.2.4)}\ .$$

With $v_{\text{CM}} = \dfrac{m}{M+m}v \Rightarrow E'_{\text{kin}}/E_{\text{kin}} \approx \dfrac{m}{M} = 0.01.$

Question: Why is the transfer of kinetic energy of the bullet into rotational energy more efficient?

5.7 $I_{CM} = \frac{1}{2}MR^2$; $D = I_{CM} \cdot \frac{d\omega}{dt}$

$$\Rightarrow \omega = \omega_0 + \frac{1}{I_{CM}} \int_0^t D\,dt'$$

$$= \omega_0 + \frac{D_0}{I_{CM}} \int_0^t e^{-at'}\,dt'$$

$$= \omega_0 + \frac{2D_0}{aMR^2}\left[1 - e^{-at}\right] .$$

For $t \to \infty \Rightarrow \omega(\infty) = \omega_0 + (2D_0/aMR^2)$.
Numerical example: $\omega(t = 10\,s) = 136.4\,s^{-1}$ (because $\omega_0 = 10\,s^{-1}$).

5.8 $E_{kin} = E_{rot} + E_{trans} = \frac{1}{2}I_{CM}\omega_0^2 + \frac{1}{2}MR^2\omega_0^2$

$$E_p = Mgh = E_{kin}$$

$$\Rightarrow h = \frac{\omega_0^2}{2Mg}(I_{CM} + MR^2) .$$

a) Full cylinder: $I_{CM} = \frac{1}{2}MR^2 \Rightarrow h_1 = \frac{3}{4}\frac{\omega_0^2 R^2}{g}$.

b) Hollow cylinder: $I_{CM} = MR^2 \Rightarrow h_2 = \frac{\omega_0^2 R^2}{g}$.

Numerical example: $h_1 = 17.2\,cm$; $h_2 = 22.9\,cm$.

14.6 Chapter 6

6.1 Tensile strength in the height z above the end of the rope:

$$\sigma = \varrho \cdot g \cdot z .$$

Relative elongation:

$$\varepsilon(z) = \frac{1}{E}\sigma(z) .$$

Total elongation

$$\Delta L = \int_0^L \varepsilon(z)\,dz = \frac{1}{E}\int_0^L \sigma(z)\,dz$$

$$= \frac{\varrho g}{E}\int_0^L z\,dz = \frac{\varrho g}{2E}L^2 .$$

a) $\varrho_{St} = 7.7 \cdot 10^3\,kg/m^3$, $E = 2 \cdot 10^{11}\,N/m^2 \Rightarrow \Delta L = 15.3\,m$.
b) $\Delta\varrho = \varrho_{St} - \varrho_W = 6.67 \cdot 10^3\,kg/m^3 \Rightarrow \Delta L = 13.3\,m$.
c) The maximum tensile stress $\sigma_{max} = \varrho \cdot g \cdot L$ appears for $z = L$ at the upper end of the rope. It should be smaller than $\sigma_{tear} = 8 \cdot 10^8\,N/m^2$

$$\Rightarrow L < \frac{\sigma_{tear}}{\varrho g} = 10^4\,m .$$

6.2 The maximum deflection is according to (6.23)

$$s = \frac{L^3}{3EI}F \quad \text{with} \quad I - \text{cross sectional moment of inertia} .$$

a) $I = \frac{1}{12}d^3 b = 4.2 \cdot 10^{-6}\,m^4 \Rightarrow s = 0.4\,m$.
b) $I = \frac{1}{12}(b_1 d_1^3 - b_2 d_2^3) = 7.8 \cdot 10^{-6}\,m^4$

$$\Rightarrow s = 0.22\,m .$$

The two cross sectional areas are
a) $5 \cdot 10^{-3}\,m^2$, b) $7.5 \cdot 10^{-3}\,m^2$.
Although the area in b) is only 1.5 times larger than in a) the double-T-profile has twice the stability in the z-direction and 10 times higher stability when bending into the y-direction.

6.3 $p(h = 10{,}000\,m) \approx 10^8\,Pa \approx 10^3\,atm$.
$F = 4\pi r^2 \cdot p = 2.8 \cdot 10^9\,N$. This force equals the weight of $2.8 \cdot 10^5$ tons.
According to (6.9) is $\Delta V = -p \cdot V/K$. After Tab. 6.1 is $K = 1/\kappa = 1.56 \cdot 10^{11}\,N/m^2$.
a)

$$\Rightarrow \frac{\Delta V}{V} = -\frac{10^{18}}{1.56 \cdot 10^{11}} = 6.4 \cdot 10^{-4}$$

$$\Rightarrow \frac{\Delta r}{r} = \frac{1}{3}\frac{\Delta V}{V} \approx 2.1 \cdot 10^{-4} .$$

The radius of the solid sphere decreases by 0.3 mm. This can be also obtained in the following way: $\Delta V/V = -p \cdot \kappa$ and $\kappa = (3/E)(1 - 2\mu) \Rightarrow \Delta r/r = -p/E(1 - 2\mu)$. Inserting the numerical values for E and μ from Tab. 6.1 one obtains the same results for $\Delta r/r$.
b) Compression of a hollow sphere with radius r and wall thickness d: Now the elastic back pressure during the compression is missing since the inner sphere with radius $(r - d)$ is a gas volume, where the compression modulus is smaller by 3 orders of magnitude. We therefore get for $d \ll r$ the pressure

$$p = -\frac{E}{1 - 2\mu}\left(\frac{\Delta r}{r} - \frac{\Delta r}{r - d}\right) \approx \frac{E}{1 - 2\mu}\frac{d\,\Delta r}{r^2}$$

$$\Rightarrow \frac{\Delta r}{r} \approx -\frac{p}{E}\frac{r}{d}(1 - 2\mu) .$$

For $d = 0.2\,m$ and $r = 1.5.\,m \Rightarrow r/d = 7.5 \Rightarrow \Delta r/r1.5 \cdot 10^{-3}$. The compression is larger by a factor 7.5 compared to the solid sphere.

6.4 The tangential force acting on the wave is

$$F = \frac{\text{power}}{\text{lenght/time}} = \frac{3 \cdot 10^5}{2\pi R \cdot 25}\,N = 3.8 \cdot 10^4\,N .$$

a) The torque acting on the axis is

$$D = FR = \frac{\pi}{2}G\frac{R^4}{L}\varphi \Rightarrow \varphi = \frac{2FL}{\pi G R^3}$$

$$= \frac{1}{G}3.87 \cdot 10^9\,rad .$$

With $G = 8 \cdot 10^{10}\,N/m^2 \Rightarrow \varphi = 5.2 \cdot 10^{-2}\,rad \approx 3°$.

6.5 From $\kappa = -(1/V)\,\mathrm{d}V/\mathrm{d}p$ (6.32) $\Rightarrow \mathrm{d}V/V = -\kappa \mathrm{d}p$. Integration yields

$$\ln V = -\kappa \cdot p + C \quad \text{with} \quad C = \ln V(p=0) = \ln V_0$$
$$\Rightarrow V = V_0 \mathrm{e}^{-\kappa p}$$
$$\Rightarrow \varrho = \varrho_0 \mathrm{e}^{+\kappa p}$$

with $\kappa = 4.8 \cdot 10^{-10}\,\mathrm{m^2/N}$ and a pressure $p = 10^8\,\mathrm{N/m^2}$ at 10^4 m water depth we get $\kappa \cdot p = 0.048$. This gives

$$\varrho = \varrho_0 \cdot \mathrm{e}^{0.048} \approx \varrho_0(1 + 0.048)\,.$$

The density rises by 4.8%.

6.6 $M = \varrho[1\,\mathrm{m}^3 - (1-2d)^2(1\,\mathrm{m} - d)] = \varrho \cdot 0.0968\,\mathrm{m}^3 = 755\,\mathrm{kg}$.
The cube immerses about 0.755 m. Its centre of gravity S_b is 0.4069 m above its lower edge, i. e. 0.348 m below the water surface. The centre of gravity of the displaced water is 0.3775 m below the water surface i. e. below S_b.
For a tilt angle $\varphi = 24°$ the deeper upper edge of the open cube comes below the water surface. The cube runs full with water. For this angle φ the meta-centre M is still above S_b, i. e. the position of a closed cube would be stable.

6.7
$$W = g\Big[(\varrho_\mathrm{b} - \varrho_\mathrm{l})a^3(h-a) + \int_0^a \big[(\varrho_\mathrm{b} - \varrho_\mathrm{l})a^2(a-z) + \varrho_\mathrm{b} a^2 z\big]\mathrm{d}z\Big]$$
$$= gha^3\,[\varrho_\mathrm{b} - \varrho_\mathrm{l}(1 - a/2h)]\,.$$

With $\varrho_\mathrm{b} = 7.8 \cdot 10^3\,\mathrm{kg/m^3} \Rightarrow W = 2.51 \cdot 6.85 \cdot 10^3\,\mathrm{N \cdot m} = 1.72 \cdot 10^4\,\mathrm{N \cdot m}$.
The lift in air would require the work $mgh = gh \cdot a^3 \varrho_\mathrm{b} = 1.96 \cdot 10^4\,\mathrm{N \cdot m}$.

6.8 $F = A\,\Delta p = \pi r^2\,\Delta p = \frac{1}{4}\pi d^2\,\Delta p = 2.5 \cdot 10^4\,\mathrm{N}$ for each of the two semi-spheres., i. e. each horse had to pull with $3.125\,\mathrm{N} \mathrel{\hat{=}} 318\,\mathrm{kp}$. If one side of the sphere had been tied to a tree, 8 horses with the pulling force 318 kp each would have been sufficient but less impressive.

6.9 a) The ratio of the two measured values is

$$\frac{\varrho_\mathrm{gold} V}{(\varrho_\mathrm{gold} - \varrho_\mathrm{l})V} = \frac{19.3}{18.3} = 1.0546\,.$$

b)
$$\frac{0.8\varrho_\mathrm{gold} + 0.2\varrho_\mathrm{copper}}{0.8\varrho_\mathrm{gold} + 0.2\varrho_\mathrm{copper} - 1} = \frac{17.2}{16.2} = 1.062\,.$$

c)
$$\frac{1.0550 - 1.0546}{1.0546} \approx 3.8 \cdot 10^{-4}\,.$$

6.10 $M_\mathrm{wood} = L\pi r^2 \varrho_\mathrm{s} = \varrho_\mathrm{l} V_\mathrm{i}$; $V_\mathrm{e}/V_\mathrm{cyl} = 0.525/1 = \varrho_\mathrm{wood}/\varrho_\mathrm{l}$, where $r = d/2$ and V_i is the immersed volume.
$\Rightarrow M = 16.5\,\mathrm{kg} \Rightarrow V_\mathrm{i} = 1.65 \cdot 10^{-2}\,\mathrm{m}^3$.
a) The immersed segment of the cylinder has the volume (see the figure)

$$V_\mathrm{i} = \tfrac{1}{2}L\big[r^2\alpha - (r-h)\sin(\alpha/2)r\big]\,,$$

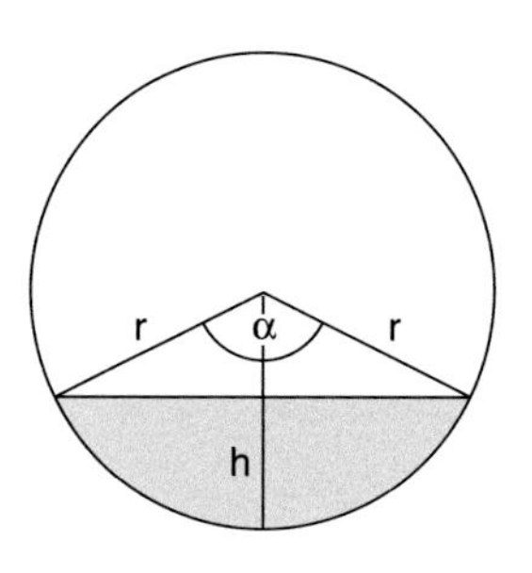

where h is the height below the water surface and α is the segment angle. With

$$h = r(1 - \cos\alpha/2) \Rightarrow V_\mathrm{i} = \tfrac{1}{2}Lr^2\left(\alpha - \tfrac{1}{2}\sin\alpha\right)\,.$$

Inserting $V_\mathrm{i} = 1.65 \cdot 10^{-2}\,\mathrm{m}^3$ and $r = 0.1$ m we get $\alpha \approx 184.5° \Rightarrow h = 0.108$ m. The cylinder immerses slightly more than half of it height.
b) The sphere has the volume $V_\mathrm{k} = \frac{4}{3}\pi R^3$ and the mass $M_\mathrm{k} = V_\mathrm{k} \cdot \varrho_\mathrm{steel}$. It experiences a buoyancy $g \cdot V_\mathrm{k} \cdot \varrho_\mathrm{l} = g \cdot M_\mathrm{k} \cdot \varrho_\mathrm{l}/\varrho_\mathrm{steel}$. At equilibrium is reached by an immersion depth where the buoyancy just compensates the weight of the sphere.

$$(M_\mathrm{cyl} + M_\mathrm{K})g = (\pi r^2 \varrho_\mathrm{l} H + M_\mathrm{K}\varrho_\mathrm{l}/\varrho_\mathrm{steel})g$$
$$\Rightarrow h = 0.553\,\mathrm{m} \text{ at } \varrho_\mathrm{steel} = 7800\,\mathrm{kg/m^3}\,.$$

14.7 Chapter 7

7.1
$$g(h) = G\frac{M_\mathrm{E}}{(R+h)^2} = G\frac{M_\mathrm{E}}{R^2(1+h/R)^2} \approx g(h=0)(1 - 2h/R)\,.$$

Inserting into (7.5a) gives

$$\frac{\mathrm{d}p}{p} = -\frac{\varrho_0}{p_0}g \cdot (1 - 2h/R)\,\mathrm{d}h\,.$$

Integration yields

$$\ln p = -\frac{\varrho_0}{p_0}gh + \frac{\varrho_0}{p_0 R}gh^2 + C$$
$$\Rightarrow p = p_0 \exp\left[-\frac{\varrho_0 g \cdot (h - h^2/R)}{p_0}\right]\,.$$

7.2 $p_0 = 1$ bar. The altitude where p=1 mbar is obtained from

$$\mathrm{e}^{-h/8.33\,\mathrm{km}} = 10^{-3}$$
$$\Rightarrow h = 8.33\ln(10^3)\,\mathrm{km} = 57.5\,\mathrm{km}\,.$$

7.3
$$p(h = 100\,\mathrm{km}) \approx 6 \cdot 10^{-6}\,\mathrm{bar} = 0.6\,\mathrm{Pa}\,.$$

For $T = 250\,\mathrm{K} \Rightarrow n = \frac{p}{kT} \Rightarrow n = 1.7 \cdot 10^{20}/\mathrm{m}^3 = 1.7 \cdot 10^{14}\,\mathrm{cm}^3$. $\varrho = nm = 1.7 \cdot 10^{20} \cdot 28\,\mathrm{AMU} = 8 \cdot 10^{-6}\,\mathrm{kg/m^3}$ (for N_2).

7.4 The buoyancy G_A = weight of the displaced air $\Rightarrow G_A = \varrho(h)gV = \varrho_0 g e^{-1/8.33} \cdot 3 \cdot 10^3\,\text{m}^3 = 3.37 \cdot 10^4\,\text{N}$.
The mass of balloon + load + gas fill can be at most $3.44 \cdot 10^3$ kg. The pressure of the fill gas is $p(h) = 0.887\,p_0 = 8.87 \cdot 10^4$ Pa.
The mass of the fill gas
a) helium: $\varrho_0 = 0.1785\,\text{kg/m}^3 \Rightarrow \varrho(h) = 0.1583\,\text{kg/m}^3 \Rightarrow m_{\text{He}} = 475\,\text{kg}$, ⇒ mass of balloon + load can be at most = 2965 kg.
b) H_2: $\varrho_0 = 0.09\,\text{kg/m}^3 \Rightarrow \varrho(h) = 0.08\,\text{kg/m}^3 \Rightarrow m_{H_2} = 240\,\text{kg}$ ⇒ mass of balloon + load should be smaller than 3200 kg.

7.5 For $h = 0$ the pressure is $p = p_0 = 10^5$ Pa and $x = x_0 = 0.2$ m. According to the Boyle–Marriott law is

$$(p_0 + \varrho g h)Ax = p_0 A x_0$$

$$\Rightarrow h = a\frac{x_0 - x}{x} \quad \text{with} \quad a = \frac{p_0}{\varrho g} = 10.2\,\text{m}$$

$$\Rightarrow \frac{\Delta h}{\Delta x} \approx -a\frac{x_0}{x^2} < 0$$

$$\Rightarrow x = \sqrt{a x_0 \left|\frac{\Delta x}{\Delta h}\right|}\,,$$

with $\Delta x = 10^{-3}$ m and $\Delta h = 1$ m. The device is usable down to a depth of 35 m with an accuracy of ± 1 m.

7.6 The number of particles that have passed at least the distance x without collisions is according to (7.33)

$$N(x) = N_0 e^{-x/\Lambda}\,.$$

a) $N(x \ge \Lambda) = N_0 e^{-1} \Rightarrow N(\Lambda)/N_0 = 0.368 = 36.8\%$.
b) $N(x \ge 2\Lambda) = N_0 e^{-2} \Rightarrow N(2\Lambda)/N_0 = 13.5\%$.

7.7 The probability W is

$$W = \int_{v_1}^{v_2} f(v)\,dv = \frac{4}{\sqrt{\pi}v_W^3}\int_{v_1}^{v_2} v^2 e^{-v^2/v_W^2}\,dv$$

$$\approx \frac{4\overline{v}^2}{\sqrt{\pi}v_W^3}\Delta v e^{-\overline{v}^2/v_W^2}\,,$$

with $\overline{v} = (v_1 + v_2)/2 = 950$ m/s and $\Delta v = (v_1 - v_2) = 100$ m/s. For N_2-molecules at $T = 300$ K is $v_W = 422$ m/s.

$$\Rightarrow W = \frac{4 \cdot 950^2 \cdot 100}{\sqrt{\pi} \cdot 422^3} \cdot e^{-5.06} = 1.7 \cdot 10^{-2}\,.$$

7.8 From (7.6) we obtain

$$\ln\frac{p_1}{p_2} = \frac{\varrho_1}{p_1} g\,\Delta h\,.$$

The density is obtained from $\varrho_1 = m \cdot p_1/(kT)$ with $m = 0.71 m_{N_2} + 0.29 m_{O_2}$ as $\varrho_1 = 1.24\,\text{kg/m}^3$

$$\Rightarrow \Delta h = 866\,\text{m}\,.$$

7.9 $\Delta\boldsymbol{v} = \boldsymbol{v}_1 - \boldsymbol{v}_2 \Rightarrow (\Delta\boldsymbol{v})^2 = v_1^2 + v_2^2 - 2v_1 v_2 \cos\alpha$.
a) Since the direction s of the velocity vectors are uniformly distributed the average values $\overline{v_1^2} = \overline{v_2^2} = \overline{v^2}$ and $\overline{\cos\alpha} = 0$:

$$\overline{(\Delta v)^2} = 2\overline{v^2} \Rightarrow \sqrt{\overline{(\Delta v)^2}} = \sqrt{2}\sqrt{\overline{v^2}}\,,$$

$$\Delta v = |\Delta\boldsymbol{v}|$$

$$= \sqrt{(v_{x1} - v_{x2})^2 + (v_{y1} - v_{y2})^2 + (v_{z1} + v_{z2})^2}$$

$$= \left(v_{x1}^2 + v_{x2}^2 + v_{y1}^2 + v_{y2}^2 + v_{z1}^2 + v_{z2}^2 - 2(v_{x1}v_{x2} + v_{y1}v_{y2} + v_{z1}v_{z2})\right)^{1/2};$$

with $\overline{v_x^2} = \overline{v_y^2} = \overline{v_z^2} = \frac{1}{3}\overline{v^2}$ and $\overline{v_x} = \overline{v_y} = \overline{v_z} = 0$

$$\Rightarrow \overline{\Delta v^2} = 6 \cdot \tfrac{1}{3}\overline{v^2} = 2 \cdot \overline{v^2}\,.$$

b) Here all absolute values v of the velocity $\boldsymbol{v}$ have the same value $\Rightarrow \overline{\Delta v} = \sqrt{2}v$.

7.10 The molecular density at $p = 10^5$ Pa and $T = 20\,°\text{C}$ is $n = 2.5 \cdot 10^{19}/\text{cm}^3$

$$\Rightarrow \sigma_{\text{Ar}} = \frac{1}{n\Lambda} = \frac{10^{-25} \cdot 10^7}{2.5 \cdot 1.5}$$

$$= 2.6 \cdot 10^{-19}\,\text{m}^2 = 26\,\text{Å}^2\,.$$

If both collision partners are moving, the mean time between two collisions is $\tau = \Lambda/\overline{\Delta v}$ where $\overline{\Delta v}$ is the mean relative velocity. We have the following numerical values:

$$\sigma_{N_2} = 31 \cdot 10^{-16}\,\text{cm}^2\,,$$

$$\overline{\Delta v}_{\text{Ar}} = \sqrt{2}\overline{v} = 565\,\text{m/s}$$

$$\Rightarrow \tau_{\text{Ar}} = \Lambda/\overline{\Delta v} = 2.6 \cdot 10^{-10}\,\text{s}\,,$$

$$\tau_{N_2} = 1.8 \cdot 10^{-10}\,\text{s}\,.$$

7.11 The density is (as in 7.10) $n = 2.5 \cdot 10^{19}/\text{cm}^3$.
a) $N = \dfrac{M}{m_{\text{He}}} = \dfrac{0.1}{6.68 \cdot 10^{-27}} = 1.5 \cdot 10^{25}$, where N is the total number of the He-atoms in the container.
b) $\sigma_{\text{He-He}} = 10 \cdot 10^{-16}\,\text{cm}^2$

$$\Rightarrow \Lambda = \frac{1}{n\sigma} = 4 \cdot 10^{-7}\,\text{m}\,.$$

c) The sum $\sum_i S_i$ is:

$$\sum_i S_i = \sum_i N_i v_i\,\Delta t = N\overline{v}\Delta t$$

$$= 1.5 \cdot 10^{25} \cdot 1260\,\text{m} \quad \text{for} \quad \Delta t = 1\,\text{s}$$

$$= 6.3 \cdot 10^{19}\ \text{light seconds}$$

$$= 2 \cdot 10^{12}\ \text{light years}\,.$$

7.12 Assume, two atoms with velocities v_1 and v_2 pass the disc at $t = 0$. Their arrival times at the detector are

$$t_1 = \frac{L}{v_1}\,; \quad t_2 = \frac{L}{v_2} \Rightarrow \Delta t = L\frac{\Delta v}{v_1 v_2}$$

with $\Delta v = v_1 - v_2$.
If one atom passes the disc at the beginning of the opening time Δt_0, the other atom at the end, the time difference between the arrival times at the detector is

$$\Delta t_{max} = \Delta t + \Delta t_0\,.$$

The time spectrum $N(t)$ of the arriving atoms with the velocity distribution $N(v) = N \cdot f(v)$ can be obtained as follows: With $v = L/t$ and $dv = -(L/t^2)dt = (v/t)dt$ we get the distribution function

$$f(v)\,dv = \frac{1}{t} v f\left(\frac{L}{t}\right) dt\,.$$

The function $f(v) \propto v^2 e^{-v^2/v_W^2}$ (see (7.30)) is then converted to

$$f(v,t) \propto \frac{L^3}{t^4} e^{-L^2/t^2 v_W^2} dt\,.$$

If the time profile of the velocity selector is $g(t)$ the time dependence of the detector signal is

$$S(t) = \int_{-\infty}^{+\infty} g(t') f(t-t')\,dt'\,.$$

If the opening time Δt_0 of the selector were infinitely short ($\Delta t_0 \to 0$), the difference Δt of the arrival times of the atoms at the detector (because of their different velocities) would be

$$\Delta t = \frac{t^2}{L}\Delta v = \frac{L}{v_W^2}\Delta v = \frac{1}{600^2} = 1.6\,\text{ms}\,.$$

Taking into account the finite opening time, the convolution $S(t)$ gives a time profile with a half width Δt which is for a rectangular opening time profile $g(t)$ with $\Delta t_0 = 1$ ms approximately $\Delta t \approx 2.5$ ms.

7.13 $\frac{m}{2}v_0^2 > G\frac{M_E m}{R+h} \Rightarrow v_0 > \sqrt{\frac{2GM_E}{R+h}}$

$$\Rightarrow v_0(h) = v_0(h=0)\sqrt{\frac{1}{1+h/R}}$$

$$\approx v_0(0)\left(1 - \frac{1}{2}h/R\right)\,.$$

For $h = 100$ km, $\Rightarrow v_0(h) = 0.992 v_0(0) = 11.1$ km/s.
a) If half of all molecules within the Maxwell distribution has a velocity $v > \overline{v}$

$$\Rightarrow v > \overline{v} = \sqrt{\frac{8kT}{\pi m}} = 11.1\,\text{km/s}$$

$$\Rightarrow T = 1.6 \cdot 10^5\,\text{K}\,.$$

7.14 The density of the outside air in a height of 50 m at $T = 300$ K is

$$\varrho = \varrho_0 e^{-\varrho_0 g h/p_0}\,,$$

with $\varrho_0 = 1.29\,\text{kg/m}^3$, $p_0 = 10^5\,\text{N/m}^2 \Rightarrow \varrho = 1.28\,\text{kg/m}^3$.
The exhaust gases must have a temperature $T > T_0$.
Because the pressure at the upper end of the smokestack is at the same temperature the same for the exhaust gases and the outside air it follows

$$\Rightarrow \varrho_1/\varrho_2 = T_2/T_1 \Rightarrow T_2 = 452\,\text{K}\,.$$

For the outside air is $p_1 = p_0 e^{-\varrho_1 g h/p_0}$ and for the exhaust gases inside of the smokestack $p_2 = p_0' \cdot e^{-\varrho_2 g h/p_0'}$. With $p_0' = p_2(h=0)$ and $p_1(h) = p_2(h)$ we obtain with the approximation $e^x \approx 1 + x$

$$p_0' - \varrho_2 g h = p_0 - \varrho_1 g h$$

$$\Rightarrow \Delta p_0 = p_0 - p_0' = \Delta\varrho g h$$

$$= (1.28 - 0.85) \cdot 9.81 \cdot 50\,\text{Pa} = 211\,\text{Pa}\,.$$

7.15 $\varrho_0(\text{He}) = 0.178\,\text{kg/m}^3$ at $p_0 = 1$ bar.
$\Rightarrow \varrho_{He}(1.5\,\text{bar}) = 0.267\,\text{kg/m}^3$.
From $m_{He} + m_{Bal} = V\varrho_{Air} \Rightarrow$

$$V = \frac{m_{Bal}}{\varrho_{Air} - \varrho_{He}(1.5\,\text{bar})}$$

$$= \frac{0.01}{1.023}\,\text{m}^3 = 9.8 \cdot 10^{-3}\,\text{m}^3\,.$$

7.16 a) $\frac{1}{2}m \cdot \langle v^2 \rangle = \frac{3}{2}kT = 3.1 \cdot 10^{-16}\,\text{J} \mathrel{\hat=} 1.9 \cdot 10^3\,\text{eV}$.
The ionization energy of the H-atom is 13.5 eV. At a density of $5 \cdot 10^{29}\,/\text{m}^3$ the mean distance between the protons is $1.25 \cdot 10^{-10}$ m. The mean potential energy, due to the Coulomb repulsion, is $E_p \approx 1.8 \cdot 10^{-18}$ J which is small compared to the mean kinetic energy at a temperature of 15 million Kelvin. This means that the matter in the central part of the sun can be safely regarded as ideal gas.

b) $\overline{v} = \sqrt{\frac{8kT}{\pi m}}\overline{v}_p = 5.6 \cdot 10^5\,\text{m/s}$;
$\overline{v}_{el} = 2.4 \cdot 10^7\,\text{m/s} = 0.08\,c$.
c) $p = nkT = 1 \cdot 10^{14}\,\text{Pa} \mathrel{\hat=} 10^9$ atm.

7.17 $M_{Atm} = \frac{4\pi R^2 \cdot 1.013 \cdot 10^5\,\text{N}}{9.81\,\text{m/s}^2} = 5.3 \cdot 10^{18}\,\text{kg}$.
The comparison with the earth mass $M_E = 6 \cdot 10^{24}$ kg shows that $M_{Atm} \approx 10^{-6} M_E$.

7.18 $M_B + \varrho_{He} \cdot V = \varrho_{Air} \cdot V$

$$\Rightarrow V = \frac{M_B}{\varrho_{Air} - \varrho_{He}}\,.$$

a) $h = 0$, $T = 300\,\text{K} \Rightarrow p_{Air} = 1$ bar, $p_{He} = 1.1$ bar $\Rightarrow$ $\varrho_{Air} = 1.23\,\text{kg/m}^3$, $\varrho_{He}(p = 1.1\,\text{bar}) = 0.196\,\text{kg/m}^3$

$$\Rightarrow V = \frac{300\,\text{kg}}{(1.23 - 0.196)\,\text{kg/m}^3} = 290\,\text{m}^3\,.$$

b) $h = 20\,\text{km}, T = 217\,\text{K}, p = 5.5 \cdot 10^{-2}\,\text{bar}$

$$\Rightarrow \varrho_{\text{Air}} = 0.9\,\text{kg/m}^3\,, \quad \varrho_{\text{He}}(p = 0.055\,\text{bar}) = 0.042\,\text{kg/m}^3$$

$$\Rightarrow V = \frac{300}{0.09 - 0.042}\,\text{m}^3 = 6250\,\text{m}^3\,.$$

The balloon has to expand considerably. On the ground it has only 5% of its maximum volume.

7.19 a) If the pressure at the upper end of the atmosphere (which is here assumed to have a sharp edge) should be $p_1 = 10\,\text{bar}$, the pressure at the bottom must be $p_0 = 11\,\text{bar}$. We assume as mean pressure

$$\Rightarrow \varrho_{\text{Air}}(\overline{p} = 10.5\,\text{bar}\,, T = 300\,\text{K}) = 1.23 \cdot 10.5\,\text{kg/m}^3 = 12.9\,\text{kg/m}^3\,,$$

$$\Rightarrow \varrho \cdot g \cdot h = 10^5\,\text{Pa}$$

$$\Rightarrow h = \frac{10^5}{12.9 \cdot 9.81\,\text{m}} = 7.9 \cdot 10^2\,\text{m}\,,$$

b) The density of solid air at $T = 0\,\text{K}$ is $\varrho = 10^3\,\text{kg/m}^3$. $\Rightarrow h = 10\,\text{m}$.

14.8 Chapter 8

8.1 a) The force acting on the area A is according to (8.41b)

$$F_{\text{w}} = c_{\text{w}} \frac{\varrho}{2} u^2 \cdot A\,.$$

Numerical values: $A = 100\,\text{m}^2$, $\varrho_{\text{L}} = 1.225\,\text{kg/m}^3$, $u = 100\,\text{km/h} = 27.8\,\text{m/s}$.
$\Rightarrow F = 5.67 \cdot 10^4\,\text{N}$. This corresponds to a weight of 5.8 tons.

b) For a simple estimation we assume that the streamlines of the wind above the roof are following the roof profile. The air above the roof then passes through a path length $S_2 = 2 \cdot 6\,\text{m} = 12\,\text{m}$. In the same time the horizontal wind flow passes only a distance $S_1 = 2 \cdot 6\,\text{m} \cdot \sin(\alpha/2) = 11.6\,\text{m}$. The velocity is then $u_2 = 100\,\text{km/h} \cdot 12/11.6 = 103.4\,\text{km/h} = 28.7\,\text{m/s}$. From the Bernoulli equation

$$p = p_0 - \tfrac{1}{2}\varrho \cdot \overline{u}_2^2\,.$$

we can determine the pressure difference $\Delta p = p - p_0$. With p_0 (pressure below the roof) $= 10^5\,\text{N/m}^2$ and $\frac{1}{2}\varrho\overline{u}_2^2 = 531\,\text{N/m}^2$ the pressure p becomes $p = (10^5 - 531)\,\text{N/m}^2$ and the difference $531\,\text{N/m}^2$. The force is $F = A \cdot \Delta p$. With $A = 2L_y \cdot 6\,\text{m} \cdot \sin(\alpha/2) = 96.7\,\text{m}^2$ effective roof area (projection onto a horizontal plane) we get

$$F = 531 \cdot 96.7\,\text{N} = 5.1 \cdot 10^4\,\text{N}\,.$$

8.2 The buoyancy depends not only on the wing profile but also on the stalling angle (Fig. 8.43). When a plane flies upside down the buoyancy is much smaller but can be still larger than zero if the stalling angle is correctly chosen.

8.3 The mean free path length $\Lambda = 1/(n \cdot \sigma)$ in liquids with typical densities $n = 3 \cdot 10^{28}\,/\text{m}^3$ and $\sigma = 10^{-19}\,\text{m}^2 \Rightarrow \Lambda = 3 \cdot 10^{-10}\,\text{m}$. The boundary layer where molecules diffuse from neighbouring layers is therefore very thin. The appearance of curls at large velocities is not caused by diffusion but by macroscopic turbulence (Convection).

8.4 The following relations apply:

$$\begin{aligned}\mathbf{grad}\,(\boldsymbol{a} \cdot \boldsymbol{b}) &= (\boldsymbol{b} \cdot \nabla)\boldsymbol{a} + (\boldsymbol{a} \cdot \nabla)\boldsymbol{b} + \boldsymbol{a} \times (\nabla \times \boldsymbol{b}) \\ &\quad + \boldsymbol{b} \times (\nabla \times \boldsymbol{a}) \\ \Rightarrow \mathbf{grad}\,(\boldsymbol{u} \cdot \boldsymbol{u}) &= \mathbf{grad}\,\overline{u}^2 \\ &= 2 \cdot (\boldsymbol{u} \cdot \nabla)\boldsymbol{u} + 2 \cdot \boldsymbol{u} \times (\nabla \times \boldsymbol{u})\,.\end{aligned}$$

The last equation can be verified in component representation. For the x-component the left hand side can be written as:

$$\begin{aligned}&\frac{\partial}{\partial x}\left(u_x^2 + u_y^2 + u_z^2\right) \\ &\quad = 2u_x \frac{\partial u_x}{\partial x} + 2u_y \frac{\partial u_y}{\partial x} + 2u_z \frac{\partial u_z}{\partial x}\,.\end{aligned} \tag{14.5}$$

For the components on the right hand side is

$$\begin{aligned}&2\left(u_x \frac{\partial}{\partial x} + u_y \frac{\partial}{\partial y} + u_z \frac{\partial}{\partial z}\right) u_x \\ &\quad + 2\left[u_y(\text{curl}\,\boldsymbol{u})_z - u_z(\text{curl}\,\boldsymbol{u})_y\right]\,.\end{aligned} \tag{14.6}$$

The second bracket [ ] in Eq. 14.6 is in the component representation

$$u_y\left(\frac{\partial}{\partial x}u_y - \frac{\partial}{\partial y}u_x\right) - u_z\left(\frac{\partial}{\partial z}u_x - \frac{\partial}{\partial x}u_z\right)\,.$$

Inserting this into (14.5) the right hand side gives the same expression as the left hand side.
Analogous results are obtained for the y- and z-component.

8.5 The pressure at the height h is

$$p(h) = \varrho \cdot g \cdot (H - h) + p_0\,.$$

At the exit of the pipe the Bernoulli equation yields

$$\begin{aligned}\Delta p = p(h) - p_0 &= \tfrac{1}{2}\varrho u_x^2 \\ u_x^2 &= 2g(H - h)\,.\end{aligned}$$

The trajectory of the liquid stream is a parabola. The initial velocity is

$$v = \{u_x, u_y = 0, u_z = 0\}\,.$$

The drop time can be obtained from $h = (1/2)\,gt^2 \Rightarrow t = \sqrt{2h/g}$.

a) The point of impinge is

$$P = \{x_i = u_x \cdot t\,;\quad y = z = 0\} = \left\{2\sqrt{h(H-h)}, 0, 0\right\}\,.$$

The velocity at P is

$$\boldsymbol{v}(P) = \{u_x, u_z = gt\}$$
$$|\boldsymbol{v}| = \sqrt{u_x^2 + u_z^2} = \sqrt{2gH}\,.$$

This is the same velocity as for body falling vertically from the height H.
b) According to the Hagen–Poiseuille Law is:

$$-\frac{dV}{dt} = -\pi R^2 \frac{dH}{dt} = \frac{\pi r^4}{8\eta L}\Delta p$$

with $\Delta p = \varrho g H + p_0 - p_0$

$$\Rightarrow \frac{dH}{dt} = -\frac{r^4}{R^2}\frac{\varrho g H}{8\eta L} \Rightarrow H = H_0 e^{-at}$$

with $H_0 = H(t=0)$ and $a = \dfrac{r^4 \varrho g}{8R^2 L \cdot \eta}$.

8.6 The probe in Fig. 8.10c measures the total pressure

$$p_0 = p + \tfrac{1}{2}\varrho u^2 = \varrho g h$$
$$= 10^3 \cdot 9.81 \cdot 1.5 \cdot 10^{-1}\,\text{Pa} = 1470\,\text{Pa}\,.$$

The results of the measurements in Fig. 8.10a give $p = 10\,\text{mbar} = 10^3\,\text{Pa}$
$\Rightarrow 0.5\,\varrho u^2 = 470\,\text{Pa} \Rightarrow u = 0.97\,\text{m/s}$.

8.7 If the funnel is filled up to the height H the radius R of the water surface is $R = H \cdot \tan(\alpha/2)$. The volume of the water is then

$$V = \tfrac{1}{3}\pi R^2 H = \tfrac{1}{3}\pi H^3 \tan^2(\alpha/2) = \tfrac{1}{9}\pi H^3\,,$$

because $\tan^2 30° = 1/3$.
a) The reduction of the water volume per time unit is

$$\frac{dV}{dt} = \frac{dV}{dH}\frac{dH}{dt} = \frac{1}{3}\pi H^2 \frac{dH}{dt}\,.$$

On the other side the Hagen–Poiseuille Law demands

$$\frac{dV}{dt} = -\frac{\pi r^4}{8\eta L}\Delta p$$

with $r = d/2$ and $\Delta p = \varrho g H$

$$\Rightarrow \frac{dH}{dt} = -\frac{3}{8}\frac{r^4 \varrho g}{\eta L H} \Rightarrow H\,dH = -a\,dt$$

with $a = \frac{3}{8}\frac{r^4 \varrho g}{\eta L} \approx 7.2 \cdot 10^{-4}\,\text{m}^2\text{s}^2$.
Integration gives:

$$H^2 = -2at + H_0^2$$

with $H_0 = H(t=0)$

$$\Rightarrow H = \sqrt{H_0^2 - 2at}\,.$$

b)

$$\frac{dM}{dt} = \varrho \frac{dV}{dt} = -\frac{1}{3}\pi a H \varrho$$
$$= -\frac{1}{3}\pi a \varrho \sqrt{H_0^2 - 2at}$$
$$\Rightarrow M(t) = \tfrac{1}{9}\pi\varrho\left(H_0^2 - 2at\right)^{3/2}\,.$$

c) The time when all of the water has streamed out of the funnel (i. e. $H(t) = 0$) is

$$T = H_0^2/2a\,.$$

With $H_0 = 0.3\,\text{m}$, $r = 2.5 \cdot 10^{-3}\,\text{m}$, $L = 0.2\,\text{m}$, $\eta = 1.0 \cdot 10^{-3}\,\text{Pa s} \Rightarrow T = 62.5\,\text{s}$.
d) With 4 litre water $\Rightarrow H_0 = (9V/\pi)^{1/3} = 0.225\,\text{m}$. The time to fill the container with $V = 4\,\text{l}$ completely is with $a = 7.2 \cdot 10^{-4}$ $T = 35\,\text{s}$. If the outflowing water in the funnel is continuously substituted by pouring water into the funnel in order to keep the water level always constant at $H = H_0$, the time to fill the 4 l container is obtained by:

$$V = \frac{1}{3}\pi a H_0 \cdot t$$
$$\Rightarrow t = \frac{3 \cdot 4 \cdot 10^{-3}}{\pi \cdot 7.2 \cdot 10^{-4} \cdot 0.225\,\text{s}} = 23.6\,\text{s}\,.$$

8.8

$$\frac{dV}{dt} = \frac{\pi R^4}{8\eta L}\Delta p \quad \text{with} \quad \Delta p = \varrho g(\Delta h + L\sin\alpha)$$
$$= 1.5 \cdot 10^{-4}(0.1 + \sin\alpha)\,\text{m}^3/\text{s}.$$

The mean flow velocity is

$$\overline{u} = \frac{1}{A}\frac{dV}{dt} = \frac{1}{\pi r^2}\frac{dV}{dt} = 7.6(0.1 + \sin\alpha)\,\text{m/s}\,.$$

The Reynold's number is

$$\text{Re} = 2300 = \frac{\varrho r u_c}{\eta}\,.$$

This gives the critical velocity

$$\overline{u}_c = \frac{\eta \text{Re}}{\varrho r} = 0.92\,\text{m/s}\,.$$

The inclination angle α is then for $\overline{u} = \overline{u}_c$

$$\sin\alpha = 0.021 \Rightarrow \alpha = 1.2°\,.$$

8.9

$$\frac{dV}{dt} = \frac{\pi R^4}{8\eta L}\varrho g\,\Delta h = 10^{-3}\,\text{m}^3; \quad \Delta h = 20\,\text{m}$$
$$\Rightarrow R = \left(\frac{10^{-3} \cdot 8\eta L}{\pi \varrho g \cdot \Delta h}\right)^{1/4} = 6 \cdot 10^{-3}\,\text{m} = 6\,\text{mm}$$
$$\Rightarrow d = 1.2\,\text{cm}$$
$$\Rightarrow u = 8.8\,\text{m/s}\,.$$

This is already above the critical velocity, which means that d has to be larger because the flow resistance is for $\overline{u} > u_c$ larger than obtained from the Hagen–Poiseuille law.

8.10 The total force acting on the ball is

$$F = am = m^* g - 6\pi\eta r v \quad \text{with} \quad m^* = (\varrho_K - \varrho_l)\frac{4}{3}\pi r^3$$

$$\Rightarrow \frac{dv}{dt} = \frac{m^*}{m} g - \frac{6\pi\eta r v}{m} .$$

Rearrangement, division by m^* and multiplication by m yields

$$\frac{dv}{g - (6\pi\eta r v / m^*)} = \frac{m^*}{m} dt ,$$

with the abbreviations

$$b = \frac{6\pi\eta r}{g m^*} \quad \text{and} \quad c = g\frac{m^*}{m}$$

$$\Rightarrow \frac{dv}{1 - bv} = c\, dt .$$

Integration gives

$$-\frac{1}{b}\ln(1 - bv) = ct + C_1$$

$$\Rightarrow v = \frac{1}{b}\left(1 - e^{-bC_1} e^{-bct}\right) .$$

Since $v(0) = v_0 \Rightarrow e^{-bC_1} = 1 - v_0 b$

$$\Rightarrow v(t) = \frac{1}{b}\left(1 + (v_0 b - 1)e^{-bct}\right)$$

$$\Rightarrow z(t) = \frac{1}{b}t - \frac{v_0 b - 1}{b^2 c} e^{-bct} .$$

8.11 Division of (8.36a)by ϱ and applying the differential operator $\mathbf{rot} = \nabla\times$ onto both sides yields

$$\frac{\partial}{\partial t}\mathbf{rot}\,\boldsymbol{u} + \nabla \times (\boldsymbol{u}\cdot\nabla)\boldsymbol{u} = -\frac{1}{\varrho}\nabla\times(\nabla p) - \nabla\times\boldsymbol{g} + \frac{\eta}{\varrho}\nabla\times \text{div}\,\mathbf{grad}\,\boldsymbol{u} .$$

Now we use the relations $\nabla\times\nabla p = 0$ and $\nabla\times\nabla\cdot(\nabla\boldsymbol{u}) = \mathbf{0}$. If the influence of gravity can be neglected ($\Rightarrow \boldsymbol{g} = \mathbf{0}$) we obtain with $\Omega = \mathbf{rot}\,\boldsymbol{u}$ the relation

$$(\boldsymbol{u}\cdot\nabla)\boldsymbol{u} = \tfrac{1}{2}\mathbf{grad}\,u^2 - \boldsymbol{u}\times\mathbf{rot}\,\boldsymbol{u} = \tfrac{1}{2}\nabla u^2 + (\Omega\times\boldsymbol{u}) .$$

Vector multiplication with ∇ gives with $\nabla\times\nabla = 0$

$$\nabla\times(\boldsymbol{u}\cdot\nabla)\boldsymbol{u} = \nabla\times(\boldsymbol{\Omega}\times\boldsymbol{u}) .$$

Then (8.36a) converts to

$$\frac{\partial\Omega}{\partial t} + \nabla\times(\boldsymbol{\Omega}\times\boldsymbol{u}) = 0 .$$

14.9 Chapter 9

9.1 Through the capillary flows per second the air mass $\varrho \cdot dV/dt \propto p \cdot dV/dt$. At the high pressure side this is $p_1 \cdot dV_1/dt$ and at the low pressure side $p_2 \cdot dV_2/dt$. It is

$$p_1\frac{dV_1}{dt} = p_2\frac{dV_2}{dt} .$$

The pumped-out volume is V_2. According to Hagen–Poiseuille we get

$$p_2\frac{dV}{dt} = \frac{\pi R^4}{8\eta L}(p_1 - p_2)\frac{p_1 + p_2}{2}$$

with $p_1 = 10^5$ Pa, $p_2 = 10^{-1}$ Pa

$$\Rightarrow p_2\frac{dV}{dt} = 4.25\cdot 10^{-3}\,\text{m}^3\,\text{Pa/s} .$$

In order to maintain a pressure of 10^{-3} hPa $= 10^{-1}$ Pa the throughput of the vacuum pump must be at least $dV/dt = 4.25\cdot 10^{-2}\,\text{m}^3/\text{s} = 42.5\,\text{l/s}$.

9.2 The force acting on each hemisphere is

$$F = \pi\cdot\left(\frac{d}{2}\right)^2 \Delta p = 2.5\cdot 10^4\,\text{N} .$$

One has to pull on each hemisphere with this force in order to separate the two hemispheres.

9.3 For $p = 10^{-5}$ hPa is $n = 2.5\cdot 10^{17}\,/\text{m}^3$.

$$\Lambda = 6\,\text{m} , \ \tau = \frac{\Lambda}{\overline{v}} \approx 1.2\cdot 10^{-2}\,\text{s} \quad \text{for} \quad \overline{v} = 500\,\text{m/s}$$

$$Z_1 = n\cdot\sigma\cdot\overline{v}_{\text{rel}} \approx n\sigma\sqrt{2}\cdot\overline{v} \quad \text{with} \quad \sigma = 10^{-14}\,\text{cm}^2$$

The number Z_1 of collisions between the molecules is

$$Z_1 \approx 180\,\text{s}^{-1} .$$

The number Z_2 of collisions per sec with the wall is

$$Z_2 = \tfrac{1}{4}n\overline{v} \approx 3\cdot 10^{19}\,\text{m}^{-2}\text{s}^{-1} .$$

Onto the whole container wall with $A = 3.26\,\text{m}^2$ impinge $9.8\cdot 10^{19}$ molecules per second $\Rightarrow Z_1/Z_2 = 1.8\cdot 10^{-18}$; $\sum s_i = n\cdot V\cdot\Lambda/\tau = n\cdot V\cdot\overline{v} = 5\cdot 10^{19}\,\text{m/s}$.

9.4 The number of collisions per second and per m^2 with the wall is $Z_2 = \frac{1}{4}n\cdot\overline{v}$. At a pressure $p = 10^{-7}$ hPa $\Rightarrow n = 2.5\cdot 10^{15}\,\text{m}^{-3}$. With $\overline{v} = 500\,\text{m/s} \Rightarrow Z_2 = 3\cdot 10^{17}\text{m}^{-2}\,\text{s}^{-1}$. A complete monolayer on the wall is achieved for Z collisions, where

$$Z = \frac{1\,\text{m}^2}{(\text{surface area per molecule})} = \frac{1}{0.15\cdot 0.2\cdot 10^{-18}} = 3.3\cdot 10^{19} .$$

Since the number of wall colissions per sec and m^2 is $Z_2 = 3\cdot 10^{17}$, it follows that after about 100 s the wall is completely covered by a monolayer.

9.5 The suction capacity $\mathrm{d}V/\mathrm{d}t$ of a mechanical pump at the pressure $p_1 = 0.1$ hPa must be equal to the suction capacity of the diffusion pump at the pressure $p_2 = 10^{-6}$ hPa.

$$p_2 \cdot 3000\,\mathrm{l/s} = p_1(\mathrm{d}V/\mathrm{d}t)_{\text{mech. pump}}$$

$$\Rightarrow \frac{\mathrm{d}V}{\mathrm{d}t} = \frac{p_2}{p_1} \cdot 3000\,\mathrm{l/s} = 3 \cdot 10^{-2}\,\mathrm{l/s} = 0.1\,\mathrm{m^3/h}\,.$$

It is, however, advisable to use a larger mechanical pump. Because the diffusion pump reaches its full suction capacity already at a pressure of 10^{-4} hPa, where the mechanical pump needs a suction capacity of $\mathrm{d}V/\mathrm{d}t = 10^{-3} \cdot 3000\,\mathrm{l/s} = 10\,\mathrm{m^3/h}$ in order to prevent the rise of the pressure p_1 above 10^{-1} hPa.

9.6 When passing through the gas the intensity of the electron beam decreases according to $I = I_0 \mathrm{e}^{-n \cdot \sigma \cdot x}$. The number of produced ions is then equal to the difference $(I_0 - I)/q$, where $q = -\mathrm{e} = -1.6 \cdot 10^{-19}$ C is the electron charge. For $n \cdot \sigma \cdot x \ll 1$ we obtain

$$\begin{aligned}\frac{I_0 - I}{\mathrm{e}} &= \frac{I_0}{\mathrm{e}} n\sigma x \\ &= \frac{10^{-2}}{1.6 \cdot 10^{-19}} \cdot 2.5 \cdot 10^{15} \cdot 10^{-18} \cdot 2 \cdot 10^{-2}\,\mathrm{s^{-1}} \\ &= 3 \cdot 10^{12}\,\text{ions/s}\,.\end{aligned}$$

The ion current is then 0.5 μA.

9.7 The mean free path length Λ at a pressure $p = 10^{-2}$ hPa is $\Lambda = 6 \cdot 10^{-3}$ m (see Tab. 9.1) and therefore comparable to the distance $d = 1$ cm between the hot wire and he wall. This case is between the limiting cases $\Lambda \gg d$ and $\Lambda \ll d$. For $\Lambda \gg d$ we get from (7.49)

$$\frac{\mathrm{d}W}{\mathrm{d}t} = \kappa F \Delta T = 52\,\mathrm{mW}\,.$$

For $\kappa = n_1 \cdot \overline{v}(f/2)k = 4.4\,\mathrm{N\,m^{-1}\,K^{-1}}$ $F = 2\pi r_1 \cdot l = 7.8 \cdot 10^{-5}\,\mathrm{m^2}$ und $\Delta T = 150$ K.
For $\Lambda \ll d$ we obtain

$$\frac{\mathrm{d}W}{\mathrm{d}t} = \lambda \cdot F \frac{\mathrm{d}t}{\mathrm{d}x} = 60\,\mathrm{mW}\,.$$

If we choose for $\Lambda \approx d$ the average of the two values we get

$$\frac{\mathrm{d}W}{\mathrm{d}t} \approx 56\,\mathrm{mW}\,.$$

The electric power input is

$$P_\mathrm{e} = U \cdot I = 1\,\mathrm{W}\,.$$

Only 5.6% of the input power are transported by heat conduction through the gas. The major part (94.4%) are lost due to heat radiation and heat conduction through the mountings.

9.8 We assume that every molecule that hits the wall sticks there for some time and then evaporates again. For a ball at rest

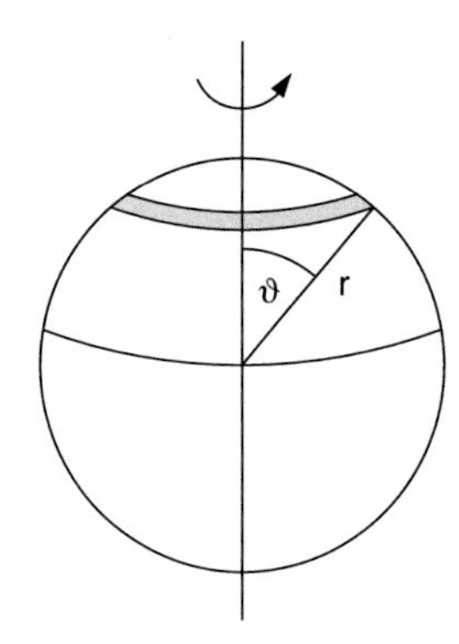

with a temperature equal to that of the gas the velocity distribution of the evaporating molecules is equal to that of the impinging molecules. Therefore there is no net angular momentum transfer. This is different for the rotating ball. Molecules that impinge onto a strip around the latitude ϑ get the rotational velocity

$$\overline{u} = \omega \cdot r \cdot \sin\vartheta\,,$$

while resting on the rotating sphere. We choose the z-axis into the direction of the rotational axis. The velocity perpendicular to the rotational axis is for the impinging molecules $v_\perp = (v_x^2 + v_y^2)^{1/2}$ and the evaporating molecules have the velocity $v'_\perp = (v_x^2 + v_y^2 + u^2)^{1/2}$.
The number of molecules impinging per sec onto the surface element $\mathrm{d}A = 2\pi r^2 \sin\vartheta\,\mathrm{d}\vartheta$ (the grey strip in the Figure) is according to (7.47)

$$\frac{\mathrm{d}N}{\mathrm{d}t} = \frac{n}{4}\overline{v}\,\mathrm{d}A\,.$$

Each evaporating molecule wins the additional momentum $m \cdot \overline{u}$ in tangential direction. The total number impinging on $\mathrm{d}A$ gets the additional momentum per sec

$$\begin{aligned}\Delta p(\vartheta) &= \frac{n}{4}\overline{v}m\overline{u}\,\mathrm{d}A \\ &= \frac{n}{4}\overline{v}m\omega r \sin\vartheta \cdot 2\pi r^2 \sin\vartheta\,\mathrm{d}\vartheta\,.\end{aligned}$$

The torque transferred to the ball by these molecules is then with $F = \mathrm{d}p/\mathrm{d}t$

$$\mathrm{d}D(\vartheta) = \frac{n}{2}\pi\overline{v}m\omega r^4 \sin^3\vartheta\,\mathrm{d}\vartheta\,.$$

Integration over all ϑ gives with

$$\int_{-\pi/2}^{+\pi/2} \sin^3\vartheta\,\mathrm{d}\vartheta = \frac{4}{3}\,,$$

the torque

$$\begin{aligned}D &= \tfrac{2}{3}\pi m\overline{v}\omega n r^4 \\ &= \tfrac{1}{2}V_{\text{sphere}} m\overline{v}\omega n r \quad \text{with} \quad V = M/\varrho\end{aligned}$$

$$D = -\frac{\mathrm{d}}{\mathrm{d}t}L = -I\frac{\mathrm{d}\omega}{\mathrm{d}t}$$

$$\rightarrow \frac{\mathrm{d}\omega}{\mathrm{d}t} = \frac{D}{(2/5)Mr^2} = a\omega$$

with

$$a = \frac{5nm\overline{v}}{4r\varrho} = \frac{10}{\pi}\frac{p}{r\varrho\overline{v}} \approx 3.18\frac{p}{r\varrho\overline{v}} ,$$

where the relations $p = (1/3)n \cdot m \cdot \overline{v}^2$ and $\overline{v}^2 = (3kT)/m$ have been used.

$$\Rightarrow \frac{\mathrm{d}\omega}{\omega} = -a\,\mathrm{d}t \Rightarrow \omega = \omega_0 \mathrm{e}^{-at} .$$

For $\omega = 0.99\,\omega_0$ we get $\mathrm{e}^{-at} = 0.99$

$$\Rightarrow t = \frac{1}{a}\ln\frac{100}{99} = \frac{0.01}{a} .$$

Numerical example:

$$r = 1 \cdot 10^{-3}\,\mathrm{m},$$
$$\varrho = 5 \cdot 10^3\,\mathrm{kg/m^3},$$
$$\overline{v} = 5 \cdot 10^2\,\mathrm{m/s},$$
$$p = 10^{-3}\,\mathrm{hPa} = 10^{-1}\,\mathrm{Pa}$$
$$\Rightarrow a = 1.3 \cdot 10^{-4}\,\mathrm{s^{-1}} \Rightarrow t = 78\,\mathrm{s}.$$

14.10 Chapter 10

10.1 The ratio of $E_{\mathrm{kin}}/E_{\mathrm{pot}}$ is generally larger for liquids than for solids. Therefore, the atoms move in the upper part of the interaction potential $V(r_{ik})$ between neighbouring atoms. Here the slope of the attractive part of the potential is smaller, therefore, the mean distance $\langle r_{ik}\rangle$ increases faster with increasing energy than in the lower part of the potential (Fig. 6.1).

10.2 a)

$$\Delta L = \alpha \cdot L \cdot \Delta T$$
$$= 16 \cdot 10^{-6} \cdot 20 \cdot 40\,\mathrm{m}$$
$$= 1.28 \cdot 10^{-2}\,\mathrm{m} = 1.28\,\mathrm{cm} .$$

b) The maximum distortion can be obtained from the figure below as

$$x = R - d = R(1 - \cos\beta) .$$

Since the length L increases by $\Delta L = 1.28\,\mathrm{cm}$, the half length of the circular arc is

$$R \cdot \beta = (10 + 0.64 \cdot 10^{-2}) = 10.0064\,\mathrm{m}$$

$$\frac{10}{R} = \sin\beta .$$

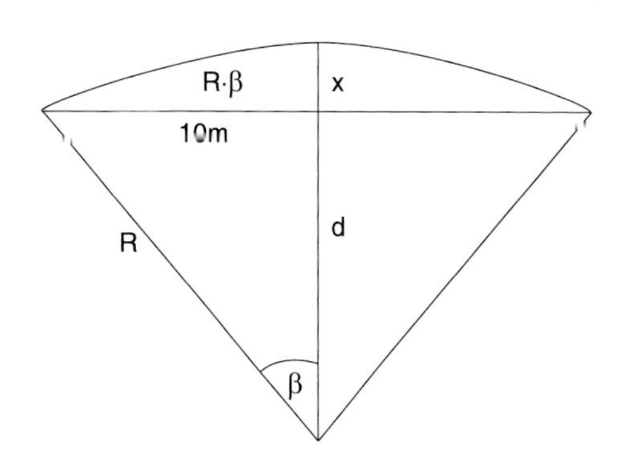

Division by $R \cdot \sin\beta$ gives

$$\Rightarrow \frac{\beta}{\sin\beta} = 1.00064 \quad \Rightarrow \frac{1}{1 - \frac{1}{6}\beta^2} = 10.0064\,\mathrm{m}$$

$$\Rightarrow \beta = 0.00623 \mathrel{\hat{=}} 3.57^\circ$$

$$\Rightarrow R = \frac{10\,\mathrm{m}}{\sin 3.57^\circ} = 160.6\,\mathrm{m}$$

$$\Rightarrow x = 160.6 \cdot (1 - \cos 3.57^\circ) = 0.31\,\mathrm{m} .$$

c) If the distortion should be prevented, the necessary pressure onto the rail in the longitudinal direction can be obtained from the relation

$$\Rightarrow \frac{F}{A} = E \cdot \frac{\Delta L}{L} = 200 \cdot 10^9\,\mathrm{N/m^2} \cdot \frac{0.0128}{20}$$
$$= 0.128\,\mathrm{GPa}$$

$$F = A \cdot E \cdot \frac{\Delta L}{L} = 2.56 \cdot 10^6\,\mathrm{N} .$$

10.3 The heat energy of one mole is

$$Q = \frac{f}{2} \cdot R \cdot T .$$

a) For helium is $f = 3 \Rightarrow Q = \frac{3}{2}R \cdot T$
The heating energy is then

$$W = 10 \cdot t\,\mathrm{W\,s}$$
$$= \tfrac{3}{2}R(100 - 20)\,\mathrm{K} + 10\,\mathrm{W\,s/K} \cdot 80\,\mathrm{K} ,$$

where the last term takes into account the heating of the container wall. With $R = 8.3\,\mathrm{J/(K \cdot mol)}$ we get

$$t = \frac{120 \cdot 8.31 + 800}{10}\,\mathrm{s} = 180\,\mathrm{s} = 3\,\mathrm{min} .$$

For N_2-molecules at the temperature $T > 300\,\mathrm{K}$ is $f = 5/2$.

$$\Rightarrow t = \frac{200 \cdot 8.31 + 800}{10}\,\mathrm{s} = 246\,\mathrm{s} = 4.1\,\mathrm{min} .$$

b) The heating up to 1000 °C takes for helium the time

$$t = \frac{980 \cdot 3R/2 + 9800}{10}\,\mathrm{s} \approx 2200\,\mathrm{s} \approx 37\,\mathrm{min} ;$$

for N_2 is f in the range from 20–500 °C $f = 5$; for $T > 500\,°\text{C}$ is $f = 7/2$.

$$dQ = U(T_2) - U(T_1) = R(\tfrac{7}{2}T_2 - \tfrac{5}{2}T_1)$$
$$= 2.89 \cdot 10^4\,\text{J}$$
$$\Rightarrow t = \frac{dQ}{10}\,\text{s} = 2.89 \cdot 10^3\,\text{s} = 48\,\text{min}\,.$$

10.4 If T is lower than the temperature T_S of the surrounding the temperature T will be approach T_S by heat conduction. After the mixing the temperature T_m is above T_S. The heat losses are proportional to the temperature difference $(T - T_S)$. Therefore the temperature decline

$$dT/dt = -a \cdot (T - T_S)\,.$$

The time-dependence of $T(t)$ after the mixture at $t = t_1$ is then

$$\frac{dT}{T - T_S} = -a(t - t_1) \Rightarrow T - T_S = C\,e^{-a(t-t_1)}\,.$$

If the mixing process occurs in a very short time at $t = t_1$ it is $T(t = t_1) = T_m \Rightarrow C = T_m - T_S \Rightarrow T(t) = T_S + (T_m - T_S) \cdot e^{-a\cdot(t-t1)}$.
If the real measured temperature curve $T(t)$ is replaced by the dashed curve in Fig. 10.12b in such a way, that the areas A_1 and A_2 are equal, the true mixing temperature is obtained. The dashed curve represents the ideal case of an infinitesimal short mixing process, where the heat losses are zero during the mixing process.

10.5 1 Mole air (N_2/O_2-mixture) has a mass of about 29 g and contains $6 \cdot 10^{23}$ molecules. For the lift of the container by 10 cm in the gravity field of the earth the energy

$$E = m \cdot g \cdot h = 0.129\,\text{kg} \cdot 9.81\,\text{m/s}^2 \cdot 0.1\,\text{m}$$
$$= 0.13\,\text{N} \cdot \text{m}$$

is required.
The thermal energy of the gas at room temperature is

$$E_{th}^{gas} = (5/2)R \cdot T = 6.2 \cdot 10^3\,\text{N} \cdot \text{m}\,,$$

and that of the container with the specific heat c is

$$E_{th}^{cont} = m_C \cdot c \cdot T = 0.1 \cdot 10^3 \cdot 300$$
$$= 3 \cdot 10^4\,\text{N} \cdot \text{m}$$
$$\Rightarrow E_{tot} = 3.6 \cdot 10^4\,\text{N} \cdot \text{m}\,.$$

The energy additionally required for the lift is therefore very small compared to the thermal energy. And the cooling after the lift would be only $\Delta T = 1 \cdot 10^{-3}$ K. Nevertheless this lift is extremely improbable, not because of energetic reasons but because of statistical reasons:
For the lift the z-component of the momentum must be at least $p_z = m \cdot v_{0z} = m \cdot \sqrt{2gh} \Rightarrow p_z > 0.18\,\text{kg} \cdot \text{m/s}$. The mean velocity component $\langle v_z \rangle$ of all molecules, which is $\langle v_z \rangle = 0$ at thermal equilibrium, must be $\langle v_z \rangle > v_0 = p_z/m_{gas} = 6.2\,\text{m/s}$. The probability that a molecule has a velocity component $v_z > v_0$ is given by the integral

$$W(v_z > v_0) = \int_{v_0}^{\infty} e^{-v_z^2 m/2kT} dv_z \Big/ \int_{-\infty}^{+\infty} e^{-v_z^2 m/2kT} dv_z\,.$$

With $x^2 = \frac{1}{2}m \cdot v_z^2/kT$ and $x(v_0) \ll 1 \Rightarrow$

$$\int_{x_0}^{\infty} = -\int_0^{x_0} + \int_0^{\infty}$$
$$\text{with} \int_0^{x_0} e^{-x^2} dx \approx \int_0^{x_0} (1 - x^2) dx = x_0 - \frac{x_0^3}{3}\,.$$

This gives with $v_0 = 6.2\,\text{m/s} \Rightarrow W(v_z > v_0) = 0.49$.
The probability that all $6 \cdot 10^{23}$ molecules have at the same time the velocity component $v_z > v_0$ is then

$$W = 0.49^{-6 \cdot 10^{23}} < 10^{-10^{-23}}\,,$$

and therefore practically zero.

10.6 a) The entropy change ΔS for an isobaric temperature rise from $T_0 = 273$ K to $T_1 = 500$ K is

$$\Delta S_{isobaric} = \nu \left(C_V \ln \frac{T_1}{T_0} + R \ln \frac{V_1}{V_0} \right)\,,$$

where $\nu = 1/22.4$ is the mole fraction.
With $V_1/V_0 = T_1/T_0$ for $p =$ const and $C_p = R + C_V$ we obtain

$$\Delta S_{isobaric} = \nu C_p \ln(T_1/T_0)$$

with $C_p = 21\,\text{J/(K} \cdot \text{mol)}$

$$\Rightarrow \Delta S_{isobaric} = \frac{21}{22.4} \ln \frac{500}{273} = 0.57\,\text{J/K}\,.$$

b) isochoric heating:

$$\Delta S_{isochoric} = \nu C_V \ln \frac{T_1}{T_0}$$

with $C_V = 12.7\,\text{J/(K} \cdot \text{mol)}$

$$\Rightarrow \Delta S_{isochoric} = 0.34\,\text{J/K}\,.$$

10.7 With $M = \varrho \cdot V$ the critical mole volume is

$$V_c = \frac{0.044\,\text{kg}}{\varrho_c} = \frac{0.044}{46}$$
$$= 9.56 \cdot 10^{-4}\,\text{m}^3 = 0.956\,\text{l}\,.$$

The mole volume is compressed from 22.4 l at standard conditions of an ideal gas to 0.956 l.

From the general gas equation of an ideal gas it follows:

$$V_c = \frac{RT_c}{p_c} = 0.33 \cdot 10^{-3}\,\text{m}^3 .$$

This shows that eigen-volume and internal pressure of a real gas around the critical point cause considerable deviations from the ideal gas. From Eq. 10.129 we obtain the van-der-Waals constants

$$b = \frac{1}{3}V_c \quad \Rightarrow \quad b = 0.32 \cdot 10^{-3}\,\text{m}^3 ,$$
$$a = 3p_c V_c^2 \quad \Rightarrow \quad a = 20.8\,\text{N} \cdot \text{m}^4 .$$

Under standard conditions ($p = 1$ bar, $T = 273$ K) the internal pressure is for 1 mole

$$\frac{a}{V^2} = 4.1 \cdot 10^4\,\text{N/m}^2$$
$$\hat{=}\ 41\%\ \text{of normal pressure!}$$

The eigen-volume of CO_2 molecules is $b/4 = 8 \cdot 10^{-5}\,\text{m}^3$ and the relative correction $b/V = \frac{0.32 \cdot 10^{-3}}{22.4 \cdot 10^{-3}} = 1.4\%$.

10.8

$$\Delta S_1 = mc_v \ln \frac{323.15}{273.15}\,\text{J/K}$$
$$= 4.18 \cdot 10^3 \ln 1.183\,\text{J/K}$$
$$= 689\,\text{J/K} ;$$

$$\Delta S_2 = mc_v \ln \frac{T_m^2}{T_1 T_2}$$
$$= 0.5 \cdot 4.18 \cdot 10^3 \cdot \ln \frac{323.15^2}{273.15 \cdot 373.15}$$
$$= 49.62\,\text{J/K} .$$

10.9 The theoretically possible maximum efficiency for $T_1 = 600\,°\text{C}$ and $T_2 = 100\,°\text{C}$ is

$$\eta = \frac{T_1 - T_2}{T_1} = \frac{500\,\text{K}}{873\,\text{K}} = 0.57 .$$

The heat delivered at 100 °C amounts therefore to 43% of the heat reveived at 600 °C. When using the technique of "cogeneration of heat and mechanical power" the heat delivered at 100 °C can be partly used for heating of buildings. The efficiency increase for cooling down to 30 °C is

$$\varepsilon = \frac{100 - 30}{373} = 18.8\% .$$

However, this saves only part of this efficiency increase, because when the additional heat energy available by cooling from 100 °C down to 30 °C can be used for driving a gas-turbine connected to an electric generator, this can deliver additional electric power. The theoretical efficiency of the power station increases then from $\eta = 57\%$ to $\eta = 570/873 = 0.65 = 65\%$. In order to prevent the water vapour to condensate at temperatures below 100 °C one has to decrease the pressure in the expansion chamber. This demands additions energy for the expansion against the external pressure.

10.10 a)

$$m_1 c_1 (T_1 - T_m) = m_2 c_2 (T_m - T_2)$$
$$\rightarrow T_m = \frac{m_1 c_1 T_1 + m_2 c_2 T_2}{m_1 c_1 + m_2 c_2} .$$

Numerical values:

$$m_1 = 1\,\text{kg}, \quad c_1 = 470\,\text{J/(kg} \cdot \text{K)},$$
$$m_2 = 10\,\text{kg}, \quad c_2 = 4.17 \cdot 10^3\,\text{J/(kg} \cdot \text{K)}$$
$$\Rightarrow T_m = 23.2\,°\text{C} \,\hat{=}\, 296.34\,\text{K} .$$

b)

$$\Delta S_1 = +mc \ln \frac{T_m}{T_1} = 1 \cdot 470 \ln \frac{296.34}{573.15}$$
$$= -310\,\text{J/K}$$
$$\Delta S_2 = 10 \cdot 4.1 \cdot 10^3 \ln \frac{296.34}{293.15}$$
$$= +445\,\text{J/K}$$
$$\Delta S = \Delta S_1 + \Delta S_2 = +135\,\text{J/K} .$$

10.11 a) A mass $M = 2 \cdot 5\,\text{kg} = 10\,\text{kg}$ exerts a pressure onto the area $0.1\,\text{m} \cdot 10^{-3}\,\text{m} = 10^{-4}\,\text{m}^2$

$$p = \frac{98}{10^{-4}}\,\text{N/m}^2 = 9.8 \cdot 10^5\,\text{Pa} .$$

Since the increase of the melting temperature T_m under the pressure p at $T = -8\,°\text{C}$ is given by $dT_m/dp = 10^{-7}\,°\text{C/Pa}$ the resulting temperature increase is $dT_m = 10^{-7} \cdot 9.8 \cdot 10^5\,°\text{C} = 0.1\,°\text{C}$. Ice at $T = -8\,°\text{C}$ therfore cannot melt solely due to the pressure.

b) The heat conduction is, according to (10.35)

$$\frac{dQ}{dt} = \lambda \cdot A \cdot \frac{dT}{dx} .$$

With $A = \pi r^2 = \pi \cdot 0.25 \cdot 10^{-6}\,\text{m}^2 = 7.8 \cdot 10^{-7}\,\text{m}^2$, $dx = 5\,\text{cm} = 0.05\,\text{m}$, $dT = 35\,°\text{C} = 35\,\text{K}$ and $\lambda = 67\,\text{W/m}^{-1} \cdot \text{K}^{-1}$, for steel we get

$$dQ/dt = 0.037\,\text{W} .$$

This heat energy flows into the horizontal part of the wire and in the surrounding volume of the ice. Since the surface $2\pi rL$ of the wire with radius r and length L is large compared to its volume, we can assume that nearly all of the heat energy flows into the ice. If the wire should melt the ice, it has to increase the temperature up to T_m and in addition it has to supply the heat of fusion. The energy balance requires:

$$\frac{dQ}{dt} = (c_{\text{Ice}} \cdot \Delta T + W_m)\varrho \cdot \frac{dV}{dt} .$$

This gives the velocity with which the wire melts through the ice block:

$$\frac{dz}{dt} = \frac{1}{L \cdot d \cdot \varrho} \cdot \frac{dQ/dt}{c \cdot \Delta T + W_m} .$$

Inserting the numerical values gives: $c = 2.1\,\text{kJ}/(\text{kg}\cdot\text{K})$, $\varrho = 0.9\cdot 10^3\,\text{kg/m}^3$, $L = 0.1\,\text{m}$, $d = 10^{-3}\,\text{m}$, $W_\text{m} = 333\,\text{kJ/kg}$, $\Delta T = 8\,\text{K}$.
This finally yields the result:

$$\frac{\mathrm{d}z}{\mathrm{d}t} = 10^{-6}\,\text{m/s}\,.$$

The observed velocity is higher by one order of magnitude. The reason for this is the heat, transferred directly from the warm wire at the edges of the horizontal part into the ice. This increases $\mathrm{d}Q/\mathrm{d}t$ considerably.

10.12 During the compression of V_1 to $V_2 < V_1$ the work supplied to the system is

$$\mathrm{d}W_1 = \int_{V_1}^{V_2} p\,\mathrm{d}V\,.$$

With $\mathrm{d}S = \mathrm{d}Q/T = 0 \Rightarrow \mathrm{d}Q = 0 \Rightarrow$

$$p\cdot V^\kappa = \text{const} = C_1 \Rightarrow$$

$$\mathrm{d}W_1 = -\int_{V_1}^{V_2} \frac{C_1}{V^\kappa}\,\mathrm{d}V = \frac{C_1}{\kappa-1}\left(\frac{1}{V_2^{\kappa-1}} - \frac{1}{V_1^{\kappa-1}}\right)$$
$$= \frac{C_1}{(\kappa-1)\cdot V_2^{\kappa-1}}\left(1 - \frac{1}{(V_1/V_2)^{\kappa-1}}\right)\,.$$

During the transition $3 \to 4$ the system delivers the work

$$\mathrm{d}W_2 = \frac{C_2}{\kappa-1}\left(\frac{1}{V_1^{\kappa-1}} - \frac{1}{V_2^{\kappa-1}}\right)\,.$$

The gain of energy during one cycle is then

$$\Delta W = \frac{C_1 - C_2}{(\kappa-1)V_2^{\kappa-1}}\left(1 - \frac{1}{(V_1/V_2)^{\kappa-1}}\right)\,.$$

For the isochoric processes $2 \to 3$ and $4 \to 1$ is $\Delta W = 0$.
The heat energy supplied to the ssterm is for 1 mole

$$Q_1 = C_V(T(3) - T(1))\,.$$

For isentropic processes is $\mathrm{d}Q = 0$, $T\cdot V^{\kappa-1} = \text{const} = C_1$ for the point 2 and C_2 for the point 3.
With $\kappa = C_p/C_V \Rightarrow (\kappa - 1)\cdot C_V = C_p - C_V = R$. The efficiency then becomes

$$\eta = \frac{\Delta W}{Q_1} = 1 - \frac{1}{(V_1/V_2)^{\kappa-1}}\,.$$

Example: $V_1 = 10\,V_2$, $\kappa = 1.4 \Rightarrow \eta = 0.6$.
Note that η does not depend on the temperature but only on the compression ration V_1/V_2.

10.13 We treat at first the stationary case with $\Delta T = 0$

$$\Rightarrow T = T_0 + (T_1 - T_0)\mathrm{e}^{-\alpha_1 x}$$
$$\Rightarrow \frac{\partial^2 T}{\partial x^2} = \alpha_1^2(T_1 - T_0)\mathrm{e}^{-\alpha_1 x}\,. \tag{14.7}$$

The stationary heat conduction equation (10.38a) gives with $\partial T/\partial t = 0$

$$\frac{\lambda}{\varrho\cdot c}\frac{\partial^2 T}{\partial x^2} = h^*\cdot(T - T_0)\,.$$

Inserting of (14.7) yields

$$\frac{\partial^2 T}{\partial x^2} = \frac{\varrho c h^*}{\lambda}(T_1 - T_0)\mathrm{e}^{-\alpha_1 x}$$
$$\Rightarrow \alpha_1^2 = \frac{\varrho c h^*}{\lambda} \Rightarrow \alpha_1 = \sqrt{\frac{\varrho c h^*}{\lambda}} = \sqrt{\frac{\varrho\cdot h}{m\cdot\lambda}}\,.$$

For $\Delta T \neq 0 \Rightarrow T = T_0 + (T_1 - T_0)\mathrm{e}^{-\alpha_1 x} + \Delta T\mathrm{e}^{-\alpha_2 x}\cos(\omega t - kx)$,

$$\frac{\partial T}{\partial t} = -\Delta T\cdot\mathrm{e}^{-\alpha_2 x}\cdot\omega\sin(\omega t - kx)\,,$$
$$\frac{\partial^2 T}{\partial x^2} = \alpha_1^2(T_1 - T_0)\mathrm{e}^{-\alpha_1 x} + (\alpha_2^2 - k^2)\Delta T\mathrm{e}^{-\alpha_2 x}$$
$$\cdot\cos(\omega t - kx) - 2\alpha_2\Delta T\mathrm{e}^{-\alpha_2 x}\cdot k\cdot\sin(\omega t - kx)\,.$$

Inserting into (10.38a) gives the stated relations by comparing the coefficients of sin and cos.

14.11 Chapter 11

11.1 $F = D\cdot x$. With $F = 1\,\text{N}$ and $x = 0.05\,\text{m} \Rightarrow D = 20\,\text{N/m}$. $\Rightarrow T = 2\pi\sqrt{m/D} = 1.4\,\text{s}$.

11.2 a) Approximation $m \ll M = \mu\cdot L$.
Velocity of the transverse wave

$$v_\text{Ph} = \sqrt{F/\mu} = \sqrt{mg/\mu}\,.$$

Running time of the wave over the distance z:

$$t_1 = \frac{z}{v_\text{Ph}} = z\sqrt{\mu/mg}\,.$$

Falling time of the ball:

$$t_2 = \sqrt{2z/g}\,.$$

For $t_1 = t_2$ the ball overtakes the wave pulse.

$$\Rightarrow z = 2m/\mu\,,$$

$\Rightarrow$ with $M = \mu\cdot L$ this gives

$$z = \frac{2m}{M}L\,.$$

This shows, that for $m > M \Rightarrow z > L$ the ball cannot overtake the wave.
b) More accurate calculation for an arbitrary ratio m/M:
For the distance z below the suspension point $z = 0$ the force

$$F = \mu(L - z)g + mg$$

acts on the rope due to the weight of rope + mass m. The phase velocity of the wave is

$$v_{Ph}(z) = [(L - z + m/\mu)g]^{1/2}$$

It decreases with increasing z (due to the decreasing weight force) according to

$$\frac{dv}{dz} = -\frac{g}{2v(z)} .$$

With

$$\frac{dv}{dt} = \frac{dv}{dz} \cdot \frac{dz}{dt} = -\frac{g}{2v} v = -\frac{g}{2}$$
$$\Rightarrow v(t) = v(z=0) - \tfrac{1}{2} gt .$$

The distance z that the wave propagates during the time t is then

$$z_1(t) = v_0 t - \tfrac{1}{4} gt^2 .$$

The ball falls in this time the distance $z_2(t) = \frac{1}{2} gt^2$. The meeting point is t at $z_1 = z_2$

$$\Rightarrow v_0 t_1 - \frac{1}{4} g t_1^2 = \frac{1}{2} g t_1^2$$

$$\Rightarrow t_1 = \frac{4}{3} v_0 / g = \frac{4}{3} \sqrt{\frac{\mu L + m}{g\mu}} .$$

This gives the meeting point

$$z_m = \frac{1}{2} g t_1^2 = \frac{8}{9}\left(L + \frac{m}{\mu}\right) = \frac{8}{9} L \left(1 + \frac{m}{M}\right) .$$

11.3 According to (11.36) the power supplied to the oscillating system is

$$P = m \cdot \gamma \omega^2 A_2^2$$
$$= \frac{(F_0^2/m) \cdot \gamma \cdot \omega^2}{\left(\omega_0^2 - \omega^2\right)^2 + (2\gamma\omega)^2} .$$

It is proportional to the square of the imaginary part b in equation (11.27c), because only the friction consumes energy. The real part a in (11.27b) determines the phase shift φ, because $\tan\varphi = b/a$, but does not consume energy. For $b = 0$ and $\omega \neq \omega_0$ is $\varphi = 0$.
For $b = 0$ and $\omega = \omega_0$ no stationary oscillation is possible. The amplitude A of the forced oscillation increases until $A = \infty$ (resonance catastrophe). In this case energy is supplied by the exciter which increases the amplitude until the oscillating system is destroyed.

11.4 The pressure at equilibrium is $p = 4\varepsilon/r$ with $\varepsilon = \sigma$. Changing the radius r changes the pressure by

$$dp = \frac{dp}{dr} dr = -\frac{4\varepsilon}{r^2} dr .$$

Therefore the restoring force is

$$dF = 4\pi r^2 \, dp = -16\pi\varepsilon \, dr \rightarrow \boldsymbol{F} = -16\pi\varepsilon \boldsymbol{r}.$$

With the restoring constant $D = 16\pi\varepsilon$ the oscillation period becomes

$$T = 2\pi \sqrt{m/D} = \tfrac{1}{2} \sqrt{\pi m/\varepsilon} = \pi r \sqrt{\varrho d/ep}$$

with the mass $m = 4\pi r^2 \cdot \varrho \cdot d$ and the thickness d of the skin of the soap bubble.
(The small change of the air pressure inside the bubble due to the small change of the volume is negligible).

11.5 The phase velocity is

$$v_{Ph} = \sqrt{E/\varrho} = 5.2 \cdot 10^3 \text{ m/s} .$$

The wavelength is

$$\lambda = v_{Ph}/\nu = 0.52 \text{ m} .$$

The maximum change of the length appears between maxima and minima of the longitudinal wave. Therefore we obtain

$$(\Delta L/L)_{max} = 2A/(\lambda/2) = 4A/\lambda$$
$$\Rightarrow \Delta L/L = 7.7 \cdot 10^{-4}$$
$$\Rightarrow \sigma_{max} = E \left(\frac{\Delta L}{L}\right)_{max} = 1.7 \cdot 10^8 \text{ N/m}^2 .$$

This is below the tensile strength by a factor 9.

11.6 The intensity of the sound wave is

$$I = \frac{1}{2} \frac{\Delta p_0^2}{\varrho v_{Ph}} \Rightarrow \Delta p_0 = \sqrt{2 \varrho v_{Ph} I} .$$

At the hearing limit is $I = 10^{-12} \text{ W/m}^2$. With $\varrho = 1.25 \text{ kg/m}^3$, $v_{Ph} = 300 \text{ m/s}$ is

$$\Delta p_0 = 2.74 \cdot 10^{-5} \text{ N/m}^2 .$$

With $\Delta p_0 = v_{Ph} \cdot \varrho \cdot \omega \cdot \xi_0$ the oscillation amplitude becomes

$$\xi_0 = 1.2 \cdot 10^{-11} \text{ m} = 0.12 \text{ Å} .$$

The amplitude is therefore smaller as one atomic diameter. The acoustic particle velocity u_0 is

$$u_0 = \omega \xi_0 = 7 \cdot 10^{-8} \text{ m/s} .$$

This is small compared to the thermal velocity $\langle v \rangle = 5 \cdot 10^2$ m/s of the molecules.

11.7 a) The surface of the liquid at rest is $z = 0$ in both sides of the U-tube. For a change $\Delta z = z_0 - z = -z$ in one side of the tube rises the liquid in the other side by $+z$. The restoring force is then for an ideal liquid (no friction)

$$F = -2z\varrho g A = m\ddot{z}$$

$$\Rightarrow z(t) = \Delta z \sin\left(\sqrt{\frac{2\varrho g A}{m}}\, t\right) = \Delta z \sin \omega t$$

$$\Rightarrow \omega = 3.5\,\mathrm{s}^{-1} \Rightarrow T = \frac{2\pi}{\omega} = 1.8\,\mathrm{s}\,.$$

The velocity v is then

$$v = \dot{z} = \omega\, \Delta z \cos \omega t\,.$$

With $\Delta z = 0.1$ m is follows: $v_{\max} = \omega \cdot \Delta z = 0.35$ m/s. The acceleration is

$$a = \ddot{z} = -\omega^2\, \Delta z \sin \omega t \Rightarrow a_{\max} = 1.23\,\mathrm{m/s^2}\,.$$

b) Taking into account friction:
According to Hagen–Poiseuille the velocity profile is

$$u(r) = \frac{\Delta p}{4\eta L}\left(R^2 - r^2\right)\,.$$

Defining a mean velocity $\overline{u}$, averaged over the cross section $\pi \cdot r^2$ we obtain with (8.31) and $F_f = \Delta p \cdot \pi r^2$ the friction force on a liquid column of length L

$$F_f = 8\pi\eta L\overline{u} \quad \text{with} \quad \overline{u} = \frac{1}{\pi R^2}\int\limits_0^R \overline{u}(r) 2\pi r\, \mathrm{d}r\,.$$

In the equation of motion

$$m \cdot \ddot{z} - b \cdot \dot{z} - 2\varrho g \pi R^2 z = 0\,.$$

With $\mathrm{d}z/\mathrm{d}t = \overline{u}$, $b = 8\pi\eta L$ the damping constant

$$\gamma = b/(2m) = 4\pi\eta L/m = 4 \cdot 10^{-2}\,\mathrm{s}^{-1}\,,$$

where

$$L = \frac{m}{\varrho \pi R^2} = 1.6\,\mathrm{m} \quad \text{and} \quad \eta = 10^{-3}\,\mathrm{Pa \cdot s}$$

have been inserted. After the time $\tau = 25$ s the oscillation amplitude has decreased to $1/e$ of its initial value.

11.8 The intensity distribution of the wave, diffracted by the slit is

$$I(\alpha) = I_0 \frac{\sin^2(\pi d \sin\alpha/\lambda)}{(\pi d \sin\alpha/\lambda)^2}\,,$$

where α is the diffraction angle.
a) For $I(\alpha)/I_0 = 0.5$ is

$$\frac{\sin^2 x}{x^2} = 0.5 \Rightarrow x \approx 1.4 \Rightarrow \sin\alpha = 1.4\lambda/(\pi d)\,.$$

With $\lambda = c/\nu = 330/(2 \cdot 10^3) = 0.165$ m und $d = 0.5$ m $\Rightarrow \sin\alpha = 0.147 \Rightarrow \alpha = 9.4°$

$$\Rightarrow \Delta s = 2 \cdot 20 \cdot \tan\alpha = 6\,\mathrm{m}\,.$$

b) For $I(\alpha)/I_0 = 0.05 \Rightarrow x \approx 2.5 \Rightarrow \alpha = 17.0°$

$$\Rightarrow \Delta s = 10.9\,\mathrm{m}\,.$$

11.9 Conservation of energy demands for the intensities of the waves:

$$I_e = I_r + I_d$$

with $I = \frac{1}{2}\varrho v_{\mathrm{Ph}} u_0^2$ (u_0 = sound particle velocity).

$$\Rightarrow \tfrac{1}{2}\varrho_1 v_{\mathrm{Ph1}} u_{0e}^2 = \tfrac{1}{2}\varrho_1 v_{\mathrm{Ph1}} u_{0r}^2 + \tfrac{1}{2}\varrho_2 v_{\mathrm{Ph2}} u_{0d}^2\,.$$

With the wave resistance $z = \varrho \cdot v_{\mathrm{Ph}}$ this can be written as

$$z_1\left(u_{0e}^2 - u_{0r}^2\right) = z_2 u_{0d}^2\,.$$

At the boundary is $u_{0e} + u_{0r} = u_{0d}$

$$\Rightarrow \overline{u}_{0r} = u_{0e}\frac{z_1 - z_2}{z_1 + z_2} \quad \text{and} \quad u_{0d} = u_{0e}\frac{2z_1}{z_1 + z_2}$$

$$\Rightarrow I_r = \frac{1}{2} z_1 u_{0r}^2 = I_e\left(\frac{z_2 - z_1}{z_2 + z_1}\right)^2$$

$$= RI_e\,, \quad R = \text{reflection coefficient}$$

$$I_d = \frac{1}{2} z_2 u_{0d}^2 = 4I_e\frac{z_1 z_2}{(z_1 + z_2)^2}$$

$$= TI_e\,, \quad T = \text{transmission coefficient}\,.$$

With the numerical values

$$\varrho_{\mathrm{Air}} = 1.29\,\mathrm{kg/m^3}\,, \quad \varrho_{\mathrm{Water}} = 10^3\,\mathrm{kg/m^3}\,,$$
$$v_{\mathrm{Ph}}^{\mathrm{Air}} = 334\,\mathrm{m/s}\,, \quad v_{\mathrm{Ph}}^{\mathrm{Water}} = 1480\,\mathrm{m/s}$$

we obtain

$$Z_1 = 1.29 \cdot 334\,\mathrm{kg/(m^2 \cdot s)}\,,$$
$$Z_2 = 10^3 \cdot 1480\,\mathrm{kg/(m^2 \cdot s)}$$
$$\Rightarrow R = 99.88\%\,; \quad T = 0.12\%\,.$$

11.10

$$\xi = \xi_1 + \xi_2$$

$$= 2A\cos\left(\frac{\Delta\omega}{2}t - \frac{\Delta k}{2}z\right)\cos\left(\omega_m t - k_m z\right)$$

$$= 2A\cos(85t - 0.25z)\cos(715t - 1.75z)\,;$$

$$v_{\mathrm{1Ph}} = \frac{\omega_1}{k_1} = \frac{800}{2}\,\mathrm{m/s} = 400\,\mathrm{m/s}\,,$$

$$v_{\mathrm{2Ph}} = \frac{\omega_2}{k_2} = \frac{630}{1.5}\,\mathrm{m/s} = 420\,\mathrm{m/s}\,,$$

$$v_G = \frac{\Delta\omega}{\Delta k} = \frac{170}{0.5}\,\mathrm{m/s} = 340\,\mathrm{m/s}\,.$$

11.11 For large values of λ the term $2\pi\sigma/(\varrho\lambda)$ in equation (11.86) can be neglected. This gives

$$v_{\text{Ph}} = \sqrt{\frac{g\lambda}{2\pi}\tanh\frac{2\pi h}{\lambda}} .$$

For $\lambda = 500$ m and $h > \lambda$ is $\tanh(2\pi h/\lambda) \approx 1$.

$$\Rightarrow v_{\text{Ph}} = 28\,\text{m/s} .$$

For $\lambda = 0.5$ m is $g \cdot \lambda/2\pi \approx 0.78\,\text{m}^2/\text{s}^2$ and with $\sigma = 7.25\,\text{J/m}^2$ is $2\pi\sigma/(\varrho\cdot\lambda) \approx 9.1\cdot 10^{-2}\,\text{m}^2/\text{s}^2$

$$\Rightarrow v_{\text{Ph}} = 0.93\,\text{m/s} .$$

11.12 The frequency of the fundamental oscillation is

$$\nu_0 = \frac{1}{2L}\sqrt{F/\mu} = \frac{1}{2L}\sqrt{\sigma/\varrho} ,$$
$$\lambda_0 = 2L \Rightarrow v_{\text{Ph}} = \nu_0\lambda_0 = \sqrt{\sigma/\varrho} .$$

With $\sigma = 3\cdot 10^{10}\,\text{N/m}^2 = 3\cdot 10^4\,\text{N/mm}^2$ und $\varrho = 7.8\cdot 10^3\,\text{kg/m}^3$, $L = 1$ m

$$\Rightarrow \nu_0 = 10^3\,\text{s}^{-1}$$
$$\Rightarrow T_0 = \frac{1}{\nu_0} = 1\,\text{ms}$$
$$\Rightarrow \lambda_0 = 2\,\text{m} , \quad v_{\text{Ph}} = 2\cdot 10^3\,\text{m/s} .$$

11.13 The weight $m\cdot g$ is compensated by the elastic restoring force $F_{\text{E}} = \pi r^2 E\frac{\Delta L}{L}$

$$m\cdot g = -k\,\Delta L \quad \text{with} \quad k = \pi r^2 E/L .$$
$$\Rightarrow k = \frac{m\cdot g}{\Delta L} .$$

The oscillation period is

$$T = 2\pi\sqrt{m/k} = 2\pi\sqrt{\Delta L/g} .$$

It is independent of m as long as the mass of the rope is negligible.
With $\Delta L = 2\cdot 10^{-3}\,\text{m} \Rightarrow T = 0.09$ s.
A pendulum with the length L has the oscillation period $T = 2\pi\sqrt{L/g}$. In our example the period would be $T = 2.84$ s, i.e. 30 times as long.

11.14 A force F acting onto the end of the flat spring bends the end by a distance (see (6.20))

$$\Delta\boldsymbol{s} = -\frac{4L^3}{Ed^3b}\boldsymbol{F}$$
$$\Rightarrow \boldsymbol{F} = -k\,\Delta\boldsymbol{s} \quad \text{with} \quad k = \frac{Ed^3b}{4L^3} .$$

The oscillation period of the spring without additional mass is

$$\omega_0 = \sqrt{4k/m_{\text{F}}} = 2\sqrt{k/m_{\text{F}}} .$$

Here the fourfold restoring force constant has to be inserted because the mean deflection of the spring is

$$\Delta s = \frac{1}{L}\int_{x=0}^{L}\Delta s(x)\,\text{d}x = \frac{1}{4}\Delta s(L) .$$

With $m_{\text{F}} = \varrho\cdot b\cdot d\cdot L$ and $k = \dfrac{Ed^3b}{4L^3}$ we obtain

$$\omega_0 = \frac{d}{L^2}\sqrt{E/\varrho} .$$

The frequency deceases with $1/L^3$!
With $\omega_0 = 2\pi\cdot 100\,\text{s}^{-1}$, $L = 0.1$ m, $E = 2\cdot 10^{11}\,\text{N/m}^2$ and $\varrho = 7.8\cdot 10^3\,\text{kg} \Rightarrow d = 0.63$ mm.
If a mass m is attached to the end of the spring, the frequency becomes

$$\omega = \sqrt{k/(m+m_{\text{F}}/4)}$$
$$\Rightarrow \frac{\omega_0}{\omega} = 2\sqrt{m+m\frac{\text{F}/4}{m_{\text{F}/4}}} = 2\sqrt{1+\frac{4m}{m_{\text{F}}}} .$$

11.15 Without waves the equilibrium position is given by: buoyancy = weight,

$$\Rightarrow aqL\varrho_1\cdot g = m\cdot g \quad \Rightarrow m = aLq\varrho_1 .$$

If the buoy is immersed by Δz below its equilibrium position and then released it performs an oscillation because of the restoring force $F = -q\varrho_1 g\Delta z$

$$z = \Delta z\sin\omega_0 t \quad \text{with} \quad \omega_0^2 = \frac{q\varrho_1 g}{m} = \frac{g}{aL} .$$

If waves are present the water surface at the location of the buoy is

$$z = z_0 + h\cdot\sin\omega_0 t \quad \text{with} \quad \omega_0 = \frac{2\pi}{T} \quad \text{and} \quad z_0 = 0 .$$

The waves generate an additional periodic buoyancy

$$\Delta F_{\text{B}} = h\cdot q\cdot g\cdot\varrho_1\cdot\sin\omega t ,$$

which results in a forced oscillation

$$\Delta z = A\sin(\omega t + \varphi) .$$

The oscillation amplitude is then, according to (11.26) when we neglect friction

$$A(\omega) = \frac{hg/(aL)}{(\omega_0^2 - \omega^2)} .$$

Numerical values: $a\cdot L = 30$ m, $h = 2$ m, $T = 5\,\text{s} \Rightarrow \omega = 1.25\,\text{s}^{-1}$, $\omega_0 = 0.57\,\text{s}^{-1} \Rightarrow A(\omega) = 0.525$ m.
Without waves (plane water surface) the fraction $(1-a)L$ of the buoy are above the water surface, at the wave peak only $x = [(1-a)L - (2m - 0.525m)]$. If the buoy should be just under water at the wave peak we must set $x = 0$. This gives $L = 32.475$ m.

11.16 The radial part of the Laplace-operator is

$$\Delta_r = \frac{2}{r}\frac{\partial}{\partial r} + \frac{\partial^2}{\partial r^2} .$$

If it is applied to $\xi = (A/r)e^{i(kr-\omega t)}$, we obtain

$$\begin{aligned}\Delta_r\xi &= \left[-\frac{2}{r^3} + \frac{ik}{r^2} + \frac{2}{r^3} - \frac{2ik}{r^2} - \frac{k^2}{r}\right]Ae^{i(kr-\omega t)}\\ &= -\frac{k^2}{r}Ae^{i(kr-\omega t)} = -k^2\xi ,\end{aligned}$$

$$\frac{\partial^2\xi}{\partial t^2} = -\omega^2\xi \quad \Rightarrow \frac{\partial^2\xi}{\partial t^2} = \frac{\omega^2}{k^2}\Delta_r\xi \quad \Rightarrow c = \frac{\omega}{k} .$$

11.17 a) $v = v_0\dfrac{1}{1-\bar{u}/v_{Ph}}$; From $v/v_0 = 1.12246$

$$\Rightarrow u_Q = v_{Ph}\left(1 - \frac{1}{1.12246}\right) = 0.1091 v_{Ph}$$

with $v_{Ph} = 330\,\text{m/s}$;

$$\bar{u}_Q \approx 36\,\text{m/s} = 130\,\text{km/h} .$$

b) $v = v_0\left(1 + \dfrac{\bar{u}_B}{v_{Ph}}\right) \Rightarrow 1 + \frac{u_B}{v_{Ph}} = 1.12246$

$$\Rightarrow \bar{u}_B = 0.12246 v_{Ph} \approx 145\,\text{km/h} .$$

11.18 We choose $x = 0$ as equilibrium position. When the block is shifted to an elongation $x_0 > 0$, is potential energy is

$$E_{pot_0} = \int_0^{x_0} 2D_0x\,dx = D_0x_0^2 .$$

a) After releasing the block it slides until the reverse point x_1 and loses on the way the energy by friction

$$E_f = f_1mg(x_0 - x_1) \quad \text{with} \quad x_1 < 0 .$$

$$\Rightarrow D_0x_1^2 = D_0x_0^2 - f_1mg(x_0 - x_1)$$

$$\Rightarrow x_1 = \frac{f_1mg}{D_0} - x_0 < 0 .$$

The absolute values of the elongations are

$$|x_1| = |x_0| - f_1mg/D_0 .$$

For the general reverse points we obtain

$$|x_n| = |x_{n-1}| - f_1mg/D_0 = |x_{n-1}| - 0.059\,\text{m} ,$$

$$|x_n| = |x_0| - nf_1mg/D_0 = |x_0| - n\cdot 0.059\,\text{m} .$$

The distances between the revers points decrease linearly with n. The motion of the block is a damped, but not harmonic oscillation.

b) The block sticks at the nth reverse point, if here the restoring force is smaller than the static friction coefficient.

$$\Rightarrow 2D_0|x_n| < f_0\cdot m\cdot g \quad \Rightarrow n > \frac{D_0x_0}{f_1mg} - \frac{f_0}{2f_1} .$$

Inserting the numerical values gives

$$n > \frac{100\cdot 0.22}{0.3\cdot 2\cdot 9.81} - \frac{0.9}{2\cdot 0.3} = 2.3 ,$$

i.e. the block sticks at least at the 3rd reverse point, if it reaches it at all. In order to check this we determine its start energy at the 2nd reverse point:

$$E_p = D_0x_2^2 = D_0(x_0 - 2f_1mg/D_0)^2 .$$

It should be larger than the friction energy $f_1\cdot m\cdot g|x_3 - x_2|$, if it should reach the reverse point x_3. Inserting the numerical values gives: $E_p(x_1) = 1.05\,\text{N}\cdot\text{m}, f_1\cdot m\cdot g|x_3 - x_2| = 0.346\,\text{N}\cdot\text{m}$. This proves that the block reaches x_3 but sticks there.

c) The total energy is

$$\begin{aligned}E &= \tfrac{1}{2}D_0x_0^2 = \tfrac{1}{2}D_0x^2 + \tfrac{1}{2}mv^2 + f_1\cdot mg\cdot(x_0 - x)\\ &\Rightarrow v^2 = (D_0/m)\left(x_0^2 - x^2\right) - 2f_1g(x_0 - x) .\end{aligned}$$

The block is released at x_0. The time T at which it reaches x_1 is

$$T = \int_{t=0}^{t_0} dt + \int_{t_0}^{t_1} dt \qquad \text{where} \qquad \begin{aligned}t_0 &= t(x=0)\\ t_1 &= t(x=x_1)\end{aligned}$$

$$-dx = v\,dt \quad \Rightarrow dt = -dx/v$$

$$\Rightarrow T = -\int_{x_0}^{0}\frac{dx}{v} - \int_0^{x_1}\frac{dx}{v} = \int_{x_1}^{x_0}\frac{dx}{v} .$$

Inserting of

$$v = \sqrt{(D_0/m)(x_0^2 - x^2) - 2f_1g(x_0 - x)}$$

gives with the substitutions

$$\begin{aligned}z &= (x_0 - x) \Rightarrow x_0^2 - x^2 = (x_0 - x)\cdot(x_0 + x)\\ &= z\cdot(2x_0 - z) ; \quad dx = -dz\end{aligned}$$

$$\Rightarrow x = x_0 \Rightarrow z = 0 ; \quad x = x_1 \Rightarrow z = z_1 = x_0 - x_1$$

$$a = D_0/m ; \quad b = 2\cdot a\cdot x_0 - 2f_1g$$

$$T = \int_{z=z_1}^{0}\frac{dz}{(-az^2 + bz)^{1/2}} .$$

The integral can be solved analytically and gives

$$T = -\frac{1}{\sqrt{a}}\left(\arcsin\frac{-2az + b}{b}\right)\Bigg|_{z1}^{0} .$$

The revers point x_1 can be obtained from

$$x_1 = \frac{f_1 mg}{D_0} - x_0 < 0 .$$

With $x_0 = 0.22\,\text{m} \Rightarrow x_1 = -0.161\,\text{m} \Rightarrow z_1 = 0.22 + 0.16 = 0.38\,\text{m}$

$$\Rightarrow T = +\frac{1}{\sqrt{D_0/m}}\left[\arcsin 1 - \arcsin\left(1 - \frac{2az_1}{b}\right)\right] .$$

14.12 Chapter 12

12.1 a)

$$m \cdot \ddot{x} + 2kx\left[1 + \frac{L}{(L^2 + x^2)^{1/2}}\right] = 0 .$$

b)

$$m \cdot \ddot{x} + 4kx - (k/L^2)x^3 = 0 .$$

c)

$$m \cdot \ddot{x} + 4kx = 0 ; \quad \omega_0 = (4k/m)^{1/2} .$$

For $x_0/L = \varepsilon$ is $\omega = \omega_0 \cdot K(\varepsilon) \approx \omega_0(1 - \varepsilon^2/2)$, where $K(\varepsilon)$ is the elliptical integral, that has been discussed in Sect. 2.9.7).

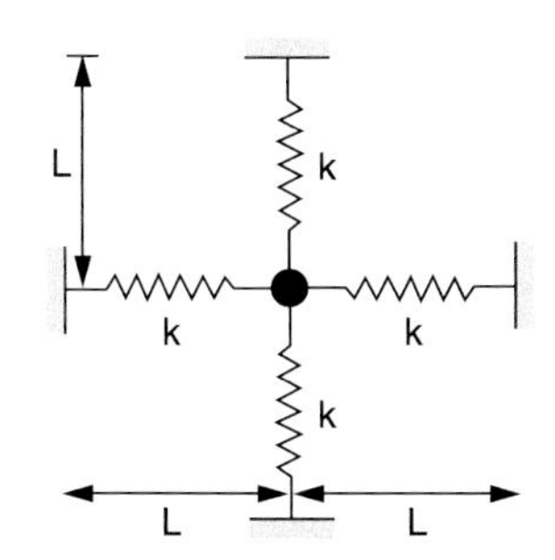

12.2

$$\ddot{x} + ax + bx^3 = 0 , \quad a, b > 0$$

$$v = \frac{\mathrm{d}x}{\mathrm{d}t} \Rightarrow \frac{\mathrm{d}^2x}{\mathrm{d}t^2} = \frac{\mathrm{d}v}{\mathrm{d}t} = \frac{\mathrm{d}v}{\mathrm{d}x} \cdot \frac{\mathrm{d}x}{\mathrm{d}t} = v\frac{\mathrm{d}v}{\mathrm{d}x}$$
$$= \frac{1}{2}\frac{\mathrm{d}(v^2)}{\mathrm{d}x} .$$

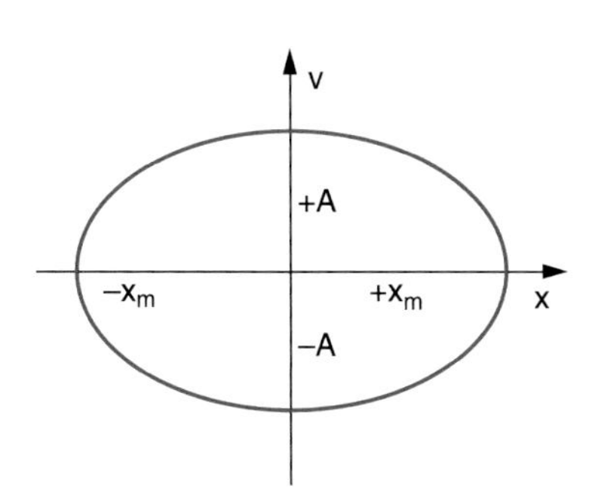

Inserting and integration over x and multiplication with the factor 2 yields the equation

$$v^2 + ax^2 + \frac{b}{2}x^4 = \text{const} = ax_0^2 + \frac{b}{2}x_0^4 = A .$$

The function $v(x)$ represents in an x-v-diagram a closed curve (called phase-space trajectory) where $x_0 = x(t = 0)$ and $v(t = 0) = 0$.

$$v(x) = \pm\left[A - ax^2 - \frac{b}{2}x^4\right]^{1/2} ,$$

which is periodically traversed. For $v = 0$ we obtain the intersection points of $v(x)$ with the x-axis

$$x_m = \pm\left[-\frac{a}{b} \pm \sqrt{\frac{a^2}{b^2} + \frac{2A}{b}}\right]^{1/2} .$$

The time for one circulation is obtained from

$$v = \frac{\mathrm{d}x}{\mathrm{d}t} \Rightarrow \mathrm{d}t = \frac{\mathrm{d}x}{v(x)}$$

$$\Rightarrow T = 4\int\limits_{x=0}^{x_m} \frac{\mathrm{d}x}{\left[A - ax^2 - \frac{b}{2}x^4\right]^{1/2}} ,$$

because the path from $x = 0$ until x_m is traversed in the time $T/4$. The elliptical integral can be found in integral tables.

12.3 a) With $y = x/x_0$ and $t^* = \omega_0 t$ we get

$$\frac{\mathrm{d}^2x}{\mathrm{d}t^2} = x_0\frac{\mathrm{d}^2y}{\mathrm{d}t^2} \quad \text{and} \quad \mathrm{d}t = \frac{1}{\omega_0}\mathrm{d}t^*$$
$$\Rightarrow \frac{\mathrm{d}^2x}{\mathrm{d}t^2} = \omega_0^2 x_0 \frac{\mathrm{d}^2y}{\mathrm{d}t^{*2}} .$$

The differential equation then transformes into

$$\frac{\mathrm{d}^2y}{\mathrm{d}t^{*2}} + \frac{k_1x_0}{\omega_0^2x_0m}y + \frac{k_2x_0^2}{\omega_0^2x_0m}y^2 = 0 .$$

This yields with $\omega_0^2 = k_1/m$ and $\varepsilon = (k_2/k_1)x_0$

$$\frac{\mathrm{d}^2y}{\mathrm{d}t^{*2}} + y + \varepsilon y^2 = 0$$

with the initial conditions: $y(0) = 1$ and $\dot{y}(0) = 0$.
b) With $t^* = \omega \cdot t$ we obtain the equations

$$\frac{\mathrm{d}^2y}{\mathrm{d}t^{*2}} + y + \varepsilon y^2 = 0 , \tag{14.8a}$$

$$\frac{\mathrm{d}^2y}{\mathrm{d}t^2} + \omega^2 y + \varepsilon\omega^2 y^2 = 0 , \tag{14.8b}$$

$$\frac{\mathrm{d}^2y}{\mathrm{d}t^2} + \omega^2 y[1 + \varepsilon y] = 0 . \tag{14.8c}$$

The last equation can be solved by series expansion with regard to ε, because the term $(1 + \varepsilon)$ can be regarded as perturbation term of the unperturbed equation with $\varepsilon = 0$.

We make the ansatz

$$y(t,\varepsilon) = y_0(t) + \varepsilon y_1(t) + \varepsilon^2 y_2(t) + \ldots$$
$$\omega = \omega_0 + \varepsilon\omega_1 + \varepsilon^2\omega_2 + \ldots .$$

Inserting this into the differential equation (14.8) and ordering the term according to the power exponents of ε, the different "coefficients" of ε^n have to be zero, if the equation should hold for arbitrary values of ε.
For the ε-free terms with ε^0 one obtains the unperturbed equation for the oscillation

$$\frac{d^2y_0}{dt^2} + \omega_0^2 y_0 = 0 \quad \Rightarrow \quad y_0 = A_0 \cdot \cos\omega_0 t .$$

For the term with ε^1 this gives

$$\frac{d^2y_1}{dt^2} + \omega_0^2(y_0^2 + y_1) + 1\omega_0\omega_1 y_0 = 0 .$$

Inserting y_0 from the previous equation we get

$$\frac{d^2y_1}{dt^2} + \omega_0^2(A_0^2\cos^2\omega_0 t + y_1) + 2\omega_0 y_1 A_0 \cos\omega t = 0 .$$

Proceeding to the quadratic term with ε^2 one obtains after some efforts

$$\omega(\varepsilon) = \omega_0\left(1 - \frac{5A^2}{12}\varepsilon^2 + O(\varepsilon^3)\right) .$$

For $\varepsilon = 0.1 \Rightarrow \omega = \omega_0(1 - 4.17 \cdot 10^{-3}A_0^2)$.

12.4 Fixpoints occur for $dx/dt = 0$. From the two equations of the problem we get for the fix points
a) $F = (x_{F1}, x_{F2}) = (0,0)$.
b) $F = (x_{F1}, x_{F2}) = (\lambda_3/\lambda_2, \lambda_1/\lambda_2)$.
They are stable, if for an elongation Δx from the fix point the conditions hold

$$\dot{x}_1 > 0 \quad \text{for} \quad \Delta x_i < 0\,; \quad i = 1,2\,,$$
$$x_1 < 0 \quad \text{for} \quad \Delta x_i > 0\,; \quad i = 1,2\,.$$

If the stability condition only holds for one direction (for instance x_1 but not for x_2) than the fix point is stable for shifts in the direction of x_1 but not for a shift in the direction of x_2.
In analogy to a saddle point on a curved surface this fix point is called saddle point. If the stability condition does not hold for both directions, the fix point is unstable.
Which of the cases applies for x_1 and x_2 can be obtained as follows:
Addition of the two equations yields

$$\dot{x}_1 + \dot{x}_2 = \lambda_1 x_1 - \lambda_3 x2 .$$

The stability depends on the sign of λ_1 and λ_3. That of λ_2 is not significant.
We distinguish between the 4 cases listed in the table. The arrows indicate the motion of the point $F = (0,0)$ at a

λ_1	λ_3	$F(0,0)$	
> 0	< 0	Saddle	↓ above, ← F → x_1, ↑ below, x_2
> 0	< 0	Unstable	↑ above, ← F →, ↓ below
< 0	> 0	Stable	↓ above, → F ←, ↑ below
< 0	< 0	Saddle	↑ above, → F ←, ↓ below

displacement. If the arrows point towards F the fix point F is stable against displacements in this direction, if they point away from F it is unstable in this direction.
For the second fix point $F = (\lambda_3/\lambda_2, \lambda_1/\lambda_2)$ no stable position exists. For $(\lambda_1 > 0, \lambda_3 < 0)$ and $(\lambda_1 < 0, \lambda_3 > 0)$ saddle-points exist. For the other two possible cases F moves on an elliptical path.

12.5 In Sect. 2.9.7 it was shown that the oscillation period T of the undamped pendulum is given by

$$T = \frac{2}{\pi}T_0 F(\varphi_0)$$

with

$$F(\varphi_0) = \int_0^{\pi/2} \frac{d\xi}{\sqrt{1 - \sin^2(\varphi_0/2)\sin^2\xi}} .$$

As can be seen in tables of elliptical integrals [2.6a, 2.6b], is

$$F\left(\varphi_0 = \frac{\pi}{4}\right) = 1.63\,; \quad F\left(\varphi_0 = \frac{\pi}{2}\right) = 1.84\,;$$
$$F\left(\varphi_0 = \frac{3}{4}\pi\right) = 2.4\,; \quad F\left(\varphi_0 = \frac{\pi}{2}\right) = \infty .$$

The oscillation periods of the undamped pendulum are then

$$T(\varphi) = a \cdot T_0 \quad \text{with} \quad \begin{array}{ll} a = 1.038 & \text{for} \quad \varphi = \pi/4\,, \\ a = 1.17 & \text{for} \quad \varphi = \pi/2\,, \\ a = 1.53 & \text{for} \quad \varphi = \frac{3}{4}\pi \\ a = \infty & \text{for} \quad \varphi = \pi\,. \end{array}$$

For the last case the pendulum reaches the metastable position in the upper reversal point where it could stay (without perturbation) infinitely long.
The equation for the damped oscillation can be solved only approximately, when using the expansion $\sin\varphi = \varphi - \frac{1}{6}\varphi^3 + \ldots$

The resulting equation

$$\ddot{\varphi} + \omega_0^2\varphi + \gamma\dot{\varphi} - \frac{\omega_0^2}{6}\varphi^3 = 0$$

can be solved for the case $\gamma/\omega_0 \ll 1$ with the ansatz $\varphi = \varphi_0 + \varepsilon\varphi_1 + \dots$
This gives the oscillation period

$$T = T_0 \frac{4F(\varphi_0)}{\sqrt{1 + \frac{1}{6}A^2}} \quad \text{with} \quad T_0 = \frac{2\pi}{\sqrt{\omega_0^2 - \gamma^2}} ,$$

where $F(\varphi_0)$ is again the elliptical integral and $A = \varphi(t = 0)$ the initial amplitude (see Probl. 12.2).

12.6

$$z(t) = \left[\frac{b}{a} + \left(\frac{1}{z_0} - \frac{b}{a}\right) \mathrm{e}^{-a(t-t_0)}\right]^{-1}$$

with $z(t = 0) = z_0$, $z(t \to \infty) = a/b$.
The doubling time for the case $a = b$:

$$T = \frac{1}{a} \ln \frac{2 - 2z_0}{1 - 2z_0} .$$

This shows that z doubles for $0 < z_0 < 0.5$ within a finite time, but for $z_0 = 0.5$ only after an infinite time. For $z_0 > 0.5$ there is no doubling at all.

12.7 The numerical analysis of the equation

$$x_{n+1} = 3.1 x_n (1 - x_n)$$

gives for $x_0 = 0.5$

$$\lim_{n\to\infty} x_{2n} = 0.5580\ldots = x_{\mathrm{F1}} ,$$

$$\lim_{n\to\infty} x_{2n+1} = 0.7646\ldots = x_{\mathrm{F2}} .$$

For $x_0 = 1/4$ the two quantities x_{F1} and x_{F2} interchange their values remain, however, independent of the initial value x_0.
For $a = 3.3$ one obtains

$$x_{\mathrm{F1}} = 0.4794\ldots , \quad x_{\mathrm{F2}} = 0.8236\ldots .$$

The Ljapunov exponent is

$$\lambda = \lim_{N\to\infty} \frac{1}{N} \sum_{i=0}^{N-1} \ln|f'(x_i)| .$$

With $f(x_n) = a x_n(1 - x_n) \Rightarrow f'(x_n) = a - 2a x_n$

$$\Rightarrow \lambda = \ln a + \tfrac{1}{2} \ln|1 - 2x_{\mathrm{F1}}| + \tfrac{1}{2} \ln|1 - 2x_{\mathrm{F2}}| .$$

For $a = 3.1 \Rightarrow \lambda = -0.264$.

For $a = 3.3 \Rightarrow \lambda = -0.619$.

12.8 For each iteration the length of a triangle side is cut in halves, but the number of sides becomes threefold larger. The total length therefore becomes

$$L_n = \left(\tfrac{3}{2}\right)^n L_0 \quad \text{and} \quad N(L/2) = 3N(L) .$$

In Eq. 12.54 is $\lambda = 1/2$ and $\lambda^\kappa = 3$.

$$\Rightarrow \kappa = -\frac{\ln 3}{\ln 2} \Rightarrow d = -\kappa = 1.58496\ldots .$$

12.9 The condition for the boundary curve is: $\dot{r} = 0$.

$$\Rightarrow (a - r^2)r = 0 \quad \Rightarrow r_1 = 0 , \quad r_2 = \sqrt{a} .$$

The first solution $r_1 = 0$ gives the stable fix point. The second solution $r_2 = \sqrt{a}$ and $\varphi = \omega_0 t$ gives as boundary curve a circle with radius $\sqrt{a}$ and centre $r = 0$, which is traversed with constant angular velocity.

12.10

$$F = -\frac{\mathrm{d}V}{\mathrm{d}x} = -2a(x - x_0) - 3b(x - x_0)^2 .$$

The equation of motion: $F = m\ddot{x}$

$$\Rightarrow \ddot{x} + \frac{2a - 6bx_0}{m} x + \frac{3b}{m} x^2 + \frac{-2ax_0 + 3bx_0^2}{m} = 0$$

$$\ddot{x} + Ax + Dx^2 + C = 0 .$$

Besides the minimum at $x = x_0$, $V(x)$ has a maximum

$$\frac{\mathrm{d}V}{\mathrm{d}x} = 2a(x - x_0) + 3b(x - x_0)^2 = 0$$

$$\Rightarrow x_{\max} = x_0 - \frac{2a}{3b} .$$

The maximum amplitude is then $2a/3b$. For $|x - x_0| > 2a/b$ no periodic motion is possible.

Index

© Springer International Publishing Switzerland 2017
W. Demtröder, *Mechanics and Thermodynamics*, Undergraduate Lecture Notes in Physics, DOI 10.1007/978-3-319-27877-3